Spectral Analysis and
Time Series

This is a volume in

PROBABILITY AND MATHEMATICAL STATISTICS

A Series of Monographs and Textbooks

Editors: Z. W. Birnbaum and E. Lukacs

A complete list of titles in this series is available from the publisher upon request.

Spectral Analysis and Time Series

VOLUME 1: UNIVARIATE SERIES
VOLUME 2: MULTIVARIATE SERIES, PREDICTION AND CONTROL

M. B. Priestley

Department of Mathematics, University of Manchester
Institute of Science and Technology

ELSEVIER
ACADEMIC
PRESS

AMSTERDAM • BOSTON • HEIDELBERG • LONDON • NEW YORK • OXFORD
PARIS • SAN DIEGO • SAN FRANCISCO • SINGAPORE • SYDNEY • TOKYO

Permissions may be sought directly from Elsevier's Science & Technology Rights
Department in Oxford, UK: phone: (+44) 1865 843830, fax: (+44) 1865 853333,
e-mail: permissions@elsevier.co.uk. You may also complete your request on-line via
the Elsevier homepage (http://www.elsevier.com), by selecting 'Customer Support'
and then 'Obtaining Permissions'

Elsevier Academic Press
84 Theobald's Road, London WC1X 8RR, UK
http://www.elsevier.com

Elsevier Academic Press
525 B Street, Suite 1900, San Diego, California 92101-4495, USA
http://www.elsevier.com

British Library Cataloguing in Publication Data
A catalogue record for this book is available from the British Library

ISBN 0-12-564922-3

Transferred to digital printing 2005
Printed and bound by Antony Rowe Ltd, Eastbourne

03 04 05 06 SE 15 14 13 12

To Nancy

Preface

Time series analysis is a very wide subject, and its historical development can be traced back to two main sources, namely, communications engineering and mathematical statistics. As a result, the subject is permeated with both engineering and statistical concepts and terminology, the former being associated with the "spectral" or "frequency domain" approach, and the latter with the "correlation" or "time domain" approach, to the analysis of time series. Many statisticians find it difficult to grasp the ideas of energy, power and frequency, while engineers may find it equally difficult to grapple with the intricacies of statistical inference based on maximum likelihood and least squares theory. However, the wide and diverse range of ideas involved make it a fascinating field of study, and the enormous growth in its use over recent years—with applications ranging from biology to astrophysics—amply demonstrate its considerable importance as a scientific tool.

In this book I have chosen to emphasize the "spectral" approach to time series, but since this is so firmly intertwined with the "time domain" approach (particularly so in view of recent developments such as "autoregressive spectral estimation"), I have included a fairly detailed treatment of time domain models so as to make the book essentially self-contained. Whenever possible I have adopted an informal style of presentation, and have not aimed at complete mathematical rigour. For the most part I have followed the style of the "applied" mathematician, but there are certain basic mathematical results (such as the spectral representation theorems) which are crucial for a proper understanding of the subject, and which, by their nature, require a careful and precise presentation. In treating such topics, I have modified the mathematical style accordingly. I have also used some elementary notions of Hilbert space theory since this "geometrical" approach does not require any great mathematical sophistication and yet it provides considerable insight into the structural properties of stationary stochastic processes. However, it is my firm belief that the most important aspects of time series analysis are related to the *ideas* rather than the techniques of the subject, and I have tried throughout to present the main ideas as fully and as clearly as possible. Where appropriate, I have tried to illuminate the theory by drawing on examples and illustrations from various physical and engineering situations.

The scope of the book is roughly as follows. Chapter 1 gives an overall view of the subject, with a discussion of the general mathematical framework and the main areas of application. Chapter 2 contains a summary of elementary probability theory (for the benefit of those readers who are not familiar with this subject), but both this chapter and the first half of Chapter 5 (which deals with statistical inference) may well be omitted by readers who have attained "intermediate level" mathematical statistics. Chapter 3 discusses the general properties of stationary processes, together with the main time domain models, and Chapter 4 deals with the basic spectral properties of these processes. Chapters 5 and 6 are both concerned with problems of estimation, Chapter 5 dealing with time domain functions and Chapter 6 with spectral functions. Chapter 7 concentrates on the more practical aspects of spectral estimation, and Chapter 8 treats the problem of "mixed spectra"—a sadly neglected topic in much of the literature. Chapter 9 extends the theory to multivariate and multidimensional processes, and Chapter 10 discusses problems of prediction, filtering and control, with particular reference to recent developments stemming from the control theory literature, such as state-space theory and Kalman filtering. Finally, Chapter 11 describes an approach to the spectral analysis of non-stationary processes, and introduces the reader to current research on non-linear time series models.

It is hoped that this book will be suitable for both post-graduate mathematicians and statisticians who are specializing in the area of time series analysis, and for research workers in applied fields such as physics, engineering, economics and biology. Although no attempt has been made to compile an exhaustive set of references, the list included should provide a reasonably comprehensive picture of the literature on the subject at the time of writing.

I gained my first knowledge of time series analysis from Professors M. S. Bartlett and H. E. Daniels, and it is a pleasure for me now to express my gratitude to them. I am extremely grateful also to my colleagues and post-graduate students at the University of Manchester Institute of Science and Technology who have read the manuscript and pointed out a number of errors and misprints. In particular, Drs T. Subba Rao, Howell Tong and Valerie Haggan have all contributed to stimulating discussions on many of the topics covered in this book, and Dr Haggan provided very substantial assistance with the numerical examples. The computations for these examples were carried out on the 1906A/7600 computer at the University of Manchester Regional Computer Centre, using the "TSALIB" (the UMIST Statistics Group Time Series Programme Library) which was constructed by Dr Haggan. The typing has been undertaken by various persons; to all of them, and to all who have contributed to the completion of this work, I express my sincere thanks.

I am grateful to the Royal Statistical Society, for their kind permission to reproduce some numerical material and figures from my papers in their Journals, Series B, Vol. 24, pp. 511–529, Vol. 27, pp. 204–237, and Series C (Applied Statistics), Vol. 30.

The work is split into two volumes; Volume 1 (Chapters 1–8) deals with the basic theory and analysis of univariate series, and Volume 2 (Chapters 9–11) with multivariate series, prediction, and control.

M. B. PRIESTLEY

Preface to Volume 2

This volume extends the theory and analysis of time series to the case of multivariate (and multi-dimensional) series, and includes a discussion of the problems of pediction, filtering, and control. The final chapter treats the analysis of non-stationary series and non-linear models.

Volume 2 uses the same notation and terminology as Volume 1, and the page numbers and chapter numbers follow on sequentially from the previous volume. Reference will occasionally be made to the material in Chapters 1–8 contained in Volume 1.

M. B. PRIESTLEY

Contents

Volume 1

Volume 2

List of Main Notation

$X(t)$	Continuous parameter time series
X_t	Discrete parameter time series
$R(r)$	Autocovariance function
$\rho(r)$	Autocorrelation function
$f(\omega)$	Normalized spectral density function
$F(\omega)$	Normalized integrated spectrum
$h(\omega)$	Non-normalized spectral density function
$H(\omega)$	Non-normalized integrated spectrum
$I_N(\omega)$	Periodogram (of a series of N observations)
$I_N^*(\omega)$	Modified periodogram
λ_s	Covariance lag window
$W(\theta)$	Spectral window
$k(u)$	Covariance lag window generator
$K(\theta)$	Spectral window generator
B	Backward shift operator
$\alpha(B)$	Autoregressive operator
$\beta(B)$	Moving average operator
$\mathrm{AR}(k)$	Autoregressive model of order k
$\mathrm{MA}(l)$	Moving average model of order l
$\mathrm{ARMA}(k, l)$	Mixed autoregressive/moving average model of order (k, l)
$R_{ij}(r)$	Cross-covariance function
$\rho_{ij}(r)$	Cross-correlation function
$h_{ij}(\omega)$	Cross-spectral density function
$\boldsymbol{h}(\omega)$	Spectral matrix
$c_{ij}(\omega)$	Co-spectrum
$q_{ij}(\omega)$	Quadrature spectrum
$\phi_{ij}(\omega)$	Phase spectrum
$\alpha_{ij}(\omega)$	Cross-amplitude spectrum
$W_{ij}(\omega)$	Complex coherency (spectrum)
$|W_{ij}(\omega)|$	Coherency (spectrum)
$g(u), g_u$	Filter impulse response function
$\Gamma(\omega)$	Filter transfer function

VOLUME 1

UNIVARIATE SERIES

Chapter 1

Basic Concepts

1.1 THE NATURE OF SPECTRAL ANALYSIS

When we look at an object we immediately notice two things. First, we notice whether or not the object is well illuminated, i.e. we observe the *strength* of the light (either emitted or reflected) from the object, and secondly we notice the *colour* of the object. In making these simple observations our eyes have, in fact, carried out a very crude form of a process termed *"spectral analysis"*. Let us examine this operation in more detail. First, what do we mean by "colour"? The usual quantitative definition of colour is expressed in terms of *"frequency"* when the light is regarded as an electromagnetic wave. Thus, we would say that blue light and red light appear to have different colours because they have different frequencies.

However, we seldom observe monochromatic light, and in general, we see a mixture of various basic colours. Now the human eye is a poor "spectrometer" in the sense that although we may be able to say whether light appears "blueish" or "reddish", we cannot say immediately how much blue or red it contains relative to other colours. We observe, rather, the over-all effect, and to obtain a more refined analysis we have to resort to optical devices such as a prism Thus, by passing the light through a prism we split it up into its various constituent colours, and determine the strength of the various components which are present. We note that by this device we have a standard method of describing the nature of any form of light—namely by stating the various colours which are present and their respective strengths, and conversely, if we know the strength of the various colours we may reconstruct the light by mixing these colours in their correct proportions. This information, namely, the colours present and their strengths, is called the *spectrum* of the light, and by using a prism we may obtain a visual display of this analysis.

1

So far we have not made use of the fact that we are discussing only electromagnetic waves in the frequency range to which the human eye is sensitive. The question therefore arises: can we apply a similar analysis to any type of "wave-like' disturbance? Clearly, if we are using the concept of "frequency", we can discuss only phenomena which possess a "wave-like" structure, i.e. which can be represented as a composition of sine and cosine waves with different amplitudes and frequencies. However, the restriction to wave-like phenomena still allows us to consider such diverse processes as radio waves, sound waves, vibration records in stress analysis, radar signals, records of "noise" in electrical apparatus, electroencephelograph traces ("brain-waves"), signals from radio stars, records of price fluctuations of commodities (e.g. fluctuations of wheat price indices), fluctuations of stocks and shares prices, records of annual unemployment figures—in fact we may usefully apply spectral analysis to any type of process which fluctuates in some form, but which exhibits a kind of "stability", i.e. tends to maintain a "steady" value, and is not obviously steadily increasing or decreasing.

1.2 TYPES OF PROCESSES

The kind of processes which we have discussed above are all examples of a general type called "time-series", i.e. the process (e.g. radio signal or price index) is varying in time, and when we refer to a record of the process we mean a graph of its values plotted against "time" as the variable. Of course, the scale of the time-axis may vary considerably: thus, for records of prices we may choose to measure time in weeks, months, or years, whereas for radio signals we would usually measure it in seconds, or possibly in milliseconds. However, we may equally well wish to study the variation of some quantity over a region of space. Suppose, for example, that we are studying variations of the thickness of cotton thread. In this case the variations occur along the length of the thread, and when we examine a record of the thickness we are studying its values at different points in space. We usually refer to such phenomena as "spatial processes". We may also generalize the examples of Section 1.1 in another direction. All the processes we have discussed so far deal with single quantities varying over time or space. Suppose, however, that we wish to study the simultaneous variation of, say, the signals received by two radio telescopes situated at different points, or simultaneous variation of the prices of bread and sugar. In such cases we would have to analyse two records, where both records cover the same period of time. Of course, in a similar way we may have to consider the simultaneous variation of two, three, four, . . . , or any number of related quantities, and such "collections" of records are called *multivariate processes*.

There is one further type of process which occurs, for example, when we are studying a quantity whose value depends on several variables, e.g. time and three space coordinates. For example, the temperature in a room will vary from point to point and also from one time instant to another. Generally, if a quantity depends on several variables it is termed a *multi-dimensional process*.

Occasionally we have to consider processes which belong to both of the above categories, i.e. processes which contain several constituent parts (or components), each part depending on several variables. Such processes may be described as *multivariate-multidimensional*. As an example of this type, let us consider the velocity field in a fluid. The velocity vector contains three components, and each component depends on three space coordinates and one time coordinate.

1.3 PERIODIC AND NON-PERIODIC FUNCTIONS

In Section 1.1 we referred to phenomena which have a "wave-like" structure. Intuitively, when we speak of a wave-like structure we usually have in mind something like the pattern of ocean waves, that is a pattern which, more or less, repeats itself after certain intervals. This idea may be expressed more precisely in terms of what is called a "periodic function". Suppose that a pattern of ocean waves happened to repeat itself perfectly at intervals of, say, p feet—or more precisely, that the section of the surface of the sea with some vertical plane repeated itself perfectly at intervals of p feet. Then if we measure distance along a horizontal line in the vertical plane, and let $f(x)$ denote the height of the surface (measured from some fixed level) at a point whose distance is x feet (from some fixed origin), we express the repetitive nature of the pattern by means of the equation

$$f(x) = f(x + kp), \qquad \text{all } x, \qquad (1.3.1)$$

where k may take any integral value, $0, \pm 1, \pm 2, \ldots$.

Generally, if a function $f(x)$ satisfies an equation of the above form it is said to be *periodic*, and if p is the smallest number such that equation (1.3.1) holds for all x, p is called the *period* of the function. If there is no value of p (other than zero) such that (1.3.1) holds for all x, the function is called non-periodic. The most familiar periodic functions which we encounter are the sine and cosine functions, since, of course, $A \sin \omega x$ and $A \cos \omega x$ are both periodic, each with period $p = (2\pi/\omega)$, see Fig. 1.1. The quantity $\omega = 2\pi/p$ is called the *angular frequency* of $\sin \omega x$ (or $\cos \omega x$) and the constant A is called the *amplitude*. These functions play a basic role in the theory of periodic functions by virtue of a most remarkable result which states that *any "well behaved" periodic function can be expressed as a*

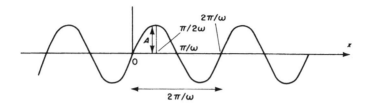

Fig. 1.1. The function $A \sin \omega x$.

(*possibly infinite*) *sum of sine and cosine functions*. This result (known as Fourier's theorem, see Chapter 4) states that virtually *any* function $f(x)$ which is periodic, with period p, may be written *as a Fourier series* in the form

$$f(x) = \sum_{r=0}^{\infty} [a_r \cos(2\pi rx/p) + b_r \sin(2\pi rx/p)] \qquad (1.3.2)$$

where $a_0, a_1, a_2, \ldots, b_0, b_1, b_2, \ldots$, are constants which may be determined from the form of $f(x)$. The various terms which appear in the summation on the right-hand side of equation (1.3.2) may be described in the following way. The first term (corresponding to $r = 0$) is simply a constant. The second term (with $r = 1$) represents cosine and sine waves with the basic period p, the next term ($r = 2$) represents cosine and sine waves with period $p/2$, the next term ($r = 3$) represents cosine and sine waves with period $p/3$, and so on. Noting that any cosine and sine wave whose period is an integral fraction of p will also repeat itself after intervals of p units, we see immediately that each of the terms in the summation repeat their values after intervals of p units, and therefore the sum is periodic with period p. This argument does not, of course, prove Fourier's theorem, but merely indicates its plausibility.

Thus, we may think of the sine and cosine functions as the basic "building bricks" in the theory of periodic functions, and we see that all periodic functions are constructed from combinations of these elements.

Let us turn now to the case of non-periodic functions. One way of looking at a non-periodic function is to regard it as a periodic function with an *infinite* period. Reasoning in this way, we might suppose that non-periodic functions could also be represented as a sum of the form (1.3.2) if we let $p \to \infty$. In other words, we might attempt to approximate to a non-periodic function by a sequence of periodic functions with longer and longer periods. This process can, in fact, be carried out in a rigorous way, and it turns out that as we increase the value of p we reduce the values of the coefficients $a_0, a_1, \ldots, b_0, b_1, \ldots$. Also, the distance between the frequencies, $2\pi r/p$, $2\pi(r+1)/p$, of neighbouring terms in (1.3.2) tends to zero, so that in the limit, the summation becomes an integral. Thus, we may expect that a non-periodic function could, perhaps, be written as

$$f(x) = \int_0^\infty \{g(\omega) \cos \omega x + k(\omega) \sin \omega x\} \, d\omega, \qquad (1.3.3)$$

where $g(\omega)$ and $k(\omega)$ are functions whose forms may be determined from the form of $f(x)$. However, in deriving equation (1.3.3) as a limiting form of (1.3.2) we have introduced a number of limiting operations and, as one may suspect, these operations will be valid only if the function $f(x)$ satisfies certain conditions. The required conditions are, roughly speaking, that $f(x)$ must "die away" as $x \to +\infty$ and as $x \to -\infty$. More precisely, if $f(x)$ is "absolutely integrable", i.e. if

$$\int_{-\infty}^\infty |f(x)| \, dx < \infty,$$

then it may be expressed as a Fourier integral of the form (1.3.3) (see Chapter 4, Section 4.5). It is often convenient to rewrite equation (1.3.2) in complex variable form, as follows.

Let the (complex valued) sequence, $\{A_r\}$, be defined by

$$A_r = \begin{cases} \frac{1}{2}(a_r - ib_r), & r > 0 \\ a_0, & r = 0 \\ \frac{1}{2}(a_{|r|} + ib_{|r|}), & r < 0 \end{cases} \qquad (1.3.4)$$

Then it is easily seen that (1.3.2) may be rewritten in the form

$$f(x) = \sum_{r=-\infty}^\infty A_r \, e^{i\omega_r x} \qquad (1.3.5)$$

where $\omega_r = 2\pi r/p$, $r = 0, \pm 1, \pm 2, \ldots$.

By analogy with the terminology used for sine and cosine functions, we call ω_r the (angular) *frequency*, and $|A_r| = (a_r^2 + b_r^2)^{1/2}$ the *amplitude*, of the complex exponential function $A_r \exp(i\omega_r x)$.

Similarly, by introducing the (complex valued) function, $p(\omega)$, defined by

$$p(\omega) = \begin{cases} \frac{1}{2}\{g(\omega) - ik(\omega)\}, & \omega > 0 \\ g(0), & \omega = 0 \\ \frac{1}{2}\{g(|\omega|) + ik(|\omega|)\}, & \omega < 0 \end{cases} \qquad (1.3.6)$$

(1.3.3) may be rewritten as

$$f(x) = \int_{-\infty}^{\infty} p(\omega)\, e^{i\omega x}\, d\omega. \tag{1.3.7}$$

The function $p(\omega)$ is called the *Fourier transform* of $f(x)$, and is of fundamental importance in spectral analysis.

Comparing (1.3.5) and (1.3.7) we see now the essential difference between periodic and non-periodic functions—namely, that whereas a periodic function can be expressed as a sum of cosine and sine terms over a *discrete set of frequencies*, $\omega_0, \omega_1, \omega_2, \omega_3, \ldots$, a non-periodic function can be expressed only in terms of cosines and sines which cover the *whole continuous range of frequencies*, i.e. from 0 to infinity. Bearing in mind that this is the only real distinction between Fourier series and Fourier integrals, we may combine both the expressions (1.3.5) and (1.3.7) in a single formula, by means of a *Fourier–Stieltjes* transform.† Thus, whether $f(x)$ is periodic (with period p), *or* non-periodic (but absolutely integrable), we may express it in the form

$$f(x) = \int_{-\infty}^{\infty} e^{i\omega x}\, dP(\omega), \tag{1.3.8}$$

where $P(\omega)$ is a (complex valued) function, called the Fourier–Stieltjes transform of $f(x)$, whose form may be determined from $f(x)$. When $f(x)$ is non-periodic the function $P(\omega)$ will, in general, be differentiable, so that (1.3.8) reduces to the same form as (1.3.7), with $dP(\omega)/d\omega \equiv p(\omega)$. On the other hand, if $f(x)$ is periodic (with period p), then $dP(\omega)$ will be of the form

$$dP(\omega) = \begin{cases} A_r, & \omega = \omega_r, r = 0, \pm 1, \pm 2, \ldots \\ 0, & \text{otherwise} \end{cases}$$

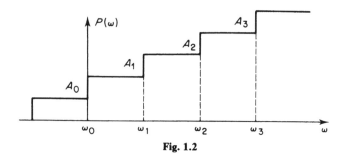

Fig. 1.2

so that in this case (1.3.8) reduces to (1.3.5). When $dP(\omega)$ is of the above form the function $P(\omega)$ has zero increments (i.e. is constant) except at the set points $\ldots \{\omega_r\}, r = 0, \pm 1, \ldots$, where it increases by amounts $\{A_r\}$. Its general form is shown in Fig. 1.2 and is given the name *step-function*.

† See Chapter 3, Section 3.6.3 for a discussion of Stieltjes integrals.

(Strictly, the graph of Fig. 1.2 refers only to the case when the A_r are real valued, i.e. when the b_r are all zero, corresponding to $f(x)$ an even function, but it indicates the general nature of $P(\omega)$ even when it takes complex values.) For comparison, a non-periodic function $f(x)$ may have a Fourier–Stieltjes transform whose typical shape is shown in Fig. 1.3, and its cor-

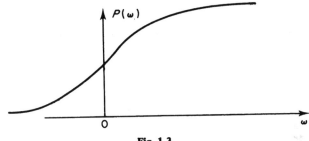

Fig. 1.3

responding derivative, $p(\omega)$, may look, typically, as shown in Fig. 1.4. (Again, it must be remembered that these graphs refer only to realvalued transforms, corresponding to even functions $f(x)$. In such cases, $p(\omega)$ is also symmetric about the origin. In general, $p(\omega)$ need not, of course, be symmetric.)

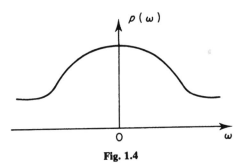

Fig. 1.4

1.4 GENERALIZED HARMONIC ANALYSIS

We have seen that the Fourier–Stieltjes transform, equation (1.3.8), includes both Fourier series and Fourier integrals as special cases, but it can be used also to represent a more general class of functions than those which are either periodic or non-periodic and absolutely integrable. In its most general form, the Fourier–Stieltjes transform was first discussed rigorously by Wiener (1930) who introduced the term "*generalized harmonic analysis*".

A very simple example of a function which can be expressed in the form (1.3.8), but which is neither periodic nor absolutely integrable is the following:

$$f(x) = \cos(x) + \cos(\sqrt{2}x) \qquad (1.4.1)$$

Clearly, $f(x)$ can be written in the required form with

$$P(\omega) = \begin{cases} \frac{1}{2}, & \omega = \pm 1, \pm\sqrt{2}, \\ 0, & \text{otherwise.} \end{cases}$$

Although $P(\omega)$ is a step-function, $f(x)$ is not periodic since the numbers 1 and $\sqrt{2}$ are not "commensurable", so that there is no value of $p(\neq 0)$ for which $f(x)$ satisfies (1.3.1) for all x. However, the above example by no means indicates the full generality of the Fourier–Stieltjes transform. In fact, the function (1.4.1) is hardly any different from a Fourier series since, although it is not strictly periodic, it may be regarded as "almost periodic", and the corresponding step-function behaviour of $P(\omega)$ is exactly the same as for Fourier series. The essentially new feature of this type of representation is that it enables us to treat functions which have a general type of "*steady-state*" form, i.e. which do not become arbitrarily large or small as $t \to \pm\infty$. In his pioneering paper, Wiener showed that functions of the "steady-state" type could also be represented in the form (1.3.8), but now the function $P(\omega)$ is no longer either a step-function or differentiable. It turns out that, for functions of this type,

$$dP(\omega) = O(\sqrt{d\omega}), \qquad (1.4.2)$$

so that the order of magnitude of the increment of $P(\omega)$ over an interval $d\omega$ is infinitesimal, but much larger than $d\omega$. Clearly, in this case $P(\omega)$ is not differentiable (i.e. its "derivative" will not be finite at any point), but may be thought of as having a form which is intermediate between a step-function (for which $dP(\omega) = O(1)$) and a differentiable function (for which $dP(\omega) = O(d\omega)$).

1.5 ENERGY DISTRIBUTIONS

In the previous section we saw that periodic and certain types of non-periodic functions could be expressed as sums or integrals of sine and cosine terms with different amplitudes and frequencies. The importance of this so-called "*spectral representation*" of a function lies in the fact that if the function represents some physical process such as a current or voltage, the total *energy* dissipated by the process in any time interval is equal to the sum of the amounts of energy dissipated by each of the sine or cosine terms (see

Chapter 4, Section 4.4). Now the energy carried by a sine or cosine term is proportional to the *square* of the amplitude. Consequently, in the case of periodic processes, the contribution of a term of the form $(a_r \cos \omega_r x + b_r \sin \omega_r x)$ to the total energy of the process is proportional to $(a_r^2 + b_r^2) = |A_r|^2$, where A_r is given by (1.3.4). If, therefore, we plot the squared amplitudes $|A_r|^2$ against the frequencies ω_r, the graph we obtain shows the relative contribution of the various sine and cosine terms to the total energy. For the moment, we will call this type of graph simply an "*energy spectrum*". (In fact, it would be more precise to describe it as a "power spectrum", but we will defer a detailed discussion of this point until Chapter 4.) Its shape shows us, for example, if some terms are contributing most of the energy with others contributing negligible energy. In other words, it tells us the *relative importance* of the various frequency components of the process. For a periodic process with, say, period p, the only frequencies of interest are those which are multiples of $(2\pi/p)$, since these are the only frequencies which enter into the representation (1.3.5), and the total energy of the process is divided up entirely among this *discrete* set of frequencies. Consequently we say that a periodic process has a *discrete spectrum*, or, in engineering terms, a *line spectrum* (see Fig. 1.5).

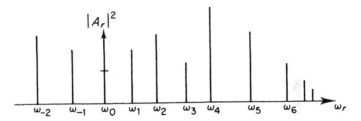

Fig. 1.5. Discrete spectrum.

On the other hand, a process which is described by a non-periodic function has a representation (1.3.7) which involves *all* frequencies, so that the total energy of the process is spread over the whole *continuous* range of frequencies. Thus, if we now plot $|p(\omega)|^2$ as a function of ω, the type of energy distribution so obtained is called a *continuous spectrum* (see Fig. 1.6).

For real-valued functions $f(x)$, $|A_{-r}|^2 = |A_r|^2$ (all r), and $|p(-\omega)|^2 = |p(\omega)|^2$ (all ω). Thus, in both cases, the energy spectra are *even* (i.e. *symmetric*) functions of the frequency variable, and need therefore be plotted only for *positive frequencies*. The forms of the graphs for negative frequencies are simply the mirror images of the graphs for positive frequencies. The physical interpretation of the function $|p(\omega)|^2$ is somewhat different from that of a discrete spectrum. In fact, $|p(\omega)|^2$ has the character of an energy *density* distribution, but again, we will defer further discussion of this point until the more detailed treatment in Chapter 4. The most general type of spectrum which occurs in practice is one which corresponds to a process composed of the sum of a periodic function and a non-periodic function. In this case the

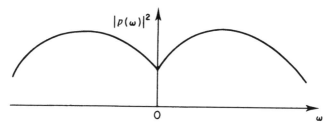

Fig. 1.6. Continuous spectrum.

spectrum is the sum of two components, a discrete distribution and a continuous distribution. Such a process is said to have a *mixed spectrum* (see Chapter 4, Section 4.9).

1.6 RANDOM PROCESSES

So far, we have discussed only *deterministic* functions, $f(x)$, i.e. functions which are such that, for each value of x, we have a rule which enables us to calculate the precise value of $f(x)$. Often, such a rule is specified in terms of a mathematical formula, such as, for example,

$$f(x) = e^{-x^2}, \qquad \text{all } x$$

or

$$f(x) = \begin{cases} e^{-x}, & x \leqslant 0 \\ e^x, & x > 0 \end{cases}$$

or,

$$f(x) = \begin{cases} 1 - |x|, & |x| \leqslant 1 \\ 0, & |x| > 1. \end{cases}$$

Deterministic functions form the whole domain of study in classical mathematical analysis, but in practical applications of spectral analysis almost all the processes which we encounter are not of this type. Let us consider, for example, a record of the variations in the thickness of a piece of yarn, plotted as a function of length (measured from some fixed point on the yarn). A typical graph might be as in Fig. 1.7 (with "thickness" and "length" measured in suitable units). If we imagine the yarn to have infinite length, so that the graph extends from zero to infinity, then it seems clear that we could not describe the complete behaviour of this graph in terms of a mathematical

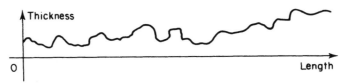

Fig. 1.7

formula which expresses thickness as a deterministic function of length. We could, perhaps, find a deterministic function which gave a good approximation to the graph over a finite interval, say, 0 to L. However, we would find that outside this interval the function was no longer a valid approximation, and this feature would persist no matter how large we made L. Why does this happen? At first, we might attempt to explain this effect by saying that the graph in Fig. 1.7 contains a large number of small "irregularities" which fail to conform to any coherent pattern. But we could then ask—what produces these irregularities? For, if we could determine the physical processes which produce the irregularities then surely we could deduce the mathematical relationship which we require. However, it turns out that the manufacturing process involves a number of factors which cannot be controlled precisely, and whose effects on the thickness of the yarn may not be fully understood. Due to these "elements of uncertainty" in the manufacturing process, the variations in thickness along the length of the yarn have a *random character* in the sense that *we cannot determine theoretically what will be the precise value of the thickness at each point of the yarn*: rather, *at each point there will be a whole range of possible values*. The only way we can describe the behaviour of the thickness (as a function of length) is to specify, at each point along the length, a *probability distribution* which describes the relative "likeliness" of each of the possible values. In the language of probability theory, we would say that, at each point of the yarn, the thickness is a "*random variable*", and the complete function (of thickness against length) is called a "*random (or stochastic) process*".

The above example is typical of an enormously wide range of observed phenomena which involve random (or stochastic) elements. Generally, a time-varying (or space-varying) quantity, $X(t)$, is called a "random process" if the situation is such that, for each t, we cannot determine theoretically a precise value for $X(t)$, but have instead a range of possible values with an associated probability distribution describing the relative likeliness of each possible value. For each individual value of t, $X(t)$ is a "random variable", and intuitively we may think of the complete process simply as a "random function". (In fact, some authors use the term "random function" as a technical expression in place of the more customary "random (or stochastic) process".)

Suppose now that $X(t)$ arises from an experiment which may be repeated under identical conditions. The first time we perform the experiment we will obtain a record of the observed value of $X(t)$, plotted as a function of t. Due to the random character of $X(t)$, the next time we perform the experiment we will almost certainly obtain a different record of observed values. Similarly, the third performance will yield yet another record of observed values, and so on. We must therefore bear in mind that *an observed record of a random process is merely one record out of a whole collection of possible records which we might have observed*. The collection of all possible records is called the "*ensemble*", and each particular record is called a "*realization*" of the process. A plot of several realizations (Fig. 1.8) illustrates clearly the

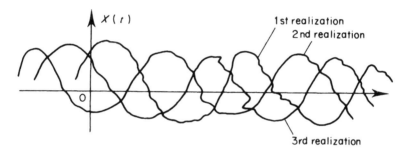

Fig. 1.8. Realizations of a random process.

point made above, namely, that for each value of t, there exists a whole range of possible values of $X(t)$. In each realization, $X(t)$ assumes one of its possible values.

In the example, on the thickness of yarn, a particular length of yarn corresponds to *one* realization; the ensemble consists of the (theoretically infinite) collection of all lengths of yarn manufactured by the process under study.

By contrast, let us consider an experiment whose outcome may be described by a deterministic function. Consider, for example, the path of a projectile moving under gravity (Fig. 1.9). As is well known, ideally the path of the projectile will be a simple parabola, which is easily described by a simple mathematical formula. Moreover, if we continue to repeat this experiment under identical conditions (i.e. with the same initial velocity and same inclination to the horizontal plane) the paths of all the other projectiles will be exactly the same as that of the first. In other words, for a "deterministic" experiment *all realizations are identical,* and we can express, for example, the relationship between height and distance by a deterministic function which associates a unique value of "height" to each value of "distance".

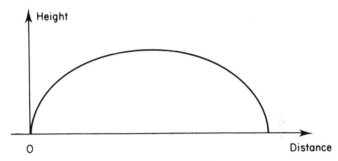

Fig. 1.9. Path of a projectile.

In the real world, the random character of most of the processes which we study arise as a result of one or more of the following situations:

(a) The physical system generating the process may itself possess "inherent" random elements, for example the random emission of electrons from the cathode of a thermionic tube (producing "noise" in an electrical current), the random emission of α-particles in radioactive material, or the Brownian motion of particles suspended in a liquid.

(b) The system may be basically deterministic but of such a complex nature that it becomes impossible for us to describe its behaviour in other than probabilistic terms as may be the case for example in econometric problems involving, say, the study of a record of the price index of a particular commodity over a period of years.

(c) Even if the system is deterministic and sufficiently simple so that we can describe it completely in terms of deterministic functions, when we *observe* the system our data may be contaminated by *errors of measurement.* By their very nature errors of measurement must be regarded as random variables, so that, in this case also, our observed records fall within the domain of random processes. As an example, consider a radio transmitter which sends out a noise-free signal. The signal is then completely deterministic, but a record of the "observed" (i.e. received) signal will, in general, contain noise due to interference in the transmission medium and due also to the imperfect nature of the receiver.

It would be futile to indulge in philosophical discussions as to which of the above factors is the major cause of "randomness" in any particular process. The important point to be noted is that *almost all quantitative phenomena occurring in science are subject to one or more of these factors, and consequently should be treated as random processes as opposed to deterministic functions.* The only case in which deterministic functions provide an adequate description is in category (c) when the errors of measurements are so small that they may be neglected—but even here the apparent determinism may disappear when the process is examined in fine detail. For example, the parabolic path of the projectile (discussed above) is based on a simplified theoretical model of the physical system. As soon as we take into account the effects of for example wind gusts and observational errors in the setting of the initial velocity, the path becomes a random process!

1.7 STATIONARY RANDOM PROCESSES

In many physical and engineering problems we encounter random processes which may be described loosely as being in a state of "statistical equilibrium". That is, if we take any realization of such a process and divide it up into a number of time intervals, the various sections of the realization look "pretty much" the same. We express this type of behaviour more precisely by saying that, in such cases, *the statistical properties of the process do not change over time, i.e. they are the same at all time points.* Random processes which possess this property are called *"stationary"*, and all processes which do not possess this property are called *"non-stationary"* (or sometimes *"evolutionary"*).

Stationary processes generally arise from any "stable" system which has achieved a "steady-state", mode of operation, and have dominated the literature on physical and engineering applications of random processes. Moreover, until recently, almost the whole of the theory of spectral analysis dealt only with processes in this category. On the other hand, non-stationary processes, such as e.g. those which exhibit a steady "trend" (increasing or decreasing with time) generally arise in the study of "unstable" systems, such as certain types of econometric models.

The mathematical theory of stationary processes is now well established, but non-stationary processes present a much more formidable mathematical problem (see Chapter 11). The basic point is that in postulating stationarity we are, in fact, making a very strong assumption regarding the structure of the process. Once this assumption is dropped, the process can become non-stationary in many different ways. This feature is in some ways similar to the situation in the study of differential equations. The theory of linear differential equations is well known, but once the assumption of linearity is removed we open the door to a large variety of modes of non-linear behaviour.

1.8 SPECTRAL ANALYSIS OF STATIONARY RANDOM PROCESSES

When we attempt to apply the ideas of spectral analysis to random processes we at once face the difficulty that the classical techniques of Fourier series and Fourier integrals apply only to deterministic functions, and cannot be applied immediately to random processes. In order to overcome this difficulty we consider, first of all, just one *realization* of the process. In each realization the process assumes a definite value at each point t, and thus each separate realization is free from random elements.

Suppose now that we are dealing with a *stationary* random process, $X(t)$. We cannot express each realization as a *Fourier series* since it will not, in general, be a periodic function of t (except in the special case of "harmonic processes"; see Chapter 3, Section 3.5.8). Similarly, we cannot express each realization as a *Fourier integral* since, by the very nature of the stationary property, it cannot be true that all realizations will decay to zero as $t \to \pm\infty$. (If this were true, clearly the process could not be stationary since its behaviour for large values of t would look quite different from its behaviour for small t.)

At this stage it might appear that we have reached an impasse, but most fortuitously, Wiener's theory of "generalized harmonic analysis" (discussed in Section 1.4) comes to our rescue. For, it turns out that each realization of a stationary process is just the type of "steady-state" function considered by Wiener. We thus have one of the most fundamental results in the theory of stationary processes, namely, that *each realization can be expressed as a Fourier–Stieltjes transform, of the form,*

$$X(t) = \int_{-\infty}^{\infty} e^{it\omega} \, dZ(\omega), \tag{1.8.1}$$

where $Z(\omega)$ is not differentiable, but has the same form as the function $P(\omega)$ discussed in Section 1.4. That is we have,

$$dZ(\omega) = \mathrm{O}(\sqrt{d\omega}) \tag{1.8.2}$$

(compare equation (1.4.2)). Owing to the peculiar nature of $dZ(\omega)$ it is difficult to visualize the form of the amplitudes $|dZ(\omega)|$ of the various frequency components in (1.8.1). However, we recall that in spectral analysis our main interest lies in the "energy distribution" of $X(t)$, i.e. we are interested in the *squared amplitudes*, $|dZ(\omega)|^2$, rather than in $|dZ(\omega)|$. According to (1.8.2) we would expect $|dZ(\omega)|^2$ to be of the order of magnitude of $(d\omega)$, so that if we now define the function $q(\omega)$ by

$$q(\omega) = |dZ(\omega)|^2/d\omega, \qquad (1.8.3)$$

we may expect $q(\omega)$ to be a reasonably "well behaved" function of ω. The function $q(\omega)$ represents, roughly speaking, the *density* of energy per unit frequency, and may thus be used as a means of describing the energy/frequency properties of $X(t)$.

So far, we have proceeded by analysing each realization of the random process separately. However, the function $Z(\omega)$ will, of course, change from realization to realization. In other words, different realizations produce different versions of $Z(\omega)$, and corresponding to the ensemble of records of $X(t)$ we have a similar ensemble of records of $Z(\omega)$. *Thus, $Z(\omega)$ (regarded as a function of ω) is itself a random process, and for each ω, $dZ(\omega)$ is a random variable.* Consequently, the function $q(\omega)$ describes the energy/frequency properties only of the particular realization from which we started, and it tells us nothing about the properties of all the other realizations. In order to construct a function which describes the properties of whole process (as opposed to merely one realization) *we now take the average value of $q(\omega)$ (for each ω) over all realizations.* We thus define a new function, $h(\omega)$, by

$$h(\omega) = \overline{q(\omega)} = \overline{|dZ(\omega)|^2}/d\omega, \qquad (1.8.4)$$

where the overbar denotes an average taken over all realizations of the process. The function $h(\omega)$ is caled the *power spectral density function* of the process, and as we shall see later, it plays a fundamental role in the analysis of stationary random processes.

We may summarize the above discussion by noting that there are two key stages in the theory of spectral analysis of stationary processes, namely,

(1) the transition from Fourier series and Fourier integrals to *generalized Fourier integrals* of the Wiener type (equation (1.8.1)),

and

(2) the operation of *averaging* the energy distributions of the individual realizations in order to obtain a "spectral" function which is characteristic of the process as a whole.

A full mathematical development of the ideas which we have sketched above requires rather careful and delicate treatment, as does the study of the precise physical interpretation of the function $h(\omega)$, given by equation

(1.8.4). These tasks will be undertaken when we return to the topic of spectral analysis with a much more detailed analysis in Chapter 4.

1.9 SPECTRAL ANALYSIS OF NON-STATIONARY RANDOM PROCESSES

We pointed out in the previous section that virtually the whole of the theory of spectral analysis of random processes applies only to stationary process, i.e. to processes whose statistical properties do not change with time. This is clearly a severe restriction since it is doubtful whether any "real-life" processes possess this property in the strict sense. In practical applications the most we could hope for is that, over the observed time interval, the process would not depart "too far" from stationarity for the results of the subsequent analysis to be invalid. (It was in this approximate sense that we previously described processes in physics and engineering as appearing to be in a state of "statistical equilibrium".)

The traditional approach to dealing with non-stationary processes has been to try to "transform" the process into a stationary form. For example, if the non-stationarity is due to the presence of a "trend", one might, perhaps, attempt to remove this trend either by fitting, say, a polynomial to the trend and then subtracting this function from the data, or by differencing the data a sufficient number of times (which will also remove a polynomial trend). This approach may be quite useful in certain cases, but there are, nevertheless, many processes which possess non-stationary characteristics of a much more complex nature, and for which the above technique is quite inadequate. The question arises, therefore, as to whether it might be possible to develop a more general theory of spectral analysis which would include non-stationary processes of a fairly general type. Recent work in this area has shown that such a generalization is indeed possible, and a new form of spectral analysis has been developed which, while not accommodating *all* non-stationary processes, does however enable us to treat a fairly large class of such processes in a unified theory which includes stationary processes as a special case. This new development is based on a further generalization of Wiener's Fourier integral (equation (1.8.1)), and introduces the interesting notion of *time-dependent spectra*, i.e. energy distributions which change from one time instant to the next. A study of the physical interpretation of these spectra involves a fundamental reappraisal of the concepts of "energy" and "frequency", and gives us a deeper insight into the (now) classical spectral theory of stationary processes. We discuss these recent developments in Chapter 11.

1.10 TIME SERIES ANALYSIS: USE OF SPECTRAL ANALYSIS IN PRACTICE

We remarked in Section 1.2 that the term *"time-series"* could logically be applied to a record of any time varying process, either deterministic or random. However, in the statistical literature this term is used exclusively in the context of random processes, and consequently the term *"time series analysis"* refers to that body of principles and techniques which deal with the analysis of observed data from random processes in which the parameter t denotes *time*.

In time series analysis the raw data with which we are presented usually takes the form of a finite number of observations on a random process $X(t)$, say for $t = 1, 2, \ldots, N$. In effect, we are given just a portion of a single realization, and from these data we try to infer as much as we can about the properties of the whole process. In the succeeding chapters we will explain how spectral analysis can provide an extremely powerful tool in the analysis of time series data. The most important applications of spectral analysis may be summarized as follows.

I. *Direct physical application*

In discussing the basic ideas of spectral analysis we have stressed throughout that the spectrum has an immediate physical interpretation as an energy/frequency distribution. In physical and engineering applications this information is, in itself, of considerable importance. For example, a mechanical structure may have to be designed to withstand the effects of certain types of random vibrations. If the structure has a number of "resonant frequencies" then it is obviously important to design it so that, if possible, these resonant frequencies are not "excited" by the random driving force. Thus, we would try to design the structure so that its resonant frequencies fall in a region where the spectrum of the random vibrations has a very low energy content. For this purpose, it is crucial to know the form of the "vibration" spectrum. (A particular example of this type of problem arises in the design of aircraft wings in relation to the effects of atmospheric turbulence.)

II. *Use in statistical model fitting*

When we are studying a process which does not represent some physical phenomenon (such as in econometrics) the spectrum no longer possesses a direct physical interpretation, but we often try to gain some insight into the probabilistic structure of the process by attempting to describe its behaviour in terms of a *statistical model*. Indeed, the problem of fitting a statistical model arises in many different types of problems (irrespective of whether or not the process is physical in origin), and one of its most important applications occurs in the analysis of certain types of stochastic control systems. Here, it turns out that the statistical model underlying the "random disturbances" which infect the control systems plays a crucial rôle in determining the optimal form of a feedback controller (see Chapter 10).

When we try to fit a statistical model to a process we usually begin by looking through a "dictionary" of standard models to see if we can find one which fits our observed data reasonably well. In compiling our dictionary we would naturally include in our list those models which, on the basis of past experience, have been found to occur most often in practical problems. Thus, in both the statistical and engineering literature one finds a wide variety of gneral types of models, the most well known being the so-called *autoregressive*, *moving-average*, and *mixed autoregressive/moving-average* schemes. These general statistical models will be discussed fully in Chapter 3, but for the moment it might be useful to mention that an *autoregressive* process $\{X_t\}$, is, for discrete time, described by an equation of the form

$$X_t + a_1 X_{t-1} + \ldots + a_k X_{t-k} = \varepsilon_t, \tag{1.10.1}$$

where $a_1, \ldots, a_k$ are constants, and ε_t is a "purely random" (or "white noise") process, whereas a *moving-average* process is described by

$$X_t = \varepsilon_t + b_1 \varepsilon_{t-1} + \ldots + b_l \varepsilon_{t-l}, \tag{1.10.2}$$

where, again, ε_t is a purely random process, and $b_1, \ldots, b_l$ are constants. The *mixed autoregressive/moving-average* model is obtained by combining both of the above equations, leading to a model of the form

$$X_t + a_1 X_{t-1} + \ldots + a_k X_{t-k} = \varepsilon_t + b_1 \varepsilon_{t-1} + \ldots + b_l \varepsilon_{t-l}. \tag{1.10.3}$$

Given a set of observations, the first step in the model fitting procedure is usually to "identify" the general type of model which would be most appropriate. Here, spectral analysis can be a very useful tool. For, as will be seen later, each type of model gives rise to a spectrum which has a characteristic shape—thus, loosely speaking, the spectrum of a typical autoregressive model will have a different general shape from that of a typical moving-average model. If, therefore, we can estimate the shape of the spectrum *directly from the observations* without assuming that the process conforms to a particular type of model (and this can, in fact, be done, see Chapter 6), then the spectral shape will provide a very useful indication of the type of model which should be fitted to the data. Crudely speaking, this approach may be described as a "spectral identikit" since it is effectively equivalent to compiling a dictionary of standard "spectral shapes", and then looking for that standard shape which best matches the form of the spectrum as estimated from the data.

However, when we have identified the general form of the model we are still faced with the formidable problem of *estimating the numerical values of the "parameters"* (i.e. the constants) which arise in the model. Again, spectral analysis could be used also in connection with this aspect of model fitting, one approach being to "adjust" the values of the parameters in the model until the spectrum as derived from the model matches the spectrum as estimated from the observations as closely as possible. This is quite a feasible method, and iterative techniques based on this approach have been developed: see Chapter 5, Section 5.4.3. However, it should be pointed out that, at least for models involving a fairly small number of parameters, there are more efficient techniques for parameter estimation based on the well established methods of "maximum likelihood" and "least squares" applied directly to the given observations. There are also alternative methods of identifying the general form of the model based on so-called "time-domain" techniques which rely on examining the form of the "autocorrelation" and "partial autocorrelation functions". See Chapter 5, Section 5.4.5.

III. *Estimation of transfer functions*

Another area where spectral analysis has important applications is the study of the behaviour of "linear time invariant" systems. Such systems occur in a multitude of practical problems, particularly in problems connected with electrical circuits, such as a simple L–R–C circuit. The basic set-up may be described in simple terms as follows. We have a device (sometimes referred to as a "black box") with two terminals, labelled "input" and "output" as shown in Fig. 1.10. Depending on the physical nature of the

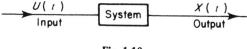

Fig. 1.10

system, the input and output may be electrical currents, voltages, mechanical displacements, rates of flow of liquids, etc., but whatever the nature of the system, if we assume that it behaves in a *linear time-invariant fashion* then this means that, at any time instant, the current value of the output process is *a linear combination of present and past values of the input process*. Thus, if we denote the values (at time t) of the input process by $U(t)$, and the output process by $X(t)$, then according to the above assumption we may express the relationship between $U(t)$ and $X(t)$ in the form

$$X(t) = \int_0^\infty g(s)U(t-s)\,ds, \tag{1.10.4}$$

where $g(s)$ is some function which does not vary with time but is determined by the properties of the particular system which we are studying. Thus, for example, if we have an R–C circuit of the form of Fig. 1.11 then the input

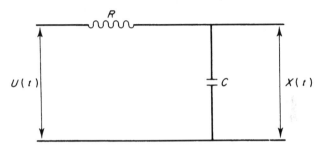

Fig. 1.11

voltage, $U(t)$, and the output voltage, $X(t)$, are related by the well known first-order linear differential equation,

$$RC\frac{dX(t)}{dt} + X(t) = U(t). \tag{1.10.5}$$

The solution of equation (1.10.5) is easily shown to be

$$X(t) = \frac{1}{RC}\int_0^\infty e^{-s/RC}U(t-s)\,ds, \tag{1.10.6}$$

which is of the form (1.10.4) if we set

$$g(s) = \frac{1}{RC}\,e^{-s/RC}. \tag{1.10.7}$$

(The quantity $T = RC$ is sometimes referred to as the "time constant" of the circuit.)

Consider now the general linear relationship as given by equation (1.10.4). If, in particular, we choose $U(t)$ to be a sine wave with frequency ω, i.e. we set $U(t) = A\exp(i\omega t)$ (using the complex form), then we find immediately that

$$X(t) = \int_0^\infty g(s)A\,e^{i\omega(t-s)}\,ds,$$

or

$$X(t) = A \cdot \Gamma(\omega) \cdot e^{i\omega t} \qquad (1.10.8)$$

where

$$\Gamma(\omega) = \int_0^\infty g(s) \, e^{-i\omega s} \, ds. \qquad (1.10.9)$$

Equation (1.10.8) illustrates a basic result in the theory of linear systems, namely, that *if the input is a sine wave the output is also a sine wave with exactly the same frequency but with modified amplitude (and phase), i.e. the original input amplitude $|A|$ becomes $\{|A \cdot \Gamma(\omega)|\}$ at the output end.*

This result clearly holds no matter what value we assign to ω, so that if the input $U(t)$ consists of a sum of sine waves with different frequencies, the output will similarly consist of a sum of sine waves with exactly the same frequencies but the amplitude (and phase) of each term will be modified as described above. If we now allow the frequency variable ω to take, in turn, all values in the range $(-\infty, \infty)$, then we may think of equation (1.10.9) as defining a complex valued *function*, $\Gamma(\omega)$, which tells us how, for each ω, the amplitude of an input sine wave of that frequency is transformed into the amplitude of the corresponding output sine wave. (For the moment, we assume that $g(s)$ is sufficiently well behaved to ensure that $\Gamma(\omega)$ exists for all ω.) The function $\{\Gamma(\omega)\}$ is then called *the transfer function of the system.*

We may also give the function $\{g(s)\}$ a physical interpretation by considering what happens to the output when we choose the input to have the form of a unit "impulse" function, that is a function which takes a very large value at just one time point, is almost zero everywhere else, and has unit area. More precisely, let $U(t)$ have the form of a Dirac δ-function, centred on the origin of the time axis, so that we may write $U(t) = \delta(t)$. Then from (1.10.4) we find that

$$X(t) = \int_0^\infty g(s)\delta(t-s) \, ds,$$

or

$$X(t) = g(t), \qquad \text{all } t(>0), \qquad (1.10.10)$$

(using the well known features of the δ-function that $\delta(t) = 0$, $t \neq 0$, and $\int_{-\varepsilon}^{\varepsilon} \delta(s) \, ds = 1$, for any $\varepsilon > 0$).

Thus the function $g(s)$ describes the form of the output of the system when the input is a unit impulse function centred on the origin. The function $g(s)$ is therefore called the *impulse response function of the system.*

It will be seen from (1.10.9) that $\Gamma(\omega)$ is the *Fourier transform* of $g(s)$. Consequently, each function uniquely determines the other, and either function may be used to describe the behaviour of the system. However, in many physical and engineering problems it turns out that the transfer function $\Gamma(\omega)$ is more useful than the impulse response function in characterizing those aspects of the system which are of prime interest. (If we know, for example, that $|\Gamma(\omega)|$ is "small" at low frequencies and "large" at high frequencies then this tells us immediately that the system will attenuate the low-frequency components in the input and accentuate the high-frequency components.) For simple systems, such as the R–C circuit described above, we may, of course, calculate the function $\Gamma(\omega)$ from purely theoretical considerations, but for larger and more complex systems this approach would, in general, be far too difficult, if not impossible. Instead, we have to resort to more empirical methods of determining the form of $\Gamma(\omega)$ which are based on the analysis of "operating records" of the system. Here, we operate the system for a finite length of time and thereby obtain sample records of the input $U(t)$ and the corresponding output $X(t)$. At this stage we may use the methods of spectral analysis to enable us to estimate the form of $\Gamma(\omega)$ from the observed data. To understand why spectral analysis is relevant to this problem, let us first recall that, for each frequency, the power spectrum is, roughly speaking, proportional to the *squared modulus* of the amplitude of that frequency component. Suppose now that both $U(t)$ and $X(t)$ are stationary random processes. In virtue of equation (1.10.8), we would expect that the squared modulus of the amplitude of each component in $X(t)$ was equal to the squared modulus of the corresponding amplitude in $U(t)$ multiplied by $|\Gamma(\omega)|^2$. We are thus led to the basic result that,

{power spectrum of the output}
= {power spectrum of the input}
× {squared modulus of the transfer function}.

(This result may be established quite rigorously; see Chapter 4, Section 4.12.) If now we are given sample records of $U(t)$ and $X(t)$ we may obtain information on the transfer function by first estimating the power spectral density functions of the two processes. Let us denote these estimated spectra by $\hat{h}_U(\omega)$, $\hat{h}_X(\omega)$. Then, by the above result, an obvious estimate of $|\Gamma(\omega)|^2$ is given by

$$|\hat{\Gamma}(\omega)|^2 = \hat{h}_X(\omega)/\hat{h}_U(\omega). \qquad (1.10.11)$$

Note, however, that this approach enables us to estimate only $|\Gamma(\omega)|^2$, rather than $\Gamma(\omega)$ itself. To obtain information on the complete function we have to introduce the notion of "cross-power spectra", and study the "cross-spectral" properties between $U(t)$ and $X(t)$. This more general approach allows us to estimate transfer functions even in cases where the observed output contains an additive "noise" disturbance, so that (1.10.4) then becomes

$$X(t) = \int_0^\infty g(s)U(t-s)\,ds + N(t),$$

where $N(t)$ denotes the noise term. This model is discussed in more detail in Chapter 9.

The method described in this section is often referred to as the "non-parametric" approach in the sense that it does not require any assumptions about the system, other than that it is linear and time invariant. Consequently, it is a suitable procedure for dealing with those cases where one has virtually no background information on the behaviour of the system under study. There are, however, other situations where one's knowledge of the physics of the system would suggest a general mathematical form for the transfer function, $\Gamma(\omega)$, but would not be sufficient to determine $\Gamma(\omega)$ in precise detail. In other words, we might be led to consider an expression for $\Gamma(\omega)$ which contained unknown "parameters" (i.e. constants). In such cases we would not usually adopt the non-parametric method (although it would still be valid) since, in general, it would be more efficient to estimate the unknown parameters by specially devised techniques. One particular case which has been studied extensively is where $\Gamma(\omega)$ may be assumed to be a *rational function* of ω (i.e. a ratio of two polynomials). Here, the unknown parameters are the coefficients of both the numerator and denominator polynomials, together with the orders of two polynomials (which usually are also unknown). In principle, the coefficients of the polynomials may be estimated by "time domain" regression analysis, but in practice it may be found useful here to combine time domain techniques together with cross-spectral analysis in order to determine appropriate values for the orders of the polynomials. We return to the discussion of this problem in Chapter 10.

IV. *Prediction and filtering problems*

One of the most important problems in the study of random processes is that of "predicting" a future value of a process, given a record of its past values. This problem is clearly of interest in the context of economic systems where, for example, one might wish to predict future values of stock market prices, sales of particular commodities, consumption of raw materials, or population growth. It is nevertheless, equally important in the analysis of many types of physical systems, and the pioneering work of Wiener on "prediction theory" was stimulated by the problem of predicting the future path of an aircraft in connection with the control of anti-aircraft guns (or "fire-control" as it has come to be known). More recently, it has been shown that prediction theory plays a fundamental rôle in the theory of stochastic control systems (see Chapter 10).

Basically, the prediction problem may be posed as follows: we are given the observed values of a random process at all past time points and wish to predict the value it will assume at some specific future time point. Thus, we are given observations up to time t, i.e. we are given the values of $\{X(s);$ $s \leq t\}$, and wish to predict the value of $X(t+\tau)$, say, where $\tau > 0$. This situation is illustrated diagrammatically in Fig. 1.12. One of the fascinating

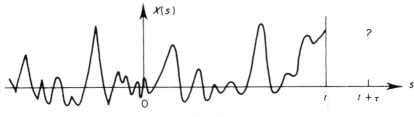

Fig. 1.12

features of this problem is the way in which the solution becomes simpler the more general the formulation! Clearly, in predicting the value of $X(t+\tau)$ we would wish to make use of the given information. In other words, we would calculate our predicted value as a "function" of the observations, $X(s)$, $s \leq t$, but some "functions" will produce more "accurate" predictors than others. To determine which function of the observations will give the most "accurate" predictor we must first decide how we will measure the accuracy of a predictor. The usual measure used in prediction theory is based on the *mean-square error* of the predictor, i.e., if we denote the predicted value of $X(t+\tau)$ by $\tilde{X}(t+\tau)$, then we measure the accuracy of $\tilde{X}(t+\tau)$ in terms of the *smallness* of the quantity

$$\mathcal{M} = \overline{\{X(t+\tau) - \tilde{X}(t+\tau)\}^2},$$

where the overbar again denotes an average taken over all realizations of the process.

Thus, the problem now is to find that "function" of the given observations which, when used as a predictor of $X(t+\tau)$, makes the value of $\mathcal{M}$ as small as possible. If we allow the free choice of *any* function, then the solution becomes almost trivial, and it is well known that the value of $\mathcal{M}$ is minimized by choosing $\tilde{X}(t+\tau)$ as the *conditional mean* of $X(t+\tau)$, given the values of $\{X(s); s \leq t\}$. However, this solution is of little use in practice since we would not be able to compute the conditional mean of $X(t+\tau)$ unless we had a very detailed knowledge of the probabilistic structure of the process. We therefore restrict ourselves to considering "simple" functions of the observations, and the classical theory developed by Wiener and Komogorov deals only with *linear* functions. (It may be shown that if $X(t)$ is what is termed a "Gaussian process", then the conditional mean of $X(t+\tau)$ is a linear function of $\{X(s); s \leq t\}$, so that in this case there is no loss of generality in restricting attention to linear functions—see Chapter 10.) To discuss this approach in a little more detail, it is simpler to consider a *discrete parameter process*, X_t, i.e. a process which is observed only at discrete values of t, say $t = 0, \pm 1, \pm 2, \ldots$. The given information now takes the form of a *sequence* of observed values, terminating at time t,

$$\ldots, X_{t-2}, X_{t-1}, X_t,$$

and a *linear* predictor of $X_{t+\tau}$ will be given by an expression of the form

$$\tilde{X}_{t+\tau} = \sum_{s=0}^{\infty} a_s X_{t-s}, \tag{1.10.12}$$

where $\{a_s\}$ is some sequence of constants which represent the relative "weighting" given to values $X_t, X_{t-1}, X_{t-2}, \ldots$. (In most cases we would clearly wish to give more weight to "fairly recent" values such as X_t, X_{t-1}, than to values observed in the "remote past" such as, e.g., X_{t-100}, X_{t-101}; we would expect therefore that the sequence $\{a_s\}$ would decrease in value as s increases.) The basic problem now is to *choose the sequence $\{a_s\}$ so as to minimize the value of $\mathcal{M}$*. This celebrated mathematical problem was first solved (almost simultaneously) by both Wiener and Kolmogorov, using related but interestingly different mathematical techniques. A more detailed account of their work (together with the work of other authors) is given in Chapter 10, but for the moment the essential point to be noted is that *the optimal form of $\{a_s\}$ is determined uniquely by the power spectral density function of $\{X_t\}$*. Thus, a knowledge of the power spectrum of $\{X_t\}$ is important if we wish to compute optimal predictors for random processes.

The problem of *linear filtering* is a more general version of the prediction problem and arises when we are unable to make accurate observations on $\{X_t\}$ directly but observe instead the process $\{Y_t\}$, where

$$Y_t = X_t + N_t,$$

$\{N_t\}$ being a "noise" disturbance. The problem now becomes: given a record of past values of $\{Y_t\}$, say $Y_t, Y_{t-1}, Y_{t-2}, \ldots$, construct a *linear filter* of the form

$$\sum_{s=0}^{\infty} b_s Y_{t-s}$$

which gives the best approximation to $X_{t+\tau}$ in the sense that the mean-square error is minimized. Note that here τ may be positive, negative, or zero, since it might well be of interest to "estimate" the (unobserved) value of X_t corresponding to a "past", "present", or "future" time point. This problem may be regarded as that of "filtering out" the noise disturbance, $\{N_t\}$, so as to reveal the underlying process, $\{X_t\}$.

Again, it turns out that the optimal choice of the sequence $\{b_s\}$ depends entirely on the spectral and cross-spectral properties of the processes $\{X_t\}$ and $\{N_t\}$. For a further discussion of filtering theory see Chapter 10.

In this chapter we have attempted to describe the main ideas underlying spectral analysis together with the range of problems to which this method is applicable. In the succeeding chapters we explore these ideas in a more detailed and systematic way, starting in Chapter 3 with the basic theory of random processes. To facilitate the discussion in Chapter 3 we first briefly review in Chapter 2 some of the elements of probability theory which form an essential framework for our subsequent study.

Chapter 2

Elements of Probability Theory

2.1 INTRODUCTION

The reader will have observed the numerous references in Chapter 1 to "randomness" and "chance phenomena". These concepts play a fundamental role in the subject of spectral analysis, but in order to deal with them in a proper scientific manner we need a mathematical theory which treats "randomness" in a quantitative way. This branch of mathematics is called *probability theory*, and in this chapter we review some of its main ideas and results. We should point out that the account given here is, of necessity, rather brief, and readers who are completely new to the subject would be well advised to refer to one of the standard texts such as Parzen (1960) or Meyer (1965) for a more comprehensive treatment. On the other hand, those who are already familiar with basic probability theory may certainly omit this chapter and proceed directly to Chapter 3, possibly referring back for references to particular results.

2.2 SOME TERMINOLOGY

Before we can give a precise meaning to the term "probability" we must first describe carefully the situations in which this term is to be applied. We therefore introduce the following terminology.

(a) *An experiment* $\mathscr{E}$ is the operation of establishing certain conditions which may produce one of several possible outcomes or results. This use of the term "experiment" is more general than its customary interpretation, and whilst it certainly includes what would normally be regarded as "experiments" (such as measuring pressures, temperatures, currents, voltages, etc.), it includes also less exciting activities such as spinning coins, throwing dice, or drawing cards from packs.

(b) *The sample space* Ω associated with an experiment $\mathscr{E}$ is the collection of *all* possible outcomes of $\mathscr{E}$. If, for example, $\mathscr{E}_1$ denotes the experiment of throwing a six-sided die, then the corresponding sample space, Ω_1, consists of the six possible scores, and this is usually written in the form $\Omega_1 = \{1, 2, 3, 4, 5, 6\}$. If $\mathscr{E}_2$ denotes the experiment of measuring, say, the value of a resistor, then (in principle) the outcome could be any positive number, so that here the sample space would be written as $\Omega_2 = \{x; 0 \leq x < \infty\}$.

(c) *An event E* is a collection of *some* of the possible outcomes of $\mathscr{E}$. In the language of set theory we would call the sample space, Ω, the *set* of all possible outcomes, and then an event, E, is a *subset* of Ω. In particular, we may regard each individual outcome as an event, since each outcome is a subset of Ω. (These are sometimes called "elementary events".) Also, the *null set* $\varnothing$ (i.e. the set which consists of no outcomes) and the complete sample space Ω are each special cases of subsets of Ω, and thus both $\varnothing$ and Ω may be regarded as events.

For the experiment $\mathscr{E}_1$ described above we might consider the following as typical events;

E_1: score is 1,
E_2: score is even   (i.e. 2, 4, or 6),
E_3: score is less than 5   (i.e. 1, 2, 3, or 4).

(d) We say that an event E has *occurred* if the outcome of the experiment belongs to the set E. For example, if the outcome of $\mathscr{E}_1$ was a score of 6 then E_2 occurred, but E_1 did not occur and E_3 did not occur.

(e) Given two subsets of Ω, say E_1, E_2, the *union* of E_1, E_2, written as $E_1 \cup E_2$, is defined as the set of outcomes which belong to *either* E_1 or E_2 or both. Thus, the *event* $(E_1 \cup E_2)$ occurs whenever E_1 or E_2 occurs (or both occur).

(f) The *intersection* of E_1, E_2, written as $E_1 . E_2$, is defined as the set of outcomes which belong to *both* E_1 and E_2. The event $(E_1 . E_2)$ occurs when both E_1 and E_2 occur.

(g) E_1 and E_2 are called *mutually exclusive* events (m.e.) if both E_1 and E_2 cannot occur simultaneously, i.e. if $E_1 . E_2 = \varnothing$ (so that the intersection contains no outcomes).

(h) A sequence of events, $E_1, E_2, \ldots, E_n$, is said to be *exhaustive* if at least one of these events *must occur*. In this case, $E_1, \ldots, E_n$, together account for all possible outcomes, so that $E_1 \cup E_2 \cup \ldots \cup E_n = \Omega$. (The definition of union may be extended in an obvious way to apply to any number of sets.)

(i)   Given an event E, the *complimentary event*, $\bar{E}$, is defined as that subset of Ω which contains all the outcomes *which do not belong to E*. Thus, the event $\bar{E}$ occurs whenever E does not occur, and vice-versa.

Some of the above definitions are illustrated diagrammatically in Fig. 2.1.

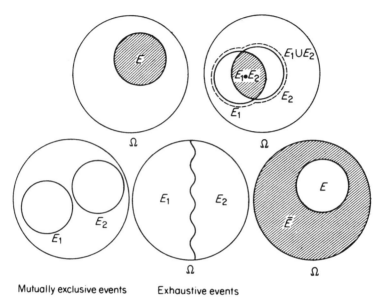

Mutually exclusive events   Exhaustive events

Fig. 2.1

2.3 DEFINITION OF PROBABILITY

The subject of probability theory is concerned with those experiments which involve "random phenomena", i.e. experiments whose outcomes cannot be predicted with certainty. Although we cannot say definitely then whether or not a particular event will occur, we may have reason to believe that some events are "more likely" to occur than others. The question now arises: how likely is the occurrence of a particular event? We attempt to answer this question by associating "probabilities" with each event.

Let E be an event associated with an experiment $\mathscr{E}$. In the *empirical approach* to probability we would define the probability of E by the following procedure. Repeat $\mathscr{E}$ n times, and let m denote the number of occurrences of E in these n repetitions. The ratio (m/n) is called the *relative frequency of occurrence* of E, and as $n \to \infty$ we observe empirically that (m/n) tends to a limiting value, p_0, say. The quantity p_0 is then called the *probability of the event E*, and we write,

$$p(E) = p_0.$$

Unfortunately, this definition is open to objections on the grounds of mathematical rigour. In particular, we must note that the limiting operation referred to above is *not* the same as that used in mathematical analysis in the sense that, given $\varepsilon (>0)$, we *cannot* find a value n_0 which is such that $|(m/n) - p_0| < \varepsilon$ for all $n \geqslant n_0$. The reason for this is simply that when we are studying random phenomena, m (the number of occurrences of the event E) is itself a random quantity, and whilst we feel sure that (m/n) will eventually settle down to a limiting value we cannot be absolutely sure that this limiting value will be approached within a certain limit after a given number of repetitions. (If we spin a coin repeatedly, it could happen that the first 100, 1000, or even 1 000 000 spins all resulted in "heads" even though the coin was a perfectly fair one.) In fact, the only way in which we could describe this limiting operation more precisely would be to say something like:

"$|(m/n) - p_0|$ will 'very probably' be less than ε for all $n \geqslant n_0$",

but this statement itself already involves the notion of probability!

In practice, however, we almost always interpret probabilities in terms of relative frequencies, and if we were told that a certain event E had a probability of 0·5 we would understand this to mean that in a "long run" of experiments E would occur on approximately half the repetitions. We would like, therefore, to preserve this interpretation of probability, and to overcome the difficulties noted above we adopt the mathematical device of defining probabilities *axiomatically*. That is, we do not state explicitly how probabilities are to be calculated but merely assert that *such quantities exist and satisfy certain laws or "axioms"*. Bearing in mind what we have said above, *we naturally choose as our axioms precisely those properties which hold for relative frequencies*. In this way we retain our physical interpretation of probability but avoid the technical deficiencies of the relative frequency definition.

2.3.1 Axiomatic Approach to Probability

Consider an experiment $\mathscr{E}$ with sample space Ω. To every event E we assign a real number $p(E)$, called the probability of E, which satisfies the following axioms:

Axiom 1 *For every event E*,

$$0 \leqslant p(E) \leqslant 1.$$

Axiom 2 *For the special event* Ω,

$$p(\Omega) = 1.$$

Axiom 3 *If* E_1, E_2 *are any two mutually exclusive events, then*

$$p(E_1 \cup E_2) = p(E_1) + p(E_2).$$

(Axiom 3 is sometimes called the "addition law".) It is easy to deduce from Axiom 3 that if E_1, E_2, . . . , E_n is a finite sequence of mutually exclusive events (i.e. E_i, E_j are mutually exclusive, for all $i \neq j$) then

$$p(E_1 \cup E_2, \ldots, \cup E_n) = p(E_1) + p(E_2) + \cdots + p(E_n).$$

The corresponding result for a countably infinite sequence of mutually exclusive events does not, however, follow from Axiom 3, and in order to deal with sample spaces which contain an infinite number of elements we must extend Axiom 3 to:

Axiom 3′ *If* E_1, E_2, . . . , E_n, . . . , *is a sequence of mutually exclusive events then*

$$p\left(\bigcup_{i=1}^{\infty} E_i\right) = \sum_{i=1}^{\infty} p(E_i).$$

The justification for the above axioms is easily seen by examining the corresponding properties for relative frequencies; thus we have,

Property 1 *For every* E, $\quad 0 \leqslant m/n \leqslant 1$.

Property 2 *The event* Ω *occurs at every repetition of* $\mathscr{E}$. *Hence in this case* $m = n$, *and* $m/n = 1$, *all* n.

Property 3 *Repeat* $\mathscr{E}$ n *times, and suppose* E_1 *occurs* n_1 *times and* E_2 *occurs* n_2 *times. Since* E_1 *and* E_2 *are m.e., the event* $(E_1 \cup E_2)$ *occurs* $(n_1 + n_2)$ *times. Hence, the relative frequency of* $(E_1 \cup E_2)$ *is,* $(n_1 + n_2)/n = $ *relative frequency of* $E_1 + $ *relative frequency of* E_2.

It may be noted at this stage that there is a simple relationship between the probability of an event E and the probability of the complimentary event $\bar{E}$. For E and $\bar{E}$ are mutually exclusive (obviously they cannot both occur together) and exhaustive (since either E or $\bar{E}$ must occur, whatever the result of the experiment). Thus, $E \cup \bar{E} = \Omega$, and we have

$$1 = p(\Omega) = p(E \cup \bar{E}) = p(E) + p(\bar{E}), \quad \text{by Axiom 3.}$$

Hence,

$$p(\bar{E}) = 1 - p(E) \qquad \text{and} \quad p(E) = 1 - p(\bar{E}). \tag{2.3.1}$$

Since the null set $\varnothing$ may be written as $\bar{\Omega}$, we have immediately that

$$p(\varnothing) = 1 - p(\Omega) = 0.$$

For this reason, $\varnothing$ is usually called the *"impossible event"*.

2.3.2 The Classical Definition

Suppose that an experiment has a finite number N of possible outcomes, say $A_1, A_2, \ldots, A_N$, (which thus constitute the "elementary events") and we are interested in an event E which is defined by the occurrence of M outcomes, say $A_1, \ldots, A_M$, i.e. E occurs if the outcome is any one of $A_1, A_2, \ldots, A_M$, so that we may write $E = A_1 \cup A_2, \ldots, \cup A_M$. If we assume that all the outcomes $A_1, \ldots, A_N$ are *equally likely*, then it is easy to show from Axioms 1, 2 and 3, that

$$p(E) = M/N. \tag{2.3.2}$$

This is a celebrated result, and is usually expressed in the form

$$p(E) = \frac{\text{number of outcomes favourable to } E}{\text{total number of outcomes}}.$$

This is the basic result which enables theoretical calculations of probabilities to be made purely on the basis of combinatorial arguments. However, its application is strictly limited to cases where the results of the experiment can be decomposed into a finite number of equally likely outcomes—such as the well known examples of spinning coins, throwing dice, and drawing balls from urns.

Equation (2.3.2.1) was at one time proposed as a *definition* of probability (the so-called "classical" definition), but this approach is open to even stronger objections than the empirical method since it requires that
(a) the number of outcomes is finite
and
(b) all outcomes are equally likely, a statement which can be made precise only by equating "equally likely" with "equal probability". Thus, as a definition it is completely "circular" since the notion of probability is inherent in one of its prior conditions.

Nevertheless, equation (2.3.2.1) is very useful for direct calculations of probabilities, and if, e.g., we spin a "fair" coin (i.e. one for which heads and tails are equally likely) then we can say immediately that

$$p(\text{heads}) = p(\text{tails}) = \tfrac{1}{2}.$$

Similarly, if we throw a fair die, then equation (2.3.2.1) tells us that the probability of each score is $1/6$, and that, e.g.,

$$p(\text{even score}) = 3/6.$$

2.3.3 Probability Spaces

In setting up the axioms of probability theory in the previous section we asserted that we would assign a probability to *every* event. As we have previously defined an event as any subset of the sample space Ω, this means that we must be able to assign a probability to *any* subset of Ω. When Ω consists of a finite number of discrete elements, such as the space Ω_1 (Section 2.2(b)), this procedure presents no difficulties. However, if Ω is a continuous space, such as Ω_2 (Section 2.2(b)), then we may no longer be able to meet this requirement in full. The point is simply that if we have a continuous space, such as the real line, R^1, we may construct some highly pathological subsets for which it is not possible to assign probabilities in a manner consistent with Axiom 3'. Such subsets are called "non-probabilizable" (or "non-measurable") and in elementary probability theory we just ignore their existence. In the more advanced theory we surmount this difficulty by restricting attention to a certain collection $\mathscr{F}$ of "well-behaved" subsets of Ω, and $\mathscr{F}$ is then called a "*σ-field*" (or "*σ-algebra*") (see Parzen (1960), pp. 148–150, or Kingman and Taylor (1966), p. 266). We then assign probabilities *only to those subsets which belong to $\mathscr{F}$*, and the term "event" then applies only to such subsets.

For a particular experiment $\mathscr{E}$ there are thus three basic ingredients in its probabilistic description: (i) the sample space Ω; (ii) the σ-field, $\mathscr{F}$, of "probabilizable" (or "measurable") subsets of Ω; (iii) the *probability measure*, $p(E)$, which assigns a probability to each subset E belonging to $\mathscr{F}$. The triplet of quantities $(\Omega, \mathscr{F}, p)$ is called a *probability space*.

2.4 CONDITIONAL PROBABILITY AND INDEPENDENCE

If we have two events, E_1, E_2, defined on the same sample space Ω (i.e. associated with the same experiment) we may ask the question: given that E_2 has occurred, what is then the probability that E_1 has also occurred? This is called the *conditional probability of E_1, given that E_2 has occurred*, and is written as $p(E_1|E_2)$.

It should be clear that, in general, the information that E_2 has occurred will affect the probability that we assign to E_1. Obviously, if we are told that E_2 has occurred then we know that, whatever the result of the experiment, it must correspond to an element of Ω which belongs to the subset E_2, and we may therefore ignore any elements of Ω which belong to E_1 but do not belong to E_2; see Fig. 2.2. As a simple example, consider the experiment $\mathscr{E}_1$ (throwing a die) and let E_1, E_2, denote the following events,

$$E_1 : \text{score is 3,} \qquad E_2 : \text{score is odd.}$$

Given no information on the result of $\mathscr{E}_1$, we have trivially that $p(E_1) = 1/6$. On the other hand, if we are told that the score is odd then there are only three remaining possibilities (i.e. score 1, 3, or 5), and assuming the die is a fair one (so that all scores are equally likely) we would expect that $p(E_1|E_2) = 1/3$.

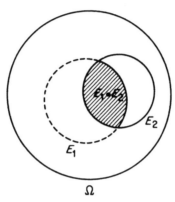

Fig. 2.2

To obtain a general formula for conditional probabilities let us consider the type of result we would expect if we go back to the "relative frequency" interpretation of probability. Suppose then that we repeat the experiment n times, and let n_2 denote the number of times E_2 occurred and m the number of times *both* E_1 and E_2 occurred (i.e. the number of times that the event $(E_1 . E_2)$ occurred). Then, the *conditional relative frequency* of E_1, given that E_2 occurred, is given by

$$\frac{m}{n_2} = \frac{m/n}{n_2/n} = \frac{\text{relative frequency of } (E_1 . E_2)}{\text{relative frequency of } (E_2)}. \qquad (2.4.1)$$

We would expect a similar result to hold for conditional probabilities, and accordingly we now *define* conditional probability by

$$p(E_1|E_2) = \frac{p(E_1 . E_2)}{p(E_2)}, \qquad (2.4.2)$$

provided, of course, that $p(E_2) > 0$. (If $p(E_2) = 0$ then effectively E_2 is an "impossible" event, and there would be little interest in considering probabilities which are conditional on the occurrence of an "impossible" event;

see, however, the section on conditional probability distributions, Section 2.12.13.)

Note that (2.4.1) is a *definition*, but in view of (2.4.1) we may interpret $p(E_1|E_2)$ (so defined) in the manner described at the beginning of this section.

Note also that if E_1 and E_2 are *mutually exclusive* events then clearly $p(E_1 . E_2) = 0$ (since the event $(E_1 . E_2)$ never occurs), and hence $p(E_1|E_2) = 0$, a result which is consistent with the definition of mutually exclusive events as events which cannot occur simultaneously.

Equation (2.4.2) may be rewritten in the slightly modified form

$$p(E_1 . E_2) = p(E_1|E_2)p(E_2) \qquad (2.4.3)$$

Our conjecture about the value of $p(\text{score } 3|(\text{score odd}))$ is now readily confirmed since in this case we have,

$$p(E_2) = \tfrac{3}{6} = \tfrac{1}{2} \qquad (\text{by } (2.3.2.1)),$$

and

$$p(E_1 . E_2) = p(\text{score } 3) = \tfrac{1}{6}.$$

Hence

$$p(E_1|E_2) = \tfrac{1}{6}/\tfrac{1}{2} = \tfrac{1}{3}.$$

2.4.1 Independent Events

We mentioned above that, in general, the occurrence of one event E_2 would affect the probability which we assign to another event E_1. However, it may turn out that, in certain cases, the conditional probability of E_1 given E_2 is exactly the same as the unconditional probability of E_1, i.e.

$$p(E_1|E_2) = p(E_1). \qquad (2.4.1.1)$$

In this situation, we say that E_1 and E_2 are *independent events*, since here knowledge of the occurrence of E_2 in no way affects the probability of the occurrence of E_1, i.e. *the event E_2 provides no information about the event E_1.*

When E_1 and E_2 are independent, in the sense that (2.4.1) holds, equation (2.4.3) takes the simpler form

$$p(E_1 . E_2) = p(E_1)p(E_2). \qquad (2.4.1.2)$$

This extremely important result is known as· the *multiplication law* for independent events.

It should be stressed that the above definition of independence is a purely mathematical one, and although it is strongly motivated by our intuitive understanding of the term "independent" it can lead to cases where two events turn out to be independent even though intuitively they would not seem to be so. The following simple example illustrates this point. Suppose that we have four cards, numbered 1, 2, 3, 4, and the experiment is to select one card "at random", i.e. so that each card has the same probability of being selected. Let E_1, E_2, denote the following events:

E_1: card selected is 1 or 2, E_2: card selected is 1 or 3.

Then

$$p(E_1) = p(E_2) = \tfrac{2}{4} = \tfrac{1}{2}.$$

Also,

$$p(E_1 \,.\, E_2) = p(\text{card 1 selected}) = \tfrac{1}{4}.$$

Hence,

$$p(E_1 | E_2) = p(E_1 \,.\, E_2)/p(E_2) = \tfrac{1}{2} = p(E_1).$$

Thus, according to our definition E_1 and E_2 are independent events. Intuitively, we might not expect this result since it could be argued that E_1 and E_2 are "linked" by the fact that both events involve card 1.

2.5 RANDOM VARIABLES

In probability theory we encounter experiments whose outcomes are described in various ways. Thus, for the experiment of spinning a coin the outcomes are described by the two words, "heads", "tails"; for the experiment of drawing a card from a standard pack the outcomes are described by a mixture of words and numbers (e.g. the card selected might be the ten of diamonds); whereas for the experiment $\mathscr{E}_1$ (throwing a die) the outcomes are described by numbers. This variation in modes of description is inconvenient, and from the point of view of mathematical development it would be much simpler if the results of all experiments could be described purely in terms of numbers. Such a scheme is easily arranged by using a numerical "code" for the outcomes. For example, in the coin spinning experiment we may simply code "heads" as 0 and "tails" as 1; in the card drawing experiment we may number the cards from 1 to 52, and then state the number of the card selected.

More generally, we may set up a numerical code for the outcomes of any experiment by assigning different (real) numbers to the possible outcomes. Putting this operation into more precise terms, we would say that we have constructed a *mapping* (or point *function*) from the original sample space Ω

onto the real line, R^1. This mapping is then called a *random variable*, a term which may seem rather strange at first but whose meaning will emerge later on. We thus make the following definition:

Definition *A random variable, $X(.)$, is a mapping (or point function) from the sample space Ω onto the real line R^1 such that to each element $\omega \in \Omega$ there corresponds a unique real number, $X(\omega)$. (See Fig. 2.3.)*

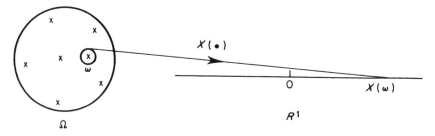

Fig. 2.3

In the coin spinning experiment we defined the random variable $X(.)$ by $X(\text{heads}) = 0$, $X(\text{tails}) = 1$. Note, however, that we could equally well have defined a different random variable, $Y(.)$, by $Y(\text{heads}) = 47$, $Y(\text{tails}) = 182$. Both $X(.)$ and $Y(.)$ would give equally valid descriptions of the outcomes, and in general there is a multitude of numerical codings which we may use for any particular experiment. In practice we would usually choose $X(.)$ rather than $Y(.)$ merely on grounds of simplicity.

In constructing a random variable we effectively replace our original (abstract) sample space Ω by a *new (concrete) sample space*, namely R^1. From this point onwards we may conduct all subsequent analysis in the new space. If the experiment consists of measuring some physical quantity (such as the value of a resistor) then the sample space Ω is itself R^1 (or an interval on R^1), so that no further mapping is required. In this case we could simply define $X(.)$ as the *identity mapping*, so that for each $\omega \in \Omega$, $X(\omega) = \omega$. In this context it is more natural to think of the random variable not as a mapping but rather *as the variable which describes the result of a random experiment*, and consequently takes different values "at random".

2.5.1 Defining Probabilities for Random Variables

So far we have defined probabilities only for subsets of the original sample space Ω. However, we frequently wish to make probability statements about the values of a random variable such as, for example, "the probability that X lies between a and b is 0.5". To make such statements meaningful we must *define* probabilities for random variables, and this is achieved by the obvious device of translating statements about the values of X into statements about corresponding subsets of Ω, as follows.

For any two numbers a, b, $(a \leqslant b)$, we *define* $p[a < X \leqslant b]$ by

$$p[a < X \leqslant b] = p[S(a, b)], \qquad (2.5.1.1)$$

where $S(a, b)$ is that subset of Ω which consists of all elements ω such that $a < X(\omega) \leqslant b$; i.e.

$$S(a, b) = \{\omega ; \omega \in \Omega \text{ and } a < X(\omega) \leqslant b\}.$$

In (2.5.1.1) we may let $a \to -\infty$ or $b \to \infty$, so that the same approach may be used also to define quantities such as $p[X \leqslant b]$ and $p[X > a]$.

More generally, for any set $B \subset R^1$ we define $p[X \in B]$ by

$$p[X \in B] = p[S(B)], \qquad (2.5.1.2)$$

where $S(B) = \{\omega ; \omega \in \Omega \text{ and } X(\omega) \in B\}$. However, if the original sample space is continuous we cannot use equations (2.5.1.1) or (2.5.1.2) immediately unless we can be sure that the set $S(a, b)$ (or the set $S(B)$) belongs to the collection $\mathcal{F}$ of "well behaved" subsets of Ω on which probabilities have been defined. Whether or not $S(a, b) \in \mathcal{F}$ depends, of course, on the form of the mapping, $X(.)$, but it may be shown that this situation will obtain if $X(.)$ satisfies the condition that, *for each x, the set* $S(x) = \{\omega ; \omega \in \Omega \text{ and } X(\omega) \leqslant x\}$ *belongs to $\mathcal{F}$*. When $X(.)$ satisfies this condition it is said to be *measurable with respect to $\mathcal{F}$*. It may then be shown further that (2.5.1.2) is valid for all *Borel sets, $B \in R^1$*. (See Parzen (1960), p. 270, Kingman and Taylor (1966), p. 284.) In the more advanced theory, the term "random variable" *is therefore restricted to measurable functions only*, and if we say that "$X(.)$ is a random variable" this statement necessarily implies that $S(x) \in \mathcal{F}$ for all x.

2.6 DISTRIBUTION FUNCTIONS

In the previous section we saw how probabilities defined on subsets of Ω could be used to define probabilities for random variables. If now we wish to describe the probabilistic properties of a random variable X in full, then in principle this means that we must specify $p[X \in B]$ for *all* Borel sets in R^1. This is clearly much too difficult an operation, and even the lesser requirement of specifying $p[a < X < b]$ for *all* intervals (a, b) is impracticable. Fortunately, we may overcome this difficulty by introducing the notion of a "*distribution function*" (sometimes called "cumulative distribution function"). Given a random variable X, the distribution function of X, $F(x)$, is defined by,

$$F(x) = p[X \leq x]. \tag{2.6.1}$$

Note that, by the discussion in Section 2.5, $p[X \leq x]$ is well defined, so that (2.6.1) defines the function F for all x.

It may be shown that the distribution function $F(x)$ uniquely determines all the probabilistic properties of the random variable. In particular, we have for any a, b $(a \leq b)$,

$$p[X \leq b] = p[X \leq a] + p[a < X \leq b] \qquad \text{(by Axiom 3)}.$$

Hence

$$p[a < X \leq b] = p[X \leq b] - p[X \leq a] = F(b) - F(a). \tag{2.6.2}$$

Thus, if we know $F(x)$ for all x then we may easily compute the probability that X lies in any given interval.

All distribution functions have the following properties:

(1) $0 \leq F(x) \leq 1$, all x, (since $F(x)$ is a "probability").

(2) $\lim_{x \to -\infty} F(x) = 0$, $\quad \lim_{x \to \infty} F(x) = 1$

   (since $S(-\infty) = \varnothing$ (null set) and $S(\infty) = \Omega$).

(3) $F(x)$ is a *non-decreasing function* in the sense that, for any $h \geq 0$ and all x,

$$F(x + h) \geq F(x).$$

   (This follows since $F(x + h) - F(x) = p[x < X \leq x + h] \geq 0$.)

(4) Further, it may be shown that, for all x, $F(x)$ is "right-continuous", i.e., for all x

$$\lim_{h \to 0+} F(x + h) = F(x)$$

(in the above, the limit of $F(x + h)$ is taken as $h \to 0$ through positive values only—see Parzen (1960), p. 173, Kingman and Taylor (1966), p. 290).

2.6.1 Decomposition of Distribution Functions

The properties of $F(x)$ listed above give us some indication of the general form of a distribution function, namely, that it starts off as zero at $-\infty$ and then rises steadily to 1 at $+\infty$, with possibly discontinuities or "jumps" at certain points. (At such a point, x_0, the value of $F(x_0)$ is defined as the limit of $F(x)$ as x approaches x_0 from the right, so as to ensure that $F(x)$ satisfies property (4).) There is, however, a very important result which enables us to decompose *any* distribution function into the sum of three terms, each term having certain characteristic properties. This result is expressed in the following theorem. (See Parzen (1960), pp. 170, 174, Kingman and Taylor (1966), p. 294.)

Theorem 2.6.1 *Lebesgue's Decomposition Theorem* *Any distribution func-tion, $F(x)$, may be written in the form*

$$F(x) = a_1F_1(x) + a_2F_2(x) + a_3F_3(x), \qquad (2.6.3)$$

where
(1) $a_i \geqslant 0, \quad i = 1, 2, 3, \quad a_1 + a_2 + a_3 = 1,$
and
(2) $F_1(x)$ *is "absolutely continuous" (continuous everywhere and differenti-able for almost all x), $F_2(x)$ is a "step-function" with a finite or countably infinite number of "jumps", and $F_3(x)$ is a "singular" function, that is, continuous with zero derivative "almost everywhere" (i.e. at almost all points).*

As we shall see, the first two terms in equation (2.6.3), $F_1(x)$ and $F_2(x)$, have important interpretations and correspond respectively to the two basic types of probability distributions which we encounter in practice, namely the "continuous" and "discrete" distributions. On the other hand, the third term, $F_3(x)$, is highly pathological and it may be safely assumed that such functions do not arise in real applications. (This is true at least for one-dimensional variables; in higher dimensions the singular term may well occur.)

In practice we therefore ignore the third term in (2.6.3) and assume that any distribution function may be written simply as the sum of the first two terms, i.e. we write

$$F(x) = a_1F_1(x) + a_2F_2(x), \qquad (2.6.4)$$

where now $a_1 \geqslant 0, a_2 \geqslant 0, a_1 + a_2 = 1$.

2.7 DISCRETE, CONTINUOUS AND MIXED DISTRIBUTIONS

Although a general distribution function may contain both an absolutely continuous term and a step-function term, as indicated by equation (2.6.4), the distribution functions which arise in the vast majority of problems contain only one or other of these terms. We therefore consider the following special cases of equation (2.6.4).

I. *The purely discrete case; $a_1 = 0$, $a_2 = 1$*

If the random variable X is such that $a_1 = 0$, $a_2 = 1$ in (2.6.4) then $F(x) \equiv F_2(x)$, so that the distribution function is a simple step-function with "jumps" $\{p_r\}$ at points $\{x_r\}$, say, $r = 1, 2, 3, \ldots$ (see Fig. 2.4). More precisely, the jump, p_r, at the point x_r is defined by,

$$p_r = F(x_r) - \lim_{x \to x_r-} F(x)$$

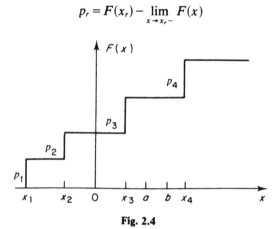

Fig. 2.4

(recall that $F(x)$ is right-continuous). If we now take any two values a, b ($a \le b$) such that the interval $(a, b]$ does not contain any of the jump points $\{x_r\}$, (Fig. 2.4), then clearly,

$$p[a < X \le b] = F(b) - F(a) = 0.$$

Thus, in effect, X cannot take any value lying between two successive jump points, i.e. *the only values which X can take are those corresponding to the "jump" points, x_1, x_2, x_3, ...* The variable X is then called a "(*purely*) *discrete random variable*", since its possible values are restricted to the *discrete* set of points x_1, x_2, x_3, ...

Discrete random variables occur in many types of problems, particularly in cases where we are interested in counting the number of times something happens. For example, if X denotes the number of defective articles in a randomly selected batch of, say, 1000 articles, then X is a discrete random variable with possible values $0, 1, 2, \ldots, 1000$. Similarly, if X denotes the number of calls which arrive at a telephone exchange in a randomly selected period of 1 hour's duration, then again X is a discrete random variable and its possible values are $0, 1, 2, \ldots, \infty$ (ideally).

We have for each r and any small $h > 0$,

$$p[x_r - h < X \le x_r + h] = F(x_r + h) - F(x_r - h) = p_r.$$

Letting $h \to 0$, we obtain

$$p[X = x_r] = p_r, \qquad r = 1, 2, 3, \ldots. \tag{2.7.1}$$

The jump p_r at the point x_r thus represents the probability that X takes the value x_r. *The set of numbers* $(p_1, p_2, p_3, \ldots)$ *is called the "discrete probability distribution" of X.* We may illustrate a discrete probability distribution graphically by constructing ordinates of heights $\{p_r\}$ at the points $\{x_r\}$, as shown in Fig. 2.5. (The ordinates $\{p_r\}$ are often joined together by a smooth curve, but it should be remembered that it is the values of this curve at the points $\{x_r\}$ only which have any probabilistic meaning.)

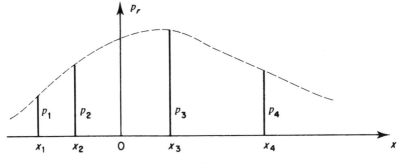

Fig. 2.5

All discrete probability distributions have the following basic properties:

(1) For each x,

$$F(x) = \sum_{r : x_r \leqslant x} p_r, \tag{2.7.2}$$

where the summation on the right-hand side extends over all values of r for which $x_r \leqslant x$. (Equation (2.7.2) follows immediately from the definition of the $\{p_r\}$, see Fig. 2.4.)

(2) $\qquad 0 \leqslant p_r \leqslant 1, \quad$ all $r \quad$ (since each p_r is a "probability"). $\tag{2.7.3}$

(3) $$\sum_r p_r = 1, \tag{2.7.4}$$

(since $1 = \lim_{x \to \infty} F(x) = \sum_r p_r$ or, equivalently, (2.7.4) follows from the fact that the events $(X = x_r)$, $r = 1, 2, \ldots$, are mutually exclusive and exhaustive).

II. *The purely continuous case; $a_1 = 1; a_2 = 0$*

If the random variable X is such that in (2.6.4) $a_1 = 1$, $a_2 = 0$, then the distribution function $F(x)$ ($\equiv F_1(x)$) is absolutely continuous and differentiable for almost all x. X is then called a (*purely*) *continuous random variable*, and $F(x)$ would typically have the form shown in Fig. 2.6.

In contrast to discrete variables, a continuous random variable can, in general, take any value either on a finite interval or on an infinite interval.

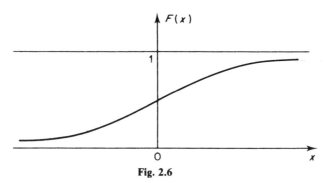

Fig. 2.6

Continuous random variables thus provide suitable (if idealized) descriptions for the measurement of physical quantities such as lengths, volumes, temperatures, pressures, voltages, etc.

If we now write

$$F(x) = \int_{-\infty}^{x} f(u) \, du, \tag{2.7.5}$$

then $f(x)$ is called the *probability density function of X*. It follows immediately from (2.7.5) that for all x s.t. $F'(x)$ exists,

$$f(x) = \frac{dF(x)}{dx}. \tag{2.7.6}$$

Now consider a small interval, $(x, x + \delta x]$. We have,

$$p[x < X \leq x + \delta x] = F(x + \delta x) - F(x) = \int_{x}^{x+\delta x} f(u) \, du,$$

or,

$$p[x < X \leq x + \delta x] = f(x)\delta x + o(\delta x), \tag{2.7.7}$$

(where $o(\delta x)$ represents a term of smaller order of magnitude than δx). Equation (2.7.7) forms the basis of the interpretation of $f(x)$ as a *density* function, namely, that according to (2.7.7) $f(x)$ represents the *density* of probability (per unit value of x) in the neighbourhood of the point x. This use of the term "probability density" is completely analogous to the use of, e.g., the term "mass density" in mechanics. For, if we have a solid body whose mass density (per unit volume) at a point P is m, then the mass in a small volume δV surrounding P is $m\delta V$. Now compare this result with (2.7.7).

Note that $f(x)$ itself does not represent a probability—rather it is $\{f(x) \cdot \delta x\}$ which has a probabilistic interpretation—but the function $f(x)$ completely determines $F(x)$ (by (2.7.6)) and thus completely specifies the properties of a continuous random variable.

Analogous to the basic properties of discrete probability distributions (2.7.3, 2.7.4), there are corresponding properties which all probability density functions satisfy, namely,

(1) $\qquad f(x) \geqslant 0, \quad$ all $x \quad$ (since $F(x)$ is non-decreasing). $\qquad$ (2.7.8)

(2) $\qquad \displaystyle\int_{-\infty}^{\infty} f(x)\, dx = 1 \quad \left(\text{since } 1 = \lim_{x \to \infty} F(x) = \int_{-\infty}^{\infty} f(x)\, dx \right).$ $\qquad$ (2.7.9)

(3) For any a, b $(a \leqslant b)$,

$$p[a < X \leqslant b] = \int_{a}^{b} f(x)\, dx. \qquad (2.7.10)$$

(Follows from (2.6.2) and (2.7.6).)

If we let $\delta x \to 0$ in (2.7.7) we find that,

$$p[X = x] = 0, \qquad \text{all } x. \qquad (2.7.11)$$

This result may seem rather surprising at first since it tells us that, if X is a continuous random variable, *the probability that X takes any particular value is zero.* However, one may again compare this with the corresponding situation for the distribution of mass in a solid body, the mass at any particular point being, ideally, zero.

In view of (2.7.11) we may include or omit either of the equality signs on the left-hand side of (2.7.10) without affecting the validity of the equation. In fact, since $p[X = a] = p[X = b] = 0$, we may write

$$p[a < X \leqslant b] = p[a \leqslant X < b] = p[a \leqslant X \leqslant b] = \int_{a}^{b} f(x)\, dx.$$

A typical probability density function may have the form shown in Fig. 2.7.

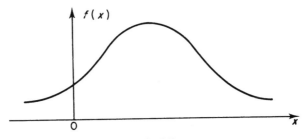

Fig. 2.7

III. *Mixed distributions; $a_1 \geq 0$, $a_2 > 0$*

We remarked previously that the random variables which arise in most practical problems are either purely discrete or purely continuous in the sense that they belong either to case I or case II. There are, nevertheless, certain situations in which we encounter random variables whose distribution functions contain *both* the continuous term $F_1(x)$ *and* the step-function term $F_2(x)$. Such a random variable, X, is said to have a *mixed* probability distribution, and its distribution function $F(x)$ will be of the form (2.6.4) in which both $a_1 > 0$ and $a_2 > 0$, and typically will have the form shown in Fig. 2.8.

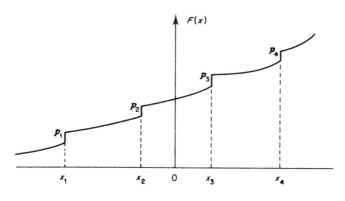

Fig. 2.8

Here, $F(x)$ will have jumps of magnitude $\{p_r\}$ at points $\{x_r\}$, $r = 1, 2, \ldots$, but between these points $F(x)$ will be continuous and differentiable. Obviously, $F(x)$ will not be differentiable at the jump points, so that we cannot describe this type of random variable completely in terms of a probability density function. We can, of course, use a density function description for the continuous term $F_1(x)$, but the step function term $F_2(x)$ cannot be treated in this way unless we introduce the notion of Dirac δ-functions. (This point is discussed further in relation to spectral analysis in Chapter 4.)

Just to convince the reader that mixed distributions are not merely the product of the mathematicians' lust for generality but do indeed arise in practice, we would mention the following well known example.

Let the random variable X denote the time which a customer has to wait for service after entering a store. If there are other people in the store when the customer enters he joins the "queue". At first sight it would seem that X

is a purely continuous variable, but if this were so it would mean that the probability that X takes any particular value is zero (cf. equation (2.7.11)). However, this result does not hold for the particular value, $X = 0$, since $X = 0$ if there is no queue when the customer enters, and the probability that there is no queue is, in general, greater than zero. One might conclude that here X has a mixed distribution with a jump point at the origin, and in fact this result can be established quite rigorously (see, e.g., Parzen (1962)).

Relationship between probability distributions and power spectra

The astute reader may have noticed a superficial similarity between the graphs used to describe various types of power spectra in Chapter 1 and the graphs of various types of distribution functions in the present chapter. This is no mere coincidence and, far from being superficial, the similarity between power spectra and probability distributions has deep significance as will emerge in the discussion in Chapter 4.

2.8 MEANS, VARIANCES AND MOMENTS

The probabilistic properties of a random variable X are completely specified by its cumulative distribution function, $F(x)$, or equivalently, by its discrete probability distribution, $\{p_i\}$, when X is discrete, or by its probability density function, $f(x)$, when X is continuous. It is often convenient, however, to summarize the main features of a discrete probability distribution or probability density function in terms of quantities called the mean, variance, and moments. In order to motivate the definitions of these quantities let us consider first the case where X is discrete valued. Suppose now that the basic experiment is repeated n times, and that in these n repetitions we observe X to take the value (x_1) n_1 times, the value (x_2) n_2 times, ..., and the value (x_k) n_k times, where we obviously have that $(n_1 + n_2 + \ldots + n_k) = n$. Then the *arithmetic mean* of these n values of X is given by

$$\bar{x} = \frac{1}{n} \sum_{i=1}^{k} n_i x_i. \tag{2.8.1}$$

In statistical inference we would regard these n values of X as a "*random sample of n observations on X*". Accordingly we call $\bar{x}$ the *sample mean*, and use it to indicate the "typical" value of X in the set of n observations.

In addition to calculating the "typical" value of X we might be interested also in measuring the "spread" (or "variability") of values in the sample.

The usual quantity used to measure "spread" is called the *variance*, and is given by,

$$s^2 = \frac{1}{n} \sum_{i=1}^{k} n_i(x_i - \bar{x})^2. \tag{2.8.2}$$

More precisely, s^2 is called the *sample variance*, and measures the *mean square deviation* of the n observations about their mean value. Clearly, s^2 has the dimension of x^2, and to obtain a measure of spread which has the same dimension as x we consider the square-root of the variance, namely,

$$s = \left(\frac{1}{n} \sum_{i=1}^{k} n_i(x_i - \bar{x})^2 \right)^{1/2}. \tag{2.8.3}$$

The quantity s measures the *root mean square deviation* about $\bar{x}$, and is called the *sample standard deviation*.

Now suppose that we let n, the number of repetitions of the experiment, tend to infinity. According to the relative frequency interpretation of probability we would expect that, as n becomes large,

$$(n_i/n) \sim p[X = x_i] = p_i,$$

say. Replacing (n_i/n) by p_i in (2.8.1) and (2.8.2), leads us to formulate the following definition.

Definition 2.8.1 *Let X be a discrete random variable with $p_i = p[X = x_i]$, $i = 1, 2, \ldots$. We define the mean, μ, and the variance, σ^2, of this discrete probability distribution by*

$$\mu = \sum_i p_i x_i, \tag{2.8.4}$$

and

$$\sigma^2 = \sum_i p_i(x_i - \mu)^2. \tag{2.8.5}$$

(The mean may be regarded as a weighted average of all the possible values of X, each value weighted by the corresponding probability.) The square-root of the variance, σ, is again called the *standard deviation* of the distribution.

If X is a continuous random variable, its probability density function is a continuous function, rather than a discrete sequence. In this case, the definitions of mean and variance follow in an obvious way from (2.8.4), (2.8.5), if we replace the summation over i by an integral. We then have the following definition:

Definition 2.8.2 *Let X be a continuous random variable with probability density function $f(x)$. The mean and variance of $f(x)$ are defined by*

$$\mu = \int_{-\infty}^{\infty} xf(x)\, dx, \qquad (2.8.6)$$

$$\sigma^2 = \int_{-\infty}^{\infty} (x - \mu)^2 f(x)\, dx, \qquad (2.8.7)$$

respectively.

The quantities μ and σ^2 are extremely useful in providing a rough summary of the form of a probability distribution (discrete or continuous), μ representing, loosely speaking, the "central value", and σ^2 the "spread" of the distribution about this "central value". However, it should be clear that we could never hope to describe fully the function $f(x)$ (or the sequence $\{p_i\}$) merely in terms of the two number μ, σ^2. Indeed, we could construct many distributions all of which have the same values of μ and σ^2, but with their shapes differing in other respects. In order to distinguish between such distributions we now introduce a further sequence of constants, called the *moments*, defined as follows.

Definition 2.8.3 The rth moment about the origin, μ'_r, is defined by

$$\mu'_r = \begin{cases} \displaystyle\sum_i x_i^r p_i, & \text{when } X \text{ is discrete,} \qquad (2.8.8) \\[2mm] \displaystyle\int_{-\infty}^{\infty} x^r f(x)\, dx, & \text{when } X \text{ is continuous,} \qquad (2.8.9) \end{cases}$$

$r = 1, 2, 3, \ldots$.

The rth moment about the mean, μ_r, is defined by

$$\mu_r = \begin{cases} \displaystyle\sum_i (x_i - \mu)^r p_i, & \text{when } X \text{ is discrete,} \qquad (2.8.10) \\[2mm] \displaystyle\int_{-\infty}^{\infty} (x - \mu)^r f(x)\, dx, & \text{when } X \text{ is continuous,} \qquad (2.8.11) \end{cases}$$

$r = 1, 2, 3, \ldots$.

Although we have defined two sets of moments, $\{\mu'_r\}$, $\{\mu_r\}$, they contain equivalent information about the shape of the distribution in the sense that the sequence $\{\mu'_r\}$, $r = 1, 2, 3, \ldots$, uniquely determines the sequence $\{\mu_r\}$, $r = 1, 2, 3, \ldots$, and vice-versa.

The first two moments (corresponding to $r = 1, 2$) add nothing to the information provided by μ, σ^2, since substituting $r = 1, 2$, in equations (2.88)–(2.8.11), we obtain for the discrete case, for example,

$$\mu_1' = \sum_i x_i p_i = \mu, \qquad\qquad \mu_1 = \sum_i (x_i - \mu) p_i \equiv 0.$$

$$\mu_2 = \sum_i (x_i - \mu)^2 p_i = \sigma^2, \qquad \mu_2' = \sum_i x_i^2 p_i = \sum_i (x_i - \mu)^2 p_i + \mu^2 \Big(\sum_i p_i \Big),$$

or,

$$\sigma^2 = \mu_2' - \mu^2. \tag{2.8.12}$$

(Similar arguments produce exactly the same results for the continuous case.)

The new information provided by the moments starts when we consider the 3rd, 4th, 5th, ... order moments. For example, it may be shown that μ_3 measures asymmetry ("skewness") about the mean, while μ_4 measures "flatness".

We may similarly define moments for a random sample of n observations, thus:

Definition 2.8.4 *The rth moment about the origin, m_r', is defined by*

$$m_r' = \frac{1}{n} \sum_i n_i x_i', \qquad r = 1, 2, 3, \dots, \tag{2.8.13}$$

and the rth moment about the mean, m_r, is defined by

$$m_r = \frac{1}{n} \sum_i n_i (x_i - \bar{x})', \qquad r = 1, 2, 3, \dots. \tag{2.8.14}$$

Corresponding to (2.8.12) we have,

$$s^2 = m_2' - \bar{x}^2, \tag{2.8.15}$$

a result which is often useful in numerical computations of sample variances.

It should be noted that:

(a) We have adopted the usual convention of using Greek letters (such as μ, σ^2, μ_r', μ_r) when referring to probability distributions, and Latin letters (such as $\bar{x}$, s^2, m_r', m_r) when referring to samples.

(b) The expressions for the mean, variance, and moments each involve (possibly) infinite sums or infinite integrals, and these may not always converge to finite values. For example, there are some distributions for which σ^2 does not exist (i.e. is not finite), such as the Cauchy distribution (Parzen (1960), p. 180).

(c) The mean is the most important and most commonly used measure of "typical" or "central" value, but alternative measures which are sometimes useful are

(i) the *median*, $\tilde{\mu}$, defined, e.g., for continuous variables by

$$\int_{-\infty}^{\tilde{\mu}} f(x)\, dx = \int_{\tilde{\mu}}^{\infty} f(x)\, dx = \tfrac{1}{2},$$

or equivalently by,

$$p[X \leq \tilde{\mu}] = p[X \geq \tilde{\mu}] = \tfrac{1}{2},$$

and

(ii) the *mode*, μ^*, defined as that value of x at which the probability density function attains its maximum value. (There are similar definitions of median and mode for discrete probability distributions and random samples.)

(d) In place of the standard deviation, σ, we sometimes use as an alternative measure of spread the *mean deviation*, defined by (in the continuous case),

$$\tilde{\sigma} = \int_{-\infty}^{\infty} |x - \mu| f(x)\, dx,$$

with similar definitions for the other cases.

2.8.1 Chebyshev's Inequality

It is intuitively obvious that the standard deviation σ provides a measure of the spread of a probability distribution in the sense that it gives us a rough scale on which we can measure the typical deviation of X from its mean value, μ. This interpretation is made more precise by the following celebrated result.

Chebyshev's inequality Let X be any random variable with finite mean, μ, and finite variance, σ^2. For any positive constant k we have,

$$p[|X - \mu| \geq k\sigma] \leq 1/k^2. \tag{2.8.16}$$

Proof. The result is valid for both discrete and continuous variables, but for simplicity we consider only the continuous case. We now have,

$$\sigma^2 = \int_{-\infty}^{\infty} (x - \mu)^2 f(x) \, dx$$

$$\geq \int_{|x - \mu| \geq k\sigma} (x - \mu)^2 f(x) \, dx \quad \text{(since the integrand is everywhere positive)}$$

$$\geq k^2 \sigma^2 \int_{|x - \mu| \geq k\sigma} f(x) \, dx$$

$$= k^2 \sigma^2 p[|X - \mu| \geq k\sigma].$$

Dividing through by $(k^2 \sigma^2)$ we obtain the required result.

Taking $k = 3$, we see that the probability that a random variable deviates from its mean by more than three standard deviations is less than $1/9$, while taking $k = 5$ shows that the probability of a deviation of more than five standard deviations is less than $0 \cdot 04$.

2.9 THE EXPECTATION OPERATOR

The formulae for the mean, variance, and moments given in the previous section may be written in a more compact form by using the "expectation operator" notation.

For any function, $g(X)$, of X, we define the *expected value of* $g(X)$, written as $E[g(X)]$, by

$$E[g(X)] = \begin{cases} \sum_i g(x_i) p_i, & \text{if } X \text{ is discrete,} & (2.9.1) \\ \int_{-\infty}^{\infty} g(x) f(x) \, dx, & \text{if } X \text{ is continuous,} & (2.9.2) \end{cases}$$

provided that the above sum or integral exists. Thus, $E[g(X)]$ may be regarded as the "mean value of $g(X)$" in the sense that it is computed as a weighted average of all possible values of $g(X)$, each value being weighted by the corresponding probability.

Using this notation, we may now rewrite the definitions of μ, σ^2, μ'_r, μ_r, in the following forms (valid for both discrete and continuous variables),

$$\mu = E[X], \qquad \sigma^2 = E[(X - \mu)^2], \qquad (2.9.3)$$

$$\mu'_r = E[X^r], \qquad \mu_r = E[(X - \mu)^r], \qquad r = 1, 2, 3, \ldots. \qquad (2.9.4)$$

The usefulness of the expectation notation lies in the fact that we may think of E as an "operator" in much the same way as that we think of $D(\equiv d/dx)$, $\int$, and Δ, as "operators". Looked at in this way, we may list the following important properties of E which hold irrespective of the nature of the particular random variable under study.

(i) For any constant c, $E[c] = c$.   (2.9.5)

(ii) E is a *linear operator* in the sense that for any set of functions $g_1(X), \ldots, g_k(X)$, each of whose expectation exists, and any set of constants $c_1, \ldots, c_k$,

$$E\left[\sum_{i=1}^{k} c_i g_i(X) \right] = \sum_{i=1}^{k} c_i E[g_i(X)]. \tag{2.9.6}$$

The first property, (2.9.5), follows trivially from the definitions (2.9.1), (2.9.2), and (2.9.6) follows simply from the corresponding linear property of sums or integrals.

Combining (2.9.5) and (2.9.6) we have that for any function $g(X)$ (whose expectation exists) and any constants b, c,

$$E[b + cg(X)] = b + cE[g(X)]. \tag{2.9.7}$$

To illustrate the above results let us now derive expressions for the mean, variance, and moments of a linear function of a random variable, X. Writing

$$Y = b + cX, \tag{2.9.8}$$

we have, with an obvious notation,

$$\mu_Y = E[Y] = E[b + cX] = b + cE[X] = b + c\mu_X, \tag{2.9.9}$$

$$\sigma_Y^2 = E[(Y - \mu_Y)^2] = E[c^2(X - \mu_X)^2] = c^2 \sigma_X^2, \tag{2.9.10}$$

$$\mu_r(Y) = E[(Y - \mu_Y)^r] = c^r \mu_r(X), \qquad r = 1, 2, \ldots. \tag{2.9.11}$$

In particular, if we take $b = -\mu_X/\sigma_X$, $c = 1/\sigma_X$, then $Y = (X - \mu_X)/\sigma_X$ has zero mean and unit variance.

2.9.1 General Transformations

We may extend (2.9.9), (2.9.10) to the case where Y is a general (i.e. not necessarily linear) function of X. Suppose that $Y = g(X)$, where g is differentiable at least twice for all x. We obtain approximate expressions for the mean and variance of Y which are valid *provided that* $\sigma_X \ll \mu_X$, i.e. provided that the variation of X about its mean is relatively *small*. With this assumption we may expand the function $g(x)$ in a Taylor series about the point $x = \mu_X$, and retain only the first two terms, i.e. we neglect terms of $O[(X - \mu_X)^2]$ and higher. (For, since σ_X is assumed small, we know from

Chebyshev's inequality that, with high probability, $(X - \mu_X)$ will also be small.) Thus, we may write,

$$Y = g(X) \doteq g(\mu_X) + (X - \mu_X)g'(\mu_X). \qquad (2.9.12)$$

By this device we have effectively "linearized" the function $g(x)$, the argument being that this will be an adequate approximation over the small range of variation of X. Equation (2.9.12) is now of the same form as (2.9.8), with $b = \{g(\mu_X) - g'(\mu_X)\}$, $c = g'(\mu_X)$, so that using (2.9.9), (2.9.10), we obtain

$$\mu_Y \doteq g(\mu_X), \qquad (2.9.13)$$

$$\sigma_Y^2 \doteq \{g'(\mu_X)\}^2 \sigma_X^2. \qquad (2.9.14)$$

As an example, suppose X takes positive values only and consider the transformation $Y = \log_e X$. Then approximately,

$$\mu_Y \doteq \log_e \mu_X, \qquad (2.9.17)$$

$$\sigma_Y^2 \doteq \sigma_X^2 / \mu_X^2. \qquad (2.9.18)$$

A better approximation for μ_Y can be obtained by retaining a further term in the Taylor series expansion of $g(x)$, giving,

$$Y \doteq g(\mu_X) + (X - \mu_X)g'(\mu_X) + \tfrac{1}{2}(X - \mu_X)^2 g''(\mu_X). \qquad (2.9.15)$$

We now have,

$$\mu_Y = E[Y] \doteq g(\mu_X) + \tfrac{1}{2}g''(\mu_X)\sigma_X^2, \qquad (2.9.16)$$

recalling that

$$E[(X - \mu_X)] = 0, \qquad E[(X - \mu_X)^2] = \sigma_X^2.$$

Note, however, that we need not bother to take account of the additional term in (2.9.15) in evaluating the variance of Y since its contribution to σ_Y^2 will be $O[(X - \mu_X)^3]$.

If we require the exact values of μ_Y, σ_Y^2, we must first find the complete probability distribution of Y. When $g(x)$ is a 1-1 function (i.e. strictly monotonic increasing or monotonic decreasing) and X is continuous with probability density function $f(x)$, the probability density function of Y, $h(y)$, may be derived as follows. Assuming that $g(x)$ is a monotonic increasing function, we may write,

$$f(x)\delta x \doteq p[x < X \leqslant x + \delta x] = p[y < Y \leqslant y + \delta y] \doteq h(y)\delta y,$$

where $y = g(x)$. Letting $\delta x \to 0$ gives,

$$h(y) = \frac{f(x)}{(dy/dx)} = \frac{f(x)}{g'(x)}. \qquad (2.9.19)$$

If $g(x)$ is a monotonic decreasing function we have,

$$f(x)\delta x \doteq p[x < X \le x + \delta x] = p[y + \delta y < Y \le y] \doteq -h(y)\delta y,$$

so that now,

$$h(y) = -f(x)/g'(x). \qquad (2.9.20)$$

Both cases are covered by the single equation,

$$h(y) = f(x)/|g'(x)|. \qquad (2.9.21)$$

Note that the expression for x in terms of y must be substituted in the right-hand side of (2.9.21), i.e. if g^{-1} denotes the inverse function of g, then (2.9.21) is more correctly written as

$$h(y) = \frac{f\{g^{-1}(y)\}}{|g'\{g^{-1}(y)\}|}. \qquad (2.9.22)$$

2.10 GENERATING FUNCTIONS

2.10.1 Probability Generating Functions

A discrete probability distribution, $p_i = p[X = x_i]$, $i = 1, 2, 3, \ldots$, may be specified simply by stating the sequence of numbers, $(p_1, p_2, \ldots)$, together with the set of possible values of X, $(x_1, x_2, \ldots)$. However, we may obtain an equivalent specification by using the number pairs (p_1, x_1), (p_2, x_2), $\ldots$, to construct a function called the *probability generating function*. Formally, we define the probability generating function (PGF), $\Pi(z)$, of a discrete random variable X by,

$$\Pi(z) = E[z^X] = \sum_i p_i z^{x_i}, \qquad (2.10.1)$$

for all z s.t. the RHS is convergent. This function is particularly useful when X takes only integer values, in which case $\Pi(z)$ becomes a polynomial function of z, and is uniformly convergent for $|z| \le 1$. Given the form of $\Pi(z)$ we can then, by expanding it as a power series in z, recover the original distribution $\{p_i\}$.

The PGF may be used to evaluate the mean and variance of a discrete random variable by noting that,

$$\Pi'(z) = \sum_i p_i x_i z^{(x_i - 1)},$$

so that, setting $z = 1$,

$$\Pi'(1) = \sum_i p_i x_i = \mu.$$

Similarly,

$$\Pi''(1) = \sum_i p_i x_i (x_i - 1) = \mu'_2 - \mu,$$

so that

$$\sigma^2 = \mu'_2 - \mu^2 = \Pi''(1) + \Pi'(1) - \{\Pi'(1)\}^2.$$

2.10.2 Moment Generating Functions

Using an approach similar to that described above we may construct a polynomial function from the moments (about the origin), $\{\mu'_r\}$, of a random variable X. The *moment generating function* (MGF), $M(t)$, is defined by

$$M(t) = E[e^{tX}] = \begin{cases} \sum_i p_i e^{tx_i}, & \text{when } X \text{ is discrete,} \\ \int_{-\infty}^{\infty} e^{tx} f(x) \, dx, & \text{when } X \text{ is continuous.} \end{cases} \quad (2.10.2)$$

(Note that whereas the PGF, $\Pi(z)$, is defined only for discrete random variables, the MGF, $M(t)$, is defined for both discrete and continuous variables.)

When X is continuous, $M(t)$ is the *(two-sided) Laplace transform of the probability density function* $f(x)$.

Assuming that $M(t)$ exists for $|t| \leq T$, say, and expanding $\exp(tX)$ as a power series (and using the linear property of E), we may rewrite (2.10.2) in the form,

$$M(t) = \sum_{r=0}^{\infty} \frac{t^r}{r!} E[X^r] = \sum_{r=0}^{\infty} \mu'_r \frac{t^r}{r!}, \quad (2.10.3)$$

(defining $\mu'_0 = 1$). Thus, if we are given the form of $M(t)$ for a particular distribution we may determine the moments up to any order by expanding $M(t)$ as a power series in t and then obtain μ'_r as the coefficient of $\{t^r/r!\}$. Alternatively, if a power series expansion of $M(t)$ is not readily obtained, μ'_r may be calculated from

$$\mu'_r = \left[\frac{d^r \{M(t)\}}{dt^r} \right]_{t=0} \quad (2.10.4)$$

When X is discrete valued there is a simple relationship between the PGF, $\Pi(z)$, and the MGF, $M(t)$. For, writing $z = e^t$ in (2.10.1) we have,

$$\Pi(e^t) \equiv E[e^{tX}] \equiv M(t). \qquad (2.10.5)$$

Setting $t = 0$ in (2.10.2) gives,

$$M(0) = E[1] = 1, \qquad \text{for all distributions,}$$

and hence

$$\Pi(1) = 1, \qquad \text{for all discrete distributions.}$$

2.10.3 Characteristic Functions

We have noted previously that not all the moments may exist for certain distributions, and in such cases $M(t)$ would not exist in any interval containing the origin. (For otherwise, the moments of all orders could be calculated from (2.10.4).) However, apart from its use in calculating moments $M(t)$ has other useful properties (particularly in regard to the study of sums of random variables), and in order to exploit these properties more generally we now introduce a modified version of $M(t)$, called the *characteristic function*, which, as will be shown, *exists for all distributions*, irrespective of whether or not some or all of the moments fail to exist.

Regard the variable t in (2.10.2) as complex-valued, so that $M(t)$ is now defined over the complex plane. If some of the moments do not exist $M(t)$ will not exist on the real axis *but will always exist on the imaginary axis*. Now set $t = iu$, where u is real, and write

$$\phi(u) \equiv M(iu) = E[e^{iuX}]. \qquad (2.10.6)$$

The function, $\phi(u)$, is called the *characteristic function of X*, and exists for all distributions as is seen by writing in, for example, the continuous case,

$$\phi(u) = \int_{-\infty}^{\infty} e^{iux} f(x)\, dx, \qquad (2.10.7)$$

so that,

$$|\phi(u)| = \left| \int_{-\infty}^{\infty} e^{iux} f(x)\, dx \right| \leq \int_{-\infty}^{\infty} |e^{iux}| f(x)\, dx$$

$$= \int_{-\infty}^{\infty} f(x)\, dx = 1, \qquad \text{all } u.$$

(Heuristically, the difference between the behaviour of $M(t)$ and $\phi(u)$ is due to the fact that the function $\exp(tx)$ changes its character completely when t is transferred from the real axis to the imaginary axis. When t is real and

$x > 0$, $\exp(tx)$ "explodes" as $t \to \infty$, so that $f(x)$ has to decay at a comparable rate for the integral in (2.10.2) to converge. On the other hand, the function $\exp(iux)$, (u real), is just the sum of sine and cosine functions, and is always bounded, having unit modulus.)

The characteristic function uniquely determines the discrete probability distribution or the probability density function, and thus completely specifies the properties of the random variable, X. To see this we note that in, e.g., the continuous case, $\phi(u)$ *is the Fourier transform of* $f(x)$, and hence $f(x)$ may be recovered from $\phi(u)$ by the usual "inversion formula" for Fourier transforms (see Chapter 4, Section 4.5) namely,

$$f(x) = \frac{1}{2\pi} \int_{-\infty}^{\infty} e^{-iux} \phi(u)\, du. \tag{2.10.8}$$

In the discrete case (2.10.6) gives

$$\phi(u) = \sum_k p_k\, e^{iux_k} \tag{2.10.9}$$

and the inversion formula may be shown to be (Parzen (1960), pp. 402, 408)

$$p_k = \lim_{U \to \infty} \frac{1}{2U} \int_{-U}^{U} e^{-iux_k} \phi(u)\, du. \tag{2.10.10}$$

The function $\phi(u)$ has the same properties as $M(t)$ (when it exists), and in particular, $\phi(u)$ still generates the moments about the origin, $\{\mu_r'\}$. Expanding $\exp(itX)$ in (2.10.6) we have,

$$\phi(u) = \sum_{r=0}^{\infty} \frac{(iu)^r}{r!} \mu_r', \tag{2.10.11}$$

whenever the moments exist.

2.10.4 Cumulant Generating Functions

Let X be a random variable with MGF $M(t)$. Write,

$$K(t) = \log_e\{M(t)\} \tag{2.10.12}$$

The function $K(t)$ is called the *cumulant generating function of X.*

When $K(t)$ may be expanded in a power series of the form

$$K(t) = \kappa_1 t + \kappa_2 \frac{t^2}{2!} + \ldots + \kappa_r \frac{t^r}{r!} + \ldots \tag{2.10.13}$$

the coefficient of $(t^r/r!)$, κ_r, is called *the rth cumulant of X.* (Note that, since $M(0) = 1$, $K(0) = 0$, so that there is no constant term in the expansion of $K(t)$.)

The cumulants, $\{\kappa_r\}$, are closely related to the moments, and by expanding $M(t)$ and comparing coefficients of powers of t in (2.10.12) we may derive expressions for each of the cumulants in terms of the moments, although such expression rapidly becomes more complicated as r increases. In particular, we have for the first four cumulants,

$$\kappa_1 = \mu, \qquad \kappa_2 = \sigma^2, \qquad \kappa_3 = \mu_3, \qquad \kappa_4 = \mu_4 - 3\sigma^4. \qquad (2.10.14)$$

2.11 SOME SPECIAL DISTRIBUTIONS

So far we have discussed discrete probability distributions and probability density functions in fairly general terms. In this section we list a number of special distributions which feature prominently in applied probability.

I. *Binomial distribution*

Consider a trial which has only two possible results, "success" and "failure", with probabilities p and $(1-p)$, respectively. Let R denote the total number of "successes" which occur in n independent repetitions of the trial. Then,

$$p_i = p[R = i] = \frac{n!}{i!(n-i)!} p^i(1-p)^{n-i}, \qquad i = 0, 1, \ldots, n. \qquad (2.11.1)$$

This result is easily verified by observing that $\{p^i(1-p)^{n-i}\}$ is the probability of i successes (and $(n-i)$ failures) in some specified order, and also that the i successes and $(n-i)$ failures could occur in $nC_i = n!/[i!(n-i)!]$ different arrangements.

The discrete probability distribution given by (2.11.1) is called the *binomial distribution with parameters n, p*. (The term "parameters" is used quite generally to refer to a set of constants in a distribution whose values may vary from one application to another.) It may be checked from (2.11.1) that

$$\sum_{i=0}^{n} p_i = \{p + (1-p)\}^n = 1, \qquad \text{for all } n, p.$$

Also, by substituting (2.11.1) in (2.8.4), (2.8.5), and observing that here $x_i = i, i = 0, \ldots, n$, we find that for the binomial distribution, mean, $\mu = np$; variance, $\sigma^2 = np(1-p)$. Typical examples of random variables having binomial distributions are
(a) the number of heads in n independent spins of a coin;
(b) the number of defective articles in a randomly selected batch of n articles; and

(c) the number of boys in a family of n children (assuming that the sexes of the children may be regarded as independent events).

II. Poisson distribution

Consider the limiting form of the binomial distribution when $n \to \infty$, and $p \to 0$, in such a way that $np = a$, a positive constant. Writing $p = a/n$ in (2.11.1) we have

$$p_i = \left\{ \frac{n(n-1)\ldots(n-i+1)}{n^i} \right\} \left(1 - \frac{a}{n}\right)^{-i} \left\{ \left(1 - \frac{a}{n}\right)^n \frac{a^i}{i!} \right\}.$$

As $n \to \infty$, we have for each fixed i,

$$p_i \to e^{-a} \frac{a^i}{i!}, \qquad i = 0, 1, 2, \ldots. \tag{2.11.2}$$

The discrete probability distribution given by (2.11.2) is called the *Poisson distribution with parameter a*. In practice, it is used as an approximation to the binomial distribution in cases where n is "large" and p is "small", i.e. in cases where we are interested in the number of occurrences of a rare event in a large number of independent trials.

Note that

$$\sum_{i=0}^{\infty} p_i = e^{-a} \left(\sum_{i=0}^{\infty} \frac{a^i}{i!} \right) = 1, \qquad \text{all } a,$$

and that, by considering the limiting forms of the expressions for the mean and variance of the binomial distribution, we find that for the Poisson distribution, mean, $\mu = a$; variance, $\sigma^2 = a$.

In addition to its use as an approximation to the binomial distribution the Poisson distribution plays an important role in the study of "*Poisson processes*", i.e. random processes which describe the pattern of occurrences of "random events". Suppose that certain events occur independently at random instants of time, at a constant mean rate of λ events per unit time interval. Well known examples of this type of phenomena are

(a) the emission of α-particles from a radioactive substance;

(b) the arrival of calls at a telephone exchange; and

(c) the arrival of cars at a particular road junction.

For each t, let $N(t)$ denote the total number of events which occur in the interval $(0, t)$. The random process, $N(t)$, is called a *Poisson process*, and a basic result states that, for each t, the random variable $N(t)$ has an *exact Poisson distribution, with parameter $a = \lambda t$*.

This result may be derived heuristically as follows: Divide the time interval $(0, t)$ into n sub-intervals, each of length $\Delta t = t/n$. Since the events occur "at random" we have,

$$p[\text{an event occurs in an interval } \Delta t] \doteq \lambda . \Delta t.$$

(If, for example, the mean rate of occurrence is 1 event every 10 seconds—so that $\lambda = 1/10$—then the probability of an event occurring in an interval of 2 seconds would be $(1/10)2 = 1/5$.)

Also, since the events occur independently,

$$p[2 \text{ events in an interval } \Delta t] \doteq \lambda^2 (\Delta t)^2 = o(\Delta t)$$

Similarly,

$$p[3 \text{ events in an interval } \Delta t] = O(\Delta t^3) = o(\Delta t),$$

and so on. We may thus ignore the possibility of more than one event occurring in an interval Δt, and consequently we may assume that for each sub-interval there are only two possibilities; either an event occurs or it does not. Hence, with this approximation, the total number of events in the interval $(0, t)$ has a binomial distribution with parameters $n, p = (\lambda \Delta t)$. To recover the original situation we must now let $\Delta t \to 0$, in which case $n \to \infty$, $p \to 0$, and $np = n . (\lambda t/n) = \lambda t$. Hence, the number of events in $(0, t)$, $N(t)$, has an exact Poisson distribution with parameter (λt). Specifically,

$$p[N(t) = i] = e^{-\lambda t} \frac{(\lambda t)^i}{i!}, \qquad i = 0, 1, \dots . \qquad (2.11.3)$$

III. *Normal (Gaussian) distribution*

The normal (or Gaussian) distribution is defined by the probability density function,

$$f(x) = \frac{1}{(2\pi\sigma^2)^{1/2}} \exp[-(x - \mu)^2/(2\sigma^2)], \qquad -\infty < x < \infty. \qquad (2.11.4)$$

A continuous random variable, X, having the above probability density function is said to have *a normal distribution (or to be "normally distributed") with parameters* μ, σ^2, and we write $X = N(\mu, \sigma^2)$.

The normal distribution is by far the most important distribution in probability theory and statistical inference, and underlies most of the established forms of tests of hypotheses. Its preeminent rôle in applied probability is usually attributed to a fundamental result, called the *Central Limit Theorem*, which states, roughly speaking, that if a random variable X is generated by adding together a large number of other independent random variables then X will always have an approximate normal distribution, irrespective of the form of the probability distribution of the individual "component" random variables. In view of this remarkable result it is reasonable to suppose that the errors involved in the measurement of most physical quantities would have distributions very close to normality since such errors would generally arise as the combined effect of a large number of independent sources of error. It then follows that the measured quantity itself (i.e. "true value" + error) would also have an approximate normal distribution.

(We have, incidentally, used above the expression "independent random variables" although as yet we have not given it a precise meaning. For the moment it may be noted simply that the term "independent" is used for random variables in a closely analogous sense to that in which it is used for "events"; see Section 2.4.1. The formal definition of independence for random variables is given in Section 2.12.)

To check that (2.11.4) represents a valid form of probability density function we must show that $I = \int_{-\infty}^{\infty} f(x)\, dx = 1$, for all μ, σ^2. To show this, change the variable from x to $u = (x - \mu)/\sigma$; so that,

$$I = \frac{1}{(2\pi)^{1/2}} \int_{-\infty}^{\infty} \exp(-\tfrac{1}{2}u^2)\, du = 1, \qquad (2.11.5)$$

using the well-known result that

$$\int_{-\infty}^{\infty} \exp(-\tfrac{1}{2}u^2)\, du = (2\pi)^{1/2}.$$

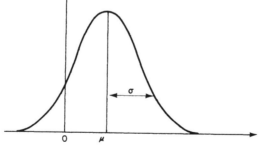

Fig. 2.9. Normal probability density function.

The general shape of the normal distribution is shown in Fig. 2.9. It is not difficult to show, by direct substitution of (2.11.4) in (2.8.6) and (2.8.7), that the *parameters* μ, σ^2, *represent the mean and variance, respectively*. (This is, of course the reason why these parameters are labelled in this way.) Thus, *if it is known that a random variable X has a normal distribution its probability density function is uniquely determined by its mean and variance.*

One of the very useful features of the normal distribution is the fact that linear transformations of normally distributed variables are also normally distributed. For, suppose $X = N(\mu, \sigma^2)$ and $Y = b + cX$, b and c being constants. Using the fact that the MGF of X is $M_X(t) = \exp(\mu t + \frac{1}{2}\sigma^2 t^2)$ (see Table 2.1), we can evaluate the MGF of Y, $M_Y(t)$, from

$$M_Y(t) = E[\exp(tY)]$$
$$= \exp(tb)E[e^{(ct)X}]$$
$$= \exp(tb)M_X(ct)$$
$$= \exp[(b + c\mu)t + \frac{1}{2}c^2\sigma^2 t^2],$$

which is the MGF of a normal distribution with mean $(b + c\mu)$ and variance $c^2\sigma^2$. This result may be summarized as:

$$\text{if } X = N(\mu, \sigma^2), \text{ then } (b + cX) = N(b + c\mu, c^2\sigma^2).$$

Setting $b = -\mu/\sigma$, $c = 1/\sigma$, it follows that

$$(X - \mu)/\sigma = N(0, 1). \qquad (2.11.6)$$

This special form (i.e. with $\mu = 0$, $\sigma = 1$), is called the *standardized normal distribution*, and its probability density function and cumulative distribution function are usually denoted by $\phi(x)$, $\Phi(x)$, respectively, i.e.

$$\phi(x) = \frac{1}{(2\pi)^{1/2}} \exp(-\tfrac{1}{2}x^2), \qquad (2.11.7)$$

$$\Phi(x) = \frac{1}{(2\pi)^{1/2}} \int_{-\infty}^{x} \exp(-\tfrac{1}{2}u^2)\, du. \qquad (2.11.8)$$

There are extensive tables of the functions $\phi(x)$, $\Phi(x)$, and these, together with the aid of (2.11.6), enable us to evaluate probabilities for normal distributions with general values of μ and σ^2, as follows.

If $X = N(\mu, \sigma^2)$, then, for any a, b, $(a \le b)$,

$$p[a \le X \le b] = p\left[\left(\frac{a - \mu}{\sigma}\right) \le \left(\frac{X - \mu}{\sigma}\right) \le \left(\frac{b - \mu}{\sigma}\right)\right]$$

$$= \Phi\left(\frac{b - \mu}{\sigma}\right) - \Phi\left(\frac{a - \mu}{\sigma}\right). \qquad (2.11.9)$$

In particular, setting $a = \mu - 1 \cdot 96\sigma$, $b = \mu + 1 \cdot 96\sigma$, we find that,

$$p[|X - \mu| > 1 \cdot 96\sigma] = 1 - p[|X - \mu| \le 1 \cdot 96\sigma]$$
$$= 1 - \{\Phi(1 \cdot 96) - \Phi(-1 \cdot 96)\} = 0 \cdot 05. \qquad (2.11.10)$$

Similarly, we find that,

$$p[|X - \mu| > 2 \cdot 57\sigma] = 1 - \{\Phi(2 \cdot 57) - \Phi(-2 \cdot 57)\} = 0 \cdot 01. \qquad (2.11.10a)$$

In other words, the ordinates at the points $(\mu \pm 1 \cdot 96\sigma)$ "cut off" 5% of the total area under the normal probability density function, and the ordinates at the points $(\mu \pm 2 \cdot 57\sigma)$ "cut off" 1% of the total area under the same curve.

As noted above, the MGF of the normal distribution, $N(\mu, \sigma^2)$, is given by

$$M(t) = \exp(\mu t + \tfrac{1}{2}\sigma^2 t^2).$$

Hence, the cumulant generating function, $K(t)$, is

$$K(t) = \log_e\{M(t)\} = \mu t + \tfrac{1}{2}\sigma^2 t^2.$$

The interesting point about this result is that here $K(t)$ is a *finite* polynomial, terminating with the term in t^2. It thus follows that, for the normal distribution,

$$\kappa_r = 0, \qquad r \geq 3. \qquad (2.11.11)$$

This result provides an important characterization of the normal distribution.

IV. *Uniform (rectangular) distribution*

A continuous random variable X whose values are restricted to a finite interval (a, b) is said to have a *uniform (or rectangular) distribution* if its probability density function $f(x)$ *is constant over the interval* (a, b). In this case,

$$f(x) = \begin{cases} c, & a \leq x \leq b \\ 0, & \text{otherwise.} \end{cases}$$

Since $\int_{-\infty}^{\infty} f(x)\, dx = 1$, we must have $c = 1/(b - a)$, so that,

$$f(x) = \begin{cases} 1/(b - a), & a \leq x \leq b \\ 0, & \text{otherwise.} \end{cases} \qquad (2.11.12)$$

(The name "rectangular distribution" is due to the fact that the shape of $f(x)$ is just a rectangle on the interval (a, b).)

Let (c, d) be a sub-interval of (a, b). Then,

$$p[c \leq X \leq d] = \int_{-\infty}^{\infty} f(x)\, dx = \frac{(d - c)}{(b - a)},$$

and we note that the above probability depends only on the length of the subinterval, and not on the location of the end points, c, d. Thus, a random variable which has a uniform distribution has no "preference" for any of the various values in its range. It is thus the continuous analogue of the discrete random variable which describes the results of an experiment with a finite number of equally likely outcomes.

The mean and variance of the uniform distribution (2.11.12) are easily shown to be

$$\mu = (a + b)/2, \qquad \sigma^2 = (b - a)^2/12. \qquad (2.11.13)$$

V. *Exponential distribution*

A continuous random variable X, taking positive values only, is said to have an *exponential distribution with parameter* $\lambda (>0)$ if its probability density function is of the form

$$f(x) = \begin{cases} \lambda \exp(-\lambda x), & x \geq 0 \\ 0, & x < 0. \end{cases} \qquad (2.11.14)$$

It is readily checked that $\int_{-\infty}^{\infty} f(x)\, dx = 1$ for all $\lambda > 0$, and that the mean and variance are,

$$\mu = 1/\lambda, \qquad \sigma^2 = 1/\lambda^2.$$

One form in which the exponential distribution arises is in the context of Poisson processes. Let $N(t)$ be a Poisson process in which the mean rate of occurrence of events is λ, and let T denote the time interval up to the occurrence of the first event. Clearly, T is a continuous positive random variable with probability density function $f(t)$, say. Let $F(t)$ denote the cumulative distribution function of T. Then, for $t \geq 0$,

$$F(t) = p[T \leq t] = 1 - p[T > t]$$
$$= 1 - p[N(t) = 0]$$
$$= 1 - \exp(-\lambda t), \qquad \text{using (2.11.3)}$$

Hence,

$$f(t) = F'(t) = \exp(-\lambda t), \qquad t \geq 0, \qquad (2.11.15)$$

so that *T has an exponential distribution with parameter* λ. More generally, it may be shown (Parzen (1962), p. 135) that *the interval between any pair of successive events in a Poisson process has the exponential distribution* (2.11.15).

As a consequence of this result the exponential distribution is often used as a *model for the life times* of items which do not exhibit any form of "ageing effect". If, whenever such an item fails it is immediately replaced by a new one of a similar type, then the assumption that there is no ageing effect means that the items fail at random time instants, so that the times at which items are replaced by new ones form a Poisson process. It then follows that the life times of the individual items, which are now the intervals between successive events in the above Poisson process, each have exponential distributions.

VI. *Chi-square* (χ^2) *distribution*

A continuous random variable X, taking positive values only, is said to have a *chi-square* (χ^2) *distribution with* ν *degrees of freedom* (ν *being a positive integer*) if its probability density function is of the form,

$$f(x) = \begin{cases} [2^{\nu/2}\Gamma(\nu/2)]^{-1}x^{\frac{1}{2}\nu-1}\exp(-x/2), & x \geq 0 \\ 0, & x < 0. \end{cases} \quad (2.11.16)$$

(Note that the parameter ν associated with this distribution is called the "degrees of freedom" (d.f.) of the distribution.) Here, Γ denotes the gamma function, defined for any $k > 0$ by

$$\Gamma(k) = \int_0^\infty x^{k-1}\exp(-x)\,dx.$$

(If k is any positive integer, we can show by repeated integration by parts that $\Gamma(k) = (k-1)!$) The constant on the right-hand side of (2.11.16) is the "normalizing factor" which ensures that $\int_{-\infty}^\infty f(x)\,dx = 1$, $\nu = 1, 2, \ldots$.

For this distribution the mean and variance turn out to be (Parzen (1960), p. 218),

$$\mu = \nu, \qquad \sigma^2 = 2\nu. \quad (2.11.17)$$

Note that the *exponential distribution* may be regarded as a special case of the *chi-square distribution in which* $\nu = 2$, so that (2.11.16) becomes

$$f(x) = \begin{cases} \frac{1}{2}\exp(-x/2), & x \geq 0 \\ 0, & x < 0, \end{cases} \quad (2.11.18)$$

corresponding to an exponential distribution with parameter $\lambda = \frac{1}{2}$.

The chi-square distribution arises most naturally as the *probability density function of the sum of squares of independent standardized normal variables*. Thus, if $X_1, \ldots, X_\nu$, are independent normal variables with $X_i = N(0, 1)$, $i = 1, \ldots, \nu$, it may be shown (Meyer (1965), p. 201) that

$$X = X_1^2 + X_2^2 + \ldots + X_\nu^2$$

has probability density function given by (2.11.16). For this reason, it is customary to write,

$$\chi_\nu^2 = \sum_{i=1}^\nu X_i^2, \quad (2.11.19)$$

meaning that the right-hand side of (2.11.19) has a chi-square distribution with ν d.f.

As $\nu \to \infty$, the chi-square distribution tends to the form of a normal distribution with the same mean and variance. Hence, if $X = \chi_\nu^2$, and ν is large, we may use the approximation,

$$X \doteq N(\nu, 2\nu). \tag{2.11.20}$$

If the $\{X_i\}$ in (2.11.19) are not independent but satisfy k linear independent "constraints", of the form

$$\sum_{j=1}^{\nu} a_{ij} X_j = 0, \qquad i = 1, \ldots, k,$$

(where the $\{a_{ij}\}$ are constants), then it may be proved that $X = \sum_{i=1}^{\nu} X_i$ *still has a χ^2 distribution, but with $(\nu - k)$ d.f. instead of ν d.f.; i.e. one degree of freedom is "lost" for each constraint.*

In the Table 2.11.1 we list the mean, variance and moment generating function for each of the above distributions. The derivation of these results will be found in Parzen (1960), Chapter 5.

Table 2.11.1. *Means, variances and moment generating functions for some special distributions*

Distribution	Mean	Variance	Moment generating function
Binomial, parameters n, p	np	$np(1-p)$	$(q + p \exp(t))^n$
Poisson, parameter a	a	a	$\exp[a(\exp t - 1)]$
Normal, parameters μ, σ^2	μ	σ^2	$\exp(\mu t + \frac{1}{2}\sigma^2 t^2)$
Uniform, on the interval (a, b)	$(a+b)/2$	$(b-a)^2/12$	$(\exp(tb) - \exp(ta))/(t(b-a))$
Exponential, parameter λ	$1/\lambda$	$1/\lambda^2$	$(1 - t/\lambda)^{-1}, \quad t < \lambda$
Chi-square, ν degrees of freedom.	ν	2ν	$(1 - 2t)^{-\nu/2}, \quad t < \frac{1}{2}$

2.12 BIVARIATE DISTRIBUTIONS

The theory of random variables discussed so far has dealt only with *univariate distributions*, that is with probability distributions which describe the properties of *single* random variables. However, there are many situations in which the result of an experiment requires several random variables for its full description. For example, if the experiment consists of selecting a person at random (from a given population) and measuring their height and weight then we require two random variables to describe the result. If we are interested in measuring the simultaneous values of pressure, temperature and volume in a gas then the description of the result would require three random variables, and so on. More generally, suppose that the experimental results are described by several random variables, $X_1, X_2, \ldots, X_p$. We can, of course, study the univariate distributions of each variable separately, but this approach would be inadequate if we wished to investigate the interrelationships between the different variables. This objective leads us to the notion of "*multivariate (or joint) probability distributions*".

Most of the new ideas involved may be illustrated by considering the case of *bivariate distributions*, i.e. distributions relating to *two* random variables, X, Y. The original sample space, Ω, is now mapped into the real plane R^2, or a subset of R^2, and as in the univariate theory it is convenient to treat the discrete and continuous cases separately.

I. *Discrete case*

Here, the random variables (X, Y) are restricted to a discrete set of possible values, $\{x_i, y_j\}$, $i = 1, 2, \ldots; j = 1, 2, \ldots$. The set of numbers,

$$p_{ij} = p[X = x_i, Y = y_j], \qquad i = 1, 2, \ldots, \qquad j = 1, 2, \ldots, \qquad (2.12.1)$$

is called the *discrete bivariate probability distribution* of (X, Y), and in all cases satisfies,

(1) $$0 \leqslant p_{ij} \leqslant 1, \qquad \text{all } i, j,$$

(2) $$\sum_i \sum_j p_{ij} = 1.$$

II. *Continuous case*

When (X, Y) are continuous variables we define the *bivariate probability density* $f(x, y)$ by,

$$p[x < X \leqslant x + \delta x, y < Y \leqslant y + \delta y] = f(x, y)\delta x \delta y. \qquad (2.12.2)$$

In all cases, $f(x, y)$ satisfies,

(1) $$f(x, y) \geqslant 0, \qquad \text{all } x, y,$$

(2) $$\int_{-\infty}^{\infty} \int_{-\infty}^{\infty} f(x, y) \, dx \, dy = 1.$$

2.12.1 Bivariate Cumulative Distribution Functions

The *bivariate cumulative distribution function* $F(x, y)$ is defined (in both the discrete and continuous cases) by

$$F(x, y) = p[X \leq x, Y \leq y]. \tag{2.12.3}$$

We then have, analogously to (2.7.2), (2.7.6),

$$F(x, y) = \sum_{i;x_i \leq x} \sum_{j;y_j \leq y} p_{ij}, \tag{2.12.4}$$

or

$$F(x, y) = \int_{-\infty}^{x} \int_{-\infty}^{y} f(u, v) \, du \, dv. \tag{2.12.5}$$

In the latter case, it follows that,

$$f(x, y) = \partial^2 F(x, y)/\partial x \, \partial y \tag{2.12.6}$$

(The function given by (2.12.4) is called a [2] *step-function*.) For bivariate distributions the result corresponding to (2.6.2) is as follows; for any constants, $a, b, c, d, (a \leq b, c \leq d)$:

$$p[a < X \leq b, c < Y \leq d] = F(b, d) - F(b, c) - F(a, d) + F(b, d). \tag{2.12.7}$$

In the continuous case (2.12.7) implies that

$$p[a < X \leq b, c < Y \leq d] = \int_{a}^{b} \int_{c}^{d} f(x, y) \, dx \, dy. \tag{2.12.8}$$

Note that for any "rectangle" (a, b, c, d), the right-hand side of (2.12.7) is always non-negative since it represents a probability. This corresponds to the "non-decreasing" property of univariate cumulative distribution functions.

2.12.2 Marginal Distributions

If we are given the bivariate distribution of (X, Y), we can determine the univariate distributions of the individual variables in the following manner. Consider first the continuous case. We have, letting $y \to \infty$ in (2.12.3),

$$F(x, \infty) = p[X \leq x, Y \leq \infty] = p[X \leq x].$$

Hence, $F(x, \infty)$ is the cumulative distribution function of X, by itself. But we may write from (2.12.5),

$$F(x, \infty) = \int_{-\infty}^{x} f_1(x) \, dx, \tag{2.12.9}$$

where $f_1(x) = \int_{-\infty}^{\infty} f(x, y)\, dy$ is thus the probability density function of X by itself; $f_1(x)$ *is called the marginal probability density function of X.* Similarly,

$$f_2(y) = \int_{-\infty}^{\infty} f(x, y)\, dx, \qquad\qquad (2.12.10)$$

the probability density function of Y by itself, is called the *marginal probability density function of Y.*

In the discrete case, the *marginal distribution* of X is given by

$$p_i = p[X = x_i] = \sum_j p_{ij}, \qquad i = 1, 2, \ldots, \qquad (2.12.11)$$

and the *marginal distribution of Y is*

$$q_j = p[Y = y_j] = \sum_i p_{ij}, \qquad j = 1, 2, \ldots \qquad (2.12.13)$$

Equations (2.12.11), (2.12.13) may be derived in the same way as (2.19.9), (2.12.10), or directly, by writing, e.g.,

$$p[X = x_i] = \sum_j p[(X = x_i) \cap (Y = y_j)].$$

2.12.3 Conditional Distributions

Suppose that after a particular experiment has been performed we are told the value of one of the variables but are not told the value of the other variable. The question then arises: what probability distribution should we assign to the unknown variable in the light of the given information about the other variable? Thus, suppose for example that in the discrete case we are given that $X = x_i$ (some fixed value). We know from the definition of conditional probability (equation (2.4.2)) that, assuming $p[X = x_i] > 0$,

$$p[Y = y_j \mid X = x_i] = \frac{p[Y = y_j, X = x_i]}{p[X = x_i]} = \frac{p_{ij}}{p_i} = \frac{p_{ij}}{\sum_j p_{ij}}.$$

For each fixed i, the sequence of numbers, $\{p_{ij}/\sum_j p_{ij}\}$, $j = 1, 2, \ldots$, is called *the conditional distribution of Y, given* $X = x_i$. Similarly, the conditional *distribution of X, given* $Y = y_j$, is $\{p_{ij}/\sum_i p_{ij}\}$, $i = 1, 2, \ldots$.

We cannot immediately apply the above reasoning to the continuous case since here the conditioning events, $(Y = y)$ or $(X = x)$, would have zero probability, and the notion of conditional probability is then undefined. We may overcome this difficulty as follows. For arbitrary small δx, δy, we have,

$$p[x < X \le x + \delta x \,|\, y < Y \le y + \delta y] = \frac{p[x < X \le x + \delta x, \, y < Y \le y + \delta y]}{p[y < Y \le y + \delta y]}$$

$$= \frac{f(x, y)\delta x \delta y}{f_2(y)\delta y} = \frac{f(x, y)}{\int_{-\infty}^{\infty} f(x, y)\, dx} \, \delta x.$$

Motivated by the limiting form of this result as $\delta y \to 0$, we thus define the *conditional probability density function of X, given* $Y = y$, by

$$f_{X|Y=y}(x) = \frac{f(x, y)}{\int_{-\infty}^{\infty} f(x, y)\, dx}. \qquad (2.12.14)$$

Similarly, the *conditional probability density function of Y, given* $X = x$, is defined by

$$f_{Y|X=x}(y) = \frac{f(x, y)}{\int_{-\infty}^{\infty} f(x, y)\, dy}. \qquad (2.12.15)$$

In, e.g., (2.12.14), the right-hand side is to be regarded *as a function of x*, with the value of y fixed. Equations (2.12.14), (2.12.15) may be rewritten in the forms,

$$f(x, y) \equiv f_{X|Y=y}(x)f_2(y) \equiv f_{Y|X=x}(y)f_1(x) \qquad (2.12.16)$$

(compare equation (2.4.3)).

2.12.4 Independent Random Variables

In Section 2.4.1 we defined independence for two events, E_1, E_2, in terms of equation (2.4.4), or equivalently, in terms of the multiplication law (2.4.5), namely,

$$p(E_1 \cdot E_2) = p(E_1)p(E_2).$$

Similarly, if we have two continuous random variables, X, Y, we say that X and Y are *independent* (or "*independently distributed*") if

$$f(x, y) = f_1(x) \cdot f_2(y), \qquad \text{all } x, y. \qquad (2.12.17)$$

In this case, comparing (2.12.16), (2.12.17), we see that

$$f_{X|Y=y}(x) \equiv f_1(x), \qquad \text{all } y$$

and

$$f_{Y|X=x}(y) \equiv f_2(y), \qquad \text{all } x.$$

Intuitively we may interpret the above results by saying that "Y provides no information regarding X", and vice-versa. In the discrete case, X, Y, are said to be independent if

$$p_{ij} = p_i \cdot q_j, \qquad \text{all } i, j. \tag{2.12.18}$$

Both cases may be covered by expressing independence in terms of the cumulative distribution function, i.e. by saying that X, Y, are independent if

$$F(x, y) = F_1(x) \cdot F_2(y), \tag{2.12.19}$$

where $F_1(x)(\equiv F(x, \infty))$, $F_2(y)(\equiv F(\infty, y))$ denote respectively the cumulative distribution functions of the marginal distributions of X and Y. It follows immediately from (2.12.7), (2.12.19) that for any a, b, c, d, $(a \leq b, c \leq d)$, the *events* $\{a < X \leq b\}$, $\{c < Y \leq d\}$ are independent in the sense that,

$$p\{a < X \leq b \cap c < Y \leq d\} = p\{a < X \leq b\} p\{c < Y \leq d\}.$$

It is often possible to establish independence for continuous variables without evaluating the precise form of $f_1(x), f_2(y)$. In fact, to prove independence it is necessary only to establish that $f(x, y)$ *admits a factorization of the form*

$$f(x, y) = u(x)v(y), \qquad \text{all } x, y, \tag{2.12.20}$$

where $u(x)$ is a function of x only, and $v(y)$ is a function of y only. When (2.12.20) holds, it follows from (2.12.9), (2.12.10) that

$$f_1(x) = \text{const} \times \{u(x)\}, \qquad f_2(y) = \text{const} \times \{v(y)\},$$

and hence (2.12.20) implies (2.12.17). Note, however, that for independence (2.12.20) must hold for *all* x, y. If, e.g., we have,

$$f(x, y) = \begin{cases} Cxy, & 0 \leq x \leq y, \quad 0 \leq y \leq 1, \\ 0, & \text{otherwise,} \end{cases}$$

(C a constant), then X, Y are *not* independent since here the range of X depends on Y. The apparent factorization of $f(x, y)$ in the required form (2.12.21) holds, therefore, only over a restricted range; the full factorization of $f(x, y)$ must be written as

$$f(x, y) = h(x, y) \cdot y, \qquad 0 \leq y \leq 1,$$

where

$$h(x, y) = \begin{cases} Cx, & 0 \leq x \leq y \\ 0, & \text{otherwise.} \end{cases}$$

2.12.5 Expectation for Bivariate Distributions

Let $g(X, Y)$ be any function of the random variables X, Y. We define the "expected value of $g(X, Y)$", written $E[g(X, Y)]$, by

$$E[g(X, Y)] = \begin{cases} \sum_i \sum_j g(x_i, y_j) p_{ij}, & \text{in the discrete case,} \\ \int_{-\infty}^{\infty} \int_{-\infty}^{\infty} g(x, y) f(x, y) \, dx \, dy, & \text{in the continuous case,} \end{cases}$$

$$(2.12.21)$$

assuming, of course, that the double sum or double integral exists. The operator E has the same linear properties as in the univariate case, and it is easily shown that the definition (2.12.21) is "consistent" with the univariate definition in the sense that, if g is a function of one variable only, say X, then $E[g(X)]$ has the same value whether it is computed from the bivariate distribution (i.e. using (2.12.21)) or from the marginal distribution of X.

A particularly important property of E arises when the variables X, Y, are independent. We then have, for any two functions $g(X)$, $h(Y)$, whose expectations exist,

$$E[g(X)h(Y)] = E[g(X)]E[h(Y)]. \qquad (2.12.22)$$

We prove this result for the continuous case only, the proof for the discrete case following in a similar manner. We have from (2.12.21),

$$E[g(X)h(Y)] = \int_{-\infty}^{\infty} \int_{-\infty}^{\infty} g(x)h(y)f(x, y) \, dx \, dy$$

$$= \int_{-\infty}^{\infty} \int_{-\infty}^{\infty} g(x)h(y)f_1(x)f_2(y) \, dx \, dy,$$

using (2.12.17) for independent variables,

$$= \left\{ \int_{-\infty}^{\infty} g(x)f_1(x) \, dx \right\} \left\{ \int_{-\infty}^{\infty} h(y)f_2(y) \, dy \right\}$$

$$= E[g(X)]E[h(Y)].$$

Note that it is *always* true that

$$E[g(X) + h(Y)] = E[g(X)] + E[h(Y)],$$

irrespective of whether or not X and Y are independent, the result following as a consequence of the linearity property of E.

2.12.6 Conditional Expectation

The *conditional expectation* of Y, given $X = x$, written as $E_Y[Y|X = x]$, is defined as the mean of the conditional distribution of Y, given $X = x$. Thus, in the continuous case we have,

$$E_Y[Y|X = x] = \int_{-\infty}^{\infty} y \cdot f_{Y|X=x}(y) \, dy, \qquad (2.12.23)$$

where $f_{Y|X=x}(y)$ is defined by (2.12.15). Similarly, the conditional expectation of X, given $Y = y$ is defined by

$$E_X[X|Y = y] = \int_{-\infty}^{\infty} x \cdot f_{X|Y=y}(x) \, dx. \qquad (2.12.24)$$

(The corresponding expressions for the discrete case are obtained with obvious modifications to (2.12.23), (2.12.24).)

The conditional expectation of Y given $X = x$, may be interpreted as follows. Suppose we repeat the basic experiment an infinite number of times and record all the different pairs of values of X and Y which occurred. We then *retain only those pairs of values in which* $X = x$, and discard all the others. Then $E_Y[Y|X = x]$ represents the *mean of the Y values, averaged only over those pairs of values which are retained.* If, for example, X, Y represent the height and weight of a person randomly selected from a given population, then, e.g., $E_Y[Y|X = 5' \, 9'']$ tells us the average weight of those persons in the population whose heights are all $5' \, 9''$.

In general, the value of $E_Y[Y|X = x]$ will vary as we vary the value of x. (The average weight for persons of height $6' \, 0''$ will obviously be greater than the average weight for persons of height $5' \, 0''$.) Thus, $E_Y[Y|X = x]$ will be a *function* of x, and we may write

$$E_Y[Y|X = x] = \phi(x), \qquad (2.12.25)$$

say, where $\phi(x)$ is called the *regression function of Y on X*. Similarly, the regression function of X on Y is defined by,

$$E_X[X|Y = y] = \psi(y), \qquad (2.12.26)$$

say. Now consider the *random variable,* $\phi(X)$, which we obtain by replacing x by X in the function $\phi(\cdot)$. The random variable $\phi(X)$ is called the *conditional expectation of Y, given X,* and we write,

$$\phi(X) = E_Y[Y|X] \qquad (2.12.27)$$

and similarly,

$$\psi(Y) = E_X[X|Y]. \qquad (2.12.28)$$

We now prove a basic result which states, roughly speaking, that if we compute means by a *"two-stage procedure"*, i.e. if we first compute the conditional mean of Y for each fixed value of X, and then average all these conditional means, we obtain simply the "over-all" mean of Y, $E[Y]$. This result may be formulated more precisely as follows:

Theorem 2.12.1

$$E_X[E_Y[Y|X]] = E[Y], \tag{2.12.29}$$

and

$$E_Y[E_X[X|Y]] = E[X]. \tag{2.12.30}$$

Proof. We consider the continuous case only. Then

$$E_X[E_Y[Y|X]] = E_X[\phi(X)] = \int_{-\infty}^{\infty} \phi(x)f_1(x)\, dx,$$

where $f_1(x)$, the marginal probability density function of X, is given by (2.12.9). Substituting the expression for $\phi(x)$ given by (2.12.23) in the above, we obtain,

$$E_X[E_Y[Y|X]] = \int_{-\infty}^{\infty} \int_{-\infty}^{\infty} y \cdot f_{Y|X=x}(y)f_1(x)\, dx\, dy$$

$$= \int_{-\infty}^{\infty} \int_{-\infty}^{\infty} yf(x, y)\, dx\, dy, \qquad \text{using (2.12.16),}$$

$$= E[Y], \qquad \text{taking } g(x, y) \equiv y \text{ in (2.12.21).}$$

Equation (2.12.30) follows similarly.

More generally, if we consider any function of Y, $g(Y)$, whose expectation exists, then the conditional expectation of $g(Y)$, given $X = x$, is given by,

$$E_Y[g(Y)|X = x] = \int_{-\infty}^{\infty} g(y)f_{Y|X=x}(y)\, dy. \tag{2.12.31}$$

Equation (2.12.29) then generalizes to,

$$E_X[E_Y[g(Y)|X]] = E[g(Y)]. \tag{2.12.32}$$

Moreover, if $h(X)$ is any function of X (such that $E[g(Y)h(X)]$ exists), then

$$E_Y[g(Y)h(X)|X = x] = h(x)E[g(Y)|X = x]$$

(since, conditional on $X = x$, $h(X)$ may be treated as a constant), and hence we have,

$$E_X[E_Y[g(Y)h(X)|X]] = E_X[h(X)E_Y[g(Y)|X]]. \tag{2.12.33}$$

A further important result is the *"least-squares" property of conditional expectation*, given in the following theorem.

Theorem 2.12.2 *For any function $u(x)$ of x write*

$$\mathscr{S}\{u(x)\} = \int_{-\infty}^{\infty} \int_{-\infty}^{\infty} \{y - u(x)\}^2 f(x, y) \, dx \, dy. \qquad (2.12.34)$$

Then, over the class of all functions $u(.)$ for which the right-hand side of (2.12.34) exists, $\mathscr{S}$ is minimized by choosing $u(x) \equiv \phi(x)$, where $\phi(x)$ is the conditional expectation of Y, given $X = x$, as given by (2.12.25).

Proof. We may write $\mathscr{S}$ in the form

$$
\begin{aligned}
\mathscr{S} &= E[\{Y - u(X)\}^2] \\
&= E[[\{Y - \phi(X)\} + \{\phi(X) - u(X)\}]^2] \\
&= E[\{Y - \phi(X)\}^2] + 2E[\{Y - \phi(X)\}\{\phi(X) - u(X)\}] \\
&\quad + E[\{\phi(X) - u(X)\}^2] \\
&= T_1 + 2T_2 + T_3, \qquad (2.12.35)
\end{aligned}
$$

say. Now $T_2 \equiv 0$, for all $u(X)$, since T_2 may be written as,

$$
\begin{aligned}
T_2 &= E_X[E_Y[\{Y - (X)\}\{\phi(X) - u(X)\}|X]] \qquad \text{(using (2.12.33))}, \\
&= E_X[\{\phi(X) - u(X)\}E_Y[\{Y - \phi(X)\}|X]].
\end{aligned}
$$

But,

$$E_Y[\{Y - \phi(X)\}|X] = E_Y[Y|X] - \phi(X) = 0, \qquad \text{all } X.$$

Thus, $T_2 = 0$, and hence,

$$\mathscr{S} = T_1 + T_3.$$

The value of T_1 is unaffected by the choice of $u(x)$, and for all $u(x)$, $T_3 \geq 0$, with equality if $u(x) \equiv \phi(x)$. Hence $\mathscr{S}$ is minimized by choosing $u(x) \equiv \phi(x)$.

The property of conditional expectation has important implications in *prediction theory*. Suppose that we are given the value of X, say $X = x$, with Y unknown, and we wish to construct some function of x, $U(x)$, which we can use *to predict* the value of Y. If we define the "best" predictor as that for which $E[\{Y - u(X)\}^2]$ is a minimum, then it follows immediately from Theorem 2.12.2 that the best choice for $u(x)$ is,

$$u(x) = E_Y[Y|X = x]. \qquad (2.12.36)$$

This point is explored in more detail in Chapter 10.

2.12.7 Bivariate Moments

The moments of a bivariate probability distribution are defined as follows:

Definition 2.12.1 *The (r, s)th moment about the origin is defined by*

$\mu'_{rs} = E[X^r Y^s]$

$$
= \begin{cases} \sum_i \sum_j x_i^r y_j^s p_{ij}, & \text{in the discrete case,} \\ \int_{-\infty}^{\infty} \int_{-\infty}^{\infty} x^r y^s f(x, y) \, dx \, dy, & \text{in the continuous case,} \end{cases} \tag{2.12.37}
$$

$$ r = 1, 2, \ldots ; s = 1, 2, \ldots . $$

In particular,

$$ \mu'_{10} = E[X] = \text{mean of } X = \mu_X, \text{ say,} $$

and

$$ \mu'_{01} = E[Y] = \text{mean of } Y = \mu_Y, \text{ say.} $$

The (r, s)th moment about the means is now defined by

$\mu_{rs} = E[(X - \mu_X)^r (Y - \mu_Y)^s]$

$$
= \begin{cases} \sum_i \sum_j (x_i - \mu_X)^r (y_j - \mu_Y)^s p_{ij}, \\ \int_{-\infty}^{\infty} \int_{-\infty}^{\infty} (x - \mu_X)^r (y - \mu_Y)^s f(x, y) \, dx \, dy \end{cases} \tag{2.12.38}
$$

$$ r = 1, 2, \ldots ; s = 1, 2, \ldots . $$

In particular,

$$ \mu_{20} = E[(X - \mu_X)^2] = \text{variance of } X = \sigma_X^2, \text{ say,} $$

and

$$ \mu_{02} = E[(Y - \mu_Y)^2] = \text{variance of } Y = \sigma_Y^2, \text{ say.} $$

On setting $r = 0$ or $s = 0$ in (2.12.37), (2.12.38), the $\{\mu'_{rs}\}$, $\{\mu_{rs}\}$ reduce to the corresponding moments of the (univariate) marginal distributions of X or Y. However, if both r and s are ≥ 1, the moments become functions of the complete bivariate distribution, and have no analogue in univariate theory.

In particular, setting $r = s = 1$ in (2.12.38) we obtain,

$$\mu_{11} = E[(X - \mu_X)(Y - \mu_Y)] = \begin{cases} \displaystyle\sum_i \sum_j (x_i - \mu_X)(y_j - \mu_Y)p_{ij}, \\[2ex] \displaystyle\int_{-\infty}^{\infty} \int_{-\infty}^{\infty} (x - \mu_X)(y - \mu_Y)f(x, y) \, dx \, dy, \end{cases}$$

which is called the *covariance between X, Y.* We sometimes write cov(X, Y) in place of μ_{11}. Analogous to (2.8.12) we have,

$$\mu_{11} = \mu'_{11} - \mu_X\mu_Y. \tag{2.12.40}$$

To prove (2.12.40) we have in, e.g., the continuous case,

$$\mu_{11} = \int_{-\infty}^{\infty} \int_{-\infty}^{\infty} (xy - \mu_X y - \mu_Y x + \mu_X\mu_Y)f(x, y) \, dx \, dy$$

$$= \int_{-\infty}^{\infty} xyf(x, y) \, dx \, dy - \mu_X \int_{-\infty}^{\infty} \int_{-\infty}^{\infty} yf(x, y) \, dx \, dy$$

$$- \mu_Y \int_{-\infty}^{\infty} \int_{-\infty}^{\infty} xf(x, y) \, dx \, dy + \mu_X\mu_Y$$

$$= \mu'_{11} - \mu_X\mu_Y - \mu_Y\mu_X + \mu_X\mu_Y = \mu'_{11} - \mu_X\mu_Y.$$

The covariance, μ_{11}, measures the degree of linear association between X, Y, in the sense that:

(a) If Y tends to *increase* from μ_Y as X *increases* from μ_X, then μ_{11} will be large and positive.

(b) If Y tends to *decrease* from μ_Y as X *increases* from μ_X then μ_{11} will be large and negative.

(c) If there is "no association" between X and Y, μ_{11} will be small.

Here, the terms "large/small" mean "large/small compared with the variation in X and Y". We therefore "standardize" μ_{11} by dividing it by $(\sigma_X^2 . \sigma_Y^2)^{1/2}$, and write

$$\rho = \mu_{11}/(\sigma_X^2 . \sigma_Y^2)^{1/2}. \tag{2.12.41}$$

The quantity ρ is called the *correlation coefficient between X and Y.* It measures linear association in the same way as μ_{11}, but we will show that ρ has the standardized range, $(-1, +1)$, so that for ρ, "large and positive" means "near $+1$"; "large and negative" means "near -1"; "small" means "near zero". To prove this result we use the following expressions for the mean and variance of linear combinations of X and Y; for any constants a, b,

$$E[aX + bY] = a\mu_X + b\mu_Y,$$
$$\text{var}(aX + bY) = a^2\sigma_X + b^2\sigma_Y^2 + 2ab\mu_{11}. \tag{2.12.42}$$

We now prove,

Theorem 2.12.3 *For all bivariate distributions with finite second order moments,*

$$|\rho| \le 1,$$

with equality if, with probability 1, there is a linear relationship between X and Y.

Proof. For any constant λ, write $Z = X + \lambda Y$. Then for all λ we must have, $0 \le \text{var}(Z) = \sigma_X^2 + \lambda^2\sigma_Y^2 + 2\lambda\mu_{11}$, by (2.12.42). The right-hand side is a quadratic in λ, and the condition that it is non-negative for all λ means that it cannot have distinct real roots, which in turn implies that

$$\mu_{11}^2 \le \sigma_X^2 \cdot \sigma_Y^2,$$

or,

$$\rho^2 = \frac{\mu_{11}^2}{\sigma_X^2 \cdot \sigma_Y^2} \le 1,$$

or,

$$|\rho| \le 1.$$

If $\rho = +1$, then $\text{var}(Z) = (\lambda\sigma_X + \sigma_Y)^2$. Hence, if we take $\lambda = -\sigma_Y/\sigma_X$, $\text{var}(Z) = 0$. It then follows, by Chebyshev's inequality, that $Z = \text{constant} = C$ (say), with probability 1, and therefore, with probability 1,

$$(-\sigma_Y/\sigma_X)X + Y = C,$$

or,

$$Y = C + (\sigma_Y/\sigma_X)X,$$

giving a linear relationship between X and Y. If $\rho = -1$, we find that, with probability 1,

$$Y = C - (\sigma_Y/\sigma_X)X.$$

If the bivariate distribution is such that the regression of, say, Y on X turns out to be *linear*, then $\phi(x) = \alpha + \beta x$, where α, β, are constants. Hence $E_Y[Y|X] = \alpha + \beta X$, and averaging over X,

$$\mu_Y = E[Y] = E_X[\alpha + \beta X] = \alpha + \beta\mu_X.$$

Also,

$$\mu'_{11} = E[XY] = E_X[X[E_Y[Y|X]]] = E_X[\alpha X + \beta X^2],$$

i.e.

$$\mu'_{11} = \alpha\mu_X + \beta\mu'_{20} = \alpha\mu_X + \beta(\sigma_X^2 + \mu_X^2).$$

Solving the above equations for α, β, gives,

$$\alpha = \mu_Y - \beta\mu_X, \qquad \beta = (\mu'_{11} - \mu_X\mu_Y)/\sigma_X^2 = \mu_{11}/\sigma_X^2.$$

Hence, if the regression of Y on X is linear, it must be of the form,

$$E_Y[Y|X = x] = \phi(x) = \mu_Y + \frac{\mu_{11}}{\sigma_X^2}(x - \mu_X). \qquad (2.12.43)$$

The *residual variance* (or "variance about the regression function") is defined by

$$\sigma_{Y.X}^2 = E[\{Y - \phi(X)\}^2].$$

When $\phi(x)$ is linear, we have,

$$\sigma_{Y.X}^2 = E\left[\left\{(Y - \mu_Y) - \frac{\mu_{11}}{\sigma_X^2}(X - \mu_X)\right\}^2\right] = \sigma_Y^2 - \frac{2\mu_{11}^2}{\sigma_X^2} + \frac{\mu_{11}^2}{\sigma_X^4} \cdot \sigma_X^2,$$

i.e.

$$\sigma_{Y.X}^2 = \sigma_Y^2 - \frac{\mu_{11}^2}{\sigma_X^2} = \sigma_Y^2(1 - \rho^2). \qquad (2.12.44)$$

Writing $\sigma_Y^2 = \sigma_{Y.X}^2 + \rho^2\sigma_Y^2$, we see that, if $\rho \sim \pm 1$, $\sigma_{Y.X}^2$ is small, and the regression line "explains" most of the variability of Y; if $\rho \sim 0$, $\sigma_{Y.X}^2$ is large, and most of the variation of Y is due to "scatter" about the regression line.

Independent random variables

If X, Y are independent variables, then,

$$\begin{aligned}\mu_{11} &= E[(X - \mu_X)(Y - \mu_Y)] \\ &= E[(X - \mu_X)]E[(Y - \mu_Y)], \qquad \text{by (2.12.21)} \\ &= 0.\end{aligned}$$

Hence, *if X, Y are independent their correlation coefficient, $\rho = 0$*. It must be stressed, however, that the converse result is not necessarily true, i.e. $\rho = 0$ *does not necessarily imply that X, Y are independent*. For, suppose, e.g., that $X = N(0, 1)$, and let $Y = X^2$. Then $\mu_{11} = E[X^3] = 0$ (since the $N(0, 1)$

distribution is symmetric about the origin) but clearly, X, Y are not independent in this case. (The conditional distribution of Y, given $X = x$, is degenerate, consisting of a unit mass of probability at the point x^2; the unconditional distribution of Y is chi-square with one degree of freedom.) There is, however, one important case where $\rho = 0$ *does* imply independence, namely, when X, Y have a *bivariate normal distribution*—see Section 2.12.9.

When $\rho = 0$, X, Y, are said to be *uncorrelated*, and in this case (2.12.42) simplifies to

$$\text{var}(aX + bY) = a^2 \sigma_X^2 + b^2 \sigma_Y^2. \qquad (2.12.45)$$

Taking $a = 1, b = \pm 1$, we have,

$$\text{var}(X \pm Y) = \sigma_X^2 + \sigma_Y^2 \qquad (2.12.46)$$

Random samples

Suppose now that we have a random sample of n pairs of observations on (X, Y) in which the pair of values (x_i, y_j) occurs n_{ij} times, $i = 1, \ldots k$; $j = 1, \ldots l$, so that, $\sum_{i=1}^{k} \sum_{j=1}^{l} n_{ij} = n$.

The sample means and variances of the X and Y are then given by applying (2.8.1), (2.8.2) to the x-values and y-values separately. The *sample covariance*, which we may denote by m_{11}, is given by replacing p_{ij} by (n_{ij}/n) in (2.12.39), leading to,

$$m_{11} = \frac{1}{n} \sum_{i=1}^{k} \sum_{j=1}^{l} (x_i - \bar{x})(y_j - \bar{y})n_{ij}, \qquad (2.12.47)$$

where $\bar{x} = (1/n) \sum_{i=1}^{k} n_i x_i$ is the sample mean of the x-values, and $\bar{y} = (1/n) \sum_{j=1}^{l} m_j y_j$ is the sample mean of the y-values, with $n_i = \sum_{j=1}^{l} n_{ij}$, $m_j = \sum_{i=1}^{k} n_{ij}$.

Usually, however, the sample would be *ungrouped*, i.e. each pair of values would be separately labelled, even though several pairs may have the same numerical values. In this case we have effectively, $n_{ii} = 1, i = 1, \ldots n, n_{ij} = 0$, $i \neq j$, so that now the sample covariance is given by

$$m_{11} = \frac{1}{n} \sum_{i=1}^{n} (x_i - \bar{x})(y_i - \bar{y}), \qquad (2.12.48)$$

where here,

$$\bar{x} = \frac{1}{n} \sum_{i=1}^{n} x_i, \qquad \bar{y} = \frac{1}{n} \sum_{i=1}^{n} y_i.$$

Similarly, for ungrouped samples the variances of the x and y values are,

$$s_X^2 = \frac{1}{n} \sum_{i=1}^{n} (x_i - \bar{x})^2, \qquad s_Y^2 = \frac{1}{n} \sum_{i=1}^{n} (y_i - \bar{y})^2,$$

and hence the *sample correlation coefficient*, r, is given by

$$r = \frac{m_{11}}{(s_X^2 \cdot s_Y^2)^{1/2}} = \frac{\sum\limits_{i=1}^{n} (x_i - \bar{x})(y_i - \bar{y})}{\left[\left(\sum\limits_{i=1}^{n} (x_i - \bar{x})\right)^2 \left(\sum\limits_{i=1}^{n} (y_i - \bar{y})^2\right)\right]^{1/2}}. \qquad (2.12.49)$$

In the same way as we showed that $-1 \leqslant \rho \leqslant 1$, for all bivariate probability distributions, we may similarly show that

$$-1 \leqslant r \leqslant 1, \qquad \text{for all bivariate samples.}$$

In fact, writing $z_i = x_i + \lambda y_i$, and using the fact that

$$\frac{1}{n} \sum_{i=1}^{n} (z_i - \bar{z})^2 \geqslant 0,$$

for all λ, we obtain

$$\left(\sum_{i=1}^{n} (x_i - \bar{x})(y_i - \bar{y})\right)^2 \leqslant \left(\sum_{i=1}^{n} (x_i - \bar{x})^2\right)\left(\sum_{i=1}^{n} (y_i - \bar{y})^2\right). \qquad (2.12.50)$$

The inequality (2.12.50) is a celebrated result, known as the *Cauchy–Schwarz inequality*.

Corresponding to (2.12.40) we have,

$$m_{11} = \left(\frac{1}{n} \sum_{i=1}^{n} x_i y_i\right) - \bar{x}\bar{y},$$

which provides a computationally more convenient formula for r, namely,

$$r = \frac{\sum\limits_{i=1}^{n} x_i y_i - \left(\sum\limits_{i=1}^{n} x_i\right)\left(\sum\limits_{i=1}^{n} y_i\right)\Big/ n}{\left\{\left[\sum\limits_{i=1}^{n} x_i^2 - \left(\sum\limits_{i=1}^{n} x_i\right)^2\Big/ n\right]\left[\sum\limits_{i=1}^{n} y_i^2 - \left(\sum\limits_{i=1}^{n} y_i\right)^2\Big/ n\right]\right\}^{1/2}}. \qquad (2.12.51)$$

2.12.8 Bivariate Moment-Generating Functions and Characteristic Functions

The bivariate MGF, $M(t_1, t_2)$, (a function of two variables) is defined by

$$M(t_1, t_2) = E[\exp(t_1 X + t_2 Y)]. \qquad (2.12.52)$$

Expanding $\exp(t_1 X + t_2 Y)$ in powers of $(t_1 X + t_2 Y)$, we have, assuming all joint moments exist,

$$M(t_1, t_2) = \sum_{r=0}^{\infty} \sum_{s=0}^{\infty} \frac{t_1^r t_2^s}{r! s!} \mu'_{rs} \qquad (2.12.53)$$

(with $\mu_{00} = 1$).

The bivariate characteristic function, $\phi(u_1, u_2)$, is defined by,

$$\phi(u_1, u_2) = E[\exp(i(u_1 X + u_2 Y))] \equiv M(iu_1, iu_2). \qquad (2.12.54)$$

Given the bivariate MGF, the MGFs of the marginal distributions of X, Y, are easily derived by noting that, e.g.,

$$M(t, 0) = E[\exp(tX + 0 \cdot Y)] = E[\exp(tX)] \qquad (2.12.55)$$

is the MGF of X. Similarly,

$$M(0, t) = E[\exp(0 \cdot X + t \cdot Y)] = E[\exp(tY)] \qquad (2.12.56)$$

is the MGF of Y. Also, writing $t_1 = t_2 = t$ in (2.12.52), we see that the MGF of $(X + Y)$ is given by

$$M_{X+Y}(t) = E[\exp(t(X + Y))] = M(t, t). \qquad (2.12.57)$$

Independent variables

When X, Y, are independent we have the following extremely important result;

$$M(t_1, t_2) = M_X(t_1) M_Y(t_2), \qquad (2.12.58)$$

where $M_X(t)$, $M_Y(t)$, denote the MGFs of the marginal distributions of X, Y, respectively. To prove (2.12.58) we write,

$$M(t_1, t_2) = E[\exp(t_1 X + t_2 Y)] = E[\exp(t_1 X) \cdot \exp(t_2 Y)]$$

$$= E[\exp(t_1 X)] E[\exp(t_2 Y)],$$

since X, Y, are independent,

$$= M_X(t_1) \cdot M_Y(t_2).$$

(Note that, since $M_X(0) = M_Y(0) = 1$, (2.12.58) is quite consistent with (2.12.55) and (2.12.56).)

Combining (2.12.57) and (2.12.58), we obtain the result that, *when and only when X and Y are independent*,

$$M_{X+Y}(t) = M_X(t) \cdot M_Y(t). \qquad (2.12.59)$$

Equation (2.12.59) is a basic result and is of fundamental importance in the study of the distributions of sums of independent random variables.

As an example of the application (2.12.59), we now show that the *sum of two independent normal variables is also a normal variable.*
Let $X = N(\mu_1, \sigma_1^2)$, $Y = N(\mu_2, \sigma_2^2)$. Then

$$M_X(t) = \exp(\mu_1 t + \tfrac{1}{2}\sigma_1^2 t^2), \qquad M_Y(t) = \exp(\mu_2 t + \tfrac{1}{2}\sigma_2^2 t^2).$$

Hence, if X, Y, are independent,

$$M_{X+Y}(t) = \exp[(\mu_1 + \mu_2) + \tfrac{1}{2}(\sigma_1^2 + \sigma_2^2)t^2],$$

which is the MGF of a normal distribution, mean $(\mu_1 + \mu_2)$, variance $(\sigma_1^2 + \sigma_2^2)$. Hence,

$$X + Y = N(\mu_1 + \mu_2, \sigma_1^2 + \sigma_2^2). \qquad (2.12.60)$$

(The above result is known generally as a "reproductive property".)

2.12.9 Bivariate Normal Distribution

The continuous random variables X, Y, are said to have a *bivariate normal distribution with parameters*, $\mu_1, \mu_2, \sigma_1^2, \sigma_2^2, \rho$, if their probability density function is of the form,

$$f(x, y) = \frac{1}{2\pi\sigma_1\sigma_2(1 - \rho^2)^{1/2}}$$

$$\times \exp\left[-\frac{1}{2(1 - \rho^2)}\left(\frac{(x - \mu_1)^2}{\sigma_1^2} - \frac{2\rho(x - \mu_1)(y - \mu_2)}{\sigma_1\sigma_2} + \frac{(y - \mu_2)^2}{\sigma_2^2}\right)\right], \quad (2.12.61)$$

$$-\infty < x < \infty, \quad -\infty < y < \infty.$$

It may be shown by direct integration that $E[X] = \mu_1$, $E[Y] = \mu_2$, $\text{var}(X) = \sigma_1^2$, $\text{var}(Y) = \sigma_2^2$, and $\text{cov}(X, Y) = \rho\sigma_1\sigma_2$, so that ρ represents the correlation coefficient between X and Y. Thus, a bivariate normal distribution is *uniquely determined by the means, variances, and covariance of the two variables.*
If X, Y are uncorrelated, i.e. if $\rho = 0$, then from (2.12.61) it is readily seen that $f(x, y)$ breaks down into the product of two univariate normal probability density functions, one a function of x only, the other a function of y only. It thus follows that, for the bivariate normal distribution, $\rho = 0$ *implies that X, Y are independent.*

To study the marginal distributions of X, Y, consider first the case where $\mu_1 = \mu_2 = 0$. We may then write,

$$f(x, y) = \left[\frac{1}{[2\pi\sigma_2^2(1-\rho^2)]^{1/2}} \exp\left(-\frac{1}{2(1-\rho^2)\sigma_2^2}\left(y - \frac{\rho\sigma_2}{\sigma_1}x\right)^2\right)\right]$$

$$\times \left[\frac{1}{(2\pi\sigma_2^2)^{1/2}} \exp\left(-\frac{x^2}{2\sigma_1^2}\right)\right] \tag{2.12.62}$$

$$= h(x, y) \cdot g(x), \text{ say.} \tag{2.12.63}$$

Now the marginal probability density function of X is given by,

$$f_1(x) = \int_{-\infty}^{\infty} f(x, y) \, dy = g(x) \int_{-\infty}^{\infty} h(x, y) \, dy.$$

Noting that, for fixed x, $h(x, y)$ is a normal probability density function, mean $\rho\sigma_2 x/\sigma_1$, variance $\sigma_2^2(1-\rho^2)$, we have

$$\int_{-\infty}^{\infty} h(x, y) \, dy = 1, \quad \text{all } x.$$

Hence,

$$f_1(x) = g(x) = \frac{1}{(2\pi\sigma_1^2)^{1/2}} \exp\left(-\frac{x^2}{2\sigma_1^2}\right), \tag{2.12.64}$$

showing that the *marginal distribution of X is $N(0, \sigma_1^2)$*. Similarly, the marginal distribution of Y is $N(0, \sigma_2^2)$.

Also, the conditional probability density function of Y, given $X = x$, is

$$f_{Y|X=x}(y) = \frac{f(x, y)}{f_1(x)} = h(x, y).$$

Hence, the *conditional distribution of Y, given $X = x$, is*

$$N((\rho\sigma_2 x/\sigma_1), \ \sigma_2^2(1-\rho^2)).$$

In the general case (where μ_1, μ_2 are not necessarily zero) a similar argument shows that:
(1) The marginal distribution of X is $N(\mu_1, \sigma_1^2)$.
(2) The marginal distribution of Y is $N(\mu_2, \sigma_2^2)$.
(3) The conditional distribution of Y, given $X = x$, is

$$N\left(\mu_2 + \frac{\rho\sigma_2}{\sigma_1}(x - \mu_1), \ \sigma_2^2(1-\rho^2)\right).$$

In particular, we have,

$$E_Y[Y|X = x] = \mu_2 + \frac{\rho\sigma_2}{\sigma_1}(x - \mu_1), \tag{2.12.65}$$

showing that, for the *bivariate normal distribution, the regression of Y on X is linear.* (Similarly, the regression of X on Y is linear.) Moreover, from the above results we see that the variance of the conditional distribution of Y, given $X = x$, is $\sigma_2^2(1 - \rho^2)$, *for all x*. Hence we have,

$$\sigma_{Y.X}^2 = \sigma_2^2(1 - \rho^2) = \sigma_Y^2(1 - \rho^2), \tag{2.12.66}$$

as in all cases of linear regression.

The moment generating function of the bivariate normal distribution is

$$M(t_1, t_2) = \exp[(t_1\mu_1 + t_2\mu_2) + \tfrac{1}{2}(\sigma_1^2 t_1^2 + \sigma_2^2 t_2^2 + 2\rho\sigma_1\sigma_2 t_1 t_2)]. \tag{2.12.67}$$

(To prove (2.12.67), write $M(t_1, t_2)$ in the form,

$$M(t_1, t_2) = E[\exp(t_1 X + t_2 Y)] = E_X[\exp(t_1 X)E_Y[\exp(t_2 Y)|X]].$$

Then use the fact that $E_Y[\exp(t_2 Y)|X]$ is the MGF of the conditional distribution of Y, given X, which by the above results, is the MGF of $N(\mu_2 + \rho(\sigma_2/\sigma_1)(X - \mu_1), \sigma_2^2(1 - \rho^2))$, together with the further fact that the marginal distribution of X is $N(\mu_1, \sigma_1^2)$.)

2.12.10 Transformations of Variables

Suppose (X, Y) are continuous variables with bivariate probability density function, $f(x, y)$.

Let

$$U = g_1(X, Y), \qquad V = g_2(X, Y) \tag{2.12.68}$$

be two new random variables defined in terms of X, Y, by the functions $g_1(x, y)$, $g_2(x, y)$. Assume that g_1, g_2 are differentiable and single-valued, so that the equations

$$u = g_1(x, y), \qquad v = g_2(x, y) \tag{2.12.69}$$

have a unique solution,

$$x = k_1(u, v), \qquad y = k_2(u, v). \tag{2.12.70}$$

Assume, further, that the Jacobian,

$$J = \frac{\partial(u, v)}{\partial(x, y)} = \begin{vmatrix} \dfrac{\partial u}{\partial x} & \dfrac{\partial u}{\partial y} \\ \dfrac{\partial v}{\partial x} & \dfrac{\partial v}{\partial y} \end{vmatrix} \neq 0. \tag{2.12.71}$$

Then the bivariate probability density function, $h(u, v)$, of (U, V) is given by

$$h(u, v) = f\{k_1(u, v), k_2(u, v)\} \cdot \frac{1}{|J|}. \qquad (2.12.72)$$

(Compare (2.12.72) with the corresponding result for univariate distributions, as given by (2.9.22)).

Equation (2.12.72) may be proved as follows. We have, for any u, v,

$$p[U \leqslant u, V \leqslant v] = p[(X, Y) \in R],$$

where R is the region in the (x, y)-plane which is such that

$$\left. \begin{array}{l} g_1(x, y) \leqslant u \\ g_2(x, y) \leqslant v \end{array} \right\} \quad \text{for all } (x, y) \in R.$$

Hence,

$$p[U \leqslant u, V \leqslant v] = \int_R \int f(x, y) \, dx \, dy.$$

Now transform the variables in the above double integral from (x, y) to (u, v), using (2.12.69). Under this transformation, the element of area $(dx \, dy)$ is transformed into $\{(1/|J|) \, du \, dv\}$. Hence,

$$p[U \leqslant u, V \leqslant v] = \int_{-\infty}^{u} \int_{-\infty}^{v} f\{k(u, v), k(u, v)\} \frac{1}{|J|} \, du \, dv,$$

and the required result follows.

2.13 MULTIVARIATE DISTRIBUTIONS

The basic ideas of bivariate probability distributions are readily extended to the general case where, instead of two, we have p random variables, $X_1, X_2, \ldots, X_p$. Thus, for p discrete valued variables we define the *multivariate discrete probability distribution* by,

$$p(i_1, i_2, \ldots, i_p) = p[X_1 = x_{i_1}, X_2 = x_{i_2}, \ldots, X_p = x_{i_p}],$$
$$i_1 = 1, 2, \ldots, \ldots, \quad i_p = 1, 2, \ldots. \qquad (2.13.1)$$

For p continuous variables, we define the multivariate probability density function by

$$p[x_1 \leqslant X_1 < x_1 + \delta x_1, \ldots, x_p \leqslant X_p < x_p + \delta x_p]$$
$$= f(x_1, \ldots, x_p)\delta x_1, \ldots, \delta x_p. \qquad (2.13.2)$$

The cumulative distribution function is defined by

$$F(x_1, x_2, \ldots, x_p) = p[X_1 \leq x_1, X_2 \leq x_2, \ldots, X_p \leq x_p]. \quad (2.13.3)$$

In the discrete case,

$$F(x_1, \ldots, x_p) = \sum_{x_{i_1} \leq x_1, \ldots, x_{i_p} \leq x_p} p(i_1, i_2, \ldots, i_p). \quad (2.13.4)$$

In the continuous case,

$$F(x_1, \ldots, x_p) = \int_{-\infty}^{x_1} \cdots \int_{-\infty}^{x_p} f(u_1, \ldots, u_p) \, du_1, \ldots, du_p, \quad (2.13.5)$$

so that,

$$f(x_1, \ldots, x_p) = \frac{\partial^p F(x_1, \ldots, x_p)}{\partial x_1, \ldots, \partial x_p}. \quad (2.13.6)$$

The random variables $X_1, \ldots, X_p$, are said to be *independent* if the *multivariate probability distribution or probability density function breaks down into the product of the p marginal distributions or density functions*. Thus, in, e.g., the continuous case, $X_1, \ldots, X_p$ are independent if $f(x_1, \ldots, x_p)$ can be written in the form

$$f(x_1, \ldots, x_p) = f_1(x_1), \ldots, f_p(x_p), \quad (2.13.7)$$

where the function f_1 involves x_1 only, $\ldots, f_p$ involves x_p only. (Compare (2.13.7) with (2.12.20).) An equivalent condition (valid for both the discrete and continuous case) is that $F(x_1, \ldots, x_p)$ be expressible in the form,

$$F(x_1, \ldots, x_p) = F_1(x_1), \ldots, F_p(x_p). \quad (2.13.8)$$

Given a suitable function, $g(X_1, \ldots, X_p)$, we define $E[g(X_1, \ldots, X_p)]$ by (in the continuous case),

$$E[g(X_1, \ldots, X_p)] = \int_{-\infty}^{\infty} \cdots \int_{-\infty}^{\infty} g(x_1, \ldots, x_p) f(x_1, \ldots, x_p) \, dx_1, \ldots, dx_p, \quad (2.13.9)$$

with a corresponding definition for the discrete case.

The definitions of marginal distributions, conditional distributions, moments, moment generating functions, and characteristic functions, are given by obvious extensions of the definitions used in the bivariate case, so that, e.g., the multivariate MGF is defined by,

$$M(t_1, \ldots, t_p) = E[\exp(t_1 X_1 + \ldots + t_p X_p)]. \quad (2.13.10)$$

Corresponding to (2.12.22), we have the result that if $X_1, \ldots, X_p$ are *independent*, then

$$E\left[\prod_{i=1}^{p} g_i(X_i)\right] = \prod_{i=1}^{p} E[g_i(X_i)], \qquad (2.13.11)$$

from which it follows that (cf. (2.12.58))

$$M_{\Sigma X_i}(t) = \prod_{i=1}^{p} M_{X_i}(t). \qquad (2.13.12)$$

2.13.1 Mean and Variance of Linear Combinations

Consider a linear combination of $X_1, \ldots, X_p$, say, $\sum_{i=1}^{p} a_i X_i$, where $a_1, \ldots, a_p$ are constant. Let

$$E[X_i] = \mu_i, \quad \text{var}(X_i) = \sigma_{ii}, \quad \text{cov}(X_i, X_j) = \sigma_{ij}, \quad i = 1, \ldots, p; \quad j = 1, \ldots, p.$$

Generalizing (2.12.42), we have,

$$E\left[\sum_{i=1}^{p} a_i X_i\right] = \sum_{i=1}^{p} a_i \mu_i, \qquad (2.13.13)$$

$$\text{var}\left[\sum_{i=1}^{p} a_i X_i\right] = \sum_{i=1}^{p} a_i^2 \sigma_{ii} + 2 \sum\sum_{i<j} a_i a_j \sigma_{ij}. \qquad (2.13.14)$$

Alternatively, (2.13.14) may be written (noting that, when $i = j$, $\text{cov}(X_i, X_j) = \text{var}(X_i) = \sigma_{ii}$),

$$\text{var}\left[\sum_{i=1}^{p} a_i X_i\right] = \sum_{i=1}^{p} \sum_{j=1}^{p} a_i a_j \sigma_{ij}. \qquad (2.13.15)$$

Using matrix notation, (2.13.15) can be written as

$$\text{var}\left[\sum_{i=1}^{p} a_i X_i\right] = a'\Sigma a, \qquad (2.13.16)$$

where $a' = (a_1, \ldots, a_p)$, and

$$\Sigma = \begin{bmatrix} \sigma_{11} & \sigma_{12} & \cdots & \sigma_{1p} \\ \sigma_{21} & \sigma_{22} & \cdots & \sigma_{2p} \\ \cdots\cdots\cdots\cdots\cdots\cdots \\ \sigma_{p1} & \sigma_{p2} & \cdots & \sigma_{pp} \end{bmatrix}$$

is called the *variance–covariance matrix* of $(X_1, \ldots, X_p)$.

If $X_1, \ldots, X_p$ are *mutually uncorrelated*, i.e. if $\sigma_{ij} = 0$, all $i \neq j$, as would certainly be the case if the variables are *independent*, then (2.13.14) simplifies to

$$\text{var}\left[\sum_{i=1}^{p} a_i X_i \right] = \sum_{i=1}^{p} a_i^2 \sigma_{ii}. \qquad (2.13.17)$$

Suppose now that we wish to find the mean and variance of a general (i.e. not necessarily linear) function, $g(X_1, \ldots, X_p)$. If the variation of each X_i about its mean μ_i is small compared with μ_i, we may, as in the univariate case, expand $g(X_1, \ldots, X_p)$ in a Taylor series about the point $(\mu_1, \ldots, \mu_p)$, and retaining only the first two terms we have, as a linear approximation,

$$g(X_1, \ldots, X_p) \sim g(\mu_1, \ldots, \mu_p) + \sum_{i=1}^{p} (X_i - \mu_i)\left[\frac{\partial g}{\partial X_i} \right], \quad (2.13.19)$$

where all the partial derivatives, $\partial g/\partial X_i$, are evaluated at the point $(\mu_1, \ldots, \mu_p)$.

We may now apply (2.13.13), (2.13.15) to the linear expression $\sum_i X_i[\partial g/\partial X_i]$, which lead to the approximate results,

$$E[g(X_1, \ldots, X_p)] \sim g(\mu_1, \ldots, \mu_p), \qquad (2.13.20)$$

$$\text{var}[g(X_1, \ldots, X_p)] \sim \sum_{i=1}^{p} \sum_{j=1}^{p} \left[\frac{\partial g}{\partial X_i} \right]\left[\frac{\partial g}{\partial X_j} \right] \text{cov}(X_i, X_j). \qquad (2.13.21)$$

2.13.2 Multivariate Normal Distribution

Let $X_1, \ldots, X_p$ be p continuous random variables with

$$E[X_i] = \mu_i, \qquad \text{var}(X_i) = \sigma_{ii}, \qquad \text{cov}(X_i, X_j) = \sigma_{ij}, \quad i, j = 1, \ldots, p.$$

We say that $(X_1, \ldots, X_p)$ have a *multivariate normal distribution* if their joint probability density function is of the form,

$$f(x_1, \ldots, x_p) = \frac{1}{(2\pi)^{p/2}\Delta^{1/2}} \exp\left[-\tfrac{1}{2}\left(\sum_{i=1}^{p} \sum_{j=1}^{p} \sigma^{ij}(x_i - \mu_i)(x_j - \mu_j) \right) \right],$$

$$(2.13.22)$$

where $\Delta = |\boldsymbol{\Sigma}| = \det(\boldsymbol{\Sigma})$, $\boldsymbol{\Sigma} = \{\sigma_{ij}\}$ being the variance–covariance matrix of $(X_1, \ldots, X_p)$, and σ^{ij} is the (i, j)th element of the inverse matrix, $\boldsymbol{\Sigma}^{-1}$.

Note that (2.13.22) may be written in matrix form as

$$f(x_1, \ldots, x_p) = \frac{1}{(2\pi)^{p/2}\Delta^{1/2}} \exp\{ -\tfrac{1}{2}[(\boldsymbol{x} - \boldsymbol{\mu})'\boldsymbol{\Sigma}^{-1}(\boldsymbol{x} - \boldsymbol{\mu})] \}, \qquad (2.13.23)$$

where $x' = (x_1, \ldots, x_p)$, $\mu' = (\mu_1, \ldots, \mu_p)$. As in the bivariate case it may be shown that:

(1) All marginal distributions are normal.
(2) The conditional distribution of each variable, given the values of the other variables, is normal.
(3) The regression function of each variable on all the other variables is *linear*.
(4) If the variables are mutually uncorrelated, i.e. if $\sigma_{ij} = 0$, $i \neq j$, then they are *independent*.

Generalizing (2.12.67), the moment generating function of the multivariate normal distribution is given by,

$$M_X(t_1, \ldots, t_p) = \exp\left[\sum_{i=1}^{p} \mu_i t_i + \tfrac{1}{2} \sum_{i=1}^{p} \sum_{j=1}^{p} \sigma_{ij} t_i t_j\right], \qquad (2.13.24)$$

or in matrix form,

$$M_X(t) = \exp[\mu' t + \tfrac{1}{2} t' \Sigma t], \qquad (2.13.25)$$

where $t' = (t_1, \ldots, t_p)$.

Using the above result we may show that linear transformations of normal variables are also normally distributed. For, let

$$Y_i = \sum_{j=1}^{p} a_{ij} X_j, \qquad i = 1, \ldots, p, \qquad (2.13.26)$$

or in matrix form,

$$Y = AX, \qquad (2.13.27)$$

where $A = \{a_{ij}\}$ is a non-singular matrix of constants. The MGF of the multivariate distribution of $Y_1, \ldots, Y_p$ is

$$M_Y(t) = E[\exp(t_1 Y_1 + \ldots + t_p Y_p)] = E[\exp(t' Y)]$$
$$= E[\exp((A't)' X)] = M_X(A't).$$

Hence,

$$M_Y(t) = \exp[(A\mu)' t + \tfrac{1}{2} t'(A\Sigma A')t] \qquad (2.13.28)$$

(using (2.13.25)), which is the MGF of a *multivariate normal distribution, with (vector) mean* $A\mu$, *and variance-covariance matrix,* $(A\Sigma A')$.

Since the marginal distributions of a multivariate normal distribution are also normal, it follows that a *linear combination of normal variables (dependent or independent) is itself normally distributed.* Thus, if $Y = \sum_{i=1}^{p} a_i X_i$, then (using (2.13.13), (2.13.15)), we have,

$$Y = N\left(\sum_{i=1}^{p} a_i \mu_i, \sum_{i=1}^{p} \sum_{j=1}^{p} a_i a_j \sigma_{ij}\right). \qquad (2.13.29)$$

If the X_i's are independent, (2.13.29) becomes,

$$Y = N\left(\sum_{i=1}^{p} a_i\mu_i, \sum_{i=1}^{p} a_i^2\sigma_{ii}\right) \qquad (2.13.30)$$

2.14 THE LAW OF LARGE NUMBERS AND THE CENTRAL LIMIT THEOREM

We conclude this chapter with a brief discussion of two limit theorems which are of fundamental importance in probability theory, namely, the *law of large numbers* and the *central limit theorem*. In essence, both these results deal with the asymptotic behaviour of sample means as the sample size tends to infinity; however, since a random sample may be regarded as a set of independent and identically distributed random variables, the results are more elegantly phrased in terms of the behaviour of *sums of independent random variables*.

It should be mentioned that, over the years, both the law of large numbers and the central limit theorem have been refined and generalized (with a view to weakening the conditions which ensure their validity), and the literature on these theorems is now extensive. Here, we discuss only the "simple" versions of these results which refer to the case of identically distributed variables with finite means and variances. (For the law of large numbers it is, in fact, sufficient to assume only the finiteness of the mean.) For a more general discussion the reader is referred to Parzen (1960), pp. 417, 430, Feller (1950), Chapter 10, and Feller (1966), Chapters 7, 8.

2.14.1 The Law of Large Numbers

Let $X_1, X_2, \ldots, X_n, \ldots$, be a sequence of independent identically distributed random variables, each having finite mean μ and finite variance σ^2. Let

$$\bar{X}_n = \frac{1}{n} \sum_{i=1}^{n} X_i, \qquad n = 1, 2, \ldots.$$

Given any ε, η, (both > 0), $\exists$ n_0 such that,

$$p[|\bar{X}_n - \mu| < \varepsilon] > 1 - \eta, \qquad for\ all\ n \geq n_0. \qquad (2.14.1)$$

Proof. We have for each n,

$$E[\bar{X}_n] = \mu. \qquad (2.14.2)$$

Also, since the X_i are independent variables,

$$\text{var}\left[\sum_{i=1}^{n} X_i\right] = \sum_{i=1}^{n} \text{var}(X_i) = n\sigma^2,$$

so that

$$\text{var}(\bar{X}_n) = \sigma^2/n. \tag{2.14.3}$$

Using Chebyshev's inequality (2.8.16), it now follows that for any $c(>0)$,

$$p[|\bar{X}_n - \mu| \ge c(\sigma/n^{1/2})] \le 1/c^2.$$

Given ε, let $c = (n)^{1/2}\varepsilon/\sigma$. Then,

$$p[|\bar{X}_n - \mu| \ge \varepsilon] < \sigma^2/n\varepsilon^2.$$

Given η, choose $n_0 = \sigma^2/(\eta\varepsilon^2)$. Then,

$$p[|\bar{X}_n - \mu| \ge \varepsilon] < \eta \qquad \text{for all } n \ge n_0,$$

and hence,

$$p[|\bar{X}_n - \mu| < \varepsilon] > 1 - \eta, \qquad \text{for all } n \ge n_0.$$

In terms of random samples we may interpret the above result as saying that, as n increases, the *mean of a random sample of n observations*, $\bar{X}_n$, gets *"closer and closer" (in the sense of (2.14.1)) to μ, the mean of the corresponding probability distribution.*

Alternatively, (2.14.1) may be written in the form

$$\lim_{n \to \infty} p[|\bar{X}_n - \mu| > \varepsilon] = 0 \tag{2.14.4}$$

and the mode of convergence implied by (2.14.4) is known generally as *"convergence in probability".*

Accordingly, we say that "$\bar{X}_n$ converges to μ in probability" as $n \to \infty$, and write,

$$p \lim_{n \to \infty} \bar{X}_n = \mu.$$

Equation (2.14.1) is known, more precisely, as the *weak law of large numbers*. We state, without proof, a deeper version of this result, called the *strong law of large numbers*, which is as follows.

Given any ε, η, (both > 0), $\exists\ n_0$ such that,

$$p[|\bar{X}_n - \mu| < \varepsilon \text{ for all } n \ge n_0] > 1 - \eta. \tag{2.14.5}$$

Note that (2.14.5) may be written in the alternative form,

$$p[\lim_{n \to \infty} \bar{X}_n = \mu] = 1, \tag{2.14.6}$$

and the mode of convergence implied by (2.14.6) is known as *"convergence with probability 1"*, or *"almost sure convergence"*. This is sometimes written as, $\bar{X}_n \overset{a.s.}{\to} \mu$.

2.14.2 Application to the Binomial Distribution

If the random variable R denotes the number of successes in n independent trials then, as explained previously, R has a binomial distribution with parameters n, p, (say). However, R may be written as

$$R = X_1 + X_2 + \ldots + X_n,$$

where the random variable X_i is defined by

$$X_i = \begin{cases} 1, & \text{if the result of the } i\text{th trial is "success",} \\ 0, & \text{if the result of the } i\text{th trial is "failure".} \end{cases}$$

The X_i are then independent identically distributed random variables, with common distribution,

$$p[X_i = 1] = p, \qquad p[X_i = 0] = 1 - p,$$

so that the common mean is

$$\mu = E[X_i] = p.$$

Applying the weak law of large numbers to the sequence $X_1, X_2, \ldots$, where now $\bar{X}_n = R/n$, we have,

$$p[|(R/n) - p| < \varepsilon] > 1 - \eta, \qquad \text{for all } n \geq n_0. \qquad (2.14.7)$$

this result being known as *Bernouilli's theorem*. Similarly, applying the strong law of large numbers we obtain,

$$p[|(R/n) - p| < \varepsilon \quad \text{for all } n \geq n_0] > 1 - \eta, \qquad (2.14.8)$$

which is known as *Borel's theorem*.

Suppose now that we have an event E associated with some general experiment. We may regard the occurrence of E as "success" and the non-occurrence of E as "failure", in which case $p = p(E)$, and (R/n) is the *relative frequency* of E in n repetitions of the experiment. Looked at in this way, the full significance of (2.14.7), (2.14.8) emerges when we realize that both these results provide *a rigorous statement of the convergence of relative frequencies to probabilities*, thereby justifying our practical interpretation of probabilities (defined axiomatically) as relative frequencies of occurrence.

2.14.3 The Central Limit Theorem

The central limit theorem was mentioned briefly in Section 2.11 when it was pointed out that, as a consequence of this theorem, we would expect the sum of a large number of independent random variables to have an approximately normal distribution. We now formulate this result more precisely, as follows.

Central limit theorem Let $X_1, X_2, \ldots, X_n, \ldots$, be a sequence of independent identically distributed random variables, each having finite mean μ and finite variance σ^2. Let

$$S_n = \sum_{i=1}^{n} X_i, \qquad n = 1, 2, \ldots.$$

Then, as $n \to \infty$, the distribution of

$$(S_n - n\mu)/(\sigma\sqrt{n}) \to N(0, 1).$$

Proof. Write

$$Y_i = X_i - \mu, \qquad i = 1, 2, \ldots,$$

$$Z_n = (S_n - n\mu)/(\sigma\sqrt{n}),$$

so that

$$Z_n = \left(\sum_{i=1}^{n} Y_i \right) \Big/ (\sigma\sqrt{n}).$$

Let $M_Y(t)$ denote the common MGF of the $\{Y_i\}$, which we assume exists. Since each Y_i has zero mean and variance σ^2 we may write,

$$M_Y(t) = 1 + \frac{t^2\sigma^2}{2!} + \frac{\mu_3 t^3}{3!} + \cdots$$

$$= 1 + \frac{t^2\sigma^2}{2!} + o(t^2). \qquad (2.14.9)$$

The MGF of $\sum_{i=1}^{n} Y_i$ is then given (using (2.13.12)) by,

$$M_{\Sigma Y_i}(t) = \{M_Y(t)\}^n = \left\{ 1 + \frac{t^2\sigma^2}{2!} + o(t^2) \right\}^n. \qquad (2.14.10)$$

Further, the MGF of Z_n is given by

$$M_{Z_n}(t) = E[\exp(tZ_n)] = E[\exp(t(\Sigma Y_i/\sigma\sqrt{n}))]$$

$$= M_{\Sigma Y_i}\left(\frac{t}{\sigma\sqrt{n}} \right)$$

$$= \left\{ 1 + \frac{t^2}{2n} + o\left(\frac{t^2}{n} \right) \right\}^n,$$

using (2.14.10). Thus, as $n \to \infty$, $M_{Z_n}(t) \to \exp(-t^2/2)$, all t, which is the MGF of the $N(0, 1)$ distribution. Since the MGF of Z_n converges, for each t, to the MGF of the $N(0, 1)$ distribution, it is reasonable to conclude that the probability distribution of Z_n converges, at each point, to the $N(0, 1)$ distribution. This conclusion, though by no means obvious, can, in fact, be proved rigorously by appealing to the "continuity theorem" (see Parzen (1960), p. 425).

It should be noted that our assumption regarding the existence of $M_Y(t)$, the MGF of each Y_i, can be relaxed without any difficulty simply by working with the *characteristic function*, $\phi_Y(u)$, of Y_i, in place of $M_Y(t)$. The proof is then virtually unaltered but has greater generality since the expansion of $\phi_Y(n)$ corresponding to (2.14.10) is valid even when the third and higher order moments of the Y_i do not exist.

Thus, for large n we may write, as an approximation,

$$(S_n - n\mu)/(\sigma\sqrt{n}) \sim N(0, 1), \tag{2.14.11}$$

or equivalently, (see (2.11.6)),

$$S_n \sim N(n\mu, n\sigma^2). \tag{2.14.12}$$

Correspondingly, for the mean, $\bar{X}_n$, of the first n variables we have for large n, writing $\bar{X}_n = (1/n)S_n$, and using (2.11.6),

$$\bar{X}_n \sim N(\mu, \sigma^2/n). \tag{2.14.13}$$

Moreover, using the approach of Section 2.14.2, it follows from the central limit theorem that if R has a binomial distribution with parameters n, p, and if n is large, then approximately,

$$(R - np)/\sqrt{npq} \sim N(0, 1), \tag{2.14.14}$$

or equivalently,

$$R \sim N(np, npq). \tag{2.14.15}$$

This result provides the basis for the use of the normal distribution as an approximation to the binomial distribution in the case where n is large.

2.14.4 Properties of Means and Variances of Random Samples

We remarked at the beginning of Section 2.14 that a random sample of n observations, $x_1, \ldots, x_n$, could be regarded as a set of n independent identically distributed random variables, and indeed this is the interpretation we must use if we wish to study the way in which quantities such as the sample mean or sample variance vary from one sample to another. Thus, in proving the law of large numbers and the central limit theorem we have incidentally proved the following results for the sample mean.

Let $\{x_i\}$ be a random sample of n observations from a distribution with mean μ, variance σ^2, and let $\bar{x} = (\sum_{i=1}^{n} x_i)/n$ denote the sample mean. Then
(1) Irrespective of the form of the distribution of the $\{x_i\}$ we always have the following exact results valid for all values of n,

$$E[\bar{x}] = \mu, \qquad \text{var}[\bar{x}] = \sigma^2/n. \qquad (2.14.16)$$

(2) Irrespective of the form of the distribution of the $\{x_i\}$, we have for *large n*,

$$\bar{x} \sim N(\mu, \ \sigma^2/n). \qquad (2.14.18)$$

(3) If the $\{x_i\}$ are *normally distributed*, then by (2.13.30) we have the *exact* result that, for all values of n,

$$\bar{x} = N(\mu, \ \sigma^2/n). \qquad (2.14.19)$$

Equation (2.14.16) tells us that the average value of $\bar{x}$ *over all samples* is equal to the "true" mean (i.e. the mean of the distribution), μ. When we think of $\bar{x}$ as an "estimate" of μ, we describe this property by the term "*unbiasedness*" and we say that "$\bar{x}$ *is an unbiased estimate of* μ".

We may now enquire whether the sample variance,

$$s^2 = \frac{1}{n} \sum_{i=1}^{n} (x_i - \bar{x})^2,$$

is similarly an unbiased estimate of σ^2. It is well known that s^2 is *not* an unbiased estimate of σ^2, this result following from the identity,

$$\sum_{i=1}^{n} (x_i - \mu)^2 \equiv \sum_{i=1}^{n} (x_i - \bar{x})^2 + n(\bar{x} - \mu)^2, \qquad (2.14.\ 20)$$

which implies that,

$$E[s^2] = E\left[\frac{1}{n} \sum_{i=1}^{n} (x_i - \mu)^2\right] - E[(\bar{x} - \mu)^2]$$

$$= (\sigma^2 - \sigma^2/n) < \sigma^2. \qquad (2.14.21)$$

However, we may easily construct an unbiased estimate of σ^2 simply by changing the divisor in s^2 from n to $(n-1)$. Writing

$$\hat{s}^2 = \frac{1}{n-1} \sum_{i=1}^{n} (x_i - \bar{x})^2, \qquad (2.14.22)$$

we now have,

$$E[\hat{s}^2] = \left(\frac{n}{n-1}\right) E[s^2] = \sigma^2. \qquad (2.14.23)$$

This correction is not particularly important for large samples (i.e. when n is large), but for small samples one would generally use $\hat{s}^2$ rather than s^2 to estimate σ^2. Further, it may be shown that if the $\{x_i\}$ are *normally distributed*, then $\{(n-1)\hat{s}^2/\sigma^2\}$ has a χ^2 *distribution with* $(n-1)$ *degrees of freedom*, i.e.,

$$(n-1)\hat{s}^2/\sigma^2 = \chi^2_{n-1}. \qquad (2.14.24)$$

A formal proof of (2.14.24) may be constructed by first showing that, for normal samples, $\hat{s}^2$ and $\bar{x}$ are independently distributed, and then obtaining the MGF of $\sum_{i=1}^{n}(x_i - \bar{x})^2$ from (2.14.20). Alternatively, the distribution of $\hat{s}^2$ may be derived directly from the multivariate normal distribution of the $\{x_i\}$ via a suitable orthogonal transformation; see, for example, Mood and Graybill (1963), p. 228. However, the plausibility of this result may be seen from the following heuristic argument.

We have, for each i,

$$(x_i - \mu)/\sigma = N(0, 1).$$

Hence, $\sum_{i=1}^{n}[(x_i - \mu)/\sigma]^2$ has a χ^2 distribution with n degrees of freedom, i.e.

$$\frac{1}{\sigma^2} \sum_{i=1}^{n}(x_i - \mu)^2 = \chi^2_n. \qquad (2.14.25)$$

Suppose now that we replace μ by $\bar{x}$ on the left-hand side of (2.14.25). We might expect that this substitution would not affect the general form of the distribution (at least, for large n), but one immediate effect is that the variables $\{(x_i - \bar{x})/\sigma\}$ now satisfy one *linear constraint*, since

$$\sum_{i=1}^{n}[(x_i - \bar{x})/\sigma] \equiv 0.$$

In view of the remarks following equation (2.11.20) we would expect that $\sum_{i=1}^{n}[(x_i - \bar{x})/\sigma]^2$ would now have a χ^2 distribution with $(n-1)$ d.f., and (2.14.24) then follows. It should be noted that (2.14.24) is an *exact result* for normal samples even though the above argument would seem to be valid only for the case of large samples.

Using the fact that the mean and variance of χ^2_{n-1} are $(n-1)$, $2(n-1)$, respectively, (2.14.24) gives for normal samples,

$$E[\hat{s}^2] = \sigma^2, \qquad (2.12.26)$$

$$\text{var}[\hat{s}^2] = 2\sigma^4/(n-1) \qquad (2.12.27)$$

$$\sim 2\sigma^4/n, \qquad \text{for large } n. \qquad (2.12.28)$$

More generally, it may be shown that, irrespective of the distribution of the $\{x_i\}$ (Parzen (1960) p. 370),

$$\text{var}(\hat{s}^2) = \frac{\mu_4}{n} - \frac{\sigma^4(n-3)}{n(n-1)},$$ (2.12.29)

where μ_4 is the fourth moment about the mean of the common distribution of the $\{x_i\}$. Recalling that, for all distributions, the fourth cumulant, κ_4, satisfies

$$\kappa_4 = \mu_4 - 3\sigma^4,$$

(2.12.29) may be written as

$$\text{var}(\hat{s}^2) = \frac{\kappa_4}{n} + \frac{2\sigma^4}{n-1}.$$ (2.12.30)

Recalling further that for normal distributions $\kappa_4 = 0$, we see that (2.12.30) reduces to (2.12.27) in that case.

Chapter 3

Stationary Random Processes

3.1 PROBABILISTIC DESCRIPTION OF RANDOM PROCESSES

In Chapter 1 we discussed the basic ideas of spectral analysis from a more or less intuitive standpoint. We are now in a position to describe these ideas more precisely using the language and framework of the theory of probability developed in the preceding chapter. As a natural starting point to this further discussion we begin by expressing our initial understanding of the term "random process" in a more formal manner.

In our previous discussion we used the term "random (or stochastic) process" to describe a fluctuating quantity, such as a voltage, displacement, or price index, etc., whose behaviour could not be described fully in terms of a deterministic equation. Thus, each time we perform the "experiment" the value of the quantity which we record at each point of time is determined (at least in part) by some random mechanism.

Let us suppose, for the moment, that the quantity which we are studying varies over time, so that we may denote it by $X(t)$. Then at each individual time point t, $X(t)$ is a *random variable*, as defined in Section 2.5, and the complete form of $X(t)$, as t varies over all its possible values, is called a *random process*. In essence, a random process is simply a *random function*, and some authors prefer this more descriptive term. All three terms, random process, stochastic process, and random function, may be regarded as completely synonymous.

We may say then that a random process is a collection, or "family", of random variables, different members of this collection being distinguished, or "*indexed*", by the different values of t to which they correspond. Typically, t will range from 0 to $+\infty$, but in some cases it may make sense to allow t to take negative values; in other cases we may be interested only in a finite time interval, $a \leqslant t \leqslant b$, say, or we may wish to study $X(t)$ only at a discrete set of time points, say every second, or every minute, in which case t takes only the values $0, \pm 1, \pm 2, \ldots$ (in suitable units). We can cover all these cases by saying that, in general, the possible values of t will form some "index set", T. Of course, when t denotes real time, T will be the real line R^1, or a subset of R^1.

We may now state the formal definition of a random process, as follows.

Definition 3.1 *A random (or stochastic) process, $\{X(t)\}$, is a family of random variables indexed by the symbol t, where t belongs to some given index set, T.*

Definition 3.2 *If t takes a continuous range of real values (finite or infinite), so that $T \subseteq R^1$, $\{X(t)\}$ is said to be a continuous parameter process. If t takes a discrete set of values, typically, $t = 0, \pm 1, \pm 2, \ldots$, then $\{X(t)\}$ is said to be a discrete parameter process.*

3.1.1 Realizations and Ensembles

As explained in Section 1.6, the fact that (for each t) $X(t)$ is a random variable means that each time we perform the experiment we will, in general, observe a different value of $X(t)$. Thus, an observed record of a random process is merely one record out of a whole collection of possible records which we might have observed. The collection of all possible records is called the *"ensemble"*, and each particular record is called a *"realization"* of the process—see Fig. 1.8.

We may think of the sample space, Ω, as consisting of a set of elements $\{\omega\}$ each of which corresponds to a particular realization, and we may then denote the various realizations by $X(t, \omega_1), X(t, \omega_2), \ldots$, and a typical realization by $X(t, \omega)$. With this notation the random process itself would be written as $X(t, .)$, but just as we find it more convenient to denote a random variable by X rather than by $X(.)$, we usually suppress the variable ω and denote a random process simply by $X(t)$.

3.1.2 Specification of a Random Process

For each t, $X(t)$ is a random variable and thus has a range of possible values, some of which may be more likely to occur than others. Accordingly, for each t, $X(t)$ will have some *probability distribution*. If $X(t)$ is a *discrete* random variable i.e. if its possible values form a discrete set, $\{x_i\}$, its properties will be described by a *discrete probability distribution*, $P_t(x_i)$, say. However, in the more usual situation $X(t)$ will be a *continuous* random variable (with a continuous range of possible values), and its properties will then be described by its *probability density function*, $f_t(x)$, say (see Section

2.7). Henceforth, unless the contrary is explicitly stated, we will assume that, for each t, $X(t)$ is a continuous random variable with probability density function $f_t(x)$ defined for all x, so that, e.g., the mean and variance of $X(t)$ will be given by

(i) $$\text{mean}\{X(t)\} = E\{X(t)\} = \int_{-\infty}^{\infty} x f_t(x)\, dx = \mu(t) \qquad (3.1.1)$$

say, and

(ii) $$\text{var}\{X(t)\} = E[\{X(t) - \mu(t)\}^2] = \int_{-\infty}^{\infty} (x - \mu(t))^2 f_t(x)\, dx = \sigma^2(t),$$
$$(3.1.2)$$

say.

At this stage we have not made any specific assumptions about the nature of the random process and so we must allow for the possibility that the probability density function of $X(t)$ will change from one time point to the next, i.e. that $f_t(x)$ will depend on both t and x. Consequently, in general, the mean and variance will change from one time point to the next, i.e. *both* $\mu(t)$ *and* $\sigma^2(t)$ *will be functions of* t.

The above approach enables us to discuss the probability distribution of $X(t)$ *for each individual value of* t. However, since we are dealing with a *collection* of random variables (corresponding to the values of the process at *all* time points) we may wish, more generally, to study the *joint variation* of the values of the process at several time points, say at times t_1, t_2, t_3. To do this we need to know the *joint probability distribution* of the three random variables $\{X(t_1), X(t_2), X(t_3)\}$, as specified by their joint probability density function, $f_{t_1, t_2, t_3}(x_1, x_2, x_3)$, say. The individual probability density functions, $f_{t_1}(x), f_{t_2}(x), f_{t_3}(x)$, represent the *marginal distributions* of $X(t_1), X(t_2), X(t_3)$, respectively, and can, of course, be calculated from $f_{t_1, t_2, t_3}(x_1, x_2, x_3)$, as explained in Section 2.12.2. However, we cannot, in general, construct the function $f_{t_1, t_2, t_3}(x_1, x_2, x_3)$ from a knowledge of $f_{t_1}(x), f_{t_2}(x), f_{t_3}(x)$, only, since these functions tell us nothing about the joint variation of the three random variables. If, for example, we wished to evaluate

$$p[X(t_1) \leq a_1, X(t_2) \leq a_2, X(t_3) \leq a_3],$$

we would require the complete form of the joint probability density function, $f_{t_1, t_2, t_3}(x_1, x_2, x_3)$, and hence we could not determine the above probability given only the functions $f_{t_1}(x), f_{t_2}(x), f_{t_3}(x)$.

More generally still we may wish to study the behaviour of the *complete process over all time points*. This situation is not merely of mathematical interest but could well arise in a practical context—for example, we might be interested in the probability that the process never exceeds the value b or falls below the value a (see Fig. 3.1). (This type of problem is frequently

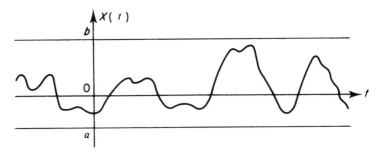

Fig. 3.1

encountered in the subject of "reliability theory".) In order to compute the above probability, namely,

$$p[a \leq X(t) \leq b, \quad \text{all } t], \tag{3.1.3}$$

we need to specify the joint probability distribution of *all* the random variables $\{X(t)\}$ which constitute the process. Now if the range of t is infinite we have to consider an infinite number of random variables; if the range of t is finite but $\{X(t)\}$ is a continuous parameter process we still have an infinite number of variables to consider. Indeed, it is only in the case where t is restricted to a discrete set of values in a finite range that we can reduce the problem to the discussion of a finite number of random variables.

Thus, it would seem that if we wish to describe the behaviour of the complete process it would, in general, be necessary to consider an *infinite dimensional probability distribution*. This presents a formidable mathematical problem which would take us outside the scope of classical probability theory, but fortunately it turns out that under fairly general conditions we may specify the probabilistic properties of the complete process by restricting attention to the behaviour of $\{X(t)\}$ at a *finite number* of time points, $t_1, t_2, \ldots t_n$. We will not attempt here to state the precise conditions under which this result holds (the interested reader will find these points discussed in Kolmogorov (1933) and Doob (1953)), but it may seem intuitively reasonable to suppose that the joint distribution of the values of the process at an arbitrarily large (but finite) number of time points will, for all practical purposes, suffice to describe the overall behaviour of the process; see Fig. 3.2. More precisely, the following theorem may be established (Doob (1953)).

Theorem 3.1 *For any positive integer n, let $t_1, t_2, \ldots, t_n$ be any admissible set of values of t. Then under general conditions the probabilistic structure of the random process $\{X(t)\}$ is completely specified if we are given the joint probability distribution of $\{X(t_1), X(t_2), \ldots, X(t_n)\}$ for all values of n and for all choices of $t_1, t_2, \ldots, t_n$.*

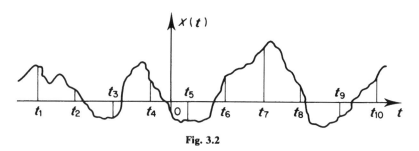

Fig. 3.2

In other words, if we are given the joint probability distribution of $\{X(t_1), X(t_2), \ldots, X(t_n)\}$ for all values of n and all choices of $t_1, \ldots, t_n$, then, in principle, we should have sufficient information to enable us to calculate the probability of any event associated with the complete overall behaviour of the process, such as, e.g., the expression (3.1.3).

3.2 STATIONARY PROCESSES

The class of all random processes is far too large to enable methods of analyses to be designed which will be suitable for all types of processes. Consequently, the development of the subject has proceeded mainly via the study of special types of random processes, such as Markov processes, branching processes, birth and death processes, diffusion processes, etc. We remarked in Section 1.7 that in the classical theory of spectral analysis of random processes we consider only the special class of processes which are termed "stationary". These processes are characterized by the feature that their *statistical properties do not change over time*, and generally arise from a stable physical system which has achieved a "steady-state" mode. (They are sometimes described as being in a state of "statistical equilibrium".) If $\{X(t)\}$ is such a process then since its statistical properties do not change with time it clearly follows that, e.g., $X(1), X(2), X(3), \ldots, X(t), \ldots$, must all have the same probability density function. However, the property of stationarity implies rather more than this since it follows also that, e.g., $\{X(1), X(4)\}, \{X(2), X(5)\}, \{X(3), X(6)\}, \ldots$, must have the same bivariate probability density function, and further that, e.g., $\{X(1), X(3), X(6)\}$,

$\{X(2), X(4), X(7)\}$, $\{X(3), X(5), X(8)\}, \ldots$, must all have the same tri-variate probability density function, and so on. We may summarize all these properties by saying that, for any set of time points $t_1, t_2, \ldots, t_n$, the *joint probability distribution of* $\{X(t_1), X(t_2), \ldots, X(t_n)\}$ *must remain unaltered if we shift each time point by the same amount.* If $\{X(t)\}$ possesses this property it is said to be "*completely stationary*". The formal definition is as follows.

Definition 3.2.1 *The process $\{X(t)\}$ is said to be completely stationary if, for any admissible $t_1, t_2, \ldots, t_n$, and any k, the joint probability distribution of $\{X(t_1), X(t_2), \ldots, X(t_n)\}$ is identical with the joint probability distribution of*

$$\{X(t_1 + k), X(t_2 + k), \ldots, X(t_n + k)\},$$

i.e. if

$$F_{X(t_1),\ldots,X(t_n)}(x_1, \ldots, x_n) \equiv F_{X(t_1+k),\ldots,X(t_n+k)}(x_1, \ldots, x_n), \quad (3.2.1)$$

where $F(.)$ denotes the distribution function of the set of random variables which appear as suffixes.

(We express the above result in terms of the distribution functions of the two sets of variables to avoid the trivial complication of distinguishing between the cases where the variables are discrete and where they are continuous.) The property described by equation (3.2.1) may be summarized by saying that *the probabilistic structure of a completely stationary process is invariant under a shift of the time origin.*

Complete stationarity is, however, a severe requirement, and we therefore relax this by introducing the notion of "*stationarity up to order m*", which is a weaker condition but nevertheless describes roughly the same type of physical behaviour. Under this weaker condition we do not insist that, e.g., the probability distribution of $X(t_1)$ must be *identical* to the probability distribution of $X(t_1 + k)$, but merely that the *main features* of these two distributions should be the same, i.e. that their *moments*, up to a certain order, should be the same. Similarly, we do not insist that the joint distribution of $\{X(t_1), X(t_2)\}$ must be identical to the joint distribution of $\{X(t_1 + k), X(t_2 + k)\}$, but merely that, up to a certain order, their joint moments are equal, and so on. These ideas lead to the following definition:

Definition 3.2.2 *The process $\{X(t)\}$ is said to be stationary up to order m if, for any admissible $t_1, t_2, \ldots, t_n$, and any k, all the joint moments up to order m of $\{X(t_1), X(t_2), \ldots, X(t_n)\}$ exist and equal the corresponding joint moments up to order m of $\{X(t_1 + k), X(t_2 + k), \ldots, X(t_n + k)\}$.*

Thus,

$$E[\{X(t_1)\}^{m_1}\{X(t_2)\}^{m_2} \ldots \{X(t_n)\}^{m_n}]$$
$$= E[\{X(t_1 + k)\}^{m_1}\{X(t_2 + k)\}^{m_2} \ldots \{X(t_n + k)\}^{m_n}], \quad (3.2.2)$$

for any k, and all positive integers $m_1, m_2, \ldots, m_n$ satisfying $m_1 + m_2 + \ldots + m_n \leq m$. In particular, setting $m_2 = m_3 = \ldots = m_n = 0$, we have that, for any t and all $m_1 \leq m$,

$$E[\{X(t)\}^{m_1}] = E[\{X(0)\}^{m_1}] \qquad \text{(taking } k = -t\text{)},$$

$$= \text{a constant, independent of } t. \qquad (3.2.3)$$

Also, for any t, s, and all m_1, m_2, satisfying $m_1 + m_2 \leq m$,

$$E[\{X(t)\}^{m_1}\{X(s)\}^{m_2}] = E[\{X(0)\}^{m_1}\{X(s-t)\}^{m_2}]$$

$$= \text{function of } (s - t) \text{ only.} \qquad (3.2.4)$$

Consider now the following special cases:

(a) *Stationarity up to order* 1 $(m = 1)$.
   If we are given that $\{X(t)\}$ is stationary up to order 1 then this implies only that $E[X(t)] = \mu$ (say), a constant independent of t.

(b) *Stationarity up to order* 2 $(m = 2)$.
   If we are given that $\{X(t)\}$ is stationary up to order 2 then we have,
   (i) $E[X(t)] = \mu$, a constant independent of t, and
   (ii) $E[X^2(t)] = \mu_2'$ (say), a constant independent of t, (using equation (3.2.3) with $m_1 = 2$).
   Hence, $\text{var}\{X(t)\} = \mu_2' - \mu^2 = \sigma^2$ (say), is also a constant, independent of t.
   Further, for any t, s,
   (iii) $E[X(t)X(s)] = \text{function of } (t - s) \text{ only}$ (using equation (3.2.4) with $m_1 = m_2 = 1$). Hence,

$$\text{cov}\{X(t), X(s)\} = E[X(t)X(s)] - \mu^2$$

$$= \text{function of } (t - s) \text{ only.}$$

To summarize, we may say that if a process is stationary up to order 2, then:
   (i) it has the same mean value, μ, at all time points;
   (ii) it has the same variance, σ^2, at all time points; and
   (iii) the covariance between the values at any two time points, s, t, depends only on $(s - t)$, the *interval* between the time points, and not on the location of the points along the time axis.

3.3 THE AUTOCOVARIANCE AND AUTOCORRELATION FUNCTIONS

Let $\{X(t)\}$ be any process which is stationary up to order 2. The properties noted above lead immediately to the introduction of two functions of fundamental importance, namely, the *autocovariance* and the *autocorrelation functions*. These functions are defined as follows:

I. *Continuous parameter processes*

Consider the quantity

$$\text{cov}\{X(t), X(t+\tau)\} = E[\{X(t)-\mu\}\{X(t+\tau)-\mu\}].$$

We know from our previous discussion that this quantity depends only on the value of τ, and does not depend on t. We may thus write

$$R(\tau) = E[\{X(t)-\mu\}\{X(t+\tau)-\mu\}]. \qquad (3.3.1)$$

For a continuous parameter process $R(\tau)$ is defined for *all values of* τ, and is called the *autocovariance function* of $\{X(t)\}$. For each τ, $R(\tau)$ measures the covariance between pairs of values of the process separated by an interval of length τ, the quantity τ usually being termed the "*lag*".

If we now write, for each τ,

$$\rho(\tau) = R(\tau)/R(0), \qquad (3.3.2)$$

then $\rho(\tau)$ is called the *autocorrelation function* of $\{X(t)\}$. Since $R(0)$ may be written as

$$R(0) = E[\{X(t)-\mu\}^2] = \text{var}\{X(t)\} = \sigma^2, \qquad (\text{all } t),$$

$\rho(\tau)$ may be written as

$$\rho(\tau) = \frac{\text{cov}\{X(t), X(t+\tau)\}}{[\text{var}\{X(t)\} \, \text{var}\{X(t+\tau)\}]^{1/2}}. \qquad (3.3.3)$$

Hence, for each τ, $\rho(\tau)$ represents the *correlation coefficient between pairs of values of* $\{X(t)\}$ *separated by an interval of length* τ. (See Section 2.12.7.) Intuitively, we may interpret $\rho(\tau)$ as a measure of the "similarity" between a realization of $X(t)$ and the same realization shifted to the left by τ units (Fig. 3.3).

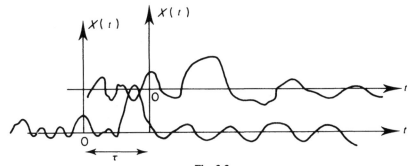

Fig. 3.3

As τ increases we would expect the correlation between $X(t)$ and $X(t+\tau)$ to decrease. (If τ is large then in general the process will, loosely speaking, have "forgotten" at time $(t+\tau)$ the value it assumed at time t.) Consequently, we would expect both $R(\tau)$ and $\rho(\tau)$ to decay to zero as $|\tau| \to \infty$. The typical form of an autocorrelation function is shown in Fig. 3.4.

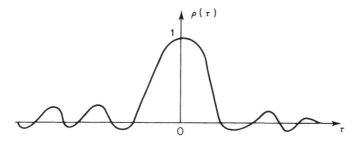

Fig. 3.4. Autocorrelation function of a continuous parameter process.

The rate at which $\rho(\tau)$ decays to zero may then be interpreted as a measure of the "memory" of the process. It should, however, be remembered that the form indicated in Fig. 3.4 is only a "typical" one; there are in fact processes for which $\rho(\tau)$ does not decay to zero as $|\tau| \to \infty$ but instead behaves in a sinusoidal manner (corresponding to a "sinusoidal memory").

II. *Discrete parameter processes*

If $X(t)$ is defined only at discrete values of t, say $t = 0, \pm 1, \pm 2, \ldots$, then both $R(\tau)$ and $\rho(\tau)$ are defined only for $\tau = 0, \pm 1, \pm 2, \ldots$. *From now on we will adopt the convention that when the parameter t is discrete valued we will use t as a suffix, so that the process will then be written as $\{X_t\}$.*

The autocovariance function, $R(\tau)$, may now be written as

$$R(\tau) = E[\{X_t - \mu\}\{X_{t+\tau} - \mu\}], \qquad \tau = 0, \pm 1, \pm 2, \ldots,$$

and the autocorrelation function, $\rho(\tau)$, as,

$$\rho(\tau) = R(\tau)/R(0), \qquad \tau = 0, \pm 1, \pm 2, \ldots.$$

It should be noted that, in this case, both $R(\tau)$ and $\rho(\tau)$ are *sequences* (see Fig. 3.5).

3.3.1 General Properties of $R(\tau)$ and $\rho(\tau)$

All autocovariance functions possess the following properties:

(1) $R(0) = \sigma^2$, (proved above).

(2) $|R(\tau)| \le R(0)$, all τ, (since $|\rho(\tau)| \le 1$, all τ, (see Section 2.12.7)).

(3) When $\{X(t)\}$ is a real valued process,
$R(-\tau) = R(\tau)$, all τ, i.e. $R(\tau)$ is an *even function.*

For, $R(-\tau) = E[\{X(t) - \mu\}\{X(t - \tau) - \mu\}]$

$$= E[\{X(t + \tau) - \mu\}\{X(t) - \mu\}],$$

(adding τ to each time point),

$$= R(\tau).$$

(The reason for the restriction that $X(t)$ should be real valued will emerge later when the autocovariance function of a complex valued process is defined.)

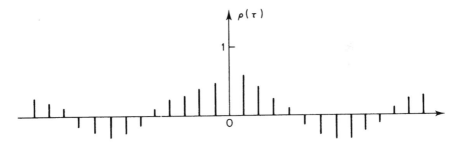

Fig. 3.5. Autocorrelation function of a discrete parameter process.

Similarly, all autocorrelation functions have the following properties:

(1) $\rho(0) = 1$ (follows from the definition).

(2) $|\rho(\tau)| \leq 1$,   all τ,

(3) $\rho(-\tau) = \rho(\tau)$,   all τ, when $X(t)$ is real-valued.

Thus, an autocorrelation function is *symmetric about the origin* $(\tau = 0)$, *and attains its maximum value of unity at the origin.*

3.3.2 Positive Semi-definite Property of $R(\tau)$ and $\rho(\tau)$

A further property of $R(\tau)$ (and $\rho(\tau)$) which it is convenient to prove at this stage is given in the following theorem:

Theorem 3.2.2 *The function $R(\tau)$ is positive semi-definite in the sense that for any set of time points $t_1, t_2, \ldots, t_n$, and all real $z_1, z_2, \ldots, z_n$,*

$$\sum_{r=1}^{n} \sum_{s=1}^{n} R(t_r - t_s) z_r z_s \geq 0. \tag{3.3.4}$$

Proof. Given any set of real numbers $z_1, z_2, \ldots, z_n$, define the random variable W by $W = \sum_{r=1}^{n} z_r X(t_r)$. Then, for all $z_1, z_2, \ldots, z_n$, we have,

$$0 \le \text{var}(W) = \sum_{r=1}^{n} \sum_{s=1}^{n} z_r z_s \, \text{cov}\{X(t_r), X(t_s)\} \qquad \text{(see (2.13.15))},$$

$$= \sum_{r=1}^{n} \sum_{s=1}^{n} z_r z_s R(t_r - t_s).$$

The corresponding result for $\rho(\tau)$ follows immediately on dividing through by $R(0)$.

This result may appear to have little significance at this stage, but at least it warns us that we cannot choose an arbitrary function (even one satisfying properties (1)–(3)) and assume that there must be some process for which it represents the autocovariance function. In order for a function to represent the autocovariance function of some process it must be positive semi-definite. Later on it will emerge that this positive semi-definite property, (3.3.4), can be used as a key result in the development of spectral analysis of random processes.

3.3.3 Complex Valued Processes

So far we have discussed only real valued processes, i.e. processes which, at each time point, assume real values. Although, of course, processes which arise in practice are all real valued it is nevertheless convenient sometimes to regard them as complex valued—just as in electrical circuit theory it is sometimes convenient to regard a voltage as a complex variable.

Suppose then that $\{X(t)\}$ is a complex valued process, of the form,

$$X(t) = U(t) + iV(t),$$

where $\{U(t)\}$, $\{V(t)\}$ are both real valued processes. The probabilistic structure of $\{X(t)\}$ is given by specifying the joint distributions of combined sets of random variables of the form

$$\{U(t_1), U(t_2), \ldots, U(t_n), V(t_1), V(t_2), \ldots, V(t_n)\},$$

for all admissible $t_1, t_2, \ldots, t_n$. The process $\{X(t)\}$ is then said to be "completely stationary" or "stationary up to order m" according as to whether the above joint distributions satisfy conditions of the form (3.2.1) or (3.2.2), respectively. (In particular, if $\{X(t)\}$ is completely stationary it follows that both $\{U(t)\}$ and $\{V(t)\}$ are each completely stationary in the sense previously defined.)

Suppose now that $\{X(t)\}$ is stationary up to order 2. The mean of $\{X(t)\}$ is defined by

$$E[X(t)] = E[U(t)] + iE[V(t)]$$

$$= \mu \text{ (say), a constant independent of } t. \qquad (3.3.5)$$

The autocovariance function of $\{X(t)\}$ is defined by

$$R(\tau) = E[\{X^*(t) - \mu^*\}\{X(t+\tau) - \mu\}], \qquad (3.3.6)$$

where * denotes the complex conjugate, so that, e.g.,

$$X^*(t) = U(t) - iV(t).$$

(Note that when, in particular, $\{X(t)\}$ is real valued, (3.3.6) reduces to the previous definitions (3.3.1).)

In general, $R(\tau)$ is complex valued, but

$$\text{var}\{X(t)\} \equiv R(0) = E[|X(t) - \mu|^2] \qquad (3.3.7)$$

is always real valued. Similarly, the autocorrelation function, defined by,

$$\rho(\tau) = R(\tau)/R(0)$$

is, in general, complex valued, and in place of property (3) we now have,

$$R(-\tau) = R^*(\tau), \qquad (3.3.8)$$

$$\rho(-\tau) = \rho^*(\tau). \qquad (3.3.9)$$

Also, in place of (3.3.4) we have,

$$\sum_{r=1}^{n} \sum_{s=1}^{n} R(t_r - t_s) z_r z_s^* \geq 0, \qquad (3.3.10)$$

for all complex numbers $z_1, z_2, \ldots, z_n$.

3.3.4 Use of the Autocorrelation Function in Practice

In the subsequent development of the theory of stationary processes we discuss the properties of certain "special models" such as the "autoregressive", "moving-average", and "mixed autoregressive/moving-average" schemes. We will see that each type of model gives rise to an autocorrelation function, $\rho(\tau)$, which has certain characteristic features. For example, in some cases $\rho(\tau)$ decays steadily to zero; in other cases it decays in a damped sinusoidal manner, or again, it may take the constant value zero after a certain point. Thus, if we are given the form of $\rho(\tau)$, or if we can estimate $\rho(\tau)$ from observational data, then we can use this information to help us to "identify" which of the special models (if any) would fit the the process under study.

3.4 STATIONARY AND EVOLUTIONARY PROCESSES

In introducing the notion of "stationarity up to order m" as a weaker version of "complete stationarity" we, in effect, replaced the requirement that two probability distributions be completely equivalent by the lesser requirement that they agree only up to their first m moments. However, we know that under general conditions the full infinite sequence of moments determines a distribution uniquely, and that the more moments we specify the more accurately we determine the distribution. Hence, we might expect that there would be some "break point", m_0, such that processes which were stationary up to order m_0 (or higher) would, for all practical purposes, be indistinguishable from processes which were completely stationary. As far as spectral analysis is concerned, it turns out that "order 2" forms the break point! In other words, processes which are stationary up to order 2 are, in this context, very similar to processes which are stationary to orders 3, 4, 5, ..., or even completely stationary processes. (This rather surprising result is due basically to the fact that "power" is a quadratic function of the values of the process, and thus the power/frequency properties depend only on second order properties. Moreover, if the process is Gaussian (see Section 3.4.1), we shall see that stationarity up to order 2 necessarily implies complete stationarity.)

We therefore divide the class of all random processes into two categories, namely:

(a) processes which are stationary up to order 2,
and

(b) all other processes.

Note that category (b) *includes, of course, all processes which are stationary up to orders* 3, 4, 5, ..., *and, in general, all completely stationary processes.* (See Fig. 3.6.)

It is tempting to assume that a completely stationary process must be stationary up to any order, but this is not necessarily true since a process may be completely stationary even though none of its moments exist. (Recall that "stationarity up to order m" implies, in particular, the existence of all joint moments up to order m.) As an example, consider a process which consists of a sequence of independent and identically distributed Cauchy variables. This process is certainly completely stationary, but since no joint moments exist it cannot be said to be stationary up to any order! However, this is a highly pathological example, and subject to the existence of the moments, a completely stationary process is, in fact, stationary up to all orders. Processes in category (a) are usually described simply as *"stationary"*, but the more precise terminology is *"weakly"* or *"wide-sense" stationary*. Processes in category (b) are called *"evolutionary"* or *non-stationary"*.

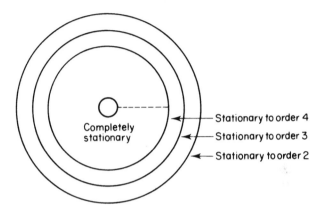

Fig. 3.6. Processes in category (*a*).

Henceforth, we will use the term *"stationary"* (*without further qualification*) *to refer to all processes in* (a), *i.e. processes which are stationary to at least order* 2.

It will be recalled that stationary processes arise from *stable* systems which have attained a *steady-state* mode of operation. By contrast, non-stationary processes describe systems which are *evolving* with time. To illustrate this distinction we list below some typical practical examples of each type of process.

Category (*a*): radio signals, radar signals, "noise in electrical circuits, vibration, turbulence, and oceanography records, economic series (e.g. price indices, stock prices, etc.) and generally any process describing a "stable" system.

Category (*b*): Processes describing the growth and decay of populations (e.g. birth and death processes), queuing systems, branching processes ("chain reactions").

3.4.1 Gaussian (Normal) Processes

$\{X(t)\}$ is called a *Gaussian* (*or normal*) *process* if, for any n and any admissible subset $t_1, t_2, \ldots, t_n$, the joint probability distribution of $\{X(t_1), X(t_2), \ldots, X(t_n)\}$ is *multivariate normal*.

Since a multivariate normal distribution is completely determined by its means, variances and covariances, it follows that the probabilistic structure of a Gaussian process is completely determined by its "mean value function" $E[X(t)]$, and its "covariance function" $\text{cov}\{X(s), X(t)\}$. Hence, if a Gaussian process is stationary up to order 2, implying that the mean and covariance function are invariant under a shift of origin, then the full multivariate distribution of $\{X(t_1), \ldots, X(t_n)\}$ is invariant under a shift or origin, and thus *the process must be completely stationary.*

3.5 SPECIAL DISCRETE PARAMETER MODELS

In our discussion of basic probability theory we highlighted certain special probability distributions (such as the binomial, Poisson, normal, etc.) on the grounds that these distributions have been found to occur frequently in practical applications. Similarly, in developing the theory of stationary random processes we build up a "dictionary" of special models which provides us with a framework for fitting models to practical data. In this section we discuss models for *discrete parameter processes*; the corresponding models for continuous parameter processes are discussed in Section 3.7.

3.5.1 Purely Random Processes; "White Noise"

The process $\{X_t\}$, $t = 0, \pm 1, \pm 2, \ldots$, is called a *purely random process* if it consists of a sequence of *uncorrelated random variables*, i.e. if for all $s \neq t$, $\text{cov}\{X_s, X_t\} = 0$. This is the simplest of all discrete parameter models and corresponds to the case where the process has "no memory", in the sense that the value of the process at time t is uncorrelated with all past values up to time $(t-1)$ (and, in fact, with all future values of the process). Graph 3.1 shows a realization of 500 observations from a Gaussian purely random process, with zero mean and unit variance.

For such a process to be stationary (up to order 2) we require only that,

$$E[X_t] = \mu \quad \text{(a constant)}, \quad \text{all } t, \tag{3.5.1}$$

and

$$E[(X_t - \mu)^2] = \sigma^2 \quad \text{(a constant } t), \quad \text{all } t. \tag{3.5.2}$$

The autocovariance function is then

$$R(s) = \text{cov}\{X_t, X_{t+s}\} = \begin{cases} 0, & s \neq 0, \\ \sigma^2, & s = 0, \end{cases} \tag{3.5.3}$$

and so is automatically a function of s only. The autocorrelation function is then given by

$$\rho(s) = \begin{cases} 1, & s = 0, \\ 0, & s = \pm 1, \pm 2, \pm 3, \ldots, \end{cases} \tag{3.5.4}$$

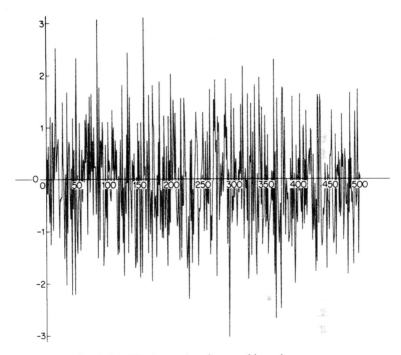

Graph 3.1. 500 observations from a white noise process.

and is shown in Fig. 3.7. From now on, we denote a "purely random process" by ε_t, and this term shall be understood to mean a *stationary* purely random process, i.e. we shall always assume that (3.5.1), (3.5.2) hold, unless the contrary is explicitly stated.

Remarks

(1) Some authors (e.g. Jenkins and Watts (1968)) use the term "purely random process" to describe, more particularly, a stationary sequence of *independent* random variables, $\{e_t\}$. Of course, if the $\{e_t\}$ are independent then they must be uncorrelated, but the converse is not necessarily true (unless $\{e_t\}$ is a Gaussian process—see Section 2.12.9). Our definition of a purely random process as a sequence of *uncorrelated* variables is, we believe,

the more widely used one, and certainly includes sequences of independent variables as a special case. As far as "second order properties" are concerned (i.e. properties of the mean, variance, and autocovariance function) there is no distinction between the two types of processes; e.g., the autocorrelation function of a stationary sequence of independent variables is as shown in Fig. 3.7.

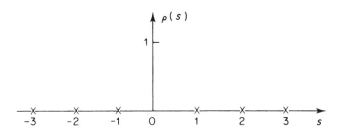

Fig. 3.7. Autocorrelation function of a purely random process.

(2) The purely random process is often called "white noise", particularly in the engineering literature. This alternative description is due to the fact that, as will be shown in the next chapter, a purely random process has a power spectrum which is "flat", i.e. has the same value at all frequencies. The term "white noise" thus arises from the analogy with "white light" in which all frequencies (i.e. "colours") are present in equal amounts.

(3) The model of a purely random process may appear to be highly artificial, and the reader may suspect that such "memoryless" processes would hardly, if ever, occur in practice. This is perfectly correct, but this model is nevertheless important in the development of the theory of stationary processes since it provides us with a basic "building block" from which we can construct more complicated models. If the idea of a sequence of uncorrelated (or independent) random variables seems too trivial, it may be useful to recall that the notion of a sequence of "independent variables" is the basic model on which virtually the whole of classical statistical inference is based.

3.5.2 First Order Autoregressive Process (Linear Markov Processes)

Following the purely random process, the next stage up in complexity is to construct a model which involves "one-step dependence", i.e. one in which the value of the process at time t depends, at least in part, on the value at time $(t-1)$. Such a model is called *first order autoregressive*, and is defined as follows.

$\{X_t\}$ is called a first order autoregressive process (denoted by AR(1)) if it satisfies the difference equation

$$X_t - aX_{t-1} = \varepsilon_t, \tag{3.5.5}$$

where a is a constant, and $\{\varepsilon_t\}$ is a (stationary) purely random process.

Rewriting (3.5.5) in the form

$$X_t = aX_{t-1} + \varepsilon_t, \tag{3.5.6}$$

we see that here X_t depends partly on X_{t-1}, and partly on the "random disturbance", ε_t. In fact, (3.5.6) may be interpreted as saying that X_t has a "linear regression" on X_{t-1}, with ε_t playing the role of the "error" term. The fact that X_t has a regression on its *own past* gives rise to the terminology "autoregressive process". However, in this case the dependence of X_t on past values extends only over one time interval. In fact, if the process is Gaussian, (so that its probabilistic structure is completely determined by its second order properties), then, given a sequence of past values, $X_{t-1}, X_{t-2}, X_{t-3}, \ldots$, the conditional distribution of X_t depends only on X_{t-1}, i.e.

$$p(X_t | X_{t-1}, X_{t-2}, X_{t-3}, \ldots) = p(X_t | X_{t-1}). \tag{3.5.7}$$

Processes satisfying the general property (3.5.7) are called *Markov processes* (see, e.g., Cox and Miller (1965)).

Here, the dependence of X_t on X_{t-1} is linear, and even though $\{X_t\}$ may not be Gaussian it is sometimes called a (wide-sense) *linear Markov process*. Graph 3.2 shows a realization of 500 observations from the AR(1) model, $X_t - 0.6X_{t-1} = \varepsilon_t$, the ε_t being normal variables with zero mean and unit variance.

To study the second order properties of this model we start by solving the difference equation (3.5.5) so as to obtain an expression for X_t in terms of $\varepsilon_t, \varepsilon_{t-1}, \varepsilon_{t-2}, \ldots$. The precise solution of this equation depends on the "initial condition", so let us suppose that initially $X_0 = 0$, (so that $X_1 = \varepsilon_1$). Then solving (3.5.5) by repeated substitution we obtain,

$$X_t = \varepsilon_t + aX_{t-1}$$

$$= \varepsilon_t + a(\varepsilon_{t-1} + aX_{t-2})$$

$$= \ldots$$

$$= \varepsilon_t + a\varepsilon_{t-1} + a^2\varepsilon_{t-2} + \ldots + a^{t-1}\varepsilon_1. \tag{3.5.8}$$

If $E[\varepsilon_t] = \mu_\varepsilon$, (all t), then by (3.5.8),

$$E[X_t] = \mu_\varepsilon(1 + a + a^2 + \ldots + a^{t-1}) = \begin{cases} \mu_\varepsilon\left(\dfrac{1 - a^t}{1 - a}\right), & a \neq 1 \\[2mm] \mu_\varepsilon t, & a = 1. \end{cases} \tag{3.5.9}$$

If $\mu_\varepsilon = 0$, $E[X_t] = 0$, all t, and $\{X_t\}$ is stationary up to order 1. If $\mu_\varepsilon \neq 0$, $E[X_t]$ is a function of t and $\{X_t\}$ is not stationary even to order 1. However, we note that if $|a| < 1$, then

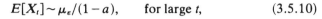

$$E[X_t] \sim \mu_\varepsilon/(1-a), \qquad \text{for large } t, \qquad (3.5.10)$$

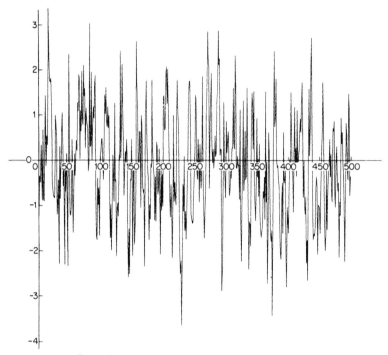

Graph 3.2. 500 observations from an AR(1) process.

and we may say that $\{X_t\}$ is "*asymptotically stationary*" to order 1.

From now on we assume that $\mu_\varepsilon = 0$, and then

$$\left.\begin{array}{l} \mathrm{var}(\varepsilon_t) = E[\varepsilon_t^2] = \sigma_\varepsilon^2, \\[4pt] \mathrm{cov}(\varepsilon_t, \varepsilon_s) = E[\varepsilon_t \varepsilon_s] = 0, \qquad s \neq t. \end{array}\right\} \qquad (3.5.11)$$

say, and

Hence,

$$\begin{aligned} \sigma_X^2 = \mathrm{var}\{X_t\} &= E[\varepsilon_t + a\varepsilon_{t-1} + \ldots + a^{t-1}\varepsilon_1]^2 \\[4pt] &= \sigma_\varepsilon^2(1 + a^2 + a^4 + \ldots + a^{2t-2}), \quad \text{by (3.5.11)} \\[4pt] &= \begin{cases} \sigma_\varepsilon^2\left(\dfrac{1 - a^{2t}}{1 - a^2}\right), & |a| \neq 1 \\[12pt] \sigma_\varepsilon^2 t, & |a| = 1. \end{cases} \end{aligned} \qquad (3.5.12)$$

Similarly, if $r \geq 0$,

$$E[X_t X_{t+r}] = E[\{\varepsilon_{t+r} + a\varepsilon_{t+r-1} + \ldots + a^{t+r-1}\varepsilon_1\}\{\varepsilon_t + a\varepsilon_{t-1} + \ldots + a^{t-1}\varepsilon_1\}]$$

$$= \sigma_\varepsilon^2 (a^r + a^{r+2} + \ldots + a^{r+2(t-1)})$$

$$= \begin{cases} \sigma_\varepsilon^2 \cdot a^r \left(\dfrac{1-a^{2t}}{1-a^2}\right), & |a| \neq 1 \\[2mm] \sigma_\varepsilon^2 t, & |a| = 1. \end{cases} \qquad (3.5.13)$$

Both var$\{X_t\}$ and cov$\{X_t, X_{t+r}\}$ are functions of t, and $\{X_t\}$ is not stationary up to order 2. However, if $|a| < 1$, then as before we may argue that *for t sufficiently large,*

$$\sigma_X^2 \sim \sigma_\varepsilon^2 / (1 - a^2) \qquad (3.5.14)$$

$$\text{cov}\{X_t, X_{t+r}\} \sim \sigma_\varepsilon^2 \cdot a^r / (1 - a^2) \qquad (3.5.15)$$

The right-hand side of (3.5.15) is now a function of r only, and we may say that $\{X_t\}$ is "*asymptotically stationary*" up to order 2.

Thus, the condition for $\{X_t\}$ to be asymptotically stationary up to order 2 is that $|a| < 1$.

It may seem at first sight that we could claim that *any* stochastic process was asymptotically stationary, since it must necessarily be true that in all cases $\lim_{t \to \infty} \text{var}\{X_t\}$ does not depend on t, and $\lim_{t \to \infty} \text{cov}\{X_t, X_{t+r}\}$ is a function of r only. However, the essential point is that the above limits may not exist, i.e. var$\{X_t\}$ and cov$\{X_t, X_{t+r}\}$ may not converge to finite values as $t \to \infty$. Only when these limits are *finite* can we say that the process is asymptotically stationary. To illustrate this point further, consider the AR(1) model with $a = 1$. In this case (3.5.12) shows that var$\{X_t\}$ increases linearly with t, and so $\to \infty$ as $t \to \infty$. Incidentally, when $a = 1$, (3.5.8) gives

$$X_t = \sum_{s=1}^{t} \varepsilon_s.$$

If the $\{\varepsilon_t\}$ are moreover independent, the above corresponds to a "*random walk*" model.

We assume, from now on, that $|a| < 1$. Equation (3.5.15) then gives the form of the autocovariance function, $R(r)$, for $r \geq 0$; since we know that $R(r)$ is an even function of r we may write

$$R(r) = \sigma_\varepsilon^2 \cdot \frac{a^{|r|}}{(1-a^2)}, \qquad r = 0, \pm 1, \pm 2, \ldots. \qquad (3.5.16)$$

The autocorrelation function is thus given by,

$$\rho(r) = R(r)/R(0) = a^{|r|}, \qquad r = 0, \pm 1, \pm 2, \ldots. \qquad (3.5.17)$$

When $a > 0$, $\rho(r) > 0$, all r, and decays to zero exponentially; when $a < 0$, $\rho(r)$ alternates in sign. These forms are shown in Figs. 3.8, 3.9. (Note that when $a \sim -1$, X_t itself will tend to alternate in sign.)

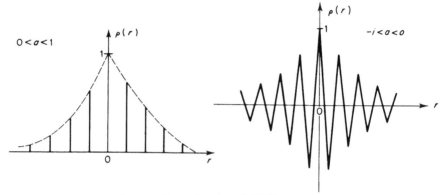

Autocorrelation function of AR(1) process.

Fig. 3.8   **Fig. 3.9**

If we *assume* that $\{X_t\}$ is stationary, the autocorrelation function may be derived by an alternative method, as follows. Multiplying both sides of (3.5.6) by X_{t-r}, $(r \geq 1)$, and taking expectations we obtain,

$$E[X_t X_{t-r}] = aE[X_{t-1}X_{t-r}] + E[\varepsilon_t X_{t-r}]. \qquad (3.5.18)$$

Now X_{t-r} is a linear function of $\varepsilon_{t-r}, \varepsilon_{t-r-1}, \ldots,$ (cf. (3.5.8)), and, for $r \geq 1$, is therefore uncorrelated with ε_t. Hence, $E[\varepsilon_t X_{t-r}] = 0$, and (3.5.18) then gives,

$$R(r) = aR(r-1),$$

so that,

$$\rho(r) = a\rho(r-1)$$
$$= a^2\rho(r-2) = \ldots = a^r\rho(0) = a^r,$$

(since $\rho(0) = 1$). Using again the fact that $\rho(r)$ is an even function of r, we then have

$$\rho(r) = a^{|r|}, \qquad r = 0, \pm 1, \pm 2, \ldots,$$

in agreement with (3.5.17). Also, multiplying both sides of (3.5.6) by X_t gives,

$$\sigma_X^2 = E[X_t^2] = aE[X_{t-1}X_t] + E[\varepsilon_t X_t].$$

But, $X_t = \varepsilon_t +$ linear function of $(\varepsilon_{t-1}, \varepsilon_{t-2}, \ldots)$. Hence

$$E[\varepsilon_t X_t] = E[\varepsilon_t^2] = \sigma_\varepsilon^2,$$

and thus,

$$\sigma_X^2 = aR(1) + \sigma_\varepsilon^2 = \sigma_\varepsilon^2/(1-a^2),$$

in agreement with (3.5.14).

It may at first seem surprising that although the AR(1) model, (3.5.5), involves only "one-stage dependence", the autocorrelation function, $\rho(r)$, does not "cut-off" (i.e. become zero) after $r = 1$, but instead tends to zero gradually as $r \to \infty$. The reason for the correlation between, e.g., X_t and X_{t-2} is because X_t is related to X_{t-1}, which in turn is related to X_{t-2}, and so on. However, if we may eliminate the effect of the "intermediate" variable by computing the *partial correlation coefficient* between X_t and X_{t-2}, given X_{t-1}, as follows. We have (see, e.g., Kendal and Stuart (1966), Chapter 27),

$$\rho_{X_t,X_{t-2}\cdot X_{t-1}} = \frac{\rho_{X_t,X_{t-2}} - \rho_{X_t,X_{t-1}}\rho_{X_{t-1},X_{t-2}}}{[\{1-\rho_{X_t,X_{t-1}}^2\}\{1-\rho_{X_{t-1},X_{t-2}}^2\}]^{1/2}}$$

$$= \frac{(a^2 - a\cdot a)}{(1-a^2)} = 0.$$

A similar argument shows that the partial correlation coefficient between X_t and X_{t-r}, given X_{t-1}, vanishes for all $r > 1$. This result provides a useful characterization of the AR(1) model.

The property of asymptotic stationarity

In the above discussion we noted that, subject to the initial condition $X_0 = 0$, X_t was not stationary (up to order 2) but was "asymptotically stationary" in the sense that it settles down to a stationary behaviour as t becomes large. On reflection, it is clear that if we fix X_0 to be zero the process cannot possibly be stationary since we have now given a special "status" to the time point $t = 0$, and the property of "invariance under a shift of the time origin" cannot hold—even for the second order moments. (Note, in particular, that with the initial condition as above, all realizations of the process must pass through the origin.) Loosely speaking, the process does not obtain a stationary character until it has "forgotten" its initial starting value, and the effect is in many ways similar to the initial "transient" behaviour of the output of a linear electrical circuit switched on at time $t = 0$.

However, instead of fixing X_0 to be zero and then studying the properties of the process as $t \to \infty$, we may alternatively take the initial condition to be of the form $X_{-N} = 0$, and then let $N \to \infty$. The solution of (3.5.5) is then

$$X_t = \sum_{s=-(N-1)}^{t} a^{t-s} \varepsilon_s,$$

and as $N \to \infty$, this becomes formally,

$$X_t = \sum_{s=-\infty}^{t} a^{t-s} \varepsilon_s = \sum_{s=0}^{\infty} a^s \varepsilon_{t-s}. \qquad (3.5.20)$$

As the right-hand side of (3.5.20) involves an infinite sum of random variables we need to discuss the sense (if any) in which this sum "converges" to a limiting random variable. It may be shown (Loeve (1963), p. 456) that if $\{U_n\}$ is a sequence of zero mean uncorrelated random variables, each with finite variance, then as $N \to \infty$, $\sum_{n=0}^{N} U_n$ converges "in mean square" (see Section 3.6) to a limiting random variable U if and only if

$$\sum_{n=0}^{\infty} E\{|U_n|^2\} < \infty,$$

in which case

$$E\left\{ \left| \sum_{n=0}^{\infty} U_n \right|^2 \right\} = \sum_{n=0}^{\infty} E\{|U_n|^2\}.$$

Thus $\sum_{n=0}^{\infty} U_n$ is convergent in mean square provided that it has finite variance. It now follows that the expression in (3.5.20) is well-defined (in the mean-square sense) provided that its variance is finite. Now,

$$\text{var}\left\{ \sum_{s=0}^{\infty} a^s \varepsilon_{t-s} \right\} = \sigma_\varepsilon^2 \left\{ \sum_{s=0}^{\infty} a^{2s} \right\} < \infty,$$

provided $|a| < 1$, the condition which we have already seen ensures the asymptotic stationarity of the process. Under this condition the right-hand side of (3.5.20) then represents a process which is stationary (up to order 2) in the conventional sense. Alternatively, if we drop the initial condition that $X_{-N} = 0$, we have

$$\left\{ X_t - \sum_{s=0}^{t+N-1} a^s \varepsilon_{t-s} \right\} = a^{t+N} X_{-N}.$$

If X_t is assumed stationary with finite variance, then

$$E\left\{ X_t - \sum_{s=0}^{t+N-1} a^s \varepsilon_{t-s} \right\}^2 = a^{2(t+N)} \sigma_X^2.$$

If $|a| < 1$, $a^{2(t+N)} \to 0$ as $N \to \infty$, and hence under this condition we see yet again that

$$\left\{ \sum_{s=0}^{t+N-1} a^s \varepsilon_{t-s} \right\}$$

converges in mean square to X_t as $N \to \infty$.

It is instructive now to consider the form of the general solution of (3.5.5). This is a linear first order difference equation with a constant coefficient, and it is well known that its general solution is made up of the sum of two parts, namely, the "complementary function" and the "particular solution".

Here, the complementary function is the solution of

$$X_t - aX_{t-1} = 0,$$

giving a function of the form Aa^t, A being an arbitrary constant. The particular solution is most conveniently computed using the "*backward difference operator*", B, defined by $BX_t = X_{t-1}$. Using this notation (3.5.5) may be written as

$$(1 - aB)X_t = \varepsilon_t,$$

and the particular solution is

$$X_t = (1 - aB)^{-1}\varepsilon_t$$

$$= \left(\sum_{s=0}^{\infty} a^s B^s \right) \varepsilon_t$$

$$= \sum_{s=0}^{\infty} a^s \varepsilon_{t-s}.$$

The general solution of (3.5.5) is thus,

$$X_t = Aa^t + \sum_{s=0}^{\infty} a^s \varepsilon_{t-s}. \tag{3.5.21}$$

In this form the constant A is arbitrary, but its value is determined once we have specified the initial condition. For example, if we determine A by the condition $X_0 = 0$, it is readily verified that (3.5.21) reduces to (3.5.8).

The second term in (3.5.21) corresponds to (3.5.20), and, with $\mu_\varepsilon = 0$, represents a zero mean stationary random process. Clearly, for (3.5.21) to represent an asymptotically stationary process the *complementary function must decay to zero as $t \to \infty$*, which shows, once again, that the condition for asymptotic stationarity is $|a| < 1$. When this condition holds, the "steady-state" behaviour of $\{X_t\}$ after it has settled down (i.e. after the complementary function has effectively decayed to zero) is described purely by the second term in (3.5.21), and this expression (or equivalently, (3.5.20)) is therefore called the *stationary solution* of (3.5.5).

3.5.3 Second Order Autoregressive Processes

An obvious extension of the AR(1) model is obtained by constructing a model which involves "two stage dependence", i.e. one in which X_t depends linearly on both X_{t-1} and X_{t-2}. Such a model is called *second order autoregressive* and is defined formally as follows.

   {X_t} *is called a second order autoregressive process* (*denoted by* AR(2)) *if it satisfies the difference equation,*

$$X_t + a_1 X_{t-1} + a_2 X_{t-2} = \varepsilon_t, \qquad (3.5.22)$$

where a_1, a_2, are constants, and {ε_t} is a purely random process. (In order to be consistent with the notation which we shall introduce later for more general models we have written the coefficients of X_{t-1} and X_{t-2} in (3.5.22) with positive signs, contrary to the convention used in the AR(1) model, (3.5.5).) We assume, without loss of generality, that $E[\varepsilon_t] = 0$, all t, so that $\text{var}\{\varepsilon_t\} = E[\varepsilon_t^2] = \sigma_\varepsilon^2$, say. Graph 3.3 shows a realization of 500 observations

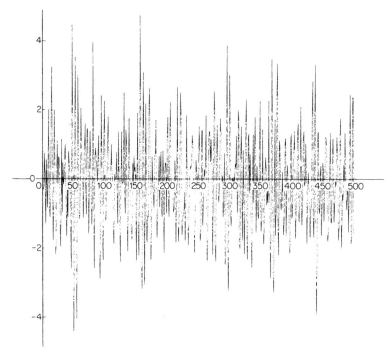

Graph 3.3. 500 observations from an AR(2) process.

from the AR(2) model, $X_t - 0 \cdot 4 X_{t-1} + 0 \cdot 7 X_{t-2} = \varepsilon_t$, the ε_t being normal variables with zero mean and unit variance. The graph exhibits a rough "periodicity", with an approximate period of five time units. (See the discussion later on the "pseudoperiodic" behaviour of certain types of AR(2) models.)

As in the previous case, we may obtain the conditions for asymptotic stationarity by studying the form of the general solution of this difference equation.

Using the backward shift operator B, (3.5.22) may be written as

$$(1 + a_1 B + a_2 B^2) X_t = \varepsilon_t,$$

or

$$(1 - \mu_1 B)(1 - \mu_2 B) X_t = \varepsilon_t, \tag{3.5.23}$$

where μ_1, μ_2, (assumed distinct) are the roots of the quadratic

$$f(z) = z^2 + a_1 z + a_2.$$

A particular solution of (3.5.23) is now given by,

$$X_t = \left(\frac{1}{(1 - \mu_1 B)(1 - \mu_2 B)} \right) \cdot \varepsilon_t$$

$$= \frac{1}{(\mu_1 - \mu_2)} \left(\frac{\mu_1}{(1 - \mu_1 B)} - \frac{\mu_2}{(1 - \mu_2 B)} \right) \cdot \varepsilon_t$$

$$= \frac{1}{(\mu_1 - \mu_2)} \left(\sum_{s=0}^{\infty} (\mu_1^{s+1} - \mu_2^{s+1}) B^s \right) \cdot \varepsilon_t$$

$$= \sum_{s=0}^{\infty} \left(\frac{\mu_1^{s+1} - \mu_2^{s+1}}{\mu_1 - \mu_2} \right) \varepsilon_{t-s}. \tag{3.5.24}$$

The general solution of (3.5.22) is obtained by adding to (3.5.24) the complementary function, i.e. by adding the solution of the homogeneous equation

$$X_t + a_1 X_{t-1} + a_2 X_{t-2} = 0. \tag{3.5.25}$$

The solution of (3.5.25) is of the form (for $\mu_1 \neq \mu_2$)

$$X_t = A_1 \mu_1^t + A_2 \mu_2^t, \tag{3.5.26}$$

A_1, A_2, being arbitrary constants, so that the general solution of (3.5.22) is

$$X_t = A_1 \mu_1^t + A_2 \mu_2^t + \sum_{s=0}^{\infty} \left(\frac{\mu_1^{s+1} - \mu_2^{s+1}}{\mu_1 - \mu_2} \right) \varepsilon_{t-s}. \tag{3.5.27}$$

Following the same argument used in the AR(1) case, it is clear that for (3.5.27) to represent an asymptotically stationary process the complementary function (3.5.26) must decay to zero as $t \to \infty$.

Thus, for asymptotically stationarity we require that

$$|\mu_1| < 1, \quad \text{and} \quad |\mu_2| < 1. \tag{3.5.28}$$

Under these conditions it is clear that the second term in (3.5.27) has finite variance, and arguing as in the AR(1) case it follows that this term is convergent (in the mean square sense) and thus represents a well defined random variable which we can call the "stationary solution" of (3.5.22). It is easy to verify that the above conditions also ensure that both $E\{X_t^2\}$ and $E\{X_t, X_{t+r}\}$ converge to finite limits as $t \to \infty$.

To express the conditions (3.5.28) in terms of the coefficients a_1, a_2, we consider the following special cases:

I. Complex or coincident roots

The condition for $f(z)$ to have complex or coincident roots is that $a_2 \geqslant \frac{1}{4}a_1^2$. In this case,

$$|\mu_1| = |\mu_2| = \sqrt{a_2},$$

and the condition for (asymptotic) stationarity becomes $a_2 < 1$. We may illustrate this graphically by representing the pair of coefficients, a_1, a_2, as a point in the "(a_1, a_2) plane", and the region corresponding to the above conditions on a_1, a_2, is shown in Fig. 3.10.

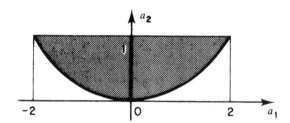

Fig. 3.10. Region of stationarity for complex or coincident roots.

II. Unequal real roots

Here, we have, $a_2 < \frac{1}{4}a_1^2$. Now $f(z)$ attains its minimum value at $z = -a_1/2$ (see Fig. 3.11). Hence, for the roots of $f(z)$ to lie between $(-1, 1)$ we must have $-1 < -a_1/2 < 1$ or $|a_1| < 2$. Also, we must have, $f(1) > 0$ and $f(-1) > 0$, i.e.

$$(1 + a_1 + a_2) > 0 \quad \text{and} \quad (1 - a_1 + a_2) > 0,$$

i.e.

$$a_1 + a_2 > -1 \quad \text{and} \quad a_1 - a_2 < 1.$$

Thus, for (asymptotic) stationarity the point (a_1, a_2) must lie *inside* the triangle defined by

$$a_2 \leq 1, \qquad a_1 + a_2 \geq -1, \qquad a_1 - a_2 \leq 1, \qquad (3.5.29)$$

as shown in Fig. 3.12.

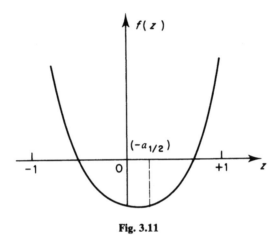

Fig. 3.11

Variance and autocorrelation function

Assuming that the conditions for asymptotic stationarity are satisfied, the variance, σ_X^2, and autocorrelation function, $\rho(r)$, of $\{X_t\}$ may be obtained as follows. We see from (3.5.24) that, for $r \geq 0$,

$$X_{t-r} = \varepsilon_{t-r} + \text{linear function of } \{\varepsilon_{t-s}; s > r\}.$$

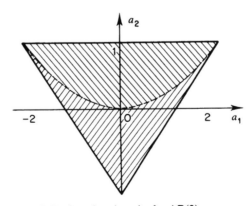

Fig. 3.12. Region of stationarity for AR(2) process.

Since $\{\varepsilon_t\}$ is an uncorrelated zero-mean sequence, we then have,

$$E[X_{t-r}\varepsilon_t] = 0, \qquad r \geq 1,$$

and

$$E[X_t\varepsilon_t] = E[\varepsilon_t^2] = \sigma_\varepsilon^2.$$

Multiplying both sides of (3.5.22) by X_{t-r} and taking expectations gives,

$$E[X_t X_{t-r}] + a_1 E[X_{t-1}X_{t-r}] + a_2 E[X_{t-2}X_{t-r}] = E[\varepsilon_t X_{t-r}], \quad r = 0, 1, 2, \ldots.$$

Using the above results we obtain,

for $r = 0$; $\qquad \sigma_X^2\{1 + a_1\rho(1) + a_2\rho(2)\} = \sigma_\varepsilon^2$ $\qquad\qquad$ (3.5.30)

for $r = 1$; $\qquad \{\rho(1) + a_1\rho(0) + a_2\rho(1)\} = 0$ $\qquad\qquad$ (3.5.31)

for $r = 2$; $\qquad \{\rho(2) + a_1\rho(1) + a_2\rho(0)\} = 0.$ $\qquad\qquad$ (3.5.32)

In general, for $r \geq 2$, we have,

$$\rho(r) + a_1\rho(r-1) + a_2\rho(r-2) = 0. \qquad\qquad (3.5.33)$$

Solving (3.5.31), (3.5.32) gives,

$$\rho(1) = -\frac{a_1}{(1+a_2)}, \qquad \rho(2) = \frac{a_1^2}{1+a_2} - a_2,$$

and substituting these values in (3.5.30) gives,

$$\sigma_X^2 = \frac{(1+a_2)\sigma_\varepsilon^2}{(1-a_2)(1-a_1+a_2)(1+a_1+a_2)}. \qquad\qquad (3.5.34)$$

(Note that the expression on the right-hand side of (3.5.34) is positive in virtue of the stationarity conditions.)

Equation (3.5.33) is a second order linear difference equation, and the general solution is of the form (for $\mu_1 \neq \mu_2$)

$$\rho(r) = B_1\mu_1^r + B_2\mu_2^r, \qquad r \geq 0,$$

where, again, μ_1, μ_2, are the roots of the quadratic $f(z) = z^2 + a_1 z + a_2$, and B_1, B_2, are constants whose values are determined by the boundary conditions $\rho(0) = 1$, $\rho(1) = -a_1/(1+a_2)$.* We then obtain,

$$\rho(r) = \frac{(1-\mu_2^2)\mu_1^{r+1} - (1-\mu_1^2)\mu_2^{r+1}}{(\mu_1-\mu_2)(1+\mu_1\mu_2)}, \qquad r \geq 0. \qquad (3.5.35)$$

(For $r < 0$, $\rho(r) = \rho(-r)$.)

To gain insight into the behaviour of the autocorrelation function, $\rho(r)$, it is instructive to consider the various cases which arise according to whether the roots of $f(z)$ are real or complex.

---

* For $\mu_1 = \mu_2 = \mu$, say, (real), the general solution of (3.5.33) takes the form $\rho(r) = (B_1 + B_2 r)\mu^r$, leading to $\rho(r) = \{1 + [(1-\mu^2)/(1+\mu^2)]r\}\mu^r, r \geq 0.$

CASE (1): REAL ROOTS, $a_2 \leqslant a_1^2/4$

If $a_1 < 0$, $a_2 > 0$, then $\mu_1 > 0$, $\mu_2 > 0$, and $\rho(r)$ decays "smoothly" to zero, as shown in Fig. 3.13. Otherwise, if $a_2 < 0$, then μ_1, μ_2, are of opposite signs,

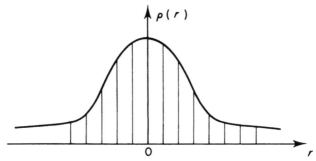

Fig. 3.13. Autocorrelation function, real positive roots.

whereas if $a_1 > 0$, $a_2 > 0$, then μ_1, μ_2, are both negative, and $\rho(r)$ alternates in sign as it decays, as shown in Fig. 3.14.

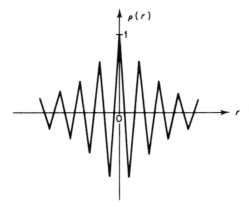

Fig. 3.14. Autocorrelation function, real negative roots.

CASE (2): COMPLEX ROOTS, $a_2 > a_1^2/4$

When the roots are real $\rho(r)$ decays exponentially to zero, the rate of decay being determined effectively by the dominant root. In this case the behaviour of $\rho(r)$ is very similar to that of the first order autoregressive (AR(1)) process, but the really interesting case occurs when the roots are complex. Here, we obtain some rather surprising features.

Since a_1, a_2, are real, μ_1, μ_2, are complex conjugates, and since further $\mu_1\mu_2 = a_2$, we may write the roots in the form

$$\mu_1 = \sqrt{a_2}\, \exp(i\theta), \qquad \mu_2 = \sqrt{a_2}\, \exp(-i\theta),$$

where θ is determined by,

$$-a_1 = \mu_1 + \mu_2 = 2\sqrt{a_2}\, \cos\theta,$$

so that

$$\cos\theta = -a_1/2\sqrt{a_2}.$$

Substituting the above values for μ_1, μ_2, in (3.5.35), we obtain,

$$\rho(r) = \frac{a_2^{r/2}\{\sin(r+1)\theta - a_2 \sin(r-1)\theta\}}{(1+a_2)\sin\theta}.$$

Setting

$$\tan\psi = \left(\frac{1+a_2}{1-a_2}\right)\tan\theta,$$

we may re-write $\rho(r)$ in the more compact form,

$$\rho(r) = a_2^{r/2}\left(\frac{\sin(r\theta + \psi)}{\sin\psi}\right). \qquad (3.5.36)$$

Although $\rho(r)$ still decays exponentially to zero (with $|a_2| < 1$, the term $a_2^{r/2}$ acts as a "damping factor"), the interesting point is that the second factor in (3.5.36) has an oscillatory form, being a periodic function with period $(2\pi/\theta)$. The autocorrelation function, $\rho(r)$, may thus be described as

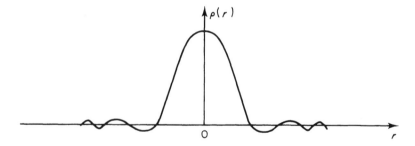

Fig. 3.15. Autocorrelation function, complex roots.

"*damped periodic*", or "*pseudo-periodic*", with period $(2\pi/\theta)$, as shown in Fig. 3.15. In fact, as $a_2 \to 1$, $\psi \to \pi/2$, and

$$\rho(r) \to \cos r\theta, \qquad (3.5.37)$$

giving an exactly periodic function. However, for stationarity we must have $a_2 < 1$, and allowing a_2 to tend to 1 produces an infinite value for σ_X^2 (unless we make $\sigma_\varepsilon^2 \to 0$).

What does the "pseudo-periodic" behaviour of $\rho(r)$ tell us about the behaviour of the process itself? First, let us consider the extreme case when $a_2 \to 1$ so that $\rho(r)$ is now exactly periodic. What does this tell us about X_t? We know, intuitively speaking, that $\rho(r)$ measures the "similarity" between original process and a copy of itself displaced by r time units. When $\rho(r)$ is of the form (3.5.37) it repeatedly attains the value $+1$ at intervals of $(2\pi/\theta)$, which means that if we displace X_t by any multiple of $(2\pi/\theta)$ we obtain a "perfect match". Expressing these ideas more precisely, we see that *an exactly periodic autocorrelation function indicates that the process itself is also exactly periodic, with the same periodicity as that of $\rho(r)$*. Reasoning in a similar manner we are led to the conclusion that a *pseudo-periodicity in $\rho(r)$ indicates some form of pseudo-periodic behaviour in X_t*, i.e. we may conclude that X_t exhibits some kind of "distorted periodicity". The damping effect in $\rho(r)$ means that if the process is displaced by just one or two periods it will still match up fairly well, whereas if it is displaced by a large number of periods the matching will be poor.

What causes the pseudo-periodic behaviour in X_t when μ_1, μ_2 are complex? At first sight there would seem to be nothing in the AR(2) model (3.5.22) which could generate this type of oscillatory behaviour. However, we gain a better understanding of this phenomenon when we consider the continuous time analogue of the AR(2) model. In continuous time the natural analogue of the linear second order *difference* equation (3.5.22) is a linear second order *differential* equation, of the form,

$$\ddot{X}(t) + \alpha \dot{X}(t) + \beta X(t) = \varepsilon(t). \tag{3.5.38}$$

In this form the AR(2) model begins to look much more familiar, and second order differential equations of the form (3.5.38) occur in a multitude of mechanical and electrical problems which involve "oscillating systems". For example, if $\alpha = 0$, $\beta > 0$, and $\varepsilon(t) \equiv 0$, (3.5.38) describes the model of *simple harmonic motion*, as typified, e.g., by a freely swinging pendulum. Here, $X(t)$ would correspond to the angular displacement of the pendulum at time t. If $\alpha > 0$ the second term on the left-hand side corresponds to a "damping term", and if $\varepsilon(t) \equiv 0$, the differential equation would represent, e.g., the motion of a pendulum swinging in a resistive medium (such as a viscous fluid) where the resistive force is proportional to velocity. Once set in motion such a pendulum will execute damped oscillations until eventually the resistive force brings the pendulum to rest. However, when the term $\varepsilon(t)$ is present it acts as a "driving force"; in the case of the AR(2) model $\varepsilon(t)$ is a random process which behaves, roughly speaking, like a series of "impulses", and it is the effect of these random impulses which sustains the motion of the pendulum. If, when the motion of the pendulum is dying out it receives an impulse, this restores the oscillatory motion until the next impulse arrives. In this way, the model (3.5.38) can still represent a stationary process even when the damping term is present.

An amusing example of this type of system was constructed by Yule (1927). Yule considered a pendulum swinging in a resistive medium, and to realize the effect of the process $\varepsilon(t)$ he imagines that there is a boy armed with a pea-shooter who shoots peas at the pendulum at random time instants. This often quoted model is known as "*Yule's pendulum*".

As another example we may consider the motion of a car travelling along a bumpy road. Assuming that the vertical displacement of the car's shock absorbers satisfy a second order differential equation (with the damping term present!), then each time the car passes over a bump its vertical displacement will oscillate until it damps out. However, if the bumps are distributed at random along the length of the road, then the motion will be sustained, and in fact, will form a stationary process.

Returning now to the original question as to the origin of the "oscillatory" behaviour of the process $\{X_t\}$, we see that, under suitable conditions, this is an inherent feature of the second order difference equation (3.5.22). The period $(2\pi/\theta)$ may be regarded as the "natural period" of the model, in the same way that a system described by a second order differential equation has a "natural period". The effect of the random process $\{\varepsilon_t\}$ (or $\varepsilon(t)$) is to cause random "phase-shifts" in the process which accounts for the "disturbed periodic" behaviour noted previously.

Before leaving this topic we may note that the transition from the discrete time model (3.5.22) to the continuous time model (3.5.38), although a natural one, is by no means a simple exercise. When we examine this point in more detail (Section 3.7.1) we will find that the construction of the process $\{\varepsilon(t)\}$ requires rather careful and delicate mathematical treatment.

3.5.4 Autoregressive Processes of General Order

Having extended the first order autoregressive model into the second order model it is now fairly obvious how to generalize the second order model into the general order autoregressive model. In effect, we simply extend the order of the linear difference equation. *Thus, we say that $\{X_t\}$ is an autoregressive process of order k (denoted by $AR(k)$) if it satisfies the difference equation,*

$$X_t + a_1 X_{t-1} + \ldots + a_k X_{t-k} = \varepsilon_t, \tag{3.5.39}$$

where $a_1, a_2, \ldots, a_k$, are constants, and $\{\varepsilon_t\}$ is a purely random process.

Equation (3.5.39) may be written more concisely in the form

$$\alpha(B)X_t = \varepsilon_t, \tag{3.5.40}$$

where

$$\alpha(z) \equiv 1 + a_1 z + \ldots + a_k z^k. \tag{3.5.41}$$

The formal solution of (3.5.40) is then given by

$$X_t = f(t) + \alpha^{-1}(B)\varepsilon_t, \tag{3.5.42}$$

where $f(t)$ (the complementary function) is the solution of the homogeneous equation $\alpha(B)X_t = 0$.

Reasoning as in the second order case it is clear that for (3.5.42) to represent an asymptotically stationary process we require that $f(t)$ *must decay to zero as $t \to \infty$*. As before, $f(t)$ will, in general, be of the form,

$$f(t) = A_1 \mu_1^t + A_2 \mu_2^t + \ldots + A_k \mu_k^t,$$

where $\mu_1, \mu_2, \ldots, \mu_k$ are the roots of the polynomial

$$g(z) = z^k + a_1 z^{k-1} + \ldots + a_k.$$

Hence, *for asymptotic stationarity we require that $|\mu_j| < 1$, all j, i.e. that all the roots of $g(z)$ must lie inside the unit circle*. Equivalently, since the roots of $g(z)$ are clearly the reciprocals of the roots of $\alpha(z)$, *the condition for asymptotic stationarity is that all the roots of $\alpha(z)$ must lie outside the unit circle*.

Moreover, writing $\alpha(B) = \prod_{i=1}^k (1 - \mu_i B)$, and expanding $\alpha^{-1}(B)$ in partial fractions, we obtain for the particular solution (assuming distinct roots),

$$X_t = \alpha^{-1}(B)\varepsilon_t = \left\{ \sum_{i=1}^k \frac{k_i}{(1 - \mu_i B)} \right\} \varepsilon_t,$$

(the $\{k_i\}$ being constants). With $|\mu_i| < 1$, all i, each term on the RHS of the above equation may now be expanded as a (mean square) convergent series in $\varepsilon_t, \varepsilon_{t-1}, \varepsilon_{t-2}, \ldots$, just as in the AR(1) case. (The case of non-distinct roots may be dealt with by a similar argument.)

Assuming that the above conditions are satisfied, and that $E[\varepsilon_t] = 0$, which, in turn, implies that $E[X_t] = 0$, multiplying both sides of (3.5.39) by X_{t-m}, $(m \ge k)$, and taking expectations shows that the autocorrelation function, $\rho(r)$, satisfies a kth order linear difference equation which is exactly the same as the left-hand side of (3.5.39), i.e. we have,

$$\rho(m) + a_1 \rho(m-1) + a_2 \rho(m-2) + \ldots + a_k \rho(m-k) = 0, \quad m = k, k+1, \ldots.$$
$$\tag{3.5.43}$$

(In fact, (3.5.43) is valid also for $m = 1, 2, \ldots (k-1)$. However, if we express these first $(k-1)$ equations in terms of $\rho(r)$ using only *positive* values of r (remembering that $\rho(r)$ is an even function), these equations do not follow the general pattern which obtains for $m \geq k$; cf. equations (3.5.30), (3.5.31), (3.5.32).) The set of equations (3.5.43) is known as the *Yule–Walker equations*. If we were given the form of the autocorrelation function, $\rho(r)$, but the coefficients $a_1, \ldots, a_k$, were unknown, we could solve the equations (3.5.43) to determine $a_1, \ldots, a_k$. In practice, however, we would not, in general, know the precise form of $\rho(r)$, and the usual procedure would be to substitute *estimates* of $\rho(r)$ (obtained from data) in (3.5.43) which could then be solved to obtain *estimates* of the coefficients $a_1, \ldots, a_k$.

The general solution of (3.5.43) is of the form (assuming $g(z)$ has distinct roots),

$$\rho(r) = B_1 \mu_1^{|r|} + B_2 \mu_2^{|r|} + \ldots + B_k \mu_k^{|r|}, \qquad (3.5.44)$$

where $B_1, B_2, \ldots, B_k$ are constants determined by the boundary conditions (i.e. by the values of $\rho(0), \rho(1), \ldots, \rho(k-1)$).

If all the μ_j's are real valued $\rho(r)$ decays to zero exponentially in a "smooth" form, the rate of decay being determined by the dominant root. If a pair of the μ_j's are complex valued $\rho(r)$ contains an oscillatory term, as in the AR(2) case.

If the roots of $\alpha(z)$ all lie strictly inside the unit circle it may still be possible to obtain a stationary solution of (3.5.39), but now we have to work in "reverse time" and express X_t as a linear function of $\varepsilon_t, \varepsilon_{t+1}, \varepsilon_{t+2} \ldots$ i.e. as a linear function of present and *future* ε_t's, of the form

$$X_t = \sum_{s=0}^{\infty} k_s \varepsilon_{t+s}.$$

For consider, e.g., the AR(1) model (3.5.5) with $|a| > 1$. We may write the particular solution as

$$X_t = -a^{-1} B^{-1} (1 - a^{-1} B^{-1})^{-1} \varepsilon_t,$$

which may be expanded in the form

$$X_t = -a^{-1} \varepsilon_{t+1} - a^{-2} \varepsilon_{t+2} - a^{-3} \varepsilon_{t+3} - \ldots.$$

The autocovariance function of this type of solution will be the same as the autocovariance function of the AR model whose roots are the reciprocals of those of $\alpha(z)$, and hence all lie *outside* the unit circle. Consequently, if we are simply seeking an AR model which generates a given autocovariance function (of suitable form), we would naturally prefer that AR model whose roots lie outside the unit circle, since in this case X_t may be expressed in terms of *present and past* ε_t's, and this description has a much more appealing physical interpretation. In the more general case, where some of the roots of $\alpha(z)$ lie inside and some outside the unit circle, the particular solution of (3.5.39) may be expressed as a linear function of past present and future ε_t's, i.e. we have a *two-sided* expression

$$X_t = \sum_{s=-\infty}^{\infty} k_s \varepsilon_{t-s}$$

(see Hannan (1970), pp. 9–14). The most difficult case to deal with is where $\alpha(z)$ has one or more roots *on* the unit circle. In this case there is no convergent expansion of X_t in terms of either past or future values of ε_t.

Henceforth, when we refer to the "condition for stationarity" for an AR model, we shall mean the condition under which a stationary solution exists with X_t expressed in terms of present and past ε_t's only.

3.5.5 Moving Average Processes

The autoregressive models discussed in the previous sections form an important class which have been found to provide useful descriptions for random processes arising in a large variety of problems. There is, however, another general class of models, calls *moving average processes*, which also occur frequently in practical applications. These are defined as follows.

$\{X_t\}$ is said to be a moving average process of order (denoted by $MA(l)$) if it may be expressed in the form

$$X_t = b_0 \varepsilon_t + b\varepsilon_{t-1} + \ldots + b_l\varepsilon_{t-l}, \tag{3.5.45}$$

where $b_0, b_1, \ldots, b_l$, are constants, and $\{\varepsilon_t\}$ is a purely random process.

Equivalently, we may write,

$$X_t = \beta(B)\varepsilon_t, \tag{3.5.46}$$

where

$$\beta(z) \equiv b_0 + b_1 z + \ldots + b_l z^l. \tag{3.5.47}$$

Note that we may, without any loss of generality, assume that $b_0 = 1$. For, if $b_0 \neq 1$ we may define a new process,

$$\varepsilon_t^* = b_0 \varepsilon_t \qquad \text{(all } t\text{)}.$$

If $\{\varepsilon_t\}$ is a purely random process then so is $\{\varepsilon_t^*\}$, and X_t can be expressed in terms of $\{\varepsilon_t^*\}$ in exactly the same form as (3.5.45) but with the coefficient of ε_t^* now unity. Alternatively, and again without loss of generality, we can always assume that ε_t has unit variance. For, if $\sigma_\varepsilon^2 \neq 1$, we simply define a new process

$$\varepsilon_t^{**} = \varepsilon_t / \sigma_\varepsilon,$$

and again, ε_t^{**} is still a purely random process. (We cannot, of course, adopt both simplifications simultaneously. If, e.g., we wish to take $\sigma_\varepsilon^2 = 1$ we cannot then assume that $b_0 = 1$.)

The purely random process, $\{\varepsilon_t\}$, forms the basic "building block" used in constructing both the autoregressive and moving average models, but the difference between the two types of models may be seen by noting that in the autoregressive case X_t is expressed as a finite linear combination of its *own* past values and the current value of ε_t, so that the value of ε_t is "drawn into" the process $\{X_t\}$ and thus influences *all* future values, $X_t, X_{t+1}, \ldots$. In the moving average case X_t is expressed directly as a linear combination of present and past values of the $\{\varepsilon_t\}$ process but of *finite extent*, so that ε_t influences only (l) future values of X_t, namely $X_{t+1}, \ldots, X_{t+l}$. This feature accounts for the fact that whereas the autocorrelation function of an autoregressive process "dies out gradually", the autocorrelation function of an MA(l) process, as we will show, "cuts off" after the point l, i.e. $\rho(r) = 0, |r| > l$.

The term "moving average" is a rather quaint one, but is now firmly established in the literature. It arose in the following way; if we were given a

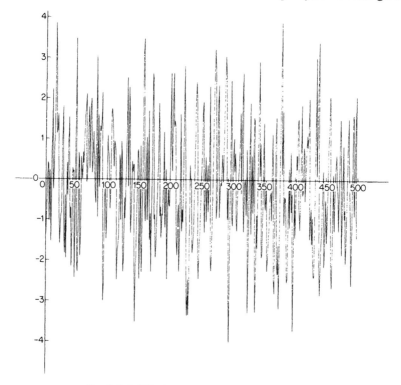

Graph 3.4. 500 observations from an MA(2) process.

complete realization of $\{\varepsilon_t\}$ we would compute X_t by constructing a weighted average of $(\varepsilon_t, \varepsilon_{t-1}, \ldots \varepsilon_{t-l})$. To compute X_{t+1} we would then take a weighted average of $(\varepsilon_{t+1}, \varepsilon_t, \ldots, \varepsilon_{t-l+1})$; in other words, we would "move on" the averaging operation by one time unit, and so on. Historically, the term was used originally to describe a rather *ad hoc* procedure for "smoothing out" irregularities in an observed process. (If we forget for the moment that $\{\varepsilon_t\}$ is a purely random process and regard it simply as just some observed process, then the process $\{X_t\}$ given by (3.5.45) will, in general, exhibit a "smoother" form than $\{\varepsilon_t\}$. This technique is, in fact, a special case of the general method of "linear filtering", and its precise effect will be discussed in more detail in Chapter 4.)

Graph 3.4 shows a realization of 500 observations from the MA(2) model, $X_t = \varepsilon_t + 1 \cdot 1\varepsilon_{t-1} + 0 \cdot 2\varepsilon_{t-2}$, the ε_t being normal variables with zero mean and unit variance.

Since $\{X_t\}$ is a linear combination of uncorrelated random variables its mean, variance, and covariances are readily obtained using the results of Chapter 2, Section 2.13.1. It is easy to see that $\{X_t\}$ is always a stationary process, irrespective of the values of $b_0, b_1, \ldots b_l$, and we have (using an obvious notation)

(i)
$$\sigma_X^2 = \sigma_\varepsilon^2 (b_0^2 + b_1^2 + \ldots + b_l^2). \tag{3.5.48}$$

(ii) Assuming $E[\varepsilon_t] = 0 (\Rightarrow E[X_t] = 0)$,

$$\rho(r) = E[X_t X_{t-r}]/\sigma_X^2$$
$$= \begin{cases} (\sigma_\varepsilon^2/\sigma_X^2)\{b_r b_0 + b_{r+1}b_1 + \ldots + b_l b_{l-r}\}, & 0 \leqslant r \leqslant l, \\ 0, & r > l. \end{cases} \tag{3.549}$$

More concisely,

$$\rho(r) = \begin{cases} \left\{ \sum_{s=r}^{l} b_s b_{s-r} \right\} \Big/ \left\{ \sum_{s=0}^{l} b_s^2 \right\}, & 0 \leqslant r \leqslant l \\ 0, & r > l, \end{cases} \tag{3.5.50}$$

$$\rho(r) = \rho(-r), \qquad\qquad r < 0.$$

Typically, $\rho(r)$ will have the form shown in Fig. 3.16.

Special case of equal weighting

When the "weights" $b_0, b_1, \ldots, b_l$ are all equal and sum to unity, i.e. when $b_0 = b_1 = \ldots = b_l = 1/(l+1)$, the above results simplify to,

$$\sigma_X^2 = \sigma_\varepsilon^2/(l+1),$$

$$\rho(r) = \begin{cases} 1 - \{|r|/(l+1)\}, & |r| \leqslant l, \\ 0, & |r| > l. \end{cases} \tag{3.5.51}$$

Equation (3.5.51) gives a "triangular" form of autocorrelation function, as shown in Fig. 3.17. (Some authors reserve the term "moving-average process" for the case of equal weights.)

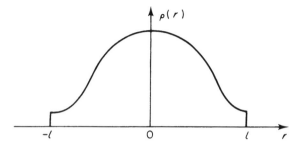

Fig. 3.16. Autocorrelation function of moving average process.

It is worth noting, in passing, that the form of the autocorrelation function, $\rho(r)$, as given by (3.5.50), has some interesting features. Apart from the fact that it "cuts-off" at the point $r = l$ (as remarked above), the numerator, which corresponds to the *autocovariance function* of $\{X_t\}$, has the form of a *convolution*—in fact, it is the convolution of the sequence $\{b_s\}$ with itself. For the moment this result may appear to have little significance, but it will be seen later (in the next chapter) that it forms a special case of an extremely important result in the context of spectral analysis.

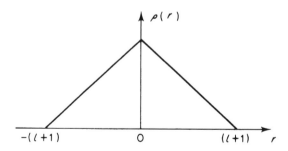

Fig. 3.17. Autocorrelation function: equal weights.

3.5.6 Mixed Autoregressive/Moving Average Processes

Since both the autoregressive and moving average models have each been found to provide useful descriptions for a large number of "real life" processes it is natural to try to combine the two in order to construct an even more general model. This combination is easily obtained as follows. We say that $\{X_t\}$ is a mixed autoregressive/moving average process of order (k, l) (denoted by $ARMA(k, l)$) if it satisfies an equation of the form,

$$X_t + a_1 X_{t-1} + \ldots + a_k X_{t-k} = b_0 \varepsilon_t + b_1 \varepsilon_{t-1} + \ldots + b_l \varepsilon_{t-l}, \quad (3.5.52)$$

where, again, $\{\varepsilon_t\}$ is a purely random process and $(a_1, \ldots, a_k, b_0, \ldots, b_l)$ are constants. (As before, we can, without loss of generality, take $b_0 = 1$.) The mixed model (3.5.52) includes, of course, both the autoregressive and moving average models as special cases, these being obtained, respectively, either by setting $b_1 = b_2 = \ldots b_l = 0$, or by setting $a_1 = a_2 = \ldots = a_k = 0$.

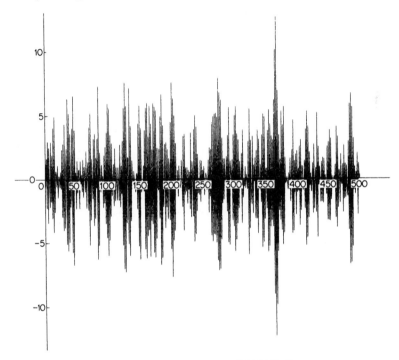

Graph 3.5. 500 observations from an ARMA(2, 2) process.

Graph 3.5 shows a realization of 500 observations from the ARMA(2,2) model,

$$X_t + 1 \cdot 4 X_{t-1} + 0 \cdot 5 X_{t-2} = \varepsilon_t - 0 \cdot 2 \varepsilon_{t-1} - 0 \cdot 1 \varepsilon_{t-2},$$

the ε_t being normal variables with zero mean and unit variance.

Using the same operator notation introduced in the two previous sections, (3.5.53) may be re-written as,

$$\alpha(B)X_t = \beta(B)\varepsilon_t. \quad (3.5.53)$$

We can regard (3.5.53) as an AR(k) model with the purely random process ε_t replaced by the MA(l) process $Z_t = \beta(B)\varepsilon_t$; equivalently, if we write $W_t = \alpha(B)X_t$, then W_t is an MA(l) process.

Equation (3.5.53) may be solved to yield an expression for X_t in terms of $\{\varepsilon_s; s \le t\}$. If we assume that the coefficients $(a_1, \ldots a_k, b_0 \ldots b_l)$ are such that $\{X_t\}$ is stationary, then we may ignore the complimentary function (which will decay to zero), and the *stationary solution* of (3.5.53) may be written formally as,

$$X_t = \alpha^{-1}(B)\beta(B)\varepsilon_t. \qquad (3.5.54)$$

If we expand $\alpha^{-1}(B)$ as a power series in B (assuming that such an expansion exists), (3.5.54) will then provide an expression for X_t as an (in general infinite) linear combination of present and past values $\{\varepsilon_t\}$. We now obtain a difference equation for the autocorrelation function, $\rho(r)$, of $\{X_t\}$, using essentialiy the same argument as used for the purely autoregressive model. Thus, if we multiply both sides of (3.5.52) by X_{t-m}, $m \geq \max(k, l+1)$, and take expectations, noting that X_{t-m} will be uncorrelated with each term on the right-hand side of (3.5.52) if $m \geq (l+1)$, we obtain,

$$\rho(m) + a_1\rho(m-1) + \ldots + a_k\rho(m-k) = 0, \qquad m \geq \max(k, l+1). \quad (3.5.55)$$

Hence, $\rho(r)$ satisfies the same type of difference equation as in the purely autoregressive case, and, for sufficiently large r, its form will be the same as that given by (3.5.44), i.e. it will, in general, contain a mixture of decaying exponential terms and damped oscillatory terms. However, the solution of (3.5.55) involves k arbitrary constants whose values are determined by $\rho(p), \rho(p+1), \ldots, \rho(p+k-1)$, where $p = \max(k, l+1)$, and these initial autocorrelations will take a different form from those corresponding to the purely autoregressive model.

Although the mixed ARMA model still belongs to the class of "*finite parameter models*", i.e. it involves only a finite number of parameters, (namely, $a_1, \ldots, a_k, b_0, \ldots b_l$), it nevertheless possesses a remarkably wide range of applicability, and most stationary processes which arise in practical problems can usually be fitted by a model of the form (3.5.52) with suitably chosen values of k and l. (In fact, the ARMA model (and its modified "integrated" form) is the basic model used by Box and Jenkins (1970) in their work on forecasting and stochastic control.)

There are obvious advantages in using a model of the form (3.5.52) which combines terms of both the autoregressive and moving average type, and in fitting models to observational data it is often possible to fit an ARMA model which involves far fewer parameters than would be required if a purely autoregressive or purely moving average model were fitted. This can be seen by noting that if $\{X_t\}$ really is an ARMA process, then attempting to express it in a purely moving average form, as in (3.5.54), leads to an *infinite* order MA model. (The expansion of $\alpha^{-1}(B)$ will, in general, lead to an infinite order power series.) Similarly, if we attempt to express it as a purely autoregressive model, by writing it in the form

$$\beta^{-1}(B)\alpha(B)X_t = \varepsilon_t,$$

we again obtain an infinite order model. In practice this means that if we try to fit either a pure autoregressive or pure moving average model to data which arise from the mixed model (3.5.52), we may find that we require an extremely large number of parameters in order to achieve a satisfactory fit.

However, it should be remembered that not *all* stationary processes can be described by ARMA models, despite the fairly general nature of this model. Just how general the ARMA model is within the class of all stationary processes is best understood by analysing the spectral properties of these models, but for the moment it may be noted that the generality of ARMA models corresponds, roughly speaking, to the generality of *rational functions* (i.e. functions which are ratios of two polynomials) within the class of all continuous integrable functions. (From an electrical engineering point of view, an ARMA process corresponds to the output obtained by passing white noise through a filter with a rational transfer function—see Chapter 4, Section 4.12.4.)

3.5.7 The General Linear Process

The various models discussed so far are all of the *"finite parameter"* type, i.e. each involves only a finite number of constants. When using one of these models to describe an observed process the parameters would invariably be unknown, and their values would have to be estimated from the given data. In such cases it is clearly imperative that the model to be fitted is of the finite parameter type since we could hardly hope to estimate an infinite number of unknown parameters from a finite amount of data. However, from the mathematical point of view it is interesting to note that *all* the previous models (i.e. the autoregressive, moving average, and ARMA) can be regarded as special cases of a yet more general model, called the *"general linear process"*, although now we have to incorporate an infinite number of parameters. The definition is as follows.

$\{X_t\}$ *is said to be a general linear process if it can be expressed in the form*

$$X_t = \sum_{u=0}^{\infty} g_u \varepsilon_{t-u}, \tag{3.5.56}$$

where $\{\varepsilon_t\}$ is a purely random process, and $\{g_u\}$ is a given sequence of constants satisfying $\sum_{u=0}^{\infty} g_u^2 < \infty$.

The condition $\sum_{u=0}^{\infty} g_u^2 < \infty$ is clearly necessary in order for X_t to have finite variance since it follows immediately from (3.5.56) that

$$\sigma_X^2 = \sigma_\varepsilon^2 \left(\sum_{u=0}^{\infty} g_u^2 \right). \tag{3.5.57}$$

Moreover, it will be recalled from our previous discussion of the AR(1) and AR(2) processes that the condition that the variance of the RHS of (3.5.56) be finite is necessary in order that, for each t, this infinite series represents (in mean square) a well defined random variable.

To obtain the form of the autocorrelation function of $\{X_t\}$ it is convenient to rewrite (3.5.56) in a slightly different form. As it stands, the right-hand side of (3.5.56) is a "one-sided" summation in the sense that it involves only present and past values of $\{\varepsilon_t\}$. However, if we define $g_u = 0$, $u < 0$, then we may re-write (3.5.56) as a "two-sided" summation, namely,

$$X \quad \sum_{u=-\infty}^{\infty} g_u \varepsilon_{t-u}. \tag{3.5.58}$$

Assuming that $E[\varepsilon_t] = 0$ (so that $E[X_t] = 0$), the autocovariance function of $\{X_t\}$ is given by,

$$R(r) = E[X_t X_{t-r}] = \sigma_\varepsilon^2 \left(\sum_{u=-\infty}^{\infty} g_u g_{u-r} \right). \tag{3.5.59}$$

Consequently, the autocorrelation function, $\rho(r)$, is

$$\rho(r) = R(r)/\sigma_X^2 = \sum_{u=-\infty}^{\infty} g_u g_{u-r} \bigg/ \sum_{u=0}^{\infty} g_u^2. \tag{3.5.60}$$

Note that, as for moving-average processes, the autocovariance function is the *convolution of the sequence* $\{g_u\}$ *with itself.* (The device of defining $g_u = 0$, $u < 0$, may have appeared to be quite pointless, but note that had we not done this, the expression for $R(r)$ could not have been written in such a neat form; the term g_{u-r} vanishes if $r > u$.)

The term "general linear process" is due to Bartlett (1955), but his definition is somewhat different in that he restricts the use of the term for the case where the $\{\varepsilon_t\}$ are *independent* random variables, as opposed to being merely uncorrelated.

Stationarity conditions for a general linear process

Initially, it would seem that no conditions on the coefficients, $\{g_u\}$, are required in order for $\{X_t\}$ to be stationary since the expression (3.5.57) for

σ_X^2 is clearly independent of t, and the expression (3.5.59) for the auto-covariance function, $R(r)$, is clearly a function of r only (and does not involve t). However, these conclusions are based on the tacit assumption that both $(\sum_{u=0}^{\infty} g_u^2)$ and $(\sum_{u=-\infty}^{\infty} g_u g_{u-r})$ *converge to finite values*—otherwise, the derivations of (3.5.57) and (3.5.59) are no longer valid.

Now we have already assumed that $\sum_{u=0}^{\infty} g_u^2 < \infty$, and the result,

$$|\text{cov}\{X_t, X_{t-r}\}| \leqslant [\text{var}\{X_t\} \, \text{var}\{X_{t-r}\}]^{1/2} = \sigma_X^2,$$

then shows that this condition ensures also that, for each r, $R(r)$ is finite.

Hence, the condition for the general linear process (3.5.56) *to be stationary is that* $\sum_{u=0}^{\infty} g_u^2 < \infty$.

Introducing the function

$$G(z) = \sum_{u=0}^{\infty} g_u z^u, \tag{3.5.61}$$

the condition $\sum_{u=0}^{\infty} g_u^2 < \infty$ then implies that the function $G(z)$ is analytic inside the unit circle, $|z| < 1$. Thus the condition for stationarity of the model $X_t = G(B)\varepsilon_t$ may be expressed by saying that $G(z)$ is analytic for $|z| < 1$, with the coefficients in its power series expansion satisfying $\sum_{0}^{\infty} g_u^2 < \infty$. Both requirements will be satisfied if $G(z)$ is analytic inside and on the unit circle, i.e. if

$$G(z) < \infty, \qquad |z| \leqslant 1. \tag{3.5.62}$$

(The function $G(z)$, or rather, the value it attains on the unit circle obtained by setting $z = e^{-i\omega}$, plays a basic role in the spectral analysis of the general linear process. We will see later that the condition (3.5.62) arises more naturally out of the requirement that the power spectrum of $\{X_t\}$ must be an integrable function.) It may be noted that if the $\{\varepsilon_t\}$ are *independent* random variables then, subject to the above condition on the $\{g_u\}$, X_t is a *completely stationary process*; see, e.g., Bartlett (1955) p. 147. As defined in (3.5.56) the general linear process appears to be simply a moving average process of *infinite order*, but in fact, it includes also the autoregressive models. Thus, if we set

$$g_u = a^u \qquad \text{(all } u\text{)},$$

(3.5.56) reduces to (3.5.20), and hence with this choice of g_u (3.5.56) generates a *first order autoregressive process*. Similarly, setting,

$$g_u = (\mu_1^{u+1} - \mu_2^{u+1})/(\mu_1 - \mu_2),$$

(3.5.56) reduces to (3.5.24), and thus generates a *second order autoregressive process*. Of course, if we make the coefficient, $\{g_u\}$, vanish beyond a certain point, i.e. if we set

$$g_u = \begin{cases} b_u, & u = 0, 1, \dots l, \\ 0, & u > l, \end{cases}$$

then (3.5.56) reduces to (3.5.45), the *moving average process of (finite) order l*.

The above results indicate that (subject, as one might expect, to certain restrictions) *an autoregressive process of finite order is equivalent to a moving average process of infinite order*, and conversely, *a moving average process of finite order is equivalent to an autoregressive process of infinite order*. For, if we consider the MA(l) model,

$$X_t = \beta(B)\varepsilon_t,$$

where β is a polynomial of degree l, then formally we may write

$$\beta^{-1}(B)X_t = \varepsilon_t,$$

and, in general, the expansion of $\beta^{-1}(B)$ will lead to a polynomial of infinite order. We may thus write, symbolically,

$$AR \text{ (finite order)} \equiv MA \text{ (infinite order)},$$

$$MA \text{ (finite order)} \equiv AR \text{ (infinite order)}.$$

However, as noted above, these relationships will hold only under certain conditions. Before considering this point further let us first consider the more general problem of *"inverting"* a general linear process into an autoregressive form. Using the function, $G(z)$, defined by (3.5.61), the model (3.5.56) may be written as

$$X_t = G(B)\varepsilon_t, \tag{3.5.63}$$

so that, formally, we may write,

$$G^{-1}(B)X_t = \varepsilon_t. \tag{3.5.64}$$

If we now expand $G^{-1}(B)$ in powers of B, (3.5.64) gives the required autoregressive form (in general, of infinite order), provided, of course, that the LHS of (3.5.64) may be expanded in a (mean-square) convergent series. This will certainly hold if $G^{-1}(z)$ admits a power series expansion,

$$G^{-1}(z) = \sum_{u=0}^{\infty} h_u z^u,$$

with $\sum_{u=0}^{\infty} |h_u| < \infty$, i.e. if $G^{-1}(z)$ is analytic inside and on the unit circle. (For, with $E(X_t) = 0$, we have

$$E\left[\left| \sum_{u=m}^{n} h_u X_{t-u} \right|^2 \right] \leq \left[\sum_{u=m}^{n} \{ \mathrm{var}(h_u X_{t-u}) \}^{1/2} \right]^2$$

$$= \sigma_X^2 \left\{ \sum_{u=m}^{n} |h_u| \right\}^2,$$

which $\to 0$ as m and $n \to \infty$, provided $\sum_{n=0}^{\infty} |h_u| < \infty$.) Thus, the model (3.5.63) is "invertible" if

$$G^{-1}(z) < \infty, \qquad |z| \leq 1. \qquad (3.5.65)$$

For the MA(l) process, $G(z) \equiv \beta(z)$ is a polynomial of degree l, and (3.5.65) is then equivalent to the condition that $\beta(z)$ has no roots inside or on the unit circle. Thus, the condition under which a finite order MA model may be inverted into (an infinite order) *AR model is that all the roots of $\beta(z)$ lie outside the unit circle.*

Imposing the condition of invertibility on an MA(l) process is one way of ensuring that there is a unique set of coefficients $b_0, b_1, \ldots, b_l$ which correspond to any given form of the autocovariance function $R(r)$. Without this condition there are, in general, 2^l different sets of coefficients all of which give rise to the same autocovariance function. To see this, consider the simple MA(1) model

$$X_t = \varepsilon_t + b\varepsilon_{t-1},$$

where $|b| < 1$. It is trivial to verify that this model has exactly the same autocovariance function as the alternative model

$$X_t = \varepsilon_t + (1/b)\varepsilon_{t-1}.$$

However, with $|b| < 1$, only the first model is invertible. In general, if we factorize $\beta(z)$ as $\beta(z) = \prod_i (1 - \lambda_i z)$, then any MA operator of the form $\prod_i (1 - \lambda_i^{\pm 1} B)$ gives rise to the same autocovariance function. Thus there are 2^l different models, but only one of these is invertible. For, introducing the "covariance generating function"

$$C(z) = \sum_{r=-\infty}^{\infty} R(r) z^r,$$

it follows from (3.5.50) that for an MA(l) process $C(z)$ may be written as

$$C(z) = \sigma_\varepsilon^2 \beta(z) \beta(z^{-1}) = \sigma_\varepsilon^2 \prod_i (1 - \lambda_i z) \prod_j (1 - \lambda_j z^{-1}).$$

If we replace any real valued λ_i by λ_i^{-1} (or any pair of complex conjugate roots by their reciprocals) it is clear that the form of $C(z)$ is unaltered (with a suitable transformation of σ_ε^2). Thus any of the 2^l different forms of $\beta(z)$, $\prod_i (1 - \lambda_i^{\pm 1} z)$, gives rise to the same autocovariance function. (For further discussion of this point see Box and Jenkins (1970), p. 195.)

The conditions for stationarity which we discussed in connection with the previous special models may now be tied together in terms of the general condition (3.5.62). For the AR(k) process, $G(z) \equiv \alpha^{-1}(z)$, and (3.5.62) reduces to the condition that *all the roots of $\alpha(z)$ must lie outside the unit circle*, as noted in Section 3.5.5. For the MA(l) process, $G(z) \equiv \beta(z)$ is a polynomial of finite order so that (3.5.62) is always satisfied, irrespective of the choice of the coefficients $b_0, b_1, \ldots, b_l$.

For the mixed ARMA model, (3.5.52),

$$G(z) = \frac{\beta(z)}{\alpha(z)}, \tag{3.5.66}$$

and (3.5.62) again reduces to the condition that all the roots $\alpha(z)$ lie outside the unit circle as in the case of the purely autoregressive model. The condition for "inverting" this model into a purely autoregressive form is obtained by applying (3.5.65) with $G(z)$ given by (3.5.66), and thus reduces to the condition that $\beta(z)$ has no roots inside the unit circle—as in the purely moving average case.

Relationship between the generating functions of the moving average and autoregressive representations

Consider the general linear process (3.5.56) (or equivalently (3.5.63)) together with its autoregressive form (3.5.64). If we assume that $G^{-1}(B)$ may be expanded in powers of B then (3.5.64) may be written specifically as,

$$\sum_{v=0}^{\infty} h_v X_{t-v} = \varepsilon_t,$$

say, or

$$H(B)X_t = \varepsilon_t, \tag{3.5.67}$$

where

$$H(z) = \sum_{v=0}^{\infty} h_v z^v. \tag{3.5.68}$$

Comparing (3.5.64), (3.5.68), we see that $H(z) \equiv G^{-1}(z)$, or

$$H(z)G(z) \equiv 1. \tag{3.5.69}$$

The functions $G(z)$, $H(z)$, are sometimes called the *generating functions* of the moving average coefficients and autoregressive coefficients, respectively, and (3.5.69) gives the basic relationship between the two generating functions.

The general linear process, as its name implies, has a fairly general structure, and as noted previously, it includes all our finite parameter models as special cases. It is interesting, therefore, to ask just how general is the general linear process? Is it "completely general" in the sense that *all* stationary processes can be represented in this form? The answer is no, i.e. not all stationary processes are of this form, but nevertheless it begins to approach the kind of representation required for full generality. To follow up this point further we have to examine the general structure of stationary processes in more depth, and we defer this discussion until we treat the topic of "prediction theory" (Chapter 10). For the moment, it may be noted that the main restriction in (3.5.56) lies in the fact that the summation on the right-hand side is "one-sided", i.e. covers only the range $u = 0$ to ∞. If we extend the range of u to $-\infty$ to ∞, so that now (3.5.56) becomes

$$X_t = \sum_{u=-\infty}^{\infty} g_u \varepsilon_{t-u} \qquad (3.5.70)$$

(with the resultant effect that now X_t depends on both past and *future* values of ε_t), then the modified model, (3.5.70), is, in fact, completely general for all stationary processes which have "*purely continuous spectra*" (see Chapter 4, Section 4.9 and Chapter 10, Section 10.1.1).

3.5.8 Harmonic Processes

The final model which we discuss in our collection of standard discrete parameter models is quite different from those which we have previously discussed. (In particular, it cannot be obtained as a special case of the general linear process). It is called the "*harmonic process*" *model*, and is defined by,

$$X_t = \sum_{i=1}^{K} A_i \cos(\omega_i t + \phi_i), \qquad (3.5.71)$$

where K, $\{A_i\}$, $\{\omega_i\}$, $(i = 1, \ldots K)$, are constants, and the $\{\phi_i\}$, $(i = 1, \ldots, K)$, are independent random variables, each having a rectangular distribution on the interval $(-\pi, \pi)$.

Equivalently, (3.5.71) may be written as

$$X_t = \sum_{i=1}^{K} (A_i' \cos \omega_i t + B_i' \sin \omega_i t), \qquad (3.5.72)$$

where $A_i' = A_i \cos \phi_i$, $B_i' = -A_i \sin \phi_i$.

(The reader may have guessed from the remarks at the end of the last section that harmonic processes do not belong to the class of stationary processes which have "purely continuous spectra". This is indeed so, and in fact, (3.5.71) corresponds to a process with a "purely discrete spectrum". We discuss these points in more detail in Chapter 4.)

From the historical point of view, models of the form (3.5.71) were probably among the first to be considered in time series analysis. They are closely related to the numerical technique of "harmonic analysis" and to the classical models of "hidden periodicities", both of which attempt to describe observational records as *sums of sine and cosine waves* whose amplitudes and frequencies are chosen so as to give the "best fit" to the data. In these earlier studies the models considered were purely *deterministic*, (i.e. did not involve any random elements) so that in (3.5.71) the *phases* $\{\phi_i\}$ would be treated as constants rather than as random variables. However, if we wish to study these models within the context of the theory of stationary processes it is essential to regard the $\{\phi_i\}$ as *random variables*. (For, if the ϕ_i are constants then the right-hand side of (3.5.71) is purely deterministic and thus represents the mean value function of $\{X_t\}$, which is clearly a function of t; $\{X_t\}$ is then a (trivial) non-stationary process.)

When the $\{\phi_i\}$ are independent rectangular random variables it is easy to show that $\{X_t\}$ is always a stationary process, irrespective of the values of the $\{A_i\}, \{\omega_i\}$. Specifically, we have,

$$E[X_t] = 0 \qquad \text{(all } t),$$

and the autocovariance function, $R(r)$, is given by,

$$R(r) = \sum_{i=1}^{K} (\tfrac{1}{2}A_i^2) \cos \omega_i r. \tag{3.5.73}$$

Hence,

$$\sigma_X^2 = R(0) = \sum_{i=1}^{K} (\tfrac{1}{2}A_i^2), \tag{3.5.74}$$

and the autocorrelation function $\rho(r)$ is

$$\rho(r) = R(r)/R(0) = \sum_{i=1}^{K} A_i^2 \cos \omega_i r \Big/ \sum_{i=1}^{K} A_i^2. \tag{3.5.75}$$

Note that both $R(r)$ and $\rho(r)$ *consist of a sum of cosine waves with exactly the same frequencies as those contained in* $\{X_t\}$; however, all the cosine terms are now "in phase", i.e. the random phases $\{\phi_i\}$ present in $\{X_t\}$ are removed from the autocovariance and autocorrelation functions.

To establish the above results we consider, for simplicity, the case $K = 1$, i.e. the case where $\{X_t\}$ contains just one cosine term. (The arguments are readily extended to the general case.)

Suppose then that

$$X_t = A \cos(\omega t + \phi).$$

Then since the probability density function of ϕ is $1/2\pi$, $-\pi \leqslant \phi \leqslant \pi$, we have,

$$E[X_t] = \frac{A}{2\pi} \int_{-\pi}^{\pi} \cos(\omega t + \phi)\, d\phi = \frac{A}{2\pi} [\sin(\omega t + \phi)]_{-\pi}^{\pi} = 0, \qquad \text{all } t.$$

Also,

$$R(r) = E[X_t X_{t-r}] = \frac{A^2}{2\pi} \int_{-\pi}^{\pi} \cos(\omega t + \phi) \cos\{\omega(t - r) + \phi\}\, d\phi$$

$$= \frac{A^2}{4\pi} \int_{-\pi}^{\pi} [\cos\{(2\omega t - \omega r) + 2\phi\} + \cos \omega r]\, d\phi.$$

The first term in the above integral vanishes, and hence,

$$R(r) = \tfrac{1}{2} A^2 \cos \omega r, \tag{3.5.76}$$

and

$$\rho(r) = \cos \omega r. \tag{3.5.77}$$

The essential feature to be noted is that *both the autocovariance function and the autocorrelation function consist of a sum of cosine terms, and hence never die out.* The form of $\rho(r)$ for the case $K = 1$ is shown in Fig. 3.18. This

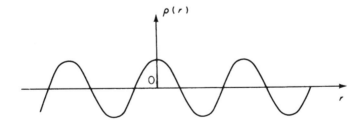

Fig. 3.18. Autocorrelation function of harmonic process.

feature contrasts sharply with the behaviour of $\rho(r)$ for autoregressive and moving-average processes. In the latter cases we have seen that, although $\rho(r)$ may oscillate, *it eventually decays to zero.* This is an extremely important distinction, and forms the basis of one method of testing for the existence of a harmonic process superimposed on, for example, an autoregressive process. (Essentially, this is the problem of detecting "signals" in the presence of "noise", and is known in the statistical literature as the problem of "mixed spectra"; see Chapter 8.)

3.6 STOCHASTIC LIMITING OPERATIONS

So far we have discussed only discrete parameter models, i.e. models in which the time parameter, t, takes integer values, $0, \pm 1, \pm 2, \ldots$. Such models provide natural descriptions of, for example, economic data where, typically, observations are made either daily, weekly, quarterly, or yearly. However, in physical applications observations are usually recorded continuously in time: e.g., a continuous record of a fluctuating voltage, and the stochastic processes which arise are then defined for all values of t. Our problem now is to construct continuous parameter analogues of the standard discrete parameter models. We have already touched on this problem in Section 3.5 when we noted that the natural continuous parameter analogue of the AR(2) model was given by the second order differential equation, (3.5.38). This is an obvious transition, and it requires little intuition to conjecture that the continuous analogue of a kth order linear *difference equation* (corresponding to the AR(k) model) is a kth order linear *differential equation*. Such differential equations are easy enough to write down, but their formulation raises a basic problem—namely, what do we mean by the derivative "$dX(t)/dt$" of a stochastic process $X(t)$? This may seem a deceptively simple question—after all, we know what we mean by the derivative of an ordinary deterministic function $f(t)$. This is defined, of course, as,

$$\lim_{h \to 0} \left(\frac{f(t+h) - f(t)}{h} \right),$$

(assuming that the limit exists). If we try to apply the same definition to a stochastic process we are faced with the problem that the value of the quantity $[\{X(t+h) - X(t)\}/h]$ is (for each t) a *random variable*, and we cannot, therefore, talk about its limiting value (as $h \to 0$) if we use the conventional definition of a limit. The basic problem then is to generalize the conventional notions of "limiting values" and "convergence" to cover the case of random variables. In other words, we need to attach a precise meaning to the statement that a sequence of *random variables*, $U_1, U_2, \ldots, U_i, \ldots$, converges to a limiting *random variable*, U. Now in discussing the "law of large numbers" (Section 2.14) we mentioned briefly two modes of convergence for random variables, namely, "*convergence in probability"* and *"convergence with probability* 1" (or *"almost sure convergence"*). However, there is a third mode of convergence, called *"mean square convergence"*, which turns out to be the most convenient form to use in the present context. The definition of "mean square convergence" is as follows.

Definition 3.6.1 *Let* $U_1, U_2, \ldots, U_i, \ldots$, *be a sequence of random variables. We say that the sequence* $\{U_i\}$ *converges in mean square iff* $\exists$ *a random variable* U *s.t.*

$$\lim_{i \to \infty} E[(U_i - U)^2] = 0.$$

We write this as

$$\text{l.i.m.}_{i \to \infty} U_i = U.$$

(Convergence in mean square is a stronger condition than convergence in probability, in the sense that the former implies the latter, see Parzen (1960), p. 416.)

The basic strategy now is to replace the notion of "ordinary convergence" by "mean square convergence", and then we may construct a whole "calculus" for stochastic processes, using essentially the same definitions of continuity, differentiation, integration, etc., as used for deterministic functions. We illustrate this approach as follows.

3.6.1 Stochastic Continuity

Definition 3.6.2 *We say that* $X(t)$ *is stochastically continuous at* $t = t_0$ *if*

$$\text{l.i.m.}_{t \to t_0} X(t) = X(t_0),$$

i.e., iff

$$\lim_{t \to t_0} E[\{X(t) - X(t_0)\}^2] = 0.$$

Now, if $\{X(t)\}$ is stationary with variance σ_X^2 and autocorrelation function, $\rho(\tau)$, then,

$$E[\{X(t) - X(t_0)\}^2] = \text{var}\{X(t)\} + \text{var}\{X(t_0)\} - 2 \, \text{cov}\{X(t), X(t_0)\}$$

$$= 2\sigma_X^2\{1 - \rho(t - t_0)\}.$$

Hence, $X(t)$ is stochastically continuous at $t = t_0$ iff

$$\lim_{t \to t_0} \rho(t - t_0) = 1,$$

i.e. iff

$$\lim_{\tau \to 0} \rho(\tau) = 1,$$

i.e. *iff* $\rho(\tau)$ *is continuous (in the ordinary sense) at* $\tau = 0$.

In fact, if $\rho(\tau)$ is continuous at $\tau = 0$ it follows that it is continuous everywhere. To prove this consider the expression (for any τ, t, h),

$$E[X(t-\tau)\{X(t+h) - X(t)\}] = \sigma_X^2 \{\rho(\tau+h) - \rho(\tau)\}.$$

By the Schwarz inequality (see proof of Theorem (2.12.3)),

$$[E[X(t-\tau)\{X(t+h) - X(t)\}]]^2 \le E[X^2(t-\tau)]E[\{X(t+h) - X(t)\}^2],$$

i.e.

$$\sigma_X^4 \{\rho(\tau+h)\} - \rho(\tau)\}^2 \le 2\sigma_X^4 (1 - \rho(h)),$$

or

$$|\rho(\tau+h) - \rho(\tau)| \le \{2(1 - \rho(h)\}^{1/2}.$$

If $\rho(\tau)$ is continuous at $\tau = 0$, given then ε we may choose δ so that

$$\{1 - \rho(h)\} \le \varepsilon^2/2 \qquad \text{if } |h| \le \delta.$$

Hence,

$$|\rho(\tau+h) - \rho(\tau)| \le \varepsilon \qquad \text{if } |h| \le \delta,$$

and the continuity of $\rho(\tau)$ is established for all τ.

It is important to remember that stochastic continuity is a property which relates only to the *average behaviour* (over all realizations) of the process *at a particular time point*. (Conditions ensuring stochastic continuity throughout an interval may be obtained but are more complicated, see, e.g., Bartlett (1955) Chapter 5.) Thus, it is possible to have a stochastically continuous process for which each realization has an infinite number of discontinuities. As an example, consider the *"random telegraph signal"* model. Here, $X(t)$ has only two possible values, say 0 or 1, and the time instants, $\{t_i\}$ at which the transitions occur form a Poisson process, rate ν, (see Section 2.11, Example (2)). A typical realization of the random telegraph signal is shown in Fig. 3.19.

It may be shown (Parzen (1962), p. 115) that $X(t)$ is a stationary process with autocorrelation function

$$\rho(\tau) = \exp(-2\nu|\tau|). \qquad (3.6.1)$$

For this process $\rho(\tau)$ is continuous at $\tau = 0$, but clearly each realization has an infinite number of discontinuities. However, as these discontinuities form a countable number of points each realization could be said to be continuous "almost everywhere".

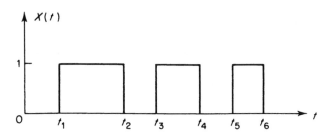

Fig. 3.19. The random telegraph signal.

3.6.2 Stochastic Differentiability

Definition 3.6.3 $X(t)$ is said to be stochastically differentiable at $t = t_0$, with derivative $\dot{X}(t_0)$, iff

$$\underset{h \to 0}{\text{l.i.m.}} \left\{ \frac{X(t_0 + h) - X(t_0)}{h} \right\}$$

exists and equals $\dot{X}(t_0)$ i.e. iff

$$\lim_{h \to 0} E\left[\left\{ \frac{X(t_0 + h) - X(t_0)}{h} \right\}^2 \right]$$

exists. Now

$$E\left[\left\{ \frac{X(t_0 + h) - X(t_0)}{h} \right\}^2 \right] = \frac{2\sigma_X^2}{h^2} \{1 - \rho(h)\}.$$

Hence, for $X(t)$ to be stochastically differentiable we require that $\lim_{h \to 0} \{[1 - \rho(h)]/h^2\}$ exists, which in turn requires at least that $\dot{\rho}(0)$ *exists and is zero.* (If $\dot{\rho}(0)$ exists it must in any case be zero by virtue of the symmetry of the function $\rho(\tau)$.) However, we may obtain a stronger condition by using the symmetry of $\rho(\tau)$ to write,

$$\frac{2(1 - \rho(h))}{h^2} \equiv -\left[\frac{\{\rho(h) - \rho(0)\} - \{\rho(0) - \rho(-h)\}}{h^2} \right].$$

The numerator of the right-hand side above represents the second difference of the three ordinates, $\rho(-h)$, $\rho(0)$, $\rho(h)$, and hence if the limit of the right hand side exists (as $h \to 0$) it must be $\ddot{\rho}(0)$, the second derivative of

$\rho(\tau)$ at $\tau = 0$. Now it can be shown (see, e.g. Yaglom (1962)) that if $\ddot{\rho}(0)$ exists then $\ddot{\rho}(\tau)$ exists for all τ. Hence, we have the result *that $\dot{X}(t)$ exists if and only if $\ddot{\rho}(\tau)$ exists.*

We show in Section 3.6.4 that, under this condition,

$$\text{cov}\{X(t), \dot{X}(t)\} = 0, \qquad \text{all } t,$$

and that

$$\text{cov}\{\dot{X}(t), \dot{X}(t+\tau)\} = -\sigma_X^2 \ddot{\rho}(\tau),$$

i.e. $\{\ddot{\rho}(\tau)/\ddot{\rho}(0)\}$ is the autocorrelation function of $\dot{X}(t)$.

The autocorrelation function of the random telegraph signal, (3.6.1), is clearly not differentiable at $\tau = 0$; consequently although this process is stochastically continuous it is not stochastically differentiable.

It may now become apparent why the mode of mean square convergence is the most convenient form for dealing with stationary processes. The point is that mean square convergence involves only quadratic functions of $\{X(t)\}$, and thus depends only on the *second order properties* of the process. Consequently, conditions for, e.g., stochastic continuity or stochastic differentiability are immediately translatable into corresponding conditions on the autocorrelation function, $\rho(\tau)$.

3.6.3 Integration

Again, using the idea of mean square convergence it is a fairly straightforward matter to extend the Riemann definition of an integral to the case where the integrand is a stochastic process. Consider, for example, an integral of the form,

$$\int_a^b g(t)X(t)\, dt,$$

where $g(t)$ is a deterministic function. As in classical analysis we start by considering an arbitrary finite set of points, $(t_0, t_1, \ldots, t_n)$ in the interval (a, b), and define the integral as

$$\text{l.i.m.} \left\{ \sum_{i=1}^{n} g(t_i)X(t_i)(t_i - t_{i-1}) \right\}$$

as $\max(t_i - t_{i-1}) \to 0$. It may be shown (see, e.g., Parzen (1962)) that this limit exists in mean square if and only if $\int_a^b \int_a^b g(s)g(t)R(t-s)\, ds\, dt$ exists as a Riemann (double) integral. (This condition arises from the fact that the Riemann integral is simply the expression for the variance of $\{\int_a^b g(t)X(t)\, dt\}$.)

We may similarly define Riemann–Stieltjes integrals which involve stochastic processes. We recall that in classical analysis, when we have two deterministic functions, $g(t)$, $F(t)$, the Riemann–Stieltjes integral

$$R = \int_a^b g(t)\, dF(t) \tag{3.6.2}$$

is defined as the limiting value of the discrete summation

$$\sum_{i=1}^n g(t_i)\{F(t_i) - F(t_{i-1})\}$$

as $\max(t_i - t_{i-1}) \to 0$. (We observe, in passing, that if $f(t) = F'(t)$ exists for all $t \in (a, b)$, then $dF(t) = f(t)\, dt$, and R may then be written in the ordinary integral form, $\int_a^b g(t)f(t)\, dt$. However, as noted in Chapter 1, Section 1.3, the advantage of introducing the Riemann–Stieltjes type of integral is that integrals of the form (3.6.2) may still be defined in a meaningful way even when the function $F(t)$ is not differentiable.)

Following the above approach, a Riemann–Stieltjes integral of the form

$$\int_a^b g(t)\, dX(t)$$

may now be defined as the mean-square limit of the discrete summation

$$\sum_{i=1}^n g(t_i)\{X(t_i) - X(t_{i-1})\}.$$

In later applications we will usually be concerned with integrals which extend over infinite ranges rather than the finite range (a, b). However, with obvious modification, the above ideas may be extended to cover these more general cases.

3.6.4 Interpretation of the Derivatives of the Autocorrelation Function

Assume, for simplicity, that $E[X(t)] = 0$. Then,

$$\sigma_X^2 \rho(\tau) = E[X(t)X(t + \tau)].$$

If $\dot{X}(t)$ exists, then differentiating the above equation w.r. to τ we obtain,

$$\sigma_X^2 \dot{\rho}(\tau) = E[X(t)\dot{X}(t + \tau)].$$

But, in view of the stationarity of $X(t)$ we may replace t by $(t - \tau)$, giving

$$\sigma_X^2 \dot{\rho}(\tau) = E[X(t - \tau)\dot{X}(t)].$$

Differentiating again w.r. to τ now gives,

$$\sigma_X^2 \ddot{\rho}(\tau) = -E[\dot{X}(t-\tau)\dot{X}(t)].$$

Hence $\{-\sigma_X^2 \ddot{\rho}(\tau)\}$ is the autocovariance function of $\dot{X}(t)$, and consequently $\{\ddot{\rho}(\tau)/\ddot{\rho}(0)\}$ is the autocorrelation function of $\dot{X}(t)$. In particular, we have that $\text{var}\{\dot{X}(t)\} = -\sigma_X^2 \ddot{\rho}(0)$, and hence $(-\sigma_X^2 \ddot{\rho}(0))^{1/2}$ gives the order of magnitude of the fluctuations of $\dot{X}(t)$. Note further that setting $\tau = 0$ in the second equation gives

$$E[X(t)\dot{X}(t)] = \sigma_X^2 \dot{\rho}(0) = 0,$$

since the existence of $\dot{X}(t)$ implies that $\dot{\rho}(0) = 0$.

Since $\dot{X}(t)$ is itself a stationary process with autocorrelation function proportional to $\ddot{\rho}(\tau)$, it is easy to derive conditions for the existence of higher derivatives. For example, $\ddot{X}(t)$ exists if and only if $\rho(\tau)$ is *four times* differentiable, and so on.

3.7 STANDARD CONTINUOUS PARAMETER MODELS

3.7.1 Purely Random Process (Continuous "White Noise")

In constructing our "dictionary" of discrete parameter models the first model we considered was that of the purely random (or "white noise") process. This process has a very simple structure but, as we have seen, it forms a basic ingredient in the construction of more complicated models such as the autoregressive, moving average, and ARMA schemes. When we come to consider continuous parameter models the natural starting point, therefore, is to define a continuous purely random process. Since, in the discrete case, this process has such a simple definition, one might imagine that it would be equally simple to define the continuous parameter version. However, far from being a simple matter, the definition of a continuous parameter purely random process poses considerable problems. In fact, it turns out that no such process exists, except in a highly degenerate sense!

To gain some insight into this rather surprising result let us try to define the continuous version by following, as closely as possible, the definition used in the discrete case. We would then define a continuous parameter purely random process, $\{\varepsilon(t)\}$, by saying that, *for any subset of times* $t_1, t_2, \ldots, t_n$, $\varepsilon(t_1), \varepsilon(t_2), \ldots, \varepsilon(t_n)$ *form a set of uncorrelated random variables.*

Assuming (as we have always done) that $E[\varepsilon(t)] = 0$, and $\text{var}\{\varepsilon(t)\} = \sigma_\varepsilon^2$, all t, it follows that the autocovariance function, $R_\varepsilon(\tau)$, has the form

$$R_\varepsilon(\tau) = \begin{cases} \sigma_\varepsilon^2, & \tau = 0 \\ 0, & \tau \neq 0, \end{cases} \tag{3.7.1}$$

and hence the autocorrelation function, $\rho(\tau)$, is

$$\rho_\varepsilon(\tau) = \begin{cases} 1, & \tau = 0 \\ 0, & \tau \neq 0 \end{cases} \qquad (3.7.2)$$

(compare equation (3.5.4)). Obviously, each realization of such a process would have an extremely erratic form since the random variables $\varepsilon(t)$, $\varepsilon(t + \Delta t)$, have to remain uncorrelated, no matter how small we make Δt. However, as defined above, $\varepsilon(t)$ *represents a degenerate process in a much stronger sense.* For, suppose we construct a random variable U by forming a linear combination of values of $\{\varepsilon(t)\}$. In continuous time a linear combination is represented by an integral so U would be of the form,

$$U = \int_a^b g(t)\varepsilon(t)\,dt, \qquad (3.7.3)$$

say. We then have (using the obvious continuous analogues of (2.13.13), (2.13.15)),

$$E(U) = 0,$$

$$\text{var}(U) = \int_a^b \int_a^b g(u)g(v)\,\text{cov}\{\varepsilon(u), \varepsilon(v)\}\,du\,dv$$

$$= \sigma_\varepsilon^2 \int_a^b \int_a^b g(u)g(v)\delta_{u,v}\,du\,dv, \qquad (3.7.4)$$

where $\delta_{u,v}$ is the Kronecker delta, i.e.,

$$\delta_{u,v} = 1, \quad u = v; \qquad \delta_{u,v} = 0, \quad u \neq v.$$

The integrand on the right-hand side of (3.7.4) vanishes everywhere except along the line $u = v$; hence the double integral is zero, and, consequently, for any choice of $g(t)$ we have,

$$\text{var}\left\{ \int_a^b g(t)\varepsilon(t)\,dt \right\} = 0, \qquad \text{if } \sigma_\varepsilon^2 \text{ is finite.}$$

In other words, if σ_ε^2 is finite, then, with probability 1, any linear combination of $\{\varepsilon(t)\}$ takes the *constant value zero.* Now it will be recalled that in the discrete models (such as the AR(k)), the process $\{X_t\}$ always turned out to be expressible as a linear combination of $\{\varepsilon_s; s \leq t\}$. Hence, the above process, $\{\varepsilon(t)\}$, would be of little value in generating, e.g., continuous parameter autoregressive processes. The only way in which we can obviate this fundamental difficulty is to allow σ_ε^2 *to be infinite.* This again leads to a degenerate process, but since in subsequent work we will usually be dealing with integrals of the form (3.7.3) rather than with $\varepsilon(t)$ itself, this particular form of degeneracy does not prove too troublesome. Note that if σ_ε^2 is infinite (3.7.1) may be written in a more informative manner as

$$R_\varepsilon(\tau) = C \cdot \delta(\tau) \qquad (3.7.5)$$

(C being a constant), i.e. *the autocovariance function now has the form of a Dirac δ-function.*

We would assure the reader (as we have done several times before) that the above points will emerge in a clearer light after we have discussed the topic of spectral analysis. In fact, the non-existence of continuous time "white-noise" is well known to engineers from power/frequency considerations. In the engineering approach such a process has to be allowed to have *infinite power*, which, as we shall see, corresponds to allowing σ_ε^2 to be infinite.

3.7.2 First Order Autoregressive Processes

We have remarked several times that in continuous time we would expect autoregressive models to be described by linear differential equations. Consequently, we would expect that the continuous parameter AR(1) would be described by a first order differential equation—corresponding to the first order difference equation (3.5.5). However, *what replaces the process $\{\varepsilon_t\}$ on the right-hand side of* (3.5.5)? The difficulties which we have encountered in discussing the continuous parameter purely random process suggest that the transition from the discrete to the continuous model is not entirely straightforward, and thus merits a fairly cautious analysis if we are not to finish up with an equation which is meaningless (both mathematically and physically). Our approach may be described as "careful" (rather than completely rigorous), but illustrates the nature of the mathematical difficulties which arise in treating continuous parameter models.

Let us start by considering again the discrete parameter model (3.5.5), which we now write as,

$$X_n - aX_{n-1} = Y_n, \qquad n = 0, \pm 1, \pm 2, \ldots,$$

or equivalently,

$$X_n(1-a) + a(X_n - X_{n-1}) = Y_n, \qquad (3.7.6)$$

where Y_n is a purely random process with variance σ_Y^2. (Note that here we denote discrete time by n, the symbol t being reserved for continuous time.) Using the difference operator Δ, defined by,

$$\Delta X_n = X_n - X_{n-1},$$

(3.7.6) may be written as

$$\Delta X_n + \left(\frac{1-a}{a}\right) X_n = \frac{Y_n}{a}. \tag{3.7.7}$$

Up to this stage the unit of time associated with the parameter n has been irrelevant, i.e. the interval between successive observations could have been 1 year, 1 hour, 1 second, or $0\cdot1$ seconds. However, in order to transform (3.7.7) into the continuous parameter form we have to consider its limiting form as this time interval tends to zero. Suppose then that the time interval between successive observations is Δt, so that we may write $t = n \cdot \Delta t$. Dividing each term in (3.7.7) by Δt gives,

$$\frac{\Delta X_n}{\Delta t} + \frac{(1-a)}{a\,\Delta t} \cdot X_n = \frac{Y_n}{a\,\Delta t}. \tag{3.7.8}$$

Now write

$$\alpha = (1-a)/a\,\Delta t, \tag{3.7.9}$$

$$\varepsilon(t) = Y_n/a\,\Delta t, \tag{3.7.10}$$

and let $\Delta t \to 0$ in (3.7.8). Clearly, the first term on the left-hand side will become the derivative, $\dot{X}(t)$, (if it exists), and formally we may write the limiting form of (3.7.8) as

$$\dot{X}(t) + \alpha X(t) = \varepsilon(t). \tag{3.7.11}$$

This is the *continuous parameter AR(1) model*, and (3.7.11) is called a *stochastic differential equation*. (The term "stochastic differential equation" is used generally to describe a differential equation in which the right-hand side (i.e. the "forcing function") is a stochastic process rather than an ordinary deterministic function.)

In writing down (3.7.11) we have tacitly assumed that the parameter α remains finite. For this to be so we obviously require that as $\Delta t \to 0$, $a \to 1$, i.e. to construct the continuous parameter model we have to consider a *sequence* of discrete parameter models in which the value of a changes as Δt changes. We recall that the parameter a represents the correlation coefficient between X_n and X_{n-1}; as the time interval Δt between these observations tends to zero we would obviously expect a to tend to unity if the process has any form of "continuity". We recall further that $|a| < 1$ is the stationarity condition for the discrete case, so that as $\Delta t \to 0$, *a moves to the boundary of the region of stationarity*.

Now consider the process $\{\varepsilon(t)\}$. This is clearly an "improper" process, taking uncorrelated values in successive infinitesimal intervals, and in this sense is consistent with our description of the *continuous time purely random process*. In the discrete case we have from (3.5.14),

$$\sigma_X^2 = \sigma_Y^2/(1-a^2).$$

From (3.7.9),

$$a = 1/(1+\alpha\,\Delta t) \sim 1 - \alpha\,\Delta t + \text{o}(\Delta t),$$

so that,

$$\sigma_X^2 = \frac{\sigma_Y^2}{2\alpha\,\Delta t}\,(1+\text{O}(\Delta t)). \tag{3.7.12}$$

Clearly, for σ_X^2 to remain finite (as $\Delta t \to 0$) we must have $\sigma_Y^2 = \text{O}(\Delta t)$, and then (3.7.10) gives,

$$\sigma_\varepsilon^2 = \text{var}\{\varepsilon(t)\} = \sigma_Y^2/[a^2(\Delta t)^2]$$
$$\sim \sigma_Y^2/(\Delta t)^2, \qquad \text{for small } \Delta t, \text{ (since } a \to 1),$$
$$= \text{O}(1/\Delta t). \tag{3.7.13}$$

Thus, $\varepsilon(t)$ is a zero mean process *whose fluctuations are* $\text{O}[1/(\Delta t)^{1/2}]$. In the limit, as $\Delta t \to 0$, $\sigma_\varepsilon^2 \to \infty$, (which as we have seen, is necessary in order to prevent complete degeneracy of the continuous purely random process).

When the $\{Y_n\}$ are *independent* (instead of being merely uncorrelated), the process

$$W(t) = \int_0^t \varepsilon(u)\,du \tag{3.7.14}$$

is called an *"additive process"* (i.e. a process with independent stationary increments). Regarding the above integral as the limiting form of a sum we see that $W(t)$ may be thought of as the limiting case of a *random walk model* in which both the step lengths and the time intervals between steps become infinitesimal.

Since $W(t)$ is the limit of a sum of independent random variables it is reasonable to suppose, by virtue of the central limit theorem (Section 2.14.3), that *for each t, $W(t)$ has a normal distribution*. However, the application here of the central limit theorem has to be treated with considerable care; although as $\Delta t \to 0$, $W(t)$ becomes the sum of an increasingly large number of independent random variables it is also true that the variance of the individual terms $\{\varepsilon(t)\Delta t\}$ is proportional to Δt and this changes as the number of random variables in $W(t)$ increases. Thus, it does not follow that any process with independent stationary increments must necessarily be Gaussian and a well known counter-example is the Poisson process, $N(t)$, discussed in Section 2.11. The increments, $\Delta N(t) = \{N(t + \Delta t) - N(t)\}$ are certainly independent (for non-overlapping intervals) and $E[\Delta N(t)] = \lambda \Delta t$, $\text{var}[\Delta N(t)] = \lambda \Delta t$, with λ the "rate" of the Poisson process. However, each $\Delta N(t)$ is, in effect, either 0 or 1, and the "jump" process $N(t)$ has, of course, a Poisson distribution, not a normal distribution. There is, in fact, a very general result which states that any additive process can be expressed as the sum of two components, namely a Gaussian process and a discontinuous or "jump" process, as typified by the Poisson process.

Wiener processes

An additive process $\{W(t)\}$ which is such that for each t, $W(t)$ has a normal distribution is called a *Wiener process*. It is not difficult to prove the more general result that $\{W(t)\}$ is then a Gaussian process in the sense of Section 3.4.1 (see Parzen (1962), p. 91).

Henceforth, we will assume that the additive process $W(t)$ defined by (3.7.14) does not contain any discontinuous or "jump" components, so that $W(t)$ is then a Wiener process, and *the process $\{\varepsilon(t)\}$ can now be described (formally) as the "derivative" of a Wiener process*. We have defined $W(t)$ above via the pathological process $\varepsilon(t)$, but the Wiener process can be introduced independently in a perfectly well defined sense: see, e.g., Parzen (1962), p. 28. The situation is, in some ways, similar to that of δ-functions; the δ-function is pathological but the integral of a δ-function is well behaved. The pathological nature of $\varepsilon(t)$ then arises from the fact that although $W(t)$ exists it is not stochastically differentiable.

The Wiener process $W(t)$ may be defined formally by the following conditions:

(a) $W(0) = 0$.

(b) $E\{W(t)\} = 0$, all $t > 0$.

(c) $W(t)$ is normal, each $t > 0$,

and

(d) $W(t)$ has stationary independent increments.

If we define the process $\{\varepsilon_h(t)\}$ by

$$\varepsilon_h(t) = \frac{1}{h}\{W(t+h) - W(t)\}, \qquad (h > 0),$$

it follows immediately from (3.7.19) that the autocovariance function of $\varepsilon_h(t)$ is given by

$$R_h(\tau) = \begin{cases} (\sigma_W^2/h)(1 - |\tau|/h), & |\tau| \leq h, \\ 0, & |\tau| > h, \end{cases} \qquad (\sigma_W^2 \text{ a constant}).$$

The process $\varepsilon(t)$ may now be defined as the limiting form of $\varepsilon_h(t)$ as $h \to 0$ and from the form of the above result as $h \to 0$ we see that $R_\varepsilon(\tau) = \sigma_W^2 \delta(\tau)$, a result which we shall confirm later.

The above discussion was based on the assumption that the $\{Y_n\}$, and hence the $\{\varepsilon(u)\}$, are independent variables. If the $\{\varepsilon(u)\}$ are merely uncorrelated but $W(t)$ is nevertheless a continuous process, we will still refer to $W(t)$ as a "Wiener process". This nomenclature is not in accord with the strict definition of the term "Wiener process", but the second order properties of $W(t)$ are, of course, the same whether the $\{\varepsilon(u)\}$ are independent or uncorrelated. Since we shall be concerned almost entirely with second order properties, the more general use of the term "Wiener process" should not lead to any confusion. In fact, the second order properties of $X(t)$ depend only on the fact that $W(t)$ has *orthogonal increments*, and hence are the same whether $W(t)$ is a strict Wiener process, a "generalized" Wiener process (in the above sense), or a discontinuous process such as a Poisson process. We shall return to this point later when we discuss "filtered Poisson processes".

It should be noted that $W(t)$ is a *non-stationary* process; in particular, its variance increases linearly with t. For, if we write $\Delta W(t) = W(t + \Delta t) - W(t)$, then for small Δt we have

$$\Delta W(t) = \int_t^{t+\Delta t} \varepsilon(u) \, du \sim \varepsilon(t) \Delta t,$$

so that,

$$\text{var}\{\Delta W(t)\} \sim (\Delta t)^2 \, \text{var}\{\varepsilon(t)\} \sim \sigma_Y^2 \sim 2\alpha \sigma_X^2 \Delta t,$$

(using (3.7.12)). Since $W(t)$ and $\Delta W(t)$ involve integrals of $\varepsilon(u)$ over non-overlapping intervals it follows that $W(t)$ and $\Delta W(t)$ are uncorrelated random variables. Hence, writing $W(t + \Delta t) = W(t) + \Delta W(t)$, we find,

$$\text{var}\{W(t + \Delta t)\} = \text{var}\{W(t)\} + \text{var}\{\Delta W(t)\}$$

$$\doteq \text{var}\{W(t)\} + 2\alpha \sigma_X^2 \Delta t.$$

If now we introduce the quantity $\sigma_W^2 = (d/dt)[\text{var}\{W(t)\}]$, then we may write,

$$\sigma_W^2 = \lim_{\Delta t \to 0} \frac{\text{var}\{(W(t + \Delta t)\} - \text{var}\{W(t)\}}{\Delta t}$$

$$= 2\alpha \sigma_X^2. \tag{3.7.15}$$

Note that σ_W^2 measures the *rate of increase of* $\text{var}\{W(t)\}$, and (3.7.15) shows that $\text{var}\{W(t)\}$ *increases with t at a constant rate*. Specifically,

$$\text{var}\{W(t)\} = \int_0^t \sigma_W^2 \, dt = t \sigma_W^2 \tag{3.7.16}$$

(a result which may be derived alternatively directly from (3.7.14)). Since $\varepsilon(t)$ itself has infinite variance the quantity σ_W^2 now becomes the basic parameter in terms of which we may express, e.g., σ_X^2; that is σ_W^2 plays the role of σ_ε^2 for the continuous case. To emphasize this point we now re-write (3.7.15) as

$$\sigma_X^2 = \sigma_W^2/2\alpha. \tag{3.7.17}$$

Also, we may now identify the constant C in the formal expression (3.7.5) for the autocovariance function of $\varepsilon(t)$. For, from (3.7.14) we have,

$$\text{var}\{W(t)\} = \int_0^t \int_0^t \text{cov}\{\varepsilon(u), \varepsilon(v)\}\, du\, dv$$

$$= C \int_0^t \int_0^t \delta(u-v)\, du\, dv$$

$$= C\left\{\int_0^t du\right\} = Ct.$$

Comparing this with (3.7.16) we see that $C = \sigma_W^2$, so that (3.7.5) may now be written,

$$R_\varepsilon(\tau) = \sigma_W^2 \delta(\tau). \tag{3.7.18}$$

In addition we have for any s, t, such that $s \leq t$,

$$\text{cov}\{W(s), W(t)\} = \text{cov}[W(s), [\{W(t) - W(s)\} + W(s)]]$$

$$= \text{cov}[W(s), \{W(t) - W(s)\}] + \text{var}\{W(s)\},$$

$$= \text{var}\{W(s)\},$$

$$= s\sigma_W^2,$$

since $\{W(t) - W(s)\}$ and $W(s)$ are uncorrelated.

We may thus write for all s, t,

$$\text{cov}\{W(s), W(t)\} = \sigma_W^2 \min(s, t). \tag{3.7.19}$$

Results (3.17.16), (3.17.17) demonstrate the non-stationary character of $\{W(t)\}$. (Note that the RHS of (3.7.17) is *not* a function of $(s - t)$ only.)

Condition for stationarity

Equation (3.7.11) is a simple first order differential equation and the particular solution may be obtained by using the "integrating factor" method. Multiplying both sides by $\exp(\alpha t)$ we may rewrite (3.7.11) as

$$\frac{d}{dt}\{\exp(\alpha t)X(t)\} = \exp(\alpha t)\varepsilon(t),$$

and hence,

$$X(t) = \exp(-\alpha t) \int_{-\infty}^{t} \exp(\alpha u)\varepsilon(u)\, du = \int_{-\infty}^{t} \exp[-\alpha(t-u)]\varepsilon(u)\, du.$$
(3.7.20)

Note that if we use the differential operator D we may write the particular solution in the alternative form

$$X(t) = \left(\frac{1}{D+\alpha}\right)\varepsilon(t),$$
(3.7.21)

and comparing (3.7.20), (3.7.21), we obtain the well known result that for any function $x(t)$,

$$\left(\frac{1}{D+\alpha}\right)x(t) \equiv \int_{-\infty}^{t} \exp[-\alpha(t-u)]x(u)\, du.$$
(3.7.22)

The general solution of (3.7.11) is obtained by adding to (3.7.20) the complementary function (CF), i.e. the solution of the homogeneous equation,

$$(D+\alpha)X(t) = 0.$$

The CF is thus of the form $A \exp(-\alpha t)$, A being an arbitrary constant, and hence the general solution of (3.7.11) is

$$X(t) = A \exp(-\alpha t) + \int_{-\infty}^{t} \exp[-\alpha(t-u)]\varepsilon(u)\, du.$$
(3.7.23)

(Compare (3.7.23) with (3.5.21).)

Following exactly the same argument as used for the discrete parameter model it is clear that for $X(t)$ to be (asymptotically) stationary the CF must decay to zero as $t \to \infty$. Hence the condition for stationarity of the continuous parameter AR(1) model (3.7.11) is

$$\alpha > 0.$$
(3.7.24)

When this condition is satisfied we can, for sufficiently large t, ignore the CF and then (3.7.20) is called the *stationary solution* of (3.7.11). However, in view of the fact that, strictly speaking, the process $\{\varepsilon(t)\}$ does not, exist, it is preferable to rewrite (3.7.20) as a Stieltjes integral in terms of $W(t)$, i.e. we write

$$X(t) = \int_{-\infty}^{t} \exp[-\alpha(t-u)]\, dW(u).$$
(3.7.25)

The autocorrelation function

The simplest way of obtaining the autocorrelation function is to start from the expression for the discrete parameter model and then construct its limiting form as $\Delta t \to 0$. In the discrete case we have, from (3.5.17),

$$\rho(r) = a^{|r|}.$$

Now write $\tau = r\Delta t$. Since $a = 1/(1 + \alpha \Delta t)$, we have,

$$\rho(\tau) = (1 + \alpha \Delta t)^{-|(\tau/\Delta t)|}.$$

As $t \to 0$ we obtain the result,

$$\rho(\tau) = \exp(-\alpha|\tau|). \tag{3.7.26}$$

The form of $\rho(\tau)$ is shown in Fig. 3.20.

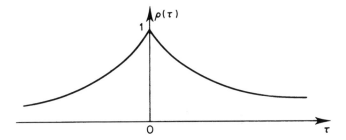

Fig. 3.20. Autocorrelation function of continuous parameter AR(1) process.

Alternatively, we can obtain a differential equation for $\rho(\tau)$ by constructing the continuous analogue of the difference equation (3.5.19). The analysis here is exactly the same as that which produced (3.7.11) from (3.5.5), and we clearly find that, for $\tau \geq 0$,

$$\dot{\rho}(\tau) + \alpha\rho(\tau) = 0. \tag{3.7.27}$$

(Here the "dot" notation refers to differentiation w.r.t. τ). The solution of (3.7.27) is

$$\rho(\tau) = B \exp(-\alpha\tau), \qquad \tau \geq 0,$$

B being an arbitrary constant. Substituting $\rho(0) = 1$ shows that $B = 1$, and using the property that $\rho(\tau)$ is an even function confirms the result (3.7.26).

A third (and more direct) method is to evaluate first the autocovariance function, $R(\tau)$, directly from (3.7.20). Thus, we have,

$$R(\tau) = \text{cov}\{X(t), X(t+\tau)\}$$

$$= \int_{-\infty}^{t} \int_{-\infty}^{t+\tau} \exp[-\alpha(t-u)] \exp[-\alpha(t+\tau-v)] \, \text{cov}\{\varepsilon(u), \varepsilon(v)\} \, du \, dv$$

$$= \sigma_W^2 \exp(-\alpha\tau) \int_{-\infty}^{t} \int_{-\infty}^{t+\tau} \exp[-\alpha(t-u)] \exp[-\alpha(t-v)] \delta(u-v) \, du \, dv,$$

(using (3.7.18))

$$= \sigma_W^2 \exp(-\alpha\tau) \int_{-\infty}^{t} \exp[-2\alpha(t-u)] \, du, \qquad \tau \geq 0.$$

But,

$$\int_{-\infty}^{t} \exp[-2\alpha(t-u)] \, du = \int_{0}^{\infty} \exp[-2\alpha v] \, dv = \frac{1}{2\alpha}.$$

Hence,

$$R(\tau) = (\sigma_W^2/2\alpha) \exp(-\alpha\tau), \qquad \tau \geq 0, \qquad (3.7.28)$$

so that, again,

$$\rho(\tau) = R(\tau)/R(0) = \exp(-\alpha\tau), \qquad \tau \geq 0.$$

(Note that $R(0) = \sigma_W^2/2\alpha$ confirms the result (3.7.17) for σ_X^2.)

Continuity and differentiability

It is clear from (3.7.26) and Fig. 3.20 that $\rho(\tau)$ *is continuous at $\tau = 0$, but $\dot{\rho}(0)$ does not exist*; (note that the "gradient" at the origin in Fig. 3.20 differs according as to whether one approaches the origin from the right or from the left). It then follows from the discussion in Section 3.6 that $\{X(t)\}$ *is stochastically continuous but is not stochastically differentiable, i.e. $\dot{X}(t)$ does not exist*. The fact that $\dot{X}(t)$ does not exist should really be obvious once we recall that the process $\{\varepsilon(t)\}$ is pathological; if the RHS of (3.7.11) does not exist neither does the LHS, and this clearly indicates the non-existence of $\dot{X}(t)$. However, despite the lack of mathematical rigour (3.7.11) still provides a very useful way of thinking of the continuous parameter AR(1) model, but it is preferable now to rewrite it using the "differential" notation, viz.,

$$dX(t) + \alpha X(t) \, dt = dW(t). \qquad (3.7.29)$$

(See, e.g., Bartlett (1955), p. 148.)

3.7.3 Examples of First Order Autoregressive Processes

To illustrate the way in which continuous parameter AR(1) processes arise we discuss briefly two physical examples of these processes.

Example 1: One dimensional Brownian motion

Brownian motion is the well known phenomenon in physics which describes the random movement of microscopic particles suspended in a liquid or gas. It was first observed experimentally by Robert Brown in 1827, and is due to the impact on the particles of randomly moving molecules of the liquid or gas. (There is a substantial literature on the subject in the context of statistical mechanics—starting with the pioneering work of Einstein (1905)—and an interesting account is given in Parzen (1962).) Here, we consider only a very simple model of one dimensional Brownian motion, as follows.

Suppose we have a particle of unit mass moving in a straight line in a resistive medium and subject to random forces. At time t, let $X(t)$ denote the velocity (or equivalently, the momentum) of the particle, $\varepsilon(t)$ the random force acting on it, and assume that the resistive force is proportional to the velocity. Then the equation of motion of the particle is of the form,

$$\dot{X}(t) = \varepsilon(t) - \alpha X(t),$$

or

$$\dot{X}(t) + \alpha X(t) = \varepsilon(t), \qquad (3.7.30)$$

where α is a positive constant.

If we now assume that the random forces, $\varepsilon(t)$, constitute a *purely random process, then the velocity* $\{X(t)\}$ *is an AR*(1) *process.*

The impulse in a small time interval Δt is $\varepsilon(t)\,\Delta t = dW(t)$, and since

$$\text{var}\{dW(t)\} \sim \sigma_W^2 \, \Delta t = \text{O}(\Delta t),$$

we see that although the impulses are small they are of order $\text{O}(\sqrt{\Delta t})$. We thus get discontinuous changes of velocity in small intervals Δt, and this gives us physical insight as to why the "acceleration", $\dot{X}(t)$, does not remain finite. The process, $X(t)$, defined by (3.7.30) is known as the "*Orstein–Uhlenbeck*" process.

Note that if $\alpha = 0$, i.e. if there is no resistance to the motion, then (3.7.30) gives

$$X(t) = \int_0^t \varepsilon(u)\, du = W(t).$$

The Wiener process thus provides a simple model for the velocity of one dimensional Brownian motion when there is no resistance, but $X(t)$ is now, of course, a non-stationary process.

There is a different (and somewhat cruder) model for Brownian motion which leads to the Wiener process as a model for the *displacement* of the particle—see, e.g., Cox and Miller (1965), p. 206. This alternative model does not provide a very accurate description of the observed phenomenon when t is small compared with the mean interval between impacts.

Example (2): R–C circuit with random input voltage

Consider the R–C circuit shown in Fig. 3.21, (i.e. a circuit containing one resistor and one capacitor). Suppose that the input voltage across AB is a

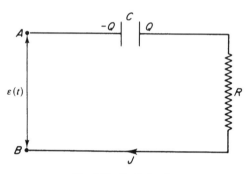

Fig. 3.21. R–C circuit.

random process whose value at time t is denoted by $\varepsilon(t)$, and let $Q(t)$ denote the charge on the capacitor at time t. Then the equation of the circuit is,

$$R\dot{Q}(t) + \frac{1}{C}Q(t) = \varepsilon(t). \qquad (3.7.31)$$

If we now assume that the random input voltage, $\varepsilon(t)$, is a *purely random process*, then it follows from (3.7.31) that *the charge, $Q(t)$, is an* AR(1) *process*. The current $J(t)$ in the circuit at time t is given by $J(t) = \dot{Q}(t)$. However, if $Q(t)$ is an AR(1) process then, as previously noted, it is not stochastically differentiable, and consequently it may be said that in this case the current does not exist! This conclusion is, of course, valid only in the "ideal" situation where $\varepsilon(t)$ is a perfect purely random process, and since such processes do not themselves exist in continuous time the above result has no real practical significance—see Doob (1949). In the more realistic case where $\varepsilon(t)$ is what is termed "band limited white noise" (see Chapter 4, Section 4.10) the current, $J(t)$, exists in a perfectly well defined sense.

3.7.4 Second Order Autoregressive Processes

To construct the continuous parameter AR(2) model we follow the same approach as for the AR(1) model, i.e. we start with the discrete parameter model and take its limiting form as Δt, the interval between successive observations, tends to zero.

For the discrete case the AR(2) model (3.5.22) may be written as,

$$X_n + a_1 X_{n-1} + a_2 X_{n-2} = Y_n, \qquad n = 0, \pm 1, \pm 2, \ldots,$$

(where again Y_n is a purely random process with variance σ_Y^2), and this may be re-written as,

$$a_2 \Delta^2 X_n - (a_1 + 2a_2)\Delta X_n + (1 + a_1 + a_2)X_n = Y_n,$$

where

$$\Delta X_n = X_n - X_{n-1}, \qquad \Delta^2 X_n = X_n - 2X_{n-1} + X_{n-2}.$$

Now introduce the time interval Δt, i.e. write $t = n \Delta t$, and divide each term in the above equation by $a_2(\Delta t)^2$, giving,

$$\frac{\Delta^2 X_n}{(\Delta t)^2} - \frac{(a_1 + 2a_2)}{a_2 \Delta t} \cdot \frac{\Delta X_n}{\Delta t} + \frac{(1 + a_1 + a_2)}{a_2(\Delta t)^2} X_n = \frac{Y_n}{a_2(\Delta t)^2}. \qquad (3.7.32)$$

Write

$$\varepsilon(t) = Y_n / [a_2(\Delta t)^2], \qquad (3.7.33)$$

$$\alpha_1 = -(a_1 + 2a_2)/(a_2 \Delta t), \qquad (3.7.34)$$

$$\alpha_2 = (1 + a_1 + a_2)/[a_2(\Delta t)^2], \qquad (3.7.35)$$

so that,

$$a_1 = -2a_2 - \alpha_1 a_2 \Delta t, \qquad (3.7.36)$$

$$1 + a_1 + a_2 = \alpha_2 a_2 (\Delta t)^2, \qquad (3.7.37)$$

$$1 - a_2 = \alpha_1 a_2 \Delta t + \alpha_2 a_2 (\Delta t)^2. \qquad (3.7.38)$$

As $\Delta t \to 0$, we obtain from (3.7.32) the *formal* differential equation,

$$\ddot{X}(t) + \alpha_1 \dot{X}(t) + \alpha_2 X(t) = \varepsilon(t), \qquad (3.7.39)$$

and *this represents the continuous parameter second order autoregressive model.*

Note, however, that for α_1 and α_2 to remain finite as $\Delta t \to 0$ we require that $a_1 \to -2$, $a_2 \to 1$. This means that the "point" (a_1, a_2) has to move to the top left-hand corner of the triangular region shown in Fig. 3.12, so that, as in the AR(1) case, the parameters of the discrete model have to converge to the boundary of the stationarity region.

Variance of X(t)

In the discrete case we have (equation (3.5.34)),

$$\sigma_X^2 = \frac{(1+a_2)\sigma_Y^2}{(1-a_2)(1+a_1+a_2)(1-a_1+a_2)}.$$

Using (3.7.36), (3.7.37), (3.7.38), and remembering that $a_1 \to -2$, $a_2 \to 1$, we then have, for small Δt,

$$\sigma_X^2 \sim \frac{2\sigma_Y^2}{\alpha_1 \Delta t \cdot \alpha_2 (\Delta t)^2 \cdot 4} = \frac{\sigma_Y^2}{2\alpha_1\alpha_2(\Delta t)^3}. \qquad (3.7.40)$$

For σ_X^2 to be finite as $\Delta t \to 0$ we require that $\sigma_Y^2 = O(\Delta t)^3$. Then,

$$\text{var}\{\varepsilon(t)\} = \frac{\sigma_Y^2}{a_2^2(\Delta t)^4} = O\left(\frac{1}{\Delta t}\right),$$

and, as before,

$$W(t) = \int_0^t \varepsilon(u)\, du$$

is a *Wiener process* (in the sense of Section 3.7.2). Again, let σ_W^2 denote the rate of increase of var$\{W(t)\}$, (see (3.7.15)). As in Section 3.7.2 we may show that

$$\sigma_W^2 = \lim_{\Delta t \to 0} \frac{\Delta \, \text{var}\{W(t)\}}{\Delta t} = \lim_{\Delta t \to 0} \frac{\text{var}\{\Delta W(t)\}}{\Delta t}.$$

But,

$$\text{var}\{\Delta W(t)\} = \text{var}\{\varepsilon(t)\Delta t\} = \frac{\sigma_Y^2}{a_2^2(\Delta t)^2} \sim \frac{\sigma_Y^2}{(\Delta t)^2} \qquad (\text{since } a_2 \to 1)$$

$$= 2\alpha_1\alpha_2\sigma_X^2\Delta t, \qquad \text{from (3.7.40).}$$

Hence,

$$\sigma_X^2 = \sigma_W^2/2\alpha_1\alpha_2. \qquad (3.7.41)$$

Conditions for stationarity

The complete solution of (3.7.39) is

$$X(t) = f(t) + (D^2 + \alpha_1 D + \alpha_2)^{-1}\varepsilon(t), \qquad (3.7.42)$$

where $f(t)$, the complementary function, is the solution of the homogeneous equation

$$(D^2 + \alpha_1 D + \alpha_2)X(t) = 0, \qquad (3.7.43)$$

and thus takes the form

$$f(t) = A_1 \exp(c_1 t) + A_2 \exp(c_2 t),$$

where A_1, A_2, are arbitrary constants and c_1, c_2 are the roots (assumed distinct) of

$$g(z) = z^2 + \alpha_1 z + \alpha_2. \tag{3.7.44}$$

For stationarity we require $f(t)$ to decay to zero as $t \to \infty$, which implies that,

$$R(c_1) < 0, \qquad R(c_2) < 0. \tag{3.7.45}$$

(The notation $R(z)$ denotes the real part of the complex variable z.) In other words, *both roots of the quadratic $g(z)$ must lie in the left-half plane.*

The conditions (3.7.45) reduce, in the case of the AR(2) model, to the equivalent conditions

$$\alpha_1 > 0 \quad \text{and} \quad \alpha_2 > 0. \tag{3.7.46}$$

In this case, the stationary solution of (3.7.39) may be written in the form,

$$X(t) = \frac{1}{(D - c_1)(D - c_2)} \cdot \varepsilon(t)$$

$$= \frac{1}{(c_1 - c_2)} \left\{ \frac{1}{D - c_1} - \frac{1}{D - c_2} \right\} \cdot \varepsilon(t)$$

$$= \int_{-\infty}^{t} \left\{ \frac{\exp[c_1(t-u)] - \exp[c_2(t-u)]}{c_1 - c_2} \right\} \varepsilon(u) \, du, \tag{3.7.47}$$

(using (3.7.22)) or, equivalently,

$$X(t) = \int_{-\infty}^{t} \left\{ \frac{\exp[c_1(t-u)] - \exp[c_2(t-u)]}{c_1 - c_2} \right\} dW(u). \tag{3.7.48}$$

(The expression (3.7.41) for σ_X^2 may be obtained directly from (3.7.47) in conjunction with (3.7.18).)

The autocorrelation function

In the discrete case the autocorrelation function satisfies the difference equation (3.5.33), namely,

$$\rho(r) + a_1 \rho(r-1) + a_2 \rho(r-2) = 0, \qquad r \geq 1.$$

In the continuous case we have analogously,

$$\ddot{\rho}(\tau) + \alpha_1 \dot{\rho}(\tau) + \alpha_2 \rho(\tau) = 0, \qquad \tau \geq 0. \tag{3.7.49}$$

The solution of (3.7.49) is of the form

$$\rho(\tau) = B_1 \exp c_1\tau + B_2 \exp c_2\tau, \qquad \tau \geq 0, \qquad (3.7.50)$$

where the constants B_1, B_2, are determined by the boundary conditions as specified by $\rho(0)$ and $\dot{\rho}(0)$. We know, of course, that $\rho(0) = 1$. To find $\dot{\rho}(0)$ we have that, in the discrete case,

$$\rho(0) = 1, \qquad \rho(1) = -a_1/(1 + a_2).$$

Hence

$$\rho(0) - \rho(1) = \frac{1 + a_1 + a_2}{1 + a_2} \sim \tfrac{1}{2}\alpha_2(\Delta t)^2,$$

using (3.7.36)–(3.7.38). Therefore, $[\rho(0) - \rho(1)]/\Delta t \to 0$ as $\Delta t \to 0$, and hence $\dot{\rho}(0) = 0$. Substituting these boundary conditions in (3.7.50) we obtain finally (recalling that $\rho(\tau)$ is an even function),

$$\rho(\tau) = \frac{c_2 \exp(c_1|\tau|) - c_1 \exp(c_2|\tau|)}{(c_2 - c_1)}. \qquad (3.7.51)$$

As in the discrete case, the detailed form of $\rho(\tau)$ depends on whether the roots c_1, c_2, are real or complex. We discuss these cases separately.

REAL ROOTS: $\alpha_2 \leq \alpha_1^2/4$

Here, $\rho(\tau)$ decays to zero in a "smooth" exponential form, as shown in Fig. 3.13.

COMPLEX ROOTS: $\alpha_2 > \alpha_1^2/4$

Set $p = (\alpha_2 - \alpha_1^2/4)^{1/2}$. Then $\rho(\tau)$ may be written as

$$\rho(\tau) = \exp(-\tfrac{1}{2}\alpha_1|\tau|)\left\{\cos p|\tau| + \frac{\alpha_1}{2p}\sin p|\tau|\right\},$$

or

$$\rho(\tau) = \exp(-\tfrac{1}{2}\alpha_1|\tau|)\left\{\frac{\cos(p|\tau| + \phi)}{\cos \phi}\right\}, \qquad (3.7.52)$$

where

$$\sin \phi = -\alpha_1/2\sqrt{\alpha_2}. \qquad (3.7.53)$$

In this case $\rho(\tau)$ decays *sinusoidally*, i.e. is *quasi-periodic with "period"*, $(2\pi/p)$, and its form is as shown in Fig. 3.15. (Alternatively, the general result (3.7.51) may be derived directly from (3.7.47), using (3.7.18); see corresponding discussion for the AR(1) model.)

Continuity and differentiability

The autocorrelation function, $\rho(\tau)$, is continuous at the origin and hence $X(t)$ is *stochastically continuous*. Moreover $\dot{X}(t)$ exists, but it may be shown that $\dot{X}(t)$ is nowhere continuous, and hence $\ddot{X}(t)$ *does not exist*. (This, again, is what we would expect in view of the fact that the process $\varepsilon(t)$ does not strictly exist.)

These results may be verified by expanding the expression for $\rho(\tau)$ given by (3.7.51) as a power series in τ, giving,

$$\rho(\tau) = 1 - \tfrac{1}{2}c_1 c_2 \tau^2 + \frac{|\tau|^3}{6} c_1 c_2 (c_1 + c_2) + \dots$$

It then follows that $\dot{\rho}(\tau)$ and $\ddot{\rho}(\tau)$ both exist (with $\dot{\rho}(0) = 0$), but the higher derivatives of $\rho(\tau)$ do not exist. In particular, the non-existence of $\rho^{(4)}(\tau)$ implies the non-existence of $\ddot{X}(t)$.

Example of an AR(2) process: L–R–C circuits

Consider the circuit shown in Fig. 3.22. Here we have a capacitor C, a resistor R, an inductor L, and a random input voltage generated, for example, by noise in a thermionic valve (see the discussion of "shot noise" in Sections 3.7.10 and 4.10).

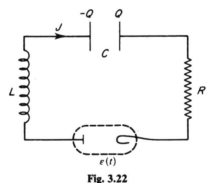

Fig. 3.22

Let the voltage at time t across the anode and cathode of the valve be denoted by $\varepsilon(t)$, and let $Q(t)$ denote the charge on the capacitor at time t. Then the equation of the circuit is

$$L\ddot{Q}(t) + R\dot{Q}(t) + C^{-1}Q(t) = \varepsilon(t).$$

Hence, if $\varepsilon(t)$ is a purely random process then $Q(t)$ is an AR(2) process. Note that here the current in the circuit, $J(t) = \dot{Q}(t)$, exists even when $\varepsilon(t)$ has the above idealized form.

3.7.5 Autoregressive Processes of General Order

Having discussed the first and second order autoregressive models in some detail we now summarize briefly the properties of the general autoregressive model. The results given below follow from fairly obvious extensions of the corresponding results for the first and second order cases.

The continuous parameter AR(k) model is described by the kth order linear differential equation

$$X^{(k)}(t) + \alpha_1 X^{(k-1)}(t) + \ldots + \alpha_k X(t) = \varepsilon(t), \qquad (3.7.54)$$

where $X^{(k)}(t) = (d/dt)^k\{X(t)\}$, etc., $\alpha_1, \ldots, \alpha_k$, are constants, and $\varepsilon(t)$ is the "derivative" of a Wiener process. Equivalently, (3.7.54) may be written formally as

$$\alpha(D)X(t) = \varepsilon_t, \qquad (3.7.55)$$

where

$$\alpha(z) = z^k + \alpha_1 z^{k-1} + \ldots + \alpha_k. \qquad (3.7.56)$$

The condition for stationarity is that *all the roots of $\alpha(z)$ have negative real parts, i.e. lie in the left-half plane.*

The autocorrelation function, $\rho(\tau)$, satisfies the differential equation

$$\alpha(D)\rho(\tau) = 0, \qquad \tau \geq 0 \qquad (3.7.57)$$

(where now D acts as a differential operator w.r.t. τ), and the solution is of the form,

$$\rho(\tau) = \sum_{i=1}^{k} B_i \exp(c_i|\tau|), \qquad (3.7.58)$$

where $B_1, \ldots, B_k$ are constants, and $c_1, \ldots, c_k$ are the roots (assumed distinct) of $\psi(z)$. As in the discrete case, if all the roots are real then $\rho(\tau)$ decays "smoothly"; a pair of complex roots gives rise to an oscillatory term in $\rho(\tau)$.

3.7.6 Moving Average Processes

The discrete parameter MA(l) model was defined in Section 3.5.5 by equation (3.5.45). Here, X_t is expressed as a weighted linear combination of present and past values of a purely random process, ε_t, the weighting extending up to l time units into the past. The natural continuous parameter analogue of this model is obtained by replacing ε_t by $\varepsilon(t)$, and the discrete summation (on the RHS of (3.5.45)) by an integral. *We thus define the continuous parameter* MA(l) *model by,*

$$X(t) = \int_0^l b(u)\varepsilon(t-u)\,du, \qquad (3.7.59)$$

where $b(u)$ is some given function and $\varepsilon(t)$ is the "derivative" of a Wiener process $W(t)$.

Note that (3.7.59) may be rewritten in the alternative forms

$$X(t) = \int_{t-l}^t b(t-u)\varepsilon(u)\,du, \qquad (3.7.60)$$

or,

$$X(t) = \int_{t-l}^t b(t-u)\,dW(u). \qquad (3.7.61)$$

If we let $l \to \infty$, (3.7.59) then becomes the "general linear process" model (to be discussed in Section 3.7.8), and the autocorrelation function of the MA(l) model may be obtained as a special case of the expression (3.7.83) for the autocorrelation function of the general linear process.

It is sometimes convenient to write the MA(l) model in terms of the differential operator D, in the same way as we expressed the AR(k) model (3.7.54) in its "operational" form (3.7.55). The "operational" form of (3.7.59), although not quite as obvious as in the autoregressive case, is readily obtained by using the formal operator relationship,

$$B^u \equiv \exp(-uD). \qquad (3.7.62)$$

This well known result (which is merely a disguised form of Taylor's theorem) follows from the fact that for any function $x(t)$ which possesses a Taylor series expansion we may write,

$$B^u \cdot x(t) = x(t-u) = x(t) - \frac{u}{1!}x'(t) + \frac{u^2}{2!}x''(t) - \ldots$$

$$= \left(1 - \frac{uD}{1!} + \frac{u^2 D^2}{2!} - \ldots\right)x(t)$$

$$= \exp(-uD) \cdot x(t).$$

The MA(l) model (3.7.59) may now be written formally as

$$X(t) = \int_0^l b(u)\{\exp(-uD) \cdot \varepsilon(t)\}\,du$$

$$= \left\{\int_0^l b(u)\exp(-uD)\,du\right\} \cdot \varepsilon(t),$$

or

$$X(t) = B(D)\varepsilon(t), \tag{3.7.63}$$

where

$$B(z) = \int_0^l b(u) \exp(-uz) \, du. \tag{3.7.64}$$

3.7.7 Mixed Autoregressive/Moving Average Processes

If we follow the same approach used in the discrete parameter case and construct a mixed ARMA model simply by combining the pure AR model (3.7.54) with the pure MA model (3.7.59) we are led to the model,

$$X^{(k)}(t) + \alpha_1 X^{(k-1)}(t) + \ldots + \alpha_k X(t) = \int_0^l b(u)\varepsilon(t-u) \, du, \tag{3.7.64}$$

or, in operational form, (using the same notation as in the previous sections),

$$\alpha(D)X(t) = B(D)\varepsilon(t). \tag{3.7.65}$$

The stationary solution (when it exists) is then given by,

$$X(t) = H(D)\varepsilon(t), \tag{3.7.66}$$

where

$$H(z) \equiv B(z)/\alpha(z). \tag{3.7.67}$$

Let us now compare (3.7.65) with the corresponding discrete parameter model (3.5.52). The stationary solution of (3.5.52) (as given by (3.5.54)) may be written as

$$X_t = K(B)Y_t, \tag{3.7.68}$$

where

$$K(z) \equiv \beta(z)/\alpha(z). \tag{3.7.69}$$

We now see an important distinction between (3.7.65) and the discrete parameter version (3.5.52). In the latter case $\beta(z)$ and $\alpha(z)$ are both finite order polynomials, so that $K(z)$ is a rational function of z. On the other hand, $B(z)$ is not a polynomial, and hence $H(z)$ is not a rational function. At this stage this distinction may not appear to be particularly important, but, as we have already hinted at the end of Section 3.5.6 and in Section 3.5.7, the functions $H(z), K(z)$, play a basic role in the spectral analysis of these models. In this context it turns out that the *rationality* of the function $K(z)$ is a very useful and important property of the ARMA scheme, and it is desirable, therefore, to try to preserve this feature in the continuous parameter case. To achieve this we have to replace the function $B(z)$ by a finite order polynomial, $B_0(z)$, where,

$$B_0(z) = \beta_0 z^l + \beta_1 z^{l-1} + \ldots + \beta_l, \tag{3.7.70}$$

say. Substituting $B_0(D)$ for $B(D)$ in (3.7.65) then leads to the new model,

$$X^{(k)}(t) + \alpha_1 X^{(k-1)}(t) + \ldots + \alpha_k X(t) = \beta_0 \varepsilon^{(l)}(t) + \beta_1 \varepsilon^{(l-1)}(t) + \ldots + \beta_l \varepsilon(t). \tag{3.7.71}$$

The stationary solution of (3.7.71) may now be written as

$$X(t) = H_0(D)\varepsilon(t), \tag{3.7.72}$$

where

$$H_0(z) = B_0(z)/\alpha(z) \tag{3.7.73}$$

now has the desired form of being a rational function of z.

From now on we will refer to (3.7.71) as the *"continuous parameter ARMA scheme"*. However, it should be noted that the expression on the RHS of (3.7.71) is of a *highly formal* nature; the process $\varepsilon(t)$ itself does not exist in a strict mathematical sense, let alone its higher derivatives!

The condition for the stationarity of (3.7.71) is that *all the roots of $\alpha(z)$ lie in the left half-plane*, cf. the discussion of the corresponding discrete parameter model in Section 3.5.6.

3.7.8 General Linear Processes

In continuous time the *general linear process* model is given by

$$X(t) = \int_0^\infty g(u)\varepsilon(t-u)\,du, \tag{3.7.74}$$

where $g(u)$ is some given function. Alternative forms of (3.7.74) are,

$$X(t) = \int_{-\infty}^t g(t-u)\varepsilon(u)\,du, \tag{3.7.75}$$

or

$$X(t) = \int_{-\infty}^t g(t-u)\,dW(u). \tag{3.7.76}$$

In "operational form" (3.7.74) may be written as

$$X(t) = G(D)\varepsilon(t), \tag{3.7.77}$$

where

$$G(z) = \int_0^\infty \exp(-uz)g(u)\, du, \qquad (3.7.78)$$

(cf. equation (3.7.63)).

As in discussing the discrete parameter version it is convenient to define $g(u) = 0$, $u < 0$, so that (3.7.74) can then be rewritten in the "two-sided" form,

$$X(t) = \int_{-\infty}^\infty g(u)\varepsilon(t-u)\, du = \int_{-\infty}^\infty g(t-u)\, dW(u).$$

The variance of $X(t)$, σ_X^2, is then given by

$$\sigma_X^2 = \int_{-\infty}^\infty \int_{-\infty}^\infty g(t-u)g(t-v)E[dW(u)\, dW(v)]. \qquad (3.7.79)$$

Using the result that

$$E[dW(u)\, dW(v)] = 0 \quad (u \neq v), \qquad E[\{dW(u)\}^2] = \sigma_W^2\, du, \qquad (3.7.80)$$

(3.7.79) reduces to

$$\sigma_X^2 = \sigma_W^2 \int_{-\infty}^\infty g^2(t-u)\, du = \sigma_W^2 \int_{-\infty}^\infty g^2(u)\, du,$$

or, since $g(u) = 0$, $u < 0$,

$$\sigma_X^2 = \sigma_W^2 \int_0^\infty g^2(u)\, du. \qquad (3.7.81)$$

Thus, the condition for σ_X^2 to be finite is that

$$\int_0^\infty g^2(u)\, du < \infty. \qquad (3.7.82)$$

This is also the *condition for the stationarity* of the general linear process model.

The autocovariance function of $X(t)$, $R(\tau)$, is given by

$$R(\tau) = E[X(t)X(t+\tau)] = \int_{-\infty}^\infty \int_{-\infty}^\infty g(t-u)g(t+\tau-v)E[dW(u)\, dW(v)]$$

$$= \sigma_W^2 \int_{-\infty}^\infty g(t-u)g(t+\tau-u)\, du, \qquad \text{(using (3.7.80))}$$

$$= \sigma_W^2 \int_{-\infty}^\infty g(u)g(u+\tau)\, du,$$

(on making the change of variable, $(t - u) \to u$). Hence, the autocorrelation function, $\rho(\tau)$, is

$$\rho(\tau) = \frac{R(\tau)}{R(0)} = \frac{\int_{-\infty}^{\infty} g(u)g(u + \tau) \, du}{\int_{0}^{\infty} g^2(u) \, du}. \qquad (3.7.83)$$

A similar discussion to that used in the discrete parameter case shows that, with suitable choices of the function $g(u)$, the general linear model includes all the previous special models. Note, in particular, that if we set

$$g(u) = \begin{cases} b(u), & 0 \leq u \leq l, \\ 0, & \text{otherwise}, \end{cases}$$

(3.7.74) reduces to the MA(l) model, whilst choosing $g(u) = \exp(-\alpha u)$ gives the AR(1) model (cf. (3.7.20)), and choosing

$$g(u) = \frac{1}{(c_1 - c_2)} \{\exp(-c_1 u) - \exp(-c_2 u)\}$$

gives the AR(2) model (cf. (3.7.48)).

3.7.9 Harmonic Processes

The discrete parameter harmonic process model (3.5.71) may be applied also to continuous parameter processes, but now the time parameter t is, of course, continuous and thus free to take any value from $-\infty$ to $+\infty$. The variance, autocovariance, and autocorrelation functions are exactly as given by (3.5.74), (3.5.73), and (3.5.75).

3.7.10 Filtered Poisson Processes, Shot Noise, and Campbell's theorem

There is a class of stochastic processes, known as "filtered Poisson processes", which have some features in common with general linear processes and it is therefore convenient, at this point, to study briefly some of their properties. The Poisson process, $N(t)$, was discussed initially in Chapter 2, Section 2.11, and it will be recalled that typically $N(t)$ denotes the total number of "events" which occur in the time interval, $(0, t)$, the time instants at which the individual events occur being purely random. For each t, the random variable $N(t)$ has a Poisson distribution with mean (λt) and variance (λt), where λ is the mean rate of occurrence per unit time. The realizations of the process $N(t)$ are all step-functions, having "jumps" of unit magnitudes at the time instants where events occur, and being constant between occurrences of successive events. The *incremental process*, $dN(t) = N(t + \delta t) - N(t)$, denotes the number of events which occur in the interval

$(t, t + \delta t)$, and, ignoring the possibility of more than one event occurring in an infinitesimal interval, $dN(t)$ takes only two values, namely 1 (with probability $\lambda \delta t$) and 0 (with probability $(1 - \lambda \delta t)$). It then follows that, to order δt,

$$E[dN(t)] = \lambda \delta t, \qquad \text{var}[dN(t)] = \lambda \delta t, \text{ and}$$

$$\text{cov}[dN(t), dN(t')] = 0, \qquad t \neq t'.$$

To illustrate the idea of a filtered Poisson process we now consider the problem of "shot noise". This is the phenomenon whereby a random fluctuating "noise" current, $X(t)$, is produced in a circuit containing a valve (or vacuum tube) due to the random emission of electrons from the cathode (see Figure 3.23) and when the circuit consists of a linear network a stochastic model for $X(t)$ may be constructed as follows.

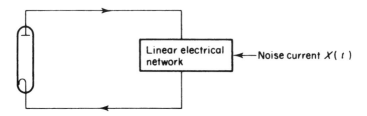

Fig. 3.23. Shot noise in a linear network.

We assume that the current at time t due to an electron emitted at time zero is described by some function $g(t)$, that the behaviour of the system is time-invariant, and that the total current produced by a sequence of electrons emitted at times $\tau_1, \tau_2, \ldots, \tau_i, \ldots$, is the sum of the currents due to the individual electrons, (this is usually referred to as the "superposition principle"). Then, assuming that the time instants at which electrons are emitted are random and conform to a Poisson process with mean rate ν, the total current at time t, $X(t)$, may be written as,

$$X(t) = \sum_{i=1}^{N(t)} g(t - \tau_i), \qquad (3.7.84)$$

where $N(t)$ is the total number of electrons emitted up to time t.

A process of the form (3.7.84), (with the $\{\tau_i\}$ as above) is generally called a "*filtered Poisson process*", and to see the connection between such processes and general linear processes we note that (3.7.84) may be re-written in the form,

$$X(t) = \int_{-\infty}^{t} g(t - \tau) \, dN(\tau). \qquad (3.7.85)$$

Comparing (3.7.85) with (3.7.76) we see that $X(t)$ has essentially the same structure as a general linear process, except that the process $W(u)$ in (3.7.76) is here replaced by $N(\tau)$. The process $N(\tau)$ has uncorrelated increments (like $W(u)$), but as noted above, has a step function form (unlike $W(u)$). The mean, variance, and covariance of $X(t)$ are easily derived as follows. Defining $g(t) = 0$, $t < 0$, we may write (3.7.85) as,

$$X(t) = \int_{-\infty}^{\infty} g(t - \tau)\, dN(\tau),$$ (3.7.86)

so that

$$E[X(t)] = \int_{-\infty}^{\infty} g(t - \tau) E[dN(\tau)] = \nu \int_{-\infty}^{\infty} g(t - \tau)\, d\tau = \nu \int_{-\infty}^{\infty} g(\tau)\, d\tau,$$ (3.7.87)

and

$$\text{cov}[X(s), X(t)] = \int_{-\infty}^{\infty} \int_{-\infty}^{\infty} g(s - \tau) g(t - \tau')\, \text{cov}[dN(\tau), dN(\tau')]$$

$$= \int_{-\infty}^{\infty} g(s - \tau) g(t - \tau)\, \text{var}[dN(\tau)]$$

(using the above properties of $dN(\tau)$),

$$= \nu \int_{-\infty}^{\infty} g(s - \tau) g(t - \tau)\, d\tau$$

$$= \nu \int_{-\infty}^{\infty} g(\tau) g(\tau + t - s)\, d\tau.$$ (3.7.88)

In particular, setting $s = t$,

$$\text{var}[X(t)] = \nu \int_{-\infty}^{\infty} g^2(\tau)\, d\tau.$$ (3.7.89)

The results (3.7.87), (3.7.88), (3.7.89), are together known as *Campbell's theorem*.

Under the condition,

$$\int_{-\infty}^{\infty} g^2(\tau)\, d\tau < \infty,$$

$X(t)$ has finite second order moments, and since $E[X(t)]$ and $\text{var}[X(t)]$ are independent of t, and $\text{cov}[X(s), X(t)]$ is a function of $(t - s)$ only, it follows that $X(t)$ *is a stationary process*.

The specific form of $g(t)$ is determined by the properties of the particular circuit under study, but when $X(t)$ represents the current at the anode of the vacuum tube a plausible form of $g(t)$ is given by,

$$g(t) = \begin{cases} 2et/T^2, & 0 \le t \le T, \\ 0, & \text{otherwise,} \end{cases} \tag{3.7.90}$$

where e is the charge on an individual electron, and T (the "transit time") is the time taken by an electron to travel from the cathode to the anode (see Parzen (1962) p. 150, and Davenport and Root (1958)). Substituting this form of $g(t)$ in (3.7.87)–(3.7.89), we obtain,

$$E[X(t)] = \nu e, \tag{3.7.91}$$

$$\text{var}[X(t)] = 4\nu e^2/3T, \tag{3.7.92}$$

and

$$R(\tau) = \text{cov}[X(t), X(t+\tau)] = \begin{cases} \dfrac{4\nu e^2}{3T}\left(1 - \dfrac{3}{2}\dfrac{|\tau|}{T} + \dfrac{1}{2}\dfrac{|\tau|^3}{T^3}\right), & |\tau| \le T \\ 0, & \text{otherwise.} \end{cases} \tag{3.7.93}$$

Note that $R(\tau)$ vanishes when $|\tau| > T$, and that $\text{var}[X(t)] = O(1/T)$. Hence, if T is sufficiently "small", $R(\tau)$ is highly concentrated around $\tau = 0$, with $R(0)$ extremely large. Thus, for T "small", $X(t)$ may be regarded as a good approximation to a purely random process. This point is discussed further in Chapter 4, Section 4.10. For a more general discussion of filtered Poisson processes see Parzen (1962), Chapter 4.

It will be seen that the autocovariance function of a filtered Poisson process has exactly the same form as that of the general linear process (3.7.76) (compare (3.7.88) with (3.7.83)) if we identify the parameter ν of the Poisson process with the parameter σ_W^2 of the Wiener process. This confirms our previous remark that the second order properties of $X(t)$ depend only on the assumption that the "residual" process has *orthogonal increments*, and thus have the same form whether the residual is a Wiener process or a Poisson process. In our discussion of "Yule's pendulum" (Section 3.5.3) we referred to the process $\{\varepsilon(t)\}$ in (3.5.38) as consisting of a series of "impulses"—in Yule's case it corresponds to the impact of "peas" at random time instants—and $\int_0^t \varepsilon(u)\,du$ is therefore more realistically modelled as a Poisson process rather than as a Wiener process, but this change does not affect the second order properties of the process. The same considerations apply to the Brownian motion model in which random forces to which the particle is subjected arise from collisions with other particles at random time instants. However, if $N(t)$ is a Poisson process with rate ν then as $\nu \to \infty$ the Poisson process becomes asymptotically a Wiener process (see

Doob (1949)) i.e. the Wiener process may be regarded as a "*continuous approximation*" to the Poisson process when the mean interval between events is small. (This follows heuristically by noting that as $\nu \to \infty$ the mean of the Poisson distribution of $N(t)$ also $\to \infty$, and hence the distribution of $N(t)$ tends asymptotically to a normal distribution.) Thus, if the number of collisions per unit time is large (as we would expect since there would be a large number of neighbouring particles), then the mean interval between collisions is small compared with t, and we may regard the random forces as arising from a Wiener process rather than a Poisson process (see also Parzen (1962) pp. 99, 129). Similar arguments apply to the models used for the random input voltages in the examples on R-C and L-R-C circuits, discussed in Sections 3.7.3 and 3.7.4.

Chapter 4

Spectral Analysis

4.1 INTRODUCTION

In this chapter we return to the main theme of this study, namely "spectral analysis". In Chapter 1 we discussed the basic ideas of spectral analysis from a more or less intuitive standpoint. However, armed with the theory of probability and the theory of random processes developed in the preceding two chapters, we are now in a position to discuss the ideas of Chapter 1 in a much more detailed and precise manner.

It will be recalled that the primary objective of spectral analysis is to decompose a time-varying quantity into a sum (or integral) of sine and cosine functions. Clearly then, the natural starting point for a discussion of this topic is the theory of *Fourier series* first mentioned in Section 1.3. In the next section we recall some standard results in Fourier series analysis.

4.2 FOURIER SERIES FOR FUNCTIONS WITH PERIODICITY 2π

Consider the sequence of functions; $\cos t$, $\sin t$, $\cos 2t$, $\sin 2t$, $\cos 3t$, $\sin 3t$, ..., $\cos nt$, $\sin nt$, ... Each of these functions is periodic, with period 2π, since for each integer n and all values of t, $\cos(nt) \equiv \cos n(t+2\pi)$, and $\sin(nt) \equiv \sin n(t+2\pi)$. It then follows that any linear combination of these functions will also be periodic, with period 2π, i.e. if we consider any function of the form

$$f(t) = \sum_{n=0}^{\infty} (a_n \cos nt + b_n \sin nt), \qquad (4.2.1)$$

where $\{a_n\}$, $\{b_n\}$ are arbitrary sequences of constants (subject to the condition that the infinite series on the RHS of (4.2.1) converges for all t), then $f(t)$

must always be a periodic function, with period 2π, whatever values we assign to the $\{a_n\}$ and $\{b_n\}$. We may thus think of the functions $\{\cos nt, \sin nt, n = 0, 1, 2, \ldots\}$ as the basic "building blocks" from which we may construct periodic functions of various forms. This observation is hardly surprising. What is much more surprising is the fact that these *cosine and sine functions are the only "building blocks" from which we can construct periodic functions*, i.e. we cannot form periodic functions by assembling mixtures of any other types of functions. This means, effectively, that *all* periodic functions are nothing more than mixtures of sine and cosine functions, and may thus be expressed in the form (4.2.1). At first sight this may not seem particularly surprising since we normally think of a periodic function as having a continuous "oscillatory" form. However, this is not necessarily the case; when we say that $f(t)$ has periodicity 2π all we mean is that $f(t) = f(t+2\pi)$, (all t), and apart from this constraint $f(t)$ can take quite an arbitrary form. In fact, $f(t)$ can have a completely arbitrary form over (say) the interval $(-\pi, \pi)$, as long as we ensure that this form is repeated exactly over all the adjacent intervals of length 2π. A simple periodic function (the so-called "saw tooth wave") is illustrated in Fig. 4.1, and it will be noted that in this case the function has no "oscillatory" features and has points of discontinuity at $t = \pm\pi, \pm3\pi, \pm5\pi, \ldots$. It may now be appreciated that the fact that (virtually) *all* periodic functions can be represented as mixtures of sine and cosine terms is indeed a most remarkable one.

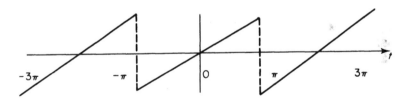

Fig. 4.1. Function with periodicity 2π.

The fundamental ideas of the above method of representing periodic functions are due to the 18th Century French mathematician J. B. J. Fourier who first developed this technique in order to solve the differential equations arising in problems of heat conduction, although D. Bernouilli in 1753 also used trigonometric series expansions to solve the differential equations for vibrating strings. Following their work the technique of "Fourier series analysis", as it has become known, has proved to be of fundamental importance in a wide range of applied mathematical fields, including, in particular, problems of heat conduction, vibrating strings, acoustics, audio-electronics, radio transmissions, and communication engineering. It has also turned out to be a rich field of research for pure mathematicians, and the basic ideas of trigonometric series expansions have been generalized and abstracted to an extent which could hardly have been envisaged by their originators. The subject of "abstract harmonic analysis" is now an important branch of functional analysis, and the results on, e.g., the "spectral resolution of unitary operators" have a direct bearing on the spectral analysis of stationary stochastic processes—see Section 4.11. (We shall return to these general considerations when we discuss the problem of spectral analysis for non-stationary processes in Chapter 11).

The ideas presented above may now be formulated more precisely as follows. Let $X(t)$ be a *deterministic* (*i.e. non-random*) function of t, which is periodic with period 2π. (Note that although we have previously used the notation, $X(t)$, to denote a stochastic process, $X(t)$ is here to be understood as an ordinary mathematical function which does not involve any random elements.) Let us suppose for the moment that $X(t)$ can, in fact, be expressed as an infinite linear combination of sine and cosine functions, so that we may write,

$$X(t) = \tfrac{1}{2}a_0 + \sum_{n=1}^{\infty} (a_n \cos nt + b_n \sin nt). \qquad (4.2.2)$$

The expression on the right-hand side of (4.2.2) is called a *Fourier series*, and the constants $\{a_n\}$, $\{b_n\}$, are called Fourier *coefficients*. The constant term, corresponding to $n = 0$, is usually denoted by $(\tfrac{1}{2}a_0)$ rather than by a_0 for reasons explained below. The next term, corresponding to $n = 1$, is called the *fundamental*, and represents a cosine/sine wave whose period exactly matches that of the given function, $X(t)$. The following term, corresponding to $n = 2$, is called the *first harmonic* and represents a cosine/sine wave whose period is exactly half that of $X(t)$, and so on. The terms corresponding to $n = 3, 4, 5, \ldots$, are called the second, third, fourth, ..., *harmonics*. For each integer m we may multiply both sides of (4.2.2) by $\cos mt$ and integrate from $-\pi$ to π. Assuming that we may integrate the right-hand side term by term we obtain,

$$\int_{-\pi}^{\pi} X(t) \cos mt \, dt = \tfrac{1}{2}a_0 \int_{-\pi}^{\pi} \cos mt \, dt$$

$$+ \sum_{n=1}^{\infty} \left\{ a_n \int_{-\pi}^{\pi} \cos nt \cos mt \, dt + b_n \int_{-\pi}^{\pi} \sin nt \cos mt \, dt \right\},$$

$$m = 0, 1, 2, \ldots. \qquad (4.2.3)$$

We now make use of one of the basic properties of the sine and cosine functions, namely, that they form a *complete orthogonal set* over the interval

$(-\pi, \pi)$, i.e.

$$\left.\begin{array}{ll} \displaystyle\int_{-\pi}^{\pi} \cos nt \cos mt \, dt = 0, & m \neq n, \\[4mm] \displaystyle\int_{-\pi}^{\pi} \sin nt \cos mt \, dt = 0, & \text{all } m, n, \\[4mm] \displaystyle\int_{-\pi}^{\pi} \sin nt \sin mt \, dt = 0, & m \neq n. \end{array}\right\} \qquad (4.2.4)$$

(These results are easily verified by direct integration.) Thus, all the terms on the right-hand side of (4.2.3) vanish except the cosine integral in which $n = m$, and (4.2.3) reduces to

$$\int_{-\pi}^{\pi} X(t) \cos mt \, dt = a_m \int_{-\pi}^{\pi} \cos^2 mt \, dt = \pi a_m,$$

i.e.,

$$a_m = \frac{1}{\pi} \int_{-\pi}^{\pi} X(t) \cos mt \, dt, \qquad m = 0, 1, 2, \ldots. \qquad (4.2.5)$$

(Writing the constant term in (4.2.2) as $(\tfrac{1}{2}a_0)$ rather than as a_0 allows us to extend the validity of (4.2.5) to $m = 0$.) Similarly, multiplying (4.2.2) by $\sin mt$, integrating from $-\pi$ to π, and using the orthogonality relations (4.2.4) we obtain,

$$\int_{-\pi}^{\pi} X(t) \sin mt \, dt = b_m \int_{-\pi}^{\pi} \sin^2 mt \, dt = \pi b_m,$$

i.e.

$$b_m = \frac{1}{\pi} \int_{-\pi}^{\pi} X(t) \sin mt \, dt, \qquad m = 1, 2, \ldots. \qquad (4.2.6)$$

Equations (4.2.5), (4.2.6), are called the *Euler–Fourier formulae*. If we are given a specific form for the function $X(t)$ (such as the saw-tooth wave illustrated in Fig. 4.1), then these equations provide us with a simple method for calculating the values of the coefficients $\{a_n\}$, $\{b_n\}$. (Some well known examples of Fourier series calculations are given in Stuart (1961) and Kufner and Kadlec (1971).) However, the derivation of (4.2.5), (4.2.6) requires further justification. We have not yet shown that any periodic function can be represented in the form (4.2.2), and have not considered the validity of the term by term integration of (4.2.3). The latter point could be covered by assuming, e.g., that the series was uniformly convergent over the interval $(-\pi, \pi)$, but the existence of the Fourier series representation

would be extremely difficult to prove formally. The usual way of circum-
venting these difficulties is to attack the problem from the "other end", i.e.
to *define* the sequences $\{a_m\}$, $\{b_m\}$ by (4.2.5), (4.2.6), and then try to prove
that, with these coefficients, the right-hand side of (4.2.2) does indeed
represent the function $X(t)$. This approach turns out to be much more
amenable to a rigorous treatment but there are still a number of analytical
points to consider, such as, (a) will (4.2.5), (4.2.6) always yield finite values
for a_m and b_m; (b) if so, will the infinite series in (4.2.2) converge for all
values of t; (c) if so, will it always converge to the value $X(t)$? The first point
is easily dealt with by imposing the condition that $X(t)$ *must be absolutely
integrable over the* interval $(-\pi, \pi)$, i.e. that

$$\int_{-\pi}^{\pi} |X(t)| \, dt < \infty. \tag{4.2.7}$$

It then follows immediately that all the a_m and b_m are finite since we have,
e.g.,

$$|a_m| \le \int_{-\pi}^{\pi} |X(t)| \, |\cos mt| \, dt \le \int_{-\pi}^{\pi} |X(t)| \, dt < \infty.$$

Points (b) and (c) require more careful thought. We note, first of all, that
each term in the Fourier series is a continuous function of t and hence we
must expect to encounter difficulties when $X(t)$ contains discontinuities (cf.
Fig. (4.1)). In fact, this does not turn out to be too serious a difficulty, and, as
one would expect, in this situation the series converges to the "average" of
the two values which form the discontinuity, i.e. if $X(t)$ has a discontinuity at
$t = t_0$, then if the Fourier series converges it will converge to the value
$\frac{1}{2}\{X(t_0-0)+X(t_0+0)\}$. (Here, $X(t_0+0)$ obviously denotes the limiting value
of $X(t_0+h)$ as $h \to 0$ through positive values, with a corresponding definition
for $X(t_0-0)$.) The question of whether a Fourier series converges at all is a
more complex one, but loosely speaking, convergence will obtain provided
that the function $X(t)$ is not pathologically "ill behaved". There are a variety
of conditions on $X(t)$ all of which ensure the convergence of its Fourier
series, and it is reasonable to suppose that these will invariably be met in
practice. (See, e.g. Titchmarsh (1939), Chapter 13, Kufner and Kadlec
(1971).) To illustrate the nature of these conditions we may refer to the
following theorem (Titchmarsh (1939)).

Jordan's Test *If $X(t)$ is of bounded variation in the neighbourhood of $t = t_0$,
then its Fourier series converges to the sum $\frac{1}{2}\{X(t_0-0)+X(t_0+0)\}$.*

(The function $X(t)$ is said to be of "bounded variation" in the interval (a, b)
if for any subdivision of this interval, $a = t_0 < t_1 < \ldots < t_n = b$,

$$\sum_{\nu=0}^{n-1} |X(t_{\nu+1})-X(t_\nu)|$$

is bounded by a constant whose value is independent of the form of the subdivision. An equivalent condition is that $X(t)$ can be represented in the form $X(t) = \phi(t) - \psi(t)$, $a < t < b$, where ϕ, ψ are each non-decreasing bounded functions.) Since it may be shown (Titchmarsh (1939)) that a function which has only a finite number of maxima and minima and a finite number of discontinuities in a given interval is of bounded variation throughout the interval, we have the following result.

If $X(t)$ has only a finite number of maxima and minima and a finite number of discontinuities in the interval $(-\pi, \pi)$, its Fourier series converges for all values of t to the value $\frac{1}{2}\{X(t-0) + X(t+0)\}$.

It follows as an immediate corollary that *if $X(t)$ is continuous and of bounded variation throughout the interval $(-\pi, \pi)$ its Fourier series converges for all values of t to $X(t)$.*

4.2.1 The L^2 Theory for Fourier Series

So far we have considered only functions which are absolutely integrable over $(-\pi, \pi)$. This class of functions is usually denoted by $L^1(-\pi, \pi)$ (or simply $L(-\pi, \pi)$), and we have studied the further conditions which are necessary for the Fourier series of such functions to be convergent. However, we may develop a more elegant approach to Fourier series by considering functions which are square integrable over $(-\pi, \pi)$, i.e. functions $X(t)$ which are such that

$$\int_{-\pi}^{\pi} \{X(t)\}^2 \, dt < \infty.$$

The class of such functions is usually denoted by $L^2(-\pi, \pi)$, and we note, in passing, that if a function belongs to $L^2(-\pi, \pi)$ it must necessarily belong to $L^1(-\pi, \pi)$. Thus, all functions in $L^2(-\pi, \pi)$ will possess Fourier series in the sense previously discussed, but we may associate with each member of $L^2(-\pi, \pi)$ a *sequence* of trigonometric series which converge (in a sense) to a limit series which we then call the "Fourier series" of that function. Of course, since the function belongs also to $L^1(-\pi, \pi)$, then this limit series will be the ordinary Fourier series, as previously defined. This approach may be formulated more explicitly as follows.

Let $X(t)$ be any function in the class $L^2(-\pi, \pi)$. Let the Fourier coefficients $\{a_n\}$, $\{b_n\}$, be defined by (4.2.5), (4.2.6). (Note that a_n, b_n, exist for all n, in virtue of the Cauchy inequality.) Write,

$$X_m(t) = \sum_{n=0}^{m} (a_n \cos nt + b_n \sin nt).$$

Then, as $m \to \infty$, $X_m(t)$ converges to $X(t)$ in the *mean square sense*, i.e.,

$$\int_{-\pi}^{\pi} \{X(t) - X_m(t)\}^2 \, dt \to 0 \qquad \text{as } m \to \infty.$$

(See, e.g. Kufner and Kadlec (1971), Ch. 3.)

4.2.2 Geometrical Interpretation of Fourier Series: Hilbert Spaces

At the beginning of this section we referred to the sine and cosine functions as the basic "building blocks" from which we could construct any type of periodic function. The idea of regarding a collection of functions as "building blocks" is a most fruitful one and, as we shall see, lends itself to a very elegant geometrical interpretation of Fourier series.

Let us first consider the exact meaning of the term "building block". It will be clear from the previous discussion that what we mean by "building blocks" is that an arbitrary periodic function can be represented as a *linear combination* of "blocks", i.e. of sines and cosines. This is intriguingly similar to the situation in vector space theory where we try to find a set of "basis" vectors such that any vector belonging to the space may be expressed as a linear combination of the basis vectors. For example, in ordinary three dimensional Euclidean space we may represent any vector, R, in the form, $R = xi + yj + zk$, where, as usual, i, j, k, denote unit vectors pointing in the directions of the three coordinate axes and x, y, z, denote the Cartesian coordinates of the point whose position vector is R. The vectors, i, j, k, thus form a "basis" for this space. As is well known, this notion may be extended to any finite dimensional vector space, with the number of (linearly independent) elements in the set of "basis" vectors corresponding to the dimension of the space. When we try to apply this approach to *functions* the first difficulty which we encounter is that of representing a function as a "vector". Consider, for the moment, a class of functions each of which is defined only at a *finite discrete set* of values of t, say $t = 1, 2, \ldots, n$. Each function in this class is then completely specified by the values which it takes at these n points on the t-axis, i.e. it is completely specified by the values, $X(1), X(2), \ldots X(n)$. Each function may thus be described by a "vector" in an n dimensional space with "coordinates" $\{X(1), X(2), \ldots, X(n)\}$. Now although a periodic function, $X(t)$, (with period 2π) is completely determined by its form over the interval $(-\pi, \pi)$, there are nevertheless an infinite number of values of t in $(-\pi, \pi)$, and to describe the function completely we must specify the value which it takes at each one of this infinite set of points in $(-\pi, \pi)$. (Note that giving a "formula" for the function, such as $X(t) = t$, $(-\pi < t < \pi)$, is merely a "shorthand" description of this procedure; in effect, the formula merely tells us how to compute each value of $X(t)$ in the interval $(-\pi, \pi)$.) However, we may still represent such a function as a "vector" provided we are prepared to consider vectors with an *infinite number of coordinates*, i.e. vectors which belong to *infinite dimensional spaces*. It turns out that infinite dimensional spaces are not all that much more difficult to treat than finite dimensional spaces, and many of the geometrical concepts, such as "orthogonality" and "orthogonal projections" carry over to the infinite dimensional case in a fairly straightforward manner. There are many different types of infinite dimensional spaces (depending on the type of mathematical structure which we impose), but for the purpose of representing functions of a continuous variable the most convenient type is that known as a *"Hilbert space"*. Basically, this is an infinite dimensional space which is such that for any two of its vectors U, V, there is defined an "inner product", usually denoted by (U, V) or $U \cdot V$. (To be technically more precise, the definition of this "inner product" has to satisfy certain fairly obvious axioms, and the space has to be "complete" in the sense that every Cauchy sequence of vectors converges to a vector contained in the space—see, e.g., Halmos (1951).) Once an inner product has been defined we may define the "length" of any vector U, usually denoted by $\|U\|$ and sometimes called the "norm" of U, by

$$\|U\|^2 = (U, U).$$

(This definition of length is clearly suggested by the corresponding property for finite dimensional spaces.) Consider now two vectors,

$$X = (x_1, \ldots x_n), \qquad Y = (y_1, \ldots y_n),$$

each belonging to R^n (the n dimensional Euclidean space). The standard definition of the "inner product" between X, Y is

$$(X, Y) = \sum_{i=1}^{n} x_i y_i, \tag{4.2.8}$$

and correspondingly, the squared lengths of X, Y, are given respectively by,

$$\|X\|^2 = \sum_{i=1}^{n} x_i^2, \qquad \|Y\|^2 = \sum_{i=1}^{n} y_i^2. \tag{4.2.9}$$

As is well known, the cosine of the angle θ between the vectors X, Y, is then given by

$$\cos \theta = (X, Y)/\|X\| \cdot \|Y\|.$$

Hence, if $(X, Y) = 0$, $\theta = \pi/2$, and the vectors X, Y, are called *"orthogonal"*. A set of linearly independent vectors, $U_1, U_2, \ldots U_n$, is said to form a *basis* for R^n if every vector $X \in R^n$ can be expressed uniquely as a linear combination of $U_1, U_2, \ldots U_n$, i.e. can be written uniquely as

$$X = \sum_{i=1}^{n} a_i U_i, \tag{4.2.10}$$

where the $\{a_i\}$ are scalars. The vectors $U_1, U_2, \ldots$ are said to form an *orthogonal basis* if, in addition, all pairs U_i, U_j, are orthogonal, i.e. if $(U_i, U_j) = 0$, all $i \neq j$. In this case the "coordinates" $a_i, \ldots a_n$, are very easily determined by taking the inner product of both sides of (4.2.10) with U_k ($k = 1, 2, \ldots n$). All the terms on the right-hand side vanish except that for which $i = k$, so that we obtain,

$$(X, U_k) = a_k \|U_k\|^2,$$

or

$$a_k = (X, U_k)/\|U_k\|^2, \qquad k = 1, 2, \ldots n. \tag{4.2.11}$$

Moreover,

$$\|X\|^2 = \left(\left(\sum_{i=1}^{n} a_i U_i \right), \left(\sum_{i=1}^{n} a_i U_i \right) \right) = \sum_{i=1}^{n} a_i^2 \|U_i\|^2. \tag{4.2.12}$$

If the U_i all have unit length they are called an *orthonormal* basis, and then (4.2.12) takes the form

$$\|X\|^2 = \sum_{i=1}^{n} a_i^2. \tag{4.2.13}$$

This result is, in fact, simply Pythagoras' theorem in n dimensions; in three dimensions it says that $R^2 = x^2 + y^2 + z^2$. We now use these results to suggest the appropriate definitions for "function spaces", i.e. infinite dimensional Hilbert spaces in which each vector represents a function of a continuous variable. Let H be a Hilbert space consisting of real valued functions which are defined on the interval $a < t < b$ and which are square integrable over this interval, i.e. H contains only those functions, $X(t)$, which belong to $L^2(a, b)$. Let $X(t)$, $Y(t)$ be any two functions in H. The inner product between $X(t)$, $Y(t)$, may be defined by noting that the obvious generalization of the discrete sum in (4.2.8) is an integral of the product of the two functions. Thus, we define the inner product by

$$(X(t), Y(t)) = \int_a^b X(t) Y(t) \, dt, \tag{4.2.14}$$

and the squared lengths of $X(t)$, $Y(t)$, are then given respectively by,

$$\|X(t)\|^2 = \int_a^b X^2(t) \, dt, \qquad \|Y(t)\|^2 = \int_a^b Y^2(t) \, dt. \tag{4.2.15}$$

(The condition that all functions in H must belong to $L^2(a, b)$ ensures that $\|X(t)\|$, $\|Y(t)\|$ are both finite, and the Cauchy inequality then ensures that

$(X(t), Y(t))$ is finite.) In accordance with the convention used for finite dimensional spaces we say that $X(t)$, $Y(t)$, are *orthogonal* if $(X(t), Y(t)) = 0$. We may call the sequence $\{U_1(t), U_2(t), \dots\}$ an *orthogonal basis* for H if

(i) $\qquad\qquad\qquad (U_i(t), U_j(t)) = 0, \qquad$ all $i \neq j$,

and

(ii) $\qquad\qquad$ every function $X(t) \in H$ can be expressed in the form

$$X(t) = \sum_{i=0}^{\infty} a_i U_i(t). \tag{4.2.16}$$

Corresponding to (4.2.11), (4.2.12), we now have,

$$a_i = \frac{(X(t), U_i(t))}{\|U_i(t)\|^2}, \qquad i = 1, 2, \dots \tag{4.2.17}$$

and

$$\|X(t)\|^2 = \sum_{i=0}^{\infty} a_i^2 \|U_i(t)\|^2. \tag{4.2.18}$$

Equation (4.2.18) is an important result usually referred to as "*Parseval's relation*". If the $U_i(t)$ form an orthonormal basis, then (4.2.18) becomes,

$$\|X(t)\|^2 = \sum_{i=0}^{\infty} a_i^2, \tag{4.2.19}$$

which is a version of Pythagoras' theorem for infinite dimensional spaces.

Now let us consider the special class of functions which are periodic, period 2π, and belong to $L^2(-\pi, \pi)$. These form a Hilbert space, H_0, of the above type, with the interval (a, b) replaced by $(-\pi, \pi)$. In particular, H_0 contains the sequence of functions $\{\cos it, \sin it, i = 0, 1, 2, \dots\}$, and the relations (4.2.4) tell us immediately that "vectors" corresponding to these functions are *orthogonal in the geometrical sense*, as defined above. (This is, of course, the reason why (4.2.4) is usually called the "orthogonality relations".) However, the basic point is that when we expand a function as a Fourier series we are, in effect, using the cosine and sine functions as a basis; thus for the class of functions which possess Fourier series the sequence $\{\cos it, \sin it, i = 0, 1, \dots\}$ *constitutes an orthogonal basis*. Equations (4.2.5), (4.2.6), for the Fourier coefficients now follow immediately from (4.2.17), noting that,

$$\left. \begin{aligned} \|\cos it\|^2 &= \int_{-\pi}^{\pi} \cos^2 it \, dt = \pi, \\[2mm] \|\sin it\|^2 &= \int_{-\pi}^{\pi} \sin^2 it \, dt = \pi. \end{aligned} \right\} \qquad i = 1, 2, \dots \tag{4.2.20}$$

4.3 FOURIER SERIES FOR FUNCTIONS OF
GENERAL PERIODICITY

The theory of Fourier series may be readily extended to functions of general periodicity (not necessarily 2π) simply by transforming the time scale so as to reduce such functions to the case previously considered.

More specifically, suppose that $X(t)$ is a periodic function with periodicity $2T$. Define a new function $Y(t)$ by $Y(t) \equiv X(tT/\pi)$. Then $Y(t)$ is periodic with period 2π and, subject to the conditions previously stated, may be expanded in a Fourier series of the form (4.2.2), i.e. we may write for any s,

$$Y(s) = \tfrac{1}{2}a_0 + \sum_{n=1}^{\infty} (a_n \cos ns + b_n \sin ns).$$

Now write $s = \pi t/T$, so that,

$$X(t) \equiv Y(\pi t/T) = \tfrac{1}{2}a_0 + \sum_{n=1}^{\infty} [a_n \cos (\pi t/T) + b_n \sin (\pi t/T)]. \qquad (4.3.1)$$

The Fourier coefficients, $\{a_n\}$, $\{b_n\}$ are now given by

$$a_n = \frac{1}{\pi} \int_{-\pi}^{\pi} Y(s) \cos ns \, ds \equiv \frac{1}{T} \int_{-T}^{T} X(t) \cos \frac{\pi nt}{T} \, dt,$$

$$b_n = \frac{1}{\pi} \int_{-\pi}^{\pi} Y(s) \sin ns \, ds = \frac{1}{T} \int_{-T}^{T} X(t) \sin \frac{\pi nt}{T} \, dt. \qquad (4.3.2)$$

We note, in passing, that the functions

$$\{\cos (\pi nt/T), \sin(\pi nt/T), \quad n = 0, 1, 2, \ldots\}$$

form a *complete orthogonal set* over the interval $(-T, T)$, i.e. the orthogonality relations (4.2.4) remain valid provided we change the limits of integration from $(-\pi, \pi)$ to $(-T, T)$.

4.4 SPECTRAL ANALYSIS OF PERIODIC FUNCTIONS

So far we have discussed Fourier series from a purely mathematical point of view. However, as indicated previously, this representation of periodic functions may be given an extremely important physical interpretation in terms of the "energy/frequency" properties of the function. Let $X(t)$ be a deterministic function, periodic with period 2π, and possessing a Fourier series expansion of the form (4.2.2). (It should be noted, once again, that we are still discussing only *deterministic* functions which do not involve any random elements.) Suppose now that $X(t)$ represents some physical time varying quantity—for example, suppose that $X(t)$ represents an electrical current. If we pass this fluctuating current through a resistance of 1 unit we will dissipate a certain amount of energy whose magnitude depends, of course, on the length of the time interval during which the current flows. In fact, it follows from the basic laws of electrical circuit theory that the total energy dissipated in the time interval $(-\pi, \pi)$ when $X(t)$ flows through a unit resistance is given by

$$\text{Total energy over interval } (-\pi, \pi) = \int_{-\pi}^{\pi} X^2(t) \, dt. \qquad (4.4.1)$$

(We are assuming that $X(t)$ is defined for both positive and negative values of t; this is purely for mathematical convenience and means simply that we regard the time instant at which the current is switched on as corresponding to $t = -\pi$, say.)

Now squaring both sides of (4.2.2), integrating from $-\pi$ to π, and using the orthogonality relations (4.2.4), we obtain

$$\int_{-\pi}^{\pi} X^2(t) \, dt = (\tfrac{1}{4}a_0^2 \cdot 2\pi) + \sum_{n=1}^{\infty} \left\{ a_n^2 \int_{-\pi}^{\pi} \cos^2 nt \, dt + b_n^2 \int_{-\pi}^{\pi} \sin^2 nt \, dt \right\}$$

$$= \pi \left[\tfrac{1}{2}a_0^2 + \sum_{n=1}^{\infty} (a_n^2 + b_n^2) \right], \qquad (4.4.2)$$

on using (4.2.20). Equation (4.4.2) is known as *Parseval's relation* for Fourier series, and follows alternatively as a special case of the more general form of Parseval's relation given by (4.2.18).

Writing

$$c_0 = a_0/2, \qquad c_n = [\tfrac{1}{2}(a_n^2 + b_n^2)]^{1/2}, \qquad n = 1, 2, \ldots,$$

we then have,

$$\text{Total energy over interval } (-\pi, \pi) = 2\pi \left\{ \sum_{n=0}^{\infty} c_n^2 \right\}. \qquad (4.4.3)$$

This result tells us the amount of energy dissipated over a "standard" time period of duration 2π. There would be little point in trying to compute the energy dissipated over the whole of the time axis (when t runs from $-\infty$ to $+\infty$) as this is clearly *infinite*, but, since the function is periodic, a description of its energy properties over the interval $(-\pi, \pi)$ is sufficient to characterize its behaviour for all time. Alternatively, instead of considering the energy dissipated over the interval $(-\pi, \pi)$ we may consider the energy dissipated *per unit time interval*. The energy per unit time is a well defined physical concept known as the "*power*". We then have from (4.4.3),

$$\text{Total power} = \frac{\text{Total energy over } (-\pi, \pi)}{2\pi} = \sum_{n=0}^{\infty} c_n^2. \qquad (4.4.4)$$

This result thus gives us a "breakdown" of the total power as an infinite sum of terms, each term c_n^2 being associated with a particular term in the Fourier series of $X(t)$. Now consider the special case when $X(t)$ consists of just *one* cosine plus sine term, i.e. is of the form,

$$X(t) = a_n \cos nt + b_n \sin nt. \qquad (4.4.5)$$

In this case all the Fourier coefficients, expect a_n and b_n, vanish, and all the c's, except c_n, vanish. Equation (4.4.4) then gives

$$\text{Total power} = c_n^2.$$

Hence, in the general case, c_n^2 *represents the contribution to the total power from the term* $(a_n \cos nt + b_n \sin nt)$.

Up till now we have distinguished the various terms in the Fourier series according to the value of n to which they correspond. However, from the physical point of view it is more informative to "label" them in terms of their *"frequencies"*. These are defined as follows; the terms $\cos nt$, $\sin nt$, each have period $(2\pi/n)$ and therefore in one time unit each term will execute $(n/2\pi)$ complete "cycles". Hence, when t is measured in seconds, we say that the term $(a_n \cos nt + b_n \sin nt)$ has a *frequency of* $(n/2\pi)$ *cycles per second*, or, as it is usually abbreviated, $(n/2\pi)$ c.p.s. In the engineering literature the terminology "cycles per second" has now been replaced by Hertz; hence we could say that the above term had a frequency of $(n/2\pi)$ Hz, this being the standard abbreviation for Hertz. (For example, the note of middle C on a piano is usually tuned to the frequency 256 c.p.s.; thus the sound wave corresponding to the "fundamental" of this note would be described by the term $\{A \cos(512\pi)t + B \sin(512\pi)t\}$, the values of A and B depending on how loud the note was played! This does not, of course, correspond to one of the terms in a Fourier series of the above type since each of these has an integer value of n—rather, it belongs to the class of Fourier series which describe functions of periodicity other than 2π.)

The quantity n itself may be interpreted as the *"angular frequency"* of the term $(a_n \cos nt + b_n \sin nt)$, and is measured in *radians per second*, assuming again that t is measured in seconds. (Note that the angle (nt) sweeps through n radians each time t moves through one unit of time.) The relationship between "frequency" (measured in c.p.s.) and "angular frequency" (measured in radians per second) is thus given by

$$\text{Angular frequency} = (\text{frequency}) \times 2\pi.$$

The quantity c_n^2 may now be interpreted as follows;

$c_n^2 =$ *contribution to the total power from the term in the Fourier series of* $X(t)$ *with frequency* $(n/2\pi)$ *c.p.s., or equivalently, with angular frequency* n *radians per second.*

.

If we plot the quantities c_n^2 against $n/2\pi$ (as illustrated in Fig. 4.2) the diagram which we obtain is called a *discrete power spectrum*, and its shape describes diagrammatically how the total power is distributed over the various frequency components of the function $X(t)$.

Fig. 4.2. Discrete power spectrum.

4.4.1 Functions of General Periodicity

Suppose now that $X(t)$ is a periodic function with period $2T$, and possessing a Fourier series expansion of the form (4.3.1), i.e. we may write,

$$X(t) = \tfrac{1}{2}a_0 + \sum_{n=1}^{\infty} [a_n \cos(\pi n t / T) + b_n \sin \pi n t / T)].$$

The Fourier coefficients, $\{a_n\}$, $\{b_n\}$ are then given by (4.3.2), and let the quantities $c_n (n = 0, 1, 2, \dots)$ be defined in terms of the $\{a_n\}$ and $\{b_n\}$ in exactly the same way as before. Then the total energy dissipated in the time interval $(-T, T)$ is given by,

$$\text{Total energy over } (-T, T) = \int_{-T}^{T} X^2(t)\, dt = 2T\left(\sum_{n=0}^{\infty} c_n^2 \right), \quad (4.4.6)$$

on using Parseval's relation (cf. (4.2.18), (4.4.2). Hence,

$$\text{Total power over } (-T, T) = \frac{\text{Total energy over } (-T, T)}{2T} = \sum_{n=0}^{\infty} c_n^2, \quad (4.4.7)$$

and we now have,

$c_n^2 =$ *contribution to the total power from the term in the Fourier series of $X(t)$ with frequency $(n/2T)$ c.p.s., or equivalently, with angular frequency $(\pi n / T)$ radians per second.*

If we plot c_n^2 against $n/2T$ we again obtain a *discrete power spectrum* whose physical interpretation is exactly the same as before except that here the discrete set of frequencies which contribute to the total power have changed from $n/2\pi$ c.p.s. to $n/2T$ c.p.s., $n = 0, 1, 2, \ldots$; as shown in Fig. 4.3.

Note that if T is much larger than π the set of frequencies, $(1/2T, 2/2T, 3/2T, \ldots)$, which form the above power spectrum is much more "closely packed" then the standard set, $(1/2\pi, 2/2\pi, 3/2\pi, \ldots)$, which form the power spectrum for functions with periodicity 2π.

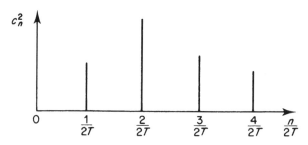

Fig. 4.3. Discrete power spectrum.

4.5 NON-PERIODIC FUNCTIONS; FOURIER INTEGRALS

We now turn our attention to the case of a real valued non-periodic function $X(t)$ (but still deterministic). Since $X(t)$ does not now possess any form of periodic structure we cannot express it in the form of a Fourier series which would be valid for *all* t. However, we can construct a Fourier series which will represent it over a *finite* interval by using the standard trick of defining a new function which is identical to $X(t)$ over a certain interval, but is periodic outside this interval. We thus choose an interval, $(-T, T)$, and define a new function, $X_T^*(t)$, by

$$
\begin{aligned}
X_T^*(t) &= X(t), & -T \leqslant t \leqslant T, \\
X_T^*(t + 2pT) &= X_T^*(t), & p = 1, 2, 3, \ldots.
\end{aligned}
\tag{4.5.1}
$$

Then $X_T^*(t)$ is certainly periodic, with period $2T$, and, subject to the usual regularity conditions, may be expressed in a Fourier series of the form (4.3.1), i.e. we may write,

$$
X_T^*(t) = \sum_{n=0}^{\infty} (a_n \cos 2\pi f_n t + b_n \sin 2\pi f_n t),
\tag{4.5.2}
$$

where, for each n, $f_n = n/2T$, and $b_0 = 0$.

Using the well known relations,

$$\cos 2\pi f_n t = \frac{1}{2}[\exp(2\pi i f_n t) + \exp(-2\pi i f_n t)],$$

$$\sin 2\pi f_n t = \frac{1}{2i}[\exp(2\pi i f_n t) - \exp(-2\pi i f_n t)],$$

we may rewrite (4.5.2) in the more convenient "complex exponential" form,

$$X_T^*(t) = \sum_{n=-\infty}^{\infty} A_n e^{2\pi i f_n t}, \tag{4.5.3}$$

where

$$A_n = \begin{cases} \frac{1}{2}(a_n - i b_n), & n \geq 1 \\ \frac{1}{2}a_0, & n = 0 \\ \frac{1}{2}(a_{|n|} + i b_{|n|}), & n \leq -1. \end{cases} \tag{4.5.4}$$

Substituting the expressions (4.3.2) for a_n, b_n, into (4.5.4) we have,

$$A_n = \frac{1}{2T} \int_{-T}^{T} X_T^*(t) e^{-2\pi i f_n t} dt$$

$$= \frac{1}{2T} \int_{-T}^{T} X(t) e^{-2\pi i f_n t} dt, \tag{4.5.5}$$

since $X(t)$ and $X_T^*(t)$ are identical over interval $(-T, T)$. Hence, for $-T \leq t \leq T$, we may write,

$$X(t) \equiv X_T^*(t) = \sum_{n=-\infty}^{\infty} \left(\int_{-T}^{T} X(t) e^{-2\pi i f_n t} dt \right) e^{2\pi i f_n t} \delta f_n, \tag{4.5.6}$$

where we have written

$$\delta f_n = f_n - f_{n-1} = 1/2T.$$

Now consider the limiting situation as we let $T \to \infty$. In this case $\delta f_n \to 0$, i.e. the discrete set of frequency points, $(\dots f_{-2}, f_{-1}, f_0, f_1, f_2, \dots)$ becomes a *continuous* set of points. The summation on the right-hand side of (4.5.6) will then become an *integral*, so that as $T \to \infty$ we obtain formally, for all t,

$$X(t) = \int_{-\infty}^{\infty} p(f) e^{2\pi i f t} df, \tag{4.5.7}$$

where

$$p(f) = \int_{-\infty}^{\infty} X(t) e^{-2\pi i f t} dt, \tag{4.5.8}$$

provided, of course, that both the above infinite integrals exist. The function, $p(f)$, is then called the *Fourier transform* of $X(t)$, and (4.5.7) is called the *Fourier integral* representation of $X(t)$. In fact, since the relationship between $X(t)$ and $p(f)$ is symmetrical, $X(t)$ and $p(f)$ are sometimes called a *"Fourier pair"*. Alternatively, (4.5.8) may be written as

$$p(f) = \int_{-\infty}^{\infty} X(t)\{\cos 2\pi ft - i \sin 2\pi ft) \, dt$$

$$= g(f) - ik(f), \qquad\qquad (4.5.8')$$

say, and then (4.5.7) may be written in the form,

$$X(t) = \int_{-\infty}^{\infty} \{g(f) \cos 2\pi ft + k(f) \sin 2\pi ft\} \, df, \qquad (4.5.7)'$$

remembering that $X(t)$ is real valued. If $X(t)$ is an even function (i.e. if $X(t) = X(-t)$, all t), then clearly $k(f) \equiv 0$, and in place of (4.5.7)' we have the *cosine transform*,

$$X(t) = \int_{-\infty}^{\infty} g(f) \cos 2\pi ft \, df.$$

Similarly, if $X(t)$ is an odd function (i.e. $X(t) = -X(-t)$, all t), we have a *sine transform*,

$$X(t) = \int_{-\infty}^{\infty} \{-k(f)\} \sin 2\pi ft \, dt.$$

(Well known examples of functions and their Fourier transforms are given in Stuart (1961), Kufner and Kadlec (1971).)

When we recall our previous discussion on the existence of Fourier series it becomes apparent that establishing the existence of the Fourier integral representations (4.5.7) as the limiting form of (4.5.6) is a step which requires some caution. In order for (4.5.7), (4.5.8), to be meaningful we require, at the very least, that the function $X(t)$ should be such that $p(f)$ exists for all f. A sufficient condition for this is that $X(t)$ *be absolutely* integrable over the infinite interval, $(-\infty, \infty)$, i.e. that

$$\int_{-\infty}^{\infty} |X(t)| \, dt < \infty. \qquad\qquad (4.5.9)$$

It then follows trivially that,

$$|p(f)| < \int_{-\infty}^{\infty} |X(t)| \, |\exp(-2ift)| \, dt = \int_{-\infty}^{\infty} |X(t)| \, dt < \infty, \qquad \text{all } f.$$

However, just as in the case of Fourier series the mere existence of $p(f)$ does not ensure that the right-hand side of (4.5.7) will converge to $X(t)$. For this to hold we require further conditions on the "good behaviour" of $X(t)$, such as, e.g., that $X(s)$ *be of bounded variation in an interval containing the point* $s = t$. It may then be shown (Titchmarsh (1948)) that the right-hand side of (4.5.7) converges to $\frac{1}{2}\{X(t+0)-X(t-0)\}$, or more simply, converges to $X(t)$ at all continuity points (cf. "Jordan's Test" for Fourier series discussed in Section 4.2).

When the above conditions are satisfied (4.5.7) provides a representation for $X(t)$ as the limiting form of a "sum" of sine and cosine functions—but the crucial point is that here the *representation involves a continuous range of frequencies*, i.e. all values of f (from $-\infty$ to $+\infty$) are present in the integral in (4.5.7). This may be contrasted with the Fourier series representation (4.3.1) for general periodic functions in which only the *discrete set* of frequencies $(\dots f_{-1}, f_0, f_1, \dots)$ play any part.

We may now rewrite (4.5.7), (4.5.8) in terms of "angular frequency" by changing the variable in (4.5.7) from f to $\omega = 2\pi f$. We then obtain,

$$X(t) = \frac{1}{\sqrt{2\pi}} \int_{-\infty}^{\infty} G(\omega) e^{i\omega t} \, d\omega, \qquad (4.5.10)$$

where

$$G(\omega) = \frac{1}{\sqrt{2\pi}} \int_{-\infty}^{\infty} X(t) e^{-i\omega t} \, dt. \qquad (4.5.11)$$

(In fact the functions $G(\omega)$, $p(f)$, are related simply by $G(\omega) \equiv \sqrt{2\pi} p(\omega/2\pi)$, and we have adopted this form of $G(\omega)$ so as to distribute the factor $(1/2\pi)$ equally between the two integrals.)

Parseval's relation

Corresponding to Parseval's relation for Fourier series (equation (4.4.2)) there is an analogous result for Fourier integrals, namely,

$$\int_{-\infty}^{\infty} X^2(t) \, dt = \int_{-\infty}^{\infty} |G(\omega)|^2 \, d\omega. \qquad (4.5.12)$$

To prove (4.5.12) we write,

$$\int_{-\infty}^{\infty} X^2(t) \, dt = \frac{1}{\sqrt{2\pi}} \int_{-\infty}^{\infty} X(t) \left\{ \int_{-\infty}^{\infty} G(\omega) e^{i\omega t} \, d\omega \right\} dt,$$

$$= \frac{1}{\sqrt{2\pi}} \int_{-\infty}^{\infty} G(\omega) \left\{ \int_{-\infty}^{\infty} X(t) e^{i\omega t} \, dt \right\} d\omega,$$

(interchanging the order of integration),

$$= \int_{-\infty}^{\infty} G(\omega)G^*(\omega)\, d\omega = \int_{-\infty}^{\infty} |G(\omega)|^2\, d\omega.$$

L^2-theory

There is also an elegant "L^2-theory" for Fourier integrals which has the advantage that the conditions for the validity of both (4.5.10) and (4.5.11) then assume a symmetrical form. Specifically, if we assume only that $X(t)$ belongs to $L^2(-\infty, \infty)$, i.e. that

$$\int_{-\infty}^{\infty} X^2(t)\, dt < \infty, \tag{4.5.13}$$

then although it does not necessarily follow that the infinite integral on the right hand side of (4.5.11) exists (since $X(t)$ need not necessarily belong to $L^1(-\infty, \infty)$) we can nevertheless define the sequence of functions,

$$G_\alpha(\omega) = \frac{1}{\sqrt{2\pi}} \int_{-\alpha}^{\alpha} X(t)\, e^{-i\omega t}\, dt,$$

and it may then be shown (Titchmarsh (1948)) that there exists a function $G(\omega)$, belonging to $L^2(-\infty, \infty)$, such that as $\alpha \to \infty$, $G_\alpha(\omega)$ *converges in mean square* to $G(\omega)$. Since $G(\omega)$ also belongs to $L^2(-\infty, \infty)$ it follows that

$$X_\alpha(t) = \frac{1}{\sqrt{2\pi}} \int_{-\alpha}^{\alpha} G(\omega)\, e^{i\omega t}\, d\omega$$

converges in mean square to $X(t)$.

This approach to Fourier integrals is due to Plancherel. Of course, if $X(t)$ belongs to both $L^1(-\infty, \infty)$ and $L^2(-\infty, \infty)$ then the function $G(\omega)$ defined above is identical with the function $G(\omega)$ defined by (4.5.11).

4.5.1 Nature of Conditions for the Existence of Fourier Series and Fourier Integrals

We have seen that periodic functions have to satisfy certain conditions in order that they should possess Fourier series and also that non-periodic functions have to satisfy certain conditions in order that they should possess Fourier integrals. However, these two sets of conditions differ in quite a fundamental way. In the case of Fourier series the conditions are essentially of a "technical" nature, i.e. they demand, roughly speaking, only that the function be reasonably "well behaved", so that, in practice, we could reasonably assume that virtually any periodic function has a Fourier series. (A function which fails to satisfy the required conditions is certainly pathological.) On the other hand, the conditions for the existence of Fourier integrals are much more stringent, and we certainly cannot assume that any reasonably "well behaved" non-periodic function possesses a Fourier

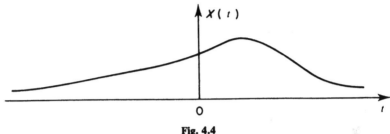

Fig. 4.4

transform. The basic condition, (4.5.9), that $X(t)$ be absolutely integrable over $(-\infty, \infty)$ implies that both $\int_{-\infty}^{-T} |X(t)|\, dt$ and $\int_{T}^{\infty} |X(t)|\, dt \to 0$ as $T \to \infty$, so that typically (but not invariably) $X(t)$ *will decay to zero as* $t \to +\infty$ *and as* $t \to -\infty$. In fact, if $\lim_{t \to -\infty} X(t)$ and $\lim_{t \to \infty} X(t)$ exist then clearly both will be zero. However, whether or not these limits exist we may still say that $X(t)$ "decays" as $|t|$ becomes large in the sense that if the set $\alpha_n(\varepsilon)$ is defined by $\alpha_n(\varepsilon) = \{t; |X(t)| > \varepsilon,\ n \le |t| < n+1\}$, then for any $\varepsilon\,(>0)$ the measure of the set $\{\alpha_n(\varepsilon)\}$ tends to zero as $n \to \infty$. Thus, the typical form of function which we might expect to possess a Fourier integral is shown in Fig. 4.4; the function illustrated in Fig. 4.5, although perfectly "well behaved", would certainly not be expected to have a Fourier integral.

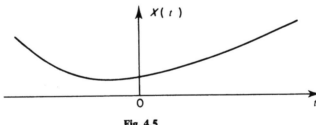

Fig. 4.5

Note that although we have imposed a similar condition of absolute integrability for the existence of Fourier series ((4.2.7)), in this case we demanded only that $X(t)$ be absolutely integrable over the *finite* interval $(-\pi, \pi)$. This is far less restrictive than the *infinite* version (4.5.9) and does not involve any conditions on the asymptotic behaviour of $X(t)$.

It is sometimes convenient, nevertheless, to treat certain functions which do not satisfy (4.5.9) as if they had Fourier transforms. To do this we need to make use of the theory of "generalized functions"—in particular, the "Dirac δ-function" is extremely useful in this context. For example, the function $X(t) = 1$, all t, clearly does not satisfy (4.5.9), and hence it does not possess a Fourier transform in the strict sense. However, it is very useful to think of its "Fourier transform" as being represented formally by the generalized function, $\{\sqrt{2\pi}\delta(\omega)\}$, i.e. a function which is infinite at $\omega = 0$ and zero everywhere else.

4.6 SPECTRAL ANALYSIS OF NON-PERIODIC FUNCTIONS

Let $X(t)$ be a non-periodic function satisfying (4.5.13), and possessing a Fourier integral of the form (4.5.10), i.e. we may write,

$$X(t) = \frac{1}{\sqrt{2\pi}} \int_{-\infty}^{\infty} G(\omega) e^{i\omega t} \, d\omega,$$

where $G(\omega)$ is given by (4.5.11). The total energy dissipated over the infinite time interval $(-\infty, \infty)$ is given by,

$$\text{Total energy over } (-\infty, \infty) = \int_{-\infty}^{\infty} X^2(t) \, dt$$

$$= \int_{-\infty}^{\infty} |G(\omega)|^2 \, d\omega, \qquad (4.6.1)$$

in virtue of Parseval's relation (4.5.12). Thus once again, we have a decomposition of the total energy as (the limiting form of) a "sum of terms", each "term" representing the contribution of a particular group of frequency components. More precisely, we see from (4.6.1) that $\{|G(\omega)|^2 \, d\omega\}$ *represents the contribution to the total energy from those components in* $X(t)$ *whose frequencies lie between* ω *and* $\omega + d\omega$. We thus have here a *continuous* distribution of energy over frequency.

The quantity $|G(\omega)|^2$ itself does not represent a measure of energy; rather, it represents an *energy density function*, i.e. $|G(\omega)|^2$ measures the *density* of energy contributed by components with frequencies in the neighbourhood of ω. If we plot $|G(\omega)|^2$ against ω the graph which we obtain may be called a (continuous) *energy spectral density function*, and its typical shape is shown in Fig. 4.6. The interpretation of $|G(\omega)|^2$ as a *density* function arises as an inevitable consequence of the fact that here we are dealing with a distribution of energy over a *continuous* range of frequencies. It will be recalled that in the case of a continuous random variable X, the *probability density function* $f(x)$ is defined analogously by $p[x < X \leqslant x + \delta x] = f(x)\delta x$. The quantity $f(x)$ does not itself represent a probability; rather it represents the density of probability in the neighbourhood of the point x—see Section 2.

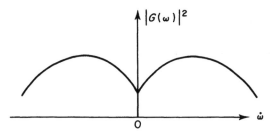

Fig. 4.6. Continuous energy spectral density function.

When $X(t)$ is real valued (as we have assumed throughout), $|G(\omega)|^2$ is an *even function of* ω, i.e. $|G(\omega)|^2 = |G(-\omega)|^2$, and we have illustrated this feature in Fig. 4.6. This follows from the fact that, by (4.5.11),

$$G(-\omega) = \frac{1}{\sqrt{2\pi}} \int_{-\infty}^{\infty} X(t)\, e^{i\omega t}\, dt = G^*(\omega).$$

Hence,

$$|G(-\omega)|^2 = G(-\omega)G^*(-\omega) = G^*(\omega)G(\omega) = |G(\omega)|^2.$$

We may summarize the above discussion by noting that there are two essential differences between the spectral properties of periodic and non-periodic functions, namely:

(a) Periodic functions are represented by a discrete set of frequency components; non-periodic functions involve a *continuous* range of frequencies.

(b) For periodic functions the total energy over the time interval $(-\infty, \infty)$ is *infinite*—consequently, we describe their spectral properties in terms of the distribution of *power* (i.e. energy per unit time) over a discrete set of frequencies. For non-periodic functions which satisfy (4.5.13) the total energy over $(-\infty, \infty)$ is *finite*—hence we describe their properties in terms of an *energy density* distribution over a continuous range of frequencies. In the latter case the total power, as given by

$$\lim_{T \to \infty} \left[\frac{1}{2T} \{\text{energy in } (-T, T)\} \right]$$

is, of course, zero.

4.7 SPECTRAL ANALYSIS OF STATIONARY PROCESSES

So far our discussion has been concerned only with *deterministic* (i.e. non-random) functions. However, this analysis was introduced mainly as background material for the main topic of this chapter, namely, the spectral analysis of *stationary stochastic processes*. When we consider stochastic processes the first problem which we encounter is the fact that a stochastic process is not just a single function, rather it represents, in general, an infinite number of different realizations of the process. To obviate this difficulty we have to begin by considering each realization separately.

Consider then a *single realization*, $X(t)$, of a continuous-parameter stationary process, and in order to avoid what is at this stage an irrelevant complication, we will assume that $E[X(t)] = 0$, all t, and that $X(t)$ is stochastically continuous. Typically, this may have the form illustrated in Fig. 4.7. (We retain the notation $X(t)$ even though, for the moment, we are discussing a realization rather than a stochastic process.)

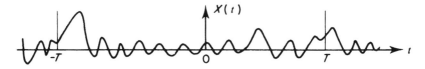

Fig. 4.7. Realization of a stationary process.

Before we can apply the ideas of Fourier series and Fourier integrals we must first ask: Can we represent a typical realization as a Fourier series? The answer to this question is clearly "No", since we have no reason to suppose that a realization of a general stationary process will be periodic in any way. (In the case of harmonic processes (see Section 3.58), almost all realizations will, in fact, be periodic if the $\{\omega_r\}$ are commensurate; however this is the only model which possesses this property, and, as we shall see later, it constitutes a very special case.)

The next question is: Can we represent a typical realization as a Fourier integral? Again, the answer is clearly "No", since we have no reason to suppose that a realization will satisfy the condition (4.5.9). Indeed, the very nature of "stationarity" suggests that a realization will almost certainly not "decay" at infinity. At this stage we seem to have reached an empass since if we cannot write $X(t)$ either as a Fourier series or as a Fourier integral there would seem to be no way in which we can meaningfully talk about a breakdown of the total energy (or power) into the various frequency components. (The very notion of "frequency" implies that we can represent the quantity as a "sum" of sine and cosine functions.) However, we may attempt to overcome this particular difficulty by using a "truncation" technique similar to that discussed in Section 4.5, i.e. we choose some arbitrary number T and "chop off" the realization at the points $t = -T$ and $t = +T$. That is, we define a new function, $X_T(t)$, by,

$$X_T(t) = \begin{cases} X(t), & -T \leq t \leq T, \\ 0, & \text{otherwise,} \end{cases} \qquad (4.7.1)$$

(see Fig. 4.7). (Note that $X_T(t)$ is *not* the same as the function, $X_T^*(t)$, defined by (4.5.1); $X_T^*(t)$ was defined so as to be periodic whereas $X_T(t)$ is non-periodic.) Then $X_T(t)$ certainly dies away at $\pm\infty$ and, being non-zero only on the finite interval $(-T, T)$, it clearly satisfies the condition (4.5.9). Hence, assuming that it is continuous, we may express $X_T(t)$ as a Fourier integral, i.e. may write it in the form (cf. (4.5.10), (4.5.11)),

$$X_T(t) = \frac{1}{\sqrt{2\pi}} \int_{-\infty}^{\infty} G_T(\omega) e^{i\omega t} d\omega, \qquad (4.7.2)$$

where

$$G_T(\omega) = \frac{1}{\sqrt{2\pi}} \int_{-\infty}^{\infty} X_T(t) e^{-i\omega t} dt$$

$$= \frac{1}{\sqrt{2\pi}} \int_{-T}^{T} X(t) e^{-i\omega t} dt, \qquad (4.7.3)$$

since on the interval $(-T, T)$, $X(t)$ and $X_T(t)$ are identical. We now know from the discussion in Section 4.6 that $|G_T(\omega)|^2$ has the following physical interpretation;

$|G_T(\omega)|^2 d\omega =$ contribution to the total energy of $X_T(t)$ contributed by those components with frequencies between $\omega, \omega + d\omega$.

It is tempting now simply to let $T \to \infty$, in which case $X_T(t)$ and $X(t)$ would become identical for *all* t, and $\{\lim_{T\to\infty} |G_T(\omega)|^2\}$ would then describe the properties of $X(t)$ itself. However, a little thought soon reveals that this approach is far too crude. For, if $G_T(\omega)$ converges to a well behaved function as $T \to \infty$ then we might just as well have started from a Fourier integral representation of $X(t)$ without introducing the function $X_T(t)$. In fact, we already know that $X(t)$ does not, in general, possess a Fourier integral and hence it is clear that $\{\lim_{T\to\infty} |G_T(\omega)|^2\}$ will *not* exist (i.e. be finite). This is equally clear from physical considerations since, due to the "steady state" nature of a stationary process, the amount of energy required to sustain the process from $t = -\infty$ to $t = +\infty$ will obviously be infinite. This situation is very similar to that encountered in our study of periodic functions where again we saw that the total energy for the infinite time interval $(-\infty, \infty)$ would be infinite, and there we surmounted this difficulty by concentrating on *power* rather than energy. This suggests that although $|G_T(\omega)|^2$ may $\to \infty$ as $T \to \infty$, the *power* contributed by the various frequency components may, in fact, converge to a finite limit as $T \to \infty$. Recalling that power is measured in terms of "energy per unit time", this suggests that, under suitable conditions,

$$\lim_{T \to \infty} \frac{|G_T(\omega)|^2}{2T} \tag{4.7.4}$$

may be finite for all ω.

We note that if the above assertion is true,

$$\lim_{T \to \infty} \frac{|G_T(\omega)|^2}{2T} \, d\omega = \text{contribution to the total power of}$$
$$X(t) \text{ contributed by components with}$$
$$\text{frequencies between } \omega, \omega + d\omega.$$

Hence, $\lim_{T \to \infty}[|G_T(\omega)|^2/2T]$ would have an interpretation as a *power density* function (see the discussion in Section 4.6). Before we formalize these ideas there is one further point to consider, namely, that the analysis above refers only to a single realization. Thus, the value of $\{|G_T(\omega)|^2/2T\}$ will depend on the particular realization we have chosen, and if the limit in (4.7.4) does exist it will relate only to the properties of the particular realization under study. We could, in principle, calculate the value of $\{|G_T(\omega)|^2/2T\}$ for every realization, but this value will change from one realization to another. If therefore, we wish to construct some quantity which will characterize the power/frequency properties of the whole sto-chastic process (as opposed to merely one realization) it is natural to *average the values of* $\{|G_T(\omega)|^2/2T\}$ *over the different realizations.* Thus, before proceeding to the limit as $T \to \infty$ we replace $\{|G_T(\omega)|^2/2T\}$ by its average value, i.e. by $E\{|G_T(\omega)|^2/2T\}$. This leads us finally to considering the quantity, $h(\omega)$, defined by,

$$h(\omega) = \lim_{T \to \infty} \left[E\left\{ \frac{|G_T(\omega)|^2}{2T} \right\} \right]. \tag{4.7.5}$$

When it exists, $h(\omega)$ has the following interpretation,

$$h(\omega) \, d\omega = \text{average (over all realizations) of the}$$
$$\text{contribution to the total power from}$$
$$\text{components in } X(t) \text{ with frequencies}$$
$$\text{between } \omega \text{ and } \omega + d\omega.$$

The function $h(\omega)$ is called the (*non-normalized*) *power spectral density function* of $X(t)$, (or more simply, the (*non-normalized*) *spectrum* of $X(t)$), and it plays a fundamental role in the spectral analysis of stationary stochastic processes.

We may introduce also two other related functions, namely, the *band spectrum*, $h(\omega_1, \omega_2)$, and the (*non-normalized*) *integrated spectrum*, $H(\omega)$. These are defined by

$$h(\omega_1, \omega_2) = \int_{\omega_1}^{\omega_2} h(\omega)\, d\omega, \tag{4.7.6}$$

= average contribution to the total power from
components with frequencies between ω_1 and ω_2,

and

$$H(\omega) = \int_{-\infty}^{\omega} h(\theta)\, d\theta \tag{4.7.7}$$

= average contribution to the total power
from all components with frequencies
less than or equal to ω.

Since $H(\omega)$ is the integral of $h(\omega)$ we may write $h(\omega)$ as the derivative of $H(\omega)$, i.e.

$$h(\omega) = dH(\omega)/d\omega. \tag{4.7.8}$$

The definition of the (non-normalized) power spectral density function $h(\omega)$ (as given by (4.7.5)) involves the limiting operation of letting $T \to \infty$, and it may be suspected therefore that this function will not exist for certain types of stationary processes. This is indeed so, but further consideration of this point is best left until the next section where we derive a much simpler alternative expression for $h(\omega)$. Meanwhile, it may be helpful to summarize the different spectral properties of periodic functions, non-periodic functions, and stationary stochastic processes, as follows.

(a) For *periodic functions* we have a distribution of *power* over a *discrete* set of frequencies.

(b) For *non-periodic functions* we have a distribution of *energy* over a *continuous* range of frequencies.

(c) For *stationary stochastic processes* for which $h(\omega)$ exists, we have a distribution of *power* over a *continuous* range of frequencies. However, as remarked above, $h(\omega)$ does not exist for all stationary processes, and we have already noted that the realizations of a harmonic process have properties very similar to those of periodic functions. We may therefore expect that harmonic processes would give rise to a power distribution over a *discrete* set of frequencies only—and this is indeed the case, as we shall see later. In fact, in the most general situation a stationary process may give rise to a *mixture* of both continuous and discrete power distributions. We consider this point in greater detail in Section 4.10.

4.8 RELATIONSHIP BETWEEN THE SPECTRAL DENSITY FUNCTION AND THE AUTOCOVARIANCE AND AUTOCORRELATION FUNCTIONS

The spectral density function $h(\omega)$ depends purely on the probabilistic properties of the process $X(t)$ so that if we know, e.g., that $X(t)$ is a kth order autoregressive process then we should be able to calculate the specific form of $h(\omega)$ for this particular model. Calculating $h(\omega)$ directly from the definition (4.7.5) would be a formidable task, but fortunately we may derive an alternative expression for $h(\omega)$ which is ideally suited for this purpose. Specifically, we will show that $h(\omega)$ is simply the *Fourier transform of the autocovariance function*; thus once we have determined the form of the autocovariance function for a particular model the calculation of $h(\omega)$ is essentially trivial. This result may seem surprising since it is difficult at first sight to see why the autocovariance function should be involved in any way with the power/frequency properties of the process. However, this difficulty is easily resolved once we realize that the *power* (at frequency ω) depends essentially on the quantity, $|G_T(\omega)|^2$, i.e. on the *squared modulus* of the Fourier transform of $X_T(t)$. Now there is a basic result in Fourier analysis which tells us that if $F(\omega)$ is the Fourier transform of a function $f(t)$, then $|F(\omega)|^2$ is the Fourier transform of the *convolution* of $f(t)$ with itself, i.e. $|F(\omega)|^2$ is the Fourier transform of the function $\{\int_{-\infty}^{\infty} f(u)f(u-t)\, du\}$. This result applies strictly to deterministic functions only, but one immediately notices the strong similarity between the above expression for a convolution and the expression $E[X(t)X(t+\tau)]$ for the autocovariance function of a zero-mean process. Thus, as soon as we start to consider "power" it is inevitable that we will be led to expressions involving the convolution of $X(t)$ with itself, and equally inevitable that this, in turn, will lead to the appearance of the autocovariance function.

Before deriving the relationship between $h(\omega)$ and the autocovariance function we first prove a lemma giving a slightly more general form of the "convolution" property referred to above.

Lemma 4.8.1 *Let* $f(t)$, $g(t)$, *be real valued functions, each possessing a Fourier transform, and let*

$$F(\omega) = \int_{-\infty}^{\infty} e^{-i\omega t} f(t)\, dt, \qquad G(\omega) = \int_{-\infty}^{\infty} e^{-i\omega t} g(t)\, dt.$$

Then,

$$F(\omega)G^*(\omega) = \int_{-\infty}^{\infty} e^{-i\omega t}k(t)\,dt, \tag{4.8.1}$$

where

$$k(t) = \int_{-\infty}^{\infty} f(u)g(u-t)\,du. \tag{4.8.2}$$

Proof. We have,

$$\int_{-\infty}^{\infty} e^{-i\omega t}k(t)\,dt = \int_{-\infty}^{\infty}\int_{-\infty}^{\infty} e^{-i\omega t}f(u)g(u-t)\,du\,dt$$

$$= \int_{-\infty}^{\infty} e^{-i\omega u}f(u)\left\{\int_{-\infty}^{\infty} e^{i\omega(u-t)}g(u-t)\,dt\right\}du.$$

But, for each u,

$$\int_{-\infty}^{\infty} e^{i\omega(u-t)}g(u-t)\,dt = \int_{-\infty}^{\infty} e^{i\omega y}g(y)\,dy, \qquad (\text{setting } y = u - t)$$

$$= G^*(\omega), \qquad \text{which is independent of } u.$$

Hence,

$$\int_{-\infty}^{\infty} e^{-i\omega t}k(t)\,dt = G^*(\omega)\left\{\int_{-\infty}^{\infty} e^{-i\omega u}f(u)\,du\right\} = G^*(\omega)F(\omega),$$

as required.

In particular, when $f(t) \equiv g(t)$ we have,

$$|F(\omega)|^2 = \int_{-\infty}^{\infty} e^{-i\omega t}k(t)\,dt, \tag{4.8.3}$$

where now,

$$k(t) = \int_{-\infty}^{\infty} f(u)f(u-t)\,du. \tag{4.8.4}$$

We now prove the following basic result relating the power spectral density function $h(\omega)$ and the autocovariance function $R(\tau)$.

Theorem 4.8.1 *Let $\{X(t)\}$ be a zero-mean continuous parameter stationary process with power spectral density function, $h(\omega)$, which exists for all ω, and autocovariance function, $R(\tau)$. Then $h(\omega)$ is the Fourier transform of $R(\tau)$, i.e.,*

$$h(\omega) = \frac{1}{2\pi}\int_{-\infty}^{\infty} e^{-i\omega\tau}R(\tau)\,d\tau. \tag{4.8.5}$$

Proof. If we apply (4.8.3) to the function $G_T(\omega)$ defined by (4.7.3) we obtain,

$$|G_T(\omega)|^2 = \int_{-\infty}^{\infty} e^{-i\omega\tau} \left\{ \int_{-\infty}^{\infty} \frac{X_T(u)}{\sqrt{2\pi}} \cdot \frac{X_T(u-\tau)}{\sqrt{2\pi}} \, du \right\} d\tau.$$

Hence we may write,

$$\frac{|G_T(\omega)|^2}{2T} = \frac{1}{2\pi} \int_{-\infty}^{\infty} e^{-i\omega\tau} \hat{R}_T(\tau) \, d\tau,$$

where we have written,

$$\hat{R}_T(\tau) = \frac{1}{2T} \int_{-\infty}^{\infty} X_T(u) X_T(u-\tau) \, du.$$

Then,

$$h(\omega) = \lim_{T\to\infty} \left[E\left\{ \frac{|G_T(\omega)|^2}{2T} \right\} \right]$$

$$= \lim_{T\to\infty} \left\{ \frac{1}{2\pi} \int_{-\infty}^{\infty} e^{-i\omega\tau} E[\hat{R}_T(\tau)] \, d\tau \right\}. \qquad (4.8.6)$$

But it follows from the definition of $X_T(t)$ ((4.7.1)) that

$$\hat{R}_T(\tau) = \begin{cases} \dfrac{1}{2T} \displaystyle\int_{-(T-|\tau|)}^{T} X(u) X(u-|\tau|) \, du, & |\tau| \leqslant 2T \\[2mm] \qquad\qquad 0, & |\tau| \geqslant 2T. \end{cases}$$

Hence, for $|\tau| \leqslant 2T$,

$$E[\hat{R}(\tau)] = \frac{1}{2T} \int_{-(T-|\tau|)}^{T} E[X(u) X(u-|\tau|)] \, du,$$

$$= \frac{1}{2T} \int_{-(T-|\tau|)}^{T} R(\tau) \, du,$$

$$= R(\tau)\{1 - (|\tau|/2T)\}.$$

(Note that since $X(t)$ has zero mean, the autocovariance function, $R(\tau)$, may be written as

$$R(\tau) = E[X(t) X(t-\tau)].)$$

On the other hand, for $|\tau| > 2T$,

$$E[\hat{R}_T(\tau)] = 0.$$

Hence we obtain from (4.8.6),

$$h(\omega) = \lim_{T \to \infty} \left\{ \frac{1}{2\pi} \int_{-2T}^{2T} [1 - (|\tau|/2T)] \, e^{-i\omega\tau} R(\tau) \, d\tau \right\}. \qquad (4.8.7)$$

To evaluate this limit we have to consider the limiting form of the function $[1 - (|\tau|/2T)]$ as $T \to \infty$. The behaviour of this function for various values of T is shown in Fig. 4.8 from which it will be seen that as T increases it tends to

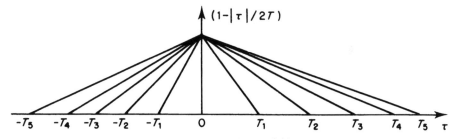

Fig. 4.8. The function $(1 - |\tau|/2T)$.

the constant values of 1 for all values of τ. Hence, if $R(\tau)$ decays to zero "fast enough" (as $\tau \to \pm \infty$) for its Fourier transform to exist, we obtain in the limit as $T \to \infty$,

$$h(\omega) = \frac{1}{2\pi} \int_{-\infty}^{\infty} e^{-i\omega\tau} R(\tau) \, d\tau. \qquad (4.8.8)$$

In deriving this extremely important result we have, incidentally, discovered the condition sufficient for the existence of $h(\omega)$, namely, that $\{X(t)\}$ should be such that its *autocovariance function* $R(\tau)$ *possesses a Fourier transform*. This will be so if the condition (4.5.9) is satisfied, i.e. *if* $R(\tau)$ *is absolutely integrable*. Hence, the power spectral density function $h(\omega)$ exists for all ω if,

$$\int_{-\infty}^{\infty} |R(\tau)| \, d\tau < \infty. \qquad (4.8.9)$$

To establish formally the existence of the limit (4.8.7) under the conditions (4.8.9) we may proceed as follows. Write

$$g_T(\tau) = \begin{cases} 1 - |\tau|/2T, & |\tau| \leq 2T \\ 0, & |\tau| > 2T. \end{cases}$$

Then $0 \leq g_T(\tau) \leq 1$, all τ, T, and for each fixed τ, $\lim_{T \to \infty} g_T(\tau) = 1$. Hence we have

$$\left| \int_{-2T}^{2T} (1 - |\tau|/2T) R(\tau) e^{-i\omega\tau} d\tau \right|$$

$$\leq \int_{-\infty}^{\infty} g_T(\tau) |R(\tau)| d\tau$$

$$\leq \int_{-\infty}^{\infty} |R(\tau)| d\tau < \infty, \qquad \text{all } T.$$

Hence the limit (as $T \to \infty$) exists, and by Lebesgue's dominated convergence theorem (see, e.g., Titchmarsh (1939), p. 345)) it follows that

$$\lim_{\tau \to \infty} \int_{-\infty}^{\infty} g_T(\tau) R(\tau) e^{-i\omega\tau} d\tau$$

$$= \int_{-\infty}^{\infty} \{ \lim_{T \to \infty} g_T(\tau) \} R(\tau) e^{-i\omega\tau} d\tau$$

$$= \int_{-\infty}^{\infty} R(\tau) e^{-i\omega\tau} d\tau$$

When (4.8.9) holds *and when $R(\tau)$ is continuous at $\tau = 0$ (so that it is continuous everywhere)* we may express $R(\tau)$ as the inverse Fourier transform of $h(\omega)$, i.e. we may write (cf. (4.5.10), (4.5.11)),

$$R(\tau) = \int_{-\infty}^{\infty} e^{i\omega\tau} h(\omega) \, d\omega. \qquad (4.8.10)$$

For real valued processes we know that $R(\tau)$ is an even function of τ, i.e. $R(-\tau) = R(\tau)$, all τ. Hence, in this case the complex term in the integral in (4.8.8) vanishes, and the expression for $h(\omega)$ reduces to

$$h(\omega) = \frac{1}{2\pi} \int_{-\infty}^{\infty} \cos \omega\tau \, R(\tau) \, d\tau. \qquad (4.8.11)$$

The power spectral density function, $h(\omega)$ is now clearly an *even function of* ω, i.e. $h(-\omega) = h(\omega)$, all ω, and hence (4.8.10) similarly reduces to a "cosine" transform, i.e.

$$R(\tau) = \int_{-\infty}^{\infty} \cos \omega\tau \, h(\omega) \, d\omega. \qquad (4.8.12)$$

The (non-normalized) integrated spectrum, $H(\omega)$, may be obtained by integrating the right-hand side of (4.8.11) with respect to ω under the integral sign, giving, for any continuity points ω_1, ω_2, s.t., $\omega_1 \le \omega_2$,

$$H(\omega_2) - H(\omega_1) = \int_{\omega_1}^{\omega_2} h(\theta) \, d\theta$$

$$= \frac{1}{2\pi} \int_{-\infty}^{\infty} \cdot \frac{\exp(-i\omega_2\tau) - \exp(-i\omega_1\tau)}{-i\tau} R(\tau) \, d\tau.$$

$$(4.8.13)$$

In particular, if $X(t)$ is real valued then $h(\omega)$ is an even function and $H(0) = \sigma_X^2/2$. Hence, for all $\omega \ge 0$,

$$H(\omega) = (\sigma_X^2/2) + \frac{1}{2\pi} \int_{-\infty}^{\infty} \frac{\exp(-i\omega\tau) - 1}{-i\tau} R(\tau) \, d\tau$$

$$= (\sigma_X^2/2) + \frac{1}{2\pi} \int_{-\infty}^{\infty} \left(\frac{\sin \omega\tau}{\tau} \right) R(\tau) \, d\tau. \qquad (4.8.13a)$$

In view of the symmetry properties which hold for real valued processes, we may, if we wish, define both the power spectral density function and integrated spectrum for *positive* frequencies only. Thus, if we define

$$h_+(\omega) = \begin{cases} h(0), & \omega = 0, \\ 2h(\omega), & \omega > 0, \end{cases}$$

$$H_+(\omega) = \int_0^{\omega} h_+(\theta) \, d\theta = 2H(\omega) - \sigma_X^2, \qquad \omega \ge 0,$$

then we may re-write (4.8.11), (4.8.12) as,

$$h_+(\omega) = \frac{2}{\pi} \int_0^{\infty} \cos \omega\tau \, R(\tau) \, d\tau, \qquad \omega > 0,$$

$$R(\tau) = \int_0^{\infty} \cos \omega\tau \, h_+(\omega) \, d\omega,$$

and (4.8.13a) as,

$$H_+(\omega) = \frac{2}{\pi} \int_0^{\omega} \frac{\sin \tau\omega}{\tau} R(\tau) \, d\tau,$$

4.8.1 Normalized Power Spectra

Setting $\tau = 0$ in (4.8.10) we have,

$$\sigma_X^2 \equiv R(0) = \int_{-\infty}^{\infty} h(\omega) \, d\omega = H(+\infty). \qquad (4.8.14)$$

Now $H(+\infty)$ represents the *total power* (i.e. the power contributed by *all* frequency components) of the process. Hence we see from (4.8.14) that the *variance* of $\{X(t)\}$, σ_X^2, is a measure of the *total power* of the process. We now define the *normalized power spectral density function* $f(\omega)$ by

$$f(\omega) = h(\omega)/\sigma_X^2. \tag{4.8.15}$$

From the above interpretation of σ_X^2 it follows that $f(\omega)$ has the interpretation;

$f(\omega)\,d\omega$ = average (over all realizations) of the *proportion* of the total power contributed by components with frequencies between ω, $\omega + d\omega$.

Dividing both sides of (4.8.8) by σ_X^2, and recalling that $R(\tau)/\sigma_X^2 \equiv \rho(\tau)$, the *autocorrelation function* of $\{X(t)\}$, we obtain,

$$f(\omega) = \frac{1}{2\pi} \int_{-\infty}^{\infty} e^{-i\omega\tau} \rho(\tau)\,d\tau, \tag{4.8.16}$$

and from (4.8.10) we obtain similarly,

$$\rho(\tau) = \int_{-\infty}^{\infty} e^{i\omega\tau} f(\omega)\,d\omega. \tag{4.8.17}$$

Thus, whereas the *non-normalized* power spectral density function $h(\omega)$, is the Fourier transform of the *autocovariance* function, $R(\tau)$, the *normalized* power spectral density function $f(\omega)$ is the Fourier transform of the *autocorrelation* function, $\rho(\tau)$.

Analogous to the definition of $H(\omega)$, we define the *normalized integrated spectrum*, $F(\omega)$, by

$$F(\omega) = \int_{-\infty}^{\omega} f(\theta)\,d\theta, \tag{4.8.18}$$

so that,

$$f(\omega) = dF(\omega)/d\omega. \tag{4.8.19}$$

The physical interpretation of $F(\omega)$ bears the same relationship to that of $H(\omega)$ as does the interpretation of $f(\omega)$ to that of $h(\omega)$.

Since $f(\omega)$ is simply a "scaled" version of $h(\omega)$ it may seem rather pointless to introduce this new function. Indeed, as far as the physical properties of the process are concerned, $f(\omega)$ and $h(\omega)$ contain essentially the same information, and in practice it makes little difference which of these two functions we choose to work with. However, as we shall see, $f(\omega)$ has rather more elegant mathematical properties than $h(\omega)$, and plays an important role in establishing the connection between spectral density functions and probability density functions.

Properties of $f(\omega)$

The normalized power spectral density function always has the following properties.

(1) $\int_{-\infty}^{\infty} f(\omega)\, d\omega = 1$

(This follows immediately on setting $\tau = 0$ in (4.8.17) and remembering that $\rho(0) = 1$.)

(2) $f(\omega) \geq 0,$ all ω
(Since $|G_T(\omega)|^2 \geq 0 \Rightarrow h(\omega) \geq 0 \Rightarrow f(\omega) \geq 0$, all ω.)

(3) For real valued processes,
$f(-\omega) = f(\omega),$ all ω
(Since in this case $\rho(-\tau) = \rho(\tau)$, and the "even" property of $f(\omega)$ follows directly from (4.8.16).)

Properties of $F(\omega)$

The normalized integrated spectrum always has the following properties.

(1) $0 \leq F(\omega) \leq 1,$ all ω
(This follows immediately from properties (1) and (2) above.)

(2) $F(-\infty) = 0,$ $F(+\infty) = 1$
(These follow from the definition of $F(\omega)$ together with property (1) above.)

(3) $F(\omega)$ is a non-decreasing function of ω, i.e.
$\omega_1 > \omega_2 \Rightarrow F(\omega_1) \geq F(\omega_2)$
(This follows from property (2) above.)

If we now compare the above properties of $F(\omega)$ with those of a probability distribution function (given in Section 2.6), we see that $F(\omega)$ does, in fact, have all the properties of a probability distribution function. (The property of "right-continuity" is relatively unimportant; we can always ensure that this holds by suitably defining the value of $F(\omega)$ at each point of discontinuity; see Section 2.6.1.) Hence, from a purely mathematical point of view, the normalized integrated spectrum is completely equivalent to a probability distribution function, and hence we may exploit all the known properties of probability distribution functions to establish further corresponding properties of the function $F(\omega)$. For this reason $F(\omega)$ is sometimes referred to as the "*spectral distribution function*". (It should be noted that the special properties of $f(\omega)$ and $F(\omega)$ which hold for real valued processes are not, of course, general properties of probability distribution functions for real valued random variables.) In particular, it follows that $f(\omega)$, being the derivative of $F(\omega)$, has *the properties of a probability density function*, and $\rho(\tau)$, being the Fourier transform of $f(\omega)$, has *the properties of a characteristic function* (except, of course, for the fact that a characteristic function of a real valued random variable is not, in general, real valued.)

Corresponding to (4.8.11), (4.8.12), (4.8.13), we have for real valued processes,

$$f(\omega) = \frac{1}{2\pi} \int_{-\infty}^{\infty} \cos \omega\tau\rho(\tau)\, d\tau \qquad (4.8.20)$$

$$\rho(\tau) = \int_{-\infty}^{\infty} \cos \omega\tau f(\omega)\, d\omega, \qquad (4.8.21)$$

$$F(\omega) = \frac{1}{2\pi} \int_{-\infty}^{\infty} \frac{\sin \tau\omega}{\tau}\rho(\tau)\, d\tau + \tfrac{1}{2}. \qquad (4.8.22)$$

Corresponding to the functions $h_+(\omega)$, $H_+(\omega)$, we may similarly define normalized spectral density functions and integrated spectra for positive frequencies only by writing

$$f_+(\omega) = \begin{cases} f(0), & \omega = 0 \\ 2f(\omega), & \omega \geqslant 0 \end{cases}$$

and

$$F_+(\omega) = \int_0^{\omega} f_+(\theta)\, d\theta = 2F(\omega) - 1, \qquad \omega \geqslant 0.$$

4.8.2 The Wiener–Khintchine Theorem

We have previously noted that there are certain types of stationary processes for which the function $h(\omega)$ does not exist (i.e. does not have a finite value for all ω) and in such cases it is clear that the function $f(\omega)$ will not exist either. However, we encountered precisely the same situation in the context of probability distributions when we saw that although the probability distribution function always exists, the probability density function need not always exist—in particular, the probability density function does not exist if the random variable is discrete valued, in which case the distribution function is a step function and is not differentiable everywhere (see Section 2.7). In view of the equivalence between the properties of integrated spectra and probability distribution functions we may expect that the *function $F(\omega)$ will always exist for all stationary processes*, but that $f(\omega)$ will exist only when $F(\omega)$ is absolutely continuous, or equivalently, when $\rho(\tau)$ satisfies the condition analogous to (4.8.9), namely that

$$\int_{-\infty}^{\infty} |\rho(\tau)| \, d\tau < \infty.$$

But, even if $\rho(\tau)$ does not die away fast enough for the above condition to hold the fundamental relationship between the properties of $\rho(\tau)$ and the "spectral properties" of the process still remains, although, of course, it can no longer be expressed in the form (4.8.17) when $f(\omega)$ does not exist. However, this difficulty is removed if we re-write (4.8.17) using the more general Stieltjes form of integral, namely,

$$\rho(\tau) = \int_{-\infty}^{\infty} e^{i\omega\tau} \, dF(\omega). \qquad (4.8.23)$$

The above integral known as a "Fourier–Stieltjes" transform, is now well defined even when $F(\omega)$ is not differentiable everywhere—e.g., if $F(\omega)$ has the form of a step-function (see Section 3.6.3.), but if $f(\omega) = F'(\omega)$ exists for all ω then $dF(\omega) = f(\omega) \, d\omega$, and (4.8.23) clearly reduces to the original form (4.8.17).

Equation (4.8.23) is of fundamental importance in the theory of stationary processes, and was first proved in a rigorous form by N. Wiener (1930) and later by A. Khintchine (1934). The celebrated result known as the "Wiener–Khintchine theorem" proves that, under general conditions, the function $F(\omega)$ always exists, and that $\rho(\tau)$ is related to $F(\omega)$ by (4.8.23). A formal statement of this result is as follows.

The Wiener–Khintchine Theorem *A necessary and sufficient condition for $\rho(\tau)$ to be the autocorrelation function of some stochastically continuous stationary process, $\{X(t)\}$, is that there exists a function, $F(\omega)$, having the properties of a distribution function on $(-\infty, \infty)$, (i.e. $F(-\infty) = 0$, $F(+\infty) = 1$, and $F(\omega)$ non-decreasing), such that, for all $\tau, \rho(\tau)$ may be expressed in the form,*

$$\rho(\tau) = \int_{-\infty}^{\infty} e^{i\omega\tau} \, dF(\omega).$$

Proof. The simplest way of establishing the "necessary" part of this result is by appealing to the "positive semi-definite" property of $\rho(\tau)$ derived in section (3.3.2). We may then make use of a general theorem due to Bochner (1936) that any positive semi-definite function which is continuous everywhere must have a Fourier–Stieltjes transform of the form (4.8.23), with

$F(\omega)$ having the properties of a distribution function. The continuity of $\rho(\tau)$ follows directly from the assumption that $X(t)$ is stochastically continuous, see Section 3.6.1. The Wiener–Khintchine theorem may thus be regarded simply as a special case of Bochner's theorem, but since we have already proved what is essentially a restricted form of this result in Theorem (4.8.1) it is of interest to see how our previous approach can be "rigorized" to cover the more general case. The following proof is due originally to Loève (1945), and we now follow the account of this given by Bartlett (1955).

Using (4.8.6) together with the subsequent evaluation of $E[\tilde{R}(\tau)]$ we may write,

$$E\left[\frac{|G_T(\omega)|^2}{2T\sigma_X^2}\right] = \frac{1}{2\pi}\int_{-2T}^{2T}\left(1-\frac{|\tau|}{2T}\right)e^{-i\omega\tau}\rho(\tau)\,d\tau$$

$$= \frac{1}{2\pi}\int_{-\infty}^{\infty}\phi_T(\tau)\,e^{-i\omega\tau}\,d\tau, \qquad (4.8.24)$$

say, where

$$\phi_T(\tau) = \begin{cases} [1-(|\tau|/2T)]\rho(\tau), & |\tau| \leqslant 2T \\ 0, & |\tau| > 2T. \end{cases}$$

Since our aim is to study the behaviour of $\rho(\tau)$ we would like to invert (4.8.24) so as to express $\phi_T(\tau)$ as the inverse transform of the left-hand side, but we cannot do this in a straightforward way since we have not, as yet, established that the left-hand side is integrable. We can, however, perform this inversion by a suitable limiting operation, as follows.

Write $f_T(\omega) = E[|G_T(\omega)|^2/2T\sigma_X^2]$, multiply both sides of (4.8.24) by $e^{i\omega t}[1-(|\omega|/\Omega_0)]$, and integrate from $\omega = -\Omega_0$ to $\omega = \Omega_0$. We then obtain,

$$\int_{-\Omega_0}^{\Omega_0} [1-(|\omega|/\Omega_0)]f_T(\omega)\,e^{i\omega t}\,d\omega$$

$$= \frac{1}{2\pi}\int_{-\infty}^{\infty}\phi_T(\tau)\left\{\int_{-\Omega_0}^{\Omega_0}[1-(|\omega|/\Omega_0)]e^{-i\omega(\tau-t)}\,d\omega\right\}d\tau$$

$$= \frac{1}{2\pi}\int_{-\infty}^{\infty}\phi_T(\tau)\left[\frac{\sin\{\tfrac{1}{2}\Omega_0(\tau-t)\}}{\tfrac{1}{2}\Omega_0(\tau-t)}\right]^2\Omega_0\,d\tau. \qquad (4.8.25)$$

Now by its definition $f_T(\omega)$ is clearly non-negative for all ω. Hence $\{[1-(|\omega|/\Omega_0)]f_T(\omega)\}$ is non-negative for all ω s.t. $|\omega| \leqslant \Omega_0$, and thus, if suitably scaled, may be regarded as a probability density function. Consequently, the left-hand side of (4.8.25) has the mathematical properties of a "scaled" characteristic function (see Section 2.10.3). Now as $\Omega_0 \to \infty$ the function

$$\frac{\Omega_0}{2\pi}\left[\frac{\sin\{\frac{1}{2}\Omega_0(\tau-t)\}}{\frac{1}{2}\Omega_0(\tau-t)}\right]^2$$

behaves like $\delta(\tau-t)$ (a Dirac δ-function), and hence as $\Omega_0 \to \infty$ the right-hand side of (4.8.25) converges uniformly to $\phi_T(t)$, provided $\phi_T(t)$ is continuous, which follows from the continuity of $\rho(\tau)$, which in turn follows from the assumed stochastic continuity of $\{X(t)\}$. Hence, by the limit theorem for sequences of characteristic functions (see Parzen (1960), p. 425) it follows that $\phi_T(t)$ is proportional to a characteristic function, and since moreover $\phi_T(0) = \rho(0) = 1$, $\phi_T(t)$ is *exactly* a characteristic function. Also, as $T \to \infty$, $\phi_T(t)$ converges uniformly to $\rho(t)$ (each t), and therefore by the same limit theorem it follows that $\rho(t)$ is itself a characteristic function. Hence, there must exist some distribution function, $F(\omega)$, which is such that $\rho(\tau)$ is its Fourier–Stieltjes transform, and the required result follows.

To prove the "sufficiency" part of the theorem we must show that given functions $\rho(\tau), F(\omega)$, satisfying (4.8.23) we can always construct a stationary process which has $\rho(\tau)$ as its autocorrelation function. Such a process is easily constructed as follows. Let U be a random variable having $F(\omega)$ as its distribution function, and let Φ be another random variable having a uniform distribution over $(-\pi, \pi)$ and distributed independently of U. Consider the (complex valued) process,

$$X(t) = \exp[i(\Phi + Ut)]. \tag{4.8.26}$$

Since Ω and Φ are independent,

$$E[X(t)] = E[\exp(i\Phi)]E[\exp(iUt)]$$
$$= 0, \quad \text{all } t,$$

since

$$E[\exp(i\Phi)] = \frac{1}{2\pi}\int_{-\pi}^{\pi}(\cos\phi + i\sin\phi)\,d\phi = 0.$$

Hence, the variance of $X(t)$ is given by,

$$\sigma_X^2 = E[|X(t)|^2] = 1, \quad \text{all } t,$$

and the autocorrelation function of $X(t)$ is thus,

$$\rho_X(\tau) = E[X^*(t)X(t+\tau)] = E[\exp(-i(\Phi + Ut))\exp(i(\Phi + Ut + U\tau))]$$

$$= E[\exp(iU\tau)] = \int_{-\infty}^{\infty} e^{i\omega\tau}\,dF(\omega) = \rho(\tau), \quad \text{all } \tau.$$

Thus, the process defined by (4.8.26) has the given function, $\rho(\tau)$, as its autocorrelation function, and the proof is complete.

Corresponding to (4.8.23) there is a similar expression for the auto-covariance function, $R(\tau)$, which (subject to the same conditions on $\{X(t)\}$) takes the form

$$R(\tau) = \int_{-\infty}^{\infty} e^{i\omega\tau} \, dH(\omega). \tag{4.8.27}$$

Equations (4.8.23), (4.8.27), are known as the "*spectral representations of the autocorrelation and autocovariance functions*", respectively.

4.8.3 Discrete Parameter Processes

So far we have considered only continuous parameter processes, i.e. processes which are observed continuously over time. We now turn our attention to discrete parameter stationary processes, i.e. processes $\{X_t\}$, which are observed only at a discrete set of time points, say at $t = 0, \pm 1, \pm 2, \ldots$. For such processes there exists a parallel spectral theory, and all the quantities previously introduced, namely the normalized and non-normalized power spectral density functions and integrated spectra, have corresponding definitions and essentially the same physical interpretation as in the continuous parameter case. In particular, we still have the fundamental relationship between the autocorrelation function and the spectral density function, but the following two points should be noted.

(a) When t is restricted to integer values the autocovariance function $R(r)$ and autocorrelation function $\rho(r)$ are similarly defined only for integer values of r; consequently, *integrals such as* (4.8.8), (4.8.16), *have to be replaced by discrete sums*.

(b) When t is restricted to integer values *all spectral functions are defined only for frequencies in the range*, $-\pi \le \omega \le \pi$. The reason for this will emerge in the following discussion.

The analogue of the Wiener–Khintchine theorem, known as Wold's theorem (Wold 1938), is as follows.

Wold's Theorem A *necessary and sufficient condition for the sequence* $\{\rho(r); r = 0, \pm 1, \pm 2, \ldots\}$ *to be the autocorrelation function for some discrete parameter stationary process, $\{X_t; t = 0, \pm 1, \pm 2, \ldots\}$, is that there exists a function $F(\omega)$, having the properties of a distribution function on the interval $(-\pi, \pi)$, (i.e. $F(-\pi) = 0$, $F(\pi) = 1$, and $F(\omega)$ is non-decreasing), such that*

$$\rho(r) = \int_{-\pi}^{\pi} e^{i\omega r} \, dF(\omega), \qquad r = 0, \pm 1, \pm 2, \ldots, \tag{4.8.28}$$

Proof. We start by converting the discrete sequence $\{\rho(r)\}$ into a continuous function by plotting $\rho(r)$ (as in Fig. (3.6)) and joining consecutive ordinates together by straight lines. We then have a function, $\rho^c(\tau)$ say, given for all τ, as illustrated in Fig. 4.9.

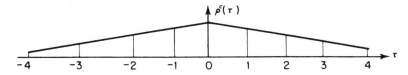

Fig. 4.9. The function $\rho^c(\tau)$.

Specifically,

$$\rho^c(\tau) = (s+1-\tau)\rho(s)+(\tau-s)\rho(s+1),$$

where s is such that, $s \leqslant \tau < s+1$, and

$$\rho^c(-\tau) = \rho^c(\tau), \qquad \text{all } \tau.$$

The function $\rho^c(\tau)$ so constructed is clearly continuous everywhere, and moreover it may be shown (Priestley (1963)) that there exists a continuous parameter stationary process which has $\rho^c(\tau)$ as its autocorrelation function. Hence, by the Wiener–Khintchine theorem there must exist a non-decreasing function, $Q(\omega)$ say, such that $\rho^c(\tau)$ may be written

$$\rho^c(\tau) = \int_{-\infty}^{\infty} e^{i\omega\tau}\, dQ(\omega).$$

Since the above holds for all τ, it must hold, in particular when τ takes integer values r, in which case $\rho^c(r)$ and $\rho(r)$ are identical. Hence, we may write

$$\rho(r) = \int_{-\infty}^{\infty} e^{i\omega r}\, dQ(\omega), \qquad r=0,\ \pm 1,\ \pm 2,\ \dots$$

If we now split up the range of integration into non-overlapping intervals, each of length 2π, then we may write,

$$\rho(r) = \sum_{s=-\infty}^{\infty} \int_{(2s-1)\pi}^{(2s+1)\pi} e^{i\omega r}\, dQ(\omega)$$

$$= \sum_{s=-\infty}^{\infty} \int_{-\pi}^{\pi} \exp[i(\omega+2s\pi)r]\, dQ(\omega+2s\pi).$$

But since both s and r take integer values only,

$$\exp(i2s\pi r) = 1, \qquad \text{all } s, r.$$

Hence, we have,

$$\rho(r) = \sum_{s=-\infty}^{\infty} \int_{-\pi}^{\pi} \exp(i\omega r) \, dQ(\omega + 2s\pi) = \int_{-\pi}^{\pi} \exp(i\omega r) \left\{ \sum_{s=-\infty}^{\infty} dQ(\omega + 2s\pi) \right\}.$$

Now write $dF(\omega) = \sum_{s=-\infty}^{\infty} dQ(\omega + 2s\pi)$, (with $F(-\pi)$ defined as zero), and (4.8.28) follows. It is easily verified that $F(\pi) = 1$ by noting that,

$$F(\pi) = \int_{-\pi}^{\pi} \left\{ \sum_{s=-\infty}^{\infty} dQ(\omega + 2s\pi) \right\} = \int_{-\infty}^{\infty} dQ(\omega) = Q(\infty) = 1.$$

The essential point which emerges from the above discussion is the fact that when t is restricted to integer values *we cannot distinguish in any way between the functions* $\exp(i\omega t)$ *and* $\exp[i(\omega + 2s\pi)t]$, s *any integer.* Thus, if the discrete parameter process $\{X_t\}$ is obtained by observing a continuous parameter process $\{X(t)\}$ at $t = 0, \pm 1, \pm 2, \ldots$, the components in $X(t)$ with frequencies, $\omega - 2\pi$, $\omega + 2\pi$, $\omega - 4\pi$, $\omega + 4\pi$, $\omega - 6\pi$, $\omega + 6\pi, \ldots$ will all appear to have frequency ω. This feature, whereby discrete sampling changes the apparent frequency, is illustrated in Fig. 4.10, and is usually referred to as the *"aliasing effect"*.

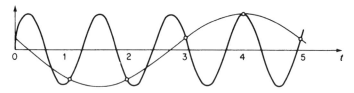

Fig. 4.10. The aliasing effect.

The frequency ω is then said to be the "alias" of each of the frequencies, $\omega \pm 2\pi$, $\omega \pm 4\pi$, $\omega \pm 6\pi, \ldots$ etc. Since every frequency outside the range $(-\pi, \pi)$ has an "alias" inside this range there is clearly no point in defining any spectral quantity outside the range $(-\pi, \pi)$. (We discuss a more general form of the aliasing effect in Chapter 7 when we consider the problem of estimating spectral density functions from experimental data.)

When $\{X_t\}$ is such that $F(\omega)$ is differentiable everywhere the (normalized) power spectral density function, $f(\omega) = dF(\omega)/d\omega$, exists for all ω, and (4.8.28) may then be written as

$$\rho(r) = \int_{-\pi}^{\pi} e^{i\omega r} f(\omega) \, d\omega, \qquad r = 0, \pm 1, \pm 2, \ldots \qquad (4.8.29)$$

Comparing (4.8.29) with (4.5.3), (4.5.5), we see that $\{\rho(r)/2\pi\}$ is the coefficient of $\exp(-i\omega r)$ in a (complex) *Fourier series* expansion of $f(\omega)$. Hence, when we invert (4.8.29) we obtain,

$$f(\omega) = \frac{1}{2\pi} \sum_{r=-\infty}^{\infty} \rho(r) e^{-i\omega r}, \qquad -\pi \le \omega \le \pi. \tag{4.8.30}$$

(The above expression for $f(\omega)$ is essentially the same as in the continuous parameter case (cf. (4.8.16)), the main differences being that, as noted above, the integral in (4.8.16) has been replaced by a *discrete sum* and the frequency range is now $-\pi$ to π.) As before, when $\{X_t\}$ is real valued $\{\rho(r)\}$ is an even sequence, and (4.8.30) reduces to a cosine transform. Thus, for real valued processes we have,

$$f(\omega) = \frac{1}{2\pi} \sum_{r=-\infty}^{\infty} \rho(r) \cos r\omega, \qquad -\pi \le \omega \le \pi, \tag{4.8.31}$$

$$= \frac{1}{2\pi} + \frac{1}{\pi} \sum_{r=1}^{\infty} \rho(r) \cos r\omega. \tag{4.8.32}$$

Integrating (4.8.32) w.r. to ω term by term we find,

$$F(\omega) = \int_{-\pi}^{\omega} f(\theta)\, d\theta = \frac{(\omega + \pi)}{2\pi} + \frac{1}{\pi} \sum_{r=1}^{\infty} \rho(r) \frac{\sin r\omega}{r}, \qquad -\pi \le \omega \le \pi. \tag{4.8.33}$$

Multiplying both sides of (4.8.29) by σ_X^2 we obtain a similar representation for the autocovariance function, $R(r)$, namely,

$$R(r) = \int_{-\pi}^{\pi} e^{i\omega r}\, dH(\omega), \qquad r = 0, \pm 1, \pm 2, \ldots \tag{4.8.34}$$

where $H(\omega) = \sigma_X^2 F(\omega)$ is the non-normalized integrated spectrum. When the non-normalized spectral density function, $h(\omega) = dH(\omega)/d\omega$, exists for all ω we have, analogously to (4.8.29)–(4.8.33),

$$R(r) = \int_{-\pi}^{\pi} e^{i\omega r} h(\omega)\, d\omega, \qquad r = 0, \pm 1, \pm 2, \ldots \tag{4.8.34}$$

$$h(\omega) = \frac{1}{2\pi} \sum_{r=-\infty}^{\infty} R(r) e^{-i\omega r}, \qquad -\pi \le \omega \le \pi. \tag{4.8.35}$$

For real valued processes,

$$h(\omega) = \frac{1}{2\pi} \sum_{r=-\infty}^{\infty} R(r) \cos r\omega, \qquad -\pi \le \omega \le \pi, \tag{4.8.36}$$

$$= \frac{\sigma_X^2}{2\pi} + \frac{1}{\pi} \sum_{r=1}^{\infty} R(r) \cos r\omega, \tag{4.8.37}$$

and

$$H(\omega) = \sigma_X^2 \left(\frac{\omega + \pi}{2\pi} \right) + \frac{1}{\pi} \sum_{r=1}^{\infty} R(r) \frac{\sin r\omega}{r}, \qquad -\pi \leq \omega \leq \pi. \quad (4.8.38)$$

4.9 DECOMPOSITION OF THE INTEGRATED SPECTRUM

We saw in Section 4.8.1 that the normalized integrated spectrum $F(\omega)$ has all the mathematical properties of a probability distribution function, and consequently we may immediately apply any general theorems on probability distribution functions to the function $F(\omega)$. Now one of the basic theorems on probability distribution functions is the "Lebesgue decomposition theorem" (discussed in Section 2.6.1) which tells us the most general form which a probability distribution function can take namely, that it can be expressed as the sum of three functions, one being absolutely continuous, another being a step-function, and the third being the so-called "singular" function. We saw further (Section 2.7) that in practice we could ignore the "singular" term, and that in most cases only one of the remaining two terms would be present—the absolutely continuous function if, typically, the underlying random variable took a continuous range of values, and the step-function if the random variable was discrete valued. It now follows that there must be a corresponding decomposition of the integrated spectrum $F(\omega)$ (for both continuous parameter and discrete parameter processes) but the more interesting problem is to try to identify the types of stochastic processes which correspond to the individual terms in the decomposition. The spectral decomposition theorem is virtually identical to Theorem 2.6.1, and takes the following form.

Theorem 4.9.1 *Any normalized integrated spectrum, $F(\omega)$, may be written in the form,*

$$F(\omega) = a_1 F_1(\omega) + a_2 F_2(\omega) + a_3 F_3(\omega), \qquad (4.9.1)$$

where

(1) $a_i \geq 0$, $i = 1, 2, 3$,

$a_1 + a_2 + a_3 = 1$,

and

(2) $F_1(\omega)$, $F_2(\omega)$, $F_3(\omega)$, are distribution functions of the following types;
  (a) $F_1(\omega)$ is absolutely continuous with derivative $F_1'(\omega)$ which exists for almost all ω, and the density function $f_1(\omega)$, which is such that $F_1(\omega) = \int_{-\infty}^{\omega} f_1(u)\, du$, exists for all ω.

(b) $F_2(\omega)$ is a step function with steps $\{p_r\}$ at points $\{\omega_r\}$, say, $r = 1, 2, 3, \ldots$, and $\sum_r p_r = 1$.

(c) $F_3(\omega)$ is a "singular" function with zero derivative almost everywhere.

As for probability distributions the "singular" function, $F_3(\omega)$, is highly pathological, (at least for one dimensional processes) and in practice we ignore the possible existence of $F_3(\omega)$. We may then write

$$F(\omega) = a_1 F_1(\omega) + a_2 F_2(\omega), \qquad (4.9.2)$$

where $a_1 \geqslant 0$, $a_2 \geqslant 0$, $a_1 + a_2 = 1$ (cf. equation (2.6.4)). Now let $\rho(\tau)$ be the autocorrelation function of the process whose normalized integrated spectrum is $F(\omega)$. Then by the Wiener–Khintchine theorem we may write (see (4.8.23)),

$$\rho(\tau) = \int e^{i\tau\omega} \, dF(\omega), \qquad (4.9.3)$$

the range of integration being either $(-\infty, \infty)$ (in the continuous parameter case), or $(-\pi, \pi)$ (in the discrete parameter case). Substituting (4.9.2) in (4.9.3) we find that we may write $\rho(\tau)$ in the form,

$$\rho(\tau) = a_1 \rho_1(\tau) + a_2 \rho_2(\tau), \qquad (4.9.4)$$

where

$$\rho_1(\tau) = \int e^{i\tau\omega} \, dF_1(\omega) = \int e^{i\tau\omega} f_1(\omega) \, d\omega, \qquad (4.9.5)$$

and

$$\rho_2(\tau) = \int e^{i\tau\omega} \, dF_2(\omega). \qquad (4.9.6)$$

But since $F_2(\omega)$ is a step function, $dF_2(\omega) = 0$ except at the "jump" points $\{\omega_r\}$, and $dF(\omega_r) = p_r$, $r = 1, 2, 3, \ldots$ Hence, (4.9.6) reduces to,

$$\rho_2(\tau) = \sum_{r=1}^{\infty} e^{i\tau\omega_r} \cdot p_r, \qquad (4.9.7)$$

which may then be written alternatively in the form,

$$\rho_2(\tau) = \sum_{r=1}^{\infty} (A_r \cos \tau\omega_r + B_r \sin \tau\omega_r),$$

where $A_r = p_r$, $B_r = ip_r$.

Thus, *the autocorrelation function may also be decomposed into two terms,* the first term, $\rho_1(\tau)$, being the Fourier transform of the continuous function, $f_1(\omega)$, and the second term, $\rho_2(\tau)$, being the sum of sine and cosine terms. Although $\rho_2(\tau)$ is not necessarily a periodic function (unless the frequencies $\{\omega_r\}$ are commensurable), it shares many of the properties of periodic functions (it is sometimes called an "almost periodic function"), and its main feature is that *it never dies away to zero no matter how large τ becomes*. On the other hand, $\rho_1(\tau)$, being the Fourier transform of a continuous integrable function, *certainly decays to zero as* $\tau \to \pm\infty$. In fact, if $f_1(\omega)$ is a function of "bounded variation", it follows from the Riemann–Lebesgue lemma (see Titchmarsh (1939), p. 403) that $\rho_1(\tau) = O(1/\tau)$ for large τ, and if $f_1'(\omega)$ is of bounded variation then $\rho(\tau) = O(1/\tau^2)$. We call $F_1(\omega)$ (or $f_1(\omega)$) the *continuous component*, and $F_2(\omega)$ the *discrete component*, of the spectrum. The typical general form of $F(\omega)$ is illustrated in Fig. 4.11.

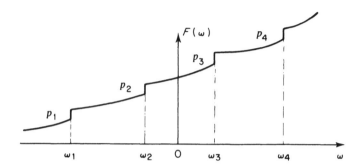

Fig. 4.11. The general form of the integrated spectrum.

This general form of integrated spectrum is exactly the same as the general form of probability distribution functions shown in Fig. 2.8 (Section 2.7). However, in the case of probability distributions it will be recalled that in the vast majority of practical problems the distribution function contained just *one* component, being either purely continuous or purely discrete according as to whether the underlying random variable was continuous or discrete valued. We shall see that a similar situation holds for spectra; in the main the integrated spectrum contains just *one* of the functions, $F_1(\omega)$, $F_2(\omega)$, although cases do arise in practice where both functions are present, rather more frequently than with probability distributions.

We may invert (4.9.5) immediately giving,

$$f_1(\omega) = \frac{1}{2\pi} \int_{-\infty}^{\infty} e^{-i\tau\omega} \rho_1(\tau)\, d\tau.$$

However, the inversion of (4.9.7) to yield an expression for the $\{p_r\}$ in terms of $\rho_2(\tau)$ is not so straightforward. At first sight the right-hand side of (4.9.7) looks like a complex Fourier series (with the p_r as coefficients) but this is not quite so since, in particular, it will not represent a periodic function unless the $\{\omega_r\}$ are commensurable. Nevertheless, we may obtain an inversion formula similar to that for Fourier series (cf. (4.5.5)) as follows. Multiply both sides of (4.9.6) by $[\exp(-i\tau\omega_s)/2T]$ and integrate from $-T$ to T. Then,

$$\frac{1}{2T}\int_{-T}^{T}p_2(\tau)\exp(-i\tau\omega_s)\,d\tau = \sum_r p_r\left\{\frac{1}{2T}\int_{-T}^{T}\exp(-i\tau(\omega_s-\omega_r))\,d\tau\right\}. \quad (4.9.8)$$

But,

$$\frac{1}{2T}\int_{-T}^{T}\exp[-i\tau(\omega_s-\omega_r)]\,d\tau = \begin{cases} \dfrac{\sin[T(\omega_s-\omega_r)]}{T(\omega_s-\omega_r)}, & \omega_r \neq \omega_s \\ 1, & \omega_r = \omega_s. \end{cases}$$

Hence, as $T \to \infty$,

$$\frac{1}{2T}\int_{-T}^{T}\exp[-i\tau(\omega_s-\omega_r)]\,d\tau \to \begin{cases} 1, & \omega_r = \omega_s \\ 0, & \omega_r \neq \omega_s, \end{cases}$$

and thus, if we let $T \to \infty$ in (4.9.8) we find that all the terms on the right-hand side vanish, except that corresponding to $r = s$. We finally obtain,

$$p_s = \lim_{T\to\infty}\left\{\frac{1}{2T}\int_{-T}^{T}p_2(\tau)\,e^{-i\tau\omega_s}\,d\tau\right\}, \qquad s = 1, 2, 3, \ldots \quad (4.9.9),$$

which enables us to compute the magnitude of the "jump" in the discrete spectrum at the point $\omega = \omega_s$. (Equation (4.9.9) is, of course, identical to that used to invert the characteristic function of a discrete probability distribution).

We now consider in more detail the following special cases of (4.9.2).

CASE 1: $a_1 = 1, a_2 = 0$

Here, $F(\omega) \equiv F_1(\omega)$, i.e. the integrated spectrum consists entirely of the absolutely continuous component, and the process $\{X(t)\}$ is then said to have *a purely continuous spectrum*. This corresponds to the case in probability theory where the distribution function is purely continuous as is typified, e.g., by the normal, rectangular, exponential, and chi-squared distributions. Just as most of the "standard" probability distributions are of the continuous type it is equally true that most of the standard models for stationary processes belong to the category of "purely continuous spectra". In fact, with one exception, *all the models discussed in Chapter 3 belong to this category*. Specifically, the following models all have (in general) purely continuous spectra;

Purely random processes
Autoregressive processes (of any order)
moving average processes (of any order)
Mixed ARMA processes (of any order)
General linear processes.

Since in this case $F(\omega) \equiv F_1(\omega)$, it follows that *the spectral density function,* $f(\omega) \equiv f_1(\omega)$ *exists for all* ω. Similarly, $\rho(\tau) \equiv \rho_1(\tau)$, and hence the *autocorrelation function of the process decays to zero as* $\tau \to \pm\infty$, following the behaviour of $\rho_1(\tau)$.

CASE 2: $a_1 = 0$, $a_2 = 1$

Here, $F(\omega) \equiv F_2(\omega)$, i.e. the integrated spectrum consists entirely of the discrete component, and the process is then said to have a *"purely discrete spectrum"*. In the engineering literature this is sometimes called a *"line spectrum"*. (In the context of probability distributions this case corresponds to a (purely) discrete distribution such as the binomial, Poisson, or geometric distributions.) The main example of a process with a purely discrete spectrum is the "harmonic process" model (see Section 3.5.8), namely,

$$X(t) = \sum_{i=1}^{K} A_i \cos(\omega_i t + \phi_i), \qquad (4.9.10)$$

where $K, \{A_i\}, \{\omega_i\}, i = 1, \ldots K$, are constants, and the $\{\phi_i\}$ are independent random variables each rectangularly distributed over $(-\pi, \pi)$. (We shall show in the next section that the spectrum for this model is purely discrete with "jumps" at the points $\{\pm\omega_i\}, i = 1, \ldots K$.) In this case, $\rho(\tau) \equiv \rho_2(\tau)$, and *consequently* $\rho(\tau)$ *never dies out*. In fact, as we have seen in Section 3.5.8, the model (4.9.10) has autocorrelation function,

$$\rho(\tau) = \sum_{i=1}^{K} (\tfrac{1}{2}A_i^2) \cos \omega_i \tau \Big/ \sum_{i=1}^{K} (\tfrac{1}{2}A_i^2), \qquad (4.9.11)$$

which, being a sum of cosine functions, is an *"almost periodic"* function. (Note that (4.9.11) is a special case of the general form given by (4.9.9).)

CASE 3: $a_1 > 0$, $a_2 > 0$

Here, $F(\omega)$ is a mixture of both the continuous and the discrete components, and consequently the process $\{X(t)\}$ is said to have a *"mixed spectrum"*. The autocorrelation function, $\rho(\tau)$, similarly contains a mixture of the two components, $\rho_1(\tau), \rho_2(\tau)$, *one of which* $(\rho_1(\tau))$ *decays to zero as* $\tau \to \pm\infty$, *whie the other* $(\rho_2(\tau))$ *never dies out*. This situation corresponds to the case of a random variable which has a "mixed" probability distribution—such as the waiting time in a queue discussed in Section 2.7. Random variables having mixed probability distributions arise very rarely indeed but processes with mixed spectra, although not particularly common, do arise in practice, and rather more frequently than mixed probability distributions. An example of such a process is easily constructed as follows. Let,

$$X(t) = Y(t) + Z(t), \qquad (4.9.12)$$

where $\{Y(t)\}$ and $\{Z(t)\}$ are uncorrelated processes (i.e. $\text{cov}\{Y(t), Z(t')\} = 0$, all t, t'), $Y(t)$ has a purely continuous spectrum (so that, e.g., $Y(t)$ may be an autoregressive, moving average, or a general linear process), and $Z(t)$ has a purely discrete spectrum (e.g. $Z(t)$ may be a harmonic process). A fairly general model for $X(t)$ would thus be (e.g. in the continuous parameter case)

$$X(t) = \int_{-\infty}^{\infty} g(u)\varepsilon(t-u)\, du + \sum_{i=1}^{K} A_i \cos(\omega_i t + \phi_i), \qquad (4.9.13)$$

where, as previously, $\varepsilon(t)$ denotes the "derivative" of a Wiener process (see Section 3.7.8). Since $Y(t)$, $Z(t)$, are uncorrelated it is trivial to verify that the autocovariance function of $X(t)$ is the sum of the autocovariance functions of $Y(t)$ and $Z(t)$, i.e., with an obvious notation,

$$R_X(\tau) = R_Y(\tau) + R_Z(\tau), \qquad \text{all } \tau, \qquad (4.9.14)$$

so that the autocorrelation function of $X(t)$ is given by,

$$\rho_X(\tau) = \frac{R_X(\tau)}{R_X(0)} = \frac{\sigma_Y^2}{\sigma_Y^2 + \sigma_Z^2}\rho_Y(\tau) + \frac{\sigma_Z^2}{\sigma_Y^2 + \sigma_Z^2}\rho_Z(\tau), \qquad (4.9.15)$$

where $\sigma_Y^2 = \text{var}\{Y(t)\}$, $\sigma_Z^2 = \text{var}\{Z(t)\}$. Hence $\rho_X(\tau)$ is of the general form (4.9.9), and it thus follows (e.g. from (4.8.22)) that the integrated spectrum of $X(t)$, $F_X(\omega)$, is the same linear combination of the integrated spectra of $Y(t)$ and $Z(t)$. Consequently, $X(t)$ has a mixed spectrum. (The above discussion shows that spectra have an "additive" property, i.e. the (non-normalized) spectrum of the sum of two uncorrelated processes is the sum of the individual (non-normalized) spectra. This property does not, of course, hold for probability distributions; the distribution function of the sum of two uncorrelated random variables is *not* the sum of the two distribution functions.) The three different cases have been classified above in terms of the behaviour of the integrated spectrum, $F(\omega)$. However, we may alternatively classify them according to the behaviour of the spectral density function $f(\omega)$, as follows.

In Case 1, $f(\omega) \equiv f_1(\omega)$ exists *everywhere* (for *all* ω).
In Case 2, $F(\omega)$ is a step function with jumps at $\{\omega_r\}$ and hence its derivative, $f(\omega)$, exists for $\omega \neq \omega_r$, $r = 1, 2, \ldots$, but is *identically zero*. At $\omega = \omega_r$ $(r = 1, 2, \ldots)$, $f(\omega)$ is *infinite*.
In Case 3, $f(\omega)$ exists for $\omega \neq \omega_r$, $r = 1, 2, \ldots$, but is *not* identically zero. At $\omega \doteq \omega_r$, $r = 1, 2, \ldots$, $f(\omega)$ is *infinite*.
In both Case 2 and Case 3 the step-function component, $F_2(\omega)$, is present, and this, of course, is the reason why $f(\omega)$ becomes infinite at the points $\{\omega_r\}$. However, it is convenient to introduce the *formal* derivative of $F_2(\omega)$ and write $f_2(\omega) = "dF_2(\omega)/d\omega"$. As noted above, $f_2(\omega)$ is zero everywhere

except at the jump points $\{\omega_r\}$ where it is infinite. It may thus be represented by a *sum of δ-functions*, i.e. we may write formally,

$$f_2(\omega) = \sum_{r=1}^{\infty} p_r \delta(\omega - \omega_r), \tag{4.9.16}$$

where p_r is the magnitude of the jump in $F_2(\omega)$ at the point $\omega = \omega_r$. We may then write the "complete" spectral density function for Case 3 in the form,

$$f(\omega) = a_1 f_1(\omega) + a_2 \sum_{r=1}^{\infty} p_r \delta(\omega - \omega_r). \tag{4.9.17}$$

Thus, for the mixed spectrum case we may think of $f(\omega)$ as consisting of the continuous function $f_1(\omega)$ superimposed on which is a series of δ-function "spikes" at the points $\{\omega_r\}$. We cannot, of course, illustrate the δ-function graphically in a precise way, but a rough idea of the general form of $f(\omega)$ is given in Fig. 4.12.

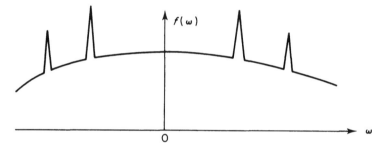

Fig. 4.12. Spectral density function for Case 3.

From the physical point of view it is clear that the presence of the harmonic process terms in (4.9.10), (4.9.13), must inevitably mean that in each case the spectral density function will contain infinite "peaks" at the points $\{\omega_r\}$. The reason for this becomes obvious when one realizes that the harmonic process model (4.9.10) is, *in itself, a spectral decomposition of the process*, i.e. (4.9.10) tells us immediately that $X(t)$ consists purely of a sum of sine and cosine functions at a *discrete set of frequencies*. Thus, there is a finite amount of power concentrated at each of three frequencies, and consequently, when we try to express the power distribution in terms of the *density* of power over *all* frequencies we naturally find that the density is *infinite* at the discrete set of frequencies which occur in the model. (This situation is exactly comparable to that of a discrete probability distribution in which there is a finite amount of probability concentrated at each of a discrete set of points. If we attempted to describe a discrete probability distribution in terms of a probability density function we would obtain, as above, a density function which consisted of a sum of δ-functions.)

The form of the spectral density function illustrated in Fig. 4.12 highlights a very important point in the physical interpretation of power spectra, namely, that if the process contains a *strictly periodic component* this will reveal itself as an *infinite "peak"* (i.e. a δ function) in the spectral density function at the appropriate frequency. If, however, the spectral density function has a *"sharp peak"* of large, but finite, magnitude, then in general this would indicate that the process contains a *"nearly periodic"* component at the appropriate frequency. This feature will be discussed in more detail in relation to the form of the spectral density function of the AR(2) model which is derived in the next section.

4.10 EXAMPLES OF SPECTRA FOR SOME SIMPLE MODELS

In this section we evaluate the spectral density functions for some of the standard models discussed in Chapter 3. We adopt the approach of assuming that the spectral density function exists, and then evaluate it directly from the autocorrelation function using (for real valued processes) the basic relationships (4.8.20) or (4.8.31). If it should turn out that a particular model has a purely discrete spectrum (so that the spectral density function does not then exist at all frequencies) this will readily become apparent in the course of attempting to evaluate the Fourier transform of the autocorrelation function. (As we have seen, it will lead to the emergence of δ-function terms). We now consider the following models for real valued processes.

I. *Purely random process; discrete parameter case*

Here, the process consists simply of a sequence of uncorrelated random variables and the autocorrelation function is given by (3.5.4), namely,

$$\rho(r) = \begin{cases} 1, & r = 0 \\ 0, & r = \pm 1, \pm 2, \ldots \end{cases} \qquad (4.10.1)$$

From (4.8.31) the (normalized) spectral density function is given by

$$f(\omega) = \frac{1}{2\pi} \sum_{r=-\infty}^{\infty} \rho(r) \cos r\omega, \qquad -\pi \leq \omega \leq \pi,$$

which, on substituting the above form of $\rho(r)$ reduces to

$$f(\omega) = 1/2\pi, \qquad \text{all } \omega \text{ in the range } (-\pi, \pi). \qquad (4.10.2)$$

The function $f(\omega)$ clearly exists for all ω in the required range and hence *the process has a purely continuous spectrum*. The fact that $f(\omega)$ takes a constant value over $(-\pi, \pi)$ means that the *total power is uniformly distributed over all frequencies*, i.e. each frequency component contributes exactly the same amount to the total power. This property is highly analogous to that of white light where, again, all frequencies (i.e. "colours") are present in equal amounts, and it is for this reason that the purely random process is often referred to as "*white noise*". The form of $f(\omega)$ is shown in Fig. 4.13, and, in probabilistic terms, corresponds to a uniform (or rectangular) probability distribution over the interval $(-\pi, \pi)$.

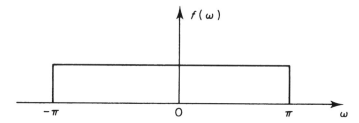

Fig. 4.13. Spectral density function of a purely random process.

II. *Purely random process; continuous parameter case*

The autocorrelation function of the process $\{\varepsilon(t)\}$ is essentially the same as above, i.e.

$$\rho(\tau) = \begin{cases} 1, & \tau = 0, \\ 0, & \tau \neq 0. \end{cases} \tag{4.10.3}$$

From (4.8.20) the (normalized) spectral density function is given by,

$$f(\omega) = \frac{1}{2\pi} \int_{-\infty}^{\infty} \rho(\tau) \cos \omega\tau \, d\tau$$

$$= 0, \qquad \text{all } \omega,$$

when $\rho(\tau)$ has the above form. Hence, $f(\omega)$ exists only in the sense *that it is identically zero at all frequencies*. (Note that $\rho(\tau)$ is discontinuous at the origin and hence the process $\{\varepsilon(t)\}$ is not stochastically continuous and therefore not covered by the general theory of the Wiener–Khintchine theorem.) However, we know from (3.7.18) that the autocovariance function of $\varepsilon(t)$ is given by

$$R(\tau) = \sigma_W^2 \delta(\tau),$$

where $\delta(\tau)$ is a Dirac δ-function and σ_W^2 is the rate of increase of variance of the Wiener process for which $\varepsilon(t)$ is the "derivative". Hence, the *non-normalized* spectral density function, $h(\omega)$, is given by (see (4.8.11)),

$$h(\omega) = \frac{\sigma_W^2}{2\pi} \int_{-\infty}^{\infty} \delta(\tau) \cos \omega\tau \, d\tau,$$

or,

$$h(\omega) = \frac{\sigma_W^2}{2\pi}, \qquad \text{all } \omega. \tag{4.10.4}$$

Thus, the process $\{\varepsilon(t)\}$ has a *uniform non-normalized spectrum* over the frequency range $(-\infty, \infty)$, and is correspondingly referred to as "*continuous white noise*".

The situation here may seem somewhat anomalous in that the non-normalized spectral density function is well behaved but the normalized spectral density function is highly degenerate, being identically zero. From a statistical point of view this is simply a consequence of the fact that $\varepsilon(t)$ has infinite variance. From the physical point of view it is clear that if a process has constant power at all frequencies over the *infinite* range $(-\infty, \infty)$ it must have infinite total power, and hence when $h(\omega)$ is normalized by dividing it by the total power the result must be zero. Now one of the basic characteristics of a purely random process is the fact that its spectral density function is uniform over frequency, and this explains why a continuous parameter purely random process must be pathological in that it has to have infinite variance.

The problem is exactly the same as that encountered in Bayesian inference when we try to construct a uniform prior probability distribution for a parameter which has an infinite range of possible values. Such a distribution does not, of course, exist, and in a similar sense the continuous parameter purely random process does not exist. In physical applications one can usually obtain a sufficiently accurate approximation to this process by using what is termed "*band limited white noise*". This is a process whose spectral density function is constant over a large, *but finite*, range, and effectively zero outside this range, i.e. $f(\omega)$ has the approximate form,

$$f(\omega) = \begin{cases} \dfrac{1}{2\omega_0}, & -\omega_0 \leqslant \omega \leqslant \omega_0, \\ 0, & \text{otherwise.} \end{cases} \tag{4.10.5}$$

Physical processes which approximate to white noise

The following are two well known examples of physical processes which give reasonable approximations to continuous parameter white noise.

(A) SHOT NOISE IN A VACUUM TUBE

This process has already been discussed in Chapter 3, Section 3.7.10, and it will be recalled that the noise current $X(t)$ at the anode of a vacuum tube has the autocovariance function given by (3.7.93). The non-normalized spectral density function $h(\omega)$ is given by the Fourier (cosine) transform of (3.7.93), and a straightforward calculation gives (see, e.g., Parzen (1962), p. 115),

$$h(\omega) = \frac{2\nu e^2}{\pi} \cdot \frac{1}{(\omega T)^4}\{(\omega T)^2 + 2(1 - \cos \omega T - \omega T \sin \omega T)\}.$$

If we expand the numerator in powers of (ωT) we find,

$$(\omega T)^2 + 2(1 - \cos \omega T - \omega T \sin \omega T) = \frac{(\omega T)^4}{4} - \frac{5(\omega T)^6}{360} + O(\omega T)^8.$$

Hence, for values of ω such that (ωT) is sufficiently small to neglect terms of order $(\omega T)^2$ and higher, $h(\omega)$ may be regarded as taking the constant value,

$$h(\omega) = \nu e^2/2\pi = eI/2\pi,$$

where $I = \nu e$ denotes the mean value of the current. Assuming that T (the time taken for the electron to travel from the cathode to the anode) is of the order of 10^{-9} seconds, then, as noted by Parzen, we may regard $h(\omega)$ as constant up to values of ω in the region of 10^9 radians per second, i.e. up to frequencies of about 100 MHz. Within this range of frequencies we may thus regard the noise current $X(t)$ as a white noise process.

(B) THERMAL NOISE

The term "thermal noise" is used to describe the small random voltage fluctuations between the ends of a resistor due to the random thermal motion of the conducting electrons inside the resistor. It may be shown that up to fairly large frequencies the non-normalized spectral density function of the voltage is approximately constant, and takes the value,

$$h(\omega) = 4RkT,$$

where k is Boltzmann's constant, T is the absolute temperature, and R is the value of the resistor. (See Doob (1949), and Middleton (1960), for a detailed discussion of the underlying physics of the problem.)

III. $AR(1)$ process; continuous parameter case

The autocorrelation function for this process is given by (3.7.26), namely,

$$\rho(\tau) = \exp(-\alpha|\tau|), \qquad (\alpha > 0). \tag{4.10.6}$$

Hence, by (4.8.20), the normalized spectral density function is,

$$f(\omega) = \frac{1}{2\pi} \int_{-\infty}^{\infty} \exp(-\alpha|\tau|) \cos \omega\tau \, d\tau$$

$$= \frac{1}{\pi} \int_{0}^{\infty} \exp(-\alpha\tau) \cos \omega\tau \, d\tau. \qquad (4.10.7)$$

The simplest way of evaluating this integral is to use the fact that $\cos \omega\tau$ is the real part of $\exp(i\omega\tau)$, so that $f(\omega)$ may be written as,

$$f(\omega) = \frac{1}{\pi} R\left\{ \int_{0}^{\infty} \exp[-\tau(\alpha - i\omega)] \, d\tau \right\},$$

$$= \frac{1}{\pi} R\left\{ \frac{1}{\alpha - i\omega} \right\} = \frac{1}{\pi}\left(\frac{\alpha}{\alpha^2 + \omega^2} \right). \qquad (4.10.8)$$

Hence, $f(\omega)$ exists for all ω and the AR(1) process thus has a purely continuous spectrum. The typical form of $f(\omega)$ is shown in Fig. 4.14 from

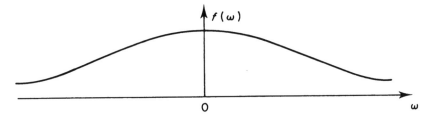

Fig. 4.14. Spectral density function of a continuous parameter AR(1) process.

which it will be seen that for this process most of the power is concentrated at the low frequency end of the spectrum, with the power at high frequencies gradually falling off as the frequency increases.

It is instructive to consider the limiting forms of $\rho(\tau)$ and $f(\omega)$ as $\alpha \to \infty$ and as $\alpha \to 0$. As $\alpha \to \infty$, $\rho(\tau)$ tends to the form of the autocorrelation function of the purely random process (4.10.3), and correspondingly, $f(\omega)$ tends to a "flat" form over $(-\infty, \infty)$, as shown in Figs. 4.15(a) and (b). On

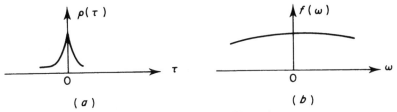

Fig. 4.15. (a) $\alpha \to \infty$; (b) $\alpha \to 0$.

the other hand, as $\alpha \to 0$, $\rho(\tau)$ tends to a "flat" form, and $f(\omega)$ tends to a δ-function form at $\omega = 0$—see Figs. 4.16(a) and (b). These features illustrate an important relationship between $\rho(\tau)$ and $f(\omega)$, namely: *the "narrower" $\rho(\tau)$, the "wider" $f(\omega)$, and vice-versa.* This follows from a basic mathematical result which states, roughly speaking, that one cannot make a function

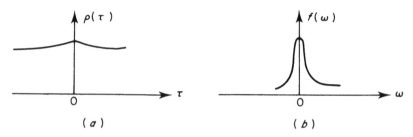

Fig. 4.16. (a) $\alpha \to 0$; (b) $\alpha \to \infty$.

and its Fourier transform both "arbitrarily narrow". We discuss this point in more detail in Chapter 8 in connection with the analysis of mixed spectra, but meanwhile it is interesting to note that the mathematical result referred to above is exactly the same as that which underlies the celebrated "uncertainty principle" in quantum mechanics.

IV. $AR(1)$ process; discrete parameter case

Here, the autocorrelation function is given by (see (3.5.17))

$$\rho(r) = a^{|r|}, \qquad (|a| < 1),$$

so that, by (4.8.32),

$$f(\omega) = \frac{1}{2\pi}\left\{1 + 2\sum_{r=1}^{\infty} a^r \cos r\omega\right\} \tag{4.10.9}$$

$$= \frac{1}{2\pi}R\left\{1 + 2\sum_{r=1}^{\infty} (a\,e^{i\omega})^r\right\}$$

$$= \frac{1}{2\pi}\left[1 + 2R\left\{\frac{a\,e^{i\omega}}{1 - a\,e^{i\omega}}\right\}\right]$$

$$= \frac{1 - a^2}{2\pi(1 - 2a\cos\omega + a^2)}, \qquad -\pi \leq \omega \leq \pi. \tag{4.10.10}$$

When $a > 0$ the typical form of $f(\omega)$ is shown in Fig. 4.17. For $a > 0$ most of the power is again concentrated at the low frequency end of the spectrum; when $a < 0$ the power is concentrated mainly at the high frequency end.

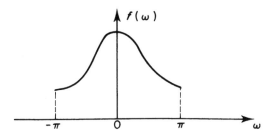

Fig. 4.17. Spectral density function of a discrete parameter AR(1) process with $a > 0$.

V. AR(2) *process; continuous parameter case*

The autocorrelation function for this model is given by (3.7.51), i.e.

$$\rho(\tau) = \frac{c_2 \exp(c_1|\tau|) - c_1 \exp(c_2|\tau|)}{(c_2 - c_1)},$$

where c_1, c_2, are the roots of $u^2 + \alpha_1 u + \alpha_2$, α_1 and α_2 are the parameters of the model (3.7.39), and for stationarity we require $R(c_1) < 0$, $R(c_2) < 0$. Hence we have, by (4.8.20),

$$f(\omega) = \frac{1}{2\pi} \int_{-\infty}^{\infty} \left\{ \frac{c_2 \exp(c_1|\tau|) - c_1 \exp(c_2|\tau|)}{c_2 - c_1} \right\} \cos \omega\tau \, d\tau$$

$$= \frac{1}{\pi} \int_0^{\infty} \left\{ \frac{c_2 \exp(c_1\tau) - c_1 \exp(c_2\tau)}{c_2 - c_1} \right\} \cos \omega\tau \, d\tau.$$

Each of the terms in the above integral is of the same form as (4.10.7) and may thus be evaluated in the same way. Recalling that $(c_1 + c_2) = -\alpha_1$, $c_1 c_2 = \alpha_2$, we obtain (after some algebra),

$$f(\omega) = \frac{1}{\pi} \left\{ \frac{\alpha_1 \alpha_2}{(\alpha_2 - \omega^2)^2 + \alpha_1^2 \omega^2} \right\}. \tag{4.10.11}$$

We see, once again, that $f(\omega)$ exists for all ω, and hence this process also has a purely continuous spectrum. The typical form of $f(\omega)$ (for $\omega \geq 0$) is shown in Fig. 4.18. The denominator of the right-hand side of (4.10.10) is a quadratic function of (ω^2) and attains its minimum value when $\omega^2 = (\alpha_2 - \alpha_1^2/2)$. This will correspond to a maximum of $f(\omega)$ provided $(\alpha_2 - \frac{1}{2}\alpha_1^2)^{1/2}$ is real valued. Hence, if $\alpha_2 > \frac{1}{2}\alpha_1^2$, $f(\omega)$ has a "peak" at $\omega = (\alpha_2 - \frac{1}{2}\alpha_1^2)^{1/2} = \omega_0$ (say). Now in our previous discussion of discrete spectra we pointed out that if $f(\omega)$ has a "peak" (at a non-zero frequency) this indicates a "pseudo-periodic" behaviour in the behaviour of the process. We may thus conclude that if $\alpha_2 > \alpha_1^2/2$, the process is pseudo-periodic, with "period" $(2\pi/\omega_0) = 2\pi/(\alpha_2 - \frac{1}{2}\alpha_1^2)^{1/2}$. This feature of the AR(2) process had been noted previously when we were studying the behaviour of its autocorrelation function

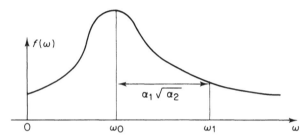

Fig. 4.18. Special density function for a continuous parameter AR(2) process.

(see Section (3.7.4)). It was found then that if $\alpha_2 > \frac{1}{4}\alpha_1^2$, the autocorrelation function had a damped oscillatory form—indicating a pseudo-periodic behaviour in the process with "period" $2\pi/(\alpha_2 - \frac{1}{4}\alpha_1^2)^{1/2}$. The apparent discrepancy between the values of the "period" as determined from the spectral density function and the autocorrelation function is due to the fact that the process is not truly periodic but merely "mimics" this type of behaviour in certain respects. We must therefore expect to obtain different estimates of the approximate periodicity if we concentrate on different features of the process. To emphasize this point we may note that if $\frac{1}{4}\alpha_1^2 < \alpha_2 < \frac{1}{2}\alpha_1^2$, the autocorrelation function oscillates (indicating pseudo-periodic behaviour) but the spectral density function has no peak at a non-zero frequency (contradicting the suggestion of pseudo-periodic behaviour).

We may obtain a rough measure of the width of the peak in $f(\omega)$ by computing the distance between ω_0 and the "half-power point" ω_1 (say). The frequency ω_1 is defined by $f(\omega_1) = \frac{1}{2}f(\omega_0)$, and the "bandwidth" of the peak is then defined as $b = 2(\omega_1 - \omega_0)$. If α_1 is small it is not difficult to show that, approximately, $b = 2\alpha_1\sqrt{\alpha_2}$. Hence, as $\alpha_1 \to 0$, $b \to 0$ and $f(\omega)$ tends to a δ-function form. In the limiting case the spectrum thus becomes discrete and the process is then strictly periodic. In fact, as $\alpha_1 \to 0$, $\rho(\tau) \to \cos(\sqrt{\alpha_2})\tau$, and now both the autocorrelation function and the spectral density function give the same value for the period. (See the discussion at the end of Section 3.5.3).

VI.   *AR(2) process; discrete parameter case*

For this process the autocorrelation function is (see (3.5.35)),

$$\rho(r) = \frac{(1-\mu_2^2)\mu_1^{|r|+1} - (1-\mu_1^2)\mu_2^{|r|+1}}{(\mu_1-\mu_2)(1+\mu_1\mu_2)},$$

where μ_1, μ_2 are the roots of $u^2 + a_1 u + a_2$, and a_1, a_2 are the parameters in the model (3.5.22). (For stationarity we require $|\mu_1| < 1$, $|\mu_2| < 1$.) The normalized spectral density function is then given by (from (4.8.32)),

$$f(\omega) = \frac{1}{2\pi}\left[1 + 2\sum_{r=1}^{\infty}\left\{\frac{(1-\mu_2^2)\mu_1^{r+1} - (1-\mu_1^2)\mu_2^{r+1}}{(\mu_1-\mu_2)(1+\mu_1\mu_2)}\right\}\cos\omega r\right].$$

Each term in the above summation is of the same form as (4.10.9) and may be evaluated by the same technique as previously used. After some reduction (and making use of the results; $(\mu_1 + \mu_2) = -a_1$, $\mu_1\mu_2 = a_2$) we finally obtain,

$$f(\omega) = \frac{(1-a_2)[(1+a_2)^2 - a_1^2]}{2\pi(1+a_2)[(1-a_2)^2 + a_1^2 + 2a_1(1+a_2)\cos\omega + 4a_2\cos^2\omega]},$$

$$-\pi \le \omega \le \pi. \qquad\qquad (4.10.12)$$

As in the continuous parameter AR(2) case, $f(\omega)$ may contain a peak at a non-zero frequency for suitable values of the parameters a_1, a_2. Specifically, if

$$a_2 > 0 \quad\text{and}\quad \left|\frac{a_1(1+a_2)}{4a_2}\right| < 1,$$

then $f(\omega)$ has a peak at $\omega = \cos^{-1}\{-a_1(1+a_2)/4a_2\}$. If $a_2 < 0$ and $|a_1(1+a_2)/4a_2| < 1$, $f(\omega)$ has a minimum at $\omega = \cos^{-1}\{-a_1(1+a_2)/4a_2\}$. For other parameter values the power may be concentrated either at the low frequency end or at the high frequency end of the spectrum.

VII.   *Harmonic processes*

The harmonic process model is (see Sections 3.5.8, 3.7.9),

$$X(t) = \sum_{i=1}^{K} A_i \cos(\omega_i t + \phi_i),$$

and from (3.5.75) the corresponding autocorrelation function is

$$\rho(\tau) = \sum_{i=1}^{K} A_i^2 \cos\omega_i\tau \Big/ \sum_{i=1}^{K} A_i^2.$$

Consider now the special case when $K = 1$, so that the expression for $\rho(\tau)$ reduces to

$$\rho(\tau) = \cos \omega_1 \tau.$$

If we try to compute the spectral density function for this form of $\rho(\tau)$ we obtain, in the continuous parameter case,

$$f(\omega) = \frac{1}{2\pi} \int_{-\infty}^{\infty} \cos \omega_1 \tau \cos \omega \tau \, d\tau$$

$$= \frac{1}{\pi} \int_0^{\infty} \cos \omega_1 \tau \cos \omega \tau \, d\tau. \qquad (4.10.13)$$

The above infinite integral cannot be evaluated in the ordinary way since when $\omega = \pm \omega_1$ it becomes infinite. However, consider a truncated form of the integral, viz.,

$$\frac{1}{\pi} \int_0^N \cos \omega_1 \tau \cos \omega \tau \, d\tau = \frac{1}{2\pi} \int_0^N \{\cos(\omega + \omega_1)\tau + \cos(\omega - \omega_1)\tau\} \, d\tau$$

$$= \left\{ \frac{1}{2\pi} \left(\frac{\sin(\omega + \omega_1)N}{(\omega + \omega_1)} \right) + \frac{1}{2\pi} \left(\frac{\sin(\omega - \omega_1)N}{(\omega - \omega_1)} \right) \right\}$$

$$= \tfrac{1}{2} \{\delta_N(\omega + \omega_1) + \delta_N(\omega - \omega_1)\}, \quad \text{say},$$

where

$$\delta_N(\theta) = \frac{1}{\pi} \left(\frac{\sin(N\theta)}{\theta} \right).$$

It is easily verified that $\int_{-\infty}^{\infty} \delta_N(\theta) \, d\theta = 1$, all N, but writing

$$\delta_N(\theta) = \frac{N}{\pi} \left(\frac{\sin(N\theta)}{N\theta} \right)$$

we see that as $\theta \to 0$ the second factor $\to 1$ and hence $\delta_N(\theta) = 0(N) \to \infty$ as $N \to \infty$. In fact, it is well known that, as $N \to \infty$, $\delta_N(\theta)$ behaves like a Dirac δ-function—as may be verified directly from the integral expression (4.10.13), (see, e.g. Rice (1944–5)). Hence we may write formally,

$$f(\omega) = \tfrac{1}{2} \{\delta(\omega + \omega_1) + \delta(\omega - \omega_1)\}. \qquad (4.10.14)$$

The normalized integrated spectrum thus takes the form

$$F(\omega) = \int_{-\infty}^{\omega} f(\omega) \, d\theta = \begin{cases} 0, & \omega < -\omega_1 \\ \tfrac{1}{2}, & -\omega_1 \le \omega < \omega_1 \\ 1, & \omega \ge \omega_1, \end{cases}$$

i.e. $F(\omega)$ has *a step-function form*, as illustrated in Fig. 4.19, and hence $X(t)$ has a *purely discrete spectrum* with "jumps" at $\omega = -\omega_1$, $\omega = \omega_1$. In the general case (i.e. when K is not necessarily 1) it follows by an obvious extension of the above argument that the spectral density function of $X(t)$ may be written formally as

$$f(\omega) = \frac{1}{2} \sum_{i=1}^{K} \{A_i^2 [\delta(\omega + \omega_i) + \delta(\omega - \omega_i)]\} / \sum_{i=1}^{K} A_i^2, \qquad (4.10.16)$$

i.e. $X(t)$ has a purely discrete spectrum, and $F(\omega)$ is a step function with "jumps" at $\omega = \pm\omega_i$, $i = 1, \ldots K$, the magnitude of the jump at $\omega = \pm\omega_i$ being $\frac{1}{2}A_i^2 / (\sum_{i=1}^{K} A_i^2)$. (This last result may be verified directly from (4.9.9).)

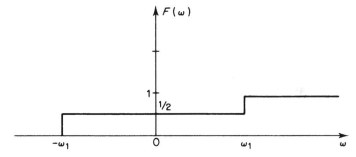

Fig. 4.19. Integrated spectrum for a harmonic process ($K = 1$).

In the discrete parameter case we obtain exactly the same formal expression for $f(\omega)$, except for the fact that here ω is limited to the range $|\omega| \leqslant \pi$ (with, of course, the same restriction on the values of the $\{\omega_i\}$, $i = 1, \ldots K$).

We could if we wished, use the same method as above (i.e. direct transformation of the autocorrelation function) to calculate the spectral density functions of the $MA(k)$ process (moving average of order k) and of the general linear process. However, these expressions may be derived by a more elegant technique using the theory of "linear filters" to be discussed in Section 4.12. (This technique provides also a simple method for calculating the spectral density function of an autoregressive process.) We therefore defer these calculations for the moment, but the reader may find it interesting to derive them independently and then to check the results with those given in Section 4.12.

4.11 SPECTRAL REPRESENTATION OF STATIONARY PROCESSES

At the outset of our discussion of the spectral analysis of stationary processes (Section 4.7) we posed the question as to whether a realization of such a process could be represented either as a Fourier series or as a Fourier integral. We concluded that neither a Fourier series nor a Fourier integral could, in general, provide a valid representation, but we did not investigate whether there was some other form of "Fourier type" expansion which could in fact be used for stationary processes. In the subsequent discussion we circumvented this problem by using a Fourier integral to represent the realization over a *finite interval* $(-T, T)$ and then defined, e.g., the non-normalized power spectral density function $h(\omega)$ by considering the limiting situation as $T \to \infty$. However, the original problem still remains to be solved, and, although it may seem a purely technical one, its solution is crucial if we wish to preserve the physical interpretation of power spectra. For, unless we can find some way of representing a *complete* realization (i.e. from $t = -\infty$ to $t = +\infty$) as a "sum" of sine and cosine terms we cannot talk in any meaningful way about the "contribution to the total power from that component with frequency ω". Indeed, the whole concept of "frequency" depends essentially on the tacit assumption that there exists a "sine and cosine" description of the underlying phenomenon. Fortunately, a suitable representation for stationary processes does, in fact, exist, but to obtain this we have to consider a new form of "Fourier expansion" which is more general than Fourier series and Fourier integrals. The nature of this more general form of Fourier expansion, due to Wiener (1930), was mentioned briefly during our preliminary discussion of spectral analysis in Chapter 1 (see Sections 1.4, 1.8) when it was pointed out that if we generalize the ordinary Fourier integral to a *Fourier–Stieltjes integral* then this form of integral can be used to represent both Fourier series and (ordinary) Fourier integrals in a unified manner, and furthermore, it enables us also to represent other functions which do not possess either Fourier series or Fourier integrals. Thus, given a realization of a continuous parameter process, $X(t)$, we are led to considering a representation of $X(t)$ of the form,

$$X(t) = \int_{-\infty}^{\infty} e^{it\omega} \, dZ(\omega). \qquad (4.11.1)$$

This representation of $X(t)$ is very similar to the Fourier–Stieltjes representation (4.8.23) of $\rho(\tau)$. However, whereas in practice $F(\omega)$ is either differentiable or a step-function, $Z(\omega)$ has a more complicated structure. Clearly, $Z(\omega)$ cannot be differentiable even when $F(\omega)$ is differentiable (i.e. when the process has a purely continuous spectrum) for then (4.11.1) would reduce to an ordinary Fourier integral. It turns out that in this case $dZ(\omega) = O(\sqrt{d\omega})$, i.e. the increment in $Z(\omega)$ over an infinitesimal interval $d\omega$ is infinitesimal, but of a much larger order of magnitude than $d\omega$ (so that the "derivative", $dZ(\omega)/d\omega$, will not, of course, exist). On the other hand, if

$F(\omega)$ is a step function (i.e. if the process has a purely discrete spectrum) then $Z(\omega)$ has essentially the same step function form (with "jumps" at the same points as $F(\omega)$). When the process has a mixed spectrum, so that $F(\omega)$ contains both discrete and continuous components, $Z(\omega)$ similarly contains components of each of the above forms.

In the above discussion we considered only the representation of a single realization. However, given that we can represent one realization in the form (4.11.1) there is no reason why we should not represent each realization in the same form, but of course, we must not expect the function $Z(\omega)$ to be the same for all realizations. In other words, we must allow the form of $Z(\omega)$ to vary from realization to realization, that is, corresponding to each realization of the process $X(t)$ there will be a realization of $Z(\omega)$. Thus, if we now think of (4.11.1) as providing a representation for the *process* (as opposed to a particular realization) *then $Z(\omega)$ is itself a complex valued stochastic process* (as a function of ω), and for each ω, $Z(\omega)$ *is a complex random variable*. The integral on the right-hand side of (4.11.1) now becomes a *stochastic integral* and is now defined in the *mean-square sense*. (See Section 3.6.3.)

In this form equation (4.11.1) is called the *"spectral representation of the process $X(t)$"* and in essence it tells us that (*virtually*) *any stationary process can be represented as (the limit of) the sum of sine and cosine functions with random coefficients $dZ(\omega)$, or more precisely, with random amplitudes $|dZ(\omega)|$ and random phases $\arg\{dZ(\omega)\}$*. The sine and cosine functions thus form the basic "building blocks" of all stationary processes, just as they do for periodic deterministic functions. Moreover, as we shall see, the stochastic process $\{Z(\omega)\}$ has the special property that its *increments at different values of ω are uncorrelated*, i.e. for any two distinct frequencies, ω, ω', the random variables, $dZ(\omega) = \{Z(\omega + d\omega) - Z(\omega)\}$, and $dZ(\omega') = \{Z(\omega' + d\omega') - Z(\omega')\}$ are uncorrelated. The process $\{Z(\omega)\}$ is called an *"orthogonal process"*. (In this respect it is similar to the Wiener process but differs in other respects; in particular, the "incremental process" $\{dZ(\omega)\}$ is not, in general, stationary.) In addition, there must obviously be some intimate relationship between the properties of the process $\{Z(\omega)\}$ and the spectral properties of $\{X(t)\}$. It turns out that this relationship is most conveniently expressed in terms of the function $H(\omega)$, the (non-normalized) integrated spectrum of $x(t)$. Specifically, we will show that, for each ω,

$$E[|dZ(\omega)|^2] = dH(\omega), \qquad (4.11.2)$$

i.e. at each frequency ω, the increment in $H(\omega)$ is equal to the mean of the squared amplitude of the corresponding component in (4.11.1). When $\{X(t)\}$ has a purely continuous spectrum, so that $dH(\omega) = h(\omega)\,d\omega$,

$h(\omega)$ being the (non-normalized) spectral density function, (4.11.2) reduces to

$$E[|dZ(\omega)|^2] = h(\omega)\, d\omega. \qquad (4.11.3)$$

The above main result, namely that (virtually) any stationary process can be represented as a *"sum"* (*of the form* (4.11.1)) *of sine and cosine functions with uncorrelated coefficients* is certainly one of the most important ones in the whole of the theory of stationary processes. (The restriction "virtually" is due to the fact that in the continuous parameter case the process has to be stochastically continuous—the same condition as was required for the spectral representation of the autocorrelation function given in the Wiener–Khintchine theorem.) Not only does it provide a "canonical form" for describing any stationary process, but, as pointed out above, it is crucial to the physical interpretation of power spectra. It is a fascinating result also in that it lends itself to a variety of different proofs which together reveal a rich collection of mathematical ideas. The proof which we present below is a heuristic one in the sense that here we do not concern ourselves overmuch with the mathematical "frills." However, the discussion will, we trust, illustrate the essential ideas involved in the derivation of (4.11.1). A formal statement of the result is as follows.

Theorem 4.11.1 *Spectral representation of continuous parameter stationary processes* Let $\{X(t)\}$, $-\infty < t < \infty$, be a zero-mean stochastically continuous stationary process. Then there exists an orthogonal process, $\{Z(\omega)\}$, such that, for all t, $X(t)$ may be written in the form,

$$X(t) = \int_{-\infty}^{\infty} e^{it\omega}\, dZ(\omega), \qquad (4.11.4)$$

the integral being defined in the mean-square sense. The process $\{Z(\omega)\}$ has the following properties;

(i) $E[dZ(\omega)] = 0$,   all ω,

(ii) $E[|dZ(\omega)|^2] = dH(\omega)$,   all ω,   (4.11.5)

where $H(\omega)$ is the (non-normalized) integrated spectrum of $X(t)$,

(iii) for any two distinct frequencies, ω, ω', $(\omega \neq \omega')$,

$$\mathrm{cov}[dZ(\omega), dZ(\omega')] = E[dZ^*(\omega)\, dZ(\omega')] = 0. \qquad (4.11.6)$$

Proof. We start by considering a single realization, $X(t)$, on a *finite* interval, $-T \le t \le T$, and then make this realization periodic outside this interval. Thus, we define a new function, $X_T^*(t)$, by,

$$X_T^*(t) = X(t), \qquad -T \le t \le T,$$

$$X_T^*(t + 2pT) = X^*(t), \qquad p = \pm 1, \pm 2, \ldots.$$

Then $X_T^*(t)$ is periodic, with period $2T$, and hence, according to Section 4.5 (equation (4.5.3)), may be written as a Fourier series in the form,

$$X_T^*(t) = \sum_{n=-\infty}^{\infty} A_n e^{2\pi i f_n t}, \tag{4.11.7}$$

where $f_n = n/2T$ and A_n is given by

$$A_n = \frac{1}{2T} \int_{-T}^{T} X_T^*(t) e^{-2\pi i f_n t}.$$

(see (4.5.5)). We may now rewrite (4.11.7) in the form

$$X(t) = \frac{1}{\sqrt{2\pi}} \sum_{n=-\infty}^{\infty} G_T(\omega_n) e^{i\omega_n t} \delta\omega_n, \tag{4.11.8}$$

where the function $G_T(\omega)$ is defined *for all* ω by,

$$G_T(\omega) = \frac{1}{\sqrt{2\pi}} \int_{-T}^{T} X_T^*(t) e^{-i\omega t} \, dt,$$

$$= \frac{1}{\sqrt{2\pi}} \int_{-T}^{T} X(t) e^{-i\omega t} \, dt, \tag{4.11.9}$$

and $\omega_n = 2\pi n/2T$, so that $\delta\omega_n = \omega_{n+1} - \omega_n = 2\pi/2T$.

Although $X_T^*(t)$ is not the same as the function $X_T(t)$ defined by (4.7.1) (recall that $X_T(t)$ was defined to be *zero* outside $(-T, T)$), the function $G_T(\omega)$ defined by (4.11.9) is exactly the same as the $G_T(\omega)$ defined by (4.7.3). Hence, if we now think of (4.11.9) as defining $G_T(\omega)$ in terms of the *process* $X(t)$, then we know from (4.7.5) that when the (non-normalized) spectral density function, $h(\omega)$, exists,

$$\lim_{T \to \infty} \left[E\left\{ \frac{|G_T(\omega)|^2}{2T} \right\} \right] = h(\omega),$$

so that, as $T \to \infty$,

$$|G_T(\omega_n)| = O(\sqrt{T}) = O(1/\sqrt{\delta\omega_n}).$$

Thus, $|G_T(\omega_n)| \to \infty$ as $T \to \infty$, *but* $\{|G_T(\omega_n)|\delta\omega_n\} = O(\sqrt{\delta\omega_n}) \to 0$ as $T \to \infty$. We now define the function

$$Z_T(\omega) = \frac{1}{\sqrt{2\pi}} \int_{-\infty}^{\omega} G_T(\theta) \, d\theta.$$

Then,

$$\Delta Z_T(\omega_n) = Z_T(\omega_{n+1}) - Z_T(\omega_n) \sim \frac{1}{\sqrt{2\pi}} G_T(\omega_n) \delta\omega_n,$$

for small $\delta\omega_n$, i.e. for large T. Hence, for large T,

$$E[|\Delta Z_T(\omega_n)|^2] \sim E\left[|G_T(\omega_n)|^2 \frac{\delta\omega_n}{2\pi}\right] \cdot \delta\omega_n,$$

$$= E\left[\frac{|G_T(\omega_n)|^2}{2T}\right] \cdot \delta\omega_n,$$

and as $T \to \infty$ (so that $\delta\omega_n \to 0$),

$$E[|\Delta Z_T(\omega_n)|^2] \sim h(\omega_n)\delta\omega_n = \Delta H(\omega_n), \qquad (4.11.10)$$

where $H(\omega) = \int_{-\infty}^{\omega} h(\theta)\, d\theta$ is the (non-normalized) integrated spectrum of $X(t)$. Now (4.11.8) provides a "spectral representation" for $X_T^*(t)$, which in turn provides a representation for $X(t)$, but one which is valid only over the interval $(-T, T)$. To obtain a representation valid for all t we should like to let $T \to \infty$, but we cannot consider the limiting form of (4.11.8) as it stands since, as noted above, $G_T(\omega_n)$ does not converge to a finite value as $T \to \infty$. However, when we combine $G_T(\omega_n)$ with $\delta\omega_n$ the product $\{G_T(\omega_n)\delta\omega_n\}$ is "well behaved". This suggests that if we re-write (4.11.8) in terms of $\Delta Z_T(\omega_n)$ rather than $G_T(\omega_n)$ we may obtain an appropriate limiting form. (This is the essential reason why the spectral representation of $X(t)$ takes a Fourier–Stieltjes integral form rather than an ordinary Fourier integral form.) We thus have, for $-T \le t \le T$,

$$X(t) \equiv X_T^*(t) = \sum_{n=-\infty}^{\infty} e^{it\omega_n} \Delta Z_T(\omega_n). \qquad (4.11.11)$$

We can now let $T \to \infty$, in which case $\delta\omega_n \to 0$, (i.e. the spacings between successive points in the sequence $\{\omega_n\}$ tends to zero) and the discrete sum on the right-hand side of (4.11.11) converges in mean square to a Stieltjes integral with respect to the limiting form of the function $Z_T(\omega)$. We then have, *for all t*,

$$X(t) = \int_{-\infty}^{\infty} e^{it\omega}\, dZ(\omega), \qquad (4.11.12)$$

and the limiting form of (4.11.10) becomes,

$$E[|dZ(\omega)|^2] = dH(\omega) = \sigma_X^2\, dF(\omega), \qquad (4.11.13)$$

where $F(\omega)$ is the normalized integrated spectrum of $X(t)$.

When the spectral density functions, $h(\omega)$ and $f(\omega)$ exist for all ω, (4.11.12) may be written as,

$$E[|dZ(\omega)|^2]/d\omega = h(\omega) = \sigma_X^2 f(\omega).$$

The inversion of (4.11.12) is not as straightforward as in the case of ordinary Fourier integrals. In particular, it should be noted that $dZ(\omega)$ is *not* the Fourier transform of $X(t)$—rather it is the "derivative", $\{dZ(\omega)/d\omega\}$, which, in a formal sense, plays the role of the Fourier transform. We may therefore think of $Z(\omega)$ as the "integral" of the Fourier transform, and then by analogy with (4.8.10) and (4.8.13), we obtain, for any ω_1, ω_2, $\omega_1 \leqslant \omega_2$,

$$Z(\omega_2) - Z(\omega_1) = \frac{1}{2\pi} \int_{-\infty}^{\infty} \frac{\exp(-i\omega_2 t) - \exp(-i\omega_1 t)}{-it} X(t)\, dt. \qquad (4.11.14)$$

(More precisely, the above infinite integral should be interpreted as the mean-square limit as $T \to \infty$ of the integral over the finite range, $(-T, T)$. Equation (4.11.14) can, in fact, be proved quite rigorously without introducing the formal "derivative" of $Z(\omega)$, see e.g. Bartlett (1955), p. 169.)

To prove that $Z(\omega)$ is an orthogonal process we first note that, since $X(t)$ has zero mean, $(E[X(t)] = 0$, all $t)$ it follows immediately from (4.11.9) that $E[G_T(\omega)] = 0$, all ω, which implies that $E[\Delta Z_T(\omega)] = 0$, all ω, which in turn implies that $E[dZ(\omega)] = 0$, all ω.

Now, taking the complex conjugate of both sides of (4.11.12) we have,

$$X^*(t) = \int_{-\infty}^{\infty} e^{-it\omega}\, dZ^*(\omega),$$

and replacing t by $(t + \tau)$ in (4.11.12) gives,

$$X(t + \tau) = \int_{-\infty}^{\infty} e^{i(t+\tau)\omega}\, dZ(\omega).$$

If we now multiply together the two equations above, and express the product of integrals on the right-hand side as a double integral, we obtain

$$X^*(t)X(t + \tau) = \int_{-\infty}^{\infty} \int_{-\infty}^{\infty} \exp(-it\omega) \exp[i(t + \tau)\omega']\, dZ^*(\omega)\, dZ(\omega').$$

Taking expectations of both sides now gives,

$$E[X^*(t)X(t + \tau)] = \int_{-\infty}^{\infty} \int_{-\infty}^{\infty} \exp[it(\omega' - \omega)] \exp(i\tau\omega') E[dZ^*(\omega)\, dZ(\omega')].$$

$$(4.11.15)$$

Since $X(t)$ is a stationary process the left-hand side of (4.11.15) is the autocovariance function $R(\tau)$, and is therefore a function of τ only and does not depend on t. Hence, the right-hand side must be independent of t. This can be so only if the *contribution to the double integral is zero when $\omega \neq \omega'$*, which tells us immediately that,

$$E[dZ^*(\omega)\, dZ(\omega')] = 0, \qquad \text{for all } \omega \neq \omega', \qquad (4.11.16)$$

and thus proves the required orthogonality property of $Z(\omega)$.

Having established this property, an alternative proof of the Wiener–Khintchine theorem follows almost immediately from (4.11.15). For, in view of (4.11.16) the double integral on the right-hand side of (4.11.15) reduces to a single integral, i.e. we may simply set $\omega' = \omega$, in which case we obtain,

$$R(\tau) = \int_{-\infty}^{\infty} e^{i\tau\omega} E[|dZ(\omega)|^2] = \int_{-\infty}^{\infty} e^{i\tau\omega} \, dH(\omega).$$

Moreover, writing, $Z(\omega) = \int_{-\infty}^{\omega} dZ(\theta)$, we have for any $\omega_1, \omega_2, (\omega_1 \leqslant \omega_2)$,

$$E[|Z(\omega_2) - Z(\omega_1)|^2] = \int_{\omega_1}^{\omega_2} \int_{\omega_1}^{\omega_2} E[dZ^*(\theta) \, dZ(\theta')]$$

$$= \int_{\omega_1}^{\omega_2} E[|dZ(\theta)|^2] = \int_{\omega_1}^{\omega_2} dH(\theta)$$

$$= H(\omega_2) - H(\omega_1). \qquad (4.11.17)$$

Setting $\omega_1 = -\infty$ and writing ω for ω_2 we then have, for all ω,

$$E[|Z(\omega)|^2] = H(\omega). \qquad (4.11.17a)$$

Also, for any two non-overlapping intervals $(\omega_1, \omega_2), (\omega_3, \omega_4)$, we have,

$$E[\{Z(\omega_4) - Z(\omega_3)\}^* \{Z(\omega_2) - Z(\omega_1)\}] = E\left[\left\{\int_{\omega_3}^{\omega_4} dZ^*(\theta)\right\}\left\{\int_{\omega_1}^{\omega_2} dZ(\theta')\right\}\right]$$

$$= \int_{\omega_3}^{\omega_4} \int_{\omega_1}^{\omega_2} E\{dZ^*(\theta) \, dZ(\theta')\} = 0,$$

$$(4.11.18)$$

since the fact that $(\omega_1, \omega_2), (\omega_3, \omega_4)$ are non-overlapping intervals means that there are no common values of θ, θ' in the range of the double integral. These results provide alternative forms of the relationship between $Z(\omega)$ and $H(\omega)$ and of the orthogonality property of $Z(\omega)$.

Spectral representation of discrete parameter stationary processes

The results of Theorem 4.11.1 carry over almost completely to the discrete parameter case, the only modifications being that, when t takes integer values only, the range of integration in (4.11.4) becomes $-\pi$ to π instead of $-\infty$ to ∞, and the condition that the process be stochastically continuous is, of course, no longer meaningful. The reason for restricting the frequency range to $(-\pi, \pi)$ is exactly the same as in the discussion in Section (4.8.3), and the spectral representation for discrete parameter processes thus takes the following form.

Let $\{X_t\}$, $t = 0, \pm1, \pm2, \ldots$, be a zero mean stationary process. Then there exists an orthogonal process, $Z(\omega)$, on the interval $(-\pi, \pi)$ such that for all integral t,

$$X_t = \int_{-\pi}^{\pi} e^{it\omega} \, dZ(\omega). \qquad (4.11.19)$$

The process $Z(\omega)$ has the same properties as in Theorem 4.11.1, and in particular,

$$E[|dZ(\omega)|^2] = dH(\omega), \qquad -\pi \leq \omega \leq \pi.$$

Spectral representation of real valued processes

The representations (4.11.4), (4.11.19), are quite general and apply to all stationary processes, both real valued and complex valued. However, when $X(t)$ (or X_t) is real valued we may re-write (4.11.4), (4.11.19), in terms of sine and cosine functions rather than complex exponentials, as follows. We note first that, e.g., in the continuous parameter case,

$$X^*(t) = \int_{-\infty}^{\infty} e^{-i\omega t} \, dZ^*(\omega) = \int_{-\infty}^{\infty} e^{i\omega t} \, dZ^*(-\omega).$$

But $X(t)$ is real valued if and only if $X(t) \equiv X^*(t)$, i.e. if and only if $dZ(\omega) = dZ^*(-\omega)$, all ω.

Hence, writing,

$$dU_0(\omega) = R[dZ(\omega)], \qquad dV_0(\omega) = -\mathscr{I}[dZ(\omega)],$$

we have,

$$dU_0(\omega) = dU_0(-\omega), \qquad dV_0(\omega) = -dV_0(-\omega), \qquad \text{all } \omega. \qquad (4.11.20)$$

Now (4.11.4) may be written as,

$$X(t) = \int_{-\infty}^{\infty} (\cos \omega t + i \sin \omega t)\{dU_0(\omega) - i \, dV_0(\omega)\},$$

i.e.

$$X(t) = \int_{-\infty}^{\infty} \cos \omega t \, dU_0(\omega) + \int_{-\infty}^{\infty} \sin \omega t \, dV_0(\omega), \qquad (4.11.21)$$

the complex terms vanishing in virtue of (4.11.20). Finally, we may write,

$$X(t) = \int_{0}^{\infty} \cos \omega t \, dU(\omega) + \int_{0}^{\infty} \sin \omega t \, dV(\omega), \qquad (4.11.22)$$

where

$$dU(\omega) = \{dU_0(\omega) + dU_0(-\omega)\} = 2\ dU_0(\omega), \qquad \omega \neq 0$$

$$dU(0) = dU_0(0),$$

and

$$dV(\omega) = \{dV_0(\omega) - dV_0(-\omega)\} = 2\ dV_0(\omega).$$

It is easily verified from their definitions that $U(\omega)$ and $V(\omega)$ are each orthogonal processes, and in addition, cross-orthogonal, i.e.

$$E[dU(\omega)\ dU(\omega')] = 0, \qquad \omega \neq \omega'$$

$$E[dV(\omega)\ dV(\omega')] = 0, \qquad \omega \neq \omega',$$

$$E[dU(\omega)\ dV(\omega')] = 0, \qquad \text{all } \omega,\ \omega',$$

while, when $\omega' = \omega$,

$$E[|dU(\omega)|^2] = E[|dV(\omega)|^2] = dH_+(\omega),$$

where $H_+(\omega)$ is the "positive frequency" integrated spectrum, defined in Section 4.8.

A similar result holds for the discrete parameter case.

Continuous and discrete spectra

We noted in Section 4.9 that the normalized integrated spectrum, $F(\omega)$, could be decomposed as a linear combination of two terms, $F_1(\omega)$, (the continuous component) and $F_2(\omega)$, (the discrete component): cf. (4.9.2). In terms of the non-normalized integrated spectrum the decomposition may be written,

$$H(\omega) = H_1(\omega) + H_2(\omega), \qquad (4.11.22)$$

where

$$H_1(\omega) \equiv a_1 \sigma_X^2 F_1(\omega), \qquad H_2(\omega) \equiv a_2 \sigma_X^2 F_2(\omega),$$

a_1, a_2, being the coefficients in (4.9.2).

There is a corresponding decomposition for the process $Z(\omega)$, i.e. we may write (see Doob (1953), p. 488)

$$Z(\omega) = Z_1(\omega) + Z_2(\omega),$$

so that, e.g., in the continuous parameter case,

$$X(t) = \int_{-\infty}^{\infty} e^{it\omega}\ dZ_1(\omega) + \int_{-\infty}^{\infty} e^{it\omega}\ dZ_2(\omega) \qquad (4.11.23)$$

$$= X_c(t) + X_d(t), \qquad (4.11.24)$$

say, where $Z_1(\omega)$, $Z_2(\omega)$, are each orthogonal processes and,

$$E[|dZ_1(\omega)|^2] = dH_1(\omega) = h(\omega)\,d\omega,$$

$$E[|dZ_2(\omega)|^2] = dH_2(\omega).$$

$$E[dZ_1^*(\omega)\,dZ_2(\omega)] = 0, \qquad \text{all } \omega, \omega'.$$

The decomposition (4.11.23) follows immediately on splitting up the range of integration in (4.11.12) into the two sets, Ω_1, Ω_2, where Ω_1 is the set of points on the ω-axis where $H(\omega)$ is continuous, and Ω_2 the set of discontinuity points of $H(\omega)$. Defining,

$$Z_i(\omega) = \begin{cases} Z(\omega), & \omega \in \Omega_i \\ 0, & \text{otherwise,} \quad i = 1, 2, \end{cases}$$

the representation (4.11.23) follows. The orthogonality of $Z_1(\omega)$, $Z_2(\omega)$, follows from the fact that Ω_1 and Ω_2 are disjoint sets. (The argument can easily be extended to the more general case where the third "singular" component, $H_3(\omega)$, is included in (4.11.22)—see Theorem 4.9.1.) The function $H_2(\omega)$ is a step-function (following the same form as $F_2(\omega)$) with jumps $\{p_r\}$ at the points $\{\omega_r\}$, say, $r = 1, 2, 3, \ldots$. Hence,

$$E[|dZ_2(\omega)|^2] = \begin{cases} 0, & \omega \neq \omega_r \\ p_r, & \omega = \omega_r \end{cases} \quad r = 1, 2, 3, \ldots,$$

and hence, when $\omega \neq \omega_r$ $(r = 1, 2, 3, \ldots)$, $dZ_2(\omega)$ is zero with probability 1. The second term in (4.11.23) thus reduces to a discrete sum, i.e. $X_d(t)$ is of the form,

$$X_d(t) = \sum_r A_r e^{i\omega_r t},$$

where $A_r = dZ_2(\omega_r)$ denotes the jump in $Z_2(\omega)$ at the discontinuity point ω_r. We may now write,

$$X(t) = \int_{-\infty}^{\infty} e^{it\omega}\,dZ_1(\omega) + \sum_r A_r e^{i\omega_r t}. \qquad (4.11.25)$$

The random variables $\{A_r\}$ are, of course, uncorrelated (this following immediately from the corresponding property of $dZ_2(\omega)$), and analogously to (4.9.9) we have,

$$A_r = \operatorname*{l.i.m.}_{T \to \infty} \left\{ \frac{1}{2\pi} \int_{-T}^{T} X(t)\, e^{-i\omega_r t}\, dt \right\}. \qquad (4.11.26)$$

Alternative proofs of the spectral representation theorem

We remarked earlier that there were a number of different proofs of the spectral representation theorem, some of which revealed interesting relationships between the spectral theory of stationary processes and other branches of pure mathematics. We now give a brief sketch of some of these alternative proofs.

(A) ANALYTICAL APPROACH

Perhaps the most straightforward way of proving (4.11.4) rigorously is to "reverse" the argument used in the heuristic proof of Theorem 4.11.1, i.e. given the process $\{X(t)\}$, we *define* $Z(\omega)$ by (4.11.14) (this defines $Z(\omega)$ up to an additive constant). We may then show that $Z(\omega)$ is an orthogonal process (this follows fairly easily from the form of the factor multiplying $X(t)$ in the integral in (4.11.14)), and then prove that, with this definition of $Z(\omega)$, the integral on the right-hand side of (4.11.4) represents $X(t)$ in mean-square, i.e. that

$$E\left[\left|X(t) - \int_{-\infty}^{\infty} \exp(it\omega)\,dZ(\omega)\right|^2\right] = 0.$$

This approach is due to Blanc-Lapierre and Fortet (1946), and further details are given in Bartlett (1955), p. 169, and Yaglom (1962), p. 36.

(B) FUNCTION-THEORY APPROACH

Cramer (1951) constructed an interesting proof using the methods of "function theory" in a Hilbert space setting. This approach has now become well established in the theory of stationary processes, and in particular Parzen (1959), developed this technique in a very lucid and elegant manner. The basic ideas may be described as follows.

We first consider the collection of all (complex valued) random variables U which have zero mean and finite variance, i.e.

$$E(U) = 0, \qquad E(|U|^2) < \infty.$$

We may show that this collection forms a Hilbert space H (see Section 4.2.2) if we define the inner-product between any two random variables U, V by

$$(U, V) = E(U^*V),$$

so that the norm of U is then given by,

$$\|U\|^2 = E(|U|^2).$$

(With this definition of inner product two random variables are "orthogonal" if they are uncorrelated, and it is thus consistent with the use of the term "orthogonal" in probability theory usage.) For each t, $X(t)$ is a random variable of the above type and hence belongs to H. As t varies from $-\infty$ to $+\infty$, $X(t)$ traces out a "curve" in H; let H_x denote the smallest subspace of H which contains this "curve".

Now we know from the Wiener–Khintchine theorem that (with $E[X(t)] = 0$),

$$E[X^*(s)X(t)] = \int_{-\infty}^{\infty} e^{i\omega(t-s)} \, dH(\omega). \qquad (4.11.27)$$

For each fixed t we may think of $\exp(i\omega t)$ as a function of ω, and henceforth we will denote this function of ω by $\phi_t(\omega)$, so that (4.11.27) can be written as,

$$E[X^*(s)X(t)] = \int_{-\infty}^{\infty} \phi_s^*(\omega)\phi_t(\omega) \, dH(\omega). \qquad (4.11.28)$$

We now introduce a second Hilbert space H_ϕ, which is the space "spanned" by the family of functions, $\{\phi_t(\omega)\}$, $-\infty < t < \infty$, i.e. H_ϕ consists of all functions ϕ which may be expressed as linear combinations of the $\{\phi_t(\omega)\}$, i.e. which may be written in the form,

$$\phi(\omega) = \sum_i c_i \phi_{t_i}(\omega),$$

(the $\{c_i\}$ being constants), together with functions which are obtained as limits of such linear combinations. The inner-product between any two functions, $\phi_1(\omega)$, $\phi_2(\omega)$ in H_ϕ is defined by,

$$(\phi_1(\omega), \phi_2(\omega)) = \int_{-\infty}^{\infty} \phi_1^*(\omega)\phi_2(\omega) \, dH(\omega). \qquad (4.11.29)$$

We now set up a mapping, M, between elements of H_x and elements of H_ϕ which is such that, for each t,

$$M[\phi_t(\omega)] = X(t),$$

and is extended to linear combinations of the $\{\phi_t(\omega)\}$ by

$$M\left[\sum_i c_i \phi_{t_i}(\omega)\right] = \sum_i c_i M[\phi_{t_i}(\omega)].$$

This mapping, M, clearly preserves inner-products since, for any s, t,

$$(X(s), X(t)) = \int_{-\infty}^{\infty} \phi_s^*(\omega)\phi_t(\omega) \, dH(\omega) = (\phi_s(\omega), \phi_t(\omega)).$$

Now for any interval, (ω_a, ω_b), define the "indicator function", $I_{\omega_a, \omega_b}(\omega)$, by

$$I_{\omega_a, \omega_b}(\omega) = \begin{cases} 1, & \omega_a \leq \omega < \omega_b, \\ 0, & \text{otherwise.} \end{cases}$$

Let $(\omega_{-n} < \omega_{-n+1} < \ldots < \omega_0 < \ldots < \omega_{n-1} < \omega_n)$ be a subset of points on the ω-axis. The crux of the proof lies in recognizing the intuitively obvious fact that we can approximate to $\phi_t(\omega)$ by sums of the form,

$$\phi_t(\omega) \sim \sum_{j=-n}^{n-1} a_j I_{\omega_j, \omega_{j+1}}(\omega),$$

where $a_j = \phi_t(\omega_j)$. In effect, we are simply forming a "step-function" approximation to $\phi_t(\omega)$, and clearly the accuracy of the approximation will increase as we increase n and decrease the interval between successive points, ω_j, ω_{j+1}. In fact, there is a well known result in the theory of functions which states that any continuous function can be constructed as the limit of a sequence of functions, starting from step functions of the above type, and this result underlies one approach to the definition of Lebesgue integration—see Riesz and Sz. Nagy (1955). Hence, we may write,

$$\phi_t(\omega) = \lim_{n \to \infty} \sum_j a_j I_{\omega_j, \omega_{j+1}}(\omega). \tag{4.11.30}$$

Now define the process $Z(\omega)$ by writing, for any ω_a, ω_b, s.t. $\omega_a \leq \omega_b$,

$$Z(\omega_b) - Z(\omega_a) = M[I_{\omega_a, \omega_b}(\omega)].$$

Then $Z(\omega)$ is clearly an orthogonal process since for any two non-overlapping intervals (ω_1, ω_2), (ω_3, ω_4),

$$E[\{Z(\omega_4) - Z(\omega_3)\}^* \{Z(\omega_2) - Z(\omega_1)\}] = (I_{\omega_4, \omega_3}(\omega), I_{\omega_2, \omega_1}(\omega))$$

$$= 0, \quad \text{by (4.11.29), (cf. (4.11.18))}.$$

Also,

$$E[|Z(\omega_b) - Z(\omega_a)|^2] = \|I_{\omega_b, \omega_a}(\omega)\|^2 = \int_{\omega_a}^{\omega_b} dH(\omega)$$

$$= H(\omega_b) - H(\omega_a). \tag{4.11.31}$$

Now applying the mapping M to each side of (4.11.30) we obtain,

$$M[\phi_t(\omega)] = \lim_{n \to \infty} \sum_j a_j M[I_{\omega_j, \omega_{j+1}}(\omega)]$$

$$= \lim_{n \to \infty} \sum_j \phi_t(\omega_j)\{Z(\omega_{j+1}) - Z(\omega_j)\}. \tag{4.11.32}$$

As $n \to \infty$ and the intervals between the $\{\omega_j\}$ decreases the right-hand side of (4.11.32) converges to the Stieltjes integral,

$$\int_{-\infty}^{\infty} \phi_t(\omega) \, dZ(\omega).$$

But, by the definition of M, $M[\phi_t(\omega)] = X(t)$, and $\phi_t(\omega)$ is, by definition, the function $\exp(it\omega)$. Hence we finally obtain,

$$X(t) = \int_{-\infty}^{\infty} e^{it\omega} \, dZ(\omega),$$

and, on writing ω for ω_a, $\omega + d\omega$ for ω_b in (4.11.31), we have,

$$E[|dZ(\omega)|^2] = dH(\omega).$$

A full account of Cramer's original proof is given in Doob (1953) and Grenander and Rosenblatt (1957a), and a somewhat more general version of essentially the same approach is given by Parzen (1959, 1961a).

The essence of the above proof may be described fairly simply in the following way. What we are doing finally (as $n \to \infty$) is using a set of δ-functions as an orthogonal basis for the space H_ϕ, and then writing,

$$\phi_t(\omega) = \int_{-\infty}^{\infty} \phi_t(\theta)\delta(\theta - \omega) \, d\theta, \qquad \text{all } \omega. \qquad (4.11.32a)$$

For each ω, the mapping of $\delta(\theta - \omega)$ is, in effect, the quantity "$\{dZ(\omega)/d\omega\}$", and applying the mapping to the above representation of $\phi_t(\omega)$ immediately gives the spectral representation of $X(t)$. In fact, these ideas can be made quite precise, and a more succinct version of the above proof (which bypasses the limiting process) can be constructed as follows. First, we re-write (4.11.32a) in a more rigorous form as,

$$\phi_t(\omega) = \int_{-\infty}^{\infty} \phi_t(\theta) \, dI(\theta - \omega), \qquad (4.11.32b)$$

where

$$I(\theta) = \begin{cases} 1, & \theta \geq 0, \\ 0, & \theta < 0, \end{cases}$$

so that $I(\theta - \omega)$ is the indicator function of the set $(-\infty, \theta)$, i.e. $I(\theta - \omega) \equiv I_{-\infty,\theta}(\omega)$, in the previous notation. (Note that if the right-hand side of (4.11.32b) is interpreted as a Lebesgue–Stieltjes integral then the result holds quite generally, we do not even require continuity of the $\phi_t(\omega)$.) Now, for each θ, define $Z(\theta)$ by

$$Z(\theta) = M[I(\theta - \omega)].$$

This is exactly the same as the previous definition of $Z(\theta)$, recalling that $I(\theta - \omega) \equiv I_{-\infty,\theta}(\omega)$. We now apply the mapping M to both sides of

(4.11.32b), and since the integral is a "linear operation" we obtain immediately,

$$X(t) = M[\phi_t(\omega)] = \int_{-\infty}^{\infty} \phi_t(\theta) \, d[M[I(\theta - \omega)]]$$

$$= \int_{-\infty}^{\infty} \phi_t(\theta) \, dZ(\theta).$$

The properties of $Z(\theta)$ follow as before.

(C) FUNCTIONAL ANALYTIC APPROACH

Perhaps the most elegant proof of the spectral representation theorem is that due to Kolmogorov (1941a). In this approach the spectral representation of stationary processes emerges as an almost trivial special case of the more general spectral theory of "unitary operators". The algebraic manipulations involved are delightfully simple, but to understand what is going on "beneath the surface" one requires a fairly deep knowledge of functional analysis. A very brief sketch of the proof for the discrete parameter case is as follows. Again, we regard X_t as an element of a Hilbert space, and define the inner product and norm as above. We now introduce the operator U defined by

$$UX_t = X_{t+1}, \qquad t = 0, \pm 1, \pm 2, \ldots \qquad (4.11.33)$$

The crucial point is that since $\{X_t\}$ is a stationary process we have, for all s, t,

$$(UX_s, UX_t) = E[X_{s+1}^* X_{t+1}] = E[X_s^* X_t] = (X_s, X_t).$$

Thus, the operator U preserves inner products, and in the language of functional analysis we call U a "*unitary operator*". We now appeal to a basic result which states that *a unitary operator has a spectral representation of the form*,

$$U = \int_{-\pi}^{\pi} e^{i\omega} \, dE(\omega), \qquad (4.11.34)$$

where the $\{dE(\omega)\}$ are the so-called "orthogonal projection operators" (see, e.g. Taylor (1964), Ch. 6).

From the "orthogonality" property of the $\{dE(\omega)\}$ it may be shown that, for all integral t, U^t has the representation,

$$U^t = \int_{-\pi}^{\pi} e^{it\omega} \, dE(\omega). \qquad (4.11.35)$$

The spectral representation of X_t is now immediate on writing

$$X_t = U^t X_0 = \int_{-\pi}^{\pi} e^{it\omega} \, dZ(\omega), \qquad (4.11.36)$$

where
$$Z(\omega) = E(\omega) \cdot X_0. \qquad (4.11.37)$$

The orthogonality property of $Z(\omega)$ follows from the corresponding property of $E(\omega)$.

A detailed and rigorous account of this proof is given by Kolmogorov (1941a) (see also Doob (1953), p. 638). Hannan (1960), p. 4, describes a simplified version of what is essentially the same proof, the simplification arising from the fact that Hannan first restricts the discussion to the case where the parameter t is restricted to a finite set of values, but the basic ideas involved are very similar.

In the "finite dimensional" case the spectral representation of the process can be derived fairly simply from the spectral resolution of the covariance matrix. Thus, we consider a finite set of random variables, $X = (X_1, X_2, \ldots X_N)$ and write $R_X = \{R(i, j)\}$, where $R(i, j) = \text{cov}\{X_i, X_j\}$. (The matrix R_X is, of course, hermitian and positive semi-definite.) Suppose R_X has N distinct eigenvalues, $\lambda_1, \lambda_2, \ldots \lambda_N$, and eigenvectors, $\mu_1, \mu_2, \ldots \mu_N$, (so that $\mu_j' \mu_k = \delta_{jk}$). Then we may express R_X in terms of its "spectral resolution" as

$$R_X = \sum_{k=1}^{N} \lambda_k \mu_k \mu_k', \qquad (4.11.38)$$

where the prime denotes both transposition and conjugation. Since R_X is hermitian, all the λ_k are real valued, and since it is positive semi-definite, $\lambda_k \geq 0$, all k. Equation (4.11.38) may be written alternatively as

$$R_X = M' \Lambda M, \qquad (4.11.39)$$

where $M = (\mu_1, \mu_2, \ldots \mu_N)$ and $\Lambda = \text{diag}(\lambda_1, \lambda_2, \ldots \lambda_N)$. Corresponding to (4.11.38), we may obtain a "spectral representation" for the $\{X_i\}$ by writing

$$Y = M'X. \qquad (4.11.40)$$

Then, (with an obvious notation)

$$R_Y = M R_X M' = \Lambda.$$

Hence R_Y is a diagonal matrix, and consequently the $\{Y_i\}$ (the components of Y) are *orthogonal* random variables. Since M is an orthogonal matrix, (4.11.40) can be inverted to give

$$X = MY, \qquad (4.11.41)$$

or

$$X_i = \mu_i' Y = \sum_j \mu_{ij} Y_j,$$

say, which gives the required "spectral representation".

To adapt the above results to the case of a "finite" stationary process we form the $(X_1, X_2, \ldots X_N)$ into a *"circular"* process (cf. Section 5.3.5), i.e. we define an autocovariance function, $R(r) = E\{X_t X_{t+r}\}$, for *all* r by using the convention that when $(t+r) > N$, we reduce its value modulo N. This means, e.g., that $X_{N+1} \equiv X_1$, $X_{N+2} \equiv X_2, \ldots$ and so on. The stationarity condition now implies that,

$$E\{X_t X_{t+r}\} = E\{X_{t+1} X_{t+1+r}\} = \ldots = E\{X_{N+1-r} X_1\},$$

i.e. we must have $R(r) = R(N-r)$, and the covariance matrix of $(X_1, X_2, \ldots X_N)$ now takes the form

$$\boldsymbol{R}_X = \begin{vmatrix} R(0) & R(1) & R(2) & \ldots & R(N-1) \\ R(N-1) & R(0) & R(1) & \ldots & R(N-2) \\ \cdots\cdots\cdots\cdots\cdots\cdots\cdots\cdots\cdots\cdots\cdots\cdots \\ R(1) & R(2) & \ldots & R(N-1) & R(0) \end{vmatrix}.$$

This matrix has the form of a *circulant*, and shows that (as far as second order properties are concerned) the $\{X_i\}$ possess a form of "circular symmetry", i.e. $\boldsymbol{R}_X$ is invariant under cyclic permutations of the $\{X_i\}$. (In fact, the term "circular process" arises from the fact that we may think of the $\{X_i\}$ as points arranged on a circle.) Thus, if we introduce the cyclic permutation matrix

$$T = \begin{bmatrix} 0 & 1 & 0 & \ldots & 0 \\ 0 & 0 & 1 & \ldots & 0 \\ \cdots\cdots\cdots\cdots\cdots\cdots \\ 0 & 0 & 0 & \ldots & 1 \\ 1 & 0 & 0 & \ldots & 0 \end{bmatrix},$$

then $Z = TX$ has the same covariance matrix as X, i.e. we have

$$T' \boldsymbol{R}_X T = \boldsymbol{R}_X, \qquad \text{or} \quad \boldsymbol{R}_X T = T \boldsymbol{R}_X.$$

(Note that since T is a permutation matrix it is orthogonal.) Hence $\boldsymbol{R}_X$ commutes with T (and with all powers of T), and therefore has the same eigenvectors as T. But the eigenvectors of T are

$$\boldsymbol{\mu}_j' = \frac{1}{\sqrt{N}} [\exp(i\omega_j), \exp(2i\omega_j), \ldots \exp(Ni\omega_j)],$$

where $\omega_j = 2\pi j / N$, and are thus the eigenvectors of $\boldsymbol{R}_X$. The spectral resolution of $\boldsymbol{R}_X$ is therefore given by (from (4.11.38)),

$$R(s, t) = \frac{1}{N} \sum_{k=1}^{N} \lambda_k \exp(is\omega_k) \exp(-it\omega_k) = \frac{1}{N} \sum_{k=1}^{N} \lambda_k \exp[i\omega_k(s-t)],$$

$$(4.11.42)$$

which is, of course, a function of $(s - t)$ only. The eigenvalues, $\{\lambda_k\}$, of $\boldsymbol{R}_X$ are given by inverting (4.11.42), yielding

$$\lambda_k = \sum_{r=0}^{N-1} R(r) \exp(-i\omega_k r), \qquad (4.11.43)$$

where we have written $R(r) = R(s, s+r)$ (independent of s). We observe from (4.11.43) the important result that *the eigenvalues of the covariance matrix $\boldsymbol{R}_X$ are proportional to the values of the (formal) "spectral density function"* at the frequencies $\{\omega_k\}$. The spectral representation of the (circular) process is, from (4.11.41), given by

$$X_t = \frac{1}{\sqrt{N}} \sum_{k=1}^{N} \exp(it\omega_k) Y_k, \qquad t = 1, \ldots N, \qquad (4.11.44)$$

where the $\{Y_k\}$ are othogonal variables. This representation involves only the discrete set of frequencies $\{\omega_k\}$, as we would expect from the essentially periodic form of the circular process. However, as $N \to \infty$, the circular process approaches the "ordinary" type of stationary process, the set of frequencies $\{\omega_k\}$ becomes dense over the interval $(0, \pi)$, and (4.11.44) becomes the integral representation (4.11.19).

General orthogonal expansions

In discussing the "function theory" approach (proof (B) above) we incidentally introduced the notation $\phi_t(\omega)$ to denote the complex exponential function $e^{it\omega}$. However, the reason for introducing the function $\phi_t(\omega)$ was not merely for ease of notation; rather we were aiming to indicate an extremely important generalization of the spectral representation theorem. The essential point is that proof (B) does not really depend on the fact that $\phi_t(\omega)$ represents the particular function $e^{it\omega}$; in fact, *it remains perfectly valid no matter what form $\phi_t(\omega)$ takes*, provided only that the "family" of functions, $\{\phi_t(\omega)\}$, $-\infty < t < \infty$, are such that the covariance function $R(s, t) = E[X^*(s)X(t)]$ can be expressed in the form (4.11.28) with respect to *some* suitable function $H(\omega)$, not necessarily the integrated spectrum. Of course, if $X(t)$ is stationary we know that (4.11.28) will be valid if, in particular, we choose $\phi_t(\omega)$ to be $e^{it\omega}$, and $H(\omega)$ to be the integrated spectrum, but there is no reason why we should not consider other choices of $\phi_t(\omega)$, provided that (4.11.28) is satisfied. These considerations assume greater significance in the case of *non-stationary* processes when $\phi_t(\omega) \equiv e^{it\omega}$ is no longer a permissible choice (for if it were it would necessarily mean that $X(t)$ were stationary), but there may well be other types of functions which will satisfy (4.11.28) even when $X(t)$ is non-stationary. We will return to this point in much greater detail in Chapter 11. Meanwhile, for future reference, we summarize the conclusion of proof (B) in its more general form—which is then known as the "*general orthogonal expansion theorem*" (see e.g. Bartlett (1955), p. 143 and Grenander and Rosenblatt (1957a), p. 27). To emphasize the fact that with a general choice of $\phi_t(\omega)$ the function $H(\omega)$ is no longer the integrated spectrum, we rewrite $dH(\omega)$ as $d\mu(\omega)$. In this context, we should refer more precisely to $\mu(\omega)$ as a "measure".

Theorem 4.11.2 General Orthogonal Expansions *Let $\{X(t)\}$ be a continuous parameter zero mean process (not necessarily stationary) with covariance function $R(s, t) = E[X^*(s)X(t)]$. If there exist a family of functions, $\{\phi_t(\omega)\}$, defined on the real line, and indexed by the suffix t, and a measure, $\mu(\omega)$, on the real line, such that for each t, $\phi_t(\omega)$ is quadratically integrable with respect to the measure μ, i.e.*

$$\int_{-\infty}^{\infty} |\phi_t(\omega)|^2 \, d\mu(\omega) < \infty, \qquad all \ t,$$

and for all s, t, $R(s, t)$ admits a representation of the form,

$$R(s, t) = \int_{-\infty}^{\infty} \phi_s^*(\omega)\phi_t(\omega) \, d\mu(\omega), \tag{4.11.45}$$

then the process $\{X(t)\}$ admits a representation of the form,

$$X(t) = \int_{-\infty}^{\infty} \phi_t(\omega) \, dZ(\omega), \tag{4.11.46}$$

where $\{Z(\omega)\}$ is an orthogonal process with

$$E[|dZ(\omega)|^2] = d\mu(\omega). \tag{4.11.47}$$

Conversely, if $X(t)$ admits a representation of the form (4.11.46) with $\{Z(\omega)\}$ an orthogonal process satisfying (4.11.47), then $R(s, t)$ admits a representation of the form (4.11.45).

Proof. The first part of the theorem has, in effect, already been proved in proof (B), we simply drop the constraint that $\phi_t(\omega) \equiv e^{it\omega}$, and proceed directly from (4.11.28). The second part follows immediately from the orthogonality of $Z(\omega)$. For, we have,

$$R(s, t) = E[X^*(s)X(t)] = \int_{-\infty}^{\infty}\int_{-\infty}^{\infty} \phi_s^*(\omega)\phi_t(\omega')E[dZ^*(\omega) \, dZ(\omega')].$$

Since $E[dZ^*(\omega) \, dZ(\omega')] = 0$ unless $\omega = \omega'$, the double integral reduces to a single integral, i.e. we have

$$R(s, t) = \int_{-\infty}^{\infty} \phi_s^*(\omega)\phi_t(\omega)E[|dZ(\omega)|^2] = \int_{-\infty}^{\infty} \phi_s^*(\omega)\phi_t(\omega) \, d\mu(\omega)$$

and the proof is complete.

Discrete parameter case

The above theorem holds equally well for discrete parameter processes, the only modification being that the range of integration in (4.11.45) and (4.11.46) is changed from $(-\infty, \infty)$ to $(-\pi, \pi)$. (Note that (4.11.45) may be regarded as the "infinite dimensional" version of (4.11.38).)

4.12 LINEAR TRANSFORMATIONS AND FILTERS

Apart from the direct physical interpretation of the spectrum, one of the main reasons for the widespread use of spectral analysis as an analytical tool is due to the fact that the spectrum provides a very simple description of the effects of linear transformations of a stationary process. The use of spectral analysis in this type of problem was discussed briefly in Chapter 1 (see Section 1.10, paragraph 3), and we now return to the study of linear transformations in a more detailed form.

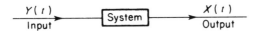

Fig. 4.20. Single input/single output system.

Suppose that we have a "system" with a single input terminal and a single output terminal. Suppose further that the "signal" applied to the input terminal is a time varying quantity $Y(t)$ and the corresponding "signal" at the output terminal is another time varying quantity $X(t)$ as described schematically in Fig. 4.20: here, we have depicted the "system" in the conventional "black-box" format. This situation arises in a multitude of practical problems, perhaps the most well known being the case where the "system" represents an electrical circuit, in which case the input and output signals will be either currents or voltages. On the other hand, the system could equally well be a mechanical device, with the input and output representing the displacements of two components, or alternatively the system could abstract in nature—such as an economic system—with the input representing, e.g., the bank rate, and the output representing the money supply. However, irrespective of the nature of the system, we can write down a general form of the relationship between input and output if we assume that the system is *linear and time-invariant*. The property of linearity means, in effect, that at any time t the value of the output is a *weighted linear combination of present, past and future values of the input*. (Although this is true only in an approximate sense for real systems it nevertheless provides a reasonably accurate description for a surprisingly large variety of cases; for physical systems the output will, of course, depend only on the present and past values of the input.) We may thus write (when t is continuous),

$$X(t) = \int_{-\infty}^{\infty} g(u) Y(t-u) \, du, \qquad (4.12.1)$$

where $g(u)$ is some (deterministic) function which depends on the detailed structure of the system but is independent of the form of the input. The property of time-invariance means that the function $g(u)$ does not itself vary over time, so that (4.12.1) holds for all t. For the purpose of the present discussion we will assume that $g(u)$ is an absolutely integrable function, i.e. that

$$\int_{-\infty}^{\infty} |g(u)| \, du < \infty. \qquad (4.12.2)$$

For physical systems which operate in "real time" the output at time t cannot, of course, depend on future values of the input. Hence in such cases we must have

$$g(u) = 0, \qquad u < 0,$$

and the range of integration in (4.12.1) is now, $(0, \infty)$. When $g(u)$ satisfies this condition the system is said to be "*physically realizable*" or "*non-anticipative*". For example, if the system is a simple R–C circuit, and $Y(t)$, $X(t)$, denote respectively the input and output voltages, then as shown in Section (1.10), (4.12.1) holds with $g(u)$ of the form

$$g(u) = \begin{cases} (1/RC) \exp(-u/RC), & u \geq 0, \\ 0, & u < 0. \end{cases}$$

More generally, (4.12.1) will hold (with $g(u) = 0$, $u < 0$) for a general electrical network provided it contains only "linear" components, i.e. consists of resistors, capacitors, and inductors, but does not contain, for example, a rectifying device such as a diode (this being a non-linear unit).

Suppose now that we feed the system with a random input. More specifically, suppose that $Y(t)$ is a stationary stochastic process with variance σ_Y^2 and autocorrelation function $\rho_Y(\tau)$. Then it follows from (4.12.1) (by an obvious generalization of (2.13.15)) that

$$\text{cov}(X(t), X(t+\tau)) = \int_{-\infty}^{\infty} \int_{-\infty}^{\infty} g(u)g(v) \, \text{cov}(Y(t-u), Y(t+\tau-v)) \, du \, dv$$

$$= \sigma_Y^2 \int_{-\infty}^{\infty} \int_{-\infty}^{\infty} g(u)g(v)\rho_Y(\tau+u-v) \, du \, dv. \qquad (4.12.3)$$

The right-hand side of (4.12.3) clearly depends only on τ (and not on t). Also, it follows immediately from (4.12.1) that if $E[Y(t)]$ is independent of t so is $E[X(t)]$. Hence, $X(t)$ is also stationary, and we may now re-write (4.12.3) as

$$\sigma_X^2 \rho_X(\tau) = \sigma_Y^2 \int_{-\infty}^{\infty} \int_{-\infty}^{\infty} g(u)g(v)\rho_Y(\tau + u - v)\, du\, dv, \qquad (4.12.4)$$

where σ_X^2, $\rho_X(\tau)$, denote respectively the variance and autocorrelation function of $X(t)$.

Now consider the case where $Y(t)$ has a purely continuous spectrum with (normalized) spectral density function, $f_Y(\omega)$, which exists for all ω. (Note that this implies, in particular, that $E[Y(t)] = 0$, which in turn implies that $E[X(t)] = 0$.) Then the (normalized) spectral density function of $X(t)$, $f_X(\omega)$, is given by (cf. (4.8.16)),

$$\sigma_X^2 f_X(\omega) = \frac{\sigma_X^2}{2\pi} \int_{-\infty}^{\infty} \rho_X(\tau)\, e^{-i\omega\tau}\, d\tau.$$

$$= \frac{\sigma_Y^2}{2\pi} \int_{-\infty}^{\infty} \int_{-\infty}^{\infty} \int_{-\infty}^{\infty} g(u)g(v)\rho_Y(\tau + u - v)\, e^{-i\omega\tau}\, du\, dv\, d\tau,$$

$$\text{(using (4.12.4))}$$

$$= \frac{\sigma_Y^2}{2\pi} \left\{ \int_{-\infty}^{\infty} g(u)\, e^{i\omega u}\, du \right\} \left\{ \int_{-\infty}^{\infty} g(v)\, e^{-i\omega v}\, dv \right\}$$

$$\times \left\{ \int_{-\infty}^{\infty} e^{-i\omega(\tau + u - v)} \rho_Y(\tau + u - v)\, d\tau \right\}. \qquad (4.12.5)$$

Writing $\tau' = \tau + u - v$ in the last integral (with u, v, fixed) this becomes,

$$\int_{-\infty}^{\infty} e^{-i\omega(\tau + u - v)} \rho_Y(\tau + u - v)\, d\tau = \int_{-\infty}^{\infty} e^{-i\omega\tau'} \rho_Y(\tau')\, d\tau'$$

$$= 2\pi f_Y(\omega), \qquad \text{all } u, v.$$

Noting that the second integral in (4.12.5) is the complex conjugate of the first integral, we thus obtain,

$$\sigma_X^2 f_X(\omega) = \sigma_Y^2 f_Y(\omega)|\Gamma(\omega)|^2, \qquad (4.12.6)$$

where

$$\Gamma(\omega) = \int_{-\infty}^{\infty} g(u)\, e^{-i\omega u}\, du, \qquad (4.12.7)$$

is called the *transfer function* of the system. The transfer function, $\Gamma(\omega)$, is the Fourier transform of $g(u)$, and if condition (4.12.2) holds this ensures that $\Gamma(\omega)$ exists (i.e. is finite) for all ω, (cf. condition (4.5.9)). Rewriting (4.12.6) in terms of the non-normalized spectral density functions, $h_X(\omega)$ $(= \sigma_X^2 f_X(\omega))$, and $h_Y(\omega)(= \sigma_Y^2 f_Y(\omega))$, we have

$$h_X(\omega) = h_Y(\omega)|\Gamma(\omega)|^2. \tag{4.12.8}$$

However, for the right-hand side of (4.12.1) to represent a "well defined" process (i.e. for the integral to converge in mean-square) we certainly require that $X(t)$ should have finite variance, or equivalently, finite total power. The condition for this is,

$$\int_{-\infty}^{\infty} h_X(\omega) \, d\omega < \infty,$$

which, in view of (4.12.8), means that $\Gamma(\omega)$ and $h_Y(\omega)$ must together satisfy the condition,

$$\int_{-\infty}^{\infty} |\Gamma(\omega)|^2 h_Y(\omega) \, d\omega < \infty. \tag{4.12.9}$$

If $h_Y(\omega)$ is a bounded function of ω, i.e. if $h_Y(\omega) \leq M$, (a constant), all ω, then a sufficient condition for (4.12.9) to hold is,

$$\int_{-\infty}^{\infty} |\Gamma(\omega)|^2 \, d\omega < \infty,$$

which, (by Parseval's theorem), is equivalent to,

$$\int_{-\infty}^{\infty} g^2(u) \, du < \infty. \tag{4.12.10}$$

Note that we do not require $Y(t)$ necessarily to have finite variance (i.e. $\int_{-\infty}^{\infty} h_Y(\omega) \, d\omega < \infty$); this need not be so as long as (4.12.9) is satisfied. Thus, we could take $Y(t)$ to be a purely random process (i.e. "white noise"), so that $h_Y(\omega)$ is a constant, and $X(t)$ is still well defined provided $g(u)$ then satisfies (4.12.10).

Equations (4.12.7), (4.12.8) are basic in the study of linear transformations, (4.12.8) giving a more precise version of the result stated in Section 1.10, namely:

Non-normalized spectral density function of the output equals (non-normalized spectral density function of the input) × (the squared modulus of the transfer function).

The important feature of this result is that the *value of the output spectral density function at frequency ω depends purely on $|\Gamma(\omega)|^2$ and the value of the input spectral density function at the same frequency ω*; it is not "contaminated" by the value of the input spectral density function at any other frequency, ω', say. This is in sharp contrast with the "time-domain" form of the input/output relationship, as given by (4.12.1), where in general the value of the output at time t depends not only on the value of the input at time t but also on the values of the input at all other times. The "frequency-domain" description, viz. (4.12.8), has, in a sense, "disentangled" the relationship, so that we may now study the properties of the system separately at each frequency.

It is instructive to re-derive (4.12.8) by an alternative method based on the spectral representation of stationary processes. Let the input process have spectral representation,

$$Y(t) = \int_{-\infty}^{\infty} e^{i\omega t} \, dZ_Y(\omega).$$ (4.12.11)

Substituting this expression in (4.12.1) we obtain,

$$X(t) = \int_{-\infty}^{\infty} g(u)\left\{\int_{-\infty}^{\infty} e^{i\omega(t-u)} \, dZ_Y(\omega)\right\} du$$

$$= \int e^{i\omega t} \Gamma(\omega) \, dZ_Y(\omega),$$

(on changing the order of integration)

$$= \int e^{i\omega t} \, dZ_X(\omega),$$ (4.12.12)

say, where

$$dZ_X(\omega) = \Gamma(\omega) \, dZ_Y(\omega) \qquad \text{(all } \omega\text{)}.$$ (4.12.13)

(If we think heuristically of $dZ_Y(\omega)$ as the "Fourier transform" of $Y(t)$ then (4.12.13) follows immediately on noting that $X(t)$ is the convolution of $g(t)$ and $Y(t)$ and hence its "Fourier transform", $dZ_X(\omega)$ is the product of $\Gamma(\omega)$, (the Fourier transform of $g(u)$) and $dZ_Y(\omega)$.) Since $Z_Y(\omega)$ is an orthogonal process it follows that $Z_X(\omega)$ is similarly an orthogonal process (recall that, for each ω, $\Gamma(\omega)$ is simply a constant), and hence (4.12.12) gives the spectral representation of the output process, $X(t)$. Hence, if we denote the (non-normalized) integrated spectra of $X(t)$, $Y(t)$, by $H_X(\omega)$, $H_Y(\omega)$, respectively, we have,

$$dH_X(\omega) = E[|dZ_X(\omega)|^2] = |\Gamma(\omega)|^2 E[|dZ_Y(\omega)|^2] = |\Gamma(\omega)|^2 \, dH_Y(\omega).$$
(4.12.14)

When the spectral density functions exist for all ω, $dH_X(\omega) = h_X(\omega) \, d\omega$, $dH_Y(\omega) = h_Y(\omega) \, d\omega$, and (4.12.14) reduces to (4.12.8). (Equation (4.12.14) is, however, valid even when the sprectral density functions do not exist.)

Discrete parameter processes

When the input and output are both discrete parameter processes, the analogue of (4.12.1) is

$$X_t = \sum_{u=-\infty}^{\infty} g_u Y_{t-u}, \qquad (4.12.15)$$

where $\{g_u\}$ is a given deterministic sequence. We then obtain, following either of the above arguments (with obvious modifications),

$$\sigma_X^2 f_X(\omega) = \sigma_Y^2 f_Y(\omega)|\Gamma(\omega)|^2, \qquad -\pi \leqslant \omega \leqslant \pi, \qquad (4.12.16)$$

where $f_X(\omega)$, $f_Y(\omega)$, denote respectively the (normalized) spectral density functions of $\{X_t\}$ and $\{Y_t\}$ (each now defined only for $|\omega| \leqslant \pi$), and

$$\Gamma(\omega) = \sum_{u=-\infty}^{\infty} g_u e^{-i\omega u}, \qquad (4.12.17)$$

is again called the *transfer function*. In terms of the non-normalized spectral density functions (4.12.16) may be written as,

$$h_X(\omega) = h_Y(\omega)|\Gamma(\omega)|^2. \qquad (4.12.18)$$

The condition analogous to (4.12.9) that $\{X_t\}$ should have finite variance is,

$$\int_{-\pi}^{\pi} |\Gamma(\omega)|^2 h_Y(\omega).d\omega < \infty. \qquad (4.12.19)$$

Examples

As illustrations of the above results we now apply them to the problem of calculating some spectral density functions.

1. *General linear processes*

The continuous parameter general linear process, as defined by (3.7.74), (Section 3.7.8), may be regarded as a special case of (4.12.1) when $Y(t) \equiv \varepsilon(t)$ (the purely random process), and $g(u) = 0$, $u < 0$. In this case, we have from (4.10.4),

$$h_Y(\omega) = \sigma_W^2/2\pi, \qquad \text{all } \omega$$

(σ_W^2 being a constant), and (4.12.8) then gives,

$$h_X(\omega) = (\sigma_W^2/2\pi) \cdot |\Gamma_0(\omega)|^2, \qquad (4.12.20)$$

where

$$\Gamma_0(\omega) = \int_0^\infty g(u)\, e^{-i\omega u}\, du.$$

Since here $h_Y(\omega)$ is a constant, the condition for $X(t)$ to have finite variance reduces to (4.12.10), which in this case becomes,

$$\int_0^\infty g^2(u)\, du < \infty. \tag{4.12.21}$$

In the discrete parameter case the general linear process is defined by (3.5.56) (Section 3.5.7), and this is a special case of (4.12.15) in which $Y_t = \varepsilon_t$ is a purely random process and $g_u = 0$, $u < 0$. We then have,

$$h_Y(\omega) = \sigma_\varepsilon^2/2\pi, \qquad -\pi \le \omega \le \pi,$$

and (4.12.18) gives,

$$h_X(\omega) = (\sigma_\varepsilon^2/2\pi)\,.\,|\Gamma_1(\omega)|^2, \qquad -\pi \le \omega \le \pi, \tag{4.12.22}$$

where

$$\Gamma_1(\omega) = \sum_{u=0}^\infty g_u\, e^{-i\omega u}.$$

Note that $\Gamma_1(\omega)$ is essentially the same as the function $G(z)$ defined by (3.5.61); in fact, $\Gamma_1(\omega)$ is the form which $G(z)$ assumes when we set $z = e^{-i\omega}$, i.e. $\Gamma_1(\omega) \equiv G(e^{-i\omega})$. The condition (3.5.62), namely that $G(z)$ be analytic inside the unit circle, is now seen to be equivalent to the condition that $h_X(\omega)$ be an integrable function, i.e. that $\{X_t\}$ has finite variance.

Similarly, the function $\Gamma(z)$ defined by (3.7.78) reduces to $\Gamma_0(\omega)$ on setting $z = i\omega$, and the condition (3.7.82) is, of course, identical to (4.12.21).

2. *Moving average processes*

If we set $Y(t) \equiv \varepsilon(t)$ and

$$g(u) = \begin{cases} 1/l, & 0 \le u \le l, \\ 0, & \text{otherwise}, \end{cases} \tag{4.12.23}$$

then $X(t)$ is a continuous parameter MA(l) process of the form (3.7.59) with uniform "weighting" (i.e. the function $b(u)$ in (3.7.59) here takes a constant value). With the above form of $g(u)$ it is easy to show that,

$$|\Gamma(\omega)|^2 = \frac{\sin^2(l\omega/2)}{(l\omega/2)^2}, \tag{4.12.24}$$

and hence,

$$h_X(\omega) = \frac{\sigma_W^2 \sin^2(l\omega/2)}{2\pi . (l\omega/2)^2}. \tag{4.12.25}$$

More generally, if $Y(t)$ is a general stationary process, and $g(u)$ is of the above form, then $X(t)$ may be said to be obtained by applying a "moving average transformation" to $Y(t)$. From the above result we see that the "*moving average*" *operation multiplies the spectral density function by the factor on the right-hand side of* (4.12.24).

In the discrete parameter case the "equal-weighting" MA(l) model corresponds to taking $Y_t = \varepsilon_t$, a purely random process, and

$$g_u = \begin{cases} 1/(l+1), & u = 0, 1, \ldots, l, \\ 0, & \text{otherwise,} \end{cases}$$

(see section 3.5.5). Then,

$$|\Gamma(\omega)|^2 = \frac{\sin^2\{(l+1)\omega/2\}}{(l+1)^2 \sin^2(\omega/2)}. \tag{4.12.26}$$

Hence,

$$h_X(\omega) = \frac{\sigma_\varepsilon^2}{2\pi} . \frac{\sin^2\{(l+1)\omega/2\}}{(l+1)^2 \sin^2(\omega/2)}, \qquad -\pi \leqslant \omega \leqslant \pi. \tag{4.12.27}$$

Both (4.12.25) and (4.12.27) can of course be derived directly by evaluating the Fourier transform of the appropriate autocovariance function.

4.12.1 Filter Terminology; Gain and Phase

Relationships of the form (4.12.1), (4.12.15), are now known, almost universally, as "*filters*", and we say for example that $X(t)$ is a "*filtered version*" of $Y(t)$. The motivation for this nomenclature (which originated in electrical engineering) is due to the fact that by designing the system so that the function $g(u)$ is such that $\Gamma(\omega)$ vanishes over a band of frequencies we can "filter out" from the input all components with frequencies in this band. For example, if we design the system so that $\Gamma(\omega) = 0$, $\omega_1 \leqslant |\omega| \leqslant \omega_2$, then it follows from (4.12.13) that $dZ_X(\omega) = 0$, $\omega_1 \leqslant |\omega| \leqslant \omega_2$. Hence, the representation (4.12.12) of $X(t)$ contains no components with frequencies in the ranges $(-\omega_2, -\omega_1)$, (ω_1, ω_2), i.e. these frequencies have been completely suppressed by the system. In practice we cannot quite achieve this degree of "perfect suppression" since the above form of $\Gamma(\omega)$ would place unrealistic demands on the form of the function $g(u)$ (that is we could not, in general, synthesize an electrical filter with the required form of $g(u)$), but there are practical designs which provide a close approximation to this type of filter. (Were this not so it would be impossible to design radio receivers—for the whole point of a radio tuner (AM or FM) is that it is able to suppress all unwanted transmissions whose frequencies lie outside a narrow band centred on the frequency of the wanted signal. The ability with which the tuner is able to perform this task determines its "selectivity".)

In the language of filter theory we call $\{g(u)\}$ the *impulse response function*, since it represents the output of the system when the input consists of a δ-function "impulse" at $t = 0$, i.e. if we set $Y(t) = \delta(t)$ in (4.12.1) then,

$$X(t) = \int_{-\infty}^{\infty} g(u)\delta(t - u) \, du = g(t), \qquad \text{all } t.$$

Similarly, in the discrete parameter case, the sequence $\{g_u\}$ represents the output of the system when the input consists of a "unit impulse" at $t = 0$; i.e. if we set $Y_t = 1$, $t = 0$, $Y_t = 0$, $t \neq 0$, then (4.12.15) gives,

$$X_t = g_t, \qquad \text{all } t.$$

Since $\Gamma(\omega)$ is, in general, complex valued we may write it in the form,

$$\Gamma(\omega) = \gamma(\omega) \exp(i\phi(\omega)), \tag{4.12.28}$$

say. Then $\gamma(\omega)$ is called the *gain* (at frequency ω), and $\phi(\omega)$ the *phase-shift* (at frequency ω), of the filter. For, in the spectral representation (4.12.11) of the input, $Y(t)$, the term $\exp(i\omega t)$ has amplitude $|dZ_y(\omega)|$ and phase $\arg\{dZ_y(\omega)\}$. In the output $X(t)$ the amplitude of this term is

$$|dZ_X(\omega)| = |\Gamma(\omega) \, dZ_y(\omega)| = \gamma(\omega)|dZ_y(\omega)|, \tag{4.12.29}$$

i.e. on passing through the filter the amplitude has been multiplied by $\gamma(\omega)$. Similarly, the phase of $\exp(i\omega t)$ in the output is

$$\arg\{dZ_X(\omega)\} = \arg\{\Gamma(\omega) \, dZ_Y(\omega)\} = \phi(\omega) + \arg\{dZ_Y(\omega)\}, \tag{4.12.30}$$

i.e. the phase has been increased or "shifted" by $\phi(\omega)$. (As noted in Section 1.10, if the input is a pure sine wave the output is also a pure sine wave with exactly the same frequency, but the amplitude and phase are modified as above.) In order to have $\phi(\omega) = 0$, all ω (i.e. zero phase-shift at all frequencies) it is clearly necessary that $\Gamma(\omega)$ be real valued for all ω. This means that $g(u)$ must be an *even function*, (i.e. $g(u) = g(-u)$), and hence the filter cannot be *physically realizable*. Thus, all physically realizable filters have to have complex valued transfer functions and hence must produce phase-shifts at some (or all) frequencies.

In order to give a visual impression of the gain and phase characteristics of a filter it is customary to plot graphs of $\gamma(\omega)$ and $\phi(\omega)$ (as functions of ω) which are then called *gain and phase plots*. Note, however, that from the nature of its definition $\phi(\omega)$ is determined only mod 2π. If $\Gamma_R(\omega)$, $\Gamma_I(\omega)$, denote respectively the real and imaginary parts of $\Gamma(\omega)$ then

$$\phi(\omega) = \tan^{-1}\left\{\frac{\Gamma_I(\omega)}{\Gamma_R(\omega)}\right\}.$$

Hence, if we always take $\phi(\omega)$ to lie in $(-\pi, \pi)$ the graph may contain spurious discontinuities. Ideally, we should plot $\phi(\omega)$ on a "cylinder" rather than on a plane!

As an example, consider the "R–C filter" discussed previously in Sections 1.10 and 4.12 for which,

$$g(u) = \begin{cases} (1/RC)\exp(-u/RC), & u \geqslant 0 \\ 0, & u < 0. \end{cases}$$

The transfer function, $\Gamma(\omega)$, is then (from (4.12.7)),

$$\Gamma(\omega) = \frac{1}{RC}\int_0^\infty \exp(-u/RC)\exp(-i\omega u)\, du = \frac{1}{1 + iRC\omega}.$$

Hence, the gain and phase characteristics of this filter are given by,

$$\gamma(\omega) = |\Gamma(\omega)| = \frac{1}{(1 + R^2 C^2 \omega^2)^{1/2}} = \frac{1}{(1 + T^2 \omega^2)^{1/2}},$$

and

$$\phi(\omega) = \arg\{\Gamma(\omega)\} = \tan^{-1}\{-RC\omega\} = \tan^{-1}\{-T\omega\},$$

where $T = RC$ is called the "time constant" of the circuit. At $\omega = 0$, $\gamma(0) = 1$, and as ω increases $\gamma(\omega)$ decreases. Thus, this filter attenuates the high frequencies relative to the low frequencies, and may be regarded as a crude approximation to the "ideal" low-pass filter. (It is, in fact, frequently used in audio amplifiers as a simple form of "scratch filter".) The forms of $\gamma(\omega)$ and $\phi(\omega)$ are shown in Fig. 4.21. Note that $\phi(\omega)$ decreases mono-

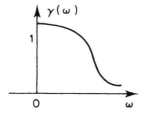
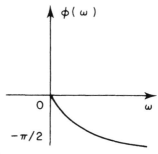

Fig. 4.21

tonically as ω increases, attaining the limiting value of $(-\pi/2)$ as $\omega \to \infty$. At very large frequencies there is therefore an approximate 90° phase shift between the input and output. Also, for large ω, $\gamma(\omega) \sim 1/\omega T$, and hence as ω doubles its value $\gamma(\omega)$ is reduced by a factor of $\frac{1}{2}$. In engineering language we would say that the limiting rate of attenuation of this R–C filter is "6dB per octave".

Cascaded filters

The basic relationships between the input and output spectral density functions extend very easily to the case where, instead of having a single filter, we have a set of n filters connected in series so that the output of the first filter becomes the input to the second filter, and so on. This arrangement is known as "cascaded filters", and is illustrated in Fig. 4.22.

Fig. 4.22. Cascaded filters.

Let $X_1(t)$ denote the input to the first filter, $X_2(t)$ the input to the second filter, . . . , and $X_{n+1}(t)$ the final output. If we denote the transfer function of the jth filter by $\Gamma_j(\omega)$, and write,

$$X_j(t) = \int_{-\infty}^{\infty} e^{i\omega t} \, dZ_j(\omega), \qquad j = 1, \ldots (n+1),$$

then, assuming that (4.12.9) is satisfied at each stage, (4.12.13) gives

$$dZ_{j+1}(\omega) = \Gamma_j(\omega) \, dZ_j(\omega)$$

$$= \ldots$$

$$= \Gamma_j(\omega)\Gamma_{j-1}(\omega) \ldots \Gamma_1(\omega) \, dZ_1(\omega).$$

Hence,

$$X_{n+1}(t) = \int_{-\infty}^{\infty} e^{i\omega t} \, dZ_{n+1}(\omega)$$

$$= \int_{-\infty}^{\infty} e^{i\omega t} \{\Gamma_n(\omega)\Gamma_{n-1}(\omega) \ldots \Gamma_1(\omega)\} \, dZ_1(\omega). \qquad (4.12.31)$$

Thus, the transfer function of the complete system, $\Gamma(\omega)$, is given by,

$$\Gamma(\omega) = \Gamma_n(\omega)\Gamma_{n-1}(\omega) \ldots \Gamma_1(\omega), \qquad (4.12.32)$$

and, with an obvious notation,

$$h_{X_{n+1}}(\omega) = |\Gamma_n(\omega)|^2 |\Gamma_{n-1}(\omega)|^2 \ldots |\Gamma_1(\omega)|^2 h_{X_1}(\omega)_1. \quad (4.12.33)$$

If $\gamma_j(\omega)$, $\phi_j(\omega)$, denote respectively the gain and phase of the jth filter, then the gain $\gamma(\omega)$ and phase $\phi(\omega)$ of the complete system are given by,

$$\gamma(\omega) = \gamma_n(\omega)\gamma_{n-1}(\omega) \ldots \gamma_1(\omega), \quad (4.12.34)$$

$$\phi(\omega) = \phi_n(\omega) + \phi_{n-1}(\omega) + \ldots + \phi_1(\omega). \quad (4.12.35)$$

Special Forms of Filters

1. *Band-pass filters*

If we choose $g(u)$ so that $|\Gamma(\omega)|^2$ has the form,

$$|\Gamma(\omega)|^2 = \begin{cases} 1, & \omega_1 \leq |\omega| \leq \omega_2, \\ 0, & \text{otherwise,} \end{cases} \quad (4.12.36)$$

(see Fig. 4.23), then the system is called a *"band-pass"* filter. The effect of this filter is to allow all frequency components in the "band" (ω_1, ω_2) to pass through unattenuated, but all other frequency components are completely suppressed. As mentioned above, this is an "ideal" form of filter in the sense that we cannot synthesize it using, for example, an electrical network. However, there are electrical filter designs which closely approximate the above form—the main difference being that in practice $|\Gamma(\omega)|^2$ would tend to have a continuous form with "rounded corners" rather than the discontinuous "ideal" form shown in Fig. 4.23.

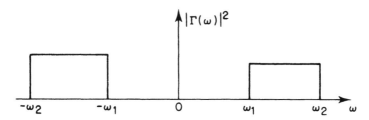

Fig. 4.23. Band-pass filter.

The impulse response function $g(u)$ is obtained by inverting (4.12.7), giving,

$$g(u) = \frac{1}{2\pi} \int_{-\infty}^{\infty} \Gamma(\omega) e^{i\omega u} \, d\omega. \quad (4.12.37)$$

In particular, when $\Gamma(\omega)$ (rather than $|\Gamma(\omega)|^2$) has the form (4.12.36), we find,

$$g(u) = \frac{1}{\pi} \int_{\omega_2}^{\omega_1} e^{i\omega u} \, d\omega,$$

i.e.

$$g(u) = \left\{ \frac{\sin \omega_2 u - \sin \omega_1 u}{\pi u} \right\}, \qquad \text{all } u. \qquad (4.12.38)$$

This does not, of course, correspond to a physically realizable filter (since here $\Gamma(\omega)$ is real valued) but we cannot use this form of $g(u)$ even in an "off-line" computational form since it has to operate on the input process over an infinite time interval.

2. Low-pass filters

If we choose $g(u)$ so that $|\Gamma(\omega)|^2$ has the form

$$|\Gamma(\omega)|^2 = \begin{cases} 1, & |\omega| \le \omega_0, \\ 0, & |\omega| > \omega_0, \end{cases} \qquad (4.12.39)$$

(see Fig. 4.24), then we obtain a *"low-pass"* filter. This filter passes all components with frequencies lower than ω_0, but completely suppresses all components with higher frequencies. (It is, in fact, a special case of the

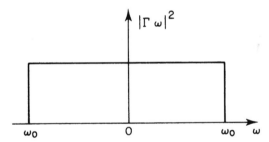

Fig. 4.24. Low-pass filter.

band-pass filter where $\omega_1 = 0$). Again, we cannot synthesize this filter exactly, but filters with approximating characteristics are commonly used in, e.g., audio amplifiers to reduce the effect of high frequency distortion.

3. High-pass filters

If we choose $g(u)$ so that

$$|\Gamma(\omega)|^2 = \begin{cases} 1, & |\omega| > \omega_0, \\ 0, & |\omega| \le \omega_0, \end{cases} \qquad (4.12.40)$$

(Fig. 4.25) we obtain a *"high-pass"* filter. Here, the filter passes all components with frequencies higher than ω_0 and suppresses all frequencies less than ω_0. Approximate versions of this type of filter are used in audio amplifiers to suppress low frequency distortion arising, e.g., from record turntable "rumble". (Bass and treble controls on audio amplifiers are simply variable filters which accentuate or attenuate the low and high frequencies (respectively) to varying degrees, depending upon the setting of the controls.)

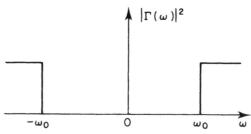

Fig. 4.25. High-pass filter.

4.12.2 Operational Forms of Filter Relationships: Calculation of Spectra

Apart from their obvious practical importance the filter relationships (4.12.6), (4.12.16), provide an elegant "short-cut" method for evaluating the spectral density functions of autoregressive and moving average models. To understand the basis of this method it is useful to re-write (4.12.1) and (4.12.15) in their "operational" forms.

I. *Differential operator form* (*continuous parameter case*)

Using the well known relationship, $B^u \equiv \exp(-uD)$, where B is the backward shift operator and D is the differential operator (see Section 3.7.6, equation (3.7.62)) we may write

$$Y(t-u) = \exp(-uD) . Y(t).$$

Substituting this expression for $Y(t-u)$ in (4.12.1) we obtain,

$$X(t) = \left\{ \int_{-\infty}^{\infty} g(u)\, e^{-uD}\, du \right\} . Y(t)$$

$$= G(D)Y(t), \qquad (4.12.41)$$

say, where

$$G(z) = \int_{-\infty}^{\infty} g(u)\, e^{-uz}\, du \equiv \Gamma(-iz). \qquad (4.12.42)$$

Hence, $\Gamma(\omega) \equiv \dot{G}(i\omega)$ and (4.12.6) may be written as

$$\sigma_X^2 f_X(\omega) = \sigma_Y^2 |G(i\omega)|^2 f_Y(\omega). \tag{4.12.43}$$

On the basis of this result we may infer more generally that if two stationary processes, $X(t)$ and $Y(t)$, are related by an equation of the form

$$X(t) = \phi(D) Y(t), \tag{4.12.44}$$

then

$$\sigma_X^2 f_X(\omega) = \sigma_Y^2 |\phi(i\omega)|^2 f_Y(\omega). \tag{4.12.45}$$

Alternatively, if we have a relationship of the form,

$$\Psi(D)X(t) = Y(t), \tag{4.12.46}$$

then, $\sigma_X^2 |\Psi(i\omega)|^2 f_X(\omega) = \sigma_Y^2 f_Y(\omega)$, and hence,

$$\sigma_X^2 f_X(\omega) = \frac{\sigma_Y^2}{|\Psi(i\omega)|^2} \cdot f_Y(\omega), \tag{4.12.47}$$

assuming that $\Psi(i\omega)$ does not vanish for any real ω. Strictly, the above justification of (4.12.45), (4.12.47) is valid only when $\phi(iz)$ and $\Psi(iz)$ each possess Fourier transforms, so that, e.g., (4.12.44) can then be transformed back into the convolution integral form (4.12.1). However, the range of validity of these results is much wider than might be expected, and it is easy to show, for example, that (4.12.45) holds even when ϕ is a polynomial of finite order. To show this, suppose first that

$$X(t) = D^k \cdot Y(t).$$

Then it follows immediately from the spectral representation of $Y(t)$ ((4.12.11)) that

$$X(t) = \int_{-\infty}^{\infty} e^{i\omega t} (i\omega)^k \, dZ_Y(\omega)$$

and hence

$$\sigma_X^2 f_X(\omega) \, d\omega = |i\omega|^{2k} E[|dZ_Y(\omega)|^2] = \sigma_Y^2 |i\omega|^{2k} f_Y(\omega) \, d\omega.$$

The corresponding result when ϕ is a polynomial follows by an obvious extension of the above argument.

Examples

We now use these results to derive the spectral density functions of the following models.

1. $AR(1)$ processes; continuous parameter

Here we have (cf. (3.7.11)),

$$\dot{X}(t) + \alpha X(t) = \varepsilon(t), \qquad (\alpha > 0).$$

This is exactly of the form (4.12.46) if we set

$$\Psi(D) = D + \alpha, \qquad Y(t) \equiv \varepsilon(t).$$

We then have immediately from (4.12.47) that,

$$\sigma_X^2 f_X(\omega) = \frac{1}{|i\omega + \alpha|^2} h_Y(\omega) = \frac{1}{(\alpha^2 + \omega^2)} \cdot \frac{\sigma_W^2}{2\pi}.$$

But, from (3.7.17), $\sigma_W^2 = 2\alpha\sigma_X^2$, and hence,

$$f_X(\omega) = \frac{1}{\pi} \cdot \frac{\alpha}{(\alpha^2 + \omega^2)},$$

in agreement with the previously derived result (4.10.8).

2. $AR(2)$ processes; continuous parameter

The AR(2) model (3.7.39) may be obtained from (4.12.46) on setting

$$\Psi(D) = D^2 + \alpha_1 D + \alpha_2, \qquad Y(t) \equiv \varepsilon(t).$$

Hence (4.12.47) now gives

$$\sigma_X^2 f_X(\omega) = \frac{1}{|-\omega^2 + i\alpha_1\omega + \alpha_2|^2} \cdot \frac{\sigma_W^2}{2\pi} = \frac{1}{(\alpha_2 - \omega^2)^2 + \alpha_1^2\omega^2} \cdot \frac{\sigma_W^2}{2\pi}.$$

From (3.7.41) we have, $\sigma_W^2 = 2\alpha_1\alpha_2\sigma_X^2$, and thus,

$$f_X(\omega) = \frac{1}{\pi} \left\{ \frac{\alpha_1\alpha_2}{(\alpha_2 - \omega^2)^2 + \alpha_1^2\omega^2} \right\},$$

in agreement with (4.10.11).

3. AR processes of general order; continuous parameter

If we set

$$\Psi(D) = D^k + \alpha_1 D^{k-1} + \ldots + \alpha_k, \qquad Y(t) \equiv \varepsilon(t),$$

then (4.12.46) gives the AR(k) model defined by (3.7.54). (We recall that for stationarity of $X(t)$ the roots of $\Psi(z)$ must all lie in the left-half plane—see Section 3.7.5.)

We now have from (4.12.47),

$$\sigma_X^2 f_X(\omega) = \frac{1}{|(i\omega)^k + \alpha_1(i\omega)^{k-1} + \ldots + \alpha_k|^2} \cdot \frac{\sigma_W^2}{2\pi}. \qquad (4.12.48)$$

For any given value of k we can, of course, simplify this expression by evaluating the denominator explicitly and calculating an expression for σ_X^2 in terms of σ_W^2 and $\alpha_1, \ldots \alpha_k$, as in the first two examples.

4. *MA(l) processes; continuous parameter*

This model has been considered previously and the spectral density function for the "uniform weighting" case is given by (4.12.25). We merely note that for all MA models the relationship between $X(t)$ and $\varepsilon(t)$ is initially given in the convolution integral form (cf. the discussion leading to (4.12.25)) and hence there is no point here in using the operational form of the filter results. The expression for $f_X(\omega)$ follows by direct application of (4.12.6), so that for the general MA(l) model, (3.7.59), we have,

$$\sigma_X^2 f_X(\omega) = (\sigma_W^2/2\pi) \cdot |B(\omega)|^2, \qquad (4.12.49)$$

where

$$B(\omega) = \int_0^l b(u) e^{-i\omega u} \, du. \qquad (4.12.50)$$

5. *Difference operator form (discrete parameter case)*

We may rewrite (4.12.15) in the form

$$X_t = \left(\sum_{u=-\infty}^{\infty} g_u B^u \right) Y_t,$$

(where, as previously, B denotes the backward shift operator), or,

$$X_t = G(B) Y_t, \qquad (4.12.51)$$

where

$$G(z) = \sum_{u=-\infty}^{\infty} g_u z^u, \qquad (4.12.52)$$

so that

$$\Gamma(\omega) = \sum_{u=-\infty}^{\infty} g_u e^{-i\omega u} \equiv G(e^{-i\omega}). \qquad (4.12.53)$$

Hence, (4.12.6) may now be written as,

$$\sigma_X^2 f_X(\omega) = \sigma_Y^2 |G(e^{-i\omega})|^2 f_Y(\omega), \qquad -\pi \leqslant \omega \leqslant \pi. \qquad (4.12.54)$$

On the basis of this result we may infer that if two stationary processes, $\{X_t\}$, $\{Y_t\}$, are related by an equation of the form

$$X_t = \phi(B) Y_t, \qquad (4.12.55)$$

then

$$\sigma_X^2 f_X(\omega) = \sigma_Y^2 |\phi(e^{-i\omega})|^2 f_Y(\omega), \qquad -\pi \leqslant \omega \leqslant \pi. \qquad (4.12.56)$$

Similarly, if we have,

$$\Psi(B) X_t = Y_t, \qquad (4.12.57)$$

then

$$\sigma_X^2 |\Psi(e^{-i\omega})|^2 f_X(\omega) = \sigma_Y^2 f_Y(\omega),$$

and therefore,

$$\sigma_X^2 f_X(\omega) = \frac{\sigma_Y^2}{|\Psi(e^{-i\omega})|^2} f_Y(\omega), \qquad -\pi \leqslant \omega \leqslant \pi, \qquad (4.12.58)$$

assuming that $\Psi(e^{-i\omega})$ does not vanish in the range $(-\pi, \pi)$.

4.12.3 Transformation of the Autocovariance Function

If we define the "autocovariance generating functions" of X_t, Y_t by

$$\tilde{R}_X(z) = \sum_{s=-\infty}^{\infty} R_X(s) z^s, \qquad \tilde{R}_Y(z) = \sum_{s=-\infty}^{\infty} R_Y(s) z^s,$$

then clearly

$$\tilde{R}_X(e^{-i\omega}) = 2\pi \sigma_X^2 f_X(\omega), \qquad \tilde{R}_Y(e^{-i\omega}) = 2\pi \sigma_Y^2 f_Y(\omega).$$

Hence, replacing $e^{-i\omega}$ by z in (14.12.56) we obtain

$$\tilde{R}_X(z) = \phi(z)\phi(z^{-1})\tilde{R}_Y(z). \qquad (4.12.58a)$$

Equating coefficients of z^s on both sides, and writing

$$\phi(z)\phi(z^{-1}) = \sum_{u=-\infty}^{\infty} \theta_u z^u,$$

say, we find

$$R_X(s) = \sum_{u=-\infty}^{\infty} \theta_u R_Y(s-u) = \left(\sum_{u=-\infty}^{\infty} \theta_u B^u \right) R_Y(s),$$

i.e.

$$R_X(s) = \phi(B)\phi(B^{-1})R_Y(s), \qquad (4.12.58b)$$

(where now the shift operator B acts on s). Similarly, the autocorrelation functions of X_t, Y_t are related by

$$\rho_X(s) = \frac{R_X(s)}{R_X(0)} = \frac{1}{K_Y} \cdot \phi(B)\phi(B^{-1})\rho_Y(s),$$

where

$$K_Y = \phi(B)\phi(B^{-1})\rho_Y(0).$$

As an illustration of the above results suppose, for example, that $X_t = \Delta Y_t = (1 - B)Y_t$. Then (4.12.58b) gives

$$R_X(s) = (1 - B)(1 - B^{-1})R_Y(s)$$
$$= 2R_Y(s) - R_Y(s - 1) - R_Y(s + 1)$$
$$= -\Delta^2 R_Y(s - 1).$$

More generally, if $X_t = \Delta^d Y_t$, then (4.12.58b) gives

$$R_X(s) = (-1)^d \Delta^{2d} R_Y(s + d).$$

Examples

1. *AR(1) processes; discrete parameter*

The discrete parameter AR(1) model is given by (see Section 3.5.2, equation (3.5.6)),

$$X_t - aX_{t-1} = \varepsilon_t,$$

where $|a| < 1$, and ε_t is a purely random process, and may be obtained from (4.12.57) by setting

$$Y_t = \varepsilon_t, \qquad \Psi(B) = 1 - aB, \qquad f_Y(\omega) = 1/2\pi, \qquad -\pi \leqslant \omega \leqslant \pi.$$

We then have from (4.12.58),

$$\sigma_X^2 f_X(\omega) = \frac{\sigma_\varepsilon^2}{|1 - a \exp(-i\omega)|^2} \cdot \frac{1}{2\pi}$$
$$= \frac{\sigma_\varepsilon^2}{2\pi} \cdot \frac{1}{(1 - 2a \cos \omega + a^2)}.$$

Recalling (from (3.5.14)) that $\sigma_X^2 = \sigma_\varepsilon^2/(1-a^2)$, we obtain,

$$f_X(\omega) = \frac{1}{2\pi} \cdot \frac{(1-a^2)}{(1-2a\cos\omega + a^2)}, \qquad -\pi \le \omega \le \pi$$

in agreement with (4.10.10).

2. $AR(2)$ processes; discrete parameter

The discrete parameter AR(2) model (cf. (3.5.22)) is

$$X_t + a_1 X_{t-1} + a_2 X_{t-2} = \varepsilon_t,$$

and corresponds to (4.12.57) on setting

$$Y_t = \varepsilon_t, \qquad \Psi(B) = 1 + a_1 B + a_2 B^2, \qquad f_Y(\omega) = 1/2\pi, \qquad -\pi \le \omega \le \pi.$$

Then (4.12.58) gives,

$$\sigma_X^2 f_X(\omega) = \frac{\sigma_\varepsilon^2}{|1 + a_1 \exp(-i\omega) + a_2 \exp(-2i\omega)|^2} \cdot \frac{1}{2\pi}.$$

i.e.

$$\sigma_X^2 f_X(\omega) = \frac{\sigma_\varepsilon^2}{2\pi} \cdot \frac{1}{\{(1-a_2)^2 + a_1^2 + 2a_1(1+a_2)\cos\omega + 4a_2\cos^2\omega\}}.$$

Substituting the expression (3.5.34) for σ_X^2, the above result then becomes identical to (4.10.12).

3. AR processes of general order; discrete parameter

Writing,

$$\Psi(B) = 1 + a_1 B + \ldots + a_k B^k, \qquad f_Y(\omega) = 1/2\pi,$$

(4.12.57) gives the discrete parameter AR(k) model defined by (3.5.39). (Note that for stationarity of X_t the roots of $\Psi(z)$ must all lie outside the unit circle—see Section 3.5.4.) Equation (4.12.58) now gives,

$$\sigma_X^2 f_X(\omega) = \frac{\sigma_\varepsilon^2}{2\pi} \cdot \frac{1}{|1 + a_1 \exp(-i\omega) + \ldots + a_k \exp(-ik\omega)|^2} \qquad (4.12.59)$$

4. MA processes; discrete parameter

We have already considered the special case of the MA(l) model with "uniform weighting", and its spectral density function is given by (4.12.27).

The general MA(l) model, (3.5.45), may be obtained from (4.12.55) by setting

$$\phi(B) = b_0 + b_1 B + \ldots + b_l B^l, \qquad f_Y(\omega) = 1/2\pi,$$

(cf. (3.5.46), (3.5.47)), and hence by (4.12.56) we find,

$$\sigma_X^2 f_X(\omega) = \frac{\sigma_\varepsilon^2}{2\pi} \cdot |b_0 + b_1 \exp(-i\omega) + \ldots + b_l \exp(-il\omega)|^2. \qquad (4.12.60)$$

This expression reduces, of course, to (4.12.27) if we take

$$b_0 = b_1 = \ldots = b_l = 1/(l+1).$$

4.12.4 Processes with Rational Spectra

When discussing the mixed ARMA model in Section 3.5.6 we observed, in passing, that the degree of generality of this model among the class of all stationary processes (with purely continuous spectra) corresponded, roughly speaking, to the degree of generality of rational functions among the class of all continuous functions. We hinted that this feature would emerge more clearly when we considered the spectral properties of this model, and we now take up this point in more detail. Consider first the discrete parameter case for which the ARMA(k, l) model is given by (3.5.52), namely,

$$X_t + a_1 X_{t-1} + \ldots + a_k X_{t-k} = b_0 \varepsilon_t + b_1 \varepsilon_{t-1} + \ldots + b_l \varepsilon_{t-l}, \qquad (4.12.61)$$

where, as usual, ε_t denotes a purely random process. The operational form of this model is given by (3.5.53), namely,

$$\alpha(B)X_t = \beta(B)\varepsilon_t, \qquad (4.12.62)$$

where

$$\alpha(z) = 1 + a_1 z + \ldots + a_k z^k, \qquad \beta(z) = b_0 + b_1 z + \ldots + b_l z^l.$$

For stationarity of X_t we require that $\alpha(z)$ has *all its roots lying outside the unit circle*—see the discussion following equation (3.5.66).

Writing $Z_t = \beta(B)\varepsilon_t$, we have from (4.12.58), (4.12.56), that (with an obvious notation),

$$\sigma_X^2 f_X(\omega) = \frac{\sigma_Z^2}{|\alpha(e^{-i\omega})|^2} \cdot f_Z(\omega)$$

$$= \sigma_\varepsilon^2 \cdot \frac{|\beta(e^{-i\omega})|^2}{|\alpha(e^{-i\omega})|^2} f_\varepsilon(\omega)$$

$$= \frac{\sigma_\varepsilon^2}{2\pi} \cdot \frac{|\beta(e^{-i\omega})|^2}{|\alpha(e^{-i\omega})|^2}. \qquad (4.12.63)$$

Since α, β, are both finite order polynomials it follows that

$$\frac{|\beta(e^{-i\omega})|^2}{|\alpha(e^{-i\omega})|^2} = \frac{\beta(e^{-i\omega})\beta(e^{i\omega})}{\alpha(e^{-i\omega})\alpha(e^{i\omega})}$$

can be expressed as the ratio of two polynomials in $(e^{-i\omega})$, i.e. is a rational function of $(e^{-i\omega})$. We thus have the important result that *the spectral density function of a discrete parameter ARMA process is a rational function of* $(e^{-i\omega})$.

Conversely, if we are given that a stationary process X_t has spectral density function of the form (4.12.63), where σ_ε^2 is a given constant and both $\alpha(z)$ and $\beta(z)$ have all their roots outside the unit circle, then we may show that X_t is an ARMA(k, l) process of the form (4.12.61). For, write

$$Z_t = \alpha(B)X_t, \qquad W_t = \beta^{-1}(B)Z_t.$$

(Note that β is "invertible" since it has no roots inside the unit circle—see Section 3.5.7.) Then by (4.12.56), (4.12.58), we have,

$$\sigma_W^2 f_W(\omega) = \frac{\sigma_Z^2}{|\beta(e^{-i\omega})|^2} f_Z(\omega) = \frac{\sigma_X^2 |\alpha(e^{-i\omega})|^2}{|\beta(e^{-i\omega})|^2} f_X(\omega) = \frac{\sigma_\varepsilon^2}{2\pi}, \qquad \text{all } \omega.$$

Hence, $f_W(\omega)$ is a constant (independent of ω) and W_t is therefore a purely random process. Also, since $\int_{-\pi}^{\pi} f_W(\omega)\, d\omega = 1$, it follows that $\sigma_W^2 = \sigma_\varepsilon^2$, and W_t may now be identified with the process ε_t appearing in (4.12.61).

Thus, the property that the *spectral density function is a rational function of* $(\exp(-i\omega))$ *completely characterizes the discrete parameter ARMA model.*

In the continuous parameter case the ARMA model is given by (3.7.71), namely,

$$\alpha(D)X(t) = B_0(D)\varepsilon(t), \tag{4.12.64}$$

where

$$\alpha(z) = 1 + \alpha_1 z + \ldots + \alpha_k z^k, \qquad B_0(z) = \beta_0 + \beta_1 z + \ldots + \beta_l z^l,$$

and for stationarity we require that $\alpha(z)$ has all its roots in the left-hand plane, see Section 3.7.7. We now obtain from (4.12.45), (4.12.47),

$$\sigma_X^2 |\alpha(i\omega)|^2 f_X(\omega) = \frac{\sigma_W^2}{2\pi} \cdot |B_0(i\omega)|^2,$$

or

$$\sigma_X^2 f_X(\omega) = \frac{\sigma_W^2}{2\pi} \cdot \frac{|B_0(i\omega)|^2}{|\alpha(i\omega)|^2}. \tag{4.12.65}$$

Thus, *the spectral density function,* $f_X(\omega)$, *is now a rational function of* ω.

Conversely, given that $X(t)$ has spectral density function of the form (4.12.65), with $\alpha(z)$ satisfying the above condition, we may show that $X(t)$ is an ARMA model of the form (4.12.64).

In engineering terminology we may express the above results by saying that an ARMA model is obtained by passing a "white-noise" process through *a filter with a rational transfer function*, e.g., in the discrete parameter case, X_t is obtained by passing ε_t through a filter with transfer function $\{\beta(e^{-i\omega})/\alpha(e^{-i\omega})\}$ as shown in Fig. 4.26.

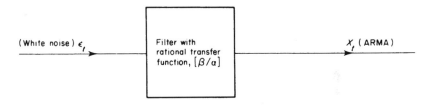

Fig. 4.26. Generation of ARMA processes.

4.12.5 Axiomatic Treatment of Filters

So far we have defined a filter as a linear transformation of the convolution integral form (4.12.1), or its discrete parameter analogue (4.12.15). However, we noted in Section 4.12.2 that the relationship between the input and output spectra was valid even when the operator $\phi(D)$ (in (4.12.44)) was a polynomial—in which case we cannot express the relationship between $X(t)$ and $Y(t)$ in the form (4.12.1) if $g(u)$ is to be a "well behaved" function. This means that if we take (4.12.1) as our definition of a filter then we cannot strictly consider relationships of the form

$$X(t) = D^k . Y(t) \qquad (4.12.66)$$

in this setting. Similarly, if we have a *"straight wire"* filter which leaves the input completely unaltered, i.e. which gives

$$X(t) = Y(t), \qquad (4.12.67)$$

or more generally, a *"pure delay"* filter, i.e.

$$X(t) = Y(t - t_0) \qquad (4.12.68)$$

(where t_0 is a positive constant), then again we cannot strictly express these in the form (4.12.1). We can "force" both (4.12.67) and (4.12.68) into the form (4.12.1) by allowing $g(u)$ to be a δ-function: if we take $g(u) = \delta(u)$ this gives (4.12.67), and $g(u) = \delta(u - t_0)$ gives (4.12.68), but then $g(u)$ is not a "well behaved" function. Now it would be highly unsatisfactory if our treatment of filters excluded a simple relationship of the form (4.12.68), and quite absurd if it excluded the simplest of all relationships, namely (4.12.67)! However, if we bypass (4.12.1) and work directly from the operational form (4.12.44), it is easily verified that a formal application of (4.12.45) is perfectly valid. That is, if we set

$$\phi(D) = D^k, \qquad\qquad \text{for (4.12.66)},$$

$$\phi(D) = 1, \qquad\qquad \text{for (4.12.67)},$$

$$\phi(D) = \exp(-t_0 D), \qquad \text{for (4.12.68)},$$

then (4.12.45) gives the correct results. Although we derived (4.12.45) *starting from* (4.12.1) the discussion suggests that (4.12.45) has a much wider range of validity, and this in turn suggests that we can treat relationships of the form (4.12.66)–(4.12.68) without any sacrifice of rigour provided we adopt a more general definition of the term "filter". We now consider this more general approach and illustrate the development for the case of continuous parameter processes. (The difficulties referred to above do not, of course, arise in the discrete parameter case, and (4.12.15) is quite general as a description of linear relationships. For example, (4.12.67) corresponds simply to taking $g_0 = 1$, $g_u = 0$, $u \neq 0$, in (4.12.15).)

The basic idea is that instead of defining a filter as an *explicit* relationship of the form (4.12.1) we now define it as an "operator" which satisfies certain axioms. More specifically, we define a *filter* as an *operator* $\mathcal{L}$ which transforms an *input function* $Y(t)$ into an *output function* $X(t)$. Symbolically, we write

$$X(t) = \mathcal{L}\{Y(t)\}. \qquad\qquad (4.12.69)$$

The filter is said to be *time-invariant* if, for any τ,

(i) $$\mathcal{L}\{Y(t+\tau)\} = X(t+\tau), \qquad\qquad (4.12.70)$$

and is said to be *linear* if, for any input functions, $X_1(t)$, $Y_2(t)$, and constants a_1, a_2,

(ii) $$\mathcal{L}\{a_1 Y_1(t) + a_2 Y_2(t)\} = a_1 \mathcal{L}\{Y_1(t)\} + a_2 \mathcal{L}\{Y_2(t)\}. \qquad (4.12.71)$$

The first property (4.12.70) says that if we translate the input by τ time units the only effect this will have will be to translate the output by the same amount. The second property (4.12.71) says that if the input is a linear combination of two functions the output will be the same linear combination of the individual outputs. (This is sometimes called the property of "*super-position*".) Equation (4.12.71) may be extended to

$$\mathscr{L}\left\{\sum_i a_i Y_i(t)\right\} = \sum_i a_i \mathscr{L}\{Y_i(t)\}. \tag{4.12.72}$$

(When the summation extends over a finite number of values of i the proof follows by induction, and it is reasonable to demand that $\mathscr{L}$ be such that (4.12.72) remains valid when the summation covers an infinite set of values of i, and in addition holds for the limiting forms of such sums.)

It is easily verified that the convolution integral (4.12.1) satisfies the above properties, as do the relationships (4.12.66)–(4.12.68). However, the interesting point is that (4.12.70) and (4.12.72) are all we need to derive the relationship between the input and output spectra. The derivation proceeds as follows. First, consider the situation when the input consists of just a single complex exponential term, i.e. we have

$$Y(t) = \exp(i\omega t),$$

say. Then the corresponding output $X(t)$ satisfies

$$\begin{aligned}
X(t+\tau) &= \mathscr{L}\{Y(t+\tau)\}, && \text{(by property (i))} \\
&= \mathscr{L}\{\exp(i\omega\tau) \cdot \exp(i\omega t)\} \\
&= \exp(i\omega\tau) \cdot \mathscr{L}\{\exp(i\omega t)\}, && \text{(by property (ii))} \\
&= \exp(i\omega\tau) \cdot X(t). && \text{(4.12.73)}
\end{aligned}$$

Now (4.12.73) holds for all τ and t and thus may be regarded as a functional equation in τ. In particular, setting $t = 0$, we have,

$$X(\tau) = X(0) \cdot \exp(i\omega\tau), \qquad \text{all } \tau.$$

This now establishes the functional form of $X(t)$, and we may write, for all t,

$$X(t) = \Gamma(\omega) \cdot \exp(i\omega t), \tag{4.12.74}$$

where $\Gamma(\omega) = X(0)$. (Note that $X(0)$ is the response of the filter at time 0 to the input $\exp(i\omega t)$, and hence will depend on the value of ω, i.e. $X(0)$ is a function of ω.) Equation (4.12.74) is of fundamental importance since it tells us that the more generally defined filter has the same basic property as the convolution integral, namely, that *if the input is a sine wave the output is also a sine wave of exactly the same frequency but with modified amplitude and phase, as determined by* $\Gamma(\omega)$. In the special case when $\mathscr{L}$ corresponds to the convolution integral,

$$X(t) = \int_{-\infty}^{\infty} g(u)\, Y(t-u)\, du,$$

we have for the input $\exp(i\omega t)$,

$$X(t) = \int_{-\infty}^{\infty} g(u) \exp[i\omega(t-u)] \, du,$$

so that

$$\Gamma(\omega) = X(0) = \int_{-\infty}^{\infty} g(u) \exp(-i\omega u) \, du.$$

Thus $\Gamma(\omega)$, being the Fourier transform of $g(u)$, is identical with the function $\Gamma(\omega)$ defined by (4.12.7). We therefore continue to call $\Gamma(\omega)$ the *transfer function* even in the more general setting.

The relationship between the input and output spectral density functions when $Y(t)$ is a general stationary process follows almost immediately from (4.12.74) by (a) using the spectral representation to express $Y(t)$ as a "linear combination" of complex exponentials, and (b) applying the linearity property of $\mathcal{L}$. The integral in the spectral representation is simply the limiting form of a sum, and hence we may appeal to (4.12.72) when applying $\mathcal{L}$ to this expression. Proceeding in this way we have,

$$X(t) = \mathcal{L}\{Y(t)\} = \mathcal{L}\left\{\int_{-\infty}^{\infty} \exp(i\omega t) \, dZ_Y(\omega)\right\}$$

$$= \int_{-\infty}^{\infty} \mathcal{L}\{\exp(i\omega t)\} \, dZ_Y(\omega), \qquad \text{by (4.12.72)}$$

$$= \int_{-\infty}^{\infty} \exp(i\omega t)\Gamma(\omega) \, dZ_Y(\omega), \qquad \text{by (4.12.74)}$$

$$= \int_{-\infty}^{\infty} \exp(i\omega t) \, dZ_X(\omega),$$

say, where

$$dZ_X(\omega) = \Gamma(\omega) \, dZ_Y(\omega), \qquad \text{all } \omega.$$

From this point onwards the argument is exactly the same as that following equation (4.12.13), and as before we obtain the result (4.12.14), which, when the spectral density functions exist, reduces to,

$$\sigma_X^2 f_X(\omega) = \sigma_Y^2 |\Gamma(\omega)|^2 f_Y(\omega). \tag{4.12.75}$$

The condition analogous to (4.12.9) is still required, i.e. we require that $\mathcal{L}$ be such that $\Gamma(\omega)$ and $h_Y(\omega)(=\sigma_Y^2 f_Y(\omega))$ together satisfy,

$$\int_{-\infty}^{\infty} |\Gamma(\omega)|^2 h_Y(\omega) \, d\omega < \infty$$

in order to ensure that $X(t)$ has finite variance.

When $\mathscr{L}$ corresponds to a differential operator, i.e. we have,

$$X(t) = \phi(D) \cdot Y(t),$$

then assuming that $\phi(z)$ has a power series expansion,

$$\phi(z) = \sum_{j=0}^{\infty} a_j z^j,$$

we may write formally,

$$\phi(D) \cdot \exp(i\omega t) = \sum_j a_j(D^j \cdot \exp(i\omega t))$$

$$= \left(\sum_j a_j(i\omega)^j \right) \cdot \exp(i\omega t)$$

$$= \phi(i\omega) \cdot \exp(i\omega t).$$

Hence, by (4.12.47) we obtain $\Gamma(\omega) \equiv \phi(i\omega)$, and when we substitute this expression for $\Gamma(\omega)$ in (4.12.75) we recover the previous result (4.12.45).

We now return to the previous examples, (4.12.66)–(4.12.68). For (4.12.66) we have $\phi(D) = D^k$ and hence,

$$\Gamma(\omega) = (i\omega)^k,$$

so that

$$\sigma_X^2 f_X(\omega) = \sigma_Y^2 \omega^{2k} f_Y(\omega), \qquad (4.12.76)$$

provided,

$$\int_{-\infty}^{\infty} \omega^{2k} f_Y(\omega) \, d\omega < \infty.$$

In particular, when $k = 1$, (4.12.76) gives the expression for the spectral density function of $X(t) = \dot{Y}(t)$, namely,

$$\sigma_X^2 f_X(\omega) = \sigma_Y^2 \omega^2 f_Y(\omega), \qquad (4.12.77)$$

provided that $\int_{-\infty}^{\infty} \omega^2 f_Y(\omega) \, d\omega < \infty$.

Thus, the effect of differentiating a stationary process is to introduce the factor (ω^2) into its spectral density function, and this, in turn, has the effect of increasing the relative contribution of the high frequency components to the total power, as we would expect on intuitive grounds. We may note fruther that the autocovariance function, $R_Y(\tau)$, of $Y(t)$ may be written (cf. (4.8.10)) as

$$R_Y(\tau) = \sigma_Y^2 \int_{-\infty}^{\infty} e^{i\omega\tau} f_Y(\omega) \, d\omega.$$

Differentiating twice w.r. to τ gives,

$$-R_Y''(\tau) = \sigma_Y^2 \int_{-\infty}^{\infty} e^{i\omega\tau} (\omega^2 f_Y(\omega))\, d\omega.$$

Hence, $\{-R_Y''(\tau)\}$ is the Fourier transform of $\{\sigma_Y^2 \omega^2 f_Y(\omega)\}$, and therefore, by (4.12.77), *it is the autocovariance function of the process* $\dot{Y}(t)$. It now follows that the autocorrelation function of $\dot{Y}(t)$ is given by,

$$\rho_{\dot{Y}}(\tau) = \frac{R_Y''(\tau)}{R_Y''(0)} = \frac{\rho_Y''(\tau)}{\rho_Y''(0)}. \qquad (4.12.78)$$

For (4.12.67) we have trivially that $\Gamma(\omega) \equiv 1$, all ω, and hence $f_X(\omega) \equiv f_Y(\omega)$. For (4.12.68) we have, $\phi(D) = \exp(-t_0 D)$, and hence,

$$\Gamma(\omega) = \exp(-it_0\omega).$$

Since $|\Gamma(\omega)|^2 = 1$, all ω, the transformation (4.12.68) leaves the spectral density function unaltered (as we would expect from the stationarity property of $Y(t)$). However, although the gain is unity at all frequencies, the phase-shift, $\phi(\omega)$, is now,

$$\phi(\omega) = \arg\{\Gamma(\omega)\} = -t_0\omega. \qquad (4.12.79)$$

Hence, *when we shift the time origin we introduce a phase-shift which increases linearly with frequency.*

Chapter 5

Estimation in the Time Domain

5.1 TIME SERIES ANALYSIS

In the previous chapters we discussed the basic properties of stationary processes, and introduced the autocovariance and correlation functions as ways of describing certain features of these processes. We then considered a number of standard models (such as the AR (autoregressive), MA (moving average), and ARMA (mixed autoregressive/moving average)), and showed that each of these models gave rise to autocorrelation functions with certain characteristic shapes. Later, we introduced the concept of power spectra, and explained how, for example, the spectral density function of a process could be determined provided we are given the precise form of the autocorrelation function—or equivalently, provided we know the precise model of the process. However, in practical problems we hardly ever have precise prior knowledge of either the autocorrelation function of a process or of its model. There are one or two very limited cases where we can deduce the appropriate model from physical considerations, such as the study of simple electrical networks like the R–C and L–R–C discussed in Chapter 3, Sections 3.7.3 and 3.7.4, but even here we have to make certain assumptions in order to derive an explicit form of the model. For example, the charge $Q(t)$ in the R–C circuit is an AR(1) process only if the input voltage $\varepsilon(t)$ is a purely random process; otherwise if $\varepsilon(t)$ is a general stationary process $Q(t)$ is no longer necessarily AR(1). In more general problems it would be virtually impossible to deduce anything about the complete autocorrelation function or the model from purely theoretical considerations, and even in those rare cases where we have sufficient insight to be able to say, e.g., that the process is AR(2), we would certainly not, in general, know the values of the parameters (α_1 and α_2) of the model, and consequently we would not be able to determine the precise form of the autocorrelation function.

291

It would, however, be quite wrong to conclude at this stage that the theory of stationary processes is of little value in practical problems; on the contrary, the concepts of autocorrelation functions and power spectra are extremely important in the analysis of experimental data because, even though we may not be able to determine these functions theoretically, the observed data themselves contain a great deal of information which we can use to *estimate* these unknown quantities. The problem of estimating unknown quantities from observational data leads us, quite naturally, into the field of *statistical inference*, and the body of statistical techniques used to analyse data from stationary processes is known as "*time series analysis*". The basic principles involved are essentially the same as those which arise in the more familiar problems of estimating the values of unknown parameters in probability distributions, and it may be useful, therefore, to summarize the main ideas of classical statistical inference before considering further the specific problem of estimating autocorrelation functions and power spectra. The discussion presented in the following section is of necessity rather brief and condensed. Readers who are unfamiliar with the theory of statistical inference and who wish to study a more detailed account may consult, e.g., Mood and Graybill (1963), Kendall and Stuart (1966), Vol. 2, Wilks (1962), or Silvey (1970). An advanced mathematical account of the subject is given by Lehmann (1959).

5.2 BASIC IDEAS OF STATISTICAL INFERENCE

The subject of statistical inference is concerned primarily with the problem of drawing inferences about the values of unknown parameters in probability distributions on the basis of observational data. The main ideas may be illustrated by considering the following simple situation. Suppose we are interested in studying the number of cars which pass a fixed point in the road in a given time interval of, say, one hour. Assuming that the time instants at which cars pass the fixed point form a Poisson process (see Chapter 2, Section 2.11), the total number X which pass during a period of 1 hour will then have a Poisson distribution, i.e. we may write,

$$P[X = r] = \exp(-a)\frac{a^r}{r!}, \qquad r = 0, 1, 2, \ldots, \qquad (5.2.1)$$

where the parameter a depends on λ, the mean rate of arrival of cars. If we have no prior knowledge about the value of λ then, of course, the value of a will be unknown, and there is no way in which we can determine this value from purely theoretical considerations. However, we may obtain some indication of "plausible" values of a by actually conducting the experiment, i.e. by counting the number of cars which pass in a particular period of one hour. Suppose we do this, and observe, say, 23 cars. This gives us one *realized value* of X (i.e. a value resulting from one performance of the experiment). We may then argue that since the parameter a represents the mean of the Poisson distribution its range of plausible values would be something like 15 to 35. We cannot, of course, definitely rule out any particular value of a simply on the basis of the single observation, $X = 23$, i.e. even if we consider an absurdly high value, say $a = 1000$, it is still just possible that we would have observed $X = 23$, in the sense that this value of X would have a non-zero probability. Nevertheless, we would argue that if the probability distribution of X really did have a mean as large as 1000 it is extremely unlikely that we would observe a value as low as 23. Our conclusions, therefore, are that the observed value of X does give us some information about the parameter a, but of a rather vague and indefinite form. We might be tempted to express this uncertainty about the value of a in terms of probability statements, but if we adopt the classical inter-pretation of probabilities as limiting relative frequencies then it soon becomes clear that this approach is quite invalid. For, when we repeat the experiment the value of X changes but the value of a (although unknown) remains the same from one repetition to the next. There is thus no sense (other than a trivial one) in which we can talk about the relative frequency of occurrence of different values of a. (Note that there is a basic difference between an "unknown" quantity and a "random" quantity.) We cannot, therefore, ask, "what is the probability that a has a particular value", and one of the main objectives of statistical inference is to formulate the types of questions about parameters which we can answer in a precise and unam-biguous manner.

Before proceeding further let us note that we would not in any case expect to gain much information from just a single observation on X. In practice, we would repeat the experiment a number of times, i.e. we would observe the number of cars in, say, 20 periods, each of one hour's duration. If we denote the number of cars in the successive periods by $X_1, X_2, \ldots, X_{20}$, then all the X_i's have the same probability distribution, namely, that given by (5.2.1), and, if the periods are non-overlapping, then $X_1, X_2, \ldots, X_{20}$ are indepen-dent random variables. The joint probability distribution of $X_1, \ldots, X_{20}$ is then the product of the marginal distributions (each of the form (5.2.1)—cf. (2.13.7)), but still involves only the single parameter, a. The observational data now consist of realized values $x_1, x_2, \ldots, x_{20}$ of $X_1, X_2, \ldots, X_{20}$ respectively, and we may use all the observations $x_1, \ldots, x_{20}$ to make inferences about the value of a.

The essential ideas contained in this problem may be given a more general formulation by noting that: (a) the assumption that the X_i are independent is typical but not crucial; we may drop this provided we have sufficient information to be able to write down the joint probability distribution of the $\{X_i\}$; (b) in the above problem the X_i are discrete valued whereas in practice we generally deal with observations on continuous variables, e.g. we may have a sequence of 50 observations, each being a realized value of a random variable having a normal distribution with unknown mean μ and unknown variance σ^2, and (c) many problems involve more than just one unknown parameter, e.g., in the example referred to in (b) above there would be two unknown parameters, namely, the mean and the variance of the normal distribution; typically, we may have several unknown parameters, $\theta_1, \theta_2, \ldots, \theta_p$, say. We may now state a fairly general form of the problem as follows.

We are given observational data $x_1, x_2, \ldots, x_n$ which constitute realized values of n random variables $X_1, X_2, \ldots, X_n$. The joint probability distribution of $X_1, X_2, \ldots, X_n$ has a known mathematical form, with joint probability density function,

$$f(x_1, x_2, \ldots, x_n, \theta_1, \theta_2, \ldots, \theta_p),$$

but the values of the p parameters $\theta_1, \theta_2, \ldots, \theta_p$ are unknown. We wish to obtain some information about the values of $\theta_1, \theta_2, \ldots, \theta_p$ from the observational data, $x_1, x_2, \ldots, x_n$.

It is interesting to compare the problem of statistical inference with that studied in probability theory. *In probability theory we assume that the parameters $(\theta_1, \ldots, \theta_p)$ are given and we use the function $f(x_1, \ldots, x_n, \theta_1, \ldots, \theta_p)$ to tell us that the "relative likeliness" of different possible values of $X_1, \ldots, X_n$. In inference, we are given the values of $X_1, \ldots, X_n$ which actually occurred, and we use the function $f(x_1, \ldots, x_n, \theta_1, \ldots, \theta_p)$ to tell us (in some sense) the "relative likeliness" of the different possible values of $\theta_1, \ldots, \theta_p$.* There is an obvious form of "duality" between the problems of statistical inference and probability theory, the problem of inference being, in effect, the "inverse" of probability theory. However, as noted previously, we cannot make probability statements about parameters, and this feature removes a good deal of the "symmetry" between the two problems.

(There is an alternative approach, known as *Bayesian inference*, in which probabilities are interpreted subjectively as "degrees of belief" rather than as limiting relative frequencies. In this approach there is no logical difficulty in making probability statements about parameters, and problems of inference can then be treated in a common framework with probability theory. From a philosophical point of view this is a considerable advantage, but the subjective interpretation of probability raises difficulties of a different kind. A lucid and comprehensive account of Bayesian inference is given by Lindley (1965).)

In the traditional treatment the subject of inference is divided into two major categories, namely (i) estimation and (ii) hypothesis testing. In "estimation" we have no preconceived ideas about the values of the parameters $\theta_1, \ldots, \theta_p$ and we try to find either the most plausible values of each parameter (called "*point estimates*") or alternatively, plausible ranges of values for each parameter (called "*interval estimates*"). In "hypothesis testing" we start with a hypothesis which specifies particular values for the parameters, and we then wish to determine whether the data are consistent with these hypothetical values. The two topics are of course closely related, and in an advanced discussion they could be treated in a unified manner within the context of a subject called "*decision theory*". However, for the purpose of the present discussion we concentrate on the "estimation" side of inference; for a treatment of hypothesis testing see, e.g., Mood and Graybill (1963).

5.2.1 Point Estimation

Again, we introduce the basic ideas by considering a simple example. Suppose we have a machine which mass-produces metal cylinders. Due to variations in the production process the diameters of the cylinders vary slightly from one cylinder to another, so that the diameter X of a randomly selected cylinder is a continuous random variable which may reasonably be supposed to have a normal distribution with a certain mean μ, and a certain variance σ^2. If the values of μ and σ^2 are both unknown we may obtain some information about them by selecting a sample of n cylinders from the output of the machine, and measuring their diameters. (This operation is, of course, exactly the same thing as repeating n times the "experiment" of selecting a single cylinder.) Suppose then that the measured diameters give values $x_1, x_2, \ldots, x_n$. A natural (point) estimate of μ is given by the *sample mean*, $\bar{x} = (1/n)(\sum_{i=1}^{n} x_i)$. This estimate has an obvious intuitive appeal, and, if we are dealing with a "random sample", has the nice property that as n increases $\bar{x}$ gets "closer and closer" to the true value of μ. For $x_1, x_2, \ldots, x_n$ are realized values of n random variables, $X_1, X_2, \ldots, X_n$. If the sample is selected "randomly" we may regard $X_1, X_2, \ldots, X_n$ as n *independent* random variables (this is, in effect, the definition of the term "random sample"), and all the X_i have the same probability distribution, namely the probability distribution of the original random variable X, i.e., $N(\mu, \sigma^2)$. It then follows from the discussion in Chapter 2, Section 2.14, that $\bar{X} = (1/n)(\sum_{i=1}^{n} X_i)$ converges in probability to μ as $n \to \infty$.

Similarly, a natural (point) estimate of σ^2 is given by the *sample variance*, $s^2 = (1/n)[\sum_{i=1}^{n} (x_i - \bar{x})^2]$, and again it may be shown from the results in Section 2.14 that as n increases s^2 gets "closer and closer" to σ^2. Moreover, the convergence of $\bar{x}$ and s^2 to μ and σ^2, respectively, holds under fairly general conditions, irrespective of the form of the probability distribution of the X_i, i.e. the assumption that the X_i above have normal distributions is not required for this result. *Thus, if we are given a random sample of n observations from any probability distribution, the sample mean $\bar{x}$ and the sample variance s^2 are, in general, the natural estimates of the mean and variance of the probability distribution* (with the exception of certain "pathological" probability distributions such as, e.g., the Cauchy distribution where the variance is infinite).

This approach generalizes fairly easily to the bivariate case. Suppose, for example, we are interested in the joint distribution of the diameters and heights of the cylinders. Then if X, Y denote respectively the diameter and height of a randomly selected cylinder, we may reasonably assume that the pair of random variables (X, Y) has a bivariate normal distribution (see Section 2.12.9) with means μ_X, μ_Y, variances σ_X^2, σ_Y^2, and covariance $\mu_{11} = \sigma_X \sigma_Y \rho$ (so that ρ denotes the correlation coefficient between X and Y). If we now select a random sample of n cylinders, and measure the diameter and height of each, we obtain n *pairs* of observations, (x_1, y_1), $(x_2, y_2), \ldots, (x_n, y_n)$. The means, μ_Y, μ_Y may again be estimated by the corresponding sample means, i.e., by

$$\bar{x} = \frac{1}{n} \sum_{i=1}^{n} x_i \qquad \text{and} \qquad \bar{y} = \frac{1}{n} \sum_{i=1}^{n} y_i$$

respectively, and similarly the variances σ_X^2, σ_Y^2 may be estimated by the sample variances,

$$s_X^2 = \frac{1}{n} \sum_{i=1}^{n} (x_i - \bar{x})^2, \qquad \text{and} \qquad s_Y^2 = \frac{1}{n} \sum_{i=1}^{n} (y_i - \bar{y})^2.$$

The natural estimate of the covariance μ_{11} is the *sample covariance* m_{11} as defined in Section 2.12.7 (cf. equation (2.12.48)), namely

$$m_{11} = \frac{1}{n} \sum_{i=1}^{n} (x_i - \bar{x})(y_i - \bar{y}). \tag{5.2.2}$$

The correlation coefficient ρ may then be estimated by the *sample correlation coefficient*,

$$r = m_{11}/(s_X^2 \cdot s_Y^2)^{1/2}, \tag{5.2.3}$$

as explained in Section 2.12.7 (cf. equation (2.12.49)).

As in the univariate case, these estimates remain valid, irrespective of the form of the underlying bivariate probability distribution. In particular, *the sample covariance m_{11} is the natural estimate of μ_{11} whatever the form of the joint distribution of the random variables X, Y.*

Sampling distributions

Although we know that the sample mean $\bar{x}$ converges in probability to μ as $n \to \infty$ it is still important to know, for any finite value of n, the "degree of accuracy" which we may ascribe to $\bar{x}$ as an estimate of μ. Of course, for some samples the value of $\bar{x}$ may turn out to be very close to μ, but for other samples $\bar{x}$ may deviate considerably from μ. Since we do not know beforehand which sample we are going to select the accuracy of $\bar{x}$ can be measured only in terms of its degree of variability over the different samples. The crucial step now is to recognize that *the variability of $\bar{x}$ over different samples is precisely what the probability distribution of the random variable,* $\bar{X} = (1/n)(\sum_{i=1}^{n} X_i)$, *describes*. The probability distribution of $\bar{X}$ is therefore called the *sampling distribution* of the estimate, $\bar{x}$. Since random sampling corresponds to independent X_i, the distribution of $\bar{X}$ follows immediately from the properties of sums of independent random variables given in section (2.14.4). In fact, we know from equation (2.14.9) that if the $\{X_i\}$ are normally distributed with mean μ and variance σ^2 then $\bar{X}$ also has an exact normal distribution with mean μ and variance (σ^2/n). Hence, in this case the sampling distribution of $\bar{x}$ is $N(\mu, \sigma^2/n)$, and we may write,

$$\bar{X} = N(\mu, \sigma^2/n).$$

This result holds in an approximate sense for large n even if the underlying distribution is not normal, by virtue of the central limit theorem: see equation (2.14.18). However, irrespective of the form of the distribution we always have the exact results (for all n) that the mean and variance of the sampling distribution of $\bar{x}$ are μ, and (σ^2/n), respectively: see equation (2.14.16). Hence, we have the general results that,

$$E[\bar{X}] = \mu, \qquad \text{var}[\bar{X}] = \sigma^2/n. \tag{5.2.4}$$

Similarly, it follows from (2.14.24) that the sampling distribution of s^2 is proportional to a χ^2 distribution with $(n-1)$ degrees of freedom, or more precisely,

$$(ns^2/\sigma^2) = \chi_{n-1}^2. \tag{5.2.5}$$

These ideas may now be stated in a more general form as follows. We are given a random sample of n observations, $x_1, x_2, \ldots x_n$, from a probability distribution which involves an unknown parameter θ. We construct an estimate $\hat{\theta}$ of θ by forming a suitable function of the observations, say, $\hat{\theta} = \hat{\theta}(x_1, x_2, \ldots, x_n)$. The *sampling distribution* of $\hat{\theta}$ is defined as the probability distribution of the *random variable,* $\hat{\theta}(X_1, X_2, \ldots, X_n)$, and it describes the variation of $\hat{\theta}$ over the different possible samples. The random variable, $\hat{\theta}(X_1, \ldots, X_n)$ is called an "*estimator*" (for θ), while $\hat{\theta}(x_1, \ldots, x_n)$, (the value of $\hat{\theta}$ calculated from the particular sample at hand) is called an *"estimate"*, or *"statistic"*. (The term "statistic" is used generally to refer to any function of the observations.)

Notation

In the preceding discussion we have used different symbols to refer to the observations $(x_1, x_2, \ldots, x_n)$ from the particular sample we have selected as opposed to the random variables $(X_1, X_2, \ldots, X_n)$, which denote general values of the first observation, second observation, ..., and so on. This convention was used in order to emphasize the logical distinction between the two ideas, but once this is understood it becomes unnecessarily cumbersome to retain the two sets of symbols. From now on we will therefore simply denote a random sample by $(X_1, \ldots, X_n)$, but it should be clear from the context whether we are discussing a particular set of numerical values or whether we are referring to a set of random variables.

Properties required for estimators

We have observed that $\bar{X}$ is the "natural" estimator for μ, and in practice we would hardly ever consider estimating μ in any other way (apart from in special cases such as the Cauchy distribution referred to previously). However, since the normal distribution is symmetric about its mean and attains its maximum value at that point it follows that, for this distribution, the mean, median, and mode all have the same value, namely μ. In view of this we might consider estimating μ by either the sample median or the sample mode. Intuitively, we would feel that $\bar{X}$ is a "better" estimator than the other two alternatives, but to establish this we need to define precisely what we mean by a "better" estimator. In a more general situation there may not be a "natural" estimator of a parameter θ, and we may then have to consider several possible estimators, with no clear intuition as to which is to be preferred. (As a simple illustration consider again the estimation of the parameter a in a Poisson distribution. Since here the mean and the variance of the distribution are both equal to a it is not intuitively clear whether we should estimate a by the sample mean or the sample variance.) In order to resolve this type of problem we must first decide on the criteria by which we judge the "goodness" of an estimator, and to this end we set out below a list of "desirable properties".

(A) UNBIASEDNESS

We say that $\hat{\theta}$ is an *unbiased estimator* for θ if the average value of $\hat{\theta}$ over all possible samples is equal to the "true value", θ, whatever value θ takes, i.e. if

$$E(\hat{\theta}) = \theta, \quad \text{all } \theta. \tag{5.2.6}$$

Here, E denotes expectation with respect to the sampling distribution of $\hat{\theta}$. If (5.2.6) does not hold then $\hat{\theta}$ is said to be a "*biased*" estimator, and

$$b(\hat{\theta}) = \{E(\hat{\theta}) - \theta\} \tag{5.2.7}$$

is called the *bias* of $\hat{\theta}$.

Since the sampling distribution of $\hat{\theta}$ will, in general, depend on n, the number of observations in the sample, $b(\hat{\theta})$ will also depend on n. If $b(\hat{\theta}) \to 0$ as $n \to \infty$ (all θ), then $\hat{\theta}$ is said to be "*asymptotically unbiased*". Unbiasedness is clearly a desirable property, but a biased estimator may still be quite useful provided it is asymptotically unbiased. On the other hand, the lack of asymptotic unbiasedness would, in general, be considered a serious defect in an estimator. (There would be little point in using $\hat{\theta}$ to estimate θ if its values over different samples are "clustering" around a mean of, say, $(\theta + 100)$.)

From (5.2.3) we can now say that $\bar{X}$ is an unbiased estimator for μ, but we know from (2.14.21) that s^2 is not an unbiased estimator for σ^2. However, we know also from (2.14.23) that $\hat{s}^2 = [n/(n-1)]s^2$ is an unbiased estimator. (Note that although s^2 is biased, it is nevertheless asymptotically unbiased.)

(B) RELATIVE EFFICIENCY

Suppose that we have a choice of two different estimators θ_1, θ_2 for the same parameter θ. If both θ_1 and θ_2 are unbiased estimators then we would prefer that estimator whose sampling distribution has the smaller variance. For, suppose that the two sampling distributions are as shown in Fig. (5.1), so

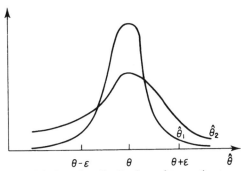

Fig. 5.1. Sampling distributions of two estimators.

that θ_1 has the smaller variance. Then the values of $\hat{\theta}_1$ (over different samples) are more highly "concentrated" around the true value θ than are the values of $\hat{\theta}_2$. Consequently, the probability that $\hat{\theta}_1$ lies in, say, the interval $(\theta - \varepsilon, \theta + \varepsilon)$ is higher than the probability that $\hat{\theta}_2$ lies in the same interval. We therefore say in this case that $\hat{\theta}_1$ is a more *efficient* estimator than $\hat{\theta}_2$ and define the *"relative efficiency"* of $\hat{\theta}_2$ as compared with $\hat{\theta}_1$ by,

$$\text{relative efficiency} = \left[\frac{\text{var}(\hat{\theta}_1)}{\text{var}(\hat{\theta}_2)} \times 100\right] \%. \qquad (5.2.8)$$

Returning now to the choice of sample mean, sample median, or sample mode, it is a well known result that, for samples from a normal distribution, the sample mean is indeed more efficient than both the sample median and sample mode. Similarly, the sample mean is a more efficient estimator of the parameter a of a Poisson distribution than is the sample variance.

Later on we shall discuss a fundamental result called the "Cramer–Rao inequality" which states that, under general conditions, *there is a lower bound to the variance of any unbiased estimate of a parameter θ.* If we are able to find an estimator $\hat{\theta}$ whose variance attains this lower bound then we know that $\hat{\theta}$ is the most efficient estimator that we can construct. We therefore say that $\hat{\theta}$ is a *fully efficient estimator* for θ, or alternatively, that $\hat{\theta}$ is an MVUE (minimum variance unbiased) estimator.

(C) MEAN SQUARE ERROR CRITERION

If we have two estimators $\hat{\theta}_1, \hat{\theta}_2$ both of which are *biased*, then the variance is no longer an appropriate measure of efficiency. For although $\hat{\theta}_1$ may have a smaller variance than $\hat{\theta}_2$, it may also have a larger bias, in which case it would not be immediately obvious which estimator is to be performed. One way of dealing with this difficulty is to construct a more general criterion, called the *mean square error*, which takes account of both the variance and the bias. For a general estimator $\hat{\theta}$ of a parameter θ, the mean square error $M^2(\hat{\theta})$ is defined by

$$M^2(\hat{\theta}) = E[(\hat{\theta} - \theta)^2]. \qquad (5.2.9)$$

We may express $M^2(\hat{\theta})$ in terms of the variance and bias of $\hat{\theta}$ by writing,

$$M^2(\hat{\theta}) = E[\{(\hat{\theta} - E(\hat{\theta})) + (E(\hat{\theta}) - \theta)\}^2]$$
$$= E[(\hat{\theta} - E(\hat{\theta}))^2] + [E(\hat{\theta}) - \theta]^2 + 2(E(\hat{\theta}) - \theta)E[\hat{\theta} - E(\hat{\theta})].$$

Noting that $E[\hat{\theta} - E(\hat{\theta})] \equiv 0$, that the first term is the variance of $\hat{\theta}$, and that the second term is the squared bias, we obtain,

$$M^2(\hat{\theta}) = \text{var}(\hat{\theta}) + b^2(\hat{\theta}). \qquad (5.2.10)$$

If now we have to choose between two biased estimators $\hat{\theta}_1, \hat{\theta}_2$ a sensible procedure would be to choose that estimator which has the smaller mean square error, i.e. we would prefer $\hat{\theta}_1$ to $\hat{\theta}_2$ if $M^2(\hat{\theta}_1) < M^2(\hat{\theta}_2)$. This procedure is, of course, quite consistent with the measure of efficiency discussed in (B), since when an estimator is unbiased the mean-square error reduces simply to the variance.

(D) CONSISTENCY

We have already observed that the sample mean $\bar{X}$ of a normal sample gets "closer and closer" to μ as n increases, i.e. the accuracy of $\bar{X}$ as an estimator if μ increases as the sample size increases. This property, called *consistency*, is clearly a desirable feature for any estimator, since it is reasonable to expect that an estimator based on, say, 100 observations should be more accurate than one based on only 50 observations.

Formally, we say that $\hat{\theta}$ is a consistent estimator for θ if $\hat{\theta} \to \theta$ in probability as the sample size, $n \to \infty$, i.e. if, given ε, η, (both >0), $\exists n_0$ such that,

$$p[|\hat{\theta} - \theta| < \varepsilon] > 1 - \eta, \quad \text{for all } n \geq n_0. \qquad (5.2.11)$$

A *sufficient condition* for consistency is that the mean square error of $\hat{\theta}$ should converge to zero as $n \to \infty$, i.e. that

$$M^2(\hat{\theta}) = E[(\hat{\theta} - \theta)^2] \to 0 \quad \text{as } n \to \infty. \qquad (5.2.12)$$

For, by analogy with Chebyshev's inequality (see Chapter 2, Section 2.8.1), we have, for any $c > 0$,

$$p[|\hat{\theta} - \theta| \geq cM(\hat{\theta})] \leq 1/c^2. \qquad (5.2.13)$$

(The inequality (5.2.13) follows in exactly the same way as the original version of Chebyshev's inequality given by (2.8.16)—the proof requires only the trivial modification of starting with the expression for $M^2(\hat{\theta})$ rather than with the expression for σ^2.) Hence we may write,

$$p[|\hat{\theta} - \theta| < cM(\hat{\theta})] \geq 1 - 1/c^2.$$

Now given ε choose c so that $cM(\theta) = \varepsilon$, then,

$$p[|\hat{\theta} - \theta| < \varepsilon] \geq 1 - \frac{M^2(\theta)}{\varepsilon^2}.$$

If $M^2(\hat{\theta}) \to 0$ as $n \to \infty$ then given η we can make $M^2(\hat{\theta})/\varepsilon^2 < \eta$ by choosing n sufficiently large, and the result follows.

Equivalently, using (5.2.10) we may say that a sufficient condition for $\hat{\theta}$ to be consistent is that

$$\text{var}(\hat{\theta}) \to 0 \text{ as } n \to \infty$$

and

$$b^2(\hat{\theta}) \to 0 \text{ as } n \to \infty. \tag{5.2.14}$$

The definition of consistency given above, based on convergence in probability, is the usual one used in estimation theory. However, some authors describe this as "weak consistency", and introduce two alternative forms, namely:

1. "mean square consistency"—when $\hat{\theta} \to \theta$ in mean square, i.e. when $M^2(\hat{\theta}) \to 0$;
2. "strong consistency"—when $\hat{\theta} \to \theta$ almost surely (i.e. with probability 1).

The result just proved shows that mean-square consistency is a stronger condition than weak consistency in the sense that the former implies the latter. (We may recall the general result that convergence in mean square always implies convergence in probability, but the converse does not hold).

(E) SUFFICIENCY

When, for example, we choose to estimate the mean μ of a distribution by the sample mean $\bar{X}$ we are, in effect, ignoring all the information in the sample observations $X_1, \ldots, X_n$ other than that contained *in a single function* of $X_1, \ldots, X_n$, namely $\bar{X} = (1/n) \sum_{i=1}^{n} X_i$. Nevertheless, we may feel that there is no loss in this procedure since it seems reasonable to suppose that the single function $\bar{X}$ contains all the information in the sample "*relevant to the estimation of μ*". More generally, if we have a parameter θ and choose as estimator some function of the observations $\hat{\theta} = \hat{\theta}(X_1, \ldots, X_n)$, we would like $\hat{\theta}$ to contain, if possible, all the information in the sample relevant to the estimation of θ. This property is a very important one, and is given the name "*sufficiency*", i.e. if $\hat{\theta}$ has this property we say that $\hat{\theta}$ is a "*sufficient statistic*" for θ. However, before we could determine whether or not an estimator has this property we must first define in precise terms what we mean by "all the information in the sample relevant to the estimation of θ"; in particular, we must consider carefully what we mean by "information". The point to be grasped is that observations yield information about an unknown parameter θ only if their probability distribution depends on θ. For otherwise inferences about plausible values of θ cannot be made on the basis of the relative likeliness of the set of observed values $X_1, \ldots, X_n$ as various values of θ are considered. (For example, we clearly cannot draw any inferences as to whether or not a particular coin is unbiased by observing the number of sixes which occur in several throws of a dice, the basic reason for this being that the distribution of the number of sixes is in no way related to the parameter p (say), the probability that a single spin of the coin will produce heads.) Now let us return to the problem of estimating μ. If someone else records the observations in the sample and discloses only the value of the sample mean $\bar{X}$ what information is lacking? We lack information on the individual observations $X_1, X_2, \ldots, X_n$ but we know that these cannot take completely arbitrary values since their mean must agree with the given value of $\bar{X}$. Hence, the uncertainty associated with the unknown observations is described by the *conditional probability distribution of* $X_1, \ldots, X_n$ *given the value of* $\bar{X}$. We can now answer the question as to whether the individual observations provide further information about μ additional to that provided by the single statistic $\bar{X}$. For, if the conditional distribution of $X_1, \ldots, X_n$ given $\bar{X}$ does not involve μ then no additional information about μ is available, and $\bar{X}$ is then a sufficient statistic for μ. On the other hand, if this conditional distribution does involve μ then further information would be available if the individual values of $X_1, \ldots, X_n$ were known, and $\bar{X}$ by itself would not be a sufficient statistic. (It might be guessed that if the observations came from a normal distribution then $\bar{X}$ is, in fact, a sufficient statistic for μ; we prove this result below.) We may now formulate a general definition of "sufficiency", as follows.

$\hat{\theta} = \hat{\theta}(X_1, \ldots, X_n)$ *is said to be a "sufficient statistic" for θ if the conditional distribution of* $X_1, \ldots, X_n$, *given the value of* $\hat{\theta}(X_1, \ldots, X_n)$, *does not involve θ.*

Let $f(x_1, \ldots, x_n, \theta)$ be the joint probability density function of $X_1, \ldots, X_n$. (For a random sample, $f(x_1, \ldots, x_n, \theta)$ will be of the form $f(x_1, \theta) \ldots f(x_n, \theta)$.) Then if $g(\hat{\theta}, \theta)$ is the probability density function of $\hat{\theta}(X_1, \ldots, X_n)$, we may in all cases (i.e. irrespective of whether or not $\hat{\theta}$ is sufficient for θ) write

$$f(x_1, \ldots, x_n, \theta) = g(\hat{\theta}, \theta) \cdot h(x_1, \ldots, x_n, \theta), \qquad (5.2.15)$$

where $h(x_1, \ldots, x_n, \theta)$ is the conditional probability density function of $X_1, \ldots, X_n$ given the value of $\hat{\theta}$ (cf. equation (2.12.16), Section (2.12.3)). The point now is that if $\hat{\theta}$ is a sufficient statistic for θ then, by definition, $h(x_1, \ldots, x_n, \theta)$ does not involve θ, and we may write,

$$f(x_1, \ldots, x_n) = g(\hat{\theta}, \theta) \cdot h(x_1, \ldots, x_n). \qquad (5.2.16)$$

This result provides the basis for a more general result, known as the "*factorization theorem*", which states that *if the joint probability density function of* $X_1, \ldots, X_n$ *splits up into the product of two functions, one of which involves some function* $\hat{\theta}(x_1, \ldots, x_n)$ *and θ, while the other involves* $x_1, \ldots, x_n$ *but not θ, then $\hat{\theta}$ is a sufficient statistic for θ.* (See, e.g., Mood and Graybill (1963).)

It now follows that if $\hat{\theta}$ is a sufficient statistic for θ, so is *any single valued function* $\phi(\hat{\theta})$, say, of $\hat{\theta}$.

The above discussion refers to the case of a single unknown parameter. A similar definition is used in the multiparameter case, namely: The *m* statistics $\hat{\theta}_1 = \hat{\theta}_1(X_1, \ldots, X_n), \ldots, \hat{\theta}_m = \hat{\theta}_m(X_1, \ldots, X_n)$ are said to be *"jointly sufficient"* for the k parameters $\theta_1, \ldots, \theta_k$ if the conditional distribution of $X_1, \ldots, X_n$, given $\hat{\theta}_1, \ldots, \hat{\theta}_m$, does not involve $\theta_1, \ldots, \theta_m$.

The usefulness of this definition applies only when $m < n$. (If we take $m = n$ then it is true trivially that all the observations are jointly sufficient for the unknown parameters.) However, m may be less than, equal to, or greater than k, the number of parameters.

We conclude this section by showing that, for a normal distribution with known σ^2, $\bar{X}$ is a sufficient statistic for μ. For, if the sample observations are chosen randomly the joint probability density function of $X_1, \ldots, X_n$, is

$$f(x_1, \ldots, x_n) = \prod_{i=1}^{n} \left(\frac{1}{(2\pi\sigma^2)^{1/2}} \exp\left[-\frac{1}{2\sigma^2} (x_i - \mu)^2 \right] \right)$$

$$= \left(\frac{1}{2\pi\sigma^2} \right)^{n/2} \exp\left[-\frac{1}{2\sigma^2} \sum_{i=1}^{n} (x_i - \mu)^2 \right]. \qquad (5.2.17)$$

But, we have the identity,

$$\sum_{i=1}^{n} (x_i - \mu)^2 \equiv \sum_{i=1}^{n} (x_i - \bar{x})^2 + n(\bar{x} - \mu)^2,$$

hence,

$$f(x_1, \ldots, x_n, \mu) = \left(\frac{1}{2\pi\sigma^2} \right)^{n/2} \exp\left[-\frac{1}{2\sigma^2} \left(\sum_{i=1}^{n} (x_i - \bar{x})^2 + n(\bar{x} - \mu)^2 \right) \right]$$

$$= g(\bar{x}, \mu) \cdot h(x_1, \ldots, x_n), \qquad (5.2.18)$$

say, where

$$g(\bar{x}, \mu) = \exp\left[-\frac{n}{2\sigma^2} (\bar{x} - \mu)^2 \right], \qquad (5.2.19)$$

$$h(x_1, \ldots, x_n) = \left(\frac{1}{2\pi\sigma^2} \right)^{n/2} \exp\left[-\frac{1}{2\sigma^2} \sum_{i=1}^{n} (x_i - \bar{x})^2 \right]. \qquad (5.2.20)$$

Since $h(x_1, \ldots, x_n)$ does not involve μ it follows that $\bar{X}$ is a sufficient statistic for μ.

5.2.2 General Methods of Estimation

The estimation of means and variances is a fairly straightforward problem in the sense that the sample analogues of these quantities suggest themselves as "natural" estimators. However, when we consider more difficult problems of estimation it may well turn out that there is no "natural" estimator of a particular parameter, and our intuition may not prove very helpful in suggesting just what function of the observations we should consider. In such situations we have to use one of a number of "general estimation procedures" which guide us towards the construction of a suitable estimator. We now discuss briefly three such general procedures.

I. *Method of maximum likelihood*

By far the most general and powerful method of estimation is that known as *"maximum likelihood"*, due to R. A. Fisher. This method produces almost all the well known estimates, and more importantly, may be applied (in principle) to any type of estimation problem provided only that we can write down the joint probability distribution of the observations. The essential ideas underlying the method may be explained as follows. Suppose, for simplicity, that we have a random sample of n observations $X_1, \ldots, X_n$ from a distribution with probability density function $f(x, \theta)$ involving just one unknown parameter θ. If we denote the complete set of observations by the vector $X = (X_1, \ldots, X_n)$, then the joint probability density function of X may be written as

$$f(x, \theta) = f(x_1, \theta)f(x_2, \theta) \ldots f(x_n, \theta). \qquad (5.2.21)$$

In probability theory we use the function $f(x, \theta)$ to evaluate the *probability* of possible sets of values of X when we know the value of θ. In particular, if

$$f(x_1, \theta) > f(x_2, \theta),$$

we may say (loosely speaking) that the value x_1 is "more likely" than the value x_2. However, in estimation problems we observe the value of X, say x, and wish to say something about θ. Now consider $f(x, \theta)$ *as a function of θ* (with x fixed at the observed values). We call this the *likelihood function* of θ, and if

$$f(x, \theta_1) > f(x, \theta_2),$$

we may say that θ_1 is a "more plausible" value of θ than θ_2 since θ_1 *ascribes a larger probability to the observed x* than does θ_2. The method of "maximum likelihood" is based on the principle that we should estimate θ by its *most plausible* values given the observations x, i.e. that we should choose that value $\hat{\theta}$ of θ which maximizes the likelihood function $f(x, \theta)$ and thus ascribes the maximum possible probability to the given observations. In many cases it turns out to be more convenient to work with the logarithm of the likelihood function, rather than with the likelihood function itself, and writing

$$L = L(x, \theta) = \log_e\{f(x, \theta)\}, \qquad (5.2.22)$$

we may equally well determine $\hat{\theta}$ by maximizing $L(x, \theta)$ rather than by maximizing $f(x, \theta)$. L is called the *log-likelihood function*, and is considered primarily as a *function of θ*. However, because L also involves the observations, x, it will vary from sample to sample, and hence, for each fixed θ, it is a *random variable*. When we wish to emphasize the latter interpretation we will write it as $L(X, \theta)$; otherwise we will denote it simply by $L(\theta)$. We now state an important result, referred to previously, which gives a lower bound for the variance of any unbiased estimate. This result arises from the celebrated *Cramer–Rao inequality* (see, e.g., Wilks (1962), p. 351), which states the following:

Cramer–Rao inequality *Let $T = T(X)$ be any unbiased estimator of θ. Then, under general conditions on $f(x, \theta)$,*

$$\text{var}(T) \geq \frac{1}{E[\partial L(X, \theta)/\partial \theta]^2}, \tag{5.2.23}$$

with equality if and only if $\partial L/\partial \theta$ may be written in the form

$$\frac{\partial L}{\partial \theta} = k(\theta)[T(x) - \theta], \tag{5.2.24}$$

where $k(\theta)$ is some positive function of θ which does not involve x.

(The main condition required on $f(x, \theta)$ for the validity of (5.2.23), (5.2.24), is that we may interchange the operations of differentiation with respect to θ and integration with respect to x; this implies, in particular, that the range of the x_i must not depend on θ.) Moreover, if there exists some function of x, $T(x)$, and some positive function $k(\theta)$ of θ, such that $\partial L/\partial \theta$ may be expressed in the form (5.2.24), then it may be shown that $T(X)$ is necessarily *an unbiased estimator of θ whose variance attains the lower bound*, (5.2.23), and further that $T(X)$ is a *sufficient statistic for θ*. In this case we say that $T(X)$ is a *fully efficient* estimator of θ.

It may be shown further that the function $k(\theta)$ has a simple interpretation, namely, that,

$$k(\theta) = 1/\text{var}(T) \tag{5.2.25}$$

To illustrate the above results we consider again the estimation of the mean of a normal distribution with known variance. We have from (5.2.18),

$$L(\mu) = \log\{f(x, \mu)\} = \log\{g(\bar{x}, \mu)\} + \log\{h(x)\},$$

where $g(\bar{x}, \mu)$ and $h(x)$ are defined by (5.2.19), (5.2.20). Hence,

$$\frac{\partial L}{\partial \mu} = \frac{\partial}{\partial \mu}[\log\{g(\bar{x}, \mu)\}] = \frac{n}{\sigma^2}(\bar{x} - \mu).$$

This is exactly of the form (5.2.24), with $T(x) \equiv \bar{x}$, and $k(\mu) = n/\sigma^2$. We can then say immediately that $\bar{X}$ is an unbiased, fully efficient estimator of μ, and is a sufficient statistic for μ. Also, we have

$$\text{var}(\bar{X}) = 1/k(\mu) = \sigma^2/n,$$

in agreement with the well known result.

An alternative expression for the lower bound for var(T) may be derived. For, under the same conditions on $f(x, \theta)$, it may be shown that

$$E\left[\frac{\partial L(\mathbf{X}, \theta)}{\partial \theta}\right]^2 = -E\left[\frac{\partial^2 L(\mathbf{X}, \theta)}{\partial \theta^2}\right]. \tag{5.2.26}$$

Hence, in place of (5.2.23) we may write,

$$\text{var}(T) \geq -\frac{1}{E[\partial^2 L(\mathbf{X}, \theta)/\partial \theta^2]}, \tag{5.2.27}$$

with equality if and only if (5.2.24) holds.

Now clearly, the smaller var(T), i.e. the larger $\{1/\text{var}(T)\}$, the larger the precision of T as an estimate of θ, and hence, the more we "know" about θ. Consequently, the Cramer–Rao upper bound for $\{1/\text{var}(T)\}$ is called the (*Fisher*) *information* (I) on θ given by the observations, and we write

$$I = I(\theta) = E[\partial L/\partial \theta]^2 = -E[\partial^2 L/\partial \theta^2]. \tag{5.2.28}$$

Properties of maximum likelihood estimators

The maximum likelihood estimate, $\hat{\theta}$, is obtained by finding that value of θ which maximizes $L(\theta)$, i.e. which ascribes the maximum possible probability to the given observations. When the derivative of $L(\theta)$ (w.r. to θ) exists and is continuous for all θ, $\hat{\theta}$ will thus be the solution of

$$\partial L/\partial \theta = 0. \tag{5.2.29}$$

Besides being intuitively reasonable, maximum likelihood estimates have, in general, three very desirable properties, as follows.

(1) First, if there exists a sufficient statistic, T, for θ then the maximum likelihood estimate, $\hat{\theta}$, will be a function of T only, and hence will also be a sufficient statistic for θ. For, if T is sufficient for θ then by (5.2.16) we may write $f(x, \theta) = g(t, \theta) \cdot h(x)$, say. Hence, $L = \log\{g(t, \theta)\} + \log\{h(x)\}$, and $\partial L/\partial \theta = (\partial/\partial \theta)[\log\{g(t, \theta)\}]$. Thus, $\hat{\theta}$, being the solution of $\partial L/\partial \theta = 0$, must depend on x through t only.

(2) Secondly, if there is a fully efficient estimate, T, of θ, then the maximum likelihood estimate will be identical to it. For, if T is fully efficient then we

know that we may write, $\partial L/\partial\theta = k(\theta)(T - \theta)$. Hence, the solution of $\partial L/\partial\theta = 0$ is $\hat{\theta} = T$.

(3) However, whether or not a fully efficient estimate exists, the outstanding property of maximum likelihood estimates is that, for large numbers of observations, they are always *approximately fully efficient*; more precisely, they are said to be "*asymptotically fully efficient*". (There are a few provisos needed, but apart from the condition on $f(x, \theta)$ previously noted, they are almost invariably satisfied in practice.) The basic reason for asymptotic efficiency may be explained heuristically as follows. Suppose we expand $(\partial L/\partial\theta)$ evaluated at $\hat{\theta}$ about the "true" value θ. Then we have

$$0 = \frac{\partial L(\hat{\theta})}{\partial\theta} = \frac{\partial L(\theta + \hat{\theta} - \theta)}{\partial\theta} = \frac{\partial L(\theta)}{\partial\theta} + (\hat{\theta} - \theta)\frac{\partial^2 L(\theta_1)}{\partial\theta^2}, \qquad (5.2.30)$$

where θ_1 lies between θ and $\hat{\theta}$. (Note that, e.g., the expression $\partial L(\hat{\theta})/\partial\theta$ means the value of $\partial L(\theta)/\partial\theta$ evaluated at $\theta = \hat{\theta}$.)

Hence,

$$\frac{\partial L(\theta)}{\partial\theta} = -\frac{\partial^2 L(\theta_1)}{\partial\theta^2}(\hat{\theta} - \theta). \qquad (5.2.31)$$

Now as n (the number of observations) $\to \infty$, it may be shown that $\partial^2 L(\theta_1)/\partial\theta^2$ may be replaced by $E[\partial^2 L(\theta)/\partial\theta^2]$. Hence, for n large we may write approximately,

$$\frac{\partial L(\theta)}{\partial\theta} = -E\left[\frac{\partial^2 L(\theta)}{\partial\theta^2}\right](\hat{\theta} - \theta). \qquad (5.2.32)$$

This is of the form

$$\partial L/\partial\theta = k(\theta)(T - \theta),$$

where

$$T = \hat{\theta} \quad \text{and} \quad k(\theta) = -E\left[\frac{\partial^2 L}{\partial\theta^2}\right].$$

Hence, by the previous result we have that $\hat{\theta}$ is asymptotically *unbiased and fully efficient*, with asymptotic variance given by

$$\text{var}(\hat{\theta}) \sim \frac{1}{k(\theta)} = -\frac{1}{E[\partial^2 L/\partial\theta^2]} = \frac{1}{I}. \qquad (5.2.33)$$

It may be shown further that, under general conditions, the asymptotic distribution of $\hat{\theta}$ is normal, and hence we may write

$$\hat{\theta} \sim N\left(\theta, \frac{1}{I}\right). \qquad (5.2.34)$$

Multiparameter case

The method extends to the multi-parameter case in a fairly straightforward way. If the joint probability distribution of the observations involves several unknown parameters, so that the joint probability density function is $f(x, \theta_1, \theta_2, \ldots, \theta_p)$, say, then the likelihood function, L, is now a function of p variables, $\theta_1, \ldots, \theta_p$, i.e.

$$L(\theta_1, \ldots, \theta_p) = \log_e\{f(x, \theta_1, \theta_2, \ldots, \theta_p)\}.$$

The maximum likelihood estimates $\hat{\theta}_1, \ldots, \hat{\theta}_p$ are obtained by maximizing L w.r. to each of the variables, and, when the partial derivatives exist, are given by the solution of the set of equations,

$$\frac{\partial L}{\partial \theta_1} = 0, \qquad \frac{\partial L}{\partial \theta_2} = 0, \ldots, \qquad \frac{\partial L}{\partial \theta_p} = 0. \qquad (5.2.35)$$

Suppose, for example, that we have a random sample of n observations, $X_1, \ldots, X_n$, from a normal distribution whose mean μ and variance σ^2 are both unknown. Then the joint probability density function of $X_1, \ldots, X_n$ is

$$f(x, \mu, \sigma^2) = \left(\frac{1}{2\pi\sigma^2}\right)^{n/2} \exp\left[-\frac{1}{2\sigma^2} \sum_{i=1}^{n} (x_i - \mu)^2\right].$$

The log-likelihood function is therefore,

$$L(x, \mu, \sigma^2) = -\frac{n}{2} \log 2\pi - \frac{n}{2} \log(\sigma^2) - \frac{1}{2\sigma^2} \sum_{i=1}^{n} (x_i - \mu)^2.$$

Hence,

$$\frac{\partial L}{\partial \mu} = 0 \Rightarrow \frac{1}{\hat{\sigma}^2} \sum_{i=1}^{n} (x_i - \hat{\mu}) = 0,$$

giving

$$\hat{\mu} = (1/n)\left(\sum_{i=1}^{n} x_i\right) = \bar{x},$$

and

$$\frac{\partial L}{\partial \sigma^2} = 0 \Rightarrow -\frac{n}{2\hat{\sigma}^2} + \frac{1}{2\hat{\sigma}^4} \sum_{i=1}^{n} (x_i - \hat{\mu})^2 = 0,$$

giving

$$\hat{\sigma}^2 = \frac{1}{n} \sum_{i=1}^{n} (x_i - \hat{\mu})^2 = \frac{1}{n} \sum_{i=1}^{n} (x_i - \bar{x})^2.$$

Thus, the maximum likelihood estimates of μ and σ^2 are just the usual sample mean and sample variance, but note that $\hat{\sigma}^2$ is the "biased" sample variance (i.e. uses devisor n instead of $(n-1)$).

In place of the information I (given by (5.2.28)) we now have an *information matrix*, $\mathbb{Z}$, with (i, j)th element,

$$I_{ij} = E\left[\frac{\partial L}{\partial \theta_i} \cdot \frac{\partial L}{\partial \theta_j}\right] = -E\left[\frac{\partial^2 L}{\partial \theta_i \, \partial \theta_j}\right]. \tag{5.2.36}$$

As in the single parameter case it may be shown that as the number of observations, $n \to \infty$, $\hat{\boldsymbol{\theta}} = (\hat{\theta}_1, \ldots, \hat{\theta}_p)$ has the asymptotic multivariate normal distribution,

$$\hat{\boldsymbol{\theta}} \sim N(\boldsymbol{\theta}, \mathbb{Z}^{-1}). \tag{5.2.37}$$

Thus, for large n,

$$\begin{aligned} \text{var}(\hat{\theta}_i) &\sim I^{ii}, \quad \text{all } i \\ \text{cov}(\hat{\theta}_i, \hat{\theta}_j) &\sim I^{ij}, \quad \text{all } i, j, \end{aligned} \tag{5.2.38}$$

where I^{ij} denotes the (i, j)th element of the matrix $\mathbb{Z}^{-1}$. The estimator $\hat{\boldsymbol{\theta}}$ is thus asymptotically unbiased and fully efficient. (See, e.g., Wilks (1962) p. 360.)

II. *Method of least squares*

The method of "least squares" is, perhaps, the oldest estimation procedure, and was first developed by Gauss in the early 19th century. Like the method of maximum likelihood it is a powerful technique which is widely used in many different types of problems. However, whereas maximum likelihood can, in principle, be applied to *any* estimation problem (provided we know the general form of the joint probability distribution of the observations), the method of least squares applies only to the case where the unknown parameters arise in expressions for the *means* of the observations, these expressions being *linear* in the parameters. Thus, for a typical observation y we would have,

$$E(y) = \theta_1 x_1 + \theta_2 x_2 + \ldots + \theta_q x_q, \tag{5.2.39}$$

where the values of $x_1, x_2, \ldots, x_q$ are *known*, and $\theta_1, \theta_2, \ldots, \theta_q$ are unknown parameters. Equivalently, we may write,

$$y = \theta_1 x_1 + \theta_2 x_2 + \ldots + \theta_q x_q + \varepsilon, \tag{5.2.40}$$

where $E(\varepsilon) = 0$. The random variable, ε, is usually regarded as a "measurement error", and (5.2.40) then gives us a *linear model* in which, apart from the error term, the value of y is a linear function of q other variables $x_1, \ldots, x_q$ whose values are assumed to be known exactly. (Note that in this model ε represents only the error in the measurement of y; the x variables are assumed to be free of error). For example, y might be the voltage measured at some point in an electrical circuit and $x_1, \ldots, x_q$ the "pre-set" values of q variable resistors in the circuit. Suppose now that we perform the experiment n times, each time using a different set of values of $x_1, \ldots, x_q$. Let y_i denote the value of y observed in the ith experiment and $x_{1i}, \ldots, x_{qi}$ the values of $x_1, \ldots, x_q$ chosen for the ith experiment. Then we may write,

$$y_i = \theta_1 x_{1i} + \theta_2 x_{2i} + \ldots + \theta_q x_{qi} + \varepsilon_i, \qquad i = 1, \ldots, n, \qquad (5.2.41)$$

where $E(\varepsilon_i) = 0$, all i, and ε_i represents the error of measurement in the ith experiment. The problem now is to estimate the unknown values of $\theta_1, \ldots, \theta_q$, on the basis of the observed values $\{y_i\}$, and the known values $\{x_{1i}\}, \ldots, \{x_{qi}\}$, $i = 1, \ldots, n$. The "*principle of least squares*" says that we should *choose as estimates of* $\theta_1, \ldots, \theta_q$ *those values*, $\hat{\theta}_1, \ldots, \hat{\theta}_q$, *which minimize the quantity*,

$$Q(\theta_1, \ldots, \theta_q) = \sum_{i=1}^{n} \{y_i - \theta_1 x_{1i} - \theta_2 x_{2i} - \ldots - \theta_q x_{qi}\}^2. \qquad (5.2.42)$$

Thus, the "least squares" estimates, $\hat{\theta}_1, \ldots, \hat{\theta}_q$, are chosen so as to minimize the sum of squares of the deviations of the observations from their theoretical means (as given by (5.2.39)).

Differentiating Q partially w.r. to each θ_j gives,

$$\frac{\partial Q}{\partial \theta_j} = (-2) \sum_{i=1}^{n} x_{ji}\{y_i - \theta_1 x_{1i} - \theta_2 x_{2i} - \ldots - \theta_q x_{qi}\}, \qquad j = 1, \ldots, q.$$

Setting these derivatives equal to zero we obtain the following equations for $\hat{\theta}_1, \ldots, \hat{\theta}_q$,

$$S_{0j} = \hat{\theta}_1 S_{1j} + \hat{\theta}_2 S_{2j} + \ldots + \hat{\theta}_q S_{qj}, \qquad j = 1, \ldots, q \qquad (5.2.43)$$

where

$$S_{0j} = \sum_{i=1}^{n} y_i x_{ji}, \qquad j = 1, \ldots, q, \qquad (5.2.44)$$

$$S_{kj} = \sum_{i=1}^{n} x_{ki} x_{ji}, \qquad k = 1, \ldots, q; \quad j = 1, \ldots, q. \qquad (5.2.45)$$

Writing,

$$\boldsymbol{\theta} = (\theta_1, \ldots, \theta_q)', \qquad \boldsymbol{y} = (y_1, \ldots, y_n)', \qquad (5.2.46)$$

$$\boldsymbol{X} = \begin{bmatrix} x_{11} & x_{12} & \cdots & x_{1n} \\ x_{21} & x_{22} & \cdots & x_{2n} \\ \vdots & \vdots & & \vdots \\ x_{q1} & x_{q2} & \cdots & x_{qn} \end{bmatrix} \qquad (5.2.47)$$

the above equation may be written in matrix form as,

$$\boldsymbol{Xy} = (\boldsymbol{XX'})\hat{\boldsymbol{\theta}} \qquad (5.2.48)$$

$$= \boldsymbol{S} \cdot \hat{\boldsymbol{\theta}}, \qquad (5.2.49)$$

say, where

$$\boldsymbol{S} = \boldsymbol{XX'}. \qquad (5.2.50)$$

The set of equations (5.2.48) for $\hat{\theta}_1, \ldots, \hat{\theta}_q$, are usually called the "*normal equations*".

If $\boldsymbol{S}$ is non-singular, then the solution for $\hat{\boldsymbol{\theta}}$ is given immediately by,

$$\hat{\boldsymbol{\theta}} = \boldsymbol{S}^{-1}\boldsymbol{Xy}. \qquad (5.2.51)$$

Moreover, if the error terms ε_i are uncorrelated, i.e. $\text{cov}(\varepsilon_i, \varepsilon_j) = 0$, all $i \neq j$, and all have the same variance, i.e. $\text{var}(\varepsilon_i) = \sigma^2$, all i, then it may be shown (see, e.g. Wilks (1962), p. 286), that each $\hat{\theta}_i$ is an *unbiased estimate* of θ_i, i.e. $E(\hat{\theta}_i) = \theta_i$, all i, and the *variance-covariance matrix of $\hat{\boldsymbol{\theta}}$ is* $(\sigma^2 \boldsymbol{S}^{-1})$, i.e.

$$\text{cov}(\hat{\theta}_i, \hat{\theta}_j) = (i, j)\text{th element of } \sigma^2 \boldsymbol{S}^{-1}, \qquad (5.2.52)$$

and

$$\text{var}(\hat{\theta}_i) = i\text{th diagonal element of } \sigma^2 \boldsymbol{S}^{-1}, \qquad (5.2.53)$$

Regression models

The model (5.2.41) is sometimes referred to as "*multiple linear regression*", $x_{1i}, \ldots, x_{qi}$ being called the "independent" or "regressor" variables. In particular, a *linear regression* between y and a single variable x, of the form,

$$y_i = \alpha + \beta x_i + \varepsilon_i, \qquad (5.2.54)$$

can be treated as a special case of (5.2.41) in which $q = 2$, $x_{1i} \equiv 1$ (all i), $x_{2i} = x_i$, and θ_1, θ_2, have been relabelled as α and β to preserve the usual notation used for this case. It is easily shown that minimizing

$$Q = \sum_{i=1}^{n} (y_i - \alpha - \beta x_i)^2$$

with respect to α and β gives the following least squares estimates,

$$\hat{\alpha} = \bar{y} - \hat{\beta}\bar{x}, \tag{5.2.55}$$

$$\hat{\beta} = \frac{\sum_{i=1}^{n} (y_i - \bar{y})(x_i - \bar{x})}{\sum_{i=1}^{n} (x_i - \bar{x})^2} \tag{5.2.56}$$

where

$$\bar{y} = \frac{1}{n} \sum_{i=1}^{n} y_i, \qquad \bar{x} = \frac{1}{n} \sum_{i=1}^{n} x_i.$$

These equations for $\hat{\alpha}$, $\hat{\beta}$, can, of course, also be obtained as a special case of (5.2.51).

A *polynomial regression* between y and a single variable x, of the form,

$$y_i = \alpha + \beta_1 x_i + \beta_2 x_i^2 + \ldots + \beta_p x_i^p + \varepsilon_i, \tag{5.2.57}$$

can be treated similarly as a special case of (5.2.41) in which $q = p + 1$, $x_{1i} \equiv 1$ (all i), $x_{ji} = x_i^{j-1}$, $j = 2, \ldots, p^{+1}$. The least squares estimates of $\alpha_1, \beta_1, \ldots, \beta_p$, are given by (5.2.51) with the matrix X taking the special form obtained by substituting the above values of x_{1i} and x_{ji}.

Gauss–Markov theorem

The method of least squares has a strong intuitive appeal, particularly in the context of polynomial regression models when it reduces to the well known device of fitting a curve to a set of points by choosing its parameters so as to minimize the sum of squares of the vertical deviations of the points from the curve. However, the outstanding property of the method is that when the ε_i satisfy the properties mentioned above, namely that they are uncorrelated and have a common variance, it produces estimates which are optimal within the class of all "*linear unbiased estimates*". To explain this more precisely let us first note that when we evaluate the right-hand side of (5.2.51) in detail, each least squares estimate $\hat{\theta}_i$ is a *linear combination* of the observations, $y_1, y_2, \ldots, y_n$, i.e. it is of the form

$$\hat{\theta}_i = a_1 y_1 + a_2 y_2 + \ldots + a_n y_n,$$

(where the a are functions of the $x_{1i}, \ldots, x_{qi}$ variables). Suppose now that we consider some other unbiased estimate of θ_i, say θ_i^*, which is also a linear combination of $y_1, \ldots, y_n$, say,

$$\theta_i^* = b_1 y_1 + b_2 y_2 + \ldots + b_n y_n.$$

Here, the b_i can be chosen in any way we please so long as θ_i^* is unbiased, i.e. the b_i must satisfy

$$E\left[\sum_{i=1}^n b_i y_i \right] = \theta_i, \qquad \text{all } i.$$

Then we may show that for all choices of $b_1, \ldots, b_n$ (subject to the above constraint),

$$\text{var}(\theta_i^*) \geq \text{var}(\hat{\theta}_i),$$

with equality if $b_i = a_i$, $i = 1, \ldots, n$. Thus, any other linear unbiased estimate is bound to be less efficient than the least squares estimate. The least squares estimates are therefore said to be the "best linear unbiased estimates" (BLUE) of the parameters. This basic result is embodied in the *Gauss–Markov Theorem* (see, e.g. Wilks (1962), p. 286) which, in fact, states a rather more general result, namely, that if we consider any given linear combination of $\theta_1, \ldots, \theta_q$, say,

$$\psi = \alpha_1 \theta_1 + \alpha_2 \theta_2 + \ldots + \alpha_q \theta_q,$$

then subject to the conditions on the $\{\varepsilon_i\}$ noted above, the BLUE of ψ is given by

$$\hat{\psi} = \alpha_1 \hat{\theta}_1 + \alpha_2 \hat{\theta}_2 + \ldots + \alpha_q \hat{\theta}_q.$$

Equivalence with maximum likelihood estimation

So far we have not made any assumptions about the distribution of the error terms $\{\varepsilon_i\}$ other than that they are uncorrelated and have a common variance, and the optimal property of the least squares estimates described above holds irrespective of the precise distribution of the $\{\varepsilon_i\}$. However, if we now introduce the much stronger additional assumptions that the $\{\varepsilon_i\}$ have a *multivariate normal* distribution, then it is easily seen that the *least squares procedure leads to exactly the same estimates as the maximum likelihood method*. For, if the $\{\varepsilon_i\}$ have a multivariate normal distribution, and are uncorrelated with common variance σ^2, the log-likelihood function of $(\theta_1, \ldots, \theta_q)$ is

$$L(\theta_1, \ldots, \theta_q) = \log\left[\left(\frac{1}{2\pi\sigma^2} \right)^{n/2} \right] - \frac{1}{2\sigma^2} \sum_{i=1}^n (y_i - \theta_1 x_{1i} \ldots - \theta_q x_{qi})^2 \qquad (5.2.58)$$

$$= \log\left[\left(\frac{1}{2\pi\sigma^2} \right)^{n/2} \right] - \frac{1}{2\sigma^2} Q(\theta_1, \ldots, \theta_q), \qquad (5.2.59)$$

where $Q(\theta_1, \ldots, \theta_q)$ is the "sum of squares" given by (5.2.42). Hence, maximizing L w.r. to $\theta_1, \ldots, \theta_q$, is exactly the same as minimizing S, and the two procedures clearly lead to the same estimates.

Weighted least squares

When $\{\varepsilon_i\}$ are uncorrelated with common variance σ^2 their variance covariance is $(\sigma^2 I)$, I being the identity matrix of order n. If we drop these assumptions and allow the $\{\varepsilon_i\}$ to have a general variance–covariance Σ, then the Gauss–Markov theorem is no longer valid and the equivalence between least squares and maximum likelihood no longer holds even under the assumption of normality. However, if the matrix Σ is known a priori, we may recover the above properties by modifying the method as follows.

Assuming that the $\{\varepsilon_i\}$ have a multivariate normal distribution, the log-likelihood function of $(\theta_1, \ldots, \theta_q)$ may be written in the form

$$L(\theta_1, \ldots, \theta_q) = -\tfrac{1}{2} \log[(2\pi)^n \det(\Sigma)] - \tfrac{1}{2}[(y - X\theta)'\Sigma^{-1}(y - X\theta)], \quad (5.2.60)$$

where y, X, θ, are defined by (5.2.46), (5.2.47) (cf. the form of the multivariate normal distribution given in Chapter 2, equation (2.12.23)). Hence, maximizing L w.r. to $\theta_1, \ldots, \theta_q$, is equivalent to minimizing the quadratic form,

$$Q^*(\theta_1, \ldots, \theta_q) = (y - X\theta)'\Sigma^{-1}(y - X\theta). \quad (5.2.61)$$

In scalar form this expression becomes,

$$Q^*(\theta_1, \ldots, \theta_q) = \sum_{i=1}^{n} \sum_{j=1}^{n} \sigma^{ij}(y_i - \theta_1 x_{1i} - \ldots \theta_q x_{qi})(y_j - \theta_1 x_{1j} - \ldots \theta_q x_{qj}),$$
$$(5.2.62)$$

where σ^{ij} is the (i, j)th element of Σ^{-1}.

In particular, if the $\{\varepsilon_i\}$ are uncorrelated but with unequal variances, i.e. $\text{var}(\varepsilon_i) = \sigma_i^2$, $i = 1, \ldots, n$, then Σ is a diagonal matrix, and hence $\sigma^{ij} = 0$, $i \neq j$, $\sigma^{ii} = 1/\sigma_i^2$. In this case, we have,

$$Q^*(\theta_1, \ldots, \theta_q) = \sum_{i=1}^{n} \frac{1}{\sigma_i^2}(y_i - \theta_1 x_{1i} - \ldots - \theta_q x_{qi})^2. \quad (5.2.63)$$

The method of estimation of $\theta_1, \ldots, \theta_q$, based on minimizing $Q^*(\theta_1, \ldots, \theta_q)$ is thus called "*weighted least squares*".

The minimization of (5.2.61) leads to the "generalized normal equations" (replacing (5.2.48)),

$$X\Sigma^{-1}y = (X\Sigma^{-1}X')\hat{\theta}. \quad (5.2.64)$$

Non-linear least squares

The method of least squares can, in principle, be extended to the case where the expressions for the means of the observations are non-linear functions of $\theta_1, \ldots, \theta_q$. Thus, suppose that we have the model,

$$y_i = g_i(\theta_1, \ldots, \theta_q) + \varepsilon_i, \qquad i = 1, \ldots, n, \qquad (5.2.65)$$

where the g_i are *known* functions of $\theta_1, \ldots, \theta_q$. Under the assumptions that the $\{\varepsilon_i\}$ are again uncorrelated with common variance, the least squares estimates of $\theta_1, \ldots, \theta_q$ are obtained by minimizing the expression

$$Q(\theta_1, \ldots, \theta_q) = \sum_{i=1}^{n} \{y_i - g_i(\theta_1, \ldots, \theta_q)\}^2. \qquad (5.2.66)$$

(Under the assumption of normality on the $\{\varepsilon_i\}$, this procedure is again equivalent to maximum likelihood estimation).

Unfortunately, we can no longer obtain a simple analytical expression for the vector of least squares estimates $\hat{\boldsymbol{\theta}}$, and in practice the computation of $\hat{\boldsymbol{\theta}}$ would usually be performed iteratively, starting with an initial rough guess, $\hat{\boldsymbol{\theta}}_0$, say, and then using an approximate "linearization" of the right-hand side of (5.2.66).

III. *Method of moments*

It is worth mentioning a third general estimation procedure called the "method of moments". It is nowhere near as general or powerful as the methods of maximum likelihood or least squares, and there is no guarantee that the estimators which it produces are asymptotically efficient, although they are in general, consistent. However, it is a useful empirical technique, and the essence of the method is as follows.

Suppose we have a random variable X whose probability distribution involves k unknown parameters, $\theta_1, \ldots, \theta_k$. Then we may express the first k moments of the distribution as functions of $\theta_1, \ldots, \theta_k$, i.e. we may write,

$$
\begin{aligned}
\mu_1' &= h_1(\theta_1, \ldots, \theta_k) \\
\mu_2' &= h_2(\theta_1, \ldots, \theta_k) \\
&\vdots \\
\mu_k' &= h_k(\theta_1, \ldots, \theta_k),
\end{aligned}
\qquad (5.2.67)
$$

say. In principle we may invert these equations to obtain expressions for

$\theta_1, \ldots, \theta_k$ in terms of $\mu'_1, \ldots, \mu'_k$, giving, say,

$$\theta_1 = p_1(\mu'_1, \mu'_2, \ldots, \mu'_k)$$
$$\theta_2 = p_2(\mu'_1, \mu'_2, \ldots, \mu'_k)$$
$$\vdots \qquad \vdots \qquad \qquad (5.2.68)$$
$$\theta_k = p_k(\mu'_1, \mu'_2, \ldots, \mu'_k).$$

If now we are given a random sample of n observations on X, then we can compute the first k sample moments $m'_1, m'_2, \ldots, m'_k$ (as described in Section 2.8). The basic idea behind the method of moments is to argue that since, under general conditions, $m'_1, m'_2, \ldots, m'_k$ are "reasonable" estimates of $\mu'_1, \mu'_2, \ldots, \mu'_k$, we may obtain "reasonable" estimates of $\theta_1, \ldots, \theta_k$ by substituting the sample moments for the distributional moments in (5.2.68). We then obtain estimates of $\theta_1, \ldots, \theta_k$, given by,

$$\hat{\theta}_1 = p_1(m'_1, m'_2, \ldots, m'_k)$$
$$\hat{\theta}_2 = p_2(m'_1, m'_2, \ldots, m'_k)$$
$$\vdots \qquad \qquad (5.2.69)$$
$$\hat{\theta}_k = p_k(m'_1, m'_2, \ldots, m'_k).$$

5.3 ESTIMATION OF AUTOCOVARIANCE AND AUTOCORRELATION FUNCTIONS

5.3.1 Form of the Data

We now return to the problem posed at the beginning of this chapter, namely that of estimating the second order properties of a stationary process from experimental data. The data almost invariably consist of the values of the process recorded over a finite time interval during one run of the experiment, i.e. we are given only *one portion of one realization* of the process. Thus, for a discrete parameter process $\{X_t\}$ we would be given the values of X_t for $t = 1, 2, \ldots, N$, say, i.e. we would be given the observed values of $X_1, X_2, \ldots, X_N$ (see Fig. 5.2). In the continuous parameter case

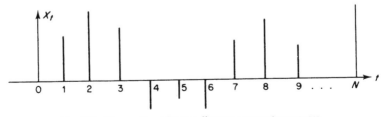

Fig. 5.2. Observations from a discrete parameter process.

the data would consist of a continuous record of $X(t)$ over the interval $(0, T)$, say, i.e. we would be given the observed values of $X(t)$ for all t in the interval $0 \leq t \leq T$ (see Fig. 5.3). However, if we wish to analyse such a record

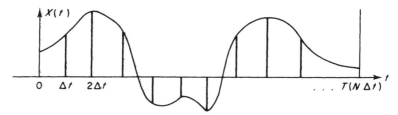

Fig. 5.3. Observations from a continuous parameter record.

by digital methods we would have to read off its values at small intervals Δt giving the discrete set of values $X(\Delta t), X(2\Delta t), \ldots, X(N\Delta t)$, where $N = T/\Delta t$. This means, of course, that we can now estimate its auto-covariance and autocorrelation functions only at the corresponding discrete set of lags, but assuming that Δt is chosen sufficiently small so as to avoid serious "aliasing" effects (see Section 4.8.3 and Section 7.1.1), this pro-cedure will usually be quite adequate. In reading a continuous record at intervals Δt we are, in effect, converting the original continuous parameter process into a discrete parameter process, and discussion which follows will therefore concentrate almost entirely on the discrete parameter case. There are, nevertheless, a limited number of cases where we might wish to analyse a continuous parameter process by analogue methods, and then we can make use of the full information in a continuous record. We will discuss appropriate estimation procedures for continuous records in a later section.

5.3.2 Estimation of the Mean

Let $\{X_t\}$ be a discrete parameter stationary process with

$$E[X_t] = \mu, \quad \text{(all } t\text{)},$$

$$\text{var}[X_t] = \sigma^2, \quad \text{(all } t\text{)}, \qquad (5.3.1)$$

$$\text{cov}[X_t, X_{t+r}] = R(r) = \sigma^2 \rho(r).$$

The mean μ, the variance σ^2, the autocovariance function $R(r)$ and the autocorrelation function $\rho(r)$ are all unknown. We are given N observations $X_1, X_2, \ldots, X_N$ and our main interest is in estimating the functions $R(r)$ and $\rho(r)$, but, as a preliminary step, we must first estimate the mean μ. We may note that this problem is logically different from the much simpler problem, discussed previously, of estimating the mean of a probability distribution on the basis of a random sample. In the first place, the observations $X_1, \ldots, X_n$ are, in general, correlated, and secondly, only when the process is completely stationary will they have a common probability distribution. If the process is stationary only up to order 2 (our standard assumption) then the distributions of $X_1, \ldots, X_N$ may all be different. Since we are not, at this stage, making any assumptions about the general form of these distributions we cannot attempt to use any general estimation methods, such as maximum likelihood, but we know that although the distributions of $X_1, \ldots, X_N$ may be different they all have the same mean value μ. This suggests that the usual estimate, namely the sample mean, may still provide a reasonable estimate of μ; whether this is true or not will emerge from its sampling properties. We therefore propose to estimate μ by the sample mean,

$$\bar{X} = \frac{1}{N} \sum_{t=1}^{N} X_t. \tag{5.3.2}$$

We have,

$$E[\bar{X}] = \frac{1}{N} \sum_{t=1}^{N} E[X_t] = \frac{1}{N} \cdot N\mu = \mu, \tag{5.3.3}$$

which tells us immediately that $\bar{X}$ is an *unbiased estimate of* μ.

Also, applying the result (2.13.15) (Chapter 2),

$$\text{var}[\bar{X}] = \frac{1}{N^2} \sum_{s=1}^{N} \sum_{t=1}^{N} \text{cov}(X_s, X_t) = \frac{\sigma^2}{N^2} \sum_{s=1}^{N} \sum_{t=1}^{N} \rho(t-s). \tag{5.3.4}$$

We now make a change of variables from s, t, to s and $r = (t - s)$. (The double summation in (5.3.4) extends over the lattice points in a square. Instead of summing first over, e.g., rows and then adding the row sums we are now summing over diagonals and adding the diagonal sums together.) The summation over r goes from $-(N-1)$ to $(N-1)$; for $r > 0$, s goes from 1 to $N - r$, while for $r < 0$, s goes from $-r$ to N. The summand in (5.3.4) is a function of r only, and hence summation over s gives the expression $(N - |r|)$ for both positive and negative r. We thus have,

$$\text{var}(\bar{X}) = \frac{\sigma^2}{N^2} \sum_{r=-(N-1)}^{(N-1)} (N - |r|)\rho(r),$$

$$= \frac{\sigma^2}{N} \sum_{r=-(N-1)}^{(N-1)} \left(1 - \frac{|r|}{N}\right)\rho(r). \tag{5.3.5}$$

This expression for $\text{var}(\bar{X})$ is exact for all values of N. In particular, it follows that if

$$\lim_{N \to \infty} \left\{ \sum_{r=-(N-1)}^{N-1} \left(1 - \frac{|r|}{N}\right)\rho(r) \right\}$$

is finite then $\text{var}(\bar{X}) \to 0$ as $N \to \infty$, and since $\bar{X}$ is unbiased it follows by (5.2.14) that $\bar{X}$ is then a consistent estimate of μ. However, if X_t has a purely continuous spectrum, with (normalized) spectral density function $f(\omega)$, then by (4.8.30),

$$f(\omega) = \frac{1}{2\pi} \sum_{r=-\infty}^{\infty} \rho(r) \exp(-i\omega r).$$

It now follows from an argument similar to the discussion in the proof of Theorem 4.8.1 (Chapter 4) that

$$\lim_{N \to \infty} \sum_{r=-(N-1)}^{(N-1)} \left(1 - \frac{|r|}{N}\right)\rho(r) = \sum_{r=-\infty}^{\infty} \rho(r) = 2\pi f(0). \tag{5.3.6}$$

(The LHS is the Caesaro sum of the sequence $\{\rho(r)\}$ and therefore converges to the "unweighted" sum, $\sum_{-\infty}^{\infty} \rho(r)$, whenever the latter is convergent.) Hence, in this case, $\text{var}(\bar{X}) \to 0$ as $N \to \infty$ and, by (5.2.14), $\bar{X}$ is an *unbiased and consistent estimate of* μ. Also, for large N we have the useful approximation,

$$\text{var}(\bar{X}) \sim (2\pi\sigma^2/N) . f(0). \tag{5.3.7}$$

This may well give a larger value for $\text{var}(\bar{X})$ than the usual expression, (σ^2/N), which applies in the case of independent observations. For example, if X_t is an AR(1) process with parameter a ($|a| < 1$) then from (4.10.10) we have

$$2\pi f(0) = \frac{1-a^2}{(1-a)^2} = \frac{1+a}{1-a}, \tag{5.3.8}$$

so that here,

$$\text{var}(\bar{X}) \sim \frac{\sigma^2}{N} . \left(\frac{1+a}{1-a}\right). \tag{5.3.9}$$

We may say, roughly speaking, that for the same accuracy in the estimation of the mean the *"equivalent number of independent observations"* is $N((1-a)/(1+a))$.

There is an interesting relationship between $\bar{X}$ and μ. For each t, μ is the average of X_t over all realizations, i.e. for each t, μ *is the ensemble average of* X_t, whereas, for each realization, $\bar{X}$ is the average of the $\{X_t\}$ over the time points $t = 1, \ldots, N$, i.e. $\bar{X}$ *is the time average of* X_t. In using $\bar{X}$ as an estimate of μ we have, in effect, replaced an ensemble average by the corresponding time average. (Since we are given only a sample from a single realization this is clearly a reasonable procedure.) We shall return to this point later when we discuss the idea of *"ergodicity"*.

5.3.3 Estimation of the Autocovariance Function

When we previously considered the problem of estimating a covariance (Section 5.2.1) we were dealing with the conventional situation where we had a random sample of pairs of observations, $(x_1, y_1), \ldots, (x_n, y_n)$, from a common bivariate distribution, and its covariance is then estimated by (5.2.2). In the present context we have only one set of observations, $X_1, \ldots, X_N$, but we may form these into the pairs,

$$(X_1, X_{r+1}), (X_2, X_{r+2}), \ldots, (X_{N-r}, X_N),$$

i.e. we pair together observations separated by r time units. In general, these pairs will all come from different bivariate distributions, so that instead of having n pairs from the same distribution we have one pair from each of $(N-r)$ different distributions. However, since $\{X_t\}$ is stationary we know that these bivariate distributions will all have the same covariance, namely $R(r)$, and arguing as in the previous section we may still use an expression of the form (5.2.2) as an estimate of $R(r)$, where now the $\{x_i\}$ correspond to $\{X_1, \ldots, X_{N-r}\}$, the $\{y_i\}$ to $\{X_{r+1}, \ldots, X_N\}$ and n, the total number of pairs, is $(N-r)$. However, since both sets of observations come from the same stationary process there is no point here in using separate estimates $(\bar{x}, \bar{y})$ for the means of the $\{x_i\}$ and $\{y_i\}$; we may instead use the previous estimate $\bar{X}$ for the common mean μ. We now obtain for each r the following estimate of $R(r)$:

$$\hat{R}^*(r) = \frac{1}{N-|r|} \sum_{t=1}^{N-|r|} (X_t - \bar{X})(X_{t+|r|} - \bar{X}), \qquad r = 0, \pm 1, \pm 2, \ldots, \pm(N-1).$$

$$(5.3.10)$$

(We attach an asterisk to $\hat{R}^*(r)$ in order to distinguish it from another estimate which we shall introduce later.)

We note that,

(1) The autocovariance, $R(r)$, and the estimate, $\hat{R}^*(r)$, have the same kind of relationship as μ and $\bar{X}$, namely, that whereas $R(r)$ is the *ensemble* average of the product $[\{X_t - \mu\}\{X_{t+|r|} - \mu\}]$, $\hat{R}^*(r)$ is the *time average* of the product $[\{X_t - \bar{X}\}\{X_{t+|r|} - \bar{X}\}]$.

(2) Since our sample covers only N time points there are no pairs of observations separated by more than $(N-1)$ time units, and hence there is no possible way of estimating the values of $R(r)$ for $|r| \geq N$.

(3) Since $R(r)$ is an even function, (i.e. $R(-r) = R(r)$, all r) for real valued processes, it would be sensible to make $\hat{R}^*(r)$ also an even function. We have ensured that this property holds by writing $\hat{R}^*(r)$ in (5.3.10) as a function of $|r|$ rather than as a function of r.

If, for the moment, we ignore the effect of estimating μ by $\bar{X}$ then it is easily seen that $R^*(r)$ *is an exactly unbiased estimate of* $R(r)$. For, replacing $\bar{X}$ by μ in (5.3.10), we have,

$$E[\hat{R}^*(r)] = \frac{1}{N - |r|} \sum_{t=1}^{N-|r|} E[\{X_t - \mu\}\{X_{t+|r|} - \mu\}]$$

$$= \frac{1}{N - |r|} \sum_{t=1}^{N-|r|} R(r)$$

$$= R(r), \quad \text{all } r. \tag{5.3.11}$$

When we allow for the effect of estimating μ by $\bar{X}$ it turns out that $\hat{R}^*(r)$ is no longer exactly unbiased but is asymptotically unbiased. The bias may be evaluated by writing,

$$\sum_{t=1}^{N-|r|} [\{X_t - \mu\}\{X_{t+|r|} - \mu\}]$$

$$\equiv \sum_{t=1}^{N-|r|} [\{(X_t - \bar{X}) + (\bar{X} - \mu)\}\{(X_{t+|r|} - \bar{X}) + (\bar{X} - \mu)\}]$$

$$\doteq \sum_{t=1}^{N-|r|} [\{X_t - \bar{X}\}\{X_{t+|r|} - \bar{X}\}] + (N - |r|)(\bar{X} - \mu)^2.$$

(Here we are ignoring the "end-effects" arising out of the summation extending only over $t = 1$ to $N - |r|$, i.e. we are assuming, e.g., that

$$\sum_{t=1}^{N-|r|} (X_t - \bar{X}) \sim \sum_{t=1}^{N} (X_t - \bar{X}) = 0.)$$

Hence, using (5.3.11), we have

$$E[\hat{R}^*(r)] = R(r) - E[(\bar{X} - \mu)^2] = R(r) - \text{var}(\bar{X}) \sim R(r) - (2\pi\sigma^2/N) \cdot f(0),$$

$$\tag{5.3.12}$$

using (5.3.7). Thus, the bias of $\hat{R}^*(r)$ is $O(1/N)$ (assuming $f(0)$ exists), and, as we shall see later, is a smaller order of magnitude than the sampling fluctuations of $\hat{R}^*(r)$. As a first approximation, we may therefore ignore the bias term.

Some authors, notably Parzen (1961b) and Schaerf (1964), suggest the alternative estimate,

$$\hat{R}(r) = \frac{1}{N} \sum_{t=1}^{N-|r|} (X_t - \bar{X})(X_{t+|r|} - \bar{X}). \tag{5.3.13}$$

This estimate, which is based on the use of the divisor N (rather than $(N - |r|)$) has a larger bias than $\hat{R}^*(r)$, but it has been asserted that, in general, $\hat{R}(r)$ has a smaller mean square error than $R^*(r)$. Ignoring the effect of estimating μ, we have,

$$E[\hat{R}(r)] = \frac{1}{N} \sum_{t=1}^{N-|r|} R(r) = \left(1 - \frac{|r|}{N}\right) R(r). \qquad (5.3.14)$$

Thus, although for fixed r the bias is $O(1/N)$, it nevertheless increases with $|r|$. When we take account of the effect of estimating μ the bias has an additional term corresponding to (5.3.12), i.e. we find,

$$E[\hat{R}(r)] = R(r) - \frac{|r|}{N} R(r) - \frac{2\pi\sigma^2(N - |r|)}{N^2} \cdot f(0). \qquad (5.3.15)$$

A plot of $\hat{R}(r)$ (or $\hat{R}^*(r)$) against r, $(r = 0, \pm 1, \ldots \pm(N-1))$ is usually called the "*sample autocovariance function*". To distinguish between $\hat{R}^*(r)$ and $\hat{R}(r)$ we will stretch the usual terminology and refer to the former as the "*unbiased estimate*", and to the latter as the "*biased estimate*".

Choice of estimates

It would be fair to say that nowadays most time series analysts prefer to use the biased estimate $\hat{R}(r)$ rather than the unbiased estimate $\hat{R}^*(r)$ and this is reflected in the fact that the majority of computer time series packages also use the biased estimate. To statisticians this may seem surprising since, in general, there is a natural tendency to use unbiased rather than biased estimates, particularly when, as in this case, an unbiased estimate can be constructed so easily. To understand the reason for this apparent departure from the usual practice we must consider the problem of estimating the autocovariance function $R(r)$ in more detail. The essential point is that we are not concerned so much with estimating $R(r)$ *for a particular value of r*. Rather, we are concerned with estimating the *function $R(r)$ for all values of r* from 0 to $\pm(N-1)$. It is in this sense that the biased estimate $\hat{R}(r)$ has a very important advantage over the unbiased estimate $\hat{R}^*(r)$ in that $\hat{R}(r)$ *shares with $R(r)$ the property that it is a positive semi-definite function*—as defined in Section 3.3.2. (The positive semi-definite property of $\hat{R}(r)$ follows from a result to be proved in Chapter 6, namely, that the finite Fourier transform of $\hat{R}(r)$ is a non-negative function called the "periodogram".) On the other hand, $\hat{R}^*(r)$ is not necessarily positive semi-definite.

Now let us consider the mean square error of $\hat{R}(r)$. Ignoring the estimation of μ, the bias is (from (5.3.14)),

$$E[\hat{R}(r) - R(r)] = \frac{|r|}{N} \cdot R(r).$$

The magnitude of the bias thus depends on two factors, (i) the value of $|r|$ relative to N, and (ii) the value of $R(r)$. When $|r|$ is relatively small compared with N the bias will be small; when $|r|$ approaches $(N-1)$ the bias is effectively $R(r)$. However, if $\{X_t\}$ has a purely continuous spectrum we know that $R(r) \to 0$ as $|r| \to \infty$. Consequently, if N is large (relative to the rate of decay of $R(r)$) then by the time $|r|$ approaches $(N-1)$, $R(r)$ itself will be small. In this case we may expect that the bias of $\hat{R}(r)$ will remain small for all r. Moreover, as will be proved shortly, $\operatorname{var}\{\hat{R}(r)\} = O(1/N)$, all r, whereas $\operatorname{var}\{\hat{R}^*(r)\} = O(1/(N-|r|))$. Consequently, as $|r|$ approaches $(N-1)$, $\operatorname{var}\{\hat{R}^*(r)\}$ increases substantially, and the "tail" of the function $\hat{R}^*(r)$ can produce a wild and erratic behaviour. The variance of $\hat{R}(r)$ remains $O(1/N)$ for all r, and hence the sampling fluctuations of $\hat{R}(r)$ remain small for all values of r.

In comparing the mean square errors of $\hat{R}(r)$ and $\hat{R}^*(r)$ we may say, in general, that,

(a) in the region where $|r|$ is small relative to N there will be little difference between the two estimates, and

(b) when $|r|$ becomes large relative to N the larger bias of $\hat{R}(r)$ is more than compensated for by its smaller variance.

For a given value of r the precise values of the two mean square errors will, of course, depend on the value of N relative to the rate of decay of $R(r)$, but it has been asserted that in most cases $\hat{R}(r)$ will have the smaller mean square error. Although this assertion does not seem to have been proved in any general form, Parzen (1961b) has shown that for a particular AR(1) process, $\hat{R}(r)$ has the smaller mean square error for all $|r| > 0$, the difference increasing substantially as $|r|$ increases.

We conclude, therefore, that $\hat{R}(r)$ is the more satisfactory estimate *and henceforth we shall use* (5.3.13) *as our standard autocovariance function estimator.*

Variances and covariances of the estimates

The second order sampling properties of the estimates $\hat{R}(r)$, $\hat{R}^*(r)$ involve the evaluation of the expectation of, e.g., the product $\{\hat{R}(r) \cdot \hat{R}(r+s)\}$. This, in turn, leads to *fourth order* functions of $\{X_t\}$, and to facilitate these calculations we now add the extra assumption that $\{X_t\}$ *is stationary up to order four.* This means, in particular, that an expression of the form

$$E[X_t X_{t+u} X_{t+v} X_{t+w}]$$

is a function of u, v, w, only, and does not depend on t.

We assume further that the mean μ is known *a priori*, so that without loss of generality we may set,

$$E[X_t] = 0, \qquad \text{all } t. \tag{5.3.16}$$

There is now no need to subtract the sample mean $\bar{X}$ from X_t in the expressions for $\hat{R}(r)$ and $\hat{R}^*(r)$, and the expression for, e.g., $\hat{R}(r)$ reduces to,

$$\hat{R}(r) = \frac{1}{N} \sum_{t=1}^{N-|r|} X_t X_{t+|r|}. \qquad (5.3.17)$$

We now have the exact result,

$$E[\hat{R}(r)] = \left(1 - \frac{|r|}{N}\right) R(r), \qquad \text{all } r,$$

and, for $r \ge 0$, $r + v \ge 0$,

$$\text{cov}\{\hat{R}(r), \hat{R}(r+v)\} = E[\hat{R}(r)\hat{R}(r+v)] - E[\hat{R}(r)]E[\hat{R}(r+v)],$$

(see (2.12.40), Chapter 2)

$$= \frac{1}{N^2} E\left[\sum_{t=1}^{N-r} \sum_{s=1}^{N-r-v} X_t X_{t+r} X_s X_{s+r+v}\right] - \left(1 - \frac{r}{N}\right)\left(1 - \frac{r+v}{N}\right) R(r)R(r+v). \qquad (5.3.18)$$

We now apply a standard result for quadravariate distributions which states that (under the condition (5.3.16)),

$$E[X_t X_{t+r} X_s X_{s+r+v}] = E[X_t X_{t+r}]E[X_s X_{s+r+v}] + E[X_t X_s]E[X_{t+r} X_{s+r+v}]$$
$$+ E[X_t X_{s+r+v}]E[X_{t+r} X_s] + \kappa_4(s-t, r, v), \qquad (5.3.19)$$

where $\kappa_4(s - t, r. v)$ is the fourth joint cumulant of the distribution of $[X_t, X_{t+r}, X_s, X_{s+r+v}]$ (see, e.g., Isserlis (1918)). Substituting (5.3.19) into (5.3.18), using the fact that $E[X_t X_{t+r}] = R(r)$, $E[X_s X_{s+r+v}] = R(r+v)$, etc., and noting that the summation (over s, t) of the first term on the RHS of (5.3.19) cancels with the second term on the RHS of (5.3.18), we obtain

$$\text{cov}\{\hat{R}(r), \hat{R}(r+v)\} =$$

$$\frac{1}{N^2} \sum_{t=1}^{N-r} \sum_{s=1}^{N-r-v} \{R(s-t)R(s+v-t)$$

$$+ R(s+r+v-t)R(s-t-r) + \kappa_4(s-t, r, v)\}. \qquad (5.3.20)$$

This expression may be simplified by noting that in each term on the RHS of (5.3.20) the variables s, t occur only in the form $(s - t)$, and to exploit this feature we rearrange the double summation (over s and t) so as to sum first over the diagonal lattice points—exactly as we did in deriving (5.3.5) from (5.3.4). Formally, we make a change of variables from s and t to $m = (s - t)$ and t. The summand depends only on m, and a careful examination of the limits of t (for fixed $m > 0$ and fixed $m < 0$) gives the result,

$\operatorname{cov}\{\hat{R}(r), \hat{R}(r + v)\} =$

$$\frac{1}{N} \sum_{m=-(N-r)+1}^{(N-r-v-1)} \left\{1 - \frac{\eta(m) + r + v}{N}\right\}\{R(m)R(m + v)$$

$$+ R(m + r + v)R(m - r) + \kappa_4(m, r, v)\} \quad (5.3.21)$$

where the function $\eta(m)$ is defined by,

$$\eta(m) = \begin{cases} m, & m > 0 \\ 0, & -v \leq m \leq 0 \\ -m - v, & -(N-r)+1 \leq m < -v. \end{cases} \quad (5.3.22)$$

Equation (5.3.21) gives an *exact* expression for $\operatorname{cov}\{\hat{R}(r), \hat{R}(r + v)\}$, and this pioneering result is due to Bartlett (1946).

When $\{X_t\}$ is a *Gaussian* process, as we shall henceforth assume, all joint distributions are multivariate normal and hence,

$$\kappa_4(m, r, v) \equiv 0,$$

so that this term no longer appears in (5.3.21). Setting $v = 0$, we then have,

$$\operatorname{var}\{\hat{R}(r)\} = \frac{1}{N} \sum_{m=-(N-r)+r}^{(N-r-1)} \left\{1 - \frac{|m| + r}{N}\right\}\{R^2(m) + R(m + r)R(m - r)\}.$$
$$(5.3.23)$$

Although (5.3.21) and (5.3.23) are exact these expressions are rather complicated but, for large N, Bartlett (1946) has given approximations which have a simpler form. The first step in the approximation is to replace the factor $\{1 - [(\eta(m) + r + v)/N]\}$ in (5.3.21) by 1 (see discussion leading from (5.3.5) to (5.3.6)); and secondly, we may replace the limits in the summation by $-\infty$ and ∞. We then have, for large N, the approximate results,

$$\operatorname{cov}\{\hat{R}(r), \hat{R}(r + v)\} \sim \frac{1}{N} \sum_{m=-\infty}^{\infty} \{R(m)R(m + v) + R(m + r + v)R(m - r)\}$$
$$(5.3.24)$$

and

$$\operatorname{var}\{\hat{R}(r)\} \sim \frac{1}{N} \sum_{m=-\infty}^{\infty} \{R^2(m) + R(m + r)R(m - r)\}. \quad (5.3.25)$$

When $\{X_t\}$ has a purely continuous spectrum we may express both the above results in terms of the spectral density function. For, the first term in (5.3.24) is the value of the convolution of $R(r)$ with itself at the point v, and the second term is the value of the same convolution at the point $(v + 2r)$. But the convolution of $R(r)$ with itself is the Fourier transform of $h^2(\omega)$, where $h(\omega) = \sigma^2 f(\omega)$ is the non-normalized spectral density function of $\{X_t\}$, (see lemma 4.8.1). Hence we may write,

$$\text{cov}\{\hat{R}(r), \hat{R}(r+v)\} \sim \frac{2\pi}{N} \int_{-\pi}^{\pi} \{e^{i\omega v} + e^{i\omega(2r+v)}\} h^2(\omega)\, d\omega \quad (5.3.26)$$

and

$$\text{var}\{\hat{R}(r)\} \sim \frac{2\pi}{N} \int_{-\pi}^{\pi} \{1 + e^{2i\omega r}\} h^2(\omega)\, d\omega. \quad (5.3.27)$$

In particular, when $r = 0$,

$$\hat{R}(0) \equiv s_x^2 = \frac{1}{N} \sum_{t=1}^{N} (X_t - \bar{X})^2$$

is the sample variance, and (5.3.25) now gives,

$$\text{var}\{\hat{R}(0)\} \sim \frac{2}{N} \sum_{m=-\infty}^{\infty} R^2(m) = \frac{2\sigma^4}{N} \sum_{m=-\infty}^{\infty} \rho^2(m). \quad (5.3.28)$$

When X_t is an AR(1) process with parameter a, we have $\rho(m) = a^{|m|}$, and (5.3.28) reduces to

$$\text{var}\{\hat{R}(0)\} \sim \frac{2\sigma^4}{N}\left(\frac{1+a^2}{1-a^2}\right). \quad (5.3.29)$$

This result may be compared with the corresponding result for the variance of the sample variance $\hat{s}^2$ when we have N' independent normal observations; here, we have (from (2.12.28), Chapter 2),

$$\text{var}(\hat{s}^2) \sim 2\sigma^4/N'.$$

Comparing this with (5.3.29) we see that, as far as the estimation of σ^2 is concerned, the "equivalent number of independent observations" is,

$$N' = N\left(\frac{1-a^2}{1+a^2}\right). \quad (5.3.30)$$

All the above results are easily modified for the "unbiased" estimate, $\hat{R}^*(r)$, given by (5.13.10). For, since

$$\hat{R}^*(r) = \left(\frac{N}{N-r}\right)\hat{R}(r), \qquad (r \geq 0),$$

all that is necessary is to replace the divisor outside the summation in

(5.3.20) by $1/\{(N-r)(N-r-v)\}$, and to replace the divisors in (5.3.24) and (5.3.25) by $(1/N-r)$. We now see, in particular, that for $r \geqslant 0$,

$$\text{var}\{\hat{R}^*(r)\} \sim \frac{1}{N-r} \sum_{m=-\infty}^{\infty} \{R^2(m) + R(m+r)R(m-r)\}, \qquad (5.3.25a)$$

which confirms the point previously noted that whereas

$$\text{var}\{\hat{R}(r)\} = O(1/N),$$

we now see

$$\text{var}\{\hat{R}^*(r)\} = O(1/(N-r)).$$

It is now possible to compute the mean square errors of $\hat{R}(r)$ and $\hat{R}^*(r)$ (using the previously derived expressions for the bias and the variance of each estimate).

Graphs 5.1 to 5.5 show the theoretical autocovariance function superimposed on the sample autocovariance function for five models: white noise, AR(1), AR(2), MA(2) and ARMA(2, 2). In each case, the sample autocovariances were computed from (5.3.13), using a stretch of $N = 500$ observations. (Numerical details of the autocovariance calculations are given in the Appendix, and graphs of the original data are given in Chapter 3, Graphs 3.1–3.4.

In each graph, the full curve denotes the theoretical autocovariance function, and the dotted curve the sample autocovariance function.

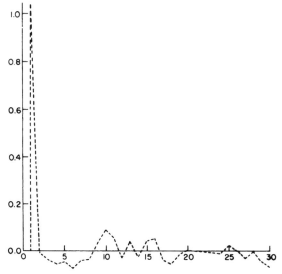

Graph 5.1. Sample and theoretical autocovariance functions for the white noise series.

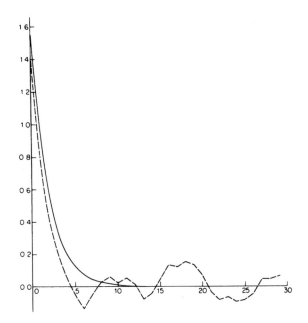

Graph 5.2. Sample and theoretical autocovariance functions for the AR(1) series.

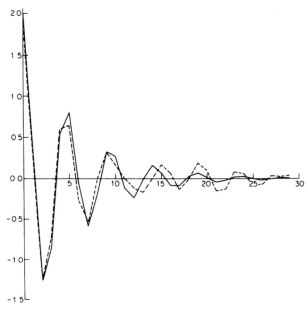

Graph 5.3. Sample and theoretical autocovariance functions for the AR(2) series.

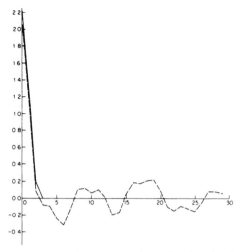

Graph 5.4. Sample and theoretical autocovariance functions for the MA(2) series.

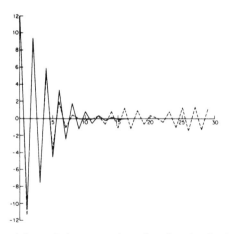

Graph 5.5. Sample and theoretical autocovariance functions for the ARMA(2, 2) series.

5.3.4 Estimation of the Autocorrelation Function

Since the autocorrelation function $\rho(r)$ is related to the autocovariance function $R(r)$ by,

$$\rho(r) = R(r)/R(0),$$

a natural estimate of $\rho(r)$ is

$$\hat{\rho}(r) = \hat{R}(r)/\hat{R}(0), \qquad r = 0, \pm 1, \ldots, \pm(N-1), \qquad (5.3.31)$$

where $\hat{R}(r)$ is given by (5.3.13). Alternatively, if we use the unbiased estimate of $R(r)$ the corresponding estimate of $\rho(r)$ is

$$\hat{\rho}^*(r) = \hat{R}^*(r)/\hat{R}(0), \qquad r = 0, \pm 1, \ldots, \pm(N-1), \qquad (5.3.32)$$

where $\hat{R}^*(r)$ is given by (5.3.10).

A plot of $\hat{\rho}(r)$ (or $\hat{\rho}^*(r)$) against r $(r = 0, \pm 1, \ldots, \pm(N-1))$ is usually called the "*sample autocorrelation function*". The choice between $\hat{\rho}(r)$ and $\hat{\rho}^*(r)$ is based on pretty much the same considerations as that between $\hat{R}(r)$ and $\hat{R}^*(r)$. The sequence $\{\hat{\rho}(r)\}$ is positive semi-definite (this following from the corresponding property of $\hat{R}(r)$), whereas this does not necessarily hold for the sequence $\{\hat{\rho}^*(r)\}$. Moreover, it is easily shown that,

$$|\hat{\rho}(r)| \leq 1, \qquad \text{all } r, \qquad (5.3.33)$$

and, again, this property does not necessarily hold for $\hat{\rho}^*(r)$. To prove (5.3.33) one notes that $\hat{\rho}(r)$ may be written as the ordinary sample correlation coefficient (corresponding to (5.2.3)) for the two sets of $(N+r)$ observations,

$$0, \ldots, 0, \quad X_1, \ldots, X_N$$

and

$$X_1, \ldots, X_N, \quad 0, \ldots, 0.$$

(In the first set there are r zeros preceding the X's and in the second set there are r zeros following the X's.)

(Another estimate which was used in the past but is hardly used at all nowadays may be obtained by applying (5.2.3) to the two sets of $(N-r)$ observations, $\{X_1, \ldots, X_{N-r}\}$ and $\{X_{r+1}, \ldots, X_N\}$. This estimate also satisfies the inequality (5.3.33), but apart from this it does not seem to have any other justification. It's disadvantages are that, for each r, it uses two different estimates of the mean μ and two different estimates of the variance σ^2 (derived separately from the first $(N-r)$ and last $(N-r)$ observations). It does not, therefore, make full use of the stationary character of $\{X_t\}$.)

The sampling properties of $\hat{\rho}(r)$ (and, for that matter, $\hat{\rho}^*(r)$) are extremely complicated, but, for large N, we may obtain approximate expressions for the second order properties by using the approximation technique discussed in Chapter 2, Section 2.13.1, leading to (2.13.20), (2.13.21). (A more rigorous version of this approach has been developed by Lomnicki and Zaremba (1957c)). Let $\delta\{\hat{R}(r)\}$, $\delta\{\hat{R}(r+v)\}$ denote the deviations of $\hat{R}(r)$ and $\hat{R}(r+v)$ from their respective means, i.e.

$$\delta\{\hat{R}(r)\} = \hat{R}(r) - E[\hat{R}(r)], \qquad \delta\{\hat{R}(r+v)\} = \hat{R}(r+v) - E[\hat{R}(r+v)].$$

Assuming that $\delta\{\hat{R}(r)\}$ and $\delta\{\hat{R}(r+v)\}$ are each small compared with

$E[\hat{R}(r)]$, $E[\hat{R}(r+v)]$, respectively, we have approximately,

$$E[\hat{\rho}(r)] \sim \frac{E[\hat{R}(r)]}{E[\hat{R}(0)]} = \left(1 - \frac{|r|}{N}\right)\rho(r), \qquad (5.3.34)$$

and the deviation of $\hat{\rho}(r)$ from $E[\hat{\rho}(r)]$ is, to the first order, given by,

$$\delta\{\hat{\rho}(r)\} = \delta\left\{\frac{\hat{R}(r)}{\hat{R}(0)}\right\} \sim \frac{\delta\{\hat{R}(r)\}}{\hat{R}(0)} - \frac{\hat{R}(r)\delta\{\hat{R}(0)\}}{\hat{R}^2(0)} \sim \frac{\delta\{\hat{R}(r)\}}{R(0)} - \frac{R(r)\delta\{\hat{R}(0)\}}{R^2(0)}.$$

Hence,

$$\delta\{\hat{\rho}(r)\}\delta\{\hat{\rho}(r+v)\} \sim \frac{\delta\{\hat{R}(r)\}\delta\{\hat{R}(r+v)\}}{R^2(0)} - \frac{R(r+v)\delta\{\hat{R}(0)\}\delta\{\hat{R}(r)\}}{R^3(0)}$$

$$- \frac{R(r)\delta\{\hat{R}(0)\}\delta\{\hat{R}(r+v)\}}{R^3(0)} + \frac{R(r)R(r+v)[\delta\{\hat{R}(0)\}]^2}{R^4(0)}.$$

Taking expectations, we have,

$$\text{cov}\{\hat{\rho}(r), \hat{\rho}(r+v)\} \sim \frac{1}{\sigma^4}[\text{cov}\{\hat{R}(r), \hat{R}(r+v)\} - \rho(r+v)\,\text{cov}\{\hat{R}(0), \hat{R}(r)\}$$

$$- \rho(r)\,\text{cov}\{\hat{R}(0), \hat{R}(r+v)\} + \rho(r)\rho(r+v)\,\text{var}\{\hat{R}(0)\}]$$

$$(5.3.35)$$

Using (5.3.24) (and assuming that $\{X_t\}$ is a Gaussian process so that the term $\kappa_4(m, r, v)$ is absent), we now obtain for $r > 0$, $r+v > 0$,

$$\text{cov}\{\hat{\rho}(r), \hat{\rho}(r+v)\}$$

$$\sim \frac{1}{N}\sum_{m=-\infty}^{\infty} [\rho(m)\rho(m+v) + \rho(m+r+v)\rho(m-r) + 2\rho(r)\rho(r+v)\rho^2(m)$$

$$- 2\rho(r)\rho(m)\rho(m-r-v) - 2\rho(r+v)\rho(m)\rho(m-r)]$$

$$(5.3.36)$$

and setting $v = 0$ gives for $r > 0$,

$$\text{var}\{\hat{\rho}(r)\} \sim \frac{1}{N}\sum_{m=-\infty}^{\infty} [\rho^2(m) + \rho(m+r)\rho(m-r)$$

$$+ 2\rho^2(r)\rho^2(m) - 4\rho(r)\rho(m)\rho(m-r)]. \quad (5.3.37)$$

These approximate expressions for the covariance and variances of the estimated autocorrelation function are basic results which are again due to Bartlett (1946).

Bartlett also proves the interesting result that the term involving the fourth cumulant $\kappa_4(m, r, v)$ is absent from (5.3.36) under more general conditions than the assumption that the process of Gaussian. He shows that if $\{X_t\}$ is a *general linear process* of the form (3.5.56) (see Section 3.5.7) but with the $\{\varepsilon_t\}$ *independent* (rather than simply uncorrelated) then the term $\kappa_4(m, r, v)$ automatically vanishes in (5.3.36).

The above results demonstrate an important feature of the estimated autocorrelation function, namely that, *in general, there will be fairly high correlation between neighbouring points of this function.* This means that when the autocorrelation function $\rho(r)$ decays to zero as $|r| \to \infty$ (as in the case of autoregressive processes) the *estimated autocorrelation function $\hat{\rho}(r)$ will be less damped and will not, therefore, decay as quickly as $\rho(r)$.* This feature applies also to the estimated autocovariance function $\hat{R}(r)$.

We now apply (5.3.36), (5.3.37) to the following special cases;

(1) *Purely random processes*

When $\{X_t\}$ is a purely random process, $\rho(0) = 1$, $\rho(r) = 0$, $r \neq 0$, and hence,

$$\text{cov}\{\hat{\rho}(r), \hat{\rho}(r+v)\} \sim 0 \qquad (v > 0), \qquad (5.3.38)$$

$$\text{var}\{\hat{\rho}(r)\} \sim \frac{1}{N} \qquad (r > 0). \qquad (5.3.39)$$

In this case, the standard deviation of $\hat{\rho}(r)$ is, to the above order of approximation, $1/\sqrt{N}$.

(2) *Processes with decaying autocorrelation functions*

If $\rho(s) \to 0$ as $|s| \to \infty$ (as when $\{X_t\}$ has a purely continuous spectrum), and if r is sufficiently large so that we may write,

$$\rho(s) \sim 0, \qquad |s| \geq r,$$

then (5.3.36), (5.3.37) reduce to,

$$\text{cov}\{\hat{\rho}(r), \hat{\rho}(r+v)\} \sim \frac{1}{N} \sum_{m=-\infty}^{\infty} \{\rho(m)\rho(m+v)\}, \qquad (5.3.40)$$

and

$$\text{var}\{\hat{\rho}(r)\} \sim \frac{1}{N} \sum_{m=-\infty}^{\infty} \rho^2(m). \qquad (5.3.41)$$

Note that the RHS of (5.3.40) is a function of v only. Hence, the *"tail" of the estimated autocorrelation function* (i.e. the region where r is "large"), *behaves approximately as a stationary process.* In this region, $\hat{\rho}(r)$ is approximately a stationary process, *but more highly autocorrelated than the original process $\{X_t\}$.*

(3) AR(1) processes

In the case when $\{X_t\}$ is an AR(1) process with $\rho(r) = a^{|r|}$, (3.5.40), (3.5.41), give

$$\text{cov}\{\hat{\rho}(r), \hat{\rho}(r+v)\} \sim \frac{a^{|v|}}{N}\left[\frac{1+a^2}{1-a^2} + |v|\right] \qquad (5.3.42)$$

and

$$\text{var}\{\hat{\rho}(r)\} \sim \frac{1}{N}\left[\frac{1+a^2}{1-a^2}\right]. \qquad (5.3.43)$$

The estimate $\hat{\rho}^*(r)$

The properties of $\hat{\rho}^*(r)$ may be obtained immediately from those of $\hat{\rho}(r)$ by using the relation,

$$\hat{\rho}^*(r) = \left(\frac{N}{N-|r|}\right)\rho(r), \qquad (5.3.44)$$

which follows from the definitions (5.3.31), (5.3.32). Approximate expressions for the variances and covariances of $\hat{\rho}^*(r)$, $(r > 0)$, are thus obtained from (5.3.36), (5.3.37) by replacing the divisor N in each case by $(N - r)$. We then find that whereas

$$\text{var}\{\hat{\rho}(r)\} = O(1/N), \qquad \text{all } r, \qquad (5.3.45)$$

$$\text{var}\{\hat{\rho}^*(r)\} = O(1/(N-r)), \qquad (5.3.46)$$

and thus increases as r approaches $(N - 1)$.

Also, from (5.3.34) we have the approximate result that for large N,

$$E[\hat{\rho}^*(r)] \sim \rho(r). \qquad (5.3.47)$$

However, Kendall and Stuart (1966) have investigated the bias in $\hat{\rho}^*(r)$ in more detail, and have shown that when $\{X_t\}$ is a purely random process,

$$E[\hat{\rho}^*(r)] = -(1/(N-r)), \qquad (5.3.48)$$

so that here there is a small but negative bias. They show further that for an AR(1) process the bias in $\hat{\rho}^*(r)$ is also negative for all r, and is $O(1/(N-r))$.

The following graphs show the theoretical versus sample autocorrelation functions for the same five models whose autocovariance functions are given in Graphs 5.1–5.5. The sample autocorrelations were each computed from $N = 500$ observations, using the estimate (5.3.31), and, as previously, the full curve denotes the theoretical function and the dotted curve the sample function.

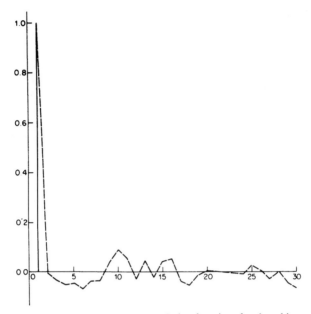

Graph 5.6. Sample and theoretical autocorrelation functions for the white noise series.

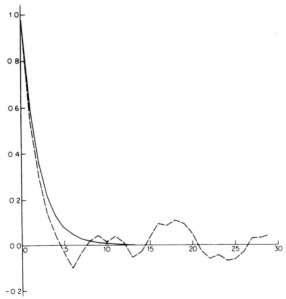

Graph 5.7. Sample and theoretical autocorrelation functions for the AR(1) series.

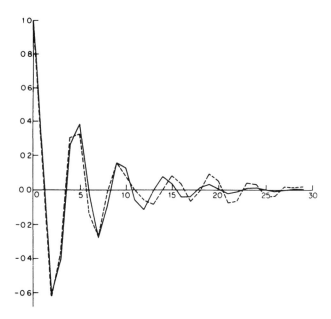

Graph 5.8. Sample and theoretical autocorrelation functions for the AR(2) series.

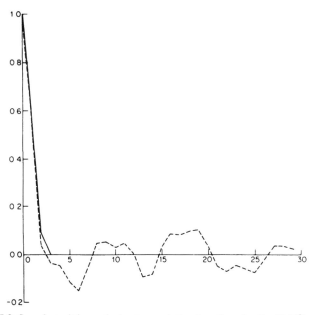

Graph 5.9. Sample and theoretical autocorrelation functions for the MA(2) series.

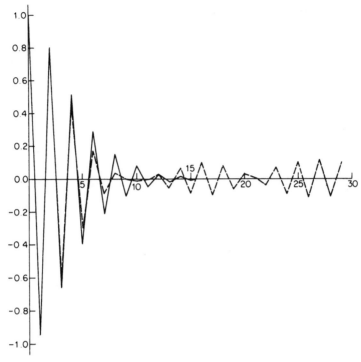

Graph 5.10. Sample and theoretical autocorrelation functions for the ARMA(2, 2) series.

5.3.5 Asymptotic Distribution of the Sample Mean, Autocovariances and Autocorrelations

We know that, under general conditions, the distribution of the mean of a sample of N independent observations, suitably scaled, tends to normality as $N \to \infty$, this result following as a consequence of the central limit theorem. In the present context the observations are, of course, correlated, but there are various generalized forms of the central limit theorem which hold for stationary processes. We may conclude, therefore, that provided $\{X_t\}$ is sufficiently "well behaved", the sample mean $\bar{X}$ defined as (5.3.2) will, when suitably scaled, also have an asymptotic normal distribution. More specifically, the following result (Anderson (1971), p. 478) may be established.

Let X_t be a general linear process of the form

$$X_t = \mu + \sum_{u=-\infty}^{\infty} g_u \varepsilon_{t-u}, \qquad (5.3.49)$$

where the ε_t are independent and identically distributed with $E(\varepsilon_t) = 0$, $E(\varepsilon_t^2) < \infty$, and $\sum_{u=-\infty}^{\infty} |g_u| < \infty$. Then $\sqrt{N}(\bar{X} - \mu)$ has (as $N \to \infty$) a limiting normal distribution with zero mean and variance $\sigma^2\{\sum_{r=-\infty}^{\infty} \rho(r)\}$. (The asymptotic mean and variance of $\sqrt{N}(\bar{X} - \mu)$ follow immediately from (5.3.3), (5.3.5).)

We remarked previously that, for finite N, the distribution theory of the sample autocovariances and autocorrelations is extremely complicated. This is particularly so for the sample autocorrelations; for each r, $\hat{\rho}(r)$ (or $\hat{\rho}^*(r)$) is a ratio of two quadratic forms in the $\{X_t\}$ and it is not difficult to understand why there are virtually no exact results on their sampling distributions. In fact, most of the work on finite sample distributions has not been concerned directly with the $\{\hat{\rho}(r)\}$ but has concentrated instead on closely related quantities called the *"circular autocorrelations"*. These are obtained by artificially extending the original set of observations, $\{X_1, \ldots, X_N\}$, to $\{X_1, \ldots, X_N, X_{N+1}, \ldots, X_{N+r}\}$ where $X_{N+1} = X_1$, $X_{N+2} = X_2, \ldots$, $X_{N+r} = X_r$, hence the use of the term "circular". With this convention we may now include the cross products, $(X_t X_{t+r})$, for *all* values of t from 1 to N, so that, e.g., in (5.3.10) the summation does not terminate at $(N - r)$ but runs from $t = 1$ to $t = N$. This device leads to "circular autocorrelations" given (assuming $E(X_t) = 0$) by

$$\hat{\rho}_c(r) = \sum_{t=1}^{N} X_t X_{t+|r|} \Big/ \sum_{t=1}^{N} X_t^2. \qquad (5.3.50)$$

This modification simplifies, to some extent, the distribution theory and provided N is large it can be argued that the distributions of the circular autocorrelations can be used as approximations to the distributions of the conventional (i.e. non-circular) autocorrelations. Most of the exact results which have been obtained refer only to the distribution of the *first* circular autocorrelation $\hat{\rho}_c(1)$ under fairly simple models for $\{X_t\}$. The distribution $\hat{\rho}_c(1)$ has been studied extensively (R. L. Anderson (1942), R. L. Anderson and T. N. Anderson (1950), Dixon (1944), Koopmans (1942), Madow (1945), and Von Neumann (1941)), and various approximations have been derived for the distributions of the higher order autocorrelations (see, e.g., T. W. Anderson (1948), Daniels (1956), Dixon (1944), Hannan (1955), Jenkins (1956), Koopmans (1942), Leipnik (1947), Rubin (1945) and Whittle (1951)).

However, as far as the asymptotic theory is concerned it would seem fairly plausible that under sufficiently strong conditions on $\{X_t\}$, both the sample autocovariances and sample autocorrelations will have a limiting normal distributions (as $N \to \infty$). This may be seen in an intuitive way by first regarding $U_t = (X_t X_{t+r})$ as a new stochastic process, and then noting, e.g., that $\hat{R}(r)$ is simply the sample mean of $U_1, U_2, \ldots, U_{N-r}$. Rigorous derivations of the asymptotic distributions of the sample autocovariances and autocorrelations have been given by Lomnicki and Zaremba (1959a) and Anderson and Walker (1964). The following theorem (Anderson (1971, p. 478)) illustrates the general nature of these results.

Let $\{X_t\}$ *be a process of the form* (5.3.49) *with* $E(\varepsilon_t) = 0$, $E(\varepsilon_t^2) < \infty$, $E(\varepsilon_t^4) < \infty$, *and with* $\sum_{u=-\infty}^{\infty} |g_u| < \infty$. *Then for any given n the limiting joint distribution of* $\sqrt{N}\{\hat{R}(0) - R(0)\}, \ldots, \sqrt{N}\{\hat{R}(n) - R(n)\}$, *(as $N \to \infty$) is multivariate normal with zero means and, for Gaussian* $\{X_t\}$, *covariances given by* (5.3.24) *with the factor* $(1/N)$ *omitted. (For non-Gaussian processes the covariances include the additional term involving* $\kappa_4(m, r, v)$.*)*

Moreover, under the additional assumption that $\sum_{u=-\infty}^{\infty} |u|g_u^2 < \infty$, *the limiting joint distribution of* $\sqrt{N}\{\hat{\rho}(1) - \rho(1)\}, \ldots, \sqrt{N}\{\hat{\rho}(n) - \rho(n)\}$ *(as $N \to \infty$) is multivariate normal with zero means and covariances given by* (5.3.36) *with the factor* $(1/N)$ *omitted. (Since the asymptotic covariance of* $\hat{\rho}(r)$, $\hat{\rho}(r+v)$, *does not involve* $\kappa_4(m, r, v)$ *when* X_t *is a linear process of the above form the condition* $E(\varepsilon_t^4) < \infty$ *may be dropped here.)*

Approximate tests for autocorrelations

We may use the asymptotic normality of the $\{\hat{\rho}(r)\}$ as the basis for approximate large sample tests and confidence intervals. For a particular value of r we have,

$$\hat{\rho}(r) \sim N(\rho(r), \sigma^2(\hat{\rho})), \tag{5.3.51}$$

where $\sigma^2(\hat{\rho})$, the asymptotic variance of $\hat{\rho}(r)$, is given by (5.3.37), or, when r is large, by (5.3.41). Unfortunately, the expression for $\sigma^2(\hat{\rho})$ involves the values of the theoretical autocorrelation function $\rho(m)$ for all m, and in general these are unknown. A crude estimate of $\sigma^2(\hat{\rho})$ can be obtained by substituting $\hat{\rho}(m)$ for $\rho(m)$ in (5.3.37) or (5.3.41). However, if we can specify a model for $\{X_t\}$ (with known parameters) then the corresponding theoretical form of $\rho(r)$ can be evaluated, and we can then test whether a given estimate $\hat{\rho}(r)$ is consistent with its theoretical value.

It is important to note, however, that this type of test is valid only as a test of $\hat{\rho}(r)$ for a *particular value of r*. In general we *cannot construct uniform critical values* or simultaneous confidence intervals for *all* the $\hat{\rho}(r)$ (extending from $r = 0$ to $r = \pm(N-1)$) since, as we have already observed, neighbouring values of $\hat{\rho}(r)$ are highly correlated. There is only one special case when neighbouring values are effectively uncorrelated, namely when $\{X_t\}$ is a purely random process—see (5.3.38). In this case we have

$$E\{\hat{\rho}(r)\} \sim 0, \qquad \text{all } r \neq 0, \tag{5.3.52}$$

and from (5.3.39),

$$\text{var}\{\hat{\rho}(r)\} \sim \frac{1}{N}, \qquad \text{all } r \neq 0. \tag{5.3.53}$$

Hence, on the basis of the asymptotic normal distribution, the approximate two-sided 5% critical values for $\hat{\rho}(r)$ are $\pm 2\sqrt{1/N}$. Similarly, an approximate 95% of confidence interval for $\rho(r)$ is $\{\hat{\rho}(r) \pm 2\sqrt{1/N}\}$. (Alternatively, if we use Kendall and Stuart's expression for the bias in $\hat{\rho}(r)$ (see (5.3.48)), the 5% critical values become $\{-(1/N) \pm 2\sqrt{1/N}\}$.) If the calculated value of $\hat{\rho}(r)$ falls outside the interval $\pm 2\sqrt{1/N}$ we may say that $\hat{\rho}(r)$ differs significantly from its theoretical value (namely zero) at roughly the 5% level, and this might lead us to suspect the hypothesis that $\{X_t\}$ is a purely random process. Although in this case the critical values are the same for all r we must nevertheless be extremely careful if we try to apply this test to the sample autocorrelation function *over a range of values of r*. We know that even if the hypothesis that $\{X_t\}$ is a purely random process is true we would still expect to find roughly 5% of the $\hat{\rho}(r)$ lying outside the critical values, and we must allow for this feature in assessing the "significance" of departures of $\hat{\rho}(r)$ from its predicted form.

5.3.6 The Ergodic Property

The methods we have used to estimate the mean and autocovariance function using observations from a single realization are all based on the basic strategy of replacing *ensemble averages* by their corresponding *time averages*. If the resulting estimates are mean square consistent then it follows that the time average over a period of N points converges in mean square, as $N \to \infty$, to the corresponding ensemble average. Processes which possess this type of property are called *"ergodic"* (a term borrowed from statistical mechanics), and the problem of determining conditions under which this property holds in a general sense has been the subject of deep mathematical study; see, e.g. Doob (1953). In the present discussion we consider only the convergence of the sample mean and sample auto-covariances.

The sample mean

Since $\bar{X}$ is an unbiased estimate of μ (by (5.3.2)) it converges to μ in mean square if

$$\lim_{N \to \infty} \text{var}(\bar{X}) = 0. \qquad (5.3.54)$$

We have already seen that a sufficient condition for (5.3.54) to hold is that

$$\lim_{N \to \infty} \left\{ \sum_{r=-(N-1)}^{(N-1)} \left(1 - \frac{|r|}{N}\right) \rho(r) \right\}$$

is finite, but we may obtain a condition which is both necessary and sufficient by examining the limiting behaviour of the RHS of (5.3.5) in more detail. For the moment we drop the assumption that the spectral density function $f(\omega)$ necessarily exists for all ω, and instead use the more general representation of $\rho(r)$ in terms of the integrated spectrum $F(\omega)$. We have from (4.8.28) (Section 4.8.3),

$$\rho(r) = \int_{-\pi}^{\pi} e^{i\omega r} \, dF(\omega),$$

and substituting this expression for $\rho(r)$ in (5.3.5) gives,

$$\text{var}(\bar{X}) = \sigma^2 \int_{-\pi}^{\pi} \left\{ \frac{1}{N} \sum_{r=-(N-1)}^{(N-1)} \left(1 - \frac{|r|}{N}\right) e^{i\omega r} \right\} dF(\omega) \qquad (5.3.55)$$

$$= \sigma^2 \int_{-\pi}^{\pi} G_N(\omega) \, dF(\omega), \qquad (5.3.56)$$

say, where

$$G_N(\omega) = \frac{1}{N^2} \cdot \frac{\sin^2(N\omega/2)}{\sin^2(\omega/2)}. \qquad (5.3.57)$$

Since $|\sin^2(N\omega/2)| \leq 1$, all N, ω, it follows that for all $\omega \neq 0$, $G_N(\omega) \to 0$ as $N \to \infty$. Also, writing $\theta = (N\omega/2)$, we have for small ω,

$$G_N(\omega) = \frac{1}{N^2} \frac{\sin^2 \theta}{\sin^2(\theta/N)} \sim \frac{\sin^2 \theta}{\theta^2} \to 1 \qquad \text{as } \theta \to 0.$$

Hence, $G_N(\omega) \to 1$ as $\omega \to 0$, all N. Thus, as $N \to \infty$, $G_N(\omega)$ vanishes everywhere except at $\omega = 0$ where it is unity. We now obtain from (5.3.36),

$$\lim_{N \to \infty} \text{var}(\bar{X}) = \sigma^2 \{F(0+) - F(0-)\}, \qquad (5.3.58)$$

where $F(0+)$, $F(0-)$ denote the limits of $F(\omega)$ as $\omega \to 0$ from the right and from the left, respectively. It now follows immediately that a necessary and sufficient condition for $\lim_{N \to \infty} \text{var}(\bar{X}) = 0$ is that $\{F(0+) - F(0-)\} = 0$, i.e. that $F(\omega)$ be continuous at $\omega = 0$. Hence, we have the result that $\bar{X}$ *converges to μ in mean square as $N \to \infty$ if and only if $F(\omega)$ is continuous at $\omega = 0$, i.e. if and only if $F(\omega)$ does not have a "jump" at $\omega = 0$.* (This result is confirmed by the approximation (5.3.7) which indicates that if $f(0)$ is finite, $\text{var}(\bar{X}) = O(1/N) \to 0$ as $N \to \infty$.)

A more elegant, but less direct, proof of the above result may be constructed from the discrete parameter version of equation (4.11.26). In the discrete parameter case, with $E(X_t) = 0$, the expression for the jump at $\omega = 0$ in the spectral function $Z(\omega)$ is given by

$$A_0 = \underset{N \to \infty}{\text{l.i.m.}} \left\{ \frac{1}{N} \sum_{t=1}^{N} X_t \right\}. \qquad (5.3.59)$$

(The limits in (4.11.6) were chosen to be symmetrical about zero, but since the process is stationary the limits in that equation could well be changed to 0 and $2T$.) It now follows that $\bar{X}$ converges in mean square to $E(X_t) = 0$ if and only if $A_0 = 0$, i.e. if and only if the integrated spectrum is continuous at $\omega = 0$.

It may, at first, seem surprising that the convergence of $\bar{X}$ to μ depends only on the behaviour of the spectrum at the origin. However, an intuitive explanation of this result may be given as follows. Suppose $F(\omega)$ has a jump at $\omega = 0$. Then $\{X_t\}$ has a "mixed" spectrum with a harmonic component at zero frequency, e.g. X_t contains a component of the form $(A \cos \phi)$, where ϕ has a rectangular distribution on $(-\pi, \pi)$ (cf. the general harmonic process model (4.9.10)). We may therefore write,

$$X_t = A \cos \phi + Y_t, \qquad (5.3.60)$$

say, where, w.l.o.g, we may take $E(Y_t) = 0$. Since ϕ is rectangularly distributed over $(-\pi, \pi)$, we have $E(\cos \phi) = 0$ and hence $E(X_t) = 0$. But, for each realization ϕ takes a particular value which it retains throughout that realization. Hence, for a given realization the term $(A \cos \phi)$ acts as a non-zero additive constant and now

$$\bar{X} = \frac{1}{N} \sum_{t=1}^{N} X_t = \frac{1}{N} \sum_{t=1}^{N} (A \cos \phi + Y_t) = A \cos \phi + \frac{1}{N} \sum_{t=1}^{N} Y_t. \qquad (5.3.61)$$

Thus, even though $(1/N) \sum_{t=1}^{N} Y_t$ may converge to zero, $\bar{X}$ always contains the constant term, $A \cos \phi$, (however large N), and cannot therefore, converge to zero.

Jumps in $F(\omega)$ at non-zero frequencies do not cause the same difficulty. For, if $F(\omega)$ has jumps at $\omega = \pm \omega_0$, say, this corresponds to a component in

X_t of the form $A \cos(\omega_0 t + \phi)$, and now,

$$\frac{1}{N} \sum_{t=1}^{N} A \cos(\omega_0 t + \phi) = \frac{1}{N} \left[A \cos\phi \sum_{t=1}^{N} \cos \omega_0 t - A \sin\phi \sum_{t=1}^{N} \sin \omega_0 t \right].$$

(5.3.62)

But,

$$\sum_{t=1}^{N} \cos \omega_0 t = \frac{1}{2} \left\{ \frac{\sin(N+\frac{1}{2})\omega_0}{\sin \frac{1}{2}\omega_0} - 1 \right\} = \frac{\sin\{\frac{1}{2}N\omega_0\} \cos\{\frac{1}{2}(N+1)\omega_0\}}{\sin \frac{1}{2}\omega_0}$$

(5.3.63)

and

$$\sum_{t=1}^{N} \sin \omega_0 t = \frac{1}{2} \left\{ \frac{\cos \frac{1}{2}\omega_0 - \cos(N+\frac{1}{2})\omega_0}{\sin \frac{1}{2}\omega_0} \right\} = \frac{\sin\{\frac{1}{2}(N+1)\omega_0\} \sin\{\frac{1}{2}N\omega_0\}}{\sin \frac{1}{2}\omega_0}.$$

(5.3.64)

For $\omega_0 \neq 0$ in the range $(-\pi, \pi)$, both these functions remain finite as $N \to \infty$, and the contribution of the term $A \cos(\omega_0 t + \phi)$ to $\bar{X}$ vanishes in the limit.

The sample covariances

For each fixed r, the sample covariance $\hat{R}(r)$ is an asymptotically un-biased estimate of $R(r)$, and it is therefore mean square consistent if $\lim_{N \to \infty} \text{var}\{\hat{R}(r)\} = 0$. We know from (5.3.27) that if $\{X_t\}$ is Gaussian and if the (non-normalized) spectral density $h(\omega)$ exists for all ω and is square integrable, then $\text{var}\{\hat{R}(r)\} = O(1/N)$, and it is then certainly true that $\lim_{N \to \infty} \text{var}\{\hat{R}(r)\} = 0$. Hence, *for Gaussian processes a sufficient condition for $\hat{R}(r)$ to converge in mean square to $R(r)$ for each fixed r is that the process $\{X_t\}$ has a purely continuous spectrum whose spectral density function is square integrable.* The same conclusions clearly hold also for the alternative estimate $\hat{R}^*(r)$.

Note that whereas the convergence of $\bar{X}$ depends only on the continuity of $F(\omega)$ at the single point, $\omega = 0$, the convergence of $\hat{R}(r)$ depends on the absolute continuity of $F(\omega)$ for all ω.

5.3.7 Continuous Parameter Processes

We remarked previously that the estimation procedures which we have discussed for discrete parameter processes could be applied also to samples from continuous parameter processes if the continuous record is first converted into a discrete set of observations by reading off its values at successive small time intervals. For most purposes this approach would be adequate, but if we wish to analyse a continuous record in its original form then, in principle, there is no difficulty in adapting the estimates we have already obtained to the continuous parameter case. The modifications required are usually fairly obvious, and consist mainly in changing discrete sums (over t) into integrals.

Suppose then that $\{X(t)\}$ is a continuous parameter stationary process, with mean μ, variance σ^2, autocovariance function $R(\tau)$, and autocorrelation function $\rho(\tau)$. Given a continuous sample recorded over the time interval $(0, T)$, i.e. given observations on $X(t)$ for $0 \le t \le T$, we wish to estimate μ, $R(\tau)$, and $\rho(\tau)$. We again estimate μ by the time-average of the observations which now takes the form,

$$\bar{X} = \frac{1}{T} \int_0^T X(t) \, dt. \tag{5.3.65}$$

Arguing exactly as in the discrete we find $E(\bar{X}) = \mu$ and

$$\text{var}(\bar{X}) = \frac{\sigma^2}{T^2} \int_0^T \int_0^T \rho(t-s) \, ds \, dt = \frac{\sigma^2}{T} \int_{-T}^T \left(1 - \frac{|\tau|}{T}\right) \rho(\tau) \, d\tau, \tag{5.3.66}$$

so that for large T,

$$\text{var}(\bar{X}) \sim \frac{\sigma^2}{T} \int_{-\infty}^{\infty} \rho(\tau) \, d\tau \tag{5.3.67}$$

$$= \frac{2\pi\sigma^2}{T} \cdot f(0), \tag{5.3.68}$$

when $f(\omega)$, the (normalized) spectral density function of $X(t)$, exists.

The estimate of $R(\tau)$ corresponding to (5.3.13) is,

$$\hat{R}(\tau) = \frac{1}{T} \int_0^{T-|\tau|} \{X(t) - \bar{X}\}\{X(t+|\tau|) - \bar{X}\} \, dt. \tag{5.3.69}$$

The "unbiased" estimate $\hat{R}^*(\tau)$ (corresponding to (5.3.10)) is obtained by changing the divisor in (5.3.69) from T to $(T-|\tau|)$. We have (assuming μ is known)

$$E[\hat{R}(\tau)] = \left(1 - \frac{|\tau|}{T}\right) R(\tau), \tag{5.3.70}$$

and when $\{X(t)\}$ is Gaussian,

$$\text{cov}\{\hat{R}(\tau), \hat{R}(\tau+v)\} \sim \frac{1}{T} \int_{-\infty}^{\infty} \{R(u)R(u+v) + R(u+\tau+v)R(u-\tau)\} \, du \tag{5.3.71}$$

$$= \frac{2\pi}{T} \int_{-\infty}^{\infty} \{e^{i\omega v} + e^{i\omega(2\tau+v)}\} f(\omega) \, d\omega, \tag{5.3.72}$$

and

$$\text{var}\{\hat{R}(\tau)\} \sim \frac{1}{T} \int_{-\infty}^{\infty} \{R^2(u) + R(u+\tau)R(u-\tau)\} \, du \qquad (5.3.73)$$

$$= \frac{2\pi}{T} \int_{-\infty}^{\infty} \{1 + e^{2i\omega\tau}\} f(\omega) \, d\omega. \qquad (5.3.74)$$

These results follow in exactly the same way as in the discrete case—cf. the derivation of (5.3.24), (5.3.25), (5.3.26), (5.3.27).

5.4 ESTIMATION OF PARAMETERS IN STANDARD MODELS

The standard finite parameter models for stationary processes discussed in Chapter 3 include three main types:
1. Autoregressive (AR) models
2. Moving average (MA) models
3. Mixed autoregressive/moving average (ARMA) models.

We now consider the problem of fitting these models to observational data. (We do not consider here the general linear model (3.5.56) since this involves an infinite number of parameters.) The harmonic model (3.5.71), although involving only a finite number of parameters, belongs to a different category since it corresponds to a process with a purely discrete spectrum. This model is treated in Chapter 6 when we study frequency domain methods. For reasons previously discussed we concentrate mainly on discrete parameter models, but some remarks on continuous parameter models are included at the end of this section.

The process of fitting any of the above models to data involves two separate stages, namely:
(a) the estimation of the *parameters* of the model;
(b) the determination of the *order* of the model.

For example, in the AR(k) model (3.5.39), viz.

$$X_t + a_1 X_{t-1} + \ldots + a_k X_{t-k} = \varepsilon_t, \qquad (5.4.1)$$

we must estimate $a_1, a_2, \ldots, a_k$ and $\sigma_\varepsilon^2 = \text{var}(\varepsilon_t)$, and also determine a suitable value for k, the order of the AR model. For the ARMA model (3.5.52) we have to estimate the parameters of both the AR and MA operators, and determine the values of both k and l, the respective orders of these operators. The logical sequence of these operators would be to determine the order (or orders) first, and then estimate the parameters. However, the methods which have been developed for order determination all involve parameter estimation as a preliminary stage. We therefore consider stage (a) first (on the assumption that the order of the model is given) and discuss stage (b) later.

5.4.1 Estimation of Parameters in Autoregressive Models

The AR(k) model (5.4.1) was first introduced in Section 3.5.4, and we saw then that, with $E(\varepsilon_t) = 0$ (our standard assumption), $E(X_t) = 0$, all t. Thus, (5.4.1) is appropriate only for an AR process with zero mean. If we wish to allow for X_t having a non-zero mean μ, then a suitable model is easily constructed by using (5.4.1) as a model for the deviations $\{X_t - \mu\}$. This leads to the slightly more general AR model,

$$X_t - \mu + a_1(X_{t-1} - \mu) + \ldots + a_k(X_{t-k} - \mu) = \varepsilon_t. \qquad (5.4.2)$$

We assume that the coefficients $a_1, \ldots, a_k$ satisfy the condition given in Section 3.5.4 which ensures that $\{X_t\}$ is (asymptotically) stationary, namely, that all the roots of $\alpha(z) = 1 + a_1 z + \ldots + a_k z^k$ lie outside the unit circle.

Given N observations $X_1, X_2, \ldots, X_N$ the problem is to estimate the unknown parameters $\mu, a_1, a_2, \ldots, a_k$ and σ_ε^2. (We assume for the present that k is known.) If we re-write (5.4.2) as

$$X_t = \mu - a_1(X_{t-1} - \mu) \ldots - a_k(X_{t-k} - \mu) + \varepsilon_t, \qquad (5.4.3)$$

then we may observe a superficial similarity between (5.4.3) and the classical multiple linear regression model, (5.2.41), in the sense that (5.4.3) expresses X_t as a linear function of $(X_{t-1} - \mu), \ldots, (X_{t-k} - \mu)$ with $\mu, (-a_1), \ldots, (-a_k)$ acting as the regression coefficients and ε_t as the "residual". Of course, (5.4.3) is not identical with the classical regression model since, as t varies from 1 to N, X_t simultaneously plays the role of both the "dependent" and "independent" variables. Also, the "dependent" variables $(X_{t-1} - \mu)$, $\ldots, (X_{t-k} - \mu)$ are themselves correlated. However, since the values of X_t, $X_{t-1}, \ldots, X_{t-k}$ are all observed there is nothing to prevent us from applying the same least squares procedure used in regression analysis (see Section 5.2.2, equation (5.2.42)). We cannot be certain that here the least squares estimates will have the same optimal properties as in the classical regression case, but in a classic paper Mann and Wald (1943) have in fact shown that the standard results for the variances and covariances of the estimates are still valid asymptotically (as $N \to \infty$). Thus, if we adopt a least squares approach then we estimate $\mu, a_1, \ldots, a_k$ by those values which minimize

$$Q(\mu, a_1, \ldots, a_k) = \sum_{t=k+1}^{N} \varepsilon_t^2 \qquad (5.4.4)$$

$$= \sum_{t=k+1}^{N} \{X_t - \mu + a_1(X_{t-1} - \mu) + \ldots + a_k(X_{t-k} - \mu)\}^2. \qquad (5.4.5)$$

Note that we cannot include the terms $\varepsilon_1^2, \varepsilon_2^2, \ldots, \varepsilon_k^2$ in (5.4.4) since these cannot be computed in terms of the observed X_t. For example, we have

$$\varepsilon_1 = X_1 - \mu + a_1(X_0 - \mu) + \ldots + a_k(X_{-(k-1)} - \mu)$$

and we cannot compute ε_1 since $X_0, X_{-1}, \ldots, X_{-(k-1)}$ are not observed. This is an unavoidable feature of the AR model, but if N is large compared with k (as is almost invariably the case in practice), the effect of ignoring the first k ε_t will be only small, and we may then regard (5.4.4) as a good approximation to the "full" sum of squares, $\{\sum_{t=1}^{N} \varepsilon_t^2\}$. This approximation is, in effect, equivalent to setting,

$$\varepsilon_1 = \varepsilon_2 = \ldots = \varepsilon_k = 0. \tag{5.4.6}$$

The least squares criterion (5.4.4) can (as usual) be justified from a maximum likelihood approach under the assumption that the residuals ε_t are Gaussian. Thus, if the ε_t have a multivariate normal distribution then, being uncorrelated, they are now independent, and the joint probability density function of $\varepsilon_{k+1}, \varepsilon_{k+2}, \ldots, \varepsilon_N$ is

$$f(\varepsilon_{k+1}, \ldots, \varepsilon_N) = \left(\frac{1}{2\pi\sigma_\varepsilon^2}\right)^{(N-k)/2} \exp\left[-\frac{1}{2\sigma_\varepsilon^2}\left(\sum_{t=k+1}^{N} \varepsilon_t^2\right)\right]. \tag{5.4.7}$$

If we make a transformation from $(\varepsilon_{k+1}, \ldots, \varepsilon_N)$ to $(X_{k+1}, \ldots, X_N)$ it is easily seen that the Jacobean is unity (see Section 2.12.10), and hence the joint probability density function of $(X_{k+1}, \ldots, X_N)$ is,

$$f(x_{k+1}, \ldots, x_n) = \left(\frac{1}{2\pi\sigma_\varepsilon^2}\right)^{(N-k)/2} \exp\left\{-\frac{1}{2\sigma_\varepsilon^2}\left[\sum_{t=k+1}^{N} \{x_t - \mu + a_1(x_{t-1} - \mu)\right.\right.$$

$$\left.\left. + \ldots + a_k(x_{t-k} - \mu)\}^2\right]\right\}. \tag{5.4.8}$$

To be more precise, the RHS of (5.4.8) does not represent exactly the joint probability density function of $X_{k+1}, \ldots, X_N$; rather, it represents the *conditional* probability density function of $X_{k+1}, \ldots, X_N$ given that the initial observations $X_1, \ldots, X_k$ remain fixed at $x_1, x_2, \ldots, x_k$, their observed values. The exact probability density function of $X_{k+1}, \ldots, X_N$ could be obtained by multiplying the conditional density function (5.4.8) by the joint probability density function of $X_1, \ldots, X_k$, but provided k is small compared with N this modification will have only a small "end effect" (see, e.g. Bartlett (1955), p. 241), and we may therefore use (5.4.8) as an adequate approximation to the likelihood function of $\mu, a_1, \ldots, a_k$ given the observations $X_1, \ldots, X_N$. With this approximation the log-likelihood function is,

$$L(\mu, a_1, \ldots, a_k) = -(N-k)\log\{\sigma_\varepsilon \sqrt{2\pi}\}$$

$$-\frac{1}{2\sigma_\varepsilon^2} \sum_{t=k+1}^{N} \{X_t - \mu + a_1(X_{t-1} - \mu)$$

$$+ \ldots + a_k(X_{t-k} - \mu)\}^2. \qquad (5.4.9)$$

If σ_ε^2 is known the (conditional) maximum likelihood estimates of $\mu, a_1, \ldots,$ a_k are clearly obtained by minimizing the second term in (5.4.9), i.e. by minimizing

$$Q(\mu, a_1, \ldots, a_k) = \sum_{t=k+1}^{N} \{X_t - \mu + a_1(X_{t-1} - \mu) + \ldots + a_k(X_{t-k} - \mu)\}^2,$$

$$(5.4.10)$$

which is identical with the sum of squares quantity (5.4.5). Thus, in this case the (conditional) maximum likelihood estimates of the parameters are identical with the least squares estimates.

Minimizing (5.4.10) w.r. to $\mu, a_1, \ldots, a_k$ gives

$$\sum_{t=k+1}^{N} \{X_t - \hat{\mu} + \hat{a}_1(X_{t-1} - \hat{\mu}) + \ldots + \hat{a}_k(X_{t-k} - \hat{\mu}) = 0, \qquad (5.4.11)$$

and

$$\sum_{t=k+1}^{N} \{X_t - \hat{\mu} + \hat{a}_1(X_{t-1} - \hat{\mu}) + \ldots + \hat{a}_k(X_{t-k} - \hat{\mu})\}(X_{t-j} - \hat{\mu}) = 0,$$

$$j = 1, \ldots, k. \qquad (5.4.12)$$

From (5.4.11) we have,

$$\hat{\mu} = \frac{\bar{X}_1 + \hat{a}_1 \bar{X}_2 + \ldots + \hat{a}_k \bar{X}_k}{1 + \hat{a}_1 + \ldots + \hat{a}_k}, \qquad (5.4.13)$$

where

$$\bar{X}_{j+1} = \frac{1}{N-k} \sum_{t=k+1-j}^{N-j} X_t, \qquad j = 0, 1, \ldots, k-1.$$

Provided again that k is small compared with N, the quantities $\bar{X}_1,$ $\bar{X}_2, \ldots, \bar{X}_k$ (which represent the means of different sets of $(N-k)$ observations) will all be close to the overall mean, $\bar{X} = (1/N) \sum_{t=1}^{N} X_t$. Making this approximation, (5.4.13) simplifies to

$$\hat{\mu} = \bar{X}, \qquad (5.4.14)$$

and (5.4.12) now gives

$$\sum_{t=k+1}^{N} \{(X_t - \bar{X}) + \hat{a}_1(X_{t-1} - \bar{X}) + \ldots + \hat{a}_k(X_{t-k} - \bar{X})\}(X_{t-j} - \bar{X}) = 0$$

$$j = 1, \ldots, k. \qquad (5.4.15)$$

As a further approximation we may write (under the same conditions on k and N),

$$\sum_{t=k+1}^{N} (X_{t-l} - \bar{X})(X_{t-j} - \bar{X}) \doteq N\hat{R}(j-l), \qquad (5.4.16)$$

where $\hat{R}(r)$ is the sample autocovariance function defined by (5.3.13). We may then write (5.4.15) as

$$\hat{R}(j) + \hat{a}_1\hat{R}(j-1) + \ldots + \hat{a}_k\hat{R}(j-k) = 0, \qquad j = 1, \ldots, k. \qquad (5.4.17)$$

In matrix form these equations may be written as,

$$\hat{R}_k \cdot \hat{a} = -\hat{r}, \qquad (5.4.18)$$

where

$$\hat{a} = (\hat{a}_1, \hat{a}_2, \ldots, \hat{a}_k)', \qquad r = (\hat{R}(1), \hat{R}(2), \ldots, \hat{R}(k))', \quad (5.4.19)$$

and

$$\hat{R}_k = \begin{bmatrix} \hat{R}(0) & \hat{R}(1) & \ldots & \hat{R}(k-1) \\ \hat{R}(1) & \hat{R}(0) & \ldots & \hat{R}(k-2) \\ \vdots & \vdots & & \vdots \\ \hat{R}(k-1) & \hat{R}(k-2) & \ldots & \hat{R}(0) \end{bmatrix}. \qquad (5.4.20)$$

Thus, approximate expressions for the maximum likelihood (or least squares) estimates $\hat{a}_1, \ldots, \hat{a}_k$ may be obtained by computing the first $(k+1)$ sample autocovariances $\hat{R}(0), \hat{R}(1), \ldots, \hat{R}(k)$ and then solving the set of linear equations (5.4.17). Note that the set of equations (5.4.17) is *identical in form to the Yule–Walker equations* (3.5.43) (Section 3.5.4) which express the *theoretical* values of the parameters $a_1, \ldots, a_k$ in terms of the *theoretical* autocorrelation function $\rho(r)$, or equivalently, in terms of the *theoretical* autocovariance function $R(r)$. The estimates (5.4.17) may therefore be obtained by writing down the expressions for $a_1, \ldots, a_k$ in terms of $R(r)$ (or in terms of $\rho(r)$) and then substituting $\hat{R}(r)$ for $R(r)$ (or $\hat{\rho}(r)$ for $\rho(r)$). This procedure is sometimes known as "parameter estimation by fitting to the sample autocovariance (or autocorrelation) function".

To summarize the above discussion we note that for an AR model the estimates $\hat{a}_1, \ldots, \hat{a}_k$ may be calculated by a number of different methods, depending on the degree of accuracy which we require. Starting with the most precise method (and continuing in decreasing order of precision) the different methods may be listed as follows.

1. *Use of the exact likelihood function*

We have already noted that (5.4.9) is only an approximation to the log-likelihood function since it does not take account of the probability distribution of the initial observations $X_1, \ldots, X_k$. If we introduce this extra factor in (5.4.8) we can write down an expression for the exact log-likelihood function, and then obtain a more accurate approximation to the maximum likelihood estimates.

Thus, under the assumption that the $\{\varepsilon_t\}$ are Gaussian, the $\{X_t\}$ are also Gaussian, and hence (with $\mu = 0$) the joint probability density function of $X_1, \ldots, X_k$ may be written (cf. (2.13.22)),

$$f(x_1, \ldots, x_k) = \left(\frac{1}{2\pi\sigma_\varepsilon^2}\right)^{k/2} |V_k|^{1/2} \exp\left(-\frac{1}{2\sigma_\varepsilon^2}\{x' V_k x\}\right),$$

where $\sigma_\varepsilon^2 V_k^{-1} (= R_k)$ denotes the variance–covariance matrix of $x = (x_1, x_2, \ldots, x_k)'$. Multiplying (5.4.8) by the above density function, we may now write the exact likelihood function as

$$f(X_1, \ldots, X_n, a_1, \ldots, a_k) = \left(\frac{1}{2\pi\sigma_\varepsilon^2}\right)^{N/2} |V_k|^{1/2} \exp[-\tilde{Q}(a_1, \ldots, a_k)/2\sigma_\varepsilon^2],$$

$$(5.4.20a)$$

where

$$\tilde{Q}(a_1, \ldots, a_k) = \sum_{i=1}^{k} \sum_{j=1}^{k} v_{ij} X_i X_j + Q(a_1, \ldots, a_k),$$

v_{ij} being (i, j)th element of V_k, and $Q(a_1, \ldots, a_k)$ the sum of squares function given by (5.4.10). (In the terminology of Box and Jenkins we would call Q the "conditional sum of squares" and $\tilde{Q}$ the "unconditional sum of squares".) The exact log-likelihood function is thus (apart from a constant),

$$L(a_1, \ldots, a_k) = -\frac{N}{2} \log \sigma_\varepsilon^2 + \tfrac{1}{2} \log |V_k| - \frac{\tilde{Q}(a_1, \ldots, a_k)}{2\sigma_\varepsilon^2}.$$

To illustrate the use of (5.4.20a), consider the AR(1) model (3.5.5). Here, the probability density function of X_1 is given by

$$f(x_1) = \frac{1}{2\pi\sigma_X^2} \exp(-x_1^2/2\sigma_X^2) = \frac{(1-a^2)}{2\pi\sigma_\varepsilon^2} \exp[-x_1^2(1-a^2)/2\sigma_\varepsilon^2],$$

(using (3.5.14)). Hence, the exact likelihood function is

$$f(X_1, \ldots, X_n, a) = \left(\frac{1}{2\pi\sigma_\varepsilon^2}\right)^{N/2} (1 - a^2) \exp\left(-\frac{1}{2\sigma_\varepsilon^2}\{(1 - a^2)X_1^2\right.$$
$$\left. + \sum_{t=2}^{N} (X_t - aX_{t-1})^2\}\right).$$

Now the matrix V_k depends, of course, on the parameters $(a_1, \ldots, a_k)$, and the derivatives of $\log|V_k|$ with respect to these parameters are complicated functions. However, it may be shown (see, e.g., Box and Jenkins (1970), p. 277) that for large N, $\log|V_k|$ is negligible compared with $\tilde{Q}$. Hence, we may ignore this term in the log-likelihood function, and to this order of approximation the exact maximum likelihood estimates of $a_1, \ldots, a_k$ are given by minimizing $\tilde{Q}(a_1, \ldots, a_k)$.

2. *Least squares estimates*

The least squares (i.e. conditional maximum likelihood) estimates are obtained by solving the normal equations (5.4.11), (5.4.12) for $\hat{a}_1, \ldots, \hat{a}_k$.

3. *Approximate least squares estimates using $\hat{\mu} = \bar{X}$*

Approximate least squares estimates are obtained by using (5.4.14) in place of (5.4.13), (i.e. estimating μ by $\hat{\mu} = \bar{X}$) and then solving the set of equations (5.4.15) for $\hat{a}_1, \ldots, \hat{a}_k$. The equations (5.4.15) may be solved by using a standard multiple regression "package"—the use of (5.4.15) in place of (5.4.12) is equivalent to first estimating μ by $\bar{X}$ and then fitting $a_1, \ldots, a_k$ by a multiple linear regression of $(X_t - \bar{X})$ on $(X_{t-1} - \bar{X}), \ldots, (X_{t-k} - \bar{X})$.

4. *Yule–Walker equation estimates*

A further approximation to the least squares estimates is obtained by using (5.4.14) together with the approximation (5.4.16), in which case the normal equations (5.4.12) reduce to the Yule–Walker equations (5.4.17).

Box and Jenkins (1970) observe that for moderate and large N the differences between the estimates obtained by the various methods will be small. However, appreciable differences can arise (even for moderately large N) if the AR operator has a characteristic root near the unit circle. In this case the Yule–Walker equations may lead to poor coefficient estimates, and one would try to obtain the closest approximation to the exact maximum likelihood estimates.

Estimation of σ_ε^2

If σ_ε^2 is unknown it may be estimated by using a standard result in linear hypothesis theory which gives an approximately unbiased estimate for the residual variance about the "regression" function, viz. (see, e.g., Scheffe (1959), p. 22),

$$\hat{\sigma}_\varepsilon^2 = \frac{1}{N-2k-1} Q(\hat{\mu}, \hat{a}_1, \ldots, \hat{a}_k) \tag{5.4.21}$$

$$\doteq \left(\frac{N-k}{N-2k-1}\right)\{\hat{R}(0) + \hat{a}_1\hat{R}(1) + \ldots + \hat{a}_k\hat{R}(k)\}. \tag{5.4.22}$$

The divisor in (5.4.21) is obtained from the general expression; (number of observations—number of parameters estimated). Here, we have used, in effect, $(N-k)$ observations (since the first k residuals are "lost" in the regression analysis) and we have estimated $(k+1)$ parameters.

Note that (5.4.21) is not the same as the maximum likelihood estimate of σ_ε^2; if we regard σ_ε^2 as an additional unknown parameter in (5.4.9) then it is easily shown that the conditional maximum likelihood estimate of σ_ε^2 is

$$\hat{\sigma}_\varepsilon^2 = \frac{1}{N-k} Q(\hat{\mu}, \hat{a}_1, \ldots, \hat{a}_k), \tag{5.4.21a}$$

whereas minimizing the exact likelihood function (but ignoring $|V_k|$) gives (with $\mu = 0$),

$$\hat{\sigma}_\varepsilon^2 = \frac{1}{N} \tilde{Q}(\hat{a}_1, \ldots, \hat{a}_k). \tag{5.4.21b}$$

Examples

1. *AR(1) model*

Here $k = 1$, and the Yule–Walker equation (5.4.17) becomes,

$$\hat{R}(1) + \hat{a}_1\hat{R}(0) = 0.$$

Hence,

$$\hat{a}_1 = -\hat{R}(1)/\hat{R}(0) = -\hat{\rho}(1). \tag{5.4.23}$$

where $\hat{\rho}(r)$ is the sample autocorrelation function given by (5.3.31). This result is obvious when one recalls that for this model, $\rho(r) = (-a_1)^{|r|}$, so that $\rho(1) = -a_1$; in particular we see that

$$E(\hat{a}_1) = a_1. \tag{5.4.24}$$

Treating μ as known, it may be shown (Bartlett (1955), p. 241) that

$$E\left[\frac{\partial^2 L}{\partial a_1^2}\right] = -\left(\frac{N}{1-a_1^2}\right).$$

Using the general result (5.2.33), we then have for large N,

$$\text{var}(\hat{a}_1) \sim (1-a_1^2)/N. \tag{5.4.25}$$

On the basis of an asymptotic normal distribution for $\hat{a}_1$, (5.4.25) together with (5.4.24) may be used to derive a (non-symmetric) confidence interval for a_1. The unbiased estimate of σ_ε^2 is,

$$\hat{\sigma}_\varepsilon^2 \doteq \left(\frac{N-1}{N-3}\right)\{R(\hat{0}) + \hat{a}_1 \hat{R}(1)\} \tag{5.4.26}$$

$$= \left(\frac{N-1}{N-3}\right) \hat{R}(0)\{1-\hat{\rho}^2(1)\} \tag{5.4.27}$$

$$= \left(\frac{N-1}{N-3}\right) s_X^2\{1-\hat{\rho}^2(1)\}, \tag{5.4.28}$$

where $s_X^2 \ (\equiv \hat{R}(0))$ is the sample variance of $X_1, \ldots, X_N$.

2. AR(2) model

We now have $k = 2$, and the Yule–Walker equations (5.4.17) give,

$$\hat{R}(1) + \hat{a}_1 \hat{R}(0) + \hat{a}_2 \hat{R}(1) = 0, \tag{5.4.29}$$

and

$$\hat{R}(2) + \hat{a}_1 \hat{R}(1) + \hat{a}_2 \hat{R}(0) = 0, \tag{5.4.30}$$

leading to,

$$\hat{a}_1 = \hat{\rho}(1)\{\hat{\rho}(2) - 1\}/(1 - \hat{\rho}^2(1)), \tag{5.4.31}$$

$$\hat{a}_2 = \{\hat{\rho}^2(1) - \hat{\rho}(2)\}/(1 - \hat{\rho}^2(1)). \tag{5.4.32}$$

Also, (5.4.22) gives,

$$\hat{\sigma}_\varepsilon^2 \doteq \left(\frac{N-2}{N-5}\right)\{\hat{R}(0) + \hat{a}_1 \hat{R}(1) + \hat{a}_2 \hat{R}(2)\}. \tag{5.4.33}$$

Asymptotic distribution of $\hat{a}_1, \ldots, \hat{a}_k$

Under the assumption that the $\{\varepsilon_t\}$ are independent and identically distributed (but not necessarily normal), with finite fourth moments,

Mann and Wald (1943) proved that for the least squares estimates $\{\sqrt{N}(\hat{a}_1 - a_1), \sqrt{N}(\hat{a}_2 - a_2), \ldots, \sqrt{N}(\hat{a}_k - a_k)\}$ have an asymptotic multivariate normal distribution with zero means and variance–covariance matrix $\sigma_\varepsilon^2 \boldsymbol{R}_k^{-1}$, where $\boldsymbol{R}_k$ is the matrix,

$$\boldsymbol{R}_k = \begin{bmatrix} R(0) & R(1) & \cdots & R(k-1) \\ R(1) & R(0) & \cdots & R(k-2) \\ \vdots & \vdots & & \vdots \\ R(k-1) & R(k-2) & \cdots & R(0) \end{bmatrix}. \tag{5.4.34}$$

As an approximation we may substitute $\hat{\sigma}_\varepsilon^2$ for σ_ε^2 and replace the elements of $\boldsymbol{R}_k$ by the corresponding sample autocovariances. In this case the matrix $\boldsymbol{R}_k$ becomes equivalent to the matrix $\hat{\boldsymbol{R}}_k$ defined by (5.4.20)—cf. the corresponding results (5.2.52), (5.2.53) for the classical regression model.

In the special case when $k = 1$, the above result gives

$$N \operatorname{var}(\hat{a}_1) \sim \sigma_\varepsilon^2 \{R(0)\}^{-1} = \sigma_\varepsilon^2 / \sigma_X^2.$$

But we know from (3.5.14) (Section 3.5.2, Chapter 3) that for this AR(1) model, $\sigma_X^2 = \sigma_\varepsilon^2 / (1 - a_1^2)$. (Note that the parameter a in (3.5.5) corresponds to $(-a_1)$.) Hence, we obtain $\operatorname{var}(\hat{a}_1) \sim (1 - a_1^2)/N$, in agreement with the previously derived result (5.4.25).

Numerical Example 1: Graph 3.2 (Chapter 3) shows 500 observations generated from the AR(1) model

$$X_t + aX_{t-1} = \varepsilon_t,$$

where $\varepsilon_t \sim N(0, 1)$, and $a = -0 \cdot 6$. Using these data (which are listed in the Appendix), the (conditional) maximum likelihood estimate of a was $\hat{a} = -0 \cdot 54994$, with estimated residual variance, $\hat{\sigma}_\varepsilon^2 = 0 \cdot 97703$. From (5.4.25) we find $\operatorname{var}(\hat{a}) \sim 0 \cdot 00128$, so that an approximate 95% confidence interval for $\hat{a}$ is $(-0 \cdot 62, -0 \cdot 48)$.

Numerical Example 2: Graph 3.3 (Chapter 3) shows 500 observations generated from the AR(2) model

$$X_t + a_1 X_{t-1} + a_2 X_{t-2} = \varepsilon_t,$$

with $\varepsilon_t \sim N(0, 1)$, $a_1 = -0 \cdot 4$, $a_2 = 0 \cdot 7$. These data (see the Appendix) gave (conditional) maximum likelihood estimates,

$$\hat{a}_1 = -0 \cdot 35308, \qquad \hat{a}_2 = 0 \cdot 6952, \qquad \hat{\sigma}_\varepsilon^2 = 0 \cdot 97948.$$

5.4.2 Estimation of Parameters in Moving Average Models

The MA(l) model was introduced in Section 3.5.5, and defined by (3.5.45). As in the AR case we now slightly modify this model to allow for X_t

to have a non-zero mean μ, the modified model being,

$$X_t = \mu + \varepsilon_t + b_1 \varepsilon_{t-1} + \ldots + b_l \varepsilon_{t-l} \qquad (5.4.35)$$

$$= \mu + \beta(B) \cdot \varepsilon_t, \qquad (5.4.36)$$

where $\beta(z) = 1 + b_1 z + \ldots + b_l z^l$. As noted in Section 3.5.5 we take the coefficients of ε_t to be unity, and we assume that $\beta(z)$ has no roots inside or on the unit circle.

Although we can write formally,

$$\varepsilon_t = \beta^{-1}(B)\{X_t - \mu\}, \qquad (5.4.37)$$

the expansion of $\beta^{-1}(B)$ will contain an infinite number of terms so that we cannot now express the "residuals" ε_t as finite linear functions of the observations $(X_1, \ldots, X_N)$. In this case it is difficult to write down an explicit expression for the likelihood function. Also the problem is no longer of the "linear hypothesis" type, and we cannot appeal to standard linear regression theory. Of course, if the $\{\varepsilon_t\}$ are normal then the observations $(X_1, \ldots, X_N)$ have a multivariate normal distribution and we can write down their joint probability density function in terms of their variance–covariance matrix, which is in fact the matrix R_N defined by (5.4.34). However, the elements of R_N (i.e. the autocovariances of X_t) are complicated functions of the coefficients $b_1, \ldots, b_l$ and this leads to difficulties in attempting to derive the exact maximum likelihood estimates of these parameters. Moreover, estimates of $b_1, \ldots, b_l$ obtained by fitting to the sample autocorrelation (or autocovariance) function are, in general, highly inefficient; see, e.g., Hannan (1960), p. 47.

However, Box and Jenkins (1970) have developed a simple but elegant procedure for evaluating an approximate likelihood function *numerically* by first determining the $\{\varepsilon_t\}$ recursively, as follows. Given observations $X_1, X_2, \ldots, X_N$ from the model (5.4.35) we first set (as starting values)

$$\varepsilon_0 = \varepsilon_{-1} = \varepsilon_{-2} = \ldots = \varepsilon_{-(l-1)} = 0. \qquad (5.4.38)$$

We then have,

$$\varepsilon_1 = X_1 - \mu,$$

$$\varepsilon_2 = X_2 - \mu - b_1 \varepsilon_1,$$

$$\varepsilon_3 = X_3 - \mu - b_1 \varepsilon_2 - b_2 \varepsilon_1,$$

$$\vdots \qquad (5.4.39)$$

$$\varepsilon_l = X_l - \mu - b_1 \varepsilon_{l-1} - b_2 \varepsilon_{l-2} - \ldots - b_{l-1} \varepsilon_1,$$

$$\varepsilon_t = X_t - \mu - b_1 \varepsilon_{t-1} - b_2 \varepsilon_{t-2} - \ldots - b_l \varepsilon_{t-l}, \qquad l \leq t \leq N.$$

Suppose for the moment that the order, l, is known. Under the assumption that the $\{\varepsilon_t\}$ are normal (and ignoring the effect of the "initial conditions" (5.4.38)) the (conditional) log-likelihood function for $(\mu, b_1, \ldots, b_l)$ can be written as

$$L(\mu, b_1, \ldots, b_l) = \text{const} - \frac{1}{2\sigma_\varepsilon^2} Q(\mu, b_1, \ldots, b_l), \qquad (5.4.40)$$

where

$$Q(\mu, b_1, \ldots, b_l) = \sum_{t=1}^{N} \varepsilon_t^2. \qquad (5.4.41)$$

The (conditional) maximum likelihood estimates of $\mu, b_1, \ldots, b_l$ are thus obtained by minimizing $Q(\mu, b_1, \ldots, b_l)$, and may therefore be regarded also as least squares estimates. (Note, however, that Q is no longer a quadratic function of the parameters.) For a given set of values of μ, $b_0, \ldots, b_l$ we may evaluate the $\{\varepsilon_t\}$ from (5.4.39), and then evaluate Q from (5.4.41). Hence we can determine the numerical value of Q for any given set of parameter values, and by plotting contours of Q over a suitable set of points in the parameter space we may determine numerically those values, $\hat{\mu}, \hat{b}_1, \ldots, \hat{b}_l$, which minimize Q. Alternatively, given initial starting values $\hat{\mu}^{(0)}, \hat{b}_1^{(0)}, \ldots, \hat{b}_l^{(0)}$ an iterative "hill-climbing" technique may be used. (The initial starting values may be obtained, for example, by fitting to the observed autocorrelation function. The resulting estimates, although inefficient, may still provide quite useful starting values.)

Exact likelihood function

As remarked previously, the basic form of the multivariate normal distribution of $X_1, \ldots, X_n$ does not provide a very convenient form of likelihood function for the estimation of $b_1, \ldots, b_l$. However, Box and Jenkins (1970), p. 269 have derived a more useable form of the exact likelihood function for an MA(l) model. The basis of their method is to derive the joint distribution of $X_1, \ldots, X_N$ by starting from the joint distribution of $(\varepsilon_{-(l-1)}, \ldots, \varepsilon_{-1}, \varepsilon_0, \ldots, \varepsilon_N)$. Thus, assuming $\mu = 0$ we consider the transformation

$$\varepsilon_{-(l-1)} = \varepsilon_{-(l-1)}$$
$$\vdots \qquad \vdots$$
$$\varepsilon_{-1} = \varepsilon_{-1}$$
$$\varepsilon_0 = \varepsilon_0$$
$$\varepsilon_1 = X_1 - b_1 \varepsilon_0 - b_2 \varepsilon_{-1} \ldots - b_l \varepsilon_{-(l-1)}$$
$$\vdots \qquad \vdots$$
$$\varepsilon_N = X_N - b_1 \varepsilon_{N-1} - b_2 \varepsilon_{N-2} - \ldots - b_l \varepsilon_{N-l},$$

which we re-write in matrix form as

$$\varepsilon = LX + M\varepsilon_*,$$

where $\varepsilon = (\varepsilon_{-(l-1)}, \ldots, \varepsilon_0, \ldots, \varepsilon_N)'$, $\varepsilon_* = (\varepsilon_{-(l-1)} \ldots \varepsilon_0)'$, L is an $(N+l) \times N$ matrix, and M is an $(N+l) \times l$ matrix. (The elements of L and M are clearly indicated by the scalar form of the above transformation, and it should be noted that they depend on the parameters $b_1, \ldots, b_l$). The joint probability density function of the elements of ε is

$$f(\varepsilon) = \left(\frac{1}{2\pi\sigma_\varepsilon^2}\right)^{(N+l)/2} \exp[-(\varepsilon'\varepsilon/2\sigma_\varepsilon^2)],$$

and since the Jacobian of the transformation is unity, the joint probability density function of X and ε_* may be written

$$f(X, \varepsilon_*) = \left(\frac{1}{2\pi\sigma_\varepsilon^2}\right)^{(N+l)/2} \exp[-\tilde{Q}(b_1, \ldots, b_l, \varepsilon_*)/2\pi\sigma_\varepsilon^2],$$

where

$$\tilde{Q}(b_1, \ldots, b_l, \varepsilon_*) = (LX + M\varepsilon_*)'(LX + M\varepsilon_*).$$

Let $\hat{\varepsilon}_*$ be that value of ε_* which minimizes $\tilde{Q}(b_1, \ldots, b_l, \varepsilon_*)$ (for fixed X). Then by standard least squares theory we have the decomposition

$$\tilde{Q}(b_1, \ldots, b_l, \varepsilon_*) = \tilde{Q}(b_1, \ldots, b_l) + (\varepsilon_* - \hat{\varepsilon}_*)'M'M(\varepsilon_* - \hat{\varepsilon}_*),$$

where

$$\tilde{Q}(b_1, \ldots, b_l) = (LX + M\hat{\varepsilon}_*)'(LX + M\hat{\varepsilon}_*).$$

Substituting the decomposition of $\tilde{Q}(b_1, \ldots, b_l, \varepsilon_*)$ into the expression for $f(X, \varepsilon_*)$ we see that $f(X, \varepsilon_*)$ may be factorized into the product of two terms, the first of which involves only X, and the second involves both X and ε_*. The first factor may therefore be identified with the joint density function of $X_1, \ldots, X_N$ and the second factor with the density function of the conditional distribution of ε_*, given $X_1, \ldots, X_N$. Specifically, the joint probability density function of $X_1, \ldots, X_N$ may be written as

$$f(X_1, \ldots, X_N) = \left(\frac{1}{2\pi\sigma_\varepsilon^2}\right)^{N/2} |M'M|^{-1/2} \exp[-\tilde{Q}(b_1, \ldots, b_l)/2\sigma_\varepsilon^2],$$

$$(5.4.41a)$$

while the conditional density of ε_*, given X, is

$$f(\varepsilon_*|X) = \left(\frac{1}{2\pi\sigma_\varepsilon^2}\right)^{l/2} |M'M|^{1/2} \exp[-\{(\varepsilon_* - \hat{\varepsilon}_*)'M'M(\varepsilon_* - \hat{\varepsilon}_*)\}/2\sigma_\varepsilon^2].$$

If, for large N, we again ignore the factor corresponding to the determinant, $|M'M|^{-1/2}$, the exact maximum likelihood estimates of $b_1, \ldots, b_l$ are obtained by minimizing $\tilde{Q}(b_1, \ldots, b_l)$. Now under the assumption that the ε_t are Gaussian, $\hat{\varepsilon}_*$ is the *conditional expectation* of ε_*, given X.

Hence, $(LX + M\hat{\varepsilon}_*) = E[\varepsilon|X]$, and, writing $\hat{\varepsilon}_t = E[\varepsilon_t|X_1, \ldots, X_N]$, $\tilde{Q}(b_1, \ldots, b_l)$ may be expressed as

$$\tilde{Q}(b_1, \ldots, b_l) = \sum_{t=-(l-1)}^{N} \hat{\varepsilon}_t^2. \qquad (5.4.41b)$$

As a computational algorithm for dealing with the exact likelihood function, Box and Jenkins (1970) discuss a procedure which they call "back forecasting" (see also Chatfield and Prothero (1973a)). This is a device whereby the given data are used to "forecast" the unobserved values of $X_0, X_{-1}, X_{-2}, \ldots$, from which one may obtain "estimates" of ε_0, $\varepsilon_{-1}, \ldots, \varepsilon_{-(l-1)}$ together with improved "estimates" of $\varepsilon_1, \ldots, \varepsilon_N$, instead of those used in (5.4.39) which are based on the rather crude initial conditions (5.4.38). In using "back forecasting" we, in effect, "estimate" $\varepsilon_{-(l-1)}, \ldots, \varepsilon_{-1}, \varepsilon_0, \ldots, \varepsilon_N$ by their conditional expectations, given the observations $X_1, \ldots, X_N$ and a set of parameter values $b_1, \ldots, b_l$ so that we can then evaluate the exponent $\tilde{Q}(b_1, \ldots, b_l)$ in the exact likelihood function (5.4.41a).

(The use of (5.4.38) amounts to estimating $\varepsilon_0, \varepsilon_{-1}, \ldots, \varepsilon_{-(l-1)}$, by their *unconditional* expectations.) This technique can lead to an appreciable increase in the accuracy of the estimates in cases where the MA operator has a characteristic root near the unit circle. For, if we regard (5.4.35) as a difference equation in $\{\varepsilon_t\}$ then the solution for ε_t (given the $\{X_t\}$) will contain a "transient" term (corresponding to the complimentary function). If the MA operator has a root near the unit circle then the transient term will decay very slowly, and the effect of the "crude" initial conditions (5.4.38) will persist through the subsequently computed ε_t for a substantial number of terms. But the method is not so computationally efficient if there is a root near the unit circle.

The (conditional) maximum likelihood estimate of the residual variance σ_ε^2 is given by

$$\hat{\sigma}_\varepsilon^2 = \frac{1}{N} Q(\hat{\mu}, \hat{b}_1, \ldots, \hat{b}_l). \qquad (5.4.42)$$

Although we are now no longer within the class of linear models (and hence cannot appeal to the general result quoted in the previous section which gives an unbiased estimate of σ_ε^2), Jenkins and Watts (1968) suggest that, by analogy with (5.4.21), a more appropriate estimate of σ_ε^2 (which takes account of the number of parameters estimated) is given by

$$\hat{\sigma}_\varepsilon^2 = \frac{1}{N-(l+1)} \cdot Q(\hat{\mu}, \hat{b}_1, \ldots, \hat{b}_l). \tag{5.4.42a}$$

An alternative approach to the problem of fitting of MA models was developed by Walker (1961). Instead of trying to obtain the likelihood function for $b_1, \ldots, b_l$ based on the distribution of the original observations, Walker considered the asymptotic multivariate normal distribution of the sample autocorrelations. (In other words, Walker considers the maximum likelihood estimation of $b_1, \ldots, b_l$, as if the raw data consisted purely of the sample autocorrelations.) The expressions for the theoretical autocorrelations in terms of $b_1, \ldots, b_l$ are easily obtained (see equation (3.5.50), Section 3.5.5), and the asymptotic variance–covariance matrix of the sample autocorrelations is given by (5.3.36).

Another approach to estimating MA parameters was discussed by Durbin (1959, 1960). Durbin's method is based on replacing the operator $\beta^{-1}(B)$ in (5.4.37) by a *finite* order polynomial. In effect, this procedure converts the MA model into a finite order AR model, and the parameter of the approximating AR model can then be estimated as described in the preceding section. A form of "likelihood" function for the parameters of the original MA model can then be obtained from the distribution of the estimated parameters of the approximating AR model. See also McClave (1973, 1974), Mentz (1977), Godolphin (1977, 1978) and Osborn (1976, 1977).

Numerical Example 3: 500 observations generated from the MA(2) model

$$X_t = \varepsilon_t + b_1 \varepsilon_{t-1} + b_2 \varepsilon_{t-2},$$

with $\varepsilon_t \sim N(0, 1)$, $b_1 = 1 \cdot 1$, $b_2 = 0 \cdot 2$, are shown in Graph 3.4. Using these data (see the Appendix), the (conditional) maximum likelihood estimates of the parameters were

$$\hat{b}_1 = 1 \cdot 05299, \qquad \hat{b}_2 = 0 \cdot 15756, \qquad \hat{\sigma}_\varepsilon^2 = 0 \cdot 97835.$$

5.4.3 Estimation of Parameters in Mixed ARMA Models

The mixed ARMA model of order (k, l) for a zero mean process is given by (3.5.52) (Section 3.5.6). Allowing for a non-zero mean μ, the model now becomes,

$$X_t - \mu + a_1(X_{t-1} - \mu) + \ldots + a_k(X_{t-k} - \mu) = \varepsilon_t + b_1 \varepsilon_{t-1} + \ldots + b_l \varepsilon_{t-l}.$$

$$\tag{5.4.43}$$

(As in the MA case we may take $b_0 = 1$ in (3.5.52).) Without further conditions the model (5.4.43) is not "identifiable" in the sense that there are many sets of coefficients, $(a_1, \ldots, a_k, b_1, \ldots, b_l)$, all of which give rise to the same autocovariance function for X_t. To ensure identifiability we impose the following conditions (cf. the discussion in Chapter 10, Section 10.5.1):

(a) The polynomials

$$\alpha(z) = 1 + \alpha_1 z + \ldots + a_k z^k \quad \text{and} \quad \beta(z) = 1 + b_1 z + \ldots + b_l z^l$$

have no common factors.

(b) All the roots of $\alpha(z)$ and $\beta(z)$ lie outside the unit circle.

(c) a_k and b_l are not both zero.

Again the $\{\varepsilon_t\}$ are not finite linear functions of the $\{X_t\}$, and the problem is not of the "linear hypothesis" type. However, as with the pure MA model it is possible to evaluate the ε_t recursively, using the following technique due to Box and Jenkins (1970). We first set, as starting values,

$$\varepsilon_1 = \varepsilon_2 = \ldots = \varepsilon_k = 0 = \varepsilon_0 = \varepsilon_{-1} = \ldots = \varepsilon_{-(l-k-1)} \qquad \text{(assuming } l > k+1\text{).}$$
$$(5.4.44)$$

Note that the starting values (5.4.44) are a combination of those previously used for the AR and MA models. We may now write,

$$\varepsilon_{k+1} = X_{k+1} - \mu + a_1(X_k - \mu) + \ldots + a_k(X_1 - \mu),$$

$$\varepsilon_{k+2} = X_{k+2} - \mu + a_1(X_{k+1} - \mu) + \ldots + a_k(X_2 - \mu) - b_1 \varepsilon_{k+1},$$

$$\vdots \qquad \qquad \vdots$$

$$\varepsilon_{k+l} = X_{k+l} - \mu + a_1(X_{k+l-1} - \mu) + \ldots + a_k(X_l - \mu)$$
$$- b_1 \varepsilon_{k+l-1} - \ldots - b_{l-1} \varepsilon_{k+1}, \qquad (5.4.45)$$

$$\varepsilon_{k+l+1} = X_{k+l+1} - \mu + a_1(X_{k+l} - \mu) + \ldots + a_k(X_{l+1} - \mu)$$
$$- b_1 \varepsilon_{k+l} - \ldots - b_l \varepsilon_{k+1},$$

$$\vdots \qquad \qquad \vdots$$

etc.

Given the order (k, l) of the model, we may evaluate the sum of squares function,

$$Q(\mu, a_1, \ldots, a_k, b_1, \ldots, b_l) = \sum_{t=k+1}^{N} \varepsilon_t^2, \qquad (5.4.46)$$

and obtain the (conditional) maximum likelihood estimates $\hat{\mu}, \hat{a}_1, \ldots, \hat{a}_k, \hat{b}_1, \ldots, \hat{b}_l$ by minimizing Q numerically, as described in the preceding section. (Again, we may use an interative approach, starting with initial estimates $\hat{\mu}^{(0)}, \hat{a}_1^{(0)} \ldots \hat{a}_k^{(0)}, \hat{b}_1^{(0)}, \ldots, \hat{b}_l^{(0)}$.) By analogy with (5.4.42a), the residual variance, σ_ε^2, may now be estimated by

$$\hat{\sigma}_\varepsilon^2 = \frac{1}{N - 2k - l - 1} Q(\hat{\mu}, \hat{a}_1, \ldots, \hat{a}_k, \hat{b}_1, \ldots, \hat{b}_l). \qquad (5.4.47)$$

(Note that the effective number of observations used is $(N - k)$, and we have estimated $(k + l + 1)$ parameters.) Alternatively, instead of using the starting values (5.4.44) we may instead set

$$(X_0 - \mu) = (X_{-1} - \mu) = \ldots = (X_{-(k-1)} - \mu) = 0 = \varepsilon_0 = \varepsilon_{-1} = \ldots = \varepsilon_{-(l-k-1)}.$$

We can then compute expressions for $\varepsilon_1, \ldots, \varepsilon_k$, and extend the summat on in (5.4.46) to include terms for $t = 1, \ldots, k$. However, if the AR operator has a root near the unit circle then the solution of the difference equation (5.4.43) for X_t will contain a slowly decaying transient, and the effect of the "crude" starting values for the X's may persist for quite some time. In this case it would be preferable to use Box and Jenkins' back forecasting technique, so that $X_0, \ldots, X_{-(k-1)}$ are estimated by their conditional expectations (given $X_1, \ldots, X_N$) rather than by their unconditional expectations (as above), and similarly for the ε's. These conditional expectations are, in practice, computed by re-writing (5.4.43) in its "forward" form (obtained by replacing B by B^{-1} in the AR and MA operators) and then assuming that the new residuals are strictly independent; see Box and Jenkins (1970). If we simply ignore the first k terms in the sum of squares function (i.e. use (5.4.46)) then the problem of transient effects due to starting values of the X's does not arise.

Exact likelihood function

Box and Jenkins (1970), p. 272 suggest that the result (5.4.41b) can be extended to ARMA models by first writing an ARMA model as a general linear process, i.e. as an infinite order MA model. This leads to an "exact" form of likelihood functions for an ARMA model in which (for $\mu = 0$) the exponent takes the form

$$\tilde{Q}(a_1, \ldots, a_k, b_1, \ldots, b_l) = \sum_{t=-\infty}^{N} \hat{\varepsilon}_t^2.$$

Box and Jenkins further suggest that, in practice, the above summation may be truncated to the finite form

$$\tilde{Q}(a_1, \ldots, a_k, b_1, \ldots, b_l) \sim \sum_{t=-(P-1)}^{N} \hat{\varepsilon}_t^2,$$

where P is a sufficiently large integer. However, Newbold (1974) has derived a precise closed form expression for the exact likelihood function of an ARMA model which avoids this type of approximation. Newbold's method is based on an extension of the "transformation" approach developed by Box and Jenkins for the finite order MA model. (Previously, Tiao and Ali (1971) had derived the exact likelihood function for an ARMA(1,1) model.) See also Ali (1977, 1978), Cooper and Thompson (1977), Kohn (1977), McLeod (1977), Pham-Dinh (1977a), Ansley (1979), Nicholls and Hall (1979) and Ljung and Box (1979).

Other methods for estimating the parameters of ARMA models have been given by Durbin (1960), Walker (1962) and more recently by Revfeim (1969), Hannan (1969b, 1970), Parzen, (1971, 1974), Akaike (1973), and Anderson (1975, 1977).

Revfeim's approach is interesting in that, in common with other recent studies, it is based essentially on a "frequency domain" analysis, and uses the characteristic property of ARMA models that their spectral density functions are *rational functions* of $\exp(-i\omega)$. Specifically, we have from (4.12.63) that (with $\mu = 0$) the (non-normalized) spectral density function of X_t is

$$h(\omega) = \frac{\sigma_\epsilon^2}{2\pi} \frac{|1 + b_1 \exp(-i\omega) + \ldots + b_l \exp(-il\omega)|^2}{|1 + a_1 \exp(-i\omega) + \ldots + a_k \exp(-ik\omega)|^2}. \qquad (5.4.48)$$

If, given observations $X_1, \ldots, X_N$, we now define a function $I_X^*(\omega)$ (called the "modified periodogram" or "sample spectral density function") by

$$I_X^*(\omega) = \frac{1}{2\pi} \sum_{r=-(N-1)}^{(N-1)} \hat{R}(r) e^{-ir\omega}, \qquad (5.4.49)$$

(where $\hat{R}(r)$ is given by (5.3.13)), then, as will be shown in Chapter 6, we have for large N,

$$E[I_X^*(\omega)] \sim h(\omega), \qquad \text{all } \omega, \qquad (5.4.50)$$

so that $I_X^*(\omega)$ is an asymptotically unbiased estimate of $h(\omega)$. Moreover, this result holds under quite general conditions and does not depend upon $\{X_t\}$ satisfying any particular finite parameter model.

Hence, we may write,

$$I_X^*(\omega) = h(\omega) + \xi(\omega), \qquad (5.4.51)$$

where, approximately, $E[\xi(\omega)] = 0$, all ω. The function $I_X^*(\omega)$ can be evaluated over any range of values of ω, and the problem of estimating $(a_1, \ldots, a_k, b_1, \ldots, b_l)$ may be viewed as equivalent to fitting the rational function (5.4.48) to values of $I_X^*(\omega)$. Revfeim used a logarithmic transformation of (5.4.51), giving,

$$\log\{I_X^*(\omega)\} = \log\{h(\omega)\} + \eta(\omega),$$

where $\eta(\omega) = \log\{1 + \xi(\omega)/h(\omega)\}$, (since $\log\{I_X^*(\omega)\}$ has more appropriate sampling properties than $I_X^*(\omega)$ itself, see Chapter 6), and then used an approximate "linearization" of the numerator and denominator in (5.4.48) by expanding each about their values corresponding to an initial set of estimates $(\hat{b}_1^{(0)}, \ldots, \hat{b}_l^{(0)})$ and $(\hat{a}_1^{(0)}, \ldots, \hat{a}_k^{(0)})$. The estimation of $(a_1, \ldots, a_k, b_1, \ldots, b_l)$ is then performed by an iterative least squares fit of $\log\{I_X^*(\omega)\}$ to $\log\{h(\omega)\}$ over a chosen set of frequencies, $\omega_1, \omega_2, \ldots, \omega_K$. The same approach can obviously be applied to the pure AR and pure MA models since each can be regarded as a special case of the general ARMA model.

Hannan (1969b, 1970), Parzen (1971), and Akaike (1973) also follow a frequency domain approach, but their estimation procedures are more sophisticated than the simple fitting technique described above. In particular, Hannan's method involves only two iterations, after which the resulting estimates are asymptotically fully efficient. Anderson (1975) discusses both frequency domain and time domain methods; see also Nicholls (1972, 1973) and Pham-Dinh (1975). We would remark that although there has recently been considerable interest in frequency domain methods, the basic development of the frequency domain approach to maximum-likelihood parameter estimation was given by Whittle in what have now become classic reference works (Whittle (1951, 1953)). (Further developments of Whittle's results are discussed by Walker (1964)).

We have seen that for linear models the log-likelihood function is dominated by the "sum of squares" function, $\sum_{t=1}^{N} \varepsilon_t^2$. The main result underlying the frequency domain approach to maximum likelihood estimation is that, for large N, this expression may be written

$$\sum_{t=1}^{N} \varepsilon_t^2 \sim \frac{N\sigma_\varepsilon^2}{2\pi} \int_{-\pi}^{\pi} \frac{I_X^*(\omega)}{h(\omega)} d\omega. \qquad (5.4.52)$$

To establish this result we introduce the modified periodogram $I_\varepsilon^*(\omega)$ of $\varepsilon_1, \ldots, \varepsilon_N$ (defined in the same way as $I_X^*(\omega)$, but with ε_t replacing X_t), and then use the asymptotic relationship (see theorem (6.2.2), Chapter 6),

$$I_X^*(\omega) \sim 2\pi h(\omega) I_\varepsilon^*(\omega)/\sigma_\varepsilon^2.$$

The above result follows immediately on noting that

$$\frac{1}{N} \sum_{t=1}^{N} \varepsilon_t^2 = \int_{-\pi}^{\pi} I_\varepsilon^*(\omega) d\omega.$$

The spectral density function $h(\omega)$ in (5.4.42) is written in the algebraic form appropriate to the particular model under consideration, and hence involves the parameters of the model. Thus, if we are fitting, e.g., an ARMA model, we may obtain approximate maximum likelihood estimates of the parameters $(a_1, \ldots, a_k, b_1, \ldots, b_l)$ by finding that rational function of the form (4.12.63) which best "approximates" $I_X^*(\omega)$ in the sense of minimizing the integral (5.4.52).

Numerical Example 4: 500 observations generated from the ARMA(2, 2) model

$$X_t + a_1 X_{t-1} + a_2 X_{t-2} = \varepsilon_t + b_1 \varepsilon_{t-1} + b_2 \varepsilon_{t-2},$$

with $\varepsilon_t \sim N(0, 1)$, $a_1 = 1 \cdot 4$, $a_2 = 0 \cdot 5$, $b_1 = -0 \cdot 2$, $b_2 = -0 \cdot 1$, are shown in Graph 3.5. Based on these data (see the Appendix), the (conditional) maximum likelihood estimates of the parameters (i.e. those values minimizing (5.4.46)) were

$$\hat{a}_1 = 1 \cdot 5236, \qquad \hat{a}_2 = 0 \cdot 62139, \qquad \hat{b}_1 = -0 \cdot 1189,$$

$$\hat{b}_2 = -0 \cdot 20355, \qquad \hat{\sigma}_\varepsilon^2 = 0 \cdot 97334.$$

5.4.4 Confidence Intervals for the Parameters

Box and Jenkins (1970) have developed a general method for obtaining simultaneous confidence intervals for the parameters AR, MA, and ARMA models. The basic results can be derived either from linear hypothesis theory or maximum likelihood theory, but there are subtle differences between the two approaches and it may be instructive, therefore, to consider them separately.

I. *Linear hypothesis: least squares approach*

We treat first the estimation of parameters in AR models. Here, the model is linear (being essentially a classical regression model) and we can appeal to the general theory of linear hypotheses. Consider a general linear hypothesis of the form (5.2.41). In vector form this may be written as,

$$H : y = X\theta + \varepsilon, \tag{5.4.53}$$

where

$$y' = (y_1, \ldots, y_N), \qquad \theta' = (\theta_1, \ldots, \theta_q), \qquad \varepsilon' = (\varepsilon_1, \ldots, \varepsilon_N),$$

and

$$X = \begin{pmatrix} x_{11} & x_{12} & \cdots & x_{1q} \\ x_{21} & x_{22} & \cdots & x_{2q} \\ \vdots & \vdots & & \vdots \\ x_{N1} & x_{N2} & \cdots & x_{Nq} \end{pmatrix} = (x_1, x_2, \ldots, x_q),$$

say. We assume that ε has the multivariate normal distribution $N(0, \sigma_\varepsilon^2 I)$. (We now denote the number of observations by N rather than n to make the notation consistent with that of the preceding sections.)

Under the general model H, θ is a vector of q unknown parameters which belongs to some parameter space Ω, (usually a subset of R^q). From the geometrical point of view, the least squares estimate, $\hat{\theta}$, of θ may be identified by the property that $X\hat{\theta}$ is the projection of y on the subspace spanned by $x_1, x_2, \ldots, x_q$, and is given by the normal equations (5.2.48), viz.

$$X'y = (X'X)\hat{\theta} \qquad (5.4.54)$$

see Fig. 5.4. (Note that the matrix (5.4.54) is the transpose of the matrix (5.2.47).)

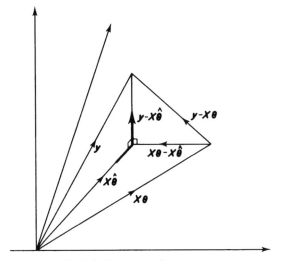

Fig. 5.4. Geometry of least squares.

Let ω be that single point in Ω which corresponds to the "true" value of θ, and consider a test of the null-hypothesis,

$$H_0: \theta \varepsilon \omega,$$

against the general model H. Let

$$Q_1 = \min_{\theta \in \Omega} \sum_{i=1}^{N} (y_i - E(y_i))^2 = \|y - X\hat{\theta}\|^2 \qquad (5.4.55)$$

and

$$Q_0 = \min_{\theta \in \omega} \sum_{i=1}^{N} (y_i - E(y_i))^2 = \|y - X\theta\|^2 \qquad (5.4.56)$$

(since ω admits only one value of the parameters, namely $\boldsymbol{\theta}$). By standard linear hypothesis theory (see Scheffe (1959)), it follows that, under H_0, Q_1 and $(Q_0 - Q_1)$ are independent, Q_1 is distributed as $\sigma_\varepsilon^2 \chi_{N-q}^2$, $(Q_0 - Q_1)$ as $\sigma_\varepsilon^2 \chi_q^2$, and hence

$$\frac{(Q_0 - Q_1)/q}{Q_1/(N-q)} = F_{q,N-q}, \tag{5.4.57}$$

where $F_{q,N-q}$ denotes, as usual, the F-distribution with q, $N-q$, degrees of freedom (note that H involves q unknown parameters while H_0 involves zero unknown parameters).

Now write for each $\boldsymbol{\theta}$,

$$Q(\boldsymbol{\theta}) = \sum_{i=1}^{N} \left\{ y_i - \sum_{j=1}^{q} x_{ij}\theta_j \right\}^2 = \|\boldsymbol{y} - \boldsymbol{X\theta}\|^2, \tag{5:4.58}$$

$Q(\boldsymbol{\theta})$ represents the sum of squares of the deviations of the y_i from their theoretical means, and we have,

$$Q_1 = Q(\hat{\boldsymbol{\theta}}), \qquad Q_0 = Q(\boldsymbol{\theta}).$$

Hence, (5.4.57) gives,

$$\frac{\{Q(\boldsymbol{\theta}) - Q(\hat{\boldsymbol{\theta}})\}}{Q(\hat{\boldsymbol{\theta}})} = \left(\frac{q}{N-q}\right) F_{q,N-q}. \tag{5.4.59}$$

Now let $F_{q,N-q}(\alpha)$ denote the upper $\alpha\%$ point of the $F_{q,N-q}$ distribution. Then a $(100-\alpha)\%$ *confidence region* for $\boldsymbol{\theta}$ may be obtained from the inequality,

$$Q(\boldsymbol{\theta}) \le Q(\hat{\boldsymbol{\theta}}) \left\{ 1 + \frac{q}{N-q} \cdot F_{q,N-q}(\alpha) \right\}. \tag{5.4.60}$$

Alternative expression

An alternative expression for the confidence region may be obtained as follows. We write,

$$\boldsymbol{y} - \boldsymbol{X\theta} = (\boldsymbol{y} - \boldsymbol{X\hat{\theta}}) + (\boldsymbol{X\hat{\theta}} - \boldsymbol{X\theta}),$$

and it is clear from the geometry of Fig. 5.6 that $(\boldsymbol{y} - \boldsymbol{X\hat{\theta}})$ and $(\boldsymbol{X\hat{\theta}} - \boldsymbol{X\theta})$ are orthogonal. Hence,

$$\|\boldsymbol{y} - \boldsymbol{X\theta}\|^2 = \|\boldsymbol{y} - \boldsymbol{X\hat{\theta}}\|^2 + \|\boldsymbol{X\hat{\theta}} - \boldsymbol{X\theta}\|^2,$$

or

$$Q_0 = Q_1 + \|\boldsymbol{X\hat{\theta}} - \boldsymbol{X\theta}\|^2,$$

so that

$$Q(\boldsymbol{\theta}) - Q(\hat{\boldsymbol{\theta}}) = Q_0 - Q_1 = \|X\hat{\boldsymbol{\theta}} - X\boldsymbol{\theta}\|^2 = [X(\hat{\boldsymbol{\theta}} - \boldsymbol{\theta})]'[X(\hat{\boldsymbol{\theta}} - \boldsymbol{\theta})]$$

$$= (\hat{\boldsymbol{\theta}} - \boldsymbol{\theta})'X'X(\hat{\boldsymbol{\theta}} - \boldsymbol{\theta}). \tag{5.4.61}$$

Substituting the above expression for $\{Q(\boldsymbol{\theta}) - Q(\hat{\boldsymbol{\theta}})\}$ in (5.4.59) we may now write the confidence region as,

$$(\hat{\boldsymbol{\theta}} - \boldsymbol{\theta})'X'X(\hat{\boldsymbol{\theta}} - \boldsymbol{\theta}) \leq Q(\hat{\boldsymbol{\theta}})\left\{\frac{q}{N-q} \cdot F_{q,N-q}(\alpha)\right\}. \tag{5.4.62}$$

Noting that $\hat{\sigma}_\varepsilon^2 = Q(\hat{\boldsymbol{\theta}})/(N-q)$ is the usual estimate of the residual variance, σ_ε^2, we may rewrite the above region as,

$$(\hat{\boldsymbol{\theta}} - \boldsymbol{\theta})'X'X(\hat{\boldsymbol{\theta}} - \boldsymbol{\theta}) \leq q\hat{\sigma}_\varepsilon^2 F_{q,N-q}(\alpha). \tag{5.4.63}$$

The confidence region for $\boldsymbol{\theta}$ may now be computed from either (5.4.60) or (5.4.63). Box and Jenkins (1970) suggest that (5.4.60) is rather easier to use since all one has to do is to plot the contours of $Q(\boldsymbol{\theta})$ which satisfy (5.4.60), and this immediately gives the region of points, $\boldsymbol{\theta}$, which constitute the $(100 - \alpha)\%$ confidence region. In fact, (5.4.63), when plotted in the $\boldsymbol{\theta}$-space, also gives the same type of result, but in order to implement (5.4.63) one must, of course, be able to compute $(X'X)$ explicitly. However, the important point to be noted is that although an inequality of the form (5.4.60) may be shown to be valid even when the model is non-linear (but the residuals are normal), (5.4.63) is valid only for linear models. Thus, while (5.4.63) would be valid (asymptotically) for the linear AR models, it would not apply to the MA or ARMA models, these being non-linear. A confidence region of the same form as (5.4.60) can be derived for *all* the standard models by appealing to maximum likelihood theory, as we now show.

II. *General models (linear and non-linear); likelihood approach*

For all the standard Gaussian time series models we may express the approximate log-likelihood function in the form

$$L(\boldsymbol{\theta}) = \text{const} - \frac{1}{2\sigma_\varepsilon^2} Q(\boldsymbol{\theta}), \tag{5.4.64}$$

where $\boldsymbol{\theta}$ denotes the set of unknown parameters (e.g. for the ARMA models, $\boldsymbol{\theta} = (\mu, a_1, \ldots, a_k, b_1, \ldots, b_l)$, $q = l + k + 1$), and $Q(\boldsymbol{\theta})$ the sum of squares function defined by (5.4.58), is now given by,

$$Q(\boldsymbol{\theta}) = \sum_t \varepsilon_t^2, \tag{5.4.65}$$

where the limits of the summation over t depend on the initial starting values appropriate to the model; see the previous sections.

Let $\hat{\boldsymbol{\theta}}$ denote the maximum likelihood (or least squares) estimate of $\boldsymbol{\theta}$. Then we may write,

$$L(\boldsymbol{\theta}) = L\{\hat{\boldsymbol{\theta}} + (\boldsymbol{\theta} - \hat{\boldsymbol{\theta}})\}$$

$$= L(\hat{\boldsymbol{\theta}}) + \sum_{i=1}^{q} (\theta_i - \hat{\theta}_i) \frac{\partial L(\hat{\theta}_i)}{\partial \theta_i} + \frac{1}{2} \sum_{i=1}^{q} \sum_{j=1}^{q} (\theta_i - \hat{\theta}_i)(\theta_j - \hat{\theta}_j) \frac{\partial^2 L(\hat{\boldsymbol{\theta}})}{\partial \theta_i \, \partial \theta_j} + \dots$$

$$\tag{5.4.66}$$

Assuming $L(\boldsymbol{\theta})$ to be locally quadratic (so that we may neglect derivatives of $L(\boldsymbol{\theta})$ higher than the second), and noting that $\partial L(\hat{\boldsymbol{\theta}})/\partial \theta_i = 0$, all i (since $\hat{\boldsymbol{\theta}}$ is the maximum likelihood estimate of $\boldsymbol{\theta}$) and writing $L_{ij} = \partial^2 L(\hat{\boldsymbol{\theta}})/\partial \theta_i \, \partial \theta_j$, we have,

$$L(\boldsymbol{\theta}) \doteq L(\hat{\boldsymbol{\theta}}) + \frac{1}{2} \sum_{i=1}^{q} \sum_{j=1}^{q} (\theta_i - \hat{\theta}_i)(\theta_j - \hat{\theta}_j) L_{ij}. \tag{5.4.67}$$

Now it may be shown that when N is large we may approximate by replacing L_{ij} by $E[\partial^2 L(\boldsymbol{\theta})/\partial \theta_i \, \partial \theta_j]$, in which case $-\{L_{ij}\}^{-1}$ is approximately the variance–covariance matrix of $\{\hat{\theta}_i\}$. (The notation $\{L_{ij}\}^{-1}$ means the inverse of the matrix whose (i, j)th element is L_{ij}.) With this approximation, the quadratic form $\{-\sum_i \sum_j L_{ij}(\theta_i - \hat{\theta}_i)(\theta_j - \hat{\theta}_j)\}$ has a χ_q^2 distribution.

Using (5.4.64), (5.4.67), we now have,

$$\{Q(\boldsymbol{\theta}) - Q(\hat{\boldsymbol{\theta}})\} \sim 2\sigma_\varepsilon^2 \cdot \tfrac{1}{2}\chi_q^2 = \sigma_\varepsilon^2 \chi_q^2. \tag{5.4.68}$$

When σ_ε^2 is unknown, we estimate it, as previously, by

$$\hat{\sigma}_\varepsilon^2 = \frac{1}{N-q} \cdot Q(\hat{\boldsymbol{\theta}}). \tag{5.4.69}$$

Then *if* the model is linear,

$$(N-q)\hat{\sigma}_\varepsilon^2 = \sigma_\varepsilon^2 \chi_{N-q}^2, \tag{5.4.70}$$

and is independent of $\{Q(\boldsymbol{\theta}) - Q(\hat{\boldsymbol{\theta}})\}$. Hence in this case,

$$\frac{Q(\boldsymbol{\theta}) - Q(\hat{\boldsymbol{\theta}})}{(N-q)\hat{\sigma}_\varepsilon^2} = \frac{q}{N-q} \cdot F_{q,N-q}, \tag{5.4.71}$$

and a $(100 - \alpha)\%$ confidence region for $\boldsymbol{\theta}$ is given by,

$$\frac{Q(\boldsymbol{\theta}) - Q(\hat{\boldsymbol{\theta}})}{Q(\hat{\boldsymbol{\theta}})} \leq \frac{q}{N-q} \cdot F_{q,N-q}(\alpha), \tag{5.4.72}$$

or

$$Q(\boldsymbol{\theta}) \leqslant Q(\hat{\boldsymbol{\theta}})\left\{1 + \frac{q}{N-q}F_{q,N-q}(\alpha)\right\}, \qquad (5.4.73)$$

in agreement with (5.4.60).

Box and Jenkins (1970) suggest that for the general case we may, for large N, treat $\hat{\sigma}_\varepsilon^2$ as a constant, and approximate to the region (5.4.73) by

$$Q(\boldsymbol{\theta}) \leqslant Q(\hat{\boldsymbol{\theta}})\left\{1 + \frac{\chi_q^2(\alpha)}{N}\right\}. \qquad (5.4.74)$$

Alternative expression

If $L(\boldsymbol{\theta})$ is locally quadratic then so is $Q(\boldsymbol{\theta})$. Hence we may write,

$$Q(\boldsymbol{\theta}) \doteq Q(\hat{\boldsymbol{\theta}}) + \tfrac{1}{2}\sum_{i=1}^{q}\sum_{j=1}^{q}(\theta_i - \hat{\theta}_i)(\theta_j - \hat{\theta}_j)Q_{ij}, \qquad (5.4.74a)$$

where

$$Q_{ij} = \frac{\partial^2 Q(\hat{\boldsymbol{\theta}})}{\partial\theta_i\,\partial\theta_j} = -2\sigma_\varepsilon^2 L_{ij}.$$

Hence (5.4.71) may be rewritten,

$$\sum_{i=1}^{q}\sum_{j=1}^{q}(\theta_i - \hat{\theta}_i)(\theta_j - \hat{\theta}_j)Q_{ij} = q\hat{\sigma}_\varepsilon^2 F_{q,N-q}. \qquad (5.4.75)$$

A $(100-\alpha)\%$ confidence region for $\boldsymbol{\theta}$ may thus be derived from,

$$\sum_{i=1}^{q}\sum_{j=1}^{q}(\theta_i - \hat{\theta}_i)(\theta_j - \hat{\theta}_j)Q_{ij} \leqslant q\hat{\sigma}_\varepsilon^2 F_{q,N-q}(\alpha). \qquad (5.4.76)$$

In practice one could approximate to Q_{ij} by using the second differences of $Q(\hat{\boldsymbol{\theta}})$, i.e. one could estimate the Q_{ij} numerically. Alternatively, in some cases one can evaluate Q_{ij} analytically from the expression for the likelihood function.

Applications to AR, MA, and ARMA models

Equation (5.4.74) (or (5.4.73)) may be applied directly to the AR, MA, and ARMA models since in each case the "sum of squares" function $Q(\boldsymbol{\theta})$ is computed directly from the residuals ε_t and the parameter values $\boldsymbol{\theta}$. However, in the AR case (but only in the AR case) one could use (5.4.63) or (5.4.76) as an alternative. The use of (5.4.63) requires the calculation of the matrix $(X'X)$. For an AR(k) model with N observations $X_1, \ldots, X_N$,

$$
X = \begin{pmatrix}
X_k & X_{k-1} & \cdots & X_1 \\
X_{k+1} & X_k & \cdots & X_2 \\
\vdots & & & \\
X_{N-1} & X_{N-2} & \cdots & X_{N-k}
\end{pmatrix}
$$

and, as an approximation (for large N) we may write $X'X \sim N . \hat{R}_k$, where $\hat{R}_k$ is the matrix defined by (5.4.20).

5.4.5 Determining the Order of the Model

In the preceding sections we discussed the problem of parameter estimation on the assumption that the orders of the autoregressive and moving-average operators were known *a priori*. In practice, the orders of these operators are almost invariably unknown and constitute, in effect, additional unknown parameters for which suitable values have to be inferred from the data. In the case of AR model fitting we may follow the analogy with multiple regression analysis and approach the problem via hypothesis testing, i.e. we could consider a test of the null hypothesis that the model is AR(k) against the alternative hypothesis that it is AR($k+1$), and continue increasing the value of k until the test gave a non significant result. A pioneering study of tests of this type was given by Whittle (1951, 1952b). However, as observed by Akaike (1974c), the problem of model fitting in the context of time series analysis is perhaps best viewed as a "multiple decision procedure" rather than as a hypothesis testing situation. To pursue the analogy with regression analysis, one may prefer a more exploratory technique such as Mallows' "C_p plots" (Mallows (1973)) rather than the more classical "analysis of variance" method which is tied down to fixed significance levels. We now describe a number of exploratory techniques of this type which have been suggested for determining the order (or orders) of standard time series models.

I. *Residual variance plots*

Suppose we are fitting an AR(k) model of the form (5.4.2), and can assume that the true model is autoregressive of finite (but unknown) order. If we choose a value of k smaller than the true order then we may expect that the (unbiased) estimated residual variance, $\hat{\sigma}_\varepsilon^2$, as given by (5.4.22) will be larger than the true residual variance, σ_ε^2, since the additional terms omitted from the model would "explain" a further part of the variance of X_t. On the other hand, once the value of k reaches the true order any further increase in

k will not significantly reduce the residual variance. Hence, if we fit a sequence of models of increasing order, evaluate $\hat{\sigma}_\varepsilon^2$ in each case, and then plot $\hat{\sigma}_\varepsilon^2$ against k, we may expect the graph to decrease at first and then "level out" at the point where k approaches the true order. This method of order determination for AR models, based on residual variance plots, was proposed by Whittle (1963), p. 35. Jenkins and Watts (1968) also use this method, and suggest that the same technique can be applied also to the determination of the order of MA models and ARMA models. Thus, in fitting an MA model, we may again fit a sequence of models of increasing order l and plot the estimated residual variance $\hat{\sigma}_\varepsilon^2$ (now given by (5.4.42a)) against l. For ARMA models the estimated residual variance $\hat{\sigma}_\varepsilon^2$ (given by (5.4.47)) is a function of both k and l (the orders of the AR and MA operators), but by plotting $\hat{\sigma}_\varepsilon^2$ on a grid of values of k and l it is possible to examine its behaviour over a set of plausible values—although, of course, the computation involved would normally be substantial.

II.   *Partial autocorrelation function*

Box and Jenkins (1970) suggest that in the case of AR models additional information on the order can be gained by examining the "*partial autocorrelation function*". Let $\pi_m = (-a_m)$, where a_m is the last coefficient in an AR(m) model. Then π_m is the partial correlation coefficient between X_t and X_{t+m} (holding $X_{t+1}, \ldots, X_{t+m-1}$ "fixed"), and a plot of π_m against m is called the partial autocorrelation function. If the true model is AR(k) then clearly $\pi_m = 0$, $m > k$, and the partial autocorrelation function vanishes after a finite number of terms. If we now estimate π_m by $\hat{\pi}_m = -\hat{a}_m$, where $\hat{a}_m$ is the estimate of the last coefficient in a fitted AR(m) model, then the graph of $\hat{\pi}_m$ against m should indicate the order of the true model. When the a_m are estimated from the Yule–Walker equations, a recursive method of calculating $\hat{\pi}_m$ has been given by Durbin (1960)—see also Whittle (1963) p. 37 and Anderson (1971), p. 188. Denoting, more explicitly, the jth coefficient of an AR model of order m by $a_{j,m} (j = 1, \ldots, m)$, Durbin's formula is

$$\hat{a}_{m+1,j} = \hat{a}_{m,j} - \hat{a}_{m+1,m+1}\hat{a}_{m,m-j+1}, \qquad j = 1, \ldots, m,$$

$$\hat{a}_{m+1,m+1} = \frac{\hat{\rho}(m+1) - \sum\limits_{j=1}^{m} \hat{a}_{mj}\hat{\rho}(m+1-j)}{1 - \sum\limits_{j=1}^{m} \hat{a}_{mj}\hat{\rho}(j)}.$$

It may be shown (Quenouille (1949)) that for an AR(k) process the $\{\pi_m\}$, $m > k$, are approximately independently distributed, each with mean zero and variance approximately $(1/N)$, N being the number of observations on which the model is fitted. Also, for large N the $\hat{\pi}_m$ may be taken to be approximately normally distributed, and hence a rough procedure for testing $[\pi_m = 0]$ is to examine whether $\hat{\pi}_m$ lies between $\pm 2\sqrt{1/N}$. (See also Hamilton and Watts (1978)).

III. *Akaike's FPE, AIC, and BIC criteria*

A more refined version of the "residual variance plot" described above was developed by Akaike (1969a) for AR order determination. In Akaike's method one again starts by fitting AR models of increasing order, k, and for each k one computes the expression,

$$\text{FPE}(k) = \frac{n+k}{n-k} \cdot \hat{\sigma}_\varepsilon^2, \tag{5.4.77}$$

where n is the number of observations to which the model is fitted, and $\hat{\sigma}_\varepsilon^2$, the maximum likelihood estimate of the residual variance, is given by

$$\hat{\sigma}_\varepsilon^2 = \{\hat{R}(0) + \hat{a}_1\hat{R}(1) + \ldots + \hat{a}_k\hat{R}(k)\}, \tag{5.4.78}$$

cf. (5.4.21a) and (5.4.22). The expression FPE stands for "final prediction error", and the motivation for this terminology arises from the following property of the AR model. If one observes the process up to time $(t-1)$ and wishes to predict the next value, X_t, by a linear combination of past values, then the "best" predictor (in the mean square error sense) is given by (with $\mu = 0$),

$$\hat{X}_t = -a_1 X_{t-1} \ldots - a_k X_{t-k}, \tag{5.4.79}$$

and the "one-step" prediction error is $X_t - \hat{X}_t = \varepsilon_t$, with mean square error,

$$E[X_t - \hat{X}_t]^2 = \sigma_\varepsilon^2. \tag{5.4.80}$$

(These results follow very simply from the general theory of prediction to be developed in Chapter 10, but the result (5.4.79) for the "best" predictor of X_t is fairly obvious from an intuitive point of view). Hence, the residual variance, σ_ε^2, may be interpreted also as the one-step prediction mean square error. However, the predictor (5.4.79) can be used only when the coefficients $a_1, \ldots, a_k$ are known exactly; if these coefficients are replaced by their least squares estimates $\hat{a}_1, \ldots, \hat{a}_k$ (obtained from an independent set of observations) then the mean square error of the *estimated* one-step predictor is no longer σ_ε^2, but Akaike showed that it may be estimated by the FPE(k) quantity given by (5.4.77). If we now plot FPE(k) against k the graph will, in general, show a definite minimum at a particular value of k, and the basis of Akaike's method is to use the value of k at which the FPE attains its minimum as the appropriate order of the model. Using this criterion, the fitted AR model has the property that it leads to the *estimated* one step predictor with the smallest mean square error.

More recently, Akaike (1971, 1974c) introduced a new expression called AIC (Akaike's information criterion). This very general criterion, which is based on information theoretic concepts, can be used for statistical model identification in a wide range of situations and is not restricted to the time series context. When a model involving q independently adjusted parameters is fitted to data, the AIC is defined by,

$$\text{AIC}(q) = (-2) \log_e[\text{maximized likelihood}] + 2q. \qquad (5.4.81)$$

We have seen that for all the standard models (AR, MA, and ARMA) the log-likelihood function for n (effective) observations is (apart from an additive constant),

$$L = -\frac{n}{2} \log \sigma_\varepsilon^2 - \frac{1}{2\sigma_\varepsilon^2} Q(\boldsymbol{\theta}), \qquad (5.4.82)$$

where Q is the sum of squares quantity previously defined (as appropriate for each class of models) and $\boldsymbol{\theta}$ denotes the set of coefficients in the model. When maximized w.r. to both $\boldsymbol{\theta}$ and σ_ε^2, we obtain, (as previously noted),

$$\hat{\sigma}_\varepsilon^2 = \frac{1}{n} Q(\hat{\boldsymbol{\theta}}), \qquad (5.4.83)$$

and the maximum of L is

$$\hat{L} = -\frac{n}{2} \log \hat{\sigma}_\varepsilon^2 - \frac{n}{2}. \qquad (5.4.84)$$

Ignoring the second term (which is a constant independent of both $\boldsymbol{\theta}$ and k), the AIC criterion (5.4.81) now becomes equivalent to

$$\text{AIC}(q) = n \log \hat{\sigma}_\varepsilon^2 + 2q \qquad (5.4.85)$$

If we plot $\text{AIC}(q)$ against q the graph will again, in general, show a definite minimum value, and the appropriate order of the model is determined by that value of q at which $\text{AIC}(q)$ attains its minimum value. The minimum value of AIC is called MAIC (minimum AIC). [In practice, for AR models $\hat{\sigma}_\varepsilon^2$ may be computed from (5.4.78)].

Although the FPE criterion was originally introduced for AR order determination only, the AIC criterion is quite general and can be applied to all the standard models (including MA and ARMA schemes). Thus, for e.g., the ARMA model (5.4.43), $\hat{\sigma}_\varepsilon^2$ is given by (5.4.83) with $\hat{\boldsymbol{\theta}} = (\hat{\mu}, \hat{a}_1, \dots, \hat{a}_k, \hat{b}_1, \dots, \hat{b}_l)$, and, with the estimation procedure described in Section 5.4.3, $Q(\boldsymbol{\theta})$ is given by (5.4.46), $q = (k + l + 1)$, and $n = (N - k)$. The orders of the

AR and MA operators are now determined by computing the AIC criterion over a selected grid of values of k and l (subject to some specified upper bounds for both these quantities) and finding those values of k and l at which AIC attains its minimum.

The AIC criterion has largely superseded the FPE method, and is now generally accepted as one of the most reliable methods for order determination. It has been widely applied to both simulated and real data (see Akaike (1974c) for references) with very successful results, and has become a firmly established tool in time series model fitting. The AIC criterion may be used also in connection with the problem of *discriminating* between different classes of models. Suppose, for example, that we wish to fit either a pure AR or pure MA model but are not sure which class to use. We may, of course, start by fitting an ARMA(k, l) model and use the AIC criterion to determine whether or not the subclass with $l = 0$ (the AR models) give a lower AIC value than the subclass with $k = 0$ (the MA models).

Relationships between the various criteria

It is easily seen that, for large N, the FPE and AIC are asymptotically equivalent procedures. For, from (5.4.77) we have,

$$\log[\text{FPE}(q)] = \log\left[\frac{(1 + q/n)}{(1 - q/n)} \cdot \hat{\sigma}_\varepsilon^2\right]$$

$$= \log(1 + q/n) - \log(1 - q/n) + \log \hat{\sigma}_\varepsilon^2$$

$$\doteq \log \hat{\sigma}_\varepsilon^2 + \frac{2q}{n}, \qquad \text{for large } n. \tag{5.4.86}$$

Hence, from (5.4.85), we have,

$$\text{AIC}(q) \doteq n \log[\text{FPE}(q)], \tag{5.4.87}$$

and asymptotically, the two criterion are equivalent as far as the determination of q is concerned.

Note, however, that the residual variance plot procedure described in § I is not equivalent to either the FPE or AIC methods. For, if we now denote the estimated residual variance $\hat{\sigma}_\varepsilon^2$ more fully by $\hat{\sigma}_\varepsilon^2(q)$, q being the number of parameters in the model, then for the AR case we have (from (5.4.21)),

$$\hat{\sigma}_\varepsilon^2(q) = \frac{1}{N - 2k - 1} Q(\hat{\mu}, \hat{a}_1, \ldots, \hat{a}_k) = \frac{1}{n - q} Q(\hat{\mu}, \hat{a}_1, \ldots, \hat{a}_k), \tag{5.4.88}$$

where $n = (N - k)$ is the (effective) number of observations used in fitting, and $q = k + 1$ is the number of parameters involved. Hence, for (5.4.83) we may write,

$$\hat{\sigma}_\varepsilon^2(q) = \frac{n}{n-q} \hat{\sigma}_\varepsilon^2,$$

where $\hat{\sigma}_\varepsilon^2$ is the maximum likelihood estimate of σ_ε^2, so that,

$$\log[\hat{\sigma}_\varepsilon^2(q)] = \log \hat{\sigma}_\varepsilon^2 - \log(1 - q/n)$$
$$\doteq \log \hat{\sigma}_\varepsilon^2 + q/n, \quad \text{for large } n. \quad (5.4.89)$$

Comparing (5.4.85) with (5.4.89) we see that $n \log[\hat{\sigma}_\varepsilon^2(q)]$ differs from the AIC criterion by a term which is approximately (q). This term is, in fact, quite important since whereas $\log[\hat{\sigma}_\varepsilon^2(q)]$ will, in general, merely "level out" after a certain point, the AIC graph, as previously noted, exhibits a *definite minimum value*. (As q increases the additional term, q, tends to "pull up" the AIC value linearly.) An incorrect application of the FPE or AIC method in which the additional term was omitted did lead to spurious results—see Akaike (1974c) for further discussion of this point.

It should be noted further that when fitting AR models, Akaike uses as starting values,

$$X_0 = X_{-1} = \ldots X_{-(k-1)} = 0,$$

so that here $n = N$, i.e. the effective number of observations to which the model is fitted is the same as the number of data points observed. In contrast, the method described in §I for fitting AR models uses (5.4.6) as starting values, and, as already noted, for this method, $n = (N - k)$.

Recently, Akaike (1978, 1979) has developed a new order determination criterion, called BIC, which is derived from a Bayesian modification of the AIC criterion. For a model involving q parameters fitted to n observations, the BIC criterion takes the form

$$\text{BIC}(q) = n \log \hat{\sigma}_\varepsilon^2 - (n-q) \log\left(1 - \frac{q}{n}\right) + q \log n + q \log\left\{q^{-1}\left(\frac{\hat{\sigma}_X^2}{\hat{\sigma}_\varepsilon^2} - 1\right)\right\},$$

where, as above, $\hat{\sigma}_\varepsilon^2$ is the maximum likelihood estimate of the residual variance based on a q-parameter model, and $\hat{\sigma}_X^2$ is the raw sample variance of the observations. When q is small relative to n we may use the approximation, $\{-(n-q) \log[1 - (q/n)]\} \doteq q$, so that

$$\text{BIC}(q) \doteq \text{AIC}(q) + q(\log n - 1) + q \log\left\{q^{-1}\left(\frac{\hat{\sigma}_X^2}{\hat{\sigma}_\varepsilon^2} - 1\right)\right\}.$$

Thus, the basic difference between BIC and AIC is that the second term in (5.4.85), $2q$, is replaced by the term $\{q + q \log n\}$. This has the effect of increasing the weight attached to the "penalty term" (which takes account of the number of parameters in the model), and consequently the minimization of BIC leads, in general, to lower model orders than those obtained by minimizing AIC. Shibata (1976) has shown that when the true model is AR(k_0), the estimate $\hat{k}$ derived from the AIC criterion is not a consistent estimate of k_0, but rather tends to overestimate k_0. On the other hand, the estimate obtained from the BIC criterion may well underestimate the true order.

Schwarz (1978) suggested the order criterion

$$S(q) = n \log \hat{\sigma}_\varepsilon^2 + q \log n,$$

which is similar to Akaike's BIC in terms of its dependence on $\log n$. In fact, if we use the above approximate expression for BIC, we may write

$$\text{BIC}(q) \doteq S(q) + q\left[\log\left\{q^{-1}\left(\frac{\hat{\sigma}_x^2}{\hat{\sigma}_\varepsilon^2} - 1\right)\right\} - 1\right] = S(q) + \text{O}(q),$$

where $\text{O}(q)$ denotes a term which is functionally independent of n. (Further discussion of order criteria is given by Kashyap (1977) and Stone (1979). Hannan's criterion, based on a "penalty term" $\text{O}(\log \log n)$, is discussed in Chapter 9, Section 9.4, in the context of multivariate model fitting, and Parzen's CAT criterion is discussed in Chapter 7, Section 7.8.)

IV. *Visual inspection of the autocorrelation function*

It is sometimes possible to gain insight into the order of an AR or MA model by examining the graphical form of the sample autocorrelation function. We saw in Chapter 3 that different orders of AR and MA models give rise to autocorrelation functions with different characteristic shapes. Typically, the AR models produce exponentially decaying autocorrelation functions whereas those for MA models drop to zero exactly after a finite number of terms. By inspecting the behaviour of the sample autocorrelation functions one can thus obtain some indication as to which class of models might be appropriate, and also some guidance as to a suitable order for the model. Box and Jenkins (1970) have developed this technique in great detail and have given extensive illustrations of its application to real data (together with a corresponding analysis of the partial autocorrelation function). However, its effective use depends on the experience and knowledge of the user, and, moreover, the behaviour of the sample autocorrelation function does not always accurately reflect the behaviour of the true autocorrelation function; see the remarks in Section 5.3.4. This approach, therefore, requires a good deal of care and caution in its application, and although it may be very useful as a preliminary guide, it would, in general, be advisable

to supplement it by one of the more quantitative techniques described in the
preceding paragraph.

V. *Use of the "inverse autocorrelation function"*

The "*inverse autocorrelation function*" was introduced by Cleveland
(1972) as a normalized form of the Fourier transform of the reciprocal of the
spectral density function. Thus, we consider a discrete parameter process
$\{X_t\}$ whose spectral density function $h(\omega)$ exists for all ω and is such that
$h(\omega) > 0$, all ω, and $\{h(\omega)\}^{-1}$ admits a Fourier series expansion. The "*inverse
autocovariance*" of lag r, Ri(r), is then defined as the rth Fourier coefficient
of $(1/4\pi^2)\{h(\omega)\}^{-1}$, i.e. we write

$$\frac{1}{4\pi^2}\{h(\omega)\}^{-1} = \frac{1}{2\pi}\sum_{r=-\infty}^{\infty} \mathrm{Ri}(r)\, e^{-i\omega r},$$

so that by the standard inversion formula,

$$\mathrm{Ri}(r) = \frac{1}{4\pi^2}\int_{-\pi}^{\pi} \{h(\omega)\}^{-1} e^{i\omega r}\, d\omega, \qquad r = 0, \pm 1, \pm 2, \ldots . \tag{5.4.90}$$

The "inverse autocorrelation function", $\rho\mathrm{i}(r)$, is now defined by

$$\rho\mathrm{i}(r) = \mathrm{Ri}(r)/\mathrm{Ri}(0), \qquad r = 0, \pm 1, \pm 2, \ldots .$$

The motivation for introducing the functions $\{\mathrm{Ri}(r)\}, \{\rho\mathrm{i}(r)\}$, is that, in a
sense, they possess the "dual" properties of R(r) and $\rho(r)$. For, as observed
by Cleveland (1972), if X_t is an ARMA(k, l) process then clearly Ri(r) has
exactly the same properties as the (ordinary) autocovariance function of the
ARMA(l, k) process obtained by interchanging the AR and MA operators.
(This property follows immediately from the form of the spectral density
function of an ARMA process given by (4.12.63).) In particular, if X_t
conforms to a pure AR(k) model then Ri(r) behaves exactly like the
autocovariance function of a pure MA(k) model, i.e. Ri(r) will be zero for all
lags greater than k. Conversely, if X_t conforms to a pure MA(l) model, then
Ri(r) behaves like the autocovariance function of a pure AR(l) model, and
decays gradually to zero.

In order to use the above properties as an aid to model identification, we
need some method of estimating the inverse covariances and correlations
from data. Cleveland (1972) suggests two possible approaches:

(a) First construct a standard non-parametric "window" estimate of
$h(\omega)$ (using the methods discussed in Chapter 6—see Section 6.2.3),
giving estimates $\hat{h}(\omega_1), \ldots, \hat{h}(\omega_M)$, say, at a discrete set of frequencies, $\omega_1, \ldots, \omega_M$. The inverse covariances can then be estimated by

approximating the integral (5.4.90) by a sum over the points $\omega_1, \ldots, \omega_M$, and replacing each $h(\omega_i)$ by $\hat{h}(\omega_i)$.

(b) Alternatively, we may fit a long order AR model to the observations, yielding, say, an $AR(p)$ model (p suitable large) with estimated coefficients $\hat{a}_1, \hat{a}_2, \ldots, \hat{a}_p$, and then estimate $h(\omega)$ by substituting these estimated coefficients in the theoretical expression for the spectral density function of an AR model, namely (4.12.59). (This is known as "autoregressive spectral estimation"—see Section 7.8.) If we adopt this alternative procedure it follows from the above discussion that the resulting estimates of the inverse covariances have the same form as the autocovariances of an $MA(p)$ model (with parameters $\hat{a}_1, \ldots, \hat{a}_p$), and hence, from (3.5.49), the estimates $\hat{R}i(r)$ take the form

$$\hat{R}i(r) = \begin{cases} \dfrac{1}{\hat{\sigma}_X^2} \sum_{j=0}^{p-r} \hat{a}_j \hat{a}_{j+r}, & r = 0, 1, \ldots, p-1, \\ 0, & r \geqslant p, \end{cases}$$

(with $\hat{a}_0 = 1$). Note that with this method $\hat{R}i(r)$ must necessarily be zero for $r \geqslant p$, and hence the need to choose p suitably large in order for $\hat{R}i(r)$ to reflect reasonably accurately the behaviour of the true inverse covariance function, $Ri(r)$.

The duality between the covariances and inverse covariances may be exploited still further by noting that for a pure MA model the inverse covariances must satisfy a set of linear equations of exactly the same form as the Yule–Walker equation (3.5.43) for a pure AR model. More generally, for an $ARMA(k, l)$ model of the form (5.4.43), the $Ri(r)$ must satisfy the dual version of the "higher order" Yule–Walker equations (3.5.55), i.e. we have (cf. Cleveland (1972), Parzen (1974)),

$$Ri(m) + b_1 Ri(m-1) + \ldots + b_l Ri(m-l) = 0, \qquad m > l.$$

Bhansali (1980) has investigated the asymptotic sampling properties of the estimated inverse covariances and correlations, and has derived conditions under which estimates produced by each of the two methods described above are consistent. Chatfield (1979) gives some practical illustrations of the use of inverse correlations, and, in particular, discusses the relative merits of partial and inverse correlations as an aid to model identification.

Numerical Examples: Using the AIC criterion (as given by (5.4.85)), the correct models for white noise, AR(1), AR(2) and MA(2) series were easily identified. Several different ARMA models were fitted to each series, but the minimum AIC value was clearly attained with the correct form of model. To illustrate the nature of the numerical results, we show below some of the AIC values for various alternative models fitted to (i) the AR(2) data, and (ii) the MA(2) data.

(i) AR(2) data:

Fitted model	AIC value
ARMA(1, 0)	322·3
ARMA(1, 1)	197·5
ARMA(2, 0)	−6·4
ARMA(2, 1)	−4·4
ARMA(3, 0)	−4·4
ARMA(3, 1)	−3·6
ARMA(4, 0)	−2·4

The minimum AIC value thus occurs with an ARMA(2, 0) (i.e. an AR(2)) model.

(ii) MA(2) data:

Fitted model	AIC value
ARMA(1, 0)	184·4
ARMA(0, 1)	1·3
ARMA(1, 1)	−6·4
ARMA(2, 0)	101·5
ARMA(0, 2)	−6·9
ARMA(3, 0)	67·2
ARMA(0, 3)	−5·0

Here, the minimum AIC value occurs with an ARMA(0, 2) (i.e. an MA(2)) model.

However, the AIC analysis of the ARMA(2, 2) series was less clear cut. The results of fitting various ARMA models to these data were as follows:

(iii) ARMA(2, 2) data:

Fitted model	AIC value
ARMA(1, 0)	259·3
ARMA(2, 0)	14·7
ARMA(2, 1)	−3·6
ARMA(2, 2)	−5·5
ARMA(3, 0)	3·4
ARMA(3, 1)	−5·8
ARMA(3, 2)	−3·8
ARMA(4, 0)	−4·2

The minimum AIC value is attained with an ARMA(3, 1) model, but it will be seen that the AIC values for the ARMA(3, 1) and ARMA(2, 2) models are very close. The coefficients of the fitted ARMA(2, 2) model are given in Numerical example 4, and the fitted ARMA(3, 1) model was

$$X_t + 0.96196\,X_{t-1} - 0.10891\,X_{t-2} - 0.24169\,X_{t-3} = \varepsilon_t - 0.68053\,\varepsilon_{t-1},$$

with $\hat{\sigma}_\varepsilon^2 = 0.97283$. Graphs 5.11 and 5.12 show the spectral density functions of the fitted ARMA(2, 2) and ARMA(3, 1) models respectively, each superimposed on the spectral density function of the true ARMA(2, 2) model. The spectra of the two fitted models are almost identical, and each provides a good fit to the spectrum of the original model. This feature may explain why: for this particular set of data, it is difficult to distinguish clearly between the two alternative models.

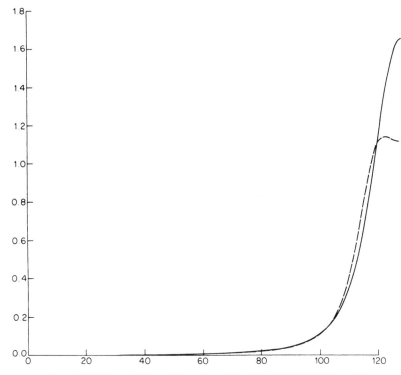

Graph 5.11. Spectral density function of the fitted ARMA(2, 2) model.

5.4.6 Continuous Parameter Models

Model aliasing

We have already noted that one way of dealing with continuous parameter processes is to "convert" them into discrete parameter processes by taking observations only at discrete intervals of time, and if we do this then we may, of course, fit any of the standard discrete parameter models which we have previously discussed. An interesting point arises here, namely that if we sample a continuous parameter AR(k) process at discrete time intervals Δt the resulting process does *not*, in general, conform to a discrete parameter AR(k) model. It has been asserted that *a sampled continuous parameter AR(k) process becomes a discrete parameter ARMA (k, $k-1$) process* (i.e. a mixed autoregressive/moving average model with the autoregressive operator of order k and the moving average operator of order ($k-1$)), but a general result of this form does not seem to have been firmly established. What is certainly clear is that the result holds for $k = 1$, i.e. that a sampled AR(1) process remains an AR(1) process. This is easily demonstrated by noting that the autocorrelation function for a continuous parameter AR(1) process is of the form,

$$\rho(\tau) = \exp(-\alpha|\tau|),$$

(where $\alpha > 0$), and the sampled process therefore has autocorrelation

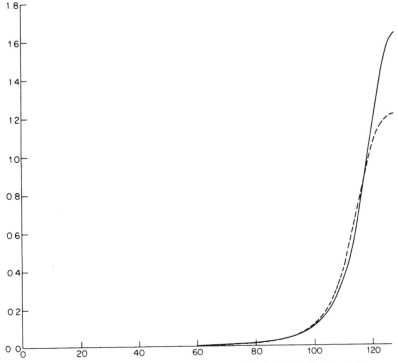

Graph 5.12. Spectral density function of the fitted ARMA(3, 1) model.

function,

$$\rho(r) = \exp(-\alpha \Delta t |r|), \qquad r = 0, \pm 1, \pm 2, \ldots$$

which can be written as, $\rho(r) = a^{|r|}$, where $a = \exp(-\alpha \Delta t)$, and satisfies $|a| < 1$ since $\alpha > 0$. Thus the sampled process has an autocorrelation function which is exactly of the form of a discrete parameter AR(1) model. It has also been established quite rigorously by Bartlett (1946) that if an AR(2) process is sampled the residuals obtained after fitting a discrete parameter AR(2) model are not uncorrelated but have non-zero first order correlations. Phadke and Wu (1969) have also studied this problem and show that when an AR(k) process is sampled the residuals, after fitting a discrete parameter AR(k) model *with specified coefficients*, have non-zero autocorrelations up to lag($k-1$) (see also Pandit and Wu (1975)). The results which have been obtained so far certainly show that when sampling a continuous parameter process the AR model structure is not preserved, and from a practical point of view, the use of an ARMA(k, $k-1$) model to represent a sampled AR(k) process may well be quite a reasonable procedure. This change in model form is, in a sense, a time domain analogue of the change in the form of the spectrum which occurs in discrete time sampling, usually termed the "aliasing effect" (see Section 4.8.3). We may therefore refer to the above feature as *"model aliasing"*.

Direct treatment of continuous parameter models

The basic estimation methods which we have discussed for discrete parameter models can, in principle, be extended to the continuous parameter case, enabling the coefficients of such models to be estimated directly from continuous records, if such are available. Although the formal treatment of, e.g., continuous AR models is fairly straightforward (as we shall indicate below), an attempt to formulate a general inference theory for continuous parameter model raises some rather deep conceptual problems. This general topic was discussed in an important paper by Grenander (1950), and has been studied also by Bartlett (1955). As soon as we start to consider sample observations which take the form of continuous records we meet the problem of specifying what we may describe loosely as the "joint probability distribution" of the observations. (This is an essential element in any attempt to develop a maximum likelihood approach to estimation of the model parameters.) The fact that any finite section of a continuous record contains, in effect, an infinite number of observations raises obvious difficulties, but in some cases it is possible to obtain an appropriate likelihood function by first considering a discrete set of observations taken at time intervals Δt, and then letting $\Delta t \to 0$ and suitably "stabilizing" the resulting function; see Bartlett (1955), p. 242. See also Durbin (1961), Phadke and Wu (1969) and Pham-Dinh (1977b).

However, the least squares estimation of the coefficients of AR models can be carried through fairly straightforwardly. Consider, for example, the continuous parameter zero-mean AR(2) model given by (3.7.39) (Section 3.7.4) namely,

$$\ddot{X}(t) + \alpha_1 \dot{X}(t) + \alpha_2 X(t) = \varepsilon(t). \qquad (5.4.91)$$

Given a continuous sample record of $X(t), 0 \le t \le T$, the least squares estimate of α_1 and α_2 are given by minimizing

$$Q(\alpha_1, \alpha_2) = \int_0^T \varepsilon^2(t)\, dt = \int_0^T \{\ddot{X}(t) + \alpha_1 \dot{X}(t) + \alpha_2 X(t)\}^2\, dt.$$

Differentiating $Q(\alpha_1, \alpha_2)$ w.r. to α_1, α_2 and setting these derivatives equal to zero gives

$$\int_0^T \dot{X}(t)\{\ddot{X}(t) + \hat{\alpha}_1 \dot{X}(t) + \hat{\alpha}_2 X(t)\}\, dt = 0, \qquad (5.4.92)$$

and

$$\int_0^T X(t)\{\ddot{X}(t) + \hat{\alpha}_1 \dot{X}(t) + \hat{\alpha}_2 X(t)\}\, dt = 0. \qquad (5.4.93)$$

Since $X(t)$ and $\dot{X}(t)$ are orthogonal we may, as an approximation valid for large T, neglect the third term in (5.4.92) and the second term in (5.4.93). This considerably simplifies the normal equations, and the solution is now immediate, namely,

$$\hat{\alpha}_1 = -\int_0^T \dot{X}(t)\ddot{X}(t)\, dt \Big/ \int_0^T \{\dot{X}(t)\}^2\, dt, \qquad (5.4.94)$$

$$\hat{\alpha}_2 = -\int_0^T X(t)\ddot{X}(t)\, dt \Big/ \int_0^T \{X(t)\}^2\, dt. \qquad (5.4.95)$$

The expression for $\hat{\alpha}_2$ can be written in the asymptotically equivalent form,

$$\hat{\alpha}_2 = \int_0^T \{\dot{X}(t)\}^2\, dt \Big/ \int_0^T \{X(t)\}^2\, dt \qquad (5.4.96)$$

(see Bartlett (1955) p. 267), and since $\ddot{X}(t)$ does not strictly exist for an AR(2) process, the expression (5.4.96) for $\hat{\alpha}_2$ is to be preferred to (5.4.95). Bartlett shows also that for large T,

$$\text{var}(\hat{\alpha}_1) \sim \sigma_W^2 \Big/ \int_0^T \{\dot{X}(t)\}^2\, dt, \qquad (5.4.97)$$

$$\text{var}(\hat{\alpha}_2) \sim \sigma_W^2 \Big/ \int_0^T \{X(t)\}^2 \, dt, \tag{5.4.98}$$

$$\text{cov}(\hat{\alpha}_1, \hat{\alpha}_2) \sim 0, \tag{5.4.99}$$

where σ_W^2 is the rate of increase of variance of the Wiener process, $W(t) = \int_0^t \varepsilon(s) \, ds$, (see Section 3.7.4).

Since $X(t)$ and $\dot{X}(t)$ are orthogonal, the estimate of α_1 is unaltered (asymptotically) if we assume *a priori* that $\alpha_2 = 0$. In this case (5.4.91) is an AR(1) model for $Z(t) \equiv \dot{X}(t)$, and hence the least squares estimate for the coefficient α in the AR(1) model,

$$\dot{X}(t) + \alpha X(t) = \varepsilon(t) \tag{5.4.100}$$

is (replacing $\dot{X}(t)$ by $X(t)$ in (5.4.94)),

$$\hat{\alpha} = -\int_0^T X(t) \dot{X}(t) \, dt \Big/ \int_0^T \{X(t)\}^2 \, dt \tag{5.4.101}$$

Also, from (5.4.97), (again replacing $\dot{X}(t)$ by $X(t)$),

$$\text{var}(\hat{\alpha}) \sim \sigma_W^2 \Big/ \int_0^T \{X(t)\}^2 \, dt. \tag{5.4.102}$$

But,

$$\int_0^T \{X(t)\}^2 \, dt \sim T\sigma_X^2 = T \cdot \frac{\sigma_W^2}{2\alpha}, \tag{5.4.103}$$

from (3.7.17) (Section 3.7.2). Hence, for large T we have,

$$\text{var}(\hat{\alpha}) \sim 2\alpha/T. \tag{5.4.104}$$

5.5 ANALYSIS OF THE CANADIAN LYNX DATA

Graph 5.13 shows the logarithms (to base 10) of the annual trappings of Canadian lynx over the period 1821–1934, giving a total of 114 observations. This is a celebrated set of data, and has been the subject of a great deal of study among time series analysts. (Campbell and Walker (1977) give an interesting review of previous analyses, and also list the original data.) The dominant feature of the graph is that the data contain persistent oscillations with a steady period of approximately ten years, but with irregular variations in amplitude. The sample autocovariance function is shown in Graph 5.14, and the strong periodic behaviour of this function confirms the "pseudo periodicity" in the data. However, the autocovariance function also shows some degree of damping, which is consistent with the irregular variations of amplitude in the data. The form of both the data and the autocovariance function suggests that either data contain a strictly periodic component corrupted by "error", or alternatively that the data conform to some "pseudo periodic" type of ARMA

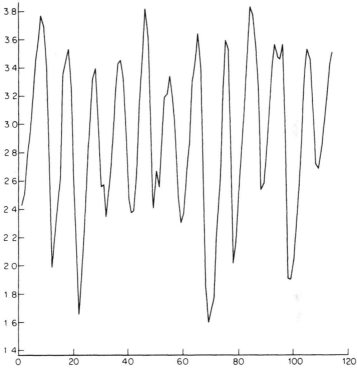

Graph 5.13. The Canadian lynx series.

model. The former type of model is that chosen by Campbell and Walker (1977), and we will discuss this approach further in Chapter 6. For the present we concentrate on the latter type of model, and one obvious candidate which comes to mind is the AR(2) model, which is well known to be capable of generating "pseudo periodic" behaviour (see Section 3.5.3). However, a fitted AR(2) model does not provide a very good match to the sample autocovariance function, the autocovariance function of the model damping much more rapidly than that computed from the data—see Hannan (1960, p. 44). It would seem reasonable, therefore, to consider the possibility of fitting higher order AR models, as suggested by

Tong (1977a). Fitting a succession of AR(k) models, with k varying from 0 to 20, gave the following AIC values.

k:	0	1	2	3	4	5	6
AIC:	$-133{\cdot}85$	$-241{\cdot}08$	$-322{\cdot}39$	$-322{\cdot}74$	$-325{\cdot}70$	$-325{\cdot}22$	$-324{\cdot}04$

k:	7	8	9	10	11	12	13
AIC:	$-327{\cdot}07$	$-326{\cdot}68$	$-325{\cdot}89$	$-327{\cdot}95$	$-337{\cdot}54$	$-336{\cdot}58$	$-335{\cdot}65$

k:	14	15	16	17	18	19	20
AIC:	$-333{\cdot}79$	$-331{\cdot}85$	$-331{\cdot}53$	$-329{\cdot}53$	$-329{\cdot}57$	$-328{\cdot}01$	$-326{\cdot}63$

The minimum value of AIC is attained with an eleventh order model, and on the basis of this criterion Tong (1977) concluded that the lynx data were best fitted by an AR(11) model. Writing this model as

$$X'_t + \sum_{i=1}^{11} a_i X'_{t-i} = \varepsilon_t,$$

where $X'_t = (X_t - \bar{X})$, $\bar{X}$ ($= 2{\cdot}9036$) denoting the sample mean, the least

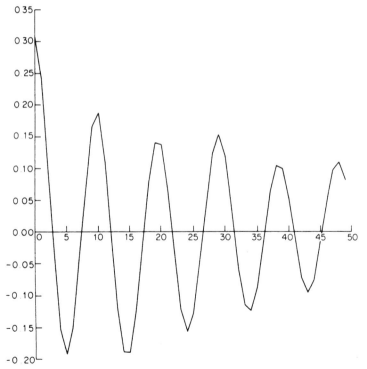

Graph 5.14. Autocovariance function of the lynx series.

squares estimates of the coefficients $\{a_i\}$ are

i	$\hat{a}_i$	i	$\hat{a}_i$
1	$-1 \cdot 13886$	7	$-0 \cdot 06819$
2	$0 \cdot 50842$	8	$0 \cdot 04039$
3	$-0 \cdot 21296$	9	$-0 \cdot 13388$
4	$0 \cdot 27003$	10	$0 \cdot 18521$
5	$-0 \cdot 11246$	11	$0 \cdot 31094$
6	$0 \cdot 12417$		

The estimated residual variance is $\hat{\sigma}_\varepsilon^2 = 0 \cdot 0423$.

The AR(11) model provides quite a good fit to the data, but we may enquire whether we could obtain a significant improvement in the fit if we widen the class of fitted models to the ARMA category. Fitting ARMA models of various orders gave the following AIC values.

Fitted model	AIC value
ARMA(1, 2)	$-295 \cdot 5$
ARMA(1, 3)	$-302 \cdot 3$
ARMA(2, 1)	$-324 \cdot 1$
ARMA(2, 2)	$-322 \cdot 5$
ARMA(3, 1)	$-322 \cdot 2$
ARMA(3, 2)	$-324 \cdot 9$
ARMA(3, 3)	$-337 \cdot 7$
ARMA(3, 4)	$-335 \cdot 8$
ARMA(4, 1)	$-324 \cdot 9$
ARMA(4, 2)	$-323 \cdot 1$
ARMA(4, 3)	$-335 \cdot 8$
ARMA(4, 4)	$-335 \cdot 3$

Over the class of models considered, the minimum AIC value is attained with the following ARMA(3, 3) model:

$$X_t' - 1 \cdot 9324 X_{t-1}' + 1 \cdot 5535 X_{t-2}' + 0 \cdot 3616 X_{t-3}'$$

$$= \varepsilon_t - 0 \cdot 7855 \varepsilon_{t-1} + 0 \cdot 0852 \varepsilon_{t-2} + 0 \cdot 4594 \varepsilon_{t-3}.$$

The corresponding estimated residual variance is $\hat{\sigma}_\varepsilon^2 = 0 \cdot 04654$. As measured by the AIC criterion, the improvement of the ARMA(3, 3) model over the AR(11) model is only marginal, and the residual variance for the ARMA(3, 3) model is, in fact, slightly larger than that of the AR(11) model (although the ARMA model contains only six parameters compared with the eleven parameters of the AR model). However, a more substantial improvement may be obtained by considering the class of "*subset autoregressive*" models. (These are constructed by searching over the class of AR(k) models, with various subsets of parameters set to zero.) Tong (1977) considered the following subset AR model:

$$X'_t - 1 \cdot 0938 X'_{t-1} + 0 \cdot 3571 X'_{t-2} + 0 \cdot 1265 X'_{t-4}$$
$$- 0 \cdot 3244 X'_{t-10} + 0 \cdot 3622 X'_{t-11} = \varepsilon_t.$$

This model has an AIC value of $-345 \cdot 96$, and gives an estimated residual variance $\hat{\sigma}_\varepsilon^2 = 0 \cdot 04405$.

In relation to the accuracy of the estimated coefficients in the above models, it is important to note that in all cases the characteristic polynomial corresponding to the AR operator has roots very close to the unit circle, as one would expect from the largely cyclical form of the data. (For the AR(11) model the smallest modulus is $0 \cdot 87$ and the largest is $0 \cdot 98$.) This means that the usual asymptotic expressions for the variances of the coefficients may no longer be reliable, but it may be of interest to note that for the coefficients of the fitted ARMA(3, 3) model these expressions give the following values.

AR coefficients	Standard deviation
$\hat{a}_1$	$0 \cdot 1598$
$\hat{a}_2$	$0 \cdot 2494$
$\hat{a}_3$	$0 \cdot 1523$

MA coefficients	Standard deviation
$\hat{b}_1$	$0 \cdot 1469$
$\hat{b}_2$	$0 \cdot 1415$
$\hat{b}_3$	$0 \cdot 0983$

An alternative approach, as mentioned above, is to "remove" the cyclical component by fitting a periodic term to the data, and then fit an AR model to the residuals. This approach has been studied in considerable detail by Campbell and Walker (1977), and we will discuss their model later in Chapter 6. Meanwhile, we may note that the "near cyclical" feature of the data is further reflected in the AR models by observing that, e.g., if we ignore the fourth order lag, the AR(11) operator may be factorized roughly in the form $(1 - B)(1 - 0 \cdot 3B^{10})$. The second factor corresponds to a "damped" periodic component with a period of ten years (see the discussion of "seasonal" models in Section 10.2). An interesting discussion of the roots of the various models described above is given by Haggan (1977).

Chapter 6

Estimation in the Frequency Domain

In this chapter we consider the estimation of the frequency domain (i.e. spectral) properties of a stationary process from observational data. From a theoretical point of view the spectral properties of all types of stationary process are completely described by the *integrated spectrum*, and it may therefore seem logical to concentrate attention entirely on the estimation of this function. However, if a process has a purely continuous spectrum its physical properties are more readily identified from a knowledge of its *spectral density function* rather than from its integrated spectrum, just as the properties of a continuous random variable are more easily recognized from the form of its probability density function than from its distribution function. Consequently, in the case of continuous spectra our main interest would usually lie in the estimation of spectral density functions. On the other hand, if the process has a purely discrete spectrum then the integrated spectrum is a step-function, and its estimation involves simply the estimation of the location and magnitude of its "jumps".

Our estimation procedures thus differ according as to whether the process has (a) a purely discrete spectrum, (b) a purely continuous spectrum, or (c) a mixed spectrum, and in the following sections we give a separate discussion for each of these cases.

For reasons previously explained (see Section 5.3.1) we will again deal mainly with discrete parameter processes. If continuous parameter processes are to be analysed by digital methods then it is necessary to convert them into discrete parameter processes by reading off their values at discrete time intervals. This operation introduces an "*aliasing effect*" (see Section 4.8.3) which we shall discuss in more detail in Chapter 7. For the moment we simply note that when a continuous parameter process is read at intervals Δt, the power outside the (angular) frequency range $(-\pi/\Delta t, \pi/\Delta t)$ is, loosely speaking, superimposed back onto this interval. Consequently, as long as we choose Δt sufficiently small so that there is negligible power at frequencies higher than $(\pi/\Delta t)$, the aliasing effect will not be serious. Alternatively, we can sometimes analyse continuous parameter processes directly (using analogue methods) and later on we will indicate how the estimation methods for discrete parameter processes can be extended to the continuous parameter case.

6.1 DISCRETE SPECTRA

The usual model for a process with a purely discrete spectrum is given by the harmonic model (4.9.10) (Section 4.9) which, in discrete time, takes the form,

$$X_t = \sum_{i=1}^{K} A_i \cos(\omega_i t + \phi_i), \tag{6.1.1}$$

where $K, \{A_i\}, \{\omega_i\}, i = 1, \ldots, K$, are constants, and the $\{\phi_i\}$ are independently and rectangularly distributed on $(-\pi, \pi)$. When t takes integer values (i.e. $t = 0, \pm 1, \pm 2, \ldots$) the $\{\omega_i\}$ are restricted to the range $(-\pi, \pi)$, i.e. $|\omega_i| \leq \pi$, all i. For this process, the (normalized) integrated spectrum $F(\omega)$ has jumps at $\omega = \pm\omega_i$ $(i = 1, \ldots, K)$, the magnitude of the jump at $\pm\omega_i$ being $(\frac{1}{2}A_i^2/C)$, where $C = \sum_{i=1}^{K} A_i^2$. Thus, to estimate the spectrum we must estimate $K, A_i(i = 1, \ldots, K), \omega_i(i = 1, \ldots, K)$. (Note that in general K, the number of terms in (6.1.1), is also an unknown parameter of the model.)

However, models of the precise form (6.1.1) rarely occur, and when they do the problem of determining the parameters is basically a mathematical rather than a statistical one. The point is that although the $\{\phi_i\}$ are defined as random variables they retain constant values throughout each realization. For any particular realization they are, in effect, simply constants, and hence each individual realization is essentially deterministic and free from random elements. The determination of the parameters is now a purely mathematical problem (which may be treated by a technique similar to harmonic analysis). Since (6.1.1) does not include any time dependent random variables there is no term in the model which could describe *errors of observation*. In other words, the model (6.1.1) is based on the unrealistic assumption that we could observe such a process with perfect accuracy. A more realistic model is obtained by assuming that there are errors of observations superimposed on the RHS of (6.1.1), which then leads to the modified model,

$$X_t = \sum_{i=1}^{K} A_i \cos(\omega_i t + \phi_i) + \varepsilon_t, \tag{6.1.2}$$

where $\{\varepsilon_t\}$ is a purely random process, independent of the $\{\phi_i\}$, with $E(\varepsilon_t) = 0$, (all t), and $E(\varepsilon_t^2) = \sigma_\varepsilon^2$, where σ_ε^2 is a further unknown parameter. (The process ε_t here plays the role of the "error term" ε in the standard regression model (5.2.40)). Since ε_t has a purely continuous uniform spectrum we should, strictly speaking, say that X_t has a *mixed spectrum*, but when the continuous component is uniform we will stretch the terminology somewhat and, following the usual convention, still refer to (6.1.2) as a "purely discrete spectrum" model. We reserve the term "mixed spectrum" for the more complex case where the continuous component is nonuniform.

6.1.1 Estimation

Given N observations $X_1, X_2, \ldots, X_N$ from the model (6.1.2) our problem now is to estimate the set of unknown parameters:

$$\{K, \quad A_i (i = 1, \ldots, K), \quad \omega_i (i = 1, \ldots, K), \quad \text{and } \sigma_\varepsilon^2\}.$$

Let us consider first the simpler situation where the frequencies $\{\omega_i\}$ are known *a priori*—so that, by implication, the value of K is also known. Then re-writing (6.1.2) as

$$X_t = \sum_{i=1}^{K} (A_i' \cos \omega_i t + B_i' \sin \omega_i t) + \varepsilon_t, \tag{6.1.3}$$

where $A_i' = A_i \cos \phi_i$, $B_i' = -A_i \sin \phi_i$, so that $A_i = \sqrt{A_i'^2 + B_i'^2}$ and $\phi_i = \tan^{-1}(-B_i'/A_i')$, we may regard (6.1.3) as a standard multiple linear regression model (cf. (5.2.41)). Since we have observations from only one realization we might as well treat the $\{\phi_i\}$ as constants (as far as estimation is concerned), in which case A_i' and B_i' may be treated as unknown parameters. We may now estimate A_i', B_i', $(i = 1, \ldots, K)$ by least squares, i.e. by minimizing

$$Q = \sum_{t=1}^{N} \left\{ X_t - \sum_{i=1}^{K} (A_i' \cos \omega_i t + B_i' \sin \omega_i t) \right\}^2. \tag{6.1.4}$$

Differentiating Q w.r. to A_i', B_i', and setting the derivatives equal to zero we obtain the normal equations,

$$\sum_{t=1}^{N} X_t \cos \omega_j t = \sum_{i=1}^{K} \hat{A}_i' c_{ij} + \sum_{i=1}^{K} \hat{B}_i' d_{ij}, \quad j = 1, \ldots, K \tag{6.1.5}$$

and

$$\sum_{t=1}^{N} X_t \sin \omega_j t = \sum_{i=1}^{K} \hat{A}_i' d_{ji} + \sum_{i=1}^{K} \hat{B}_i' s_{ij}, \quad j = 1, \ldots, K \tag{6.1.6}$$

where

$$c_{ij} = \sum_{t=1}^{N} \cos \omega_i t \cos \omega_j t,$$

$$s_{ij} = \sum_{t=1}^{N} \sin \omega_i t \sin \omega_j t, \qquad (6.1.7)$$

$$d_{ij} = \sum_{t=1}^{N} \sin \omega_i t \cos \omega_j t.$$

The solution of these equations is considerably simplified if all the ω_i are multiples of $(2\pi/N)$, i.e. if the data length N is an integral multiple of the period of each of the sine and cosine terms in (6.1.3). (This would arise, e.g., if the trigonometric terms in (6.1.3) represent a "seasonal" component (i.e. a function with a period of one year) in an economic series, in which case all the frequencies (ω_i) will be harmonics of the fundamental frequency corresponding to the period of one year. The ω_i will then have the required form provided the observation period covers an integral number of years.) In this case we may write,

$$\omega_i = 2\pi p_i/N, \qquad i = 1, \ldots, K, \qquad (6.1.8)$$

where $p_1, p_2, \ldots, p_K$ are all integers satisfying $0 \leq p_i \leq [N/2]$, ($[N/2]$ being $N/2$ when N is even or $(N-1)/2$ when N is odd), and then make use of the well known orthogonality relations for the functions $\cos(2\pi pt/N)$, $\sin(2\pi pt/N)$, (p taking integer values in the range $0 \leq p \leq [N/2]$), namely,

$$\sum_{t=1}^{N} \cos\left(\frac{2\pi pt}{N}\right) \cos\left(\frac{2\pi qt}{N}\right) = \begin{cases} 0, & 0 \leq p \neq q \leq [N/2], \\ N/2, & 0 < p = q < N/2, \\ N, & p = q = 0 \quad \text{or } N/2, N \text{ even.} \end{cases}$$

$$\sum_{t=1}^{N} \sin\left(\frac{2\pi pt}{N}\right) \cos\left(\frac{2\pi qt}{N}\right) = 0, \qquad \text{all } p, q, \qquad (6.1.9)$$

$$\sum_{t=1}^{N} \sin\left(\frac{2\pi pt}{N}\right) \sin\left(\frac{2\pi qt}{N}\right) = \begin{cases} 0, & 0 \leq p \neq q \leq [N/2] \\ N/2, & 0 < p = q < N/2 \\ 0, & p = q = 0 \quad \text{or } N/2, N \text{ even.} \end{cases}$$

Apart from being well known these relations follow immediately from (5.3.63), (5.3.64) on writing, e.g.,

$$\sum_{t=1}^{N} \cos\left(\frac{2\pi pt}{N}\right) \cos\left(\frac{2\pi qt}{N}\right) = \tfrac{1}{2} \sum_{t=1}^{N} \left\{ \cos\left(\frac{2\pi t}{N}\right)(p+q) + \cos\left(\frac{2\pi t}{N}\right)(p-q) \right\}.$$

Thus, when the $\{\omega_i\}$ are of the form (6.1.8), and assuming that no p_i is either 0 or $N/2$ when N is even, we have,

$$c_{ij} = s_{ij} = 0, \qquad \text{all } i \neq j,$$
$$d_{ij} = 0, \qquad \text{all } i, j, \qquad (6.1.10)$$
$$c_{ii} = s_{ii} = N/2, \qquad \text{all } i.$$

Substituting (6.1.10) into (6.1.5), (6.1.6), we obtain for $i = 1, \ldots, K$,

$$\hat{A}_i' = \frac{2}{N} \sum_{t=1}^{N} X_t \cos \omega_i t, \qquad (6.1.11)$$

$$\hat{B}_i' = \frac{2}{N} \sum_{t=1}^{N} X_t \sin \omega_i t. \qquad (6.1.12)$$

If, e.g., $p_1 = 0$ then $c_{11} = N$ and now,

$$\hat{A}_1' = \frac{1}{N} \sum_{t=1}^{N} X_t \cos \omega_1 t = \frac{1}{N} \sum_{t=1}^{N} X_t = \bar{X}, \qquad (6.1.13)$$

the sample mean. This result is obvious when it is realized that with $\omega_1 = 0$ the term $(A_1' \cos \omega_1 t + B_1' \sin \omega_1 t)$ becomes simply the constant, A_1'. We have

$$E(\hat{A}_i') = \frac{2}{N} \sum_{t=1}^{N} E(X_t) \cos \omega_i t.$$

Recalling that $E(\varepsilon_t) = 0$ it follows that (conditional on the ϕ_i fixed),

$$E(X_t) = \sum_{i=1}^{K} (A_i' \cos \omega_i t + B_i' \sin \omega_i t),$$

and the orthogonality relations (6.1.10) now yield,

$$E(\hat{A}_i') = A_i' \qquad \text{(all } i\text{)}, \qquad (6.1.14)$$

so that $\hat{A}_i'$ is an unbiased estimate of A_i'. Similarly,

$$E(\hat{B}_i') = B_i' \qquad \text{(all } i\text{)}, \qquad (6.1.15)$$

so that $\hat{B}_i'$ is an unbiased estimate of B_i'. Also, from (6.1.11),

$$\text{var}(\hat{A}_i') = \left(\frac{2}{N}\right)^2 \sigma_\varepsilon^2 \sum_{t=1}^{N} \cos^2 \omega_i t = \frac{2\sigma_\varepsilon^2}{N}. \qquad (6.1.16)$$

Similarly,

$$\text{var}(\hat{B}_i') = 2\sigma_\varepsilon^2/N, \qquad (6.1.17)$$

and from (6.1.10),

$$\text{cov}(\hat{A}'_i, \hat{B}'_i) = 0 \qquad \text{(all } i\text{).} \qquad (6.1.18)$$

An estimate of σ_ε^2 may be obtained by using the usual expression for the unbiased estimate of the residual variance in a regression model. In this case (remembering that (6.1.3) involves $2K$ parameters) we obtain,

$$\hat{\sigma}_\varepsilon^2 = \frac{1}{N - 2K} \sum_{t=1}^{N} \left\{ X_t - \sum_{i=1}^{K} (\hat{A}'_i \cos \omega_i t + \hat{B}'_i \sin \omega_i t) \right\}^2. \qquad (6.1.19)$$

When the $\{\omega_i\}$ are not all of the form $(2\pi p_i/N)$, the expressions (6.1.11) for $\hat{A}'_i, \hat{B}'_i$ hold as approximations which are valid to $O(1/N)$. For we now have for large N,

$$c_{ii} \sim N/2, \qquad s_{ii} \sim N/2,$$

whereas $c_{ij}, s_{ij} (i \neq j)$ and d_{ij} are all $O(1)$ (i.e. remain finite as $N \to \infty$) provided $|\omega_i \pm \omega_j| \gg 2\pi/N$, i.e. provided the $\{\omega_i\}$ are spaced sufficiently wide apart, and no ω_i is within π/N of 0 or π. In place of (6.1.14), (6.1.15), we now find,

$$E(\hat{A}'_i) = A'_i + O(1/N), \qquad E(\hat{B}'_i) = B'_i + O(1/N). \qquad (6.1.20)$$

6.1.2 Periodogram Analysis

In the previous section we considered the estimation of $\{A'_i\}$, $\{B'_i\}$, on the assumption that the $\{\omega_i\}$ were known *a priori*. However, as previously mentioned, the $\{\omega_i\}$ are, in general, unknown, as is the value of K, the number of trigonometric terms in (6.1.3), and the estimation procedure has to be extended to cover these additional unknown parameters. With the $\{\omega_i\}$ unknown the model (6.1.3) is no longer of the "linear hypothesis" form and we cannot use the standard least squares approach. Also, it is clear from the form of (6.1.11), that we cannot estimate $\{A'_i\}$ and $\{B'_i\}$ until we have first estimated the $\{\omega_i\}$. The usual procedure adopted here is to locate the $\{\omega_i\}$ by using a "*search*" technique based on a function called the *periodogram*. This function was first introduced by Schuster (1898) in connection with what was then called the model of "*hidden periodicities*". (The residual ε_t in (6.1.3) was regarded as masking the periodic trigonometric terms, these periodicities being thought of as "hidden" beneath the noise process.) The basic ideas underlying periodogram analysis may be explained heuristically as follows. Suppose we guess the value of ω_1 as $\hat{\omega}_1$, say. We can then compute the estimates $\hat{A}'_1$, $\hat{B}'_1$, from (6.1.11) using $\hat{\omega}_1$ in place of ω_1. If we have been lucky and the guessed value $\hat{\omega}_1$ is exactly equal to ω_1 (or close to it) then $\hat{A}'_1$, $\hat{B}'_1$, will be close to A'_1, B'_1, and hence the squared amplitude $(A'^2_1 + B'^2_1)$ will be non-zero. On the other hand, if $\hat{\omega}_1$ is substantially removed from ω_1

(or any of the other frequencies present in the model), then, in effect, we are estimating the coefficients of a term which does not exist in the model, and consequently $(A_1'^2 + B_1'^2)$ will be close to zero. If we now choose a sufficiently fine set of trial frequencies $\hat{\omega}_1, \hat{\omega}_2, \hat{\omega}_3, \ldots$, and plot the squared amplitudes $(\hat{A}_p'^2 + \hat{B}_p'^2)$ against $\hat{\omega}_p$, the ordinates will be non-zero when $\hat{\omega}_p$ is near one of the $\{\omega_i\}$ but will be close to zero otherwise. We can therefore locate the values of the $\{\omega_i\}$ by inspecting the squared amplitudes $(A_p'^2 + B_p'^2)$ and selecting those whose values are "appreciably" greater than zero. This is the basis of the method of periodogram analysis, but in this technique we actually plot the "normalized" amplitudes, $\{(N/2)(A_p'^2 + B_p'^2)\}$ (rather than the squared amplitudes themselves), the effect of the factor $(N/2)$ being to magnify the differences between the "large" and "small" ordinates. We now give a more formal description of the method.

Given N observations $X_1, \ldots, X_N$ the function $I_N(\omega)$, called the *periodogram*, is defined for all ω in the range $-\pi \leq \omega \leq \pi$ by

$$I_N(\omega) = \{A(\omega)\}^2 + \{B(\omega)\}^2, \tag{6.1.21}$$

where

$$A(\omega) = \sqrt{\frac{2}{N}} \sum_{t=1}^{N} X_t \cos \omega t, \tag{6.1.22}$$

$$B(\omega) = \sqrt{\frac{2}{N}} \sum_{t=1}^{N} X_t \sin \omega t. \tag{6.1.23}$$

Alternatively, we may write $I_N(\omega)$ in the form,

$$I_N(\omega) = \frac{2}{N} \left| \sum_{t=1}^{N} X_t e^{-i\omega t} \right|^2. \tag{6.1.24}$$

Although $I_N(\omega)$ is defined for *all* ω in $(-\pi, \pi)$ we cannot, of course, evaluate it numerically as a continuous function of ω; we can compute it only at a discrete set of frequencies. We therefore evaluate $I_N(\omega)$ at the set of frequencies $0, 2\pi/N, 4\pi/N, 6\pi/N, \ldots$ (the reason for choosing this set of frequencies will become apparent later), i.e. we evaluate

$$I_p \equiv I_N(\omega_p), \qquad \omega_p = 2\pi p/N, \qquad p = 0, 1, \ldots, [N/2],$$

and then plot I_p against p (see Fig. 6.1)

Now if ω_p coincides exactly with one of the frequencies ω_i in the model (6.1.3), then comparing (6.1.11), (6.1.12) with (6.1.22), (6.1.23), we have

$$A(\omega_p) = \sqrt{\frac{N}{2}} \hat{A}_i', \qquad B(\omega_p) = \sqrt{\frac{N}{2}} \hat{B}_i'. \tag{6.1.25}$$

Hence, when $\omega_p = \omega_i$,

$$E[I_p] = \frac{N}{2}[E(\hat{A}_i'^2) + E(\hat{B}_i'^2)].$$

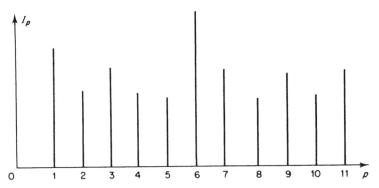

Fig. 6.1. A periodogram plot.

But,

$$E(\hat{A}_i'^2) = \{E(\hat{A}_i')\}^2 + \text{var}(\hat{A}_i')$$

$$\doteq A_i'^2 + \frac{2\sigma_\varepsilon^2}{N},$$

using (6.1.14), (6.1.16). A similar result holds for $E(\hat{B}_i'^2)$, and hence,

$$E[I_p] \doteq \frac{N}{2}(A_i'^2 + B_i'^2) + 2\sigma_\varepsilon^2$$

$$= \frac{N}{2}A_i^2 + 2\sigma_\varepsilon^2, \qquad \text{if } \omega_p = \omega_i. \qquad (6.1.26)$$

On the other hand, when ω_p does not fall near one of the ω_i we may evaluate $E[I_p]$ simply by inserting an additional term in (6.1.3) with frequency ω_p and amplitude $A_{\omega_p} = 0$. Thus, setting $A_i = 0$ in (6.1.26) we find

$$E[I_p] \doteq 2\sigma_\varepsilon^2, \qquad \text{if } |\omega_p - \omega_i| \gg 0, \quad i = 1, \ldots, K. \qquad (6.1.27)$$

Hence, as ω_p approaches each of the ω_i the ordinate is $O(N)$, and the *periodogram exhibits a large peak*; otherwise the ordinates are $O(1)$. We may therefore locate the true frequencies $\{\omega_i\}$ by noting the positions of the peaks in the graph of the periodogram ordinates.

The results (6.1.26), (6.1.27) are exact if all the $\{\omega_i\}$ are of the form (6.1.8), but hold only approximately otherwise. We need to know more precisely, therefore, how the $\{I_p\}$ behave when the $\{\omega_i\}$ are not restricted to multiples of $2\pi/N$. In particular, we need to know what happens when ω_p is close to (but not exactly equal to) ω_i—will the periodogram still exhibit a well defined peak, and if so, how close must ω_p be to ω_i for this effect to occur? To answer these questions we have to study the sampling properties of the periodogram in a more precise form.

6.1.3 Sampling Properties of the Periodogram

We consider first the "null" case of the model (6.1.2) in which $A_i = 0$, all i (or equivalently, $K = 0$) so that $X_t(\equiv \varepsilon_t)$ is a purely random process and has no discrete component in its spectrum. If we now assume further that $\{X_t\}$ is a *Gaussian* process then we can obtain the exact sampling distribution of the $I(\omega_p)$, $p = 0, 1, \ldots, N/2$. For, under the above assumptions $\{X_t\}$ is a sequence of independent random variables, each normally distributed with zero mean and variance, σ_X^2 say, i.e. for each t,

$$X_t = N(0, \sigma_X^2).$$

It now follows that $A(\omega_p)$, being a linear combination of the $\{X_t\}$, is also normally distributed with zero mean and variance given by,

$$\text{var}\{A(\omega_p)\} = \frac{2\sigma_X^2}{N} \left(\sum_{t=1}^{N} \cos^2 \omega_p t \right)$$

$$= \begin{cases} \sigma_X^2, & \omega_p = 2\pi p/N, \quad p \neq 0, \quad N/2, \\ 2\sigma_X^2, & \omega_p = 0, \end{cases}$$

in virtue of (6.1.9). Hence,

$$A(\omega_p) = \begin{cases} N(0, \sigma_X^2), & p \neq 0, \dfrac{N}{2} \quad (N \text{ even}) \\[2em] N(0, 2\sigma_X^2), & p = 0 \quad \text{or} \quad \dfrac{N}{2}. \end{cases} \tag{6.1.28}$$

A similar argument shows that,

$$B(\omega_p) = \begin{cases} N(0, \sigma_X^2), & p \neq 0, \dfrac{N}{2} \quad (N \text{ even}) \\[2em] 0, & p = 0 \quad \text{or} \quad \dfrac{N}{2}. \end{cases} \tag{6.1.29}$$

In addition, using the orthogonality relations (6.1.9) we have,

$$\text{cov}\{A(\omega_p), B(\omega_q)\} = \frac{2\sigma_X^2}{N} \left(\sum_{t=1}^{N} \cos \omega_p t \sin \omega_q t \right) = 0, \qquad \text{all } p, q.$$

Similarly,

$$\text{cov}\{A(\omega_p), A(\omega_q)\} = \text{cov}\{B(\omega_p), B(\omega_q)\} = 0, \qquad p \neq q.$$

But the $\{A(\omega_p)\}$ and $\{B(\omega_p)\}$ are linear combinations of the $\{X_t\}$ and hence have a multivariate normal distribution (see Section 2.13.2). Since they are mutually uncorrelated they must all be *independent* random variables. It now follows immediately that for $p \neq 0$, $N/2$, $I_p (\equiv I_N(\omega_p))$, being the sum of the squares of two independent zero mean normal variables, has a distribution which is proportional *to chi-squared on 2 degrees of freedom* (or, equivalently, an exponential distribution, see (2.11.19), Section 2.11). More specifically,

$$I_p = A^2(\omega_p) + B^2(\omega_p) = \sigma_X^2 \chi_2^2, \qquad p \neq 0, N/2. \qquad (6.1.30)$$

For $p = 0$ or $N/2$ (N even),

$$I_p = A^2(\omega_p) = (2\sigma_X^2)\chi_1^2.$$

Also, since for different p the $\{A(\omega_p)\}$ and $\{B(\omega_p)\}$ are independent, it follows further that the $\{I_p\}$, $p = 0, 1, \ldots, [N/2]$, are *independently distributed*. We thus have the following theorem.

Theorem 6.1.1 *If $\{X_t\}$ is a Gaussian purely random process with zero mean and variance σ_X^2 then the $\{I_p\}$, $p = 0, 1, \ldots, [N/2]$, are independently distributed, and for each p,*

$$I_p = \begin{cases} \sigma_X^2 \cdot \chi_2^2, & p \neq 0, \quad N/2 \ (N \text{ even}) \\ 2\sigma_X^2 \cdot \chi_1^2, & p = 0, \quad N/2. \end{cases}$$

In particular, using the result that the mean and variance of the χ_ν^2 distribution are ν, 2ν, respectively, we have

$$E[I_p] = 2\sigma_X^2, \qquad \text{all } p, \qquad (6.1.31)$$

$$\text{var}[I_p] = \begin{cases} 4\sigma_X^4, & p \neq 0, \quad N/2, \\ 8\sigma_X^4, & p = 0, \quad N/2. \end{cases} \qquad (6.1.32)$$

Next, we obtain expressions for $\text{cov}\{I_N(\omega), I_N(\omega')\}$ when ω, ω', are not necessarily of the form $2\pi p/N$, and for $E[I_N(\omega)]$ in the general case when the $A_i \neq 0$. To facilitate the derivation of these results we first apply a simple algebraic transformation to $I_N(\omega)$ to enable us to write it in a more convenient form.

Lemma 6.1.1 *Let $I_N(\omega)$ be defined by (6.1.24). Then for each ω we may write $I_N(\omega)$ in the alternative form,*

$$I_N(\omega) = 2 \sum_{s=-(N-1)}^{(N-1)} \hat{R}(s) \cos s\omega, \qquad (6.1.33)$$

where

$$\hat{R}(s) = \frac{1}{N} \sum_{t=1}^{N-|s|} X_t X_{t+|s|}. \tag{6.1.34}$$

Proof. From (6.1.24) we have,

$$I_N(\omega) = \frac{2}{N} \left| \sum_{t=1}^{N} X_t e^{-i\omega t} \right|^2$$

$$= \frac{2}{N} \left[\sum_{t=1}^{N} X_t e^{-i\omega t} \right] \left[\sum_{r=1}^{N} X_r e^{i\omega r} \right]$$

$$= \frac{2}{N} \left[\sum_{t=1}^{N} \sum_{r=1}^{N} X_t X_r \cos(t-r)\omega \right].$$

(The sine terms vanish due to the symmetry of the factor $(X_t X_r)$.) Now transform the indices t and r to t and $s = (t-r)$. (This device simply transforms the double summation over t and r into a summation over the diagonal points of the lattice; cf. the argument leading from (5.3.4) to (5.3.5).) Then s goes from $-(N-1)$ to $(N-1)$ and, for fixed s, t goes from 1 to $N-|s|$. We may now write,

$$I_N(\omega) = 2 \sum_{s=-(N-1)}^{(N-1)} \left\{ \frac{1}{N} \sum_{t=1}^{N-|s|} X_t X_{t+|s|} \right\} \cos s\omega$$

$$= 2 \sum_{s=-(N-1)}^{(N-1)} \hat{R}(s) \cos s\omega,$$

$$= 2 \left\{ \hat{R}(0) + 2 \sum_{s=1}^{N-1} \hat{R}(s) \cos s\omega \right\}, \tag{6.1.35}$$

where $\hat{R}(s)$ is defined by (6.1.34). Note that $\hat{R}(s)$ is the (biased) *sample autocovariance function* of X_t, as given by (5.3.17). (For the model (6.1.2) $E[X_t] = 0$ so that there is no need here to subtract the sample mean $\bar{X}$ from X_t in computing $\hat{R}(s)$.)

The form of $E[I_N(\omega)]$ for the general case is now given by the following theorem.

Theorem 6.1.2 *Let X_t be given by (6.1.2). Then for all ω,*

$$E[I_N(\omega)] = 2\sigma_\varepsilon^2 + \frac{1}{2N} \sum_{i=1}^{K} A_i^2 \left\{ \frac{\sin^2(\frac{1}{2}N(\omega + \omega_i))}{\sin^2(\frac{1}{2}(\omega + \omega_i))} + \frac{\sin^2(\frac{1}{2}N(\omega - \omega_i))}{\sin^2(\frac{1}{2}(\omega - \omega_i))} \right\}.$$

$$\tag{6.1.36}$$

Proof. From (6.1.33) we have,

$$E[I_N(\omega)] = 2 \sum_{s=-(N-1)}^{(N-1)} E[\hat{R}(s)] \cos s\omega$$

$$= 2 \sum_{s=-(N-1)}^{(N-1)} \left(1 - \frac{|s|}{N}\right) R_X(s) \cos s\omega, \qquad (6.1.37)$$

using (5.3.14), where $R_X(s)$ is the theoretical autocovariance function of X_t. Now from (6.1.2) we may write X_t in the form

$$X_t = Z_t + \varepsilon_t, \qquad (6.1.38)$$

where

$$Z_t = \sum_{i=1}^{K} A_i \cos(\omega_i t + \phi_i).$$

Since ε_t is independent of the $\{\phi_i\}$, $\{Z_t\}$ and $\{\varepsilon_t\}$ are independent processes, and hence,

$$R_X(s) = R_Z(s) + R_\varepsilon(s), \qquad (6.1.39)$$

where $R_Z(s)$, $R_\varepsilon(s)$ are the autocovariance functions of Z_t, ε_t, respectively. The autocovariance function of Z_t is (according to (3.5.73), Section 3.5.8),

$$R_Z(s) = \frac{1}{2} \sum_{i=1}^{K} A_i^2 \cos \omega_i s, \qquad (6.1.40)$$

while ε_t, being a purely random process, has autocovariance function,

$$R_\varepsilon(s) = \sigma_\varepsilon^2 \delta_{0,s} \qquad (6.1.41)$$

where $\delta_{0,s} = 1$, $s = 0$, and $\delta_{0,s} = 0$, $s \neq 0$. Substituting (6.1.39), (6.1.40) into (6.1.37) we obtain,

$$E[I_N(\omega)] = \sum_{i=1}^{K} A_i^2 \left\{ \sum_{s=-(N-1)}^{(N-1)} \left(1 - \frac{|s|}{N}\right) \cos s\omega_i \cos s\omega \right\} + 2\sigma_\varepsilon^2. \quad (6.1.42)$$

We now use the well known result,

$$\sum_{s=-(N-1)}^{(N-1)} \left(1 - \frac{|s|}{N}\right) \cos s\theta = \frac{1}{N} \frac{\sin^2(N\theta/2)}{\sin^2(\theta/2)} = 2\pi F_N(\theta), \text{ say,} \quad (6.1.43)$$

$F_N(\theta)$ being the "*Fejer kernel*". (To prove (6.1.43) we may use the identity that for any sequence $a(s)$

$$\sum_{s=-(N-1)}^{(N-1)} (N - |s|)a(s) \equiv \sum_{u=0}^{N-1} \sum_{s=-u}^{u} a(s),$$

together with the further results,

$$\sum_{s=-u}^{u} \cos s\theta = \frac{\sin((u+\frac{1}{2})\theta)}{\sin(\frac{1}{2}\theta)}, \tag{6.1.44}$$

and

$$\sum_{u=0}^{N-1} \frac{\sin((u+\frac{1}{2})\theta)}{\sin(\frac{1}{2}\theta)} = \frac{\sin^2(N\theta/2)}{\sin^2(\theta/2)}, \tag{6.1.45}$$

(6.1.44), (6.1.45) following directly from (5.3.63), (5.3.64). Alternatively, we may write the LHS of (6.1.43) as

$$\sum_{s=-(N-1)}^{(N-1)} \cos s\theta - \frac{1}{N} \sum_{s=-(N-1)}^{(N-1)} |s| \cos s\theta,$$

evaluate the first term from (5.3.63), and evaluate the second term by differentiating $\Sigma \cos s\theta$ term by term w.r. to θ.) The final result (6.1.36) now follows from (6.1.42) on writing,

$$\sum_{s=-(N-1)}^{(N-1)} \left(1 - \frac{|s|}{N}\right) \cos s\omega_i \cos s\omega$$

$$= \frac{1}{2} \sum_{s=-(N-1)}^{(N-1)} \left(1 - \frac{|s|}{N}\right) \{\cos s(\omega + \omega_i) + \cos s(\omega - \omega_i)\},$$

and applying (6.1.43) to each term.

The Fejer kernel $F_N(\theta)$, whose form is shown in Fig. 6.2, plays a central role in Fourier analysis (see, e.g. Lancos (1956)), and we will meet this function many times in our study of spectral analysis.

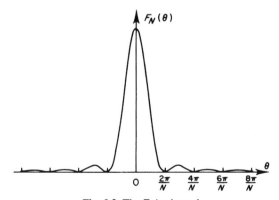

Fig. 6.2. The Fejer kernel.

It will be seen from Fig. 6.2 that $F_N(\theta)$ has a large peak at $\theta = 0$, of magnitude $N/2\pi$, and then decays as $|\theta|$ increases, with subsidiary peaks at (approximately) $\theta = \pm 3\pi/N, \pm 5\pi/N, \pm 7\pi/N, \ldots$. The zeros of the function occur at $\pm 2\pi/N, \pm 4\pi/N, \pm 6\pi/N, \ldots$. The form of $F_N(\theta)$ near $\theta = 0$ follows by setting $\phi = N\theta/2$, so that

$$2\pi F_N(\theta) = \frac{1}{N}\frac{\sin^2\phi}{\sin^2(\phi/N)} \sim \frac{N\sin^2\phi}{\phi^2}, \qquad \text{for small } \phi.$$

Hence,

$$\lim_{\theta \to 0} 2\pi F_N(\theta) = N \lim_{\phi \to 0} \left(\frac{\sin\phi}{\phi}\right)^2 = N.$$

The form of $E[I_N(\omega)]$ may now be visualized by superimposing $2K$ Fejer kernels "centered" on the points $\omega \pm \omega_i$, $i = 1, \ldots, K$, together with the constant term $2\sigma_\varepsilon^2$, as typified by Fig. 6.3.

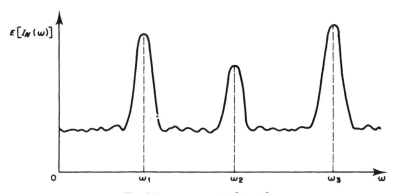

Fig. 6.3. Behaviour of $E[I_N(\omega)]$.

For $|\theta| \gg 2\pi/N$, $F_N(\theta) = O(1/N)$, and hence if the ω_i are spaced sufficiently wide apart (i.e. if $|\omega_i \pm \omega_j| \gg 2\pi/N$, $i \neq j$) the behaviour of the periodogram in the neighbourhood of $\omega = \omega_i$ is determined essentially by the term,

$$\frac{1}{N}\left\{\frac{\sin^2(\frac{1}{2}N(\omega + \omega_i))}{\sin^2(\frac{1}{2}(\omega + \omega_i))} + \frac{\sin^2(\frac{1}{2}N(\omega - \omega_i))}{\sin^2(\frac{1}{2}(\omega - \omega_i))}\right\}.$$

As ω approaches ω_i the second term gives a peak of magnitude N, but we must recall that in searching for periodicities the periodogram is computed only for $\omega = \omega_p = 2\pi p/N$, $p = 0, 1, \ldots, [N/2]$. This amounts to placing a "grid", with intervals $2\pi/N$, on top of the graph in Fig. 6.3, giving a figure of the form of Fig. 6.4. If ω_i is exactly of the form $2\pi p_i/N$ then $I_N(\omega_p)$ will exhibit a peak of height $(NA_i^2/2)$ at $p = p_i$, in agreement with (6.1.26). However, if ω_i is not of the form $2\pi p_i/N$ the height of the observed peak may be substantially reduced. The worst case occurs when ω_i falls midway between two neighbouring periodogram ordinates, say at $2\pi p/N$ and $2\pi(p+1)/N$. In this case the height of the peak is reduced by the factor $(4/\pi^2)$ (Whittle (1952a)) and, depending on the magnitude of A_i^2, this effect

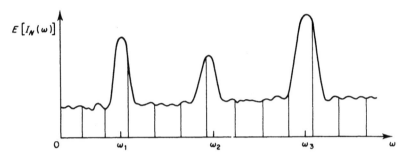

Fig. 6.4. The standard periodogram ordinates.

could be quite serious in terms of our ability to detect the presence of this frequency component. We are, of course, at liberty to compute $I_N(\omega)$ for any values of ω we choose, and one way of trying to surmount the above difficulty is to compute $I_N(\omega)$ at a finer grid of points. We might, for example, compute $I_N(\omega)$ for $\omega = \pi p/N$, or even for $\omega = \pi p/2N$, $p = 0, 1, 2, \ldots$. Alternatively, we could retain the standard grid, $\omega = 2\pi p/N$, $p = 0, 1, \ldots$, and compute additional values of $I_N(\omega)$ at a fine grid of points between two particular neighbouring ordinates, say between $2\pi p/N$ and $2\pi(p+1)/N$, if we suspect that one of the ω_i lies between these points.

This device is quite feasible from a computational point of view (and indeed has proved quite useful in practical applications), but it introduces complications in the distributional properties of the periodogram ordinates. For, as soon as we introduce values of $I_N(\omega)$ at frequencies other than the standard set $\{2\pi p/N\}$ *we lose the crucial property of independence between the different ordinates.* (This is a key property in the derivation of the test for periodogram ordinates to be discussed in the next section.) It could be argued that since $I_N(\omega)$ is a trigonometric polynomial of degree $(N-1)$ it is uniquely determined by its values at N points; in fact, it is easily shown (see Section 7.6) that by using "harmonic interpolation" we may, for any ω, express $I_N(\omega)$ purely in terms of $\{I_N(\omega'_p), \; p = 0, 1, \ldots, N-1\}$, where $\omega'_p = 2\pi p/(2N-1)$. This is certainly true, but all it means is that given the values of the $\{I_N(\omega'_p)\}$ we may compute the value of any other periodogram ordinate without recourse to the original observations, $X_1, \ldots, X_N$. It does not help us a great deal with the problem of *searching* for periodic components since our search method is essentially a graphical technique which is based on our ability to spot and identify abnormally large values in the *computed* set of periodogram ordinates.

Variance and covariance

It is convenient at this point to derive more general expressions for the variance and covariance of periodogram ordinates which will be required at a later stage. We consider only the "null" case of (6.1.2) (i.e. where $A_i = 0$, all i) and again assume that $\{X_t\}$ is a sequence of strictly *independent* random variables (not merely uncorrelated). However, whereas previously we assumed further that $\{X_t\}$ was a Gaussian process we now drop this assumption and allow the X_t to be non-normal.

From (6.1.35) we have for any ω_1, ω_2,

$$\text{cov}\{I_N(\omega_1), I_N(\omega_2)\} = 4 \sum_{s=0}^{N-1} \sum_{r=0}^{N-1} a(s)a(r) \cos s\omega_1 \cos r\omega_2 \, \text{cov}\{\hat{R}(s), \hat{R}(r)\}$$

(6.1.46)

where $a(0) = 1$, $a(s) = 2$, $s \geq 1$. To evaluate the above expression we use the exact result for $\text{cov}\{\hat{R}(r), \hat{R}(s)\}$, the general form of which is given by (5.3.21). In this case the assumption that the $\{X_t\}$ are independent leads to a considerable simplification since this implies that

(a) $\quad \kappa_4(s - t, r, v) = \begin{cases} \kappa_4 = \mu_4 - 3\sigma_X^4, & s - t = r = v = 0, \\ 0, & \text{otherwise,} \end{cases}$

(6.1.47)

(b) $\qquad\qquad\qquad R(s) = \begin{cases} \sigma_X^2, & s = 0, \\ 0, & s \neq 0, \end{cases}$

(6.1.48)

where κ_4 is the 4th cumulant, μ_4 the 4th moment, and σ_X^2 the variance of X_t. Using these results (5.3.21) reduces to,

$$\text{cov}\{\hat{R}(s), \hat{R}(r)\} = 0, \qquad r \neq s,$$

(6.1.49)

$$\text{var}\{\hat{R}(s)\} = \frac{(N-s)}{N^2}\sigma_X^4(1 + \delta_{0,s}) + \delta_{0,s}\frac{\kappa_4}{N}, \qquad s \geq 0.$$

(6.1.50)

Substituting these expressions in (6.1.46) we find,

$$\text{cov}[I_N(\omega_1), I_N(\omega_2)] = \frac{4\kappa_4}{N} + \frac{4\sigma_X^4}{N}\left[2 + 4\sum_{s=1}^{N-1}\left(1 - \frac{s}{N}\right)\cos s\omega_1 \cos s\omega_2\right]$$

(6.1.51)

$$= \frac{4\kappa_4}{N} + \frac{4\sigma_X^2}{N}\left[\sum_{s=-(N-1)}^{(N-1)}\left(1 - \frac{|s|}{N}\right)\{\cos s(\omega_1 + \omega_2) + \cos s(\omega_1 - \omega_2)\}\right].$$

(6.1.52)

Applying (6.1.43) we now obtain:

Theorem 6.1.3 *If the $\{X_t\}$ are independent with finite fourth order cumulant κ_4,*

$$\text{cov}\{I_N(\omega_1), I_N(\omega_2)\} = \frac{4\kappa_4}{N} + \frac{4\sigma_X^4}{N^2}\left[\frac{\sin^2(\tfrac{1}{2}N(\omega_1+\omega_2))}{\sin^2(\tfrac{1}{2}(\omega_1+\omega_2))} + \frac{\sin^2(\tfrac{1}{2}N(\omega_1-\omega_2))}{\sin^2(\tfrac{1}{2}(\omega_1-\omega_2))}\right]$$

$$(6.1.53)$$

This result, which was given by Bartlett (1955), p. 278, leads immediately to a number of interesting features, namely:

(1) Setting $\omega_1 = \omega_2 = \omega$, we have,

$$\text{var}\{I_N(\omega)\} = \begin{cases} 4\sigma_X^4 + \dfrac{4\kappa_4}{N} + O\left(\dfrac{1}{N^2}\right), & \omega \neq 0, \pm\pi \\[2ex] 8\sigma_X^4 + \dfrac{4\kappa_4}{N}, & \omega = 0, \pm\pi. \end{cases} \qquad (6.1.54)$$

In particular, when X_t is normal (so that $\kappa_4 = 0$), and ω is of the form $2\pi p/N$ (so that the term $O(1/N^2)$ is exactly zero), (6.1.54) reduces exactly to (6.1.32). However, (6.1.54) shows that (6.1.32) is still valid up to $O(1/N)$ even when X_t is non-normal and ω is not of the form $2\pi p/N$.

(2) Also, for $\omega_1 \neq \pm\omega_2$,

$$\text{cov}\{I_N(\omega_1), I_N(\omega_2)\} = 0, \text{ if } X_t \text{ normal (i.e. } \kappa_4 = 0) \text{ and } \omega_1, \omega_2,$$
both multiples of $2\pi/N$.

$$= O(1/N^2), \text{ if } X_t \text{ normal and } |\omega_1 \pm \omega_2| \gg 2\pi/N.$$

$$= O(1/N), \text{ if } X_t \text{ non-normal } (\kappa_4 \neq 0) \text{ and}$$
$|\omega_1 \pm \omega_2| \gg 2\pi/N, \text{ or } \omega_1, \omega_2, \text{ both multiples}$
of $2\pi/N$. $\qquad (6.1.55)$

The first statement in (6.1.55) is, of course, already implicit in the previously derived result that, for normal X_t, I_p and I_q are independently distributed $(p \neq q)$. The further statements in (6.1.55) show that even for non-normal X_t the periodogram ordinates are asymptotically uncorrelated if ω_1, ω_2, are both multiples of $(2\pi/N)$ or are spaced sufficiently far apart. If ω_1, ω_2, are fixed frequencies (independent of N) then the condition on the spacing between ω_1 and ω_2 is automatically satisfied.

We note that (6.1.53) can be re-written more concisely in the form,

$$\text{cov}\{I_N(\omega_1), I_N(\omega_2)\} = \frac{4\kappa_4}{N} + \frac{8\pi\sigma_X^4}{N}\{F_N(\omega_1+\omega_2) + F_N(\omega_1-\omega_2)\}, \qquad (6.1.56)$$

where $F_N(\theta)$ is the Fejer kernel defined by (6.1.43).

6.1.4 Tests for Periodogram Ordinates

If we observe that the periodogram contains a number of peaks we cannot conclude immediately that each of these corresponds to a genuine periodic component in X_t. For, even in the "null" case (i.e. when $A_i = 0$, all i) it is still possible that peaks may occur in the periodogram ordinates due merely to random sampling fluctuations. We must therefore apply a test to a periodogram peak to determine whether its value is significantly larger than that which would be likely to arise if there were no genuine periodic components in the model.

The usual procedure is to start by plotting the periodogram ordinates at the "standard frequencies" $\omega_p = 2\pi p/N$, $p = 0, 1, \ldots, [N/2]$, and then test the value of the *largest* observed peak. We consider the general model (6.1.2) and set up the null hypothesis, $H_0: A_i = 0$, all i (or equivalently, $K = 0$). Assuming that the $\{X_t\}$ are *normal* we know from theorem (6.1) that, under H_0, the ordinates $I_p (\equiv I_N(2\pi p/N))$, $p = 0, 1, \ldots, [N/2]$ are independently distributed, and for $p = 1, \ldots, [N/2]$, (N odd), each I_p has a distribution proportional to χ_2^2. Specifically, we have for each $p \neq 0$,

$$I_p/\sigma_X^2 = \chi_2^2.$$

Since χ^2 on two degrees of freedom is equivalent to an exponential distribution (see Section (2.11)), it follows that (I_p/σ_X^2) has probability density function,

$$f(x) = \tfrac{1}{2}\exp(-x/2), \qquad 0 \leq x < \infty. \tag{6.1.57}$$

Hence, for any $z\,(\geq 0)$,

$$p[(I_p/\sigma_X^2) \leq z] = \int_0^z \tfrac{1}{2}\exp(-x/2)\,dx = 1 - \exp(-z/2). \tag{6.1.58}$$

This distribution was used by Schuster (1898) to test the significance of a periodogram ordinate selected at random. Now let

$$\gamma = \left\{ \max_{1 \leq p \leq [N/2]} (I_p) \right\} \Big/ \sigma_X^2, \tag{6.1.59}$$

Under H_0, γ is the maximum of $[N/2]$ independent identically distributed exponential variables, and we have for any z,

$$
\begin{aligned}
p[\gamma > z] &= 1 - p[\gamma \leq z] \\
&= 1 - p[(I_p/\sigma_X^2) \leq z, \text{ all } p] \\
&= 1 - (1 - \exp(-z/2))^{[N/2]}, \tag{6.1.60}
\end{aligned}
$$

using (6.1.58) together with the property that the $\{I_p\}$ are independent.

Equation (6.1.60) gives us the distribution of γ under the null hypothesis, and *if the value σ_X^2 were known a priori* we could use this distribution to construct an exact test for the value of the maximum ordinate. Under the alternative hypothesis (not all $A_i = 0$) γ will be "large", and hence we use a one-sided test with critical region of the form, $\{\gamma > z_0\}$, where z_0 is chosen so that the RHS of (6.1.60) is equal to α, the significance level of the test.

In practice, however, σ_X^2 is almost invariably unknown and we then have to estimate it from the data. Since, under H_0, $E[I_p] = 2\sigma_X^2$ (all $p \geq 1$), we have,

$$E\left[\sum_{p=1}^{[N/2]} I_p\right] = 2\left[\frac{N}{2}\right]\sigma_X^2, \tag{6.1.61}$$

and hence

$$v = \frac{1}{2[N/2]} \cdot \sum_{p=1}^{[N/2]} I_p \tag{6.1.62}$$

is an unbiased estimate of σ_X^2. (In fact, it follows from (6.1.33) and the orthogonality relations (6.1.9) that,

$$\sum_{p=1}^{[N/2]} I_p \sim N\hat{R}(0) = \sum_{t=1}^{N} X_t^2, \tag{6.1.63}$$

so that v is essentially the usual expression for the sample variance of $\{X_t\}$.)

We may now construct a large sample test for $\max(I_p)$ by replacing σ_X^2 by v in (6.1.59), leading to the test statistic,

$$g^* = \frac{\max(I_p)}{\{1/2[N/2]\} \sum_{p=1}^{[N/2]} I_p}. \tag{6.1.64}$$

If, for large N, we neglect the sampling fluctuations of the denominator (i.e. treat the denominator as having the constant value σ_X^2) then g^* will have the same distribution as γ, so that asymptotically we have (from (6.1.60)),

$$p[g^* > z] \sim 1 - (1 - \exp(-z/2))^{[N/2]}. \tag{6.1.65}$$

This asymptotic distribution of g^* forms the basis of Walker's large sample test for $\max(I_p)$ (Walker (1914)).

However, Fisher (1929), in a celebrated paper, derived an *exact test* for $\max(I_p)$, based on the statistic,

$$g = \frac{\max(I_p)}{\sum_{p=1}^{[N/2]} I_p}, \tag{6.1.66}$$

the above expression being known as "*Fisher's g statistic*". Since σ_X^2 acts as a "proportionality" factor in the distributions of both I_p and $\sum_p I_p$, it is clear that the distribution of $I_p/(\sum_p I_p)$ will not involve σ_X^2. This feature is exactly the same as that which arises in Students' t-statistic and the device of dividing I_p by $(\sum_p I_p)$ to remove the dependence on σ_X^2 is sometimes called "*Studentization*".

Fisher showed that (for the case N odd) the exact distribution of g under H_0 is given by,

$$p[g > z] = n(1-z)^{n-1} - \frac{n(n-1)}{2}(1-2z)^{n-1}$$

$$+ \ldots + (-1)^a \frac{n!}{a!(n-a)!}(1-az)^{n-1}, \qquad (6.1.67)$$

where $n = [N/2]$, and a is the largest integer less than $(1/z)$. (Fisher derived this distribution by ingenious geometrical arguments, but Whittle (1952a) and Grenander and Rosenblatt (1957a) have given alternative analytical derivations.) The asymptotic distribution of g^* given by (6.1.65) amounts, effectively, to using only the first term of the exact distribution of g. To see this, we first note that since $g^* \equiv 2ng$, the first term of the exact distribution (6.1.66) gives,

$$p[g^* > z] = p\left[g > \frac{z}{2n}\right] \sim n\left(1 - \frac{z}{2n}\right)^{n-1} \sim n \exp(-z/2).$$

On the other hand, the asymptotic distribution (6.1.65) gives,

$$p[g^* > z] \sim 1 - (1 - \exp(-z/2))^n \sim n \exp(-z/2),$$

(using only the first term in the expansion of $(1 - \exp(-z/2))^n$).

The test procedure is now completed by choosing z_α so that, under the null distribution (6.1.66),

$$p[g > z_\alpha] = \alpha,$$

α being the chosen significance level of the test. Alternatively, when N is large we may use the asymptotic distribution (6.1.65) and choose z_α so that

$$p[g^* > z] = \alpha.$$

If the calculated value of g(or g^*) exceeds z_α, then max(I_p) is significant at $(100\alpha)\%$ and we reject H_0 and conclude that X_t contains a periodic component.

Mention may be made also of an alternative approach due to Hartley (1949) who treats the problem from an "analysis of variance" viewpoint. The main difference between Hartley's test and Fisher's test is the method by which max(I_p) is "Studentized" to enable an exact test to be constructed from the distribution (6.1.60) of γ. In Hartley's approach a specified number of periodic terms are fitted by least squares and an estimate of $\sigma_\varepsilon^2 (= \sigma_X^2$ under H_0) is then derived in the usual way from the residual sum of squares. An analysis of variance table is then formed in which the total sum of squares of the observations (about their mean) is decomposed into the sums of squares due to the "regression" on each of the periodic terms, plus the residual sum of squares. The test is then based on the "F_{max} ratio", i.e. the ratio of the largest "regression" sum of squares to the residual sum of squares.

Should the test of max(I_p) give a significant result, all that we may strictly infer is that the null hypothesis is discredited, and consequently that there exists a periodic component at *some* frequency ω. However, we usually wish to conclude more specifically that there exists a periodic component at the frequency corresponding to max(I_p). Suppose max(I_p) occurs at $p = p'$, corresponding to the frequency,

$$2\pi p'/N = \Omega,$$

say. Then Hartley (1949) has shown that the probability of a "misleading result" i.e. the probability that the significance of max(I_p) is due to a periodic component at some other frequency ω (where $|\omega - \Omega| \gg 2\pi/N$) is less than α, the significance level of the test. Thus, if max(I_p) is significant we may safely estimate the frequency, ω_1 (say), of the first detected periodic component by the location of the peak, i.e. we estimate ω_1 by

$$\hat{\omega}_1 = 2\pi p'/N = \Omega. \tag{6.1.68}$$

We then estimate the corresponding coefficients, A_1', B_1', by

$$\hat{A}_1' = \sqrt{\frac{2}{N}} A(\hat{\omega}_1), \qquad \hat{B}_1' = \sqrt{\frac{2}{N}} B(\hat{\omega}_1), \tag{6.1.69}$$

where $A(\omega)$, $B(\omega)$ are given by (6.1.22), (6.1.23) (cf. (6.1.25)), and the squared amplitude A_1^2 is estimated by

$$\hat{A}_1^2 = \hat{A}_1'^2 + \hat{B}_1'^2 = \frac{2}{N} \cdot I_{p'}, \tag{6.1.70}$$

in agreement with the form of estimate suggested by (6.1.26). If we replace $\hat{\omega}_1$ by its true value ω_1 in (6.1.69) then, as noted previously, A_1' and B_1' become the least squares estimates of A_1', B_1', when ω_1 is itself of the form $2\pi p/N$. If ω_1 is not of this form, $\hat{A}_1'$ and $\hat{B}_1'$ are equivalent to the least squares estimates to O($1/N$), provided Ω and $(\pi - \Omega)$ are $\gg 2\pi/N$ (see the discussion leading to equations (6.1.11), (6.1.12) and (6.1.20)). Whittle

(1952a) has shown that (on the assumption that the $\{\omega_i\}$ are spaced sufficiently well apart),

$$E(\hat{\omega}_i) = \omega_i + O(1/N), \qquad (6.1.71)$$

$$\text{var}(\hat{\omega}_i) = \frac{24\sigma_\varepsilon^2}{N^3(A_i'^2 + B_i'^2)} + o(1/N^3). \qquad (6.1.72)$$

The expression for $\text{var}(\hat{\omega}_i)$ is worthy of note since it shows that $\text{var}(\hat{\omega}_i) = O(1/N^3)$, in contrast with the usual situation in linear models where the variance of an estimated parameter is $O(1/N)$; see also Walker (1971).

Fisher's test deals only with the largest periodogram ordinate, but Whittle (1952a) suggested that it may be extended to provide a test for the second largest ordinate by omitting the term $I_{p'}$ from the denominator of the g statistic, and substituting $(n-1)$ for n in the distribution of g. Thus, if the largest ordinate $I_{p'}$ turns out to be significant, we may test the second largest ordinate $I_{p''}$ say, by using the statistic,

$$g' = \frac{I_{p''}}{\left\{\sum_{p=1}^{[N/2]} I_p\right\} - I_{p'}},$$

and referring to Fisher's g distribution with n replaced by $(n-1)$. If the second largest ordinate turns out to be significant we continue with the above procedure to test the third largest ordinate, and so on, until we fail to obtain a significant result. In this way, we obtain an estimate of K, the number of periodic components in the model (6.1.3).

This test procedure works well if all the true frequencies $\{\omega_i\}$ are exactly (or very close to) integer multiples of $(2\pi/N)$. However, if some of the ω_i are not of this form the power of the test will certainly be affected. The worst case occurs if one of the frequencies ω_k falls midway between two periodogram ordinates, say between $2\pi j/N$ and $2\pi(j+1)/N$. In this case it follows from (6.1.36) that, as mentioned previously, the height of the corresponding periodogram peak is reduced by the factor $4/\pi^2$ (or approximately 0·4), but Fisher's test will still be reasonably powerful provided what might be termed the *"signal to noise ratio"* $(A_k^2/\sigma_\varepsilon^2)$ is fairly large. If the signal to noise ratio $A_k^2/\sigma_\varepsilon^2$ is critical it is possible that the factor $(4/\pi^2)$ might cause this component to remain undetected. One way of attempting to remove this difficulty is to compute the periodogram ordinates at a smaller spacing than $2\pi/N$. This device will help the user to identify those situations when this difficulty is present, but, as has been pointed out, the correlation between the ordinates at such a smaller spacing will invalidate Fisher's test.

However, the main difficulty arises in the test of the second largest ordinate. For, if ω_k falls roughly midway between $2\pi j/N$ and $2\pi(j+1)/N$ the periodic component will provide substantial contributions to *both* these ordinates, and possibly also to one of the adjacent pair of ordinates at $2\pi(j-1)/N$ and $2\pi(j+2)/N$. Thus, if we suspect the existence of a second periodic component, and wish to test the second largest periodogram ordinate, it might be advisable to omit all ordinates which fall within $2\pi/N$ of the largest peak.

Grenander and Rosenblatt (1957a) have derived the distribution of the ratio of the rth largest periodogram ordinate to the sum of the periodogram ordinates, under the null hypothesis H_0 that X_t is a Gaussian purely random process. Denoting the rth largest periodogram ordinate by $I_{(r)}$, Grenander and Rosenblatt have shown that the null distribution of $g_{(r)} = I_{(r)}/\{\sum_p I_p\}$ is given by

$$p[g_{(r)} > z] = \frac{n!}{(r-1)!} \sum_{j=1}^{a} \frac{(-1)^{j-r}(1-jz)^{n-1}}{j(n-j)!(j-r)!}, \qquad (6.1.73)$$

where, as before, a is the largest integer less than $(1/z)$. (When $r=1$, (6.1.73) reduces, of course, to (6.1.67), the distribution of $g \equiv g_{(1)}$.) This distribution provides an appropriate test for the situation where we know that if the null hypothesis is false, X_t *must contain exactly r periodic terms, the values of r being known a priori.* (Percentage points for the distributions of the $\{g_{(r)}\}$ have been tabulated by Shimshoni (1971), for $r = 1, 2, 5, 7, 10, 25, 50$.) However, if the number of periodic terms is unknown we cannot simply apply a sequence of tests based on $g \ (= g_{(1)})$, $g_{(2)}$, $g_{(3)}$, . . . since *the distribution (6.1.73) holds only in the null case* and if the maximum periodogram ordinate turns out to be significant the null distributions for $g_{(2)}$, $g_{(3)}$, . . . are no longer appropriate. It may be noted further that if X_t contains a number of periodic terms, all of comparable amplitudes, then the power of the g-test to detect these may be substantially reduced. For, if, e.g., X_t contains three terms with equal squared amplitudes then the denominator in (6.1.66) will include three periodogram ordinates each of which is comparable in value to $\max(I_p)$, and consequently the value of g will be substantially smaller than it would have been if there was only one periodic term present. Hartley's alternative test (referred to above) does not suffer from this disadvantage, since the denominator here (which estimates the residual variance) is based on the residual sum of squares and thus all the periodogram ordinates which one wishes to test are excluded from this expression. This means that when testing a particular ordinate the denominator is not "contaminated" by the possible presence of periodic terms at other frequencies. However, in Hartley's test the total number of periodogram ordinates to be tested has to be specified *a priori* (and must be considerably smaller than $n = [N/2]$ in order to leave sufficient degrees of freedom for the residual sum of squares).

Canadian lynx data

The Canadian lynx data were discussed in Section 5.5 where we fitted various types of ARMA models. However, we noted that these data showed a strong cyclical behaviour, and we now examine the form of the periodogram, shown in Graph 6.1. The periodogram (computed from 114

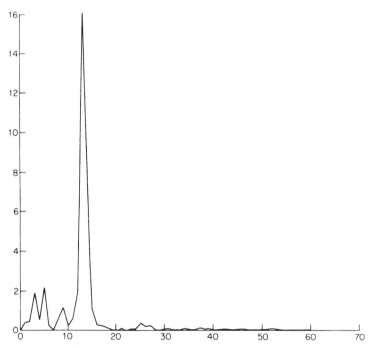

Graph 6.1. Periodogram of the Canadian lynx series.

observations) is dominated by a very large peak occurring at $p = 12$ (corresponding to the frequency $2\pi \times (12/114)$), and it is clear that if we were to test the magnitude of this peak on the null hypothesis of a purely random process we would obtain a highly significant result. (Note that the periodogram also contains minor peaks at the harmonics of the above frequency—although, due to the scaling of the graph, these do not stand out as clearly as the main peak. However, the harmonic frequencies are clearly visible in the estimated spectral density function when this is plotted on a logarithmic scale; see Graph 7.13. The presence of these harmonics is due to the asymmetrical shape of the individual cycles in the data.) The periodogram ordinates shown in the graph were computed using the "fast Fourier transform" algorithm (see Section 7.6) and are consequently evaluated at the frequencies $(2\pi p/128)$. The peak ordinate shown in the graph is therefore slightly smaller than that obtained at the frequency $(24\pi/114)$. Carrying out the formal g-test, we have

$$I_{12} = 21 \cdot 0246, \qquad \sum_{p=1}^{57} I_p = 35 \cdot 23,$$

giving the value of Fisher's g-statistic as

$$g = 0 \cdot 5968.$$

This value is, as expected, highly significant. (Using just the first term in (6.1.67), we find that, with $N = 114$, the upper 5%, 1% and 0.1% points of the g-distribution are approximately $0 \cdot 1181$, $0 \cdot 1431$ and $0 \cdot 1776$ respectively.) It is important to remember that this test confirms the presence of a periodic component *only under the null hypothesis of a purely random process*. If we consider the more general model in which we have a periodic component superimposed on a residual process having a general continuous spectrum, then the g-test is no longer valid and we have to treat the problem by methods appropriate for analysing *"mixed spectra"*. We examine this type of problem in Chapter 8.

Some further examples of the application of Fisher's g-test are given in Chapter 8 when we discuss the more general problem of "mixed spectra".

Secondary analysis

If the true frequencies $\{\omega_i\}$ are not all multiples of $(2\pi/N)$ it is possible to improve on the estimate (6.1.68) by using a technique called "secondary analysis". This technique, which is well known to meteorologists and oceanographers, was developed initially for the purely deterministic case (where the residual ε_t is excluded), and a description appropriate to this case is given by Brunt (1917), pp. 196–199. To illustrate the basic ideas, let us consider the case where X_t contains just one periodic term and no residual, so that we may write,

$$X_t = A \cos(\omega_0 t + \phi),$$

say. Suppose that the maximum periodogram ordinate leads to an estimated frequency ω_1 and let $T_1 = 2\pi/\omega_1$ denote the corresponding estimated period. We now divide up the N observations into a number of groups, each group containing mT_1 observations, say, and perform a separate harmonic analysis for each group. For the sth group the estimated amplitudes are given by,

$$\hat{A}'_s = \left(\frac{2}{mT_1}\right) \sum_{t=(s-1)mT_1}^{smT_1} A \cos(\omega_0 t + \phi) \cos(\omega_1 t),$$

$$\hat{B}'_s = \left(\frac{2}{mT_1}\right) \sum_{t=(s-1)mT_1}^{smT_1} A \cos(\omega_0 t + \phi) \sin(\omega_1 t).$$

Writing $t' = smT_1 - t$,

$$\left(\frac{mT_1}{2}\right)\hat{A}'_s = \frac{A}{2} \sum_{t'=0}^{mT_1} [\cos\{(\omega_1 + \omega_0)(smT_1 - t') + \phi\}$$

$$+ \cos(\omega_1 - \omega_0)(smT_1 - t') - \phi\}],$$

with a similar expression for $(mT_1/2)\hat{B}'_s$. Hence,

$$(mT_1/2)(\hat{A}'_s + i\hat{B}'_s) = \tfrac{1}{2}A\left[\sum_{t'} \exp[i(\omega_1 + \omega_0)(smT_1 - t') + \phi] \right.$$

$$+ \exp[i(\omega_1 - \omega_0)(smT_1 - t') - \phi]\Big]$$

$$= \tfrac{1}{2}A|G(\omega_1 + \omega_0)| \exp[i(\omega_1 + \omega_0)smT_1 + \phi + \psi_1]$$

$$+ \tfrac{1}{2}A|G(\omega_1 - \omega_0)| \exp[i(\omega_1 - \omega_0)smT_1 - \phi + \psi_2],$$

where

$$G(\theta) = \sum_{t'=0}^{mT_1} \exp(-it'\theta), \quad \text{and} \quad \psi_1 = \arg\{G(\omega_1 + \omega_0)\}, \quad \psi_2 = \arg\{G(\omega_1 - \omega_0)\}.$$

Now if mT_1 is sufficiently large (and ω_0 sufficiently removed from zero),

$$|G(\omega_1 + \omega_0)| = \frac{\sin(\tfrac{1}{2}mT_1(\omega_1 + \omega_0))}{\sin(\tfrac{1}{2}(\omega_1 + \omega_0))}$$

will be small and hence we can neglect the first term above. We then have, as an approximation,

$$\phi_s = \tan^{-1}(\hat{B}'_s/\hat{A}'_s) \doteq (\omega_1 - \omega_0)smT_1 - \phi + \psi_2 \qquad (\text{mod } 2\pi).$$

If we now compute $\hat{A}'_s$, $\hat{B}'_s$ for each group, evaluate ϕ_s, and plot ϕ_s against s, then the graph should be roughly linear with slope.

$$\beta \doteq (\omega_1 - \omega_0)mT_1 = 2\pi m(1 - (\omega_0/\omega_1)),$$

so that

$$\omega_0 \doteq \omega_1(1 - (\beta/2\pi m)).$$

The value of β may be estimated by regression analysis (or "by eye"), and, when substituted in the above expression for ω_0, leads to an improved estimate of the true frequency. Alternatively, if we let $T_0 = 2\pi/\omega_0$ denote the true period, then,

$$T_0 \sim T_1/(1 - (\beta/2\pi m)).$$

This method works quite well in practice (see, e.g., Campbell and Walker (1977) who applied it to the estimation of the period of the Canadian lynx data), but there is a snag in the estimation of the slope β in that the phases ϕ_s are defined only mod 2π by the values of $\hat{A}'_s$ and $\hat{B}'_s$. This means that some values of ϕ_s may have to be "shifted" by multiples of 2π in order to obtain an approximately linear relationship between ϕ_s and s.

The method of "secondary analysis" is closely related to a more general technique due to Tukey called "*complex demodulation*"; see, e.g. Tukey (1961), Bingham, Godfrey and Tukey (1967), and Section 11.2.2.

Canadian lynx data: Campbell and Walker (1977) used the method of secondary analysis to obtain a more accurate estimate of the frequency of a fitted periodic component in the Canadian lynx data. Here, the maximum periodogram ordinate occurs at the frequency $2\pi(12/114)$, and taking $T_1 = 28/3$ (years), $m = 3$, and using the four groups of data given by the first 112 observations, they found that $\beta \doteq 33°$, giving a revised estimated period, $T_0 \doteq 9.63$ years.

6.2 CONTINUOUS SPECTRA

We now consider the case where the observed process $\{X_t\}$ has a purely continuous spectrum so that its spectral density function exists for all ω. We recall from Chapter 4 that most of the standard models (including the AR, MA, and ARMA) do, in fact, have purely continuous spectra, the only exception being the harmonic process model which, as we have just seen, has a purely discrete spectrum. Now the AR, MA, and ARMA models may be regarded as special cases of the *general linear* process model (see Section 3.5.7), which, in the discrete parameter case, takes the form,

$$X_t = \sum_{u=0}^{\infty} g_u \varepsilon_{t-u}, \tag{6.2.1}$$

where $\{\varepsilon_t\}$ *is a purely random process.* If we now generalize this model one stage further by extending the summation on the RHS of (6.2.1) to the "two-sided" form, i.e. we allow u to run from $-\infty$ to $+\infty$, then we obtain,

$$X_t = \sum_{u=-\infty}^{\infty} g_u \varepsilon_{t-u}, \tag{6.2.2}$$

and this may be taken as the prototype model for processes with purely continuous spectra (see the discussion following equation (3.5.70), Section 3.5.7). We will therefore take (6.2.2) as our standard reference model for continuous spectra processes, and assume that $\sum_{u=-\infty}^{\infty} g_u^2 < \infty$ (the condition required for X_t to be stationary with finite variance), and that $E(\varepsilon_t) = 0$, so that $E(X_t) = 0$, and $E(\varepsilon_t^2) = \sigma_\varepsilon^2$. Given a sample of N observations $X_1, X_2, \ldots, X_N$ the problem now is to estimate, for all ω in $(-\pi, \pi)$, either the non-normalized spectral density function $h(\omega)$ or the normalized spectral density function $f(\omega)$ of $\{X_t\}$. Since $f(\omega) \equiv h(\omega)/\sigma_X^2$ (cf. Section 4.8.1), it makes very little difference whether we choose to estimate $h(\omega)$ or $f(\omega)$; once we have constructed an estimate $\hat{h}(\omega)$ of $h(\omega)$ we can easily obtain an estimate of $f(\omega)$ simply by dividing $\hat{h}(\omega)$ by s_X^2, where s_X^2 is the variance of the sample observations. However, the sampling properties of estimates of $h(\omega)$ are somewhat simpler than those of estimates of $f(\omega)$, and consequently we shall study mainly the estimation of $h(\omega)$.

For real valued discrete parameter processes we have from (4.8.36) (Section 4.8.3),

$$h(\omega) = \frac{1}{2\pi} \sum_{s=-\infty}^{\infty} R(s) \cos s\omega, \qquad -\pi \le \omega \le \pi, \qquad (6.2.3)$$

where $R(s) = E(X_t X_{t+s})$ (with $E(X_t) = 0$) is the autocovariance function of $\{X_t\}$.

We assume throughout (unless otherwise stated) that $h(\omega)$ is continuous for all ω; this will certainly hold if $R(s)$ is absolutely summable, i.e. if $\sum_{s=-\infty}^{\infty} |R(s)| < \infty$.

We saw in Chapter 5 (Section 5.3.3) that we can estimate $R(s)$ for $s = 0, \pm 1, \ldots, \pm(N-1)$ (but *not* for $|s| \ge N$), and hence it would seem that the natural way to estimate $h(\omega)$ would be to replace $R(s)$ in (6.2.3) by its sample estimate, $\hat{R}(s)$, and truncate the summation over s at the points $-(N-1)$ and $(N-1)$. The interesting point is that *we have already met a function which is almost exactly of this form* (albeit in a different context), *namely, the periodogram.* In fact, if we write,

$$I_N^*(\omega) = \frac{1}{4\pi} I_N(\omega), \qquad (6.2.4)$$

where $I_N(\omega)$ is the periodogram of $X_1, X_2, \ldots, X_N$, as defined by (6.1.24), then it follows from (6.1.33) that,

$$I_N^*(\omega) = \frac{1}{2\pi} \sum_{s=-(N-1)}^{(N-1)} \hat{R}(s) \cos s\omega, \qquad (6.2.5)$$

where

$$\hat{R}(s) = \frac{1}{N} \sum_{t=1}^{N-|s|} X_t X_{t+|s|}, \qquad (6.2.6)$$

is the sample estimate of $R(s)$ appropriate to the case where $E(X_t) = 0$ (cf. (5.3.17)). Comparing (6.2.5) with (6.2.3) we see that $I_N^*(\omega)$ is just the sample version of $h(\omega)$, and for this reason $I_N^*(\omega)$ is sometimes referred to as the "*sample spectral density function*". However, because of its close relationship with the periodogram we shall call $I_N^*(\omega)$ the "*modified periodogram*", or (when it is unlikely to cause confusion) we shall refer to it simply as the "*periodogram*". If it is not known *a priori* that $E(X_t) = 0$ we would compute $\hat{R}(s)$ from the more general expression (5.3.13), which is equivalent to applying (6.2.6) to $X_t' = (X_t - \bar{X})$ $(t = 1, \ldots, N)$, $\bar{X}$ being the sample mean. The RHS of (6.2.5) then becomes identical to the (modified) periodogram of X_t'. (It may be noted that the effect of subtracting $\bar{X}$ to correct for a non-zero mean affects mainly the values of $I_N^*(\omega)$ in the neighbourhood of $\omega = 0$. Using an obvious notation we have, $I_{N,X'}^*(2\pi p/N) \equiv I_{N,X}^*(2\pi p/N)$, $(2\pi p/N) \neq 0, \pi$, by virtue of the orthogonality relations (6.1.9) applied to the functions $A_{X'}(\omega)$, $B_{X'}(\omega)$.)

It follows immediately from (6.2.5) that,

$$E[I_N^*(\omega)] = \frac{1}{2\pi} \sum_{s=-(N-1)}^{(N-1)} E[\hat{R}(s)] \cos s\omega,$$

$$= \frac{1}{2\pi} \sum_{s=-(N-1)}^{(N-1)} \left(1 - \frac{|s|}{N}\right) R(s) \cos s\omega, \qquad (6.2.7)$$

using (5.3.14) (which is exact when $E(X_t) = 0$). As $N \to \infty$ the factor $(1 - |s|/N) \to 1$ for each s (cf. the discussion leading from (5.3.5) to (5.3.6)) and hence,

$$E[I_N^*(\omega)] \to \frac{1}{2\pi} \sum_{s=-\infty}^{\infty} R(s) \cos s\omega = h(\omega) \qquad (6.2.8)$$

(since h is assumed to be continuous), showing that, for each ω, $I_N^*(\omega)$ *is an asymptotically unbiased estimate of* $h(\omega)$. This result may be established more rigorously by writing (using the real form of (4.8.34))

$$R(s) = \int_{-\pi}^{\pi} h(\theta) \cos s\theta \, d\theta, \qquad (6.2.9)$$

which, when substituted in (6.2.7) gives,

$$E[I_N^*(\omega)] = \int_{-\pi}^{\pi} h(\theta) \left[\frac{1}{4\pi} \sum_{s=-(N-1)}^{(N-1)} \left(1 - \frac{|s|}{N}\right) \{\cos s(\theta + \omega) + \cos s(\theta - \omega)\}\right] d\theta,$$

$$= \int_{-\pi}^{\pi} h(\theta) \{\tfrac{1}{2} F_N(\theta + \omega) + \tfrac{1}{2} F_N(\theta - \omega)\} d\theta,$$

where, from (6.1.43),

$$F_N(\theta) = \frac{1}{2\pi} \cdot \frac{\sin^2(N\theta/2)}{N\sin^2(\theta/2)},$$
(6.2.10)

is the Fejer kernel. Changing the variable of integration from θ to $(-\theta)$ in the first term and noting that both $h(\theta)$ and $F_N(\theta)$ are even functions of θ, we finally obtain,

$$E[I_N^*(\omega)] = \int_{-\pi}^{\pi} h(\theta) F_N(\theta - \omega) \, d\theta.$$
(6.2.11)

Now it is well known that as, $N \to \infty$, $F_N(\theta) \to \delta(\theta)$ (the Dirac δ-function). This is easily verified by noting (a) that for all $\theta \neq 0$, $F_N(\theta) \to 0$ as $N \to \infty$, and (b) that $\int_{-\pi}^{\pi} F_N(\theta) \, d\theta = 1$, (see, e.g., Titchmarsh (1939), p. 413). Thus, since $h(\theta)$ is assumed continuous at $\theta = \omega$ the RHS of (6.2.11) converges, as $N \to \infty$, to $h(\omega)$, in agreement with (6.2.8).

Moreover, if $h(\omega)$ is sufficiently "smooth" we can determine the order of magnitude of the bias in $I_N^*(\omega)$ as an estimate of $h(\omega)$. Specifically, *if $h(\omega)$ satisfies a Lipshitz condition of order 1*, i.e. if $|h(\omega_1) - h(\omega_2)| < K|\omega_1 - \omega_2|$ as $|\omega_1 - \omega_2| \to 0$, K being a constant, then it follows from a further result due to Fejer (1910) that

$$E[I_N^*(\omega)] = \int_{-\pi}^{\pi} h(\theta) F_N(\theta - \omega) \, d\theta = h(\omega) + O\left(\frac{\log N}{N}\right).$$
(6.2.12)

The bias in $I_N^*(\omega)$ is thus $O((\log N)/N)$ if $h(\omega)$ satisfies the above condition. Note that the condition that $h(\omega)$ *has a bounded first derivative* is certainly sufficient since (by the mean value theorem) this implies that $h(\omega)$ satisfies a Lipshitz condition of order 1.

6.2.1 Finite Fourier Transforms

It is instructive to consider an alternative derivation of (6.2.11), based on the spectral representation of X_t. We first introduce a function called the *"finite Fourier transform"* of the observations, which is defined by

$$\zeta_X(\omega) = \frac{1}{\sqrt{2\pi N}} \sum_{t=1}^{N} X_t e^{-i\omega t}, \qquad -\pi \leq \omega \leq \pi.$$
(6.2.13)

(Note that $\zeta_X(\omega)$ may be expressed in terms of the finite cosine and sine transforms, $A(\omega)$, $B(\omega)$, (6.1.22), (6.1.23), by

$$\zeta_X(\omega) = (1/\sqrt{4\pi})\{A(\omega) - iB(\omega)\}.)$$

Using the spectral representation of X_t, viz. (cf. (4.11.19), Section 4.11),

$$X_t = \int_{-\pi}^{\pi} e^{it\theta} \, dZ(\theta),$$

where the process $Z(\theta)$ has orthogonal increments, (6.2.13) may be written as,

$$\zeta_X(\omega) = \int_{-\pi}^{\pi} \left(\frac{1}{\sqrt{2\pi N}} \sum_{t=1}^{N} e^{it(\theta - \omega)} \right) dZ(\theta). \tag{6.2.14}$$

But, from (5.3.63), (5.3.64),

$$\sum_{t=1}^{N} e^{it\phi} = e^{i(N+1)\phi/2} \cdot \frac{\sin(N\phi/2)}{\sin(\phi/2)}. \tag{6.2.15}$$

Hence we may write

$$\zeta_X(\omega) = \int_{-\pi}^{\pi} F_N^{1/2}(\theta - \omega) \exp[i(N+1)(\theta - \omega)/2] \, dZ(\theta), \tag{6.2.16}$$

where $F_N(\theta)$ is the Fejer kernel given by (6.2.10). Now it follows immediately from (6.1.24) and (6.2.4), that the periodogram $I_N^*(\omega)$ may be written as,

$$I_N^*(\omega) = |\zeta_X(\omega)|^2 = \zeta_X(\omega)\zeta_X^*(\omega) \tag{6.2.17}$$

(* denoting the complex conjugate), and hence from (6.2.16),

$$E[I_N^*(\omega)] = \int_{-\pi}^{\pi} \int_{-\pi}^{\pi} F_N^{1/2}(\theta - \omega) \exp[i(N+1)(\theta - \omega)/2] F_N^{1/2}(\theta' - \omega)$$

$$\times \exp[-i(N+1)(\theta' - \omega)/2] E[dZ(\theta) \, dZ^*(\theta')]$$

Since $Z(\theta)$ has orthogonal increments, $E[dZ(\theta) \, dZ^*(\theta')] = 0$, $\theta \neq \theta'$, while $E[|dZ(\theta)|^2] = h(\theta) \, d\theta$ (cf. (4.11.13), Section 4.11). The above double integral thus reduces to a single integral (obtained by setting $\theta' = \theta$) and we obtain,

$$E[I_N^*(\omega)] = \int_{-\pi}^{\pi} F_N(\theta - \omega) E[|dZ(\theta)|^2] = \int_{-\pi}^{\pi} F_N(\theta - \omega) h(\theta) \, d\theta, \tag{6.2.18}$$

in agreement with (6.2.11).

However, (6.2.16) is an interesting equation and contains rather more information than we have so far exploited in deriving (6.2.18).

Although the function $F_N^{1/2}(\theta)$ does not strictly tend to a δ-function as $N \to \infty$ it does, nevertheless, behave in a similar way to a δ-function. In particular we have (as with $F_N(\theta)$) that $F_N^{1/2}(\theta) \to 0$ as $N \to \infty$ for all $\theta \neq 0$, while as $\theta \to 0$, $F_N^{1/2}(\theta) \to \sqrt{N/2\pi}$. Hence, as $N \to \infty$, $F_N^{1/2}(\theta)$ vanishes everywhere except at the origin.

If we now consider the set of finite Fourier transforms at the frequencies $\omega_p = 2\pi p/N$, $p = 0, 1, \ldots, [N/2]$, then for N large we may write from (6.2.16),

$$\zeta_X(\omega_p) \sim \sqrt{\frac{N}{2\pi}} \Delta Z(\omega_p) = \frac{\Delta Z(\omega_p)}{(\Delta \omega_p)^{1/2}}, \qquad (6.2.19)$$

where $\Delta Z(\omega_p) = Z(\omega_p) - Z(\omega_{p-1})$, and $\Delta \omega_p = \omega_p - \omega_{p-1} = 2\pi/N$. Equation (6.2.19) is interesting in that it shows the close relationship between the finite Fourier transforms $\zeta_X(\omega_p)$ and the increments $\Delta Z(\omega_p)$. That such a relationship exists is fairly obvious when we realize that the spectral representation tells us, loosely speaking, that the $\{X_t\}$ are the Fourier coefficients of the function "$dZ(\theta)/d\theta$", so that "$dZ(\theta)/d\theta$" can, in a sense, be recovered by constructing the Fourier series in $\{X_t\}$. (A more precise formulation of this idea is given (for the continuous parameter case) by the result (4.11.14), Section 4.11.) This is, in effect, just what we have done in constructing the function $\zeta_X(\omega)$. Another way of seeing this result is to note that $\zeta_X(\omega_p)$ is simply the discrete parameter analogue of the function $\{G_T(\omega_n)/\sqrt{2T}\}$ introduced in the derivation of the spectral representation in Section 4.11—see (4.11.9).

Equation (6.2.19) now gives us immediately that,

$$E[I_N^*(\omega_p)] = E[|\zeta_X(\omega_p)|^2] \sim E[|\Delta Z(\omega_p)|^2]/\Delta \omega_p,$$

$$= \Delta H(\omega_p)/\Delta \omega_p, \qquad \text{(cf. 4.11.17)},$$

$$\to h(\omega_p),$$

as $N \to \infty$, and correspondingly $\Delta \omega_p \to 0$. (Above, $H(\omega) = \int_{-\infty}^{\omega} h(\theta)\, d\theta$ is the non-normalized integrated spectrum.) This result provides yet another derivation of (6.2.8).

6.2.2 Properties of the Periodogram of a Linear Process

We have seen (6.2.8) that the periodogram $I_N^*(\omega)$ is, for each ω, an asymptotically unbiased estimate of the (non-normalized) spectral density function $h(\omega)$. However, despite this quite appealing property it turns out that, in its "raw state", *the periodogram is an extremely poor (if not a useless) estimate of the spectral density function.* The reasons for this rather surprising result are,

(a) $I_N^*(\omega)$ *is not a consistent estimate of* $h(\omega)$*, in the sense that* var$\{I_N^*(\omega)\}$ *does not tend to zero as* $N \to \infty$.

(b) *As a function of ω, $I_N^*(\omega)$ typically has an erratic and wildly fluctuating form*—this feature is a consequence of the fact that for any two fixed neighbouring frequencies, ω_1, ω_2, $\text{cov}\{I_N^*(\omega_1), I_N^*(\omega_2)\}$ *decreases as N increases*. (The magnitude of the correlation between neighbouring ordinates determines the general overall "smoothness" of the function–the lower the correlation the more erratic the behaviour of the function).

To establish the above properties we need to extend the results of Section 6.1.3 on the second order moments of the periodogram to the case when $\{X_t\}$ is a general linear process of the form (6.2.2). (The results of Section 6.1.3 apply only when $\{X_t\}$ is an independent sequence.) In extending these results we rely heavily on an extremely important relationship between the periodogram of $\{X_t\}$ and the periodogram of the residual process $\{\varepsilon_t\}$ in (6.2.2). The general form of this relationship may be obtained heuristically in the following way.

From (6.2.2) we may think of $\{X_t\}$ as a "filtered version" of $\{\varepsilon_t\}$, and it then follows from the discrete parameter analogue of (4.12.13) that (with an obvious notation),

$$dZ_X(\omega) = \Gamma(\omega)\,dZ_\varepsilon(\omega), \qquad (6.2.20)$$

where

$$\Gamma(\omega) = \sum_{u=-\infty}^{\infty} g_u\,e^{-i\omega u}, \qquad (6.2.21)$$

(cf. (4.12.17)), and we now denote the orthogonal processes which appear in the spectral representations of $\{X_t\}$, $\{\varepsilon_t\}$ by $dZ_X(\omega)$, $dZ_\varepsilon(\omega)$ respectively. Using the asymptotic relationship (6.2.19) between $\zeta_X(\omega)$ and $dZ_X(\omega)$ we may deduce from (6.2.20) that,

$$\zeta_X(\omega) \sim \Gamma(\omega)\zeta_\varepsilon(\omega), \qquad (6.2.22)$$

where $\zeta_X(\omega)$ is given by (6.2.13), and $\zeta_\varepsilon(\omega)$ is defined in exactly the same way for $(\varepsilon_1, \ldots, \varepsilon_N)$. We now have (using (6.2.17)),

$$I_{N,X}^*(\omega) \sim |\Gamma(\omega)|^2 I_{N,\varepsilon}^*(\omega), \qquad (6.2.23)$$

where $I_{N,X}^*(\omega)$, $I_{N,\varepsilon}^*(\omega)$ now denote the periodograms of $(X_1, \ldots, X_N)$, $(\varepsilon_1, \ldots, \varepsilon_N)$ respectively. In fact, if we regard $I_{N,X}^*(\omega)$ as the sample version of $h_X(\omega)$ (cf. (6.2.3), (6.2.5)) then (6.2.23) may be thought of simply as the sample version of the basic filter relationship (4.12.18). Now since $\{\varepsilon_t\}$ is a purely random process its (non-normalized) spectral density function is

$$h_\varepsilon(\omega) = \sigma_\varepsilon^2/2\pi, \qquad -\pi \leqslant \omega \leqslant \pi \qquad (6.2.24)$$

and (4.12.18) gives

$$h_X(\omega) = (\sigma_\varepsilon^2/2\pi) \cdot |\Gamma(\omega)|^2, \qquad (6.2.25)$$

where $\Gamma(\omega)$ is again given by (6.2.21). We may now re-write (6.2.23) as,

$$I_{N,X}^*(\omega) \sim 2\pi h_X(\omega) \cdot \frac{1}{\sigma_\varepsilon^2} \cdot I_{N,\varepsilon}^*(\omega). \qquad (6.2.26)$$

The asymptotic relationship (6.2.26) was given by Bartlett (1955) (cf. also Sargen (1953)) and is a key result. It enables us to derive approximate expressions for the sampling properties of $I_{N,X}^*(\omega)$ immediately from the corresponding properties of $I_{N,\varepsilon}^*(\omega)$; under the additional assumption that the $\{\varepsilon_t\}$ are independent (not merely uncorrelated) the properties of $I_{N,\varepsilon}^*(\omega)$ are, in turn, covered by the results of Section 6.1.3. However, our heuristic derivation of (6.2.26) does not indicate the order of magnitude of the error incurred if we treat (6.2.26) as an equality for finite N. To investigate this point further we now adopt a more rigorous approach and reformulate (6.2.22) and (6.2.26) in the following more precise forms.

Theorem 6.2.1 *Let $\{X_t\}$ be a general linear process of the form* (6.2.2) *in which the $\{\varepsilon_t\}$ are uncorrelated with $E(\varepsilon_t) = 0$, $E(\varepsilon_t^2) = \sigma_\varepsilon^2$, and $\sum_{u=-\infty}^\infty |g_u||u|^{1/2} < \infty$. Then,*

$$\zeta_X(\omega) = \Gamma(\omega)\zeta_\varepsilon(\omega) + r_N(\omega), \qquad (6.2.27)$$

where $E[|r_N(\omega)|^2] = O(1/N)$, uniformly in ω. Moreover, if the $\{\varepsilon_t\}$ are independent with $E(\varepsilon_t^4) < \infty$, then $E[|r_N(\omega)|^4] = O(1/N^2)$, uniformly in ω.

Proof. We have,

$$\zeta_X(\omega) = \frac{1}{\sqrt{2\pi N}} \sum_{t=1}^N X_t \exp(-i\omega t)$$

$$= \frac{1}{\sqrt{2\pi N}} \sum_{u=-\infty}^\infty \sum_{t=1}^N \exp[-i\omega(t-u)]\varepsilon_{t-u} \exp(-i\omega u)g_u$$

$$= \frac{1}{\sqrt{2\pi N}} \sum_{u=-\infty}^\infty \exp(-i\omega u)g_u\left(\sum_{t=1-u}^{N-u} \varepsilon_t \exp(-i\omega t)\right)$$

$$= \frac{1}{\sqrt{2\pi N}} \sum_{u=-\infty}^\infty \exp(-i\omega u)g_u\left\{\left(\sum_{t=1}^N - \sum_t{}' + \sum_t{}''\right)\varepsilon_t \exp(-i\omega t)\right\}, \qquad (6.2.28)$$

where $\sum'$ runs from

$$t = (N-u+1) \text{ to } t = N, \qquad u > 0,$$

or

$$t = 1 \text{ to } -u, \qquad u < 0,$$

and $\sum''$ runs from

$$t = 1 - u \text{ to } 0, \qquad\qquad u > 0,$$

or

$$t = N + 1 \text{ to } N - u, \qquad\quad u < 0.$$

Hence we may write $\zeta_X(\omega)$ in the form (6.2.27) where

$$r_N(\omega) = \frac{1}{\sqrt{2\pi N}} \sum_{u=-\infty}^{\infty} \exp(-i\omega u) g_u \theta(u),$$

with

$$\theta(u) = \left(\sum_t'' - \sum_t'\right) \varepsilon_t \exp(-i\omega t).$$

Since the number of terms in both $\sum''$ and $\sum'$ is $O(|u|)$, and since the ε_t are uncorrelated, we have,

$$E[|\theta(u)|^2] \leq K|u|\sigma_\varepsilon^2,$$

(K a constant independent of ω), and hence,

$$E[|r_N(\omega)|^2] \leq \frac{1}{2\pi N} \left[\sum_{u=-\infty}^{\infty} |g_u| \{E(|\theta(u)|^2\}^{1/2} \right]^2$$

$$\leq \frac{K\sigma_\varepsilon^2}{2\pi N} \left[\sum_{u=-\infty}^{\infty} |g_u| |u|^{1/2} \right]^2. \tag{6.2.29}$$

Hence, provided $\sum_{u=-\infty}^{\infty} |g_u| |u|^{1/2} < \infty$, $E\|r_N(\omega)|^2\| = O(1/N)$, uniformly in ω. The "remainder term", $r_N(\omega)$, thus has mean square which is $O(1/N)$, and this is sometimes expressed by writing,

$$\zeta_X(\omega) = \Gamma(\omega)\zeta_Y(\omega) + O_R(1/N^{1/2}), \tag{6.2.30}$$

where $O_R(1/N^{1/2})$ denotes a random variable which is $O(1/N^{1/2})$ in the sense that its mean square is $O(1/N)$.

The fourth absolute moment of $r_N(\omega)$ depends on the fourth moments of $\theta(u)$ and the proof of the second part of the theorem follows by an obvious modification of the above discussion under the additional assumption that the $\{\varepsilon_t\}$ are independent with finite fourth order moments.

The condition $\sum_{u=-\infty}^{\infty} |g_u| |u|^{1/2} < \infty$ is required in order to ensure that the coefficients $\{g_u\}$ in the model (6.2.2) decay to zero "sufficiently fast". The order of magnitude of $|r_N(\omega)|$ clearly depends on the rate at which the $\{g_u\}$ decay to zero; if the $\{g_u\}$ decay "instantaneously", so that $g_0 = 1$ (say), $g_u = 0$, $u \neq 0$, then $X_t \equiv \varepsilon_t$, and $r_N(\omega) \equiv 0$ (all N, ω), and (6.2.22) is then exact for all N. If we impose stronger conditions on the rate of decay of the $\{g_u\}$ then we can obtain correspondingly stronger results on the order of magnitude of the moments of $|r_N(\omega)|$. Specifically, it may be shown that if the $\{\varepsilon_t\}$ are independent with finite mth order moments, then if,

$$\sum_{u=-\infty}^{\infty} |g_u||u|^\alpha < \infty \qquad (\alpha > 0), \qquad (6.2.31)$$

we have,

$$E[|r_N(\omega)|^m] = O(1/N^{m\alpha}), \qquad (6.2.32)$$

uniformly in ω, (see Walker (1965)). Note that (with $\alpha > 0$),

$$\sum_{u=-\infty}^{\infty} |g_u||u|^\alpha < \infty \Rightarrow \sum_{u=-\infty}^{\infty} |g_u| < \infty \Rightarrow \sum_{u=-\infty}^{\infty} g_u^2 < \infty, \qquad (6.2.33)$$

the last condition being the one originally imposed on the $\{g_u\}$ to ensure that $\{X_t\}$ is stationary with finite variance.

Theorem 6.2.2 *Let $\{X_t\}$ be a general linear process of the form* (6.2.2) *in which the $\{\varepsilon_t\}$ are independent with $E(\varepsilon_t) = 0$, $E(\varepsilon_t^2) = \sigma_\varepsilon^2$, $E(\varepsilon_t^4) < \infty$, and $\sum_{u=-\infty}^{\infty} |g_u||u|^\alpha < \infty$, $\alpha > 0$. Then,*

$$I_{N,X}^*(\omega) = 2\pi h(\omega) \cdot \frac{1}{\sigma_\varepsilon^2} \cdot I_{N,\varepsilon}^*(\omega) + R_N(\omega), \qquad (6.2.34)$$

where,

$$E[|R_N(\omega)|^2] = O(1/N^{2\alpha}),$$

uniformly in ω.

Proof.

$$I_{N,X}^*(\omega) = \zeta_X(\omega)\zeta_X^*(\omega)$$

$$= \{\Gamma(\omega)\zeta_\varepsilon(\omega) + r_N(\omega)\}\{\Gamma^*(\omega)\zeta_\varepsilon^*(\omega) + r_N^*(\omega)\},$$

where $r_N(\omega)$ is defined as in (6.2.27). Hence we may write,

$$I_{N,X}^*(\omega) = |\Gamma(\omega)|^2 I_{N,\varepsilon}^*(\omega) + R_N(\omega),$$

$$= 2\pi h(\omega) \cdot \frac{1}{\sigma_\varepsilon^2} \cdot I_{N,\varepsilon}^*(\omega) + R_N(\omega),$$

where

$$R_N(\omega) = r_N(\omega)\Gamma^*(\omega)\zeta_\varepsilon^*(\omega) + r_N^*(\omega)\Gamma(\omega)\zeta_\varepsilon(\omega) + |r_N(\omega)|^2.$$

$$= U_1 + U_2 + U_3,$$

say. Now,

$$E(|U_1|^2) = |\Gamma(\omega)|^2 E(|r_N(\omega)|^2|\zeta_\varepsilon(\omega)|^2)$$

$$\leqslant |\Gamma(\omega)|^2 \{E(|r_N(\omega)|^4 E(|\zeta_\varepsilon(\omega)|^4)\}^{1/2},$$

by the Schwarz inequality. But

$$E(|\zeta_\varepsilon(\omega)|^4) = E(|I^*_{N,\varepsilon}(\omega)|^2) = O(1),$$

and

$$E(|r_N(\omega)|^4) = O(1/N^{4\alpha}),$$

uniformly in ω, by (6.2.32). Hence,

$$E(|U_1|^2) = O(1/N^{2\alpha}).$$

Similarly,

$$E(|U_2|^2) = O(1/N^{2\alpha}),$$

and

$$E(|U_3|^2) = E(|r_N(\omega)|^4) = O(1/N^{4\alpha}),$$

as already noted. Finally, we write,

$$E(|R_N(\omega)|^2) \leq [\{E(|U_1|^2)\}^{1/2} + \{E(|U_2|^2)^{1/2} + E(|U_3|^2)\}^{1/2}]^2,$$

and the result follows.

The asymptotic relationship (6.2.26), or its more precise form (6.2.34), is of considerable importance since it allows us to obtain asymptotic expressions for the sampling properties of $I^*_{N,X}(\omega)$ directly from the known results on $I^*_{N,\varepsilon}(\omega)$. Thus, if the $\{\varepsilon_t\}$ are normal and the $\{g_u\}$ satisfy the required conditions we may deduce from Theorem 6.1.1 that, *asymptotically*, *the set of random variables* $\{I_{N,X}(\omega_p)/2\pi h(\omega_p)\}$, $p = 0, 1, \ldots, [N/2]$, $(\omega_p = 2\pi p/N)$, *are independently distributed*, and for $p \neq 0$, $N/2$ (N even), each is asymptotically distributed as χ^2_2, i.e. we may write,

$$I_{N,X}(\omega_p) \sim \begin{cases} 2\pi h(\omega_p) \cdot \chi^2_2, & p \neq 0, N/2, \\ 4\pi h(\omega_p) \cdot \chi^2_1, & p = 0, N/2. \end{cases} \tag{6.2.35}$$

In particular, (6.2.35) gives (for $p \neq 0$, $N/2$),

$$E[I_{N,X}(\omega_p)] \sim 4\pi h(\omega_p), \qquad \mathrm{var}[I_{N,X}(\omega_p)] \sim 16\pi^2 h^2(\omega_p),$$

and hence, since $I^*_{N,X}(\omega) = (1/4\pi)I_{N,X}(\omega)$, we have, $E[I^*_{N,X}(\omega_p)] \sim h(\omega_p)$ (which is, of course, already covered by the more general result (6.2.8)), and

$$\mathrm{var}[I^*_{N,X}(\omega_p)] \sim h^2(\omega_p). \tag{6.2.36}$$

This last result, (6.2.36), shows that $\mathrm{var}\{I^*_{N,X}(\omega_p)\} = O(1)$ (as $N \to \infty$), and confirms the assertion made at the beginning of this section that $\mathrm{var}\{I^*_{N,X}(\omega_p)\} \nrightarrow 0$, and showing that *it is not a consistent estimate of* $h(\omega_p)$. The restriction that ω_p be of the form $(2\pi p/N)$ is not important as far as the mean and variance of $I_{N,X}(\omega_p)$ are concerned; in fact, the above results hold for general ω, as we now show by generalizing (6.1.56) (cf. theorem 6.1.3), as follows.

Theorem 6.2.3 *Let $\{X_t\}$ be a general linear process of the form (6.2.3), with the $\{\varepsilon_t\}$ and $\{g_u\}$ satisfying the conditions of theorem 6.2.2. Then,*

$$\text{cov}\{I_{N,X}^*(\omega_1), I_{N,X}^*(\omega_2)\}$$

$$= \left[\frac{e}{n} + \frac{2\pi}{N}\{F_N(\omega_1 + \omega_2) + F_N(\omega_1 - \omega_2)\}\right] h(\omega_1)h(\omega_2) + O(1/N^\alpha),$$

(6.2.37)

where $e = (\kappa_4(\varepsilon)/\sigma_\varepsilon^4) = \{E(\varepsilon_t^4)/\sigma_\varepsilon^4 - 3\}$, $F_N(\theta)$ is the Fejér kernel given by (6.1.43), and the remainder term is $O(1/N^\alpha)$ uniformly in ω_1, ω_2.

Proof. From Theorem (6.2.2) we may write,

$$\text{cov}\{I_{N,X}^*(\omega_1), I_{N,X}^*(\omega_2)\} = \frac{4\pi^2}{\sigma_\varepsilon^4} h(\omega_1)h(\omega_2) \, \text{cov}\{I_{N,\varepsilon}^*(\omega_1), I_{N,\varepsilon}^*(\omega_2)\}$$

$$+ V_1 + V_2 + V_3,$$

(6.2.38)

where $V_1 \propto \text{cov}\{I_{N,\varepsilon}^*(\omega_1), R_N(\omega_2)\}$, V_2 is similarly defined with ω_1, ω_2, interchanged, and $V_3 = \text{cov}\{R_N(\omega_1), R_N(\omega_2)\}$. The main expression in (6.2.37) follows immediately from the first term in (6.2.38), using (6.1.56) to evaluate $\text{cov}\{I_{N,\varepsilon}^*(\omega_1), I_{N,\varepsilon}^*(\omega_2)\}$ (remembering that, for example, $I_{N,\varepsilon}^*(\omega_1) = (1/4\pi)I_{N,\varepsilon}(\omega_1)$). It remains to prove that the remainder terms, $V_1 + V_2 + V_3 = O(1/N^\alpha)$. Now,

$$|\text{cov}\{I_{N,\varepsilon}^*(\omega_1), R_N(\omega_2)\}| \leq [\text{var}\{I_{N,\varepsilon}^*(\omega_1)\} \, \text{var}\{R_N(\omega_2)\}]^{1/2},$$

and

$$\text{var}\{I_{N,\varepsilon}^*(\omega_1)\} = O(1), \qquad \text{var}\{R_N(\omega_2)\} = O(1/N^{2\alpha}).$$

Hence, $V_1 = O(1/N^\alpha)$, and similarly $V_2 = O(1/N^\alpha)$. Finally,

$$|V_3| \leq [\text{var}\{R_N(\omega_1)\} \, \text{var}\{R_N(\omega_2)\}]^{1/2} = O(1/N^{2\alpha}),$$

and the result follows.

Setting $\omega_1 = \omega_2 = \omega$, (6.2.37) gives for *all* ω ($\neq 0, \pm\pi$)

$$\text{var}\{I_{N,X}^*(\omega)\} = h^2(\omega)(1 + e/N) + O(1/N^\alpha) \to h^2(\omega), \qquad \text{as} \quad N \to \infty.$$

(6.2.38)

Before concluding our discussion of the sampling properties of the periodogram it is convenient at this stage to derive some important general results (which we shall need later) concerning weighted integrals of the periodogram. These results are contained in the following theorem.

Theorem 6.2.4 *Let $\phi_1(\omega)$, $\phi_2(\omega)$, be two real valued functions defined on $-\pi \leq \omega \leq \pi$, each of which has at most a finite number of discontinuities and is both absolutely integrable and square integrable, i.e. for $i = 1, 2$,*

$$\int_{-\pi}^{\pi} |\phi_i(\omega)| \, d\omega < \infty \qquad \text{and} \qquad \int_{-\pi}^{\pi} \phi_i^2(\omega) \, d\omega < \infty.$$

Let $\{X_t\}$ be a general linear process of the form (6.2.2) with the $\{\varepsilon_t\}$ satisfying the conditions of theorem (6.2.2) and $g_u = O(1/|u|^k)$ for some $k > 2$. Write, for $i = 1, 2$,

$$\hat{\psi}_i = \int_{-\pi}^{\pi} \phi_i(\omega) I^*_{N,X}(\omega) \, d\omega, \tag{6.2.39}$$

$$\psi_i = \int_{-\pi}^{\pi} \phi_i(\omega) h(\omega) \, d\omega. \tag{6.2.40}$$

Then,

(1) $\quad \lim_{N \to \infty} E(\hat{\psi}_i) = \psi_i, \qquad i = 1, 2,$ \hfill (6.2.41)

(2) $\quad \lim_{N \to \infty} N \, \text{cov}(\hat{\psi}_1, \hat{\psi}_2) = e\psi_1\psi_2 + 4\pi \int_{-\pi}^{\pi} \phi_1(\omega)\bar{\phi}_2(\omega)h^2(\omega) \, d\omega,$ \hfill (6.2.42)

where

$$e = \{E(\varepsilon_t^4)/\sigma_\varepsilon^4 - 3\} \qquad \text{and} \qquad \bar{\phi}_2(\omega) = \tfrac{1}{2}\{\phi_2(\omega) + \phi_2(-\omega)\}.$$

In particular, when $\phi_1(\omega) \equiv \phi_2(\omega) = \phi(\omega)$, say,

(3) $\quad \lim_{N \to \infty} N \, \text{var}(\psi) = e\psi^2 + 4\pi \int_{-\pi}^{\pi} \phi(\omega)\bar{\phi}(\omega)h^2(\omega) \, d\omega,$ \hfill (6.2.43)

where $\bar{\phi}(\omega) = \tfrac{1}{2}\{\phi(\omega) + \phi(-\omega)\}.$

Proof. Under the stated conditions on $\phi_1(\omega)$, $\phi_2(\omega)$, we have,

$$E(\hat{\psi}_i) = \int_{-\pi}^{\pi} \phi_i(\omega) E[I^*_{N,X}(\omega)] \, d\omega, \qquad i = 1, 2, \tag{6.2.44}$$

and

$$\text{cov}(\hat{\psi}_1, \hat{\psi}_2) = \int_{-\pi}^{\pi} \int_{-\pi}^{\pi} \phi_1(\omega)\phi_2(\omega') \, \text{cov}\{I^*_{N,X}(\omega), I^*_{N,X}(\omega')\} \, d\omega \, d\omega'. \tag{6.2.45}$$

Part (1) of the theorem follows immediately from (6.2.44) since by (6.2.10) $E[I^*_{N,X}(\omega)]$ converges uniformly to $h(\omega)$ as $N \to \infty$. To prove part (2) we first note that if $g_u = O(1/|u|^k)$, some $k > 2$, then $\exists \alpha$ satisfying $1 < \alpha < k - 1$, s.t. $\sum_u |g_u| |u|^\alpha < \infty$. With this value of α the remainder term in (6.2.37) is $o(1/N)$, uniformly in ω, ω', and substituting (6.2.37) into (6.2.45) we now have,

$$\text{cov}(\hat{\psi}_1, \hat{\psi}_2) = \int_{-\pi}^{\pi} \int_{-\pi}^{\pi} \phi_1(\omega)\phi_2(\omega')h(\omega)h(\omega')$$

$$\times \left[\frac{e}{N} + \frac{2\pi}{N}\{F_N(\omega + \omega') + F_N(\omega - \omega')\} + o\left(\frac{1}{N}\right) \right] d\omega \, d\omega',$$

$$= \frac{e}{N} \cdot \psi_1\psi_2 + \frac{2\pi}{N} \int_{-\pi}^{\pi} \int_{-\pi}^{\pi} \phi_1(\omega)\phi_2(\omega')h(\omega)h(\omega')$$

$$\times \{F_N(\omega + \omega') + F_N(\omega - \omega')\} \, d\omega \, d\omega' + o\left(\frac{1}{N}\right).$$

$$(6.2.46)$$

Now as $N \to \infty$, $F_N(\theta) \to \delta(\theta)$ (the δ-function). Hence,

$$\int_{-\pi}^{\pi} \phi_2(\omega')h(\omega')\{F_N(\omega + \omega') + F_N(\omega - \omega')\} \, d\omega' \to h(\omega)\{\phi_2(\omega) + \phi_2(-\omega)\},$$

$$(6.2.46a)$$

almost everywhere, i.e. at all continuity points of $\phi_2(\omega)$. (Recall that $h(\omega)$ is an even function of ω.) Part (2) now follows, as does part (3) (being a special case of (2)).

The expression for $\text{cov}\{\hat{\psi}_1, \hat{\psi}_2\}$ must obviously be symmetric in $\phi_1(\omega)$, $\phi_2(\omega)$ and hence the second term in (6.2.42) could equally well be written as

$$4\pi \int_{-\pi}^{\pi} \bar{\phi}_1(\omega)\phi_2(\omega)h^2(\omega) \, d\omega,$$

where $\bar{\phi}_1(\omega) = \frac{1}{2}\{\phi_1(\omega) + \phi_1(-\omega)\}$. (This alternative expression can be derived directly from (6.2.46) by integrating first w.r. to ω.) Hence, *if either $\phi_1(\omega)$ or $\phi_2(\omega)$ is an even function of ω*, then (6.2.42) simplifies to

$$\lim_{N \to \infty} N \, \text{cov}(\hat{\psi}_1, \hat{\psi}_2) = e\psi_1\psi_2 + 4\pi \int_{-\pi}^{\pi} \phi_1(\omega)\phi_2(\omega)h^2(\omega) \, d\omega. \qquad (6.2.47)$$

The results of theorem 6.2.4 are important, and their generalization to the case where the functions $\phi_1(\omega)$, $\phi_2(\omega)$ depend on N (the number of observations) plays a central role in the estimation of spectral density functions. There are, in fact, various alternative proofs of theorem 6.2.4, some of which hold under somewhat weaker conditions than those we have assumed. One alternative approach is based on re-writing the expression for $\hat{\psi}_i$, (6.2.39), as a linear combination of the sample autocovariances, $\hat{R}(s)$

(using (6.2.5)), and then using (5.3.21) to evaluate the covariances of the sample autocovariances. Proofs along these lines are given by Grenander and Rosenblatt (1957a), Parzen (1957a); in particular, Parzen's proof does not assume that $\{X_t\}$ is a general linear process and replaces this assumption by weaker conditions on the fourth moments.

Discussion

(a) The result (6.2.38) shows that $\text{var}\{I^*_{N,X}(\omega)\} \not\to 0$ as $N \to \infty$, and hence $I^*_{N,X}(\omega)$ *is not a consistent estimate of* $h(\omega)$ *in the sense that it does not converge to* $h(\omega)$ *in mean square.* (In fact, Grenander (1951) has shown that $I^*_{N,X}(\omega)$ does not converge even to a limiting random variable as $N \to \infty$.) Moreover, it is clear from (6.2.37) that for any two fixed neighbouring frequencies, ω_1, ω_2, $\text{cov}\{I^*_{N,X}(\omega_1), I^*_{N,X}(\omega_2)\}$ *decreases as N increases.* These two features confirm the statement made at the beginning of this section that, as a function of ω, $I^*_{N,X}(\omega)$ *has an erratic and wildly fluctuating form.* The typical form of the periodogram of a process with continuous spectrum is shown in Fig. 6.5, from which it will be seen that it is of little use as an estimate of $h(\omega)$.

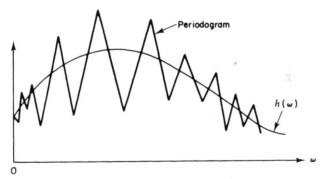

Fig. 6.5. Periodogram of a process with a continuous spectrum.

(b) We have already noted that $I^*_{N,X}(\omega)$ is the "natural" estimate of $h(\omega)$, the expression for $I^*_{N,X}(\omega)$ being simply the sample version of the expression for $h(\omega)$ (see the remarks following equation (6.2.5)). We thus have a case where the "natural" estimate is quite unsatisfactory, and we may therefore enquire whether one of the general estimation procedures, such as the method of maximum likelihood discussed in Section 5.2.2, could be used to obtain a more satisfactory estimate. However, we cannot immediately apply the method of maximum likelihood since here the problem is to estimate an unknown continuous *function* rather than a finite set of parameters. Nevertheless, we may argue that if we can estimate the values of $h(\omega)$ at the discrete set of points, $\omega_p = 2\pi p/N$, $p = 0, 1, \ldots, [N/2]$, then in practice this would probably suffice, and now we have only a finite number of parameters to estimate. Interestingly, it turns out that under the assumption that the observations are normally distributed the maximum likelihood estimate of $h(\omega_p)$ is effectively, $I^*_{N,X}(\omega_p)$, the corresponding periodogram ordinate! Moreover, it may be shown further that the set of periodogram ordinates, $\{I^*_{N,X}(\omega_p); p = 0, 1, \ldots, [N/2]\}$, are (essentially) a set of sufficient statistics for the parameters $\{h(\omega_p); p = 0, 1, \ldots, [N/2]\}$: see Whittle (1957).

These results present us with an unusual situation; the "natural" estimate of $h(\omega)$ is unsatisfactory and yet classical statistical estimation theory fails to suggest any alternative form of estimate. Of course, the problem is different from that encountered in the usual applications of maximum likelihood theory since here the observations are correlated, but quite apart from this, the basic difficulty stems from the fact that, as mentioned above, we are trying to estimate a *continuous function* rather than a set of parameters. The erratic nature of $I^*_{N,X}(\omega)$ would not in itself be too serious a defect if we were concerned only with estimating $h(\omega)$ *at one particular value of* ω (although the lack of consistency would still apply). It is only when we think of estimating the *function* $h(\omega)$ by the *function* $I^*_{N,X}(\omega)$ that its erratic form becomes a serious fault—because if we know *a priori* that $h(\omega)$ is a smooth continuous function then we would naturally require any reasonable estimating function to share the same property. However, the property of "smoothness" is a concept which is not normally met in classical estimation theory, and the construction of estimating functions which possess this characteristic requires special "smoothing" techniques which are designed specifically to take account of this feature. We discuss these techniques in the next section, but meanwhile we may take particular note of the "sufficiency" property of the periodogram ordinates—this gives us a strong indication that in devising more appropriate estimates of $h(\omega)$ we need consider only functions of the periodogram ordinates. As we shall see, although in its "raw" state the periodogram is not a satisfactory estimate of $h(\omega)$, a suitably "smoothed" version of $I^*_{N,X}(\omega)$ can produce quite a good estimate of $h(\omega)$.

(c) The difficulties associated with estimating density functions are not unique to the problem of estimating spectral density functions, they arise also in, e.g., the estimation of *probability density functions*. In fact, the two problems are very closely linked, and the basic "smoothing" techniques which we discuss in the next section can be applied equally well to the estimation of probability density functions; see, Rosenblatt (1956a), Whittle (1958), Parzen (1962a), Bartlett (1963), Boneva, *et al.* (1971), Rosenblatt (1971), and Silverman (1978). From the physical point of view the fundamental difficulty in estimating the spectral density function stems from the fact that when the process has a purely continuous spectrum the power at a particular frequency ω is *zero*, just as when a random variable X has an absolutely continuous probability distribution the probability that X takes any particular value is zero. The best we can hope to achieve, in physical terms, is to measure the power in a small band of frequencies centred on ω, say from $(\omega - \varepsilon)$ to $(\omega + \varepsilon)$, and this is exactly analogous to the well known device of estimating a probability density function by constructing a histogram. (The simple "histogram" estimate of a probability density function corresponds to the use of a rectangular Daniell "window" in spectral estimation; see Section 6.2.3.)

The problem of estimating "non-parametric regression functions" is also of the same general type, and again the same basic ideas of using smoothing functions can be applied to the study of this area of function fitting; see Priestley and Chao (1972) and Nadaraya (1970).

(d) As a point of historical interest we would mention that it has been fashionable in the past to refer to the so-called "failure" of periodogram analysis to deal satisfactorily with processes which possess continuous spectra. Presumably, the use of this type of phraseology is intended as a colourful description of the following features: (1) the periodogram is an unsatisfactory estimate of the spectral density function; and (2) if a process with a purely continuous spectrum is *incorrectly* modelled as one with a discrete spectrum, then the use of the periodogram tests described in Section 6.1.4 may indicate the existence of a large number of spurious periodic components. (This latter feature is due to the fact that when the spectrum is continuous the mean and variance of $I_{N,X}^*(\omega)$ are proportional to $h(\omega)$, $h^2(\omega)$, respectively, and hence $I_{N,X}^*(\omega)$ may produce spurious peaks in regions where $h(\omega)$ is relatively large.) However, neither of these results is, in itself, a valid criticism of periodogram analysis which remains entirely appropriate for the purpose for which it was designed, namely, the analysis of processes *with discrete spectra*. It is hardly fair to criticise the method of periodogram analysis on the grounds that it fails to produce sensible results when *misapplied* to processes with continuous spectra. Schuster's pioneering paper of 1896 contains remarks which suggest that Schuster was, in fact, aware of the dangers of applying this technique in the case of continuous spectra, and bearing in mind that the probabilistic theory of processes with continuous spectra did not exist at that time, Schuster's observations show extraordinary insight. The main conclusion to be drawn from the above discussion is that in spectral analysis, as in most other branches of statistics, there is no such thing as a "universal data analysis technique" which can be safely applied without due consideration of the basic underlying statistical model.

6.2.3 Consistent Estimates of the Spectral Density Function; Spectral Windows

Before discussing the problem of constructing consistent estimates of the spectral density function $h(\omega)$, it is instructive to consider in more detail the basic reason why the periodogram $I_{N,X}^*(\omega)$ is an inconsistent estimate of $h(\omega)$. As noted above, this is indeed a surprising result since $I_{N,X}^*(\omega)$ is essentially the same function of the sample autocovariances $\{\hat{R}(s)\}$ as $h(\omega)$ is of the theoretical autocovariances $\{R(s)\}$, and we know from the results of Chapter 5 (Section 5.3.6) that for each fixed s, $\hat{R}(s)$ is (under general conditions) a consistent estimate of $R(s)$, being asymptotically unbiased and having a variance which is $O(1/N)$. We might therefore expect that a linear combination of the $\{\hat{R}(s)\}$ would be a consistent estimate of the same linear combination of the $\{R(s)\}$. Some authors try to explain the inconsistency of $I_{N,X}^*(\omega)$ by pointing out that $I_{N,X}^*(\omega)$ includes all the sample autocovariance extending from $s = 0$ to $s = (N - 1)$ (cf. (6.1.35)) and hence, no matter how large N becomes, $I_{N,X}^*(\omega)$ always involves the "tail" of the sample autocovariance function, which is a poor estimate of the corresponding theoretical autocovariances since in this region $\hat{R}(s)$ is based on just a small number of pairs of observations. This is certainly true, and if we were considering the unbiased estimate $\hat{R}^*(s)$ (cf. (5.3.10)), we could say further that $\text{var}\{\hat{R}^*(s)\}$, being $O(1/(N - |s|))$, increases substantially as s approaches $(N - 1)$. (This effect is sometimes described by saying that the tail of the function $\hat{R}^*(s)$ exhibits a "wild" behaviour.) However, $I_{N,X}^*(\omega)$ involves the biased estimate $\hat{R}(s)$ rather than the unbiased version, and we know that $\text{var}\{\hat{R}(s)\}$ remains $O(1/N)$ even when s approaches $(N - 1)$ (cf. the remarks following (5.3.25a)). Of course, the bias of $\hat{R}(s)$ may increase over a certain range as s increases, but we are not concerned here with the bias, we are simply trying to explain the fact that $\text{var}\{I_{N,X}^*(\omega)\}$ does not tend to zero, and the fact that as s increases $\hat{R}(s)$ is based on fewer and fewer observations does not explain this effect. The reason for the behaviour of $\text{var}\{I_{N,X}^*(\omega)\}$ lies rather in the fact that, according to (6.1.35), $I_{N,X}^*(\omega)$ involves N sample autocovariances, and although the variance of each is $O(1/N)$, the cumulative effect of the N terms produces a variance which is $O(1)$. As a simple illustration of this result consider a random variable U which is of the form $U = \sum_{i=1}^{N} U_i$ where the $\{U_i\}$ are uncorrelated and each has variance $1/N$. Then it follows trivially that $\text{var}(U) = 1$. Of course, the $\{\hat{R}(s)\}$ are not, in general, uncorrelated, but the basic effect is the same, as verified by the result (6.2.38). We thus see that the reason why $\text{var}\{I_{N,X}^*(\omega)\}$ does not tend to zero is because, loosely speaking, it contains "too many" sample autocovariances.

One way of obtaining an expression with a reduced variance is simply to omit some of the terms in (6.1.35). If we do this we will certainly reduce the variance, but on the other hand the terms omitted will affect the expected value of the new expression, and the general effect will be to increase the bias. However, we know that if the process has a purely continuous spectrum then $R(s) \to 0$ as $|s| \to \infty$, and hence if we omit only those terms which correspond to the "tail" of the sample autocovariance function then hopefully the bias will not be affected too seriously. These ideas suggest that we might consider as an estimate of $h(\omega)$ an expression of the form

$$\hat{h}_0(\omega) = \frac{1}{2\pi} \sum_{s=-M}^{M} \hat{R}(s) \cos s\omega, \qquad (6.2.48)$$

where M is some integer $< (N-1)$ whose precise value is as yet unspecified. An estimate of the form (6.2.48) is called a "*truncated periodogram*", and M is called the "*truncation point*". Later, we will study the variance and bias of $\hat{h}_0(\omega)$ in some detail, but for the moment we may obtain some insight into the properties of this type of estimate by arguing heuristically that since $\hat{h}_0(\omega)$ contains $(M+1)$ sample autocovariances, whereas $I_{N,X}^*(\omega)$ contains N sample autocovariances, we would expect $\hat{h}_0(\omega)$ to have a variance which is roughly $(M/N) \times \text{var}\{I_{N,X}^*(\omega)\} = O(M/N)$. (In fact, the result that $\text{var}\{\hat{h}_0(\omega)\} = O(M/N)$ can be established quite rigorously, as we will show later.) Also, using (5.3.14) we have,

$$E[\hat{h}_0(\omega)] = \frac{1}{2\pi} \sum_{s=-M}^{M} \left(1 - \frac{|s|}{N}\right) R(s) \cos s\omega \qquad (6.2.49)$$

$$\to h(\omega), \qquad \text{as } M \to \infty \qquad (6.2.50)$$

(cf. the derivation of (6.2.8)). Hence, if we make the value of M depend on N in such a way that $M \to \infty$ as $N \to \infty$, then $\hat{h}_0(\omega)$ will be an asymptotically unbiased estimate of $h(\omega)$. Thus, we may conclude tentatively that *provided we make $M \to \infty$ as $N \to \infty$, but sufficiently slowly (relative to N) so that $(M/N) \to 0$ as $N \to \infty$,* then both the bias and variance of $\hat{h}_0(\omega)$ will tend to zero as $N \to \infty$, and $\hat{h}_0(\omega)$ *will then be, for each ω, a consistent estimate of* $h(\omega)$. It is easy to find values of M which satisfy the above requirements; for example, we could choose $M = \sqrt{N}$, or more generally, we could choose $M = N^\alpha$, for any α such that $0 < \alpha < 1$.

Now the estimate $\hat{h}_0(\omega)$ can be regarded as a special case of the more general form of estimate,

$$\hat{h}(\omega) = \frac{1}{2\pi} \sum_{s=-(N-1)}^{(N-1)} \lambda(s) \hat{R}(s) \cos s\omega, \qquad (6.2.51)$$

where, to recover (6.2.48), we define the function $\lambda(s)$ to be,

$$\lambda(s) = \begin{cases} 1, & |s| \leq M, \\ 0, & |s| > M. \end{cases} \tag{6.2.52}$$

Thus, $\hat{h}_0(\omega)$ corresponds to a special case of (6.2.51) in which we assign unit "weight" to the first M sample autocovariances and zero "weight" to the remainder. However, once we have written $\hat{h}_0(\omega)$ in the form (6.2.51) we see that there is no reason why we should not consider other estimates of the general form (6.2.51) in which, for example, $\lambda(s)$ decreases *gradually*, rather than having the discontinuous form (6.2.52). For such estimates the contribution of the "tail" of the sample autocovariance function would be reduced (rather than eliminated), but provided $\lambda(s)$ decreases at a "suitable rate" (relative to N) we may expect that expressions of the form (6.2.51) will still provide consistent estimates of $h(\omega)$ under quite a wide choice of functions $\lambda(s)$. The same idea can be used equally well to estimate the normalized spectral density function, $f(\omega) = h(\omega)/\sigma_X^2$. The estimate of $f(\omega)$ corresponding to (6.2.51) is,

$$\hat{f}(\omega) = \frac{\hat{h}(\omega)}{s_X^2} = \frac{1}{2\pi} \sum_{s=-(N-1)}^{(N-1)} \lambda(s)\hat{\rho}(s) \cos s\omega, \tag{6.2.53}$$

where $\hat{\rho}(s) = \hat{R}(s)/\hat{R}(0)$ is the sample autocorrelation function. Estimates of the general form (6.2.51) were first introduced by Grenander and Rosenblatt (1953) and, as we shall see shortly, there are many different forms of $\lambda(s)$ which we can use, all of which lead to consistent estimates of $h(\omega)$ if the function $\lambda(s)$ decreases at the appropriate rate. To establish these properties we need to obtain expressions for the mean and variance of the general estimate $\hat{h}(\omega)$ given by (6.2.51), but before doing so we will first show that $\hat{h}(\omega)$ *can be re-written in an alternative form, namely, as a weighted integral of the periodogram* $I_{N,X}^*(\omega)$. This result can be seen quite easily by noting that (6.2.51) is the (discrete) Fourier transform of the product of $\lambda(s)$ and $\hat{R}(s)$ and hence, by the property of Fourier transforms (cf. the discrete version of lemma 4.8.1), $\hat{h}(\omega)$ can be expressed as the convolution of the Fourier transform of $\lambda(s)$ with $I_N^*(\omega)$, the Fourier transform of $\hat{R}(s)$. The result can, however, be established directly as follows.

Since both $\hat{R}(s)$ and $\cos s\omega$ are even functions of s we may as well take $\lambda(s)$ also to be a (real) even function of s, in which case we may re-write (6.2.51) as

$$\hat{h}(\omega) = \frac{1}{2\pi} \sum_{s=-(N-1)}^{(N-1)} \lambda(s)\hat{R}(s) e^{-is\omega}. \tag{6.2.54}$$

Similarly, we may re-write (6.2.5) as

$$I_{N,X}^*(\omega) = \frac{1}{2\pi} \sum_{s=-(N-1)}^{(N-1)} \hat{R}(s) \, e^{-is\omega}, \qquad (6.2.54a)$$

which, when inverted, gives for $|s| \leq (N-1)$,

$$\hat{R}(s) = \int_{-\pi}^{\pi} I_{N,X}^*(\theta) \, e^{is\theta} \, d\theta. \qquad (6.2.55)$$

Substituting (6.2.55) in (6.2.54) we now obtain,

$$\hat{h}(\omega) = \int_{-\pi}^{\pi} I_{N,X}^*(\theta) \left\{ \frac{1}{2\pi} \sum_{s=-(N-1)}^{(N-1)} \lambda(s) \, e^{-is(\omega-\theta)} \right\} d\theta,$$

or

$$\hat{h}(\omega) = \int_{-\pi}^{\pi} I_{N,X}^*(\theta) W(\omega - \theta) \, d\theta, \qquad (6.2.56)$$

where

$$W(\theta) = \frac{1}{2\pi} \sum_{s=-(N-1)}^{(N-1)} \lambda(s) \, e^{-is\theta}. \qquad (6.2.57)$$

Hence, $\hat{h}(\omega)$ may be expressed as a weighted integral of the periodogram, $I_{N,X}^*(\theta)$, where the weight function, $W(\theta)$, is the (discrete) Fourier transform of the sequence $\{\lambda(s)\}$. For most of the sequences $\{\lambda(s)\}$ which we shall use the function $W(\theta)$ typically is highly concentrated in the region of $\theta = 0$; the more slowly $\lambda(s)$ decays to zero the more concentrated is $W(\theta)$. The RHS of (6.2.56) thus produces a weighted average of the values of $I_{N,X}^*(\theta)$ with the largest weights attached to ordinates in the neighbourhood of $\theta = \omega$; in other words, $\hat{h}(\omega)$ corresponds to a "locally" weighted average of periodogram ordinates in the neighbourhood of the frequency ω.

Quite apart from its relationship with (6.2.53), the estimate (6.2.56) has a strong intuitive appeal in its own right. For, one of the deficiencies of $I_{N,X}^*(\omega)$ as an estimate of $h(\omega)$ is due to the fact that $I_{N,X}^*(\omega)$ is a highly erratic function, and a natural procedure for obtaining an improved estimate is to "smooth" $I_{N,X}^*(\omega)$ locally in the neighbourhood of the required frequency; this is exactly what the expression (6.2.56) computes. *Thus, the device of weighting the sample autocovariance function so as to reduce the contribution from the "tail" has exactly the same effect as smoothing the periodogram by a weighted integral of the form* (6.2.56). Moreover, if we start with an estimate of the form (6.2.56) we can easily re-write it in the form (6.2.54), as follows. Consider the expression (6.2.56) where now the $W(\theta)$ is an arbitrary periodic function, period 2π, integrable over $(-\pi, \pi)$. Substituting (6.2.54a) in (6.2.56) we obtain,

$$\hat{h}(\omega) = \frac{1}{2\pi} \sum_{s=-(N-1)}^{(N-1)} e^{-is\omega} \hat{R}(s) \left\{ \int_{-\pi}^{\pi} e^{is(\omega-\theta)} W(\omega-\theta) \, d\theta \right\}$$

$$= \frac{1}{2\pi} \sum_{s=-(N-1)}^{(N-1)} \lambda(s) \hat{R}(s) e^{-i\omega s}, \qquad (6.2.58)$$

where

$$\lambda(s) = \int_{-\pi}^{\pi} e^{is(\omega-\theta)} W(\omega-\theta) \, d\theta$$

$$= \int_{-\pi}^{\pi} e^{is\theta} W(\theta) \, d\theta, \qquad (s = -(N-1), \ldots, (N-1)) \quad (6.2.59)$$

(remembering that $W(\theta)$ is periodic, period 2π). Note that (6.2.58) involves the Fourier coefficients $\lambda(s)$ of $W(\theta)$ only for $|s| \le N-1$, even though $W(\theta)$ may have non-zero Fourier coefficients for $|s| > N-1$. Hence, if we write,

$$W^*(\theta) = \frac{1}{2\pi} \sum_{s=-(N-1)}^{(N-1)} \lambda(s) e^{-is\theta}, \qquad (6.2.60)$$

with $\lambda(s)$ defined by (6.2.59), then irrespective of the values of the higher order Fourier coefficients of $W(\theta)$ we always have,

$$\int_{-\pi}^{\pi} I_{N,X}^*(\theta) W(\omega-\theta) \, d\theta \equiv \int_{-\pi}^{\pi} I_{N,X}^*(\theta) W^*(\omega-\theta) \, d\theta. \quad (6.2.61)$$

The basis of this result lies in the fact that $I_{N,X}^*(\theta)$ is a trigonometric polynomial of finite degree $(N-1)$ and the same result would, of course, hold if we replace $I_{N,X}^*(\theta)$ by any other trigonometric polynomial of the same degree.

We may note also that if we choose $\lambda(s)$ to be a real even function of s (as is invariably the case) then $W(\theta)$, as defined by (6.2.57), is also a real even function θ, so that (6.2.57) may be written alternatively in "real" form as,

$$W(\theta) = \frac{1}{2\pi} \sum_{s=-(N-1)}^{(N-1)} \lambda(s) \cos s\theta. \qquad (6.2.62)$$

In referring to the estimate $\hat{h}(\omega)$ given by (6.2.53) we call $\{\lambda(s)\}$ the "*lag window*", and $W(\theta)$ the "*spectral window*". The reason for this rather unusual terminology stems from the fact that we often think of $W(\theta)$ as being effectively zero outside a small interval, say $(-\varepsilon, \varepsilon)$, and hence the integral (6.2.56) may be regarded as giving a "view" of $I_{N,X}^*(\theta)$ through a narrow "slit" or "window" from $(\omega - \varepsilon)$ to $(\omega + \varepsilon)$. The term "window" was first introduced by Blackman and Tukey (1959) in connection with the function $W(\theta)$, and has since become firmly established in the literature. (In the earlier papers on spectral estimation $\lambda(s)$ and $W(\theta)$ were referred to simply as the "weight sequence" and "weight function".)

Some special windows

Before deducing the asymptotic sampling properties of the general form of estimate it may be helpful at this point to consider the following specific forms of $\lambda(s)$, together with the corresponding forms of $W(\theta)$, which have been suggested by various authors.

I. The "truncated periodogram" window

The initial estimate which we considered, namely the function $\hat{h}_0(\omega)$ given by (6.2.48), corresponds to taking the lag window, $\lambda(s)$, to be of the form (6.2.52), viz,

$$\lambda(s) = \begin{cases} 1, & |s| \leq M, \\ 0, & |s| > M, \end{cases} \qquad (6.2.63)$$

where $M(< N-1)$, the "window parameter", denotes the truncation point in the sum (6.2.48). The form of $\lambda(s)$ is shown in Fig. 6.6. The corresponding

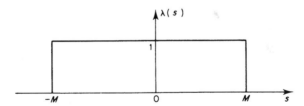

Fig. 6.6. The lag window for the truncated periodogram.

spectral window $W(\theta)$ is from (6.2.62) given by,

$$W(\theta) = \frac{1}{2\pi} \sum_{s=-M}^{M} \cos s\theta = \frac{1}{2\pi}\left\{\frac{\sin[(M+\frac{1}{2})\theta]}{\sin(\theta/2)}\right\} = D_M(\theta), \quad (6.2.64)$$

say (cf. 6.1.44)). The function $D_M(\theta)$ is known as the "*Dirichlet kernel*" and, in common with the Fejer kernel $F_N(\theta)$ (defined by (6.1.43)), plays an important role in Fourier analysis (see Lanczos (1956)). Its form is illustrated in Fig. 6.7 from which it will be noted that $D_M(\theta)$ has a large peak at $\theta = 0$, of magnitude $(2M+1)/2\pi$, together with subsidiary peaks at (approximately)

$$\theta = \pm \frac{5\pi}{2M+1}, \ \pm \frac{9\pi}{2M+1}, \ \pm \frac{13\pi}{2M+1}, \dots$$

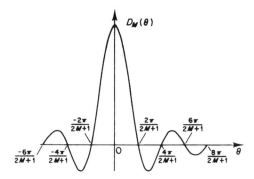

Fig. 6.7. The Dirichlet kernel; the spectral window for the truncated periodogram.

The zeros occur at

$$\pm \frac{2\pi}{2M+1}, \ \pm \frac{4\pi}{2M+1}, \ \pm \frac{6\pi}{2M+1}, \dots$$

The form of $D_M(\theta)$ near $\theta = 0$ is seen by setting $\phi = (M + \frac{1}{2})\theta$, so that

$$2\pi D_M(\theta) = \frac{\sin \phi}{\sin(\phi/2M+1)} \sim (2M+1)\left(\frac{\sin \phi}{\phi}\right), \qquad \text{for small } \phi.$$

Comparing Figs. 6.2 and 6.7, it will be seen that the Fejer kernel and the Dirichlet kernel share the common feature of having one large peak, or "main lobe", at the origin together with subsidiary peaks of decreasing magnitude, or "side lobes", on both sides of the origin. However, there is one important difference between the two kernels, namely that the Dirichlet kernel takes negative values for certain values of θ whereas the Fejer kernel is non-negative everywhere. Now it follows immediately from (6.2.56) that if the spectral window, $W(\theta)$, is non-negative for all θ then the estimate $\hat{h}(\omega)$ is also non-negative for all ω. (We recall that the periodogram $I_N^*(\theta)$ is itself non-negative for all θ.) We know that the true spectral density function $h(\omega)$ is a non-negative function, and it is desirable, therefore, that the estimate $\hat{h}(\omega)$ should share this property. The truncated periodogram estimate, which uses the Dirichlet kernel as its spectral window, may take negative values for certain values of ω, but an estimate based on the Fejer kernel as its spectral window will always be non-negative.

II. *Bartlett's window*

Bartlett (1950) proposed an estimate of $h(\omega)$ which is essentially equivalent to (6.2.53) with the lag window $\lambda(s)$ having the form,

$$\lambda(s) = \begin{cases} 1 - |s|/M, & |s| \leq M, \\ 0, & |s| > M. \end{cases} \qquad (6.2.65)$$

Here, $M(<N)$ is again the "window parameter", and as in the previous example, the value of M determines the point at which the sample auto-covariance functions is truncated. However, whereas in the previous example all the autocovariances up to lag M were given equal weight, we now apply linearly decreasing weights to the autocovariances up to lag M and zero weights thereafter. The form of $\lambda(s)$ for the Bartlett estimate is shown in Fig. 6.8.

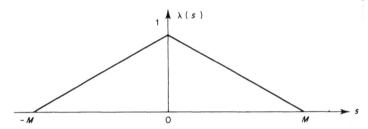

Fig. 6.8. The lag window for the Bartlett estimate.

The corresponding spectral window is (from (6.2.62)),

$$W(\theta) = \frac{1}{2\pi} \sum_{s=-M}^{M} \left(1 - \frac{|s|}{M}\right) \cos s\theta = \frac{1}{2\pi M} \left\{ \frac{\sin(M\theta/2)}{\sin(\theta/2)} \right\}^2 = F_M(\theta), \qquad (6.2.66)$$

cf. (6.1.43). Thus, *the spectral window for the Bartlett estimate is the Fejer kernel of order M, $F_M(\theta)$*, whose general form is illustrated in Fig. 6.2. Since $F_M(\theta)$ is non-negative everywhere it follows that *the Bartlett estimate is similarly non-negative everywhere.*

It is interesting to recall the reasoning which led Bartlett to propose the above form of estimate. Initially, Bartlett argued that one could reduce the large sampling fluctuations of the periodogram by first splitting up the original sample of N observations into (N/M) subsamples, each subsample containing M observations, and then, for each ω, average the corresponding periodogram ordinates computed separately from each subsample. Bartlett worked with the unbiased autocovariance estimates,

$$\hat{R}^*(s) = \left(\frac{N}{N - |s|}\right) \hat{R}(s)$$

(cf. (5.3.10), (5.3.13)), in terms of which the periodogram for each sub-sample can be written as,

$$I_M^*(\omega) = \frac{1}{2\pi} \sum_{s=-M}^{M} \left(1 - \frac{|s|}{M}\right) \tilde{R}^*(s) \, e^{-i\omega s}, \qquad (6.2.67)$$

where $\tilde{R}^*(s)$ denotes the unbiased autocovariance estimate computed from the subsample. When we average the periodograms over the subsamples we obtain an expression which is of the same form as (6.2.67) but with $\tilde{R}^*(s)$ replaced by its "averaged" value. Neglecting the "end-effects" associated with each subsample, we may effectively replace the "averaged" value of $\tilde{R}^*(s)$ by $\hat{R}^*(s)$, the unbiased estimate computed from the full sample of N observations. We thus finally obtain as an estimate of $h(\omega)$,

$$\hat{h}_B(\omega) = \frac{1}{2\pi} \sum_{s=-M}^{M} \left(1 - \frac{|s|}{M}\right) \hat{R}^*(s) \, e^{-i\omega s} \qquad (6.2.68)$$

$$= \frac{1}{2\pi} \sum_{s=-M}^{M} \left(1 - \frac{|s|}{M}\right)\left(1 - \frac{|s|}{N}\right)^{-1} \hat{R}(s) \, e^{-i\omega s}. \qquad (6.2.69)$$

The estimate (6.2.68) is the one originally proposed by Bartlett, and we see from (6.2.69) that strictly the lag window for this estimate should be written as

$$\lambda(s) = \begin{cases} (1 - |s|/M)/(1 - |s|/N), & |s| \leq M, \\ 0, & |s| > M. \end{cases} \qquad (6.2.70)$$

However, as previously explained, the value of M is usually chosen to be much smaller than N, and hence over the range, $0 \leq |s| \leq M$, we can treat the factor $(1 - |s|/N)^{-1}$ as unity, and adequately approximate the lag window (6.2.70) by the simpler form (6.2.65). The estimate of $h(\omega)$ corresponding to the lag window (6.2.65) is the one now commonly known as the "Bartlett estimate", and its sampling properties are somewhat simpler than those for the lag window (6.2.70).

III. *The Daniell (or "rectangular") window*

Daniell (1946) was one of the first authors to consider the problem of estimating the spectral density function and suggested that one may estimate $h(\omega)$ by averaging the periodogram over a small interval centred on the frequency ω, say from $(\omega - \pi/M)$ to $(\omega + \pi/M)$, i.e. Daniell's estimate is,

$$\hat{h}_D(\omega) = \left(\frac{M}{2\pi}\right) \int_{\omega-(\pi/M)}^{\omega+(\pi/M)} I_N^*(\theta) \, d\theta. \qquad (6.2.71)$$

(For ω near $\pm\pi$ we extend the range of $I_N^*(\theta)$ by making it periodic, with period 2π.) This estimate is thus a particular case of (6.2.56) in which the spectral window $W(\theta)$ is periodic, period 2π, and on $(-\pi, \pi)$ has the *rectangular form*,

$$W(\theta) = \begin{cases} M/2\pi, & -\pi/M \leq \theta \leq \pi/M, \\ 0, & \text{otherwise,} \end{cases} \qquad (6.2.72)$$

as shown in Fig. 6.9. (Note that $W(\theta) \geq 0$, all θ, and hence the Daniell

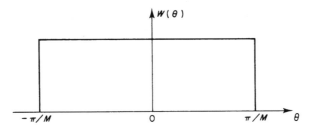

Fig. 6.9. The Daniell spectral window.

estimate, $\hat{h}_D(\omega)$, is always non-negative.) The corresponding lag window is, from (6.2.59), given by,

$$\lambda(s) = \frac{M}{2\pi} \int_{-\pi/M}^{\pi/M} e^{is\theta} \, d\theta = \frac{\sin(\pi s/M)}{(\pi s/M)}, \qquad \text{all } s. \qquad (6.2.73)$$

Here, the $\{\lambda(s)\}$ do not vanish after a certain value of $|s|$, so that there is no "truncation point". The spectral window, $W(\theta)$, has non-zero Fourier coefficients of arbitrarily high order, but the spectral estimate $\hat{h}_D(\omega)$ involves only the first N values of $\lambda(s)$, i.e. we may write,

$$\hat{h}_D(\omega) = \frac{1}{2\pi} \sum_{s=-(N-1)}^{(N-1)} \frac{\sin(\pi s/M)}{(\pi s/M)} \hat{R}(s) \, e^{-is\omega}. \qquad (6.2.74)$$

Note that the equivalence between (6.2.74) and (6.2.71) is *exact* even though the Fourier transform of the first N values of $\lambda(s)$ does not recover the exact rectangular form of $W(\theta)$ given by (6.2.72); see the remarks following (6.2.59).

The parameter M which appears in the Daniell window does not correspond to a "truncation point" (as in the two previous examples), but determines rather the "degree of smoothing" which is applied to the periodogram—the smaller M the greater the smoothing. Also, as M decreases, the lag window $\lambda(s)$ becomes more concentrated around $s = 0$, and decays more rapidly to zero. In this sense, M plays the same role here as in the truncated periodogram and Bartlett windows in that it controls the "*bandwidth*" of the spectral window. (We shall discuss the idea of "bandwidth" in much greater detail in Chapter 7 when we explain how this property of the spectral window effects the behaviour of $\hat{h}(\omega)$.)

Loosely speaking, the Daniell window may be regarded as the "dual" of the truncated periodogram window; the spectral window for the former having the same shape as the lag window for the latter, and the lag window for the Daniell estimate has a similar (but not identical) form to the spectral window for the truncated periodogram.

IV. *The general Tukey window*

Tukey, who together with Bartlett and Daniell was one of the pioneers of spectral estimation theory, suggested as estimate based on the lag window,

$$\lambda(s) = \begin{cases} 1 - 2a + 2a\cos(\pi s/M), & |s| \leq M, \\ 0, & |s| > M. \end{cases} \qquad (6.2.75)$$

(See Tukey (1949) and Blackman and Tukey (1959).) This lag window is also of the "truncated" type, the parameter M determining the truncation point, while the constant a is an additional parameter lying in the range $0 < a \leq \frac{1}{4}$ (so that $\lambda(s) \geq 0$, all s). The corresponding spectral window is a weighted linear combination of Dirichlet kernels, viz.,

$$W(\theta) = \frac{1}{2\pi} \sum_{s=-M}^{M} \{(1-2a) + a(\exp(i\pi s/M) + \exp(-i\pi s/M))\} \exp(-is\theta)$$

$$= aD_M\left(\theta - \frac{\pi}{M}\right) + (1-2a)D_M(\theta) + aD_M\left(\theta + \frac{\pi}{M}\right), \qquad (6.2.76)$$

where the Dirichlet kernel $D_M(\theta)$ is defined by (6.2.64). If we substitute (6.2.76) in (6.2.56) we see that Tukey's estimate $\hat{h}_T(\omega)$, say, can be written as a weighted linear combination of the values of the truncated periodogram estimate (example I) at the frequencies, $(\omega - \pi/M)$, ω, and $(\omega + \pi/M)$, i.e.,

$$\hat{h}_T(\omega) = a\hat{h}_0(\omega - \pi/M) + (1-2a)\hat{h}_0(\omega) + a\hat{h}_0(\omega + \pi/M), \qquad (6.2.77)$$

where $\hat{h}_0(\omega)$ is given by (6.2.48).

V. *The Tukey–Hamming window*

Tukey (1949) originally suggested taking $a = 0 \cdot 23$ in (6.2.75), in which case the lag window becomes

$$\lambda(s) = \begin{cases} 0 \cdot 54 + 0 \cdot 46 \cos(\pi s/M), & |s| \leq M, \\ 0, & |s| > M. \end{cases} \qquad (6.2.78)$$

This value of a was apparently chosen so as to reduce the magnitude of the peak in the first side lobe of the spectral window (6.2.76) relative to the magnitude of the peak in the main lobe, and the form of $\lambda(s)$ given by (6.2.78) is known as the "*Hamming*", or "*Tukey–Hamming*", window.

VI. *The Tukey–Hanning window*

A slightly more convenient computational form of the lag window (6.2.78) is given by taking $a = 0\cdot25$ in (6.2.75), so that now $\lambda(s)$ may be written as

$$\lambda(s) = \begin{cases} \frac{1}{2}\{1 + \cos(\pi s/M)\}, & |s| \le M, \\ 0, & |s| > M. \end{cases} \qquad (6.2.79)$$

This form is known as the "*Tukey–Hanning*" window, and is the one commonly referred to as the "*Tukey window*". The spectral window is now (from (6.2.76)),

$$W(\theta) = \tfrac{1}{4}D_M\!\left(\theta - \frac{\pi}{M}\right) + \tfrac{1}{2}D_M(\theta) + \tfrac{1}{4}D_M\!\left(\theta + \frac{\pi}{M}\right), \qquad (6.2.80)$$

and the corresponding spectral estimate, $\hat{h}_{T/H}(\omega)$, say, can be written as,

$$\hat{h}_{T/H}(\omega) = \tfrac{1}{4}\hat{h}_0\!\left(\omega - \frac{\pi}{M}\right) + \tfrac{1}{2}\hat{h}_0(\omega) + \tfrac{1}{4}\hat{h}_0\!\left(\omega + \frac{\pi}{M}\right), \qquad (6.2.81)$$

where, again, $\hat{h}_0(\omega)$ is the truncated periodogram defined by (6.2.48).

VII. *The Parzen window*

Parzen (1961b) suggested the lag window,

$$\lambda(s) = \begin{cases} 1 - 6(s/M)^2 + 6(|s|/M)^3, & |s| \le M/2, \\ 2(1 - |s|/M)^3, & M/2 \le |s| \le M, \\ 0, & |s| > M. \end{cases} \qquad (6.2.82)$$

The general forms of the Tukey–Hanning and Parzen lag windows are compared in Fig. 6.10.

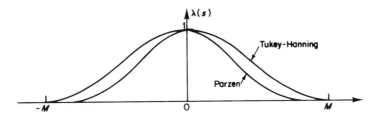

Fig. 6.10. The Parzen and Tukey–Hanning lag windows.

The Parzen spectral window corresponding to (6.2.82) is given by (assuming M even),

$$W(\theta) = \frac{1}{2\pi} \left[\sum_{s=-M/2}^{M/2} \{1 - 6(s/M)^2 + 6(|s|/M)^3\} \cos s\theta \right.$$

$$\left. + 2 \sum_{M/2 < |s| \le M} \left(1 - \frac{|s|}{M}\right)^3 \cos s\theta \right]$$

$$= \frac{3}{8\pi M^3} \left(\frac{\sin(M\theta/4)}{\frac{1}{2}\sin(\theta/2)}\right)^4 \{1 - \tfrac{2}{3}\sin^2(\theta/2)\}, \qquad (6.2.83)$$

(cf. Parzen (1963a)).

For large M the second term is small compared with the first in the neighbourhood of $\theta = 0$ (cf. the derivation of the form of the Dirichlet kernel near the origin) and as a useful approximation we may set

$$W(\theta) \sim \frac{3}{8\pi M^3} \left(\frac{\sin(M\theta/4)}{\frac{1}{2}\sin(\theta/2)}\right)^4. \qquad (6.2.84)$$

The Parzen lag window (6.2.82) may be derived by taking the Bartlett lag window, (6.2.65) with parameter $M/2$ (M even), treating it as a continuous function of s, and then forming the convolution of this function with itself. This explains why the *Parzen spectral window with parameter M has effectively the same form as the square of the Bartlett spectral window with parameter $M/2$*. The Parzen window thus shares with the Bartlett and Daniell windows the property that it always produces non-negative estimates of the spectral density function.

VIII. *The Bartlett–Priestley window*

The quadratic spectral window,

$$W(\theta) = \begin{cases} \dfrac{3M}{4\pi}\left\{1 - \left(\dfrac{M\theta}{\pi}\right)^2\right\}, & |\theta| \le \pi/M, \\[2mm] 0, & |\theta| > \pi/M, \end{cases} \qquad (6.2.85)$$

corresponding to the lag window,

$$\lambda(s) = \frac{3M^2}{(\pi s)^2}\left\{\frac{\sin(\pi s/M)}{\pi s/M} - \cos(\pi s/M)\right\}, \qquad (6.2.86)$$

was introduced by Priestley (1962c), and independently by Bartlett (1963). This form of spectral window was derived by minimizing an approximate expression for the relative mean square error of the estimated spectral density function with respect to the functional form of the window. Priestley showed that the form (6.2.85) was optimal within a somewhat restricted class of windows, but the optimality of the quadratic window with respect to the relative mean square error was later established more generally by Epanechnikov (1969). We will not elaborate further on this point at the moment since we return to the discussion of mean square errors (and other optimality criteria) in a more detailed form in Chapter 7.

IX. *The Lomnicki–Zaremba window*

Lomnicki and Zaremba (1957a) considered spectral estimates of the form,

$$\hat{h}(\omega) = \frac{1}{2\pi} \sum_{s=-(N-1)}^{(N-1)} \lambda^*(s)\hat{R}^*(s) \cos s\omega, \qquad (6.2.87)$$

where $\hat{R}^*(s)$ is the unbiased estimate of $R(s)$ (see (5.3.10)), and showed that the form of $\lambda^*(s)$ which minimizes the integrated mean square error of $\hat{h}(\omega)$ is given by,

$$\lambda^*(s) = \frac{R^2(s)}{\text{var}\{\hat{R}^*(s)\} + R^2(s)}. \qquad (6.2.88)$$

Unfortunately, this form of $\lambda^*(s)$ depends on the unknown "true" auto-covariance function $R(s)$ but Lomnicki and Zaremba suggested that one may approximate to the above form by

$$\lambda^*(s) = \frac{\rho^{|s|}}{\dfrac{1}{N-|s|}\left(\dfrac{1+\rho}{1-\rho}\right) + \rho^{|s|}}, \qquad (6.2.89)$$

where $0 < \rho < 1$. (This approximate form of $\lambda^*(s)$ is obtained by assuming that the observations conform to an AR(1) model with parameter ρ.) The parameter ρ corresponds, in a more general setting, to the rate of decay of the autocovariance function $R(s)$ and Lomnicki and Zaremba suggested that, without any prior information, one may take $\rho = \frac{1}{2}$. Lomnicki and Zaremba's estimate can, of course, be put in the general form (6.2.51) by setting $\lambda(s) = \lambda^*(s)/(1 - |s|/N)$.

X. *The Whittle window*

Whittle (1957) approached the problem of spectral estimation by intro-ducing a prior distribution over the space of all spectral density functions, and defined the optimal estimate as the one which minimizes the mean square error, evaluated with respect the prior distribution at all the frequencies $\{2\pi p/N\}$, $p = 0, 1, \ldots, [N/2]$. Although this procedure does not, in general, lead to an explicit form for the lag window, $\lambda(s)$, Whittle made certain simplifying assumptions and devised the form,

$$\lambda(s) = 1/[1 + a(s^2/N)], \qquad (6.2.90)$$

a being a constant.

XI. *The Daniells window*

Daniells (1957, 1962) suggested a novel approach by treating the estimation of $h(\omega)$ as a curve-fitting problem, given the periodogram values. He chooses a basic weight function $\alpha(\theta)$ (highly peaked in the region of $\theta = 0$), fits a polynomial of order k by minimizing with respect to $\alpha_0(\omega), \ldots, \alpha_k(\omega)$,

$$\int_{-\pi}^{\pi} \alpha(\theta)\{I_{N,X}^*(\omega + \theta) - \alpha_0(\omega) - \alpha_1(\omega)\theta - \ldots - \alpha_k(\omega)\theta^k\}^2 \, d\theta$$

and then estimates $h(\omega)$ by $\tilde{h}(\omega) = \alpha_0(\omega)$. If

$$\alpha(u) = \frac{1}{\sqrt{2\pi\sigma^2}} \exp(-u^2/2\sigma^2) \qquad (\sigma \text{ small}),$$

$\tilde{h}(\omega)$ can be written as,

$$\tilde{h}(\omega) = \frac{1}{2\pi} \sum_{s=-(N-1)}^{(N-1)} \lambda^{(k)}(s)\hat{R}(s) \cos s\omega,$$

where,

$$\lambda^{(2k)}(s) = \left(1 + \frac{\sigma^2 s^2}{2 \cdot 1!} + \ldots + \frac{\sigma^{2k} s^{2k}}{2^k \cdot k!}\right) \exp(-\tfrac{1}{2}\sigma^2 s^2). \qquad (6.2.91)$$

Scale parameter windows; window generators

It will be observed that for most (but not all) of the lag windows described above variation of the window parameter M simply "stretches" or "contracts" the function $\lambda(s)$, i.e. M acts as a *scale parameter*. For such windows we may write $\lambda(s)$ in the form,

$$\lambda(s) = k(s/M), \qquad (6.2.92)$$

where $k(u)$ is a fixed continuous even functions of u, with $k(0) = 1$. The specific forms of $k(u)$ for some of the above examples are as follows.

(a) Truncated periodogram window; $k(u) = \begin{cases} 1, & |u| \leq 1, \\ 0, & |u| > 1, \end{cases}$

(b) Bartlett window; $k(u) = \begin{cases} 1 - |u|, & |u| \leq 1, \\ 0, & |u| > 1, \end{cases}$

(c) Daniell window; $k(u) = (\sin \pi u)/(\pi u),$

(d) General Tukey window; $k(u) = \begin{cases} 1 - 2a + 2a \cos(\pi u), & |u| \leq 1, \\ 0, & |u| > 1, \end{cases}$

(g) Parzen window; $k(u) = \begin{cases} 1 - 6u^2 + 6|u|^3, & |u| \leq \frac{1}{2}, \\ 2(1 - |u|)^3, & \frac{1}{2} \leq |u| \leq 1, \\ 0, & |u| > 1. \end{cases}$

(h) Bartlett–Priestley window; $k(u) = \dfrac{3}{(\pi u)^2} \left(\dfrac{\sin \pi u}{\pi u} - \cos \pi u \right).$

Note, however, that in particular the Lomnicki–Zaremba window, the Whittle window, the Daniells window, and the "exact" form of the Bartlett window (6.2.70) *cannot* be written in the form (6.2.92), so that these windows are not of the "scale parameter" form.

The function $k(u)$ is called a *"lag window generator"*, and its Fourier transform,

$$K(\theta) = \frac{1}{2\pi} \int_{-\infty}^{\infty} k(u) \, e^{-iu\theta} \, du, \qquad (6.2.93)$$

is called a *"spectral window generator"*. (This terminology is due to Parzen (1957a, 1963a) who first introduced lag windows of the form (6.2.92).) The spectral window $W(\theta)$ corresponding to $\lambda(s)$ can be expressed in terms of $K(\theta)$ by writing,

$$W(\theta) = \frac{1}{2\pi} \sum_{s=-(N-1)}^{(N-1)} \lambda(s) \, e^{-is\theta}$$

$$= M \left\{ \frac{1}{2\pi M} \sum_{s=-\infty}^{\infty} k(s/M) \exp[-i(s/M)M\theta] \right\}.$$

For large M the above sum may be approximated by the integral (6.2.93), and hence we may write,

$$W(\theta) \sim M . K(M\theta). \qquad (6.2.94)$$

The precise relationship between $W(\theta)$ and $K(\theta)$ is given by

$$W(\theta) = M \sum_{j=-\infty}^{\infty} K\{M(\theta + 2\pi j)\}, \qquad (6.2.95)$$

(see Parzen 1963a). The result (6.2.95) is of exactly the same form as the relationship between the spectral density functions of a continuous parameter process and the discrete parameter process obtained by sampling the former at unit time intervals. This relationship was derived in the proof of Wold's theorem (equation (4.8.28)), and underlies the "aliasing effect" referred to in Section 4.8.3. The proof of (6.2.95) follows along exactly the same lines if we identify $k(u)$ and $K(\theta)$ with the autocovariance functions and spectral density function (respectively) of the continuous parameter process, and relate $\lambda(s)$ and $W(\theta)$ similarly with the discrete parameter process. The approximation (6.2.94) corresponds to using only the term for $j = 0$ in (6.2.95).

Some specific forms of $K(\theta)$ are;

(a) Truncated periodogram; $K(\theta) = \dfrac{1}{\pi} \dfrac{\sin\theta}{\theta}.$

(b) Bartlett window; $K(\theta) = \dfrac{1}{2\pi} \left(\dfrac{\sin(\theta/2)}{\theta/2} \right)^2.$

(c) Daniell window; $K(\theta) = \begin{cases} 1/2\pi, & |\theta| \le \pi, \\ 0, & |\theta| > \pi. \end{cases}$

(g) Parzen window; $K(\theta) = \dfrac{3}{8\pi} \left(\dfrac{\sin(\theta/4)}{\theta/4} \right)^4.$

(h) Bartlett–Priestley window; $K(\theta) = \begin{cases} \dfrac{3}{4\pi}(1 - (\theta/\pi)^2), & |\theta| \le \pi, \\ 0, & |\theta| > \pi. \end{cases}$

To illustrate the approximation (6.2.94) let us consider, for example, the Bartlett window for which the exact form of $W(\theta)$ is given by (6.2.66), namely,

$$W(\theta) = \frac{1}{2\pi M} \left(\frac{\sin(M\theta/2)}{\sin(\theta/2)} \right)^2 = M\tilde{K}(M\theta),$$

where,

$$\tilde{K}(\theta) = \frac{1}{2\pi M^2} \left(\frac{\sin(\theta/2)}{\sin(\theta/2M)} \right)^2 \sim \frac{1}{2\pi} \left(\frac{\sin(\theta/2)}{\theta/2} \right)^2$$

for large M, in agreement with the form of $K(\theta)$ given above.

When $\lambda(s)$ is of the form (6.2.92) we shall call both $\lambda(s)$ and $W(\theta)$ "scale parameter windows".

Exponential type windows

Parzen (1958, 1961b) introduced a further class of windows which he called the "exponential type". This class is characterized by lag windows of the form,

$$\lambda(s) = k[\exp(\alpha|s|)/M], \qquad (6.2.96)$$

where $M \to \infty$ as $N \to \infty$ in such a way that $(\log M)/N \to 0$, and α is a positive constant. (To distinguish these windows from those of the form (6.2.92), Parzen later referred to (6.2.92) as windows of the *"algebraic type".*) The motivation for introducing windows of the exponential type is that they represent a more general version of the Lomnicki–Zaremba window, which does not have a scale parameter form. Thus, if we replace $(1/(N - |s|))$ by $1/N$ in (6.2.89) we can rewrite this lag window in the form (6.2.96), with $k(u) = (1/(1 + |u|))$, $\alpha = \log(1/\rho)$, and $M = N/a$, a being a positive constant.

The reader may find the collection of windows listed in examples I–XI somewhat bewildering, but there are, in fact, many more special forms of $\lambda(s)$ and $W(\theta)$ which have appeared from time to time in the literature. (There is an interesting paper by Neave (1972) which discusses a variety of different functions.) There is really no limit to the number of different windows which we could construct; in fact, virtually any even function $\lambda(s)$ which is such that $\lambda(0) = 1$ and $\lambda(s)$ decays smoothly to zero and involves a parameter which controls the rate of decay, will provide a "reasonable" window. We defer further discussion of window "shapes" until Chapter 7 when we will consider in more detail the relative merits of the various standard windows, and examine the considerations which led to some of the standard forms. For the moment we will simply remark that, in general, the precise "shape" of the lag or spectral window is relatively unimportant as far as the sampling properties of the spectral estimate are concerned; a much more crucial decision is the choice of the window parameter (i.e. the parameter M in examples I–VII) which controls the rate of decay of $\lambda(s)$, or equivalently, the "width" of $W(\theta)$. We will discuss this question also in Chapter 7.

6.2.4 Sampling Properties of Spectral Estimates

The sampling properties of spectral estimates are extremely complicated and exact results can be obtained only in a few very special cases. However, we can obtain fairly simple asymptotic expressions for the mean, variance, and covariance of spectral estimates. These asymptotic results provide valid approximations when the sample size N is large, and are the ones which are most often used in practice.

We consider an estimate of the spectral density function $\hat{h}(\omega)$ of the form (6.2.54), namely,

$$\hat{h}(\omega) = \frac{1}{2\pi} \sum_{s=-(N-1)}^{(N-1)} \lambda_N(s)\hat{R}(s) e^{-is\omega}, \qquad (6.2.97)$$

which, according to (6.2.56), may be alternatively written as,

$$\hat{h}(\omega) = \int_{-\pi}^{\pi} I_N^*(\theta) W_N(\omega - \theta) \, d\theta, \qquad (6.2.98)$$

where,

$$W_N(\theta) = \frac{1}{2\pi} \sum_{s=-(N-1)}^{(N-1)} \lambda_N(s) \, e^{-is\theta}.$$

We now attach the suffix N to both $\lambda(s)$ and $W(\theta)$ to emphasize the dependence (via the window parameter) of both these functions on N. For example, in the case of the truncated periodogram estimate (6.2.48) we noted previously that we must have $M \to \infty$ as $N \to \infty$ in order for the estimate to be consistent; this means that M must be chosen as a function of N, and consequently the associated forms of $\lambda(s)$ and $W(\theta)$ also depend on N. (Also, since we shall be concerned only with the single process $\{X_t\}$, we now drop the suffix X in the notation for the periodogram and return to our previous convention of denoting it simply by $I_N^*(\theta)$.) We assume that $\lambda_N(s)$ is an *even sequence*, i.e. $\lambda_N(s) = \lambda_N(-s)$, all N, s, so that $W_N(\theta)$ *is a real valued even function of* θ, and, being a finite sum of cosines, is continuous for all θ. We assume further that $\lambda_N(s)$ is such that the following conditions hold;

(i) $W_N(\theta) \geq 0$, all N, θ,

(ii) $\displaystyle\int_{-\pi}^{\pi} W_N(\theta) \, d\theta = 1$, $(\Leftrightarrow \lambda_N(0) = 1)$, all N,

(iii) $\displaystyle\int_{-\pi}^{\pi} W_N^2(\theta) \, d\theta < \infty$, all N,

(iv) For any $\varepsilon \ (>0)$, $W_N(\theta) \to 0$ uniformly as $N \to \infty$ for $|\theta| > \varepsilon$,

(v) $\displaystyle\frac{\sum_{s=-(N-1)}^{(N-1)} \left(\frac{|s|}{N}\right) \lambda_N^2(s)}{\sum_{s=-(N-1)}^{(N-1)} \lambda_N^2(s)} \to 0$ as $N \to \infty$.

Note that condition (v) may be expressed alternatively in the form,

(v') $\left| \dfrac{V_N(\theta)}{V_N(0)} - 1 \right| \to 0$ as $N \to \infty$ for $|\theta| \leq \dfrac{A}{N}$,

(A a positive constant), where $V_N(\theta)$ denotes the convolution of $W_N(\theta)$ with itself.

Condition (iv) ensures that as N increases $W_N(\theta)$ becomes more and more concentrated around $\theta = 0$, so that (in view of (ii)) *as $N \to \infty$, $W_N(\theta)$ has the limiting form of a δ-function.* Condition (v) ensures that $W_N(\theta)$ is not too "narrow" in relation to $(1/N)$, i.e. that its "width" is much greater than $O(1/N)$. This will be true for example for lag windows of the "truncated" type (examples I, II, IV, VII) if the truncation point, M, is such that $(M/N) \to 0$ as $N \to \infty$, i.e. if $M = o(N)$.

It is useful to note at this stage that, under conditions (ii), (iii), (iv), $W_N(\theta)$ has the property that,

(vi) $\left\{ \int_{-\pi}^{\pi} W_N^2(\theta) \, d\theta \right\} \to \infty$ as $N \to \infty$.

This result is intuitively obvious from the fact that $W_N(\theta)$ has the limiting form of a δ-function, but a formal proof is easily constructed, as follows. Let $g(\theta)$ be any continuous bounded square integrable function defined on $(-\infty, \infty)$; then by (ii) and (iv) we have,

$$\lim_{N \to \infty} \int_{-\pi}^{\pi} W_N(\theta) g(\theta) \, d\theta = g(0).$$

Now by the Cauchy–Schwarz inequality,

$$\left| \int_{-\pi}^{\pi} W_N(\theta) g(\theta) \, d\theta \right| \leq \left\{ \int_{-\pi}^{\pi} W_N^2(\theta) \, d\theta \right\}^{1/2} \left\{ \int_{-\pi}^{\pi} g^2(\theta) \, d\theta \right\}^{1/2}.$$

Suppose that $\int_{-\pi}^{\pi} W_N^2(\theta) \, d\theta$ remains bounded as $N \to \infty$, i.e. suppose that $\exists$ a constant K such that $\int_{-\pi}^{\pi} W_N^2(\theta) \, d\theta \leq K^2$, all N. Then,

$$\left| \int_{-\pi}^{\pi} W_N(\theta) g(\theta) \, d\theta \right| \leq K \left\{ \int_{-\pi}^{\pi} g^2(\theta) \, d\theta \right\}^{1/2}, \qquad \text{all } N,$$

and letting $N \to \infty$ we obtain

$$g(0) \leq K \left\{ \int_{-\pi}^{\pi} g^2(\theta) \, d\theta \right\}^{1/2}.$$

If our assumption about $\int_{-\pi}^{\pi} W_N^2(\theta) \, d\theta$ is correct, the above inequality must hold for *all* suitable choices of $g(\theta)$. However, let $g(\theta) = \exp(-\theta^2/4\sigma^2)$, in which case $g(0) = 1$, (all σ), and

$$\int_{-\pi}^{\pi} g^2(\theta) \, d\theta < \int_{-\infty}^{\infty} g^2(\theta) \, d\theta = \int_{-\infty}^{\infty} \exp(-\theta^2/2\sigma^2) \, d\theta = \sigma\sqrt{2\pi}.$$

If σ is sufficiently small we can make $K\{\int_{-\pi}^{\pi} g^2(\theta) \, d\theta\}^{1/2}$ arbitrarily small, and hence we obtain a contradiction—which thus proves the result (vi).

A rigorous derivation of the asymptotic mean, variance and covariance of $\hat{h}(\omega)$ under the above conditions was first given by Grenander and Rosenblatt (1953) (see also Grenander and Rosenblatt (1957a, 1957b)) and Parzen (1957a), who dealt with the case where $W_N(\theta)$ is of the "scale parameter form"). The standard method is to start from the expression (6.2.97) for $\hat{h}(\omega)$, and then use the results given in Chapter 5 (cf. (5.3.21)) for the sampling properties of the samples autocovariances. However, this approach is extremely complicated, and the limiting operations involved in obtaining the asymptotic results need to be handled with considerable care and delicacy, for as N increases it affects not only the number of terms in (6.2.97) but also the weights $\lambda_N(s)$ attached to each term. Here, we adopt a different approach and work directly from (6.2.98) which, we note, is a weighted integral of the periodogram and thus has the same structure as the statistics $\hat{\psi}_i$ considered in theorem 6.2.4. At first sight it would appear that we could obtain asymptotic expressions for $E[\hat{h}(\omega)]$ and $\mathrm{cov}\{\hat{h}(\omega_1), \hat{h}(\omega_2)\}$ immediately from theorem 6.2.4 by setting,

$$\phi_1(\theta) = W_N(\omega_1 - \theta), \qquad \phi_2(\theta) = W_N(\omega_2 - \theta), \qquad (6.2.100)$$

assuming, of course, that the process $\{X_t\}$ from which the observations come satisfies the conditions of this theorem. As we shall see, this device does in fact yield the correct results, but the above substitution for ϕ_1, ϕ_2, requires further justification since in the definition of $\hat{\psi}_i$ (equation (6.2.39)) the weight function ϕ_i was a fixed function, independent of N. However, the essential point in the proof of theorem 6.2.4 is that as $N \to \infty$ the Fejer kernel $F_N(\theta)$ behaves as a δ-function with respect to the function $\phi_2(\theta)$, and we can see intuitively that this property will remain true when we replace $\phi_2(\theta)$ by $W_N(\omega_2 - \theta)$ provided $W_N(\theta)$ satisfies condition (v'); recall that the "width" of $F_N(\theta)$ is $O(1/N)$. More formally, we can show that the results of theorem 6.2.4 remain valid when $W_N(\theta)$ satisfies (v') (or equivalently, when $\lambda_N(s)$ satisfies (v)) by the following argument. Consider the LHS of (6.2.46a) and let us suppose, for simplicity, that the spectral density function $h(\omega)$ is constant for all ω; $h(\omega)$ is, in any case, a fixed continuous function, independent of N, and does not give rise to any of the difficulties referred to above. When we replace both $\phi_1(\theta)$ and $\phi_2(\theta)$ by $W_N(\theta)$ (supposing, again for simplicity, that $\omega_1 = \omega_2 = 0$), the LHS of (6.2.46a) then becomes

$$\int_{-\pi}^{\pi} W_N(\theta')\{F_N(\theta - \theta') + F_N(\theta + \theta')\} \, d\theta'. \qquad (6.2.101)$$

(Note the change of notation: ω and ω' in (6.2.46a) correspond to θ, θ', in (6.2.101).) Since both terms in the integral in (6.2.101) are of the same form it suffices to consider just the first; let us write,

$$\int_{-\pi}^{\pi} W_N(\theta') F_N(\theta - \theta') \, d\theta' = \tilde{W}_N(\theta),$$

say. Then substituting this expression into the second term of (6.2.46) gives us a term which (apart from the factor $(2\pi/N)$) has the form,

$$\int_{-\pi}^{\pi} W_N(\theta) \tilde{W}_N(\theta) \, d\theta.$$

To establish the validity of, e.g., (6.2.43) we now want to show that this is asymptotically equivalent to $\int_{-\pi}^{\pi} W_N^2(\theta) \, d\theta$ (in the sense that, as $N \to \infty$, the limit of the ratio of the two expressions is unity). Now $W_N(\theta)$ and $F_N(\theta)$ have Fourier coefficients $\lambda_N(s)$ and $(1 - |s|/N)$ $(|s| \le N - 1)$, respectively, and $\tilde{W}_N(\theta)$, being the convolution of $W_N(\theta)$ with $F_N(\theta)$, therefore has Fourier coefficients $\{(1 - |s|/N)\lambda_N(s)\}$, $(|s| \le N - 1)$. Hence, using Parseval's relation, we may write

$$\int_{-\pi}^{\pi} W_N(\theta) \tilde{W}_N(\theta) \, d\theta = \frac{1}{2\pi} \sum_{s=-(N-1)}^{(N-1)} \left(1 - \frac{|s|}{N}\right) \lambda_N^2(s).$$

Similarly,

$$\int_{-\pi}^{\pi} W_N^2(\theta) \, d\theta = \frac{1}{2\pi} \sum_{s=-(N-1)}^{(N-1)} \lambda_N^2(s),$$

and consequently,

$$\lim_{N \to \infty} \left\{ \int_{-\pi}^{\pi} W_N(\theta) \tilde{W}_N(\theta) \, d\theta \bigg/ \int_{-\pi}^{\pi} W_N^2(\theta) \, d\theta \right\} = \lim_{N \to \infty} \left\{ 1 - \frac{\sum_{s=-(N-1)}^{(N-1)} \frac{|s|}{N} \lambda_N^2(s)}{\sum_{s=-(N-1)}^{(N-1)} \lambda_N^2(s)} \right\}$$

$$= 1, \quad \text{by condition (v)}.$$

The validity of the other parts of theorem (6.2.4) may be established by a similar argument. Thus, *provided the observations come from a general linear process satisfying the conditions of theorem* (6.2.4), *and provided* $W_N(\theta)$ *satisfies the above conditions*, we may safely apply the results of this theorem when $\phi_1(\theta)$, $\phi_2(\theta)$, are given by (6.2.100), even though these functions now depend on N. However, because of the dependence of these functions on N we now have to express the results in the form of asymptotic relationships rather than as the strict limits.

We then obtain immediately from (6.2.41),

$$E\{\hat{h}(\omega)\} \sim \int_{-\pi}^{\pi} h(\theta) W_N(\omega - \theta) = \tilde{h}(\omega), \qquad (6.2.102)$$

say. If $h(\omega)$ has a bounded first derivative we can strengthen the above result by using (6.2.12), from which it follows that,

$$E[\hat{h}(\omega)] = \int_{-\pi}^{\pi} h(\theta) W_N(\omega - \theta) \cdot d\theta + O\left(\frac{\log N}{N}\right). \qquad (6.2.103)$$

Since $h(\theta)$ is assumed continuous for all θ it now follows from conditions (ii) and (iv) on $W_N(\theta)$ that,

$$\lim_{N \to \infty} E\{\hat{h}(\omega)\} = h(\omega), \qquad \text{all } \omega, \qquad (6.2.104)$$

showing that $\hat{h}(\omega)$ is an asymptotically unbiased estimate of $h(\omega)$. Also, from (6.2.43), (6.2.42),

$$N \, \mathrm{var}\{\hat{h}(\omega)\} \sim e\{\tilde{h}(\omega)\}^2$$

$$+ 2\pi \int_{-\pi}^{\pi} h^2(\theta) W_N(\omega - \theta)\{W_N(\omega - \theta) + W_N(\omega + \theta)\} \, d\theta \qquad (6.2.105)$$

and

$$N \, \mathrm{cov}\{\hat{h}(\omega_1), \hat{h}(\omega_2)\} \sim e\tilde{h}(\omega_1)\tilde{h}(\omega_2)$$

$$+ 2\pi \int_{-\pi}^{\pi} h^2(\theta) W_N(\omega_1 - \theta)\{W_N(\omega_2 - \theta) + W_N(\omega_2 + \theta)\} \, d\theta, \qquad (6.2.106)$$

where e (the fourth cumulant of the residual process) is as defined in theorem (6.2.4). We can simplify (6.2.105) by observing that the first term is $O(1)$ and is therefore negligible compared with the second term which $\to \infty$ as $N \to \infty$. (We know that $W_N(\theta) \to \delta(\theta)$ as $N \to \infty$, and hence, by condition (vi), $\int_{-\pi}^{\pi} W_N^2(\omega - \theta) \, d\theta \to \infty$.) In fact, if $\{X_t\}$ is a Gaussian process the first term vanishes automatically. Note, however, an important distinction between the case where the weight function is fixed (independent of N) and the case where it tends to a δ-function; in the former case both terms in (6.2.105) are important, and the variance differs according to whether or not $\{X_t\}$ is Gaussian, whereas in the latter case the first term is negligible compared with the second, and the asymptotic variance is the same whether $\{X_t\}$ is Gaussian or not. Also, since $W_N(\omega - \theta) \to \delta(\omega - \theta)$ and $W_N(\omega + \theta) \to \delta(\omega + \theta)$, the term

$$\int_{-\pi}^{\pi} h^2(\theta) W_N(\omega - \theta) W_N(\omega + \theta) \, d\theta,$$

will vanish as $N \to \infty$ unless $\omega = 0$ or $\pm \pi$. (Recall that $W_N(\theta)$ is even and periodic, period 2π, so that, e.g., $W_N(\pi + \theta) \equiv W_N(\theta - \pi) \equiv W_N(\pi - \theta)$.) On the other hand, when $\omega = 0$ or $\pm \pi$ the second part of the second term in (6.2.105) becomes identical to the first part of the second term, so that we may now write,

$$\text{var}\{\hat{h}(\omega)\} \sim (1 + \delta_{\omega,0,\pi}) \frac{2\pi}{N} \int_{-\pi}^{\pi} h^2(\theta) W_N^2(\omega - \theta) \, d\theta, \quad (6.2.107)$$

where

$$\delta_{\omega,0,\pi} = \begin{cases} 1, & \omega = 0, \pm \pi, \\ 0, & \omega \neq 0, \pm \pi. \end{cases}$$

Finally, since, for large N, $W_N^2(\omega - \theta)$ is highly concentrated about $\theta = \omega$ (cf. condition (iv)) and since $h(\theta)$ is assumed continuous, we may write,

$$\int_{-\pi}^{\pi} h^2(\theta) W_N^2(\omega - \theta) \, d\theta \sim h^2(\omega) \int_{-\pi}^{\pi} W_N^2(\omega - \theta) \, d\theta = h^2(\omega) \int_{-\pi}^{\pi} W_N^2(\theta) \, d\theta,$$

since $W_N(\theta)$ has period 2π. We now have from (6.2.107),

$$\text{var}\{\hat{h}(\omega)\} \sim (1 + \delta_{\omega,0,\pi}) \frac{2\pi}{N} h^2(\omega) \int_{-\pi}^{\pi} W_N^2(\theta) \, d\theta. \quad (6.2.108)$$

But, by Parseval's theorem,

$$2\pi \int_{-\pi}^{\pi} W_N^2(\theta) \, d\theta = \sum_{s=-(N-1)}^{(N-1)} \lambda_N^2(s), \quad (6.2.109)$$

so that we may re-write the asymptotic variance of $\hat{h}(\omega)$ in the alternative form,

$$\text{var}\{\hat{h}(\omega)\} \sim (1 + \delta_{\omega,0,\pi}) h^2(\omega) \cdot \frac{1}{N} \left\{ \sum_{s=-(N-1)}^{(N-1)} \lambda_N^2(s) \right\}. \quad (6.2.110)$$

The above expression (or its equivalent form for scale parameter windows to be derived below) is the standard one used to evaluate the asymptotic variance of estimates of the spectral density function. It is easily computed for any of the standard forms of lag windows $\lambda_N(s)$ discussed previously, and we will later tabulate the asymptotic variances of some of the standard estimates.

Turning to the covariances we see that the first term is (6.2.106) is always $O(1)$ (or exactly zero if $\{X_t\}$ is Gaussian); the second term involves the integral of products of functions which tend to δ-functions centred on ω_1, ω_2, and $(-\omega_2)$. Hence, if ω_1, ω_2, are fixed frequencies such that $\omega_1 \neq \pm \omega_2$, the second term $\to 0$, and

$$\lim_{N \to \infty} \text{cov}\{\hat{h}(\omega_1), \hat{h}(\omega_2)\} = 0, \quad \omega_1 \neq \pm \omega_2. \quad (6.2.111)$$

This result tells us that the spectral estimates at any two distinct fixed frequencies are asymptotically uncorrelated, as we observed in the case of the periodogram. However, one of the objectives in smoothing the periodogram was to inject more correlation between neighbouring ordinates, and we would therefore expect such ordinates of $\hat{h}(\omega)$ to have greater covariance than the corresponding ordinates of the periodogram. This is indeed correct, but it is not revealed by the "crude" asymptotic result (6.2.111) since the interval between any two fixed frequencies ω_1, ω_2 becomes "large" (relative to $1/N$) as $N \to \infty$. It is more informative, therefore, to consider the covariance between two neighbouring frequencies when N is large but finite. Thus, consider (6.2.106) when ω_1, ω_2, are neighbouring frequencies (neither of which is close to zero or $\pm \pi$) and N is large. As noted above, the first term on the RHS is always $O(1)$ (giving a contribution $O(1/N)$ to the covariance), while the second part of the second term involved the integral of the product of two functions tending to δ-functions centred on ω_1 and $(-\omega_2)$, and so tends to zero even when ω_1 and ω_2 are close. The dominant term in the expression for the covariance is therefore the first part of the second part in (6.2.106), and we may therefore use the approximation,

$$\text{cov}\{\hat{h}(\omega_1), \hat{h}(\omega_2)\} \sim \frac{2\pi}{N} \int_{-\pi}^{\pi} h^2(\theta) W_N(\omega_1 - \theta) W_N(\omega_2 - \theta) \, d\theta \qquad (6.2.112)$$

The integrand in (6.2.112) involves the product of two functions tending to δ-functions centred on ω_1, ω_2; for N large (but finite) the integral will be small *unless ω_1 and ω_2 are sufficiently close together so that $|\omega_1 - \omega_2|$ is the same order of magnitude as the "width" of the function $W_N(\theta)$*. To illustrate this point consider, for example, the Daniell window (cf. (6.2.72)) for which

$$W_N(\theta) = \begin{cases} M/2\pi, & |\theta| \le \pi/M, \\ 0, & \text{otherwise.} \end{cases}$$

The product $\{W_N(\omega_1 - \theta) . W_N(\omega_2 - \theta)\}$ will be zero for all θ if $|\omega_1 - \omega_2| > 2\pi/M$, but if $|\omega_1 - \omega_2| < 2\pi/M$ there will be a range of values of θ for which the product is non-zero. *The spectral estimates corresponding to this window will thus be effectively uncorrelated only if the separation between the frequencies ω_1, ω_2, is greater than the "bandwidth", $2\pi/M$, of the window,* but for frequencies whose separation lies within the bandwidth of the window the correlation will be non-zero, and will increase as $|\omega_1 - \omega_2|$ decreases.

The main conclusion of this result, namely that for two spectral estimated to be effectively uncorrelated the separation between their frequencies must be appreciably greater than the "bandwidth" of the spectral window, holds quite generally for most of the standard windows, and is one of the reasons why the concept of bandwidth is particularly important. We will return to this point again when we discuss the general notion of bandwidth in Chapter 7.

Results for scale parameter windows

In the particular case when the lag window is of the scale parameter form, i.e. when $\lambda_N(s)$ is of the form, $\lambda_N(s) = k(s/M)$, we can express the above results in the more precise form of "strict limits" since now the function $k(u)$ does not depend on N. We assume as before that $k(u)$ is an even continuous function with $k(0) = 1$, and now add the additional assumptions that $K(\theta)$ is positive and $\int_{-\infty}^{\infty} k^2(u)\, du < \infty$. Conditions (i)–(v) above will be automatically satisfied provided that as $N \to \infty$, $M \to \infty$ and $M/N \to 0$. We now have,

$$\frac{1}{M} \sum_{s=-(N-1)}^{(N-1)} \lambda_N^2(s) = \frac{1}{M} \sum_{s=-(N-1)}^{(N-1)} k^2(s/M) \to \int_{-\infty}^{\infty} k^2(u)\, du, \qquad \text{as } N \to \infty.$$

Hence, in place of (6.2.110) we may now write,

$$\lim_{N \to \infty} \left[\frac{N}{M} \text{var}\{\hat{h}(\omega)\} \right] = (1 + \delta_{\omega,0,\pi}) h^2(\omega) \int_{-\infty}^{\infty} k^2(u)\, du, \quad (6.2.113)$$

and in place of (6.2.111) we have (cf. Parzen (1957a)),

$$\lim_{N \to \infty} \left[\frac{N}{M} \text{cov}\{\hat{h}(\omega_1), \hat{h}(\omega_2)\} \right] = 0, \qquad \omega_1 \neq \pm\omega_2. \quad (6.2.114)$$

Equation (6.2.113) is the standard expression used to compute the asymptotic variance of a spectral estimate based on a scale parameter window. The result (6.2.113) can be used also for windows of the *exponential type* (i.e. where $\lambda(s)$ is of the form (6.2.96)) if we replace M by $(\log M/\alpha)$ and replace $\{\int_{-\infty}^{\infty} k^2(u)\, du\}$ by the number 2 (see Parzen (1961b)).

Approximate expressions for the bias

We have seen that the periodogram is an asymptotically unbiased estimate of $h(\omega)$, but it will nevertheless be biased for N finite. The introduction of the smoothing window means that the estimate $\hat{h}(\omega)$ is, in effect, a weighted average of neighbouring periodogram ordinates, and this will, in general, further increase the bias—although provided $M \to \infty$ as $N \to \infty$ $\hat{h}(\omega)$ will also be asymptotically unbiased.

Let

$$b(\omega) = [E\{\hat{h}(\omega)\} - h(\omega)] \quad (6.2.115)$$

denote the bias in $\hat{h}(\omega)$ at frequency ω. From (6.2.103) we have (recalling that $\int_{-\pi}^{\pi} W_N(\theta)\, d\theta = 1$),

$$b(\omega) = \int_{-\pi}^{\pi} \{h(\theta) - h(\omega)\} W_N(\omega - \theta)\, d\theta + \mathrm{O}\left(\frac{\log N}{N}\right). \quad (6.2.116)$$

The second term, $\mathrm{O}((\log N)/N)$, is the bias due to the periodogram itself, and in general this will be much smaller than the first term which represents the bias due to the smoothing. We therefore ignore the second term, and concentrating on the first term only we find

$$b(\omega) \sim \int_{-\pi}^{\pi} \{h(\omega - \theta) - h(\omega)\} W_N(\theta)\, d\theta. \quad (6.2.117)$$

Suppose now that $h(\omega)$ is "locally quadratic", i.e. is twice differentiable with a bounded second derivative, and for small θ,

$$h(\omega - \theta) \sim h(\omega) - \theta h'(\omega) + \frac{\theta^2}{2} h''(\omega) + \mathrm{o}(\theta^2). \quad (6.2.118)$$

Substituting (6.2.118) in (6.2.117) we obtain

$$b(\omega) \sim \tfrac{1}{2} h''(\omega) \int_{\pi}^{\pi} \theta^2 W_N(\theta)\, d\theta. \quad (6.2.119)$$

(Note that $W_N(\theta)$ is an even function of θ and hence $\int_{\pi}^{\pi} \theta W_N(\theta)\, d\theta = 0$.) Equation (6.2.119) gives us only a rather crude expression for the bias, but it illustrates a very important feature of the smoothing procedure. The second factor in (6.2.119) is a measure of the squared *width* of the window, $W_N(\theta)$; if we think of $W_N(\theta)$ as a probability density function then this term would represent its variance. On the other hand, the first factor $h''(\omega)$ depends on the *curvature* of the true spectral density function $h(\omega)$ in the sense that $\max_\omega |h''(\omega)|$ will be large if $h(\omega)$ has a sharp peak, but will be small if $h(\omega)$ is a fairly "flat" function. Thus, the magnitude of the bias depends basically on two factors, (a) the width of the window and (b) the degree of curvature of the spectral density function. In order to keep the magnitude of the bias low, the width of the window should ideally be "tailored" to the curvature properties of the spectral density function. We will explore this point more fully in Chapter 7.

A more precise and elegant treatment of the bias is possible for windows of the scale parameter form, following an approach due to Parzen (1957a).

Writing $\lambda_N(s) = k(s/M)$ in (6.2.97) and taking expectations we have,

$$b(\omega) = \frac{1}{2\pi} \sum_{s=-(N-1)}^{(N-1)} \left\{ k(s/M)\left(1 - \frac{|s|}{N}\right) - 1 \right\} R(s) \, e^{-is\omega}$$

$$-\frac{1}{2\pi} \sum_{|s|\geqslant N} R(s) \, e^{-is\omega}$$

$$= \frac{1}{2\pi} \sum_{s=-(N-1)}^{(N-1)} \{ k(s/M) - 1 \} R(s) \, e^{-is\omega}$$

$$-\frac{1}{2\pi N} \sum_{s=-(N-1)}^{(N-1)} |s| k(s/M) R(s) \, e^{-is\omega}$$

$$-\frac{1}{2\pi} \sum_{|s|\geqslant N} R(s) \, e^{-is\omega}. \tag{6.2.120}$$

Now let $r(>0)$ be the largest integer such that

$$k^{(r)} = \lim_{u \to 0} \left\{ \frac{1 - k(u)}{|u|^r} \right\} \tag{6.2.121}$$

exists, is finite, and is non-zero. Parzen calls r the "*characteristic exponent*" of the function $k(u)$. Assume that $\sum_{s=-\infty}^{\infty} |s|^q |R(s)| < \infty$ for $q \leqslant r$, and that $(N/M^r) \to \infty$ as $N \to \infty$. Then taking the limit of the above expression for $b(\omega)$ as $N \to \infty$ we see that the second and third terms $\to 0$ as $o(M^{-r})$, uniformly in ω, whilst the first term is asymptotically

$$-k^{(r)} M^{-r} h^{(r)}(\omega),$$

where

$$h^{(r)}(\omega) = \frac{1}{2\pi} \sum_{s=-\infty}^{\infty} |s|^r R(s) \, e^{-is\omega}, \tag{6.2.122}$$

is called by Parzen the "*generalized r^{th} derivative*" of $h(\omega)$. We finally obtain

$$\lim_{N \to \infty} \{ M^r b(\omega) \} = -k^{(r)} h^{(r)}(\omega), \tag{6.2.123}$$

so that an asymptotic expression for the bias is

$$b(\omega) \sim -M^{-r} k^{(r)} h^{(r)}(\omega). \tag{6.2.124}$$

Note that $h^{(r)}(\omega)$ is *not* the same as the (ordinary) rth derivative of $h(\omega)$; if r is even then clearly

$$h^{(r)}(\omega) = (-1)^{r/2} \left(\frac{d}{d\omega} \right)^r . \, h(\omega),$$

but when r is odd there is no simple relationship between the two quantities. For r even $k^{(r)}$ may similarly be related to the rth derivative of $k(u)$ at the origin. For, from the definition of $k^{(r)}$ we may write for small u,

$$k(u) \sim 1 - k^{(r)}|u|^r.$$

Hence, for r even,

$$k^{(r)} = -(r!)^{-1} \left[\frac{d^r(k(u))}{du^r} \right]_{u=0}.$$

Now the behaviour of $k(u)$ near the origin is closely related to the behaviour of its Fourier transform $K(\theta)$, when θ is large, and hence $k^{(r)}$ is, in effect, a measure of the "width" of $K(\theta)$. In particular, when $r = 2$ (as would be the case, e.g., with Parzen's window, example VII), then,

$$k^{(2)} = -\tfrac{1}{2}k''(0) = \tfrac{1}{2} \int_{-\infty}^{\infty} \theta^2 K(\theta) \, d\theta,$$

(inverting (6.2.93) and differentiating twice w.r. to u) and $h^{(2)}(\omega) = -h''(\omega)$. Hence in this case (6.2.123) reduces to,

$$b(\omega) \sim \frac{1}{2M^2} \cdot h''(\omega) \int_{-\infty}^{\infty} \theta^2 K(\theta) \, d\theta, \qquad (6.2.125)$$

which is virtually the same as the previously derived "crude" expression (6.2.119). (Recall that $W_N(\theta) \sim M \cdot K(M\theta)$; cf. (6.2.95)).

Let $Q(>0)$ be such that $\sum_{s=-\infty}^{\infty}|s|^Q|R(s)| < \infty$. The larger Q the faster do the covariances decay to zero. On the other hand, the larger r the slower does the function $k(u)$ decay. In the above discussion we have assumed that $Q \geqslant r$, i.e. that $k(u)$ decays as fast as or faster than the autocovariance function $R(s)$. If $r > Q$, i.e. if $k(u)$ decays more slowly than $R(s)$, then clearly the bias will now be dominated by the second and third terms in (6.2.120) and the above asymptotic expression is no longer valid. We now obtain instead (Parzen (1957a)),

$$b(\omega) \sim o(M^{-Q}),$$

provided $M/N \to 0$ (when $Q \leqslant 1$) or $M^Q/N \to 0$ ($Q > 1$). The latter situation will arise, in particular, in the case of the "truncated periodogram" estimate (example (I)) for which $k(u) = 1$, $|u| \leqslant 1$, $k(u) = 0$, $|u| > 1$. Here, $\lim_{u \to 0}[(1 - k(u))|u|^r]$ exists and is zero for all integers r, and thus for this window we may say that r is *infinite*. For this estimate all one can say about the bias is that, provided N satisfies the above conditions, $b(\omega) = o(M^{-Q})$, where Q is as defined above.

The asymptotic bias of a spectral estimate corresponding to any of the standard scale parameter windows is easily evaluated from (6.2.124), and the asymptotic variance is also easily evaluated from (6.2.113) or (if the window is not of the scale parameter form) from (6.2.110) or (6.2.108). To illustrate the typical form of the calculations we consider the following specific windows.

Bartlett window

Here,

$$k(u) = \begin{cases} 1 - |u|, & |u| \le 1, \\ 0, & |u| > 1, \end{cases}$$

so that

$$\int_{-\infty}^{\infty} k^2(u)\, du = 2 \int_0^1 (1-u)^2\, du = 2/3.$$

Hence, (6.2.113) gives,

$$\mathrm{var}\{\hat{h}(\omega)\} \sim \frac{2M}{3N} \cdot h^2(\omega) \qquad (\omega \ne 0, \pm\pi).$$

For this form of $k(u)$, $r = 1$, and

$$k^{(1)} = \lim_{u \to 0} \left(\frac{1 - k(u)}{|u|} \right) = 1.$$

Hence, (6.2.124) gives,

$$b(\omega) = [E\{\hat{h}(\omega)\} - h(\omega)] \sim -\frac{1}{M} h^{(1)}(\omega),$$

where

$$h^{(1)}(\omega) = \frac{1}{2\pi} \sum_{s=-\infty}^{\infty} |s| R(s)\, e^{-i\omega s}.$$

Daniell window

Here, $k(u) = \sin \pi u / (\pi u)$, and

$$\int_{-\infty}^{\infty} k^2(u)\, du = \frac{2}{\pi} \int_0^{\infty} \frac{\sin^2 x}{x^2}\, dx = 1.$$

Hence, by (6.2.113),

$$\mathrm{var}\{\hat{h}(\omega)\} \sim \frac{M}{N} h^2(\omega) \qquad (\omega \ne 0, \pm\pi).$$

(For this window it is, in fact, simpler to evaluate the variance directly from

(6.2.108).) Also, for small u, $k(u) \sim 1 - (\pi u)^2/6 + O(|u|^3)$, so that $r = 2$, and $k^{(2)} = \pi^2/6$. Hence, by (6.2.124),

$$b(\omega) \sim -\frac{\pi^2}{6M^2} h^{(2)}(\omega) = \frac{\pi^2}{6M^2} h''(\omega).$$

Parzen window

Here,

$$k(u) = \begin{cases} 1 - 6u^2 + 6|u|^3, & |u| \le \frac{1}{2} \\ 2(1 - |u|)^3, & \frac{1}{2} \le |u| \le 1 \\ 0, & |u| > 1. \end{cases}$$

$$\int_{-\infty}^{\infty} k^2(u)\, du = \frac{151}{280} \doteq 0 \cdot 539285.$$

Also, $r = 2$, and $k^{(2)} = 6$. Hence, by (6.2.113),

$$\text{var}\{\hat{h}(\omega)\} \sim 0 \cdot 539285 \frac{M}{N} h^2(\omega) \qquad (\omega \ne 0, \pm\pi),$$

and by (6.2.124),

$$b(\omega) \sim -\frac{6}{M^2} h^{(2)}(\omega) = \frac{6}{M^2} h''(\omega).$$

Truncated periodogram

Here,

$$k(u) = \begin{cases} 1, & |u| \le 1 \\ 0, & |u| > 1. \end{cases}$$

$\int_{-\infty}^{\infty} k^2(u)\, du = 2$, and hence by (6.2.113),

$$\text{var}\{\hat{h}(\omega)\} \sim \frac{2M}{N} \cdot h^2(\omega).$$

For this form of $k(u)$, r is *infinite* and, as noted above, we cannot now apply (6.2.124) to evaluate the asymptotic bias.

The asymptotic bias and variances for the standard windows are summarized in Table 6.1. These results are, of course, derived under the assumption that the conditions on M, Q and r required for the validity of (6.2.113) and (6.2.124) hold. The asymptotic variances of all the estimates involve the factor $\{(M/N)h^2(\omega)\}$ and thus differ only with regard to the numerical coefficient of this factor. We have therefore tabulated only this numerical coefficient, i.e. the quantity given in the table is

$$\int_{-\infty}^{\infty} k^2(u)\, du \sim N \, \text{var}\{\hat{h}(\omega)\}/\{Mh^2(\omega)\}.$$

Table 6.1. *Asymptotic bias and variances*

Estimate	r	Bias	$N \, \text{var}\{\hat{h}(\omega)\}/\{Mh^2(\omega)\}$
Truncated periodogram	∞	See remarks in text.	2
Bartlett window	1	$-\dfrac{1}{M} h^{(1)}(\omega)$	2/3
Daniell window	2	$\dfrac{\pi^2}{6M^2} h''(\omega)$	1
Parzen window	2	$\dfrac{6}{M^2} h''(\omega)$	0·539285
General Tukey window	2	$\dfrac{\pi^2 a}{M^2} h''(\omega)$	$2(1 - 4a + 6a^2)$
Tukey–Hanning window	2	$\dfrac{\pi^2}{4M^2} h''(\omega)$	3/4
Tukey–Hamming window	2	$\dfrac{0·23\pi^2}{M^2} h''(\omega)$	0·7948
Bartlett–Priestley window	2	$\dfrac{\pi^2}{10M^2} h''(\omega)$	6/5

Note: The above numerical coefficients for the variance apply when $\omega \neq 0, \pm\pi$. When $\omega = 0, \pm\pi$, all the above values should be doubled.

Modified expression for the asymptotic variance

Neave (1970c) derived a modified expression for the asymptotic variance of $\hat{h}(\omega)$ under the more general limiting conditions that as $N \to \infty$, $M \to \infty$ and $(M/N) \to \gamma$, where γ is a constant satisfying $0 \leq \gamma < 1$. For estimates based on scale parameter windows Neave showed that

$$\lim_{n \to \infty} \frac{N}{M} \, \text{var}\{\hat{h}(\omega)\} = (1 + \delta_{0,\omega,\pi}) h^2(\omega) \int_{-1/\gamma}^{1/\gamma} k^2(u)(1 - \gamma|u|)\, du.$$

$$(6.2.126)$$

(When $\gamma = 0$, (6.2.126) reduces, of course, to the standard expression (6.2.113). Also, when $k(u)$ vanishes for $|u| > 1$, as is usually the case, then the limits in the above integral become independent of γ and are simply -1 and $+1$.) Neave (1970b) evaluated the exact variance of $\hat{h}(\omega)$ when the process is a Gaussian white noise, and showed that, with γ replaced by M/N, (6.2.126) provides a slightly better approximation than (6.2.113) when applied to the Parzen and Tukey windows. In a subsequent paper, Neave (1971) evaluated the exact variance of $\hat{h}(\omega)$ for general Gaussian processes by writing (from (6.2.51)),

$$\text{var}\{\hat{h}(\omega)\}$$

$$= \frac{1}{4\pi^2} \sum_{s=-(N-1)}^{(N-1)} \sum_{r=-(N-1)}^{(N-1)} k\left(\frac{s}{M}\right) k\left(\frac{r}{M}\right) \cos s\omega \cos r\omega \, \text{cov}\{\hat{R}(s), \hat{R}(r)\},$$

and then using (5.2.31) (with $\kappa_4 = 0$) to evaluate $\text{cov}\{\hat{R}(s), \hat{R}(r)\}$. This leads to a rather cumbersome expression for $\text{var}\{\hat{h}(\omega)\}$, but it can be evaluated numerically for specified forms of $R(s)$ (or equivalently, $h(\omega)$), $k(u)$, M and N. Neave considered four different forms of spectral shapes, and concluded that, in general, (6.2.126) slightly underestimates the exact value of $\text{var}\{\hat{h}(\omega)\}$, whereas the standard expression (6.2.113) slightly overestimates it.

Neave (1971) further pointed out that his exact calculations show that, at a given frequency ω, the value of $\text{var}\{\hat{h}(\omega)\}$ may be influenced by the values of $h(\omega)$ at neighbouring frequencies. (This is contrary to the behaviour indicated by (6.2.113) or (6.2.136), according to which $\text{var}\{\hat{h}(\omega)\}$ depends purely on the value of $h(\omega)$.) Neave describes this effect as "*variance leakages*", and the underlying reason for this feature can be seen by comparing the more accurate expression (6.2.107) for the variance with the standard approximation (6.2.108). (The phenomenon of "leakage" is better known in the context of bias studies, and is discussed in more detail in Section 7.5.)

Consistency of spectral estimates

Under the conditions (i)–(v) on $W_N(\theta)$ we know that $\hat{h}(\omega)$ is an asymptotically unbiased estimate of $h(\omega)$, and hence, for each fixed ω, $\hat{h}(\omega)$ will then be a consistent estimate if its variance tends to zero as $N \to \infty$. Using (6.2.108) it is clear that $\hat{h}(\omega)$ is consistent if

$$\lim_{N \to \infty} \left\{ \frac{1}{N} \int_{-\pi}^{\pi} W_N^2(\theta) \, d\theta \right\} = 0,$$

i.e. if

$$\lim_{N \to \infty} \left\{ \frac{1}{N} \sum_s \lambda_N^2(s) \right\} = 0.$$

From (6.2.113) we see that for scale parameter windows the condition for consistency reduces simply to $M \to \infty$, $(M/N) \to 0$, as $N \to \infty$.

Moreover, under general conditions we may show that $\hat{h}(\omega)$ is a *uniformly consistent* estimate, in the sense that for any $\varepsilon > 0$,

$$p\left\{\max_{\omega} |\hat{h}(\omega) - h(\omega)| < \varepsilon\right\} \to 1, \qquad \text{as } N \to \infty.$$

Grenander and Rosenblatt (1957a), p. 262 establish this result under the assumption that X_t is a linear process, $h(\omega) > 0$, all ω, its second derivative $h''(\omega)$ is bounded, $W_N(\theta) \to \delta(\theta)$, and $\sum_s |\lambda_N(s)| = \mathrm{o}(N^{1/2})$. However, under the slightly more restrictive assumptions that $\hat{h}(\omega)$ is based on a scale parameter window with $\int_{-\infty}^{\infty} |k(u)|\, du < \infty$, $M = \mathrm{o}(N^{1/2}) \to \infty$, $\sum_{s=-\infty}^{\infty} s^2 |R(s)| < \infty$, and X_t is a Gaussian process, a proof is easily constructed as follows. Write

$$\{\hat{h}(\omega) - h(\omega)\} = [\hat{h}(\omega) - E\{\hat{h}(\omega)\}] + [E\{\hat{h}(\omega)\} - h(\omega)].$$

We know that the second term (i.e. the bias) is either $\mathrm{O}(M^{-r}h^{(r)}(\omega))$ or $\mathrm{o}(M^{-2})$, depending on whether r (the characteristic exponent) satisfies $r \leqslant 2$ or $r > 2$. If $r \leqslant 2$ then $h^{(r)}(\omega)$ is certainly bounded, and hence in all cases the second term converges uniformly to zero. As far as the first term is concerned we have,

$$2\pi \max_{\omega} |\hat{h}(\omega) - E\{\hat{h}(\omega)\}| \leqslant \sum_{s=-(N-1)}^{(N-1)} |k(s/M)| \cdot |\hat{R}(s) - E\{\hat{R}(s)\}|.$$

The above condition on $R(s)$ implies that $R(s) = \mathrm{o}(|s|^{-3})$, and hence is square summable, in which case (5.3.23) gives $\mathrm{var}\{\hat{R}(s)\} \leqslant K/N$, where K is a constant which is independent of s. We thus have

$$2\pi E\left[\max_{\omega} |\hat{h}(\omega) - E\{\hat{h}(\omega)\}| \right] \leqslant \sqrt{\frac{K}{N}} \left\{ \sum_{s=-(N-1)}^{(N-1)} |k(s/M)| \right\}.$$

But as $M \to \infty$,

$$\left\{ \frac{1}{M} \sum_s |k(s/M)| \right\} \to \int_{-\infty}^{\infty} |k(u)|\, du = \mathrm{O}(1),$$

and hence, provided $(M/\sqrt{N}) \to 0$, $E[\max_{\omega}|\hat{h}(\omega) - E\{\hat{h}(\omega)\}|] \to 0$, and the required result follows.

Hannan (1961) derives the same result (for the case of scale parameter windows) under slightly weaker conditions on $k(u)$, but, like Grenander and Rosenblatt, takes X_t to be a linear process of the form (6.2.2), with the residuals ε_t independent and normal, and with the coefficients $\{g_u\}$ decaying at a suitable rate. (Grenander and Rosenblatt do not state any specific assumptions on the distribution of the ε_t, but since they use the result, $\mathrm{var}(\hat{R}(s)) = \mathrm{O}(1/N)$, they are presumably assuming implicitly that the ε_t have at least finite fourth order moments.) Brillinger (1975), p. 265 gives a somewhat stronger form of the above result under correspondingly stronger conditions, and closely related results are given also by Lomnicki and Zaremba (1957a). Lomnicki and Zaremba again assume that X_t is a linear process with the ε_t (effectively) independent and having finite fourth order moments, and show that $\hat{h}(\omega)$ is what they term a "uniformly strongly consistent" estimate, i.e. $\max_\omega [E\{\hat{h}(\omega) - h(\omega)\}^2] \to 0$, as $N \to \infty$, if and only if $\lim_{N\to\infty} \lambda_N^*(s) = 1$, all s, and $\lim_{N\to\infty} \sum_{s=-(N-1)}^{(N-1)} \{\lambda_N^*(s)\}^2/(N - |s|) = 0$, where $\lambda_N^*(s) = \lambda_N(s)(1 - |s|/N)$ (cf. (6.2.87)).

Approximate distributions of spectral estimates

Consider the general estimate $\hat{h}(\omega)$ given by (6.2.98). If we approximate the integral in (6.2.98) by a discrete sum over the ordinates $\omega_p = 2\pi p/N$, then we may write

$$\hat{h}(\omega) \sim \left(\frac{2\pi}{N}\right) \sum_p W_N(\omega - \omega_p) I_N^*(\omega_p). \tag{6.2.127}$$

Suppose now that the observations $\{X_t\}$ come from a linear process of the form (6.2.2) in which the residuals $\{\varepsilon_t\}$ are Gaussian. Then we know from (6.2.35) that, asymptotically, the set of random variables, $\{I_N^*(\omega_p)/2\pi h(\omega_p)\}$ are independently distributed, and $I_N^*(\omega_p) \sim \frac{1}{2}h(\omega_p)\chi_2^2$, $p \neq 0$, $N/2$, and $I_N^*(\omega_p) \sim h(\omega_p)\chi_1^2$, $p = 0$, $N/2$ (N even). In fact, these asymptotic distributions still hold for any *fixed* number of frequencies if the residuals $\{\varepsilon_t\}$ are independent but not necessarily Gaussian, since any finite set of the random variables $A_\varepsilon(\omega_p)$, $B_\varepsilon(\omega_p)$ (cf. (6.1.22)) being linear combinations of the $\{\varepsilon_t\}$, will still be asymptotically normal by the central limit theorem. Hence for large N the estimate $\hat{h}(\omega)$, approximated by (6.2.127), may be regarded as a weighted linear combination of independent χ^2 variables. If the weights were all equal over a band of frequencies and zero elsewhere it would then follow that asymptotically the sum also had a χ^2 distribution, but when the weights are unequal (as is generally the case, except for the Daniell window) the distribution of the sum is no longer χ^2. However, it is reasonable to adopt the well known device of *approximating the distribution of* $\{\hat{h}(\omega)/h(\omega)\}$ *by a distribution of the form* $\{a\chi_\nu^2\}$, where the constants a and ν are chosen so that the mean and variance of $\{a\chi_\nu^2\}$ are the same as the asymptotic mean and variance of $\{\hat{h}(\omega)/h(\omega)\}$. (This approximation for the distribution of $\hat{h}(\omega)/h(\omega)$ was suggested by Tukey (1949).)

Using (6.2.104) and (6.2.110), and recalling that the mean and variance of χ_ν^2 are ν and 2ν, respectively (cf. Section 2.11), we now set

$$1 = a\nu,$$

$$\frac{1}{N}\left\{\sum_s \lambda_N^2(s)\right\} = \begin{cases} 2a^2\nu, & \omega \neq 0, \pm\pi, \\ a^2\nu, & \omega = 0, \pm\pi. \end{cases}$$

Hence we obtain,

$$a = 1/\nu$$

and

$$\nu = 2N \Big/ \Big\{ \sum_s \lambda_N^2(s) \Big\}, \qquad (\omega \neq 0, \pm\pi). \tag{6.2.128}$$

Equivalently, for estimates which use scale parameter windows we have,

$$\nu = 2N \Big/ \Big\{ M \int_{-\infty}^{\infty} k^2(u) \, du \Big\}. \tag{6.2.129}$$

(Note that these approximations do not hold for the periodogram ordinates themselves, i.e. if we set $\lambda_N(s) = 1$, $s = 0, \pm 1, \ldots \pm (N-1)$ in (6.2.128) we apparently find that $I_N^*(\omega)/h(\omega) \sim \chi_1^2$, whereas the correct result is, $I_N^*(\omega)/h(\omega) \sim \frac{1}{2}\chi_2^2$. The reason for this is that the expression (6.2.110) for the asymptotic variance does not hold for the periodogram; if we consider, e.g., the truncated periodogram estimate we know that (6.2.110) is valid only when $M/N \to 0$, so that we cannot substitute $M = N$. If we do substitute $M = N$ in the top line of Table 6.1 the expression we obtain for the variance of the periodogram is twice as large as the correct value.)

The parameter ν is called that "*equivalent degrees of freedom*" of the spectral estimate, and from (6.2.129) we find the results given in Table 6.2.

Table 6.2

Estimate	Equivalent Degrees of freedom (ν)
Truncated periodogram	N/M
Bartlett window	$3N/M$
Daniell window	$2N/M$
Parzen window	$3 \cdot 708614 \, (N/M)$
Tukey–Hanning window	$8N/3M$
Tukey–Hamming window	$2 \cdot 5164 \, (N/M)$
Bartlett–Priestley window	$1 \cdot 4 \, (N/M)$

The equivalent number of degrees of freedom for the Daniell window can be seen intuitively by recalling that this spectral window has a rectangular form over the interval $(-\pi/M, \pi/M)$, and thus includes N/M periodogram ordinates at the standard spacing of $2\pi/N$. The spectral estimate is, in this case, an unweighted average of N/M ordinates, each distributed proportional to χ_2^2, and hence its distribution is proportional to a χ^2 variable with $(2N/M)$ degrees of freedom. (We assume, as usual, that the spectral density function $h(\omega)$ may be regarded as approximately constant over the width of the window.)

It will be seen that in each case in Table 6.2, ν, the equivalent number of degrees of freedom, is proportional to N/M whereas the variance is proportional to M/N. Hence, as M *decreases* (relative to N), i.e. as the degree of smoothing *increases*, the equivalent number of degrees of freedom *increases* and the variance *decreases*. In other words, the higher the degree of freedom the larger the degree of smoothing and the smaller the variance. However, if we increase the degrees of freedom too much we may introduce a substantial bias in the estimate (in general the bias increases as M decreases) and thus invalidate the χ^2 approximation which was based on the assumption that the estimate was approximately unbiased.

Approximate confidence intervals for spectral ordinates

The χ^2 approximation for the distribution of $\{\hat{h}(\omega)/ah(\omega)\}$ may be used to construct an approximate confidence interval for the value of the spectral density function $h(\omega)$ at a particular frequency ω. Let $a_\nu(\alpha)$, $b_\nu(\alpha)$ denote respectively the lower and upper $\{100(\alpha/2)\}\%$ points of the χ^2 distribution with ν degrees of freedom, i.e. $a_\nu(\alpha)$, $b_\nu(\alpha)$ are defined by

$$p[\chi^2_\nu \leqslant a_\nu(\alpha)] = p[\chi^2_\nu \geqslant b_\nu(\alpha)] = \frac{\alpha}{2}.$$

Then we have approximately (recalling that $a = 1/\nu$),

$$p\left[a_\nu(\alpha) < \frac{\nu\hat{h}(\omega)}{h(\omega)} < b_\nu(\alpha)\right] = 1 - \alpha, \qquad (6.2.130)$$

and hence an approximate $\{100(1-\alpha)\}\%$ confidence interval for $h(\omega)$ is given by (cf. Section 5.2.3),

$$\left[\frac{\nu\hat{h}(\omega)}{b_\nu(\alpha)}, \frac{\nu\hat{h}(\omega)}{a_\nu(\alpha)}\right]. \qquad (6.2.131)$$

It must be stressed that the above interval provides an approximate $\{100(1-\alpha)\}\%$ confidence interval for $h(\omega)$ *only at a particular frequency* ω. If (6.2.131) is computed for several values of ω the set of intervals obtained will *not* represent a simultaneous $\{100(1-\alpha)\}\%$ confidence interval for all the values of $h(\omega)$ involved. An asymptotic confidence band for $h(\omega)$ over the entire frequency range can, however, be constructed by considering the distribution of the *maximum* deviation of $\hat{h}(\omega)$ from $h(\omega)$, and we will describe this approach later when we consider Grenander and Rosenblatt's goodness-of-fit tests (Section 6.2.6).

Asymptotic normality of spectral estimates

In Section 5.3.5 we saw that if the observations came from a linear process whose coefficients satisfy suitable conditions then, as the sample size $N \to \infty$, the sample autocovariances are asymptotically normal. Since $\hat{h}(\omega)$ is a linear combination of the sample autocovariances it is reasonable to suppose that $\hat{h}(\omega)$ will also have an asymptotic normal distribution, although it must be remembered that in the expression for $\hat{h}(\omega)$ the number of sample auto-covariances involved also changes with N, and hence a strict derivation of the asymptotic normality of $\hat{h}(\omega)$ based on the asymptotic distribution of the autocovariances is by no means straightforward. However, we know that for the standard spectral estimates considered in Table 6.2, $N/M \to \infty$ as $M \to \infty$, and hence the degrees of freedom, $\nu \to \infty$ as $N \to \infty$. Since the χ_ν^2 distribution (suitably scaled) tends to a normal distribution as $\nu \to \infty$, it follows also that on the basis of the above χ^2 approximation one would expect $\hat{h}(\omega)$ to be asymptotically normal. It is, in fact, possible to prove quite rigorously that, under fairly general conditions, $\hat{h}(\omega)$ does have an asymp-totic normal distribution and proofs of these results have been given by Lomnicki and Zaremba (1957b, 1959a,b), Rosenblatt (1959), Brillinger (1965, 1969a), Brillinger and Rosenblatt (1967), Hannan (1970), and Anderson (1971). The following theorem (Anderson (1971), pp. 539, 545) illustrates the general nature of these results.

Suppose that the observations $\{X_t\}$ come from a general linear process of the form (6.2.2), viz., $X_t = \sum_{u=-\infty}^{\infty} g_u \varepsilon_{t-u}$, with $\sum_{u=-\infty}^{\infty} |g_u| < \infty$, the $\{\varepsilon_t\}$ being independent random variables with $E[\varepsilon_t] = 0$, $E[\varepsilon_t^2] < \infty$, and $E[\varepsilon_t^4] < \infty$. Let $\hat{h}(\omega)$ be an estimate of the general form (6.2.97) with $\lambda_N(s)$ of scale parameter form derived from a lag window generator, $k(u)$, (cf. (6.2.92)), $k(u)$ being an even continuous function with $k(0) = 1$, $k(u) = 0$, $|u| > 1$. Let M be chosen so that, as $N \to \infty$, $M \to \infty$ and $(M/N) \to 0$. Then $\sqrt{N/M}[\hat{h}(\omega) - E\{\hat{h}(\omega)\}]$ has a limiting normal distribution with zero mean and variance given by (6.2.113). Moreover, the estimates, $\hat{h}(\omega_1), \ldots, \hat{h}(\omega_k)$, at any fixed number of frequencies have a limiting joint normal distribution with variances and covariances given by (6.2.113), (6.2.114).

Using the asymptotic normality of $\hat{h}(\omega)$ we may now construct an alternative form of confidence interval for $h(\omega)$. Thus, approximate $\{100(1 - \alpha)\}\%$ probability limits for $\hat{h}(\omega)$ are,

$$h(\omega) \pm c(\alpha)\sqrt{\text{var } \hat{h}(\omega)},$$

where $c(\alpha)$ is the two-sided $(100\alpha)\%$ point of the standardized normal distribution $(c(\alpha) = 1 \cdot 96$ for a 95% interval, and $c(\alpha) = 2 \cdot 58$ for a 99% interval). Using (6.2.110) for the asymptotic variance of $\hat{h}(\omega)$ an approximate confidence interval may be written (for $\omega \neq 0, \pm \pi$) as

$$\hat{h}(\omega) \left\{ 1 \pm c(\alpha) \left(\frac{\sum_s \lambda_N^2(s)}{N} \right)^{1/2} \right\}^{-1}, \qquad (6.2.132)$$

or alternatively (by (6.2.128)) as,

$$\hat{h}(\omega)\{1 \pm c(\alpha)\sqrt{2/\nu}\}^{-1}, \qquad (6.2.133)$$

where ν is the equivalent degrees of freedom of $\hat{h}(\omega)$. (The form of (6.2.133) follows directly from (6.2.131) on replacing $a_\nu(\alpha)$, $b_\nu(\alpha)$ by the lower and upper $(100\alpha/2)\%$ points of the $N(\nu, 2\nu)$ distribution.)

Logarithmic transformation of $\hat{h}(\omega)$

The confidence intervals (6.2.131), (6.2.132), are both "multiplicative" in form, i.e. in both cases the width of the interval is proportional to $\hat{h}(\omega)$ and thus varies with frequency. However, if we make a logarithmic trans- formation then the corresponding confidence intervals for $\log h(\omega)$ have *uniform width* (i.e. are independent of ω). From (6.2.131) we obtain a $\{100(1-\alpha)\%\}$ confidence interval for $\log h(\omega)$ as

$$\left[\log \hat{h}(\omega) + \log\left(\frac{\nu}{b_\nu(\alpha)}\right), \ \log \hat{h}(\omega) + \log\left(\frac{\nu}{a_\nu(\alpha)}\right) \right], \qquad (6.2.134)$$

while (6.2.133) gives,

$$[\log \hat{h}(\omega) - \log\{1 + c(\alpha)\sqrt{2/\nu}\}, \ \log \hat{h}(\omega) - \log\{1 - c(\alpha)\sqrt{2/\nu}\}]. \qquad (6.2.135)$$

Once the form of the interval has been computed it can then be applied at any frequency point.

The logarithmic transformation is indeed a very natural one from the point of view of both statistical and engineering considerations. Since the asymptotic mean and variance of $\hat{h}(\omega)$ are proportional to $h(\omega)$, $h^2(\omega)$, respectively, a logarithmic transformation is clearly suggested as a *variance stabilizing technique* (cf. Section 2.9.1). Using (2.9.17), (2.9.18) we find,

$$E[\log_e \hat{h}(\omega)] \sim \log_e h(\omega), \qquad (6.2.136)$$

and

$$\mathrm{var}[\log_e \hat{h}(\omega)] \sim \frac{1}{N}\left\{ \sum_s \lambda_N^2(s) \right\}, \qquad (6.2.137)$$

so that (asymptotically) the *variance of $\log_e \hat{h}(\omega)$ is independent of ω*. Also, since variance stabilizing transformations generally have the property of bringing the distribution of the statistic closer to normality (cf. Kendall and Stuart (1966), Vol. 3, p. 93), it may be expected that the distribution of $\log_e \hat{h}(\omega)$ will be closer to normality than that of $\hat{h}(\omega)$; see Jenkins and Priestley (1957), Jenkins (1961). (It may be proved that $\log_e \hat{h}(\omega)$ has an asymptotic normal distribution under the same conditions as those which apply to $\hat{h}(\omega)$; see, e.g., Anderson (1971), p. 541.)

Plotting $\hat{h}(\omega)$ on a logarithmic scale is also in accord with the engineering use of the dB (decibel) scale for measuring power levels. In engineering usage a power level P_1 is compared with a reference power level $P_0(<P_1)$ by saying that "P_1 is n dB above P_0", where

$$n = 10 \log_{10}(P_1/P_0).$$

(For example, doubling a power level corresponds to an increase of approximately 3 dB.)

6.2.5 Estimation of the Integrated Spectrum

If we wish to estimate the (non-normalized) integrated spectrum, $H(\omega) = \int_{-\pi}^{\omega} h(\phi) \, d\phi$, it would seem natural to take as an estimate,

$$H^*(\omega) = \int_{-\pi}^{\omega} \hat{h}(\phi) \, d\phi, \qquad (6.2.138)$$

where $\hat{h}(\omega)$ is of the general form (6.2.98), i.e. $\hat{h}(\omega)$ is a smoothed version of the periodogram based on a smoothing window $W_N(\theta)$. However, as it turns out we can obtain a perfectly satisfactory estimate of $H(\omega)$ by integrating the raw periodogram $I_N^*(\omega)$ directly, i.e. there is no need here to smooth $I_N^*(\omega)$ first before integrating. This may seem a curious result since we already know that $I_N^*(\omega)$ is not a consistent estimate of $h(\omega)$ and it is therefore surprising that the integral of $I_N^*(\omega)$ turns out to be a consistent estimate of the integral of $h(\omega)$. It may be helpful at this point to consider the closely analogous problem of estimating the probability distribution of a continuous random variable X from a random sample of observations on X. If we denote the distribution function of X by $F(x)$ then it is well known that the empirical distribution function $\hat{F}(x)$ (a step function defined by $\hat{F}(x) =$ proportion of sample observations $\leq x$) is, for each x, a consistent estimate of $F(x)$, but the derivative of $\hat{F}(x)$ (which is zero everywhere except at the jump points of $\hat{F}(x)$, and non-existent at the jump points) is clearly not a consistent estimate of $f(x) = F'(x)$, the probability density function of X. To obtain a consistent estimate of $f(x)$ we have to smooth the function "$d\hat{F}(x)/dx$" by, e.g., constructing a histogram. (As pointed out previously, the histogram estimate of $f(x)$ is equivalent to smoothing $\{d\hat{F}(x)/dx\}$ by a rectangular "window", and is thus the analogue of the Daniell estimate of the spectral density functions.) To pursue this point in more detail let us now consider the slightly more general problem of estimating a weighted integral of $h(\omega)$, say,

$$H_A = \int_{-\pi}^{\pi} A(\phi)h(\phi)\,d\phi, \tag{6.2.139}$$

where $A(\phi)$ is some given fixed bounded function, independent of the sample size, N. The quantity, H_A, reduces to the integrated spectrum, $H(\omega)$, if, in particular, we choose $A(\phi)$ as

$$A_\omega(\phi) = \begin{cases} 1, & \phi \leq \omega, \\ 0, & \phi > \omega. \end{cases} \tag{6.2.140}$$

On the other hand, if we take $A(\phi)$ as,

$$A_{\omega_1,\omega_2}(\phi) = \begin{cases} 1, & \omega_1 \leq \phi \leq \omega_2, \\ 0, & \text{otherwise} \end{cases} \tag{6.2.141}$$

then H_A becomes the *band spectrum* $\{H(\omega_2) - H(\omega_1)\}$. Suppose now that we consider as an estimate of H_A,

$$H_A^* = \int_{-\pi}^{\pi} A(\phi)\hat{h}(\phi)\,d\phi, \tag{6.2.142}$$

where $\hat{h}(\phi)$ is given by (6.2.98). Substituting (6.2.98) in (6.2.142) and changing the order of integration we may write,

$$H_A^* = \int_{-\pi}^{\pi} I_N^*(\theta)A_N(\theta)\,d\theta, \tag{6.2.143}$$

say, where,

$$A_N(\theta) = \int_{-\pi}^{\pi} W_N(\phi - \theta)A(\phi)\,d\phi.$$

But $A(\phi)$ is a fixed bounded function (independent of N) whereas $W_N(\theta)$ tends to a δ-function as $N \to \infty$. Hence, for sufficiently large N, we have,

$$A_N(\theta) \sim A(\theta),$$

(at all continuity points of $A(\theta)$), and the estimate (6.2.143) is essentially the same as

$$\hat{H}_A = \int_{-\pi}^{\pi} I_N^*(\theta)A(\theta)\,d\theta. \tag{6.2.144}$$

In other words, once we introduce a fixed function $A(\phi)$ in the integral (6.2.142), the effect of the window $W_N(\theta)$ becomes "submerged" inside the function $A(\phi)$. It is easy to see that H_A and H_A^* are equivalent as far as their asymptotic sampling properties are concerned (cf. Parzen (1957a), and the only advantage of H_A^* would be its use in conjunction with an estimate $\hat{h}(\phi)$ of the "truncated" type. In this case H_A^* would involve only $M(\ll N)$ sample autocovariances whereas $\hat{H}_A$ involves the complete sample autocovariance function. (However, this point is significant only when $\hat{h}(\phi)$ and $I_N^*(\phi)$ are computed by first computing the sample autocovariances and then taking Fourier transforms; if, as is now common, $I_N^*(\theta)$ is computed directly from the data via the "fast Fourier transform" algorithm and $\hat{h}(\phi)$ is computed directly from $I_N^*(\theta)$, then the above advantage disappears.) The asymptotic sampling properties of $\hat{H}_A$ follow immediately from the results of theorem 6.2.4.

Returning now to the problem of estimating the integrated spectrum, $H(\omega)$, we choose the function A as in (6.2.140), and the estimate corresponding to (6.2.144) becomes,

$$\hat{H}(\omega) = \int_{-\pi}^{\omega} I_N^*(\theta)\, d\theta. \tag{6.2.145}$$

Assuming that the observations come from a general linear process satisfying the conditions of theorem (6.2.4) we now obtain from (6.2.41), (6.2.42),

$$\lim_{N \to \infty} E[\hat{H}(\omega)] = H(\omega), \tag{6.2.146}$$

and,

$$\lim_{N \to \infty} N \operatorname{cov}[\hat{H}(\omega_1), \hat{H}(\omega_2)] = eH(\omega_1)H(\omega_2)$$
$$+ 4\pi \int_{-\pi}^{\pi} A_{\omega_1}(\phi)\bar{A}_{\omega_2}(\phi)h^2(\phi)\, d\phi, \tag{6.2.147}$$

where

$$\bar{A}_{\omega_2}(\phi) = \tfrac{1}{2}\{A_{\omega_2}(\phi) + A_{\omega_2}(-\phi)\},$$

and $e = \{E(\varepsilon_t^4)/\sigma_\varepsilon^4 - 3\}$, ε_t being the residual process.

Since $A_\omega(\phi)$ is a fixed bounded function it is clear that $\operatorname{var}[\hat{H}(\omega)] = O(1/N) \to 0$ as $N \to \infty$. Since $\hat{H}(\omega)$ is also an asymptotically unbiased estimate of $H(\omega)$, it follows that $\hat{H}(\omega)$ *is a consistent estimate of* $H(\omega)$. For Gaussian processes the first term in (6.2.147) vanishes $(e = 0)$, and the expression for the covariance reduces simply to the second term. Note that, for computational purposes, $\hat{H}(\omega)$ can be written (for real valued processes) as,

$$\hat{H}(\omega) = \frac{(\pi + \omega)}{2\pi} s_X^2 + \frac{1}{\pi} \left[\sum_{s=1}^{(N-1)} \hat{R}(s) \frac{\sin s\omega}{s} \right], \qquad (6.2.148)$$

where s_X^2 is the sample variance of the observations (cf. (4.8.34)).

If we wish to estimate $H_+(\omega)(= 2 \int_0^\omega h(\phi) \, d\phi = 2H(\omega) - \sigma_X^2)$, the integrated spectrum defined for positive frequencies only (cf. Section 4.8), then we choose $A(\phi)$ as

$$A_\omega(\phi) = \begin{cases} 2, & 0 \le \phi \le \omega, \\ 0, & \text{otherwise,} \end{cases}$$

and the corresponding estimate becomes,

$$\hat{H}_+(\omega) = 2 \int_0^\omega I_N^*(\theta) \, d\theta \quad (= 2\hat{H}(\omega) - s_X^2). \qquad (6.2.149)$$

We now have,

$$\lim_{N \to \infty} E[\hat{H}_+(\omega)] = H_+(\omega), \qquad (6.2.150)$$

and,

$$\lim_{N \to \infty} N \operatorname{cov}[\hat{H}_+(\omega_1), \hat{H}_+(\omega_2)]$$

$$= eH_+(\omega_1)H_+(\omega_2) + 8\pi \int_0^{\min(\omega_1, \omega_2)} h^2(\phi) \, d\phi. \qquad (6.2.151)$$

For Gaussian processes the first term in (6.2.151) vanishes and we have,

$$\lim_{N \to \infty} N \operatorname{cov}[\hat{H}_+(\omega_1), \hat{H}_+(\omega_2)] = 8\pi \int_0^{\min(\omega_1, \omega_2)} h^2(\phi) \, d\phi, \qquad (6.2.152)$$

and in particular, setting $\omega_1 = \omega_2 = \omega$,

$$\lim_{N \to \infty} N[\operatorname{var} \hat{H}_+(\omega)] = 8\pi \int_0^\omega h^2(\phi) \, d\phi. \qquad (6.2.153)$$

If we compare (6.2.152), (6.2.153), with (3.7.19), (3.7.16), we observe an important feature of the process, $\sqrt{N}\{\hat{H}_+(\omega) - H_+(\omega)\}$, namely that *its asymptotic second order properties are closely related to those of a Wiener process* $W(t)$. However, whereas $W(t)$ has constant variance per unit time, $\sqrt{N}\hat{H}_+(\omega)$ does not, in general, have constant variance per unit frequency, unless the observations come from a purely random process, in which case $h(\phi)$ is a constant and $\sqrt{N}\hat{H}_+(\omega)$ does then have constant variance per unit frequency. (In the general case $\sqrt{N}\hat{H}_+(\omega)$ can be reduced to a variable having a constant increase of variance by a suitable transformation of the frequency scale.) This property of $\hat{H}_+(\omega)$ can be seen intuitively by approximating the integral in (6.2.149) by a discrete sum and writing,

$$\hat{H}_+(\omega_p) \sim \frac{4\pi}{N} \sum_{q=0}^{p} I_N^*\left(\frac{2\pi q}{N}\right). \qquad (6.2.154)$$

For a normal process the $\{I_N^*(2\pi q/N)\}$ $(q = 0, 1, 2, \ldots)$ are asymptotically independent random variables (each proportional to χ_2^2), and hence $\hat{H}_+(\omega_p)$, being the limiting form of a sum of independent random variables, *may be regarded as the limiting form of a random walk.* Also, since the $\{I_N^*(2\pi q/N)\}$ are asymptotically independent, $\hat{H}_+(\omega_p)$ will (when suitably normalized) have an asymptotic normal distribution. These properties of $\hat{H}_+(\omega)$ underlie the "goodness-of-fit" test proposed by Grenander and Rosenblatt (1953), which we discuss below.

6.2.6 Goodness-of-Fit Tests

Suppose we wish to test the hypothesis that the observed process conforms to a particular type of model, such as an autoregressive or moving average model. If the model is fully specified (i.e. if the order and parameters are given) then we can test the hypothesis by comparing the estimated spectral density function or estimated integrated spectrum with their theoretical forms, and if there is good agreement we would accept the hypothesis; otherwise we would reject the hypothesis. This type of analysis is called a *"goodness-of-fit test"*, and various tests may be constructed depending on (a) the type of function selected (i.e. the spectral density function or integrated spectrums) and (b) the criterion used to measure the overall agreement between the estimated and theoretical forms of the function. In principle, these tests can be used also to *construct confidence intervals for the complete function* under test. We discuss below some specific tests which have been proposed.

I. *Grenander and Rosenblatt's test for the non-normalized integrated spectrum*

Grenander and Rosenblatt (1953, 1957a) proposed a goodness of fit test for the (positive) non-normalized integrated spectrum $H_+(\omega)$ based effectively on the maximum absolute deviation of $\hat{H}_+(\omega)$ from $H_+(\omega)$ over the range $0 \leqslant \omega \leqslant \pi$, i.e. they considered the statistic,

$$\max_{0 \leqslant \omega \leqslant \pi} \left[\sqrt{N}|\hat{H}_+(\omega) - H_+(\omega)|\right], \qquad (6.2.155)$$

N being the number of observations in the sample from which $\hat{H}_+(\omega)$ is computed. (In their original study Grenander and Rosenblatt actually considered the statistic, $\max_{0 \le \omega \le \pi} [\sqrt{N}|\frac{1}{2}\hat{H}_+(\omega) - \frac{1}{2}H_+(\omega)|]$, but it is clear that this statistic leads to exactly the same test as (6.2.155).) On the hypothesis that the specified $H_+(\omega)$ is the true integrated spectrum of the process, we know from the above discussion that for normal processes $\sqrt{N}\{\hat{H}_+(\omega) - H_+(\omega)\}$ behaves asymptotically like a continuous random walk (over the range $0 \le \omega \le \pi$) with covariance structure (6.2.152), and hence an event of the form,

$$\max_{0 \le \omega \le \pi} [\sqrt{N}|\hat{H}_+(\omega) - H_+(\omega)|] \le a,$$

say, may be identified with the event that the random walk starting at $\omega = 0$ does not cross boundaries placed at $(-a)$ and a before it reaches $\omega = \pi$. The probability of this event can now be derived by considering a random walk with "absorbing barries" at $(-a)$ and a, and using well known methods to find the probability that the walk is not absorbed in the barries over the interval $(0, \pi)$. These are the basic ideas underlying the development of Grenander and Rosenblatt's test. Formally, they show that for a *normal* general linear process of the form (6.2.2), with $g_u = o(1/u^{3/2})$,

$$\lim_{N \to \infty} p\left[\max_{0 \le \omega \le \pi} \sqrt{N}|\hat{H}_+(\omega) - H_+(\omega)| \le a \right] = p\left[\max_{0 \le \omega \le \pi} |\eta(\omega)| \le a \right], \quad (6.2.156)$$

where $\eta(\omega)$ is a normal process with,

$$E[\eta(\omega)] = 0, \quad \text{all } \omega,$$

$$\text{cov}[\eta(\omega_1), \eta(\omega_2)] = 8\pi G\{\min(\omega_1, \omega_2)\},$$

where

$$G(\omega) = \int_0^\omega h^2(\phi)\, d\phi,$$

(cf. (6.2.152)). The RHS of (6.2.156) can be evaluated by first transforming the frequency scale so as to obtain a constant rate of increase of variance, i.e. if we set $t = 8\pi G(\omega)$, then,

$$p\left[\max_{0 \le \omega \le \pi} |\eta(\omega)| \le a \right] = p\left[\max_{0 \le t \le 8\pi G(\pi)} |W(t)| \le a \right],$$

where $W(t)(= \eta\{G^{-1}(t/8\pi)\})$ is now a standard Wiener process with $\sigma_W^2 = 1$. The probability that a Wiener process starting at $t = 0$ is not absorbed by barriers at $-a, a$, can be derived by using the method of "images" (see, e.g., Sommerfeld (1949)), and we obtain (cf. Grenander and

Rosenblatt (1957a), p. 195),

$$p\left[\max_{0\leqslant t\leqslant 8\pi G(\pi)} |W(t)|\leqslant a\right]=\Delta^{[1]}\left(\frac{a}{\sqrt{8\pi G(\pi)}}\right),$$

where,

$$\Delta^{[1]}(x)=\sum_{k=-\infty}^{\infty}(-k)^k[\Phi\{(2k+1)x\}-\Phi\{(2k-1)x\}], \quad (6.2.157)$$

and $\Phi(x)$ denotes the distribution function of the standardized normal distribution. (The deviation of (6.2.157) is considered in more detail in Chapter 8 when we discuss a similar test procedure for analysing processes with "mixed spectra". Meanwhile we attach the superfix [1] to $\Delta(x)$ to distinguish it from a different quantity, $\Delta^{[2]}(x)$, which is introduced later.) Putting the above results together we have,

$$\lim_{N\to\infty} p\left[\frac{\max_{0\leqslant\omega\leqslant\pi} \sqrt{N}|\hat{H}_+(\omega)-H_+(\omega)|}{\sqrt{8\pi G(\pi)}}\leqslant a\right]=\Delta^{[1]}(a). \quad (6.2.158)$$

In principle, this result can be used directly to test a specified form of $H_+(\omega)$; the chosen probability level $\Delta^{[1]}(a)$ determines the value of a, and the specified form of $H_+(\omega)$ determines $G(\pi)$. (Grenander and Rosenblatt (1957a), p. 196, have given a table of values of a for various values of $\Delta^{[1]}(a)$, e.g., for $\Delta^{[1]}(a)=0\cdot99$, $a=2\cdot8070$, while for $\Delta^{[1]}(a)=0\cdot95$, $a=2\cdot2414$.) However, in practice it is more convenient to replace $G(\pi)$ in (6.2.158) by a consistent estimate computed from the sample observations. Noting that $G(\pi)=(1/4\pi)\sum_{s=-\infty}^{\infty} R^2(s)$, Grenander and Rosenblatt (1957a) suggest the estimate,

$$\hat{G}(\pi)=\frac{1}{4\pi}\sum_{s=-M}^{M}\hat{R}^2(s), \quad (6.2.159)$$

where $\hat{R}(s)$ is the usual sample autocovariance function and $M(<N-1)$ is chosen to be of the form $M=kN^\theta$, k a constant and $0<\theta<1$. (If we wish to avoid the somewhat arbitrary choice of the "truncation point", M, we may, as suggested by Lomnicki and Zaremba (1959a) and Hannan (1960), consider instead,

$$G^*(\pi)=\tfrac{1}{2}\int_0^\pi \{I_N^*(\theta)\}^2\, d\theta=\frac{1}{8\pi}\left\{\sum_{s=-(N-1)}^{(N-1)}\hat{R}^2(s)\right\}, \quad (6.2.160)$$

which is based on the full sample autocovariance function and is also a consistent estimate of $G(\pi)$. Note that $E[\{I_N^*(\theta)\}^2]=$ var$[I_N^*(\theta)]+[E\{I_N^*(\theta)\}]^2\sim 2h^2(\theta)$, $\theta\neq 0$, $\pm\pi$, (cf. (6.2.36)) so that asymptotically the expected value of $\int_0^\pi\{I_N^*(\theta)\}^2\, d\theta$ is *twice* that of $\int_0^\pi h^2(\theta)\, d\theta$.) Replacing $G(\pi)$ by $\hat{G}(\pi)$ (or $G^*(\pi)$) in (6.2.158) enables us to test the goodness of fit of a specified $H_+(\omega)$ simply by examining the maximum absolute deviation of $\hat{H}_+(\omega)$ from $H_+(\omega)$. This substitution also allows us to construct an approximate confidence band for the complete function, $H_+(\omega)$. Thus, if we set $\Delta^{[1]}(a) = \alpha$ (and determine the value of a accordingly), then an approximate $(100\alpha)\%$ confidence band for $H_+(\omega)$ is given by

$$\hat{H}_+(\omega) - a\sqrt{\frac{8\pi\hat{G}(\pi)}{N}} \leqslant H_+(\omega) \leqslant \hat{H}_+(\omega) + a\sqrt{\frac{8\pi\hat{G}(\pi)}{N}} \qquad (6.2.161)$$

Test for white noise with given power

To illustrate the above test let us consider the problem of testing the hypothesis H_0 that the observations come from a white noise (i.e. purely random) process with given variance (i.e. total power) σ_X^2. In this case the hypothesized form of $H_+(\omega)$ is

$$H_+(\omega) = \sigma_X^2 \omega / \pi, \qquad 0 \leqslant \omega \leqslant \pi,$$

and since $h(\omega) = \sigma_X^2 / 2\pi$, $G(\pi) = \sigma_X^4 / 4\pi$. Hence the test corresponding to (6.2.158) is now; reject H_0 if,

$$\max_{0 \leqslant \omega \leqslant \pi} \left| \frac{\hat{H}_+(\omega)}{\sigma_X^2} - \frac{\omega}{\pi} \right| > a\sqrt{\frac{2}{N}}, \qquad (6.2.162)$$

where a is determined by setting $\Delta^{[1]}(a) = 1 - \alpha$, α being the significance level chosen for the test. The test may thus be performed graphically by constructing a straight line through the points $(0, 0)$, $(\pi, 1)$ (representing the function $\{H_+(\omega)/\sigma_X^2\}$), superimposing the function $\hat{H}_+(\omega)/\sigma_X^2$, and checking that throughout the range $(0, \pi)$, $\hat{H}_+(\omega)/\sigma_X^2$ always lies between the two parallel lines which lie $(a\sqrt{2/N})$ above and below the line representing $H_+(\omega)/\sigma_X^2$; see Fig. 6.11. It is important to note that:

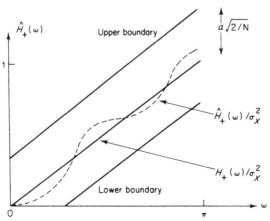

Fig. 6.11. Test for white noise with given power.

(a) Grenander and Rosenblatt's test is appropriate only when the (positive) *non-normalized* integrated spectrum $H_+(\omega)$ is *fully specified*. Thus, in the above test for white noise we require that the value of σ_X^2 be specified *a priori* as part of the hypothesis being tested. We could, of course, estimate σ_X^2 by s_X^2 (the sample variance) in computing $G(\pi)$, but even if we do this, so that the test becomes

$$\max_{0 \leqslant \omega \leqslant \pi} \left| \frac{\hat{H}_+(\omega)}{s_X^2} - \frac{\sigma_X^2 \omega}{s_X^2 \pi} \right| > a \sqrt{\frac{2}{N}}, \qquad (6.2.163)$$

the value of σ_X^2 is still required in order to evaluate the test statistic. (If we wished to test merely for white noise, without specifying σ_X^2, the appropriate procedure would be to test the form of the *normalized* integrated spectrum, as in Bartlett's test described below.)

(b) The above form of Grenander and Rosenblatt's test is valid only for *normal* processes. In the non-normal case the covariance properties of $\hat{H}_+(\omega)$ involve the quantity e (cf. (6.2.151)), and the test is therefore sensitive to departures from normality. Although e can be estimated from the observations the estimate suggested by Grenander and Rosenblatt converges rather slowly. (For an interesting discussion of the problems involved in extending the test to the non-normal case see Grenander and Rosenblatt (1957a), pp. 198–201.)

Grenander and Rosenblatt (1957a) have given a number of numerical illustrations of their test applied to various simulated series.

II. *Bartletts' tests for the normalized integrated spectrum*

Bartlett (1954, 1955) proposed two tests, both of which are similar in form to Grenander and Rosenblatt's test, but designed to test the goodness of fit of the (positive) *normalized* integrated spectrum, $F_+(\omega)(= H_+(\omega)/\sigma_X^2)$, (cf. Section 4.8.1). A natural estimate of $F_+(\omega)$ is

$$\hat{F}_+(\omega) = \hat{H}_+(\omega)/s_X^2,$$

where $\hat{H}_+(\omega)$ is given by (6.2.149), and as above, s_X^2 denotes the sample variance. Recalling that $s_X^2 \equiv \hat{R}(0) = 2 \int_0^\pi I_N^*(\theta) \, d\theta$, $\hat{F}_+(\omega)$ may be written,

$$\hat{F}_+(\omega) = \frac{\int_0^\omega I_N^*(\theta) \, d\theta}{\int_0^\pi I_N^*(\theta) \, d\theta}. \qquad (6.2.164)$$

This suggests that if we wish to test a specified form of $F_+(\omega)$ we might consider as a suitable test statistic,

$$\max_{0 \leqslant \omega \leqslant \pi} \sqrt{N} |\hat{F}_+(\omega) - F_+(\omega)|. \qquad (6.2.165)$$

However, although it might be suspected that the process $\sqrt{N}\{\hat{F}_+(\omega) - F_+(\omega)\}$, $0 \le \omega \le \pi$, behaves very much like $\sqrt{N}\{\hat{H}_+(\omega) - H_+(\omega)\}$, it differs from the latter in one very important respect, namely, that when $\omega = \pi$, $\sqrt{N}\{\hat{F}_+(\pi) - F_+(\pi)\} \equiv 0$ (since both $\hat{F}_+(\omega)$ and $F_+(\omega)$ are normalized so that $\hat{F}_+(\pi) = F_+(\pi) = 1$). Thus, if we try to interpret $\sqrt{N}\{\hat{F}_+(\omega) - F_+(\omega)\}$ as a random walk the walk is constrained not only to start at zero *but also to terminate at zero*. This means that the increments of $\sqrt{N}\{\hat{F}_+(\omega) - F_+(\omega)\}$ are no longer independent, and the asymptotic distribution theory of the statistic (6.2.165) is quite different from the Grenander and Rosenblatt statistic which corresponds to the "open ended" walk, $\sqrt{N}\{\hat{H}_+(\omega) - H_+(\omega)\}$. Bartlett, however, introduced the ingenious device of relating the asymptotic distribution of (6.2.165) to the *Kolmogorov–Smirnov statistics* (used for testing the goodness of fit of an empirical probability distribution function) by first considering the case of normal white noise and then replacing each of the integrals in (6.2.164) by discrete sums over the frequency points $\omega_p = 2\pi p/N$, $p = 1, 2, \ldots, [N/2]$. Thus, suppose that the observations $\{X_t\}$ come from a general linear process of the form (6.2.2), i.e.,

$$X_t = \sum_{u=-\infty}^{\infty} g_u \varepsilon_{t-u},$$

with $\{\varepsilon_t\}$ a normal process, so that the $\{\varepsilon_t\}$ are then independent random variables. Let $I_{N,X}^*(\omega)$, $I_{N,\varepsilon}^*(\omega)$ denote the periodograms of X_t, ε_t respectively. We know from theorem 6.1.1 that for $p = 1, 2, \ldots, [N/2]$ (assuming, for convenience, that N is odd), $I_{N,\varepsilon}^*(\omega_p) = (\sigma_\varepsilon^2/4\pi)\chi_2^2$, and for different values of p the $I_{N,\varepsilon}^*(\omega_p)$ are independent. Hence, if we approximate the integrals in (6.2.164) by sums, and consider the modified estimate (corresponding to the $\{\varepsilon_t\}$),

$$\tilde{F}_+^{(\varepsilon)}(\omega_p) = \frac{\sum\limits_{q=1}^{p} I_{N,\varepsilon}^*(\omega_q)}{\sum\limits_{q=1}^{m} I_{N,\varepsilon}^*(\omega_q)}, \tag{6.2.166}$$

(where $m = [N/2]$), then the $\tilde{F}_+^{(\varepsilon)}(\omega_p)$ have the same distribution as the variables,

$$W_p = \frac{\sum\limits_{q=1}^{p} H_q}{\sum\limits_{q=1}^{m} H_q}$$

where $H_1, H_2, \ldots, H_m$, are independent identically distributed exponential variables. Now let $x_{(1)} \leq x_{(2)} \leq \ldots \leq x_{(m)} = 1$ denote an *ordered* random sample from a uniform distribution on the interval $(0, 1)$. It is well known that the joint distribution of the $\{x_{(i)}\}$ is the same as the joint distribution of $W_1, W_2, \ldots, W_m$. (This result may be seen intuitively by thinking of $x_{(1)}$, $x_{(2)}, \ldots, x_{(m)}$ as the times of events in a Poisson process, conditional on there being m events in the interval $(0, 1)$, and then recalling that for an unconditional Poisson process the intervals between events are exponentially distributed.) Now consider the empirical distribution function $\hat{\psi}(x)$ computed from $x_{(1)}, x_{(2)}, \ldots, x_{(m)}$. We have,

$$\hat{\psi}(x) = k/m, \qquad x_{(k)} \leq x < x_{(k+1)},$$

i.e. $\hat{\psi}(x)$ is a step function with jumps of magnitude $1/m$ at the points $x_{(k)}$. The theoretical distribution function for a uniform distribution on $(0, 1)$ is,

$$\psi(x) = x, \qquad 0 \leq x \leq 1,$$

and it follows from the theory of the well known Kolmogorov–Smirnov test (see, e.g., Feller (1948)) that

$$\lim_{m \to \infty} p[\max_x \sqrt{m} |\hat{\psi}(x) - \psi(x)| \leq a] = \Delta^{[2]}(a), \qquad (6.2.167)$$

where

$$\Delta^{[2]}(a) = \sum_{j=-\infty}^{\infty} (-1)^j \exp(-2a^2 j^2). \qquad (6.2.168)$$

Thus, with asymptotic probability $\Delta^{[2]}(a)$, the function $\hat{\psi}(x)$ will lie completely in the region R bounded by two parallel lines lying at a distance $a/\sqrt{m}$ above and below the line $\psi(x) = x$, as shown in Fig. 6.12. But, if we

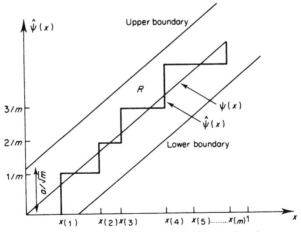

Fig. 6.12. Test for the empirical distribution function.

plot (p/m) against W_p (with p/m on the *vertical* axis and W_p on the *horizontal* axis), the configuration of points which we obtain has exactly the same probabilistic properties as the set of jump points of the function $\hat{\psi}(x)$; this follows immediately from the equivalence between the distributions of the $\{x_{(i)}\}$ and the $\{W_p\}$. If we ignore the probability that $\hat{\psi}(x)$ will cross either boundary at some point even though all the jump points remain within R (which is negligible for large m since the height of each jump is $1/m$ and the distance between the boundaries is $O(1/\sqrt{m})$), we may equate

$$\lim_{m \to \infty} p\left[\max_{1 \le p \le m} \sqrt{m} \left| W_p - \frac{p}{m} \right| \le a \right]$$

with the LHS of (6.2.167). Recalling that the $\tilde{F}_+^{(\varepsilon)}(\omega_p)$ (as given by (6.2.166)) have the same distribution as the W_p, and that $m = [N/2]$, we finally obtain the asymptotic result that for a normal purely random process,

$$\lim_{N \to \infty} p\left[\max_{p} \sqrt{\frac{N}{2}} \left| \tilde{F}_+^{(\varepsilon)}(\omega_p) - \frac{p}{m} \right| \le a \right] = \Delta^{[2]}(a), \qquad (6.2.169)$$

where $\Delta^{[2]}(a)$ is given by (6.2.168). Note that the theoretical form of $F_+(\omega)$ for the process $\{\varepsilon_t\}$ is

$$F_+^{(\varepsilon)}(\omega) = \frac{\omega}{\pi}, \quad \text{so that} \quad F_+^{(\varepsilon)}(\omega_p) = \frac{2p}{N} \sim \frac{p}{m}.$$

For $a = 1 \cdot 36$, $\Delta^{[2]}(a) = 0 \cdot 95$, while for $a = 1 \cdot 63$, $\Delta^{[2]}(a) = 0 \cdot 99$. (It may be noted that the argument which we have used here to derive (6.2.169) is, in effect, the reverse of the argument used by Bartlett in his original derivation of this result. Bartlett derived (6.2.169) directly from the equivalence between $\tilde{F}_+^{(\varepsilon)}(\omega_p)$ and W_p, treating the W_p as a random walk constrained to terminate at the value 1 when $p = m$, and obtained the limiting probability that this conditional walk is not absorbed in the boundaries at $\pm a\sqrt{2/N}$. Bartlett then used this result, together with the equivalence between the W_p and the $\{x_{(i)}\}$, to provide an alternative derivation of the Kolmogorov–Smirnov result (6.2.167). (See Bartlett (1955), pp. 92–94, 284–287.)

Test for white noise (unspecified power)

Using the result (6.2.169) we may now construct a test for the hypothesis that the observations, $\{X_t\}$, come from a white noise process (with unspecified variance) by plotting $\tilde{F}_+^{(X)}(\omega_p)$ against p, and checking that the absolute deviations of $\tilde{F}_+^{(X)}(\omega_p)$ from the straight line, p/m, (i.e. the line joining the points $(0, 0)$ and $(m, 1)$) do not exceed $\pm a\sqrt{2/N}$. Here, a is determined by setting $\Delta^{[2]}(a) = 1 - \alpha$, α being the significance level of the test. Thus, $a = 1 \cdot 36$ or $1 \cdot 63$ corresponding to 5% or 1% significance levels, respectively.

Bartlett's T_p test

The above test for white noise is readily extended to the more general case where we wish to test the goodness of fit of a specified $F_+(\omega)$ of general form. The basic idea, due to Bartlett (1954), is to scale the $I_{N,X}^*(\omega_p)$ by dividing each ordinate by $f(\omega_p)$ ($f(\omega)$ being the spectral density function corresponding to the given form of $F_+(\omega)$) so that the scaled ordinates are equivalent to the $I_{N,\varepsilon}^*(\omega)$ and may then be tested as above. Under the conditions of theorem 6.2.2 (with $\alpha > \frac{1}{2}$) we have,

$$I_{N,X}^*(\omega) \sim 2\pi \frac{\sigma_X^2}{\sigma_\varepsilon^2} f(\omega) I_{N,\varepsilon}^*(\omega) + O\left(\frac{1}{N^{1/2}}\right),$$

and hence the statistic

$$T_p = \frac{\sum\limits_{q=1}^{p} I_{N,X}^*(\omega_q)/f(\omega_q)}{\sum\limits_{q=1}^{m} I_{N,X}^*(\omega_q)/f(\omega_q)} \tag{6.2.170}$$

is asymptotically equivalent to $\tilde{F}_+^{(\varepsilon)}(\omega_p)$. Hence we have

$$\lim_{N \to \infty} p \left| \max_p \sqrt{\frac{N}{2}} \left| T_p - \frac{p}{m} \right| \le a \right| = \Delta^{[2]}(a), \tag{6.2.171}$$

and we may test the given form of $F_+(\omega)$ (or equivalently, the given form of $f(\omega)$) by checking that the absolute deviations of T_p from the line p/m do not exceed $\pm a\sqrt{2/N}$.

Bartlett's U_p test

The result (6.2.171) can, in principle, be used to construct a set of simultaneous asymptotic confidence intervals for the values of $f(\omega)$ at the points $\{\omega_p\}$, $p = 1, 2, \ldots, m$, but the evaluation of such intervals would be rather complicated since the statistic T_p is a complicated function of the $\{f(\omega_p)\}$. Bartlett (1954) therefore suggested that one might consider instead the "unscaled" statistic

$$U_p = \frac{\sum\limits_{q=1}^{p} I_{N,X}^*(\omega_q)}{\sum\limits_{q=1}^{m} I_{N,X}^*(\omega_q)}. \tag{6.2.172}$$

Unfortunately, the asymptotic distribution of U_p is not related in any simple way with that of $\tilde{F}_+^{(\varepsilon)}(\omega)$, and the analogy with the empirical distribution function breaks down completely. The general "random walk" approach is no longer valid (unless we consider only the distribution of U_p conditional on the denominator fixed), and little seems to be known about the strict asymptotic properties of U_p. However, we know that $E[U_p] \sim F_+(\omega_p)$, and Bartlett suggested that, as an approximate procedure, one might test the statistic

$$\frac{\{U_p - F_+(\omega_p)\}}{(4\pi \int_0^\pi f^2(\theta)\, d\theta)^{1/2}} \qquad (6.2.173)$$

by treating it as if it had the same asymptotic distribution as $\{T_p - (p/m)\}$, viz., that given by (6.2.171). This approach leads to a test which accepts the hypothesized form of $F_+(\omega)$ if,

$$\max_p |U_p - F_+(\omega_p)| \le a\left(\frac{8\pi}{N}\int_0^\pi f^2(\theta)\, d\theta\right)^{1/2}. \qquad (6.2.174)$$

If we replace the denominator in (6.2.173) by the sample estimate suggested by Bartlett, namely,

$$\left[\frac{4\pi^2}{N} \cdot \frac{1}{s_X^4} \cdot \sum_{p=1}^m \{I_{N,X}^*(\omega_p)\}^2\right]^{1/2} = v^{1/2}, \qquad (6.2.175)$$

say (cf. (6.2.160)), we obtain a set of simultaneous confidence intervals for the $\{F_+(\omega_p)\}$ of the form,

$$U_p - a\sqrt{\frac{2v}{N}} \le F_+(\omega_p) \le U_p + a\sqrt{\frac{2v}{N}}, \qquad (6.2.176)$$

with asymptotic confidence level $\{100\Delta^{[2]}(a)\}\%$.

It is interesting to compare Bartlett's U_p test with Grenander and Rosenblatt's test for $\hat{H}_+(\omega)$. We can rewrite U_p as,

$$U_p = \left(\sum_{q=1}^p I_{N,X}^*(\omega_q)\right)\bigg/\left(\frac{Ns_X^2}{4\pi}\right) \sim \frac{2}{s_X^2}\int_0^{\omega_p} I_{N,X}^*(\theta)\, d\theta = \frac{\hat{H}_+(\omega_p)}{s_X^2},$$

where $\hat{H}_+(\omega_p)$ is given by (6.2.149). Hence, from (6.2.174) the U_p test is based effectively on the acceptance region,

$$\left|\frac{\hat{H}_+(\omega_p)}{s_X^2} - F_+(\omega_p)\right| \le a\left(\frac{8\pi}{N}\int_0^\pi f^2(\theta)\, d\theta\right)^{1/2},$$

a being determined by the required probability level $\Delta^{[2]}(a)$. On the other hand, the acceptance region for Grenander and Rosenblatt's test can be re-written (from (6.5.158)) as

$$\left|\frac{\hat{H}_+(\omega_p)}{\sigma_X^2} - F_+(\omega_p)\right| \leq a\left(\frac{8\pi}{N}\int_0^\pi f^2(\theta)\,d\theta\right)^{1/2},$$

a now being determined by setting the probability level equal to $\Delta^{[1]}(a)$. The two test statistics appear to be very similar, the only difference being that $\hat{H}_+(\omega)$ is divided by s_X^2 in the former and by σ_X^2 in the latter, but with the crucial effect that in the U_p test the statistic vanishes identically at $\omega_p = \pi$. This is reflected in the fact that for the same significance level the values of a differ in the two cases; e.g. for a 5% significance level a takes the value 1·36 for the U_p test and the value 2·2414 for the Grenander and Rosenblatt test.

Both the T_p and U_p tests have the advantage that, being based on estimates of the *normalized* integrated spectra, they can be expressed as functions of the sample *autocorrelations*, and hence are relatively insensitive to departures from normality, as opposed to the Grenander and Rosenblatt statistic which is a function of the sample autocovariances and is dependent on the assumption of normality. Moreover, the T_p test remains asymptotically valid even if $f(\omega)$ is not completely specified but is only partially specified as having the form of, e.g., an autoregressive spectral density function, with the parameter estimated from the observations; see Hannan (1960). This modification does not, however, hold for the Grenander and Rosenblatt test. Finally, as observed by Bartlett, in practice one is usually more interested in testing the form of the normalized spectra (rather than the non-normalized versions which require more information to be specified) and this also indicates the merit of the T_p and U_p tests.

Bartlett (1954) has given numerical illustrations of both the T_p and U_p tests applied to the Canadian lynx data, and Gower (1955) has applied these tests to the Beveridge Wheat Price Index series. See also Campbell and Walker (1977) for a further study of the application of the T_p and U_p tests to the Canadian lynx data.

III. *Grenander and Rosenblatt's test for the non-normalized spectral density function*

Grenander and Rosenblatt (1957a) have also proposed an asymptotic test for the non-normalized spectral density function $h(\omega)$ which, as in the previous cases, can be used also to construct asymptotic confidence bands for an unknown form of $h(\omega)$. They show, using a heuristic argument, that,

$$\lim_{l\to 0}\lim_{N\to\infty} p\left[\max_{k\cdot}\sqrt{\frac{Nl}{2\pi}}\left|\log\int_{kl}^{(k+1)l} I_N^*(\theta)\,d\theta - \log\int_{kl}^{(k+1)l} h(\theta)\,d\theta\right|\right.$$

$$\left. \leq \Phi^{-1}\left(1 - \frac{\alpha l}{\pi}\right)\right] = \exp(-2\alpha). \qquad (6.2.177)$$

(Note, once again, the used of a logarithmic transformation in deriving confidence bands for $h(\omega)$, cf. the remarks preceding equation (6.2.136).) This result does not involve the parameter e, and hence it remains valid asymptotically even for non-normal processes, in agreement with the feature previously noted that the asymptotic variance of a consistent spectral estimate does not depend on e. For the validity of this result we must have that, as $N \to \infty$, $1/l \to \infty$ more slowly than N. If we assume that $h(\theta)$ is roughly constant over the interval $\{kl, (k+1)l\}$, then, with a suitable choice of α, (6.2.177) provides an asymptotic confidence band for the values of $h(\omega)$ at all the points $\omega = l/2, 3l/2, 5l/2, \ldots$, of the form

$$\exp\left\{-\sqrt{\frac{2\pi}{Nl}}\Phi^{-1}\left(1-\frac{\alpha l}{\pi}\right)\right\}\frac{1}{l}\int_{kl}^{(k+1)l} I_N^*(\theta)\, d\theta \leq h\{(k+\tfrac{1}{2})l\}$$

$$\leq \exp\left\{\sqrt{\frac{2\pi}{Nl}}\Phi^{-1}\left(1-\frac{\alpha l}{\pi}\right)\right\}\frac{1}{l}\int_{kl}^{(k+1)l} I_N^*(\theta)\, d\theta.$$

$$(6.2.178)$$

Woodroofe and Van Ness (1967) obtained confidence bands of a similar form by considering $\max_{0 \leq k \leq M} |\hat{h}(\pi k/M) - h(\pi k/M)|$, where $\hat{h}(\omega)$ is based on a general scale parameter window with parameter M. Their results are derived under general conditions which assume, in particular, that X_t is a linear process of the form (6.2.2) with the $\{\varepsilon_t\}$ independent and having finite eighth order moments. (Hannan (1970), p. 292 derived the same type of results under slightly different conditions.) Woodroofe and Van Ness showed also that,

$$\lim_{N \to \infty} \left(\frac{N}{M \log M}\right)^{1/2} \max_{\omega} [|\hat{h}(\omega) - h(\omega)|/h(\omega)] = \left\{4\pi \int_{-\infty}^{\infty} K^2(\theta)\, d\theta\right\}^{1/2},$$

in probability, $K(\theta)$ being the spectral window generator. An important feature of this result is that it shows that the width of a confidence band which holds simultaneously for all ω is $O\{(\log M)^{1/2}\}$ times the width of a confidence interval at a single frequency.

Further asymptotic properties of the periodogram have been derived by Whittle (1959), Parthasarathy (1960) and Walker (1965).

IV. *Bartlett's homogeneity of chi-squared test*

Bartlett (1954) considered a different type of test for the goodness of fit of a specified normalized spectral density function $f(\omega)$. This test is based on the result that for linear processes satisfying the conditions of theorem (6.2.2), the random variables $\{I_N^*(\omega_p)/f(\omega_p)\}$, $p = 1, 2, \ldots, m = [N/2]$, ($N$ odd) are asymptotically independent and identically distributed, each being

$\frac{1}{2}\sigma_X^2 \chi_2^2$ (cf. (6.2.35)). Suppose now that we divide up the frequency range $(0, \pi)$ into a number of disjoint sub intervals $I_1, I_2, \ldots, I_k$, say, and let n_j denote the number of the $\{\omega_p\}$ points which fall in I_j. If we sum the scaled periodogram ordinates over each sub interval we have, for each j,

$$\sum_{\omega_p \in I_j} \{I_N^*(\omega_p)/f(\omega_p)\} \sim (\tfrac{1}{2}\sigma_X^2)\chi_{2n_j}^2,$$

and hence the asymptotic distribution of each of the quantities,

$$S_j^2 = \frac{1}{2n_j} \sum_{\omega_p \in I_j} \{I_N^*(\omega_p)/f(\omega_p)\}, \qquad j = 1, \ldots, k, \qquad (6.2.179)$$

is the same as that of the variance of a normal sample with $2n_j$ degrees of freedom. Hence, if the specified form of $f(\omega)$ is the correct one, the S_j^2 may be regarded as independent estimates of the same variance (viz., $\frac{1}{2}\sigma_X^2$), and we may test the goodness of fit of $f(\omega)$ by applying Bartlett's "homogeneity of χ^2 test" (usually referred to as the "homogeneity of variances test") to the S_j^2. The test statistic is (see, e.g. Fraser (1958), p. 260)

$$M = C\left[2\left(\sum_j n_j\right)\log_e\left\{\frac{\sum_j n_j s_j^2}{\sum_j n_j}\right\} - 2\sum_j n_j \log_e s_j^2\right], \qquad (6.2.180)$$

where

$$C = \left[1 + \frac{1}{3(k-1)}\left\{\left(\sum_j \frac{1}{2n_j}\right) - \frac{1}{2\sum_j n_j}\right\}\right]^{-1},$$

and on the null hypothesis that the specified form of $f(\omega)$ is the true one, M is asymptotically distributed as χ_{k-1}^2.

The grouping of the scaled periodogram ordinates into the sub-intervals $I_1, I_2, \ldots, I_k$ must, of course, be determined *a priori* (i.e. the choice of $I_1, \ldots, I_k$ must be made without reference to the computed values of the periodogram ordinates). One would usually choose a wide grouping for frequency ranges which are of little interest (i.e. where the spectral density function is likely to retain the same form under the null and alternative hypotheses), and a narrow grouping over ranges where the shape of the spectral density function is more critical. This feature endows the test with a considerable degree of flexibility, and the use of wide grouping intervals for frequency ranges of little interest produces a more sensitive test for the ordinates in the more critical regions. The test is also asymptotically valid even when $f(\omega)$ is not fully specified but involves parameters which have to be estimated from the data, as would arise, for example, if we wish to test the goodness of fit of an AR model of known order but with unknown parameters. In this case, $f(\omega)$ would have a known functional form, although it would involve the AR parameters whose values would then have to be estimated, but asymptotically this would not affect the validity of the test; see Hannan (1960). Bartlett (1954) argues that the test is relatively insensitive to departures from normality, but for this to hold it would seem necessary to restrict the choice of subintervals so that the proportion of periodogram ordinates in each subinterval is small. A numerical illustration of the test is given by Bartlett (1954) who used it to test the fit of an AR(2) model to the Canadian lynx data.

V. *Quenouille's goodness of fit test for autoregressive models*

The tests which we have described above are all designed to test the goodness of fit of a specified form of spectrum, and in principle they could be used to test the fit of a fully specified finite parameter model, such as an AR, MA, or ARMA scheme. Some of the above tests can be applied even when the model is not fully specified (so that its parameters are estimated from the data), but if we specifically wish to test the fit of, say, an AR model, it would seem more natural to base the test on the behaviour of the sample covariance or autocorrelation function rather than on the sample spectral functions, i.e. it would seem more natural to use a time domain rather than a frequency domain approach. The test which we describe below was proposed by Quenouille (1947) as a method of testing the fit of AR models; it is based on the sample autocorrelation function and, as such, it would have followed on more logically after our discussion of the time domain estimation in Chapter 5. However, the approach which we use to derive Quenouille's test is essentially a frequency domain one, and it is convenient, therefore, to discuss the detailed analysis at this point.

Suppose then that we wish to test the hypothesis H_0 that the observations come from an AR(k) model, where the order k is known *a priori* and we will suppose initially that the parameters are also fully specified. The basic idea behind Quenouille's test is to construct a sequence of linear functions of the sample autocorrelations such that, under H_0, they are asymptotically distributed with zero mean, common variance and zero covariance. Since, under general conditions, the sample autocorrelations are asymptotically normal, it follows that such linear functions will be asymptotically independent, and each normally distributed with zero mean. The sum of squares of these functions (suitably scaled) will then, under H_0, have an asymptotic χ^2 distribution, and this can be used as the test statistic.

Let $\hat{R}(s)$ denote the sample autocovariance of lag s (as given by (6.2.6)), and consider a linear function of the form,

$$Q(s) = \sum_u A_u \hat{R}(s-u).$$

Inverting (6.2.5) we have,

$$\hat{R}(s) = \int_{-\pi}^{\pi} e^{i\omega s} I_N^*(\omega) \, d\omega,$$

so that $Q(s)$ may be re-written as

$$Q(s) = \int_{-\pi}^{\pi} e^{i\omega s} A(\omega) I_N^*(\omega) \, d\omega,$$

where

$$A(\omega) = \sum_u A_u e^{-i\omega u}.$$

We now have from theorem 6.2.4,

$$E[Q(s)] \sim \int_{-\pi}^{\pi} \exp(i\omega s) A(\omega) h(\omega) \, d\omega = \bar{Q}(s), \qquad \text{say, all } s,$$

and for any s, t,

$$\text{cov}[Q(s), Q(t)] \sim \frac{2\pi}{N} \int_{-\pi}^{\pi} \exp(i\omega s) A(\omega)$$

$$\times \{\exp(-i\omega t) A^*(\omega) + \exp(i\omega t) A^*(-\omega)\} h^2(\omega) \, d\omega$$

$$+ \frac{e}{N} \cdot \bar{Q}(s) \cdot \bar{Q}(t). \quad (6.2.181)$$

(Note that here the function $A(\omega)$ may be complex valued and hence the second term in the above integral must be conjugated; also $A(\omega)$ is independent of N and does not have a limiting δ-function form as with spectral windows.) Since $A^*(-\omega) \equiv A(\omega)$, the first term in (6.2.181) can be written (ignoring for the moment the factor $2\pi/N$) as

$$\int_{-\pi}^{\pi} \exp[i\omega(s-t)] A(\omega) A^*(\omega) h^2(\omega) \, d\omega + \int_{-\pi}^{\pi} \exp[i\omega(s+t)] A^2(\omega) h^2(\omega) \, d\omega$$

$$= U_1 + U_2,$$

say. Now under H_0, the observations $\{X_t\}$ come from an AR(k) model, say,

$$X_t + a_1 X_{t-1} + \ldots + a_k X_{t-k} = \varepsilon_t,$$

so that by (4.12.59) we may write $h(\omega)$ in the form

$$h(\omega) = \frac{\sigma_\epsilon^2}{2\pi} \cdot \frac{1}{\alpha(\omega)\alpha^*(\omega)},$$

where $\alpha(\omega) = \{1 + a_1 \exp(-i\omega) + \cdots + a_k \exp(-ik\omega)\}$. With this substitution U_1 becomes,

$$U_1 = \frac{\sigma_\epsilon^4}{4\pi^2} \int_{-\pi}^{\pi} \exp[-i\omega(s-t)] \frac{A(\omega)A^*(\omega)}{\alpha^2(\omega)\{\alpha^*(\omega)\}^2} \, d\omega.$$

The problem is to choose the coefficients A_u, or equivalently the function $A(\omega)$, so that $Q(s)$ and $Q(t)$ are asymptotically uncorrelated. If, for the moment, we consider just the term U_1 it is clear that there are two obvious alternative choices for $A(\omega)$, namely

(1) $A(\omega) = \alpha^2(\omega) = A^{(1)}(\omega),$

say, or

(2) $A(\omega) = \alpha(\omega)\alpha^*(\omega) = A^{(2)}(\omega),$

say. With either choice, U_1 becomes zero since $\int_{-\pi}^{\pi} \exp[i\omega(s-t)] \, d\omega = 0$, $s \neq t$. Similarly, it is clear that with the second choice U_2 also vanishes for all s, t. With the first choice U_2 becomes

$$U_2 = \frac{\sigma_\epsilon^4}{4\pi^2} \int_{-\pi}^{\pi} \exp[i\omega(s+t)] \frac{\alpha^2(\omega)}{\{\alpha^*(\omega)\}^2} \, d\omega.$$

Now $\alpha^2(\omega)$ is a finite polynomial in $\exp(-i\omega)$ of degree $2k$; on the other hand, assuming that the AR(k) model is invertable into an infinite order MA form, $\{\alpha^*(\omega)\}^{-2}$ contains only positive powers of $\exp(i\omega)$. The above integral for U_2 will therefore vanish for all s, t, if both s and t are each $\geq k$. The same argument shows immediately that $\bar{Q}(s) = 0$

(1)   for all $s > 0$ with $A(\omega) = A^{(2)}(\omega)$,

or

(2)   for all $s > k$ with $A(\omega) = A^{(1)}(\omega)$.

Thus, if we set $A(\omega) = A^{(2)}(\omega)$ the required conditions on the $Q(s)$ will hold for $s > 0$, while if we set $A(\omega) = A^{(1)}(\omega)$ the conditions will hold for $s > k$. Either form of $A(\omega)$ may be used to construct a test, *but the advantage of $A^{(1)}(\omega)$ is that the resulting test remains valid asymptotically even when the coefficients, $a_1, \ldots, a_k$, of the AR model are not specified but are estimated from the data* (see Bartlett (1955), p. 261), whereas it is not clear that this property holds for the test based on $A^{(2)}(\omega)$. Since, in practice, we are usually more interested in testing whether the observations fit a *general* AR(k) model (rather than one with coefficients specified *a priori*), the test based on $A^{(1)}(\omega)$ is obviously the more useful one, and this leads to the form of the test originally proposed by Quenouille (1947). (Note that the coefficients $a_1, \ldots, a_k$, would be estimated from the first k sample auto-covariances and asymptotically this will not affect the $\{Q(s)\}$ for $s \ge k+1$.) The alternative form of test based on $A^{(2)}(\omega)$ was considered by Bartlett and Diananda (1950).

To obtain an explicit expression for Quenouille's test statistic we first note that the frequency domain function $\alpha(\omega)$ corresponds to the time domain operator $\psi(B) = (1 + a_1 B + \ldots + a_k B^k)$. Hence the function $A^{(1)}(\omega) = \alpha^2(\omega)$ corresponds to the operator $\psi^2(B) = (1 + a_1 B + \ldots + a_k B^k)^2$, and we may therefore write $Q(s)$ in the form,

$$Q(s) = \psi^2(B) \cdot \hat{R}(s)$$

(where now the shift operator B acts on s, i.e. $B\hat{R}(s) = \hat{R}(s-1)$, etc.). The variance of $Q(s)$ is given by substituting $s = t$ in the expression (6.2.181) for the covariance (recalling that for $s > k$ the term U_2 vanishes) giving,

$$\text{var}[Q(s)] \sim \frac{2\pi}{N} \cdot \frac{\sigma_\varepsilon^4}{4\pi^2} \cdot 2\pi = \frac{\sigma_\varepsilon^4}{N}.$$

We now scale $Q(s)$ by dividing it by $\hat{R}(0) = s_X^2$ (and then, for large N, treat s_X^2 as asymptotically equivalent to the constant σ_X^2), so that writing

$$\tilde{Q}(s) = Q(s)/s_X^2 = \psi^2(B) \cdot \hat{\rho}(s)$$

(where $\hat{\rho}(s)$ is the sample autocorrelation of lag s) we have that asymptotically the $Q(s)$ are independent ($s > k$), each normally distributed with zero mean and variance given by,

$$\text{var}[\hat{Q}(s)] \sim \frac{1}{N} \cdot \frac{\sigma_\varepsilon^4}{\sigma_X^4}.$$

If we now standardize the $\tilde{Q}(s)$ so as to have unit asymptotic variance and choose a set of values of s, say $s = k+1, k+2, \ldots, k+m$, we obtain the test statistics

$$Q = \sum_{s=k+1}^{k+m} \left\{ \sqrt{N} \frac{\sigma_X^2}{\sigma_\varepsilon^2} \cdot \tilde{Q}(s) \right\}^2 = \sum_{s=k+1}^{k+m} \left[\sqrt{N} \frac{\sigma_X^2}{\sigma_\varepsilon^2} \{ \psi^2(B) \cdot \hat{\rho}(s) \} \right]^2, \qquad (6.2.182)$$

which, under H_0, is asymptotically distributed as χ^2 with m degrees of freedom. In his original derivation of the test Quenouille used the "unbiased" sample autocorrelation,

$$\hat{\rho}^*(s) = \left(\frac{N}{N - |s|} \right) \hat{\rho}(s)$$

(cf. (5.3.44)), in terms of which Q may be written,

$$Q = \sum_{s=k+1}^{k+m} \left[(N-s)^{1/2} \left(1 - \frac{|s|}{N}\right)^{1/2} \frac{\sigma_X^2}{\sigma_\varepsilon^2} \{\psi^2(B)\hat{\rho}^*(s)\} \right]^2$$

$$\sim \sum_{s=k+1}^{k+m} \left[(N-s)^{1/2} \frac{\sigma_X^2}{\sigma_\varepsilon^2} \{\psi^2(B)\hat{\rho}^*(s)\} \right]^2,$$

but Quenouille suggested that the factor $(N-s)^{1/2}$ should be changed to $(N-s+k)^{1/2}$ to improve the accuracy of the asymptotic approximation. With this modification we finally obtain the statistic,

$$Q = \sum_{s=k+1}^{k+m} \left[(N-s+k)^{1/2} \frac{\sigma_X^2}{\sigma_\varepsilon^2} \{\psi^2(B)\hat{\rho}^*(s)\} \right]^2. \qquad (6.2.183)$$

In computing Q we would evaluate the ratio $(\sigma_X^2/\sigma_\varepsilon^2)$ in terms of the coefficients $a_1, \ldots, a_k$ (or their estimated values). For example, in the case $k = 2$ (corresponding to an AR(2) model) we have (from (3.5.34)),

$$\frac{\sigma_X^2}{\sigma_\varepsilon^2} = \frac{(1+a_2)}{(1-a_2)(1-a_1+a_2)(1+a_1+a_2)},$$

and $\psi(B) = 1 + a_1 B + a_2 B^2$, so that,

$$\psi^2(B)\hat{\rho}^*(s) = \{\hat{\rho}^*(s) + 2a_1\hat{\rho}^*(s-1) + (a_1^2 + 2a_2)\hat{\rho}^*(s-2)$$

$$+ 2a_1 a_2\hat{\rho}^*(s-3) + a_2^2\hat{\rho}^*(s-4)\}.$$

Quenouille's test is equivalent to a test based on the partial autocorrelations; in fact it is easy to show (Bartlett (1955), p. 264) that

$$\psi^2(B)\hat{\rho}^*(s) \sim \frac{\sigma_\varepsilon^2}{\sigma_X^2} \cdot \hat{\pi}_s \qquad (s > k),$$

where $\hat{\pi}_s$ is the sample partial correlation of lag s ($\hat{\pi}_s = -\hat{a}_s$), where $\hat{a}_s$ is the least-squares estimate of the last coefficient in a fitted AR(s) model, and is the (sample) partial correlation coefficient between X_t and X_{t-s}, holding $X_{t-1}, \ldots, X_{t-s+1}$ fixed; see Section 5.4.5. Hence, the χ^2 test based on Q is asymptotically equivalent to a χ^2 test based on $N(\sum_{s=k+1}^{k+m} \hat{\pi}_s^2)$, in accordance with the result noted in Section 5.4.5 that for an AR(k) model, $\hat{\pi}_j \sim N(0, 1/N), j > k$.

On a more general note, Walker (1952) has shown that Quenouille's test is asymptotically equivalent to the likelihood ratio test for testing the hypothesis of an AR(k) model against the alternative hypothesis of an AR($k+m$) model, as proposed by Whittle (1951). Walker (1952) and Hannan (1958b) have investigated the asymptotic power of both the Quenouille and Bartlett and Diananda tests.

Numerical illustrations of Quenouille's test are given by Hannan (1960), Bartlett (1954), and Campbell and Walker (1977). Chanda (1964) has studied modifications of the test statistic aimed at improving the accuracy of the χ^2 approximation for finite N.

Although the Bartlett and Diananda tests (based on $A^{(2)}(\omega)$) is not obviously "robust" with respect to the estimation of the parameters, the resulting statistic has the attractive property that it can be expressed very neatly in terms of the sample autocorrelations of the residual process, ε_t. For, setting $A(\omega) = A^{(2)}(\omega) = \alpha(\omega)\alpha^*(\omega)$, we have (with an obvious notation),

$$Q(s) = \int_{-\pi}^{\pi} \exp(i\omega s)\alpha(\omega)\alpha^*(\omega)I_{N,X}^*(\omega)\,d\omega,$$

$$\sim \int_{\pi}^{\pi} \exp(i\omega s)I_{N,\varepsilon}^*(\omega)\,d\omega,$$

(using theorem 6.2.2 and (4.12.59))

$$= \hat{R}_\varepsilon(s), \qquad s > 0.$$

Hence, the χ^2 test now based on (a standardized form of) $\sum_{s=1}^{m} \tilde{Q}^2(s)$ is asymptotically equivalent to a χ^2 test based $\{\sum_{s=1}^{m} \hat{\rho}_\varepsilon^2(s)\}$, which is exactly the same as the test statistic suggested by Box and Pierce (1970) in the context of what they termed a "portmanteau lack of fit test". Box and Pierce (see also Box and Jenkins (1970), p. 290) propose more generally that the statistic $Q = N\{\sum_{s=1}^{m} \hat{\rho}_\varepsilon^2(s)\}$ may be used to test the fit of a general ARMA model, treating its distribution under the null hypothesis as approximately χ^2 on $(m - k - l)$ degrees of freedom, where k, l denote respectively the orders of the AR and MA operators. (The Bartlett–Diananda test was designed only for pure AR models). Davies *et al.* (1977) studied the sampling properties of the Q statistic for moderate sample sizes and concluded that in such cases the mean and variance of Q can differ substantially from the values predicted by the asymptotic χ^2 approximation, the mean being far too low. In view of this, the true significance level may be much lower than that given by the χ^2 distribution. Prothero and Wallis (1976) and Ljung and Box (1978) suggest that Q be modified to

$$Q' = N(N+2) \sum_{s=1}^{m} (N-s)^{-1}\hat{\rho}_\varepsilon^2(s)$$

on the grounds that Q' has a distribution which is closer to the asymptotic χ^2 form. Some power studies of the Q' statistic are given by Davies and Newbold (1979).

Hosking (1978) has given a derivation of Quenouille's test and the "portmanteau" test via the use of Silvey's "Lagrange multiplier test" for suitable alternative hypotheses. Godfrey (1979) uses the same approach.

A goodness of fit test for pure MA models has been constructed by Wold (1949); see, e.g., Bartlett (1955) p. 262.

6.2.7 Continuous Parameter Processes

We mentioned at the beginning of this chapter that the methods of spectral analysis appropriate to discrete parameter processes could be applied equally well to continuous parameter processes if we first convert a continuous parameter record into discrete parameter form by reading it at small intervals of time Δt. Although this procedure introduces the "aliasing effect" (which we will discuss in more detail in Chapter 7), this will not be serious if Δt is chosen sufficiently small, and if we wish to use digital methods of spectral analysis the above device has to be adopted. However, if we wish to analyse continuous parameter records directly there is no difficulty in constructing a theory of spectral estimation which parallels that of the discrete parameter case very closely. In fact, with the obvious modifications (namely, changing sums into integrals and altering the frequency range from $(-\pi, \pi)$ to $(-\infty, \infty)$), the two approaches are virtually identical, see, e.g., Parzen (1957a) who treats both the discrete and continuous cases together.

Suppose then that we have a continuous parameter stationary process $\{X(t)\}$ whose spectrum is purely continuous with (non-normalized) spectral density function $h(\omega)$ and autocovariance function $R(\tau)$, and assume for simplicity that $E[X(t)] = 0$, all t. Given a finite sample record of $X(t)$ of length T, say, $0 \leq t \leq T$, we may estimate $R(\tau)$ for $0 \leq |\tau| \leq T$ as described in Section 5.3.7, i.e. we set

$$\hat{R}(\tau) = \frac{1}{T} \int_0^{T-|\tau|} X(t) X(t + |\tau|) \, dt. \qquad (6.2.184)$$

(If $E[X(t)]$ is not assumed zero we of course estimate it by $\bar{X} = (1/T) \int_0^T X(t) \, dt$ and use the more general estimator (5.3.69).)

To estimate $h(\omega)$ we consider the continuous analogue of (6.2.97), viz.

$$\hat{h}(\omega) = \frac{1}{2\pi} \int_{-T}^{T} \lambda_T(\tau) \hat{R}(\tau) e^{-i\omega\tau}, \qquad (6.2.185)$$

where now the covariance lag window $\{\lambda_T(\tau)\}$ is defined for all τ in $(-T, T)$. Defining the (modified) *periodogram* by,

$$I_T^*(\theta) = \frac{1}{2\pi T} \left| \int_0^T X(t) e^{-i\theta t} \, dt \right|^2 \qquad (6.2.186)$$

(cf. (6.1.24)), we may write,

$$I_T^*(\theta) = \frac{1}{2\pi} \int_{-T}^{T} \hat{R}(\tau) e^{-i\theta t} d\tau$$

(cf. lemma 6.1.1) so that,

$$\hat{R}(\tau) = \int_{-\infty}^{\infty} I_T^*(\theta) e^{i\theta t} d\theta, \qquad (6.2.187)$$

and (6.2.185) can now be re-written as,

$$\hat{h}(\omega) = \int_{-\infty}^{\infty} I_T^*(\theta) W_T(\omega - \theta) d\theta, \qquad (6.2.188)$$

where $W_T(\theta)$, the spectral window, is given by,

$$W_T(\theta) = \frac{1}{2\pi} \int_{-T}^{T} \lambda_T(\tau) e^{-i\theta t} d\tau \qquad (6.2.189)$$

(cf. (6.2.97), (6.2.98), (6.2.99)).

The lag window $\{\lambda_T(\tau)\}$ can be chosen exactly as in the discrete parameter case, and any of the standard forms discussed in Section 6.2.3 can be used, provided, of course, we remember that $\lambda_T(\tau)$ is now defined for *all* τ in $(-T, T)$, and that $W_T(\theta)$ is now the integral transform of $\lambda_T(\tau)$. In particular, if we choose a lag window of scale parameter form then $\lambda_T(\tau)$ takes the form,

$$\lambda_T(\tau) = k(\tau/M),$$

where $k(u)$ is a fixed continuous even function, with $k(0) = 1$, and M is the window parameter which controls the "width" of the spectral window which is now,

$$W_T(\theta) = M \cdot K(M\theta) \qquad (6.2.190)$$

where

$$K(\theta) = \frac{1}{2\pi} \int_{-\infty}^{\infty} k(u) e^{-iu\theta} du. \qquad (6.2.191)$$

Note that (6.2.190) is an exact result whereas the corresponding result for the discrete parameter case (6.2.95) is an approximation. The standard form of $k(u)$ and $K(\theta)$ listed in Section 6.2.3 thus apply exactly to the continuous parameter case. Assuming that $\int_{-\infty}^{\infty} k^2(u) du < \infty$, and that as $T \to \infty$,

$M \to \infty$, $(M/T) \to 0$, we have analogously to (6.2.102), (6.2.113), (6.2.114),

$$\lim_{T \to \infty} E[\hat{h}(\omega)] = h(\omega),$$

$$\lim_{T \to \infty} \left[\frac{T}{M} \operatorname{var}\{\hat{h}(\omega)\} \right] = (1 + \delta_{\omega,0}) h^2(\omega) \int_{-\infty}^{\infty} k^2(u) \, du, \quad (6.2.192)$$

and

$$\lim_{T \to \infty} \left[\frac{T}{M} \operatorname{cov}\{\hat{h}(\omega_1), \hat{h}(\omega_2)\} \right] = 0, \qquad \omega_1 \neq \pm \omega_2,$$

(cf. Parzen (1957a)). The expression for the asymptotic variance is the same as in the discrete case, and the results previously derived for the standard windows (including the asymptotic expressions for the bias) hold equally well in the present case.

Spectral estimation via wave analysers

When dealing with continuous records we may use an alternative method of spectral estimation based on what is known as a "wave analyser". This is very much an engineering orientated approach, and it is particularly suited to cases where the process $\{X(t)\}$ is recorded in the form of an electrical signal. The basic idea is to pass the record through a narrow band filter, centred on frequency ω, say which (ideally) will remove all components except those with frequencies in the neighbourhood of ω. The "power" in $\{X(t)\}$ at frequency ω is then estimated by estimating the total power (i.e. the variance) of the filtered output. Analytically, the procedure may be described as follows. Let $g_\omega(u)$ denote the impulse response function of a filter, centred on frequency ω, and let $\Gamma_\omega(\theta)$ denote the corresponding transfer function,

$$\Gamma_\omega(\theta) = \int_{-\infty}^{\infty} g_\omega(u) \, e^{-iu\theta} \, du.$$

If we pass $X(t)$ through this filter, the output $Y(t)$ is given by

$$Y(t) = \int_{-\infty}^{\infty} g_\omega(u) X(t-u) \, du,$$

and the estimate of $h(\omega)$ is obtained by estimating the variance of $Y(t)$ from a record of length T, i.e. we estimate $h(\omega)$ by

$$\hat{h}(\omega) = s_Y^2 = \frac{1}{T} \int_0^T |Y^2(t)| \, dt. \qquad (6.2.193)$$

In electrical engineering terminology the technique may be explained diagrammatically as in Fig. 6.13. Assuming that $g_\omega(u)$ satisfies conditions analogous to those of theorem 6.2.2, the periodograms of $Y(t)$ and $X(t)$ are,

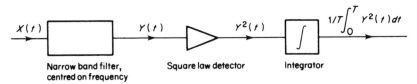

Fig. 6.13

with an obvious notation, related by (cf. (6.2.23)),

$$I^*_{T,Y}(\theta) \sim |\Gamma_\omega(\theta)|^2 I^*_{T,X}(\theta),$$

while setting $\tau = 0$ in (6.2.181) gives,

$$s^2_Y \equiv R_Y(0) = \int_{-\infty}^\infty I^*_{T,Y}(\theta)\, d\theta.$$

Hence, we find,

$$\hat h(\omega) = \int_{-\infty}^\infty I^*_{T,Y}(\theta)\, d\theta \sim \int_{-\infty}^\infty |\Gamma_\omega(\theta)|^2 I^*_{T,X}(\theta)\, d\theta. \qquad (6.2.194)$$

Thus, although the estimate (6.2.193) was motivated mainly by physical considerations, the result (6.2.194) shows that it is, in fact, of the same basic form as the more conventional estimate (6.2.188) in the sense that both are weighted integrals of the periodogram; *in* (6.2.194) *the function* $|\Gamma_\omega(\theta)|^2$ *plays the role of the spectral window*, $W_T(\omega - \theta)$, and, like $W_T(\omega - \theta)$, $|\Gamma_\omega(\theta)|^2$ is also highly concentrated in the neighbourhood of $\theta = \omega$. We can obtain an even closer relationship between (6.2.188) and (6.2.194) if, as would usually be the case, we work with a *fixed* filter, $\Gamma(\omega)$, and "tune" it to the required frequency ω by multiplying X_t by $\exp(i\omega t)$. (This technique is well known in the radio engineering context where it is known as "heterodyning"; the multiplication of $X(t)$ by $\exp(i\omega t)$ is here affected by "mixing" $X(t)$ with the output of a local oscillator.) Thus, let $g_0(u)$, $\Gamma_0(\omega)$, denote the impulse response function and transfer function of a fixed narrow band filter centred on *zero* frequency. Multiplying $X(t)$ by $\exp(i\omega t)$ and passing the product through the filter gives the output,

$$Y(t) = \int_{-\infty}^\infty g_0(u) X'(t - u)\, du,$$

where

$$X'(t) \equiv \exp(i\omega t) \cdot X(t).$$

But it follows from the definition of the periodogram, (6.2.186), that,

$$I_{T,X'}^*(\theta) = I_{T,X}^*(\theta - \omega).$$

Hence from (6.2.194) we obtain,

$$
\begin{aligned}
\hat{h}(\omega) &= \int_{-\infty}^{\infty} I_{T,Y}^*(\theta)\, d\theta \\
&\sim \int_{-\infty}^{\infty} |\Gamma_0(\theta)|^2 I_{T,X'}^*(\theta)\, d\theta \\
&= \int_{-\infty}^{\infty} |\Gamma_0(\theta)|^2 I_{T,X}^*(\theta - \omega)\, d\theta \\
&= \int_{-\infty}^{\infty} I_{T,X}^*(\theta) |\Gamma_0(\omega - \theta)|^2\, d\theta \quad\quad (6.2.195)
\end{aligned}
$$

(replacing $(\theta - \omega)$ by $-\theta$ and recalling that $I_{T,X}^*(\theta)$ is an even function of θ). Equation (6.2.195) is now of exactly the same form as (6.2.188), with $|\Gamma_0(\theta)|^2$ playing the role of $W_T(\theta)$. Of course, to obtain a consistent estimate of $h(\omega)$ we would, in principle, have to ensure that the "width" of $|\Gamma_0(\theta)|^2$ tends to zero as $T \to \infty$, but this is merely an academic point. In practice we would be operating with a fixed record length T and the filter "bandwidth" would be set in relation to T by the same criterion which affect the choice of the window parameter M and which we shall discuss in detail in the next chapter.

In the above discussion we have tacitly ignored the problem due to "transients" in the filter output. If the filtering is performed "on-line" by a physical device (such as an electrical circuit) then it naturally will have to be physically realizable, so that $g_0(u) = 0$, $u < 0$, and $Y(t)$ can be written

$$Y(t) = \int_0^\infty g_0(u) X'(t - u)\, du = \int_{-\infty}^t g_0(t - u) X'(u)\, du.$$

If $g_0(u)$ is non-zero for all $u > 0$, the above integral cannot be evaluated (for any t) from only a finite length of $X'(t)$. However, if we suppose, more realistically, that $g_0(u)$ decays sufficiently fast so that $g_0(u) \sim 0$ for $u > u_0$, say, then the full filter output $Y(t)$, can be evaluated for $t \geq u_0$ from a record of $X(t)$ starting at $t = 0$. For $t < u_0$, the output of the filter will not have reached its "steady state" form, and this initial portion of the output is usually referred to as a "transient". Consequently, the lower limit in the integral in (6.2.193) should be set as u_0 rather than zero (and the divisor changed to $(T - u_0)$).

In the previous analysis we have, in effect, assumed that $T \gg u_0$, i.e. we have assumed that the total record length is much larger than the duration of the transients, and hence the effect of changing the lower limit in (6.2.193) will be negligible. (Note, however, that the smaller we make the width of $|\Gamma_0(\theta)|^2$, the more slowly will $g_0(u)$ decay.) For further discussion of the wave-analyser approach to spectral estimation see Priestley and Gibson (1965).

6.3 MIXED SPECTRA

So far we have considered the problem of spectral estimation for (a) processes with purely discrete spectra, and (b) processes with purely continuous spectra. However, we know from our discussion of spectral theory (Chapter 4) that stationary processes can arise in which the spectra are neither purely discrete nor purely continuous but have a "mixed" structure, i.e. contain both discrete and continuous components. A typical model for a process with a mixed spectrum is (in the discrete parameter case) (cf. (4.9.12), (4.9.13)),

$$X_t = Y_t + Z_t, \tag{6.3.1}$$

where $\{Y_t\}$, $\{Z_t\}$ are uncorrelated processes, Y_t being a general linear process with (non-normalized) spectral density function $h(\omega)$ and Z_t a harmonic process of the form,

$$Z_t = \sum_{i=1}^{K} A_i \cos(\omega_i t + \phi_i) = \sum_{i=1}^{K} (A_i \cos \omega_i t + B_i' \sin \omega_i t). \tag{6.3.2}$$

The analysis of processes with mixed spectra presents some formidable problems, and as far as spectral analysis is concerned this case is by far the most difficult one to deal with. Fortunately, most stationary processes which are encountered in practice belong to the category of "purely continuous spectra", but processes with mixed spectra do occur in certain types of practical problems and their study is by no means of theoretical interest only. In the language of communications engineering we may think of the model (6.2.195) as representing a *"signal"* Z_t (consisting of a sum of periodic terms) superimposed on which there is a *noise* process Y_t. The analysis of mixed spectra may be regarded, therefore, as the *problem of detecting signals in the presence of "coloured" (i.e. autocorrelated) noise*. By contrast, the model (6.1.2) for which the classical periodogram tests were designed corresponds to the much simpler problem of detecting signals in the presence of *white noise*. Examples of processes with mixed spectra occur in radar, stress analysis, oceanography, and communications engineering, and we will discuss some of these applications in Chapter 8.

If, in (6.3.2), the frequencies $\{\omega_i\}$ and the value of K are known *a priori*, we can estimate the coefficients A_i', B_i' by regression analysis and then estimate the continuous spectral density function $h(\omega)$ from the "residuals",

$$\left\{ X_t - \sum_{i=1}^{K} (\hat{A}_i' \cos \omega_i t + \hat{B}_i' \sin \omega_i t) \right\}$$

by the methods discussed in the previous sections. However, if the $\{\omega_i\}$ (and K) are unknown, we must first "detect" the presence and location of the harmonic components. The periodogram tests for detecting harmonic components which we discussed in Section 6.1 (such as Fisher's g-test) are valid only when $\{Y_t\}$ is a purely random process (so that $h(\omega)$ is then constant), and we know that even if $Z_t \equiv 0$, the sampling properties of the periodogram of $\{X_t\}$ depend on the form of $h(\omega)$. Thus, if we simply ignore the fact that $\{Y_t\}$ may have a non-uniform spectral density function and apply the g-test we are likely to reach misleading conclusions. (If $Z_t \equiv 0$, so that there are no genuine harmonic components present, we are still likely to find large periodogram ordinates in regions where $h(\omega)$ is large.) On the other hand, if the complete form of $h(\omega)$ is known *a priori*, we can reduce the problem to the case where $\{Y_t\}$ is a purely random process by dividing the periodogram ordinates of X_t by $h(\omega)$ (cf. theorem 6.2.2), but if $h(\omega)$ is unknown we cannot obviously follow the same approach using an estimated form of $h(\omega)$ since the estimation of $h(\omega)$ would seem to require, as a first step, the removal of the harmonic terms, Z_t. For, if we ignore the presence of Z_t and try to estimate $h(\omega)$ by smoothing the periodogram of X_t each term in Z_t will produce a Fejer kernel component in the periodogram (cf. (6.1.36)), and even after smoothing by a spectral window we will still be left with a number of spurious peaks in the estimated spectral density function. This feature raises another fundamental difficulty in the analysis of mixed spectra, namely, that of distinguishing between a strict harmonic component and a narrow peak in the continuous spectrum, given only a finite number of observations. On the basis of a sample of N observations a strict harmonic component will produce a Fejer kernel term in the periodogram which has the form of a continuous peaked function with finite width $O(1/N)$, and in a purely empirical analysis it would be impossible to determine whether this continuous peaked function is due to a harmonic component or whether it simply reflects a narrow peak in the continuous spectrum.

At this stage it might appear that the problem of analysing mixed spectra is an intractable one in that the estimation of any one of the two components in the spectrum seems to require *a priori* information on the other component. However, the situation is not quite as hopeless as it may seem, and we can, in fact, devise a method of detecting and locating the harmonic components without using specific *a priori* information on the form of the continuous spectral density function $h(\omega)$. Also, we can modify the standard methods of estimating spectral density functions so as to allow us to construct a reliable estimate of $h(\omega)$ even though harmonic components may be present. However, these methods require the development of more refined analytical techniques than those which we have so far considered, and we postpone further discussion of mixed spectra until Chapter 8 when we shall re-examine this problem in much greater detail.

Chapter 7

Spectral Analysis in Practice

7.1 SETTING UP A SPECTRAL ANALYSIS

In Chapter 6 we considered the theoretical aspects of the problem of estimating the spectral density function of a stationary process from a finite sample of observations. Let $\{X_t\}$ be the (discrete parameter) stationary process under study and let $f(\omega)$, $h(\omega)$ denote respectively its normalized and non-normalized spectral density functions. Given a sample of N observations $X_1, X_2, \ldots, X_N$, the general class of estimators of $h(\omega)$ considered in Chapter 6 is given by (6.2.97), namely,

$$\hat{h}(\omega) = \frac{1}{2\pi} \sum_{s=-(N-1)}^{(N-1)} \lambda_N(s)\hat{R}(s) e^{-i\omega s}, \qquad (7.1.1)$$

where

$$\hat{R}(s) = \frac{1}{N} \sum_{t=1}^{N-|s|} (X_t - \bar{X})(X_{t+|s|} - \bar{X}), \qquad (7.1.2)$$

$(\bar{X} = (1/N) \sum_{t=1}^{N} X_t$ being the sample mean) denotes the sample auto-covariance function, and $\{\lambda_N(s)\}$ is a sequence of decreasing weights which we called the *"covariance lag window"*. The corresponding class of estimators of $f(\omega)(=h(\omega)/\sigma_X^2)$ is given by,

$$\hat{f}(\omega) = \frac{\hat{h}(\omega)}{s_X^2} = \frac{1}{2\pi} \sum_{s=-(N-1)}^{(N-1)} \lambda_N(s)\hat{\rho}(s) e^{-i\omega s}, \qquad (7.1.3)$$

where $\hat{\rho}(s) = \hat{R}(s)/\hat{R}(0)$ is the sample autocorrelation function, and s_X^2 $(\equiv \hat{R}(0))$ is the sample variance (cf. (6.2.53)). We showed that (7.1.1) could be written alternatively in the algebraically equivalent form,

$$\hat{h}(\omega) = \int_{-\pi}^{\pi} I_{N,X-\bar{X}}^*(\theta) W_N(\omega - \theta) \, d\theta, \qquad (7.1.4)$$

502

where

$$I^*_{N,X-\bar{X}}(\theta) = \frac{1}{2\pi N} \left| \sum_{t=1}^{N} (X_t - \bar{X}) \, e^{-i\theta t} \right|^2, \qquad (7.1.5)$$

denotes the periodogram of $\{(X_t - \bar{X})\}$, and

$$W_N(\theta) = \frac{1}{2\pi} \sum_{s=-(N-1)}^{(N-1)} \lambda_N(s) \, e^{-is\theta}, \qquad (7.1.6)$$

is the *spectral window* corresponding to the covariance lag window $\{\lambda_N(s)\}$; see Section 6.2.3. In the case of continuous parameter processes we noted that we could still use estimators of the above form provided, of course, that we first convert the continuous sample record into discrete parameter form by reading off its values at discrete time intervals Δt. (Throughout most of Section 6.2 we assumed that the value of $E\{X_t\}$ was known and hence, without loss of generality, could be assumed zero, in which case $\hat{R}(s)$ takes the simpler form, (6.2.6). Here, we consider the more realistic case where $E\{X_t\}$ is unknown, so that we now use the more general form of $\hat{R}(s)$, as given by (5.3.13), and consequently the definition of the periodogram is modified from (6.1.24) and (6.2.4) to (7.1.5). This means, in effect, that we are now working with the values $\{X_t - \bar{X}\}$ rather than the $\{X_t\}$ themselves, but this change will not affect the asymptotic sampling properties of $h(\omega)$ or $\hat{f}(\omega)$. If, as we will later discuss, we consider a method of computing $h(\omega)$ which is based on replacing (7.1.4) by a discrete sum over the set of frequencies, $\omega_p = 2\pi p/N$, $p = 1, 2, \ldots$, we may note that (with an obvious notation), $I^*_{N,X-\bar{X}}(2\pi p/N) \equiv I^*_{N,X}(2\pi p/N)$, $p \neq 0$, i.e. these periodogram ordinates are not affected by subtracting the mean $\bar{X}$, except at zero frequency where $I^*_{N,X-\bar{X}}(0) = 0$; see the discussion in Section 6.2.)

The sequence $\{\lambda_N(s)\}$ typically involves a parameter M, controlling its rate of decay; in particular, if, as is usually the case, $\lambda_N(s)$ is of "scale parameter form" (see Section 6.2.3) i.e. if $\lambda_N(s) = k(s/M)$, where $k(u)$ is some fixed even continuous square integrable function of u with $k(0) = 1$, then we know from the results of Section 6.2.4 that, irrespective of the precise form of $k(u)$, $\hat{h}(\omega)$ will always be a consistent estimate of $h(\omega)$ if, as $N \to \infty$, $M \to \infty$ in such a way that $M/N \to 0$. When $k(u)$ is such that, $k(u) = 0$, $|u| > 1$, then $\lambda_N(s) = 0$, $|s| > M$, and the effective limits of the summation over s in (7.1.1) are $-M$ to M. This occurs, for example, with the "truncated periodogram" and Bartlett windows (see Section 6.2.3), and M is then called the "*truncation point*".

However, when we start to think about the problem of *computing* spectral estimates we must first decide on the following;

(1) the mathematical form of the lag window $\lambda_N(s)$ (or $k(u)$), or equivalently, of the spectral window $W_N(\theta)$;

(2) the value of the parameter M (the above conditions on M determine merely its order of magnitude in relation to N, they certainly do not determine the precise value of M for a given value of N);

(3) the value of N, (the sample size) when an arbitrarily large amount of data is available;

(4) the value of Δt, (the time domain sampling interval) when analysing continuous parameter processes;

(5) the set of values of ω for which $\hat{h}(\omega)$ (or $\hat{f}(\omega)$) is to be computed; if these are to be equally spaced then we must choose $\Delta\omega$, the frequency interval between adjacent estimates;

(6) the most efficient way of computing $\hat{h}(\omega)$ (or $\hat{f}(\omega)$).

In the following sections we shall try to provide answers to these questions by examining how changes in the above quantities affect the statistical properties of the spectral estimates. In so doing, we are led to introduce certain criteria which, in principle, determine "optimal" forms of estimates, but typically involve properties of the true spectral density function (or the true autocovariance function) which may not be known *a priori*. For example, the notion of the "bandwidth" of the spectral density function (effectively, the width of its narrowest peak) plays a major role in determining the appropriate value of M. However, even if these properties are not known completely the "optimality" theory is still useful in that it focuses attention on those features which are particularly important when we consider the problem of "designing" a spectral estimation procedure.

The problem of choosing Δt, the time domain sampling interval, falls into a rather different category from the others listed above, and its choice is determined mainly by considering a result known as the "aliasing theorem". We have already referred to this result in our previous discussions, particularly in Section 4.8.3 when we were considering the proof of Wold's theorem. We now examine this result in a slightly more general form.

7.1.1 The Aliasing Effect

If we sample a continuous parameter process $\{X(t)\}$ at time intervals Δt it is clear that we will lose some information on its spectral properties. For, when we look at $X(t)$ only at a discrete set of equally spaced time points we miss detailed information on its high frequency (i.e. rapidly oscillating) components, and this will obviously affect our ability to estimate the high frequency end of the spectrum. To study this effect in more detail let us suppose that $\{X(t)\}$ admits a spectral representation of the form (cf. theorem 4.11.1, Section 4.11),

$$X(t) = \int_{-\infty}^{\infty} e^{it\omega} \, dZ_X(\omega). \tag{7.1.7}$$

Denoting the sampled process by $\{Y_t\}$ i.e. writing $Y_t \equiv X(t . \Delta t)$, $t = 0, \pm 1,$ $\pm 2, \ldots,$ we have from (7.1.7),

$$Y_t = \int_{-\infty}^{\infty} \exp(it . \Delta t . \omega) \, dZ_X(\omega)$$

$$= \sum_{k=-\infty}^{\infty} \int_{(2k-1)\pi/\Delta t}^{(2k+1)\pi/\Delta t} \exp(it . \Delta t . \omega) \, dZ_X(\omega)$$

$$= \sum_{k=-\infty}^{\infty} \int_{-\pi/\Delta t}^{\pi/\Delta t} \exp\left[it . \Delta t\left(\omega + \frac{2k\pi}{\Delta t}\right)\right] dZ_X\left(\omega + \frac{2k\pi}{\Delta t}\right).$$

But since t and k take only integer values, $\exp(i . 2k\pi t) = 1$, all k, t, and hence we may write,

$$Y_t = \sum_{k=-\infty}^{\infty} \int_{-\pi/\Delta t}^{\pi/\Delta t} \exp(it . \Delta t . \omega) \, dZ_X\left(\omega + \frac{2k\pi}{\Delta t}\right)$$

$$= \int_{-\pi/\Delta t}^{\pi/\Delta t} \exp(it . \Delta t . \omega) \, dZ_Y(\omega), \qquad (7.1.8)$$

say, where

$$dZ_Y(\omega) = \sum_{k=-\infty}^{\infty} dZ_X\left(\omega + \frac{2k\pi}{\Delta t}\right). \qquad (7.1.9)$$

The orthogonality of $dZ_Y(\omega)$ follows from

$$E[dZ_Y(\omega) \, dZ_Y^*(\omega')] = \sum_k \sum_l E\left[dZ_X\left(\omega + \frac{2k\pi}{\Delta t}\right) dZ_X^*\left(\omega' + \frac{2l\pi}{\Delta t}\right)\right] = 0,$$

$$\omega \neq \omega',$$

since both ω, ω', lie in the range $(-\pi/\Delta t, \pi/\Delta t)$ and hence, if $\omega \neq \omega'$, there are no values of k, l, for which $\omega + 2k\pi/\Delta t$ and $\omega' + 2l\pi/\Delta t$ are equal. Thus, we have shown that Y_t *has a spectral representation which extends only over the frequency range,* $(-\pi/\Delta t, \pi/\Delta t)$, the basic reason for this being that when t is restricted to integer multiples of Δt *we cannot distinguish between the frequency components* $\exp(i\omega t)$ *and* $\exp[i(\omega + 2k\pi/\Delta t)t]$. The components in $X(t)$ with frequencies $\omega - 2\pi/\Delta t$, $\omega + 2\pi/\Delta t$, $\omega - 4\pi/\Delta t$, $\omega + 4\pi/\Delta t$, $\omega - 6\pi/\Delta t$, $\omega + 6\pi/\Delta t, \ldots$, will all *appear* to have frequency ω; these frequencies are said to be "aliases" of ω, and every frequency outside the range $(-\pi/\Delta t, \pi/\Delta t)$ has an "alias" inside this range. (The term "alias" is due to Tukey.) The above analysis is virtually the same as that used in Section 4.8.3, except that there we were dealing with the case $\Delta t = 1$. We may define the (non-normalized) integrated spectrum of Y_t by,

$$dH_Y(\omega) = E[|dZ_Y(\omega)|^2], \qquad |\omega| \le \pi/\Delta t, \qquad (7.1.10)$$

and it follows from (7.1.9) and the orthogonality properties of $dZ_X(\omega)$ that,

$$dH_Y(\omega) = \sum_{k=-\infty}^{\infty} E\left[\left|dZ_X\left(\omega + \frac{2k\pi}{\Delta t}\right)\right|^2\right],$$

or,

$$dH_Y(\omega) = \sum_{k=-\infty}^{\infty} dH_X\left(\omega + \frac{2k\pi}{\Delta t}\right), \qquad |\omega| \le \frac{\pi}{\Delta t}, \qquad (7.1.11)$$

where $H_X(\omega)$ is the (non-normalized) integrated spectrum of $\{X(t)\}$ (cf. (4.11.5)). If $X(t)$ has (non-normalized) spectral density function $h_X(\omega)$ then (7.1.11) becomes,

$$h_Y(\omega) = \sum_{k=-\infty}^{\infty} h_X\left(\omega + \frac{2k\pi}{\Delta t}\right), \qquad |\omega| \le \frac{\pi}{\Delta t}. \qquad (7.1.12)$$

Also, if $E[X(t)] = 0$ then $E[Y_t] = 0$ and we have,

$$R_Y(s) = \text{cov}\{Y_t, Y_{t+s}\} = \int_{-\pi/\Delta t}^{\pi/\Delta t} \exp(i\omega s \Delta t) E[|dZ_Y(\omega)|^2]$$

$$= \int_{-\pi/\Delta t}^{\pi/\Delta t} \exp(i\omega s \Delta t)\, dH_Y(\omega), \qquad (7.1.13)$$

with,

$$H_Y\left(\frac{\pi}{\Delta t}\right) = \int_{-\pi/\Delta t}^{\pi/\Delta t} dH_Y(\omega) = \int_{-\pi/\Delta t}^{\pi/\Delta t} \sum_k dH_X\left(\omega + \frac{2k\pi}{\Delta t}\right)$$

$$= \int_{-\infty}^{\infty} dH_X(\omega) = H_X(\infty) = \text{var}\{X(t)\}.$$

When $h_Y(\omega)$ exists,

$$R_Y(s) = \int_{-\pi/\Delta t}^{\pi/\Delta t} \exp(i\omega s \Delta t) h_Y(\omega)\, d\omega,$$

and hence,

$$h_Y(\omega) = \frac{\Delta t}{2\pi} \sum_{s=-\infty}^{\infty} R_Y(s) \exp(-i\omega s \Delta t). \qquad (7.1.14)$$

Equation (7.1.12) is the fundamental result known as the "aliasing theorem". It tells us that the spectral density function of the sampled process, $h_Y(\omega)$, extends only from $-\pi/\Delta t$ to $\pi/\Delta t$, and that at each point, ω, within this interval its value is a superposition of the values of $h_X(\omega)$ at the points ω, $\omega \pm 2\pi/\Delta t$, $\omega \pm 4\pi/\Delta t$, ... (The range $(-\pi/\Delta t, \pi/\Delta t)$ corresponds to the *angular* frequency scale; if we set $f = \omega/2\pi$, so that f denotes frequency measured in Hz (or cycles per second), then the spectral density function of Y_t extends from $-1/2\Delta t$ Hz. to $1/2\Delta t$ Hz.) In graphical terms we can describe the construction of $h_Y(\omega)$ as follows; we take the graph of $h_X(\omega)$ over $(-\infty, \infty)$, mark vertical "fold" lines at all the points $\omega = \pm \pi/\Delta t$, $\pm 3\pi/\Delta t$, $\pm 5\pi/\Delta t$, ..., and then fold back the graph "concertina fashion" onto the interval $(-\pi/\Delta t, \pi/\Delta t)$. (Recall that, for real valued process, $h_X(\omega)$ is an even function of ω.) The graph of $h_Y(\omega)$ is then given by superimposing all the folded portions additively, as shown in Fig. 7.1. The frequency $\omega_N = 2\pi/\Delta t$ is called the *Nyquist* or *folding frequency*.

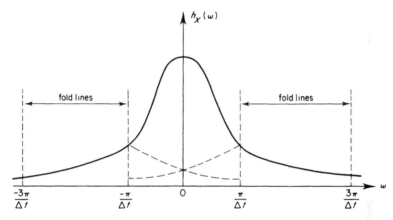

Fig. 7.1. Aliasing effect.

Choosing the interval Δt

Clearly, we cannot recover the form of $h_X(\omega)$ completely from a knowledge of $h_Y(\omega)$ unless $h_X(\omega)$ *is exactly zero for all ω outside the interval* $(-\pi/\Delta t, \pi/\Delta t)$. In this exceptional case we have immediately from (7.1.12),

$$h_Y(\omega) \equiv h_X(\omega), \qquad |\omega| \leqslant \pi/\Delta t, \tag{7.1.15}$$

and the spectral properties of $X(t)$ are completely determined by its sampled values. If $h_X(\omega)$ has any "*band limited*" form, i.e. if $\exists\, \omega_0$ such that $h_X(\omega) = 0$, $|\omega| > \omega_0$, then we can always choose Δt sufficiently small for the above condition to hold, namely by choosing $\Delta t < \pi/\omega_0$. (This result is a special case of a more general "sampling theorem" for band-limited functions, see, e.g. Whittaker (1935).) However, in the more realistic case where $h_X(\omega)$ does not vanish outside a certain interval (but tends to zero gradually as $\omega \to \pm\infty$) we can reduce the aliasing effect to negligible proportions by *choosing Δt sufficiently small so that $h_X(\omega)$ is negligible when $|\omega| \geq \pi/\Delta t$*, i.e. *by choosing Δt so that $X(t)$ has negligible power at frequencies higher than $\pi/\Delta t$.*

In practice it is sometimes possible to determine the frequency limit beyond which $X(t)$ has negligible power from a knowledge of the frequency range of the recording mechanism. Alternatively, if we know that we are interested in estimating $h_X(\omega)$ only over a restricted range, say for $|\omega| \leq \omega_0$, then we can pass the observed record of $X(t)$ through a *band pass filter* so as to remove (effectively) all frequencies higher than ω_0 (see Section 4.12.1), and then Δt is determined by

$$\omega_0 \leq \pi/\Delta t,$$

i.e.

$$\Delta t \leq \pi/\omega_0. \tag{7.1.16}$$

Equivalently, if t is measured in seconds and if $f_0 = \omega_0/2\pi$ denotes the upper frequency limit measured in "cycles per second" (or Hz), then Δt is determined by,

$$f_0 \leq 1/2\Delta t,$$

i.e.

$$\Delta t \leq 1/2f_0. \tag{7.1.17}$$

For example, if we wish to estimate the spectral density function only up to 5 Hz then we must choose a sampling interval of $(1/10)$th second, or smaller. On the other hand, if we wish to estimate the spectral density function up to 20 000 Hz we must sample at intervals of $(1/40000)$th second, or smaller.

Transformation of the frequency scale

Once we have sampled the process $\{X(t)\}$ we would usually treat the $\{Y_t\}$ simply as observations from a discrete parameter process with a spectral density function defined on the standard range $(-\pi, \pi)$. In other words, we would treat the $\{Y_t\}$ as a standard set of discrete time observations, which amounts, in effect, to taking Δt as unity. However, once we have estimated the spectral density function of Y_t in this way we can easily convert it back into the original frequency range $(-\pi/\Delta t, \pi/\Delta t)$ as follows. If we write $\omega' = \Delta t \cdot \omega$ in (7.1.8) we can rewrite the spectral representation of Y_t in the standard form as

$$Y_t = \int_{-\pi}^{\pi} e^{it\omega'} \, dZ_Y\left(\frac{\omega'}{\Delta t}\right).$$

Denoting the spectral density function of Y_t *defined over the range* $(-\pi, \pi)$ by $h_Y^*(\omega)$, we now have,

$$h_Y^*(\omega) \, d\omega = E\left[\left|dZ_Y\left(\frac{\omega}{\Delta t}\right)\right|^2\right] = h_Y\left(\frac{\omega}{\Delta t}\right) \frac{d\omega}{\Delta t}.$$

Hence, $h_Y^*(\omega)$ and $h_Y(\omega)$ are related by,

$$h_Y(\omega) = \Delta t \cdot h_Y^*(\omega \cdot \Delta t), \qquad |\omega| \leq \frac{\pi}{\Delta t}. \tag{7.1.18}$$

This relationship can be obtained alternatively by noting that

$$h_Y^*(\omega) = \frac{1}{2\pi} \sum_{s=-\infty}^{\infty} R_Y(s) \, e^{-i\omega s}, \tag{7.1.19}$$

and comparing (7.1.19) with (7.1.14).

Thus, to convert the spectral density function, $h_Y^*(\omega)$, back to the original scale we multiply it by Δt and scale the frequency range accordingly.

Random sampling intervals

We have observed that sampling a continuous parameter process inevitably results in a loss of some form of "information"; in particular when we sample at equal time intervals we lose some information on the spectral properties of the process. However, if we sample instead at random time instants we will still lose "information" but this may not be necessarily of a "systematic" nature provided the random mechanism is suitably chosen. In particular, it has been shown by Shapiro and Silverman (1960) that if we sample the process $\{X(t)\}$ at time instants $\{t_i\}$ which correspond *to the points of a Poisson process* (see Section 2.11), then we may recover the full spectral properties of $\{X(t)\}$ from a knowledge of the spectral properties of $\{Y_t\}$. In this sense, "Poisson process" sampling may be said to be "*alias free*". Shapiro and Silverman (1960) considered also the more general case where the time instants $\{t_i\}$ correspond to the points of a *renewal process* (i.e. the intervals $(t_i - t_{i-1})$ are independent identically distributed positive random variables), and show that this scheme will also be alias free provided that the probability distribution of the intervals $(t_i - t_{i-1})$ satisfies certain conditions, which include, as a special case, the Poisson process model for which the distribution of the intervals is exponential. Brillinger (1972) studied the case where the $\{t_i\}$ form a *general point process*. Gaster and Roberts (1977) discuss how such random sampling schemes occur quite naturally in certain physical applications, such as a laser anemometer. Akaike (1960) and Shapiro and Silverman (1960) have investigated the properties of "*jittered sampling*", i.e.

where the $\{t_i\}$ are of the form $t_i = i\Delta t + \varepsilon_i$, the $\{\varepsilon_i\}$ being independent and identically distributed random variables. This form of sampling might arise, for example, in the case where we intend to sample at equal fixed time intervals but the sampling mechanism involves "jitter" which produces small random perturbations from the assigned time points.

7.2 MEASURES OF PRECISION OF SPECTRAL ESTIMATES

A natural approach to the problem of determining the parameters (such as N and M) of the estimate $\hat{h}(\omega)$ is to choose these so that the resulting estimate has a prescribed degree of "precision". To do this we must first decide how we are to measure the "precision" of the estimate. Bearing in mind that the standard deviation of $\hat{h}(\omega)$ is (approximately) proportional to $h(\omega)$ (cf. (6.2.108)), Parzen (1957b) suggested two alternative measures, namely the "$p\%$ *Gaussian range of percentage error*" and the "*mean square percentage error*", defined as follows.

(a) The $p\%$ *Gaussian range of percentage error*, $\Delta_p(\omega)$, is defined by,

$$\Delta_p(\omega) = \gamma_p \cdot \frac{v(\omega)}{h(\omega)} + \frac{|b(\omega)|}{h(\omega)}, \qquad (7.2.1)$$

where $v^2(\omega) = \mathrm{var}\{\hat{h}(\omega)\}$, $b(\omega) = [E\{\hat{h}(\omega)\} - h(\omega)]$ denotes the bias of $\hat{h}(\omega)$, and γ_p is the two sided $p\%$ point of the standardized normal distribution, i.e.

$$\frac{1}{\sqrt{2\pi}} \int_{\gamma_p}^{\infty} \exp(-x^2/2) \, dx = \{(100-p)/2\}\%.$$

The motivation for this definition can be seen by noting that,

$$|\hat{h}(\omega) - h(\omega)| \le |\hat{h}(\omega) - E\{\hat{h}(\omega)\}| + |E\{\hat{h}(\omega)\} - h(\omega)|,$$

and hence

$$p\left[\frac{|\hat{h}(\omega) - h(\omega)|}{h(\omega)} \le \delta\right] \ge p\left[\frac{|\hat{h}(\omega) - E\{\hat{h}(\omega)\}|}{h(\omega)} \le \delta - \frac{|b(\omega)|}{h(\omega)}\right].$$

Using the asymptotic normality of $\hat{h}(\omega)$, the RHS above may, for large N, be set to be $p\%$ by choosing

$$\delta = \gamma_p \cdot \frac{v(\omega)}{h(\omega)} + \frac{|b(\omega)|}{h(\omega)}.$$

Hence, if we define $\Delta_p(\omega)$ as in (7.2.1), we have (asymptotically),

$$P\left[\frac{|\hat{h}(\omega) - h(\omega)|}{h(\omega)} \leqslant \Delta_p(\omega)\right] \geqslant \frac{p}{100}. \tag{7.2.2}$$

Thus, $h(\omega)\Delta_p(\omega)$ may be regarded as an upper bound for a $p\%$ confidence interval for $h(\omega)$, based on the asymptotic normal distribution and taking due account of the bias. The confidence intervals (6.2.131), (6.2.132) take no account of the bias, but it should be remembered that, as with (6.2.131), (6.2.132), a confidence interval based on $\Delta_p(\omega)$ applies only to a particular frequency.

(b) *The mean square percentage error,* $\eta(\omega)$, is defined by,

$$\eta^2(\omega) = [E\{\hat{h}(\omega) - h(\omega)\}^2]/h^2(\omega) = \{v^2(\omega) + b^2(\omega)\}/h^2(\omega). \tag{7.2.3}$$

Parzen notes that these two measures are essentially equivalent; in fact it follows by Chebycheff's inequality that,

$$P\left[\frac{|\hat{h}(\omega) - h(\omega)|}{h(\omega)} \leqslant \Delta_p(\omega)\right] \geqslant 1 - \left\{\frac{\eta(\omega)}{\Delta_p(\omega)h(\omega)}\right\}^2.$$

In particular, if $b(\omega) \equiv 0$, then $\Delta_p(\omega) \equiv \gamma_p . \eta(\omega)$. We may therefore work with either quantity as a measure of the precision of $\hat{h}(\omega)$, and choose whichever is the more convenient to the context of the subsequent analysis.

Note that if we use the asymptotic expressions for $v^2(\omega)$ and $b(\omega)$ given by (6.2.113), (6.2.124), appropriate for the usual case of scale parameter windows, we may write (for $\omega \neq 0, \pi$),

$$\Delta_p(\omega) = \gamma_p . \left\{\frac{M}{N} \int_{-\infty}^{\infty} k^2(u) \, du\right\}^{1/2} + \frac{k^{(r)}}{M^r} \cdot \frac{1}{\{\lambda^{(r)}(\omega)\}^r} \tag{7.2.4}$$

where r is the "characteristic exponent" of $k(u)$, $k^{(r)}$ is defined by (6.2.121), and

$$\lambda^{(r)}(\omega) = \left|\frac{h(\omega)}{h^{(r)}(\omega)}\right|^{1/r}, \tag{7.2.5}$$

$h^{(r)}(\omega)$ being the "rth generalized derivative" of $h(\omega)$ defined by (6.2.127). (For reasons which will be discussed in the next section, Parzen (1957b) calls $\lambda^{(r)}(\omega)$ the "*spectral bandwidth of order r*" *at frequency* ω.) Similarly, we may write $\eta^2(\omega)$ as,

$$\eta^2(\omega) = \frac{M}{N}\left\{\int_{-\infty}^{\infty} k^2(u) \, du\right\} + \left[\frac{k^{(r)}}{M^r\{\lambda^{(r)}(\omega)\}^r}\right]^2 \tag{7.2.6}$$

We thus see that both $\Delta_p(\omega)$ and $\eta(\omega)$ depend on the properties of the true spectral density function, $h(\omega)$, *purely in terms of the quantity*, $\lambda^{(r)}(\omega)$, and depend on the covariance lag window via the quantities $\{\int_{-\infty}^{\infty} k^2(u)\,du\}$ and $k^{(r)}$.

(c) *The "signal to noise ratio"*

Parzen (1961b) introduced a third measure of precision which he called the "signal to noise ratio" of the estimate and which is defined by,

$$\mathrm{SNR}\{\hat{h}(\omega)\} = \frac{E\{\hat{h}(\omega)\}}{[\mathrm{var}\{\hat{h}(\omega)\}]^{1/2}}. \tag{7.2.7}$$

(An estimate with a high SNR may be said to have high "precision" in the sense that its *standard* deviation is small compared with its mean.) Using the asymptotic normality of $\hat{h}(\omega)$ we have, for large N,

$$P\left[\left|\frac{\hat{h}(\omega) - E\{\hat{h}(\omega)\}}{E\{\hat{h}(\omega)\}}\right| \leq \delta\right] \doteq \frac{p}{100}$$

if we set $\delta = \gamma_p . [\mathrm{var}\{\hat{h}(\omega)\}]^{1/2}/E\{\hat{h}(\omega)\}$, where γ_p is, as above, the two-sided $p\%$ point of the standardized normal distribution. Hence we have,

$$\delta = \gamma_p \cdot \frac{1}{\mathrm{SNR}\{\hat{h}(\omega)\}}, \tag{7.2.8}$$

so that, in effect, the SNR measure is a simplified version of Δ_p *in which the bias of $\hat{h}$ is ignored*. In fact, if $b(\omega) = 0$, then $\Delta_p(\omega) = \gamma_p/\mathrm{SNR}\{\hat{h}(\omega)\}$. The "signal to noise ratio" may be expressed also in terms of ν, *the equivalent number of degrees of freedom of $\hat{h}(\omega)$* defined by (6.2.129), Section 6.2.4. For, writing $\hat{h}(\omega) \doteq h(\omega) . (1/\nu) . \chi_\nu^2$ (and recalling that the mean and variance of χ_ν^2 are ν, 2ν, respectively) we have immediately that,

$$\mathrm{SNR}\{\hat{h}(\omega)\} \doteq \left(\frac{\nu}{2}\right)^{1/2}, \tag{7.2.9}$$

or

$$\nu \doteq 2[\mathrm{SNR}\{\hat{h}(\omega)\}]^2. \tag{7.2.10}$$

The use of the SNR as a measure of precision is very closely related to the approach adopted by Blackman and Tukey (1959). They measure the precision of a spectral estimate effectively by the width of the $p\%$ confidence interval for $\log h(\omega)$ computed on the basis of the approximating χ^2 distribution for $\hat{h}(\omega)$. However, the approximating χ^2 distribution takes no account of the bias in $\hat{h}(\omega)$ (cf. Section 6.2.4, equations (6.2.128), (6.2.129)), and the resulting confidence interval for $\log h(\omega)$ depends purely on ν (and the percentage points of the χ^2 distribution), thus making it equivalent to $\mathrm{SNR}\{\hat{h}(\omega)\}$.

A common feature of both the SNR measure and the Blackman–Tukey measure is that they attempt to assess the accuracy of the estimate $\hat{h}(\omega)$ purely in terms of its variance *and ignore the bias*. Blackman and Tukey (1959) appear to justify this step by regarding $\hat{h}(\omega)$ as an estimate of $E\{\hat{h}(\omega)\}$ rather than as an estimate of $h(\omega)$ itself. Now from (6.2.102) we see that $E\{\hat{h}(\omega)\}$ is a weighted integral of $h(\omega)$, and if the problem was really to estimate a weighted integral of this form the basic considerations would be quite different. For, in that case the precise form of the function $W_N(\theta)$ would be determined purely by the *form of the quantity we wish to estimate*, and not by considerations of the statistical properties of the estimate. In any case, we surely would not be interested in estimating simply *any* weighted integral of $h(\omega)$, but if the main interest were the estimation of a function of this type then the most plausible form of $W_N(\theta)$ would be,

$$W_N(\theta) = \begin{cases} 1/2h, & |\theta| \le h, \\ 0, & |\theta| > h, \end{cases}$$

(corresponding to the rectangular window, but now having a fixed width, independent of N). With this choice of $W_N(\theta)$, $E\{\hat{h}(\omega)\}$ has a simple physical interpretation, namely that it represents the power contributed by components with frequencies between $(\omega - h)$ and $(\omega + h)$. However, a general weighted integral of $h(\omega)$ has no simple physical interpretation.

7.3 RESOLVABILITY AND BANDWIDTH

7.3.1 The Role of Spectral Bandwidth

We see from (7.2.4), (7.2.6), that the basic spectral quantity which affects both $\Delta_p(\omega)$ and $\eta(\omega)$ is the expression $\lambda^{(r)}(\omega)$, defined by (7.2.5). In order, therefore, to be able to use either $\Delta_p(\omega)$ or $\eta(\omega)$ we should, ideally, know the value of $\lambda^{(r)}(\omega)$ for all ω. Now $\lambda^{(r)}(\omega)$ is a complicated function of the true spectral density function, $h(\omega)$, and it would be quite unrealistic to suppose that we could have prior knowledge of the precise form of $\lambda^{(r)}(\omega)$ as a function of ω. However, in some situations we can obtain a rough idea of the order of magnitude of $\inf_\omega\{\lambda^{(r)}(\omega)\}$ by relating this quantity to the very important physical notion of "*bandwidth*". Consider, for simplicity, the case where the true spectral density function, $h(\omega)$, is unimodal and contains just one dominant peak at $\omega = \omega_0$, say, as illustrated in Fig. 7.2. If we wish to introduce a measure of the "width" of the peak we may follow the standard engineering practice and define the *spectral bandwidth* B_h as the distance between the *half-power points* ω_1, ω_2, where $\omega_1(<\omega_0)$ and $\omega_2(>\omega_0)$ are defined by $h(\omega_1) = h(\omega_2) = \frac{1}{2}h(\omega_0)$. The points ω_1, ω_2, correspond to the frequencies at which the power density drops to half the value which it

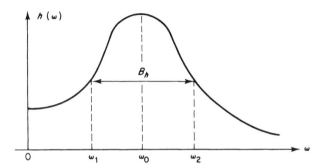

Fig. 7.2. Spectral bandwidth.

attains at the peak frequency, ω_0. Thus, the spectral bandwidth is defined by,

$$B_h = (\omega_2 - \omega_1). \qquad (7.3.1)$$

(Here we have assumed that ω_0 is not near $\pm\pi$. If $h(\omega)$ has a peak near or at $\pm\pi$—as would occur, e.g., in the case of the discrete parameter AR(1) model (3.5.5) with a close to -1 (cf. (4.10.10)), then we may still apply the definition (7.3.1) provided we use the periodic extension of $h(\omega)$ outside the interval $(-\pi, \pi)$.)

Suppose now that B_h is small (relative to ω_0); then we may expand $h(\omega)$ locally about $\omega = \omega_0$ and write,

$$h(\omega) \sim h(\omega_0) + \frac{1}{2}(\omega - \omega_0)^2 h''(\omega_0) + O\{(\omega - \omega_0)^3\},$$

(remembering that $h'(\omega_0) = 0$). When $\omega = \omega_2$, $h(\omega_2) = \frac{1}{2}h(\omega_0)$, and neglecting terms $O\{(\omega - \omega_0)^3\}$ we have, as a rough approximation,

$$\tfrac{1}{2}h(\omega_0) \doteq h(\omega_0) + \tfrac{1}{2}(\omega_2 - \omega_0)^2 h''(\omega_0),$$

which gives,

$$(\omega_2 - \omega_0) \doteq |h(\omega_0)/h''(\omega_0)|^{1/2}, \qquad (7.3.2)$$

with a similar expression for $(\omega_0 - \omega_1)$. Hence, to this order of approximation,

$$B_h \doteq 2|h(\omega_0)/h''(\omega_0)|^{1/2}. \qquad (7.3.3)$$

Now for lag windows with characteristic exponent $r = 2$ we have from

(6.2.122), $h^{(2)}(\omega) = h''(\omega)$, so that,

$$\lambda^{(2)}(\omega) = |h(\omega)/h''(\omega)|^{1/2}. \qquad (7.3.4)$$

We thus see that there is a close relationship between $\lambda^{(2)}(\omega)$ and B_h; in fact, at the peak frequency ω_0,

$$B_h \doteq 2\lambda^{(2)}(\omega_0). \qquad (7.3.5)$$

Also, it is clear that $|h''(\omega)|$ will be large in the neighbourhood of the peak and relatively small elsewhere, so that it is reasonable to regard $\frac{1}{2}B_h$ *as a rough lower bound for* $\lambda^{(2)}(\omega)$.

For windows with characteristic exponents other than 2 there is no simple general relationship between $\lambda^{(r)}(\omega)$ and the "half-power" spectral bandwidth B_h, but Parzen (1957b) has shown that for fairly typically shaped spectral density functions $\lambda^{(r)}(\omega)$ has the dimensions of "bandwidth" and is, in fact, proportional to B_h at a peak. Parzen therefore calls $\lambda^{(r)}(\omega)$ the *spectral bandwidth of order r* (*at frequency* ω), but it would be dangerous to regard $\lambda^{(r)}(\omega)$ as *equivalent* to B_h except in the special case $r = 2$ when we have the simple relationship (7.3.5).

When $h(\omega)$ has a multipeaked form the physical bandwidth would usually be defined as the half-power bandwidth of the narrowest peak. However, the notion of a "well defined peak" does not lend itself to a precise analytical description, and we prefer therefore to define the *overall spectral bandwidth* B_h by,

$$B_h = 2\{\inf_{\omega}|h(\omega)/h''(\omega)|^{1/2}\}, \qquad (7.3.6)$$

so that, by definition, $\frac{1}{2}B_h$ provides a lower bound for $\lambda^{(2)}(\omega)$.

For a typical multipeaked spectrum B_h will be roughly the same as the physical bandwidth, i.e. B_h will correspond roughly to the width of the

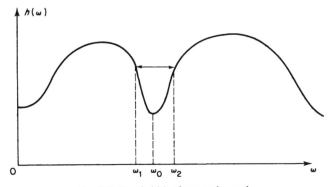

Fig. 7.3. Bandwidth of spectral trough.

narrowest peak. If, however, in an exceptional case $h(\omega)$ contains a sharp "trough" between two fairly broad peaks, and if the trough is sufficiently narrow so that it represents the "sharpest feature" in $h(\omega)$, then B_h, as defined by (7.3.6), will now correspond to the width of the trough. To see this, let us use the form of $h(\omega)$ illustrated in Fig. (7.3). The bandwidth of the trough may be measured as the distance between the "*double-power*" *points*, i.e. if ω_0 denotes the frequency at which the trough attains its minimum, and ω_1, ω_2 are such that $h(\omega_1) = h(\omega_2) = 2h(\omega_0)$, then the bandwidth is $|\omega_2 - \omega_1|$. If $|\omega_2 - \omega_1|$ is sufficiently small we can use exactly the same analysis as that leading to (7.3.2) and we find immediately that

$$(\omega_2 - \omega_0) \doteq \{h(\omega_0)/h''(\omega_0)\}^{1/2}.$$

Hence, in this case the value of $\inf_\omega |h(\omega)/h''(\omega)|^{1/2}$ would be dominated by the behaviour of $h(\omega)$ in the neighbourhood of the trough.

Relationship between spectral bandwidth and the rate of decay
of the autocovariance function

In Section 4.10 we derived the spectral density function of a continuous parameter AR(1) process and noted, in passing, that the "narrower" the autocorrelation (or autocovariance) function, the "wider" the spectral density function, and vice versa. Although we demonstrated this only for the AR(1) case we noted further that this feature would hold for general forms of spectral density functions since it follows as a consequence of a basic mathematical result that we cannot make a function and its Fourier transform both arbitrarily narrow. We may therefore expect that *the spectral bandwidth B_h will be "small" if the autocovariance function $R(s)$ decays "slowly", whereas B_h will be "large" if $R(s)$ decays "quickly"*. As a simple illustration let us consider the two extreme cases: (1) a purely random (i.e. "white noise") process, and (2) a harmonic process with a single periodic component. In (1), $R(s)$ decays immediately to zero and $h(\omega)$, being a constant, may be said to have infinite bandwidth. In (2), $R(s)$ is periodic and so never decays, while $h(\omega)$ has the form of a δ-function and so has zero bandwidth.

Thus, *in principle, information on the rate of decay of $R(s)$ (or $\rho(s)$) is equivalent to information on the spectral bandwidth*. Various measures of the rate of decay of $R(s)$ have been suggested but it is difficult to obtain precise relationships between these and B_h. However, we may note that

$$h''(\omega) = -\frac{1}{2\pi} \sum_{s=-\infty}^{\infty} s^2 R(s) e^{-i\omega s},$$

so that

$$|h''(\omega)| \leq \frac{1}{2\pi} \left\{ \sum_{s=-\infty}^{\infty} s^2 |R(s)| \right\}. \qquad (7.3.7)$$

Now $\{\sum_{s=-\infty}^{\infty} s^2 |R(s)|\}^{1/2}$ is analogous to the "standard deviation" of the function $|R(s)|$, and hence if $R(s)$ decays quickly $|h''(\omega)|$ will be small, and consequently B_h will be large. In the AR(1) case we can obtain a more precise result as follows. Consider, e.g., the discrete parameter model, $X_t - aX_{t-1} = \varepsilon_t$ ($|a| < 1$), for which $R(s)$ is exponentially damped and has the form,

$$R(s) = \frac{\sigma_\varepsilon^2}{(1-a^2)} \cdot a^{|s|}$$

(cf. (3.5.16)). The rate of decay of $R(s)$ is here determined by the magnitude of a; a small value of a corresponds to a high rate of decay while $|a| \sim 1$ corresponds to a slow rate of decay. The corresponding spectral density function is given by (cf. (4.10.10)),

$$h(\omega) = \frac{\sigma_X^2}{2\pi} \cdot \frac{(1-a^2)}{(1 - 2a \cos \omega + a^2)}.$$

If $a > 0$, $h(\omega)$ has a single peak at $\omega = 0$ and setting $h(\omega_0) = \frac{1}{2}h(0)$ and assuming ω_0 is small we find that approximately,

$$\omega_0 \doteq (1-a)/\sqrt{a}$$

so that

$$B_h \doteq 2\omega_0 \doteq 2[(1/\sqrt{a}) - \sqrt{a}]. \qquad (7.3.8)$$

7.3.2 The Role of Window Bandwidth

We know from the results of Section 6.2.4 that, for scale parameter windows, var$\{\hat{h}(\omega)\} = O(M/N)$, and bias $\{\hat{h}(\omega)\} = O(1/M')$ (cf. (6.2.113), (6.2.124)). Hence, in general, the effect of increasing the parameter M is to increase the variance and decrease the bias, while decreasing M decreases the variances and increases the bias, i.e.,

as $M\uparrow$, variance $\uparrow$, bias $\downarrow$,

as $M\downarrow$, variance $\downarrow$, bias $\uparrow$.

To be specific, let us consider the Daniell window ((6.2.72)) for which $W_N(\theta)$ has the rectangular form,

$$W_N(\theta) = \begin{cases} M/2\pi, & |\theta| \leq \pi/M, \\ 0, & |\theta| > \pi/M. \end{cases} \qquad (7.3.9)$$

With this window we have (Table 6.1, Section 6.2.4)

$$v^2(\omega) = \text{var}\{\hat{h}(\omega)\} \sim \frac{M}{N} h^2(\omega), \tag{7.3.10}$$

$$b(\omega) = \text{bias}\{\hat{h}(\omega)\} \sim \frac{\pi^2}{6M^2} h''(\omega). \tag{7.3.11}$$

Suppose now that we have an unlimited amount of data available so that the values of both M and N are at our disposal, and that we decide to use $\Delta_p(\omega)$ as our measure of precision. Given the value of $\lambda^{(2)}(\omega)$, equation (7.2.4) contains two unknowns, N and M, and if we were interested only in estimating $h(\omega)$ *at one particular frequency* ω, we could choose N and M in any way so long as (7.2.4) is satisfied, i.e. *we could choose any value of M we like* and, provided there is a corresponding value of N which satisfies (7.2.4), the resulting estimate will always have the required precision. This apparent paradox is due to the fact that we are considering the statistical properties of $\hat{h}(\omega)$ only at a particular frequency. However, in practice we are hardly ever interested in estimating just one value of $h(\omega)$; in general we would wish to estimate $h(\omega)$ over a range of frequencies, i.e. we would wish to estimate the *overall shape* of $h(\omega)$, and now we have to introduce the concepts of "resolvability" which soon reveals the crucial importance of the parameter M.

"Resolvability" is a well known term in optics and other branches of physics, and it is used, e.g., to describe the degree of "fine detail" which an optical device (such as telescope) can reproduce. Its use in the context of spectral analysis has a similar purpose, and can be illustrated as follows. Suppose that the true spectral density function $h(\omega)$ has two distinct peaks at $\omega = \omega_1, \omega = \omega_2$, say. If the estimate $\hat{h}(\omega)$ is to reproduce accurately the shape of $h(\omega)$, then $\hat{h}(\omega)$ should also have two distinct peaks at ω_1 and ω_2. Now from (6.2.102) we have,

$$E\{\hat{h}(\omega)\} \sim \int_{-\pi}^{\pi} h(\theta) W_N(\omega - \theta) \, d\theta \tag{7.3.12}$$

$$= \frac{M}{2\pi} \int_{\omega-(\pi/M)}^{\omega+(\pi/M)} h(\theta) \, d\theta, \tag{7.3.13}$$

when $W_N(\theta)$ is given by (7.3.9). Thus, $E\{\hat{h}(\omega)\}$ is the average value of $h(\theta)$ over an interval of width $2\pi/M$ centred on ω, and if M is small enough for $2\pi/M > |\omega_2 - \omega_1|$ then the two peaks at ω_1 and ω_2 will be "merged together" in $\hat{h}(\omega)$; see Fig. 7.4. In order to separate or "*resolve*" the values of $h(\omega)$ at $\omega = \omega_1, \omega = \omega_2$, we must clearly choose M sufficiently large so that *the width of $W_N(\theta)$ is not greater than the distance between* ω_1 *and* ω_2, i.e. so that,

$$2\pi/M \leq |\omega_2 - \omega_1|. \tag{7.3.14}$$

Another way of looking at it is to say that in order to resolve the values of $h(\omega)$ at ω_1 and ω_2 we want the estimates $\hat{h}(\omega_1)$ and $\hat{h}(\omega_2)$ to be approximately *uncorrelated*, so that, e.g., $\hat{h}(\omega_1)$ is not unduly influenced by the value of $\hat{h}(\omega_2)$. We know from the discussion following equation (6.2.112) that, in

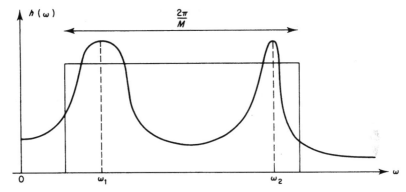

Fig. 7.4. Effect of window width.

the case of the Daniell window, $\hat{h}(\omega_1)$ and $\hat{h}(\omega_2)$ will be approximately uncorrelated provided $|\omega_2 - \omega_1| \geqslant 2\pi/M$, which leads to the same condition as (7.3.14).

Now in addition to distinguishing the peaks at ω_1, ω_2 we may also want $\hat{h}(\omega)$ to reproduce, as far as possible, the correct overall shape of $h(\omega)$. This means, e.g., that, for each peak, we want to be able to resolve the values of $h(\omega)$ at, say, each of the two *half-power points*, the distance between which is (by definition) the bandwidth of the peak; see Fig. 7.5. To reproduce each of the peaks (and troughs) in $h(\omega)$ we must therefore choose M *so that the width*

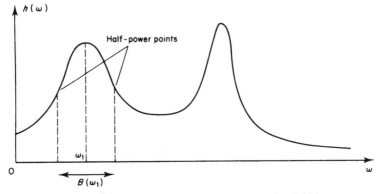

Fig. 7.5. Width of $W_N(\theta)$ in relation to spectral bandwidth.

of $W_N(\theta)$ is not greater than the bandwidth of the narrowest peak (or trough) in $h(\omega)$, or more precisely, so that

$$\text{width of } W_N(\theta) \leq B_h, \tag{7.3.15}$$

where B_h is the overall spectral bandwidth defined by (7.3.6). In practice it would be advisable to leave some margin of safety and choose the width of $W_N(\theta)$ to be somewhat smaller than B_h. However, if we try to make the width of $W_N(\theta)$ very small we will have to make M very large, which in turn means that N, the number of observations, will also have to be very large in order to satisfy (7.2.4). We would suggest, therefore, that as a general practical procedure we should make,

$$\text{width of } W_N(\theta) = \tfrac{1}{2} B_h. \tag{7.3.16}$$

This choice agrees very closely with that by Parzen (1957b) (although Parzen's approach is based on slightly different considerations), and the use of (7.3.16) (as opposed to treating (7.3.15) as an equality) makes it rather safer to assume that estimates whose frequencies are separated by an interval B_h are approximately uncorrelated even when windows other than the rectangular Daniell one are used.

So far, our discussion of "resolvability" has been centred on the Daniell window but it is fairly obvious that similar considerations apply to all the standard spectral windows. However, the Daniell window has a simple rectangular form and it is clear what we mean by the "width" of this function. The definition of "width", or to be more precise, "*bandwidth*", for windows with more general shapes is not so clear, but we clearly need such a definition before we can apply (7.3.16) to windows of general form. We now discuss a number of different definitions of window bandwidth which have been suggested.

Various definitions of window bandwidth

(1) Half-power points definition

Assume that $W_N(\theta)$ is an even function of θ attaining its maximum at $\theta = 0$. Then the "half-power" bandwidth of $W_N(\theta)$ is defined as the distance between the "half-power" points on the main lobe $W_N(\theta)$, i.e. the "half-power" bandwidth, B_{HP}, is defined by

$$B_{HP} = 2\theta_1, \tag{7.3.17}$$

where θ_1 is such that $W_N(\pm\theta_1) = \tfrac{1}{2} W_N(0)$. If θ_2 is the first zero of $W_N(\theta)$, then linear interpolation between 0 and θ_2 gives, as a rough approximation, $\theta_1 \doteq \tfrac{1}{2}\theta_2$, so that $B_{HP} \doteq \theta_2$, see Fig. 7.6.

(2) *Parzen's definition*

Parzen's (1957b) measure of window bandwidth, B_p, is defined as the width of the rectangular window which has the same area as $W_N(\theta)$ (viz., unity) and the same height as $W_N(\theta)$ at $\theta = 0$. Thus,

$$B_p = 1/W_N(0). \tag{7.3.18}$$

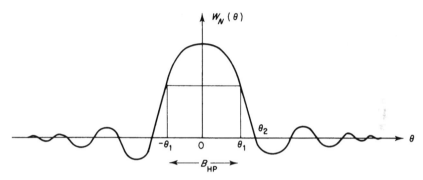

Fig. 7.6. Half-power points bandwith.

(3) *Jenkins' definition*

Jenkins (1961) defines the bandwidth as the width of that rectangular window which produces an estimate with the same asymptotic variance as that corresponding to $W_N(\theta)$. Now the asymptotic variance corresponding to $W_N(\theta)$ is

$$\frac{1}{N}\left\{\sum_{s=-(N-1)}^{(N-1)} \lambda_N^2(s)\right\}h^2(\omega),$$

while that corresponding to a rectangular window on the interval $(-\pi/M, \pi/M)$ is $(M/N)h^2(\omega) = (2\pi/BN)h^2(\omega)$, where $B = 2\pi/M$ is the width of the rectangular window. Hence, Jenkins' bandwidth, B_J, is defined by

$$\frac{1}{N}\left\{\sum_{s=-(N-1)}^{(N-1)} \lambda_N(s)\right\} = \frac{2\pi}{B_J N}, \tag{7.3.19}$$

i.e.

$$B_J = 2\pi\Big/\left\{\sum_s \lambda_N^2(s)\right\} \sim \frac{2\pi}{M\{\int_{-\infty}^{\infty} k^2(u)\,du\}}, \tag{7.3.20}$$

when $\lambda_N(s) = k(s/M)$ is of scale parameter form.

(4) Grenander's definition

Grenander (1951) defined the bandwidth B_G by regarding $W_N(\theta)$ as a probability distribution and then evaluating its "standard deviation", i.e. B_G is defined by,

$$B_G = \left\{ \int_{-\pi}^{\pi} \theta^2 W_N(\theta) \, d\theta \right\}^{1/2}. \qquad (7.3.21)$$

Criticisms of the above definitions

The above definitions are all reasonable interpretations of the intuitive notion of "width". The half-power points definition follows the analogy with the spectral bandwidth, and both Jenkins' and Parzen's definitions derive from an attempt to "force" a general window $W_N(\theta)$ into an equivalent rectangular form for which the width is uniquely determined. However, the definitions of both B_{HP} and B_p may be criticized on the grounds that they characterize only one particular geometrical feature of the window. B_{HP} does not take account of the shape of $W_N(\theta)$ outside the interval $(-\theta_1, \theta_1)$, while B_p does not take any account of the shape of $W_N(\theta)$ except at the single point $\theta = 0$. On the other hand, Jenkins' definition B_J does take account of the complete shape of $W_N(\theta)$ but ties down the bandwidth to the *variance* properties of the estimate, so that both the bandwidth and the variance then depend on the same functional of $W_N(\theta)$. Grenander's definition B_G again takes account of the complete window shape but, as we shall see later, is based on a measure which is not entirely appropriate for functions defined only on $(-\pi, \pi)$.

Of course, which definition we regard as the most appropriate depends on how we wish to use it. If we require a definition of bandwidth so as to make use of (7.3.16), i.e. so as to choose M so that $W_N(\theta)$ is sufficiently "narrow" relative to the spectral bandwidth B_h to ensure that the estimate $\hat{h}(\omega)$ does not "blur" the shape of $h(\omega)$, then, in effect, we are trying to control the *degree of smoothing* produced by $W_N(\theta)$ so that

$$E[\hat{h}(\omega)] \sim \int_{-\pi}^{\pi} h(\theta) W_N(\omega - \theta) \, d\theta$$

does not differ too much from $h(\omega)$. This means that we are trying to control the *bias* (in the neighbourhood of peaks in $h(\omega)$). Consequently, we want to define the bandwidth of $W_N(\theta)$ in such a way that it is directly related to the *bias* (rather than the variance) of $h(\omega)$. These considerations suggest a new definition of window bandwidth, as follows.

(5) A new definition

Brillinger (1975) defines the bandwidth of a general kernel, and in the context of spectral windows Brillinger's definition takes the form,

$$B_B = \left\{ \int_{-\pi}^{\pi} (1 - \cos \theta) W_N(\theta) \, d\theta \right\}^{1/2} . \qquad (7.3.22)$$

(Treating $W_N(\theta)$ as a "circular" probability distribution of a random variable X defined in $(-\pi, \pi)$, B_B can be expressed as $B_B = 1 - E(\cos X)$ which is a natural definition of the "variance" of a circular distribution.) Since

$$\lambda_N(s) = \int_{-\pi}^{\pi} e^{i\theta s} W_N(\theta) \, d\theta = \int_{-\pi}^{\pi} \cos s\theta W_N(\theta) \, d\theta,$$

($W_N(\theta)$ being a real even function), we may write,

$$B_B^2 = \lambda_N(0) - \lambda_N(1).$$

If $W_N(\theta)$ is of scale parameter form, i.e. if $\lambda_N(s) = k(s/M)$, then,

$$B_B^2 = k(0) - k(1/M).$$

Now if $k(u)$ has characteristic exponent r then for small u,

$$k(u) \sim k(0) - k^{(r)}|u|^r,$$

and hence,

$$k(0) - k(1/M) \sim k^{(r)}(1/M)^r.$$

Thus,

$$B_B \sim \{(1/M^r)k^{(r)}\}^{1/2}. \qquad (7.2.23)$$

This result shows that B_B is close to the type of measure we want since it depends essentially on $k^{(r)}$, and $k^{(r)}$ is just that functional of $k(u)$ which enters into the expression for the asymptotic bias of $\hat{h}(\omega)$; cf. (6.2.124). Unfortunately, however, we see from (7.2.23) that B_B does not have the correct dimensions for a "bandwidth", as may be seen alternatively by noting that if r is even, say $r = 2s$, then

$$k^{(2s)} = -\frac{1}{(2s)!} \left[\frac{d^{2s}\{k(u)\}}{du^{2s}} \right]_{u=0} = \frac{-(i)^{2s}}{(2s)!} \int_{-\infty}^{\infty} \theta^{2s} K(\theta) \, d\theta.$$

Hence, B_B has the dimension of θ^s, and the correct dimension for a bandwidth measurement is obtained by taking the sth root of B_B. We now propose the following definition for the bandwidth of a scale parameter window.

Definition 7.3.1 *Let $W_N(\theta)$ be a scale parameter spectral window whose corresponding lag window has characteristic exponent r, and let $k^{(r)}$ be defined by (6.2.121). We define the bandwidth of $W_N(\theta)$ by,*

$$B_W = C\left\{\frac{1}{M^r}k^{(r)}\right\}^{1/r} \tag{7.3.24}$$

where C is a constant to be determined.

We now determine C by requiring that for the rectangular window on $(-\pi/M, \pi/M)$ B_W should be exactly equal to the width of this window, $2\pi/M$. (The bandwidth of the rectangular window on $(-\pi/M, \pi/M)$ is uniquely given as $2\pi/M$ according to the half-power points, Parzen, or Jenkins definitions, and any reasonable definition of bandwidth must give the value $2\pi/M$ for this window.) Alternatively, we can say that C is chosen so that for scale parameter windows $W_N(\theta)$ with $r = 2$, B_W is the width of the rectangular window which produces the same bias as $W_N(\theta)$. (The rectangular window corresponds to $r = 2$ and hence the bias of the rectangular window, being $O(1/M^2)$, can be compared only with other windows for which $r = 2$ also.) Now for the rectangular window, $r = 2$ and $k^{(2)} = \pi^2/6$. Hence, for this window,

$$\frac{C\{k^{(2)}\}^{1/2}}{M} = \frac{C}{M}\cdot\frac{\pi}{\sqrt{6}}.$$

We require this to be $2\pi/M$ and hence we set, $C = 2\sqrt{6}$. We may now define the bandwidth of $W_N(\theta)$ explicitly by the following.

Definition 7.3.2

$$B_W = \frac{2\sqrt{6}}{M}\{k^{(r)}\}^{1/r}. \tag{7.3.25}$$

(When $r = 2$, (7.3.25) coincides with an alternative definition of bandwidth given by Parzen (1961b). However, Parzen's alternative definition was given only for windows with $r = 2$.) Using (6.2.124) we may now express the bias in terms of B_W thus,

$$\text{bias}\{\hat{h}(\omega)\} \sim -\frac{1}{M^r}k^{(r)}h^{(r)}(\omega) = -\left\{\frac{B_W}{2\sqrt{6}}\right\}^r h^{(r)}(\omega). \tag{7.3.26}$$

In particular, when $r = 2$, $h^{(2)}(\omega) = -h''(\omega)$, and

$$\text{bias}\{\hat{h}(\omega)\} \sim (B_W^2/24)h''(\omega). \tag{7.3.27}$$

In the neighbourhood of the narrowest peak in $h(\omega)$, $2|h(\omega)/h''(\omega)|^{1/2} \sim B_h$

(the spectral bandwidth), and hence in this region,

$$|\text{bias}\{\hat{h}(\omega)\}| \sim \frac{1}{6}\left\{\frac{B_W}{B_h}\right\}^2 h(\omega). \tag{7.3.28}$$

This result shows very clearly how the relative magnitude of B_W to B_h controls the bias; the smaller the ratio, B_W/B_h, the smaller the bias.

CASE $r = 2$

Most of the standard windows (other than the Bartlett window) fall in this category. In this case B_W and B_B are identical (apart from the factor $2\sqrt{6}$) and also,

$$\begin{aligned}
B_B &\sim \frac{1}{M}\{k^{(2)}\}^{1/2} = -\frac{1}{2M}\{k''(0)\}^{1/2} = \frac{1}{2M}\left\{\int_{-\infty}^{\infty} \theta^2 K(\theta)\, d\theta\right\}^{1/2} \\
&\sim \frac{1}{2M}\left\{\int_{-M\pi}^{M\pi} \theta^2 \cdot \frac{1}{M} \cdot W_N\left(\frac{\theta}{M}\right) d\theta\right\}^{1/2} \\
&= \frac{1}{2}\left\{\int_{-\pi}^{\pi} \theta^2 W_N(\theta)\, d\theta\right\}^{1/2}, \tag{7.3.29}
\end{aligned}$$

which is $\frac{1}{2} \times B_G$, where B_G is Grenander's definition of bandwidth. However, this definition is now seen to be appropriate only for windows with $r = 2$, i.e. for windows such that $\int_{-\infty}^{\infty} \theta^2 K(\theta)\, d\theta < \infty$ (which rules out, e.g., the Bartlett window).

Returning to the general case we have,

$$B_W = \frac{2\sqrt{6}}{M}\{k^{(r)}\}^{1/r} = O(1/M),$$

while (for $\omega \neq 0, \pi$),

$$\text{var}\{\hat{h}(\omega)\} \sim \frac{M}{N}h^2(\omega)\left\{\int_{-\infty}^{\infty} k^2(u)\, du\right\} = O(M).$$

Hence, we find that,

$$\text{var}\{\hat{h}(\omega)\} \times B_W = \left(\frac{2\sqrt{6}}{N}\right)\{k^{(r)}\}^{1/r}\left\{\int_{-\infty}^{\infty} k^2(u)\, du\right\} = \text{constant},$$

independent of M, i.e. we have,

$$\text{variance} \times \text{bandwidth} = \text{constant}. \tag{7.3.30}$$

The basic result was first derived by Grenander (1951) using his own definition (B_G) of bandwidth. It is easy to see that the same result holds for Jenkins' definition (this follows directly from the definition of B_J), and also for Parzen's definition, since we have,

$$B_p = \frac{1}{W_N(0)} \sim \frac{1}{MK(0)} = O(1/M).$$

Bandwidth of general windows

If $W_N(\theta)$ is not of scale parameter form we need something to replace r, the characteristic component, since this parameter is defined only for scale parameter windows. Nevertheless, we need some way of determining the power of $\{\int_{-\pi}^{\pi} (1 - \cos \theta) W_N(\theta) \, d\theta\}$ which gives the correct dimensions for bandwidth. Now we certainly wish to preserve the basic relationship,

variance × bandwith = constant.

Hence, we now define r to be such that

$$\lim_{N \to \infty} \left\{ \int_{-\pi}^{\pi} W_N^2(\theta) \, d\theta \right\} \left\{ \int_{-\pi}^{\pi} (1 - \cos \theta) W_N(\theta) \, d\theta \right\}^{1/r} \qquad (7.3.31)$$

exists and is non-zero. (If no finite value of r satisfies this condition we take $r = \infty -$ as for the truncated periodogram window.) The bandwidth of $W_N(\theta)$ is now defined by,

$$B_W = 2\sqrt{6} \left\{ \int_{-\pi}^{\pi} (1 - \cos \theta) W_N(\theta) \, d\theta \right\}^{1/r} \qquad (7.3.32)$$

If, in particular, $W_N(\theta)$ is of scale parameter form, then,

$$\int_{-\pi}^{\pi} W_N^2(\theta) \, d\theta \sim M^2 \int_{-\pi}^{\pi} K^2(M\theta) \, d\theta \sim M \int_{-\infty}^{\infty} K^2(\theta) \, d\theta = O(M),$$

and

$$\int_{-\pi}^{\pi} (1 - \cos \theta) W_N(\theta) \, d\theta \sim \frac{1}{M^r} k^{(r)},$$

so that in this case the value of r determined by (7.3.31) is exactly the same as the characteristic exponent of the window. Note that $\int_{-\pi}^{\pi} (1 - \cos \theta) W_N(\theta) \, d\theta \to 0$ as $N \to \infty$ since $W_N(\theta) \to \delta(\theta)$ as $N \to \infty$. On the other hand, $\int_{-\pi}^{\pi} W_N^2(\theta) \, d\theta \to \infty$ as $N \to \infty$ by property (vi) of Section 6.2.4.

Comparison of bandwidths

We show in Table 7.1 expressions for the bandwidths of some of the standard windows according to the various definitions discussed above. The value of B_W is easily obtained from a knowledge of r and $k^{(r)}$, using (7.3.26).

Table 7.1. *Bandwidths of standard windows.*

Window	r	$k^{(r)}$	B_W	B_{HP} (approx.)	B_P	B_J
Bartlett	1	1	$4\cdot9/M$	$2\pi/M$†	$2\pi/M$	$3\pi/M$
Daniell (rectangular)	2	$\pi^2/6$	$2\pi/M$	$2\pi/M$	$2\pi/M$	$2\pi/M$
Parzen	2	6	$12/M$	$4\pi/M$†	$8\pi/3M$	$3\cdot72\pi/M$
Tukey–Hanning	2	$\pi^2/4$	$2\cdot45\pi/M$	$2\pi/M$	$2\pi/M$	$8\pi/3M$
Tukey–Hamming	2	$0\cdot23\pi^2$	$2\cdot35\pi/M$	$2\pi/M$	$2\pi/M$	$2\cdot52\pi/M$
Bartlett–Priestley	2	$\pi^2/10$	$1\cdot55\pi/M$	$1\cdot41\pi/M$	$4\pi/3M$	$5\pi/3M$

†Note: These half power points were computed by linear interpolation between
$\theta = 0$ and $\theta = 2\pi/M$ (Bartlett), and between $\theta = 0$ and $\theta = 4\pi/M$ (Parzen).
More accurate half-power bandwidths for these windows are; Bartlett:
$5\cdot56/M$, Parzen: $8/M$.

We see from the above table that all the four definitions of bandwidth lead
to comparable expressions for the standard windows. Moreover, even if the
spectral bandwidth B_h is known exactly, the choice of the bandwidth of
$W_N(\theta)$ in relation to B_h is to some extent arbitrary, except for the main
condition that the bandwidth of $W_N(\theta)$ must not be greater than B_h. We may
therefore conclude that the particular definition of bandwidth which we
adopt is relatively unimportant, but in view of our observations on the
relative merits of the various definitions we would naturally prefer the
definition B_W, as given by (7.3.24). *We propose, therefore, to adopt B_W as our
standard measure of window bandwidth*, and we shall use this quantity as the
basis of the "design relations" for spectral estimation which we discuss in the
next section.

Grenander's uncertainty principle

The relationship (7.3.30) tells us that for fixed N we cannot obtain
simultaneously arbitrarily low values for both variance and bandwidth.
Thus, if we choose M small so as to make the variance small we must be
prepared to accept a large bandwidth; on the other hand, if we make M large
so as to give us a small bandwidth we must accept a large variance.
Grenander (1951, 1958) noted the close similarity between the nature of
this result and the famous Heisenberg *"uncertainty principle"* in quantum
mechanics. Relating bandwidth to "resolvability" and variance to "reliability" he then formulated his uncertainty principle of spectral estimation
in the form; *"reliability and resolvability are antagonistic".*

As remarked above, Grenander first proved (7.3.30) using his own
definition (B_G) of window bandwidth, and although there is no doubt about
the mathematical validity of this result (which, as we have seen, holds

equally well with the B_J, B_P, and B_W definitions), Lomnicki and Zaremba (1957a, 1959b) have disputed the interpretation which Grenander gave. Lomnicki and Zaremba argue that, in a more precise sense, "resolvability" is the ability to distinguish between the values of $h(\omega)$ at neighbouring frequencies, and, as such, it depends not only on the bias but also on the variance. Similarly, they argue that the "reliability" of an estimate depends not only on its variance but also on its bias. They prefer, therefore, to define "resolvability" in terms of the accuracy with which one is able to estimate the derivative $h'(\omega)$ and they show that, according to their own criterion of integrated mean square error, the "optimal" form of $\lambda_N(s)$ for estimating $h(\omega)$ (namely, (6.2.88)) is "optimal" also for estimating $h'(\omega)$. They thus conclude that resolvability and reliability are not antagonistic, but on the contrary, that estimate which has maximum reliability also has maximum resolvability.

This apparent contradiction arises, of course, from the use of different interpretations of the terms "resolvability" and "reliability". Although there is clearly some justification for the point of view adopted by Lomnicki and Zaremba there is no doubt about the implication of (7.3.30), namely, that "*bandwidth and variance are antagonistic*". Using either B_G or B_W as measures of bandwidth this then implies that "*bias and variance are antagonistic*", and further elaboration on this point seems irrelevant and merely complicates the issue.

7.4 DESIGN RELATIONS FOR SPECTRAL ESTIMATION: CHOICE OF WINDOW PARAMETERS, RECORD LENGTH AND FREQUENCY INTERVAL

We now return to the questions posed at the beginning of this chapter concerning the choice of the parameters involved in the computation of spectral estimates. We will consider separately the cases of (a) unlimited record length, and (b) fixed record length. In case (a) the parameters to be determined are N (the total number of observations), M and $\Delta\omega$ (the frequency interval between adjacent estimates); in case (b) N is fixed and only M and $\Delta\omega$ have to be determined. We treat first the case of discrete parameter processes, and later extend the discussion to data sampled from continuous parameter processes. (In the latter case there is, of course, a further parameter to be selected, namely Δt, the time sampling interval.).

I. *Unlimited record length*

Here, the values of both N and M are at our disposal, and we can deal with the problem in two stages. The basic strategy is first to select M so that the estimate has a prescribed degree of *resolution*, and then determine N so that, according to one of the measures discussed in Section 7.2, the estimate has a prescribed degree of *precision*.

Let us consider first the question of resolution. The ability of the estimate to reveal the "fine detail" of the true spectral density function $h(\omega)$ depends on the magnitude of the window bandwidth B_W in relation to what might be termed the *"critical bandwidth"* β_h of $h(\omega)$. Following the discussion in Section 7.3.2 we may distinguish between the following cases.

(i) If we wish simply to be able to *separate* two peaks in $h(\omega)$ at frequencies ω_1, ω_2, then the critical bandwidth is $\beta_h = |\omega_2 - \omega_1|$. More generally, if $h(\omega)$ is known to have several peaks, say at frequencies ω_1, ω_2, $\omega_3, \ldots$, then β_h would be the smallest interval between adjacent peaks, i.e. $\beta_h = \min_i |\omega_{i+1} - \omega_i|$.

(ii) If, as is usually the case, $h(\omega)$ has a number of peaks (and troughs) and we require the estimate $\hat{h}(\omega)$ to reproduce fairly accurately the *overall shape* of $h(\omega)$, then the critical bandwidth $\beta_h = B_h$, where B_h is the overall spectral bandwidth, as defined by (7.3.6). (As remarked in Section 7.3.1, B_h would usually be determined by the width of the narrowest peak in $h(\omega)$, but in exceptional cases B_h may be determined rather by the width of a sharp trough.)

Suppose now that we have sufficient prior information about the shape of $h(\omega)$ to enable us to obtain an approximate value for the critical bandwidth β_h as described above. Then, following the approach discussed in Section 7.3.2, we set the required degree of resolution by making the window bandwidth B_W a specified fraction of β_h, i.e. we choose M so that,

$$B_W = \alpha \cdot \beta_h,$$

where α is a specified constant lying in the range $0 < \alpha < 1$. In most practical situations we would usually set $\alpha = \frac{1}{2}$ (cf. the discussion leading to (7.3.16)), i.e. we would choose the window bandwidth to be half the critical bandwidth, in which case the above equation becomes,

$$B_W = \tfrac{1}{2}\beta_h. \tag{7.4.1}$$

However, if we were particularly concerned with estimating the shape of $h(\omega)$ in the region of a peak and wish to achieve a high degree of resolution we might set $\alpha = \frac{1}{3}$ or even $\alpha = \frac{1}{4}$. Once β_h is specified, the value of M corresponding to (7.4.1) is easily found from Table 7.1. For example, we have:

$$\text{Daniell window} \qquad M = 4\pi/\beta_h,$$

$$\text{Parzen window} \qquad M = 24/\beta_h, \tag{7.4.2}$$

$$\text{Tukey–Hanning window} \qquad M = 4{\cdot}9\pi/\beta_h.$$

Note that the smaller the value of β_h, the larger the corresponding value of M. Also, for given β_h, the Parzen window requires a larger value of M than either the Daniell or the Tukey–Hanning windows.

Having chosen M as above, we now determine N so that $\hat{h}(\omega)$ attains a specified degree of precision. In Section 7.2 we discussed a number of measures of precision, but for our present purposes the most natural measure would seem to be $\Delta_p(\omega)$, defined by (7.2.1), and evaluated in (7.2.4). For a given choice of window (i.e. for a given form of $k(u)$), and fixed values of $\Delta_p(\omega)$, γ_p, and M, (7.2.4) determines N as a function of $\lambda^{(r)}(\omega)$. Thus, according to (7.2.4) the appropriate value of N would vary with frequency (since $\lambda^{(r)}(\omega)$ varies with ω), but this is of little practical significance since we would hardly ever know $\lambda^{(r)}(\omega)$ as a function of ω, and in any case, if we wish to estimate the complete form of $h(\omega)$ the total record length required is determined by the number of observations needed to estimate $h(\omega)$ (with the specified precision) at that value of ω which requires the largest number of observations. Hence, if we replace $\lambda^{(r)}(\omega)$ in (7.2.4) by its lower bound, *the value of N so obtained will provide the required degree of precision over the most "difficult" portions of the spectrum* (typically, the region where $h(\omega)$ has its narrowest peak), and will provide a higher degree of precision at all other frequencies (except, perhaps, for the isolated points, $\omega = 0$, $\pm \pi$, where the variance is doubled). In particular, for windows with $r = 2$, the lower bound for $\lambda^{(2)}(\omega)$ is $\frac{1}{2}B_h$, B_h being the spectral bandwidth as defined by (7.3.6).

For a general scale parameter window $W_N(\theta)$ with characteristic exponent r, write

$$I_W = \int_{-\infty}^{\infty} k^2(u)\, du, \qquad (7.4.3)$$

$$B'_W = 2\sqrt{6}\{k^{(r)}\}^{1/r}. \qquad (7.4.4)$$

(From (7.3.25) it will be seen that B'_W is the value which B_W takes when we set $M = 1$; B'_W may be called the "standardized" window bandwidth.) Now let us consider windows with $r = 2$. From (7.2.4), $\Delta_p(\omega)$ may be written as,

$$\gamma_p \sqrt{\left(\frac{I_W M}{N}\right) + \frac{1}{M^2}\left\{\frac{B'_W}{2\sqrt{6}\lambda^{(2)}(\omega)}\right\}^2} = \Delta_p(\omega). \qquad (7.4.5)$$

Replacing $\lambda^{(2)}(\omega)$ by its lower bound, $\frac{1}{2}B_h$, we now obtain,

$$\gamma_p \sqrt{\left(\frac{I_W M}{N}\right) + \frac{1}{6M^2}\left\{\frac{B'_W}{B_h}\right\}^2} = \Delta_p(\omega). \qquad (7.4.6)$$

For example, suppose we decide to use the Daniell window (7.3.9), and choose $\Delta_p(\omega) = 0\cdot25$, with $p = 0\cdot95$, i.e. we require that the estimate should have not more than 25% proportional error at the 95% confidence level. Then $\gamma_p = 1\cdot96$, and from Tables 6.1 and 7.1 we find that for this window, $I_W = 1$, $B'_W = 2\pi$. Hence, (7.4.6) gives,

$$1\cdot96\sqrt{\left(\frac{M}{N}\right) + \frac{2\pi^2}{3M^2}\cdot\frac{1}{B_h^2}} = 0\cdot25. \tag{7.4.7}$$

With the value of M determined by (7.4.2), (7.4.7) now determines the value of N required for an estimate based on the Daniell window to have the required precision.

However, it is important to note that *this procedure is valid only if the second term in (7.4.6) (which represents the absolute value of the relative bias) is small compared with the chosen value of* $\Delta_p(\omega)$. If β_h is fairly large relative to B_h, the value of M given by (7.4.2) will be small, and the second term in (7.4.6) may well by itself exceed the value of $\Delta_p(\omega)$, in which case there will be no value of N satisfying (7.4.6). (Even with $\beta_h = B_h$, and B_W chosen according to (7.4.1), the second term in (7.4.6) is $\frac{1}{24} = 0\cdot042$, (cf. (7.3.29))). However, the essential point is that if the second term in (7.4.6) is only slightly less than $\Delta_p(\omega)$, the solution of (7.4.6) could lead to an unnecessarily large value of N. To see this, we write the solution of (7.4.6) as

$$N = \frac{\gamma_p^2 I_W M}{[\Delta_p - (1/6M^2)\{B'_W/B_h\}^2]^2}, \tag{7.4.8}$$

then if the denominator in (7.4.8) is very small, the resulting value of N will be very large. In this case, choosing M *larger* than the value given by (7.4.2) will reduce the bias term, and so increase the denominator in (7.4.8), and could well result in a smaller value of N for the same level of precision. The best procedure is to minimize (7.4.6) w.r. to M, i.e. to find that value of M, $M_{\min}$, say, which results in the smallest value of N for fixed $\Delta_p(\omega)$. If $M_{\min}$ turns out to be as large as, or larger than, the value of M determined by (7.4.2), then we retain the value $M_{\min}$ (since this will provide more than adequate resolution) and determine the value of N (which is now the smallest possible) by substituting $M_{\min}$ for M in (7.4.6). On the other hand, if $M_{\min}$ turns out to be smaller than the value of M determined by (7.4.2), we have to retain the latter value of M, and then determine N from (7.4.6).

Minimizing (7.4.6) w.r. to M gives the following expression for $M_{\min}$,

$$M_{\min} = \left\{\frac{5}{6\Delta_p(\omega)}\right\}^{1/2}\frac{B'_W}{B_h}. \tag{7.4.9}$$

and the minimum value of N is obtained by substituting (7.4.9) in (7.4.8), giving,

$$N_{\min} = \left(\frac{25\sqrt{5}}{16\sqrt{6}}\right)\frac{\gamma_p^2}{\{\Delta_p(\omega)\}^{5/2}B_h}\cdot(B'_W I_W) \tag{7.4.9a}$$

The procedure may be summarized as follows. We first determine the value of M, say M_β, according to (7.4.2); M_β is then the smallest value of M which provides the required degree of resolution. We then evaluate $M_{\min}$ from (7.4.9). If $M_{\min} \geq M_\beta$, we ignore M_β and use the value $M_{\min}$. The value of N is then given by solving (7.4.6) with $M = M_{\min}$. If $M_{\min} < M_\beta$, we ignore $M_{\min}$ and use M_β; N is then determined by solving (7.4.6) with $M = M_\beta$.

In fact, recalling that $B_W = B'_W/M$, we can write (7.4.1) as,

$$M_\beta = 2B'_W/\beta_h.$$

Comparing the above equation with (7.4.9) we see that,

$$M_\beta > \text{ or } < M_{\min} \text{ according to whether } \Delta_p(\omega) \gtrless \tfrac{5}{24}(\beta_h/B_h)^2.$$

If we write $\beta_h = 2\alpha B_h$ then,

$$M_\beta > \text{ or } < M_{\min} \text{ according to whether } \Delta_p(\omega) \gtrless 5\alpha^2/6.$$

In the usual case, $\beta_h = B_h$, corresponding to $\alpha = \tfrac{1}{2}$, and then, $M_\beta >$ or $< M_{\min}$ according to whether $\Delta_p(\omega) \gtrless \tfrac{5}{24} = 0\cdot2083$. Thus, if $\Delta_p(\omega) = 0\cdot25$, $M_\beta > M_{\min}$ and we use the value of M_β in determining N. If $\Delta_p(\omega) = 0\cdot1$, $M_{\min} > M_\beta$, and we use the value of $M_{\min}$. However, if we set $\alpha = \tfrac{1}{4}$, so that (7.4.1) gives $B_W = \tfrac{1}{4}B_h$ (for high resolution), then $M_\beta > M_{\min}$ if $\Delta_p(\omega) > 5/(6 \times 16) = 0\cdot052083$. In this case, even if $\Delta_p(\omega) = 0\cdot1$ (which is about the smallest value we would consider in practice), we have $M_\beta > M_{\min}$.

When $\Delta_p(\omega) > \tfrac{5}{24}(\beta_h/B_h)^2$ (so that we use the value M_β), the second term in (7.4.6) satisfies,

$$\frac{1}{6M^2}\left\{\frac{B'_W}{B_h}\right\}^2 = \frac{1}{24}\left\{\frac{\beta_h}{B_h}\right\}^2 < \frac{1}{5}\Delta_p(\omega).$$

Hence, determining N by substituting M_β in (7.4.6) is valid when the second term in (7.4.6) is smaller than $\tfrac{1}{5}\Delta_p(\omega)$.

To illustrate the operation of these design criteria we now consider the following examples.

EXAMPLE 1

We are given that the spectral density function $h(\omega)$ has a number of well separated peaks, the narrowest having a bandwidth of $\pi/20$ radians per unit time (on the frequency scale $-\pi$ to π). Since the peaks are well separated the width of the troughs will be larger than the width of the narrowest peak and hence we may set $B_h = \pi/20$. Our objective is to estimate the overall shape of $h(\omega)$, so that following the discussion in paragraph (ii) above, we set $\beta_h = B_h = \pi/20$. Using the Daniell window and substituting this value of β_h in (7.4.2) gives $M = 80$.

Suppose now that we set $\Delta_p(\omega) = 0 \cdot 25$, with $p = 0 \cdot 95$, so that $\gamma_p = 1 \cdot 96$. For the Daniell window, $B'_W = 2\pi$, and (7.4.9) gives $M_{\min} = 73$, which is smaller than the value of M derived above, and thus (strictly) would not provide the required degree of resolution, as we would expect since here $\Delta_p(\omega) > \frac{5}{24}$. We therefore retain the value $M = 80$, and substituting this value in (7.4.7) we find that the value of N required for the specified precision is given by,

$$1 \cdot 96\sqrt{80/N} + \tfrac{1}{24} = 0 \cdot 25,$$

giving $N = 7081$. (This value of N is quite large by normal standards, but is due to the fact that we are here trying to estimate a spectral density function with a very small bandwidth.)

EXAMPLE 2

The spectral density function $h(\omega)$ again has bandwidth $B_h = \pi/20$ (as in Example 1), but now our aim is merely to separate the peaks, the minimum distance between which is given as $\pi/10$. In this case we set $\beta_h = \pi/10$ (see paragraph (i) above), and again using the Daniell window we find from (7.4.2) that $M = 40$. For $\Delta_p(\omega) = 0 \cdot 25$, ($p = 0 \cdot 95$), $M_{\min}$ has the same value as in Example 1, namely 73, but this is now larger than the value of M determined by β_h. (Here, $\frac{5}{24}(\beta_h/B_h)^2 = 0 \cdot 8333 > \Delta_p(\omega)$.) Hence, we now retain the value $M = 73$, and (7.4.7) gives,

$$1 \cdot 96\sqrt{\frac{73}{N}} + \frac{800}{3 \times 73^2} = 0 \cdot 25,$$

giving $N = 7014$.

Note that if we had used the first value of M, namely, $M = 40$, then N would be given by,

$$1 \cdot 96\sqrt{40/N} + \tfrac{1}{6} = 0 \cdot 25,$$

giving $N = 22\,127$. Thus, increasing M from 40 to 73 reduces the number of observations required by a factor of 3.

A similar analysis could be carried through using the mean square percentage error $\eta(\omega)$ (defined by (7.2.3)) in place of $\Delta_p(\omega)$. However, it would seem more natural for the user to specify a value for the percentage error at a particular confidence level rather than to specify a value for the mean square percentage error.

Remark

It may seem, at first sight, that the procedure described above for choosing B_W leads to a value of M which depends only on the constant, β_h (cf.

(7.4.2.)) and hence does not conform with the condition required for the consistency of $\hat{h}(\omega)$, namely that $M \to \infty$ as $N \to \infty$. The point here is that (7.4.2) really gives the *smallest* value of M required for the specified degree of resolution; if we increase M above the value given by (7.4.2) we reduce the window bandwidth and the resolution is improved. Similarly, for a given value of M, (7.4.5) gives the *smallest* value of N required for the specified degree of precision. If we are given a larger number of observations than the value of N indicated by (7.4.5) we could then increase M (so as to improve the resolution) without reducing the precision.

Design criteria based on "degrees of freedom"

Some authors have suggested that the precision of $\hat{h}(\omega)$ should be measured by ν, the equivalent number of degrees of freedom, defined by (6.2.129) or, equivalently, by the SNR, defined by (7.2.7). As noted previously, this measure ignores the bias in $\hat{h}(\omega)$, and moreover it is hard to see how one could specify an appropriate value for ν without first considering the magnitude of the variance. In other words ν would not seem to be the natural parameter to specify when one first starts to think about the design of spectral estimates. Typical values of ν which have been suggested in the literature range from about 10 to 50, but choosing "degrees of freedom" in this arbitrary fashion would seem to be a rather dubious procedure.

However, it should be noted that in passing from (7.4.5) to (7.4.6) we have replaced the bias term by its maximum value. In so doing we are erring on the side of caution, and ensuring that the proportional error never exceeds its specified value at *any* frequency. In regions where $h(\omega)$ is fairly "flat" the bias will be substantially smaller than at a peak frequency, and it could be argued that replacing $\lambda^{(2)}(\omega)$ by its lower bound, $\frac{1}{2}B_h$, leads to an inflated estimate of $\Delta_p(\omega)$ over this region. Since we would virtually never know $\lambda^{(2)}(\omega)$ for each ω, the only alternative to the above approach is to ignore the bias completely (as suggested, e.g. by Blackman and Tukey (1959)) and measure the precision purely in terms of the variance. The resulting criterion can now be expressed purely in terms of ν, the degrees of freedom. For, setting $b(\omega) = 0$ in (7.2.1) we have,

$$\gamma_p \frac{v(\omega)}{h(\omega)} = \Delta_p(\omega),$$

or, using (6.2.129),

$$\gamma_p \sqrt{2/\nu} = \Delta_p(\omega).$$

This equation determines ν for given values of p and $\Delta_p(\omega)$. For example, with $p = 0.95$, and $\Delta_p(\omega) = 0.1$, $\gamma_p = 1.96$ and we have, $\nu \doteq 768$. For the more typical values of ν, say $\nu = 50$, the value of $\Delta_p(\omega)$ (with $p = 0.95$) is $1.96\sqrt{2/50} \doteq 0.39$, leading to a proportional error of at least 39%, while for $\nu = 10$, $\Delta_p(\omega) \doteq 0.8766$, corresponding to a proportional error of at least 87%. These results are by no means atypical, and show *that spectral estimates may often have proportional errors of the order of 60%–80%*. We know from (6.2.129) that ν is proportional to N/M; for example, for the Daniell window, $\nu = 2N/M$ (see Table 6.2). Thus, $\nu = 50$ corresponds to $M = N/25$, and even here the proportional error is approximately 40% (with no account being taken of the bias). We may note the salutary warning that to obtain a small proportional error we require careful design and large amounts of data!

If we are prepared to measure the precision of $\hat{h}(\omega)$ purely in terms of its variance, the design procedure would be as follows. First, we determine the window bandwidth B_W from (7.4.1), which then determines M from (7.4.2). We then determine ν as above, or alternatively specify a value of ν directly. Finally, we determine the value of N by using Table 6.2 to express ν in terms of N and M.

EXAMPLE 3

Consider the spectral density function described in Example 1. If we again choose $M = 80$, and select $\nu = 50$, corresponding to $\Delta_p(\omega) \doteq 0.39$, then, using the Daniell window for which $\nu = 2N/M$, we require $N = \nu M/2 = 2000$ observations.

Since this approach ignores the bias we could, as a slight variation, use the approximate χ^2 distribution of $\hat{h}(\omega)$ in place of the asymptotic normal distribution. Thus, instead of determining ν by specifying a fixed value for $\Delta_p(\omega)$ we could specify a fixed length for the $p\%$ confidence interval of the χ_ν^2 distribution. This then determines ν from (6.2.131).

Note, however that confidence intervals constructed without due consideration of the bias (whether based on the normal or χ^2 distributions) are likely to be quite unreliable in regions where $h(\omega)$ has peaks or troughs. (For windows with $r = 2$ the asymptotic bias is proportional to the second derivative $h''(\omega)$, which is negative at peaks and positive at troughs. Hence $\hat{h}(\omega)$ will tend to underestimate $h(\omega)$ at peaks and overestimate it at troughs—as we would expect intuitively from the local "smoothing" effect of the window $W_N(\theta)$.)

Variations of window bandwidth with frequency

The selection of the window bandwidth B_W according to (7.4.1) is aimed, typically, at ensuring that we have adequate resolvability in the region of the narrowest peak in $h(\omega)$, and if we use the same value of B_W at all frequencies we will, of course, have more than adequate resolvability in regions where

$h(\omega)$ is fairly "flat". However, in principle, there is no reason why we should not allow B_W to vary with frequency, i.e. we may use a narrow window bandwidth, corresponding to a large value of M, in the neighbourhood of peaks (assuming that their locations are known approximately), and a wide window bandwidth, corresponding to a small value of M, in regions where $h(\omega)$ relatively is "flat". In the "flat" regions a high degree of smoothing will not affect the bias too severely, and will have the beneficial effect of reducing the variance of the estimate. To implement this approach successfully one must naturally have prior knowledge as to the regions where $h(\omega)$ is likely to be "peaked" or "flat", but given this information there is no reason why we should not exploit the opportunity to increase the precision of the estimate in the "flat" regions.

If we wished to make B_W completely frequency dependent then, we would replace B_h in (7.4.1) by the "*bandwidth at frequency ω*", so that, e.g. for windows with $r = 2$ we replace B_h by $2|h(\omega)/h''(\omega)|^{1/2}$, and (7.4.1) then becomes,

$$B_W = \left| \frac{h(\omega)}{h''(\omega)} \right|^{1/2}. \tag{7.4.10}$$

In this case B_W would vary continuously with ω, but to implement (7.4.10) we would need to know $|h(\omega)/h''(\omega)|^{1/2}$ for all ω. A more practical procedure would be to choose B_W as in (7.4.1) in the region of the peaks, and to set $B_W = \frac{1}{2} \times$ "local bandwidth" in the "flat" regions.

From the computational point of view the evaluation of $\hat{h}(\omega)$ using different values of M for different values of ω would be rather tedious if we compute $\hat{h}(\omega)$ in the form (7.1.1), namely, as a weighted sum of sample autocovariances. (Most computer programs work with a fixed value of M for all frequencies.) However, if, as we will later discuss, we evaluate $\hat{h}(\omega)$ directly from the periodogram (by replacing the integral in (7.1.4) by a discrete sum) there is little difficulty in varying the window bandwidth with frequency, particularly if a simple shaped window such as the Daniell is used.

II. *Fixed record length*

If only a limited length of record is available, i.e. if N is fixed, then we have only one parameter (viz., M) to adjust, and we can no longer construct $\hat{h}(\omega)$ so as to have prescribed resolution and precision. We could, of course, still choose M so as to set the resolution to a required level (provided the appropriate value was not greater than $N - 1$), but then we would have no control over the degree of precision. In statistical terms this is equivalent to fixing the bias of an estimator with no control over its variance. Now we have already seen that, in general, as the bias decreases the variance increases (and vice versa), so that it would be more sensible to choose M so as to effect a compromise between bias and variance. We would therefore choose M so as to minimize either the $p\%$ Gaussian range of percentage error $\Delta_p(\omega)$ or the relative mean square error $\eta^2(\omega)$. Since these measures of precision are effectively equivalent it will not matter a great deal which one we choose, but in this case it is more convenient (from the algebraic point of view) to work with $\eta^2(\omega)$. For a given choice of window we can easily evaluate $\eta^2(\omega)$ from (7.2.3) using the standard asymptotic values for the bias and variance, and then minimize the resulting expression with respect to M.

For example, in the case of the Daniell window we have, substituting (7.3.10), (7.3.11) in (7.2.3),

$$\eta^2(\omega) = \frac{M}{N} + \left(\frac{\pi^2}{6M^2}\right)^2 \left|\frac{h''(\omega)}{h(\omega)}\right|^2. \tag{7.4.11}$$

The second term (which represents the squared bias) will be largest in the region where $h(\omega)$ contains its narrowest peak; we may therefore follow the same approach as above and, as an approximation, replace $|h''(\omega)/h(\omega)|$ by $(4/B_h^2)$, where B_h is the spectral bandwidth defined by (7.3.6). With this substitution (7.4.11) becomes

$$\eta^2(\omega) = \frac{M}{N} + \frac{4\pi^4}{9M^4 B_h^4}. \tag{7.4.12}$$

Differentiating $\eta^2(\omega)$ w.r. to M and setting the derivative equal to zero we find that $\eta^2(\omega)$ is minimized by setting,

$$M = \left\{\frac{16\pi^4 N}{9B_h^4}\right\}^{1/5}. \tag{7.4.13}$$

Thus, the value of M which minimizes the relative mean square error is $O(N^{1/5})$, and this result clearly holds for all windows with characteristic exponent $r = 2$. A similar analysis can be carried through for windows with other values of r, and we find that the optimal value of M is $O(N^{(1/2r+1)})$. (See the more general discussion of relative mean square errors in Section 7.5.) However, the precise value of M now involves $\lambda^{(r)}(\omega)$, and for practical purposes we would have to replace $\lambda^{(r)}(\omega)$ by its approximate lower bound expressed in terms of B_h. (As a very rough empirical rule we may take B_h/r as a lower bound for $\lambda^{(r)}(\omega)$.) Alternatively, if we follow the "degrees of freedom" approach, then, with N fixed, the specification of ν (the equivalent number of degrees of freedom), completely determines M. Thus, if N is fixed at 1000, say, and we select ν to be 50, then, for example, if we are using the Daniell window we choose M so that $\nu = 2N/M = 50$, giving $M = N/25 = 40$. However, this approach is undesirable in that, since it ignores the bias, it determines M as a fixed proportion of N without taking into account the magnitude of B_h.

Estimation procedures when there is no prior information on spectral bandwidth

The design relations which we have discussed above all require a knowledge of the spectral bandwidth B_h before they can be applied in practice. Indeed, a knowledge of B_h is crucial in order to determine a suitable value for the window bandwidth B_W from (7.4.1), and the value of N from (7.4.6), (in case (I)), and to determine M as a function of N from (7.4.13) (in case (II)). Without such knowledge there is no method of constructing estimates which satisfy well defined objective criteria. It might appear that we are being over optimistic in expecting to possess this type of knowledge before we have estimated $h(\omega)$, but in many physical and engineering applications it is possible to gain some idea of the magnitude of B_h, or a lower bound to this quantity, by considering the physical properties of the uinderlying mechanism. For example, if we are studying a problem in vibration analysis in which the observations represent the discplacement of a vibrating system it may be possible to fix a lower bound for B_h by arguing that if the spectral bandwidth was lower than some critical value the mechanism would fracture due to resonance effects. (A lower bound for B_h would certainly suffice since replacing B_h by its lower bound in the design equations would not decrease the precision of the estimates, but may indicate that a larger number of observations are needed than are, in fact, required.)

However, having made these observations it must be admitted that there are many situations in which we have no prior knowledge whatsoever on the spectral bandwidth. Unfortunately, these occur only too frequently in practice; for example, in the analysis of economic time series where the concept of bandwidth does not have any obvious physical interpretation. In such cases one may still attempt to estimate spectral density functions but now we have to adopt more of a "trial and error" approach to the selection of the estimation parameters. Although this may seem unscientific it should be remembered that spectral analysis has been applied with considerable success to many economic time series, and it is now one of the major tools used in analysing this type of data; see, e.g. Granger and Hatanaka (1964), and Gudmundsson (1971).

Perhaps the obvious question which arises at this stage is; why not estimate the spectral bandwidth from the data? Unfortunately, no completely satisfactory method of estimating B_h has been discovered, and it is debatable as to whether any satisfactory method really exists. The amount of information which we can extract from any set of data depends, of course, on the number of assumptions which we are prepared to make, and assuming only second order stationarity it is doubtful whether there is enough information in the data to enable us to construct a reliable estimate of B_h. Suppose, for example, we choose an arbitrary window bandwidth, construct an estimate $\hat{h}_0(\omega)$, and then try to estimate B_h by using the bandwidth of $\hat{h}_0(\omega)$. A little thought will show us that this method is futile, for "bandwidth" is a measure of smoothness and the smoothness of $\hat{h}_0(\omega)$ will depend just as much on the bandwidth of the window used as on the bandwidth of $h(\omega)$, the true spectral density function. If we choose a large window bandwidth, $\hat{h}_0(\omega)$ will be much smoother than $h(\omega)$; on the other hand, if we choose a narrow window bandwidth $\hat{h}_0(\omega)$ will exhibit erratic fluctuations (rather like the raw periodogram). In neither case will the smoothness properties of $\hat{h}_0(\omega)$ bear any relation to smoothness properties of $h(\omega)$.

There are, however, a number of semi-empirical procedures which we can use when prior information on spectral bandwidth is lacking, and we describe some of these below. We consider only the case where N, the number of observations, is fixed, so that the basic problem is to choose the window bandwidth B_W or equivalently the window parameter M.

(1)   *Using the sample autocovariance function*

We saw in Section 7.3.1 that, roughly speaking, the spectral bandwidth was inversely proportional to the rate of decay of the autocovariance function $R(s)$. Hence, in principle, we can extract the information we seek from the autocovariance (or autocorrelation) function, but, of course, this function is unknown just as the spectral density function is unknown. Now even with zero prior knowledge we can estimate $R(s)$ by the estimator $\hat{R}(s)$ given by (7.1.2), and this does not involve the same difficulties which occur in the estimation of $h(\omega)$, i.e. there are no estimation parameters to be selected. We can therefore construct the sample autocovariance function $\hat{R}(s)$, plot it as a function of s, and try to determine a suitable value for M from its observed rate of decay. Relating the rate of decay of $\hat{R}(s)$ (however measured) to the spectral bandwidth would be quite a difficult exercise, but if we consider spectral estimates of the *truncated* type (such as the truncated periodogram, the Bartlett, Parzen, and Tukey estimates) then we can obtain a rough idea of the appropriate value of M by reasoning as follows. For these estimates M represents the point at which the sample autocovariance function is truncated, and following our first ideas on spectral estimation (see the discussion leading to (6.2.48)) a fairly natural procedure would be to choose M so that

$$R(s) \sim 0, \qquad |s| > M,$$

i.e. so that the theoretical autocovariance function is effectively zero for all lags greater than M. To put this approach into practice we replace $R(s)$ by $\hat{R}(s)$, and select M as that value of s beyond which $\hat{R}(s)$ appears to be effectively zero, making due allowance for sampling fluctuations in $\hat{R}(s)$. If $\hat{R}(s)$ decays to zero in a fairly obvious way (as in Fig. 7.7) then the

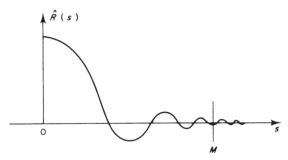

Fig. 7.7. Determination of truncation point from the sample autocovariance function.

appropriate value of M is easily determined. However, needless to say, the analysis will rarely be as clear cut as this in practice, and in order to make this technique more precise we should ideally have some idea of the standard deviation of $\hat{R}(s)$. (However, we know from the results of Section 5.3.3 that the sampling properties of $\hat{R}(s)$ depend on the theoretical autocovariance function.)

Although rather crude, this method can be quite effective and has been applied with reasonable success in a number of problems. It is certainly preferable to choosing a completely arbitrary value for M, and, in cases where we have no information at all on the order of magnitude of the spectral bandwidth, it at least enables us to attempt to "match" the value of M to the properties of the process.

There are, of course, a number of dangers inherent in this approach, the main ones being:

(a) The results of Section 5.3.3 show that the sample autocovariances are themselves autocorrelated, and consequently $\hat{R}(s)$ *will, in general, decay more slowly than the theoretical autocovariance function, $R(s)$.* Also, the sampling fluctuations of the individual $\hat{R}(s)$ will further distort the form of this function unless N is large compared with the rate of decay of $R(s)$.

(b) If, for example, $h(\omega)$ contains a large peak with a wide bandwidth and a much smaller peak with a narrow bandwidth, then, as pointed out by Tukey (1961), the effect of the small peak will be obscured in the sample autocovariance function and the value of M selected will reflect only the bandwidth of the large peak. This warns us that we must not expect $\hat{R}(s)$ to reveal all the "fine detail" present in $R(s)$, but limitations of this type are bound to affect any empirical method of extracting information on spectral bandwidth.

(2) *"Window closing"*

The idea behind the technique of "window closing" is to start with a arbitrary but large window bandwidth B_W (corresponding to a small value of M), and compute the resulting spectral estimate $\hat{h}(\omega)$. We then compute a further sequence of spectral estimates by "closing" the window, i.e. by progressively decreasing B_W (or equivalently, by increasing M). The initial estimate, based on a large B_W, will be a very smooth function, but as we decrease B_W the estimates will progressively exhibit more detail and become more "erratic" in form. By examining the manner in which the estimates change as we decrease B_W we may be able to detect the point at which the smoothing has been relaxed "too far", and this will then enable us to select a suitable value for B_W.

Jenkins (1965) suggests that we might implement this approach by initially choosing three values of B_W, a "large" value $B_W^{(1)}$, a "medium" value $B_W^{(2)}$, and a "small" value $B_W^{(3)}$, corresponding to three values of M, M_1, M_2 and M_3, with, e.g., $M_3/M_1 = 4$. The forms of these estimates can then be examined to see whether a value of M intermediate between M_1 and M_3 should be chosen, or whether M should be even smaller than M_1 or larger than M_3.

In essence, the technique of "window closing" does, in fact, depend on prior knowledge of the properties of the spectral density function. For, if we reject some estimates as being "too smooth" and others as being "too rough", then we are saying, in effect that we know the degree of smoothness which the true spectral density function possesses. However, it may be that in some situation our prior knowledge is not sufficiently precise to allow us to express it quantitatively (in terms of B_h), but we can express it graphically by indicating whether a particular estimate looks "too smooth" or "too rough". In such cases "window closing" may be a very useful technique.

(3) *Choosing M as a fixed proportion of N*

One of the many "folklores" which bedevil the subject of spectral estimation is the recommendation that for estimates of the truncated type the truncation point M should be chosen as a fixed proportion of N, say 20% or 30% of N. We mention this suggestion merely to dismiss it, since it is clearly unwise to adopt a general rule of always choosing M to be (say) 30% of N irrespective of the properties of the underlying process. One of the most important features to emerge from our previous discussion is that, ideally, M should be chosen so as to "match" the properties of the true spectral density function $h(\omega)$ and we can see this in a very striking way if we consider the extreme case where $h(\omega)$ corresponds to a uniform (i.e. "white") spectrum. In this case we would naturally choose $M = 0$ (whatever the value of N) since here all we need do is to estimate the variance of the process. (This value of M is quite in accord with both (7.4.2) and (7.4.13) since for a white spectrum we may take B_h to be infinite.) Thus, although we may have no prior knowledge of the spectral bandwidth, it is obviously desirable to try to "learn" something about the properties of the process before choosing a value for M, and this is precisely the aim of the two methods described above.

The choice of M as a fixed proportion of N has little intuitive appeal, and its only theoretical justification is that it controls the value of the relative variance (i.e. the ratio of $\text{var}\{\hat{h}(\omega)\}$ to $h^2(\omega)$). In this sense it is closely related to the "degrees of freedom" approach, but it should be noted that the equivalent number of degrees of freedom ν depends not only on the ratio N/M but also on the form of window used; see Table 6.2. For example, for the Bartlett window $\nu = 3N/M$, so that choosing M as 30% of N will give $\nu = 10$ degrees of freedom, a very low value. Even if we neglect the bias, the proportional error at the 95% confidence level will then be approximately 87%.

The above choice of M might, however, be suitable as an upper limit to be used in conjunction with a "window closing" procedure.

Choice of the frequency interval $\Delta\omega$

Having determined the values of N and M the only remaining parameter to be selected is $\Delta\omega$, the frequency interval at which the estimate $\hat{h}(\omega)$ is to be evaluated. For estimates of the truncated type the summation in (7.1.1) terminates at $s = \pm M$, and $\hat{h}(\omega)$ is a trigonometric polynomial of degree M. It then follows from the "sampling theorem" (see Section 7.1.1) that the complete function $\hat{h}(\omega)$ is determined by its values at the discrete set of points, $\omega_r = 2\pi r/M'$, $r = 0, 1, \ldots, [M'/2]$, where M' is any integer $\geq (2M+1)$. In particular, if $k(1) = 0$ (as holds for the Bartlett, Tukey-Hanning, and Parzen windows) then $\hat{h}(\omega)$ is a trigonometric polynomial of degree $(M-1)$, and the condition on M' becomes, $M' \geq 2M - 1$. Thus, $\hat{h}(\omega)$ is completely determined by its values at the points, $\omega_r = \pi r/M$, $r = 0, 1, 2, \ldots, M$, corresponding to a frequency interval $\Delta\omega = \pi/M$.

More generally, Parzen (1957a) has shown that if we set $\Delta\omega = \frac{1}{2}B_W$, (where B_W is the window bandwidth) and evaluate $\hat{h}(\omega)$ for $\omega = \Delta\omega, 2(\Delta\omega)$, $3(\Delta\omega), \ldots, k(\Delta\omega)$ (where $k = [\pi/\Delta\omega]$), then the values of $\hat{h}(\omega)$ at all other frequencies may be computed using (effectively) linear interpolation, and the resulting estimates will have the same asymptotic bias and variance as those computed directly from (7.1.1). If we follow (7.4.1) and choose B_W to be half the bandwidth of the narrowest peak in $h(\omega)$, then the above value for $\Delta\omega$ will ensure that we have at least four or five computed values of $\hat{h}(\omega)$ in the region of a peak.

For the Daniell window, $B_W = 2\pi/M$, and the above rules gives $\Delta\omega = \pi/M$, leading to the evaluation of $\hat{h}(\omega)$ at the points $\omega_r = \pi r/M$, $r = 0, 1, \ldots, M$, exactly as for the truncated type estimates.

The above discussion should be regarded as providing only a guideline for the number of spectral estimates which would normally be computed, and it should be emphasized that there is no reason why we should not evaluate $\hat{h}(\omega)$ (directly) for any values of ω we like, and for any number of points we like. If we particularly wished to study the form of $h(\omega)$ over a certain range of frequencies it would be quite in order (and indeed, very sensible) to evaluate $\hat{h}(\omega)$ over a much finer set of points over this range.

Designs for continuous parameter processes

We now discuss the estimation of the spectral density function of a continuous parameter process $X(t)$ based on sampled observations. If $X(t)$ is sampled at intervals Δt, giving observations $Y_t = X(t \cdot \Delta t)$, then, ignoring aliasing errors, the spectral density functions of $\{Y_t\}$ and $\{X(t)\}$ are, from (7.1.15) and (7.1.18), related by,

$$h_X(\omega) = \Delta t \cdot h_Y^*(\omega \Delta t), \qquad |\omega| \le \pi/\Delta t, \qquad (7.4.14)$$

where $h_Y^*(\omega)$ is given by (7.1.19). Given N observation on Y_t we may construct an estimate, $\hat{h}_Y^*(\omega)$, of $h_Y^*(\omega)$ in the usual way (i.e. by using an expression of the form (7.1.1)), and the natural estimate of $h_X(\omega)$ is then,

$$\hat{h}_X(\omega) = \Delta t \cdot \hat{h}_Y^*(\omega \Delta t). \qquad (7.4.15)$$

The relative bias and relative variance of $\hat{h}_X(\omega)$ are exactly the same as for $\hat{h}_Y^*(\omega)$, since we have immediately from (7.4.14), (7.4.15), writing $\omega' = \omega \Delta t$,

$$[E\{\hat{h}_X(\omega)\} - h_X(\omega)]/h_X(\omega) = [E\{\hat{h}_Y^*(\omega')\} - h_Y^*(\omega')]/h_Y^*(\omega'),$$

and

$$\text{var}\{\hat{h}_X(\omega)\}/h^2(\omega) = \text{var}\{\hat{h}_Y^*(\omega')\}/\{h_Y^*(\omega')\}^2.$$

Hence, the measures of precision $\Delta_p(\omega)$, $\eta(\omega)$ and SNR, (defined respectively by (7.2.1), (7.2.3), (7.2.7)) take the same values for $\hat{h}_X(\omega)$ as for $\hat{h}_Y^*(\omega)$, and the design relations for $\hat{h}_X(\omega)$ are therefore the same as for $\hat{h}_Y^*(\omega)$.

The only problem is to translate information on the bandwidth of $h_X(\omega)$, which, with t measured in seconds, would usually be measured in Hz (cycles per second) on a frequency range $-1/2\Delta t$ to $1/2\Delta t$, into the equivalent information on the bandwidth of $h_Y^*(\omega)$ measured in radians per second on a frequency range $-\pi$ to π, which is the standard form we have used throughout our previous discussions on bandwidth. Now if $b_h^{(x)}$ denotes the bandwidth of $h_X(\omega)$ measured in Hz, the bandwidth of $h_X(\omega)$ measured in radians per second is given by,

$$B_h^{(X)} = 2\pi b_h^{(X)}.$$

The corresponding bandwidth for $h_Y^*(\omega)$ is now given by,

$$B_h^{(Y)} = \Delta t B_h^{(X)} = 2\pi(\Delta t)b_h^{(X)}. \qquad (7.4.16)$$

The routine for estimating $h_X(\omega)$ can be summarized as follows,

(1) Select the sampling interval Δt (in seconds) according to the rule discussed in Section 7.1.1, i.e. select Δt so that $h_X(\omega)$ contains negligible power at frequencies higher than $\pi/\Delta t$ radians per sec ($= 1/2\Delta t$ Hz), or, if pre-filtering is used, so that $\pi/\Delta t = \omega_0$, where ω_0 is the highest frequency at which we require an estimate of $h_X(\omega)$. (Tukey suggests that, in order to leave a suitable margin of safety, one should choose Δt so that $\pi/\Delta t = 1\cdot5\omega_0$, or, if $f_0 = \omega_0/2\pi$, so that $\Delta t = 1/3f_0$.)

(2) Given the value of $b_h^{(X)}$ compute $B_h^{(Y)}$ from (7.4.16). We may now apply any of the design procedures which we have previously discussed for the case of discrete parameter processes.

EXAMPLE 4

We wish to estimate the spectral density function of a continuous parameter process over the range 0 to 25 Hz, the spectral bandwidth being given as $0\cdot5$ Hz.

To estimate the spectral density function up to 25 Hz, we must choose the sampling interval Δt so that $1/2\Delta t \geq 25$, i.e. so that $\Delta t \leq 1/50$ sec. Setting $\Delta t = 1/50$, and using the fact that $b_h^{(X)} = 0\cdot5$, the value of $B_h^{(Y)}$ is now given by (7.4.11) as,

$$B_h^{(Y)} = 2\pi \cdot \frac{1}{50} \cdot \frac{1}{2} = \frac{\pi}{50} \text{ (radians/second)}.$$

Suppose now that an unlimited length of record is available and that we wish to design a spectral estimate, based on the Daniell window, whose proportional error is not more than 25% at the 95% confidence level. Using the above value for $B_h^{(Y)}$, the value of M given by (7.4.2) is, (setting $\beta_h = B_h^{(Y)}$),

$$M = 4\pi/B_h^{(Y)} = 200.$$

(The value of M_{min} (from (7.4.9)) is 183, which is smaller than the above value. Hence we retain the value $M = 200$.) The number of observations required for $\Delta_p(\omega) = 0.25$ with $p = 0.95$ is now given by (7.4.7), which in this case becomes

$$1.96\sqrt{200/N} + \tfrac{1}{24} = 0.25,$$

giving $N = 17\ 702$, corresponding to a recording time of (approximately) 354 seconds (or roughly 6 minutes). Thus, the process should be recorded over a period of 354 seconds and sampled at intervals of $1/50$ second.

Following the above discussion on the choice of $\Delta\omega$, the estimates would usually be computed at the frequencies $\omega_r = \pi r/200$, $r = 0, 1, \ldots, 200$. This then gives 200 spectral ordinates over the frequency range 0 to 25 Hz with a frequency interval of 0.125 Hz.

Discussion

We would warn the reader not to read too much into the discussion of design procedures given in this section. The discussion shows that, in principle, spectral estimates can be designed so as to achieve given levels of resolution and precision provided we know their sampling properties and have some prior knowledge of the shape of the spectral density function to be estimated. However although these "design relations" may appear to have been constructed in a fairly precise way, their practical usefulness is limited by the following considerations.

(1) The expressions we have used for the bias and variance are valid only in an *asymptotic* sense and their use with finite N and M is at best an approximation; similarly, the use of normal or χ^2 distributions for spectral estimates is also valid only asymptotically. Quite apart from these points, we have made further approximations in e.g., replacing the quantity $\lambda^{(2)}(\omega)$ by the spectral bandwidth B_h.

(2) To implement the design relations in an exact form we need precise prior knowledge of the spectral bandwidth.

(3) For fairly low proportional errors (e.g., 25%) the design relations lead to quite large values of N (the total number of observations). In Example 1 we found that we required approximately 7000 observations, and in many practical situations it would be quite unrealistic to expect to have available time series of this length.

Nevertheless, the study of "design relations" is valuable in that, (a) as mentioned previously, it directs our attention to the important features of the spectral density function on which, ideally, we should have prior knowledge; in particular, it encourages us to think hard about obtaining, whenever possible, some idea of the order of magnitude of the spectral bandwidth, and (b) it warns us that for typical lengths of series (and in this context, 1000 observations would usually be regarded as a fairly long series) we must be prepared to accept quite high proportional errors, typically of the order of 60%–80%.

In dealing with practical problems of spectral estimation the "design relations" should be regarded as providing guidelines rather than as procedures to be followed strictly. Bearing in mind the limitations imposed by the lengths of series normally available, a more empirical approach, based, e.g. on the "window closing" technique described above, may well be a more suitable form of analysis.

Numerical Examples

We now apply the spectral estimation techniques discussed in the preceding sections to series generated from four standard models, namely AR(1), AR(2), MA(2) and ARMA(2, 2). The series and their models are shown in Graphs 3.2–3.5, Chapter 3, and the data are listed in the Appendix. The sample autocovariance and autocorrelation functions are shown in Graphs 5.2–5.5, 5.7–5.10, Chapter 5, and the autocovariances are also listed in the Appendix. For these examples the theoretical spectral density functions are known, and hence we may determine appropriate values for the window bandwidth (or truncation point) according to the criteria discussed above. In the following calculations approximate values of the (half power points) spectral bandwidths were obtained from the graphs of the theoretical spectral density functions. (More accurate values could have been obtained by numerical interpolation, but since the formulae which determine window bandwidth are themselves only approximate it was not felt necessary to calculate the spectral bandwidths with great accuracy.) The estimated density functions are all computed from series of length 500 using the *Parzen window*, and the details of the calculations are as follows.

AR(1) series. The spectral density function has a single peak at $\omega = 0$ with approximate bandwidth $B_h = 46\pi/128$. Setting $B_w = \frac{1}{2}B_h$, and recalling that for the Parzen window $B_w = 12/M$ (see Table 7.1), the appropriate value of M, say M_1, is given by

$$\frac{12}{M_1} = \frac{23\pi}{128} \quad \text{or} \quad M_1 = 21 \cdot 26.$$

Alternatively, we may determine that value of M, say M_2, which minimizes the relative mean square error for the given number of observations, viz. $N = 500$. The general expression for that value of M which minimizes the relative mean square error is given by (7.5.8) (see Section 7.5), and, for the Parzen window, is given by

$$M_2 = \left[4 \times \left(\frac{12}{2\sqrt{6} \times \frac{1}{2}B_h} \right)^4 \times \frac{N}{0.54} \right]^{1/5}.$$

Substituting $N = 500$, $B_h = 46\pi/128$ gives

$$M_2 = 16.74.$$

Carrying out similar calculations for the other series we obtain the following table of results.

Series	Spectral bandwidth	M_1	M_2
AR(1)	$46\pi/128$	21·26	16·74
AR(2)	$15\pi/128$	65·19	41·02
MA(2)	$105\pi/128$	9·31	8·65
ARMA(2, 2)	$23\pi/128$	42·52	29·14

It will be seen that the values of M_1 and M_2 vary considerably over the four series. The AR(2) series has the smallest bandwidth and consequently requires the largest truncation points; the MA(2) series has a relatively large bandwidth and correspondingly requires a fairly low truncation point. However, remembering that the values of M_1 and M_2 are based on formulae which are intended to serve mainly as a *guide* to the choice of parameter in spectral estimation, it was decided to compute three spectral estimates for each series, using in all cases *the same three values of M_1*, namely, $M = 12, 25, 50$. (These correspond to window bandwidths of 1, 0·48 and 0·24 respectively, on an angular frequency range of $-\pi$ to π.) These three truncation points cover the range of values of M_1 and M_2 shown in the above table reasonably well, and the resulting spectral estimates are shown below in Graphs 7.1–7.12. For each graph the spectral ordinates were calculated at a frequency spacing of $(\pi/128)$; thus the indexing variable, r, on the frequency axis runs from 0 to 128. The full curve represents the theoretical density function, and the dotted curve the estimate. Note that both curves correspond to the *normalized* spectral density function, i.e. the spectral estimates were computed as (smoothed) Fourier transforms of the sample *autocorrelation* function, using (7.1.3) with $\lambda_N(s)$ having the form of the Parzen lag window, (6.2.82).

These results agree well with what one would expect from the asymptotic theory. Thus, with $M = 12$ the spectral estimates show an overall "smooth" form, but with ordinates in the neighbourhood of the peak being, in general, substantially underestimated. With the larger truncation

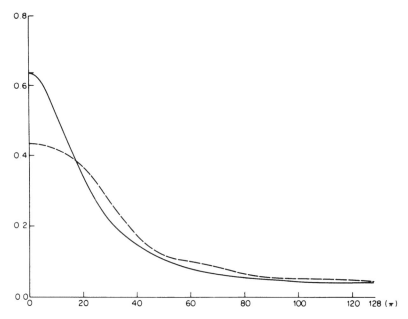

Graph 7.1. Theoretical and estimated spectral density functions for the AR(1) series, $M = 12$.

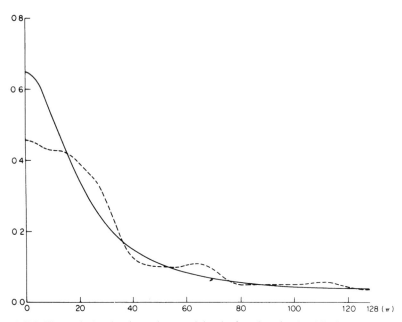

Graph 7.2. Theoretical and estimated spectral density functions for the AR(1) series, $M = 25$.

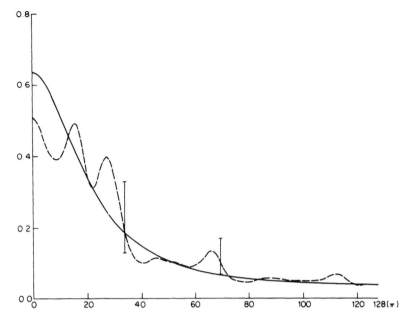

Graph 7.3. Theoretical and estimated spectral density functions for the AR(1) series, $M = 50$.

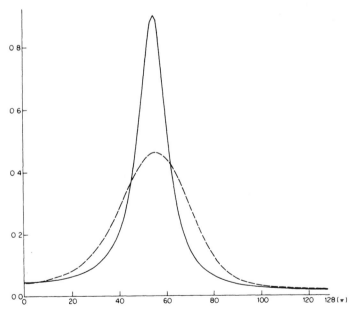

Graph 7.4. Theoretical and estimated spectral density functions for the AR(2) series, $M = 12$.

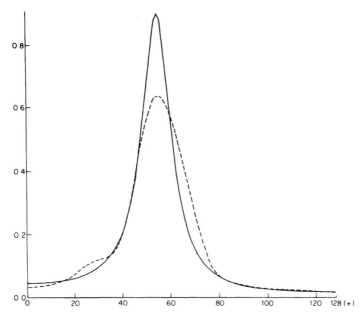

Graph 7.5. Theoretical and estimated spectral density functions for the AR(2) series, $M = 25$.

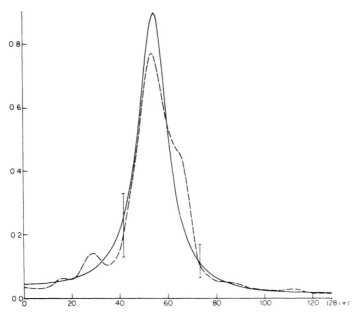

Graph 7.6. Theoretical and estimated spectral density functions for the AR(2) series, $M = 50$.

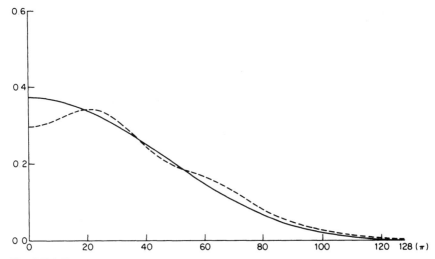

Graph 7.7. Theoretical and estimated spectral density functions for the MA(2) series, $M = 12$.

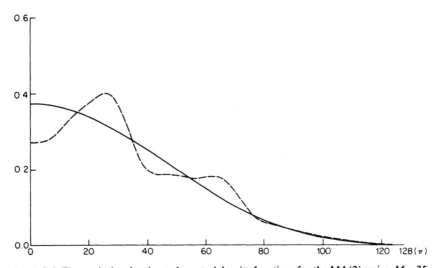

Graph 7.8. Theoretical and estimated spectral density functions for the MA(2) series, $M = 25$.

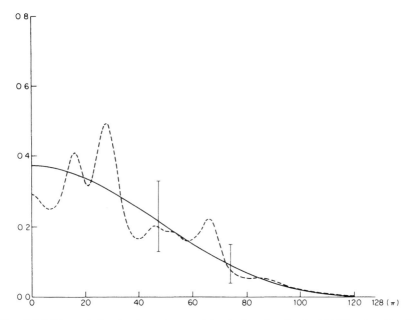

Graph 7.9. Theoretical and estimated spectral density functions for the MA(2) series, $M = 50$.

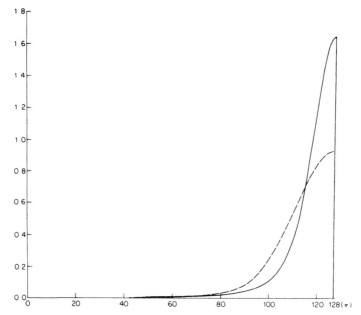

Graph 7.10. Theoretical and estimated spectral density functions for the ARMA(2, 2) series, $M = 12$.

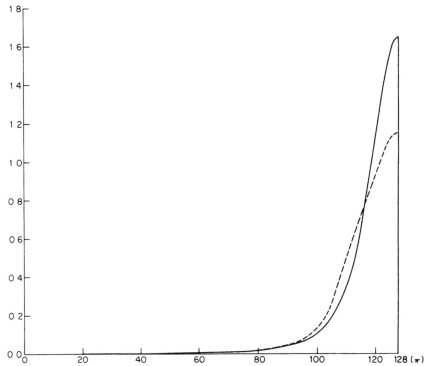

Graph 7.11. Theoretical and estimated spectral density functions for the ARMA(2, 2) series, $M = 25$.

point, $M = 50$, the estimates exhibit a more erratic form (since, for most of the series, $M = 50$ is unnecessarily large), but in the case of the AR(2) series, the "peak" ordinates are still underestimated. Again, this is what we would expect since, as seen above, for good resolution of the peak, the AR(2) series requires a truncation point of about 65. It is interesting to note that, judged from Graph 5.3, the theoretical autocovariance function of the AR(2) series appears to have effectively damped down to zero at about lag 25, and the sample autocovariance seems to have similarly decayed at about lag 30. If we were choosing the truncation point on the basis of the behaviour of the sample autocovariance function, we would probably have selected $M = 30$ (or smaller); however, as noted above, we require M to be about 65 for good resolution, and even with $M = 50$ the peak ordinates are underestimated. Similar remarks may be made about the ARMA(2, 2) series: the theoretical autocovariance function appears to have decayed at about lag 15, but the value of M_1 is as high as 42·52.

Approximate values for the relative bias and relative standard deviation may be computed from the expressions given in Table 6.1. For the

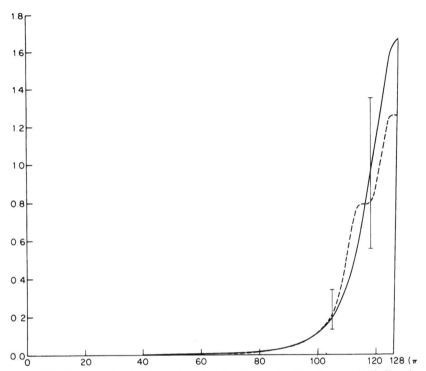

Graph 7.12. Theoretical and estimated spectral density functions for the ARMA(2, 2) series, $M = 50$.

Parzen window, we have

$$\text{relative standard deviation} \doteq \left\{0.54 \frac{M}{N}\right\}^{1/2},$$

$$\text{relative bias} \doteq \frac{6}{M^2}\left|\frac{h''(\omega)}{h(\omega)}\right|^{1/2} = \frac{24}{M^2 B_h^2} \quad \text{(at the peak).}$$

For a given value of M (and with $N = 500$), the relative standard deviation takes the same value for each of the four series. These values are as follows.

	Relative standard deviation (all series)
$M = 50$	0·2322
$M = 25$	0·1643
$M = 12$	0·1138

The relative bias depends on the spectral bandwidth, and thus varies from series to series. The table below shows the (approximate) relative bias *at the peak* ordinate for each series.

		Relative bias (at peak)
AR(1) series:	$M = 50$	0·0075
	$M = 25$	0·0301
	$M = 12$	0·1308
AR(2) series:	$M = 50$	0·0783
	$M = 25$	0·2833
	$M = 12$	1·2297
MA(2) series:	$M = 50$	0·0014
	$M = 25$	0·0058
	$M = 12$	0·0251
ARMA(2, 2) series:	$M = 50$	0·0313
	$M = 25$	0·1205
	$M = 12$	0·5230

Confidence intervals

Ignoring the bias, approximate confidence intervals for individual spectral ordinates can be computed either from (6.2.131) (based on the χ^2 approximation) or from (6.2.133) (based on the normal approximation). For a given value of M the "scaling factors", i.e. the factors multiplying $h(\omega)$ in the confidence limits, are the same for all series, and the factors for a 95% confidence interval are as shown below.

95% confidence interval scaling factors

(I) *Normal Approximation* (cf. (6.2.133))
   $M = 50$: (0·6870, 1·8365)
   $M = 25$: (0·7564, 1·4751)
   $M = 12$: (0·8176, 1·2872)

(II) χ^2 *approximation* (cf. (6.2.131))
   $M = 50$ ($\nu = 37$):   (0·6645, 1·6735)
   $M = 25$ ($\nu = 74$):   (0·7424, 1·4201)
   $M = 12$ ($\nu = 155$): (0·8342, 1·3136)

(Confidence intervals based on the χ^2 approximation are shown on the graphs by vertical lines at certain frequencies.)

Taking the bias into account, we may compute for each series the 95%
"Gaussian range of percentage error", $\Delta_p(\omega)$ (defined by (7.2.1)), by
multiplying the relative standard deviation by 1·96 and adding the
"peak" relative bias. This gives us a rough upper bound to the overall
relative error of spectral estimates, and for $M = 25$ the values of
$\max\{\Delta_p(\omega)\}$ for the various series are given below.

Series	Δ_p
AR(1)	0·3522
AR(2)	0·6054
MA(2)	0·3279
ARMA(2, 2)	0·4426

7.4.1 Prewhitening and Tapering

"Prewhitening" and "tapering" are techniques which aim to improve the
accuracy of spectral estimates by making certain preliminary transformations to the data before the spectral estimation procedures are
applied. We now discuss briefly the basic ideas involved.

Prewhitening

The most difficult form of spectral density function to estimate is one
which contains sharp peaks. For, in this case we have to choose a very small
window bandwidth, leading to a large value of M, and consequently N would
have to be very large in order to keep the variance at a reasonable level.
Alternatively, if N and M are fixed, the bias would be substantial in the
region of the peaks. On the other hand, if the process has a "white"
spectrum, i.e. if $h(\omega)$ is a constant, then we can choose any value of M we
like (including zero) without affecting the bias, since here,

$$E[\hat{h}(\omega)] \sim \int_{-\pi}^{\pi} h(\theta) W_N(\omega - \theta)\, d\theta = h(\omega),$$

irrespective of the form of $W_N(\theta)$, so long as $\int_{-\pi}^{\pi} W_N(\theta)\, d\theta = 1$. The method
of "prewhitening" (or "pre-filtering", as it is sometimes called) was sugges-
ted by Press and Tukey (1956), and its objective is to "flatten" the spectral
density function by passing the original (mean corrected) process X_t through
a linear filter constructed so that its output Y_t has a spectral density function
which is "nearly white". If we denote the impulse response function of the
filter by $\{g_u\}$ then $\{X_t\}$ and $\{Y_t\}$ are related (see Section 4.12) by

$$Y_t = \sum_u g_u X_{t-u} \tag{7.4.17}$$

and the corresponding spectral density functions satisfy (cf. (4.12.18)),

$$h_Y(\omega) = |\Gamma(\omega)|^2 h_X(\omega),\qquad(7.4.18)$$

where $\Gamma(\omega)$, the transfer function of the filter, is given by,

$$\Gamma(\omega) = \sum g_u\, e^{-i\omega u}.$$

If the filter has been appropriately chosen, and $h_Y(\omega)$ is, in fact, nearly white (i.e. if the bandwidth of $h_Y(\omega)$ is much larger than the bandwidth of $h_X(\omega)$), then we can estimate $h_Y(\omega)$ from the $\{Y_t\}$ values using standard methods, and now the choice of the window parameter M is not critical. Denoting the estimate of $h_Y(\omega)$ by $\hat{h}_Y(\omega)$, we can now estimate $h_X(\omega)$ by inverting (7.4.18), i.e. we estimate $h_X(\omega)$ by,

$$\hat{h}_X(\omega) = |\Gamma(\omega)|^{-2}\hat{h}_Y(\omega).\qquad(7.4.19)$$

Although this is an ingenious idea, its application is limited by the following points.

(a) If $h_Y(\omega)$ is to be "nearly white", then ideally $\{g_u\}$ should be chosen so that

$$|\Gamma(\omega)|^2 \propto h_X^{-1}(\omega),$$

which means that we must have prior knowledge of the approximate shape of $h_X(\omega)$. In particular we must know the approximate location and bandwidths of the peaks.

(b) If $h_X(\omega)$ has some very sharp peaks then in order for $|\Gamma(\omega)|^2$ to suppress these the impulse response function, $\{g_u\}$, will have to have a very long "tail", which means that Y_t cannot be evaluated from (7.4.17) unless we have a very large number of observations on X_t.

Blackman and Tukey (1959) suggest that the first difficulty may be overcome by performing a pilot spectral analysis (on the X_t observations) using a fairly broad spectral window, but the success of the subsequently designed filter will then depend on just how well the pilot analysis has determined the position and bandwidths of the spectral peaks.

It is interesting to note that, using the relationship (6.2.23) between the periodograms of X_t and Y_t, the estimate (7.4.19) can be written as,

$$\hat{h}_X(\omega) \doteq |\Gamma(\omega)|^{-2} \int_{-\pi}^{\pi} |\Gamma(\theta)|^2 I_{N,X}^*(\theta) W_N(\omega - \theta)\, d\theta.$$

Although this is not quite of the same form as the standard estimate (7.1.4), the prewhitening operation can be regarded as a method of modifying the window $W_N(\theta)$ so as to "match" it to the shape of $h_X(\omega)$.

Tapering

Tapering is a different technique from prewhitening, its purpose being to reduce the bias in spectral estimates by reducing the bias due to the periodogram. Now, from (7.1.4) the expected value of $\hat{h}(\omega)$ can be written as

$$E[\hat{h}(\omega)] = \int_{-\pi}^{\pi} E[I_{N,X-\bar{X}}^{*}(\theta)] W_N(\omega - \theta)\, d\theta, \qquad (7.4.20)$$

and in our discussion of the asymptotic bias in section (6.2.4) in effect we ignored the bias in the periodogram and replaced (7.4.20) by

$$E[\hat{h}(\omega)] \sim \int_{-\pi}^{\pi} h(\omega) W_N(\omega - \theta)\, d\theta. \qquad (7.4.21)$$

The argument used is that the periodogram bias being $O((\log N)/N)$ (cf. (6.2.12)), is, in general, negligible compared with the bias due to the smoothing window $W_N(\theta)$, which according to (6.2.124) is $O(1/M')$. However, if the window has discontinuities, the periodogram bias may be worth correcting by "tapering" the data, as follows. Given the observations $X_1, X_2, \ldots, X_N$, we form a new set of observations $\{Y_t\}$ by writing,

$$Y_t = h_t \cdot X_t, \qquad t = 1, \ldots, N, \qquad (7.4.22)$$

where $\{h_t\}$ is a suitable sequence of constants called a "*taper*" (or "*fader*" or "*data window*"). To estimate the spectral density function of $\{X_t\}$ we replace the periodogram, $I_{N,X-\bar{X}}^{*}(\theta)$, by $|d_N(\theta)|^2$, where,

$$d_N(\theta) = \frac{1}{H^{1/2}} \sum_{t=1}^{N} (Y_t - h_t \tilde{X}) e^{-i\theta t}, \qquad (7.4.23)$$

with

$$\tilde{X} = \left(\sum_{t=1}^{N} Y_t \right) \Big/ \left(\sum_{t=1}^{N} h_t \right) \qquad \text{and} \quad H = 2\pi \sum_{i=1}^{N} h_t^2.$$

(Note that if $h_t = 1$, all t, then $Y_t \equiv X_t$, and $|d_N(\theta)|^2$ reduces to the standard definition of the periodogram, as given by (7.1.5).) The spectral estimate is now evaluated from (7.1.4), with $I_{N,X-\bar{X}}^{*}(\theta)$ replaced by $|d_N(\theta)|^2$.

To see the effect of the taper on the bias we may assume that $E[X_t] = 0$ so that the term $\tilde{X}$ is dropped from $d_N(\omega)$, and using the spectral representation of X_t (cf. (4.11.19)) write,

$$X_t = \int_{-\pi}^{\pi} e^{i\lambda t}\, dZ(\lambda).$$

We now have,

$$d_N(\theta) = \int_{-\pi}^{\pi} H(\lambda - \theta)\, dZ(\lambda), \qquad (7.4.24)$$

where,

$$H(\theta) = \frac{1}{H^{1/2}} \sum_{t=1}^{N} h_t\, e^{i\theta t}. \qquad (7.4.25)$$

Hence, using the orthogonality of $dZ(\omega)$,

$$E[|d_N(\theta)|^2] = \int_{-\pi}^{\pi} |H(\lambda - \theta)|^2 E[|dZ(\lambda)|^2]$$

$$= \int_{-\pi}^{\pi} |H(\lambda - \theta)|^2 h(\lambda)\, d\lambda. \qquad (7.4.26)$$

Writing the estimate of $h(\omega)$ as,

$$\hat{h}^*(\omega) = \int_{-\pi}^{\pi} |d_N(\theta)|^2 W_N(\omega - \theta)\, d\theta, \qquad (7.4.27)$$

we find

$$E[\hat{h}^*(\omega)] = \int_{-\pi}^{\pi} W_N^*(\omega - \lambda) h(\lambda)\, d\lambda, \qquad (7.4.28)$$

where

$$W_N^*(\phi) = \int_{-\pi}^{\pi} |H(\theta - \phi)|^2 W_N(\theta)\, d\theta. \qquad (7.4.29)$$

Thus, $E[\hat{h}^*(\omega)]$ is a weighted integral of $h(\lambda)$, the weight function $W_N^*(\phi)$ being the convolution of the spectral window $W_N(\theta)$ with the function $H(\theta)$. Note that $\hat{h}^*(\omega)$ itself cannot be written as a weighted integral of the periodogram of the form (7.1.4) with $W_N(\theta)$ replaced by $W_N^*(\theta)$. However, we can write $\hat{h}^*(\omega)$ in an analogous form to (7.1.1) by introducing the "modified sample autocovariances",

$$\hat{R}^*(s) = \left(\frac{1}{\sum\limits_{t=1}^{N} h_t^2} \right) \sum_{t=1}^{N-|s|} (Y_t - h_t \tilde{X})(Y_{t+|s|} - h_{t+|s|} \tilde{X})$$

and then we have,

$$\hat{h}^*(\omega) = \frac{1}{2\pi} \sum_{s=-(N-1)}^{(N-1)} \lambda_N(s)\hat{R}^*(s)\, e^{-i\omega s}.$$

The standard estimate $\hat{h}(\omega)$ computed from the raw data corresponds to taking $h_t = 1$, $t = 1, \ldots, N$, in which case,

$$|H(\theta)|^2 = \frac{1}{2\pi N}\left\{\frac{\sin(N\theta/2)}{\sin(\theta/2)}\right\}^2 = F_N(\theta), \qquad (7.4.30)$$

the Fejér kernel (cf. (6.2.15)) and (7.4.26) now becomes identical with (6.2.8). The exact value of $E[\hat{h}(\omega)]$ is thus given by (7.4.28) with $W_N^*(\phi)$ given by,

$$W_N^*(\phi) = \int_{-\pi}^{\pi} W_N(\theta)F_N(\theta-\phi)\, d\theta. \qquad (7.4.31)$$

Now the width of $W_N(\theta)$ is $O(1/M)$ while the width of $F_N(\theta)$ is $O(1/N)$. Thus, if $M = o(N)$ (as we have assumed throughout) then for large N, $F_N(\theta)$ acts as a δ-function w.r. to $W_N(\theta)$. Hence, if $W_N(\theta)$ is continuous everywhere we may write, $W_N^*(\phi) \sim W_N(\phi)$, and the expression for $E[\hat{h}(\omega)]$ given by (7.4.28) reduces to (7.4.21). In this case, if N is sufficiently large relative to M, the periodogram bias has a negligible effect on the bias of $\hat{h}(\omega)$, and there would be little to be gained in "tapering" the data by a transformation of the form (7.4.22).

However, if $W_N(\theta)$ has sharp discontinuities the effect of the function $F_N(\theta)$ on the bias of $\hat{h}(\omega)$ may become noticeable. Consider, e.g. the case of the Daniell window; this has discontinuities at $\theta = \pm 2\pi/M$, and when this window is convoluted with $F_N(\theta)$ the "side lobes" of $F_N(\theta)$ (see Fig. 6.2) will produce a similar "ripple" in the function $W_N^*(\phi)$ near $\phi = \pm 2\pi/M$. This means that, as far as the computation of the *exact* value of $E[\hat{h}(\omega)]$ is concerned, the Daniell window no longer has the strict rectangular form but has "ripples" at each end of its range, whose effect will be to increase the bias.

The result of tapering the data, as in (7.4.22), is to replace the Fejér kernel by the more general function $|H(\theta)|^2$, and the objective is to choose the sequence $\{h_t\}$ in such a way that $|H(\theta)|^2$ has much smaller side lobes than $F_N(\theta)$. The effects of tapering can be seen in a more intuitive way by thinking of the sample observations as having been selected from the complete realization by multiplying $\{X_t\}$ by the function,

$$h_t^{(0)} = \begin{cases} 1, & t = 1, \ldots, N, \\ 0, & \text{otherwise.} \end{cases} \qquad (7.4.32)$$

The Fourier transform of the data is then the convolution of the Fourier transform of $h_t^{(0)}$ with the "Fourier transform" of $\{X_t\}$ (viz. $dZ(\lambda)$), and this is exactly the effect which is revealed by equation (6.2.16). Since $h_t^{(0)}$ changes "sharply" from zero to 1 at $t = 1$ and $t = N$, its Fourier transform (and the squared modulus of its Fourier transform) will exhibit side lobes. However, if we replace $h_t^{(0)}$ by a "taper" h_t which changes "smoothly" from zero to 1, then its Fourier transform will, in general, have much smaller side lobes.

This feature is, in fact, a particular case of the well known "*Gibbs phenomenon*" (see, e.g., Kufner and Kadlec (1971), p. 271), which refers to the ripple effect (or "overshoot") in the partial sums of a Fourier series at a point where the function is discontinuous. To see this we first note that since $W_N^*(\phi)$ is the convolution of $W_N(\theta)$ and $|H(\theta)|^2$, its Fourier coefficients are the product of the Fourier coefficients of these functions. Hence we may write

$$W_N^*(\phi) = \frac{1}{2\pi} \sum_{s=-(N-1)}^{(N-1)} \lambda_N(s) \tilde{h}_s \, e^{-i\phi s},$$

where the $\{\tilde{h}_s\}$ are the Fourier coefficients of $|H(\theta)|^2$. In the "untapered" case $|H(\theta)|^2 = F_N(\theta)$, $\tilde{h}_s = (1 - |s|/N)$, and $W_N^*(\phi)$ is then the nth partial Cesaro sum of the Fourier series for $W_N(\phi)$. The ripples in such partial sums are smaller than they would have been had we used the "raw" partial sums corresponding to $\tilde{h}_s = 1$, all s, but then can be suppressed further by the use of more general "*convergence factors*" (or "*σ-factors*"); see Lanczos (1956). This is just what we are doing by introducing the general factor $\tilde{h}_s$ in place of $(1 - |s|/N)$. (Kaiser (1966) suggests the use of a "*prolate spheroidal*" factor which is particularly effective in suppressing the "ripples".) If the lag window $\lambda_N(s)$ is of the truncated type, with truncation point $M \ll N$, then in the untapered case the above expression for $W_N^*(\phi)$ becomes,

$$W_N^*(\phi) = \frac{1}{2\pi} \sum_{s=-M}^{M} \lambda_N(s) \left(1 - \frac{|s|}{N}\right) e^{-is\phi},$$

which is effectively the same as,

$$W_N^*(\phi) \doteq \frac{1}{2\pi} \sum_{s=-M}^{M} \lambda_N(s) \, e^{-is\phi} = W_N(\phi),$$

and we see, once again, that in this case $W_N^*(\phi)$ and $W_N(\phi)$ are virtually identical. However, for the Daniell or Bartlett–Priestley windows, where $\lambda_N(s)$ is not of the truncated type, the factor $(1 - |s|/N)$ will have a noticeable effect.

As examples of particular tapers we could choose h_t to be a *trapezium* shaped function, i.e. $h_t = 1$, over the central portion of the data and changes linearly from 1 to zero at each end of the data stretch. Alternatively, we could consider a *"cosine" taper* in which h_t again takes the value 1 over the central portion but varies as a cosine function over the initial and final stretches. If we give h_t the form of a cosine function over the complete range of the data, i.e. if we set

$$h_t = \tfrac{1}{2}\{1 - \cos(2\pi t/N)\}, \qquad t = 1, \ldots, N, \qquad (7.4.33)$$

the resulting form of h_t is known as a *Hanning taper* and it is, of course, closely related to the Tukey–Hanning lag window (6.2.79). The function $|H(\theta)|^2$ is then given by the square of (6.2.80), with M replaced by $N/2$, and its side lobes are considerably smaller than those of the Fejer kernel. In fact, any lag window of the truncated type can be used as a taper if we change M to $N/2$ and shift the function so that it attains its maximum value of 1 at (approximately) $t = N/2$. Thus, the Bartlett lag window corresponds to a *triangular* taper, which is a special case of the trapezium taper. Thomson (1977) describes the use of the *"prolate spheroidal"* taper.

Since the result of tapering is, roughly speaking, to reduce the effective length of the data, it is not surprising that the corresponding spectral estimate has a larger variance than one using the same window but based on untapered data.

Specifically, if h_t is of the form $h_t = a(t/N)$, $t = 1, \ldots, N$, where $a(u)$ is a bounded continuous function which vanishes outside the interval $(0, 1)$, then it may be shown that the asymptotic variance of $\hat{h}(\omega)$ is given by the standard expression, (6.2.108), multiplied by the factor

$$\left\{\int_0^1 a^4(u)\, du\right\} \bigg/ \left\{\int_0^1 a^2(u)\, du\right\}^2$$

(Brillinger (1975), p. 151). (It is easily shown by the Schwarz inequality that this factor is always greater than or equal to unity.) Thus, if, e.g., we use a cosine taper on the initial one-tenth and final one-tenth portions of the data, the above result shows that the variance is increased by the factor $1 \cdot 116$ (Brillinger, 1975), but hopefully this would be compensated by a corresponding reduction in the bias of the estimate.

If the true spectral density function $h(\omega)$ has a very wide dynamic range, with (say) the low frequency power exceeding the high frequency power by a factor of 10^4 or more (as may well happen in certain types of physical applications: see, e.g., Thomson (1977)), then the "ripples" in an untapered Daniell estimate could introduce severe bias in the high frequency region, and clearly it would be advisable to taper the data in this situation. However, if we know, *a priori*, that the spectral density function is of this form (or has a very large peak at a known frequency) then, as observed by Tukey (1967), it would be more sensible to "pre-whiten" the data rather than to search for "superefficient" tapers. The technique of "pre-whitening" may thus be used as an alternative to "tapering" in cases where one has some prior knowledge of the general shape of the spectral density function.

The idea of using tapering to reduce bias was suggested originally by Cooley and Tukey (1965), and is discussed further by Jones (1971) and Cooley *et al.* (1967).

7.5 CHOICE OF WINDOW

So far, the discussion has concentrated on the problem of choosing the parameters of a spectral estimate once the general form of the window has been selected. However, in practice, the first step is to choose a particular form of window (Bartlett, Daniell, Parzen, Tukey–Hanning, etc.). We have pointed out previously that the properties of the estimate depend more critically on the choice of the window bandwidth (or equivalently, on the choice of M) than on the form of the window, but nevertheless, a choice of window has to be made before an estimate can be computed, and it is not surprising that, even allowing for a suitable choice of M, different windows lead to estimates with different statistical properties. Various criteria have been suggested for judging the merit of different windows, and we now discuss critically some of the more important ones.

I. *Leakage*

We know from (6.2.102) that, for a general window $W_N(\theta)$, $E\{\hat{h}(\omega)\}$ is a weighted average of $h(\theta)$, weighted by $W_N(\omega - \theta)$. Now consider, e.g., the truncated periodogram window, (6.2.64), shown in Fig. 6.7. This window has a main lobe centred on $\theta = 0$, but in addition has a substantial negative side lobe centred on $\theta \doteq 3\pi/(2M + 1)$, together with subsidiary side lobes (alternating between positive and negative values) at $\theta \doteq 5\pi/(2M + 1)$, $7\pi/(2M + 1), \ldots$. With this window, there will be a substantial contribution to $E\{\hat{h}(\omega)\}$ from the values of $h(\theta)$ at $\theta \doteq \omega \pm 3\pi/(2M + 1)$, $\omega \pm 5\pi/(2M + 1), \ldots$, and if, for example, $\omega \pm 3\pi/(2M + 1)$ happens to coincide with a peak in $h(\theta)$ then this could seriously affect the value of $E\{\hat{h}(\omega)\}$. This effect is known as "*leakage*" (the idea being that the window allows values of $h(\theta)$, for θ outside the main lobe, to "leak" through into the expression for $E\{\hat{h}(\omega)\}$), and it has been suggested that this may be used as a basis for window selection, i.e., if we have a choice of several windows we should choose the one with the smallest leakage. Figure 7.8 shows the relative shapes of the truncated periodogram, Bartlett, Parzen, and Tukey-Hanning windows. (The graphs are based on the approximation (6.2.94), viz., $W_N(\theta) \sim MK(M\theta)$, where, for each window, $K(\theta)$ is the window

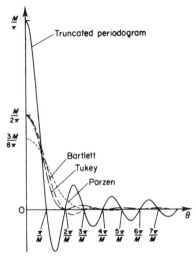

Fig. 7.8. Comparison of various spectral windows.

generator defined by (6.2.93).) It will be seen that, of these, the truncated periodogram window has by far the largest side lobes, and that the side lobes of the Bartlett window are larger than those of both the Parzen and Tukey–Hanning windows. (The Parzen window has slightly smaller side lobes than the Tukey–Hanning window). On this basis one would certainly reject the truncated periodogram window, and choose either the Parzen or Tukey–Hanning window in preference to the Bartlett window. (In fact, the selection of a as 0.23 in the general Tukey window, (6.2.75)—which leads to the Tukey–Hanning window, (6.2.78)—was motivated by an attempt to superimpose suitable combinations of truncated periodogram windows so as to obtain a near cancellation of the positive and negative lobes.) The phenomenon of leakage has been discussed extensively in the literature, *but we would argue that, judged in isolation, it does not provide a very sound basis for window selection*, for the following reasons.

(i) If side lobe leakage is to be avoided, a near ideal window is the rectangular (Daniell) form which has no side lobes at all. The estimate corresponding to the rectangular window can be computed exactly from $(N-1)$ sample autocovariances, even though the Fourier coefficients of this window do not drop to zero after $s = (N-1)$; see the discussion following (6.2.59). The rectangular estimate suffers from a form of "*periodogram-leakage*" due to the bias in the periodogram, as mentioned in the discussion on tapering in Section 7.4.1, but in general this is of a much lower order of magnitude than the effects described above, and is, in any case, a somewhat different phenomenon. (The windows shown in Fig. 7.8 suffer from "*window-leakage*", i.e. the estimate $\hat{h}(\omega)$ is contaminated by values of $I_N(\theta)$ at frequencies far removed from ω, and the "periodogram-leakage" is superimposed on this effect, although here the "window-leakage" far outweighs the "periodogram-leakage". For the rectangular window, there is no "window-leakage", and the "periodogram-leakage" is the sole effect.) In the past the rectangular estimate has been avoided since its evaluation from (7.1.1) involves the computation of all the $(N-1)$ sample autocovariances; however, since the advent of the "fast Fourier transform", which we discuss in the next section, the evaluation of this estimate directly from (7.1.4) presents no computational difficulties. Nevertheless, as we shall see later, this window is not optimal according to a relative mean square error criterion.

(ii) The degree of distortion due to side lobes can be controlled by the choice of M. As we increase M we contract the window and thus reduce the distance between the first side lobe and the main lobe. Of course, if M is inappropriately chosen and is made excessively small, so that B_W, the bandwidth of $W_N(\theta)$, is large relative to B_h, the spectral bandwidth, then the leakage effect can be troublesome. For suppose $h(\omega)$ has a very narrow peak at $\omega = \omega_0$. If, for example, we use the Bartlett window we will obtain subsidiary peaks in $E\{\hat{h}(\omega)\}$ at those values of ω for which $(\omega \pm \omega_0)$ coincides with a side lobe, leading to a "ripple" effect in the form of $E\{\hat{h}(\omega)\}$.

(If B_W is large relative to B_h then instead of $W_N(\theta)$ acting as a "pseudo" δ-function with respect to $h(\theta)$, $h(\theta)$ will act as a "pseudo" δ-function with respect to $W_N(\theta)$. In this case, the convolution of $h(\theta)$ with $W_N(\theta)$ will reproduce, in effect, a "smoothed" version of $W_N(\theta)$ in the region of the peak; hence the "ripple effect" in $E(\hat{h}(\omega)\}$.)

(iii) However, the most important point is that *leakage is merely one aspect of window performance*. It is, in fact, a relevant consideration only in as much as it affects the magnitude of the *bias*.

Similar criticisms apply to any method of judging the merits of windows which is based on *ad hoc* geometrical features such as continuity, or continuity of derivatives, etc. If we wish to formulate an objective approach to window selection then obviously we must construct a criterion of "efficiency" which takes account of both the *bias and the variance* of the estimate, and then try to determine the functional form of the window which optimizes this criterion. This approach has been criticized on the grounds that any such criterion is, in a sense, "arbitrary", but before we can compute a spectral estimate we have to select a particular window, and in so doing we are bound to be guided by some "criterion", even if it amounts to nothing more than following current fashion or using whatever window happens to be available on the local computer! We feel, therefore, that there is considerable merit in the construction of a sensible criterion, which we may then use either to compare the performance of the standard windows or to derive the form of the "optimal" window. In following this approach we are in accord with well established principles of general statistical inference where we define measures of efficiency for estimators and then derive the form of estimator which has maximum efficiency. However, the problem of spectral analysis involves the estimation of a complete *function* (as opposed to the estimation of a single parameter) and consequently there are various possible criteria which may be used to measure the overall efficiency of the various spectral estimates. Those which have been suggested so far may be summarized as follows.

II. *The integrated mean square error*

The first attempt to derive an optimum estimate is due to Lomnicki and Zaremba (1957a) who considered estimates of the form (7.1.1) (but with $\hat{R}(s)$ replaced by the "unbiased" estimate $\hat{R}^*(s)$, given by (5.3.10)), and proposed as a criterion the integrated mean square error, defined by,

$$M_I = \int_{-\pi}^{\varepsilon} E\{\hat{h}(\omega) - h(\omega)\}^2 \, d\omega. \tag{7.5.1}$$

They showed that the lag window $\lambda_N(s)$ which minimizes M_I is that given by (6.2.88), but unfortunately this function depends on the unknown form of the theoretical autocovariance function, and hence cannot be implemented in practice. Lomnicki and Zaremba suggested that, as a practical device, one could approximate to the lag window (6.2.88) by a function of the form (6.2.89), but this form of $\lambda_N(s)$ is no longer optimal.

Lomnicki and Zaremba's criterion has been criticized on the grounds that it gives equal weight to all frequencies (cf. Bartlett (1957), Jenkins (1961)). However, without any prior information on $h(\omega)$ it would be difficult to suggest any obvious alternative weighting.

III. *The expected maximum squared error*

Parzen (1961b) suggested the following criterion,

$$M_M = E[\max_{\omega} \{\hat{h}(\omega) - h(\omega)\}^2]. \tag{7.5.2}$$

Parzen points out that this is a relevant quantity to consider when constructing confidence bands for the complete spectral density function (cf. the discussion on confidence bands for the integrated spectrum in Section 6.2.6), and shows that, using a form of Chebysheff's inequality,

$$p[\max_{\omega}|\hat{h}(\omega) - h(\omega)| > \varepsilon] \leqslant \frac{1}{\varepsilon^2} M_M, \qquad (7.5.3)$$

for any $\varepsilon > 0$. Thus, given the value of M_M for a particular estimate we can construct confidence bands for the complete spectral functions. However, the evaluation of M_M for estimates based on different windows would seem to be quite a difficult problem, and, as yet, no results have been reported on the derivation of the functional form of $W_N(\theta)$ which optimizes this criterion.

IV.   *The maximum relative mean square error*

Grenander and Rosenblatt (1957a) considered the mean square error at a particular frequency, viz.,

$$E\{\hat{h}(\omega) - h(\omega)\}^2 = \text{var}\{\hat{h}(\omega)\} + [\text{bias}\{\hat{h}(\omega)\}]^2,$$

but since both the (asymptotic) variance and squared bias are proportional to $h^2(\omega)$ it would seem more suitable to consider the *relative mean square error* $\eta^2(\omega)$ defined by (7.2.3), namely,

$$\eta^2(\omega) = \{v^2(\omega) + b^2(\omega)\}/h^2(\omega), \qquad (7.5.4)$$

where $v^2(\omega)$, $b(\omega)$, denote respectively the variance and bias of $\hat{h}(\omega)$. We may now construct an overall criterion by considering the maximum value of $\eta^2(\omega)$, i.e. we set

$$M_R = \max_{0 < \omega < \pi} \{\eta^2(\omega)\}. \qquad (7.5.5)$$

Using the asymptotic expressions (6.2.113), (6.2.124), for the variance and bias of scale parameter windows we may write,

$$\eta^2(\omega) \sim \frac{M}{N} \int_{-\infty}^{\infty} k^2(u)\, du + \frac{1}{M^{2r}} \{k^{(r)}\}^2 \left\{\frac{1}{\lambda^{(r)}(\omega)}\right\}^{2r}, \qquad (7.5.6)$$

where r is the characteristic exponent of $k(u)$, $k^{(r)}$ is defined by (6.2.121), and $\lambda^{(r)}(\omega) = |h(\omega)/h^{(r)}(\omega)|^{1/r}$ is the "spectral bandwidth of order r" at frequency ω, defined by (7.2.5). Since the first term is independent of ω, the maximum value of $\eta^2(\omega)$ will, to this order of approximation, be determined by the minimum value of $\lambda^{(r)}(\omega)$. Thus, writing $\lambda_0^{(r)} = \inf_{\omega}\{\lambda^{(r)}(\omega)\}$ (assumed >0), the maximum value of $\eta^2(\omega)$ is given by replacing $\lambda^{(r)}(\omega)$ by $\lambda_0^{(r)}$

in (7.5.6). Now let I_W, B'_W be defined as in (7.4.3), (7.4.4), so that M_R can then be written as,

$$M_R \sim \frac{M}{N} I_W + \frac{1}{M^{2r}} \left\{ \frac{B'_W}{2\sqrt{6}\lambda_0^{(r)}} \right\}^{2r}. \tag{7.5.7}$$

We now minimize (7.5.7) w.r to M. Differentiating this expression and setting the derivative to zero gives,

$$M = \left[2r \left\{ \frac{B'_W}{2\sqrt{6}\lambda_0^{(r)}} \right\}^{2r} \frac{N}{I_W} \right]^{1/(2r+1)}, \tag{7.5.8}$$

and substituting this value of M in (7.5.7) gives,

$$\min_{M}\{M_R\} \sim N^{-2r/(2r+1)} \left\{ \frac{B'_W I_W}{2\sqrt{6}\lambda_0^{(r)}} \right\}^{2r/(2r+1)} \{(2r)^{1/(2r+1)} + (2r)^{-2r/(2r+1)}\}. \tag{7.5.9}$$

We see from this result that $\min\{M_R\}$ depends on the choice of window only through r and the product $(B'_W . I_W)$. Since $\min\{M_R\} = O(N^{-2r/(2r+1)})$ it would seem sensible to choose a window with as large a value of r as possible, so as to make the order of magnitude of $\min\{M_R\}$ as small as possible. However, *no window with $r > 2$ can be non-negative*, since $r > 2 \Rightarrow k^{(2)} = 0 \Rightarrow \int_{-\infty}^{\infty} \theta^2 K(\theta) \, d\theta = 0$. Hence, it would seem that *the most useful class of windows to consider is that for which $r = 2$.* (Parzen (1957b) reached the same conclusion but for somewhat different reasons.) Fortunately, almost all the windows listed in Table 7.1 have characteristic exponent $r = 2$, the one exception being the Bartlett window for which $r = 1$. Thus, for the Daniell, Parzen, Tukey–Hanning, Tukey–Hamming, and Bartlett–Priestley windows we find that (setting $r = 2$ in (7.5.9)), $\min\{M_R\} = O(N^{-4/5})$, whereas for the Bartlett window, $\min\{M_R\} = O(N^{-2/3})$, which is a larger order of magnitude than in the other cases. On the basis of the asymptotic value of $\min\{M_R\}$ we would therefore reject the Bartlett window in favour of the others.

If we now restrict attention to windows with $r = 2$, then $\lambda_0^{(r)} = \inf_{\omega}\{\lambda^{(2)}(\omega)\} = \frac{1}{2}B_h$, where B_h is the spectral bandwidth defined by (7.3.6). Setting $r = 2$ in (7.5.9) now gives,

$$\min\{M_R\} \sim 0.4626 N^{-4/5} \left\{ \frac{B'_W I_W}{\frac{1}{2}B_h} \right\}^{4/5}, \tag{7.5.10}$$

and the value of $\min\{M_R\}$ depends on the form of the window only through the product $(B'_W . I_W)$. We may thus regard the product $(B'_W . I_W)$ as an index of the "efficiency" of a window, in the sense that the smaller this product, the smaller the minimum relative mean square error. Below, we compare the values of this product for the standard scale parameter windows with $r = 2$.

Table 7.2 *Comparative efficiencies of windows*

Window	B'_W	I_W	$B'_W . I_W$
Daniell	2π	1	6·2832
Parzen	12	0·54	6·4800
Tukey–Hanning	$2·45\pi$	0·75	5·7715
Tukey–Hamming	$2·35\pi$	0·7948	5·8665
Bartlett–Priestley	$1·55\pi$	1·2	5·8403

The most striking feature of these results is that, on the basis of $B'_W I_W$, there is not a great deal of difference between the various windows, the values of $(B'_W I_W)$ being quite close to each other. The Tukey–Hanning window achieves the smallest value, but it must be remembered that this window is *not* non-negative. Of the remainder, the Bartlett–Priestley window (which is non-negative and has a quadratic shape) achieves the smallest value. In fact, it may be shown (as below) that, in the above sense, *this window is optimal over the class of non-negative windows with $r = 2$.* That is, if we consider only non-negative windows, the functional form of $W_N(\theta)$ which minimizes (7.5.10) is the quadratic window (6.2.85). This window was first proposed on these grounds by Priestley (1962c), and later by Bartlett (1965), and its optimality properties were noted also by Epanechnikov (1969). See also Priestley (1965a).

Of those considered in Table 7.2, the Parzen window has the largest value of the $(B'_W I_W)$ index, being comparable with the Daniell window. It may be noted that the Daniell window, which according to the "side lobe leakage" philosophy has a near-ideal shape, performs slightly worse than most of the other windows according to the above criterion.

It is interesting to note also that, on the above criterion, the performance of the Tukey–Hamming window is marginally inferior to that of the Tukey–Hanning window. The Tukey–Hamming window corresponds to setting $a = 0·23$ in the general Tukey window (6.2.75), and this value of a was chosen specifically to reduce side lobe leakage. The Tukey–Hanning window corresponds to setting a at the "non-optimal" value of 0·25.

A formal proof of the optimality of the quadratic window is easily constructed using the following variational argument.

Optimality property of the quadratic window

We first observe that for windows with $r = 2$,

$$B'_W \propto \{k^{(2)}\}^{1/2} = \left\{ \frac{1}{2} \int_{-\infty}^{\infty} \theta^2 K(\theta)\, d\theta \right\}^{1/2}$$

(cf. (7.3.29)), where

$$K(\theta) = \frac{1}{2\pi} \int_{-\infty}^{\infty} k(u) \, e^{-iu\theta} \, du \qquad (7.5.11)$$

is the spectral window generator. Also, $I_W = 2\pi \int_{-\infty}^{\infty} K^2(\theta) \, d\theta$, and hence to find the functional form of $K(\theta)$ which minimizes $(B'_W I_W)$ it suffices to find that form which minimizes,

$$C_K = \left\{ \int_{-\infty}^{\infty} K^2(\theta) \, d\theta \right\} \left\{ \int_{-\infty}^{\infty} \theta^2 K(\theta) \, d\theta \right\}^{1/2}. \qquad (7.5.12)$$

(Note that C_K is essentially equivalent to the criterion considered by Grenander in the formulation of his "uncertainty principle"; see Section 7.3.2.)

Now let,

$$K_0(\theta) = \begin{cases} \dfrac{3}{4\pi} \left\{ 1 - \left(\dfrac{\theta}{\pi} \right)^2 \right\}, & |\theta| \leqslant \pi, \\[2mm] 0, & |\theta| > \pi, \end{cases} \qquad (7.5.13)$$

which is the form of the Bartlett–Priestley window generator (see Section 6.2.3), and let $K(\theta)$ be any other non-negative function satisfying

$$\int_{-\infty}^{\infty} K(\theta) \, d\theta = \int_{-\infty}^{\infty} K_0(\theta) \, d\theta = 1, \qquad (7.5.14)$$

and

$$\int_{-\infty}^{\infty} \theta^2 K(\theta) \, d\theta = \int_{-\infty}^{\infty} \theta^2 K_0(\theta) \, d\theta. \qquad (7.5.15)$$

We will show that,

$$\int_{-\infty}^{\infty} K^2(\theta) \, d\theta \geqslant \int_{-\infty}^{\infty} K_0^2(\theta) \, d\theta. \qquad (7.5.16)$$

Write,

$$K(\theta) = K_0(\theta) + \varepsilon(\theta).$$

Then by (7.5.14), (7.5.15), we must have,

$$\int_{-\infty}^{\infty} \varepsilon(\theta) \, d\theta = 0, \qquad (7.5.17)$$

and

$$\int_{-\infty}^{\infty} \theta^2 \varepsilon(\theta) \, d\theta = 0. \qquad (7.5.18)$$

and since $K(\theta)$ is non-negative we must also have,

$$\varepsilon(\theta) \geqslant 0, \qquad |\theta| > \pi. \tag{7.5.19}$$

Now,

$$\int_{-\infty}^{\infty} K^2(\theta) \, d\theta = \int_{-\infty}^{\infty} K_0^2(\theta) \, d\theta + 2 \int_{-\infty}^{\infty} \varepsilon(\theta) K_0(\theta) \, d\theta + \int_{-\infty}^{\infty} \varepsilon^2(\theta) \, d\theta$$

$$= \int_{-\infty}^{\infty} K_0^2(\theta) \, d\theta + 2U_1 + U_2,$$

say, and

$$U_1 = \frac{3}{4\pi} \int_{-\pi}^{\pi} \left\{ 1 - \left(\frac{\theta}{\pi}\right)^2 \right\} \varepsilon(\theta) \, d\theta$$

$$= \frac{3}{4\pi} \left[\int_{-\infty}^{\infty} \left\{ 1 - \left(\frac{\theta}{\pi}\right)^2 \right\} \varepsilon(\theta) \, d\theta - \int_{|\theta|>\pi} \left\{ 1 - \left(\frac{\theta}{\pi}\right)^2 \right\} \varepsilon(\theta) \, d\theta \right].$$

But the first term above vanishes by (7.5.17), (7.5.18), and therefore,

$$\frac{4\pi U_1}{3} = \int_{|\theta|>\pi} \left\{ \left(\frac{\theta}{\pi}\right)^2 - 1 \right\} \varepsilon(\theta) \, d\theta \geqslant 0,$$

since, by (7.5.19), $\varepsilon(\theta) \geqslant 0$ on $|\theta| > \pi$. Also, it is obvious that $U_2 \geqslant 0$, and hence we have proved the inequality (7.5.15), with equality iff $\varepsilon(\theta) \equiv 0$.

We have thus established the result that the quadratic function $K_0(\theta)$, as given by (7.5.13), is the functional form of $K(\theta)$ which minimizes C_K, and hence minimizes the index $(B'_W I_W)$.

V. *Minimizing the record length*

In discussing the maximum relative mean square error criterion we first minimized $\eta^2(\omega)$ w.r. to M by setting M according to (7.5.9). Jenkins (1961) pointed out that this approach may not be suitable if, e.g., we are particularly concerned with estimating $h(\omega)$ in the neighbourhood of a peak, since then the value of M would be determined rather by resolution considerations, as discussed in Section 7.4. To cater for this situation we now adopt a slightly different approach, and instead of seeking the window which minimizes M_R we now seek the window which minimizes the number of observations required to construct an estimate with a given level of resolution and precision.

Suppose then that we are given the spectral bandwidth B_h and that we set the resolution by choosing M so that,

$$B_W = \alpha \cdot B_h, \tag{7.5.20}$$

where α is a prescribed constant. (This represents a slightly more general version of (7.4.1).) If we use $\Delta_p(\omega)$ as our measure of precision, then for given p, $\Delta_p(\omega)$, the required number of observations N is given by solving (7.2.1) for that frequency ω_0 which corresponds to the minimum value of $\lambda^{(2)}(\omega)$ (see the discussion following (7.4.2)). When $\omega = \omega_0$, we have from (7.3.28),

$$|\text{bias}\{\hat{h}(\omega_0)\}| \sim \tfrac{1}{6}\{B_W/B_h\}^2 h(\omega_0) = \tfrac{1}{6}\alpha^2 \cdot h(\omega_0),$$

by (7.5.20). This determines the value of the second term in (7.2.1), and using the fact that $\text{var}\{\hat{h}(\omega)\}/h^2(\omega) \sim (I_W \cdot M)/N$, (7.2.1) becomes

$$\gamma_p \cdot \sqrt{\frac{I_W M}{N}} = \{\Delta_p(\omega_0) - \tfrac{1}{6}\alpha^2\}. \tag{7.5.21}$$

Assuming that α and $\Delta_p(\omega_0)$ are chosen so that $\Delta_p(\omega_0) > \tfrac{5}{6}\alpha^2$, the value of M determined by (7.5.20) will exceed the value of $M_{\min}$ as given by (7.4.9) (see Section 7.4), and the required value of N is given by,

$$N = \frac{\gamma_p^2 I_W M}{\{\Delta_p(\omega_0) - \tfrac{1}{6}\alpha^2\}^2}, \tag{7.5.22}$$

with M given by (7.5.20).

Now, by the definition of B'_W, it follows immediately that $B_W = B'_W/M$, and hence (7.5.20) gives,

$$M = B'_W/\{\alpha B_h\}.$$

Substituting this value of M in (7.5.22) we obtain,

$$N = \left[\frac{\gamma_p^2}{\alpha B_h \{\Delta_p(\omega_0) - \tfrac{1}{6}\alpha^2\}^2}\right] \cdot (B'_W I_W). \tag{7.5.23}$$

Since γ_p, $\Delta_p(\omega_0)$, α and B_h are constants which are quite independent of the window form, we see from (7.5.23) that *the value of N depends on the choice of window through exactly the same index, $(B'_W I_W)$, as determines the magnitude of the maximum relative mean square error, M_R*. On the other hand, if $\Delta_p(\omega_0) < \tfrac{5}{6}\alpha^2$, then $M_{\min}$ would be the appropriate value of M to use (see Section 7.4), and the corresponding value of N, $N_{\min}$ (as given by (7.4.9a)), still depends on the window only through $(B'_W I_W)$. Hence, it follows that *the window which achieves the smallest value of $(B'_W I_W)$ will not only minimize M_R but will also have the property that it requires the smallest number of observations for a given degree of resolution and precision*. Thus, from Table 7.2 we see that, among the windows considered, the Tukey–Hanning window requires the smallest number of observations, while within the class of non-negative windows the Bartlett–Priestley window requires the smallest number of observations.

To illustrate these results let us consider again the spectral density function discussed in Example 1 of Section 7.4. Here, B_h is given as $\pi/20$ (radians per unit time). If we set $\alpha = \frac{1}{2}$, $p = 0 \cdot 95$, $\Delta_p(\omega_0) = 0 \cdot 25$, then $\gamma_p = 1 \cdot 96$, and the numbers of observations appropriate for the various windows are found from (7.5.23). For all the windows, the value of M given by (7.5.20) with $\alpha = \frac{1}{2}$ will exceed $M_{\min}$ since here $\Delta_p(\omega_0) > \frac{5}{24}$; see Section 7.4. Using Table 7.2 for the values of $(B'_W I_W)$ we obtain:

Daniell:	7081	(as in example 1).
Parzen:	7303	
Tukey–Hanning:	6504	
Tukey–Hamming:	6611	
Bartlett–Priestley:	6582.	

We may note that, irrespective of the values of α, γ_p, $\Delta_p(\omega_0)$ and B_h, the required numbers of observations (as determined from (7.5.23)) will always be in the above ratios, since these ratios are determined by the relative values of $(B'_W I_W)$.

Comparison of equibandwidth variances

As a slight variation on the above theme, we may compare the performances of two windows (corresponding to the same value of r) by first adjusting the values of M so as to give them equal bandwidths, and then compare their variances. For example, if we make the values of B_W the same for the Parzen and Tukey–Hanning windows, the respective values of M, M_P and M_T, say, must satisfy (see Table 7.1)),

$$12/M_P = 2 \cdot 45 \pi/M_T, \qquad \text{or} \quad M_P \doteq 1 \cdot 56 M_T.$$

Thus, for equal bandwidths (as measured by B_W), *and hence for equal asymptotic bias*, the truncation point for the Parzen window should be approximately $1 \cdot 56$ times that for the Tukey–Hanning window. Now, the asymptotic variance of the Parzen estimate is $0 \cdot 54 M_P/N$, while that of the Tukey estimate is $0 \cdot 75 M_T/N$. When M_P and M_T are related as above, the variance of the Parzen estimate is approximately $1 \cdot 123$ times that of the Tukey estimate. This ratio is just the ratio of the values of the index $(B'_W I_W)$ for the two windows, and we can see that this result holds generally since we can write, $(B'_W I_W) = N(B'_W/M)((M/N)I_W)$, the first factor being the bandwidth B_W and the second factor the variance of the estimate. (In fact, we see from (7.3.31) that $(1/N)(B'_W I_W)$ is just the "constant" in the equation, bandwidth $\times$ variance = constant.)

If the same procedure is followed using other definitions of window bandwidth then, of course, the ratio of the variances will change. Thus, equalizing bandwidths on the basis of the Parzen definition B_P gives the variance of the Parzen estimate as 0.96 times the variance of the Tukey estimate, while equalizing bandwidths using the Jenkins' definition B_J leads (by definition of B_J) to exactly equal variances. However, these other definitions of window bandwidth do not provide a valid basis for comparison of windows since equalizing bandwidths according to either the B_P or B_J definitions *does not give equal asymptotic bias* in the two estimates. In fact, *any* two windows whose bandwidths are equated according to the B_J definition are bound to lead to equal variances, but the bias of the two estimates may be quite different.

Discussion

As with the discussion of design relations in Section 7.4 we would advise the reader not to read too much into the discussion on optimality criteria. In particular, it must be remembered that throughout we have used asymptotic expressions for the bias and variance, and the resulting expression for e.g., the relative mean square error is only approximate in the case of finite N. Nevertheless, the broad conclusions of this analysis are certainly interesting, and may be summarized as follows.

(a) The most useful class of windows seems to be that with characteristic exponent $r = 2$. This class produces a relative mean square error of order $O(1/N^{4/5})$ whereas windows with $r = 1$ lead to a relative mean square error of order $O(1/N^{2/3})$. On the other hand, windows with $r > 2$ can never be non-negative.

(b) According to the $(B'_W I_W)$ index, there is not a great deal of difference between the efficiencies of the windows listed in Table 7.2 (all of which have $r = 2$), although small differences do exist. However, bearing in mind the approximate nature of this analysis, it would be unwise to attach too much importance to very small differences in the values of this index.

(c) As a point of theoretical interest mainly, the functional form of the non-negative window which minimizes the index $(B'_W I_W)$ is the quadratic window whose generator is given by (7.5.13). To the order of the approximations used, this window also minimizes the maximum relative mean square error, but this is due essentially to the fact that the expression used for the bias is equivalent to using the *quadratic* approximation (6.2.118). If we had included higher order terms, the form of the optimal window would no doubt have been different. (This approximate expression for bias is not sufficiently accurate to take account of, e.g., "window-leakage" effects.)

7.6 COMPUTATION OF SPECTRAL ESTIMATES: THE FAST FOURIER TRANSFORM

In previous sections we dealt with the "design" of spectral estimates, i.e. the problem of choosing the form of the spectral window and the values of the associated parameters. Once the window and its parameters have been selected we are left with the task of computing the estimates, and we now turn our attention to the study of efficient methods of computation.

Prior to about 1965 the standard method of evaluating a spectral estimate $\hat{h}(\omega)$ was via (7.1.1), i.e. as a weighted sum of sample autocovariances $\hat{R}(s)$. The bulk of these calculations lies in the computation of the $\{\hat{R}(s)\}$, and consequently it was customary to use a lag window of the truncated type (such as the Bartlett, Parzen, or Tukey windows), so that only $(M-1)$ $(\ll N-1)$ autocovariances need to be computed. By contrast, the Daniell window requires all $(N-1)$ autocovariances. However, the introduction by Cooley and Tukey (1965) of the "fast Fourier transform" algorithm has virtually revolutionized the computational aspects of spectral analysis. This algorithm provides a general numerical technique for computing finite Fourier transforms in an extremely fast and efficient manner, and although it has a wide range of applications outside the context of spectral analysis it is particularly suited to the computation of spectral estimates. Using this algorithm we can quite easily compute the periodogram $I_{N,X-\bar{X}}^*(\theta)$, directly from the data (i.e. from (7.1.5)), and then evaluate $\hat{h}(\omega)$ from (7.1.4) by replacing the integral by a sum over a discrete set of periodogram ordinates. It turns out that this method is generally more efficient (in terms of computing time) than the use of (7.1.1), even for estimates using truncated lag windows, and this is now the standard method used in computer programs for spectral analysis. Before considering the details of this method we first discuss briefly the basic ideas underlying the algorithm. Our discussion deals only with the most simple form of the algorithm; for more comprehensive accounts the reader is referred to Cooley and Tukey (1965), Gentleman and Sande (1966), Bingham *et al.* (1967) Brigham and Morrow (1967), Cooley *et al.* (1967, 1970) and Brigham (1974).

The fast Fourier transform

Suppose that we are given N numbers $x_0, x_1, \ldots, x_{N-1}$ and wish to compute the finite Fourier transforms,

$$d(\omega_p) = \sum_{t=0}^{N-1} x_t \exp(i\omega_p t), \qquad p = 0, 1, \ldots, (N-1), \qquad (7.6.1)$$

where $\omega_p = 2\pi p/N$. For each p the evaluation of $d(\omega_p)$ requires N (complex) multiplications and additions, and hence the evaluation of the complete set of the $\{d(\omega_p)\}$ requires N^2 operations. However, if N can be factorized in the form $N = r \cdot s$, $(r, s,$ integers), then we can effect a substantial reduction in the arithmetic by computing $d(\omega_p)$ in two stages, as follows. First we note that each t in the range $(0, N-1)$ may be written,

$$t = rt_1 + t_0,$$

where, as t goes from 0 to $(N-1)$, t_1 goes from 0 to $(s-1)$ and t_0 goes from 0 to $(r-1)$. Similarly, each p in the range $(0, N-1)$ may be written,

$$p = sp_1 + p_0,$$

where p_1 goes from 0 to $(r-1)$ and p_0 from 0 to $(s-1)$. The expression (7.6.1) for $d(\omega_p)$ may now be re-written as

$$d(\omega_p) = \sum_{t_0} \sum_{t_1} x_{(rt_1+t_0)} \exp[2\pi i p_{(rt_1+t_0)}/N]$$

$$= \sum_{t_0} \exp(2\pi i p t_0/N)\left\{\sum_{t_1} x_{(rt_1+t_0)} \exp(2\pi i p t_1/s)\right\}. \tag{7.6.2}$$

But, writing $p = sp_1 + p_0$,

$$\exp(2\pi i p t_1/s) \equiv \exp(2\pi i p_1 t_1) \cdot \exp(2\pi i p_0 t_1/s) \equiv \exp(2\pi i p_0 t_1/s)$$

(since p_1, t_1, are both integers). Hence, for each p, the expression resulting from the summation over t_1 in (7.6.2) does not depend on the value of p_1, and is thus a function of p_0 and t_0 only. We may therefore write,

$$\sum_{t_1} x_{(rt_1+t_0)} \exp(2\pi i p t_1/s) = a(p_0, t_0), \tag{7.6.3}$$

say, and evaluate $d(\omega_p)$ in the form,

$$d(\omega_p) = \sum_{t_0} a(p_0, t_0) \exp(2\pi i p t_0/N). \tag{7.6.4}$$

The evaluation of $d(\omega_p)$ may thus be broken down into two stages, namely (i) the evaluation (for each p_0, t_0) of the transforms (7.6.3), which run over s points, and (ii) the evaluation of the transforms (7.6.4), which run over r points. Now there are rs pairs of values of (p_0, t_0) and each $a(p_0, t_0)$ requires s additions and multiplications giving, in all, rs^2 operations. Also, (7.6.4) requires r additions and multiplications for each p, and there are N values of p to consider. Hence, the total number of operations required is,

$$rs^2 + Nr = N(r+s),$$

which will generally be substantially smaller than the N^2 operations required for a direct evaluation of (7.6.1). For example, if $N = 437$ then·a direct evaluation of (7.6.1) requires 190 969 operations. On the other hand, writing $437 = 19 \times 23$, the above algorithm requires only $437 \times 42 = 18\ 354$ operations, which is approximately 10% of the previous figure.

The above device can be extended considerably if N may be factorized in the form

$$N = r_1 . r_2 . r_3 \ldots r_p,$$

where $r_1, r_2, \ldots, r_p$ are prime numbers (not necessarily distinct). In this case we may start by applying the above algorithm with $r = r_1$, $s = (r_2 . r_s \ldots r_p)$, and then successively reduce the evaluation of (7.6.3) by taking $r = r_2$, $s = (r_3 \ldots r_p), \ldots$, and so on. In this way we reduce the problem of evaluating the "N point" transform (7.6.1) to that of evaluating p relatively short transforms, and the total number of operations required is now $N(r_1 + r_2 + \ldots + r_p)$. In particular, if N happens to be of the form $N = 2^p$, the number of operations required is $2pN$, i.e. $[N . (2 \log N / \log 2)]$, which is $O(N \log N)$, as compared with N^2 operations required for direct evaluation. The transforms in this case are very easy to evaluate (since they are all "2 point" transforms and thus entail essentially only simple addition and subtraction), and computer programs for fast Fourier transforms are often based on this form of the algorithm. Although N will rarely be of the form 2^p initially one may add zeros to the data so as to increase the (apparent) number of data points until it attains the required form. Adding zeros to the original data will not, of course, affect the numerical values of the finite Fourier transforms but it will affect the frequencies at which the transforms are evaluated. Thus, if we add zeros so as to increase N to N' (say) then the transforms will be evaluated at the frequencies $\omega'_p = 2\pi p / N'$, rather than at $\omega_p = 2\pi p / N$. Some authors have suggested that whenever the record length is artificially increased by the addition of zeros the modified record should be "*tapered*" at both ends (see Section 7.4.1) so as to "smooth" the transition from non-zero to zero values; see, e.g., Tukey (1967), Godfrey (1974), Brillinger (1975). Computer programs are now available for the more general forms of the algorithm (in which N is assumed to be highly composite but not necessarily a power of 2); a general program has been given by Singleton (1969).

Calculation of covariances by inverting the periodogram

The fast Fourier transform algorithm provides an efficient method for calculating the periodogram, $I^*_{N, X - \bar{X}}(\theta)$, directly from the "mean-corrected" data. Thus, writing $X'_t = X_t - \bar{X}$, we apply the algorithm to compute the finite Fourier transforms (cf. Section 6.2.1),

$$\zeta_{X'}(\omega_p) = \frac{1}{\sqrt{2\pi N}} \sum_{t=1}^{N} X_t' e^{-i\omega_p t}, \qquad p = 0, 1, \ldots, [N/2], \qquad (7.6.5)$$

and then evaluate the periodogram ordinates from,

$$I^*_{N, X - \bar{X}}(\omega_p) = |\zeta_{X'}(\omega_p)|^2. \qquad (7.6.6)$$

This technique gives us the periodogram ordinates at the frequencies $\omega_p = 2\pi p/N$ (or at $\omega_p' = 2\pi p/N'$ if zeros have been added to the data), and we can now evaluate the spectral estimates $\hat{h}(\omega)$ by for example approximating the integral in (7.1.4) by a sum over the discrete set of frequencies $\{\omega_p\}$ (or $\{\omega_p'\}$). However, we noted previously that it was sometimes useful to examine the behaviour of the sample autocovariance function, $\hat{R}(s)$, so as to obtain a rough idea of an appropriate value for the window bandwidth (or truncation point) when such information was not available *a priori*. If the periodogram ordinates have already been computed we can evaluate the sample autocovariances by finding the inverse Fourier transform of the periodogram. This operation involves the computation of two Fourier transforms, but if we use the fast Fourier transform in both cases it is often faster (for series of, say, 1000 observations or longer) to evaluate the $\{\hat{R}(s)\}$ in this way rather than computing them directly from (7.1.2). To recover the $\{\hat{R}(s)\}$ from the periodogram ordinates we note first that from (6.2.5) we may write,

$$I^*_{N, X - \bar{X}}(\omega) = \frac{1}{2\pi} \sum_{s=-(N-1)}^{(N-1)} \hat{R}(s) e^{-is\omega} \qquad (7.6.7)$$

(where $\hat{R}(s)$ is here evaluated from the mean corrected observations X_t' so that (6.2.6) now becomes identical with the general expression (5.3.13) for $\hat{R}(s)$). Treating (7.6.7) as a Fourier series, the inversion formula is,

$$\int_{-\pi}^{\pi} I^*_{N, X - \bar{X}}(\omega) e^{is\omega} \, d\omega = \begin{cases} \hat{R}(s), & s = -(N-1), \ldots, (N-1), \\ 0, & |s| \geq N. \end{cases}$$
$$(7.6.8)$$

We cannot, of course, evaluate the LHS of (7.6.8) since the periodogram ordinates are computed only at a discrete set of frequencies. However, since (7.6.7) is a *finite* trigonometric polynomial we can obtain the $\{\hat{R}(s)\}$ from the periodogram ordinates at a discrete set of points. Let us then try to obtain the $\{\hat{R}(s)\}$ from the ordinates at frequencies $\omega_p = 2\pi p/N$, $p = 0, \pm 1, \ldots, \pm[N/2]$. Writing (for ease of notation) $I^*(\omega_p)$ in place of $I^*_{N, X - \bar{X}}(\omega_p)$, (7.6.7) gives

$$I^*(\omega_p) = \frac{1}{2\pi} \sum_{s=-(N-1)}^{(N-1)} \hat{R}(s) e^{-i\omega_p s}, \qquad (7.6.9)$$

and multiplying (7.6.9) by $\exp(i\omega_p r)$ and summing over p we obtain,

$$\sum_{p=-[N/2]}^{[N/2]} I^*(\omega_p) \exp(i\omega_p r) = \frac{1}{2\pi} \sum_s \hat{R}(s)\left\{\sum_p \exp[i\omega_p(r-s)]\right\}. \qquad (7.6.10)$$

Now,

$$\sum_{p=-[N/2]}^{[N/2]} \exp[i\omega_p(r-s)] = \begin{cases} 0, & (r-s) \neq 0, \pm N, \\ N, & (r-s) = 0, \pm N. \end{cases}$$

Hence the LHS of (7.6.10) is $(N/2\pi)\{\hat{R}(r) + \hat{R}(N-|r|)\}$ and we clearly *cannot* recover $\hat{R}(s)$ using the frequencies $\{\omega_p\}$. This feature becomes obvious once we realize that (for real valued processes) the set of frequencies $\{\omega_p\}$ provide only $[N/2]$ periodogram ordinates from which we obviously cannot determine the N unknowns $\hat{R}(0), \hat{R}(1), \ldots, \hat{R}(N-1)$. It suggests, however, that if we use the frequencies $\omega'_p = 2\pi p/(2N-1)$, $p = 0, \pm 1, \ldots,$ $\pm(N-1)$, then we will have just the right number of ordinates. Using these frequencies we now have,

$$\sum_{p=-(N-1)}^{(N-1)} I^*(\omega'_p) \exp(i\omega'_p r) = \frac{1}{2\pi} \sum_s \hat{R}(s)\left\{\sum_p \exp[i\omega'_p(r-s)]\right\}, \qquad (7.6.11)$$

and now,

$$\sum_{p=-(N-1)}^{(N-1)} \exp[i\omega'_p(r-s)] = \begin{cases} 0, & (r-s) \neq 0, \pm(2N-1), \\ (2N-1), & (r-s) = 0, \pm(2N-1). \end{cases}$$

Since there are no values of r, s, such that $(r-s) = \pm(2N-1)$, (7.6.11) gives,

$$\hat{R}(r) = \frac{2\pi}{2N-1} \sum_{p=-(N-1)}^{(N-1)} I^*(\omega'_p) \exp(i\omega'_p r), \qquad (7.6.12)$$

$r = 0, \pm 1, \ldots, \pm(N-1)$. Equation (7.6.12) gives an *exact* expression for $\hat{R}(r)$; it is not merely an approximation to (7.6.8). In fact, if we define $\hat{R}(s)$ to be zero for $|s| \geq N$ we can regard (7.6.7) as a trigonometric polynomial of degree P for any integer $P \geq (2N-1)$. It then follows that (7.6.12) will provide an exact expression for $\hat{R}(r)$ if we choose the $\{\omega'_p\}$ to be of the form $\omega'_p = 2\pi p/P$, $p = 0, \pm 1, \ldots,$ for any integer $P \geq (2N-1)$.

It should be noted that the fast Fourier transform algorithm produces the periodogram ordinates only at frequencies $2\pi p/N$, where N is the number of data points. To obtain the values of $I^*(\omega)$ at $\omega'_p = 2\pi p/(2N-1)$ we must therefore add $(N-1)$ zeros to the data so that the "apparent" number of data points becomes $(2N-1)$.

Evaluation of weighted integrals of the periodogram

The standard forms of spectral estimates are all based on weighted integrals of the periodogram, and the relationship (7.6.12) enables us to express such integrals *exactly* as sums over discrete sets of frequencies. Consider an expression of the form,

$$I = \int_{-\pi}^{\pi} I^*(\omega) W(\omega) \, d\omega. \tag{7.6.13}$$

Using (7.6.9) we may re-write this as,

$$I = \frac{1}{2\pi} \sum_{s=-(N-1)}^{(N-1)} \hat{R}(s) \left\{ \int_{-\pi}^{\pi} e^{-is\omega} W(\omega) \, d\omega \right\}$$

$$= \frac{1}{2\pi} \sum_{s=-(N-1)}^{(N-1)} \lambda(s) \hat{R}(s), \tag{7.6.14}$$

say, where $\lambda(s)$ is the sth coefficient of the Fourier series expansion of $W(\omega)$ over the interval $(-\pi, \pi)$. Substituting (7.6.12) in (7.6.14) we now obtain,

$$I = \frac{2\pi}{2N-1} \sum_{p=-(N-1)}^{(N-1)} I^*(\omega_p') \left\{ \frac{1}{2\pi} \sum_{s=-(N-1)}^{(N-1)} \lambda(s) \exp(i\omega_p' s) \right\}$$

i.e. $\tag{7.6.15}$

$$I = \frac{2\pi}{2N-1} \sum_{p=-(N-1)}^{(N-1)} I^*(\omega_p') W(\omega_p'),$$

where, as above, $\omega_p' = 2\pi p/(2N-1)$, $p = 0, \pm 1, \ldots, \pm(N-1)$. Note, once again, that the result (7.6.15) is *exact*, i.e. (7.6.15) is not merely a discrete approximation to (7.6.13). Also, (7.6.15) remains exact if we set $\omega_p' = 2\pi p/P$, where P is any integer satisfying $P \geq 2N-1$. (In deriving (7.6.15) we have tacitly assumed that the Fourier coefficients of $W(\omega)$ vanish for $|s| \geq N$. However, even if $W(\omega)$ has non-zero coefficients for $|s| \geq N$ the argument is still valid since, by (7.6.14), the value of I depends only on the values of $\lambda(s)$ for $|s| \leq N-1$. See the discussion following (6.2.59).)

Evaluation of spectral estimates by sums over discrete frequencies

Using the result (7.6.15) we can now write a general spectral estimate $\hat{h}(\omega)$ as a weighted sum of periodogram ordinates over the frequencies $\{\omega_p'\}$. Thus, if $\hat{h}(\omega)$ is any estimate of the form (7.1.4), namely,

$$\hat{h}(\omega) = \int_{-\pi}^{\pi} I^*(\theta) W_N(\omega - \theta) \, d\theta, \tag{7.6.16}$$

then we can compute $\hat{h}(\omega)$ in the form,

$$\hat{h}(\omega) = \frac{2\pi}{2N-1} \sum_{p=-(N-1)}^{(N-1)} I^*(\omega_p') W_N(\omega - \omega_p'). \qquad (7.6.17)$$

The estimate $\hat{h}(\omega)$ can be computed exactly from (7.6.17), using the fast Fourier transform algorithm to evaluate the periodogram ordinates $\{I^*(\omega_p')\}$. As with the evaluation of the autocovariances, it will be necessary to add $(N-1)$ zeros to the data in order to obtain the periodogram ordinates at frequencies $\omega_p' = 2\pi p/(2N-1)$.

Alternatively, if we use the standard form of the algorithm and thus obtain the periodogram ordinates at frequencies $\omega_p = 2\pi p/N$, we can approximate the integral (7.6.16) by a sum over the $\{\omega_p\}$ and compute $\hat{h}(\omega)$ in the form,

$$\hat{h}(\omega) = \frac{2\pi}{N} \sum_{p=-[N/2]}^{[N/2]} I^*(\omega_p) W_N(\omega - \omega_p). \qquad (7.6.18)$$

This would usually provide a sufficiently accurate approximation for the computation of $\hat{h}(\omega)$. In fact, if the window $W_N(\theta)$ is of the "truncated type", then (7.6.16) reduces to a trigonometric polynomial of degree $(M-1)$, (where M is the truncation point) and now (7.6.16) can be written exactly as a sum over any set of frequencies of the form $2\pi p/P$, provided $P \geq 2M-1$. Thus, if $M \leq \frac{1}{2}(N+1)$ then (7.6.18) will give an exact expression for $\hat{h}(\omega)$.

Some authors (see, e.g. Hannan (1970), Brillinger (1975)) suggest an even simpler approximation for $\hat{h}(\omega)$, namely,

$$\hat{h}(\omega) = \sum_j w_j I^* \left[\frac{2\pi\{p(\omega)+j\}}{N} \right]. \qquad (7.6.19)$$

where, for each frequency ω, $p(\omega)$ is chosen as that integer such that $\{2\pi p(\omega)/N\}$ is closest to ω, and $\{w_j\}$ is a fixed sequence of weights (i.e. independent of ω) satisfying $\sum_j w_j = 1$, and concentrated around the region of $j = 0$. If we write $w_j = (2\pi/N)W_N\{\omega - 2\pi(p(\omega)+j)/N\}$ then clearly (7.6.19) and (7.6.18) are very close in form, the only difference being that in (7.6.18) the weights w_j vary with the frequency ω whereas in (7.6.19) they remain independent of ω. For those frequencies ω which are exactly of the $2\pi p/N$ (so that here $\omega \equiv 2\pi p(\omega)/N$), (7.6.18) and (7.6.19) become identical in form. Now we saw in Section 7.4 that we would usually evaluate $\hat{h}(\omega)$ at frequency intervals (π/M), where $M(\ll N)$ is the window parameter. Hence, if we evaluate $\hat{h}(\omega)$ only at the frequencies $\omega_p = 2\pi p/N$ this would, in general, provide more than an adequate number of points over the interval $(-\pi, \pi)$. (If N (or $N' = N + $ number of zeros added) is a multiple of $2M$ then the frequencies $\{\omega_p\}$ contain, as a subset, the frequencies $\{\pi p/M\}$.)

In particular, if the window $W_N(\theta)$ is such that $W_N(\theta)$ vanishes outside a small interval centred on $\theta = 0$ (as occurs, e.g. with the Daniell and Bartlett–Priestley windows) then the corresponding form of (7.6.19) may be written,

$$\hat{h}(\omega) = \sum_{j=-m}^{m} w_j I^* \left[\frac{2\pi\{p(\omega)+j\}}{N} \right], \qquad (7.6.20)$$

where m is such that $(2m + 1)$ periodogram ordinates fall within the interval over which $W_N(\theta)$ is non-zero. This form of $\hat{h}(\omega)$ corresponds to a window bandwidth $(2m \times (2\pi/N))$ so that in the case of, e.g., the Daniell window m is related to the window parameter M by $m = N/2M$, and the equivalent number of degrees of freedom is $4m$ (see Table 6.2). The form of estimate (7.6.19) will often be sufficiently accurate even when ω is not exactly a multiple of $(2\pi/N)$. In fact, if we take the original form of $\hat{h}(\omega)$ as given by (7.6.16) and define a new estimate $\hat{h}^*(\omega)$ by,

$$\hat{h}^*(\omega) = \hat{h}\{2\pi p(\omega)/N\}, \qquad (7.6.21)$$

where $p(\omega)$ is as defined above, then Parzen (1957a) has shown that $\hat{h}(\omega)$ and $\hat{h}^*(\omega)$ have the same asymptotic sampling properties. It thus follows that the estimate (7.6.20) has the same asymptotic properties as an estimate based on the Daniell window with $M = N/2m$.

Spectral estimation by averaging over time blocks

Certain types of "spectral analysers" (i.e. special purpose computers designed for analysing continuous parameter records) compute estimates of spectral density functions by dividing up the sample record into a number of "blocks" covering different time intervals, and then averaging the squared modulus of the finite Fourier transform (at a given frequency) over the different blocks. This procedure is effectively the same as estimating the spectral density function by Bartlett's original method of averaging periodograms computed from sub-intervals; the only difference between this method and that of applying a Bartlett window to the "complete" periodogram is that the amount of data to be stored at each stage is considerably smaller. The user selects the length T' of each block, which is equivalent to selecting the bandwidth of the corresponding Bartlett window (the bandwidth being $\propto 1/T'$), and the user also selects the time sampling interval Δt for analog–digital conversion.

However, most spectral analysers also allow for "tapering" of each block, which in this case is equivalent to using a different spectral window (i.e. other than the Bartlett window). If tapering is used over the *whole* record length it is not equivalent to filtering (see Section 7.4.1); however, if tapering is applied to each block (i.e. sub-interval) then it becomes equivalent simply to filtering the observations. To see this, suppose we digitize the record and obtain a total of T observations which we then divide up into M blocks, each containing T' observations (so that $T/T' = M$), as illustrated in Fig. 7.9.

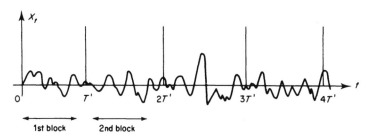

Fig. 7.9. Dividing a record up into time blocks.

Let,

$$Y_t(\omega) = \sum_{u=0}^{T'-1} g_u X_{t-u} \exp[-i\omega(t-u)] \qquad (t \geq T'), \qquad (7.6.22)$$

where $\{X_t\}$ denotes a digitized version of the continuous parameter record. Then the original Bartlett estimate of the spectral density function $h(\omega)$ is

$$\hat{h}_0(\omega) = \frac{1}{M} \sum_{m=1}^{M} |Y_{mT'}^{(0)}(\omega)|^2, \qquad (7.6.23)$$

where $Y_t^{(0)}(\omega)$ is obtained from (7.6.22) by setting

$$g_u = \frac{1}{\sqrt{2\pi T'}}, \qquad \text{all } u.$$

(Note that $|Y_{mT'}^{(0)}(\omega)|^2$ is then the periodogram for the mth block.) However, $\hat{h}_0(\omega)$ is not strictly equivalent to

$$\hat{h}_B(\omega) = \int_{-\pi}^{\pi} I_T^*(\theta) W_B(\omega - \theta) \, d\theta,$$

($W_B(\theta)$ being the Bartlett window). To recover $\hat{h}_B(\omega)$ we must compute,

$$\hat{h}_1(\omega)(\equiv \hat{h}_B(\omega)) = \frac{1}{T} \sum_{t=1}^{T} |Y_t^{(0)}(\omega)|^2,$$

i.e. we must average *all* the $|Y_t^{(0)}(\omega)|^2$, not merely those corresponding to $t = T', 2T', \ldots, MT'$. (We cannot, of course, compute $Y_t^{(0)}(\omega)$ exactly for small t.)

Suppose now that instead of computing the "raw" Fourier transform, $Y_t^{(0)}(\omega)$, for each block we taper the observations in each block by multiplying each by a sequence of constants, $h_1, h_2, \ldots, h_{T'}$, i.e. we compute,

$$Y_{T'}^{(1)}(\omega) = \sum_{t=1}^{T'} h_t X_t e^{-i\omega t},$$

$$Y_{2T'}^{(1)}(\omega) = \sum_{t=T'+1}^{2T'} h_{t-T'} X_t e^{-i\omega t},$$

$$\vdots$$

This is equivalent to writing

$$Y_t^{(1)}(\omega) = \sum_{u=0}^{T'-1} g_u X_{t-u} \exp[-i\omega(t-u)], \qquad t = T', 2T', \ldots \qquad (7.6.24)$$

(which is now of the same form as (7.6.22)) with

$$g_u = h_{T'-u}, \qquad u = 0, 1, \ldots, (T'-1).$$

However, if we consider (7.6.24) as defined for *all* t (not merely $t = T', 2T', 3T', \ldots$) then $Y_t^{(1)}(\omega)$ may be regarded as a filtered version of X_t, with filter impulse response function $\{g_u\}$. *Hence the taper $\{h_t\}$ acts now just like a filter* $\{g_u\}$. If we were to compute,

$$\hat{h}_g(\omega) = \frac{1}{T} \sum_{t=1}^{T} |Y_t^{(1)}(\omega)|^2, \qquad (7.6.25)$$

this would be equivalent (apart from "end effects") to smoothing the complete periodogram of X_t by the window,

$$W_g(\theta) = \left| \sum_{u=1}^{T'-1} g_u e^{-iu\theta} \right|^2. \qquad (7.6.26)$$

(cf. Section (6.2.7) and note the discrete parameter analogue of equation (6.2.195)).

Alternatively, we may write

$$W_g(\theta) = \left| \sum_{u=0}^{T'-1} h_{T'-u} \exp(-iu\theta) \right|^2$$

$$= \left| \sum_{v=1}^{T'} h_v \exp[-i(T'-v)\theta] \right|^2$$

$$= \left| \sum_{v=1}^{T'} h_v \exp(-iv\theta) \right|^2. \qquad (7.6.27)$$

Thus, by applying a taper we can obtain spectral estimates corresponding to an arbitrary window by averaging the $|Y_t^{(1)}(\omega)|^2$ over blocks. Of course, a spectral analyser would not usually compute (7.6.25), rather it would compute

$$\hat{h}_g(\omega) = \frac{1}{M} \sum_{m=1}^{M} |Y_{mT'}^{(1)}(\omega)|^2, \qquad (7.6.28)$$

but this expression can be regarded as an approximation to (7.6.25), the approximation involved being the same as that used in expressing the original Bartlett estimate as a smoothed version of the complete periodogram.

As specific examples of this method of computation we may consider the following.

(a) If we take $h_t = 1/\sqrt{(2\pi T')}$, all t, we obtain the spectral estimate corresponding to the Bartlett window.

(b) If we take $\{h_t\}$ as a Hanning taper we obtain an estimate whose window is the squared modulus of the Tukey–Hanning window.

(c) If we take a triangular taper of the form,

$$h_t = 1 - \frac{2|\tfrac{1}{2}T' - t|}{T'}$$

(with T' even) we obtain an estimate corresponding to the Parzen window.

In general, to obtain an estimate corresponding to a (non-negative) window $W(\theta)$ we have to choose the taper so that the $\{h_t\}$ are the Fourier coefficients of $\{W(\theta)\}^{1/2}$.

Missing observations

It sometimes happens that observations corresponding to certain time points fail to be recorded, or are recorded with obvious gross errors. In this situation we have only a partial set of observations from which to estimate the spectral density function. Parzen (1963b) has developed an approach to the problem of missing observations based on the following device. Let the function $g(t)$ be defined by,

$$g(t) = \begin{cases} 1, & \text{if an observation is recorded at time } t, \\ 0, & \text{if an observation is missing at time } t. \end{cases}$$

Denoting the original process by $\{X_t\}$, the given set of observations may be written as,

$$Y_t = g(t) \cdot X_t, \qquad t = 1, 2, \ldots, N.$$

(The function $g(t)$ plays the same analytical role as the function h_t introduced in the discussion of "tapering"; see Section 7.4.1). Suppose now that
there is a "regular" pattern of missing observations, so that the function $g(t)$
is known for all t and is such that

$$R_g(s) = \lim_{N \to \infty} \left\{ \frac{1}{N} \sum_{t=1}^{N-|s|} g(t)g(t+|s|) \right\} \qquad (7.6.29)$$

exists for all s. Assuming that $\{X_t\}$ is a zero mean stationary process with
autocovariance function, $R_X(s)$, Parzen suggests that $R_X(s)$ may be estimated by

$$\hat{R}_X(s) = \hat{R}_Y(s)/R_g(s), \qquad (7.6.30)$$

(for all s such that $R_g(s) \neq 0$), where

$$R_Y(s) = \frac{1}{N} \sum_{t=1}^{N-|s|} Y_t Y_{t+|s|}.$$

It is easily seen $\hat{R}_X(s)$ is an asymptotically unbiased estimate of $R_X(s)$, and
may, moreover, be shown to be a consistent estimate (Parzen (1963b)). (If
we omit the limit in (7.6.29) and define $R_g(s)$ simply as

$$\left\{ \frac{1}{N} \sum_{t=1}^{N-|s|} g(t)g(t+|s|) \right\}$$

then, in effect, $\hat{R}_X(s)$ is obtained by summing the products $\{X_t . X_{t+|s|}\}$ only
over those pairs of time points, $(t, t+|s|)$, for which both the corresponding
observations have been recorded, and then dividing this sum by the number
of terms included.)

The spectral density function of $\{X_t\}$, $h_X(\omega)$, may now be estimated using
standard expressions of the form (7.1.1) provided we replace the usual
sample autocovariances by estimates of the form (7.6.30). Alternatively, we
can compute a modified form of finite Fourier transform $d_N(\theta)$ as in (7.4.23)
with $g(t)$ replacing h_t, and then form a modified version of the periodogram
$|d_N(\theta)|^2$. We can then use spectral estimates of the form (7.6.16) (or their
discrete sum analogues) if we replace $I^*(\theta)$ by its modified version $|d_N(\theta)|^2$.
Parzen (1963b) has derived the asymptotic variance of spectral estimates of
this form and, as we would expect, one of the effects of missing observations
is to increase the variance of these estimates—in much the same way as
"tapering" also increases the variance.

The case where the pattern of missing observations has a periodic
structure has been discussed by Jones (1962, 1971) and Parzen (1963b). See
also Bloomfield (1970), Neave (1970), and Clinger and Van Ness (1976).

7.7 TREND REMOVAL AND SEASONAL ADJUSTMENT; REGRESSION ANALYSIS AND DIGITAL FILTERS

So far in our discussion of spectral analysis we have assumed that the observations come from a stationary process $\{X_t\}$ whose mean (possibly unknown) is *constant*. However, there are certain situations where the observations exhibit a "*trend*", e.g., they may show a tendency to increase (or decrease) steadily over time, or may fluctuate in periodic manner. This type of behaviour occurs typically with economic time series where, for example, a record of a price index may show a steadily increasing trend over a number of years; one may also observe fluctuations which recur periodically from year to year and are due to "*seasonal effects*". In such cases it would no longer be sensible to treat the observations as coming from a stationary process, but we may instead consider a more general model of the form,

$$X_t = \mu_t + Y_t, \qquad (7.7.1)$$

where μ_t denotes the time dependent mean of X_t (i.e. the "trend") and the "residual" Y_t is a zero mean stationary process. Although $\{X_t\}$ would not, in general, be stationary (and hence would not possess a "spectrum") we can nevertheless consider the problem of estimating the spectral density function $h_Y(\omega)$ of Y_t (assuming that Y_t has an absolutely continuous spectrum), given observations on X_t.

There are two distinct approaches to this problem. If we are interested in studying not only the spectral density function of Y_t but also the form of the trend then we would try to construct an estimate of μ_t, $\hat{\mu}_t$ (say), from the data, and then estimate $h_Y(\omega)$ from the estimated residuals $\{X_t - \hat{\mu}_t\}$. With no information on the form of μ_t we obviously cannot construct a consistent estimate of it; however, if we are able to assume that μ_t has a known functional form (e.g., that μ_t is a polynomial in t) then we can estimate the unknown parameter of this function, in which case the problem becomes essentially that of "*regression analysis*" (see Section 5.2.2). On the other hand, if we are not interested in the form of the trend then we may attempt to "remove" μ_t by applying suitable linear filtering operations to the data. For example, if we know that μ_t is a polynomial of degree k in t then differencing the data $(k-1)$ times will reduce μ_t to a constant, while differencing k times will reduce it to zero. We can then estimate the spectral density function of the differenced data (which will correspond to the differenced values of Y_t rather than Y_t itself), and obtain $h_Y(\omega)$ by applying the appropriate "inverse filter", as we will describe below. The theory underlying the use of these filtering operations is exactly the same as that described in Section 4.12, but in this context the application of linear filters to discrete parameter processes is known as "*digital filtering*" (in contrast to the application of linear filters to continuous parameter processes which would normally be performed by analog methods). We now consider these two approaches in more detail.

Regression analysis of time series

Let us suppose that we can express μ_t as a linear combination of known functions, $\phi_1(t), \phi_2(t), \ldots, \phi_q(t)$, so that the model (7.7.1) takes the form,

$$X_t = \theta_1\phi_1(t) + \theta_2\phi_2(t) + \ldots + \theta_q\phi_k(t) + Y_t, \qquad (7.7.2)$$

where q is known, $\phi_1(t), \ldots, \phi_q(t)$ are known functions, but $\theta_1, \ldots, \theta_q$ are unknown parameters and Y_t is a zero mean process with unknown spectral density function $h_Y(\omega)$. Given n observations on $\{X_t\}$, say for $t = 1, 2, \ldots, n$, the first problem is to estimate the parameters $\theta_1, \ldots, \theta_q$. Since the model (7.7.2) is linear in $\theta_1, \ldots, \theta_q$, one obvious approach is to estimate these parameters by "least squares" (cf. Section 5.2.2). However, we must take note of the fact that, contrary to the assumptions made in the standard linear model, the residuals $\{Y_t\}$ in (7.7.2) will not, in general, be uncorrelated (unless $\{Y_t\}$ happens to be a purely random process). In this case the maximum likelihood estimates of $\boldsymbol{\theta} = (\theta_1, \theta_2, \ldots, \theta_q)'$ (under the assumption that $\{Y_t\}$ is a Gaussian process) corresponds to using a "*weighted least squares*" criterion (of the form (5.2.63)), and the resulting estimates $\hat{\boldsymbol{\theta}}$ are given by (5.2.64). These estimates are then the "*best linear unbiased estimates*" (BLUE) of $\theta_1, \ldots, \theta_q$ (in the sense of the Gauss–Markov theorem) irrespective of the form of the distribution of $\{Y_t\}$. However, the evaluation of the weighted least squares estimates (5.2.64) requires a knowledge of $\boldsymbol{\Sigma}_Y$, the variance–covariance matrix of $(Y_1, Y_2, \ldots, Y_n)$, and since the spectral density function of Y_t is unknown it follows that the matrix $\boldsymbol{\Sigma}_Y$ will also be unknown. Thus, in general, we cannot compute the weighted least squares estimates directly, although we could consider an iterative procedure in which we start with some estimate $\hat{\boldsymbol{\theta}}_0$, estimate the residuals, construct the corresponding $\hat{\boldsymbol{\Sigma}}_Y$, and then compute a revised estimate $\hat{\boldsymbol{\theta}}_1$ using weighted least squares based on $\hat{\boldsymbol{\Sigma}}_Y$, and so on. This would be quite a lengthy procedure, and it is therefore of interest to enquire whether, under general conditions, the *unweighted* least squares estimate $\hat{\boldsymbol{\theta}}^*$ (say)—which can be computed directly from (5.2.51)—would be comparable in efficiency with the *weighted* least squares estimate $\hat{\boldsymbol{\theta}}$.

This question was first considered in a general form by Grenander (1954), and subsequently by Rosenblatt (1956b) and Grenander and Rosenblatt (1957a), Chapter 7. They proved a fundamental theorem which gives a general set of conditions under which $\hat{\boldsymbol{\theta}}^*$ is *asymptotically efficient*, in the sense that the limiting forms of the (suitably normalized) variance–covariance matrices of $\hat{\theta}^*$ and $\hat{\theta}$ are identical. Under these conditions, which hold for a surprisingly wide range of regression models, we can use the simple unweighted least squares estimate $\hat{\theta}^*$ and be assured that, for large samples, $\hat{\theta}^*$ has virtually the same efficiency as the weighted least squares estimate $\hat{\theta}$. To explain the nature of these conditions we must first introduce the notion of the "*regression spectrum*", which is defined as follows. Following Grenander (1954) we assume that the regression functions, $\{\phi_r(t)\}$, possess the following properties;

(i) $\Phi_r(n) = \sum_{t=1}^{n} |\phi_r(t)|^2 \to \infty$ as $n \to \infty$, for each r,

(ii) $\lim_{n \to \infty} \dfrac{\Phi_r(n+1)}{\Phi_r(n)} = 1,$ for each r,

(iii) $\lim_{n \to \infty} \dfrac{\sum_{t=1}^{n} \phi_r(t+h)\phi_s^*(t)}{\{\Phi_r(n) \cdot \Phi_s(n)\}^{1/2}} = R_{rs}(h),$

say, exists for each r, s, and all integers $h \geq 0$. Defining $\phi_r(t) = 0$, $t < 0$, (each r), the above definition of $R_{rs}(h)$ can now be extended to all integer values of h, and we can write

$$R(h) = \{R_{rs}(h)\}, \qquad h = 0, \pm 1, \pm 2, \ldots,$$

i.e. $R(h)$ is a $q \times q$ matrix with entry $R_{rs}(h)$ in the rth row and sth column.

(iv)   The matrix $R(0)$ is non-singular.

Property (i) is imposed to ensure that the contribution of each of the functions $\phi_r(t)$ increases unboundedly as $n \to \infty$. Property (ii) is to ensure that each of the functions $\phi_r(t)$ is, in a sense, a "slowly increasing" function of t (i.e. $\phi_r(n+1)$ will not "dominate" the sum of squares of $\phi_r(t)$ over $t = 1, \ldots, n+1$). Property (iii) allows us to introduce "correlation functions" for the $\{\phi_r(t)\}$, defined as limits of time averages, and property (iv) ensures that, in a sense, the $\{\phi_r(t)\}$ are not asymptotically "linear dependent".

It is not difficult to show that $R(h)$ is (for each h) a Hermitian matrix and that, for any $q \times 1$ vector α, $\alpha^* R(h)\alpha$ is a non-negative sequence (cf. Grenander and Rosenblatt (1957a), p. 234). It then follows that $\exists$ a Hermitian matrix function, $M(\lambda) = \{M_{rs}(\lambda)\}$, such that $R_{rs}(h)$ admits a "spectral representation" of the form,

$$R_{rs}(h) = \int_{-\pi}^{\pi} e^{ih\lambda}\, dM_{rs}(\lambda),$$

or, in matrix form,

$$\boldsymbol{R}(h) = \int_{-\pi}^{\pi} e^{ih\lambda} \, d\boldsymbol{M}(\lambda), \qquad h = 0, \pm 1, \pm 2, \dots \qquad (7.7.3)$$

(cf. Section 4.8.3), where $\boldsymbol{M}(\lambda)$ has positive semi-definite matrix increments, i.e. for every interval $(\lambda_1, \lambda_2), (\lambda_1 < \lambda_2), \Delta\boldsymbol{M}(\lambda_1, \lambda_2) = \{\boldsymbol{M}(\lambda_2) - \boldsymbol{M}(\lambda_1)\}$ is a positive semi-definite matrix. The matrix function $\boldsymbol{M}(\lambda)$ is called the "*spectral distribution function*" of the regression functions (cf. the spectral representation of the autocovariance function of a stationary process, as given by (4.8.34)).

Now let λ be a point such that for any interval (λ_1, λ_2) containing λ (i.e. $\lambda_1 < \lambda < \lambda_2$), $\Delta\boldsymbol{M}(\lambda_1, \lambda_2) = \boldsymbol{M}(\lambda_2) - \boldsymbol{M}(\lambda_1) > 0$, i.e. $\Delta\boldsymbol{M}(\lambda_1, \lambda_2)$ is a positive semi-definite matrix and not the null matrix. Grenander calls the set of all such points λ the "*regression spectrum*" and we denote it by S. The set S thus consists of all the *points of increase* of the function $\boldsymbol{M}(\lambda)$. (As we shall see below, in the case of polynomial or trigonometric regression functions S consists simply of a discrete set of $p(\le q)$ points $\lambda_1, \lambda_2, \dots, \lambda_p$.) Grenander and Rosenblatt (1957a, p. 244) show that, in general, the regression spectrum S can be decomposed into disjoint sets $E_1, E_2, \dots, E_p$ $(p \le q)$ such that

$$\boldsymbol{M}(E_i) > 0, \qquad i = 1, \dots, p,$$

and

$$\boldsymbol{M}(E_i)\boldsymbol{M}^{-1}\boldsymbol{M}(E_j) = \delta_{ij}\boldsymbol{M}(E_i), \qquad \text{all } i, j, \qquad (7.7.4)$$

where, for any set E, $\boldsymbol{M}(E) = \int_E d\boldsymbol{M}(\lambda)$, and

$$\boldsymbol{M} = \int_{-\pi}^{\pi} d\boldsymbol{M}(\lambda) = \boldsymbol{M}(\pi) - \boldsymbol{M}(-\pi) = \boldsymbol{R}(0),$$

is non-singular. The sets $\{E_i\}$ are called the "*elements of the regression spectrum*".

We are now in a position to state the conditions under which the unweighted least squares estimate $\hat{\boldsymbol{\theta}}^*$ is asymptotically efficient. First, we construct normalized forms of the asymptotic variance–covariance matrices of $\hat{\boldsymbol{\theta}}^*$ and $\hat{\boldsymbol{\theta}}$ as follows. Let

$$\boldsymbol{D}_n = \text{diag}\{\Phi_1^{1/2}(n), \dots, \Phi_q^{1/2}(n)\},$$

i.e. $\boldsymbol{D}_n$ is a diagonal matrix with entry $\Phi_r^{1/2}(n)$ in the rth row and column. The normalized variance–covariance matrices of $\hat{\boldsymbol{\theta}}^*$, $\hat{\boldsymbol{\theta}}$ are then defined as,

$$\boldsymbol{V}_n(\hat{\boldsymbol{\theta}}) = \boldsymbol{D}_n[E\{(\hat{\boldsymbol{\theta}} - \theta)(\hat{\boldsymbol{\theta}} - \theta)'\}]\boldsymbol{D}_n \qquad (7.7.5)$$

and

$$V_n(\hat{\boldsymbol{\theta}}^*) = \boldsymbol{D}_n[E\{(\hat{\boldsymbol{\theta}}^* - \theta)(\hat{\boldsymbol{\theta}}^* - \boldsymbol{\theta})'\}]\boldsymbol{D}_n. \qquad (7.7.6)$$

(The reason for the pre and post multiplication by $\boldsymbol{D}_n$ is to prevent the matrices from degenerating as $n \to \infty$.) We now say that the unweighted least squares estimate $\hat{\boldsymbol{\theta}}^*$ is *asymptotically efficient* (as compared with the weighted least squares estimate $\hat{\boldsymbol{\theta}}$) if

$$\lim_{n \to \infty} V_n(\hat{\boldsymbol{\theta}}^*) = \lim_{n \to \infty} V_n(\hat{\boldsymbol{\theta}}).$$

Grenander and Rosenblatt's basic theorem can now be stated as follows:

The unweighted least squares estimate, $\hat{\boldsymbol{\theta}}^*$, is asymptotically efficient if and only if the spectral density function of Y_t, $h_Y(\omega)$, is constant on each of the elements, E_i, of the regression spectrum S, (i.e. if $h_Y(\omega) = c_i$, say, for all $\omega \in E_i$).

It is clear that the above condition will hold for all positive piecewise continuous spectral density functions $h_Y(\omega)$ if and only if each element E_i consists of a single point λ_i and the $\{\lambda_i\}$ are distinct, i.e. *if and only if the elements of S are $p(\leq q)$ distinct points, $\lambda_1, \lambda_2, \ldots, \lambda_p$.* (In this case $\boldsymbol{M}(\lambda)$ increases only at the p points, $\lambda_1, \lambda_2, \ldots, \lambda_p$.) For real valued processes $h_Y(\omega)$ is an even function and hence each E_i can here consist of a *pair* of points, $\{\lambda_i, -\lambda_i\}$.

A rigorous proof of the general form of Grenander and Rosenblatt's theorem has been given by Grenander and Rosenblatt (1957a), pp. 241–245, under the assumption that $h_Y(\omega)$ is positive and piecewise continuous. The corresponding problem for vector valued processes has been studied by Rosenblatt (1956b), and a very full discussion of the vector case is given by Hannan (1970), Chapter 8. (See also Watson (1967) for a comprehensive survey of regression problems.) However, we can gain useful insight into the general nature of Grenander and Rosenblatt's result from the following heuristic proof of the "sufficiency" part of their theorem.

We first re-write the model (7.7.2) in the matrix form,

$$\boldsymbol{X} = \boldsymbol{\phi}\boldsymbol{\theta} + \boldsymbol{Y}, \qquad (7.7.7)$$

where,

$$\boldsymbol{X} = \begin{pmatrix} X_1 \\ \vdots \\ X_n \end{pmatrix}, \qquad \boldsymbol{Y} = \begin{pmatrix} Y_1 \\ \vdots \\ Y_n \end{pmatrix}, \qquad \boldsymbol{\theta} = \begin{pmatrix} \theta_1 \\ \vdots \\ \theta_q \end{pmatrix}, \qquad \boldsymbol{\phi}_r = \begin{pmatrix} \phi_r(1) \\ \vdots \\ \phi_r(n) \end{pmatrix}$$

and $\boldsymbol{\phi} = (\boldsymbol{\phi}_1, \boldsymbol{\phi}_2, \ldots, \boldsymbol{\phi}_q)$. Then from (5.2.51), the unweighted least squares estimate $\hat{\theta}^*$ can be written,

$$\hat{\theta}^* = (\phi'\phi)^{-1}\phi'X, \tag{7.7.8}$$

or, using (7.7.7),

$$\hat{\theta}^* = \theta + (\phi'\phi)^{-1}\phi'Y = \theta + AY, \tag{7.7.9}$$

say. The variance–covariance matrix of $\hat{\theta}^*$ is thus given by

$$E[(\hat{\theta}^* - \theta)(\hat{\theta}^* - \theta)'] = A\Sigma_Y A' = (\phi'\phi)^{-1}\phi'\Sigma_Y\phi(\phi'\phi)^{-1}. \tag{7.7.10}$$

(Note that (5.2.52) is a special case of this result when $\Sigma_Y = \sigma_Y^2 I$.) Also, the weighted least squares estimate $\hat{\theta}$ is (from (5.2.64)) given by

$$\hat{\theta} = (\phi'\Sigma_Y^{-1}\phi)^{-1}\phi'\Sigma_Y^{-1}X, \tag{7.7.11}$$

$$= \theta + (\phi'\Sigma_Y^{-1}\phi)^{-1}\phi'\Sigma_Y^{-1}Y, \tag{7.7.12}$$

and by a similar argument we find that,

$$E[(\hat{\theta} - \theta)(\hat{\theta} - \theta)'] = (\phi'\Sigma_Y^{-1}\phi)^{-1} \tag{7.7.13}$$

The normalized variance–covariance matrix of $\hat{\theta}^*$, $V_n(\hat{\theta}^*)$, (given by (7.7.6)) can thus be written,

$$V_n(\hat{\theta}^*) = [D_n(\phi'\phi)^{-1}D_n][D_n^{-1}\phi'\Sigma_Y\phi D_n^{-1}][D_n(\phi'\phi)^{-1}D_n]$$

$$= B_n C_n B_n,$$

say. We now consider the limiting of C_n as $n \to \infty$. Since $\{Y_t\}$ is a stationary process the (i, j)th element of Σ_Y is $R_Y(i-j)$, where $R_Y(s)$ is the auto-covariance function of Y_t. Hence the (i, j)th element of C_n is

$$\frac{\sum_{u=1}^{n}\sum_{v=1}^{n}\phi_i(u)\phi_j(v)R_Y(u-v)}{\Phi_i^{1/2}(n)\Phi_j^{1/2}(n)}$$

$$= \sum_{s=0}^{n-1} R_Y(s)\left\{\frac{\left[\sum_{v=1}^{n-s}\phi_i(v+s)\phi_j(v)\right]}{\Phi_i^{1/2}(n)\Phi_j^{1/2}(n)}\right\} + \sum_{s=-(n-1)}^{-1} R_Y(s)\left\{\frac{\left[\sum_{u=s+1}^{n}\phi_i(u)\phi_j(u+s)\right]}{\Phi_i^{1/2}(n)\Phi_j^{1/2}(n)}\right\}$$

$$\sim \sum_{s=-(n-1)}^{(n-1)} R_Y(s)R_{ij}(s)$$

$$= \int_{-\pi}^{\pi}\left\{\sum_{s=-(n-1)}^{(n-1)} R_Y(s)\, e^{is\lambda}\right\} dM_{ij}(\lambda)$$

$$\sim 2\pi\int_{-\pi}^{\pi} h_Y(\lambda)\, dM_{ij}(\lambda).$$

Hence we have,

$$\lim_{n \to \infty} C_n = 2\pi \int_{-\pi}^{\pi} h_Y(\lambda) \, dM(\lambda). \tag{7.7.14}$$

Also, it is easily seen that,

$$\lim_{n \to \infty} B_n^{-1} = \lim_{n \to \infty} [D_n^{-1}(\phi'\phi)D_n^{-1}] = R(0) = M.$$

Hence we finally obtain the result,

$$\lim_{n \to \infty} V_n(\hat{\theta}^*) = M^{-1}\left\{ 2\pi \int_{-\pi}^{\pi} h_Y(\lambda) \, dM(\lambda) \right\} M^{-1}. \tag{7.7.15}$$

(Grenander and Rosenblatt prove this result rigorously by bounding $h_Y(\lambda)$ above and below by finite order trigonometric polynomials, the bounding polynomials corresponding to the spectral density functions of finite order MA models for which only a finite number of the autocovariances are non-zero.)

A similar argument shows that,

$$\lim_{n \to \infty} V_n(\hat{\theta}) = \left\{ \frac{1}{2\pi} \int_{-\pi}^{\pi} \frac{1}{h_Y(\lambda)} \, dM(\lambda) \right\}^{-1}, \tag{7.7.16}$$

and hence $\hat{\theta}^*$ will be asymptotically efficient if,

$$M^{-1}\left\{ 2\pi \int_{-\pi}^{\pi} h_Y(\lambda) \, dM(\lambda) \right\} M^{-1} \left\{ \frac{1}{2\pi} \int_{-\pi}^{\pi} \frac{1}{h_Y(\lambda)} \, dM(\lambda) \right\} = I. \tag{7.7.17}$$

Now suppose that $h_Y(\omega)$ does, in fact, take the constant value c_i on the element, E_i, of the regression spectrum S $(i = 1, \ldots, p)$. Then,

$$\int_{-\pi}^{\pi} h_Y(\lambda) \, dM(\lambda) = \sum_i c_i \int_{E_i} dM(\lambda) = \sum_i c_i M(E_i).$$

Using the orthogonality of the $M(E_i)$, the LHS of (7.7.17) can be written,

$$M^{-1}\left\{ \sum_i c_i M(E_i) M^{-1} \frac{1}{c_i} M(E_i) \right\} = M^{-1}\left\{ \sum_i M(E_i) \right\}$$

(using the properties (7.7.4)). But

$$M = \int_{-\pi}^{\pi} dM(\lambda) = \sum_i \int_{E_i} dM(\lambda) = \sum_i M(E_i),$$

and hence we have shown that the LHS of (7.7.17) is I, and consequently that $\hat{\theta}^*$ is asymptotically efficient if $h_Y(\omega)$ is constant on the elements of S.

There are two cases of considerable interest in which the elements of S are, in fact, distinct points, namely the cases of *polynomial regression* and *trigonometric regression*, so that for these models the unweighted least squares estimate $\hat{\theta}^*$ is always asymptotically efficient irrespective of the form of $h_Y(\omega)$ (provided only that it is positive and piecewise continuous). In the case of polynomial regression the mean μ_t takes the form,

$$\mu_t = \theta_1 + \theta_2 t + \ldots + \theta_q t^{q-1}, \tag{7.7.18}$$

so that here $\phi_r(t) = t^{r-1}$, $r = 1, 2, \ldots, q$. We now have that,

$$\Phi_n(r) = \sum_{t=1}^{n} t^{2(r-1)} \sim \frac{n^{2r-1}}{2r-1},$$

so that properties (i) and (ii) required for the regression functions are certainly satisfied. Also,

$$R_{rs}(h) = \lim_{n \to \infty} \left[\Phi_n^{-1/2}(r) \Phi_h^{-1/2}(s) \right] \sum_{t=1}^{n} t^{r-1}(t+h)^{s-1}$$

$$= \frac{\{(2r-1)(2s-1)\}^{1/2}}{(r+s-1)}, \quad \text{(independent of } h\text{)}.$$

Hence we may write,

$$R_{rs}(h) = \int_{-\pi}^{\pi} e^{ih\lambda} \, dM_{rs}(\lambda),$$

where, for each r, s, $M_{rs}(\lambda)$ has a single "jump" point at $\lambda = 0$ and is constant elsewhere. Thus, $M(\lambda)$ increases only at $\lambda = 0$ and hence the elements of the spectrum S consists of just one point, namely the origin. It therefore follows that *the unweighted least squares estimate $\hat{\theta}^*$ is asymptotically efficient for the polynomial regression model.*

In the case of trigonometric regression we have,

$$\mu_t = \theta_1 \exp(i\lambda_1 t) + \theta_2 \exp(i\lambda_2 t) + \ldots + \theta_q \exp(i\lambda_q t), \tag{7.7.19}$$

where $\lambda_1, \lambda_2, \ldots, \lambda_q$ are specified frequencies. Here, $\phi_r(t) = \exp(i\lambda_r t)$, and again it is easily verified that these functions satisfy the required properties with

$$\Phi_n(r) = n \qquad \text{(all } r\text{)}$$

and

$$R_{rs}(h) = \lim_{n \to \infty} \left[\frac{1}{n} \sum_{t=1}^{n} \exp[i(\lambda_r - \lambda_s)t] \cdot \exp(i\lambda_s h) \right] = \delta_{rs} \exp(i\lambda_s h).$$

Hence, we may write

$$R(h) = \int_{-\pi}^{\pi} e^{it\lambda} \, dM(\lambda),$$

where now $M(\lambda)$ is constant except at the "jump" points $\lambda_1, \lambda_2, \ldots, \lambda_q$. Thus, if the $\{\lambda_i\}$ are distinct the elements of S are distinct points, and again *the unweighted least squares estimate $\hat{\theta}^*$ is asymptotically efficient.*

Note that (7.7.19) can be written in "real" form as,

$$\mu_t = \sum_{r=1}^{q} (A_r \cos \lambda_r t + B_r \sin \lambda_r t), \tag{7.7.20}$$

and thus μ_t can be regarded as a stationary process of the "harmonic model" form (cf. Section 4.10) with a purely discrete spectrum provided the phases of the q frequency components are interpreted as independent random variables, uniformly distributed over $(-\pi, \pi)$. Looking at μ_t in this way, we can regard X_t itself as a stationary process with a "mixed spectrum". We consider the general analysis of processes with spectra in Chapter 8, but if the frequencies $\{\lambda_i\}$ are known *a priori* it would, in general, be best to treat μ_t as a deterministic trend and then to estimate the "amplitudes" θ_i by least squares regression analysis. The continuous component in the spectrum (namely, the spectral density function of Y_t) can then be estimated from the residuals of the regression model. The fact that μ_t, regarded as a stationary process, has a purely discrete spectrum with "jumps" at the points $\{\lambda_i\}$ $(i = 1, \ldots, q)$ ties in completely with the form of $M(\lambda)$ corresponding to the deterministic interpretation of μ_t. This equivalence is, of course, due to the fact that if $\theta_r \exp(i\lambda_r t)$ is regarded as a stationary process then $R_{rr}(h)$ is the limiting form of the sample autocorrelation function of this process (as $n \to \infty$), and by the ergodic property $R_{rr}(h)$ will be identical with the theoretical autocorrelation function (see Section 5.3.6).

Estimating the spectral density function of the residuals

Once we have estimated the parameters θ of the regression function we can construct the estimated mean $\hat{\mu}_t$ and estimate $h_Y(\omega)$ from the estimated residuals $\{X_t - \hat{\mu}_t\}$ by the standard methods which we have previously discussed. If the regression functions $\{\phi_r(t)\}$ satisfy the properties (i)–(iv) specified above, and the parameters θ are estimated by unweighted or weighted least squares, then the removal of the "trend" μ_t does not affect the asymptotic sampling properties of $\hat{h}_Y(\omega)$, i.e. *the asymptotic sampling properties of $\hat{h}_Y(\omega)$ are the same as if $\hat{h}_Y(\omega)$ was constructed directly from observations on $\{Y_t\}$* (Grenander and Rosenblatt (1957a), p. 202, Hannan (1970), pp. 450–452; see also Hannan (1958a), Hannan (1960), Chapter 5).

Lomnicki and Zaremba (1957c) considered the particular case of poly-nomial regression of the form (7.7.18) and showed that if we estimate the parameters by least squares and replace X_t by $\{X_t - \hat{\mu}_t\}$ in the usual autocovariance estimate (7.1.2), then the resulting estimate of $R_Y(s)$, the autocovariance function of Y_t, has the same asymptotic variance as in the standard case (cf. (5.3.25)).

However, the removal of a trend may affect the bias of $\hat{h}_Y(\omega)$ at certain frequencies, and for small samples this effect may be worth correcting. It is fairly easy to examine the main effect of the regression analysis on the bias of $\hat{h}_Y(\omega)$, and the effect can then be approximately corrected by multiplying $\hat{h}_Y(\omega)$ by a suitable factor; see, e.g. Hannan (1960), Chapter 5, Hannan (1970), p. 451. A simple example of this phenomenon occurs when X_t has a constant (but unknown) mean μ. In our previous discussion of this case we suggested that μ be estimated by the sample mean $\bar{X}$ and that the spectral density function could then be estimated from the "mean corrected" observations $\{X_t - \bar{X}\}$, $t = 1, \ldots, N$. This is simply a special case of the general trend removal procedure, since the case of a constant mean μ can be regarded either as a special case of the regression model (7.7.18) with $q = 1$ (so that $\mu \equiv \theta_1$), or as a special case of (7.7.19) with $q = 1$ and $\lambda_1 = 0$. In either case the (unweighted) least squares estimate of μ is $\bar{X}$, and hence working with $\{X_t - \bar{X}\}$ is exactly the same as subtracting the estimated "trend". Now we noted previously that the periodogram ordinate of $\{X_t - \bar{X}\}$ always vanishes at zero frequency. Hence, if we form an estimate of the spectral density function by (say) using a discrete version of the Daniell window in which $(2m + 1)$ consecutive periodogram ordinates are averaged (i.e. using (7.6.20) with $w_j = 1/(2m + 1)$, all j), then the spectral estimate at zero frequency will, in fact, be based only on an average of $2m$ periodogram ordinates. Hence, at this frequency it would be more appropriate to divide the sum of periodogram ordinates by $(2m)$ rather than $(2m + 1)$. (As observed by Hannan (1960), this modification is closely analogous to the familiar device of using the divisor $(n - 1)$ (rather than n) when constructing an unbiased estimate of the variance from n independent observations.)

When we remove an unknown constant mean we are, in effect, also removing the component in the residual Y_t at zero frequency. Similarly, if we remove a more general trigonometric function of the form (7.7.19) we will also remove part of Y_t, namely those components at the corresponding frequencies $\lambda_1, \lambda_2, \ldots, \lambda_q$. In general, we may expect the removal of a trend to affect the bias of $\hat{h}_Y(\omega)$ at those frequencies corresponding to the location of the "jumps" in $M(\lambda)$.

If we do not wish to estimate the detailed form of $h_Y(\omega)$ but wish merely to test whether the residuals Y_t are autocorrelated (i.e. to test whether the $h_Y(\omega)$ is uniform or non-uniform) then we can use a test statistic devised by

Durbin and Watson (1950, 1951, 1971) which is based, effectively, on the first sample autocorrelation of the estimated residuals.

Digital filtering

As mentioned previously, we can sometimes remove trends by passing the observations $\{X_t\}$ through a suitable linear filter. Special types of filters may be constructed for dealing with specific functional forms of μ_t, and we now describe some standard forms.

Removing polynomial trends by differencing

If μ_t is a $(q-1)$th degree polynomial in t (of the form (7.7.18)) then it is well known that the qth differences of μ_t will be zero. Hence if we form $X_t' = \nabla^q X_t$ (where $\nabla \equiv (1-B)$ is the standard difference operator) we have,

$$X_t' = \nabla^q X_t = \nabla^q Y_t. \qquad (7.7.21)$$

It follows from the theory of Section 4.12 (cf. equation (4.12.56)) that the spectral density functions of X_t' and Y_t are related by,

$$h_{X'}(\omega) = |1 - \exp(-i\omega)|^{2q} h_Y(\omega) = 2^{2q}(\sin^2(\omega/2))^q h_Y(\omega). \qquad (7.7.22)$$

Hence, we may estimate $h_Y(\omega)$ by first forming an estimate $\hat{h}_{X'}(\omega)$ (using standard methods applied to the differenced data), and then computing,

$$\hat{h}_Y(\omega) = 2^{-2q}(\sin^2(\omega/2))^{-q} \hat{h}_{X'}(\omega) \qquad (\omega \neq 0). \qquad (7.7.23)$$

Note that to implement this approach it is necessary that the *order* of the polynomial be known *a priori*. Also, it is not possible to apply the inverse transformation at $\omega = 0$.

Seasonal adjustment filters

Many economic time series exhibit a fairly regular pattern of behaviour over each year. Thus, if we consider a series describing the number of sales of a particular item we may find that, e.g., the sales tend to be low during the winter but high during the summer. In such cases we say that the observations $\{X_t\}$ contain a *seasonal component* and we may write

$$X_t = S_t + Y_t, \qquad (7.7.24)$$

where S_t denotes the seasonal component and Y_t denotes, as before, the residual. When our interest centres mainly on the study of Y_t we may attempt to remove S_t, this operation being known as "*seasonal adjustment*". The residuals, Y_t, are then called the "*seasonally adjusted*" data.

If the seasonal component S_t maintains a constant pattern from year to year then S_t will be a periodic function of t with a period of one year. Specifically, suppose that we measure t in months, then S_t will be a periodic function with period 12, and hence (for t integer valued) may be represented as a harmonic series, i.e. we may write

$$S_t = \sum_{p=0}^{6} \left[A_p \cos\left(\frac{2\pi pt}{12}\right) + B_p \sin\left(\frac{2\pi pt}{12}\right) \right]. \qquad (7.7.25)$$

If the phases of the harmonic terms are regarded as independent uniform random variables then S_t is a stationary process with a purely discrete spectrum (the "jumps" occurring at frequencies 0, $\pm 2\pi/12$, $\pm 4\pi/12, \ldots$, $\pm\pi/2$), and X_t is then a stationary process with a "mixed spectrum". However, the seasonal component S_t may be removed by operating on X_t with any of the following filters.

A. THE MOVING AVERAGE FILTER

Suppose, more generally, that S_t is periodic with period s. If s is odd, say $s = 2r + 1$, we may remove S_t by operating on X_t with the filter,

$$X_t' = \frac{1}{(2r+1)}\{X_{t-r} + \ldots + X_t + \ldots + X_{t+r}\} = \phi(B)X_t, \qquad (7.7.26)$$

say, i.e. X_t' is formed by taking an unweighted moving average of X_t over $(2r+1)$ points. Since S_t is periodic with period $(2r+1)$ the filter $\phi(B)$ will clearly reduce S_t to a constant value, and hence we may write

$$X_t' = \mu + \phi(B)Y_t.$$

The spectral density function of Y_t can now be estimated as described above, i.e. by estimating the spectral density function of X_t' and applying the inverse filter transformation. The squared modulus of the transfer function of the moving average filter $\phi(B)$ is (from (4.12.26)),

$$|\phi(e^{-i\omega})|^2 = \frac{\sin^2\{(2r+1)\omega/2\}}{(2r+1)^2\sin^2(\omega/2)}. \qquad (7.7.27)$$

Note that since $\phi(B)$ is symmetric about the time point t, $\phi(\exp(-i\omega))$ is real valued, and hence this filter does not introduce any phase shifts.

If s is even we cannot construct a "symmetric" filter of the above form, and the usual procedure is to modify $\phi(B)$ slightly and write, if $s = 2r$, say,

$$X_t' = \frac{1}{2r}\{\tfrac{1}{2}X_{t-r} + X_{t-r+1} + \ldots + X_t + \ldots + X_{t+r-1} + \tfrac{1}{2}X_{t+r}\}. \qquad (7.7.28)$$

Note that in applying these filters to a finite stretch of data we "lose" r points at the beginning and at the end of the stretch; thus, if we start with N values of X_t we can compute only $(N - 2r)$ values of X'_t. Similarly, in applying the qth order difference filter we lose $(q - 1)$ points at the beginning of the record.

B. THE s-STEP DIFFERENCE FILTER

Another way of removing S_t is to use the filter $(1 - B^s)$, i.e. to form

$$X'_t = (1 - B^s)X_t = X_t - X_{t-s}. \tag{7.7.29}$$

This filter has been used extensively by Box and Jenkins (1970) as a method of modelling seasonal variation, and it is clear that it will remove any component in X_t which is periodic with period s. We thus have,

$$X'_t = (1 - B^s)Y_t = Y_t - Y_{t-s},$$

and again the spectral density function of Y_t may be estimated by estimating the spectral density function of X'_t and applying the inverse filter transformation. For this filter the squared modulus of the transfer function is,

$$|1 - \exp(-is\omega)|^2 = \{2(1 - \cos s\omega)\} = 4\sin^2(s\omega/2). \tag{7.7.30}$$

Note that this transfer function vanishes at $\omega = 2\pi p/s$, $p = 0, 1, \ldots, [s/2]$, and hence the filter "kills" all components in X_t at these frequencies.

C. THE s-STEP MOVING AVERAGE FILTER

Consider the simple moving average filter,

$$X'_t = \frac{1}{(2r + 1)} \sum_{u=-r}^{r} X_{t-u}. \tag{7.7.31}$$

If we let $r \to \infty$ (and ignore, for the moment, the fact that the weights $(1/2r + 1) \to 0$), we see that, for all t, X'_t is simply an average of X_t over a very long time interval, and hence X'_t loses its dependence on t and becomes, in effect, a constant. (This feature is reflected in the fact that as $r \to \infty$ the transfer function of the moving average filter (viz., (7.7.27)) $\to 0$ everywhere except at $\omega = 0$.)

If we now consider a more general version of (7.7.31), namely,

$$X'_t = \frac{1}{(2r + 1)} \sum_{u=-r}^{r} X_{t-su}, \tag{7.7.32}$$

then it is clear that, as $r \to \infty$, $X'_t, X'_{t+s}, X'_{t+2s}, \ldots$, are, in effect, all averages of the same set of $\{X_t\}$. In other words, as $r \to \infty$, X'_t becomes periodic with period s. This can be seen also from the form of the squared modulus of the transfer function corresponding to (7.7.32), namely,

$$\frac{1}{(2r+1)^2}\left|\sum_{u=-r}^{r}\exp(-i\omega us)\right|^2 = \frac{\sin^2\{(2r+1)s\omega/2\}}{(2r+1)^2\sin^2(s\omega/2)}. \qquad (7.7.33)$$

As $r \to \infty$, this function vanishes except when $s\omega = 2\pi p$, i.e. except when $\omega = 2\pi p/s$, $p = 0, 1, 2, \ldots, [s/2]$, where it tends to unity. Hence, when we apply this filter with a large value of r we effectively remove all frequency components except those at $\omega = \{2\pi p/s\}$ which are passed with unit gain and zero phase shift. Since S_t contains components only at these frequencies it is passed (substantially) unchanged through the filter, while Y_t is (substantially) suppressed. (The filter will pass the components in Y_t at frequencies $\{2\pi p/s\}$, but since Y_t has a continuous spectrum these will be negligible compared with S_t.)

Thus, for large r, the output of the filter X'_t will approximate S_t, and we can recover Y_t by computing,

$$\hat{Y}_t = X_t - X'_t. \qquad (7.7.34)$$

When operating with a finite stretch of data, say from $t = 1, \ldots, N$, we cannot allow r to exceed $[N/s]$, and with this value we can compute only s values of X'_t. However, since X'_t is (ideally) periodic, period s, we can extend X'_t outside this range of s values by making it a strictly periodic function with period s.

It is possible to develop more general models of seasonal behaviour in which, e.g., the seasonal component is multiplicative rather than additive, so that (7.7.24) becomes

$$X_t = S_t . Y_t.$$

If we know a priori that the model is of this form we can reduce it to an additive form by means of a logarithmic transformation on X_t. However, there are cases where it might be more realistic to drop the assumption that S_t is strictly periodic and replace it by a more flexible model in which S_t is, roughly speaking, a stationary process with a continuous spectrum "concentrated" around the frequencies $\{2\pi p/s\}$. Such a model might arise, for example, by allowing each of the coefficients, $\{A_p\}$, $\{B_p\}$ in (7.7.25) to be time dependent. An interesting discussion of these more general seasonal models is given by Durbin (1963) and Durbin and Murphy (1975).

7.8 AUTOREGRESSIVE, ARMA, AND MAXIMUM ENTROPY SPECTRAL ESTIMATION; CAT CRITERION

The spectral density function estimates which we have considered so far are all based on the technique of smoothing the periodogram via a suitable spectral window. This method is valid for all types of stationary processes with purely continuous spectra and it is "non-parametric" in the sense that it does not assume a specific parametric model for the observed series. However, if we know, for example, that the series conforms to an AR model we could estimate its spectral density function by the alternative method of first estimating the parameters of the AR model from the data and then substituting these estimates in the theoretical expression for the spectral density function.

For example, suppose we know that the series satisfies a discrete parameter AR(k) model of the form (3.5.39), i.e.

$$X_t + a_1 X_{t-1} + \ldots + a_k X_{t-k} = \varepsilon_t, \tag{7.8.1}$$

ε_t being a purely random process with zero mean and variance σ_ε^2. Given a set of observations $(X_1, \ldots, X_N)$, the coefficients $(a_1, \ldots, a_k)$ and σ_ε^2 can be estimated by least squares, as described in Section 5.4.1, yielding estimates $\hat{a}_1, \ldots, \hat{a}_k, \hat{\sigma}_\varepsilon^2$, say. If we now substitute these estimates in the theoretical expression for the spectral density function of an AR(k) process (viz., (4.12.59)) we obtain the estimate,

$$\hat{h}(\omega) = \frac{\hat{\sigma}_\varepsilon^2}{2\pi} \cdot \frac{1}{|1 + \hat{a}_1 \exp(-i\omega) + \ldots + \hat{a}_k \exp(-ik\omega)|^2}. \tag{7.8.2}$$

In practice the order k of the model would also be unknown, but we may estimate k by e.g., using Akaike's AIC criterion, as described in Section 5.4.5.

This method is known as "*autoregressive spectral estimation*", and has been advocated in particular by Akaike (1969b) and Parzen (1974). In fact, Parzen does not assume that the series necessarily satisfies an *exact* AR model; rather, he assumes that we can obtain an *adequate approximation* by using a finite order AR model (i.e. we regard (7.8.1) as an approximation to the possibly infinite order AR representation arising from a general linear process model). We may then use this approximate AR form to estimate the spectral density function, as above. With this formulation we still have to estimate a suitable value for k, but k is no longer the order of the "true" AR model—rather, it is the order of the AR model which, in a sense, provides the best approximation to the observed series.

Parzen (1974, 1977a,b) has developed a special technique for determining k based on a quantity which he calls CAT ("criterion for autoregressive transfer functions") which is similar in spirit to (but not identical with) Akaike's AIC criterion. The CAT criterion involves considering, for each k, the integrated relative mean square error of the spectral estimate, viz.,

$$I = \int_{-\pi}^{\pi} E\left\{ \frac{\hat{h}(\omega) - h(\omega)}{h(\omega)} \right\}^2 d\omega, \tag{7.8.3}$$

and then determines k as that value which, in effect, minimizes I. By splitting up I into two terms, the first of which depends on k while the second is independent of k, Parzen shows that the above procedure is equivalent to computing, for each k, the quantity,

$$
\mathrm{CAT}(k) = \begin{cases} \left(\dfrac{1}{N}\sum_{j=1}^{k}\dfrac{1}{\hat{\sigma}_j^2}\right) - \dfrac{1}{\hat{\sigma}_k^2}, & k = 1, 2, 3, \ldots, \\ -(1 + (1/N)), & k = 0, \end{cases} \tag{7.8.4}
$$

where $\hat{\sigma}_j^2$ is the "unbiased" estimate of the residual variance when an AR model of order j is fitted to the data (see (5.4.21)), and then choosing k as that value for which $\mathrm{CAT}(k)$ attains its minimum value.

This technique provides a "semi-automatic" method of spectral estimation in the sense that it leads to a unique form of $\hat{h}(\omega)$ based purely on the data and apparently does not require any prior information on the form of the true spectral density function. (As we have seen, in the "window" method one ideally requires prior information on the spectral bandwidth in order to make an optimal choice of the window bandwidth.) However, it should be remembered that the above spectral estimate is optimal (in the integrated relative mean square error sense) only within the class of estimates of the autoregressive form.

ARMA type estimates

Akaike (1974c, 1976) has also considered a parametric method of spectral estimation but uses an ARMA model (of the form (3.5.52)) in place of the AR model (7.8.1). The theoretical spectral density function for the ARMA(k, l) model is given by (4.12.63), and may be written in the form,

$$
h(\omega) = \frac{\sigma_\varepsilon^2}{2\pi} \frac{\left|1 + b_1 \exp(-i\omega) + \ldots + b_l \exp(-il\omega)\right|^2}{\left|1 + a_1 \exp(-i\omega) + \ldots + a_k \exp(-ik\omega)\right|^2}, \tag{7.8.5}
$$

and the estimate of $h(\omega)$ is obtained by substituting estimates of the parameters $(b_1, \ldots, b_l, a_1, a_2, \ldots, a_k)$ and σ_ε^2 in (7.8.5). (For methods of estimating parameters in ARMA models see Section 5.4.3). The orders k, l of the AR and MA operators may be determined by again applying Akaike's AIC criterion.

Beamish (1977) has studied the properties of window estimates and AR estimates (and a closely related form called "maximum entropy" estimates which we discuss below) and has applied these techniques to a large number of series simulated from known models. Not surprisingly, it turns out that all these techniques lead to comparable results, no one method being uniformly superior to the others. As one might expect, the AR estimate works well if the series is generated by an exact AR model, but if the series is generated by, e.g., an MA(1) model with $b_0 = 1$, $b_1 \sim 1$, then the AR estimate is relatively poor. The window method provides reasonable estimates for most series provided the window bandwidth is suitably selected. Although it has sometimes been asserted that the AR and ARMA methods generally lead to "smoother" estimates than the window method, Beamish found that this was not necessarily the case. In fact, in quite a number of cases (such as the MA(1) example referred to above) the window estimate was considerably smoother than the AR estimate; see Beamish and Priestley (1981). (Of course, the degree of smoothness of the window estimate depends on the magnitudes of both the window bandwidth and the spectral bandwidth; in general, for a given spectrum, the wider the window bandwidth the smoother the estimate.)

The character of the AR and ARMA spectral estimates is quite different from that of the window estimates, the basic distinction being that all window estimates are *quadratic functions* of the observations whereas the AR and ARMA estimates are very much more complicated functions of the data. The sampling properties of AR and ARMA estimates are correspondingly more complicated than those of the window estimates, although Hannan (1970), Parzen (1974) and Berk (1974) have derived some asymptotic results for their second order properties. Hannan (1970), p. 334 observes that these complicated expressions "reflect the use of a method of estimation not basically designed for spectral estimation".

It is interesting to recall that although AR and ARMA models would generally be fitted by time domain methods, they can alternatively be fitted by frequency domain methods; see, e.g. Hannan (1969a,b). To fit an ARMA model using a frequency domain approach one, in effect, fits a function of the form (7.8.5) to the periodogram ordinates (see the remarks at the end of Section 5.4.3). If the parameters of an ARMA model are, in fact, estimated in this way then the ARMA method of spectral estimation may be viewed basically as just another way of "smoothing" the periodogram. Thus, whereas the window technique smooths the periodogram by "local averaging", the ARMA method smooths it "globally" by fitting a rational function over the complete frequency range.

It is also interesting to note that the AR and ARMA methods can be combined with the window method if we regard the fitting of a parametric model simply as a "*prewhitening technique*" (see Section 7.4.1). In this approach we would fit, e.g., an AR(k) model but instead of automatically assuming that the residuals ε_t constitute a purely random process we would evaluate the residuals from the fitted model and compute their estimated spectral density function $\hat{h}_\varepsilon(\omega)$, say, by the standard window method. The final estimate of $h(\omega)$, the spectral density function of X_t, would then be given by replacing $(\hat{\sigma}_\varepsilon^2/2\pi)$ in (7.8.2) by $\hat{h}_\varepsilon(\omega)$.

This approach would seem to provide a very powerful method since even if the fitted model does not describe the observed series with very high accuracy it should, nevertheless, provide residuals which are much closer to "white noise" than the original series. Consequently, one could estimate $h_\varepsilon(\omega)$ using a window with a relatively large bandwidth, and hence derive an estimate of $h_\varepsilon(\omega)$ with fairly low bias and variance.

Maximum entropy estimates

The method of "maximum entropy" spectral estimation was proposed by Burg (1967, 1972), and although the basic ideas underlying this method are quite different from those discussed above it turns out that the "maximum entropy" estimates are, in fact, identical in form with the AR estimates. The method of maximum entropy is based on choosing a spectral estimate $\hat{h}(\omega)$ which is such that the "entropy",

$$E = \int_{-\pi}^{\pi} \log_e\{\hat{h}(\omega)\}\, d\omega, \tag{7.8.6}$$

is maximized, subject to the constraints,

$$\int_{-\pi}^{\pi} \hat{h}(\omega)\, e^{i\omega r}\, d\omega = \hat{R}(r), \qquad r = 0, \pm 1, \dots, \pm k, \tag{7.8.7}$$

where k is a given integer and the $\{\hat{R}(r)\}$ are the sample autocovariances. (For a general discussion of the concept of "entropy" in statistical inference see, e.g., Akaike (1977).) Thus, we seek that functional form of $\hat{h}(\omega)$ which maximizes E subject to the constraints that the first k Fourier coefficients of $\hat{h}(\omega)$ must match exactly the first k sample autocovariances.

The following argument (due to Akaike) shows that the resulting form of $\hat{h}(\omega)$ is exactly the same as for the AR estimate. Given $\hat{R}(0), \hat{R}(\pm 1), \dots, \hat{R}(\pm k)$, let $\{W_t\}$ be *any* stationary process whose autocovariance function $R_W(r)$ satisfies $R_W(r) = \hat{R}(r)$, $r = 0, \pm 1, \dots, \pm k$. Now consider a linear predictor $\hat{W}_{t+1}$ of W_{t+1}, given $W_t, W_{t-1}, W_{t-2}, \dots$, i.e. consider

$$\hat{W}_{t+1} = \sum_{u=0}^{\infty} b_u W_{t-u},$$

and choose the $\{b_u\}$ so as to minimize $M = E[W_{t+1} - \hat{W}_{t+1}]^2$. It is well known that this minimum value of M is given by

$$2\pi \exp\left\{\frac{1}{2\pi} \int_{-\pi}^{\pi} \log h_W(\omega)\, d\omega\right\},$$

where $h_W(\omega)$, the spectral density function of $\{W_t\}$, is an arbitrary function subject only to the constraints,

$$\int_{-\pi}^{\pi} e^{i\omega r} h_W(\omega)\, d\omega = \hat{R}(r), \qquad r = 0, \pm 1, \ldots, \pm k \qquad (7.8.8)$$

(see Chapter 10, equation (10.1.41)).

We thus clearly have that,

$$E\left[W_{t+1} - \sum_{u=0}^{k-1} a_{u+1} W_{t-u}\right]^2 \geq E[W_{t+1} - \hat{W}_{t+1}]^2$$

$$= 2\pi \exp\left[(1/2\pi) \int_{-\pi}^{\pi} \log h_W(\omega)\, d\omega\right],$$

for all sequences $\{a_u\}$ and all $\{W_t\}$. But the minimum value of

$$M' = E\left[W_{t+1} - \sum_{u=0}^{k-1} a_{u+1} W_{t-u}\right]^2$$

w.r. to $\{a_u\}$ is the same as the one step prediction error in the case where W_t is exactly an AR(k) process. In fact, the minimum value of M' w.r. to $\{a_u\}$ is given by the ratio of the following determinants:

$$\frac{\begin{vmatrix} R_W(0) & R_W(1) & \cdots & R_W(k-1) \\ R_W(-1) & R_W(0) & \cdots & R_W(k-2) \\ \vdots & \vdots & & \vdots \\ R_W(-k+1) & R_W(-k+2) & \cdots & R_W(0) \end{vmatrix}}{\begin{vmatrix} R_W(0) & R_W(1) & \cdots & R_W(k) \\ R_W(-1) & R_W(0) & \cdots & R_W(k-1) \\ \vdots & \vdots & & \vdots \\ R_W(-k) & R_W(-k+1) & \cdots & R_W(0) \end{vmatrix}}, \qquad (7.8.9)$$

which is a function only of $R_W(0), R_W(\pm 1), \ldots, R_W(\pm k)$, i.e. a function only of $\hat{R}(0); \hat{R}(\pm 1), \ldots, \hat{R}(\pm k)$, and thus the minimum value of M' is the same for *all* processes $\{W_t\}$ whose autocovariance functions satisfy the required constraints. Moreover, the values of $\{a_u\}$ which minimize M' are given by the Yule–Walker equations,

$$R_W(s+1) = \sum_{u=0}^{k-1} a_{u+1} R_W(s-u), \qquad s = 0, \ldots, (k-1), \quad (7.8.10)$$

(cf. (5.4.17)). But if, in particular, W_t is an AR(k) process then,

$$\min M' = 2\pi \exp\left[(1/2\pi) \int_{-\pi}^{\pi} \log h_{AR}(\omega)\, d\omega\right],$$

where $h_{AR}(\omega)$ denotes the spectral density function of the AR(k) model. We therefore obtain,

$$\int_{-\pi}^{\pi} \log h_{AR}(\omega)\, d\omega \geq \int_{-\pi}^{\pi} \log h_W(\omega)\, d\omega, \qquad (7.8.11)$$

where $h_W(\omega)$ is an arbitrary function subject only to the constraints (7.8.8).

We have thus shown that the functional form of $h(\omega)$ which maximizes the entropy (subject to the given constraints) corresponds exactly to the spectral density function of an AR(k) process with the coefficients $\{a_u\}$ determined by the Yule–Walker equations. Since $R_W(0), \ldots, R_W(\pm k)$ are identical with $\hat{R}(0), \ldots, \hat{R}(\pm k)$, the values of $\{a_u\}$ determined by the Yule–Walker equations are (effectively) the same as the least squares estimates of the AR parameters. Thus, for given k, the maximum entropy spectral estimate is the same as that obtained by fitting an AR(k) model.

The maximum entropy method of spectral estimation was originally proposed by Burg in 1967 but the first published paper describing this approach is due to Lacoss (1971). In this paper Lacoss discusses also another "non-quadratic" spectral estimate, called the "maximum likelihood estimate", developed originally by Capon (1969). The maximum likelihood estimate is based on the idea of constructing a filter which passes a particular frequency component in an undistorted form and rejects all other frequency components in an "optimal" manner and, given K sample autocovariances, leads to an estimate of the form,

$$\hat{h}(\omega) = \frac{1}{l(\omega) R_K^{-1} l^*(\omega)}, \qquad (7.8.12)$$

where $l(\omega) = \{1, \exp(i\omega), \exp(2i\omega), \ldots, \exp(Ki\omega)\}$, R_K is the $K \times K$ matrix with (i, j)th element $\hat{R}(i - j)$ ($\hat{R}(s)$ being the sample autocovariance function of the data), and $*$ denotes transposition and conjugation. Burg (1972) established a relationship between the maximum entropy and maximum likelihood estimates by showing that the reciprocals of the maximum likelihood estimate based on K autocovariances is equal to the average of the reciprocals of the maximum entropy estimates based on $k = 1, 2, \ldots, K$ autocovariances.

It should be noted that Burg's method for estimating the coefficients in the maximum entropy estimate is somewhat different from that based on the Yule–Walker equations, although it uses a similar type of recursion. Burg's method is described by Ulrych and Bishop (1975) and Anderson (1974).

Tukey (1978) remarks that AR spectral estimation is suitable only for relatively "smooth" spectra, and performs badly for spectra with very erratic and very small high frequency power, such as the wave-guide spectra studied by Thomson (1977) and the spectra of stars and planets studied by Comes and Michel (1974). However, Akaike has suggested that in such situations one could filter the series into a number of different frequency bands, shift the output of each band down to an interval centred on zero frequency, and then fit separate AR models to each output.

Martin (1979) has also reported poor results with AR spectral estimates. He considered a simulated AR(6) process whose spectral density function had two well defined peaks. Both peaks were easily captured via standard "window" estimates, but the AR estimates produced only one peak.

Kleiner *et al.* (1979) have discussed the problem of *"robust spectral estimation"*. They point out that conventional spectral estimates can be severely distorted by the presence of "outliers" in the data, and such outliers need not be large relative to the scale of the observations; they need only be large relative to the scale of the innovations process. These authors propose two techniques for constructing spectral estimates which are robust with respect to outliers. Both techniques make use of AR model fitting, but here the AR model is used as a *pre-whitening device* rather than as a direct spectral estimate. In their first method they fit an AR model by least squares and then subject the observations to "robust filtering"—this device, in effect, compares each observation with its value predicted from the AR model, and then "corrects" any observation which deviates substantially from its predicted value. A new AR model is then fitted to the "corrected" observations, and the process is repeated until convergence is obtained. The final AR model is used as a pre-whitening filter, and the spectrum of the original series is obtained by applying the inverse of this filter to a "windowed estimate" of the spectrum of the residuals. Their second method is similar in nature, but uses a robust method of estimating the AR coefficients (similar to Huber's "*M*-estimates" for regression parameters). The paper cited above describes the application of both methods to a number of series, involving both real and simulated data.

Gasser (1974) has proposed a method for constructing *"resistent spectral estimates"* based on smoothing the logarithm of the periodogram (via a conventional window), and then estimating the spectrum by an exponential transformation. A similar approach, which is also based on smoothing the log periodogram, is discussed by Wahba (1978).

7.9 SOME EXAMPLES

(a) Canadian lynx data

These data were first discussed in Section 5.5, and it was found that among the class of AR models the AIC criterion picked out an eleventh order model. We noted also that the data exhibited a strong cyclical behaviour, and this feature was confirmed by the very large peak in the periodogram; see Graph 6.1 (Chapter 6).

Graph 7.13 shows the form of the logarithm of the (non-normalized) spectral density functions over the interval $(0, \pi)$ based on (i) a windowed

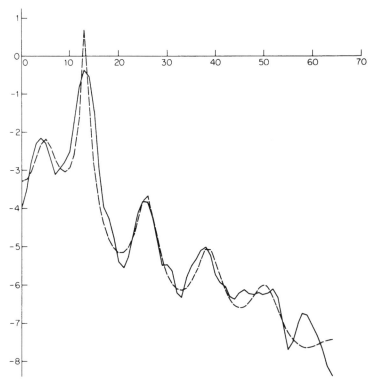

Graph 7.13. Log of the estimated spectral density functions of the lynx series based on "window" and AR model.

estimate using a Parzen window with truncation point $M = 64$, and (ii) the fitted AR(11) model. The full curve shows the windowed estimate, and the dotted curve the AR estimate. (Plotting the *logarithm* of the spectra is effectively the same as displaying the spectra on a "dB" scale.) Over most of the frequency range the agreement between the two forms of spectral estimates is good, the main distinction being that the windowed estimate produces a lower peak. A larger truncation point would increase the height of the peak in the windowed estimate—although it would also increase the "roughness" of the windowed estimate. Note the presence of subsidiary peaks at the harmonics of the main peak. This is due to the asymmetry of the cycles in the original data; see the discussion on the periodogram of the lynx data in Chapter 6.

(b) The following examples are taken from Beamish and Priestley (1981).

Using a series of 100 observations generated from the AR(4) model

$$X_t - 2.7607X_{t-1} + 3.8106X_{t-2} - 2.6535X_{t-3} + 0.9238X_{t-4} = \varepsilon_t,$$

with $\varepsilon_t \sim N(0, 1)$, two different AR spectral estimates were computed, one using coefficient estimates derived from the Yule–Walker equations, and the other using Burg coefficient estimates. For the Yule–Walker estimates, both the AIC and CAT criteria selected an AR model of order 7; with the Burg estimates both criteria selected the correct order, 4. Graph 7.14 shows the two AR spectral estimates superimposed on the

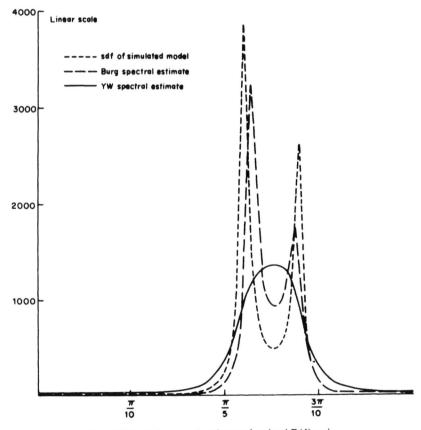

Graph 7.14. AR spectral estimates for the AR(4) series.

theoretical spectral density function. (Note that here the finely dotted curve denotes the *theoretical* form.) The theoretical spectral density function is double-peaked, and the AR/Burg spectral estimate captures this feature quite well. However, the AR/Yule–Walker spectral estimate is poor, and contains just a single peak.

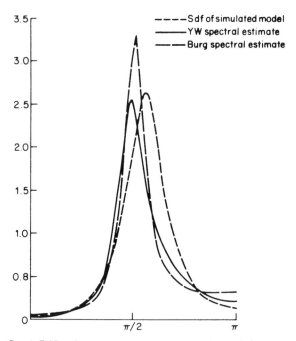

Graph 7.15. AR spectral estimates for the ARMA(2, 2) series.

Graph 7.15 shows the Yule–Walker and Burg AR spectral estimates computed from 100 observations generated from the ARMA(2, 2) model,

$$X_t + 0 \cdot 25 X_{t-1} + 0 \cdot 5 X_{t-2} = \varepsilon_t - \varepsilon_{t-1} - 0 \cdot 75 \varepsilon_{t-2},$$

with $\varepsilon_t \sim N(0, 1)$. (The AIC and CAT criteria selected a fourth order AR model with both the Yule–Walker and Burg coefficient estimates.) Here, both forms of AR spectral estimates give reasonably good results, although the AR estimate based on the Yule–Walker coefficients has an excessively high peak. For comparison, Graph 7.16 shows a windowed spectral estimate computed from the same data, using a Parzen window with truncation point $M = 30$. The windowed estimate is comparable with the AR/Burg estimate, apart from the slight "roughness" in the windowed estimate at the higher frequencies.

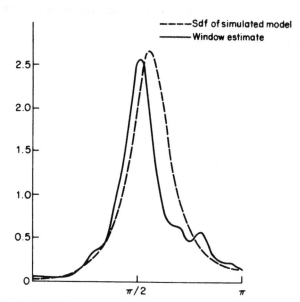

Graph 7.16. Window spectral estimate for the ARMA(2, 2) series.

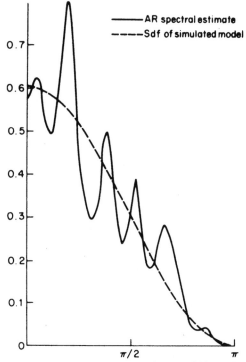

Graph 7.17. AR spectral estimate for the MA(1) series.

Graph 7.17 shows an AR/Yule–Walker spectral estimate computed from 500 observations generated from the MA(1) model,

$$X_t = \varepsilon_t + 0 \cdot 95 \varepsilon_{t-1},$$

with $\varepsilon_t \sim N(0, 1)$. Since the coefficient of ε_{t-1} is very close to 1, the coefficients in the AR representation of this model decay very slowly, and the first 13 coefficients are numerically greater than 0·5. This model thus provides a searching test for the AR method of spectral estimation. With the Yule–Walker coefficients, the AIC and CAT criteria selected a sixteenth order AR model, and the resulting spectral estimate is that shown in Graph 7.17. It will be seen that here the AR estimate is highly irregular, with quite marked oscillations over the whole frequency range. On the other hand, the windowed estimate computed from the same data (using a Parzen window with truncation point $M = 10$), and shown in Graph 7.18, gives a very smooth and accurate reproduction of the theoretical spectral density function.

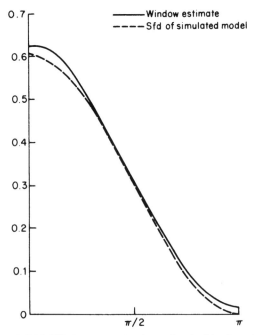

Graph 7.18. Window spectral estimate for the MA(1) series.

Chapter 8

Analysis of Processes with Mixed Spectra

8.1 NATURE OF THE PROBLEM

In our first discussions on spectral analysis (Chapter 6) it was pointed out that the analysis takes different forms depending on whether the process has (a) a purely discrete spectrum (b) a purely continuous spectrum, or (c) a mixed spectrum. Cases (a) and (b) were discussed in detail in Chapter 6 (with Chapter 7 concentrating on the practical aspects of case (b)), but case (c) was touched on only briefly in Section 6.3. We now return to the mixed spectrum case and examine it in more detail.

We noted in Section 6.3 that the mixed spectrum model is by far the most difficult one to deal with since here the spectrum contains discrete and continuous components, both of which are, in general, of unknown form. For such a process the (non-normalized) integrated spectrum $H(\omega)$ may be written as (cf. (4.11.2)),

$$H(\omega) = H_1(\omega) + H_2(\omega), \qquad (8.1.1)$$

where $H_1(\omega)$, $H_2(\omega)$ denote (respectively) the continuous and discrete components. The function $H_1(\omega)$ is absolutely continuous, and its derivative, $h_1(\omega) = H_1'(\omega)$ represents the corresponding (non-normalized) spectral density function. On the other hand, $H_2(\omega)$ is a step function, with jumps $\{p_r\}$ at, say, frequencies $\{\omega_r\}$, $r = 1, 2, \ldots, K$. The general form of $H(\omega)$ is similar to that illustrated in Fig. 4.11, and a full spectral analysis involves the estimation of K (the number of jumps), their frequencies $\{\omega_r\}$ and magnitudes $\{p_r\}$, $r = 1, \ldots, K$, and the continuous spectral density function $h_1(\omega)$.

Here, we are allowing $h_1(\omega)$ to have a general unknown form. The special case where $h_1(\omega)$ is assumed to be constant was included under the category of "purely discrete spectra", and the analysis was given in Section 6.1, cf. the remarks following equation (6.1.2).

613

8.2 TYPES OF MODELS

Let $\{X_t\}$ denote a discrete parameter zero mean stationary process with a mixed spectrum. Then we know from the discussion following equation (4.11.22) that, corresponding to the decomposition (8.1.1), X_t may be similarly decomposed in the form,

$$X_t = Y_t + Z_t, \tag{8.2.1}$$

where $\{Y_t\}$, $\{Z_t\}$, are uncorrelated processes, with Y_t corresponding to the continuous spectral component, $H_1(\omega)$, and Z_t to the discrete component, $H_2(\omega)$. We may therefore write,

$$Y_t = \int_{-\pi}^{\pi} e^{it\omega}\, dZ_1(\omega), \tag{8.2.2}$$

$$Z_t = \sum_r A_r e^{i\omega_r t}, \tag{8.2.3}$$

where $E[|dZ_1(\omega)|^2] = dH_1(\omega)$, and $E[|A_r|^2] = p_r$, cf. (4.11.25). (A similar result holds for the continuous parameter with the limits of integration in (8.2.2) changed to $(-\infty, \infty)$.)

We thus see that any process with a mixed spectrum can be split up into two parts, namely Z_t, a sum of periodic terms (with its power concentrated at the discrete set of frequencies $\{\omega_r\}$) and Y_t a process with a continuous spectrum (and hence with a continuous distribution of power over the whole frequency range). In the language of communications engineering we would call Z_t the "*signal*" and Y_t the "*noise*", this terminology arising from the fact that, for example, an (unmodulated) radio signal will consist of a sine wave transmitted at a single fixed frequency, whereas a "noise disturbance" (such as "hiss") will usually have a continuous spectrum. The process $\{X_t\}$ consists of "signal" plus "noise", and one of our objectives in analysing a record of $\{X_t\}$ is to "detect", or separate, the "signal" from the "noise". Since we are here allowing the noise spectrum to have a general form the problem may be described as the detection of signals in the presence of "*coloured*" noise. The simple case treated in Section 6.1 (where Y_t was assumed to be a purely random process) corresponds to the detection of signals in the presence of *white* noise.

The structure of the "signal" Z_t merits further consideration. When expressed in the general form (8.2.3), the coefficients $\{A_r\}$ of the harmonic terms are simply uncorrelated complex valued random variables. However, this is hardly a realistic model since, in general, we will have observations from only a single realization of $\{X_t\}$, and thus could not hope to infer anything about the variation of the $\{A_r\}$ over different realizations. All we can hope to do is to estimate the particular values which the $\{A_r\}$ take for the given realization; in other words we might just as well treat the $\{A_r\}$ as constants. But if the $\{A_r\}$ are taken as strict constants then Z_t is a deterministic function and so acts as a time dependent mean. Consequently X_t is no longer a stationary process—unless we adopt the device (used in defining the harmonic process model) of giving each of the terms in (8.2.3) a *random phase* but treating the amplitudes as constants. Thus, for each r we will treat $|A_r|$ as a constant and $\arg\{A_r\}$ as a random variable. (Note that the spectrum of Z_t does not involve the phases but depends on only the $|A_r|$.) Assuming that $\{X_t\}$ is a real valued process we are now led to using the standard harmonic process model for the signal (cf. Sections 3.5.8, 4.10), viz.,

$$Z_t = \sum_{r=1}^{K} A_r \cos(\omega_r t + \phi_r). \tag{8.2.4}$$

(For convenience we have used the same symbol A_r to denote the coefficients in both (8.2.3) and (8.2.4), but these are, of course, different quantities.)

Examples of processes with mixed spectra occur, e.g. in radar, stress analysis, oceanography, and communication engineering. In investigating, for example, the effect of gust loads on aircraft structures the stress record is usually composed of two components due to (i) the interaction of the aircraft structure with the velocity field of atmospheric turbulence, and (ii) the fundamental frequencies of the wing and tail of the aircraft. The term produced by (i) gives rise to the continuous spectrum while (ii) corresponds to the discrete spectrum (see, e.g., Press and Houlbout (1955), Press and Tukey (1956)). The mixed spectrum model has also been employed in oceanography by Whittle (1954b), who used it in the analysis of a Seiche record.

Basic assumptions

We now restate formally the precise model for mixed spectra which has been outlined above. We consider a discrete parameter zero mean stationary process of the form (8.2.1), viz.,

$$X_t = Y_t + Z_t \tag{8.2.5}$$

where

(a) Y_t is a general linear process of the form,

$$Y_t = \sum_{u=0}^{\infty} g_u \varepsilon_{t-u}, \tag{8.2.6}$$

the $\{\varepsilon_t\}$ being independent random variables with $E(\varepsilon_t) = 0$, $E(\varepsilon_t^2) = 1$,

$E(\varepsilon_t^4) < \infty$, and $\sum_{u=0}^{\infty} g_u^2 < \infty$, $\sum_{u=0}^{\infty} u|g_u| < \infty$. We denote the autocovariance function of Y_t by

$$R^{(1)}(s) = E[Y_t Y_{t+s}], \qquad (8.2.7)$$

and the corresponding spectral density function by

$$h_1(\omega) = \frac{1}{2\pi} \sum_{s=-\infty}^{\infty} R^{(1)}(s) e^{-i\omega s}. \qquad (8.2.8)$$

(b) Z_t is a harmonic process of the form,

$$Z_t = \sum_{r=1}^{K} A_r \cos(\omega_r t + \phi_r), \qquad (8.2.9)$$

the $\{A_r\}$, $\{\omega_r\}$, $r = 1, \ldots, K$, being unknown constants, and the $\{\phi_r\}$ are independent random variables, each rectangularly distributed over $(-\pi, \pi)$.

8.3 TESTS BASED ON THE PERIODOGRAM

The basic problem in the analysis of mixed spectra is to infer the complete spectral form, given observations on $\{X_t\}$. This analysis is usually carried out in two stages,
(a) the estimation of the discrete spectrum, i.e. the estimation of the amplitudes $\{A_r\}$ and frequencies $\{\omega_r\}$;
(b) the estimation of the continuous spectral density function $h_1(\omega)$.
If the frequencies, $\{\omega_r\}$, are known a priori then, as mentioned in Section 6.3, we can regard Z_t as a "regression" term and estimate the amplitudes by ordinary least squares. (The fact that the $\{\phi_r\}$ are treated as random variables does not really matter here, we simply estimate the values of $(A_r \cos \phi_r)$ and $(A_r \sin \phi_r)$ appropriate to the given realization. Also, the fact that the "residuals" Y_t are correlated does not affect the asymptotic efficiency of the least squares estimates for trigonometric polynomials; see Section 7.7.) The spectral density function $h_1(\omega)$ can then be estimated from the residuals obtained from the regressions analysis. However, apart from cases where Z_t represents, e.g., a "seasonal factor" (so that the $\{\omega_r\}$ then correspond to multiples of the basic period of one year), the values of the $\{\omega_r\}$ would generally be unknown, and consequently the value of K (the number of harmonic components) would also be unknown. In this situation we have to "search" for the presence of harmonic components, as in the periodogram analysis of Section 6.1 when Y_t was assumed to have a "white" spectrum. In Section 6.1 we proceeded by setting up the null hypothesis that there were no harmonic components present, and determined the value of K by testing

the periodogram ordinates (in order of magnitude) until we failed to obtain a significant result. We now adopt a similar approach for the mixed spectrum case, i.e. we construct a suitable test to detect the presence of harmonic components, and by repeated use we can determine the value of K. Thus, for processes with mixed spectra the various stages involved in a complete spectral analysis may be set out as follows.

(i) The first step is to test for the *existence* of the discrete spectrum, i.e. to detect the presence of a "signal".

(ii) If a discrete spectral component is detected we then estimate its unknown parameters, namely, the amplitude and frequencies of the harmonic terms.

(iii) The autocovariance function corresponding to the continuous spectrum, (i.e. the autocovariance function of Y_t) is then estimated by removing the contribution of the harmonic terms from the sample autocovariances of X_t.

(iv) The continuous spectrum may then be estimated from the "corrected" autocovariances by the same method as used in the case of purely continuous spectra. Alternatively, as we shall see later, the continuous spectrum may be estimated directly from the sample autocovariances of X_t provided that we use a specially designed form of spectral window.

We must however remember that the tests discussed in Section 6.1 (such as that based on Fisher's g-statistic) are no longer valid in the present context when the spectral density function of Y_t is allowed to have a general form. More specifically, we know that even in the null case where $A_r = 0$, all r, the sampling properties of the periodogram ordinates of $\{X_t\}$ depend on $h_1(\omega)$ (cf. (6.2.35)), and any test based on these ordinates must take account of this result. We may, nevertheless, attempt to modify the standard periodogram tests so as to allow for the general form of $h_1(\omega)$, and we now describe various tests of this type.

I. *Whittle's test*

The first study of the problem of mixed spectra is due to Whittle (1952a, 1954b). Whittle's approach is based on an ingenious attempt to reduce the problem to the classical periodogram analysis by using the asymptotic relationship between the periodogram of the general linear process Y_t, and that of the residual process ε_t. Thus, given N observations $X_1, X_2, \ldots, X_N$, let us define the periodogram $I_{N,X}(\omega)$ as in (6.1.24), i.e. write

$$I_{N,X}(\omega) = \frac{2}{N} \left| \sum_{t=1}^{N} X_t e^{-i\omega t} \right|^2, \qquad (8.3.1)$$

and let $I_{N,Y}(\omega)$, $I_{N,\varepsilon}(\omega)$, be similarly defined for the processes $\{Y_t\}$, $\{\varepsilon_t\}$.

Then with the coefficients $\{g_u\}$ in (8.2.6) satisfying the stated conditions we have the well known asymptotic relationship between $I_{N,Y}(\omega)$ and $I_{N,\varepsilon}(\omega)$ (cf. theorem 6.2.2), viz.

$$I_{N,Y}(\omega) \sim 2\pi h_1(\omega) I_{N,\varepsilon}(\omega). \tag{8.3.2}$$

Hence, for large N, $I_{N,Y}(\omega)/\{2\pi h_1(\omega)\}$ may be regarded as the periodogram of an independent process. Suppose now that we wish to test the null hypothesis,

$$H_0: A_r = 0, \quad \text{all } r,$$

i.e. that there are no harmonic components present. Under H_0, $X_t \equiv Y_t$, and consequently $I_{N,X}(\omega)/\{2\pi h_1(\omega)\}$ behaves asymptotically like the periodogram of an independent process. Thus, in principle, we may test for the existence of harmonic components by applying Fisher's g-test (Section 6.1.4) to the "standardized" periodogram ordinates, $I_{N,X}(\omega)/\{2\pi h_1(\omega)\}$. If the function $h_1(\omega)$ were known *a prior* this would be the natural way to deal with the problem of mixed spectra. We would then evaluate the quantity $I_{N,X}(\omega)/\{2\pi h_1(\omega)\}$ at the standard frequencies $\omega_p = 2\pi p/N$, $p = 0, 1, \ldots, [N/2]$ and test H_0 by referring the statistic

$$g_0^{(W)} = \frac{\max\limits_{p} [I_p/2\pi h_1(\omega_p)]}{\sum\limits_{p=1}^{[N/2]} [I_p/2\pi h_1(\omega_p)]}, \tag{8.3.3}$$

where $I_p \equiv I_{N,X}(\omega_p)$, to Fisher's g-distribution with $[N/2]$ degrees of freedom. However, as we have already noted, in general $h_1(\omega)$ is unknown, and this is where the difficulties arise. For, if we wish to use a statistic of the form $g_0^{(W)}$ in this case we have to estimate $h_1(\omega)$ from the observed process X_t. Whittle suggested a truncated periodogram estimate (corresponding to the window (6.2.63)) of the form,

$$\hat{h}_1(\omega) = \frac{1}{2\pi} \sum_{s=-(m-1)}^{(m-1)} \hat{R}_X(s) e^{-i\omega s}, \tag{8.3.4}$$

where

$$\hat{R}_X(s) = \frac{1}{N} \sum_{t=1}^{N-|s|} X_t X_{t+|s|} \tag{8.3.5}$$

is the sample autocovariance function of X_t and m denotes the truncation point. If we now substitute $\hat{h}_1(\omega)$ for $h_1(\omega)$ in (8.3.3) we obtain the test statistic,

$$g^{(W)} = \frac{\max_{p} [I_p/2\pi\hat{h}_1(\omega_p)]}{\sum_{p=1}^{N/2} [I_p/2\pi\hat{h}_1(\omega_p)]}. \tag{8.3.6}$$

Under H_0, $g^{(W)}$ is still asymptotically distributed as Fisher's g. However, if there exists a harmonic component at frequency ω_r it will contribute the term $(\frac{1}{2}A_r^2 \cos s\omega_r)$ to $E[\hat{R}(s)]$, and it is then easy to see that the contribution of the harmonic component to $E[\hat{h}_1(\omega_r)]$ is (approximately) $mA_r^2/4\pi$. Thus, in this case $\hat{h}_1(\omega)$ contains a spurious peak at $\omega = \omega_r$ with height $O(m)$, and the effect of replacing $h_1(\omega)$ by $\hat{h}_1(\omega)$ is to reduce the height of the maximum ordinate by a factor $O(1/m)$. Consequently, the power of the test is substantially reduced. Whittle noted this result and recommended that we should use a modified estimate of $h_1(\omega)$ obtained from $\hat{h}_1(\omega)$ by omitting all peaks in $\hat{h}_1(\omega)$ which we suspect are due to the presence of harmonic components (Whittle (1954b)). This is a rather dangerous procedure, for by omitting peaks in $\hat{h}_1(\omega)$ we are assuming that these are due to harmonic components *before* the existence of such terms has been established by a test on $g^{(W)}$. If, for example, Y_t is an AR(2) process then it is quite possible that $h_1(\omega)$ itself contains peaks of arbitrarily small width, and the application of the above procedure in such a case could possibly lead to the conclusion that there exists a harmonic component even when X_t has a purely continuous spectrum.

The main difficulty in this approach lies in distinguishing between peaks in the continuous spectrum and peaks due to harmonic components. However, this difficulty is more fundamental than it would at first appear. For, as observed by Bartlett (1957), given only a finite number of observations and no further information it is impossible to distinguish strict harmonic components from peaks in the continuous spectrum of arbitrarily small width. (As the noise spectrum becomes sharper and more peaked the noise component begins to look just like a "signal".) Bartlett therefore suggested that the problem of mixed spectra can be made tractable only in terms of the separation of spectral peaks of broad and narrow width. This feature is clearly indicated by the behaviour of the periodogram for a mixed process. For, the contribution of the harmonic component $(A_r \cos \omega_r t)$ to $E[I_{N,X}(\omega)]$ is (cf. theorem 6.1.2),

$$U = \frac{A_r^2}{2N} \left\{ \frac{\sin^2[\frac{1}{2}N(\omega - \omega_r)]}{\sin^2[\frac{1}{2}(\omega - \omega_r)]} + \frac{\sin^2[\frac{1}{2}N(\omega + \omega_r)]}{\sin^2[\frac{1}{2}(\omega + \omega_r)]} \right\}. \tag{8.3.7}$$

In the region $0 \leq \omega \leq \pi$, U has a maximum peak at $\omega = \omega_r$, whose shape follows that of the Fejer kernel (cf. Fig. 6.2), with bandwidth $2\pi/N$. Now consider a process with a purely continuous spectrum whose spectral density function $h(\omega)$ has bandwidth B_h. We know from (6.2.11) that (apart from a constant factor) $E[I_{N,X}(\omega)]$ is the convolution of $h(\omega)$ with the Fejer kernel, $F_N(\theta)$. Hence, if B_h is substantially larger than $2\pi/N$, the bandwidth of $E[I_{N,X}(\omega)]$ will also be very close to B_h. If B_h is comparable with (or smaller than) $2\pi/N$, $E[I_{N,X}(\omega)]$ will have a bandwidth of (approximately) $2\pi/N$. Since the whole basis of periodogram tests for harmonic components lies in the fact that harmonic components produce sharp peaks in the periodogram, it is clear that, given a sample of N observations, it will be impossible to distinguish strict harmonic components from peaks in the continuous spectrum whose widths are $\leqslant 2\pi/N$. Thus, in order to make the problem tractable we make the assumption that *the continuous spectral density function, $h_1(\omega)$, has bandwidth $B_h \gg 2\pi/N$.*

(The point discussed above is one of many where arguments based purely on asymptotic theory can give quite misleading results. We could, of course, argue that since B_h is a fixed quantity there will always be a sufficiently large value of N for which $B_h \gg 2\pi/N$, and hence that *asymptotically* we can always distinguish between harmonic components and peaks in the continuous spectrum. However, this argument is totally irrelevant as far as practical analysis is concerned. In practice we are usually given a fixed number of observations, and the question then is whether, *for that particular value of N, B_h* is sufficiently large to enable us to identify the peaks due to harmonic components.)

II. *The grouped periodogram test*

With the above restriction on B_h, an alternative test to Whittle's, suggested by Bartlett, may be constructed by grouping the periodogram ordinates. Let k be some integer $\leqslant B_h$, and divide the periodogram ordinates into $[N/2k]$ sets, each set containing k ordinates. We thus obtain the $[N/2]$ periodogram ordinates subdivided as follows:

$$I_1, I_2, \ldots, I_k; \quad I_{k+1}, \ldots, I_{2k}; \quad I_{2k+1}, \ldots, I_{3k}; \quad \ldots$$

Consider

$$\gamma_k = \frac{I_{p'}/\{2\pi h_1(\omega_{p'})\}}{\sum\limits_{p=(l-1)k+1}^{lk} I_p/\{2\pi h_1(\omega_p)\}}, \tag{8.3.8}$$

where

$$I_{p'}/\{2\pi h_1(\omega_{p'})\} = \max_{(l-1)k+1 \leqslant p \leqslant lk} [I_p/\{2\pi h_1(\omega_p)\}]$$

Under H_0, the statistic γ_k has asymptotically the same distribution as Fisher's g with k degrees of freedom, and may be used as the basis of a test for mixed spectra by employing the following device. Because of the restriction on k with respect to the bandwidth of $h_1(\omega)$, we may write $h_1(\omega) \sim$ constant, in the frequency region considered. Hence, we may approximate to γ_k by $g_k^{(B)}$, where

$$g_k^{(B)} = I_{p'} / \sum_p I_p. \tag{8.3.9}$$

Thus, even when $h_1(\omega)$ is unknown, we may test for the existence of harmonic components by referring $g_k^{(B)}$ to Fisher's g-distribution with k degrees of freedom.

The relationship between γ_k and $g_k^{(B)}$ may be examined in more detail by writing

$$\gamma_k = g_k^{(B)}[1/(1+\mu)],$$

where

$$\mu = \left(\sum_p I_p \delta_p \right) \bigg/ \left(\sum_p I_p \right),$$

and

$$\delta_p = \{h_1(\omega_{p'}) - h_1(\omega_p)\}/h(\omega_p).$$

Assume that in the region considered $h_1(\omega)$ is differentiable and has at most one zero. Let $M(h)$, $m(h)$ be the upper and lower bounds of $h_1(\omega)$ in this region. Then, since we have chosen $k \leq B_h$, $m(h)/M(h) \geq \frac{1}{2}$, and it follows that

$$\gamma_k\{2 - M(h)/m(h)\} \leq g_k^{(B)} \leq \gamma_k M(h)/m(h). \tag{8.3.10}$$

Hence, even though $M(h)/m(h) \leq 2$, it is still possible for $g_k^{(B)}$ to differ considerably from γ_k, and it may happen that $g_k^{(B)}$ is almost twice as large as γ_k. The only way to obtain a closer approximation to γ_k would be to make k much smaller than B_h, and thereby reduce the ratio $M(h)/m(h)$. However, unless $h_1(\omega)$ has a large bandwidth this procedure would leave $g_k^{(B)}$ with a very small number of degrees of freedom. There does not appear to be any systematic way of choosing k, but the value finally chosen must be a compromise between reducing the ratio $M(h)/m(h)$ and retaining sufficient degrees of freedom to give the test reasonable power.

Assuming that we may, to a sufficient degree of accuracy, test $g_k^{(B)}$ using the null distribution of γ_k, it remains only to adjust the significance level of the test to allow for the choice of subinterval whose ordinates are being tested. If we could choose a significance level α when applying Fisher's classical periodogram test to k ordinates, then the approximate significance level for a test based on $g_k^{(B)}$ becomes $\alpha' = \alpha k/[\tfrac{1}{2}N]$.

Canadian lynx data

In Section 6.1.4 we applied Fisher's g-test to the complete set of periodogram ordinates for the Canadian lynx data, and found that the large peak at ordinate $p = 12$ was highly significant according to this test. We now consider the application of the grouped periodogram test to these ordinates. Taking groupings of $k = 5$, 10 and 20 ordinates gives the following results.

Grouping (k):	5	10	20
$g_k^{(B)}$:	0·8523	0·8343	0·6371

Here $N = 114$, so that using only the first term in the g-distribution (cf. (6.1.67)), the critical value for $g_k^{(B)}$ based on a significance level α is approximately given by

$$\frac{\alpha k}{57} = k(1-z)^{k-1}.$$

Taking $\alpha = 0·05$, we obtain the following approximate critical values.

k:	5	10	20
z:	0·8279	0·5425	0·1917

The value of $g_5^{(B)}$ is thus significant at 5%, while the values of $g_{10}^{(B)}$ and $g_{20}^{(B)}$ are highly significant. The result for $k = 20$ should, perhaps, be viewed with some suspicion since its validity rests on the assumption that the spectral density function of the residual process is roughly constant over a fairly broad band of frequencies (and the same may possibly apply to the $k = 10$ grouping), but the overall nature of the above results certainly supports the hypothesis that the Canadian lynx data have a "mixed spectrum" structure. Campbell and Walker (1977) adopted a "mixed spectrum" model, and, having estimated the period of the harmonic component at 9·63 years, they fitted an AR(2) model to the residuals. Their final model is

$$X_t = 2·9036 + 0·0895 \cos(2\pi t/9·63) - 0·6249 \sin(2\pi t/9·63) + Y_t,$$

where

$$Y_t - 0·9717 Y_{t-1} + 0·2654 Y_{t-2} = \varepsilon_t,$$

with $\hat{\sigma}_\varepsilon^2 = 0·041984$.

III. *Hannan's test*

Hannan (1961) considered a test statistic which is essentially the same as Whittle's statistic $g^{(W)}$, but introduced an important modification in the estimation of $h_1(\omega)$. Instead of using the usual "window" type of estimator (of which (8.3.4) is typical) Hannan uses an estimate of the form

$$h_1^*(\omega) = \left[\hat{h}_1(\omega) - \frac{2\pi}{N} W_m(0) I_{N,X}^*(\omega)\right] \bigg/ \left(1 - \frac{2\pi}{N} W_m(0)\right), \quad (8.3.11)$$

where

$$\hat{h}_1(\omega) = \int_{-\pi}^{\pi} I_{N,X}^*(\theta) W_m(\omega - \theta)\, d\theta \qquad (8.3.12)$$

is a conventional window estimate computed from a spectral window $W_m(\theta)$, m being the window parameter. (We have written $I_{N,X}^*(\omega) = (1/4\pi) I_{N,X}(\omega)$, consistent with the notation used in previous chapters, but the notation for windows has been changed slightly, the suffix now denoting the window parameter.) If we choose $W_m(\theta)$ as the truncated periodogram window, then $\hat{h}_1(\omega)$ is, of course, identical with the estimate (8.3.4) suggested by Whittle, but the importance of the second term in (8.3.11) is that it removes the contribution which might arise from a harmonic component at frequency ω. Hannan derived the estimate $h_1^*(\omega)$ by removing the regression of X_t on the terms $\sin \omega t$ and $\cos \omega t$, and it may be shown that, under the usual conditions on the window $W_m(\theta)$, $h_1^*(\omega)$ is an asymptotically unbiased estimate of $h_1(\omega)$ *whether or not there exists a harmonic component at frequency ω.* The estimate $h_1^*(\omega)$ thus removes the difficulty of having to decide whether observed peaks in the conventional estimate, $\hat{h}_1(\omega)$, are due to the continuous or discrete spectrum, and the resulting test statistic is obtained by substituting $h_1^*(\omega)$ for $h_1(\omega)$ in (8.3.3).

We will not, at this point, discuss the properties of $h_1^*(\omega)$ further since it is a special case of a more general class of estimates based on the "*double window*" *technique* introduced by Priestley (1964a). This general class of estimates will be examined in detail later on, but for the moment we may note that $h_1^*(\omega)$ is a special case of estimates $\hat{\hat{h}}_1(\omega)$ of the form

$$(1 - c)\hat{\hat{h}}_1(\omega) = \int_{-\pi}^{\pi} I_{N,X}^*(\theta)\{W_m^{(1)}(\omega - \theta) - c W_n^{(2)}(\omega - \theta)\}\, d\theta \qquad (8.3.13)$$

where $W_m^{(1)}(\theta)$, $W_n^{(2)}(\theta)$, are general windows (with parameters m, n respectively), $n > m$, and $c = W_m^{(1)}(0)/W_n^{(2)}(0)$. The particular estimate $h_1^*(\omega)$ results effectively from (8.3.13) on setting $n = N$ and choosing

$$W_n^{(2)}(\theta) = \begin{cases} N/2\pi, & |\theta| < \pi/N \\ 0, & \text{otherwise,} \end{cases} \qquad (8.3.14)$$

so that here,

$$c = \frac{W_m^{(1)}(0)}{W_n^{(2)}(0)} = \frac{2\pi}{N} W_m^{(1)}(0). \qquad (8.3.15)$$

The basic idea underlying the construction of $\hat{h}_1(\omega)$ is that $W_m^{(1)}(\theta)$ acts as a conventional spectral window, while $W_n^{(2)}(\theta)$ removes the contribution from a harmonic component which may be presented at frequency ω. Let $B_w^{(1)}$, $B_w^{(2)}$ denote respectively the bandwidth of $W_m^{(1)}(\theta)$, $W_n^{(2)}(\theta)$. It will be shown in Section 8.6 that, whether or not there exists a harmonic component at frequency ω, $\hat{h}_1(\omega)$ is an approximately unbiased estimate of $h_1(\omega)$ provided

(i) $B_w^{(1)} \ll B_h$ (the usual requirement),

and

(ii) $2\pi/N \leqslant B_w^{(2)} < B_w^{(1)}$.

(Note that the bandwidth of the Fejer kernel generated by a harmonic component is $2\pi/N$.) Hannan chooses $W^{(2)}$ so that its bandwidth has the minimum value $2\pi/N$, but in general we may allow $W^{(2)}$ to have a larger bandwidth, subject to the condition (ii). However, conditions (i) and (ii) together clearly indicate that in order to construct an estimate of the form $h_1(\omega)$ we must have $B_h \gg 2\pi/N$, so that the condition previously noted on the bandwidth of $h_1(\omega)$ is still required for the successful application of Hannan's test. If $B_h = O(1/N)$ we would have to choose $B_w^{(1)} > B_h$, in which case $\hat{h}_1(\omega)$ could be substantially biased and, in general, would underestimate $h_1(\omega)$ at a peak frequency. This would inflate the value of $\{I_{N,X}(\omega)/2\pi\hat{h}_1(\omega)\}$ and may lead to the conclusion that harmonic components exist even when the spectrum is purely continuous.

We will see also in Section 8.6 that although $E[\hat{h}_1(\omega)]$ is essentially unaffected by the presence of a harmonic component at frequency ω, it will be affected by a harmonic component at a neighbouring frequency (i.e. within the bandwidth of $W_n^{(2)}(\omega)$) unless the windows satisfy the additional condition

$$W_m^{(2)}(\theta) - c W_n^{(1)}(\theta) = 0 \qquad (8.3.16)$$

for all θ in the interval, $|\theta| \leqslant B_w^{(2)}$. Unless $W_m^{(2)}(\theta)$ is chosen to be a rectangular window the estimate (8.3.11) will not possess this property, in which case $h_1^*(\omega)$ will be biased by the presence of a harmonic component whose frequency lies in the interval $(\omega - 2\pi/N, \omega + 2\pi/N)$. MacNeill (1974, 1977) has extended Hannan's test to the multivariate case where one observes a vector process $X_t = (X_t^{(1)}, X_t^{(2)}, \ldots, X_t^{(s)})$, say. In particular, MacNeill (1977) has constructed a test for the hypothesis that several processes share common periodic components. Nicholls (1967) discusses the power of Hannan's test when the frequency of the harmonic component is not a multiple of $(2\pi/N)$, and some discussion of the bivariate case is given by Nicholls (1969).

IV. *Bartlett's test*

Bartlett (1967, 1978) proposed a method of detecting harmonic components which, although not based directly on the periodogram ordinates, uses the closely related statistics

$$A_X(\omega_p) = \sqrt{\frac{2}{N}} \sum_{t=1}^{N} X_t \cos \omega_p t, \tag{8.3.17}$$

$$B_X(\omega_p) = \sqrt{\frac{2}{N}} \sum_{t=1}^{N} X_t \sin \omega_p t. \tag{8.3.18}$$

(We recall from (6.1.22) that $I_p = [\{A_X(\omega_p)\}^2 + \{B_X(\omega_p)\}^2].$) Let $J_X(\omega_p) = A_x(\omega_p) + iB_X(\omega_p)$, and let $J_Y(\omega_p), J_Z(\omega_p)$ be similarly defined with respect to the processes Y_t, Z_t given by (8.2.6), (8.2.9). Then clearly,

$$J_X(\omega_p) = J_Y(\omega_p) + J_Z(\omega_p). \tag{8.3.19}$$

Suppose now that Z_t contains a harmonic component with frequency ω_r, amplitude A_r, and phase ϕ_r. Bartlett (1978), p. 341 shows that, for large N,

$$J_Z(\omega_p) \sim \sqrt{\frac{N}{2}} \cdot A_r \exp(-i\phi_r) \frac{\exp(-2\pi\varepsilon) - 1}{2\pi i(\varepsilon + s)}, \tag{8.3.20}$$

where ε, s are defined by writing $\omega_r = \omega_{p-s} + (2\pi\varepsilon/N)$, $(|\varepsilon| \ll \frac{1}{2})$. Bartlett's method for detecting harmonic components is to construct a graph using $A_X(\omega_p)$, $B_x(\omega_p)$, $(p = 0, 1, \ldots, [N/2])$ as Cartesian coordinates. From (8.3.20) it will be seen that the presence of a harmonic component gives rise to a term $J_Z(\omega_p)$, whose real and imaginary components are *large* (in fact, $O(N^{1/2})$), and concentrated in *a particular direction* when ω_p lies in the neighbourhood of ω_r. Thus, by inspecting the $\{A(\omega_p), B(\omega_p)\}$ plot one should be able to detect visually the presence of harmonic terms. This technique has been illustrated numerically by Bartlett (1967) in relation to the analysis of some traffic data, but no test is as yet available for assessing the "directional" properties of such plots.

Unfortunately, all the tests based on the periodogram suffer from a number of disadvantages which may be summarized as follows:
  (a) Whittle's test is severely limited in that it requires the estimation of the complete spectral density function. Harmonic components produce peaks in this estimated function which, if not removed, may reduce the power of the test so greatly that it fails to detect harmonic

components even with fairly large amplitudes. On the other hand, the removal of these peaks before the test is applied is equivalent to the assumption that the peaks are due to harmonic components, thus begging the whole question, since the main purpose of the test is to detect harmonic components.

(b) Hannan's test also requires the estimation of the complete spectral density function. Although Hannan's estimate $h_1^*(\omega)$ is constructed in a much more subtle way (so that it does not suffer from the disadvantages noted above) it may still be biased by the presence of a harmonic component whose frequency lies between two neighbouring periodogram ordinates.

(c) The grouped periodogram test does not require the estimates of the spectral density function but introduces the difficulty of choosing k. There does not appear to be any systematic method of choosing k, and as noted previously, such a choice must be a compromise between reducing the factor $\{M(h)/m(h)\}$ and retaining sufficient degrees of freedom. Also, it must be remembered that even when k has been chosen, the error introduced by replacing γ_k by $g_k^{(B)}$ may be quite serious.

(d) Bartlett's approach is certainly an interesting one, but it relies purely on a visual inspection of the $\{A(\omega_p), B(\omega_p)\}$ plots and, as yet, no formal associated test has been developed.

In the next section, we consider a different approach to the mixed spectra problem, developed by Priestley (1962a,b, 1964a), which is not based on the periodogram but uses instead a more direct investigation of the auto-covariance function.

8.4 THE $P(\lambda)$ TEST

Consider a process X_t of the form (8.2.5). The autocovariance function of X_t, $R_X(s)$, may be written as

$$R_X(s) = R^{(1)}(s) + R^{(2)}(s), \qquad (8.4.1)$$

where $R^{(1)}(s)$, $R^{(2)}(s)$ denote the autocovariance functions of Y_t, Z_t, respectively. Since Y_t has a purely continuous spectrum we know that $R^{(1)}(s) \to 0$ as $|s| \to \infty$ (cf. Section 4.9). On the other hand, $R^{(2)}(s)$ consists of a number of sine waves with the same frequencies as Z_t (Section 3.5.8), and will never die out. Consequently, $R_X(s)$ may oscillate with varying amplitude over the initial portion of the function, but will settle down to a steady oscillation of the same form as $R^{(2)}(s)$ as $|s| \to \infty$. Thus, the feature which distinguishes a process with a mixed spectrum from one with a purely continuous spectrum is the behaviour of $R_X(s)$ for large s.

We now recall the result that the rate of decay of $R^{(1)}(s)$ is related to the bandwidth of $h_1(\omega)$ (the "wider" $h_1(\omega)$ the more quickly will $R^{(1)}(s)$ decay; cf. Section 4.10). In the previous section we showed that to make the problem tractable it was necessary to restrict the bandwidth of $h_1(\omega)$ in relation to the number of observations. However, we may equally well specify this restriction in terms of the rate of decay of the autocovariance function $R^{(1)}(s)$. Accordingly, given N observations, we may assume that there exists an integer m ($\ll N$) such that

$$R^{(1)}(s) \sim 0, \qquad |s| \geq m. \tag{8.4.2}$$

If, therefore, we are searching for the presence of harmonic components we may do this by performing a harmonic analysis of the "tail" of the auto-covariance function, i.e. we examine the function

$$P(\lambda) = \frac{1}{2\pi} \sum_{m < |s| \leq n} \hat{R}_X(s) \cos s\lambda, \qquad 0 \leq \lambda \leq \pi, \tag{8.4.3}$$

where $\hat{R}_X(s)$ is the sample autocovariance function of X_t, as given by (8.3.5), and $n \leq N$. Under the null hypothesis H_0 (that X_t has a purely continuous spectrum) we may set $R^{(2)}(s) \equiv 0$ (all s), and it then follows immediately from the condition (8.4.2) that

$$E[P(\lambda)] \sim 0, \qquad \text{all } \lambda. \tag{8.4.4}$$

However, if X_t contains a harmonic component with frequency ω_r, then it will be shown later that

$$E[P(\lambda)] \sim \begin{cases} 0, & |\lambda - \omega_r| \gg O(1/m). \\ O(n-m), & \lambda \sim \omega_r. \end{cases} \tag{8.4.5}$$

Hence, the presence of harmonic terms may be detected by investigating the peaks of this function. There is, however, no need to restrict the discussion to a function of the precise form (8.4.3), and subsequently we consider a more general expression of the form

$$P(\lambda) = \frac{1}{2\pi} \sum_{s = -(N-1)}^{(N-1)} \{w_n^{(1)}(s) - w_m^{(2)}(s)\} \hat{R}_X(s) e^{-i\omega s}, \tag{8.4.6}$$

where $w_n^{(1)}(s)$, $w_m^{(2)}(s)$ are two general "covariance lag windows", with parameters n, m respectively. Note that (8.4.3) may be written in the above form by setting

$$w_n^{(1)}(s) = \begin{cases} 1, & |s| \leq n, \\ 0, & |s| > n, \end{cases} \tag{8.4.7}$$

and

$$w_m^{(2)}(s) = \begin{cases} 1, & |s| \leq m, \\ 0, & |s| > m. \end{cases} \tag{8.4.8}$$

In general $w_n^{(1)}(s)$, $w_m^{(2)}(s)$ may both be chosen from the general class of covariance lag windows discussed in Section 6.2.3. Specifically, we assume that $w_n^{(1)}(s)$, $w_m^{(2)}(s)$ are chosen from a class of lag windows, $\{w_\nu(s)\}$, which satisfy the following conditions.

(1) For each ν, $w_\nu(s)$ is an *even* sequence, i.e.

$$w_\nu(-s) = w_\nu(s), \qquad \text{all } s, \nu.$$

(2) The corresponding spectral window,

$$W_\nu(\theta) = \frac{1}{2\pi} \sum_{s=-\infty}^{\infty} w_\nu(s) \, e^{-is\theta},$$

satisfies conditions (i)–(v) of Section 6.2.4.

(3) For any two spectral windows $W_{\nu_1}^{(1)}(\theta)$, $W_{\nu_2}^{(2)}(\theta)$ write

$$V_{1,2}(\theta) = W_{\nu_1}^{(1)}(\theta) * W_{\nu_2}^{(2)}(\theta),$$

(* denoting convolution). Then,

$$V_{1,2}(\theta) = O(1), \qquad \text{as } \nu_1, \nu_2 \to \infty,$$

for $|\theta| \gg 1/\min(\nu_1, \nu_2)$.

(This condition states merely that if both $W^{(1)}$ and $W^{(2)}$ are highly concentrated at distinct points $\theta = \theta_1$, $\theta = \theta_2$, then the integral of the product of these functions will be small if the separation between θ_1, θ_2 is much greater than the bandwidth of the "wider" of the two functions. In particular, if $W^{(1)}$, $W^{(2)}$ are each rectangular windows, with $\nu_1 > \nu_2$, then $V_{1,2}(\theta) = 0$ for $|\theta| > 2\pi/\nu_2$.)

We may now write $P(\lambda)$ in the alternative form,

$$P(\lambda) = \int_{-\pi}^{\pi} I_{N,X}^*(\theta)\{W^{(1)}(\lambda - \theta) - W^{(2)}(\lambda - \theta)\} \, d\theta \qquad (8.4.9)$$

$$= \hat{h}^{(1)}(\lambda) - \hat{h}^{(2)}(\lambda), \qquad (8.4.10)$$

say, where

$$\hat{h}^{(1)}(\lambda) = \int_{-\pi}^{\pi} I_{N,X}^*(\theta) W^{(1)}(\lambda - \theta) \, d\theta \qquad (8.4.11)$$

and $\hat{h}^{(2)}(\lambda)$ is similarly defined. (For brevity we now drop the suffixes n, m attached to $w^{(1)}(s)$, $w^{(2)}(s)$, and $W^{(1)}(\theta)$, $W^{(2)}(\theta)$, but the dependence of all these functions on the parameters n, m should be noted.) As mentioned above, the test for harmonic components is based on the existence of peaks in the function $P(\lambda)$, and we see from (8.4.5) that the height of such peaks is proportional to $(n - m)$. Hence, for the test to work efficiently, we obviously require $(n - m)$ to be large, subject, of course, to the restrictions that $n \leqslant N$

and m is large enough for (8.4.2) to hold. Accordingly, let the parameters n, m be chosen so that, as $N \to \infty$, both m and $n \to \infty$, but $n \leq N$ and $m = o(n)$. We proceed to investigate the asymptotic sampling of $P(\lambda)$ under the null hypothesis H_0 (i.e. that X_t has a purely continuous spectrum), and assuming that X_t is a *Gaussian process*.

Theorem 8.1 *Let $\{X_t\}$ be a Gaussian process, and suppose that both $h_1(\omega)$ and $h'_1(\omega)$ are bounded functions. On the null hypothesis that $Z_t \equiv 0$,*

(a) $\lim_{N \to \infty} E[P(\lambda)] = 0,$

(b) $\text{var}[P(\lambda)] \sim \dfrac{2\pi}{N} h_1^2(\lambda) \displaystyle\int_{-\pi}^{\pi} \{W^{(1)}(\theta) - W^{(2)}(\theta)\}^2 \, d\theta + o\left(\dfrac{1}{N}\right)$ $\qquad (\lambda \neq 0, \pi)$

(c) $\text{cov}[P(\lambda), P(\mu)] \sim O(1/N), \qquad$ if $|\lambda - \mu| \gg 1/m.$

Proof. If $Z_t \equiv 0$, $X_t(\equiv Y_t)$ has a purely continuous spectrum with spectral density function $h_1(\omega)$. Part (a) now follows immediately on writing (using (6.2.103)),

$$E[P(\lambda)] = \int_{-\pi}^{\pi} h_1(\theta) W^{(1)}(\lambda - \theta) \, d\theta - \int_{-\pi}^{\pi} h_1(\theta) W^{(2)}(\lambda - \theta) \, d\theta + O\left(\frac{\log N}{N}\right)$$

and noting that both the above integrals converge to $h_1(\lambda)$ since both $W^{(1)}(\theta)$ and $W^{(2)}(\theta)$ converge to δ-functions. To prove (b) and (c) we use (8.4.10) and write,

$$\begin{aligned} \text{cov}[P(\lambda), P(\mu)] = \text{cov}[\hat{h}^{(1)}(\lambda), \hat{h}^{(1)}(\mu)] - \text{cov}[\hat{h}^{(1)}(\lambda), \hat{h}^{(2)}(\mu)] \\ - \text{cov}[\hat{h}^{(1)}(\mu), \hat{h}^{(2)}(\lambda)] + \text{cov}[\hat{h}^{(2)}(\lambda), \hat{h}^{(2)}(\mu)]. \end{aligned} \quad (8.4.12)$$

From theorem 6.2.4 we have (for $\lambda \neq 0, \pi$; $\mu \neq 0, \pi$),

$$\text{cov}[\hat{h}^{(i)}(\lambda), \hat{h}^{(j)}(\mu)] = T_{ij} + o(1/N) \qquad (8.4.13)$$

where,

$$T_{ij} = \frac{2\pi}{N} \int_{-\pi}^{\pi} h_1^2(\theta) W^{(i)}(\lambda - \theta) W^{(j)}(\mu - \theta) \, d\theta.$$

If $\lambda = \mu \neq 0$, substituting (8.4.13) into (8.4.12) gives,

$$\text{var}[P(\lambda)] = \frac{2\pi}{N} \int_{-\pi}^{\pi} h_1^2(\theta) \{W^{(1)}(\lambda - \theta) - W^{(2)}(\lambda - \theta)\}^2 \, d\theta + o\left(\frac{1}{N}\right). \quad (8.4.14)$$

Since $W^{(1)}(\lambda - \theta)$, $W^{(2)}(\lambda - \theta)$ are both highly concentrated in the region $\theta = \lambda$, we may write (8.4.14) as,

$$\text{var}[P(\lambda)] = \frac{2\pi}{N} h_1^2(\lambda) \int_{-\pi}^{\pi} \{W^{(1)}(\theta) - W^{(2)}(\theta)\}^2 \, d\theta + o\left(\frac{1}{N}\right).$$

(For $\lambda = 0$ or π the above expression should be doubled.) On the other hand, if $\lambda \neq \mu$, we have,

$$|T_{ij}| \leqslant \frac{2\pi}{N} \left\{ \sup_{0 < \theta < \pi} h_1^2(\theta) \right\} \left| \int_{-\pi}^{\pi} W^{(i)}(\lambda - \theta) W^{(j)}(\mu - \theta)\, d\theta \right|$$

$$= \frac{2\pi}{N} \left\{ \sup_{0 < \theta < \pi} h_1^2(\theta) \right\} |V_{ij}(\lambda - \mu)|.$$

By condition (3), if $|\lambda - \mu| \gg 1/m$, all the V_{ij} are bounded as $N \to \infty$, and hence

$$T_{ij} = O(1/N), \qquad i, j = 1, 2.$$

Part (c) now follows from (8.4.12).

It might be remarked that if $W^{(1)}(\theta)$, $W^{(2)}(\theta)$ are both rectangular windows, with bandwidths $(2\pi/n)$, $(2\pi/m)$ respectively, then $V_{ij}(\lambda - \mu)$ is *exactly* zero if $|\lambda - \mu| \geqslant 2\pi/m$. In this case we have, to $O(1/N)$,

$$\mathrm{cov}[P(\lambda), P(\mu)] = 0, \qquad \text{if } |\lambda - \mu| \geqslant 2\pi/m.$$

Define

$$\Lambda(n, m) = 2\pi \int_{-\pi}^{\pi} \{W^{(1)}(\theta) - W^{(2)}(\theta)\}^2\, d\theta,$$

so that we may write

$$\mathrm{var}\{P(\lambda)\} \sim \frac{\Lambda(n, m)}{N} h_1^2(\lambda). \qquad (8.4.15)$$

For most choices of windows $\Lambda(n, m)$ is most conveniently evaluated by using Parseval's theorem to re-write it as

$$\Lambda(n, m) = \sum_s [\{w^{(1)}(s)\}^2 - 2w^{(1)}(s)w^{(2)}(s) + \{w^{(2)}(s)\}^2]. \qquad (8.4.16)$$

We know from condition (vi) of Section 6.2.4 that $\Lambda(n, m) \to \infty$ as $N \to \infty$; in fact, in general $\Lambda(n, m) = O(n)$ (cf. Table 6.1), so that for $|\lambda - \mu| \gg 1/m$, the correlation between $P(\lambda)$, $P(\mu)$ is $O(1/n)$. Further, $P(\lambda)$, $P(\mu)$ being linear functions of the sample autocovariances, are asymptotically normally distributed under general conditions on the moments of $\{X_t\}$ (Lomnicki and Zaremba (1959a)). Hence, if $|\lambda - \mu| \gg 1/m$, $P(\lambda)$, $P(\mu)$ may, for large N, be treated as independent random variables with zero means.

The basic ideas underlying the construction of the test may now be described intuitively as follows. Consider the values of $P(\lambda)$ at a set of points spaced sufficiently well apart along the interval $(0, \pi)$. Under H_0 these values are (for large N) approximately independent random variables, each having zero mean. If we try to test the value of $P(\lambda)$ at a particular point λ_0, say (e.g. at the point where it attains its maximum value), we run into the difficulty that var$\{P(\lambda_0)\}$ depends on the (unknown) value of $h_1^2(\lambda_0)$, and to form a suitable test statistic we would first have to estimate $h_1(\lambda_0)$, as in the periodogram tests discussed in the previous section. However, when dealing with the $P(\lambda)$ statistic we can obviate this difficulty to a large extent by basing the test on *cumulative sums* rather than on particular values. More precisely, let the integer m' be chosen so that $m' = o(m) \to \infty$ as $N \to \infty$ (e.g. we may choose $m' = m^\alpha$, $0 < \alpha < 1$), and subdivide the interval $(0, \pi)$ into subintervals of length $2\pi/m'$. If we now form the cumulative sums

$$\sum_{p=1}^{q} P\left(\frac{2\pi p}{m'}\right),$$

then, since the terms in these sums are asymptotically independent, we may derive the distributional properties of such sums by analogy with the theory of random walks. (Here we are following closely the approach used by Grenander and Rosenblatt (1957a) and Bartlett (1954) for deriving confidence bands for the integrated spectrum; see Section 6.2.6.) We now define the "standardized" cumulative sums by

$$J_q = \left\{\frac{N}{m'\Lambda(n, m)}\right\}^{1/2} \sum_{p=1}^{q} P\left(\frac{2\pi p}{m'}\right), \qquad q = 0, 1, \ldots, [m'/2]. \qquad (8.4.17)$$

Then on H_0 (writing P_p for $P(2\pi p/m')$, etc.),

$$\mathrm{cov}\{J_q, J_r\} = \frac{N}{m'\Lambda(n, m)} \sum_{p=1}^{q} \sum_{p'=1}^{r} \mathrm{cov}\{P_p, P_{p'}\}$$

$$= \frac{N}{m'\Lambda(n, m)} \left\{\sum_{p=1}^{\min(q,r)} \mathrm{var}(P_p) + \sum_{p \neq p'}\sum \mathrm{cov}(P_p, P_{p'})\right\}$$

$$= \frac{1}{m'} \left\{\sum_{p=1}^{\min(q,r)} h_1^2\left(\frac{2\pi p}{m'}\right)\right\} + T_N,$$

say. Since $\mathrm{cov}(P_p, P_{p'}) = O(1/N)$, $p \neq p'$, we have

$$|T_N| \leq K \frac{N}{m'\Lambda(n, m)} \frac{\frac{1}{2}m'(m' - 1)}{N} \qquad (K \text{ a constant}),$$

$$= O\left(\frac{m'}{\Lambda(n, m)}\right).$$

But $\Lambda(n, m) = O(n)$, and hence, since $m = o(n)$, $T_N \to 0$ as $N \to \infty$. We thus have

$$\lim_{N \to \infty} \text{cov}\{J_q, J_r\} = \lim_{N \to \infty} \left\{ \frac{1}{m'} \sum_{p=1}^{\min(q,r)} h_1^2 \left(\frac{2\pi p}{m'} \right) \right\}$$

$$\sim \frac{1}{2\pi} \int_0^{\min[(2\pi q/m'),(2\pi r/m')]} h_1^2(\omega) \, d\omega. \qquad (8.4.18)$$

We may thus, for large N, treat J_q as a diffusion process, and following Grenander and Rosenblatt (1957a) we infer (cf. Priestley (1962a)),

$$\lim_{N \to \infty} p \left[\max_{0 \leq 2\pi q/m' \leq \pi} (J_q) \leq a \right] = p \left[\max_{0 \leq q \leq \pi} (\eta(q)) \leq a \right], \qquad (8.4.19)$$

where $\eta(q)$ is a Wiener process with

$$E[\eta(q)] = 0,$$

and

$$\text{cov}\{\eta(q), \eta(r)\} = \frac{1}{2\pi} G\{\min(q, r)\},$$

with

$$G(x) = \int_0^x h_1^2(\omega) \, d\omega. \qquad (8.4.20)$$

(We may note that $p[\max_q(\eta(q)) < a]$ is the probability that the "random walk" $\eta(q)$ is not absorbed in a barrier at a). As $N \to \infty$, the distribution of J_q satisfies the diffusion equation (Bartlett (1955), p. 47),

$$\frac{\partial p(x, t)}{\partial t} = \tfrac{1}{2}\sigma^2(t) \frac{\partial^2 p(x, t)}{\partial x^2}, \qquad (8.4.21)$$

where $t = 2\pi q/m'$, and $\sigma^2(t) = (1/2\pi)h_1^2(t)$. Making a change of "time" scale by writing

$$\tau = \frac{1}{2\pi} \int_0^t h_1^2(\omega) \, d\omega,$$

we obtain an unrestricted solution of (8.4.21) in the form,

$$p_1(x, t) = \left\{ \frac{1}{2\pi\gamma(t)} \right\}^{1/2} \exp\left(-\frac{1}{2} \frac{x^2}{\gamma(t)} \right), \qquad (8.4.22)$$

where

$$\gamma(t) = \frac{1}{2\pi} G(t) = \frac{1}{2\pi} \int_0^t h_1^2(\omega) \, d\omega,$$

the general solution of (8.4.21) being a linear combination of such solutions.

If we set up an absorbing barrier at a the solution is obtained by combining functions of the type (8.4.22) to ensure that $p(x, t)$ vanishes on the boundary. The appropriate form may be obtained by the method of "images" (cf. Section 6.2.6 and Bartlett (1955)), and is given by

$$p_2(x, t) = \left\{\frac{1}{2\pi\gamma(t)}\right\}^{1/2}\left[\exp\left(-\frac{x^2}{2\gamma(t)}\right) - \exp\left(-\frac{(x-2a)^2}{2\gamma(t)}\right)\right].$$

Accordingly, we obtain

$$\lim_{N\to\infty} p\left[\max_{0\leq 2\pi q\leq \pi/m'} (J_q)\leq a\right] = \int_{-\infty}^{a} p_2(x, \pi)\, dx = 2\Phi\left\{\frac{a}{\sqrt{\gamma(\pi)}}\right\} - 1,$$

where $\Phi(x)$ is the distribution function of the standardized $N(0, 1)$ distribution. Hence, we may write

$$\lim_{N\to\infty} p\left[\frac{\max_{0<q<[m'/2]} (J_q)}{\{(1/2\pi)G(\pi)\}^{1/2}}\leq a\right] = 2\Phi(a) - 1 = \Delta^{[3]}(a), \qquad (8.4.23)$$

say. (We attach the superfix [3] to $\Delta(a)$ to distinguish it from related quantities $\Delta^{[1]}$, $\Delta^{[2]}$ introduced in Section 6.2.6.) We may use this result to determine which, if any, of the amplitudes A_r are non-zero, and the corresponding frequencies ω_r. To do so, however, we require as estimate of $G(\pi)$. Note that $G(\pi)$ *is the only functional of $h_1(\omega)$ which enters into the test based on the cumulative sums J_q*, i.e. we do not have to estimate $h_1(\omega)$ itself but only the integral of its square. Now we have

$$G(\pi) = \int_0^\pi h_1^2(\omega)\, d\omega = \frac{1}{4\pi}\sum_{s=-\infty}^{\infty}\{R^{(1)}(s)\}^2.$$

If all the $A_r = 0$ we may obtain a consistent estimate of $G(\pi)$ by

$$\hat{G}(\pi) = \frac{1}{4\pi}\sum_{s=-m}^{m}\hat{R}_X^2(s),$$

(cf. (6.2.159) and subsequent discussion.)

However, if $A_r > 0$ $(r = r_1, r_2, \ldots, r_j)$, then

$$E\left[\sum_{s=-m}^{m}\hat{R}_X^2(s)\right] \sim \sum_s\left\{R^{(1)}(s) + \frac{1}{2}\sum_{\alpha=1}^{j} A_{r_\alpha}^2\cos\omega_{r_\alpha}\right\}^2$$

$$\sim \sum_s\{R^{(1)}(s)\}^2 + 2\pi\sum_\alpha A_{r_\alpha}^2 h_1(\omega_{r_\alpha}) + \sum_s\left\{\sum_\alpha\frac{1}{2}A_{r_\alpha}^2\cos\omega_{r_\alpha}\right\}^2$$

$$\sim 4\pi G(\pi) + U_2 + U_3,$$

say. Now U_3 may be approximately removed by considering

$$U_3^* = 2 \sum_{s=m+1}^{2m} \hat{R}_X^2(s) \sim \sum_{s=-m}^{m} \left\{ \sum_\alpha \frac{1}{2} A_{r_\alpha}^2 \cos \omega_{r_\alpha} \right\}^2,$$

since $R^{(1)}(s)$ dies out for large s. The effect of including U_2, which is always positive, is slightly to overestimate $G(\pi)$, thus tending to make the subsequent test over-cautious. (The quantity U_3 is, in general, $O(m)$, and hence U_2 is negligible compared with U_3.) Our final estimate of $G(\pi)$ is

$$\hat{G}^*(\pi) = \frac{1}{4\pi} \left[\sum_{s=-m}^{m} \hat{R}_X^2(s) - 2 \sum_{s=m+1}^{2m} \hat{R}_X^2(s) \right]. \qquad (8.4.24)$$

Moreover, $\hat{G}^*(\pi)$ may be shown to be a consistent estimate of $G(\pi)$, on the null hypothesis that $Z_t \equiv 0$, by writing

$$\hat{G}^*(\pi) = \frac{1}{4\pi} \left[2 \sum_{s=-m}^{m} \hat{R}_X^2(s) - \sum_{s=-2m}^{2m} \hat{R}_X^2(s) \right] = V_1 - V_2,$$

say. Since $\hat{G}(\pi)$ is a consistent estimate of $G(\pi)$ it follows that, in probability, $V_1 \to 2G(\pi)$ and $V_2 \to G(\pi)$, as $m \to \infty$.

We now construct a test of the null hypothesis,

$$H_0: A_r = 0, \qquad \text{all } r,$$

against the alternative hypothesis

$$H_1: A_r > 0, \qquad \text{for at least one } r,$$

by using the asymptotic distribution of J_q derived above. For under H_0, J_q satisfies equation (8.4.23), for large N. However, on the alternative hypothesis

$$J_q = O\{(Nn/m')^{1/2}\},$$

when $(2\pi q/m') \sim \omega_r$ (assuming $\Lambda(n, m) = O(n)$). Then J_q will, with large probability, cross the boundary at a. Note that we need here consider only the *single* boundary case since, on H_1, J_q will tend to be large and positive. Hence, a significance test may be constructed by plotting J_q against q and determining whether J_q crosses the boundary at a, where a is determined by the significance level of the test $[1 - \Delta^{[3]}(a)]$. (In effect, a is determined by the usual two-sided percentage points of the standardized normal distribution; if $\Delta^{[3]}(a) = 0.95$ then $a = 1.96$, if $\Delta^{[3]}(a) = 0.99$, $a = 2.58$, etc.)

Some special windows

Almost all the standard windows discussed in Section 6.2.3 would serve as choices for $W^{(1)}(\theta)$ and $W^{(2)}(\theta)$, typical choices being the Dirichlet kernel,

$D_\nu(\theta)$, defined by (6.2.64), and the Fejer kernel, $F_\nu(\theta)$, defined by (6.2.66). (The Dirichlet kernel does not satisfy properties (i) and (iv) of Section 6.2.4, but in view of the Reimann–Lebesgue lemma it still behaves asymptotically like a δ-function, and since integrals of the form

$$\int_{-\pi}^{\pi} D_n(\lambda - \theta)D_m(\mu - \theta)\, d\theta$$

remain bounded as n and $m \to \infty$ provided $m|\lambda - \mu| \gg 1$, the general results remain valid.) However, there is no reason why we should choose $W^{(1)}(\theta)$ and $W^{(2)}(\theta)$ from the same class of windows. In fact, from the point of view of reducing the "bias" in $P(\lambda)$ a good choice is

$$w^{(1)}(s) = \begin{cases} (1 - |s|/n), & |s| \leq n, \\ 0, & |s| > n, \end{cases} \tag{8.4.25}$$

corresponding to

$$W^{(1)}(\theta) = F_n(\theta),$$

and

$$w^{(2)}(s) = \begin{cases} (1 - |s|/n), & |s| \leq m, \\ 0, & |s| > m, \end{cases} \tag{8.4.26}$$

which, since $m = o(n)$, is essentially the same as

$$w^{(2)}(s) = \begin{cases} 1, & |s| \leq m, \\ 0, & |s| > m, \end{cases} \tag{8.4.27}$$

corresponding to $W^{(2)}(\theta) = D_m(\theta)$. With these windows $P(\lambda)$ may be written as

$$P(\lambda) = \frac{1}{\pi} \sum_{s=m+1}^{n-1} (1 - |s|/n)\hat{R}_X(s) \cos s\lambda, \tag{8.4.28}$$

and $P(\lambda)$ is independent of the first m autocovariances. To evaluate $\Lambda(n, m)$ for the above choice of windows we have

$$\sum_s \{w^{(1)}(s)\}^2 = \sum_{s=-(n-1)}^{(n-1)} \left(1 - \frac{|s|}{n}\right)^2 \sim \frac{2n}{3},$$

$$\sum_s \{w^{(2)}(s)\}^2 \sim \sum_{s=-m}^{m} 1 \sim 2m,$$

$$\sum_s w^{(1)}(s)w^{(2)}(s) \sim \sum_{s=-m}^{m} \left(1 - \frac{|s|}{n}\right) \sim \left(2m - \frac{2m^2}{n}\right),$$

so that, for $\lambda \neq 0, \pi$,

$$\Lambda(n, m) \sim \left(\frac{2n}{3} - 2m + \frac{2m^2}{n}\right). \tag{8.4.29}$$

One important feature of the above windows is that in this case we may remove the restriction $m' = o(m)$. For now we have $T_{11} = O(1/N)$ even when $m' = O(m) = o(n)$, and $T_{12} \sim (2\pi/N)h_1^2(\lambda)\{D_m(\lambda - \mu)\} + o(1/N)$. The first term is zero if both λ and μ are of the form $\pi p/(m + \frac{1}{2})$. Similarly, $T_{21} = o(1/N)$ if λ and μ are of this form, as does T_{22}. Thus we may now subdivide the range $(0, \pi)$ into intervals of length $2\pi/(m + \frac{1}{2})$ ($\sim 2\pi/m$), and the limiting distribution of J_q remains valid, i.e. for this particular choice of windows we may take $m' = m$. This property holds irrespective of the choice of $W^{(1)}(\theta)$, provided we take $W^{(2)}(\theta) = D_m(\theta)$. Clearly, the same property holds also if we choose $W^{(1)}(\theta)$, $W^{(2)}(\theta)$, to be rectangular windows with bandwidths $2\pi/n$, $2\pi/m$, respectively.

The value of m is determined by the rate of decay of $R^{(1)}(s)$, (cf. (8.4.2)), or equivalently, by the bandwidth of $h_1(\omega)$, but so far we have said little about the value of n, except to impose the constraint $m \ll n < N$. Now the magnitude of the peak in $P(\lambda)$ due to a harmonic component will depend on both the amplitude of the component and the number of autocovariances used in forming $P(\lambda)$. If all the harmonic components have reasonably large amplitude (relative to σ_Y^2) then we may choose n fairly small relative to N and the test should still detect their presence, i.e. in this case we do not need to base the harmonic analysis of the full "tail" of the autocovariance function. However, if there exists a harmonic component with a small amplitude it may be necessary to use the full autocovariance function to detect this component. In this situation, we may take $n = N$ and write

$$\begin{aligned}
P_N(\lambda) &= I^*_{N,X}(\lambda) - \frac{1}{2\pi} \sum_{s=-m}^{m} \hat{R}_X(s) \cos \lambda \\
&= I^*_{N,X}(\lambda) - \hat{h}^{(2)}(\lambda),
\end{aligned} \tag{8.4.30}$$

corresponding to the windows

$$W^{(1)}(\theta) = D_N(\theta), \tag{8.4.31}$$

$$W^{(2)}(\theta) = D_m(\theta). \tag{8.4.32}$$

(Since $I^*_{N,X}(\theta)$ is a trigonometric polynomial of degree $(N-1)$, $D_N(\theta)$ acts exactly like a δ-function with respect to $I^*_{N,X}(\theta)$.) The derivation of the formula for var$\{P(\lambda)\}$ is not strictly valid for the case $n = N$, and the expression for var$\{P(\lambda)\}$ requires slight modification. However, it is easy to show that $P_N(\lambda)$, $P_N(\mu)$ are asymptotically independent if $m|\lambda - \mu| \gg 1$, or if both λ and μ are of the form $2\pi p/m$, and also that for $\lambda \neq 0$,

$$\text{var}[P_N(\lambda)] \sim h_1^2(\lambda)\{1 - (2m/N)\}, \qquad (8.4.33)$$

so that

$$\Lambda(N, m) \sim (N - 2m).$$

Table 8.1 gives the values of $\Lambda(n, m)$ for various choices of windows. For $\lambda = 0, \pi$ the values shown should be doubled.

Table 8.1. *Values of $\Lambda(n, m)$*

$W^{(1)}(\theta)$	$W^{(2)}(\theta)$	$\Lambda(n, m)\ (\lambda \neq 0, \pi)$
$F_n(\theta)$	$D_m(\theta)$	$\frac{2}{3}n - 2m + \frac{2m^2}{n}$
$D_n(\theta)$	$D_m(\theta)$	$2n - 2m$
$F_n(\theta)$	$F_m(\theta)$	$\frac{2}{3}n - \frac{4}{3}m + \frac{2m^2}{3n}$
$D_n(\theta)$	$F_m(\theta)$	$2n - \frac{4}{3}m$
$D_N(\theta)$	$D_m(\theta)$	$N - 2m$
$D_N(\theta)$	$F_m(\theta)$	$N - \frac{4}{3}m$

The magnitude of the peak in $P(\lambda)$ due to a harmonic component with amplitude A_r and frequency ω_r is easily evaluated by noting that such a component contributes the term $(\frac{1}{2}A_r^2 \cos s\omega_r)$ to the autocovariance function, $R^{(2)}(s)$. Since $R^{(1)}(s) \sim 0$ for $|s| > m$, we have $E\{\hat{R}_X(s)\} \sim (1 - |s|/N)R^{(2)}(s)$, and hence the contribution to $E\{P(\lambda)\}$ is given by

$$E\{P(\lambda)\} = \frac{A_r^2}{4\pi} \sum_s \{w^{(1)}(s) - w^{(2)}(s)\}\left(1 - \frac{|s|}{N}\right) \cos s\omega_r \cos s\lambda$$

$$= \tfrac{1}{4}A_r^2\{\check{W}^{(1)}(\lambda + \omega_r) + \check{W}^{(1)}(\lambda - \omega_r) - \check{W}^{(2)}(\lambda + \omega_r) - \check{W}^{(2)}(\lambda - \omega_r)\}, \qquad (8.4.32)$$

where $\check{W}^{(1)}(\theta)$, $\check{W}^{(2)}(\theta)$ are the (discrete) Fourier transforms of $(1 - |s|/N)w^{(1)}(s)$ and $(1 - |s|/N)w^{(2)}(s)$ (i.e. $\check{W}^{(1)}(\theta)$ is the convolution of $W^{(1)}(\theta)$ with $F_N(\theta)$, and similarly for $\check{W}^{(2)}(\theta)$). Since $m = O(N)$, we may set $\check{W}^{(2)}(\theta) \sim W^{(2)}(\theta)$. Similarly if $n \ll N$ we may set $\check{W}^{(1)}(\theta) \sim W^{(1)}(\theta)$.

Thus, for the choice of windows suggested above, namely $W^{(1)}(\theta) = F_n(\theta)$, $W^{(2)}(\theta) = D_m(\theta)$,

$$E[P(\lambda)] \sim \tfrac{1}{4}A_r^2[F_n(\lambda + \omega_r) + F_N(\lambda - \omega_r) - D_m(\lambda + \omega_r) - D_m(\lambda - \omega_r)]. \qquad (8.4.35)$$

Recalling that $F_n(0) = n/2\pi$, $D_m(0) = (2m+1)/2\pi$, we see that when $\lambda \sim \omega_r$,

$$E\{P(\omega_r)\} \sim \frac{A_r^2}{8\pi}(n-2m). \qquad (8.4.36)$$

The magnitude of the peak thus depends on the value of $(n-2m)$, and clearly we must choose $n > 2m$. Similarly, if we take $n = N$ and set, e.g., $W^{(1)}(\theta) = D_N(\theta)$, $W^{(2)}(\theta) = D_m(\theta)$, then

$$E[P(\lambda)] = E[I_{N,X}^*(\lambda)] - E[\hat{h}^{(2)}(\theta)]$$

$$= \frac{A_r^2}{4\pi}[F_N(\lambda + \omega_r) + F_N(\lambda - \omega_r) - D_m(\lambda + \omega_r) - D_m(\lambda - \omega_r)]$$

$$(8.4.37)$$

(cf. (6.1.36)), and when $\lambda \sim \omega_r$,

$$E[P(\omega_r)] \sim \frac{A_r^2}{8\pi}(N-2m). \qquad (8.4.38)$$

Bandwidth considerations

So far we have considered only the asymptotic properties of $P(\lambda)$ and we have shown that, under H_0,

$$\lim_{N\to\infty} E[P(\lambda)] = 0.$$

However, when we consider the $P(\lambda)$ test for a finite sample it becomes necessary to investigate the magnitude of $E\{P(\lambda)\}$ in relation to m, and the restrictions which this imposes on the bandwidth of $h_1(\omega)$. For example, if we choose $w^{(1)}(s)$, $w^{(2)}(s)$ as in (8.4.25), (8.4.27), then $P(\lambda)$ is given by (8.4.28) and under H_0,

$$E\{P(\lambda)\} = \frac{1}{\pi} \sum_{s=m+1}^{n-1} \left(1 - \frac{|s|}{n}\right)\left(1 - \frac{|s|}{N}\right)R^{(1)}(s)\cos s\lambda.$$

If $R^{(1)}(s)$ is dominated by an exponentially damped factor (as would be the case if, e.g. Y_t is an AR process), then we may write

$$|R^{(1)}(s)| \le K\rho^{|s|} \qquad (0 < \rho < 1, K \text{ a constant})$$

so that

$$|E\{P(\lambda)\}| \le K(\rho^{m+1} - \rho^n)/\pi(1-\rho). \qquad (8.4.39)$$

For example, if Y_t is an AR(1) process, with parameter ρ, then $R^{(1)}(s) = K\rho^{|s|}$

(where $K = \sigma_Y^2$). The spectral density function of Y_t is

$$h_1(\omega) = \frac{\sigma_Y^2(1-\rho^2)}{2\pi(1-2\rho\cos\omega+\rho^2)}$$

with

$$h_1(0) = \sigma_Y^2(1-\rho^2)/2\pi(1-\rho)^2.$$

If ω_0 is such that $h_1(\pm\omega_0) = \frac{1}{2}h_1(0)$, then $2(1-\rho^2) = (1-\rho)^2 + 2\rho(1-\cos\omega_0)$. If ω_0 is small, $\omega_0 \sim (1-\rho)/\sqrt{\rho}$. Hence the spectral bandwidth at $\omega = 0$ is approximately given by,

$$B_h = 2\omega_0 \sim 2\left(\frac{1}{\sqrt{\rho}} - \sqrt{\rho}\right). \tag{8.4.40}$$

When evaluating $P(\lambda)$ we choose m so that

$$R^{(1)}(s) \sim 0, \qquad |s| \geq m. \tag{8.4.41}$$

If Y_t is an MA(k) process then clearly we would choose $m = k$, in which case (8.4.41) holds exactly. In general, however, $R^{(1)}(s)$ is never exactly zero but tends to zero as $|s| \to \infty$. In such cases we may choose m so that

$$\{R^{(1)}(s)\}^2 \leq \mathrm{var}\{\hat{R}_X(s)\}, \qquad |s| \geq m. \tag{8.4.42}$$

(With this value of m the "best" estimate (in the mean square error sense) of $R^{(1)}(s)$, for $|s| > m$, is zero.) Using the asymptotic formula for var $\hat{R}_X(s)$ (cf. (5.3.41)), we may write

$$\mathrm{var}\{\hat{R}_X(s)\} \sim \frac{1}{N}\sum_{q=-\infty}^{\infty}\{R^{(1)}(q)\}^2 \sim \frac{K^2(1+\rho^2)}{N(1-\rho^2)}.$$

According to (8.4.42), m is now given approximately by

$$\rho^{2m} = (1+\rho^2)/N(1-\rho^2),$$

or

$$m = \frac{\log N}{\log(1/\rho^2)} + \frac{\log\{(1-\rho^2)/(1+\rho^2)\}}{\log(1/\rho^2)}. \tag{8.4.43}$$

Inverting (8.4.40) we obtain, for B_h small, $1-\rho \sim \frac{1}{2}B_h$. Substituting this in (8.4.43) gives

$$m \sim B_h^{-1} \log(\tfrac{1}{2}B_h . N). \tag{8.4.44}$$

In general, if we are given the value of ρ we may obtain the appropriate value of m from (8.4.43). If, instead, we are given B_h, or an upper bound to B_h, we may find m directly from (8.4.44), or more accurately, solve (8.4.40) iteratively and then use (8.4.43). Although the relation (8.4.40) holds only for an AR(1) process, (8.4.43) holds approximately for any process with an exponentially damped autocovariance function. We may now find the "bias" (i.e. the value of $E\{P(\lambda)\}$), corresponding to the value of m given by (8.4.43). We have, from (8.4.43)

$$\rho^m = \left\{ \frac{1+\rho^2}{N(1-\rho^2)} \right\}^{1/2},$$

and hence by (8.4.39)

$$|E[P(\lambda)]| = O(1/N^{1/2}). \tag{8.4.45}$$

Note that equation (8.4.43) provides merely a *lower bound* for the value of m. If N, the number of observations, is large (so that n can be made large) then we may well choose m larger than the value given by (8.4.43). This will have the effect of further reducing the "bias" in $P(\lambda)$.

Practical analysis

Given a set of observations $X_1, X_2, \ldots, X_N$, the first step is to compute the sample autocovariances $\hat{R}_X(s)$, $s = 0, \pm, \ldots, \pm(N-1)$. The next step is to determine the value of m. If we may assume that $R^{(1)}(s)$ is exponentially damped, and are given the value of the damping factor ρ, then a lower bound for m may be determined from (8.4.43). Alternatively, if we have no prior knowledge on the rate of decay of $R^{(1)}(s)$ we may obtain a rough value for m by a visual inspection of the autocovariance function. Thus, if we plot $\hat{R}_X(s)$ against s we may choose m as that point where the autocovariances appear to settle down to steady oscillations with constant amplitudes.

Once the value of m has been selected we choose $n(>2m)$ and, using (e.g.) the windows $W^{(1)}(\theta) = F_n(\theta)$, $W^{(2)}(\theta) = D_m(\theta)$, we compute $P(\lambda)$ from (8.4.28) and evaluate it at the points $\lambda_p = 2\pi p/N$, $p = 0, 1, \ldots, [N/2]$. (Although the test is constructed from values of $P(\lambda)$ separated by intervals $2\pi/m$, in general it will be useful first to evaluate $P(\lambda)$ at all points on the "finer" grid $\{2\pi/N\}$ in order to locate the peaks of this function more accurately.) If X_t has a mixed spectrum then the $\{P(\lambda_p)\}$ will contain several well defined peaks. We select the *first* peak (in order of frequency), occurring at $\lambda = \omega_0$, say, and subdivide the frequency range $(0, \pi)$ at intervals $2\pi/m$ on both sides of $\lambda = \omega_0$. These points are then used in forming the cumulative sums $\Sigma P(\lambda)$. We now compute

$$\tilde{J}_q = \left\{ \frac{N}{m\Lambda(n,m)} \right\}^{1/2} \sum_{p=0}^{q} P\left(\frac{2\pi p}{m} + \delta \right) \Big/ \left\{ \frac{1}{2\pi} \hat{G}^*(\pi) \right\}^{1/2}, \tag{8.4.46}$$

$q = 0, 1, \ldots, [m/2]$, and test whether $\max(\tilde{J}_q) \leq a$. The constant δ

$(0 \le \delta \le 2\pi/m)$ is chosen so that $((2\pi p/m) + \delta) = \hat{\omega}_0$ for some integral value of p.

If the first peak at $\hat{\omega}_0$ is shown to be significant, i.e. if $\tilde{J}_q$ crosses the boundary at a, then the amplitude, A_0, of this harmonic component may be estimated by using (8.4.36), i.e. we may construct an asymptotically unbiased estimate of A_0^2 by

$$\hat{A}_0^2 = \frac{8\pi}{(n - 2m)} \cdot P(\hat{\omega}_0). \qquad (8.4.47)$$

The corresponding frequency ω_0 is, of course, estimated by $\hat{\omega}_0$, i.e. we estimate the frequencies of the harmonic components by the location of the peaks in $P(\lambda)$. Having estimated A_0^2 we may remove the contribution of this component by correcting the sample autocovariances by computing

$$\hat{R}_{X,1}(s) = \hat{R}_X(s) - \tfrac{1}{2}\hat{A}_0^2 \cos s\hat{\omega}_0. \qquad (8.4.48)$$

We may now re-compute $P(\lambda)$ using $\hat{R}_{X,1}(s)$ in place of $\hat{R}_X(s)$, and test for the presence of further harmonic components. If another component is detected we repeat this procedure until no further components are detected.

The advantage of testing peaks in order of frequency is that as each peak is found to be significant its effect is removed before the next peak is tested. If we test peaks in order of magnitude, say, then the test may be complicated by an "interaction" of two harmonic components. Thus, if we are testing a peak at $\hat{\omega}_1$, one of the points of subdivision on either side of $\hat{\omega}_1$ may fall within $2\pi/m$ of another peak, say at $\hat{\omega}_0$, and then the peak at $\hat{\omega}_0$ would contribute appreciably to $\Sigma P(\lambda)$. If $\Sigma P(\lambda)$ contains k peaks, we must reduce the significance level α by the factor $(1/k)$. More precisely, if we would normally use a significance level α when testing a peak selected at random, we should use a significance level (α/k) when testing eah peak in order of frequency.

It has been suggested that the *selection* of the "coarse" grid, $\{2\pi p/m) + \delta\}$, so that one of its points coincides with a peak in the "finer" grid, $\{\lambda_p\}$, should also be taken into account when assessing the significance level of the test. However, in the null case this selection of grid does not maximize the quantities $\tilde{J}_q$ (over the class of all possible choices), it merely maximizes the contribution to $\tilde{J}_q$ from the neighbourhood of the peak. Consequently, as the test involves values of $P(\lambda)$ spread over the whole frequency range it seems reasonable to suppose that the test procedure will not be seriously affected; see also Bartlett (1966), p. 323 on this point.

The effect of non-normality has been discussed by Priestley (1962a). It is not difficult to show that, to $O(1/N)$, the asymptotic variance of $P(\lambda)$ is unaffected by non-normality, as is the asymptotic expression for $\text{cov}\{P(\lambda), P(\mu)\}$. Thus, the significance test is still asymptotically valid.

Priestley (1962b) has given a comparison of the asymptotic powers of Whittle's test, the grouped periodogram test, and the $P(\lambda)$ test.

Bhansali (1979) has given an interesting application of the $P(\lambda)$ test in connection with a mixed spectrum analysis of the Canadian lynx data. An application of the $P(\lambda)$ test to tide heights is described in Bhansali (1977b).

Multidimensional mixed spectra

The problem of mixed spectra can arise also in the study of multi-dimensional processes, but here the spectral structure may be much more complicated than in the one-dimensional case. We will discuss this aspect further when we consider the spectral analysis of multidimensional stationary processes in Chapter 9, but meanwhile we note that the $P(\lambda)$ test may be extended to the two-dimensional case (Priestley (1964b)), and can be used to detect various forms of discontinuities in two-dimensional spectra.

8.5 ANALYSIS OF SIMULATED SERIES

To compare the relative performances of the different tests for mixed spectra described in the preceding sections, Priestley (1962b) constructed a number of simulated series from models of the form

$$X_t = \beta(Y_t + A_0 \cos \omega_0 t).   \tag{8.5.1}$$

For each series, Y_t was generated from the AR(2) model,

$$Y_t - 1 \cdot 2 Y_{t-1} + 0 \cdot 4 Y_{t-2} = \varepsilon_t,   \tag{8.5.2}$$

ε_t being an independent normal process with $E\{\varepsilon_t\} = 0$, $E\{\varepsilon_t^2\} = 10^4$, but the values of A_0, ω_0 and β were varied, as indicated below. (The scaling factor β was introduced to keep the range of the observations in each series within specified limits.) The number of observations generated for each series was $N = 992$.

Series no.	A_0	$\omega_0/2\pi$	β
1	500	1/20	1/4
2	250	1/20	1/8
3	150	1/16	1/2
4	100	1/16	1/2
5	75	1/16	1/2

Using the result (3.5.34) for the variance of an AR(2) process, we find that $\sigma_Y^2 = 44\,870$, and hence $\sigma_Y = 211 \cdot 83$. The "signal to noise" ratio, $\{A/\sigma_Y\}$, thus varies from $2 \cdot 36$ (for series 1) to $0 \cdot 35$ (for series 5).

The $P(\lambda)$ test

The autocorrelation function for the model (8.5.2) is (cf. Section 3.5.3)

$$\rho^{(1)}(s) = \rho^{|s|} \sin(s\theta + \psi)/\sin \psi,$$

where

$$\rho = \sqrt{0 \cdot 4} = 0 \cdot 6325,$$

$$\theta = \cos^{-1}\{1 \cdot 2/3\sqrt{0 \cdot 4}\} = 0 \cdot 3221 \text{ radians},$$

$$\psi = \tan^{-1}\{(1 \cdot 4/0 \cdot 6) \tan \theta\} = 0 \cdot 6609 \text{ radians}.$$

The autocovariance function is

$$R^{(1)}(s) = K\rho^{|s|} \sin(s\theta + \psi),$$

where

$$K = \sigma_Y^2/\sin \psi = 73101 \cdot 99.$$

We apply the $P(\lambda)$ test with the usual choice of weight functions

$$W_n^{(1)}(\theta) = F_n(\theta), \qquad W^{(2)}(\theta) = D_m(\theta),$$

so that $P(\lambda)$ may be computed in the form

$$P(\lambda) = \hat{h}^{(1)}(\lambda) - \hat{h}^{(2)}(\lambda) \tag{8.5.3}$$

where $\hat{h}^{(1)}(\lambda)$ is the Bartlett type estimate,

$$\hat{h}^{(1)}(\lambda) = \frac{1}{2\pi} \sum_{s=-n}^{n} \left(1 - \frac{|s|}{n}\right) \hat{R}_X(s) \cos s\lambda, \tag{8.5.4}$$

and $\hat{h}^{(2)}(\lambda)$ is the truncated periodogram,

$$\hat{h}^{(2)}(\lambda) = \frac{1}{2\pi} \sum_{s=-m}^{m} \hat{R}_x(s) \cos s\lambda. \tag{8.5.5}$$

A lower bound for m may be obtained from (8.4.43), and substituting $N = 992$, $\rho = 0 \cdot 6325$, this gives $m > 7 \cdot 7$. Since in this case we have fairly long series we can choose m substantially larger than its lower bound, and in the subsequent analysis we take $m = 32$, $n = 128$. For computational convenience $P(\lambda)$ was evaluated in terms on the autocorrelation function instead of the autocovariance function, i.e. we have replaced $\hat{R}_X(s)$ in (8.5.4) by $\hat{\rho}_X(s) = \hat{R}_X(s)/\hat{R}_X(0)$. The only modification to the test is that we replace $\hat{G}^*(\pi)$ by

$$\hat{g}^*(\pi) = \frac{1}{4\pi}\left[\sum_{s=-m}^{m} \hat{\rho}_X^2(s) - 2\sum_{s=m+1}^{n} \hat{\rho}_X^2(s)\right]. \qquad (8.5.6)$$

Writing $P^*(\lambda) = P(\lambda)/\hat{R}_X(0)$ the test statistic now takes the form

$$\tilde{J}_q = \frac{\{N/m\Lambda(n,m)\}^{1/2} \sum_{p=0}^{q} P^*[(2\pi p/m)+\delta]}{\{\hat{g}^*(\pi)/2\pi\}^{1/2}} \qquad (8.5.7)$$

$$= \frac{1\cdot29\left\{2\pi \sum_{p=0}^{q} P^*[(2\pi p/m)+\delta]\right\}}{\{4\pi g^*(\pi)\}^{1/2}}. \qquad (8.5.8)$$

(For the above analysis, $\Lambda(n,m) = 37\cdot3$.) We recall that to apply the test we choose the first peak (in order of frequency) of $P(\lambda)$, sat at $\lambda = \lambda_0$, and select the ordinates of $P(\lambda)$ at the points $\lambda = (2\pi p/m + \delta)$, $p = 0, 1, \ldots, [m/2]$, where δ is chosen so that one of these points coincides with λ_0. These ordinates are then used to form the cumulative sums. For a given significance level the critical value for $\max(\tilde{J}_q)$ is the corresponding two-sided percentage point of the standardized normal distribution. The results of applying the $P(\lambda)$ test to the five series are summarized below.

SERIES 1

$P^*(\lambda)$ has one peak occurring at $\lambda = 48\pi/496$, and

$$2\pi P^*(48\pi/496) = 20\cdot0, \qquad 2\pi P^*(17\pi/496) = -2\cdot06, \qquad 4\pi \hat{g}^*(\pi) = 2\cdot91.$$

Hence, from (8.5.8)

$$\tilde{J}_2 = \frac{1\cdot29 \times 17\cdot94}{\sqrt{2\cdot91}} = 13\cdot6^{***}.$$

The $0\cdot1\%$ level for $\max(\tilde{J}_q)$ is $3\cdot3$, and hence the result is highly significant at this level. The frequency of the harmonic component is estimated by the location of the peak in $P(\lambda)$. Thus we obtain

$$\hat{\omega} = 48\pi/496 \qquad (\omega = \pi/10),$$

and the amplitude is estimated by (cf. (8.4.36)),

$$\hat{A}^2 = \frac{8\pi}{(n-2m)}P(\hat{\omega}) = \frac{8\pi}{64}\hat{R}_X(0)P^*(\hat{\omega}),$$

giving $\hat{A} = 115\cdot5$ $(A = 125)$.

SERIES 2

$P^*(\lambda)$ has one peak at $\lambda = 48\pi/496$, and

$$2\pi P^*(48\pi/496) = 8\cdot57, \qquad 2\pi P^*(17\pi/496) = -0\cdot802, \qquad 4\pi g^*(\pi) = 3\cdot67.$$

Hence,

$$\tilde{J}_2 = 5\cdot23^{***}.$$

This result is, again, highly significant (certainly at $0\cdot1\%$).

$$\hat{\omega} = 48\pi/496 \quad (\omega = \pi/10), \qquad \hat{A} = 24\cdot1 \quad (A = 31\cdot25).$$

SERIES 3

$P^*(\lambda)$ has one peak at $\lambda = 62\pi/496$ $(= \pi/8)$, and

$$2\pi P^*(62\pi/496) = 7\cdot17, \qquad 2\pi P^*(31\pi/496) = 0\cdot972, \qquad 4\pi\hat{g}^*(\pi) = 3\cdot90.$$

Hence,

$$\tilde{J}_2 = 5\cdot32^{***},$$

significant at $0\cdot1\%$.

$$\hat{\omega} = \pi/8 \quad (\omega = \pi/8), \qquad \hat{A} = 79\cdot4 \quad (A = 75).$$

SERIES 4

$P^*(\lambda)$ has one peak at $\lambda = 62\pi/496$, and

$$2\pi P^*(62\pi/496) = 3\cdot84, \qquad 2\pi P^*(31\pi/496) = 0\cdot536, \qquad 4\pi\hat{g}^*(\pi) = 3\cdot31.$$

Hence,

$$\tilde{J}_2 = 3\cdot19^{***}$$

which is significant certainly at 1%.

$$\hat{\omega} = \pi/8 \quad (\omega = \pi/8), \qquad \hat{A} = 51\cdot3 \quad (A = 50).$$

SERIES 5

$P^*(\lambda)$ has one peak at $\lambda = 62\pi/496$, and

$$2\pi P^*(62\pi/496) = 1\cdot93, \qquad 2\pi P^*(31\pi/496) = 0\cdot997, \qquad 4\pi\hat{g}^*(\pi) = 3\cdot89.$$

Hence,

$$\tilde{J}_2 = 1\cdot91.$$

This result is almost (but not quite) significant at 5%.

$$\hat{\omega} = \pi/8 \quad (\omega = \pi/8), \qquad \hat{A} = 37 \cdot 7 \quad (A = 35).$$

The grouped periodogram test

We now apply the grouped periodogram test to the same series. For each series we have applied the test with three different values of k (the grouping interval), namely $k = 5, 10, 20$. The test statistic is (cf. (8.3.9)),

$$g_k^{(B)} = \max_k (I_p) \Big/ \sum_k I_p. \tag{8.5.9}$$

When $g_k^{(B)}$ has been evaluated we may then correct for the factor $\{m(h)/M(h)\}$ (cf. Section 8.3) by referring $\{m(h)/M(h)\}g_k^{(B)}$ to Fisher's g-distribution with k degrees of freedom. The results are summarized in Table 8.2 below.

Table 8.2. *Grouped periodogram tests.*

Series	k	$g_k^{(B)}$	$(m/M)g_k^{(B)}$
	5	0·852	0·801
1	10	0·797**	0·646*
	20	0·542**	0·434*
	5	0·798	0·790
2	10	0·719**	0·662*
	20	0·520**	0·442**
	5	0·944**	0·802
3	10	0·780**	0·492
	20	0·718**	0·251
	5	0·844	0·743
4	10	0·613	0·478
	20	0·541**	0·384*
	5	0·823	0·699
5	10	0·496	0·313
	20	0·414*	0·145

To determine the critical value of $g_k^{(B)}$ for a significance level α, we must first allow for the choice of subset of k ordinates by reducing the significance level by the factor $k/[N/2]$. Thus, to determine the α critical level of $g_k^{(B)}$ we find the $\alpha k/[N/2]$ level of Fisher's g with k degrees of freedom, the distribution of which is given by (6.1.67). Using only the first term of (6.1.67) we find, as an approximation, that the α critical level for $g_k^{(B)}$ is given by,

$$\frac{\alpha k}{[N/2]} = k(1-z)^{k-1}.$$

For $N = 992$, $k = 5, 10, 20$, $\alpha = 0\cdot05, 0\cdot01$, the critical values are shown in Table 8.3. The values of $g_k^{(B)}$ may now be assessed for significance. Values which are significant at 5% are marked *, and those significant at 1% are marked ** in Table 8.2.

Table 8.3. *Critical values of $g_k^{(B)}$.*

		k	
α	5	10	20
0·05	0·900	0·641	0·381
0·01	0·933	0·699	0·434

It will be observed that almost all of the uncorrected values of $g_k^{(B)}$ are significant at 1%, but when the correction factor is applied only one value is significant at 1%, few are significant at 5%, the remainder not being significant at 5%. Although at least one of the uncorrected values of $g_k^{(B)}$ is significant for all series, the corrected value fails to give significant results for series 3 and 5. In both cases a certain amount of ambiguity arises when, as for series 4, $g_5^{(B)}$ and $g_{10}^{(B)}$ are not significant at 5%, but $g_{20}^{(B)}$ is significant at 5%. One notes also the importance of the correction factor, $\{m(h)/M(h)\}$. Although the "true" value of the test statistic will, in general, be greater than $\{m(h)/M(h)\}g_k^{(B)}$, it would appear dangerous to base the test purely on the value of $g_k^{(B)}$.

Whittle's test

Here, we apply the statistic (8.3.6), viz.,

$$g^{(w)} = \max \{I_p/2\pi\hat{h}(\omega_p)\} \Big/ \sum_p \{I_p/2\pi\hat{h}(\omega_p)\}. \tag{8.5.10}$$

(Although we refer to this as "Whittle's test" it should be noted that we have not followed the precise procedure suggested by Whittle, i.e. we have not removed the apparent peak in $\hat{h}(\omega)$ before evaluating the standardized periodogram ordinates. The reason for not removing this peak was explained in Section 8.3.) The estimate $\hat{h}(\omega)$ used was the function $\hat{h}^{(2)}(\omega)$ given by (8.5.5) with again $m = 32$.

SERIES 1

The maximum standardized periodogram ordinate occurs at $p = 50$, corresponding to $\omega_p = 100\pi/992$, with

$$I_{50} = 4278 \times 10^3, \qquad 2\pi\hat{h}(\omega_{50}) = 256 \times 10^3,$$

so that

$$\max I_p/2\pi\hat{h}(\omega_p) = 16\cdot7.$$

We may now compute $g^{(w)}$ and refer to Fisher's g distribution with 496 degrees of freedom. Alternatively, since N is large, we may test $\max I_p/2\pi\hat{h}(\omega_p)$ according to the asymptotic distribution (6.1.60). The critical α value for this distribution is given approximately by $z = 2\log_e(N/2\alpha)$. When $N = 992$, $\alpha = 0\cdot05$, this gives $z = 18\cdot4$. Thus, the result is not significant at 5%.

Since this test fails to detect the harmonic component even in the strongest case (i.e. when the "signal to noise" ratio is largest), we have not applied it to the other series. (The failure is no doubt due partly to the fact that the frequency of the harmonic component here falls approximately between the two periodogram ordinates at $p = 49$, $p = 50$, thereby reducing the height of the maximum ordinate by $(4/\pi^2) \doteq 0\cdot4$.)

8.6 ESTIMATION OF THE SPECTRAL DENSITY FUNCTION

Once the harmonic components are detected (or not detected, as the case may be), we then have to deal with the problem of estimating the continuous spectral density function $h_1(\omega)$. Our previous studies of estimating spectral density functions (Chapters 6 and 7) dealt only with the case where X_t is known *a priori* to have a purely continuous spectrum, but the general method of using a weighted sum of autocovariances can still be applied in the more general case of mixed spectra, once the autocovariance function of Y_t, $R^{(1)}(s)$, has been estimated. The problem is essentially the same as that of removing regression functions (see Section 7.7), and if we write Z_t in the form

$$Z_t = \sum_{r=1}^{K} (A_r' \cos \omega_r t + B_r' \sin \omega_r t) \qquad (8.6.1)$$

(so that $A_r^2 = (A_r')^2 + (B_r')^2$), and estimate A_r', B_r', as in (6.1.69), then $R^{(1)}(s)$ may be estimated from the residual process in the obvious way. Alternatively, we may, for large N, construct an estimate of $R^{(1)}(s)$ by recalling that each term in Z_t contributes the term $(\frac{1}{2}A_r^2 \cos \omega_r s)$ to $R_X(s)$, so that a natural estimate of $R^{(1)}(s)$ is given by

$$\hat{R}^{(1)}(s) = \hat{R}_x(s) - \frac{1}{2}\sum_{r=1}^{K}\hat{A}_r^2 \cos s\omega_r. \tag{8.6.2}$$

If the $\{\omega_r\}$ are known *a priori* (as would be the case if, e.g., Z_t represents a "seasonal component") then it may be shown (Priestley (1964a)) that $\hat{R}^{(1)}(s)$ is a consistent estimate of $R^{(1)}(s)$, being asymptotically unbiased and having the same asymptotic variance as the standard autocovariance estimator (see (5.3.25)) in the case when there are no harmonic components present.

Direct estimation of $h_1(\omega)$ via the "double window" technique

The spectral density function $h_1(\omega)$ may be estimated from $\hat{R}^{(1)}(s)$ by the usual technique of constructing a smoothed Fourier transform, as in the standard case of processes with purely continuous spectra. However, Priestley (1964a) developed an alternative approach in which $h_1(\omega)$ is estimated directly from $\hat{R}_X(s)$, the sample autocovariance function of $\{X_t\}$. The basic idea is to construct a suitable window which automatically removes the effect of the harmonic components, so that it is no longer necessary to remove the contribution of such terms from $\hat{R}_X(s)$.

Let $W_n^{(1)}(\theta)$, $W_m^{(2)}(\theta)$ be two windows of the type discussed in Section 8.4, where now $m < n = o(N)$. Suppose further that the windows are chosen so that they satisfy

$$W_m^{(2)}(\theta) - cW_n^{(1)}(\theta) = 0, \tag{8.6.3}$$

for *all* θ in the range $|\theta| < \pi/n$, where

$$c = W_m^{(2)}(0)/W_n^{(1)}(0). \tag{8.6.4}$$

For example, we could choose two rectangular windows of the form

$$W_n^{(1)}(\theta) = \begin{cases} n/2\pi, & |\theta| \le \pi/n \\ 0, & \text{otherwise,} \end{cases} \tag{8.6.5}$$

and

$$W_m^{(2)}(\theta) = \begin{cases} m/2\pi, & |\theta| \le \pi/m \\ 0, & \text{otherwise.} \end{cases} \tag{8.6.6}$$

Since $m < n$, condition (8.6.3) is clearly satisfied
With the above choice of $W_n^{(1)}(\theta)$, $W_m^{(2)}(\theta)$ write

$$\hat{h}^{(1)}(\omega) = \int_{-\pi}^{\pi} I_{N,X}^*(\theta)W_n^{(1)}(\omega - \theta)\,d\theta \tag{8.6.7}$$

and let $\hat{h}^{(2)}(\omega)$ be similarly defined with respect to $W_m^{(2)}(\theta)$. Consider, for simplicity, the case where X_t contains just one harmonic component (the argument being easily extended to the general case) so that we may write

$$X_t = A\cos(\omega_0 t + \phi) + Y_t. \tag{8.6.8}$$

Murthy (1962) has shown that, if $n = o(N)$, $\hat{h}^{(1)}(\omega)$ is a consistent estimate of $h_1(\omega)$ except in the neighbourhood of ω_0. However, we will show that $\hat{h}^{(1)}(\omega)$ may be used to estimate $h_1(\omega)$ at all frequencies ω such that $|\omega - \omega_0| > \pi/n$, while for $|\omega - \omega_0| \leq \pi/n$, $h_1(\omega)$ may be estimated by $\{\hat{h}^{(2)}(\omega) - c\hat{h}^{(1)}(\omega)\}/(1-c)$. Thus, the estimate which we propose may be written as,

$$\hat{h}_1(\omega) = \begin{cases} \hat{h}^{(1)}(\omega), & |\omega - \omega_0| > \pi/n, \\ \{\hat{h}^{(2)}(\omega) - c\hat{h}^{(1)}(\omega)\}/(1-c), & |\omega - \omega_0| \leq \pi/n, \end{cases} \tag{8.6.9}$$

and we will show that $\hat{h}_1(\omega)$ is a consistent estimate of $h_1(\omega)$ for all ω.

If X_t is of the form (8.6.8) it is easy to prove that (cf. (6.1.36))

$$E[I_{N,X}^*(\theta)] = E[I_{N,Y}^*(\theta)] + \tfrac{1}{4}A^2\{F_N(\theta + \omega_0) + F_N(\theta - \omega_0)\} \tag{8.6.10}$$

Substituting (8.6.10) in (8.6.7) gives,

$$E[\hat{h}^{(1)}(\omega)] = \int_{-\pi}^{\pi} E[I_{N,Y}^*(\theta)] W_n^{(1)}(\omega - \theta)\, d\theta$$

$$+ \tfrac{1}{2}A^2 \int_{-\pi}^{\pi} W_n^{(1)}(\omega - \theta) F_N(\theta - \omega_0)\, d\theta$$

$$= S_1 + S_2,$$

say. We know from (6.2.103) that

$$S_1 = \int_{-\pi}^{\pi} h_1(\theta) W_n^{(1)}(\omega - \theta)\, d\theta + O\left(\frac{\log N}{N}\right),$$

while, since $n = o(N)$, $F_N(\theta)$ acts like a δ-function with respect to $W_n^{(1)}(\theta)$, so that

$$2S_2/A^2 = W_n^{(1)}(\omega - \omega_0) + O((\log N)/N).$$

Thus, to $O((\log N)/N)$,

$$E[\hat{h}^{(1)}(\omega)] = \int_{-\pi}^{\pi} h_1(\theta) W_n^{(1)}(\omega - \theta)\, d\theta + \tfrac{1}{2}A^2 W_n^{(1)}(\omega - \omega_0). \tag{8.6.11}$$

If $|\omega - \omega_0| > \pi/n$, the second term in (8.6.11) vanishes. Also, the bias of the rectangular estimate is $O(1/n^2)$ (see Table 6.1), and hence

$$E[\hat{h}^{(1)}(\omega)] = h_1(\omega) + O(1/n^2), \qquad |\omega - \omega_0| > \pi/n. \tag{8.6.12}$$

Thus, for $|\omega - \omega_0| > \pi/n$, the harmonic component does not affect $E[\hat{h}^{(1)}(\omega)]$, and $\hat{h}^{(1)}(\omega)$ is asymptotically unbiased. Similarly, we may show that, to $O((\log N)/N)$,

$$E[\hat{h}^{(2)}(\omega)] = \int_{-\pi}^{\pi} h_1(\omega) W_m^{(2)}(\omega - \theta) \, d\theta + \tfrac{1}{2} A^2 W_m^{(2)}(\omega - \omega_0).$$

We thus find,

$$E[\hat{h}^{(2)}(\omega) - c\hat{h}^{(1)}(\omega)] \sim \int_{-\pi}^{\pi} h_1(\theta)\{W_m^{(2)}(\omega - \theta) - c W_n^{(1)}(\omega - \theta)\} \, d\theta$$

$$+ \tfrac{1}{2} A^2 \{W_m^{(2)}(\omega - \omega_0) - c W_n^{(1)}(\omega - \omega_0)\}. \quad (8.6.13)$$

If $|\omega - \omega_0| \leq \pi/n$, then by (8.6.3) the second term in (8.6.13) is zero, so that at these frequencies,

$$E[\hat{h}^{(2)}(\omega) - c\hat{h}^{(1)}(\omega)] \sim \int_{-\pi}^{\pi} h_1(\theta)\{W_m^{(2)}(\omega - \theta) - c W_n^{(1)}(\omega - \theta)\} \, d\theta$$

$$= h_1(\omega)(1 - c) + O(1/m^2). \quad (8.6.14)$$

Consequently, $\{\hat{h}^{(2)}(\omega) - c\hat{h}^{(1)}(\omega)\}/(1 - c)$ is an asymptotically unbiased estimate of $h_1(\omega)$ for $|\omega - \omega_0| \leq \pi/n$.

Further, it may be shown (Priestley, 1962b) that when X_t is of the form (8.6.8),

$$\text{var}[\hat{h}^{(1)}(\omega)] \sim \frac{2\pi}{N}\left[\int_{-\pi}^{\pi} h_1^2(\theta)\{W_n^{(1)}(\omega - \theta)\}^2 \, d\theta \right.$$

$$\left. + A^2 h_1(\omega_0)\{W_n^{(1)}(\omega - \omega_0)\}^2\right]. \quad (8.6.15)$$

Again, if $|\omega - \omega_0| > \pi/n$, the second term in (8.6.15) is zero, and (8.6.15) becomes (for $\omega \neq 0, \pi$),

$$\text{var}[\hat{h}^{(1)}(\omega)] \sim \frac{2\pi}{N} h_1^2(\omega) \int_{-\pi}^{\pi} \{W_n^{(1)}(\theta)\}^2 \, d\theta = \frac{n}{N} h_1^2(\omega),$$

which $\to 0$ as $N \to \infty$ since $n = o(N)$.

Also (Priestley, 1962b), we have

$$\text{cov}[\hat{h}^{(1)}(\omega), \hat{h}^{(2)}(\omega)] \sim \frac{2\pi}{N}\left[\int_{-\pi}^{\pi} h_1^2(\theta) W_n^{(1)}(\omega - \theta) W_m^{(2)}(\omega - \theta) \, d\theta \right.$$

$$\left. + A^2 h_1(\omega_0) W_n^{(1)}(\omega - \omega_0) W_m^{(2)}(\omega - \omega_0)\right].$$

$$(8.6.16)$$

Combining (8.6.15), (8.6.16) (and using a similar result for $\text{var}[\hat{h}^{(2)}(\omega)]$), we find,

$$\text{var}[\hat{h}^{(2)}(\omega) - c\hat{h}^{(1)}(\omega)] \sim \frac{2\pi}{N}\left[\int_{-\pi}^{\pi} h_1^2(\theta)\{W_m^{(2)}(\theta) - cW_n^{(1)}(\theta)\}^2\,d\theta \right.$$

$$\left. + A^2 h_1(\omega_0)\{W_m^{(2)}(\omega - \omega_0) - cW_n^{(1)}(\omega - \omega_0)\}^2\right].$$

(8.6.17)

As before, if $|\omega - \omega_0| \leq \pi/n$, the second term in (8.6.17) is zero, and evaluating the first term we obtain (noting that here $c = m/n$),

$$\text{var}[\{\hat{h}^{(2)}(\omega) - c\hat{h}^{(1)}(\omega)\}/(1-c)] \sim \left\{\frac{mn}{N(n-m)}\right\} h_1^2(\omega) \qquad (\omega \neq 0, \pi),$$

(8.6.18)

which also $\to 0$ as $N \to \infty$.

It will be observed that, for $|\omega - \omega_0| < \pi/n$, both the mean and variance of $\hat{h}_1(\omega)$ do not depend on A, and consequently take the same values *whether or not* there exists a harmonic component at frequency ω_0. Thus, even if the $P(\lambda)$ test gives a misleading result, i.e. if it indicates the presence of a spurious harmonic component, the estimate defined above will still be asymptotically unbiased in the neighbourhood of ω_0.

It is also worth noting that $\hat{h}_1(\omega_0)$ is still an asymptotically unbiased estimate of $h_1(\omega_0)$, even if $W_n^{(1)}(\theta)$, $W_m^{(2)}(\theta)$ do not satisfy (8.6.3), but this form of estimate will not, in general, be asymptotically unbiased at other frequencies in the neighbourhood of ω_0 unless (8.6.3) holds.

In practice, the value of m is determined by B_h, the bandwidth of $h_1(\omega)$, and we would usually choose m so that

$$2\pi/m = \tfrac{1}{2}B_h \qquad (8.6.19)$$

(see Section 7.4), i.e. so that

$$m = 4\pi/B_h.$$

Having chosen m we must choose n, noting that as n increases var $\{\hat{h}^{(1)}(\omega)\}$ increases but var$[\{\hat{h}^{(2)}(\omega) - c\hat{h}^{(1)}(\omega)\}/(1-c)]$ decreases.

There is, however, the possibility that the transition from $\hat{h}^{(1)}(\omega)$ to $\{\hat{h}^{(2)}(\omega) - c\hat{h}^{(1)}(\omega)\}/(1-c)$ might produce discontinuities in $\hat{h}_1(\omega)$ at the frequencies $\omega = \omega_0 \pm (\pi/n)$. It would seem reasonable, therefore, to choose $n = 2m$, in which case $\hat{h}^{(1)}(\omega)$ and $\{\hat{h}^{(2)}(\omega) - c\hat{h}^{(1)}(\omega)\}/(1-c)$ both have the same asymptotic variance. With this value of n the effect of any discontinuities should not be serious since the bias (for any ω) is at most $O(1/m^2)$.

Relationship between $\hat{\hat{h}}_1(\omega)$ and $P(\lambda)$

We mentioned previously (Section 8.4) that the estimate of $h_1(\omega)$ proposed by Hannan for standardizing the periodogram ordinates is a special case of the general "double window" estimate $\hat{\hat{h}}_1(\omega)$. In fact, any estimate of the "double window" form could be used to standardize the periodogram ordinates, and since $\hat{\hat{h}}_1(\omega_0)$ is asymptotically unbiased *whether or not there exists a harmonic component* at ω_0, the resulting test statistic could be referred to Fisher's g-distribution, as in Hannan's test.

Thus, while the final tests differ, Hannan's statistic is very closely related to $P(\lambda)$, as may be seen by noting that $\hat{\hat{h}}_1(\lambda)$ can be expressed in terms of $P(\lambda)$ by the relation

$$\hat{\hat{h}}_1(\lambda) \equiv \hat{h}^{(2)}(\lambda) - c'P(\lambda), \tag{8.6.20}$$

where $c' = c/(1-c)$. (To prove (8.6.20) we write $P(\lambda) = \hat{h}^{(1)}(\lambda) - h^{(2)}(\lambda)$, and $(1-c)\hat{\hat{h}}(\lambda) = \hat{h}^{(2)}(\lambda) - c\hat{h}^{(1)}(\lambda)$.)

We may also see this relationship by considering the expression

$$K(\omega) = \hat{h}^{(1)}(\omega) / [\{\hat{h}^{(2)}(\omega) - c\hat{h}^{(1)}(\omega)\}/(1-c)],$$

which is a more general version of the standardized periodogram ordinates used in Hannan's test. ($K(\omega_p)$ reduces to the standardized periodogram ordinate on setting $n = N$, $W^{(1)}(\theta) = D_N(\theta)$.) Writing $k(\omega) = \hat{h}^{(1)}(\omega)/\hat{h}^{(2)}(\omega)$, $K(\omega)$ may be written as

$$K(\omega) = \frac{(1-c)k(\omega)}{1-ck(\omega)},$$

while $P(\omega)$ can be written as

$$P(\omega) = \hat{h}^{(1)}(\omega) - \hat{h}^{(2)}(\omega) = \hat{h}^{(2)}(\omega)[1 - k(\omega)].$$

It may be noted that the roles of $W^{(1)}$, $W^{(2)}$ in $\hat{\hat{h}}_1(\omega)$ are reversed in the construction of $P(\omega)$. In the former case we eliminate the contribution of the harmonic components in order to estimate $h_1(\omega)$; in the latter case we eliminate the contribution of $h_1(\omega)$ in order to detect the harmonic components and to estimate their amplitudes and frequencies.

VOLUME 2

MULTIVARIATE SERIES,
PREDICTION AND CONTROL

Chapter 9

Multivariate and Multidimensional Processes

In Chapter 1 (Section 1.2) we touched briefly on the notions of *multivariate* and *multidimensional* processes. In this chapter we develop the theory of stationary processes for both these cases, although for the most part we shall be concerned with multivariate processes, multi-dimensional processes being treated towards the end of the chapter.

Multivariate processes arise when instead of observing just a single process $X(t)$ we observe (simultaneously) several processes, $X_1(t)$, $X_2(t)$, ..., $X_p(t)$, say. Thus, in an engineering context we may wish to study the simultaneous variations (over time) of current and voltage, or pressure, temperature and volume, or seismic records taken at a number of different geographical locations. In economics we may be interested in studying inflation rates and money supply, unemployment and interest rates, or the supply and demand of a particular commodity. We could, of course, study each quantity on its own, and treat each as a separate univariate process. Although this would give us some information about each quantity it could never reveal what might, in fact, be the most important feature of the study, namely, the *interrelationships* between the various quantities. Just as in probability theory we cannot examine relationships between random variables knowing only their marginal distributions—we need to know their *joint* probability distribution—so, in dealing with multivariate processes we need a framework for describing not only the properties of the individual processes but also the "cross links" which may exist between them. We thus require something anal-ogous to the concept of a "joint probability distribution" to describe the joint second order properties of a multivariate process. This is achieved by introducing the notions of *cross-covariance* (or *cross-correlation*) functions and *cross-spectra*, and the definitions and properties of these functions are given below.

To avoid unnecessary repetition we consider only *discrete parameter* processes, but, as in the univariate case, most of the theory applies equally well to the continuous parameter case and the modifications required should by now be obvious.

9.1 CORRELATION AND SPECTRAL PROPERTIES OF MULTIVARIATE STATIONARY PROCESSES

To introduce the new ideas involved in the study of multivariate processes we consider first the case of *bivariate* processes.

Suppose we are given two stochastic processes, $\{X_{1,t}\}$, $\{X_{2,t}\}$, $t = 0, \pm 1, \pm 2, \ldots$. We say that $\{X_{1,t}; X_{2,t}\}$ is a *stationary bivariate process* (or tnat $\{X_{1,t}\}$, $\{X_{2,t}\}$ are *jointly stationary*) if,

(a) $\{X_{1,t}\}$ and $\{X_{2,t}\}$ are each (univariate) stationary processes in the sense of Section 3.4, and

(b) $\mathrm{cov}\{X_{1,t}, X_{2,s}\}$ is a function of $(s - t)$ only.

Assuming that the above conditions hold, we may define the auto-covariance functions of $\{X_{1,t}\}$, $\{X_{2,t}\}$ in the usual way, namely,

$$R_{11}(s) = E[\{X_{1,t} - \mu_1\}^* \{X_{1,t+s} - \mu_1\}], \qquad (9.1.1)$$

$$R_{22}(s) = E[\{X_{2,t} - \mu_2\}^* \{X_{2,t+s} - \mu_2\}], \qquad (9.1.2)$$

where $\mu_1 = E[X_{1,t}]$, $\mu_2 = E[X_{2,t}]$. (We take the complex conjugate of the first factor in (9.1.1), (9.1.2), to allow for the possibility that $X_{1,t}$, $X_{2,t}$ may be complex valued.)

The corresponding autocorrelation functions are then

$$\rho_{11}(s) = R_{11}(s)/R_{11}(0), \qquad (9.1.3)$$

$$\rho_{22}(s) = R_{22}(s)/R_{22}(0). \qquad (9.1.4)$$

Cross-covariance and cross-correlation functions

The above functions describe the *auto*correlation properties, i.e. the correlation structure *within* each process. We now introduce a new function, called the *cross-covariance function*, which describes the correlation structure *between* the processes, and is defined by,

$$R_{21}(s) = \mathrm{cov}\{X_{1,t}, X_{2,t+s}\} = E[\{X_{1,t} - \mu_1\}^* \{X_{2,t+s} - \mu_2\}]. \quad (9.1.5)$$

Note that, by condition (b), $R_{21}(s)$ depends only on s and is independent of t. The normalized version of $R_{21}(s)$, called the *cross-correlation function*, is then given by,

$$\rho_{21}(s) = \frac{R_{21}(s)}{(R_{11}(0)R_{22}(0))^{1/2}}, \qquad (9.1.6)$$

We will say that $R_{21}(s)$ denotes the cross-covariance function with "$X_{1,t}$ leading $X_{2,t}$". For the sake of symmetry we may also define the cross-covariance function with "$X_{2,t}$ leading $X_{1,t}$", viz.,

$$R_{12}(s) = E[\{X_{2,t} - \mu_2\}^* \{X_{1,t+s} - \mu_1\}], \qquad (9.1.7)$$

with $\rho_{12}(s)$ defined analogously to (9.1.6). We may note, however, that the functions $R_{12}(s)$, $R_{21}(s)$, contain equivalent information since, for all s,

$$R_{12}(s) = R_{21}^*(-s). \qquad (9.1.8)$$

The complete covariance properties of the bivariate process $\{X_{1,t}, X_{2,t}\}$ are then summarized by the sequence of matrices:

$$\boldsymbol{R}(s) = \begin{bmatrix} R_{11}(s) & R_{12}(s) \\ R_{21}(s) & R_{22}(s) \end{bmatrix}, \qquad (9.1.9)$$

$\boldsymbol{R}(s)$ being called the *covariance matrix* of lag s. The *correlation matrix* of lag s, denoted $\boldsymbol{\rho}(s)$, is defined similarly.

Since $\rho_{12}(s)$ is a correlation coefficient we obviously have,

$$|\rho_{12}(s)| \leq 1, \qquad \text{all } s, \qquad (9.1.10)$$

but whereas, for real valued processes, $\rho_{11}(s)$, $\rho_{22}(s)$ are symmetric and attain their maximum value of unity at $s = 0$, in general $\rho_{12}(s)$ is not symmetric and may attain its maximum anywhere with $\max|\rho_{12}(s)| \leq 1$. Fig. 9.1 illustrates the typical form of a cross-correlation function for real valued processes.

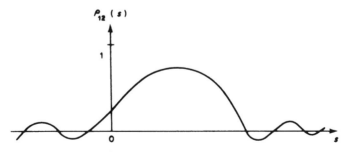

Fig. 9.1. A cross-correlation function.

Since both $X_{1,t}$, $X_{2,t}$ have constant means we may (without loss of generality) set $\mu_1 = \mu_2 = 0$, and henceforth we will assume that both $X_{1,t}$ and $X_{2,t}$ are zero mean processes.

Auto and cross-spectra

Suppose $\{X_{1,t}\}$, $\{X_{2,t}\}$ each have purely continuous spectra with (non-normalized) spectral density functions $h_{11}(\omega)$, $h_{22}(\omega)$, respectively. Then $h_{11}(\omega)$, $h_{22}(\omega)$ are the Fourier transforms of the respective autocovariance functions, i.e. we may write (cf. (4.8.35)),

$$h_{11}(\omega) = \frac{1}{2\pi} \sum_{s=-\infty}^{\infty} R_{11}(s) e^{-is\omega}, \qquad (9.1.11)$$

$$h_{22}(\omega) = \frac{1}{2\pi} \sum_{s=-\infty}^{\infty} R_{22}(s) e^{-is\omega}. \qquad (9.1.12)$$

We may now introduce the Fourier transform of the cross-covariance function, $R_{21}(s)$, and write,

$$h_{21}(\omega) = \frac{1}{2\pi} \sum_{s=-\infty}^{\infty} R_{21}(s) e^{-is\omega}. \qquad (9.1.13)$$

The function $h_{21}(\omega)$ defined by (9.1.13) is called the (non-normalized) *cross-spectral density function* or simply, the *"cross-spectrum"*, and exists for all ω provided $R_{21}(s)$ is absolutely summable, that is provided

$$\sum_{s=-\infty}^{\infty} |R_{21}(s)| < \infty. \qquad (9.1.14)$$

To see the physical interpretation of the function $h_{21}(\omega)$, we make use of the spectral representation of each process. Since $\{X_{1,t}\}$, $\{X_{2,t}\}$ are each stationary, we know from Section 4.11 that they admit spectral representations of the form (cf. (4.11.19)),

$$X_{1,t} = \int_{-\pi}^{\pi} e^{it\omega} \, dZ_1(\omega), \qquad (9.1.15)$$

$$X_{2,t} = \int_{-\pi}^{\pi} e^{it\omega} \, dZ_2(\omega), \qquad (9.1.16)$$

where $\{dZ_1(\omega)\}$, $\{dZ_2(\omega)\}$ are each orthogonal processes, i.e. for $\omega \neq \omega'$,

$$E[dZ_1^*(\omega) \, dZ_1(\omega')] = E[dZ_2^*(\omega) \, dZ_2(\omega')] = 0.$$

Substituting (9.1.15), (9.1.16) in (9.1.5) (with $\mu_1 = \mu_2 = 0$) gives

$$R_{21}(s) = \int_{-\pi}^{\pi} \int_{-\pi}^{\pi} e^{-it\omega} e^{i(s+t)\omega'} E[dZ_1^*(\omega) \, dZ_2(\omega')]. \qquad (9.1.17)$$

Now the left-hand side is a function of s only; hence the right-hand side must be a function of s only and must not depend on t. The same argument as that following (4.11.15) now shows that this can be true only if

$$E[dZ_1^*(\omega)\,dZ_2(\omega')] = 0, \qquad \omega \neq \omega',$$

i.e. $\{dZ_1(\omega)\}$, $\{dZ_2(\omega)\}$ are not only orthogonal but also *cross-orthogonal*. Using this property (9.1.17) reduces to

$$R_{21}(s) = \int_{-\pi}^{\pi} e^{is\omega} E[dZ_1^*(\omega)\,dZ_2(\omega)], \qquad (9.1.18)$$

and similarly,

$$R_{11}(s) = \int_{-\pi}^{\pi} e^{is\omega} E[|dZ_1(\omega)|^2], \qquad (9.1.19)$$

$$R_{22}(s) = \int_{-\pi}^{\pi} e^{is\omega} E[|dZ_2(\omega)|^2]. \qquad (9.1.20)$$

But, inverting (9.1.13), (9.1.11), (9.1.12), gives,

$$R_{21}(s) = \int_{-\pi}^{\pi} e^{is\omega} h_{21}(\omega)\,d\omega, \qquad (9.1.21)$$

$$R_{11}(s) = \int_{-\pi}^{\pi} e^{is\omega} h_{11}(\omega)\,d\omega, \qquad (9.1.22)$$

$$R_{22}(s) = \int_{-\pi}^{\pi} e^{is\omega} h_{22}(\omega)\,d\omega, \qquad (9.1.23)$$

and comparing (9.1.18), (9.1.19), (9.1.20) with (9.1.21), (9.1.22), (9.1.23), we now see that,

$$\begin{aligned} h_{11}(\omega)\,d\omega &= E[|dZ_1(\omega)|^2], \\ h_{22}(\omega)\,d\omega &= E[|dZ_2(\omega)|^2], \end{aligned} \qquad (9.1.24)$$

and

$$h_{21}(\omega)\,d\omega = E[dZ_1^*(\omega)\,dZ_2(\omega)].$$

Thus, the interpretation of the (non-normalized) cross-spectral density function is that $\{h_{21}(\omega)\,d\omega\}$ *represents the average value of the product of the coefficients of* $e^{it\omega}$ *in* $X_{1,t}$ *and* $X_{2,t}$. This may be compared with the interpretation, for example, of $h_{11}(\omega)$, namely that $\{h_{11}(\omega)\,d\omega\}$ represents the average value of the *square* of the coefficient of $e^{it\omega}$ in $X_{1,t}$—or the "power" due to the component with frequency ω. Alternatively, we may say that, whereas $h_{11}(\omega)\,d\omega$, $h_{22}(\omega)\,d\omega$, represent the *variances* of

$dZ_1(\omega)$, $dZ_2(\omega)$, respectively, $h_{21}(\omega)\,d\omega$ represents the *covariance* between $dZ_1(\omega)$ and $dZ_2(\omega)$.

To emphasize the distinction between the various spectra we will sometimes refer to $h_{11}(\omega)$, $h_{22}(\omega)$ as the *auto-spectral density functions* (or simply, as the "*auto-spectra*").

Introducing the Fourier transform of $R_{12}(s)$,

$$h_{12}(\omega) = \frac{1}{2\pi} \sum_{s=-\infty}^{\infty} R_{12}(s)\, e^{-is\omega}, \tag{9.1.25}$$

we may now describe the complete spectral properties of the bivariate process by the *spectral matrix*,

$$\boldsymbol{h}(\omega) = \begin{bmatrix} h_{11}(\omega) & h_{12}(\omega) \\ h_{21}(\omega) & h_{22}(\omega) \end{bmatrix} \tag{9.1.26}$$

Corresponding to the normalized auto-spectral density functions $f_{11}(\omega)$, $f_{22}(\omega)$ (the Fourier transforms of $\rho_{11}(s)$, $\rho_{22}(s)$), we may define also the *normalized cross-spectral density function*,

$$f_{12}(\omega) = \frac{1}{2\pi} \sum_{s=-\infty}^{\infty} \rho_{12}(s)\, e^{-is\omega}, \tag{9.1.27}$$

with $f_{21}(\omega)$ similarly defined.

Co-spectra and quadrature spectra

The cross-spectral density function is, in general, complex valued (since $R_{12}(s)$ need not be symmetric), and we may therefore write it in the form,

$$h_{12}(\omega) = c_{12}(\omega) - iq_{12}(\omega), \tag{9.1.28}$$

where $c_{12}(\omega)$ and $\{-q_{12}(\omega)\}$ denote respectively the real and imaginary parts of $h_{12}(\omega)$. The function $c_{12}(\omega)$ is called the *co-spectrum*, and $q_{12}(\omega)$ is called the *quadrature spectrum*, of $X_{1,t}$ and $X_{2,t}$. (This terminology is due to Tukey.) It follows immediately from (9.1.13) that, for real valued processes,

$$c_{12}(\omega) = \frac{1}{2\pi} \sum_{s=-\infty}^{\infty} \frac{1}{2}\{R_{12}(s) + R_{12}(-s)\} \cos s\omega, \tag{9.1.29}$$

and

$$q_{12}(\omega) = \frac{1}{2\pi} \sum_{s=-\infty}^{\infty} \frac{1}{2}\{R_{12}(s) - R_{12}(-s)\} \sin s\omega. \tag{9.1.30}$$

Note that $c_{12}(\omega)$ is an even function of ω, whereas q_{12} is an odd function. (Correspondingly, $\frac{1}{2}\{R_{12}(s)+R_{12}(-s)\}$ is the even part of $R_{12}(s)$, while $\frac{1}{2}\{R_{12}(s)-R_{12}(-s)\}$ is the odd part.) If we express $X_{1,t}$ and $X_{2,t}$ in terms of sines and cosines rather than complex exponentials and write (cf. (4.11.22)),

$$X_{i,t} = \int_{-\pi}^{\pi} \cos \omega t \, dU_i(\omega) + \int_{-\pi}^{\pi} \sin \omega t \, dV_i(\omega), \qquad i = 1, 2,$$

then $c_{12}(\omega)$ represents the covariance between $dU_1(t)$, $dU_2(t)$ and $dV_1(t)$, $dV_2(t)$, i.e. the covariance between the coefficients of the *"in-phase"* components in the two processes. On the other hand, $q_{12}(\omega)$ represents the covariance between $dU_1(t)$, $dV_2(t)$ and $dU_2(t)$, $dV_1(t)$, corresponding to the *"out of phase"* or "quadrature" components.

For real valued processes we may observe that at $\omega = 0$, $h_{12}(0)$ and $h_{21}(0)$ are both real valued. (This follows immediately from their definitions.) Consequently we have,

$$h_{12}(0) = c_{12}(0), \qquad q_{12}(0) = 0.$$

Cross-amplitude spectra, phase spectra and coherency

An alternative way of expressing $h_{12}(\omega)$ is to write in the "polar" form,

$$h_{12}(\omega) = \alpha_{12}(\omega) \, e^{i\phi_{12}(\omega)}, \qquad (9.1.31)$$

where

$$\alpha_{12}(\omega) = |h_{12}(\omega)| = \{c_{12}^2(\omega) + q_{12}^2(\omega)\}^{1/2}, \qquad (9.1.32)$$

and

$$\phi_{12}(\omega) = \tan^{-1}\{-q_{12}(\omega)/c_{12}(\omega)\}. \qquad (9.1.33)$$

The function $\alpha_{12}(\omega)$ is called the *cross-amplitude spectrum*, and $\phi_{12}(\omega)$ is called the *phase spectrum*. To gain an insight into the interpretations of these functions let us write,

$$dZ_i(\omega) = |dZ_i(\omega)| \, e^{i\phi_i(\omega)}, \qquad i = 1, 2,$$

so that $|dZ_i(\omega)|$ is the amplitude, and $\phi_i(\omega)$ the phase, of the component in $X_{i,t}$ with frequency ω. Now suppose, for simplicity, that $|dZ_i(\omega)|$ and $\phi_i(\omega)$ are independent random variables, $i = 1, 2$. Then,

$$\alpha_{12}(\omega) e^{i\phi_{12}(\omega)} \, d\omega = h_{12}(\omega) \, d\omega$$

$$= E[dZ_1(\omega) \, dZ_2^*(\omega)]$$

$$= E[|dZ_1(\omega)| \, |dZ_2(\omega)|]E[e^{i\{\phi_1(\omega)-\phi_2(\omega)\}}] \qquad (9.1.34)$$

Hence, we may interpret $\alpha_{12}(\omega)\,d\omega$ as the average value of the *product of the amplitudes* of the components with frequency ω in $X_{1,t}$ and $X_{2,t}$, and $\phi_{12}(\omega)$ represents the "average value" (in the sense of (9.1.34)) of the phase-shift, $\{\phi_1(\omega) - \phi_2(\omega)\}$, between the components in $X_{1,t}$ and $X_{2,t}$ at frequency ω.

According to (9.1.33) the phase spectrum is defined only mod 2π. However, $\phi_{12}(\omega)$ is conventionally taken to be in the interval $(-\pi, \pi)$, and its value is made unique by defining $\phi_{12}(\omega)$ to be the angle (in the range $(-\pi, \pi)$) between the positive half of the $c_{12}(\omega)$ axis and the line joining the origin to the point $\{c_{12}(\omega), -q_{12}(\omega)\}$. Thus, $\phi_{12}(\omega)$ is given the same sign as $\{-q_{12}(\omega)\}$.

We now define the *complex coherency* (at frequency ω) by,

$$w_{12}(\omega) = \frac{h_{12}(\omega)}{\{h_{11}(\omega)h_{22}(\omega)\}^{1/2}}. \qquad (9.1.35)$$

The *coherency* (at frequency ω) is defined as $|w_{12}(\omega)|$, and the graph of $|w_{12}(\omega)|$ as a function of ω is called the *coherency spectrum*, this term being due to Wiener. (Some authors call $|w_{12}(\omega)|^2$ the *coherence*.) Using (9.1.24), we may re-write $w_{12}(\omega)$ as,

$$w_{12}(\omega) = \frac{\text{cov}\{dZ_1(\omega), dZ_2(\omega)\}}{[\text{var}\{dZ_1(\omega)\,\text{var}\{dZ_2(\omega)\}]^{1/2}} \qquad (9.1.36)$$

so that $w_{12}(\omega)$ may be interpreted as the *correlation coefficient between the random coefficients of the components in $X_{1,t}$ and $X_{2,t}$ at frequency ω*. It follows immediately that, for all ω,

$$0 \leqslant |w_{12}(\omega)| \leqslant 1, \qquad (9.1.37)$$

for any two jointly stationary processes. As in the case of ordinary correlation coefficients the closeness of $|w_{12}(\omega)|$ to unity indicates the extent to which the random coefficients (at frequency ω) are linearly related. Later, we will see that the form of $|w_{12}(\omega)|$ over all ω determines the extent to which the processes $X_{1,t}$ and $X_{2,t}$ are *linearly related*. Roughly speaking, we may refer to $w_{12}(\omega)$ as a "correlation coefficient in the frequency domain".

Invariance of coherency under linear transformations

It is well known that the correlation coefficient between two random variables remains unchanged when we make linear transformations of the variables. Similarly, the coherency $w_{12}(\omega)$ is invariant under linear transformations of $X_{1,t}$ and $X_{2,t}$. For, suppose we write,

$$X'_{1,t} = \sum_{u=-\infty}^{\infty} a(u)X_{1,t-u} = \alpha(B)X_{1,t},$$

say, and

$$X'_{2,t} = \sum_{u=-\infty}^{\infty} b(u)X_{2,t-u} = \beta(B)X_{2,t},$$

say. Then by (4.12.54) we have (with an obvious notation),

$$dZ'_1(\omega) = \alpha(e^{-i\omega}) dZ_1(\omega), \qquad dZ'_2(\omega) = \beta(e^{-i\omega}) dZ_2(\omega),$$

so that,

$$
\begin{aligned}
h'_{11}(\omega) &= |\alpha(e^{-i\omega})|^2 h_{11}(\omega) \\
h'_{22}(\omega) &= |\beta(e^{-i\omega})|^2 h_{22}(\omega), \\
h'_{12}(\omega) &= \alpha(e^{i\omega})\beta(e^{-i\omega})h_{12}(\omega).
\end{aligned}
\tag{9.1.38}
$$

It now follows immediately from (9.1.35) that

$$|w'_{12}(\omega)| = |w_{12}(\omega)|, \tag{9.1.39}$$

for all ω such that $\alpha(e^{-i\omega})$ and $\beta(e^{-i\omega})$ are non-zero.

This result is indeed apparent when we note that the effect of the transformations is to change $dZ_1(\omega)$ to $\alpha(e^{-i\omega}) dZ_1(\omega)$ and $dZ_2(\omega)$ to $\beta(e^{-i\omega}) dZ_2(\omega)$, thus leaving the correlation coefficient between these random variables unaltered. However, it is important to remember that this result was derived on the assumption that both $X_{1,t}$ and $X_{2,t}$ are stationary process whose spectral and cross-spectral density functions exist at all frequencies. If either process contains a "trend" or a "seasonal component" (which the linear transformations were designed to remove) the above result will not, in general, be valid.

Examples

1. *Uncorrelated processes*

The simplest example of a bivariate stationary process occurs when $X_{1,t}$ and $X_{2,t}$ are *uncorrelated* processes, i.e. when $\text{cov}\{X_{1,t}, X_{2,s}\} = 0$, all s, t. (This would certainly be the case if, e.g. the two processes were *independent*.) We then have,

$$R_{12}(s) = 0, \qquad \text{all } s,$$

so that, by (9.1.13),

$$h_{12}(\omega) = 0, \qquad \text{all } \omega.$$

Consequently,

$$c_{12}(\omega) = 0, \qquad q_{12}(\omega) = 0, \qquad \alpha_{12}(\omega) = 0, \qquad w_{12}(\omega) = 0, \qquad \text{all } \omega,$$

i.e. all the cross-spectral functions are zero at all frequencies.

2. *Linear regression*

Suppose $X_{1,t}$ and $X_{2,t}$ satisfy a linear regression relationship of the form,

$$X_{1,t} = aX_{2,t} + \varepsilon_t,$$

where ε_t is a white noise (i.e. purely random) process, uncorrelated with $X_{2,t}$. Here,

$$R_{12}(s) = aR_{22}(s),$$

so that,

$$h_{12}(\omega) = ah_{22}(\omega).$$

Also,

$$h_{11}(\omega) = a^2 h_{22}(\omega) + h_\varepsilon(\omega).$$

Hence, for this model,

$$c_{12}(\omega) = \alpha h_{22}(\omega), \qquad q_{12}(\omega) = 0, \qquad \alpha_{12}(\omega) = ah_{22}(\omega), \qquad \phi_{12}(\omega) = 0,$$

and

$$w_{12}(\omega) = [1 + \{h_\varepsilon(\omega)/a^2 h_{22}(\omega)\}]^{-1/2}.$$

Note that if $\varepsilon_t \equiv 0$, then $w_{12}(\omega) = 1$, all ω, i.e. the coherency between $X_{1,t}$ and $X_{2,t}$ is unity at all frequencies—as we would expect since $X_{1,t}$ and $X_{2,t}$ satisfy an exact linear relationship when $\varepsilon_t \equiv 0$.

3. *Linear regression with delay*

Suppose now that $X_{1,t}$ and $X_{2,t}$ satisfy a linear regression relationship but with a "delay" of d time units, i.e.

$$X_{1,t} = aX_{2,t-d} + \varepsilon_t.$$

In this case,

$$R_{12}(s) = aR_{22}(s - d),$$

giving (from (9.1.13)),

$$h_{12}(\omega) = \frac{a}{2\pi} \sum_{s=-\infty}^{\infty} R_{22}(s-d) e^{-i\omega s}$$

$$= a e^{-i\omega d} h_{22}(\omega).$$

Also (as in the previous example),

$$h_{11}(\omega) = a^2 h_{22}(\omega) + h_\varepsilon(\omega).$$

Hence,

$$c_{12}(\omega) = (a \cos \omega d) h_{22}(\omega), \qquad q_{12}(\omega) = (a \sin \omega d) h_{22}(\omega)$$

$$\alpha_{12}(\omega) = a h_{22}(\omega), \qquad \phi_{12}(\omega) = -\omega d,$$

and $w_{12}(\omega)$ is the same as in the previous example.

The form of the phase spectrum $\phi_{12}(\omega)$ is particularly important in this example. It tells us that *when there is a time delay the phase spectrum is a linear function of frequency, the slope representing the magnitude of the delay.*

More generally, for any bivariate stationary process we may define,

$$d(\omega) = -\{d\phi_{12}(\omega)/d\omega\}$$

as the *"envelope delay"* (or *"group delay"*) (cf. Parzen (1967)). This function measures, in effect, the time delay between the components in the two processes at frequency ω. For the simple model considered above the time delay d is the same for all frequencies and hence in this case $d(\omega)$ takes a constant value. However, in general $d(\omega)$ will vary with frequency. The estimation of "group delay" is discussed by Hannan and Thomson (1971).

Above, we derived the cross-spectral density function $h_{12}(\omega)$ directly from (9.1.13). However, it is instructive to note a more elegant method of deriving $h_{12}(\omega)$, using (9.1.24). Using (9.1.15) we may write,

$$X_{2,t-d} = \int_{-\pi}^{\pi} e^{i(t-d)\omega} dZ_2(\omega) = \int_{-\pi}^{\pi} e^{it\omega} \{e^{-i\omega d} dZ_2(\omega)\},$$

so that (with an obvious notation) we have,

$$dZ_1(\omega) = a e^{-i\omega d} dZ_2(\omega) + dZ_\varepsilon(\omega),$$

where, since $X_{2,t}$ and ε_t are uncorrelated process, $dZ_2(\omega)$ and $dZ_\varepsilon(\omega)$ are also uncorrelated. Equation (9.1.24) now gives immediately,

$$h_{12}(\omega) = E[dZ_1(\omega) dZ_2^*(\omega)]/d\omega$$

$$= a e^{-i\omega d} E[|dZ_2(\omega)|^2]/d\omega$$

$$= a e^{-i\omega d} h_{22}(\omega),$$

as above.

4. *A model for price and supply*

Grenander and Rosenblatt (1957a, p. 36) discuss a simple form of a well known econometric model relating the price and supply of a commodity. Let $X_{1,t}$ and $X_{2,t}$ denote respectively the price and supply at time t. The model is,

$$X_{2,t} = aX_{1,t-1} + \varepsilon_t, \qquad X_{1,t} = -bX_{2,t} + \eta_t,$$

where $a > 0$, $b > 0$, and ε_t, η_t are both white noise processes, ε_t being uncorrelated with η_t. Thus, $X_{2,t}$ is linearly related to $X_{1,t}$ with a delay of one time unit, but here $X_{1,t}$ is also linearly related to $X_{2,t}$. (If, in the language of systems theory, we regard $X_{1,t}$ as the "input" and $X_{2,t}$ as the "output", then the second equation corresponds to a "feedback" loop connecting the output back to the input.) Following the second approach used in Example (3), we may write for this model,

$$dZ_2(\omega) = a\, e^{-i\omega}\, dZ_1(\omega) + dZ_\varepsilon(\omega),$$
$$dZ_1(\omega) = -b\, dZ_2(\omega) + dZ_\eta(\omega).$$

Solving these equations for $dZ_1(\omega)$, $dZ_2(\omega)$, we obtain,

$$dZ_1(\omega) = \frac{dZ_\eta(\omega) - b\, dZ_\varepsilon(\omega)}{(1 + ab\, e^{-i\omega})},$$

$$dZ_2(\omega) = \frac{a\, e^{-i\omega}\, dZ_\eta(\omega) + dZ_\varepsilon(\omega)}{(1 + ab\, e^{-i\omega})}.$$

Substituting these expressions in (9.1.24) and remembering that $dZ_\eta(\omega)$, $dZ_\varepsilon(\omega)$ are uncorrelated, we obtain (writing $\sigma_\eta^2 = \text{var}\{\eta_t\}$, $\sigma_\varepsilon^2 = \text{var}\{\varepsilon_t\}$),

$$h_{11}(\omega) = \left\{\frac{\sigma_\eta^2}{2\pi} + b^2\frac{\sigma_\varepsilon^2}{2\pi}\right\}\Big/|1 + ab\, e^{-i\omega}|^2,$$

$$h_{22}(\omega) = \left\{\frac{a^2\sigma_\eta^2}{2\pi} + \frac{\sigma_\varepsilon^2}{2\pi}\right\}\Big/|1 + ab\, e^{-i\omega}|^2,$$

and

$$h_{12}(\omega) = \left\{\frac{a\, e^{-i\omega}\sigma_\eta^2}{2\pi} - b\frac{\sigma_\varepsilon^2}{2\pi}\right\}\Big/|1 + ab\, e^{-i\omega}|^2.$$

(Here, we have used the result that, for example, since η_t is a white noise process, $E[|dZ_\eta(\omega)|^2]/d\omega = h_\eta(\omega) = \sigma_\eta^2/2\pi$, etc.)

Since $|1 + ab\,e^{-i\omega}|^2 = 1 + a^2 b^2 + 2ab\,\cos\omega$, the above expressions can be re-written as,

$$h_{11}(\omega) = \frac{(\sigma_\eta^2 + b^2 \sigma_\varepsilon^2)}{2\pi(1 + a^2 b^2 + 2ab\,\cos\omega)},$$

$$h_{22}(\omega) = \frac{(a^2 \sigma_\eta^2 + \sigma_\varepsilon^2)}{2\pi(1 + a^2 b^2 + 2ab\,\cos\omega)},$$

$$h_{12}(\omega) = \frac{(a\,e^{-i\omega}\sigma_\eta^2 - b\sigma_\varepsilon^2)}{2\pi(1 + a^2 b^2 + 2ab\,\cos\omega)}.$$

We now find

$$c_{12}(\omega) = \frac{a\sigma_\eta^2\,\cos\omega - b\sigma_\varepsilon^2}{2\pi(1 + a^2 b^2 + 2ab\,\cos\omega)}$$

$$q_{12}(\omega) = \frac{a\sigma_\eta^2\,\sin\omega}{2\pi(1 + a^2 b^2 + 2ab\,\cos\omega)}$$

$$\alpha_{12}(\omega) = \frac{[(a\sigma_\eta^2\,\cos\omega - b\sigma_\varepsilon^2)^2 + a^2 \sigma_\eta^4\,\sin^2\omega]^{1/2}}{2\pi(1 + a^2 b^2 + 2ab\,\cos\omega)}$$

$$\phi_{12}(\omega) = \tan^{-1}\left\{ \frac{a\sigma_\eta^2\,\sin\omega}{a\sigma_\eta^2\,\cos\omega - b\sigma_\varepsilon^2} \right\}.$$

Extension to general multivariate processes

The above discussion of bivariate processes is readily extended to the general multivariate case. Thus, if we have p discrete parameter processes, $X_{1,t}$, $X_{2,t}$, ... $X_{p,t}$, each having zero mean, we define the covariance matrix at lag s by

$$\mathbf{R}(s) = \{R_{ij}(s)\}, \qquad i = 1, \ldots p, \quad j = 1, \ldots p, \qquad (9.1.40)$$

where,

$$R_{ij}(s) = E[X_{j,t}^* X_{i,t+s}]. \qquad (9.1.41)$$

For $i = j$, $R_{ii}(s)$ denotes the autocovariance function of $X_{i,t}$, while for $i \neq j$, $R_{ij}(s)$ denotes the cross-covariance function between $X_{i,t}$ and $X_{j,t}$ (with $X_{j,t}$ leading $X_{i,t}$). We assume that the p processes are jointly stationary, i.e. that for all i, j, $R_{ij}(s)$ is a function of s only and does not depend on t. We then have the spectral representations,

$$X_{i,t} = \int_{-\pi}^{\pi} e^{it\omega}\,dZ_i(\omega), \qquad i = 1, \ldots p, \qquad (9.1.42)$$

with the $\{dZ_i(\omega)\}$ orthogonal and cross-orthogonal. Substituting (9.1.42) in (9.1.41) gives the spectral representation of the covariance functions,

$$R_{ij}(s) = \int_{-\pi}^{\pi} e^{is\omega} \, dH_{ij}(\omega), \qquad (9.1.43)$$

where

$$dH_{ij}(\omega) = E[dZ_i(\omega) \, dZ_j^*(\omega)].$$

The matrix $H(\omega) = \{H_{ij}(\omega)\}$ may be termed the *integrated spectral matrix* (or the *spectral distribution matrix*). The diagonal elements $H_{ii}(\omega)$ are the integrated spectra of the $\{X_{i,t}\}$, while $H_{ij}(\omega)$ is the *integrated cross-spectrum* between $X_{i,t}$ and $X_{j,t}$. When each $H_{ij}(\omega)$ is differentiable, with $dH_{ij}(\omega) = h_{ij}(\omega) \, d\omega$, (all i, j), (9.1.43) becomes,

$$R_{ij}(s) = \int_{-\pi}^{\pi} e^{is\omega} h_{ij}(\omega) \, d\omega, \qquad (\text{all } i, j). \qquad (9.1.44)$$

Inverting (9.1.44) we have,

$$h_{ij}(\omega) = \frac{1}{2\pi} \sum_{s=-\infty}^{\infty} R_{ij}(s) \, e^{-is\omega}, \qquad (9.1.45)$$

so that for $i = j$, $h_{ii}(\omega)$ is the (auto)spectral density function of $X_{i,t}$, while for $i \neq j$, $h_{ij}(\omega)$ is the cross-spectral density function of $X_{i,t}$ and $X_{j,t}$. The spectral matrix is now

$$h(\omega) = \{h_{ij}(\omega)\}, \qquad i = 1, \ldots p, \quad j = 1, \ldots p. \qquad (9.1.46)$$

The co-, quadrature, cross-amplitude, phase and coherency spectra may now be defined exactly as above for any pair of processes $X_{i,t}$ and $X_{j,t}$.

The Fourier relations (9.1.44), (9.1.45), may be written more concisely in the form,

$$R(s) = \int_{-\pi}^{\pi} e^{is\omega} h(\omega) \, d\omega, \qquad (9.1.47)$$

and

$$h(\omega) = \frac{1}{2\pi} \sum_{s=-\infty}^{\infty} R(s) \, e^{-is\omega}. \qquad (9.1.48)$$

(In (9.1.47) the integral is evaluated with respect to each element of $h(\omega)$.) Moreover, if we let X_t denote the column vector,

$$X_t = \begin{bmatrix} X_{1,t} \\ X_{2,t} \\ \vdots \\ X_{p,t} \end{bmatrix},$$

then we may write,

$$R(s) = E[X_{t+s}X_t^*],\qquad(9.1.49)$$

where the asterisk now denotes both conjugation *and* transposition, and $R(s)$ clearly has the property,

$$R^*(s) = R(-s).\qquad(9.1.50)$$

Similarly, (9.1.42) can be written as,

$$X_t = \int_{-\pi}^{\pi} e^{it\omega}\,dZ(\omega),\qquad(9.1.51)$$

where the column vector $dZ(\omega)$ has elements $dZ_1(\omega)$, $dZ_2(\omega)$, $\dots dZ_p(\omega)$. The spectral matrix $h(\omega)$ can be expressed in the form,

$$h(\omega)\,d\omega = E[dZ(\omega)\,dZ^*(\omega)],\qquad(9.1.52)$$

from which it is easily seen that $h(\omega)$ is a *Hermitian* matrix in the sense that,

$$h^*(\omega) = h(\omega),\qquad(9.1.53)$$

i.e.

$$h_{ij}(\omega) = h_{ji}^*(\omega),\qquad\text{all } i, j,$$

but more importantly, $h(\omega)$ is also a *positive semi-definite Hermitian matrix*. To prove this let $\lambda = (\lambda_1, \lambda_2, \dots \lambda_p)$ be a row vector of arbitrary (complex valued) constants, and consider the univariate process

$$Y_t = \lambda X_t.\qquad(9.1.54)$$

The autocovariance function of Y_t is

$$R_Y(s) = E[Y_{t+s}Y_t^*] = \lambda E[X_{t+s}X_t^*]\lambda^* = \lambda R(s)\lambda^*,\qquad(9.1.55)$$

and hence the spectral density function of Y_t is

$$h_Y(\omega) = \frac{1}{2\pi}\sum_{s=-\infty}^{\infty} R_Y(s)\,e^{-is\omega}$$

$$= \lambda h(\omega)\lambda^*.\qquad(9.1.56)$$

Since $h_Y(\omega) \geqslant 0$ (all ω) for all choices of $\lambda_1, \lambda_2, \dots \lambda_p$, it follows that $h(\omega)$ is a positive semi-definite matrix.

9.2 LINEAR RELATIONSHIPS

In Section 4.12 we discussed linear relationships between univariate processes of the form

$$X_t = \sum_{u=-\infty}^{\infty} g_u Y_{t-u}, \qquad (9.2.1)$$

and showed that the spectral density functions of X_t, Y_t are related by (cf. (4.12.18)),

$$h_X(\omega) = |\Gamma(\omega)|^2 h_Y(\omega), \qquad (9.2.2)$$

where

$$\Gamma(\omega) = \sum_{u=-\infty}^{\infty} g_u e^{-i\omega u}. \qquad (9.2.3)$$

The condition for X_t to have finite variance is (cf. (4.12.19)),

$$\int_{-\pi}^{\pi} |\Gamma(\omega)|^2 h_Y(\omega) \, d\omega < \infty. \qquad (9.2.4)$$

Relationships of the form (9.2.1) are most conveniently interpreted in the context of linear systems, in which case Y_t, X_t, become, respectively, the "input" and "output" (see Fig. 4.20), and $\Gamma(\omega)$ is then the system's *transfer function*, and $\{g_u\}$ the *impulse response function*. In the multivariate analogue of (9.2.1) we would have (say) q input processes, $Y_{1,t}$, $Y_{2,t}$, ... $Y_{q,t}$ and p output processes, $X_{1,t}$, $X_{2,t}$, ... $X_{p,t}$. In general, there will be "cross-links" between the various inputs and outputs (see Fig. 9.2), so that each output will be the sum of linear functions of *all* the input processes.

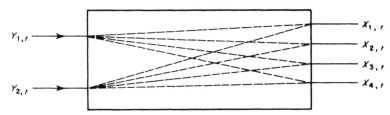

Fig. 9.2. Multivariate linear system (2 inputs, 4 outputs).

Thus, we may write the ith output as,

$$X_{i,t} = \sum_{u=-\infty}^{\infty} g_{i1}(u) Y_{1,t-u} + \sum_{u=-\infty}^{\infty} g_{i2}(u) Y_{2,t-u} + \ldots + \sum_{u=-\infty}^{\infty} g_{iq}(u) Y_{q,t-u},$$

$$i = 1, 2, \ldots p, \qquad (9.2.5)$$

or, in matrix form,

$$X_t = \sum_{u=-\infty}^{\infty} g(u) Y_{t-u}, \qquad (9.2.6)$$

where X_t is a column vector with elements $X_{1t} \dots X_{p,t}$, we define Y_t similarly, and $g(u) = \{g_{ij}(u)\}$, $u = 0, \pm 1, \pm 2, \dots$, are called the "*impulse response matrices*". (So far, we are not restricting the system to be "physically realizable", so that the summations over u in (9.2.5), (9.2.6) are two-sided.)

Introducing the spectral representations,

$$X_{i,t} = \int_{-\pi}^{\pi} e^{it\omega} dZ_i^{(X)}(\omega), \qquad i = 1, \dots p, \qquad (9.2.7)$$

$$Y_{j,t} = \int_{-\pi}^{\pi} e^{it\omega} dZ_j^{(Y)}(\omega), \qquad j = 1, \dots q, \qquad (9.2.8)$$

the jth terms on the RHS of (9.2.5) can be written as (cf. (4.12.13))

$$\int_{-\pi}^{\pi} e^{it\omega} \Gamma_{ij}(\omega) \, dZ_j^{(Y)}(\omega),$$

where

$$\Gamma_{ij}(\omega) = \sum_{u=-\infty}^{\infty} g_{ij}(u) \, e^{-i\omega u} \qquad (9.2.9)$$

represents the *transfer function between the jth input and the ith output*. Equation (9.2.5) now gives, for each ω,

$$dZ_i^{(X)}(\omega) = \Gamma_{i1}(\omega) \, dZ_1^{(Y)}(\omega) + \dots + \Gamma_{iq}(\omega) \, dZ_q^{(Y)}(\omega), \qquad i = 1, \dots p. \qquad (9.2.10)$$

This equation is of considerable importance. In the "time domain" description (9.2.5), the relationship between the ith output at time t involves weighted linear combinations of past, present and future values of all the input processes. However, the "frequency domain" form (9.2.10) has a much simpler structure. In fact (9.2.10) is simply the classical multiple linear regression model, and, as in the single input/single output case, has the feature that the spectral properties of the output at frequency ω depend only on the spectral properties of the input *at the same frequency* ω. Thus, by transforming (9.2.5) into the frequency domain form (9.2.10) we have, in a sense, "disentangled" the equation, and replaced the "lagged" relationship by a sequence of simple multiple regression equations (one for each frequency).

Writing (9.2.10) in matrix form we have,

$$dZ^{(X)}(\omega) = \Gamma(\omega)dZ^{(Y)}(\omega), \qquad (9.2.11)$$

where $\Gamma(\omega) = \{\Gamma_{ij}(\omega)\}$ is called the *transfer function matrix*. The system is thus described completely by the transfer function matrix $\Gamma(\omega)$ which, when written out in full, takes the form,

$$\Gamma(\omega) = \begin{vmatrix} \Gamma_{11}(\omega) & \Gamma_{12}(\omega) & \cdots & \Gamma_{1q}(\omega) \\ \Gamma_{21}(\omega) & \Gamma_{22}(\omega) & \cdots & \Gamma_{2q}(\omega) \\ \cdots\cdots\cdots\cdots\cdots\cdots\cdots\cdots\cdots \\ \Gamma_{p1}(\omega) & \Gamma_{p2}(\omega) & \cdots & \Gamma_{pq}(\omega) \end{vmatrix}, \qquad (9.2.12)$$

the entry in the ith row and jth column being the transfer function relating the jth input to the ith output. Equation (9.2.11) gives us immediately the relationship between the spectral matrices of the input and output. For we have

$$E[dZ^{(X)}(\omega)\, dZ^{(X)*}(\omega)] = \Gamma(\omega)E[dZ^{(Y)}(\omega)\, dZ^{(Y)*}(\omega)]\Gamma^*(\omega), \qquad (9.2.13)$$

which, on using (9.1.52), gives,

$$h_X(\omega) = \Gamma(\omega)h_Y(\omega)\Gamma^*(\omega), \qquad (9.2.14)$$

as the multivariate analogue of (9.2.2).

Noting that the variance of $X_{i,t}$ is given by integrating the ith diagonal element of $h_X(\omega)$, the condition (analogous to (9.2.4)) for each output to have finite variance is,

$$\text{tr}\left\{\int_{-\pi}^{\pi} \Gamma(\omega)h_Y(\omega)\Gamma^*(\omega)\, d\omega\right\} < \infty \qquad (9.2.15)$$

(where, for any square matrix A, $\text{tr}(A)$ denotes the "trace" of A, namely, the sum of the diagonal elements of A).

9.2.1 Linear Relationships with added Noise

Consider a single input/single output system in which the output is corrupted by a "noise" disturbance N_t as shown in Fig. 9.3. In this case

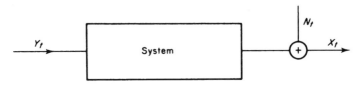

Fig. 9.3. Single input/single output system with noise.

we have in place of (9.2.1),

$$X_t = \sum_{u=-\infty}^{\infty} g_u Y_{t-u} + N_t, \qquad (9.2.16)$$

and (9.2.2) is no longer valid since $h_X(\omega)$ now depends on both Y_t and N_t. Consequently we cannot determine the transfer function gain $|\Gamma(\omega)|$ from a knowledge of $h_X(\omega)$ and $h_Y(\omega)$ only. However, if we assume that N_t is *uncorrelated* with Y_t then, as we shall see, we can determine the complete transfer function $\Gamma(\omega)$ by studying the *cross-spectrum* between Y_t and X_t.

For, assuming that N_t and Y_t are real-valued zero mean uncorrelated processes we have,

$$R_{XY}(s) = E[Y_t X_{t+s}]$$

$$= \sum_{u=-\infty}^{\infty} g_u E[Y_t Y_{t+s-u}]$$

$$= \sum_{u=-\infty}^{\infty} g_u R_{YY}(s-u). \qquad (9.2.17)$$

Taking Fourier transforms of both sides of (9.2.17) we obtain

$$h_{XY}(\omega) = \sum_{s=-\infty}^{\infty} e^{-i\omega s} \left\{ \sum_{u=-\infty}^{\infty} g_u R_{YY}(s-u) \right\}$$

$$= \sum_{u=-\infty}^{\infty} g_u e^{-i\omega u} \left\{ \sum_{s=-\infty}^{\infty} e^{-i\omega(s-u)} R_{YY}(s-u) \right\},$$

or

$$h_{XY}(\omega) = \Gamma(\omega) h_{YY}(\omega) \qquad (9.2.18)$$

(where we now denote the spectral density function of Y_t by $h_{YY}(\omega)$ rather than $h_Y(\omega)$, etc.).

We thus have one of the basic results in the study of linear systems, namely,

$$\Gamma(\omega) = h_{XY}(\omega)/h_{YY}(\omega), \qquad (9.2.19)$$

(for all ω such that $h_{YY}(\omega) \neq 0$), i.e. *the transfer function $\Gamma(\omega)$ is the ratio of the cross-spectral density function between the input and output to the spectral density function of the input.*

In place of (9.2.2) we now have,

$$h_{XX}(\omega) = |\Gamma(\omega)|^2 h_{YY}(\omega) + h_{NN}(\omega). \qquad (9.2.20)$$

We may note from (9.2.18) that (since $h_{YY}(\omega)$ is real valued),

$$\phi_{XY}(\omega) = \arg\{h_{XY}(\omega)\} = \arg\{\Gamma(\omega)\}, \qquad (9.2.21)$$

so that the *phase spectrum depends purely on the phase shifts in the transfer function* $\Gamma(\omega)$ (see Section 4.12.1). We also have

$$\alpha_{XY}(\omega) = |h_{XY}(\omega)| = |\Gamma(\omega)|h_{YY}(\omega), \qquad (9.2.22)$$

showing that the *"gain"* $|\Gamma(\omega)|$ *is the ratio of the cross-amplitude spectrum to the input spectrum.* The function, $\{\alpha_{XY}(\omega)/h_{YY}(\omega)\}$, is sometimes called the *"gain spectrum"*.

The "delay" model considered in Example (3) may be regarded as a special case of the general linear model (9.2.16) in which

$$g_u = \begin{cases} a, & u = d, \\ 0, & u \neq d. \end{cases}$$

With this form of $a(u)$,

$$\Gamma(\omega) = a\, e^{-i\omega d},$$

so that the "gain" takes the constant value a at all frequencies. Also,

$$\phi_{XY}(\omega) = \arg\{\Gamma(\omega)\} = -\omega d,$$

as shown previously.

We may gain further insight into the result (9.2.18) by re-deriving it from a frequency domain approach, as follows. Using the spectral representations of Y_t, X_t, and N_t, we may transform (9.2.16) into the corresponding frequency domain relationship,

$$dZ_X(\omega) = \Gamma(\omega)\, dZ_Y(\omega) + dZ_N(\omega), \qquad (9.2.23)$$

where $dZ_Y(\omega)$ and $dZ_N(\omega)$ are uncorrelated. As observed previously, (9.2.23) has precisely the form of a *simple linear regression* between $dZ_X(\omega)$ and $dZ_Y(\omega)$ (for each ω), and multiplying both sides of (9.2.23) by $dZ_Y^*(\omega)$ and taking expectations, we obtain,

$$E[dZ_X(\omega)\, dZ_Y^*(\omega)] = \Gamma(\omega)E[|dZ_Y(\omega)|^2], \qquad (9.2.24)$$

which immediately gives (9.2.18) on using (9.1.24). However, we may observe another interesting feature of (9.2.24), namely that the expression for $\Gamma(\omega)$ may be rewritten as,

$$\Gamma(\omega) = \operatorname{cov}\{dZ_X(\omega), dZ_Y(\omega)\}/\operatorname{var}\{dZ_Y(\omega)\},$$

which is the standard expression for the "least squares" estimate of $\Gamma(\omega)$ when (9.2.23) is regarded as a linear regression model. Thus, if we are simply given two processes, X_t and Y_t, then (9.2.19) gives the transfer function which results in the "best" linear approximation to X_t (in the least squares sense). Specifically, if we choose the sequence $\{\hat{g}_u\}$ so as to minimize

$$E[N_t^2] = E\left[X_t - \sum_{u=-\infty}^{\infty} \hat{g}_u Y_{t-u}\right]^2, \qquad (9.2.25)$$

then the minimum value of (9.2.25) is obtained by choosing the $\{\hat{g}_u\}$ as the Fourier coefficients of the function,

$$\hat{\Gamma}(\omega) = h_{XY}(\omega)/h_{YY}(\omega). \qquad (9.2.26)$$

It is easy to prove this result directly by noting that the "normal equations" arising from the minimization of (9.2.25) are identical with (9.2.17). Alternatively, the result follows from the fact that,

$$E[N_t^2] = \int_{-\pi}^{\pi} h_{NN}(\omega)\, d\omega = \int_{-\pi}^{\pi} [|dZ_N(\omega)|^2],$$

so that minimizing $E[N_t^2]$ is equivalent to minimizing $E[|dZ_N(\omega)|^2]$ for each ω.

Residual variance bound

From (9.2.20) we have,

$$\sigma_N^2 = \int_{-\pi}^{\pi} h_{NN}(\omega)\, d\omega$$

$$= \int_{-\pi}^{\pi} \{h_{XX}(\omega) - |\Gamma(\omega)|^2 h_{YY}(\omega)\}\, d\omega$$

$$= \int_{-\pi}^{\pi} \left[1 - \frac{|h_{XY}(\omega)|^2}{h_{XX}(\omega)h_{YY}(\omega)}\right] h_{XX}(\omega)\, d\omega$$

$$= \int_{-\pi}^{\pi} [1 - |w_{XY}(\omega)|^2] h_{XX}(\omega)\, d\omega, \qquad (9.2.27)$$

where $|w_{XY}(\omega)|$ is the coherency (at frequency ω) between X_t and Y_t. Hence, the degree of linear association between Y_t and X_t (as measured by the magnitude of the residual variance σ_N^2) depends on the closeness of $|w_{XY}(\omega)|^2$ to unity at all frequencies. If $|w_{XY}(\omega)|^2$ is close to 1 at all frequencies we would expect to obtain a close linear fit between the two series, thereby strengthening the interpretation of $w_{XY}(\omega)$ as a "correlation coefficient in the frequency domain". Accordingly, the expression on the RHS of (9.2.27), namely

$$\int_{-\pi}^{\pi} [1 - |w_{XY}(\omega)|^2] h_{XX}(\omega)\, d\omega, \qquad (9.2.28)$$

may be called the "*residual variance bound*", since it gives the value of the residual variance after fitting the "best possible" linear relationship between Y_t and X_t, allowing an infinite number of parameters. If we fit a "finite parameter" linear relationship the residual variance can never be smaller than (9.2.28), and in general it will be larger.

Analysis of variance

Re-writing (9.2.27) as

$$\sigma_N^2 = \int_{-\pi}^{\pi} h_{XX}(\omega)\, d\omega - \int_{-\pi}^{\pi} h_{XX}(\omega) |w_{XY}(\omega)|^2\, d\omega,$$

$$= \sigma_X^2 - \int_{-\pi}^{\pi} h_{XX}(\omega) |w_{XY}(\omega)|^2\, d\omega,$$

we have,

$$\sigma_X^2 = \int_{-\pi}^{\pi} h_{XX}(\omega) |w_{XY}(\omega)|^2\, d\omega + \sigma_N^2. \qquad (9.2.29)$$

Equation (9.2.29) may be interpreted thus,

Total variance of X_t = variance due to "regression on Y_t"
+ residual variance.

Note that if we re-write (9.2.20) as,

$$h_{NN}(\omega) = h_{XY}(\omega)\{1 - |w_{XY}(\omega)|^2\}, \qquad (9.2.30)$$

we obtain an exact analogue of the standard result in the linear regression analysis between two random variables, X, Y,

$$\sigma_{\text{resid}}^2 = \sigma_X^2(1 - \rho^2),$$

where ρ is the correlation coefficient between X and Y.

Multivariate case

When the output contains an additive noise disturbance, (9.2.6) becomes

$$X_t = \sum_{u=-\infty}^{\infty} g(u) Y_{t-u} + N_t, \qquad (9.2.31)$$

where $N'_t = (N_{1,t}, N_{2,t}, \ldots, N_{p,t})$ is a vector of p components, each $N_{i,t}$ being uncorrelated with each $Y_{j,t}$, as illustrated in Fig. 9.4. Equation (9.2.11) now becomes,

$$dZ^{(X)}(\omega) = \Gamma(\omega)\, dZ^{(Y)}(\omega) + dZ^{(N)}(\omega), \tag{9.2.32}$$

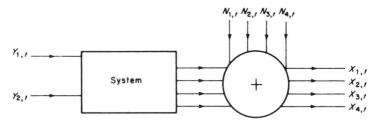

Fig. 9.4. Multivariate linear system with noise.

and multiplying both sides by $dZ^{(Y)*}(\omega)$ and taking expectations yields, (remembering that $dZ^{(Y)}(\omega)$ and $dZ^{(N)}(\omega)$ are uncorrelated).

$$E[dZ^{(X)}(\omega)\, dZ^{(Y)*}(\omega)] = \Gamma(\omega)E[dZ^{(Y)}(\omega)\, dZ^{(Y)*}(\omega)],$$

or

$$h_{XY}(\omega) = \Gamma(\omega)h_{YY}(\omega), \tag{9.2.33}$$

where $h_{YY}(\omega)$ is the *spectral matrix of* Y_t, and the $p \times q$ matrix $h_{XY}(\omega)$ may be termed the *cross-spectral matrix between* Y_t *and* X_t. (The (i, j)th element of $h_{XY}(\omega)$ is the cross-spectral density function between $X_{i,t}$ and $Y_{j,t}$). We now have, analogously to (9.2.19),

$$\Gamma(\omega) = h_{XY}(\omega)h^{-1}_{YY}(\omega), \tag{9.2.34}$$

(assuming that $h_{YY}(\omega)$ is non-singular). From (9.2.32) we have,

$$h_{XX}(\omega) = \Gamma(\omega)h_{YY}(\omega)\Gamma^*(\omega) + h_{NN}(\omega),$$

$$= h_{XY}(\omega)h^{-1}_{YY}(\omega)h^*_{XY}(\omega) + h_{NN}(\omega),$$

by (9.2.34). Hence the spectral matrix of the "residual" process N_t is given by

$$h_{NN}(\omega) = h_{XX}(\omega) - h_{XY}(\omega)h^{-1}_{YY}(\omega)h^*_{XY}(\omega). \tag{9.2.35}$$

9.2.2 The Box–Jenkins "Transfer Function" Models

When we consider the problem of fitting a linear relationship to a finite number of observations on Y_t, X_t, we realize that the general model (9.2.16) is hardly in a suitable form since it involves an arbitrarily large number of parameters. Instead, we would usually prefer to fit a finite parameter model of the form,

$$X_t + a_1 X_{t-1} + \ldots + a_k X_{t-k} = b_0 Y_t + b_1 Y_{t-1} + \ldots + b_l Y_{t-l} + N'_t, \quad (9.2.36)$$

which may be written in operator form as

$$\alpha(B)X_t = \beta(B)Y_t + N'_t, \quad (9.2.37)$$

where

$$\alpha(z) = 1 + a_1 z + \ldots + a_k z^k, \qquad \beta(z) = b_0 + b_1 z + \ldots + b_l z^l.$$

Note that (9.2.37) can be written formally as

$$X_t = \alpha^{-1}(B)\beta(B)Y_t + \alpha^{-1}(B)N'_t, \quad (9.2.38)$$

so that (9.2.35) is a special case of (9.2.16) in which $\Gamma(\omega)$ is a *rational function of* $e^{-i\omega}$, i.e.

$$\Gamma(\omega) = \beta(e^{-i\omega})/\alpha(e^{-i\omega}), \quad (9.2.39)$$

and

$$N_t = \alpha^{-1}(B)N'_t.$$

Box and Jenkins (1970) refer to (9.2.36) as a *"transfer function model"* and discuss, in particular, how such models can be used to improve the forecasts of future values of X_t by basing the forecasts not only on past values of X_t but also on past values of the related series, Y_t. (We will consider the general problem of forecasting in Chapter 10.)

Identification of $\alpha(B)$ and $\beta(B)$ by "covariance contraction"

When appropriate values of k and l are known the coefficients $(a_1, \ldots a_k, b_0, \ldots b_l)$ can be estimated by a standard least squares approach. However the main difficulty is the determination of suitable values of k and l, i.e. we require some method of identifying the structure of $\alpha(B)$ and $\beta(B)$ before carrying through the least-square estimation of their coefficients. Box and Jenkins (1970) proposed a method of identifying $\alpha(B)$ and $\beta(B)$ by examining the corresponding impulse response function g_u, which, in turn, is determined by first "pre-whitening" the input Y_t. Priestley (1971a) suggested an alternative approach based on a technique called *"covariance contraction"*. This technique may be summarized briefly as follows. We first pre-whiten both Y_t and X_t by fitting univariate ARMA models to each process, leading to, say,

$$\theta_1(B)X_t = \phi_1(B)\eta_t, \quad (9.2.40)$$

$$\theta_2(B)Y_t = \phi_2(B)\varepsilon_t, \quad (9.2.41)$$

η_t and ε_t being (univariate) white noise processes. If we can now fit a "transfer function" model to the residuals,

$$\eta_t + p_1\eta_{t-1} + \ldots + p_n\eta_{t-n} = q_0\varepsilon_t + q_1\varepsilon_{t-1} + \ldots + q_m\varepsilon_{t-m} + \xi_t, \qquad (9.2.42)$$

say, or,

$$P(B)\eta_t = Q(B)\varepsilon_t + \xi_t, \qquad (9.2.43)$$

then the corresponding "transfer function" model for (Y_t, X_t) is given by

$$P(B)\phi_1^{-1}(B)\theta_1(B)X_t = Q(B)\phi_2^{-1}(B)\theta_2(B)Y_t + \xi_t,$$

or,

$$\alpha(B)X_t = \beta(B)Y_t + N'_t,$$

where

$$\alpha(B) = P(B)\phi_2(B)\theta_1(B),$$

$$\beta(B) = Q(B)\phi_1(B)\theta_2(B),$$

$$N'_t = \phi_1(B)\phi_2(B)\xi_t.$$

It may at first appear that the fitting of a model to (η_t, ε_t) involves exactly the same difficulties as those associated with the fitting of a model to the original processes, Y_t, X_t. However, the basic strategy of this approach is based on the consideration that the structure of the operators $\alpha(B)$, $\beta(B)$ depends on both the autocorrelation and cross-correlation structure of Y_t and X_t. In fitting individual models to Y_t, X_t, we are essentially removing the autocorrelation structure and it may be expected, therefore, that the general form of the operators, $P(B)$, $Q(B)$, will be much simpler than the form of $\alpha(B)$, $\beta(B)$. Moreover we may exploit the fact that η_t and ε_t are white noise processes when using the cross-covariance function to indicate the forms of $P(B)$ and $Q(B)$, as follows. If we write,

$$\chi_t = \eta_t + p_1\eta_{t-1} + \ldots + p_n\eta_{t-n} = P(B)\eta_t, \qquad (9.2.44)$$

then we may write,

$$\chi_t = q_0\varepsilon_t + q_1\varepsilon_{t-1} + \ldots + q_m\varepsilon_{t-m} + \xi_t. \qquad (9.2.45)$$

The cross-covariance function between χ_t and ε_t is given by,

$$R_{\chi\varepsilon}(s) = E[\varepsilon_t\chi_{t+s}] = \begin{cases} q_s, & s = 0, 1, \ldots, m, \\ 0, & \text{otherwise}. \end{cases} \qquad (9.2.46)$$

Thus, the cross-covariance function at lag s is simply the coefficient q_s, and when $Q(B)$ contains only a finite number of terms, $R_{\chi\varepsilon}(s)$ *will be zero, except for lags* $s = 0, 1, \ldots, m$. On the other hand, the cross-covariance function between η_t and ε_t, $R_{\eta\varepsilon}(s)$, will not in general, vanish after a finite number of terms since, in general, the operator $\{P^{-1}(B)Q(B)\}$ will produce an infinite series when expanded in powers of B. However, there is a simple relationship between $R_{\eta\varepsilon}(s)$ and $R_{\chi\varepsilon}(s)$, namely,

$$R_{\chi\varepsilon}(s) = P(B)R_{\eta\varepsilon}(s) \qquad (9.2.47)$$

(where the shift operator B now acts on the variable s). Hence, we may regard $P(B)$ as that operator *which "contracts" the cross-covariance function,* $R_{\eta\varepsilon}(s)$, *into a function of the form* (9.2.46). In practice, we would have to work with the estimated cross-covariance function, $\hat{R}_{\eta\varepsilon}(s)$, and a suitable form for $P(B)$ can be inferred by seeking the appropriate "filter" which causes the function $\hat{R}_{\eta\varepsilon}(s)$ to decay "rapidly" to zero. In particular if $\hat{R}_{\eta\varepsilon}(s)$ decays slowly and smoothly (i.e. contains essentially "low frequency" variation) then it may be "contracted" by repeated application of the difference operator $(1-B)$. In this case $P(B)$ will be of the form

$$P(B) = (1-B)^d. \qquad (9.2.48)$$

If $\hat{R}_{\eta\varepsilon}(s)$ has an "oscillatory" form (i.e. contains essentially "high frequency" variation) it may be "contracted" by, for example, using a moving average filter, so that $P(B)$ would be of the form,

$$P(B) = \frac{1}{(2d+1)}(B^{-d} + B^{-d-1} + \ldots 1 + B + \ldots + B^d). \qquad (9.2.49)$$

In either case, a suitable value for the parameter d may be inferred by starting off with a small value (typically, $d = 1$), and increasing d until the transformed cross-covariance function exhibits the required degree of "contraction".

Once the form of $P(B)$ has been determined, the form of $Q(B)$ may be determined immediately from the contracted cross-covariance function, $\hat{R}_{\chi\varepsilon}(s)$, using (9.2.46). (In practice, the terms required in $Q(B)$ would be found by determining those lags for which $\hat{R}_{\chi\varepsilon}(s)$ differs significantly from zero.)

It should be emphasized that the above technique is suggested as a method of identifying only the *structure* of the operators, $\alpha(B)$, $\beta(B)$. The associated parameters $(a_1, \ldots, a_k, b_0, b_1, \ldots, b_l)$ can then be estimated by a full least squares approach—see Priestley (1971a).

Lead–Lag relationships

The most useful application of "transfer function" models to forecasting problems arises when one series "leads" the other series. Specifically, we will say that ε_t *leads* η_t *by d time units* if (9.2.42) reduces to

$$\eta_t + p_1\eta_{t-1} + \ldots + p_n q_{t-n} = q_d\varepsilon_{t-d} + \ldots + q_m\varepsilon_{t-m} + \xi_t, \quad (9.2.50)$$

i.e. if $q_0 = q_1 = \ldots = q_{d-1} = 0$. (More generally, we may still refer to "a lead of d units" if (9.2.50) is modified to

$$\eta_t + p_1\eta_{t-1} + \ldots + p_n\eta_{t-n} = q_{d-r}\varepsilon_{t-d+r} + \ldots + q_d\varepsilon_{t-d} + \ldots + q_m\varepsilon_{t-m} + \xi_t,$$
$$(9.2.51)$$

in which $q_{d-r}, \ldots, q_{d-1}, q_{d+1}, \ldots, q_m$, are all relatively small compared with q_d.)

This situation would correspond to the case where the "contracted" cross-covariance function, $\hat{R}_{x\varepsilon}(s)$, exhibits a well defined "peak", and the value of the lead time d may be estimated by that value of s at which $\hat{R}_{x\varepsilon}(s)$ attains its maximum. The estimated value of d, say $\hat{d}$, may be confirmed by computing the estimated phase spectrum $\hat{\phi}_{\eta\varepsilon}(\omega)$, say, of η_t and ε_t. In this case, $\hat{\phi}_{\eta\varepsilon}(\omega)$ should exhibit a strong linear "trend" with gradient $(-d)$; see Example 3 in Section 9.1. (We discuss the estimation of cross-covariances and cross-spectra in Section 9.5, but for the moment we may note that some caution is required in interpreting plots of phase spectra, since $\hat{\phi}(\omega)$, for each ω, *is determined modulo* 2π *only*. Consequently, it may be worthwhile to shift the series ε_t by $\hat{d}$ time units and recompute the phase-spectrum, which should then fail to exhibit any linear trend.)

Intervention analysis

The technique of "intervention analysis", due to Box and Tiao (1975), is designed to deal with situations where the observations $\{X_t\}$ are affected by the occurrence of some unusual event at a known time point t_0. For example, the sales of a commodity may be affected by a sudden change in advertizing policy, by a strike, or by a change in tax laws. Suppose now that we are interested in forecasting future values of the series, and to this purpose wish to fit a suitable model to the data. If the given observations include the time point t_0 we could argue that in the model fitting stage the observations prior to time t_0 should be ignored on the grounds that the occurrence of the unusual event may have changed the structure of the series. However, this would be wasteful of data (particularly with short series), and it would be more sensible to try to remove (as far as possible) the effect of the "event", so that the model can then be fitted to the full set of observations.

The approach of Box and Tiao is to characterize the occurrence of the event by introducing an artificial "indicator" series which changes its form abruptly at time t_0. For example, we may introduce the series

$$Y_t^{(1)} = \begin{cases} 0, & t \neq t_0, \\ 1, & t = t_0, \end{cases}$$

which consists of a single "spike" at $t = t_0$, or alternatively, the series

$$Y_t^{(2)} = \begin{cases} 0, & t < t_0, \\ 1, & t \geq t_0, \end{cases}$$

which takes the form of a step-function with a "jump" at $t = t_0$. If we assume now that the effect of the "event" can be described by a *linear function* of $\{Y_t^{(1)}\}$ or $\{Y_t^{(2)}\}$, we can allow for this effect by fitting a "transfer function" model between X_t and $Y_t^{(1)}$ or $Y_t^{(2)}$. Thus, we may write

$$X_t = \alpha_1^{-1}(B)\beta_1(B)Y_t^{(1)} + N_t^{(1)},$$

or

$$X_t = \alpha_2^{-1}(B)\beta_2(B)Y_t^{(2)} + N_t^{(2)},$$

and fit the polynomials $\alpha(B)$, $\beta(B)$, by the methods described above. In practice, the nature of the event will usually determine whether we use an artificial series of the form $Y_t^{(1)}$ or $Y_t^{(2)}$, but, in principle, these forms are interchangeable since we can write $Y_t^{(1)} \equiv (1 - B)Y_t^{(2)}$, and hence

$$\alpha_2^{-1}(B)\beta_2(B) \equiv \alpha_1^{-1}(B)\beta_1(B)(1 - B).$$

Although neither $Y_t^{(1)}$ nor $Y_t^{(2)}$ can be regarded as a realization of a stationary process, the above regression method would still be applicable, and the technique of "intervention analysis" has been applied quite successfully to various sets of data—see, e.g. Tiao, Box and Hamming (1975).

9.3 MULTIPLE AND PARTIAL COHERENCY

In the discussion of linear relationships in Section 9.2 it was pointed out that linear models involving lagged values of the "dependent" processes, $Y_{1,t}, Y_{2,t}, \ldots, Y_{q,t}$, are equivalent to standard multivariate linear regression models for the corresponding "Fourier transforms" $dZ_1^{(Y)}(\omega), \ldots dZ_q^{(Y)}(\omega)$. Once this point is grasped it becomes apparent that the whole "apparatus" of multivariate linear regression theory can be taken over (almost unchanged) and applied to the study of multivariate spectral relationships. In particular, the ideas of "multiple correlation" and "partial correlation" (see, e.g., Kendall and Stuart (1966), Ch. 27) have immediate analogues in the frequency domain, where they become "multiple coherency" and "partial coherency". We now discuss these notions in more precise terms.

Multiple coherency

When studying a linear regression relationship between a random variable X and several other random variables $Y_1, Y_2, \ldots, Y_q$, one may introduce the multiple correlation coefficient R which is defined so that $(1 - R^2)$ represents the proportion of var$\{X\}$ which is "explained" by the regression of X on $Y_1, \ldots, Y_q$. If there is only one Y variable (i.e. if $q = 1$) then the multiple correlation coefficient is, of course, the same as the ordinary correlation coefficient. Similarly, in studying a linear relationship between a process X_t and several other processes $Y_{1,t}, \ldots, Y_{q,t}$ one may define the "multiple coherency" (at frequency ω) as the proportion of the power of X_t at frequency ω which is "explained" by the relationship with $Y_{1,t}, \ldots, Y_{q,t}$. Thus, suppose that in the model (9.2.31) there is just one output process X_t (i.e. $p = 1$), and q input processes, so that (9.2.31) takes the form

$$X_t = \sum_{u=-\infty}^{\infty} g_1(u) Y_{1,t-u} + \ldots + \sum_{u=-\infty}^{\infty} g_q(u) Y_{q,t-u} + N_t. \qquad (9.3.1)$$

Then from (9.2.35) we can write the spectral density function of N_t as,

$$h_{NN}(\omega) = h_{XX}(\omega) - \boldsymbol{h}_{XY}(\omega) \boldsymbol{h}_{YY}^{-1} \boldsymbol{h}_{XY}^*(\omega)$$
$$= h_{XX}(\omega)[1 - W_{XY_1 Y_2 \ldots Y_q}^2(\omega)], \qquad (9.3.2)$$

where

$$W_{XY_1, \ldots, Y_q}^2(\omega) = \frac{\boldsymbol{h}_{XY}(\omega) \boldsymbol{h}_{YY}^{-1}(\omega) \boldsymbol{h}_{XY}^*(\omega)}{h_{XX}(\omega)}, \qquad (9.3.3)$$

is called the *squared multiple coherency* between X_t and $Y_{1,t}, \ldots, Y_{q,t}$ (at frequency ω). (Note that since X_t is univariate here, $\boldsymbol{h}_{XY}(\omega)$ is a $1 \times p$ matrix, $\boldsymbol{h}_{YY}(\omega)$ is $p \times p$, and $\boldsymbol{h}_{XY}^*(\omega)$ is $p \times 1$.) We may now re-write (9.3.2) as

$$h_{XX}(\omega) = W_{XY_1 \ldots Y_q}^2(\omega) h_{XX}(\omega) + h_{NN}(\omega), \qquad (9.3.4)$$

which gives a decomposition of the power (density) of X_t as the sum of the "power due to regression on $Y_{1,t}, \ldots, Y_{q,t}$" plus the "residual power". The first term thus represents the proportion of the power "explained" by $Y_{1,t}, \ldots, Y_{q,t}$.

Note that (9.3.5) is the multivariate version of (9.2.29), and, as before, the form of $W^2_{XY_1,\ldots,Y_q}(\omega)$ determines the extent to which we may fit X_t by a linear function of $Y_{1,t}, \ldots, Y_{q,t}$. If $W^2_{XY_1,\ldots,Y_q}(\omega) = 1$, all ω, then there is a perfect linear relationship between X_t and $Y_{1,t}, \ldots, Y_{q,t}$. On the other hand, if $W^2_{XY_1,\ldots,Y_q}(\omega) = 0$, all ω, then all the power resides in the residual, N_t.

When X_t is univariate, (9.2.32) may be written,

$$dZ^{(X)}(\omega) = \Gamma_1(\omega)\, dZ_1^{(Y)}(\omega) + \ldots + \Gamma_q(\omega)\, dZ_q^{(Y)}(\omega) + dZ_N(\omega),$$

where $\Gamma_i(\omega)$ $(i = 1, \ldots, q)$ is the transfer function corresponding to the ith term in (9.3.1). Using the standard expression for the "regression" sum of squares in a multiple regression model, we may, from (9.3.4) express the multiple coherency in the alternative form,

$$W_{XY_1,\ldots,Y_q}(\omega) = [\Gamma_1(\omega)h_{XY_1}(\omega) + \ldots + \Gamma_q(\omega)h_{XY_q}(\omega)]/h_{XX}(\omega). \quad (9.3.5)$$

Partial coherency

In dealing with relations between random variables one may enquire whether a high correlation between (say) X and Y_1 is due to an intrinsic association between these variables, or whether it is due merely to the fact that X and Y_1 are each highly correlated with some other variable, Y_2 say. To distinguish between these two cases one introduces the "partial correlation coefficient" which measures the correlation between X and Y_1 after the influence of Y_2 on each of these variables has been removed.

To define the "partial coherency" between X_t and $Y_{1,t}$ allowing for $Y_{2,t}$, we first remove the influence of $Y_{2,t}$ on X_t and $Y_{1,t}$ by considering the processes,

$$\eta_{1,t} = X_t - \sum_{u=-\infty}^{\infty} b_1(u)\, Y_{2,t-u}, \quad (9.3.6)$$

$$\eta_{2,t} = Y_{1,t} - \sum_{u=-\infty}^{\infty} b_2(u)\, Y_{2,t-u}, \quad (9.3.7)$$

where $\{b_1(u)\}$, $\{b_2(u)\}$, are determined by minimizing $E[\eta_{1,t}^2]$ and $E[\eta_{2,t}^2]$, respectively. Using (9.2.26), the corresponding transfer functions are given by

$$B_1(\omega) = \sum_{u=-\infty}^{\infty} b_1(u)\, e^{-i\omega u} = h_{XY_2}(\omega)/h_{Y_2 Y_2}(\omega), \quad (9.3.8)$$

$$B_2(\omega) = \sum_{u=-\infty}^{\infty} b_2(u)\, e^{-i\omega u} = h_{Y_1 Y_2}(\omega)/h_{Y_2 Y_2}(\omega). \quad (9.3.9)$$

Hence, in the spectral representations of $\eta_{1,t}$ and $\eta_{2,t}$ we have (with an obvious notation),

$$dZ_{\eta_1}(\omega) = dZ_X(\omega) - B_1(\omega)\, dZ_{Y_2}(\omega), \qquad (9.3.10)$$

$$dZ_{\eta_2}(\omega) = dZ_{Y_1}(\omega) - B_2(\omega)\, dZ_{Y_2}(\omega). \qquad (9.3.11)$$

Evaluating $E[dZ_{\eta_1}(\omega)\, dZ_{\eta_2}^*(\omega)]$ we find that the cross spectral density function of $\eta_{1,t}$ and $\eta_{2,t}$ is

$$h_{\eta_1\eta_2}(\omega) = h_{XY_1}(\omega) - B_1(\omega)h_{Y_2Y_1}(\omega) - B_2^*(\omega)h_{XY_2}(\omega)$$
$$+ B_1(\omega)B_2^*(\omega)h_{Y_2Y_2}(\omega). \qquad (9.3.12)$$

Substituting the expressions for $B_1(\omega)$, $B_2(\omega)$, given by (9.3.8) and (9.3.9), (9.3.12) reduces to

$$h_{\eta_1\eta_2}(\omega) = h_{XY_1}(\omega) - \frac{h_{XY_2}(\omega)h_{Y_2Y_1}(\omega)}{h_{Y_2Y_2}(\omega)}. \qquad (9.3.13)$$

The function $h_{\eta_1\eta_2}(\omega)$ is called the *partial cross-spectral density function* of X_t and $Y_{1,t}$ (allowing for $Y_{2,t}$), and is sometimes denoted by $h_{XY_1.Y_2}(\omega)$. The partial (complex) coherency, $w_{XY_1.Y_2}(\omega)$, is now defined as the (complex) coherency of $\eta_{1,t}$ and $\eta_{2,t}$, and is thus given by,

$$W_{XY_1.Y_2}(\omega) = \frac{h_{\eta_1\eta_2}(\omega)}{\{h_{\eta_1\eta_1}(\omega)h_{\eta_2\eta_2}(\omega)\}^{1/2}}, \qquad (9.3.14)$$

where $h_{\eta_1\eta_1}(\omega)$ and $h_{\eta_2\eta_2}(\omega)$, the spectral density functions of $\eta_{1,t}$ and $\eta_{2,t}$, are given by applying the result (9.2.32) to (9.3.6) and (9.3.7). This gives,

$$h_{\eta_1\eta_1}(\omega) = h_{XX}(\omega)\{1 - |W_{XY_2}(\omega)|^2\}, \qquad (9.3.15)$$

and

$$h_{\eta_2\eta_2}(\omega) = h_{Y_1Y_1}(\omega)\{1 - |W_{Y_2Y_1}(\omega)|^2\}. \qquad (9.3.16)$$

Substituting these expressions in (9.3.14), and using (9.3.13) finally gives,

$$W_{XY_1.Y_2}(\omega) = \frac{W_{XY_1}(\omega) - W_{XY_2}(\omega)W_{Y_2Y_1}(\omega)}{\{(1 - |W_{XY_2}(\omega)|^2)(1 - |W_{Y_2Y_1}(\omega)|^2)\}^{1/2}}. \qquad (9.3.17)$$

The function $|W_{XY_1.Y_2}(\omega)|$ measures the coherency (at frequency ω) between X_t and $Y_{1,t}$, after removing the common influence of $Y_{2,t}$ and we may note that (9.3.17) is an obvious analogue of the standard expression for the partial correlation between two random variables X, Y_1, allowing for a third random variable Y_2, namely,

$$\rho_{XY_1.Y_2} = \frac{\rho_{XY_1} - \rho_{XY_2}\rho_{Y_2Y_1}}{\{(1 - \rho_{XY_2}^2)(1 - \rho_{Y_2Y_1}^2)\}^{1/2}}. \qquad (9.3.18)$$

The *partial phase spectrum* is similarly defined as

$$\phi_{XY_1.Y_2}(\omega) = \arg\{h_{n_1 n_2}(\omega)\}, \tag{9.3.19}$$

(where $h_{n_1 n_2}(\omega)$ is given by (9.3.13)), and measures the phase shift (at frequency ω) between X_t and $Y_{1,t}$, after allowing for phase shifts in each of these processes induced by their common association with $Y_{2,t}$.

Expressions for the partial coherency may be derived also for the more general case where we wish to remove the common influence of *several* processes, $Y_{2,t}, Y_{3,t}, \ldots, Y_{q,t}$. These expressions follow the standard general formulae for partial correlations—see, e.g. Kendall and Stuart (1966, Ch. 27).

Note that although we have found it convenient to think of X_t as the "output" and $Y_{1,t}, Y_{2,t}, \ldots, Y_{q,t}$ as "inputs", this distinction is really an artificial one. In fact, given any multivariate stationary process $X_t' = \{X_{1,t}, X_{2,t}, \ldots, X_{p,t}\}$, we can choose any $X_{i,t}$ as the "output" and any subset $\{X_{j,t}, X_{k,t}, X_{l,t}, X_{m,t}, \ldots\}$ as the "inputs". In this way we can define, e.g. the partial coherency, $W_{ij.klm\ldots}(\omega)$, between any pair, $X_{i,t}$ and $X_{j,t}$, allowing for any subset $\{X_{k,t}, X_{l,t}, X_{m,t}, \ldots\}$. Partial coherencies may be computed by means of an iterative formula similar to the one used for evaluating partial autocorrelations. The iteration is based on increasing by one the number of variables in the subset $\{X_{k,t}, X_{l,t}, X_{m,t}, \ldots\}$; see Parzen (1967).

The notion of "multiple coherency" was introduced by Goodman (1963) and Koopmans (1964a, 1964b). Partial cross-spectra and coherency were introduced by Tick (1963), and developed by Koopmans (1964a, 1974), Akaike (1965), Goodman (1965), and Parzen (1967).

9.4 MULTIVARIATE AR, MA, AND ARMA MODELS

In Section 3.5 we discussed the three main types of univariate models, namely the AR (autoregressive), MA (moving average) and ARMA (mixed autoregressive/moving average). Each of these models has its corresponding multivariate extension which is obtained, essentially, by replacing the scalar parameters in the univariate model by matrix parameters.

Autoregressive models

The p-variate AR(k) model is given by

$$X_t + a_1 X_{t-1} + \ldots + a_k X_{t-k} = \varepsilon_t, \tag{9.4.1}$$

where $X_t = \{X_{1,t}, \ldots, X_{p,t}\}'$, $a_1, \ldots, a_k$ are $p \times p$ matrices, and $\varepsilon_t = \{\varepsilon_{1,t}, \ldots, \varepsilon_{p,t}\}'$ is a multivariate zero mean white noise process, i.e.,

$$E[\varepsilon_t \varepsilon_s'] = 0, \qquad s \neq t, \tag{9.4.2}$$

$$E[\varepsilon_t \varepsilon_t'] = \Sigma_\varepsilon, \qquad \text{say}. \tag{9.4.3}$$

The components of ε_t are univariate white noise processes, uncorrelated with each other at different time points, but possibly cross-correlated at common time points, so that Σ_ε is a general variance–covariance matrix—not necessarily of diagonal form. In the univariate case the condition for (asymptotic) stationarity is that the characteristic polynomial (corresponding to the AR operator) has all its zeros outside the unit circle (see Section 3.5.4). The corresponding condition for the multivariate case is (Hannan (1970), p. 326),

$$\text{all the zeros of } |\alpha(z)| \text{ lie outside the unit circle}, \tag{9.4.4}$$

where the matrix polynomial $\alpha(z)$ is given by,

$$\alpha(z) = \sum_{u=0}^{k} a_u z^u \qquad (a_0 = I), \tag{9.4.5}$$

and, for any square matrix A, we write $|A|$ for the determinant of A.

Assuming stationarity, we may derive the analogue of the Yule–Walker equations (cf. (3.5.43)) by multiplying both sides of (9.4.1) by X_{t-m}' and taking expectations. Denoting the covariance matrix (of lag s) for X_t by $R_X(s)$, this gives,

$$R_X(m) + a_1 R_X(m-1) + \ldots + a_k R_X(m-k) = 0, \qquad m = k, k+1, \ldots. \tag{9.4.6}$$

For the first order AR(1) model, (9.4.6) has the solution (for $s \geq 0$),

$$R_X(s) = a_1^s R_X(0). \tag{9.4.7}$$

(Compare with (3.5.16).)

Although the matrix form of the multivariate AR model looks very similar to the univariate form, the multivariate model has a more complicated structure. In particular, it should be noted that, when written out in scalar form, the equation for each $X_{i,t}$ involves not only lagged values of $X_{i,t}$ but also lagged values of all the other variables $\{X_{j,t}\}$. For example, a bivariate AR(2) model, when written out in full, takes the form,

$$X_{1,t} + \theta_{11}X_{1,t-1} + \theta_{12}X_{2,t-1} + \phi_{11}X_{1,t-2} + \phi_{12}X_{2,t-2} = \varepsilon_{1,t},$$
$$X_{2,t} + \theta_{21}X_{1,t-1} + \theta_{22}X_{2,t-1} + \phi_{21}X_{1,t-2} + \phi_{22}X_{2,t-2} = \varepsilon_{2,t} \tag{9.4.8}$$

In the above,

$$a_1 = \begin{bmatrix} \theta_{11} & \theta_{12} \\ \theta_{21} & \theta_{22} \end{bmatrix}, \qquad a_2 = \begin{bmatrix} \phi_{11} & \phi_{12} \\ \phi_{21} & \phi_{22} \end{bmatrix}.$$

Moving average models

The *p*-variate MA(*l*) model is,

$$X_t = \varepsilon_t + b_1 \varepsilon_{t-1} + b_2 \varepsilon_{t-2} + \ldots + b_l \varepsilon_{t-l}. \tag{9.4.9}$$

The condition analogous to that of Section 3.5.7 for the invertibility of the MA model (into an AR form) is,

$$\text{all the zeros of } |\boldsymbol{\beta}(z)| \text{ lie outside the unit circle,} \tag{9.4.10}$$

where

$$\boldsymbol{\beta}(z) = \sum_{u=0}^{l} b_u z^u \qquad (b_0 = I). \tag{9.4.11}$$

(As in the univariate case, there is no loss of generality in taking the coefficient of ε_t to be the identity matrix as long as Σ_ε is allowed to be a general variance–covariance matrix—see the discussion in Chapter 10 on the "identifiability" of multivariate ARMA models). Corresponding to (3.5.49), the covariance matrix of X_t is given by,

$$R_X(s) = \begin{cases} b_0 \Sigma_\varepsilon b_s' + b_1 \Sigma_\varepsilon b_{1+s}' + \ldots + b_{l-s} \Sigma_\varepsilon b_l', & 0 \leqslant s \leqslant l \\ 0, & s > l \end{cases} \tag{9.4.12}$$

As an illustration of the multivariate MA model, we write out in full a bivariate MA(2) model, as follows:

$$\begin{aligned} X_{1,t} &= \varepsilon_{1,t} + \gamma_{11} \varepsilon_{1,t-1} + \gamma_{12} \varepsilon_{2,t-1} + \delta_{11} \varepsilon_{1,t-2} + \delta_{12} \varepsilon_{2,t-2}, \\ X_{2,t} &= \varepsilon_{2,t} + \gamma_{21} \varepsilon_{1,t-1} + \gamma_{22} \varepsilon_{2,t-1} + \delta_{21} \varepsilon_{1,t-2} + \delta_{22} \varepsilon_{2,t-2}. \end{aligned} \tag{9.4.13}$$

For the above case,

$$b_1 = \begin{bmatrix} \gamma_{11} & \gamma_{12} \\ \gamma_{21} & \gamma_{22} \end{bmatrix}, \qquad b_2 = \begin{bmatrix} \delta_{11} & \delta_{12} \\ \delta_{21} & \delta_{22} \end{bmatrix}.$$

Autoregressive/Moving average models

The *p*-variate ARMA (k, l) model is,

$$X_t + a_1 X_{t-1} + \ldots + a_k X_{t-k} = \varepsilon_t + b_1 \varepsilon_t + b_1 \varepsilon_{t-1} + \ldots + b_l \varepsilon_{t-l}, \tag{9.4.14}$$

or, in operator form,

$$\boldsymbol{\alpha}(B)\boldsymbol{X}_t = \boldsymbol{\beta}(B)\boldsymbol{\varepsilon}_t, \tag{9.4.15}$$

where the matrix polynomials $\boldsymbol{\alpha}(z)$, $\boldsymbol{\beta}(z)$ are as defined by (9.4.5), (9.4.11). The model (9.4.14) is *stationary and invertible provided* $\boldsymbol{\alpha}(z)$ *and* $\boldsymbol{\beta}(z)$ *satisfy conditions* (9.4.4) *and* (9.4.10).

We saw in Section 3.5.7 that the univariate AR, MA, and ARMA models are all special cases of the general linear process. Similarly, the above models may be regarded as special cases of the *multivariate linear process*,

$$\boldsymbol{X}_t = \sum_{u=0}^{\infty} \boldsymbol{g}(u)\boldsymbol{\varepsilon}_{t-u}, \tag{9.4.16}$$

which, in turn, is a special case of the general linear relationship (9.2.6) in which $\boldsymbol{g}(u) = 0$, $u < 0$, and $\boldsymbol{Y}_t$ is replaced by the white noise process $\boldsymbol{\varepsilon}_t$.

If we write (9.4.16) as,

$$\boldsymbol{X}_t = \boldsymbol{G}(B) \cdot \boldsymbol{\varepsilon}_t, \tag{9.4.17}$$

where

$$\boldsymbol{G}(z) = \sum_{u=0}^{\infty} \boldsymbol{g}(u)z^u, \tag{9.4.18}$$

then we see that:

(a) for the AR(k) model, $\boldsymbol{G}(z) = [\boldsymbol{\alpha}(z)]^{-1}$,

(b) for the MA(l) model, $\boldsymbol{G}(z) = \boldsymbol{b}(z)$,   (9.4.19)

(c) for the ARMA(k, l) model, $\boldsymbol{G}(z) = [\boldsymbol{\alpha}(z)]^{-1}\boldsymbol{\beta}(z)$.

Note that for each of the above models the matrix polynomial, $\boldsymbol{G}(z)$, has a "*rational*" form.

The spectral matrix of $\boldsymbol{\varepsilon}_t$ takes the constant value,

$$\boldsymbol{h}_\varepsilon(\omega) = \frac{1}{2\pi}\boldsymbol{\Sigma}_\varepsilon \qquad (\text{all } \omega) \tag{9.4.20}$$

(this result following immediately on substituting (9.4.2), (9.4.3), in (9.1.48)). Also, the transfer function matrix for (9.4.16) is

$$\boldsymbol{\Gamma}(\omega) = \sum_{u=0}^{\infty} \boldsymbol{g}(u) \, e^{-i\omega u} = \boldsymbol{G}(e^{-i\omega}). \tag{9.4.21}$$

Hence, from (9.2.14) we obtain the following expression for the spectral matrix of the general linear process (9.4.16),

$$\boldsymbol{h}_X(\omega) = \frac{1}{2\pi}\boldsymbol{G}(e^{-i\omega})\boldsymbol{\Sigma}_\varepsilon\boldsymbol{G}^*(e^{-i\omega}). \tag{9.4.22}$$

Substituting the form of $G(z)$ given above, the spectral matrix of the AR(k) model (9.4.1) is thus,

$$h_X(\omega) = \frac{1}{2\pi} \alpha^{-1}(e^{-i\omega}) \Sigma_\varepsilon \{\alpha^{-1}(e^{-i\omega})\}^*. \qquad (9.4.23)$$

Similarly, the spectral matrix of the MA(l) model (9.4.9) is,

$$h_X(\omega) = \frac{1}{2\pi} \beta(e^{-i\omega}) \Sigma_\varepsilon \beta^*(e^{-i\omega}). \qquad (9.4.24)$$

(We recall that an asterisk attached to a matrix denotes both conjugation and transposition.)

Relationship with "transfer function" models

The multivariate ARMA model represents a more general version of the "transfer function" models discussed in Section 9.2.2. Consider, for example, the bivariate ARMA(k, l) model,

$$X_{1,t} + a_{11}X_{1,t-1} + a_{12}X_{2,t-1} + \ldots + a_{k1}X_{1,t-k} + a_{k2}X_{2,t-k}$$
$$= \varepsilon_{1,t} + b_{11}\varepsilon_{1,t-1} + b_{12}\varepsilon_{2,t-1} + \ldots + b_{l1}\varepsilon_{1,t-l} + b_{l2}\varepsilon_{2,t-l}, \qquad (9.4.25)$$

$$X_{2,t} + a'_{11}X_{1,t-1} + a'_{12}X_{2,t-1} + \ldots + a'_{k1}X_{1,t-k} + a'_{k2}X_{2,t-k}$$
$$= \varepsilon_{2,t} + b'_{11}\varepsilon_{1,t-1} + b'_{12}\varepsilon_{2,t-1} + \ldots + b'_{l1}\varepsilon_{1,t-l} + b'_{l2}\varepsilon_{2,t-l}. \qquad (9.4.26)$$

Writing $X_{1,t} = X_t$, $X_{2,t} = Y_t$, we now see that the transfer "function" model (9.2.36) corresponds only to the *first half* of the ARMA model, viz. (9.4.25). The full bivariate ARMA model corresponds, in effect, to a "double" transfer function model in which $X_{1,t}$ is first expressed as the output of a rational system with $X_{1,t}$ as input, and then $X_{2,t}$ is expressed as the output of a second rational system with $X_{1,t}$ as input. The basic distinction is that whereas the transfer function model represents an *"open-loop"* system, the full ARMA model represents a *"closed-loop"* system, the second equation, (9.4.26), corresponding to the *"feedback loop"*.

It would be appropriate, therefore, to use a transfer function model only when we know *a priori* that there exists a "causal" relationship between $X_{1,t}$ and $X_{2,t}$ which operates *in a particular direction* (i.e. we can identify $X_{2,t}$ say, as the "input" and $X_{1,t}$ as the "output"). However, if $X_{1,t}$ and $X_{2,t}$ are of "equal importance", and we simply wish to fit a linear relationship to them then the ARMA model would be the more appropriate one. We will return to this point in Chapter 10 when we discuss "closed-loop" systems.

Estimation of parameters

If the values of k and l (the orders of the models) are specified *a priori* then, assuming that ε_t has a multivariate normal distribution we can estimate the parameters of the multivariate AR, MA, and ARMA models by maximum likelihood, using a multivariate extension of the iterative technique discussed in Section 5.4. This approach has been developed by Akaike (1973) and Tunnicliffe-Wilson (1973)—see also Osborn (1977), Nichols (1976), and Phadke and Kedem (1978) for the multivariate MA case. However, although these models bear a *superficial* resemblance to the corresponding univariate ones, their structure is, in fact, much more complicated, and gives rise to quite deep "*identifiability*" problems. In the univariate case we can impose fairly simple conditions which ensure that a given covariance function determines a unique ARMA model. In the multivariate case these simple conditions no longer suffice, and the task of finding conditions under which a given covariance function "identifies" a unique multivariate ARMA model is one of the major problems of multivariate time series analysis.

Recently, there has been an intensive study of the structure of multivariate ARMA models (principally by Hannan and Akaike) using concepts such as "state-space representations" which are derived from linear systems theory. We therefore defer further discussion of the estimation and "identifiability" problems until Chapter 10, where we shall consider the "systems theoretic" approach in more detail.

A very general treatment of the estimation of parameters in multivariate linear models is given by Dunsmuir and Hannan (1976) and Deistler, Dunsmuir and Hannan (1976); see also Dunsmuir (1979) and Rissanen and Caines (1979). (A general treatment of univariate linear models is given in Hannan (1973).) The likelihood function of multivariate ARMA models is discussed by Hillmer and Tiao (1979), and a multivariate extension of the Box–Jenkins approach to ARMA model fitting is described by Tiao and Box (1979).

Multivariate order determination procedures

Akaike's AIC criterion may be extended to the case of order determination for multivariate linear models. Thus, if a p-variate AR(k) model is fitted to N observations, the AIC takes the form (see, e.g., Jones (1974)),

$$\text{AIC}(k) = N \log|\hat{\mathbf{\Sigma}}_\varepsilon^{(k)}| + 2p^2 k, \qquad (9.4.27)$$

where $\hat{\mathbf{\Sigma}}_\varepsilon^{(k)}$, the estimated variance–covariance matrix of the residuals, ε_t, may be computed as

$$\hat{\boldsymbol{\Sigma}}_\varepsilon^{(k)} \doteq \sum_{i=0}^{k} \hat{a}_i \hat{\boldsymbol{R}}_X(i) \qquad (9.4.28)$$

(cf. (5.4.22)), $\hat{\boldsymbol{R}}_X(i)$ being the sample and autocovariance matrices (as given by (9.5.3)), and $\hat{a}_i$, $(\hat{a}_0 = \boldsymbol{I})$, the estimated coefficients obtained by substituting $\hat{\boldsymbol{R}}_X(i)$ for $\boldsymbol{R}_X(i)$ in the Yule–Walker equations (9.4.6).

The multivariate form of Parzen's CAT criterion is given (Parzen (1977b)) by,

$$\text{CAT}(k) = \text{tr}\left[\left\{\frac{p}{N}\sum_{j=1}^{k} (\tilde{\boldsymbol{\Sigma}}_\varepsilon^{(j)})^{-1}\right\} - (\tilde{\boldsymbol{\Sigma}}^{(k)})^{-1}\right], \qquad (9.4.29)$$

where

$$\tilde{\boldsymbol{\Sigma}}_\varepsilon^{(j)} = \left(\frac{N}{N-pj}\right)\hat{\boldsymbol{\Sigma}}_\varepsilon^{(j)},$$

$\hat{\boldsymbol{\Sigma}}_\varepsilon^{(j)}$ being the estimated variance–covariance matrix of ε_t based on a fitted AR(j) model. Parzen (1977b) gives a comparative study of the multivariate AIC and CAT criteria.

As noted in Section 5.4.5, Shibata (1976) considered the properties of the estimated order derived from the AIC criterion applied to univariate processes, and showed that if the true model is AR(k_0), the estimate $\hat{k}$ determined by minimizing AIC(k), is not a consistent estimate of k_0, but tends to overestimate k_0 for large samples. (Similarly, the estimated order derived from the CAT criterion is not a consistent estimate of k_0.) Hannan (1979a, 1980) therefore proposed a modified form of Akaike's criterion which, for a fitted multivariate ARMA(k, l) model, takes the form

$$\phi = \log|\hat{\boldsymbol{\Sigma}}_\varepsilon| + (k+l)C_N/N, \qquad (9.4.30)$$

where C_N is a sequence which $\to \infty$ as $N \to \infty$. Assuming that the true model is a (multivariate) ARMA(k_0, l_0), Hannan shows that, under general conditions on the model, the estimated order $(\hat{k}, \hat{l})$ obtained by minimizing ϕ converges in probability to (k_0, l_0) provided only that the sequence C_N is chosen so that $C_N \to \infty$ and $N \to \infty$. (Dividing the AIC criterion by N, we obtain an equivalent criterion which corresponds to Hannan's ϕ with $C_N = 2p^2$, which does not, of course, satisfy the above condition.) Moreover, under general conditions on the model, Hannan shows further that $(\hat{k}, \hat{l})$ converges almost surely to (k_0, l_0) if

$$\liminf_{N \to \infty} \{C_N/\log N\} > 0,$$

while under somewhat stronger conditions on the model, almost sure convergence holds if

$$\liminf_{N \to \infty} \{C_N / 2 \log \log N\} > 1.$$

Hannan and Quinn (1979) consider the univariate AR(k) case, and use the criterion

$$\phi = \log \hat{\sigma}_\epsilon^2 + (2kc \log \log N)/N, \qquad (9.4.31)$$

and show that this leads to a strongly consistent estimate of k, provided that the constant c is chosen so that $c > 1$. They show that asymptotically this procedure underestimates the true order to a lesser degree than other procedures based on $C_N = O(\log N)$, and this result is confirmed by their analysis of some simulated AR(1) models based on the above form of ϕ with $c = 1$.

9.5 ESTIMATION OF CROSS-SPECTRA

We now consider the estimation of auto- and cross-spectra from finite samples. Thus, suppose we are given N observations on the (real valued) multivariate stationary process, $\boldsymbol{X}_t = \{X_{1,t}, X_{2,t}, \ldots, X_{p,t}\}'$, say for $t = 1, \ldots, N$. We may, of course, estimate $h_{ii}(\omega)$, the (auto) spectral density function of $X_{i,t}$, by the general methods described in Chapters 6 and 7, i.e. we construct an estimate of the form (cf. (7.1.1)),

$$\hat{h}_{ii}(\omega) = \frac{1}{2\pi} \sum_{s=-(N-1)}^{(N-1)} \lambda_N(s) \hat{R}_{ii}(s) e^{-i\omega s}, \qquad (9.5.1)$$

where $\lambda_N(s)$ is a suitable covariance lag window, of the type discussed in Section 6.2.3, and $\hat{R}_{ii}(s)$, the sample autocovariance function of $X_{i,t}$, is given by (cf. (7.1.2)),

$$\hat{R}_{ii}(s) = \frac{1}{N} \sum_{t=1}^{N-|s|} (X_{i,t} - \bar{X}_i)(X_{i,t+|s|} - \bar{X}_i), \qquad s = 0, \pm 1, \ldots \pm (N-1),$$

$$(9.5.2)$$

with

$$\bar{X}_i = \frac{1}{N} \sum_{t=1}^{N} X_{i,t}$$

denoting the sample mean of $X_{i,t}$. If we use the same approach to estimate $h_{ij}(\omega)$, the cross-spectral density function between $X_{i,t}$ and $X_{j,t}$, then we must construct an estimate of the cross-covariance function

$R_{ij}(s)$. A natural extension of (9.5.2) gives,

$$\hat{R}_{ij}(s) = \frac{1}{N} \sum_t (X_{j,t} - \bar{X}_j)(X_{i,t+s} - \bar{X}_i), \qquad s = 0, \pm 1, \dots \pm(N-1), \quad (9.5.3)$$

where the summation goes from $t = 1$ to $(N - s)$, $s \geq 0$, and from $t = (1 - s)$ to N, $s < 0$. It is easily seen that $\hat{R}_{ij}(s)$ is an asymptotically unbiased estimate of $R_{ij}(s)$, with

$$E[\hat{R}_{ij}(s)] = R_{ij}(s) + O(1/N), \qquad (9.5.4)$$

and a similar argument to that used in Section 5.3.3 shows that, when X_t is Gaussian process, we have for large N,

$$\text{cov}[\hat{R}_{ij}(s), \hat{R}_{ij}(u)] \sim \frac{1}{N} \left[\sum_{r=-\infty}^{\infty} R_{ii}(r)R_{jj}(r + u - s) + R_{ij}(r + u)R_{ji}(r - s) \right]$$

$$(9.5.5)$$

(compare with (5.3.24)).

We may now estimate the cross-spectral density function by

$$\hat{h}_{ij}(\omega) = \frac{1}{2\pi} \sum_{s=-(N-1)}^{(N-1)} \lambda_N(s)\hat{R}_{ij}(s)\, e^{-i\omega s}. \qquad (9.5.6)$$

At this stage we may observe that there is really no reason why the covariance lag window $\lambda_N(s)$ in (9.5.6) should be the same as that in (9.5.1), or why $\lambda_N(s)$ should be the same for all the auto-spectra, $h_{ii}(\omega)$. We recall from the discussion of Section 7.3.2 that in (9.5.1) $\lambda_N(s)$ should ideally, be chosen to "match" the rate of decay of $R_{ii}(s)$, or equivalently, so that the bandwidth of the corresponding spectral window $W_N(\theta)$ "matches" the bandwidth of $h_{ii}(\omega)$. Now for a general multivariate process there is no reason why the auto-spectra of the various components should all have equal (or comparable) bandwidth, and we certainly have no grounds for assuming that the "bandwidths" of the cross-spectra are comparable with the bandwidths of the auto-spectra. (As we shall see later when we consider the technique of "alignment", this latter assumption would be highly dubious.) Thus, even if we decide to retain the same mathematical form of $\lambda_N(s)$ for all the spectral estimates, it would, in general, be sensible to allow, e.g., different truncation points for the different auto- and cross-spectral estimates. In other words, we should allow $\lambda_N(s)$ to depend on i and j, and in (9.5.6) write more precisely, $\lambda_N(s) = \lambda_N^{(i,j)}(s)$. Whilst this is quite a feasible procedure in practice, it simplifies the study of the sampling properties of the spectral estimates if we assume that all estimates are based on a common lag window, and in the following discussion we take $\lambda_N(s)$ to be the same (in terms of both its mathematical form and truncation point) for all the auto- and cross-spectra.

Noting that (9.5.6) reduces to (9.5.1) when $i = j$, we may take (9.5.6) as the general form of estimate for all elements of the spectral matrix, $h(\omega)$, and write it more concisely as,

$$\hat{h}(\omega) = \frac{1}{2\pi} \sum_{s=-(N-1)}^{(N-1)} \lambda_N(s)\hat{R}(s)\, e^{-i\omega s}, \qquad (9.5.7)$$

where $\hat{R}(s) = \{\hat{R}_{ij}(s)\}$ is the *sample covariance matrix of lag s*. The treatment of multivariate spectral estimation now follows the univariate case very closely. Many of the univariate formulae have obvious multivariate versions, the detailed derivations being almost the same (with the familiar change of scalar quantities to matrices, where appropriate). Whenever possible we will therefore omit the details of the new derivations and simply refer back to the corresponding univariate results. Thus, we introduce the (modified) *periodogram matrix*, $I_N(\omega)$, defined by,

$$I_N(\omega) = \{I_{N,ij}(\omega)\}, \qquad i, j = 1, \ldots p,$$

where

$$I_{N,ij}(\omega) = \zeta_{X_i}(\omega)\zeta^*_{X_j}(\omega), \qquad (9.5.8)$$

with

$$\zeta_{X_i}(\omega) = \frac{1}{\sqrt{2\pi N}} \sum_{t=1}^{N} X_{i,t}\, e^{-i\omega t}, \qquad i = 1, \ldots p,$$

denoting the *finite Fourier transform* of $X_{i,t}$. For $i = j$, $I_{N,ii}$ is simply the periodogram of $X_{i,t}$; for $i \neq j$, $I_{N,ij}$ is the *cross-periodogram* between $X_{i,t}$ and $X_{j,t}$. (Note that in the univariate case we denoted the modified periodogram by $I^*_N(\omega)$—cf. (6.2.4). However, in the multivariate case we drop the asterisk since this might be erroneously interpreted as indicating conjugation and transposition.) For computational purposes the cross-periodogram may be written as,

$$I_{N,ij}(\omega) = \{A_{X_i}(\omega)A_{X_j}(\omega) + B_{X_i}(\omega)B_{X_j}(\omega)\}$$
$$- i\{B_{X_i}(\omega)A_{X_j}(\omega) - A_{X_i}(\omega)B_{X_j}(\omega)\}, \qquad (9.5.9)$$

where, for $i = 1, \ldots p$,

$$A_{X_i}(\omega) = \frac{1}{\sqrt{2\pi N}} \sum_{t=1}^{N} X_{i,t} \cos \omega t,$$

$$B_{X_i}(\omega) = \frac{1}{\sqrt{2\pi N}} \sum_{t=1}^{N} X_{i,t} \sin \omega t.$$

In the same way that $I_{N,ii}(\omega)$ represents the "raw" (i.e. unsmoothed) sample version of the auto-spectrum of $X_{i,t}$, so $I_{N,ij}(\omega)$ represents the "raw" sample cross-spectrum between $X_{i,t}$ and $X_{j,t}$. Similarly, the real part and the negative of the imaginary parts of $I_{N,ij}(\omega)$ represent the "raw" sample co-spectrum and quadrature spectrum, respectively.

We may re-write $I_N(\omega)$ in the form (cf. (6.2.5)),

$$I_N(\omega) = \frac{1}{2\pi} \sum_{s=-(N-1)}^{(N-1)} \hat{R}(s) e^{-i\omega s}, \qquad (9.5.10)$$

so that,

$$\hat{R}(s) = \int_{-\pi}^{\pi} I_N(\omega) e^{i\omega s} d\omega, \qquad (9.5.11)$$

and (9.5.7) can now be written in the alternative form (cf. (6.2.56)),

$$\hat{h}(\omega) = \int_{-\pi}^{\pi} I_N(\theta) W_N(\omega - \theta) d\theta, \qquad (9.5.12)$$

where

$$W_N(\theta) = \frac{1}{2\pi} \sum_{s=-(N-1)}^{(N-1)} \lambda_N(s) e^{-i\omega s} \qquad (9.5.13)$$

is the spectral window corresponding to $\lambda_N(s)$. By analogy with (7.6.18) we may, for large N, replace (9.5.12) by the discrete approximation,

$$\hat{h}(\omega) = \frac{2\pi}{N} \sum_p I_N(\omega_p) W_N(\omega - \omega_p), \qquad (9.5.14)$$

where $\omega_p = 2\pi p/N$. The elements of $\hat{h}(\omega)$ can now be evaluated by using the "*fast Fourier transform*" algorithm (Section 7.6) to evaluate the elements of $I_N(\omega_p)$. In particular, if $W_N(\theta)$ is a Daniell (i.e. rectangular) window with bandwidth $4\pi m/N$, then (9.5.14) reduces effectively to

$$\hat{h}(\omega) = \frac{1}{(2m+1)} \sum_{j=-m}^{m} I_N\left[\frac{2\pi\{p(\omega)+j\}}{N} \right], \qquad (9.5.15)$$

where $p(\omega)$ is that integer such that $2\pi p(\omega)/N$ is closest to ω. (Compare with (7.6.20)). Here, each diagonal element of $h(\omega)$ is computed by averaging $(2m+1)$ periodogram ordinates, so that the equivalent number of "degrees of freedom" is (approximately) $4m$; see Section 7.6. If $h(\omega)$ may be assumed to be roughly constant over the bandwidth of $W_N(\theta)$ then, from (9.5.15), $(2m+1)\hat{h}(\omega)$ is asymptotically equivalent to the sum of $(2m+1)$ independent matrices, $I_N(\omega_p) = \zeta(\omega_p)\zeta^*(\omega_p)$, where $\zeta(\omega_p)$ has a multivariate complex normal distribution with zero mean and variance–covariance matrix $h(\omega_p)$. Hence, $(2m + 1)\hat{h}(\omega)$ has a distribution which is closely related to the distribution of a sample variance–covariance matrix. Now the sample variance–covariance matrix formed from real valued normal variables has a known distribution called the "*Wishart distribution*"—see, e.g. Kendall and Stuart (1966). (Strictly, the Wishart distribution refers to the joint distribution of the distinct elements of the variance–covariance matrix, and is, in effect, the multivariate analogue of the chi-squared distribution.) Goodman (1957), in a pioneering study of multivariate spectral estimates, introduced a more general version called the "*complex Wishart distribution*", and used it as an approximation for the distribution of the estimated spectral matrix. Specifically, Goodman suggested that, for $\omega \neq 0$, π, the distribution of $(2m + 1)\hat{h}(\omega)$ may be approximated by the complex Wishart distribution with parameters $(2m + 1)$, $h(\omega)$, denoted by $W^c(2m + 1, h(\omega))$. More generally, we may apply the same approximate distribution for the estimate (9.5.14) where the weighting is non-uniform, the approximation being similar in nature to that involved in using a chi-squared distribution for general univariate spectral estimators; see Section 6.2.4 and the discussion following equation (6.2.127). Rigorous studies of the use of the complex Wishart distribution as an approximation to the distribution of $(2m + 1)\hat{h}(\omega)$ have been given by Wahba (1968) and Brillinger (1969a).

Sampling properties of spectral estimates

From (9.5.10) and (9.1.48) we have, for large N,

$$E[I_N(\omega)] \sim h(\omega),$$

so that (9.5.12) gives,

$$E[\hat{h}(\omega)] \sim \int_{-\pi}^{\pi} h(\theta) W_N(\omega - \theta) \, d\theta. \tag{9.5.16}$$

Making the usual assumption that $W_N(\theta) \to \delta(\theta)$ as $N \to \infty$, it follows from (9.5.16) that at all continuity points,

$$E[\hat{h}(\omega)] \sim h(\omega), \tag{9.5.17}$$

i.e. each element of $\hat{h}(\omega)$ is an asymptotically unbiased estimate of the corresponding element of $h(\omega)$.

Expressions for the asymptotic variances and covariances of the elements of $\hat{h}(\omega)$ can be derived by following the same approach as used in the univariate case; see Section 6.2.4. Here, we assume that X_t is a general linear process of the form (9.4.16) and use the asymptotic relationship between the periodogram matrices of X_t and ε_t. Thus, assuming that ε_t is an independent process with finite fourth order moments, and that the coefficients in (9.4.16) satisfy the condition,

$$\sum_{u=0}^{\infty} \|\mathbf{g}(u)\| \cdot |u|^{\alpha} < \infty \tag{9.5.18}$$

(where, for any matrix A, $\|A\|$ denotes the "norm" of A, which may be taken as the square root of the largest eigenvalue of AA^*), a similar argument to that of Theorem 6.2.2 gives (cf. Hannan (1970, p. 248)),

$$\mathbf{I}_{N,X}(\omega) = \mathbf{G}(e^{-i\omega})\mathbf{I}_{N,\varepsilon}(\omega)\mathbf{G}^*(e^{-i\omega}) + \mathbf{R}_N(\omega), \tag{9.5.19}$$

where each element of $\mathbf{R}_N(\omega)$ has mean square $O(1/N^{2\alpha})$. The second order properties of $\mathbf{I}_{N,X}(\omega)$ can now be derived from the corresponding properties of $\mathbf{I}_{N,\varepsilon}(\omega)$, and using (9.5.12), the properties of $\hat{\mathbf{h}}(\omega)$ follow from those of $\mathbf{I}_{N,X}(\omega)$. Denoting the (i, j)th element of $\mathbf{I}_{N,\varepsilon}(\omega)$ by $I_{ij}^{(\varepsilon)}(\omega)$, a straightforward calculation similar to that of Theorem 6.1.3 shows that

$$\mathrm{cov}\{I_{ij}^{(\varepsilon)}(\omega_1), I_{kl}^{(\varepsilon)}(\omega_2)\}$$

$$= \frac{1}{2\pi N}\sigma_{ik}\sigma_{jl}F_N(\omega_1 - \omega_2) + \frac{1}{2\pi N}\sigma_{il}\sigma_{jk}F_N(\omega_1 + \omega_2) + \frac{\kappa_{ijkl}}{N}, \tag{9.5.20}$$

where $F_N(\theta)$ is the Fejer kernel defined by (6.1.43), σ_{ij} is the (i, j)th element of Σ_ε, i.e. $\sigma_{ij} = \mathrm{cov}\{\varepsilon_{i,t}, \varepsilon_{j,t}\}$, and κ_{ijkl} is the fourth cumulant between $\varepsilon_{i,t}$, $\varepsilon_{j,t}$, $\varepsilon_{k,t}$, and $\varepsilon_{l,t}$. (Compare (9.5.20) with (6.1.53).) If $\omega_1 \neq \pm\omega_2$ then all the above terms converge to zero as $N \to \infty$. If $\omega_1 = \omega_2$, the first term is $\sigma_{ik}\sigma_{jl}/4\pi^2$, while the other two terms $\to 0$ as $N \to \infty$, while if $\omega_1 = \omega_2 = 0$ or $\pm\pi$, both the first two terms give non-zero contributions. Neglecting the third term in (9.5.20) we have, as $N \to \infty$,

$$\mathrm{cov}\{I_{ij}^{(\varepsilon)}(\omega_1), I_{kl}^{(\varepsilon)}(\omega_2)\}$$

$$\sim \begin{cases} 0, & \omega_1 \neq \omega_2, \\ (1/4\pi)\sigma_{ik}\sigma_{jl}, & \omega_1 = \omega_2 \neq 0, \pm\pi, \\ (1/4\pi)(\sigma_{ik}\sigma_{jl} + \sigma_{il}\sigma_{jk}), & \omega_1 = \omega_2 = 0, \pm\pi. \end{cases} \tag{9.5.21}$$

Note that, apart from the third term in (9.5.20), the above results are exact if both ω_1 and ω_2 are of the form $2\pi p/N$. The corresponding result for the elements of $\mathbf{I}_{N,X}(\omega)$ can now be derived as follows. Writing $\gamma_{ij}(\omega)$ for the (i, j)th element of $\mathbf{G}(e^{-i\omega})$, we have from (9.5.19) (neglecting the remainder term, $\mathbf{R}_N(\omega)$),

$$I_{ij}^{(X)}(\omega) \sim \sum_p \sum_q \gamma_{ip}(\omega)\gamma_{jq}^*(\omega)I_{pq}^{(\varepsilon)}(\omega).$$

Hence,

$$\text{cov}\{I_{ij}^{(X)}(\omega_1), I_{kl}^{(X)}(\omega_2)\} \sim \sum_p \sum_q \sum_m \sum_n \gamma_{ip}(\omega_1)\gamma_{jq}^*(\omega_1)\gamma_{km}^*(\omega_2)\gamma_{ln}(\omega_2)$$
$$\times \text{cov}\{I_{pq}^{(\varepsilon)}(\omega_1), I_{mn}^{(\varepsilon)}(\omega_2)\}.$$

Substituting the expression for $\text{cov}\{I_{pq}^{(\varepsilon)}(\omega_1), I_{mn}^{(\varepsilon)}(\omega_2)\}$ given by (9.5.20) (and neglecting the third term) yields,

$$\text{cov}\{I_{ij}^{(X)}(\omega_1), I_{kl}^{(X)}(\omega_2)\}$$

$$\sim \frac{2\pi}{N}F_N(\omega_1-\omega_2)\left\{\frac{1}{2\pi}\sum_p \sum_m \gamma_{ip}(\omega_1)\sigma_{pm}\gamma_{mk}^*(\omega_2)\right\}\left\{\frac{1}{2\pi}\sum_q \sum_n \gamma_{jq}^*(\omega_1)\sigma_{qn}\gamma_{nl}(\omega_2)\right\}$$

$$+\frac{2\pi}{N}F_N(\omega_1+\omega_2)\left\{\frac{1}{2\pi}\sum_p \sum_n \gamma_{ip}(\omega_1)\sigma_{pn}\gamma_{nl}^*(\omega_2)\right\}\left\{\frac{1}{2\pi}\sum_q \sum_m \gamma_{jq}^*(\omega_1)\sigma_{qm}\gamma_{mk}(\omega_2)\right\}.$$

$$(9.5.22)$$

Now if ω_1 and ω_2 are sufficiently close together so that $\gamma_{mk}(\omega_2) \doteq \gamma_{mk}(\omega_1)$, the first factor in the first term in (9.5.22) is the (i, k)th element of the matrix product,

$$\frac{1}{2\pi}\boldsymbol{G}(e^{-i\omega_1})\boldsymbol{\Sigma}_\varepsilon \boldsymbol{G}^*(e^{-i\omega_1}),$$

and from (9.4.22) this is $h_{ik}(\omega_1)$, the (i, k)th element of $\boldsymbol{h}_X(\omega_1)$. Evaluating the remaining terms in (9.5.22) in a similar way we obtain for $|\omega_1-\omega_2|$ small,

$$\text{cov}\{I_{ij}^{(X)}(\omega_1), I_{kl}^{(X)}(\omega_2)\}$$

$$\sim \frac{2\pi}{N}F_N(\omega_1-\omega_2)h_{ik}(\omega_1)h_{jl}^*(\omega_1)+\frac{2\pi}{N}F_N(\omega_1+\omega_2)h_{il}(\omega_1)h_{jk}^*(\omega_1).$$

$$(9.5.23)$$

This result shows, in particular, that

$$\text{cov}\{I_{ij}^{(X)}(\omega_1), I_{kl}^{(X)}(\omega_2)\} \sim 0, \qquad \omega_1 \neq \pm\omega_2, \qquad (9.5.24)$$

and

$$\text{cov}\{I_{ij}^{(X)}(\omega), I_{kl}^{(X)}(\omega)\}$$

$$\sim \begin{cases} h_{ik}(\omega)h_{jl}^*(\omega), & \omega \neq 0, \pm\pi, \\ h_{ik}(\omega)h_{jl}^*(\omega)+h_{il}(\omega)h_{jk}^*(\omega), & \omega = 0, \pm\pi. \end{cases} \qquad (9.5.25)$$

Having established (9.5.22), the remainder of the analysis is almost an exact duplicate of the treatment for the univariate case (see the proof of Theorem 6.2.4). Thus, we write from (9.5.12),

$$\text{cov}\{\hat{h}_{ij}(\omega_1), \hat{h}_{kl}(\omega_2)\}$$

$$= \int_{-\pi}^{\pi}\int_{-\pi}^{\pi} W_N(\omega_1-\theta)W_N(\omega_2-\phi)\,\text{cov}\{I_{ij}^{(X)}(\theta), I_{kl}^{(X)}(\phi)\}\,d\theta\,d\phi,$$

and substitute (9.5.22) for the covariance term.

Provided the spectral window $W_N(\theta)$ satisfies the conditions (i)–(v) *of Section* 6.2.4, the effect of the Fejer kernel in the first term (9.5.22) will be to "collapse" the double integral into a single integral along the line $\theta = \phi$, and we can then use (9.5.25) for the covariance between $I_{ij}^{(X)}(\theta)$, $I_{kl}^{(X)}(\theta)$. Treating the second term in (9.5.22) in a similar way, and assuming that all the elements of $\boldsymbol{h}_X(\omega)$ are approximately constant over the bandwidth of $W_N(\theta)$ (so that they may be taken outside the integrals) we finally obtain for $|\omega_1 - \omega_2|$ small,

$$\text{cov}\{\hat{h}_{ij}(\omega_1), \hat{h}_{kl}(\omega_2)\}$$

$$\sim \frac{2\pi}{N} h_{ik}(\omega_1)h_{jl}^*(\omega_1) \int_{-\pi}^{\pi} W_N(\omega_1-\theta)W_N(\omega_2-\theta)\,d\theta$$

$$+ \frac{2\pi}{N} h_{il}(\omega_1)h_{jk}^*(\omega_1) \int_{-\pi}^{\pi} W_N(\omega_1-\theta)W_N(\omega_2+\theta)\,d\theta. \qquad (9.5.26)$$

Since $W_N(\omega_1-\theta)$, $W_N(\omega_2-\theta)$, converge to δ-functions centred on ω_1, ω_2 respectively, each term in (9.5.26) tends to zero if ω_1 and ω_2 are distinct fixed frequencies, i.e.

$$\text{cov}\{\hat{h}_{ij}(\omega_1), \hat{h}_{kl}(\omega_2)\} \sim 0, \qquad \omega_1 \neq \pm\omega_2.$$

On the other hand, if $\omega_1 = \omega_2 = \omega \neq 0, \pm\pi$, the second term is negligible, and writing,

$$C_W = 2\pi \int_{-\pi}^{\pi} W_N^2(\omega-\theta)\,d\theta = 2\pi \int_{-\pi}^{\pi} W_N^2(\theta)\,d\theta, \qquad (9.5.27)$$

we have,

$$\text{cov}\{\hat{h}_{ij}(\omega), \hat{h}_{kl}(\omega)\} \sim \frac{C_W}{N} h_{ik}(\omega)h_{jl}^*(\omega), \qquad \omega \neq 0, \pm\pi. \qquad (9.5.28)$$

When $\omega_1 = \omega_2 = \omega = 0, \pm\pi$, both the integrals in (9.5.26) are the same, and (9.5.28) becomes,

$$\text{cov}\{\hat{h}_{ij}(\omega), \hat{h}_{kl}(\omega)\} \sim \frac{C_W}{N}\{h_{ik}(\omega)h_{jl}^*(\omega) + h_{il}(\omega)h_{jk}^*(\omega)\}, \qquad \omega = 0, \pm\pi.$$

$$(9.5.29)$$

The following special cases of (9.5.28) are of interest.

(i) Setting $i = k$, $j = l$, we obtain the following expression for the variance of a cross-spectral estimate,

$$\text{var}\{\hat{h}_{ij}(\omega)\} \sim \frac{C_W}{N} h_{ii}(\omega) h_{jj}(\omega), \qquad \omega \neq 0, \pm \pi. \qquad (9.5.30)$$

(ii) Setting $i = j$, $k = l$, we obtain the covariance between two different auto-spectral estimates, namely,

$$\text{cov}\{\hat{h}_{ii}(\omega), \hat{h}_{kk}(\omega)\} \sim \frac{C_W}{N} h_{ik}(\omega) h_{ik}^*(\omega)$$

$$= \frac{C_W}{N} |h_{ik}(\omega)|^2, \qquad \omega \neq 0, \pm \pi. \qquad (9.5.31)$$

(iii) Setting $i = j = k = l$, we obtain the variance of a single auto-spectral estimate,

$$\text{var}\{\hat{h}_{ii}(\omega)\} \sim \frac{C_W}{N} h_{ii}^2(\omega), \qquad (9.5.32)$$

while setting $i = j = k = l$ in (9.5.29) gives,

$$\text{var}\{\hat{h}_{ii}(\omega)\} \sim \frac{2C_W}{N} h_{ii}^2(\omega), \qquad \omega = 0, \pm \pi.$$

These results are, of course, identical with the results previously derived in Chapter 6—see equation (6.2.108). Note that C_W may be written alternatively as (cf. (6.2.109))

$$C_W = \sum_{s=-(N-1)}^{(N-1)} \{\lambda_N(s)\}^2, \qquad (9.5.33)$$

or, for scale parameter windows for which $\lambda_N(s) = k(s/M)$ (see Section 6.2.4), we have,

$$C_W = M \int_{-\infty}^{\infty} k^2(u) \, du = M I_W, \qquad (9.5.34)$$

where I_W is as defined in (7.4.3). (Values of I_W for the standard windows are given in Table 6.1.) For the estimate (9.5.15), we have,

$$(C_W/N) = 1/(2m+1).$$

The above results may be summarized by setting out the asymptotic variance–covariance matrix of $\hat{h}_{ii}(\omega)$, $\hat{h}_{jj}(\omega)$, $\hat{h}_{ij}(\omega)$, $\hat{h}_{ji}(\omega)$, viz. (for

$\omega \neq 0, \pm \pi)$,

	$\hat{h}_{ii}(\omega)$	$\hat{h}_{jj}(\omega)$	$\hat{h}_{ij}(\omega)$	$\hat{h}_{ji}(\omega)$
$\hat{h}_{ii}(\omega)$	$h_{ii}^2(\omega)$	$\|h_{ij}(\omega)\|^2$	$h_{ii}(\omega)h_{ij}^*(\omega)$	$h_{ii}(\omega)h_{ji}^*(\omega)$
$\hat{h}_{jj}(\omega)$	$\|h_{ij}(\omega)\|^2$	$h_{jj}^2(\omega)$	$h_{jj}(\omega)h_{ij}^*(\omega)$	$h_{jj}(\omega)h_{ji}^*(\omega)$
$\hat{h}_{ij}(\omega)$	$h_{ii}(\omega)h_{ij}^*(\omega)$	$h_{jj}(\omega)h_{ij}(\omega)$	$h_{ii}(\omega)h_{jj}(\omega)$	$\{h_{ij}(\omega)\}^2$
$\hat{h}_{ji}(\omega)$	$h_{ii}(\omega)h_{ji}(\omega)$	$h_{jj}(\omega)h_{ji}(\omega)$	$\{h_{ij}^*(\omega)\}^2$	$h_{ii}(\omega)h_{jj}(\omega)$

The matrix is premultiplied by $\dfrac{C_W}{N}$.

$$(9.5.35)$$

9.5.1 Estimation of Co-spectra and Quadrature Spectra

Once the spectral matrix $h(\omega) = \{h_{ij}(\omega)\}$ has been estimated we can construct estimates of all the associated spectral functions (such as co- and quadrature spectra) by replacing the elements of $h(\omega)$ by the corresponding elements of $\hat{h}(\omega)$ in the definitions of these functions. Thus, we may estimate the co-spectrum $c_{ij}(\omega)$ and the quadrature spectrum $q_{ij}(\omega)$, by

$$\hat{c}_{ij}(\omega) = \mathcal{R}\{\hat{h}_{ij}(\omega)\} = \frac{1}{4\pi} \sum_{s=-(N-1)}^{(N-1)} \lambda_N(s)\{\hat{R}_{ij}(s) + \hat{R}_{ij}(-s)\} \cos s\omega, \quad (9.5.36)$$

$$\hat{q}_{ij}(\omega) = -\mathcal{I}\{\hat{h}_{ij}(\omega)\} = \frac{1}{4\pi} \sum_{s=-(N-1)}^{(N-1)} \lambda_N(s)\{\hat{R}_{ij}(s) - \hat{R}_{ij}(-s)\} \sin s\omega, \quad (9.5.37)$$

where $\lambda_N(s)$ is the same covariance lag window used in (9.5.6), and $\hat{R}_{ij}(s)$ is the sample cross-covariance function given by (9.5.3). Alternatively, $\hat{c}_{ij}(\omega)$ may be computed from (9.5.14) by writing

$$\hat{c}_{ij}(\omega) = \sum_p \mathcal{R}\{I_{N,ij}(\omega_p)\} W_N(\omega - \omega_p), \quad (9.5.38)$$

the real part of $I_{N,ij}(\omega_p)$ being given by (9.5.9). Replacing $\mathcal{R}\{I_{N,ij}(\omega_p)\}$ by $-\mathcal{I}\{I_{N,ij}(\omega_p)\}$ in (9.5.38) then gives $\hat{q}_{ij}(\omega)$.

The variances and covariances of $\hat{h}_{ii}(\omega)$, $\hat{h}_{jj}(\omega)$, $\hat{c}_{ij}(\omega)$, $\hat{q}_{ij}(\omega)$, are easily derived from (9.5.28) by writing,

$$\hat{c}_{ij}(\omega) = \frac{1}{2}\{\hat{h}_{ij}(\omega) + \hat{h}_{ij}^*(\omega)\} = \frac{1}{2}\{\hat{h}_{ij}(\omega) + \hat{h}_{ji}(\omega)\} \quad (9.5.39)$$

and

$$\hat{q}_{ij}(\omega) = \frac{1}{2i}\{\hat{h}_{ij}^*(\omega) - \hat{h}_{ij}(\omega)\} = \frac{1}{2i}\{\hat{h}_{ji}(\omega) - \hat{h}_{ij}(\omega)\}. \quad (9.5.40)$$

This gives, for example,

$$\operatorname{cov}\{\hat{h}_{ii}(\omega), \hat{q}_{ij}(\omega)\} = \frac{1}{2i} \operatorname{cov}\{\hat{h}_{ii}(\omega), \hat{h}_{ij}(\omega)\} - \frac{1}{2i} \operatorname{cov}\{\hat{h}_{ii}(\omega), \hat{h}_{ji}(\omega)\}$$

$$\sim \frac{1}{2i}[h_{ii}(\omega)h_{ij}^*(\omega) - h_{ii}(\omega)h_{ji}^*(\omega)],$$

$$\omega \neq 0, \pm \pi \text{ (from (9.5.35))},$$

$$= h_{ii}(\omega)q_{ij}(\omega). \tag{9.5.41}$$

Evaluating the other terms by similar methods, we obtain the asymptotic variance–covariance matrix of $\hat{h}_{ii}(\omega)$, $\hat{h}_{jj}(\omega)$, $\hat{c}_{ij}(\omega)$, $\hat{q}_{ij}(\omega)$, as (for $\omega \neq 0, \pm \pi$),

	$\hat{h}_{ii}(\omega)$	$\hat{h}_{jj}(\omega)$	$\hat{c}_{ij}(\omega)$	$\hat{q}_{ij}(\omega)$
$\hat{h}_{ii}(\omega)$	$h_{ii}^2(\omega)$	$\lvert h_{ij}(\omega)\rvert^2$	$h_{ii}(\omega)c_{ij}(\omega)$	$h_{ii}(\omega)q_{ij}(\omega)$
$\hat{h}_{jj}(\omega)$		$h_{jj}^2(\omega)$	$h_{jj}(\omega)c_{ij}(\omega)$	$h_{jj}(\omega)q_{ij}(\omega)$
$\hat{c}_{ij}(\omega)$			$\frac{1}{2}\{h_{ii}(\omega)h_{jj}(\omega) + c_{ij}^2(\omega) - q_{ij}^2(\omega)\}$	$c_{ij}(\omega)q_{ij}(\omega)$
$\hat{q}_{ij}(\omega)$				$\frac{1}{2}\{h_{ii}(\omega)h_{jj}(\omega) + q_{ij}^2(\omega) - c_{ij}^2(\omega)\}$

with overall factor $\dfrac{C_W}{N}$.

$$\tag{9.5.42}$$

(Since $h_{ii}(\omega)$, $h_{jj}(\omega)$, $c_{ij}(\omega)$, $q_{ij}(\omega)$ are all real valued variables, the above variance–covariance matrix is symmetric, and hence we have omitted the terms below the diagonal). Also it should be emphasized that the above expressions do not hold for $\omega = 0, \pm \pi$. In particular, when $\omega = 0$ both $\hat{q}_{ij}(\omega)$ and $q_{ij}(\omega)$ are zero. The above variance–covariance matrix was first derived by Goodman (1957); see also Jenkins (1963).

9.5.2 Estimation of Cross-amplitude and Phase Spectra

Following the definitions (see (9.1.32), (9.1.33)), we may estimate the cross-amplitude spectrum $\alpha_{ij}(\omega)$ and the phase spectrum $\phi_{ij}(\omega)$ (between $X_{i,t}$ and $X_{j,t}$) by,

$$\hat{\alpha}_{ij}(\omega) = [\{\hat{c}_{ij}(\omega)\}^2 + \{\hat{q}_{ij}(\omega)\}^2]^{1/2}, \tag{9.5.43}$$

and

$$\hat{\phi}_{ij}(\omega) = \tan^{-1}\{-\hat{q}_{ij}(\omega)/\hat{c}_{ij}(\omega)\}, \tag{9.5.44}$$

where $\hat{c}_{ij}(\omega)$, $\hat{q}_{ij}(\omega)$ are given by (9.5.36), (9.5.37). Both these quantities are non-linear functions of $\hat{c}_{ij}(\omega)$ and $\hat{q}_{ij}(\omega)$, but assuming that $\hat{c}_{ij}(\omega)$ and $\hat{q}_{ij}(\omega)$ are approximately unbiased estimates of $c_{ij}(\omega)$ and $q_{ij}(\omega)$, with variances small relative to their respective means, we may obtain approximate expressions for the means and variances of $\hat{\alpha}_{ij}(\omega)$ and $\hat{\phi}_{ij}(\omega)$ by using the δ-*technique* described in Section 2.9.1. Applying this technique we find, e.g.,

$$E\{\hat{\alpha}_{ij}(\omega)\} \sim \alpha_{ij}(\omega),$$

$$\text{var}\{\hat{\alpha}_{ij}(\omega)\} \sim [c_{ij}^2(\omega) \, \text{var}\{\hat{c}_{ij}(\omega)\} + q_{ij}^2(\omega) \, \text{var}\{\hat{q}_{ij}(\omega)\}$$
$$+ 2c_{ij}(\omega)q_{ij}(\omega) \, \text{cov}\{\hat{c}_{ij}(\omega), \hat{q}_{ij}(\omega)\}]/\alpha_{ij}^2(\omega).$$

Using the variance–covariance matrix (9.5.42) to evaluate the variance and covariance in the above expression, we obtain,

$$\text{var}\{\hat{\alpha}_{ij}(\omega)\} \sim \frac{C_W}{2N} \alpha_{ij}^2(\omega)\left\{1 + \frac{1}{|w_{ij}(\omega)|^2}\right\}, \qquad (9.5.45)$$

where $|w_{ij}(\omega)| = \alpha_{ij}(\omega)/\{h_{ii}(\omega)h_{jj}(\omega)\}^{1/2}$ is the coherency spectrum, as defined by (9.1.35). Using the same method we find similarly from (9.5.44) that,

$$\text{var}\{\hat{\phi}_{ij}(\omega)\} \sim \frac{C_W}{2N}\left\{\frac{1}{|w_{ij}(\omega)|^2} - 1\right\}, \qquad (9.5.46)$$

and

$$\text{cov}\{\hat{\alpha}_{ij}(\omega), \hat{\phi}_{ij}(\omega)\} \sim 0. \qquad (9.5.47)$$

These results were given by Jenkins (1963), who pointed out the crucial role played by the coherency, $|w_{ij}(\omega)|$, in the sampling properties of $\hat{\alpha}_{ij}(\omega)$ and $\hat{\phi}_{ij}(\omega)$. In particular, we see from (9.5.46) that the asymptotic variance of the phase estimate, $\hat{\phi}_{ij}(\omega)$, is *zero* when $|w_{ij}(\omega)| = 1$, but becomes *infinite* when $|w_{ij}(\omega)| = 0$. Similarly the asymptotic variance of the cross-amplitude estimate, $\hat{\alpha}_{ij}(\omega)$, becomes infinite when $|w_{ij}(\omega)| = 0$. *Thus, at frequencies where the coherency is low, the estimates of the cross-amplitude and phase spectra may have extremely large variances— despite the "smoothing" effect of the spectral window.* The behaviour of the estimated phase spectrum may be explained from the fact that when $|w_{ij}(\omega)|$ is low we are trying to estimate the average difference between the phases of two (effectively) independent complex valued random variables.

Confidence intervals for phase spectra

When $|w_{ij}(\omega)| = 0$, so that $h_{ij}(\omega) = h_{ji}(\omega) = 0$, it follows from (9.5.42) that $\hat{c}_{ij}(\omega)$ and $\hat{q}_{ij}(\omega)$ are uncorrelated. Assuming that N is sufficiently large for $\hat{c}_{ij}(\omega)$ and $\hat{q}_{ij}(\omega)$ to be treated as bivariate normal variables, they are then (asymptotically) independent, each having zero mean and with common variance. The distribution of the ratio Z, of two independent $N(0, \sigma^2)$ variables is Cauchy, in which case the principal value of $\tan^{-1} Z$ has a uniform distribution on $(-\pi/2, \pi/2)$. Hence, *when the coherency is zero, the principal value of $\hat{\phi}_{ij}(\omega)$ is approximately uniformly distributed over $(-\pi/2, \pi/2)$.*

If $|w_{ij}(\omega)| > 0$, then the distribution of $\hat{\phi}_{ij}(\omega)$ will no longer be approximately uniform; it is now the ratio of two (asymptotically) normal variables with *non-zero means* (and, provided C_W/N is small, relatively small variances), and hence the distribution of $\hat{\phi}_{ij}$ will be "peaked" over the interval $(-\pi/2, \pi/2)$. An approximate confidence interval for $\phi_{ij}(\omega)$ can be constructed by using an approximating normal distribution for the ratio $\{-\hat{q}_{ij}(\omega)/\hat{c}_{ij}(\omega)\}$. Using the δ-technique again we find,

$$\text{var}\{-\hat{q}_{ij}(\omega)/\hat{c}_{ij}(\omega)\} = \text{var}\{\tan \hat{\phi}_{ij}(\omega)\}$$

$$\sim \{\sec^2 \phi_{ij}(\omega)\}^2 \frac{C_W}{2N}\left(\frac{1}{|w_{ij}(\omega)|^2} - 1\right). \qquad (9.5.48)$$

An approximate confidence interval for $\phi_{ij}(\omega)$ can now be derived by treating $\tan \hat{\phi}_{ij}(\omega)$ as approximately normally distributed with mean $\tan \phi_{ij}(\omega)$ and variance given by (9.5.48). Since (9.5.48) involves the true coherency, $|w_{ij}(\omega)|^2$, this must be replaced by its estimate, $|\hat{w}_{ij}(\omega)|$ (see below). Also, the factor $\sec^4 \phi_{ij}(\omega)$ may be replaced by $\sec^4 \hat{\phi}_{ij}(\omega)$ to obtain a numerical value for the variance—alternatively, by writing $\sec^2 \phi_{ij}(\omega) = 1 + \tan^2 \phi_{ij}(\omega)$, a confidence interval for $\tan \phi_{ij}(\omega)$ can be computed by iterative techniques.

Alternatively, if the spectral matrix is estimated by (9.5.15), then a confidence interval for $\phi_{ij}(\omega)$ can be constructed from (Hannan (1970), p. 257),

$$|\sin\{\hat{\phi}_{ij}(\omega) - \phi_{ij}(\omega)\}| \leq \left\{\frac{1 - |\hat{w}_{ij}(\omega)|^2}{4m|\hat{w}_{ij}(\omega)|^2}\right\}^{1/2} t_{4m}(\alpha), \qquad (9.5.49)$$

where $t_{4m}(\alpha)$ is the two-sided $(100\alpha)\%$ point of the t-distribution with $4m$ degrees of freedom.

9.5.3 Estimation of Coherency

Replacing $h_{ij}(\omega)$, $h_{ii}(\omega)$, and $h_{jj}(\omega)$ by their estimates in (9.1.35), we obtain the following estimate for the coherency spectrum,

$$|\hat{w}_{ij}(\omega)| = \frac{|\hat{h}_{ij}(\omega)|}{\{\hat{h}_{ii}(\omega)\hat{h}_{jj}(\omega)\}^{1/2}} = \left\{\frac{\hat{c}_{ij}^2(\omega) + \hat{q}_{ij}^2(\omega)}{\hat{h}_{ii}(\omega)\hat{h}_{jj}(\omega)}\right\}^{1/2}. \qquad (9.5.50)$$

Using the δ-technique (as in the previous section) we obtain,

$$\text{var}\{|\hat{w}_{ij}(\omega)|\} \sim \frac{C_W}{2N}(1 - |w_{ij}(\omega)|^2).$$ (9.5.51)

The variance of the squared coherency estimate is similarly found to be,

$$\text{var}\{|\hat{w}_{ij}(\omega)|^2\} \sim \frac{C_W}{2N} 4|w_{ij}(\omega)|^2(1 - |w_{ij}(\omega)|^2).$$ (9.5.52)

(We recall that for the spectral estimate (9.5.15), which is based on averaging $(2m + 1)$ periodogram ordinates with uniform weighting, $C_W/N = 1/(2m + 1)$.)

Since $w_{ij}(\omega)$ is the correlation coefficient between $dZ_i(\omega)$ and $dZ_j(\omega)$ we may expect the distribution of $|\hat{w}_{ij}(\omega)|$ to be closely related to that of an ordinary sample correlation coefficient. This suggests that it would be useful to apply Fisher's z-transformation to $|\hat{w}_{ij}(\omega)|$, and if we write

$$\hat{z}_{ij}(\omega) = \tanh^{-1}(|\hat{w}_{ij}(\omega)|),$$ (9.5.53)

then from (9.5.51), we have,

$$\text{var}\{\hat{z}_{ij}(\omega)\} \sim C_W/2N,$$ (9.5.54)

i.e. the transformation (9.5.53) has "stabilized" the variance of $|\hat{w}_{ij}(\omega)|$ and rendered it independent of frequency. An approximate confidence interval for the coherency $|w_{ij}(\omega)|$, can now be obtained by taking $Z_{ij}(\omega)$ to be approximately normally distributed with mean $z_{ij}(\omega) = \tanh^{-1}(|w_{ij}(\omega)|)$ and variance $(C_W/2N)$.

Enochson and Goodman (1965) suggest an improved approximation by taking $\hat{z}_{ij}(\omega)$ to be normal with mean $\{z_{ij}(\omega) + (1/4m)\}$ and variance $1/4m$ in the case where $\hat{h}(\omega)$ is given by (9.5.15).

Test for zero coherency

A test for zero coherency (at frequency ω) can be derived from the analogy between (9.2.29) and the analysis of variance for the linear regression model. Thus, if we write,

$$X_{i,t} = \sum_{u=-\infty}^{\infty} g_{ij}(u)X_{j,t-u} + N_t,$$

then

$$dZ_i(\omega) = \Gamma_{ij}(\omega)\,dZ_j(\omega) + dZ_N(\omega),$$

where

$$E[|dZ_N(\omega)|^2] = (1 - |w_{ij}(\omega)|^2)h_{ii}(\omega)\,d\omega$$

(cf. (9.2.27)), and $\Gamma_{ij}(\omega) = h_{ij}(\omega)/h_{ij}(\omega)$. If we now use an analogy of the standard F-test for the ratio of the "sum of squares due to regression" to the "residual sum of squares", we have that, under the hypothesis that $|w_{jk}(\omega)| = 0$,

$$\frac{2m|\hat{w}_{ij}(\omega)|^2}{(1 - |\hat{w}_{ij}(\omega)|^2)} = F_{2,4m}. \tag{9.5.55}$$

Here, m is the parameter associated with the spectral estimate (9.5.15)— or more generally, is obtained by equating C_W/N with $1/(2m + 1)$.

Tables and graphs for constructing confidence intervals for coherencies have been given by Amos and Koopmans (1963) and Groves and Hannan (1968) have given tables for constructing confidence intervals for the multiple coherency, $W_{XY_1...Y_q^{(\omega)}}$. Tables of the cumulative distribution function of the estimated multiple coherency have been given by Alexander and Vok (1963). Confidence intervals for partial coherencies can be obtained by using the fact that the distribution of the sample partial coherency is the same as that of the sample coherency provided the equivalent number of "degrees of freedom" of the spectral estimates is reduced by $(q - 1)$ where q is the number of Y_t's "removed" in evaluating the (true) partial coherency (see Hannan (1970), p. 262).

9.5.4 Practical Considerations: Alignment Techniques

In the previous sections we discussed general spectral estimates of the form (9.5.7), or equivalently, of the form (9.5.12), and derived their asymptotic means, variances and covariances. In particular, if $W_N(\theta)$ satisfies the usual condition for spectral windows (especially $W_N(\theta) \to \delta(\theta)$ as $N \to \infty$), then it follows from (9.5.17), (9.5.30), (9.5.32) that, provided $C_W/N \to 0$ as $N \to \infty$, *each element of $\hat{h}(\omega)$ is a consistent estimate of the corresponding element of $h(\omega)$*. However, before we can compute $\hat{h}(\omega)$ we must, as in the univariate case, first select suitable values for the parameter involved in $\hat{h}(\omega)$. These are:

(i) the functional form of the covariance lag window, $\lambda_N(s)$ (or equivalently, of the spectral window, $W_N(\theta)$);

(ii) the bandwidth parameter of the window;

(iii) the sample size, N;

(iv) the frequency domain interval, $\Delta\omega$, at which values of $\hat{h}(\omega)$ are computed;

and, in the case of continuous parameter processes,

(v) the time domain interval, Δt, at which the multivariate process is sampled.

In Chapter 7 we gave a detailed discussion on the choice of these quantities for univariate spectral analysis, and the general considerations described there apply equally well to the multivariate case. Thus, the choice of the general form of the window is governed by the same considerations as those discussed in Section 7.5. For a given form of window, the parameter M (typically, the truncation point) may be chosen so as to achieve a specified degree of "resolvability" in the spectral estimates, as described in Section 7.4. In the multivariate case we may, of course, choose different window bandwidths for the different auto-spectra and cross-spectra, in which case we could determine a separate value for M for each element of $\hat{h}(\omega)$. If we use the same window bandwidth for all elements of $\hat{h}(\omega)$, then, ideally, we choose M so that the window bandwidth B_W is a specified fraction of the narrowest spectral peak (or trough) in all the elements of $h(\omega)$. For example, if we have sufficient prior information to be able to say that over the complete set of auto- and cross-spectra the "finest detail" which we wish to resolve is represented by a "critical (spectral) bandwidth" β_h, then, following the general principles of Section 7.4, we set

$$B_W = \alpha \beta_h,$$

α being a prescribed fraction. If, for example, we take $\alpha = \frac{1}{2}$, then the corresponding value of M may be computed from (7.4.2).

When an unlimited amount of data is available then, for having chosen M, we may choose the sample size N so as to achieve some specified level of "precision" in the estimates. This "precision" might refer to a particular auto-spectral estimate, to a particular cross-spectral estimate, or to particular estimates of phase and coherency spectra. For example, we may wish to choose N so that a certain coherency estimate attains a specified relative error, or has, say, a 95% confidence interval of specified length. Once the measure of "precision" has been selected, the determination of N (for given M) follows in an obvious way from the formulae for the asymptotic variances of the various spectral functions given in the previous sections, in much the same way as we determine sample size for univariate spectral estimates. (The detailed calculations for the univariate case were described in Section 7.4.)

The frequency domain interval $\Delta\omega$ is determined by the same considerations as those discussed in Section 7.4, and the choice of the time domain sampling interval Δt is determined by ω_0, the highest frequency of importance in *all* the spectral functions. For given ω_0, we choose Δt so that $\Delta t \leq \pi/\omega_0$; see the discussion of the "aliasing effect" in Section 7.1.1.

The technique of "pre-whitening" and "tapering" (see Section 7.4.1) can, of course, also be applied in the multivariate case. (Here, the "pre-whitening" operation would, in its most general form, be accomplished by applying a multivariate linear transformation of the form (9.2.6).)

Choice of bandwidth in the estimation of coherency

Although the general principles underlying the choice of window bandwidth are essentially the same as in the univariate case, the estimation of a coherency spectrum raises special features. Let us first see what happens if we do not apply any smoothing at all in computing the spectral est mates, i.e. if we take $\lambda_N(s) = 1$, all s, in (9.5.7). In this case, $\hat{h}(\omega) \equiv I_N(\omega)$, i.e. the estimated spectral matrix is simply the periodogram matrix (compare (9.5.7) with (9.5.10)), and hence, from (9.5.50),

$$|\hat{w}_{ij}(\omega)| = \frac{|\hat{h}_{ij}(\omega)|}{\{\hat{h}_{ii}(\omega)\hat{h}_{jj}(\omega)\}^{1/2}} = \frac{|I_{N,ij}(\omega)|}{\{I_{N,ii}(\omega)I_{N,jj}(\omega)\}^{1/2}}$$

$$= \frac{|\zeta_{X_i}(\omega)||\zeta_{X_j}(\omega)|}{\{|\zeta_{X_i}(\omega)|^2|\zeta_{X_j}(\omega)|^2\}^{1/2}} \equiv 1, \qquad \text{all } \omega,$$

where $\zeta_{X_i}(\omega)$ is the finite Fourier transform for $X_{i,t}$—see (9.5.8). *Hence, if we estimate the coherency spectrum from the "raw" auto- and cross-periodograms, we obtain a function which is unity at all frequencies*, irrespective of the form of the true coherency spectrum. This result may at first seem surprising, but it becomes obvious when we realize that using the raw periodogram matrix is equivalent to estimating the correlation coefficient between $\{dZ_i(\omega), dZ_j(\omega)\}$ from the *single* pair of observations, $\{\zeta_{X_i}(\omega), \zeta_{X_j}(\omega)\}$, which is bound to lead to the value of unity. When we apply smoothing, we are assuming, in effect, that the correlation coefficient (i.e. the coherency) has the same value for a number of neighbouring frequencies, and hence can be estimated by treating $\{\zeta_{X_i}(\omega_p), \zeta_{X_j}(\omega_p)\}$, (for neighbouring values of p) as observations from the same bivariate distribution. (This is, in fact, exactly what we are doing when we use the spectral estimate (9.5.15).)

We thus see that the estimated coherency may be highly inaccurate if we apply "too little" smoothing. On the other hand, if we apply "too much" smoothing, the estimates of $h_{ij}(\omega)$, $h_{ii}(\omega)$, and $h_{jj}(\omega)$ could be seriously biased and this could also lead to an inaccurate estimate of the coherency spectrum. The selection of the "optimal" degree of smoothing for estimating coherencies is indeed a difficult problem, but provided the window bandwidth is chosen appropriately (in relation to β_h), and the resulting value of M is small compared with N, the standard coherency estimate (9.5.50) should be reasonably efficient.

Plotting phase spectra

According to (9.1.33) the phase angle $\phi_{ij}(\omega)$ is defined only mod 2π, but, as previously noted, we usually take its range as $(-\pi, \pi)$ and, by giving $\phi_{ij}(\omega)$ the same sign as $\{-q_{ij}(\omega)\}$, each pair $\{c_{ij}(\omega), q_{ij}(\omega)\}$ determines a unique value of $\phi_{ij}(\omega)$ in the interval $(-\pi, \pi)$. However, when studying phase spectra we often look for linear trends to see if there is any "lead–lag" relationship between the processes $X_{i,t}$ and $X_{j,t}$ (see Example (3) of Section 9.1). If the values of the estimated phase spectrum, $\hat{\phi}_{ij}(\omega)$, are evaluated only in the interval $(-\pi, \pi)$ such linear trends might well fail to show up on a visual inspection of the graph since, e.g., values of $\hat{\phi}_{ij}(\omega)$ just greater than $+\pi$ would appear as values near $(-\pi)$ and hence produce apparent discontinuities in the graphical form of $\hat{\phi}_{ij}(\omega)$. Ideally, we should join together the end points, $-\pi$ and π of the ϕ-axis and plot the values of $\hat{\phi}_{ij}(\omega)$ on a cylinder, but since this is somewhat impracticable we may achieve roughly the same effect by plotting, for each ω, three values of $\hat{\phi}_{ij}(\omega)$, namely, one in the range $(-3\pi, \pi)$, one in the range $(-\pi, \pi)$, and one in the range $(\pi, 3\pi)$. This leads to a graph of the form shown in Fig. 9.5, and now the presence of any linear trends should be more readily detected.

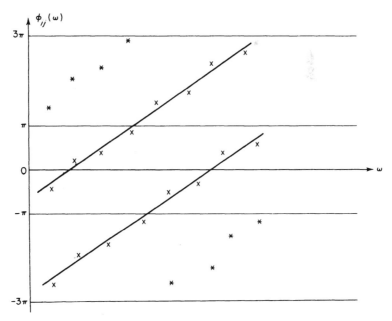

Fig. 9.5. Phase spectrum plots.

Alignment techniques

The result (9.5.17) tells us that, under general conditions, $\hat{h}_{ij}(\omega)$ is an asymptotically unbiased estimate of $h_{ij}(\omega)$ provided that $W_N(\theta) \to \delta(\theta)$ as $N \to \infty$, which typically is ensured by making the window parameter $M \to \infty$ as $N \to \infty$. Also, we know from (9.5.30) that in order for var$\{\hat{h}_{ij}(\omega)\}$ to tend to zero, we need $(C_W/N) \to 0$ as $N \to \infty$, which, by (9.5.34), typically reduces to the requirement that $(M/N) \to 0$ as $N \to \infty$. Although both requirements are easily met (for example by choosing $M = N^{\alpha}$, $0 < \alpha < 1$), for finite N, $\hat{h}_{ij}(\omega)$ will in general be a biased estimate, and to keep the bias low we must (as in the univariate case) choose the bandwidth of $W_N(\theta)$ sufficiently small, i.e. we must choose M sufficiently large in relation to the rate of decay of the cross-covariance function $R_{ij}(s)$.

So far, these considerations are exactly the same as those which arise in the estimation of auto-spectra, but we must now recall the fact that the shape of a general cross-covariance function is quite different from that of an autocovariance function. Unlike an autocovariance function, a cross-covariance function is not, in general, an even function, and its maximum value may occur anywhere. Suppose, for example, that the cross-covariance function $R_{ij}(s)$ has the form illustrated in Fig. 9.6, with its maximum

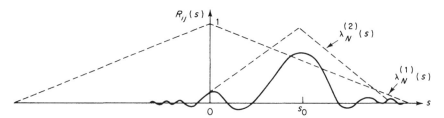

Fig. 9.6. A windowed cross-covariance function.

occurring at $s = s_0$. Suppose further that we have decided to use, say, the Bartlett lag window for which $\lambda_N(s) = (1 - |s|/M)$, $|s| \leq M$, and $\lambda_N(s) = 0$, $|s| > M$. The value of M should be chosen so that although $\lambda_N(s)$ attenuates the contribution of the "tail" of the (sample) cross-covariance function it does not severely distort the shape of $R_{ij}(s)$ in the "important" region, i.e. in the neighbourhood of $s = s_0$. We must therefore choose a fairly large value of M (certainly, substantially larger than s_0), leading to the lag window, $\lambda_N^{(1)}(s)$, say. However, this large value of M will result in a fairly large variance, and in fact, the lag window $\lambda_N^{(1)}(s)$ is "wasteful" in the sense that, since it is symmetric, it includes all the cross-covariances for $s = -(M-1), -(M-2), \ldots, 0$, which contribute little to the reduction of bias. This feature is due, of course, to the fact that we are trying to estimate a *cross-spectrum* by means of a *symmetric* lag window designed primarily for the estimation of *auto-spectra*. For the present case there is really no reason why $\lambda_N(s)$ should be symmetric about $s = 0$ and have its maximum at $s = 0$; indeed, it would be more sensible to "centre" $\lambda_N(s)$ on the point s_0, so that we could then use a much smaller value of M, as in the window $\lambda_N^{(2)}(s)$. When the value of s_0 is known *a priori* this leads to general estimates of $h_{ij}(\omega)$ of the form

$$\hat{\hat{h}}_{ij}(\omega) = \frac{1}{2\pi} \sum_{s=-(N-1)}^{(N-1)} \lambda_N(s-s_0)\hat{R}_{ij}(s) \, e^{-i\omega s}, \qquad (9.5.56)$$

where $\lambda_N(s)$ is the standard type of lag window used for the estimation of auto-spectra. Alternatively, (9.5.56) may be written

$$\hat{\hat{h}}_{ij}(\omega) = e^{-i\omega s_0}\hat{h}'_{ij}(\omega), \qquad (9.5.57)$$

where

$$\hat{h}'_{ij}(\omega) = \frac{1}{2\pi} \sum_{s=-(N-1)-s_0}^{(N-1)-s_0} \lambda_N(s)\hat{R}_{ij}(s+s_0) \, e^{-i\omega s}. \qquad (9.5.58)$$

The form (9.5.57) gives us another way of looking at this estimation procedure. For $\hat{h}'_{ij}(\omega)$ may be regarded as an estimate of

$$h'_{ij}(\omega) = \frac{1}{2\pi} \sum_{s=-\infty}^{\infty} R_{ij}(s+s_0) \, e^{-i\omega s}, \qquad (9.5.59)$$

which is related to $h_{ij}(\omega)$ by

$$h_{ij}(\omega) = h'_{ij}(\omega) \, e^{-i\omega s_0}. \qquad (9.5.60)$$

Now the function $R_{ij}(s+s_0)$ has its maximum at $s = 0$, and hence its Fourier transform, $h'_{ij}(\omega)$ will be a reasonably "smooth" function. By contrast, $h_{ij}(\omega)$ has the form of a sine wave of "frequency" s_0, modulated by $h'_{ij}(\omega)$, i.e. $h'_{ij}(\omega)$ may be thought of as the "*envelope*" of $h_{ij}(\omega)$. If we try to estimate $h_{ij}(\omega)$ directly we must choose $\lambda_N(s)$ so that the spectral window, $W_N(\theta)$, is "sharp" with respect to $h_{ij}(\omega)$, i.e. is such that $h_{ij}(\omega)$ does not vary significantly over the bandwidth B_W of $W_N(\theta)$. But if $h_{ij}(\omega)$ is of the above form (with s_0 fairly large) then B_W will have to be chosen so small that the effect of the smoothing may well be negligible. We may thus argue that since it is the function $h'_{ij}(\omega)$ which is "smooth" (*not*

$h_{ij}(\omega)$)), it will be much easier to estimate the "envelope" of the cross-spectrum rather than the cross-spectrum itself, and then convert the estimate of $h'_{ij}(\omega)$ into an estimate of $h_{ij}(\omega)$ via (9.5.57). Of course, the estimate (9.5.56) cannot be evaluated unless s_0 is known, but in cases where s_0 is unknown it would be reasonable to estimate it by $\hat{s}_0$, the location of the maximum of the *sample* cross-covariance function. This estimated value of s_0 can be checked by studying the estimated phase spectrum corresponding to $\hat{h}'_{ij}(\omega)$. If this phase spectrum shows a linear trend (over frequency) with slope $\hat{d}$, the estimate $\hat{s}_0$ may be adjusted so as to remove the trend.

If the cross-spectra are estimated directly from the periodogram matrix (using (9.5.14)) then the operation of "shifting" the cross-covariance function (so that its maximum occurs at $s = 0$) can be effected simply by translating the time-axis of one process relative to the other. Thus, if we write $X'_{i,t} \equiv X_{i,t+s_0}$ then $R_{ij}(s + s_0)$ is the cross-covariance function between $X'_{i,t}$ and $X_{j,t}$, and $h'_{ij}(\omega)$ is the corresponding cross-spectrum. Hence, $\hat{h}'_{ij}(\omega)$ can be computed by smoothing the cross-periodogram between $X'_{i,t}$ and $X_{j,t}$.

When we translate the time-axis of one process relative to the other so as to obtain the maximum cross-correlation at zero lag, the two processes are said to be "*aligned*", and hence the term "*alignment*" is used to describe the above technique. The advantages of the alignment technique were first pointed out by Akaike and Yamanouchi (1962) and by Priestley (1965a). Akaike and Yamanouchi discuss also the bias in cross-spectral estimates, and asymptotic expressions for the bias are given by Parzen (1967). (These are similar to the corresponding results for auto-spectra—see Section 6.2.4.)

Nettheim (1966) has given a treatment of the bias, and has also considered the possibility of reducing the bias by estimating cross-spectra via fitted finite order multivariate AR models. See also Parzen (1969), Cleveland and Parzen (1975), and Tick (1967).

9.6 NUMERICAL EXAMPLE

Graphs 9.1 and 9.2 show records of "input 1" and "output 1" taken from a distillation column in the Chemical Engineering Department at the University of Manchester Institute of Science and Technology. The distillation column was used to study the fractionation process, and in it two liquids, A and B, are blended to form a mixture which is then separated by fractional distillation in a multi-stage column. A simplified schematic diagram of the system is shown in Fig. 9.7. (A more detailed

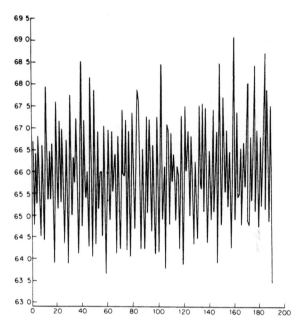

Graph 9.1. Input 1: chemical engineering data.

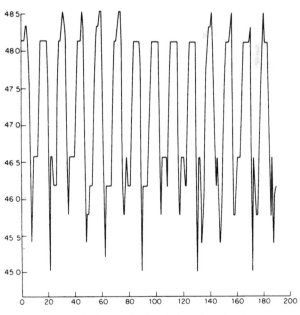

Graph 9.2. Output 1: chemical engineering data.

description of this system is given in Haggan (1975).) Input 1 represents
the steam flow rate (in kg/hour), and output 1 the flow rate of liquid B (in
kg/hour), at the points indicated in the diagram. Although the system is
usually operated in "closed-loop" form, the cycle time is of the order of 7
or 8 hours. The data analysed consisted of 190 observations at 60-second
intervals, and hence we may regard the system as effectively "open-loop".

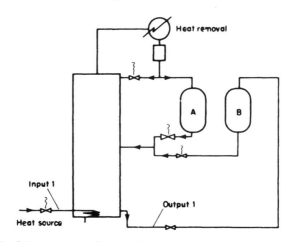

Fig. 9.7. A diagram of the distillation column under observation.

Graphs 9.3 to 9.9 show estimates of the (non-normalized) auto-spectra
of input 1 and output 1, the real and imaginary parts of the (non-
normalized) cross-spectrum between input 1 and output 1, the (non-
normalized) cross-amplitude spectrum, phase spectrum, and the
coherency spectrum. These spectra were calculated from 190 observations, using (in all cases) a Parzen window with truncation point
$M = 19$. The variable r indexing the frequency axis, denotes the values of
$(\pi\omega/128)$, so that r runs from 0 to 128 over the interval $(0, \pi)$. The
spectrum of input 1 shows that most of its power is concentrated at the
high frequency end—with a major spectral peak in the neighbourhood of
$\omega = 3\pi/4$. On the other hand, the spectrum of output 1 shows a high
concentration of power at low frequencies, roughly in the neighbourhood
of $\omega = \pi/6$. However, the coherency spectrum shows well-defined peaks
at each of these frequencies, indicating that there may well be a fairly
strong linear relationship between these two processes (see Haggan
(1975)). The plot of the phase spectrum shows a sharp discontinuity
around the frequency $\omega = \pi/2$, giving an example of the effect illustrated
in Fig. 9.5. If we adopt the device of re-plotting the phases over the three
intervals, the estimated phase spectrum would then show a smooth
variation over the whole frequency range.

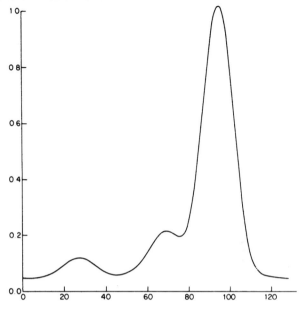

Graph 9.3. Spectrum: input 1.

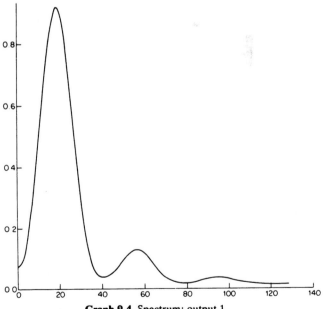

Graph 9.4. Spectrum: output 1.

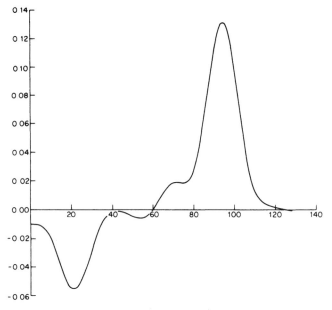

Graph 9.5. Cross-spectrum, real part, input 1/output 1.

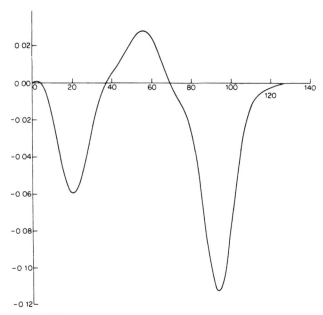

Graph 9.6. Cross-spectrum, imaginary part, input 1/output 1.

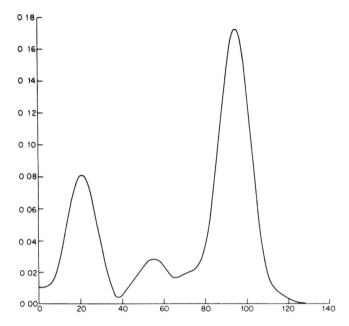

Graph 9.7. Cross-amplitude spectrum, input 1/output 1.

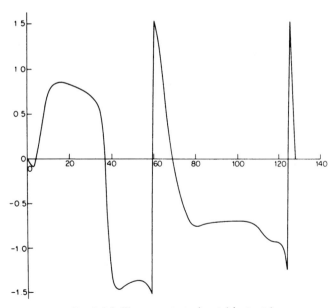

Graph 9.8. Phase spectrum, input 1/output 1.

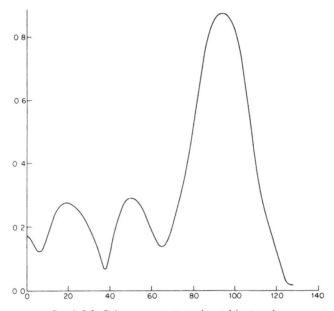

Graph 9.9. Coherency spectrum, input 1/output 1.

9.7 CORRELATION AND SPECTRAL PROPERTIES OF MULTIDIMENSIONAL PROCESSES

A process is called *multidimensional* when the parameter t which indexes its values has several components, t_1, t_2, ..., t_p, say. In this case the parameter t is vector valued and of course it can no longer represent "time", although one of its components could represent a time parameter. For example, if we observe the heights of a sea wave at different points in a given area (at a fixed time instant) then we have a process, $X(x, y)$, which depends on two spatial parameters, x and y. If we record the heights at different points over an interval of time then we have a process $X(x, y, t)$ which depends on two spatial parameters x and y, and a time parameter t. (Note the distinction between multivariate processes and multidimensional processes; in a multivariate process $\boldsymbol{X}$ is a vector and t is a scalar, whereas in a multidimensional process X is a scalar and t is a vector.) Multidimensional processes arise naturally when we consider "fields" and wish to study their spatial as well as their temporal variations. Thus, in the statistical theory of turbulence each component of the velocity vector may be regarded as a four-dimensional process, depending on three spatial coordinates and one time coordinate

(Bartlett (1955), p. 193), while Longuett-Higgins (1957) used a two-dimensional process to describe the behaviour of sea waves, as indicated above. Pierson and Tick (1957) considered two-dimensional processes in meteorology and oceanography, and a similar model has been suggested for the analysis of waves in a paper mill. Other models for two-dimensional processes have been discussed by Whittle (1954c), Walker and Young (1955) and Heine (1955).

To illustrate the basic correlation and spectral theory of multidimensional processes we will consider a two-dimensional process, $X_{t,\tau}$, which depends on two discrete parameters, t and τ, $t = 0, \pm 1, \pm 2, \ldots, \tau = 0, \pm 1, \pm 2, \ldots$ (The extension of the theory to two-dimensional continuous parameter processes should again be obvious, as should the extension to higher dimensional stationary processes.) We call $\{X_{t,\tau}\}$ a (weakly) *stationary process* if,

(i) $E\{X_{t,\tau}\} = \mu$, a constant (all t, τ);

(ii) $\text{var}\{X_{t,\tau}\} = \sigma^2$, a constant (all t, τ);

(iii) $\text{cov}\{X_{t,\tau}, X_{t',\tau'}\}$ is a function only of the "displacements", $(t' - \tau)$, $(\tau' - \tau)$.

Assuming that these conditions hold, and that $\mu = 0$, we define the (two-dimensional) *autocovariance function* by

$$R(s, u) = E[X_{t,\tau}^* X_{t+s, \tau+u}], \qquad (9.7.1)$$

(where * denotes the complex conjugate). Analogously to (4.8.34) there exists a spectral representation of $R(s, u)$ of the form

$$R(s, u) = \int_{-\pi}^{\pi} \int_{-\pi}^{\pi} e^{is\omega + iu\nu} \, dH(\omega, \nu), \qquad (9.7.2)$$

where $H(\omega, \nu)$ (suitably scaled) has the properties of a bivariate probability distribution on the region $-\pi \leq \omega \leq \pi$, $-\pi \leq \nu \leq \pi$. Corresponding to (9.7.2), there exists a spectral representation of $X_{t,\tau}$ of the form

$$X_{t,\tau} = \int_{-\pi}^{\pi} \int_{-\pi}^{\pi} e^{it\omega + i\tau\nu} \, dZ(\omega, \nu), \qquad (9.7.3)$$

where $dZ(\omega, \nu)$ is a doubly orthogonal process in ω and ν, i.e.

$$E[dZ(\omega, \nu) \, dZ^*(\omega', \nu')] = 0, \qquad \omega \neq \omega', \nu \neq \nu',$$

and is related to $H(\omega, \nu)$ by

$$dH(\omega, \nu) = E[|dZ(\omega, \nu)|^2]. \qquad (9.7.4)$$

(The orthogonality properties of $dZ(\omega, \nu)$ follow from substituting (9.7.3) in (9.7.1) and using the condition that the RHS must be a function of s and u only; the argument is virtually the same as in the one-dimensional case—see the discussion following equation (4.11.15).) Once the ortho-gonality of $dZ(\omega, \nu)$ has been established, (9.7.4) follows immediately by comparing the result of the above substitution with (9.7.2). The argument is again almost identical with that of the one-dimensional case. The function $H(\omega, \nu)$ is called the (two-dimensional) *integrated spectrum*, and the *spectral density function* (when it exists) is defined by

$$h(\omega, \nu) = \frac{\partial^2 H(\omega, \nu)}{\partial \omega \, \partial \nu}. \tag{9.7.5}$$

In this case (9.7.2) becomes

$$R(s, u) = \int_{-\pi}^{\pi} \int_{-\pi}^{\pi} e^{is\omega + iu\nu} h(\omega, \nu) \, d\omega \, d\nu \tag{9.7.6}$$

and the inverse formula is

$$h(\omega, \nu) = \left(\frac{1}{2\pi}\right)^2 \sum_{s=-\infty}^{\infty} \sum_{u=-\infty}^{\infty} R(s, u) e^{-is\omega - iu\nu}. \tag{9.7.7}$$

For real value processes we have

$$R(s, u) = R(-s, -u), \qquad h(\omega, \nu) = h(-\omega, -\nu).$$

The isotropic model

If we may assume that the statistical properties of $X_{t,\tau}$ have a "circular symmetry" (the so-called "isotropic" case) then $R(s, u)$ will depend only on $r = (s^2 + u^2)^{1/2}$, and following Bartlett (1955), p. 192, we may write

$$R(s, u) = \rho\{(s^2 + u^2)^{1/2}\} \tag{9.7.8}$$

where $\rho(r)$ is some given function of r. The properties of this model are most conveniently discussed in terms of continuous parameters, and if we now assume that t and τ (and, hence, s and u) take a continuous range of values then in place of (9.7.7) we have

$$h(\omega, \nu) = \left(\frac{1}{2\pi}\right)^2 \int_{-\infty}^{\infty} \int_{-\infty}^{\infty} R(s, u) e^{-is\omega - iu\nu} \, ds \, du$$

$$(-\infty < \omega < \infty, \quad -\infty < \nu < \infty). \tag{9.7.9}$$

Substituting (9.7.8) in (9.7.9), and changing to polar coordinates, $s = r \cos \theta$, $u = r \sin \theta$, gives

$$h(\omega, \nu) = \left(\frac{1}{2\pi}\right)^2 \int_{0}^{\infty} \int_{0}^{2\pi} \rho(r) \exp[-ir(\omega \cos \theta + \nu \sin \theta)] r \, dr \, d\theta$$

$$= \frac{1}{2\pi} \int_{0}^{\infty} J_0(r\eta) \rho(r) r \, dr \tag{9.7.10}$$

where $\eta = (\omega^2 + \nu^2)^{1/2}$, and J_0 is the Bessel function of the first kind of zero order (see, e.g., Watson (1944)). In particular, if we consider the case where

$$\rho(r) = Ce^{-\alpha r}, \tag{9.7.11}$$

C and α positive constants, then (9.7:10) becomes

$$h(\omega, \nu) = \frac{C}{2\pi} \int_0^\infty r\, e^{-\alpha r} J_0(r\eta)\, dr. \tag{9.7.12}$$

Using the general result (Watson (1944), p. 386),

$$\int_0^\infty e^{-at} J_\nu(bt) t^{\nu+1}\, dt = \frac{2a(2b)^\nu \Gamma(\nu + 3/2)}{(a^2 + b^2)^{\nu+3/2} \sqrt{\pi}},$$

valid for $\mathscr{R}(\nu) > -1$, (9.7.12) reduces to

$$h(\omega, \nu) = \frac{C'}{\{1 + (\omega^2 + \nu^2)/\alpha^2\}^{3/2}} \tag{9.7.13}$$

where $C' = \{C\Gamma(3/2)\}\{2\pi^{5/2}\alpha^2\}$.

As we would expect, the spectral density function also has circular symmetry (this holds even in the general isotropic case since, by (9.7.10), $h(\omega, \nu)$ depends only on η). The form of $\rho(r)$ used in (9.7.11) is similar to the autocovariance function of a (one-dimensional) AR(1) process, but there is no two-dimensional analogue of the AR(1) model which gives rise to this form of $R(s, u)$; see Bartlett (1955), p. 193.

Estimation of covariances and spectra

Given a sample $\{X_{t,\tau}\}$, $t = 1, 2, \ldots, N_1$, $\tau = 1, 2, \ldots, N_2$, we may estimate $R(s, u)$ by

$$\hat{R}(s, u) = \frac{1}{N_1 N_2} \sum_{t=1}^{N_1-s} \sum_{=1}^{N_2-u} X_{t,\tau} X_{t+s,\tau+u},$$

$$0 \leqslant s \leqslant N_1 - 1, \qquad 0 \leqslant u \leqslant N_2 - 1 \tag{9.7.14}$$

and

$$\hat{R}(s, u) = \hat{R}(-s, -u),$$

$$-N_1 - 1 \leqslant s < 0, \qquad -N_2 - 1 \leqslant \tau < 0$$

(assuming, of course, that the observations are real valued and are already adjusted to have zero mean).

It follows from (9.7.14) that

$$\lim_{N_1, N_2 \to \infty} E\{\hat{R}(s, u)\} = R(s, u),$$

and, when $\{X_{t,\tau}\}$ is a normal process, a similar calculation to that of Section 5.3.3 gives (cf. (5.3.25)),

$$\lim_{N_1,N_2\to\infty} N_1N_2 \, \mathrm{var}\{\hat{R}(s,u)\} = \sum_{p=-\infty}^{\infty} \sum_{q=-\infty}^{\infty} \{R(p-s,q-u)R(p+s,q+u)$$
$$+ R^2(p,q)\}, \qquad (9.7.15)$$

provided the RHS is convergent (see Priestley (1962d)).

The spectral density function, $h(\omega, \nu)$, is now estimated by

$$\hat{h}(\omega,\nu) = \left(\frac{1}{2\pi}\right)^2 \sum_{s=-(N_1-1)}^{(N_1-1)} \sum_{u=-(N_2-1)}^{(N_2-1)} \lambda(s,u)\hat{R}(s,u)\, e^{-is\omega - iu\nu} \qquad (9.7.16)$$

where $\lambda(s, u)$ is an even two-dimensional covariance lag window, i.e. $\lambda(-s, -u) = \lambda(s, u)$, all s, u. (For ease of notation we do not specifically indicate the dependence of $\lambda(s, u)$ on N_1 and N_2.)

Introducing the *two-dimensional periodogram* defined by

$$I(\omega,\nu) = \frac{1}{(2\pi)^2 N_1 N_2} \left| \sum_{t=1}^{N_1} \sum_{\tau=1}^{N_2} X_{t,\tau}\, e^{-it\omega - i\tau\nu} \right|^2 \qquad (9.7.17)$$

we may re-write $\hat{h}(\omega, \nu)$ in the alternative form

$$\hat{h}(\omega,\nu) = \int_{-\pi}^{\pi}\int_{-\pi}^{\pi} I(\lambda,\mu)W(\omega-\lambda, \nu-\mu)\, d\lambda\, d\mu \qquad (9.7.18)$$

where the *spectral window* $W(\lambda, \mu)$ is given by

$$W(\lambda,\mu) = \left(\frac{1}{2\pi}\right)^2 \sum_s \sum_u \lambda(s,u)\, e^{-is\lambda - iu\mu}. \qquad (9.7.19)$$

In order for $\hat{h}(\omega, \nu)$ to be a consistent estimate of $h(\omega, \nu)$ we obviously require $W(\lambda, \mu)$ to become "highly concentrated" about $\lambda = 0$, $\mu = 0$, as $N_1, N_2 \to \infty$. We therefore impose the following conditions on $W(\lambda, \mu)$ (which correspond in an obvious way to the conditions required in the one-dimensional case, see Section 6.2.4);

(i) $\int_{-\pi}^{\pi}\int_{-\pi}^{\pi} W(\lambda, \mu)\, d\lambda\, d\mu = 1$, all N_1, N_2;

(ii) $\int_{-\pi}^{\pi}\int_{-\pi}^{\pi} W^2(\lambda, \mu)\, d\lambda\, d\mu < \infty$, all N_1, N_2;

(iii) for any $\varepsilon_1, \varepsilon_2 > 0$,

$$W(\lambda, \mu) \to 0 \text{ as } N_1, N_2 \to \infty$$

uniformly for $|\lambda| \geq \varepsilon_1$, $|\mu| \geq \varepsilon_2$;

(iv) if $V(\omega, \mu) = W(\lambda, \mu)^* W(\lambda, \mu)$ (where * denotes convolution) then

$$\left| \frac{V(\lambda, \mu)}{V(0, 0)} - 1 \right| \to 0 \text{ as } N_1, N_2 \to \infty,$$

for $|\lambda| \leq A_1/N_1$, $|\mu| \leq A_2/N_2$, where A_1, A_2 are arbitrary constants.

Under these conditions, Grenander and Rosenblatt (1957b) have shown that

$$E\{\hat{h}(\omega, \nu)\} \sim \int_{-\pi}^{\pi} \int_{-\pi}^{\pi} h(\lambda, \mu) W(\omega - \lambda, \nu - \mu) \, d\lambda \, d\mu$$

$$\to h(\omega, \nu) \qquad \text{as} \quad N_1, N_2 \to \infty \text{ (by (iii))}, \qquad (9.7.20)$$

and, for normal processes,

$$\text{var}\{\hat{h}(\omega, \nu)\} \sim \frac{(2\pi)^2}{N_1 N_2} h^2(\omega, \nu) \left\{ \int_{-\pi}^{\pi} \int_{-\pi}^{\pi} W^2(\lambda, \mu) \, d\lambda \, d\mu \right\},$$

$$\omega \neq 0, \pm \pi, \qquad \nu \neq 0, \pm \pi. \qquad (9.7.21)$$

(For $\omega, \nu = 0, \pm \pi$, the above expression should be doubled.) Thus, provided that

$$\frac{1}{N_1 N_2} \left\{ \int_{-\pi}^{\pi} \int_{-\pi}^{\pi} W^2(\lambda, \mu) \, d\lambda \, d\mu \right\} \to 0 \qquad \text{as } N_1, N_2 \to \infty,$$

$\hat{h}(\omega, \nu)$ is a consistent estimate of $h(\omega, \nu)$.

Examples of windows satisfying the required conditions are easily obtained by considering the usual class of one-dimensional windows, $W(\lambda)$, satisfying similar conditions, and forming two-dimensional windows from the product, $W(\lambda)W(\mu)$. Thus, for example, we may take $W(\lambda)$ to be the Dirichlet kernel, $D_M(\lambda)$, or the Fejer kernel, $F_M(\lambda)$. In the latter case we have

$$W(\lambda, \mu) = F_{M_1}(\lambda) F_{M_2}(\mu)$$

$$= \left(\frac{1}{2\pi} \right)^2 \left\{ \frac{1}{M_1} \left(\frac{\sin^2 M_1 \lambda/2}{\sin^2 \lambda/2} \right) \frac{1}{M_2} \left(\frac{\sin^2 M_2 \mu/2}{\sin^2 \mu/2} \right) \right\}, \qquad (9.7.22)$$

corresponding to

$$\lambda(s, u) = \begin{cases} (1 - |s|/M_1)(1 - |s|/M_2), & |s| < M_1, \; |u| < M_2, \\ 0, & \text{otherwise}. \end{cases} \qquad (9.7.23)$$

(Note that the window parameters M_1 and M_2 will, in general, differ. The appropriate values for M_1 and M_2 will depend on the "bandwidths" of $h(\omega, \nu)$ along the ω- and ν-axes, respectively: see Priestley (1964b).)

With $W(\lambda, \mu)$ chosen as in (9.7.22), we find (cf. Section 6.2.4)

$$\int_{-\pi}^{\pi} \int_{-\pi}^{\pi} W^2(\lambda, \mu)\, d\lambda\, d\mu \sim (\tfrac{2}{3}M_1)(\tfrac{2}{3}M_2),$$

so that here (for $\omega \neq 0$, $\pm \pi$, $\nu \neq 0$, $\pm \pi$),

$$\mathrm{var}\{\hat{h}(\omega, \nu)\} \sim \frac{4}{9} \frac{M_1 M_2}{N_1 N_2} h^2(\omega, \nu). \qquad (9.7.24)$$

(The asymptotic variance of $\hat{h}(\omega, \nu)$ can be similarly evaluated for other choices of windows.)

9.7.1 Two-dimensional Mixed Spectra

In Chapter 8 we referred briefly to the multidimensional version of the mixed spectra problem, and pointed out that the types of discontinuities which could arise were more complicated than in the one-dimensional case. We now consider the two-dimensional case in more detail.

If the integrated spectrum $H(\omega, \nu)$ takes the form

$$H(\omega, \nu) = H_1(\omega, \nu) + H_2(\omega, \nu), \qquad (9.7.25)$$

where $H_1(\omega, \nu)$ is absolutely continuous and $H_2(\omega, \nu)$ contains discontinuities, we will say that $\{X_{t,\tau}\}$ possesses a *two-dimensional mixed spectrum*, and $H_2(\omega, \nu)$ will be called the *discrete spectrum*. However, by the analogy with bivariate probability distributions, it will be appreciated that the nature of the discontinuities in $H_2(\omega, \nu)$ may be more complex than those for the corresponding one-dimensional model. For, apart from the case where $H_2(\omega, \nu)$ is discontinuous only at a countable number of points, (ω_1, ν_1), (ω_2, ν_2), ... (corresponding to a discrete bivariate probability distribution), the discontinuities may form a "*ridge*", that is $H_2(\omega, \nu)$ may be discontinuous at all points along a line in the (ω, ν)-plane (corresponding to a "singular" bivariate probability distribution where there exists a linear relationship between the two random variables). In the latter case we have, in effect, a one-dimensional spectrum in the sense that the harmonic components which generate the discrete spectrum consist of waves all travelling in the same direction in the (t, τ)-plane. For example, in the region $\omega \geqslant 0$, $\nu \geqslant 0$, suppose that $H_2(\omega, \nu)$ consists of a single ridge along the line $\omega = \omega_0$, say. Then the corresponding density function may be written formally as

$$h_2(\omega, \nu) = \tfrac{1}{2}\{\delta(\omega - \omega_0) + \delta(\omega + \omega_0)\} g(\nu),$$

where $g(\nu)$ is some continuous function of ν. In this case it may be shown

(Priestley 1964b) that an appropriate model for $X_{t,\tau}$ is

$$X_{t,\tau} = Z_\tau \cos(\omega_0 t + \phi_\tau) + Y_{t,\tau},$$

where Z_τ is a stationary process with spectral density function $g(\nu)$, and $Y_{t,\tau}$ denotes the component corresponding to the continuous spectrum, $H_1(\omega, \nu)$. The discrete component of $\{X_{t,\tau}\}$ thus consists of a system of waves all travelling in the direction parallel to the t-axis, and modulated by the process Z_τ.

Further possibilities are that the discontinuities form *sets of ridges*, or occur in *rings*, i.e. at points on circles in the (ω, ν)-plane. (See Priestley (1964b) for a discussion of the physical interpretation of these models.)

In analysing a two-dimensional process we may therefore wish to test for (i) "point" discontinuities, (ii) ridges parallel to the ω-axis, (iii) ridges parallel to the ν-axis, (iv) ridges in a general direction, and (v) "circle" discontinuities, and it turns out that the $P(\lambda)$ test described in Section 8.4 can be adapted in quite an elegant way to cater for all these possibilities. This two-dimensional version of the $P(\lambda)$ test was developed by Priestley (1964b), and the basic ideas may be described briefly as follows.

Corresponding to the decomposition (9.7.25), we may similarly decompose the autocovariance function of $\{X_{t,\tau}\}$ and write

$$R(s, u) = R_1(s, u) + R_2(s, u), \tag{9.7.26}$$

where $R_1(s, u)$ corresponds to $H_1(\omega, \nu)$ and $R_2(s, u)$ to $H_2(\omega, \nu)$. Since R_1 may be expressed as the Fourier transform of the bounded function $h_1(\omega, \nu) = H_1'(\omega, \nu)$, it follows that $R_1(s, u) \to 0$ as s and $u \to \infty$. On the other hand, $R_2(s, u)$ will, in general, have a periodic structure and consequently will never die out. Thus, we may detect the presence of a discrete spectrum by examining the form of $R(s, u)$ for large values of s and u. Let m_1, m_2 be chosen so that

$$R_1(s, u) \sim 0,$$

outside the region $|s| < m_1$, $|u| < m_2$.

Given observations on $\{X_{t,\tau}\}$ for $t = 1, \ldots, N_1$, $\tau = 1, \ldots, N_2$, we may perform a harmonic analysis of the "tail" of the sample autocovariance function by writing

$$P(\lambda, \mu) = \left(\frac{1}{2\pi}\right)^2 \sum_s{}^* \sum_u{}^* \hat{R}(s, \mu) e^{-is\lambda - iu\mu}, \tag{9.7.27}$$

where $\sum_s^* \sum_u^*$ denotes a summation over values of s and u which lie outside the region $|s| < m_1$, $|u| < m_2$. More generally, we may consider the function

$$P(\lambda, \mu) = \left(\frac{1}{2\pi}\right)^2 \sum_{s=-(N_1-1)}^{(N_1-1)} \sum_{u=-(N_2-1)}^{(N_2-1)} \{w^{(1)}(s, u) - w^{(2)}(s, u)\} \hat{R}(s, u) e^{-is\lambda - iu\mu},$$

(9.7.28)

where $w^{(1)}(s, u)$, $w^{(2)}(s, u)$ are two lag windows chosen so that

$$\{w^{(1)}(s, u) - w^{(2)}(s, u)\} \sim 0 \qquad \text{for} \quad |s| < m_1, \quad |u| < m_2.$$

It is easy to see that if $X_{t,\tau}$ has a purely continuous spectrum, then

$$E[P(\lambda, \mu)] \sim 0, \qquad \text{all } \lambda, \mu,$$

but if the spectrum has a discrete component then $E[P(\lambda, \mu)]$ will be large in the neighbourhood of the discontinuities. If we now plot $P(\lambda, \mu)$ as a function of λ and μ, then the graph will indicate the form of the discrete spectrum, and may be used as a basis of a test for detecting a discrete spectral component.

Under the null hypothesis that the spectrum is purely continuous the asymptotic sampling properties of $P(\lambda, \mu)$ can be derived by methods similar to those used in the one-dimensional case, and a test constructed from cumulative sums of the $P(\lambda, \mu)$. Thus, we evaluate $P(\lambda, \mu)$ over the two-dimensional "grid", $(2\pi r/N_1, 2\pi s/N_2)$, $r = 0, 1, \ldots, [N_1/2]$, $s = 0, 1, \ldots, [N_2/2]$, and form the two-dimensional array (writing $\lambda_r = 2\pi r/N_1$, $\mu_s = 2\pi s/N_2$),

$$P(\lambda_1, \mu_1), \qquad P(\lambda_1, \mu_2), \qquad \ldots, P(\lambda_1, \mu_{[N_2/2]}),$$
$$P(\lambda_2, \mu_1), \qquad P(\lambda_2, \mu_2), \qquad \ldots, P(\lambda_2, \mu_{[N_2/2]})$$
$$\cdots \cdots \cdots \cdots \cdots \cdots \cdots \cdots \cdots \cdots \cdots \cdots \cdots \cdots$$
$$P(\lambda_{[N_1/2]}, \mu_1), P(\lambda_{[N_1/2]}, \mu_2), \ldots, P(\lambda_{[N_1/2]}, \mu_{[N_2/2]}).$$

The interesting feature is that, *simply by altering the order in which these values are formed into cumulative sums* we can construct tests for each of the different types of discontinuities mentioned above. For example, in the case of "point" discontinuities, we "diagonalize" the two-dimensional array to form a one-dimensional sequence, and test the resulting cumulative sums as in the one-dimensional case. In the case of "ridges" parallel to the ν-axis we first sum the rows of the above array and base the test on cumulative sums of the row totals. Ridges parallel to the ν-axis (and ridges in a general direction) can be treated similarly, and by using an "analysis of variance" type of decomposition we can test for "ridges versus points". The full details of these various tests, together with a discussion of the physical models underlying the different forms of discrete spectra, are given in Priestley (1964b).

Chapter 10

Prediction, Filtering and Control

10.1 THE PREDICTION PROBLEM

In Chapter 1 (Section 1.10) we discussed briefly one of the most celebrated problems in time series analysis, namely that of "*predicting*" (or "*forecasting*") a future value of a process, given a set of its past values. Suppose, for example, that we are studying a (univariate) discrete parameter stationary process $\{X_t\}$ and, having observed its values at a number of time points up to and including time t, we wish to predict its value at the time point, $t+m$ $(m>0)$. Thus, we are given $(n+1)$ past values, $X_t, X_{t-1}, X_{t-2}, \ldots, X_{t-n}$, say, and wish to predict the unknown value of X_{t+m}. Clearly, we would wish to use the information contained in the past values in predicting the future value X_{t+m}, so that if we denote the predicted value by $\tilde{X}_{t+m}$ then we would compute $\tilde{X}_{t+m}$ as some function θ of $X_t, X_{t-1}, X_{t-2}, \ldots, X_{t-n}$, i.e. we would write

$$\tilde{X}_{t+m} = \theta\{X_t, X_{t-1}, X_{t-2}, \ldots, X_{t-n}\}.$$

The question now is how do we choose the function θ? To answer this we must first decide on the "criterion" which we will use to measure the accuracy of $\tilde{X}_{t+m}$ as a predictor of X_{t+m}. The simplest and most widely used measure of accuracy is the "mean square error",

$$\mathcal{M}(m) = E[\{X_{t+m} - \tilde{X}_{t+m}\}^2], \tag{10.1.1}$$

and if we adopt this as our criterion, then the problem is to find that form of θ which minimizes $\mathcal{M}(m)$. If we allow a completely free choice of θ (i.e. if we do not restrict the form of θ in any way) then the solution is extremely simple. For we know from Section 2.12.6 that when we have just two random variables, X and Y, and wish to predict Y from a function $u(X)$, say, of X then the "best" choice of u (in the sense of minimizing the mean square error, $E[\{Y - u(X)\}^2])$ is the *conditional expectation* of Y, given X, i.e. we set (cf. (2.12.36))

$$u(X) = E[Y|X].$$

It follows in exactly the same way that the form of X_{t+m} which minimizes $\mathcal{M}(m)$ is,

$$\tilde{X}_{t+m} = E[X_{t+m}|X_t, X_{t-1}, \ldots, X_{t-n}]. \tag{10.1.2}$$

where the RHS of (10.1.2) denotes the conditional expectation of X_{t+m}, given $X_t, X_{t-1}, \ldots, X_{t-n}$. The proof of this slightly more general result is virtually the same as that of Theorem 2.12.2, and its intuitive interpretation may be seen as follows. Suppose we consider all the different possible realizations of the process $\{X_t\}$, as indicated in Fig. 1.8. Typically, two different realizations will take the same values only at isolated set of time points but, as we are considering all possible realizations, there will be some which take the same values throughout an interval of time points. Since we know the values of X_t at time $t, t-1, \ldots, t-n$, we can immediately remove all realizations which do not agree with these values (since none of these could possibly be the realization with which we are dealing). Those remaining will all take the same values at times $t, t-1, \ldots, t-n$, but will take different values outside this interval, as illustrated in Fig. 10.1. Each alternative realization in this sub-collection

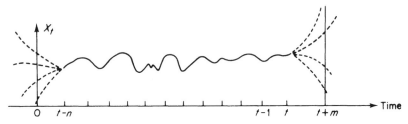

Fig. 10.1. Realizations of the process $\{X_t\}$.

will give rise to a different value of X_{t+m}, and since we have no way of knowing which particular realization to choose, the obvious device is to take the *average* of the different values of X_{t+m} generated by this sub-collection, and use this as our "best" prediction of X_{t+m}. The average value of X_{t+m} *over this sub-collection of realizations* is precisely what we mean when we refer to the *conditional* expectation of X_{t+m}, given $X_t, X_{t-1}, \ldots, X_{t-n}$, and thus the expression (10.1.2) has a very strong intuitive appeal.

Linear predictors

Although (10.1.2) is an important theoretical result it is of little practical use, since in order to evaluate this conditional expectation we would need to know the precise form of the joint probability distribution $(X_{t+m}, X_t, X_{t-1}, \ldots, X_{t-n})$. In practice we hardly ever possess such detailed knowledge of the structure of the process, and the nearest we can approach this situation is to argue (from the usual considerations) that in many cases we would expect such joint distributions to be approximately multivariate normal. However, if the joint distributions were normal then we know from Section 2.13.2 that the conditional expectation (10.1.2) is a *linear* function of $X_t, X_{t-1}, \ldots, X_{t-n}$, in which case we can write the predictor $\tilde{X}_{t+m}$ more explicitly as

$$\tilde{X}_{t+m} = a_0 X_t + a_1 X_{t-1} + \ldots + a_n X_{t-n}, \qquad (10.1.3)$$

where $a_0, a_1, \ldots, a_n$ are constants (dependent on m and n). Given that $\tilde{X}_{t+m}$ has this linear form, the only remaining step is to determine the values of the coefficients $a_0, a_1, \ldots, a_n$ which minimize the mean square error (10.1.1), and since this is a quadratic function of the $\{X_t\}$, the values of these coefficients may be determined from a knowledge only of the autocovariance function (or equivalently, the spectrum) of the process.

We can, of course, decide to consider only *linear* predictors, whether or not the process is Gaussian, the argument being that if the process is Gaussian then the linear predictor is optimal (in the mean square error sense), whereas if the process is non-Gaussian then, in general, we would be unable to evaluate the expression (10.1.2) and so we might as well seek the "best" linear predictor. In fact, the conventional theory of prediction is concerned almost entirely with linear predictors, and to emphasize this point it is sometimes referred to as "*linear least squares prediction theory*".

As a point of historical interest, we would mention that the pioneering work on linear predictors was developed at about the same time, but quite independently, by Kolmogorov (1941b) and Wiener (1949). Wiener's interest in this problem arose from his study of the prediction of the flight paths of aircraft (in connection with the problem of "fire control") which was carried out during the early 1940s, and his work was not available for general publication until the appearance of his celebrated monograph in 1949. It then emerged that Wiener and Kolmogorov had solved the same problem, and although the analytical techniques used by them appear to be quite different, we shall see later that they can be regarded as merely two different ways of "coordinatizing" the same basic geometrical problem.

The form of the prediction problem studied by Kolmogorov and Wiener is a slightly modified version of that described above in that they considered only the limiting case where n (the number of past values) $\to \infty$. This corresponds to the situation where we have observed *all* the past values of the process up to time t (i.e. we have observed a semi-infinite record). Clearly, this is unrealistic, but it turns out, as we would expect, that the coefficients a_j in (10.1.3) become negligible, in general, for sufficiently large values of j. Thus, in general, X_{t+m} does not depend critically on the "remote past history" of the process, and the solution to the "semi-infinite" version of the problem may therefore be used as a good approximation to the more realistic case where we have a large (but finite) number of past values. The reason for letting $n \to \infty$ is that it simplifies considerably the mathematical treatment of the problem, and we may note that the general solution (10.1.1) is still valid even in this limiting case.

Before describing the Kolmogorov and Wiener solutions we first discuss a technique called "spectral factorization" since this is a basic ingredient of both methods.

10.1.1 Spectral Factorization and Linear Representations

There is one case where the solution of the prediction problem is entirely trivial, namely, when the process consists of a sequence of *independent* random variables, $\{\varepsilon_t\}$. Here, the past provides no information on the future, and it follows from (10.1.2) that, irrespective of the number of past observations, the best predictor of ε_{t+m} is simply its (unconditional) mean. (When the variables are independent the conditional expectation reduces, of course, to the unconditional expectation.) Thus, if ε_t is a zero mean process, then the best predictor of ε_{t+m} is always *zero* (for all m). Suppose now that the $\{\varepsilon_t\}$ are merely *uncorrelated* (not necessarily independent), i.e. $\{\varepsilon_t\}$ is a "white noise" process. Then although it may be possible to obtain useful information on ε_{t+m} from a non-linear function of $\{\varepsilon_t, \varepsilon_{t-1}, \varepsilon_{t-2}, \ldots\}$, any *linear* function $(a_0\varepsilon_t + a_1\varepsilon_{t-1} + a_2\varepsilon_{t-2} + \ldots)$ of past values contains no useful information on ε_{t+m} (since ε_{t+m} is clearly uncorrelated with all such linear functions), and hence the best *linear* predictor of ε_{t+m} is still *zero* (for all m). The basic idea in Kolmogorov's solution is first to express the given process X_t as *a general linear process* of the form (cf. (3.5.56)),

$$X_t = \sum_{u=0}^{\infty} g_u \varepsilon_{t-u}, \qquad (10.1.4)$$

$\{\varepsilon_t\}$ being a white noise process, and then use the above result for a white noise process to obtain the best linear predictor for X_{t+m}.

However, before we can develop this approach we first need to know how to construct a general linear representation for a given stationary process $\{X_t\}$, or indeed whether such representations always exist. We know from the general discussion of stationary processes in Chapter 3 that harmonic processes constitute a special category, not included in the general class of linear processes (see Section 3.5.8). Hence, we exclude all processes where spectra contain discrete components, i.e. *we assume that* $\{X_t\}$ *has a purely continuous spectrum with (non-normalized) spectral density function* $h(\omega)$. Since $h(\omega) \geq 0$, all ω, we can always find some (possibly complex valued) function $\phi(\omega)$ such that $h(\omega)$ can be written as

$$h(\omega) = |\phi(\omega)|^2 = \phi(\omega)\phi^*(\omega) \qquad (10.1.5)$$

(* denoting complex conjugate). Also, assuming that var$\{X_t\}$ is finite, so that $h(\omega)$ is integrable, it follows that $\phi(\omega)$ is quadratically integrable (over $-\pi, \pi$) and hence can be expanded as a Fourier series which represents it at least in the "mean-square" sense (cf. Section 4.2.1), i.e. we may write

$$\phi(\omega) = \sum_{u=-\infty}^{\infty} \gamma_u e^{-i\omega u}. \qquad (10.1.6)$$

Now, using the spectral representation of $\{X_t\}$ we may write

$$X_t = \int_{-\pi}^{\pi} e^{it\omega} \, dZ(\omega), \qquad (10.1.7)$$

where $E[|dZ(\omega)|^2] = h(\omega)\, d\omega$. Assuming now that $h(\omega)$ is strictly positive, so that, $|\phi(\omega)| > 0$, all ω, we set

$$dz(\omega) = dZ(\omega)/\phi(\omega),$$

so that (10.1.7) can be re-written as

$$X_t = \int_{-\pi}^{\pi} e^{it\omega}\phi(\omega) \, dz(\omega), \qquad (10.1.8)$$

and substituting (10.1.6) in (10.1.8) (and changing the order of summation and integration) gives

$$X_t = \sum_{u=-\infty}^{\infty} \gamma_u \left\{ \int_{-\pi}^{\pi} e^{i(t-u)\omega} \, dz(\omega) \right\},$$

or

$$X_t = \sum_{u=-\infty}^{\infty} \gamma_u \varepsilon_{t-u}, \qquad (10.1.9)$$

where

$$\varepsilon_t = \int_{-\pi}^{\pi} e^{it\omega} \, dz(\omega). \qquad (10.1.10)$$

The RHS of (10.1.9) exists as a mean-square limit since,

$$2\pi \sum_{u=-\infty}^{\infty} \gamma_u^2 = \int_{-\pi}^{\pi} |\phi(\omega)|^2 \, d\omega = \int_{-\pi}^{\pi} h(\omega) \, d\omega = \text{var}\{X_t\} < \infty.$$

But ε_t is an *uncorrelated*, (i.e. *white noise*) process since the spectral density function of ε_t satisfies

$$h_\varepsilon(\omega) \, d\omega = E[|dz(\omega)|^2] = \frac{1}{|\phi(\omega)|^2} E[|dZ(\omega)|^2] = \frac{h(\omega) \, d\omega}{|\phi(\omega)|^2} = d\omega,$$

using (10.1.5). Thus, $h_\varepsilon(\omega)$ takes the constant value unity for all ω, and hence,

$$\text{var}\{\varepsilon_t\} = 2\pi.$$

We have thus shown that any stationary process with a purely continuous spectrum can be representated as a linear combination of the terms of an uncorrelated process ε_t. (Although we derived this result under the assumption that $h(\omega) > 0$, all ω, this assumption is not an intrinsic one, and the result is still valid even when $h(\omega)$ is allowed to take zero values—see, e.g. Koopmans (1974), p. 252.)

However, (10.1.9) is not quite of the desired form since it involves a two-sided summation over u (i.e. it involves both positive and negative values of u). Consequently, X_t here depends on past, present and *future* values of ε_t, whereas in the general linear process model, (10.1.4), X_t depends only on past and present values of ε_t (the summation in (10.1.4) involving only positive values of u). This may appear to be only a minor distinction, but it is, in fact, an extremely important point since the presence of future values of ε_t in (10.1.9) introduces considerable complications in the analysis of the prediction problem. (We will return to this point later.) To express X_t in the form (10.1.4) we have to choose the function $\phi(\omega)$ in (10.1.5) so that it has a *one-sided* Fourier series with only positive powers of $(e^{-i\omega})$ appearing. Now there are many ways in which we can choose a function $\phi(\omega)$ which satisfies (10.1.5). (Note that if $\phi(\omega)$ satisfies (10.1.5) then so does $\{\phi(\omega) e^{i\psi(\omega)}\}$, for arbitrary real $\psi(\omega)$.) Whether we can, in particular, choose $\phi(\omega)$ so that it has the required one-sided form depends on the form of $h(\omega)$, but we can appeal to a well known theorem in analysis which states, effectively, that if $h(\omega)$ satisfies the condition

$$\int_{-\pi}^{\pi} \log\{h(\omega)\} \, d\omega > -\infty, \tag{10.1.11}$$

then $\phi(\omega)$ can always be chosen in the required form. The precise statement of the theorem is as follows (see Doob (1953), pp. 160, 569):

If $h(\omega)$ satisfies (10.1.11) then $\exists$ a unique sequence $g_0, g_1, g_2, \ldots$, with g_0 real and positive, and $\sum_{u=0}^{\infty} |g_u|^2 < \infty$, such that the function

$$G(z) = \sum_{u=0}^{\infty} g_u z^u, \tag{10.1.12}$$

has no zeros in $|z| < 1$, i.e. inside the unit circle, and provides a factorization for $h(\omega)$ in the form

$$h(\omega) = |G(e^{-i\omega})|^2. \tag{10.1.13}$$

Note that the condition $\sum_{u=0}^{\infty} |g_u|^2 < \infty$ implies that $G(z)$ is *analytic inside the unit circle*, $|z| < 1$.

If we normalize the function $G(z)$ so that $G(0) = 1$ ($\Leftrightarrow g_0 = 1$), then $h(\omega)$ can be factorized in the alternative form,

$$h(\omega) = C|G(e^{-i\omega})|^2, \tag{10.1.14}$$

C being a constant.

The condition (10.1.11) plays a fundamental role in linear prediction theory. Since $\log\{h(\omega)\} \leq h(\omega)$, (all ω), and $\int_{-\pi}^{\pi} h(\omega) \, d\omega = \text{var}\{X_t\} < \infty$, it follows that $\int_{-\pi}^{\pi} \log\{h(\omega)\} \, d\omega$ cannot diverge to $+\infty$ and therefore must either be finite or diverge to $-\infty$. This integral will diverge to $-\infty$ if, e.g., $h(\omega)$ vanishes at every point throughout an interval, but this certainly represents a highly pathological case. In fact, (10.1.11) is a fairly weak condition, and we may expect it to hold in the vast majority of cases (certainly, in any situations of practical interest). However, if we impose a stronger condition on $h(\omega)$ then we can obtain explicit expressions for C and $G(z)$ by the following construction due to Whittle (1963). Let $R(s)$ denote the autocovariance function of $\{X_t\}$, and write

$$h(z) = \frac{1}{2\pi} \sum_{s=-\infty}^{\infty} R(s) z^s, \tag{10.1.15}$$

so that $h(z)$ reduces to the spectral density function $h(\omega)$ on setting $z = e^{-i\omega}$. (To simplify the notation we use the same symbol to denote both the spectral density function $h(\omega)$ and the function (10.1.15). However, it should be noted that, strictly, these are different functions.) We now assume that $\log\{h(z)\}$ is analytic in an annulus, $\rho < |z| < 1/\rho$, ($\rho < 1$), so that it has a Laurent expansion,

$$\log h(z) = \sum_{u=-\infty}^{\infty} c_u z^u. \tag{10.1.16}$$

Then, in this region,

$$
\begin{aligned}
h(z) &= \exp\left\{ \sum_{u=-\infty}^{\infty} c_u z^u \right\} \\
&= e^{c_0} \cdot \exp\left\{ \sum_{u=-\infty}^{-1} c_u z^u \right\} \exp\left\{ \sum_{u=1}^{\infty} c_u z^u \right\}. \tag{10.1.17}
\end{aligned}
$$

But, since the above region includes the unit circle, the $\{c_k\}$ are the Fourier coefficients of $\log\{h(\omega)\}$, and since this is an even real valued function, the $\{c_k\}$ are similarly real valued, with $c_{-k} = c_k$. Hence, if we write

$$G(z) = \exp\left\{ \sum_{u=1}^{\infty} c_u z^u \right\} \tag{10.1.18}$$

then (10.1.17) can be re-written as

$$h(z) = e^{c_0} G(z) G(z^{-1}), \tag{10.1.19}$$

and setting $z = e^{-i\omega}$ we obtain,

$$h(\omega) = e^{c_0} |G(e^{-i\omega})|^2. \tag{10.1.20}$$

Equation (10.1.20) is called the *canonical factorization of the spectral density function*, and it gives us a factorization of the required form since $G(z)$, being analytic in $|z| < \rho^{-1}$, has a Taylor series expansion in (positive) powers of z, and thus $G(e^{-i\omega})$ has a one-sided Fourier series involving only positive powers of $e^{-i\omega}$. Moreover, it is clear from (10.1.18) that $\{G(z)\}^{-1}$ is also analytic for $|z| < \rho^{-1}$, and hence $G(z)$ has no zeros in $|z| < 1$. Finally, from (10.1.18) we have $G(0) = 1$, and hence,

$$g_0 = 1 \tag{10.1.21}$$

as required.

Since the $\{c_k\}$ are the Fourier coefficients of $\log\{h(\omega)\}$ we have, by the standard inversion formula,

$$c_k = \frac{1}{2\pi} \int_{-\pi}^{\pi} e^{i\omega k} \log\{h(\omega)\} \, d\omega, \tag{10.1.22}$$

and, in particular,

$$c_0 = \frac{1}{2\pi} \int_{-\pi}^{\pi} \log\{h(\omega)\} \, d\omega. \tag{10.1.23}$$

The factorization (10.1.20) is slightly different from (10.1.5) since (10.1.20) has the constant e^{c_0} as a factor. However, if we follow through the argument following (10.1.5) we obtain in a similar manner,

$$X_t = \sum_{u=0}^{\infty} g_u \varepsilon_{t-u}, \qquad (10.1.24)$$

where the $\{g_u\}$ are the Fourier coefficients of $G(e^{-i\omega})$ and

$$\varepsilon_t = \int_{-\pi}^{\pi} e^{it\omega} \, dz(\omega), \qquad \text{with} \quad dz(\omega) = \{dZ(\omega)/G(e^{-i\omega})\}.$$

The process ε_t now has spectral density function,

$$h_\varepsilon(\omega) = \frac{1}{|G(e^{-i\omega})|^2} h(\omega) = e^{c_0},$$

so that $\{\varepsilon_t\}$ is, as before, an uncorrelated process, with

$$\text{var}\{\varepsilon_t\} = 2\pi e^{c_0}. \qquad (10.1.25)$$

Note that since $\{G(z)\}^{-1}$ is analytic for $|z| \leqslant 1$ it follows from the discussion of Section 3.5.7 that the linear representation (10.1.24) is "invertible" into an (infinite order) AR form, i.e. we may re-write (10.1.24) as

$$\sum_{u=0}^{\infty} g'_u X_{t-u} = \varepsilon_t, \qquad (10.1.26)$$

where

$$\sum_{u=0}^{\infty} g'_u z^u = \{G(z)\}^{-1}.$$

Equivalently, we may write (10.1.24) as

$$X_t = G(B)\varepsilon_t, \qquad (10.1.27)$$

and (10.1.26) as

$$\varepsilon_t = G^{-1}(B)X_t. \qquad (10.1.28)$$

If we do not assume that $\log\{h(z)\}$ is analytic in $\rho < |z| < 1/\rho$, $(\rho < 1)$, but assume merely that $\log\{h(\omega)\}$ satisfies (10.1.11), then the linear representation (10.1.24) is still valid but the AR representation may not exist.

Continuous parameter processes

For a continuous parameter process $X(t)$ with spectral density function $h(\omega)$ $(-\infty < \omega < \infty)$, the role of the condition (10.1.11) is played by the

Paley–Wiener condition,

$$\int_{-\infty}^{\infty} \frac{\log\{h(\omega)\}}{1+\omega^2}\, d\omega > -\infty. \qquad (10.1.29)$$

Under this condition it may be shown (Paley and Wiener (1934), Doob (1953), p. 584) that $h(\omega)$ can be factorized in the form

$$h(\omega) = |G(i\omega)|^2, \qquad (10.1.30)$$

where $G(z)$ has a one-sided Laplace transform, viz.,

$$G(z) = \int_0^{\infty} g(u)\, e^{-uz}\, du, \qquad (10.1.31)$$

with $\int_0^{\infty} |g(u)|^2\, du < \infty$, and $G(z)$ *is analytic and has no zeros in the right-half plane* (so that $G(i\omega)$ is analytic and has no zeros in the lower-half plane). A similar argument to that used in the discrete parameter case then shows that $X(t)$ admits the linear representation

$$X(t) = \int_0^{\infty} g(u)\varepsilon(t-u)\, du, \qquad (10.1.32)$$

where $\varepsilon(t)$ is a continuous parameter white noise process (cf. Section 3.7.1), with spectral density function given (formally) by $h_\varepsilon(\omega) = 1$, (all ω). (Compare (10.1.28) with (4.12.43).) Writing $dW(t) = \varepsilon(t)\, dt$, and comparing the above form of $h_\varepsilon(\omega)$ with (4.10.4), we see that here $\sigma_w^2 = 2\pi$, i.e.

$$E[\{dW(t)\}^2] = 2\pi\, dt.$$

Since $\varepsilon(t-u) = e^{-uD}\varepsilon(t)$, (10.1.32) may be expressed in the alternative form,

$$X(t) = \left\{ \int_0^{\infty} g(u)\, e^{-uD}\, du \right\} \varepsilon(t)$$

$$= G(D)\varepsilon(t). \qquad (10.1.33)$$

10.1.2 Geometrical Representation of Linear Prediction

If we use the Hilbert space description of a stochastic process (see Sections 4.11, 4.2.2) we may obtain a very neat geometrical representation of the linear prediction problem. Consider a Hilbert space H consisting of all complex random variables U with $E(U) = 0$, $E(|U|^2) < \infty$. We define the inner product between any two random variables U, V by

$$(U, V) = E(U^*V),$$

so that two vectors in H are *orthogonal* if they are uncorrelated. The "squared length", or "norm" of U is then given by

$$\|U\|^2 = E(|U|^2),$$

and the "squared distance" between U and V is,

$$\|U - V\|^2 = E(|U - V|^2).$$

Now consider the subspace H_t of H spanned by the random variables, $\{X_s; s \le t\}$, i.e. H_t consists of all linear combinations of $X_t, X_{t-1}, X_{t-2}, \ldots$ (together with the limits of Cauchy sequences of such linear combinations). The vector representing the random variable X_{t+m} will, in general, lie outside the subspace H_t, and the essence of the linear prediction problem is to find that vector $\tilde{X}_{t+m}$ in H_t which is "closest" to X_{t+m}. If we adopt the mean square error as our measure of "closeness" then we seek that vector $\tilde{X}_{t+m}$ such that $\|X_{t+m} - \tilde{X}_{t+m}\|^2$ is a minimum, i.e. such that the "distance" between, X_{t+m} and $\tilde{X}_{t+m}$ is a minimum. Expressed in this way the problem has a very well known and simple geometrical solution, namely, that $\tilde{X}_{t+m}$ is the *orthogonal projection* of X_{t+m} on the subspace H_t, as illustrated in Fig. 10.2. More generally, let $\mathcal{H}_0$ be a subspace of a

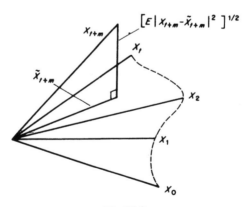

Fig. 10.2

general Hilbert space $\mathcal{H}$, and Z a vector $\in \mathcal{H}$ but $\notin \mathcal{H}_0$. Let Y be any vector $\in \mathcal{H}_0$. Then we have the *projection theorem*:

There exists a unique vector $Y^* \in \mathcal{H}_0$ such that $\|Z - Y\|^2$ is a minimum when $Y = Y^*$, and Y^* is uniquely determined by the property,

$$(Z - Y^*) \perp \text{ every vector } \in \mathcal{H}_0.$$

(See, e.g., Halmos (1951).)

The calculation of orthogonal projections is greatly simplified if we have an *orthogonal basis* for $\mathcal{H}$. For example, in the finite-dimensional case let $\{A_1, \ldots, A_p\}$ be an orthogonal basis for $\mathcal{H}_0$, so that every vector Y can be expressed as $Y = \sum_i \alpha_i A_i$. Now extend this basis to $\{A_1, \ldots, A_p, B_1, \ldots, B_q\}$ to form an orthogonal basis for $\mathcal{H}$. Then Z can be written

$$Z = \sum_i \alpha_i A_i + \sum_j \beta_j B_j.$$

The orthogonal projection of Z on $\mathcal{H}_0$ is then

$$Y^* = \sum_i \alpha_i A_i,$$

i.e. we obtain Y^* simply by setting to zero the coordinates corresponding to the basis vectors which lie outside $\mathcal{H}_0$. The proof of this result follows immediately on noting that

$$Z - Y^* = \sum_j \beta_j B_j,$$

which is clearly orthogonal to every vector $\in \mathcal{H}_0$ since $A_i \perp B_j$, all i, j.

The above discussion indicates that we can obtain a simple solution to the linear prediction problem provided we can find an orthogonal basis for the Hilbert space H generated by the process $\{X_t\}$. Since orthogonality here means "uncorrelated", all we require is a linear representation of X_t in terms of an uncorrelated process. This type of representation is a basic feature of both the Kolmogorov and Wiener solutions; the Kolmogorov approach is based on the general linear representation (10.1.23), whereas the Wiener approach is based on the spectral representation (10.1.7) using the $\{dZ(\omega)\}$ as the orthogonal basis.

10.1.3 The Kolmogorov Approach

We consider first the case of a discrete parameter process, and wish to construct the linear least squares predictor of X_{t+m}, given $X_t, X_{t-1}, \ldots$. We assume that $h(\omega)$, the spectral density function of X_t, satisfies the condition (10.1.11), so that X_t may be written as a (one-sided) linear process, as in (10.1.24). The linear least squares predictor $\hat{X}_{t+m}$ has the form

$$\tilde{X}_{t+m} = \sum_{u=0}^{\infty} a_u X_{t-u}, \qquad (10.1.34)$$

where $\{a_u\}$ is a sequence of constants to be determined. However, since each X_t is a linear combination of $\varepsilon_t, \varepsilon_{t-1}, \ldots$, we may equally well write $\tilde{X}_{t+m}$ as,

$$\tilde{X}_{t+m} = \sum_{u=0}^{\infty} b_u \varepsilon_{t-u}, \qquad (10.1.35)$$

where $\{b_u\}$ is another sequence of constants, and if we determine either the $\{a_u\}$ or the $\{b_u\}$ sequence the problem will be solved.

We may note that since $\text{var}\{\hat{X}_{t+m}\} \leqslant \text{var}\{X_{t+m}\} < \infty$, the coefficients in (10.1.35) must satisfy

$$\sum_{u=0}^{\infty} b_u^2 < \infty.$$

Now we may express X_{t+m} as

$$X_{t+m} = \sum_{u=0}^{\infty} g_u \varepsilon_{t+m-u}$$

$$= \sum_{u=0}^{m-1} g_u \varepsilon_{t+m-u} + \sum_{u=m}^{\infty} g_u \varepsilon_{t+m-u}. \qquad (10.1.36)$$

The first term above represents that part of X_{t+m} which involves future ε_t's, and so represents the "unpredictable" part of X_{t+m}. (Since the $\{\varepsilon_t\}$ are uncorrelated, past observations have no linear predictive value for future ε_t's.) On the other hand, the second term depends only on present and past values of ε_t, and thus represents the "predictable" past of X_{t+m}. It is therefore fairly clear that the linear least squares predictor of X_{t+m} is given simply by the second term in (10.1.36), i.e.

$$\tilde{X}_{t+m} = \sum_{u=m}^{\infty} g_u \varepsilon_{t+m-u} = \sum_{u=0}^{\infty} g_{u+m} \varepsilon_{t-u}. \qquad (10.1.37)$$

In fact, since the $\{\varepsilon_t\}$ form an orthogonal "basis" for $\{X_t\}$, it follows immediately from the geometrical approach discussed above that $\tilde{X}_{t+m}$ is obtained from X_{t+m} by setting the first term in (10.1.36) to zero, but a direct proof of (10.1.37) is easily constructed as follows.

From (10.1.35), (10.1.36), we have

$$\mathcal{M}(m) = E[\{X_{t+m} - \tilde{X}_{t+m}\}^2] = E\left[\left\{\sum_{u=0}^{m-1} g_u \varepsilon_{t+m-u} + \sum_{u=0}^{\infty} (g_{u+m} - b_u)\varepsilon_{t-u}\right\}^2\right]$$

$$= \sigma_\varepsilon^2\left\{\left(\sum_{u=0}^{m-1} g_u^2\right) + \sum_{u=0}^{\infty} (g_{u+m} - b_u)^2\right\}.$$

The first term above is independent of the choice of the $\{b_u\}$, and the second term is clearly minimized by choosing $b_u = g_{u+m}$ ($u = 0, 1, 2, \ldots$), leading to (10.1.37).

With this choice of $\{b_u\}$ the second term vanishes, and we have

$$\sigma_m^2 = E\{X_{t+m} - \tilde{X}_{t+m}\}^2 = \sigma_\varepsilon^2\left(\sum_{u=0}^{m-1} g_u^2\right), \qquad (10.1.38)$$

this expression being known as the *"m-step prediction variance"*.

In particular, when $m = 1$,

$$X_{t+1} - \tilde{X}_{t+1} = \varepsilon_{t+1}, \qquad (10.1.39)$$

(recalling, from (10.1.21), that $g_0 = 1$). Hence, ε_{t+1} represents the *"one-step prediction error"*, and if we re-write (10.1.39) as

$$X_{t+1} = \tilde{X}_{t+1} + \varepsilon_{t+1} \qquad (10.1.40)$$

we see that ε_{t+1} *is the essentially "new" part of* X_{t+1}, in the sense that it represents that part of X_{t+1} which is not linearly dependent on past observations. For this reason, $\{\varepsilon_t\}$ is sometimes called the *"innovations process"* of $\{X_t\}$. In geometrical terms, it denotes that part of X_{t+1} which is orthogonal to the sub-space spanned by $\{X_t, X_{t-1}, \ldots\}$, as illustrated in Fig. 10.3.

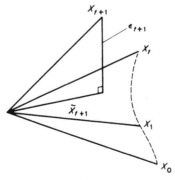

Fig. 10.3. The innovations process.

We have further, from (10.1.39),

$$\sigma_1^2 = E[\{X_{t+1} - \tilde{X}_{t+1}\}^2] = \sigma_\varepsilon^2$$

$$= 2\pi \exp\left\{\frac{1}{2\pi}\int_{-\pi}^{\pi} \log h(\omega)\, d\omega\right\}, \qquad (10.1.41)$$

on using (10.1.25), (10.1.23). This result, which is quite a basic one, gives an expression for the one-step prediction variance in terms of the spectral density function $h(\omega)$.

If we wish to write $\tilde{X}_{t+m}$ explicitly as a function of $X_t, X_{t-1}, \ldots$ (i.e. in the form (10.1.34)), then we first write (10.1.37) as

$$\tilde{X}_{t+m} = G^{(m)}(B)\varepsilon_t, \qquad (10.1.42)$$

where

$$G^{(m)}(z) = \sum_{u=0}^{\infty} g_{u+m} z^u. \qquad (10.1.43)$$

Assuming that $G(z)$ is analytic in $|z| \leq 1$, we have from (10.1.28),

$$\varepsilon_t = G^{-1}(B)X_t, \qquad (10.1.44)$$

and substituting this in (10.1.42) gives,

$$\tilde{X}_{t+m} = A(B)X_t, \qquad (10.1.45)$$

with

$$A(z) = G^{(m)}(z)G^{-1}(z). \qquad (10.1.46)$$

Expanding $A(z)$ as a power series in z now gives the required form. Note that since $\sum_{u=0}^{\infty} g_u^2 < \infty$, $G^{(m)}(z)$ is analytic in $|z| < 1$, and hence, since $G(z)$ has no zeros in $|z| < 1$, $A(z)$ is analytic in $|z| < 1$.

It is useful to introduce a special notation to describe the relationship between $G^{(m)}(z)$ and $G(z)$. For any function, $K(z)$, with Laurent expansion,

$$K(z) = \sum_{u=-\infty}^{\infty} k_u z^u,$$

we define (following Whittle (1963)),

$$[K(z)]_+ = \sum_{u=0}^{\infty} k_u z^u, \qquad (10.1.47)$$

and

$$[K(z)]_- = \sum_{u=-\infty}^{-1} k_u z^u. \qquad (10.1.48)$$

Then, from (10.1.43) we may write $G^{(m)}(z)$ as,

$$G^{(m)}(z) = [z^{-m}G(z)]_+.$$ (10.1.49)

The expression for $A(z)$ can now be written as

$$A(z) = [z^{-m}G(z)]_+/G(z).$$ (10.1.50)

If we consider the sequence of predictors $\{\tilde{X}_{t+m}\}$ for different values of t (with m fixed), then this forms a new process, which, from (10.1.45), may be regarded as the output of *a linear filter* operating on $\{X_t\}$. The transfer function of this filter is, of course, $A(e^{-i\omega})$.

The function $A(e^{-i\omega})$ is therefore called the *predictor transfer function*, and $A(z)$ is called the *predictor generating function*.

Example 1

Consider the AR(1) model,

$$X_t - aX_{t-1} = \varepsilon_t, \qquad |a| < 1.$$ (10.1.51)

Writing this as

$$X_t = (1 - aB)^{-1}\varepsilon_t,$$ (10.1.52)

we see that here,

$$G(z) = (1 - az)^{-1} = 1 + az + a^2 z^2 + \dots,$$

i.e.

$$X_t = \varepsilon_t + a\varepsilon_{t-1} + a^2\varepsilon_{t-2} + \dots$$

(cf. (3.5.21)). Thus, $g_u = a^u$ and (10.1.37) gives

$$\tilde{X}_{t+m} = a^m\varepsilon_t + a^{m+1}\varepsilon_{t-1} + a^{m+2}\varepsilon_{t-2} + \dots$$
$$= a^m(\varepsilon_t + a\varepsilon_{t-1} + a^2\varepsilon_{t-2} + \dots).$$

Hence we may write

$$\tilde{X}_{t+m} = a^m X_t.$$ (10.1.53)

For this example, once we have expressed $\tilde{X}_{t+m}$ in terms of the $\{\varepsilon_t\}$ the transformation to (10.1.53) is immediate. We can, of course, derive the same result more formally from (10.1.50) by noting that here,

$$[z^{-m}G(z)]_+ = [z^{-m}(1 - az)^{-1}]_+$$
$$= (a^m + a^{m+1}z + a^{m+2}z^2 + \dots)$$
$$= a^m(1 - az)^{-1}.$$

Hence,

$$A(z) = [z^{-m}G(z)]_+ / G(z) = a^m,$$

which means that, $a_0 = a^m$, $a_u = 0$, $u \geq 1$, leading again to (10.1.53).

The m-step prediction variance is given by

$$\sigma_m^2 = \sigma_\varepsilon^2 \left(\sum_{u=0}^{m-1} a^{2u} \right) = \sigma_\varepsilon^2 \left(\frac{1 - a^{2m}}{1 - a^2} \right).$$

This result shows that for the AR(1) model the linear least squares predictor of X_{t+m} depends only on the most recent observation X_t and does not involve X_{t-1}, X_{t-2}, ..., a feature which we would expect bearing in mind the Markov nature of the AR(1) model. We note, for future reference, that *the solution* (10.1.53) *can be derived directly from the difference equation* (10.1.51) *by setting future ε_t's to zero*, i.e. by setting

$$\varepsilon_{t+1} = \varepsilon_{t+2} = \ldots = \varepsilon_{t+m} = 0. \tag{10.1.54}$$

We note further that as $m \to \infty$, $\tilde{X}_{t+m} \to 0$ (the unconditional mean of X_{t+m}), and

$$E\{X_{t+m} - \tilde{X}_{t+m}\}^2 \to \text{var}\{X_t\}.$$

In other words, as we try to predict further and further ahead into the future, the information provided by past observations becomes negligible, the predictor then becomes simply the unconditional mean, and the predictor variance becomes identical with var$\{X_t\}$.

For the more general AR(k) model it turns out, not surprisingly, *that $\tilde{X}_{t+m}$ depends only on the last k observed values of X_t*, and may be obtained by solving the AR(k) difference equation with the future $\{\varepsilon_t\}$ set to zero, as in (10.1.54). Writing the AR(k) model as

$$X_t + a_1 X_{t-1} + \ldots + a_k X_{t-k} = \varepsilon_t,$$

we see that, in particular for $m = 1$ (setting $\varepsilon_{t+1} = 0$),

$$\tilde{X}_{t+1} = -a_1 X_t - \ldots - a_k X_{t-k+1}.$$

Thus the AR(k) model provides in itself a natural one-step predictor.

Example 2

We now consider the MA(1) model,

$$X_t = \varepsilon_t - b\varepsilon_{t-1} \qquad (|b| < 1).$$

Here,

$$g_0 = 1, \qquad g_1 = -b, \qquad g_u = 0, \qquad u \geq 2, \qquad \text{and} \quad G(z) = (1 - bz).$$

Hence,

$$[z^{-m}G(z)]_+ = \begin{cases} -b, & m = 1 \\ 0, & m \geq 2, \end{cases}$$

and consequently,

$$\tilde{X}_{t+1} = -b\varepsilon_t, \qquad \tilde{X}_{t+m} = 0, \qquad m \geq 2,$$

(as may be verified directly from (10.1.37)).

From (10.1.50) we now have, for $m = 1$,

$$A(z) = -b(1-bz)^{-1} = -b(1+bz+b^2z^2+\ldots),$$

so that $a_u = (-b)^{u+1}$, and thus,

$$\tilde{X}_{t+1} = \sum_{u=0}^{\infty} (-b)^{u+1} X_{t-u}.$$

For the general MA(l) model,

$$X_t = \varepsilon_t + b_1\varepsilon_{t-1} + \ldots + b_l\varepsilon_{t-l},$$

it follows exactly as above that (using (10.1.37)),

$$\tilde{X}_{t+m} = \begin{cases} b_m\varepsilon_t + \ldots + b_l\varepsilon_{t+m-l}, & m \leq l, \\ 0, & m > l. \end{cases}$$

Continuous parameter processes

The analogous results for continuous parameter processes may be derived as follows. We assume that $h(\omega)$ satisfies (10.1.29), so that $h(\omega)$ admits the factorization (10.1.30), and correspondingly $X(t)$ admits the linear representation (10.1.32). We now write

$$X(t+m) = \int_0^\infty g(u)\varepsilon(t+m-u)\,du$$

$$= \int_0^m g(u)\varepsilon(t+m-u)\,du + \int_m^\infty g(u)\varepsilon(t+m-u)\,du. \qquad (10.1.55)$$

Given $\{X(s); s \leq t\}$, we may argue as in the discrete case that the first term in (10.1.55) represents the "unpredictable" part of $X(t+m)$, and it follows, as previously, that the linear least squares predictor of $X(t+m)$ is given by the second term in (10.1.55), i.e.

$$\tilde{X}(t+m) = \int_m^\infty g(u)\varepsilon(t+m-u)\,du = \int_0^\infty g(u+m)\varepsilon(t-u)\,du. \qquad (10.1.56)$$

The prediction error is,

$$\{X(t+m)-\tilde{X}(t+m)\}=\int_{0}^{m}g(u)\varepsilon(t+m-u)\,du,$$

and the m-step prediction variance is,

$$\sigma_{m}^{2}=E\{X(t+m)-\tilde{X}(t+m)\}^{2}=\sigma_{w}^{2}\left\{\int_{0}^{m}g^{2}(u)\,du\right\} \qquad (10.1.57)$$

(where $\sigma_{w}^{2}\,dt=\text{var}\{dW(t)\}$, with $dW(t)=\varepsilon(t)\,dt$). We may formally express $\varepsilon(t)$ in terms of $X(t)$ by inverting (10.1.33), giving

$$\varepsilon(t)=G^{-1}(D)X(t), \qquad (10.1.58)$$

(where $G(z)$ is given by (10.1.31)). In particular, if $G^{-1}(D)$ is a finite order polynomial in D, as is the case when $X(t)$ is a finite order AR process, then (10.1.58) gives a simple expression for $\varepsilon(t)$ in terms of $X(t)$ and its derivatives. Writing $\varepsilon(t-u)=e^{-uD}\varepsilon(t)$ in (10.1.56) and substituting (10.1.58) gives,

$$\tilde{X}(t+m)=\left\{\int_{0}^{\infty}g(u+m)\,e^{-uD}G^{-1}(D)\,du\right\}X(t),$$

or

$$\tilde{X}(t+m)=A(D)X(t), \qquad (10.1.59)$$

where

$$A(z)=G^{-1}(z)\left\{\int_{0}^{\infty}g(u+m)\,e^{-uz}\,dz\right\}.$$

We now extend the $[\ ]_{+}$ and $[\ ]_{-}$ notation as follows: for any function $K(z)$ with two-sided Laplace transform,

$$K(z)=\int_{-\infty}^{\infty}k(u)\,e^{-uz}\,du,$$

we write

$$[K(z)]_{+}=\int_{0}^{\infty}k(u)\,e^{-uz}\,du,\qquad [K(z)]_{-}=\int_{-\infty}^{0}k(u)\,e^{-uz}\,du.$$

Then,

$$[e^{mz}G(z)]_{+}=\left[\int_{0}^{\infty}g(u)\,e^{-z(u-m)}\,du\right]_{+}=\left[\int_{-m}^{\infty}g(u+m)\,e^{-uz}\,du\right]_{+}$$

$$=\int_{0}^{\infty}g(u+m)\,e^{-uz}\,du.$$

Hence, $A(z)$ can be written as

$$A(z) = G^{-1}(z)[e^{mz}G(z)]_+. \qquad (10.1.60)$$

Note that although we started out with an integral representation of $\tilde{X}(t+m)$, (namely (10.1.56)) we must allow the operator $A(D)$ in (10.1.59) to be interpreted fairly widely. In particular, we must allow $A(D)$ to be, e.g., a finite polynomial in D—corresponding to the case where $\tilde{X}(t+m)$ is a linear combination of the derivatives of $X(t)$—since, as we shall see, this form arises in the study of AR processes.

Regarding $\tilde{X}(t+m)$ as the output of a *linear filter* operating on $X(t)$, we see from (10.1.59) that this filter has transfer function $A(i\omega)$. Thus, in the continuous parameter case, $A(i\omega)$ is called the *predictor transfer function*. If we wish to compute $\tilde{X}(t+m)$ by synthesizing this filter via, say, an electrical network, then it would usually be more convenient to characterize the filter in terms of its transfer function rather than its impulse function, i.e. it would be more useful to know the form of $A(i\omega)$ rather than its Fourier transform $a(u)$.

Example 3

For the continuous parameter AR(1) model we have

$$\dot{X}(t) + \alpha X(t) = \varepsilon(t) \qquad (\alpha > 0), \qquad (10.1.61)$$

the stationary solution of which is,

$$X(t) = (D + \alpha)^{-1}\varepsilon(t)$$

$$= \int_0^\infty e^{-\alpha u}\varepsilon(t - u)\, du,$$

(cf. (3.7.20)). Thus here, $g(u) = e^{-\alpha u}$, and (10.1.56) gives

$$\tilde{X}(t+m) = \int_0^\infty e^{-\alpha(u+m)}\varepsilon(t-u)\, du$$

$$= e^{-\alpha m}\int_0^\infty e^{-\alpha u}\varepsilon(t-u)\, du$$

$$= e^{-\alpha m}X(t). \qquad (10.1.62)$$

As in the discrete parameter AR(1) model, we see that $\tilde{X}(t+m)$ depends only on the most recent observation. Also, we note once again that, for the AR(1) model, once $\tilde{X}(t+m)$ has been expressed in terms of the $\{\varepsilon(t)\}$ process the transformation to (10.1.62) is immediate. However, as in the discrete case, we can obtain the result more formally by observing that, with $G(z) = 1/(z + \alpha)$,

$$\left[e^{mz} \frac{1}{z+\alpha} \right]_+ = \frac{e^{-\alpha m}}{z+\alpha},$$

and hence, from (10.1.60),

$$A(z) = (z + \alpha) \frac{e^{-\alpha m}}{z + \alpha} = e^{-\alpha m}.$$

The application of (10.1.59) now gives the result (10.1.63).
The m-step prediction variance is

$$\sigma_m^2 = \sigma_w^2 \left(\int_0^m e^{-2\alpha u} \, du \right) = \frac{\sigma_w^2}{2\alpha} (1 - e^{-2m\alpha}).$$

Example 4

Consider the continuous parameter AR(2) model,

$$\ddot{X}(t) + \alpha_1 \dot{X}(t) + \alpha_2 X(t) = \varepsilon(t) \qquad (\alpha_1 > 0, \alpha_2 > 0).$$

This can be written as

$$\alpha(D) X(t) = \varepsilon(t)$$

where

$$\alpha(z) = z^2 + \alpha_1 z + \alpha_2 = (z - c_1)(z - c_2),$$

say, and the conditions on α_1, α_2 ensure that the roots, c_1, c_2, both lie in the left-half plane. Thus, for this model,

$$G(z) = \alpha^{-1}(z) = (z - c_1)^{-1} (z - c_2)^{-1},$$

and $X(t)$ may be written (cf. (3.7.47)),

$$X(t) = \frac{1}{(c_1 - c_2)} \int_0^\infty (e^{c_1 u} - e^{c_2 u}) \varepsilon(t - u) \, du.$$

Hence,

$$g(u) = (c_1 - c_2)^{-1} \{ e^{c_1 u} - e^{c_2 u} \},$$

and therefore,

$$\int_0^\infty g(u+m) e^{-uz} \, dz = \frac{1}{(c_1 - c_2)} \int_0^\infty \{ e^{c_1(u+m)} - e^{c_2(u+m)} \} e^{-uz} \, du$$

$$= \frac{1}{(c_1 - c_2)} \left[\frac{e^{c_1 m}}{z - c_1} - \frac{e^{c_2 m}}{z - c_2} \right].$$

We thus obtain,

$$A(z) = \frac{1}{(c_1 - c_2)}[z(e^{c_1 m} - e^{c_2 m}) - (c_2 e^{c_1 m} - c_1 e^{c_2 m})],$$

and the corresponding form of $\tilde{X}(t + m)$ is

$$\tilde{X}(t + m) = A(D)X(t) = \frac{1}{(c_1 - c_2)}[(c_1 e^{c_2 m} - c_2 e^{c_1 m})X(t)$$
$$+ (e^{c_1 m} - e^{c_2 m})\dot{X}(t)].$$

If we now write $c_1 = -a + ib$, $c_2 = -a - ib$ $(a > 0)$, then

$$\frac{e^{c_1 m} - e^{c_2 m}}{c_1 - c_2} = \frac{1}{b} e^{-am} \sin bm,$$

and

$$\frac{c_1 e^{c_2 m} - c_2 e^{c_1 m}}{c_1 - c_2} = \frac{a}{b} e^{-am} \sin bm + e^{-am} \cos bm.$$

Hence, $\tilde{X}(t + m)$ may be written

$$\tilde{X}(t + m) = e^{-am}\left(\cos bm + \frac{a}{b} \sin bm\right)X(t) + \frac{e^{-am}}{b} \sin bm \dot{X}(t).$$

For the AR(1) model $A(z)$ turned out to be a constant, while for the AR(2) model $A(z)$ is a linear function of z. It is not difficult to see that for the general continuous parameter AR(k) model, $A(z)$ is a polynomial of degree $(k - 1)$ (see, e.g., Whittle (1963), p. 39). Thus, for the AR(k) case, $\tilde{X}(t + m)$ is a linear function of $X(t)$ and its first $(k - 1)$ derivatives.

In fact, $\tilde{X}(t + m)$ may be obtained by solving the differential equation for the AR(k) model with $\varepsilon(t)$ set to zero $(s > t)$, and using the values of $X(t), \dot{X}(t), \ddot{X}(t), \ldots, X^{(k-1)}(t)$ as the "initial conditions".

10.1.4 The Wiener Approach

The Wiener approach is based on a more direct line of attack in which we write $\tilde{X}_{t+m}$ in the form (10.1.34), namely,

$$\tilde{X}_{t+m} = \sum_{u=0}^{\infty} a_u X_{t-u}, \tag{10.1.63}$$

substitute this directly into (10.1.1), and minimize the resulting expression for $\mathcal{M}(m)$ with respect to the $\{a_u\}$. Following this procedure we obtain

$$
\begin{aligned}
\mathcal{M}(m) &= E[\{X_{t+m} - \tilde{X}_{t+m}\}^2] \\
&= \left[R(0) - 2 \sum_{u=0}^{\infty} a_u R(m+u) + \sum_{u=0}^{\infty} \sum_{v=0}^{\infty} a_u a_v R(u-v) \right],
\end{aligned}
$$

where, as previously, $R(s)$ denotes the autocovariance function of X_t. Differentiating $\mathcal{M}(m)$ with respect to a_u and setting the derivative equal to zero gives,

$$
\sum_{v=0}^{\infty} a_v R(u-v) = R(m+u), \qquad u = 0, 1, 2, \ldots \qquad (10.1.64)
$$

(It is easy to check, by the usual variational argument, that the stationary value of $\mathcal{M}(m)$ determined by (10.1.64) is indeed a minimum—see, e.g., Bartlett (1955), p. 200.) This equation (or rather, its continuous parameter analogue), is known as the *Wiener–Hopf equation*, and it has exactly the same form as the "normal equations" which we would obtain from a least squares regression analysis of X_{t+m} on $X_t, X_{t-1}, \ldots$. However, in the usual regression situation we have only a finite number of parameters to determine, in which case the "normal equations" form a finite set of linear equations which can be solved by standard methods. If we restricted $\tilde{X}_{t+m}$ to be a linear function of a finite number of past observations, say $X_t, X_{t-1}, \ldots, X_{t-n}$, (giving the so-called "finite memory" predictor) then (10.1.64) would similarly reduce to a finite number of linear equations which could be solved directly. In the general case, however, (10.1.64) involves an infinite number of unknowns $\{a_v\}$, and its solution requires some care. Since the LHS of (10.1.64) has the form of a convolution between the $\{a_v\}$ and $\{R(s)\}$ sequences, it may appear, at first sight, that we could solve this equation fairly easily by taking the Fourier transform of each side. Unfortunately, this simple approach does not work, the difficulty being that (10.1.64) *is valid only for $u = 0, 1, 2, \ldots$, and does not hold for negative u.* We therefore have to adopt a different approach, and we start by re-writing the expression for $\mathcal{M}(m)$ in its frequency domain form. Using the spectral representation of X_t (viz., (10.1.7)), we may write

$$
X_{t+m} = \int_{-\pi}^{\pi} e^{i(t+m)\omega} \, dZ(\omega),
$$

while, from (10.1.63), we may also write,

$$
\tilde{X}_{t+m} = \int_{-\pi}^{\pi} e^{it\omega} A(e^{-i\omega}) \, dZ(\omega),
$$

where, as in Section 10.1.3, $A(z) = \sum_{u=0}^{\infty} a_u z^u$. We now have

$$X_{t+m} - \tilde{X}_{t+m} = \int_{-\pi}^{\pi} e^{it\omega} \{ e^{im\omega} - A(e^{-i\omega}) \} \, dZ(\omega).$$

Using the fact that the $\{ dZ(\omega) \}$ are orthogonal, with $E\{ |dZ(\omega)|^2 \} = h(\omega) \, d\omega$, we obtain,

$$\mathcal{M}(m) = \int_{-\pi}^{\pi} |e^{im\omega} - A(e^{-i\omega})|^2 h(\omega) \, d\omega. \tag{10.1.65}$$

The problem of finding that sequence $\{ a_u \}$ which minimizes $\mathcal{M}(m)$ now becomes one of finding that function, $A(e^{-i\omega})$, which minimizes (10.1.65). Since (10.1.65) is a weighted average (over frequency) of the squared difference between $e^{im\omega}$ and $A(e^{-i\omega})$ (with weighting proportional to $h(\omega)$), we may regard that form of $A(e^{-i\omega})$ which minimizes $\mathcal{M}(m)$ as providing the *"best mean approximation"* to the function $e^{im\omega}$. However, we must remember that $A(e^{-i\omega})$ cannot be chosen freely; since $\tilde{X}_{t+m}$ has to be a function of present and past observations only, $A(e^{-i\omega})$ must correspondingly be a *"backward transform"*, i.e. it must have a *one-sided* Fourier series involving only *negative powers* of $e^{i\omega}$. To determine the optimal form of $A(e^{-i\omega})$ within this class of functions we use the canonical factorization of $h(\omega)$ (as given by (10.1.20)), and substituting in (10.1.65) we obtain,

$$\mathcal{M}(m) = e^{c_0} \int_{-\pi}^{\pi} |e^{im\omega} G(e^{-i\omega}) - A(e^{-i\omega}) G(e^{-i\omega})|^2 \, d\omega. \tag{10.1.66}$$

By the construction of $G(z)$, $G(e^{-i\omega})$ is clearly a backward transform, and it is then easily seen that the product, $A(e^{-i\omega}) G(e^{-i\omega})$ must also be a backward transform. It is now intuitively obvious that $\mathcal{M}(m)$ is minimized by choosing $A(e^{-i\omega})$ so that $\{ A(e^{-i\omega}) G(e^{-i\omega}) \}$ is the "backward part" of $\{ e^{im\omega} G(e^{-i\omega}) \}$. To establish this result formally we decompose $e^{im\omega} G(e^{-i\omega})$ as

$$e^{im\omega} G(e^{-i\omega}) = G_1(e^{-i\omega}) + G_2(e^{-i\omega}), \tag{10.1.67}$$

where, using the notation of (10.1.47), (10.1.48),

$$G_1(z) = [z^{-m} G(z)]_+, \qquad G_2(z) = [z^{-m} G(z)]_-.$$

(The function $G_1(e^{-i\omega})$ is called the *backward transform*, and $G_2(e^{-i\omega})$ the *forward transform*, of $e^{im\omega} G(e^{-i\omega})$.) Substituting (10.1.67) into (10.1.66), and using the result that for any backward transform $a(e^{-i\omega})$ and any forward transform $b(e^{-i\omega})$,

$$\int_{-\pi}^{\pi} a(e^{-i\omega}) b(e^{i\omega}) \, d\omega = 0$$

(which follows immediately from the orthogonality properties of the sine and cosine functions; cf. (4.2.4)), we finally obtain,

$$e^{-c_0}\mathcal{M}(m) = \int_{-\pi}^{\pi} |G_2(e^{-i\omega})|^2 \, d\omega + \int_{-\pi}^{\pi} |G_1(e^{-i\omega}) - A(e^{-i\omega})G(e^{-i\omega})|^2 \, d\omega.$$

$$(10.1.68)$$

The first term is independent of $A(e^{-i\omega})$, and the second term is clearly minimized by choosing

$$A(e^{-i\omega}) = G_1(e^{-i\omega})/G(e^{-i\omega}).$$ (10.1.69)

Writing $z = e^{-i\omega}$ and using the above expression for $G_1(z)$, we may write (10.1.69) as

$$A(z) = [z^{-m}G(z)]_+/G(z),$$ (10.1.70)

which is, of course, identical to the Kolmogorov solution (10.1.50). Since the Fourier coefficients of $\{e^{im\omega}G(e^{-i\omega})\}$ are given by

$$\frac{1}{2\pi}\int_{-\pi}^{\pi} e^{i\omega'(m+u)}G(e^{-i\omega'}) \, d\omega', \qquad u = 0, \pm 1, \pm 2, \dots,$$

(10.1.69) may be written explicitly in the form

$$A(e^{-i\omega}) = \sum_{u=0}^{\infty} e^{-i\omega u}\left\{\int_{-\pi}^{\pi} e^{i\omega'(m+u)}G(e^{-i\omega'}) \, d\omega'\right\}\Big/2\pi G(e^{-i\omega}).$$ (10.1.71)

With this choice of $A(e^{-i\omega})$ the second term in (10.1.69) vanishes, and the m-step predictor variance is given by

$$\sigma_m^2 = E[\{X_{t+m} - \tilde{X}_{t+m}\}^2] = e^{c_0}\int_{-\pi}^{\pi} |G_2(e^{-i\omega})|^2 \, d\omega.$$ (10.1.72)

Continuous parameter processes

Here, we write the predictor as,

$$\tilde{X}(t+m) = \int_0^{\infty} a(u)X(t-u) \, du,$$ (10.1.73)

and substituting this in (10.1.1) and minimizing the resulting expression with respect to $a(u)$ we find, analogous to (10.1.64),

$$\int_0^{\infty} a(v)R(u-v) \, dv = R(u+m), \qquad u > 0.$$ (10.1.74)

This is known as the *Wiener–Hopf integral equation*, and, as in the discrete parameter case, it cannot be solved by simple transform techniques since it holds only over the "half-range", $0 \le u < \infty$. Proceeding as in the previous case we may write $\mathcal{M}(m)$ as

$$\mathcal{M}(m) = \int_{-\infty}^{\infty} |e^{im\omega} - A(i\omega)|^2 h(\omega) \, d\omega,$$

where

$$A(z) = \int_0^{\infty} a(u) e^{-uz} \, du.$$

(Again, we must interpret the RHS of (10.1.73) fairly widely as a general "linear operator" acting on $X(t)$ (cf. Section 4.12.5), and not restrict $a(u)$ to be a "well behaved" function. Thus, where appropriate, this expression could represent, for example, a linear function of the derivatives of $X(t)$, corresponding to which $A(z)$ would have a polynomial form.) Assuming that $h(\omega)$ satisfies (10.1.29), and using the canonical factorization of $h(\omega)$ given by (10.1.30) we obtain,

$$\mathcal{M}(m) = \int_{-\infty}^{\infty} |e^{im\omega} G(i\omega) - G(i\omega) A(i\omega)|^2 \, d\omega.$$

Decomposing $e^{im\omega} G(i\omega)$ into the sum of its backward and forward transforms, viz.,

$$e^{im\omega} G(i\omega) = [e^{im\omega} G(i\omega)]_+ + [e^{im\omega} G(i\omega)]_-$$

and substituting this into $\mathcal{M}(m)$ we find, as previously,

$$\mathcal{M}(m) = \int_{-\infty}^{\infty} |[e^{im\omega} G(i\omega)]_-|^2 \, d\omega + \int_{-\infty}^{\infty} |[e^{im\omega} G(i\omega)]_+ - A(i\omega) G(i\omega)|^2 \, d\omega.$$

Hence, $\mathcal{M}(m)$ is minimized by choosing

$$A(z) = [e^{mz} G(z)]_+ / G(z), \tag{10.1.75}$$

in agreement with (10.1.60). Corresponding to (10.1.71), $A(i\omega)$ may be written explicitly as,

$$A(i\omega) = \frac{1}{2\pi G(i\omega)} \int_0^{\infty} e^{-i\omega u} \left\{ \int_{-\infty}^{\infty} e^{i\omega'(u+m)} G(i\omega') \, d\omega' \right\} du \tag{10.1.76}$$

Example 5

Let $\{X_t\}$ be a discrete parameter process with spectral density function,

$$h(\omega) = \frac{\sigma_\varepsilon^2}{2\pi} \frac{\{(1+b^2) - 2b \cos \omega\}}{\{(1+a^2) - 2a \cos \omega\}} \qquad (|a| < 1, \quad |b| < 1, \quad a \ne b).$$

Then $h(\omega)$ may be written as

$$h(\omega) = \frac{\sigma_\varepsilon^2}{2\pi} \left| \frac{1 - b e^{-i\omega}}{1 - a e^{-i\omega}} \right|^2 = \frac{\sigma_\varepsilon^2}{2\pi} G(e^{-i\omega})G(e^{i\omega}), \qquad (10.1.77)$$

where

$$G(z) = (1 - bz)/(1 - az).$$

Clearly $G(z)$ has no poles or zeros in $|z| \leq 1$ (by the conditions on a and b), and $G(0) = 1$. Hence (10.1.76) provides the canonical factorization of $h(\omega)$, and from (10.1.70) we have

$$A(z) = [z^{-m} G(z)]_+ / G(z).$$

But, we may expand $G(z)$ as

$$G(z) = 1 + (a - b)z(1 - az)^{-1}$$

$$= 1 + (a - b)z \left\{ \sum_{u=0}^\infty a^u z^u \right\}.$$

Hence,

$$[z^{-m} G(z)]_+ = (a - b)z^{-(m-1)} \sum_{u=m-1}^\infty a^u z^u$$

$$= \frac{(a - b)a^{m-1}}{1 - az}.$$

Thus,

$$A(z) = a^{m-1}(a - b)(1 - bz)^{-1},$$

and the coefficient of z^u in the power series expansion of $A(z)$ is

$$a_u = a^{m-1}(a - b)b^u.$$

We may therefore write the least squares predictor $\tilde{X}_{t+m}$ as

$$\tilde{X}_{t+m} = a^{m-1}(a - b) \sum_{u=0}^\infty b^u X_{t-u}.$$

Moreover, the m-step predictor variance is given by (10.1.38) as

$$\sigma_m^2 = E\{X_{t+m} - \tilde{X}_{t+m}\}^2 = \sigma_\varepsilon^2 \left\{ \sum_{u=0}^{m-1} g_u^2 \right\}.$$

From (10.1.78) we have

$$g_u = \begin{cases} (a - b)a^{u-1}, & u \neq 0 \\ 1, & u = 0. \end{cases}$$

Hence,

$$\sigma_m^2 = \sigma_\varepsilon^2 \left[1 + \sum_{u=1}^{m-1} (a-b)^2 a^{2(u-1)} \right] = \sigma_\varepsilon^2 \left[1 + (a-b)^2 \left\{ \frac{1 - a^{2(m-1)}}{1 - a^2} \right\} \right].$$

We may note that the time domain description corresponding to the spectral density function (10.1.77) is the ARMA(1, 1) model,

$$X_t - aX_{t-1} = \varepsilon_t - b\varepsilon_{t-1}.$$

Use of Kolmogorov–Wiener predictors in practice

The practical usefulness of the Kolmogorov–Wiener predictors depends very much on our ability to factorize the spectral density function in the canonical form (10.1.20) or (10.1.30). In general this is quite a tricky operation, particularly if we are simply given a set of observations on $\{X_t\}$ and have no prior knowledge as to the theoretical form of $h(\omega)$. However, Bhansali (1974, 1977a) has constructed a numerical procedure for obtaining approximations to the Kolmogorov–Wiener predictors, based on estimating $h(\omega)$ from the data and then carrying out (effectively) the analysis following (10.1.16). (See also Whittle (1963), p. 35.)

For discrete parameter processes there is one case where the factorization of $h(\omega)$ can be effected quite simply, namely when $h(\omega)$ *is a rational function of* $e^{-i\omega}$. We have already met this situation in Example 5 where, as we saw, the factorization could be obtained by inspection. The same technique can be applied to general rational functions. For, suppose $h(\omega)$ has the form

$$h(\omega) = P(e^{-i\omega})/Q(e^{-i\omega}),$$

where $P(z)$, $Q(z)$ are finite order polynomials. Since $h(z)$ has a symmetric Laurent expansion when $z = e^{-i\omega}$, if follows that if z_0 is a zero of $P(z)$ so is $1/z_0$, and similarly for $Q(z)$. We may therefore write $h(z)$ as

$$h(z) = \frac{C \prod_k (1 - b_k z)(1 - b_k z^{-1})}{\prod_k (1 - a_k z)(1 - a_k z^{-1})},$$

where $|b_k| \leq 1$, $|a_k| < 1$, all k (and C is a constant). The function $G(z)$ is now clearly given by

$$G(z) = \frac{\prod_k (1 - b_k z)}{\prod_k (1 - a_k z)}.$$

However, if $h(\omega)$ is known to be a rational function of $e^{-i\omega}$ then X_t conforms to an ARMA model (cf. Section 4.12.4). In this case we can fit an ARMA model to X_t, and obtain the predictors directly from the model via a recursive technique which we describe in Section 10.2.

10.1.5 The Wold Decomposition

So far in our discussion of the prediction problem we have assumed that the process has a purely continuous spectrum with spectral density function $h(\omega)$ satisfying (10.1.11), viz.,

$$P = \int_{-\pi}^{\pi} \log\{h(\omega)\}\, d\omega > -\infty \qquad (10.1.79)$$

(or satisfying (10.1.29) in the continuous parameter case). The question now arises as to what happens when (10.1.79) does not hold. We know from (10.1.41) that, when $P > -\infty$, the one-step prediction variance is $2\pi e^{P/2\pi}$. Thus, we would expect that *when $P = -\infty$ the one-step prediction variance will be zero, i.e. we will be able to predict X_{t+1} perfectly from a linear function of $X_t, X_{t-1}, X_{t-2}, \ldots$*.

For a general stationary discrete parameter process $\{X_t\}$ (whose spectrum need not necessarily be purely continuous) let us consider the sequence of prediction variances,

$$\sigma_m^2 = E[\{X_{t+m} - \tilde{X}_{t+m}\}^2], \qquad m = 1, 2, 3, \ldots,$$

where $\tilde{X}_{t+m}$ is the projection of X_{t+m} on H_t, the space spanned by X_t, $X_{t-1}, \ldots$. Note that, by the stationarity of X_t, σ_m^2 does not depend on t. Clearly, $H_s \subseteq H_t$, $s \leq t$, and hence the $\{\sigma_m^2\}$ form a non-decreasing sequence, i.e.

$$\sigma_1^2 \leq \sigma_2^2 \leq \ldots \leq \sigma_m^2 \leq \ldots$$

(Obviously, we would expect the prediction variances to increase as we predict further ahead into the future.)

If $\sigma_1^2 > 0$ the process is called *"regular"* (or *"non-deterministic"*), whereas if $\sigma_1^2 = 0$, the process is called *"singular"* (or *"deterministic"*). If $\sigma_1^2 > 0$, if follows from the above inequality that $\sigma_m^2 > 0$, all m, i.e. all the prediction variances are positive. On the other hand, if $\sigma_1^2 = 0$, then X_{t+1} is its own projection on H_t, i.e. $H_{t+1} \equiv H_t$ (all t) and all the H_t spaces are the same. Thus, X_{t+m} is its own projection on H_t (all $m \geq 1$), and hence

$$0 = \sigma_1^2 = \sigma_2^2 = \ldots = \sigma_m^2 = \ldots$$

We therefore see that the $\{\sigma_m^2\}$ are either all positive (corresponding to X_t regular) or all zero (corresponding to X_t singular). In the latter case all future values of X_t can be predicted "perfectly", i.e. with zero prediction variance, from its past values.

Wold (1938) proved a fundamental theorem which states that any stationary process can be decomposed into the sum of a regular process and a singular process, these two processes being uncorrelated with each other. More precisely, Wold proved the following result.

Any stationary process X_t can be expressed in the form,

$$X_t = U_t + V_t,$$ (10.1.80)

where

(a) U_t *and V_t are uncorrelated processes.*

(b) U_t *is regular with a one-sided linear representation.*

$$U_t = \sum_{u=0}^{\infty} g_u \varepsilon_{t-u},$$

with $g_0 = 1$, $\sum_{u=0}^{\infty} g_u^2 < \infty$, and ε_t a white noise process uncorrelated with V_t, i.e. $E[\varepsilon_t V_s] = 0$, all s, t. The sequence $\{g_u\}$ and the process $\{\varepsilon_t\}$ are uniquely determined.

(c) V_t *is singular, i.e. can be predicted from its own past with zero prediction variance.*

Wold's theorem may be established as follows. Let $\tilde{X}_t$ denote, as above, the projection of X_t on H_{t-1}, i.e. $\tilde{X}_t$ denotes the linear least squares predictor of X_t, given $X_{t-1}, X_{t-2}, \ldots$, and define the process $\{\varepsilon_t\}$ by

$$\varepsilon_t = X_t - \tilde{X}_t.$$

Clearly, $\varepsilon_t \perp H_{t-1}$, i.e. ε_t is orthogonal to every element of H_{t-1}. For any pair, $\{\varepsilon_s, \varepsilon_t\}$ with, say, $s < t$, we have

$$\varepsilon_t \perp H_s \subseteq H_{t-1},$$

and

$$\varepsilon_s \in H_s.$$

Hence ε_s and ε_t are orthogonal, i.e. $E[\varepsilon_s, \varepsilon_t] = 0$, all $s \neq t$.

Now consider the projection of X_t on $H_t(\varepsilon)$, the subspace spanned by $\varepsilon_t, \varepsilon_{t-1}, \ldots$. We may write this projection as

$$U_t = \sum_{u=0}^{\infty} g_u \varepsilon_{t-u},$$ (10.1.81)

where, since the $\{\varepsilon_t\}$ are orthogonal,

$$g_u = (X_t, \varepsilon_{t-u})/\sigma_\varepsilon^2$$ (10.1.82)

with

$$\sigma_\varepsilon^2 = \|\varepsilon_t\|^2 = E[\varepsilon_t^2].$$

Thus,

$$g_0 = (X_t, \varepsilon_t)/\sigma_\varepsilon^2 = ((\tilde{X}_t + \varepsilon_t), \varepsilon_t)/\sigma_\varepsilon^2$$

$$= (\varepsilon_t, \varepsilon_t)/\sigma_\varepsilon^2 \qquad (\text{since } \varepsilon_t \perp \tilde{X}_t)$$

$$= 1.$$

Now write

$$V_t = X_t - U_t.$$

Then, for $s \leq t$,

$$(V_t, \varepsilon_s) = (X_t - U_t, \varepsilon_s) = (X_t, \varepsilon_s) - (U_t, \varepsilon_s) = 0,$$

since from (10.1.81), $(U_t, \varepsilon_s) = \sigma_\varepsilon^2 g_{t-s}$, while from (10.1.82) (X_t, ε_s) has the same value. For $s > t$, $V_t \in H_t$ and $\varepsilon_s \perp H_t$; hence $\varepsilon_s \perp V_t$. We have now shown that $\varepsilon_s \perp V_t$ for all s, t, and hence $U_s \perp V_t$ for all s, t. We can therefore write

$$\|X_t\|^2 = \|U_t\|^2 + \|V_t\|^2,$$

and then we obtain

$$\sigma_\varepsilon^2 \left(\sum_{u=0}^{\infty} g_u^2 \right) = \|U_t\|^2 \leq \|X_t\|^2 < \infty.$$

Finally, let $H_t(V)$ be the space spanned by V_t, V_{t-1}, ... Then

$$V_t \in H_t = H_{t-1} \oplus \{\varepsilon_t\},$$

where $\{\varepsilon_t\}$ denotes the [1] space spanned by ε_t. Thus, $V_t \in H_{t-1}$, since $V_t \perp \varepsilon_t$, and by repeating the same argument we find that $V_t \in H_{t-s}$, all $s \geq 0$. Hence,

$$V_t \in H_{-\infty} = \bigcap_{t=-\infty}^{\infty} H_t,$$

and is therefore determined by the "remote past" of $\{X_t\}$. But the projection of V_t on H_{t-1} is the same as the projection of V_t on $H_{t-1}(V)$, since we may write

$$H_{t-1} = H_{t-1}(V) \oplus H_{t-1}(\varepsilon).$$

Hence, $V_t \in H_{t-1}(V)$,   i.e. $H_t(V) \equiv H_{t-1}(V)$ (all t),   and   clearly

$V_t \in H_{t-s}(V)$, all $s \geq 0$. We may therefore write

$$V_t \in H_{-\infty}(V) = \bigcap_{t=-\infty}^{\infty} H_t(V),$$

i.e. V_t is determined by its own "remote past" and can be predicted with zero prediction variance.

If $H_{-\infty} = \{0\}$, then $V_t \equiv 0$, and $X_t \equiv U_t$ is called "*purely non-deterministic*". If $\sigma_\varepsilon^2 = 0$, then $U_t \equiv 0$, and $X_t \equiv V_t$ is then a deterministic (i.e. singular) process.

The corresponding result for continuous parameter processes is given by Doob (1953), p. 588.

There is a close relationship between the Wold decomposition and the Lebesgue decomposition of the integrated spectrum, given by (4.9.1). If $H(\omega)$ denotes the integrated spectrum of X_t, then writing

$$H(\omega) = H_1(\omega) + H_2(\omega) + H_3(\omega),$$

(where H_1, H_2, H_3 denote, respectively, the absolutely continuous, discrete and singular components of $H(\omega)$), we may correspondingly write X_t as (cf. (4.11.24)),

$$X_t = X_t^{(1)} + X_t^{(2)} + X_t^{(3)},$$

where $X_t^{(1)}$, $X_t^{(2)}$, $X_t^{(3)}$ are mutually uncorrelated with integrated spectra $H_1(\omega)$, $H_2(\omega)$, $H_3(\omega)$ respectively. The process $X_t^{(2)}$ is clearly singular since, by (4.11.25) it may be written

$$X_t^{(2)} = \sum_r A_r e^{i\omega_r t} = \sum_r (A_r' \cos \omega_r t + B_r' \sin \omega_r t).$$

As we are assuming that the spectrum of X_t is known, the $\{\omega_r\}$ are known, as are the squared amplitudes $|A_r|^2$, and it remains only to find the coefficients $\{A_r'\}$, $\{B_r'\}$ in order to determine the complete evolution of $X_t^{(2)}$. Given a semi-infinite set of observations we can obviously determine these coefficients with complete accuracy (from linear operations on the data), and hence $X_t^{(2)}$ is deterministic. The process $X_t^{(3)}$ is also singular (since $H_3'(\omega)$ is zero almost everywhere), and thus both $X_t^{(2)}$ and $X_t^{(3)}$ may be identified with the component V_t in (10.1.80). On the other hand, $X_t^{(1)}$ is either regular or singular, according as to whether or not $h(\omega) = H_1'(\omega)$ satisfies (10.1.79). If $P > -\infty$, then $X_t^{(1)}$ is regular; otherwise it is singular. To summarize: if $P > -\infty$ then in (10.1.80),

$$U_t = X_t^{(1)}, \qquad V_t = X_t^{(2)} + X_t^{(3)} \, ;$$

if $P = -\infty$,

$$U_t \equiv 0, \qquad V_t = X_t^{(1)} + X_t^{(2)} + X_t^{(3)}.$$

We may gain further insight into the nature of the condition (10.1.79) by considering the following analogy due to Whittle (1954a). If $h(\omega)$ vanishes over an interval then clearly (10.1.79) cannot hold. Now $h(\omega)$ may be regarded as the (infinite) set of eigenvalues of the (infinite order) variance–covariance matrix of the whole process, and the case where $h(\omega)$ vanishes over an interval is similar to that of a finite set of random variables whose variance–covariance matrix has a zero eigenvalue. In the latter case it is well known that there exists a linear relationship between the variables, and consequently in the former case one would also expect to find a linear relationship between the values of X_t at different time points.

We may note, in particular, that any process with a "band-limited" spectrum, i.e. one whose spectral density function vanishes outside a restricted band of frequencies, must be singular. However, it is not necessary for $h(\omega)$ to vanish over an interval in order for the process to be singular. Consider the continuous parameter process $X(t)$ whose spectral density function has the Gaussian form,

$$h(\omega) = \frac{\sigma_X^2}{\sqrt{2\pi}} e^{-\omega^2/2},$$

with corresponding autocovariance function,

$$R(\tau) = \sigma_X^2 e^{-\tau^2/2}.$$

It is clear that $h(\omega)$ does not satisfy (10.1.29), and consequently $X(t)$ is a singular process. This result can be seen in an alternative way by noting that, since $R(\tau)$ has derivatives of all orders, $X(t)$ has stochastic derivatives of all orders. Following Bartlett (1955), p. 203, we may use the analogy with Taylor series expansions of deterministic functions and write $X(t+m)$ as

$$X(t+m) = X(t) + m\dot{X}(t) + \frac{m^2}{2}\ddot{X}(t) + \ldots + \frac{m^r}{r!}X^{(r)}(t) + \ldots \quad (10.1.83)$$

Regarding the RHS of (10.1.83) as a "linear predictor" for $X(t+m)$, we see at once that $X(t+m)$ can be predicted perfectly from its past values. If we denote the RHS of (10.1.83) by $\tilde{X}(t+m)$ then we may write

$$\tilde{X}(t+m) = e^{mD}X(t),$$

and comparing this expression with (10.1.59) we see that here

$$A(z) = e^{mz}.$$

The function e^{mz} is a *forward transform* (its two-sided Laplace transform being formally $\delta(u + m))$, and as such it is strictly inadmissible as a choice for $A(z)$. The fact that in this case we are able to express the result of applying the operator e^{mD} to $X(t)$ in terms of present and past values is due simply to the fact that here $X(t + m)$ is itself a linear function of present and past values, i.e. in geometrical terms, $X(t + m) \in H_t$, the space spanned by $\{X(s); s \leq t\}$.

It is important to note that the terms "deterministic" and "non-deterministic" which we have used above must be interpreted in a strictly *linear* context. Thus, when we say, for example, that a process X_t is deterministic we mean that X_{t+m} can be expressed as a *linear* function of $X_t, X_{t-1}, \ldots$. It is quite possible for a process to be deterministic in a general sense (by which we mean that X_{t+m} can be determined as a non-linear function of $X_t, X_{t-1} \ldots$) and yet non-deterministic in a linear sense. For, consider the continuous parameter process defined by (4.8.26), namely,

$$X(t) = e^{i(\Phi + \Omega t)},$$

where Φ and Ω are independent random variables, Φ being uniformly distributed over $(-\pi, \pi)$ and Ω having distribution function $H(\omega)$. We know from the discussion following (4.8.26) that $X(t)$ has integrated spectrum $H(\omega)$, and thus, by a suitable choice of Ω, we can arrange for $X(t)$ to have any form of spectrum we like, and, in particular, for the condition (10.1.29) to be satisfied, so that $X(t)$ is non-deterministic. However, irrespective of the form of the distribution function of Ω, it is clear that $X(t)$ depends only on knowing the values of Φ and Ω for that realization, and these are easily determined from past observations. The essential point is that, although the values of Φ and Ω may be found, they cannot be determined by *linear* operations on the past data.

10.1.6 Prediction for Multivariate Processes

The theory of univariate prediction can be extended in a formal way to multivariate processes provided we start with a linear representation or, equivalently, with the canonical factorization of the spectral matrix. In the multivariate case the construction of the canonical factorization is a very deep problem; the technique described in Section 10.1.1 (based on an expansion of $\log h(z)$) is no longer valid since, due to the non-commutivity of matrix multiplication, the identity $\exp(A + B) = \exp A \exp B$ breaks down when A and B are matrices. However, if we *assume* that the spectral matrix possesses the required factorization, we can readily obtain the multivariate analogue of the result (10.1.70). Consider a discrete parameter multivariate stationary process X_t and suppose that its spectral matrix $h(\omega)$ admits the factorization

$$h(\omega) = G(e^{-i\omega})CG^*(e^{-i\omega}), \qquad (10.1.84)$$

where the matrix valued function $G(z)$ is analytic for $|z| \le 1$, $|G(z)|$ has no zero inside $|z| < 1$, and $G(0) = I$. (As in Chapter 9, the asterisk denotes both transposition and conjugation, and $|A|$ denotes the determinant of A.) Then, as in equation (10.1.1), we may derive the linear representation of X_t,

$$X_t = \sum_{u=0}^{\infty} g(u)\varepsilon_{t-u}, \qquad (10.1.85)$$

where the $\{g(u)\}$ are the Fourier coefficients of $G(e^{-i\omega})$, and ε_t is a multivariate white noise process (cf. (9.4.16)). Now let

$$\tilde{X}_{t+m} = \sum_{u=0}^{\infty} a(u)X_{t-u} \qquad (10.1.86)$$

denote a linear predictor of X_{t+m}, and suppose that the $\{a(u)\}$ are to be chosen so as to minimize

$$E[\{X_{t+m} - \tilde{X}_{t+m}\}'\{X_{t+m} - \tilde{X}_{t+m}\}] = \mathrm{tr}(\Sigma_m),$$

where Σ_m denotes the variance–covariance matrix of the prediction errors, $\{X_{t+m} - \tilde{X}_{t+m}\}$. Using the linear representation (10.1.85), a similar argument to that of the univariate case gives (cf. (10.1.70)),

$$A(z) = [z^{-m}G(z)]_+ G^{-1}(z), \qquad (10.1.87)$$

where $A(z) = \sum_{n=0}^{\infty} a(u)z^n$, and $[K(z)]_+$ denotes the matrix formed by taking the backward transform of each element of $K(z)$. As in the univariate case, ε_t represents the one-step prediction errors and we have

$$\Sigma_1 \equiv \Sigma_\varepsilon = 2\pi C, \qquad (10.1.88)$$

where C is defined by (10.1.84). (Compare (10.1.84) with (9.4.22).)

The determinant of the variance–covariance matrix of the one-step prediction errors is given by (Whittle, 1953),

$$|\Sigma_1| = \exp\left[\frac{1}{2\pi} \int_{-\pi}^{\pi} \log\{|2\pi h(\omega)|\} \, d\omega\right], \qquad (10.1.89)$$

this result being the multivariate extension of (10.1.42).

For a full discussion of the multivariate prediction problem, see Wiener and Masani (1957, 1958), Helson and Lowdenslager (1958), Masani (1966) and Rozanov (1967); see also Doob (1953), p. 594, Hannan (1970), p. 157, and Section 10.6.

10.2 THE BOX–JENKINS APPROACH TO FORECASTING

The Kolmogorov–Wiener approach to prediction provides an explicit expression for $\tilde{X}_{t+m}$, but this has to be completely recomputed if, e.g., we change m to $(m+1)$. However, Box and Jenkins (1970) have developed a recursive algorithm for computing the predictors of discrete parameter processes which satisfy finite parameter linear models. (Following current fashion, we refer to the Box–Jenkins technique as a "forecasting method", although the terms "forecasting" and "prediction" are synonymous.)

Suppose that X_t satisfies the stationary invertible ARMA(k, l) model,

$$X_t + a_1 X_{t-1} + \ldots + a_k X_{t-k} = \varepsilon_t + b_1 \varepsilon_{t-1} + \ldots + b_l \varepsilon_{t-l}, \quad (10.2.1)$$

where we assume here that $\{\varepsilon_t\}$ is a sequence of *independent* random variables, each having zero mean and variance σ_ε^2. In describing the Box–Jenkins approach it is convenient to adopt a slightly different notation for predictors, and given $X_t, X_{t-1}, X_{t-2}, \ldots$, we now denote the predictor of X_{t+m} by $\tilde{X}_t(m)$ (instead of the previous notation, $\tilde{X}_{t+m}$). The advantage of the new notation is that it makes explicit both the "*lead time*" m, and the "*origin*" t, from which we are predicting. Now the (unconstrained) least squares predictor $\tilde{X}_t(m)$ is given by (10.1.2) as the conditional expectation of X_{t+m}, given $X_t, X_{t-1}, X_{t-2}, \ldots$, i.e.

$$\tilde{X}_t(m) = E[X_{t+m} | X_t, X_{t-1}, X_{t-2}, \ldots]. \quad (10.2.2)$$

Re-writing (10.2.1) with t replaced by $(t+1)$ we have

$$X_{t+1} + a_1 X_t + \ldots + a_k X_{t-k+1} = \varepsilon_{t+1} + b_1 \varepsilon_t + \ldots + b_l \varepsilon_{t-l+1}, \quad (10.2.3)$$

and taking conditional expectations of both sides gives

$$\tilde{X}_t(1) + a_1 X_t + \ldots + a_k X_{t-k+1} = E^*(\varepsilon_{t+1}) + b_1 E^*(\varepsilon_t) + \ldots + b_l E^*(\varepsilon_{t-l+1})$$
$$(10.2.4)$$

where E^* denotes expectation conditional on $X_t, X_{t-1}, X_{t-2}, \ldots$ given. (Note that, by definition of E^*, $E^*(X_{t-s}) = X_{t-s}$, $s > 0$.) Now for the model (10.2.1) we may express X_t as a linear function of $\varepsilon_t, \varepsilon_{t-1}, \ldots$, and vice-versa. Hence, conditional expectations given $X_t, X_{t-1}, X_{t-2}, \ldots$ are the same as conditional expectations given $\varepsilon_t, \varepsilon_{t-1}, \varepsilon_{t-2}, \ldots$. Thus we have

$$E^*(\varepsilon_{t+1}) = E[\varepsilon_{t+1} | \varepsilon_t, \varepsilon_{t-1}, \varepsilon_{t-2}, \ldots] = 0 \quad (10.2.5)$$

(since the $\{\varepsilon_t\}$ are independent), and

$$E^*(\varepsilon_{t-s}) = \varepsilon_{t-s}, \quad s \geq 0.$$

Using these results in (10.2.4) we obtain

$$\tilde{X}_t(1) + a_1 X_t + \ldots + a_k X_{t-k+1} = b_1 \varepsilon_t + \ldots + b_l \varepsilon_{t-l+1} \quad (10.2.6)$$

and comparing (10.2.3), (10.2.5), we see that

$$X_{t+1} - \tilde{X}_t(1) = \varepsilon_{t+1} \quad (10.2.7)$$

(cf. (10.1.39)), and similarly,

$$\varepsilon_t = X_t - \tilde{X}_{t-1}(1), \qquad \varepsilon_{t-1} = X_{t-1} - \tilde{X}_{t-2}(1), \qquad \varepsilon_{t-2} = X_{t-2} - \tilde{X}_{t-3}(1),$$

etc. Substituting these expressions for ε_t, ε_{t-1}, ε_{t-2}, ... in (10.2.6) we now obtain

$$\tilde{X}_t(1) + a_1 X_t + \ldots + a_k X_{t-k+1}$$
$$= b_1\{X_t - \tilde{X}_{t-1}(1)\} + \ldots + b_l\{X_{t-l+1} - \tilde{X}_{t-l}(1)\},$$

or

$$\tilde{X}_t(1) + b_1 \tilde{X}_{t-1}(1) + \ldots + b_l \tilde{X}_{t-l}(1) = (b_1 - a_1)X_t + \ldots + (b_l - a_l)X_{t-l+1}$$
$$- a_{l+1}X_{t-l} - \ldots - a_k X_{t-k+1},$$
$$(10.2.8)$$

(assuming, e.g., that $k \geq l$). Equation (10.2.8) enables us to compute $\tilde{X}_t(1)$ recursively, given the values of $\tilde{X}_{t-1}(1)$, $\tilde{X}_{t-2}(1), \ldots, \tilde{X}_{t-l}(1)$, X_t, X_{t-1}, $\ldots, X_{t-k+1}$. This is a particularly useful expression if we are interested in computing one-step ahead predictors "continuously" over time, and (10.2.8) allows us to "update" the one-step ahead predictor computed at time $(t-1)$ into the one-step ahead predictor at time t as soon as the "new" observation X_t becomes available.

Similarly, using (10.2.1) with t replaced by $(t+2)$, and taking conditional expectations (conditional on X_t, X_{t-1}, X_{t-2}, ...) of both sides gives

$$\tilde{X}_t(2) + a_1 \tilde{X}_t(1) + a_2 X_t + \ldots + a_k X_{t-k+2}$$
$$= b_2 \varepsilon_t + \ldots + b_l \varepsilon_{t-l+2}$$
$$= b_2\{X_t - \tilde{X}_{t-1}(1)\} + \ldots + b_l\{X_{t-l+2} - \tilde{X}_{t-l+1}(1)\},$$

or

$$\tilde{X}_t(2) + a_1 \tilde{X}_t(1) + b_2 \tilde{X}_{t-1}(1) + \ldots + b_l \tilde{X}_{t-l+1}(1)$$
$$= (b_2 - a_2)X_t + \ldots + (b_l - a_l)X_{t-l+2} - \ldots - a_k X_{t-k+2}.$$
$$(10.2.9)$$

From (10.2.9) we can compute $\tilde{X}_t(2)$ recursively, given $\tilde{X}_t(1)$, $\tilde{X}_{t-1}(1)$, $\ldots, \tilde{X}_{t-l+1}(1)$, X_t, X_{t-1}, $\ldots, X_{t-k+2}$, and so on. In general, we can compute $\tilde{X}_t(m)$ recursively from $\tilde{X}_t(1)$, $\tilde{X}_t(2)$, ..., $\tilde{X}_t(m-1)$, and the data, X_t, X_{t-1}, ...

A "*prediction interval*" for an individual value X_{t+m} can be constructed by assuming that $\tilde{X}_t(m)$ is normally distributed (as would certainly be the case if $\{\varepsilon_t\}$ is a normal process), with mean X_{t+m} and variance given by (10.1.38). To evaluate (10.1.38) for the model (10.2.1) we obtain the coefficients $\{g_u\}$ by writing

$$\alpha(z) = \sum_{u=0}^{k} a_u z^u, \qquad \beta(z) = \sum_{u=0}^{l} b_u z^u \qquad (a_0 = b_0 = 1),$$

and equating coefficients of powers of z in the identity, $\alpha(z)G(z) \equiv \beta(z)$.

For $m > k(\geq l)$, we obtain similarly,

$$\tilde{X}_t(m) + a_1\tilde{X}_t(m-1) + \ldots + a_k\tilde{X}_t(m-k) = 0, \qquad (10.2.10)$$

which shows that $\tilde{X}_t(m)$ satisfies the difference equations corresponding to the AR operator of (10.2.1). (The same result clearly holds for a pure AR model, which confirms the comment made in Example 1 of Section 10.1.3.)

Regarding $\tilde{X}_t(m)$ as a function of m (with t fixed), Box and Jenkins call the asymptotic form of this function the "*eventual forecast function*". It follows from (10.2.10) that *the eventual forecast function for an AR or ARMA model has exactly the same form as the complementary function*, corresponding to the solution of the homogeneous difference equations.

To illustrate the above method let us consider the ARMA (3, 3) model,

$$X_t + a_1X_{t-1} + a_2X_{t-2} + a_3X_{t-3} = \varepsilon_t + b_1\varepsilon_{t-1} + b_2\varepsilon_{t-2} + b_3\varepsilon_{t-3}.$$

The recursive equation for the one-step predictors is (from (10.2.8)),

$$\tilde{X}_t(1) + b_1\tilde{X}_{t-1}(1) + b_2\tilde{X}_{t-2}(1) + b_3\tilde{X}_{t-3}(1)$$
$$= (b_1 - a_1)X_t + (b_2 - a_2)X_{t-1} + (b_3 - a_3)X_{t-2},$$

and, from (10.2.9), the equation for the two-step predictors is

$$\tilde{X}_t(2) + a_1\tilde{X}_t(1) + b_2\tilde{X}_{t-1}(1) + b_3\tilde{X}_{t-1}(1) = (b_2 - a_2)X_t + (b_3 - a_3)X_{t-1}.$$

These recursive techniques require suitable *starting values*, and Box and Jenkins suggest that any ε_t which cannot be evaluated from the observations (i.e. those for which the corresponding values of X_t and $\tilde{X}_{t-1}(1)$ are unavailable) may be set equal to their unconditioned expectations, namely, zero.

Graph 10.1 shows the forecasted values (superimposed on the actual values) of the final ten observations from a sample of 500 observations generated from the ARMA(2, 2) model,

$$X_t + 1 \cdot 4 X_{t-2} + 0 \cdot 5 X_{t-2} = \varepsilon_t - 0 \cdot 2 \varepsilon_{t-1} - 0 \cdot 1 \varepsilon_{t-2}$$

(with $\varepsilon_t \sim N(0, 1)$), based on the same order ARMA model fitted to the first 490 observations.

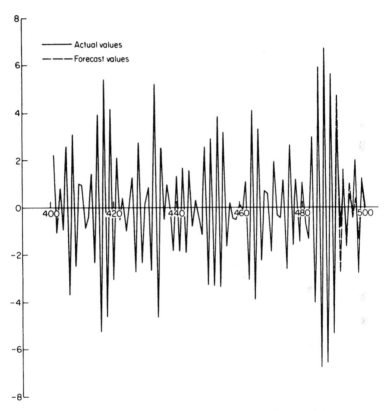

Graph 10.1. Forecasts computed from ARMA(2, 2) model.

Discussion

(1) Recalling that the Box–Jenkins method is based on the general expression for predictors as conditional expectations, it may, at first, seem rather surprising that the recursive relations all turn out to be *linear*, leading to expressions for the predictors as *linear* functions of past observations. This feature is, of course, due to the fact that we have assumed that the $\{\varepsilon_t\}$ are *independent*; had we assumed merely that the $\{\varepsilon_t\}$ are uncorrelated, then (10.2.5) would no longer be valid. Naturally, if

we assume that $\{\varepsilon_t\}$ is a *normal* process then the distinction between independence and uncorrelatedness disappears.

(2) At a basic level, the Kolmogorov, Wiener, and Box–Jenkins approaches are all equivalent, since each leads to the linear least-squares predictor of X_{t+m}. The Kolmogorov and Wiener methods apply to any stationary process with a purely continuous spectrum, whereas the Box–Jenkins method applies only to processes which satisfy finite parameter linear models. However, by restricting the processes to this class Box and Jenkins are able to exploit the linearity of the model to derive neat recursive relations *which enjoy the very considerable advantage of computational simplicity*. In the practical situation where we start with *data* (rather than a specified spectrum or model), we can fit an ARMA model by the methods described in Section 5.4.3, and the computation of predictors based on the fitted model is then an almost trivial exercise.

(3) Since the Box–Jenkins approach is based on the assumption of a finite parameter linear model, it may be regarded as a special case of the more general (and more powerful) *"Kalman filter"* algorithm (see, e.g. Caines (1972)). The Kalman algorithm is similarly based on a finite parameter linear time domain model, but it uses the more powerful "state-space" description of linear models and can therefore handle multivariate processes with hardly any additional effort. (Box and Jenkins (1970) discuss the use of "transfer function" models as one method of incorporating information from ancilliary processes in the prediction of X_t, but they do not discuss prediction based on a full multivariate model. See the discussion in Section 9.2.2 on the relationship between "transfer function" models and full multivariate models.) We discuss the topic of "Kalman filtering" (and the associated technique known as "state-space forecasting") in Section 10.6.

ARIMA models

Box and Jenkins (1970) discuss a particular type of non-stationary process in which X_t itself is non-stationary but its dth differences, $\Delta^d X_t$, follow a stationary ARMA model. Thus, writing $Y_t = \Delta^d X_t$ (where $\Delta = (1 - B)$ denotes the difference operator), we may write,

$$\alpha(B) Y_t = \beta(B)\varepsilon_t, \qquad (10.2.11)$$

where

$$\alpha(z) = \sum_{u=0}^{k} a_k z^k, \qquad \beta(z) = \sum_{u=0}^{l} \beta_l z^l \qquad (\alpha_0 = \beta_0 = 1),$$

have no zeros inside the unit circle, $|z| \leq 1$. The corresponding model for X_t is,

$$(1 - B)^d \alpha(B) X_t = \beta(B) \varepsilon_t, \tag{10.2.12}$$

or

$$\alpha^*(B) X_t = \beta(B) \varepsilon_t,$$

where $\alpha^*(z) = (1 - z)^d \alpha(z)$. Clearly (10.2.12) represents a non-stationary model *since $\alpha^*(z)$ has a d-fold root on the unit circle.* (The model (10.2.12) may be interpreted as the result of passing the white noise process, ε_t, through the "unstable" linear filter whose transfer function, $\beta(e^{-i\omega})/\alpha^*(e^{-i\omega})$, has a dth order pole on the unit circle.) Box and Jenkins call such models *"autoregressive integrated moving/average"* (hence the abbreviation, "ARIMA"), although it may be noted that this type of non-stationary process was studied also by Yaglom (1955), and sometimes described as an *"accumulated process"*. Box and Jenkins refer to (10.2.12) as an ARIMA (k, d, l) model. Once the value of d is known (or estimated), the parameters of the model may be estimated by fitting the stationary ARMA model (10.2.11) to Y_t, and recursive relations for the predictors of X_t can then be derived exactly as in the previous case. Thus, if we expand $\alpha^*(z)$ and write

$$\alpha^*(z) = 1 + a_1^* z + \ldots + a_{k+d}^* z^{l+d},$$

then (10.2.12) can be written

$$X_t + a_1^* X_{t-1} + \ldots + a_{k+d}^* X_{t-k-d} = \varepsilon_t + b_1 \varepsilon_{t-1} + \ldots + b_l \varepsilon_{t-l} \tag{10.2.13}$$

and the recursive expressions for the predictors can be obtained by taking conditional expectations of (10.2.13). This leads to equations of exactly the same form as (10.2.8), (10.2.9), with the a's replaced by the a^*'s, and k replaced by $(k + d)$.

It should be noted that although (10.2.12) is certainly a non-stationary model, it describes only one type of non-stationary behaviour, namely where the process is of the "explosive" type. Here, the second order properties vary over time, but the complete evolution of the process is determined (for all time) by a finite set of fixed parameters. Thus, if we fit this type of model to a set of observations, $X_1, X_2, \ldots, X_N$, say, the complete "future" of the process is determined by the values of the parameters fitted to the given stretch of data. Later, in Chapter 11, we will consider more general types of non-stationary processes which are governed by continuously varying time-dependent parameters.

The three-term predictor

Consider the non-stationary process,

$$X_t = \varepsilon_t + \gamma_{-1}\varepsilon_{t-1} + \gamma_0\{S\varepsilon_{t-1}\} + \gamma_1\{S^2\varepsilon_{t-1}\}, \qquad (10.2.14)$$

where $\{\varepsilon_t\}$ is an independent white noise process, γ_{-1}, γ_0, γ_1 are constants, and $S = \Delta^{-1} = (1-B)^{-1}$ denotes the summation operator. Equivalently, (10.2.14) may be written,

$$\begin{aligned}
\Delta^2 X_t &= \Delta^2\varepsilon_t + \gamma_{-1}\Delta^2\varepsilon_{t-1} + \gamma_0\Delta\varepsilon_{t-1} + \gamma_1\varepsilon_{t-1}\\
&= (\varepsilon_t - 2\varepsilon_{t-1} + \varepsilon_{t-2}) + \gamma_{-1}(\varepsilon_{t-1} - 2\varepsilon_{t-2} + \varepsilon_{t-3})\\
&\quad + \gamma_0(\varepsilon_{t-1} - \varepsilon_{t-2}) + \gamma_1\varepsilon_{t-1}\\
&= \varepsilon_t + b_1\varepsilon_{t-1} + b_2\varepsilon_{t-2} + b_3\varepsilon_{t-3}, \qquad (10.2.15)
\end{aligned}$$

say, where

$$\begin{aligned}
b_1 &= \gamma_{-1} + \gamma_0 + \gamma_1 - 2,\\
b_2 &= 1 - 2\gamma_{-1} - \gamma_0, \qquad (10.2.16)\\
b_3 &= \gamma_{-1}.
\end{aligned}$$

Thus, $\Delta^2 X_t$ satisfies an MA(3) model, and X_t is an ARIMA process of the form (10.2.12) with $d = 2$, $\alpha(z) \equiv 1$, $\beta(z) = 1 + b_1 z + b_2 z^2 + b_3 z^3$.

From (10.2.14), the one-step predictor is given (replacing t by $t+1$ and setting $\varepsilon_{t+1} = 0$), by

$$\tilde{X}_t(1) = \gamma_{-1}\varepsilon_t + \gamma_0\{S\varepsilon_t\} + \gamma_1\{S^2\varepsilon_t\}, \qquad (10.2.17)$$

and hence

$$\Delta\tilde{X}_t(1) = \gamma_{-1}\{\Delta\varepsilon_t\} + \gamma_0\varepsilon_t + \gamma_1\{S\varepsilon_t\}. \qquad (10.2.18)$$

Equation (10.2.18) may be termed a *"three-term predictor"*, by analogy with the form of "three term controllers" used in the study of linear control systems—see Section 10.4.1.

Exponentially weighted moving-average predictors

Suppose that in (10.2.14), $\gamma_{-1} = \gamma_1 = 0$, $0 < \gamma_0 < 2$, so that the model becomes

$$X_t = \varepsilon_t + \gamma_0\{S\varepsilon_{t-1}\},$$

or

$$\Delta X_t = \Delta\varepsilon_t + \gamma_0\varepsilon_{t-1} = \varepsilon_t - \theta\varepsilon_{t-1}, \qquad (10.2.19)$$

where $\theta = 1 - \gamma_0$, so that $|\theta| < 1$. Thus, in this case ΔX_t is an MA(1) process. The one-step predictor is now (from (10.2.17)),

$$\tilde{X}_t(1) = \gamma_0\{S\varepsilon_t\},$$

or

$$\Delta\tilde{X}_t(1) = \gamma_0\varepsilon_t = \gamma_0(1 - \theta B)^{-1}\Delta X_t,$$

or

$$(1 - \theta B)\tilde{X}_t(1) = \gamma_0 X_t, \tag{10.2.20}$$

or

$$\tilde{X}_t(1) = \theta\tilde{X}_{t-1}(1) + (1 - \theta)X_t. \tag{10.2.21}$$

This result shows that *the one-step predictor at time t is a weighted linear combination of the one-step predictor at time $(t-1)$ and the "new" observation, X_t.* This form of predictor is widely used in practice on an *ad hoc* basis, but as we have seen, *it is optimal (in the mean square error sense) only when X_t conforms to the model* (10.2.19). It is generally referred to as the *"exponentially weighted moving average"* predictor since, from (10.2.20), $\tilde{X}_t(1)$ may be written explicitly as

$$\tilde{X}_t(1) = \gamma_0(1 - \theta B)^{-1}X_t$$
$$= (1 - \theta)(X_t + \theta X_{t-1} + \theta^2 X_{t-2} + \dots). \tag{10.2.22}$$

Thus, $\tilde{X}_t(1)$ has the form of an (infinite order) moving average, with exponentially decaying weights, $(1 - \theta)$, $(1 - \theta)\theta$, $(1 - \theta)\theta^2$, ...

Seasonal models

In Section 7.7 we observed that the filter $(1 - B^s)$ will remove a periodic, or "seasonal", component with period s (cf. (7.7.29)). Consequently, if X_t contains such a component, $Y_t = (1 - B^s)X_t$ will have a "deseasonalized" form, and thus it is reasonable to assume that it may be represented by an ARMA model

$$\alpha(B)Y_t = \beta(B)\varepsilon_t.$$

The corresponding model for X_t is now,

$$\alpha(B)(1 - B^s)X_t = \beta(B)\varepsilon_t.$$

To allow for the possibility of non-stationarity in both the $\{Y_t\}$ model and the seasonal component, Box and Jenkins (1970) generalize the above model to

$$\alpha_1(B)\alpha_2(B^s)\Delta^d\Delta_s^D X_t = \beta_1(B)\beta_2(B^s)\varepsilon_t, \tag{10.2.23}$$

where $\alpha_1(z)$, $\alpha_2(z)$, $\beta_1(z)$, $\beta_2(z)$ are polynomials of degrees, k, K, l, L, respectively, (having all their roots outside the unit circle) and $\Delta_s = (1 - B^s)$ is the s-step difference operator, i.e. $\Delta_s X_t = X_t - X_{t-s}$. Box and Jenkins refer to (10.2.23) as a "*multiplicative model of order* $(k, d, l) \times (K, D, L)$".

Again, once the values of d and D have been determined, and suitable values specified for the orders k, K, l, L, the further parameters of the model may be estimated by fitting the model $\alpha_1(B)\alpha_2(B^s)Z_t = \beta_1(B)\beta_2(B^s)\varepsilon_t$ to $Z_t = \Delta^d \Delta_s^D X_t$, using the methods of Section 5.4.3. Recursive relations for the predictors of seasonal models are then constructed by expanding the product of operators in (10.2.23), and proceeding as in the previous cases.

Box and Jenkins derive the model (10.2.23) by arguing that the seasonal component of X_t may be modelled by

$$\alpha_2(B^s)\Delta_s^D X_t = \beta_2(D^s)e_t, \qquad (10.2.24)$$

while the "non-seasonal" component may be modelled by assuming that the "residuals" from the above model, e_t, satisfy

$$\alpha_1(B)\Delta^d e_t = \beta_1(B)\varepsilon_t. \qquad (10.2.25)$$

Substituting (10.2.25) into (10.2.24) then gives the multiplicative model, (10.2.23). Box and Jenkins maintain that (10.2.23) is a fairly general model for (non-stationary) series which contain a seasonal component of period s, and it should be noted that (10.2.23) implies rather more than the conventional additive model

$$X_t = f_t + P_t + Y_t, \qquad (10.2.26)$$

where f_t denotes a deterministic polynomial of degree $(d-1)$ (representing the "trend"), P_t is a periodic component, (period s), and Y_t is a stochastic process, with $E(Y_t) = 0$. For, if the model really is of the form (10.2.26) and we fit a model of the form (10.2.23), then this implies that Y_t is a *pathological* process, in the sense that Y_t itself has a form of "periodic" structure. To see this, we apply the operator $\Delta^d \Delta_s^D$ to both sides of (10.2.26), and noting that Δ^d "annihilates" f_t and Δ_s^D "annihilates" P_t, we obtain,

$$\Delta^d \Delta_s^D X_t = \Delta^d \Delta_s^D Y_t.$$

Consequently, Y_t also satisfies (10.2.23), i.e.

$$\alpha_1(B)\alpha_2(B^s)\Delta^d \Delta_s^D Y_t = \beta_1(B)\beta_2(B^s)\varepsilon_t,$$

and the presence of the "singular" factors $(1 - B^s)^D$, $(1 - B)^d$ (both of which have roots *on* the unit circle) mean that (a) Y_t is non-stationary (if

$d > 0$), and (b) it has a form of periodic structure, in the sense that its "spectral density function" is given by

$$h_Y(\omega) = \frac{\sigma_\varepsilon^2}{2\pi} \left| \frac{\beta_1(e^{-i\omega})\beta_2(e^{-is\omega})}{\alpha_1(e^{-i\omega})\alpha_2(e^{-is\omega})} \right|^2 \left| \frac{1}{(1 - e^{-i\omega})^d (1 - e^{-is\omega})^D} \right|^2$$

and thus has infinite "peaks" at frequencies $\omega = 2\pi k/s$, $k = 0, 1, \ldots$, $[s/2]$, *unless, of course,* $\beta_1(B)$, $\beta_2(B^s)$ *contain as factors* $(1-B)^d$, $(1-B^s)^D$. Thus, for $h_Y(\omega)$ to be bounded; it is necessary for $\beta_1(B)$, $\beta_2(B^s)$ to be of the form $\beta_1(B) = (1-B)^d \tilde{\beta}_1(B)$, $\beta_2(B^s) = (1-B^s)^D \tilde{\beta}_2(B)$ where $\tilde{\beta}_1(B)$, $\tilde{\beta}_2(B)$ have their roots outside $|z| \leqslant 1$.

In this case, we have for the solution of (10.2.23),

$$X_t = f_t + P_t + k_t + Y_t, \tag{10.2.27}$$

where k_t is the (decaying) solution of $\alpha_1(B)\alpha_2(B^s)X_t = 0$, and Y_t (the particular solution of (10.2.23)) is the stationary process,

$$Y_t = \alpha_1^{-1}(B)\alpha_2^{-1}(B)\tilde{\beta}_1(B)\tilde{\beta}_2(B^s)\varepsilon_t.$$

However, if $\beta_1(B)$, $\beta_2(B^s)$ do not contain factors of the above form then Y_t is non-stationary and has itself a periodic form.

The same feature applies also to the non-seasonal ARIMA model, (10.2.12) (which is simply a special case of (10.2.23) with $D = 0$, $\alpha_2(B^s) \equiv \beta_2(B^s) \equiv 1$). Thus, if we write the solution of (10.2.12) as

$$X_t = f_t + Y_t, \tag{10.2.28}$$

(f_t being, as above, a polynomial of degree $(d-1)$), then Y_t is a non-stationary process unless $\beta(B)$ contains $(1-B)^d$ as a factor.

Abraham and Box (1975, 1978) distinguish these two cases by saying that when $\beta_1(B)$, $\beta_2(B^s)$ contain $(1-B)^d$, $(1-B^s)^D$ as factors then the seasonal component has a strictly periodic stable "deterministic" form (i.e. it has a Fourier series representation with constant amplitude and phases), and the trend has a stable "deterministic" polynomial form, with constant coefficients. On the other hand, when $\beta_1(B)$, $\beta_2(B^s)$ do not contain these factors the seasonal component has an "adaptive" form (with the amplitudes and phases possibly changing over time) and similarly the trend has an "adaptive" polynomial form. Abraham and Box therefore argue that it is always safe to fit a model of the form (10.2.23) to non-stationary seasonal data; if the trend and seasonal components are both "stable", this will be revealed by a near cancellation of the operators $(1-B)^d$, $(1-B^s)^D$ on both sides of the fitted model.

For data with an annual periodicity of twelve months, and with t measured in months, the "seasonal" operator takes the form $(1 - B^{12})$. As observed by Abraham and Box, the operator $(1 - B^{12})$ has roots $e^{2\pi i j/12}$, $j = 1, 2, \ldots, 12$, which are evenly distributed over the unit circle. The complementary function corresponding to this operator therefore has the form

$$A_0 + \sum_{j=1}^{6} \left(A_j \cos \frac{2\pi jt}{12} + B_j \sin \frac{2\pi jt}{12} \right). \qquad (10.2.29)$$

Now, $(1 - B^{12})$ can be factorized as

$$(1 - B^{12}) = (1 - \sqrt{3}B + B^2)(1 - B + B^2)(1 + B^2)$$
$$\times (1 + B + B^2)(1 + \sqrt{3}B + B^2)(1 + B)(1 - B). \qquad (10.2.30)$$

The successive factors in (10.2.30) correspond to the successive terms in (10.2.29). Thus, $(1 - \sqrt{3}B + B^2)$, (which has roots $(\sqrt{3} \pm i)/2$), corresponds to the term $j = 1$ (the term with period twelve months), $(1 - B + B^2)$ corresponds to $j = 2$ (period six months), $(1 + B^2)$ corresponds to $j = 3$ (period four months), and so on.

The operator $\alpha_2(B^s)(1 - B^s)^D$ can describe quite a wide range of different types of seasonal behaviour. Thus, the complementary function (or equivalently, the "eventual forecast function") corresponding to $(1 - aB^s)$, $(|a| < 1)$, has the form of a "damped" periodic function, (period s); the operator $(1 - B^s)$ corresponds to a strictly periodic function; $(1 - B)(1 - aB^s)$ corresponds to a damped periodic function superimposed on a linear trend, and so on—see Box and Jenkins (1970), p. 326.

A simple but useful operator for monthly data with a yearly period is,

$$(1 - B)(1 - B^{12}).$$

This operator incorporates *a periodicity of twelve months* together with a *linear trend*. (Note that $(1 - B)^d$ and $(1 - B^{12})$ share the common root, $B = 1$; thus the complementary function corresponding to the product of these operators contains a polynomial of degree d, rather than $(d - 1)$.) For example, Box and Jenkins (1970), p. 305, fit the model,

$$(1 - B)(1 - B^{12})X_t = (1 - b_1 B)(1 - b_2 B^{12})\varepsilon_t,$$

to the logarithms of the "international airline passengers" series, and, assuming the $\{\varepsilon_t\}$ normal, obtain maximum likelihood estimates $\hat{b}_1 = 0\cdot 4$, $\hat{b}_2 = 0\cdot 6$. The model can then be written explicitly as

$$X_t - X_{t-1} - X_{t-12} + X_{t-13} = \varepsilon_t - 0\cdot 4\varepsilon_{t-1} - 0\cdot 6\varepsilon_{t-12} + 0\cdot 24\varepsilon_{t-13},$$

and recursive relations for the predictors can be derived as in (10.2.9).

We find, e.g.,

$$\tilde{X}_t(1) = X_t + X_{t-11} - X_{t-12} - 0{\cdot}4\varepsilon_t - 0{\cdot}6\varepsilon_{t-11} + 0{\cdot}24\varepsilon_{t-12},$$

with $\varepsilon_t = (X_t - \tilde{X}_{t-1}(1))$, etc. If, as often happens, the behaviour of such series is dominated by the trend and seasonal components then the "AR" operator $(1 - B)(1 - B^{12})$ will by itself account for most of the variation in the series, and consequently the predictors will not depend critically on the choice of the "MA" operator.

Box and Jenkins (1970) discuss a large number of practical applications of their forecasting and model fitting method. Further discussion of their approach is given by Box and Newbold (1971), Chatfield and Prothero (1973a, b), Newbold and Granger (1974), Chatfield (1975), Box, Hillmer and Tiao (1976), and Cleveland and Tiao (1976). Cleveland and Tiao (1976) explain how the seasonal model (10.2.23) may be used as a basis for the "*seasonal adjustment*" of time series. This approach to seasonal adjustment is based on using the model to construct the seasonal component, the "seasonally adjusted" series then being recovered by subtracting the seasonal component from the original series. See also Tiao and Hillmer (1978).

10.3 THE FILTERING PROBLEM

Suppose that we wish to predict a process X_t but X_t itself is not observable. Instead, we observe X_t with added "noise", i.e. we observe the process,

$$Y_t = X_t + N_t, \qquad (10.3.1)$$

where N_t (the noise) and X_t are jointly stationary processes. Given a semi-infinite set of observations on $\{Y_t\}$, say $Y_t, Y_{t-1}, Y_{t-2}, \ldots$, we wish to "estimate" X_{t+m} by a linear function,

$$\tilde{X}_{t+m} = \sum_{u=0}^{\infty} b_u Y_{t-u}, \qquad (10.3.2)$$

where the sequence $\{b_u\}$ is chosen so that

$$\mathcal{M}(m) = E[\{X_{t+m} - \tilde{X}_{t+m}\}^2] \qquad (10.3.3)$$

is minimized.

We call this the "*filtering problem*", the basic idea being that we wish to "filter out" (as far as possible) the noise component N_t. In the prediction problem we naturally considered only the case $m > 0$, but in the present problem it makes sense to consider all three cases, $m < 0$, $m = 0$, $m > 0$.

The case $m \leq 0$ corresponds to the situation where we wish to "estimate" the value of X_t at a time point for which an observation on Y_t has been made, while the case $m > 0$ corresponds to predicting a future value of X_t.

In engineering language we would call X_t the "*signal*", and the problem could then be described as "signal extraction from noisy data". The mixed spectrum problem (discussed in Chapter 8) also deals with the analysis of data consisting of "signal plus noise", but in that case the "signal" has a rather special form—it consists of a sum of harmonic components and so has a purely discrete spectrum, but of unknown form. Here, the "signal" X_t is assumed to have a purely continuous spectrum of known form, as is the "noise" process N_t.

Before discussing the solution of the above problem let us first consider the (unrealistic) case where a complete realization of Y_t is available. Then we can consider a "two-sided" estimate,

$$\tilde{X}_{t+m} = \sum_{u=-\infty}^{\infty} c_u Y_{t-u} \qquad (10.3.4)$$

and the problem now is to choose the $\{c_u\}$ so as to minimize $\mathcal{M}(m)$. In this form, the problem is exactly the same as that considered in Section 9.2.1, and the solution is already known. Writing

$$C(z) = \sum_{u=-\infty}^{\infty} c_u z^u, \qquad (10.3.5)$$

we know from (9.2.26) that $\mathcal{M}(m)$ is minimized by choosing

$$C(e^{-i\omega}) = h_{X'Y}(\omega)/h_{YY}(\omega), \qquad (10.3.6)$$

where $h_{X'Y}(\omega)$ is the cross-spectral density function between $X'_t \equiv X_{t+m}$ and Y_t, and $h_{YY}(\omega)$ is the spectral density function of Y_t. Assuming that X_t and N_t are uncorrelated processes, we have,

$$h_{X'Y}(\omega) = e^{itm} h_{XX}(\omega),$$

$h_{XX}(\omega)$ being the spectral density function of X_t, and hence we find

$$C(e^{-i\omega}) = e^{itm} h_{XX}(\omega)/h_{YY}(\omega). \qquad (10.3.7)$$

The minimized value of $\mathcal{M}(m)$ corresponds to the "residual variance", and hence, by (9.2.27), is given by

$$\mathcal{M}_{\min}(m) = \int_{-\pi}^{\pi} \left\{ 1 - \frac{h_{XX}(\omega)}{h_{YY}(\omega)} \right\} h_{XX}(\omega) \, d\omega. \qquad (10.3.8)$$

10.3.1 The Wiener Filter

We now return to the more realistic case where Y_t is observed only up to time t, and $\tilde{X}_{t+m}$ is restricted to the "one-sided" form (10.3.2). Writing,

$$B(z) = \sum_{u=0}^{\infty} b_u z^u \qquad (10.3.9)$$

we will determine that form of $B(z)$ which minimizes $\mathcal{M}(m)$ by using an extension of the Wiener–Hopf technique described in Section 10.1.4. Let X_t, N_t, Y_t have spectral representations,

$$X_t = \int_{-\pi}^{\pi} e^{it\omega}\, dZ_X(\omega), \qquad N_t = \int_{-\pi}^{\pi} e^{it\omega}\, dZ_N(\omega), \qquad Y_t = \int_{-\pi}^{\pi} e^{it\omega}\, dZ_Y(\omega).$$

Then $\tilde{X}_{t+m}$ may be written

$$\tilde{X}_{t+m} = \int_{-\pi}^{\pi} e^{it\omega} B(e^{-i\omega})\, dZ_Y(\omega),$$

and hence

$$\mathcal{M}(m) = E\left[\left| \int_{-\pi}^{\pi} e^{i(t+m)\omega}\, dZ_X(\omega) - \int_{-\pi}^{\pi} e^{it\omega} B(e^{-i\omega})\, dZ_Y(\omega) \right|^2 \right].$$

Assuming, as above, that X_t and N_t are uncorrelated processes, it follows that $dZ_X(\omega)$ and $dZ_N(\omega)$ are uncorrelated, and hence,

$$E[dZ_X(\omega)\, dZ_Y^*(\omega)] = E[|dZ_X(\omega)|^2] = h_{XX}(\omega)\, d\omega,$$

while

$$E[|dZ_Y(\omega)|^2] = h_{YY}(\omega)\, d\omega.$$

Hence, using the orthogonality of $dZ_X(\omega)$, $dZ_Y(\omega)$ (for different values of ω), we evaluate $\mathcal{M}(m)$ as

$$\mathcal{M}(m) = \int_{-\pi}^{\pi} \{ h_{XX}(\omega) - e^{i\omega m} B(e^{+i\omega}) h_{XX}(\omega) - e^{-i\omega m} B(e^{-i\omega}) h_{XX}(\omega)$$

$$+ |B(e^{-i\omega})|^2 h_{YY}(\omega) \}\, d\omega. \qquad (10.3.10)$$

We now assume that $h_{YY}(\omega)$ satisfies the condition (10.1.11), so that it admits a canonical factorization of the form (10.1.13), viz.,

$$h_{YY}(\omega) = |G_Y(e^{-i\omega})|^2,$$

where $G_Y(e^{-i\omega})$ is a "backward transform", i.e. has a one-sided Fourier series involving only positive powers of $(e^{-i\omega})$. Substituting this expression in (10.3.10), we may rewrite $\mathcal{M}(m)$ in the form

$$\mathcal{M}(m) = \int_{-\pi}^{\pi} |e^{i\omega m} G_0(e^{-i\omega}) - B(e^{-i\omega}) G_Y(e^{-i\omega})|^2 \, d\omega$$

$$+ \int_{-\pi}^{\pi} \left\{ 1 - \frac{h_{XX}(\omega)}{h_{YY}(\omega)} \right\} h_{XX}(\omega) \, d\omega, \tag{10.3.11}$$

where $G_0(e^{-i\omega}) = h_{XX}(\omega)/G_Y^*(e^{-i\omega})$. Decomposing $e^{i\omega m} G_0(e^{-i\omega})$ into the sum of its backward and forward transforms, i.e. writing,

$$e^{i\omega m} G_0(e^{-i\omega}) = [e^{i\omega m} G_0(e^{-i\omega})]_+ + [e^{i\omega m} G_0(e^{-i\omega})]_-$$

$$= G_1(e^{-i\omega}) + G_2(e^{-i\omega}),$$

say (cf. (10.1.67)), and using the orthogonality of backward and forward transforms with respect to ω-integration, we obtain from (10.3.11) (recalling that $B(e^{-i\omega}) G_Y(e^{-i\omega})$ is also a backward transform)

$$\mathcal{M}(m) = \int_{-\pi}^{\pi} |G_1(e^{-i\omega}) - B(e^{-i\omega}) G_Y(e^{-i\omega})|^2 \, d\omega + \int_{-\pi}^{\pi} |G_2(e^{-i\omega})|^2 \, d\omega$$

$$+ \int_{-\pi}^{\pi} \left\{ 1 - \frac{h_{XX}(\omega)}{h_{YY}(\omega)} \right\} h_{XX}(\omega) \, d\omega. \tag{10.3.12}$$

The second and third terms in (10.3.12) are independent of the choice of $B(e^{-i\omega})$, and the first term is clearly minimized by choosing

$$B(e^{-i\omega}) = G_1(e^{-i\omega})/G_Y(e^{-i\omega}), \tag{10.3.13}$$

or equivalently,

$$B(z) = [z^{-m} G_0(z)]_+ / G_Y(z). \tag{10.3.14}$$

(When $N_t \equiv 0$, $h_{XX}(\omega) = h_{YY}(\omega) = G_Y(e^{-i\omega}) G_Y^*(e^{-i\omega})$; hence $G_0(z) = G_Y(z)$, (10.3.13) reduces to (10.1.69), and (10.3.14) reduces to (10.1.70).) Equation (10.3.13) (or (10.3.14)) provides a formal solution to the filtering problem, but we may write $B(e^{-i\omega})$ explicitly as

$$B(e^{-i\omega}) = \sum_{u=0}^{\infty} e^{-i\omega u} \left[\int_{-\pi}^{\pi} e^{i\omega'(m+u)} \{ h_{XX}(\omega')/G_Y^*(e^{-i\omega}) \} \, d\omega' \right] \bigg/ 2\pi G_Y(e^{-i\omega}). \tag{10.3.15}$$

(Compare (10.3.15) with (10.1.71).) This result is due to Wiener (1949), and the filter with transfer function $B(e^{-i\omega})$ is called the *Wiener filter*.

With the above choice for $B(e^{-i\omega})$, the minimum mean square filtering error is

$$\mathcal{M}_{\min}(m) = \int_{-\pi}^{\pi} |G_2(e^{-i\omega})|^2 \, d\omega + \int_{-\pi}^{\pi} \left\{ 1 - \frac{h_{XX}(\omega)}{h_{YY}(\omega)} \right\} h_{XX}(\omega) \, d\omega. \tag{10.3.16}$$

Comparing (10.3.16) with (10.3.8), we see that the second term in (10.3.16) represents the minimum mean square error for the case where Y_t is observed for all t, and the first term represents the additional error due to the restriction that $B(e^{-i\omega})$ has to be a backward transform. Now, the second term in (10.3.16) may be written (using $h_{YY}(\omega) = h_{XX}(\omega) + h_{NN}(\omega)$),

$$\int_{-\pi}^{\pi} \left\{ \frac{h_{XX}(\omega) h_{NN}(\omega)}{h_{XX}(\omega) + h_{NN}(\omega)} \right\} d\omega, \qquad (10.3.17)$$

and hence, as observed by Yaglom (1962), p. 133, it is impossible to recover X_t with perfect precision (even when a complete realization of Y_t is available) unless the product of the "signal" and "noise" spectra vanishes at all frequencies, i.e. *unless the "signal" and "noise" have non-overlapping spectra*. In this case we can, in principle, recover X_t perfectly by passing Y_t through an ideal band-pass filter which completely suppresses those frequencies for which $h_N(\omega)$ is non-zero. In general, the effect of the Wiener filter is to attenuate those frequency components where the signal to noise power ratio is low.

Continuous parameter processes

The corresponding result for the continuous parameter case follows in much the same way. Thus, we observe $Y(s) = X(s) + N(s)$, $s \le t$ (with $X(s)$, $N(s)$ uncorrelated), and "estimate" $X(t+m)$ by

$$\check{X}(t+m) = \int_0^{\infty} b(u) Y(t-u) \, du. \qquad (10.3.18)$$

Writing

$$B(z) = \int_0^{\infty} e^{-zu} b(u) \, du, \qquad (10.3.19)$$

it follows by a very similar argument to that used above that the mean square error is minimized by choosing,

$$B(z) = [e^{mz} G_0(z)]_+ / G_Y(z), \qquad (10.3.20)$$

where now $G_0(z)$ is defined by $G_0(i\omega) = h_{XX}(\omega)/G_Y^*(i\omega)$, and $G_Y(z)$ is defined by the canonical factorization of $h_{YY}(\omega)$ in the form $h_{YY}(\omega) = |G_Y(i\omega)|^2$.

Corresponding to (10.3.15), $B(i\omega)$ may be written explicitly as

$$B(i\omega) = \frac{1}{2\pi G_Y(i\omega)} \int_0^{\infty} e^{-i\omega u} \left[\int_{-\infty}^{\infty} e^{i\omega'(u+m)} \{ h_{XX}(\omega')/G_Y^*(i\omega') \} \, d\omega' \right] du. \qquad (10.3.21)$$

(Compare with (10.1.76).)

Example

Consider the case where

$$h_{YY}(\omega) = \frac{1}{2\pi} \frac{|1 - \beta e^{-i\omega}|^2}{|1 - \alpha e^{-i\omega}|^2} \qquad (0 < |\alpha| < 1, \quad 0 < |\beta| < 1, \quad \alpha \neq \beta),$$

and

$$h_{XX}(\omega) = \frac{k}{2\pi} \frac{1}{|1 - \alpha e^{-i\omega}|^2},$$

with $k = (\alpha - \beta)(1 - \alpha\beta)/\alpha$.

For this model it is easily verified that the spectral density function of N_t is given by

$$h_{NN}(\omega) = h_{YY}(\omega) - h_{XX}(\omega) = \frac{\beta}{2\pi\alpha} \qquad \text{(all } \omega\text{)}.$$

Thus, N_t is a white noise process with variances (β/α). Writing $z = e^{-i\omega}$, the function $h_{YY}(\omega)$ factorizes as,

$$h_{YY}(z) = \left(\frac{1}{\sqrt{2\pi}} \frac{1 - \beta z}{1 - \alpha z}\right)\left(\frac{1}{\sqrt{2\pi}} \frac{1 - \beta z^{-1}}{1 - \alpha z^{-1}}\right),$$

so that here,

$$G_Y(z) = \frac{1}{\sqrt{2\pi}} \left(\frac{1 - \beta z}{1 - \alpha z}\right).$$

(Note that $G_Y(z)$ has no poles or zeros inside $|z| \leq 1$ by the conditions on α and β.) We now have,

$$G_0(z) = \frac{h_{XX}(z)}{G_Y^*(z)} = \frac{k}{2\pi(1 - \alpha z)(1 - \alpha z^{-1})} \cdot \sqrt{2\pi}\left(\frac{1 - \alpha z^{-1}}{1 - \beta z^{-1}}\right)$$

$$= \frac{k}{\sqrt{2\pi}(1 - \alpha z)(1 - \beta z^{-1})}.$$

Suppose now that we wish to construct $\tilde{X}_{t+m}$. We need to evaluate

$$[z^{-m}G_0(z)]_+ = \frac{k}{\sqrt{2\pi}}\left[\frac{z^{-m}}{(1 - \alpha\beta)}\left\{\frac{1}{1 - \alpha z} + \frac{\beta}{z - \beta}\right\}\right]_+.$$

The second term, $\beta/(z - \beta)$, has a pole at $z = \beta$, which lies inside the unit circle. Hence this term is clearly a forward transform and may be ignored.

Consequently,

$$[z^{-m}G_0(z)]_+ = \frac{k}{\sqrt{2\pi}(1-\alpha\beta)}[z^{-m}(1+\alpha z+\alpha^2 z^2+\alpha^3 z^3+\ldots)]_+$$

$$= \frac{k}{\sqrt{2\pi}(1-\alpha\beta)}(\alpha^m+\alpha^{m+1}z+\alpha^{m+2}z^2+\ldots)$$

$$= \frac{k\alpha^m}{\sqrt{2\pi}(1-\alpha\beta)(1-\alpha z)}.$$

We now obtain from (10.3.14),

$$B(z) = \frac{k\alpha^m}{(1-\alpha\beta)(1-\beta z)} = \frac{(\alpha-\beta)\alpha^{m-1}}{(1-\beta z)}$$

$$= (\alpha-\beta)\alpha^{m-1}(1+\beta z+\beta^2 z^2+\beta^2 z^3+\ldots).$$

Thus, $\tilde{X}_{t+m}$ is given by

$$\tilde{X}_{t+m} = (\alpha-\beta)\alpha^{m-1}\sum_{u=0}^{\infty}\beta^u Y_{t-u}.$$

The Kalman filter

As with the Wiener predictor, the output of the Wiener filter has to be completely recalculated if we change t or m. However, if we assume that X_t is generated by an ARMA model we can evaluate $\tilde{X}_{t+m}$ by a recursive technique due to Kalman (1960, 1963) and Kalman and Bucy (1961). This approach is based on the "state-space" representation of X_t discussed in Section 10.4.4, and we therefore defer a detailed discussion of the Kalman filter until Section 10.6. However, it is interesting to note that there is an important distinction between the Wiener and Kalman solutions, which, to some extent, reflects the change in computational technology between the 1940s and the 1960s. In Wiener's formulation the solution is expressed in terms of the *transfer function*, $B(e^{-i\omega})$, of the optimal filter, and, as previously mentioned, this is appropriate to the situation where the filtering is to be performed in an *analogue* fashion using a "physical" filter constructed, say, from an electrical network. By contrast, the Kalman solution is expressed in terms of a recursive algorithm, which is ideally suited to the use of *digital* computers.

In the Wiener model the "signal" X_t is allowed to be a general stationary process with a purely continuous spectrum. The Kalman model is somewhat more restrictive in that X_t is (effectively) assumed to follow a finite parameter ARMA model (as in the Box–Jenkins approach), but is also more general in the sense that the parameters of the model can be allowed to be time-dependent (corresponding to X_t being non-stationary). The Wiener approach can, however, be extended to a similar class of non-stationary processes via the application of the theory of "evolutionary spectra". We discuss this extension briefly in Chapter 11.

10.4 LINEAR CONTROL SYSTEMS

Consider a single input/single output system in which U_t denotes the input, X_t the true (unobservable) output, N_t an additive noise disturbance, and Y_t the observed output—as shown schematically in Fig. 10.4. If the system is physically realizable and U_t and Y_t are observed

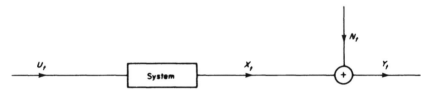

Fig. 10.4. Single input/single output system.

only at discrete time instants (say $t = 0, \pm 1, \pm 2, \ldots$), the relationships between $U_t, X_t, N_t, Y_t,$ may be expressed as,

$$X_t = h\{U_t, U_{t-1}, U_{t-2}, \ldots\}, \tag{10.4.1}$$

$$Y_t = X_t + N_t, \tag{10.4.2}$$

where the functional $h\{\cdot\}$ describes the "dynamics" of the system, i.e. the relationship between the present true output X_t and the present and past inputs $U_t, U_{t-1}, U_{t-2}, \ldots$. However, this formulation is far too general if we wish to apply statistical methods to the study of such systems. Thus if the functional $h\{\cdot\}$ is unknown we cannot attempt to "identify" its form given a finite set of observations on U_t and Y_t unless we are prepared to make much more specific assumptions about its structure. We now introduce the crucial assumption which underlies virtually all statistical studies of control systems, namely, that the functional $h\{\cdot\}$ is a *linear time invariant filter*, i.e. we assume that, for all t, the present output is a fixed linear combination of present and past inputs. With this assumption we can rewrite (10.4.1) in the form,

$$X_t = \sum_{u=0}^{\infty} a_u U_{t-u}, \tag{10.4.3}$$

where the sequence of constants $\{a_u\}$ is the impulse response function of the system. Assuming further that $\sum_{u=0}^{\infty} |a_u| < \infty$ (the condition for the system to be "stable"), we may define the function,

$$A(z) = \sum_{u=0}^{\infty} a_u z^u, \qquad (10.4.4)$$

and $A(e^{-i\omega})$ is then the transfer function of the system. The properties of the system are now completely characterized either by the sequence $\{a_u\}$ or by the function $A(z)$, and we can write (10.4.3) more concisely as,

$$X_t = A(B)U_t. \qquad (10.4.5)$$

10.4.1 Minimum Variance Control

Suppose we wish to construct a feedback controller to compensate for the disturbance N_t so that the output Y_t follows, as far as possible, some predetermined form, say we wish Y_t to follow the deterministic function V_t. Then, writing $U_t = A^{-1}(B)V_t + U_t^*$, we have from (10.4.2), (10.4.5),

$$Y_t = A(B)U_t + N_t = V_t + A(B)U_t^* + N_t,$$

and the problem is reduced to choosing U_t^* so that $Y_t^* = A(B)U_t^* + N_t$ remains as close as possible to zero. Without loss of generality we may therefore take the original predetermined form to be *zero*, all t (which is then called the "set point"), and the problem now is to design a controller which, for each t, computes U_t as a functional F of $Y_t, Y_{t-1}, \ldots$ so that

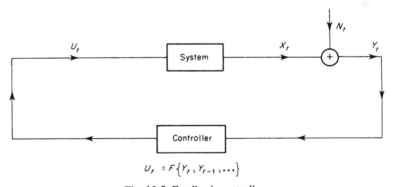

Fig. 10.5. Feedback controller.

the output remains as close as possible to zero. The optimal form of F is then determined by minimizing some "cost function" V which measures the "cost" of deviations of Y_t from its set point zero. In "*minimum variance*" control we choose as our cost function $V = E[Y_t^2]$ and then choose F to minimize this form of V.

However, in practice the inertia of the system may be such that the instantaneous effect of U_t on X_t is negligible, so that in effect $a_0 = 0$ in (10.4.3). More generally, suppose there is a delay (or "dead time") of d time units between input and output, i.e. in (10.4.3), $a_0 = a_1 = \ldots = a_{d-1} = 0$, $a_d \neq 0$. Then the input at time t does not affect the output until time $(t + d)$, and at time t the optimal choice of U_t will be that for which $V = E[Y_{t+d}^2]$ is a minimum. Thus, writing,

$$U_t = F\{Y_t, Y_{t-1}, \ldots\}$$

the problem is to choose the form of F so as to minimize $E[Y_{t+d}^2]$. The optimal form of F is easily found by noting that since, for each t, U_t is chosen as a function of $Y_t, Y_{t-1}, \ldots$, and since Y_t is itself a function of past values of U_t and N_t, we may equally well express U_t as a function of $N_t, N_{t-1}, \ldots$. The corresponding true output X_{t+d} may also be expressed as a function of $N_t, N_{t-1}, \ldots$, say,

$$X_{t+d} = G\{N_t, N_{t-1}, \ldots\}. \tag{10.4.6}$$

We now have

$$V = E[Y_{t+d}^2] = E[\{X_{t+d} + N_{t+d}\}^2]$$
$$= E[[N_{t+d} + G\{N_t, N_{t-1}, \ldots\}]^2]. \tag{10.4.7}$$

The problem of determining the functional form of G which minimizes V is now seen to be formally equivalent to the problem of *least squares prediction*. In the latter problem we are given the observed values of $N_t, N_{t-1}, \ldots$ and we seek a predictor of N_{t+d} of the form,

$$\tilde{N}_{t+d} = H\{N_t, N_{-1}, \ldots\}$$

such that $E[\{N_{t+d} - \tilde{N}_{t+d}\}^2]$ is minimized. The solution is given by (10.1.2), namely,

$$\tilde{N}_{t+d} = E[N_{t+d} | N_t, N_{t-1}, \ldots], \tag{10.4.8}$$

where the expression on the right-hand side of (10.4.8) denotes, as previously, the conditional expectation of N_{t+d}, given $N_t, N_{t-1}, \ldots$. Thus, the optimal choice of G to minimize (10.4.7) is,

$$G\{N_t, N_{t-1}, \ldots\} = -E\{N_{t+d} | N_t, N_{t-1}, \ldots\} = -\tilde{N}_{t+d}.$$

Hence, the optimal choice of the input, U_t, is that for which the corresponding true output satisfies,

$$X_{t+d} = -\tilde{N}_{t+d}. \tag{10.4.9}$$

This result is a special case of a more general result known as the *"principle of certainty equivalence"* in the statistical literature (see, e.g., Whittle (1963)), and as the *"separation principle"* in the control literature. As noted by Box and Jenkins (1970), it tells us that to keep Y_{t+d} as close as possible to zero we should choose the input so that the corresponding true output X_{t+d} exactly cancels the best predictor of the noise disturbance at time $(t+d)$, i.e. we compute $\tilde{N}_{t+d}$ on the basis of the information available up to time t, and then proceed as if the value of N_{t+d} were known "with certainty".

Now from (10.4.5)

$$X_{t+d} = B^{-d}A(B)U_t.$$

Hence, the optimal input corresponding to (10.4.9) is given by

$$U_t = -\{B^d A^{-1}(B)\}\tilde{N}_{t+d}. \tag{10.4.10}$$

(We assume that the operator $A(B)$ is "invertible", i.e. that $A(z)$ has no zeros inside or on the unit circle; in engineering language this corresponds to the system being of "minimum phase" type.) If we restrict the feedback controller to be linear in $Y_t, Y_{t-1}, \ldots$, (10.4.10) remains valid provided we now interpret $\tilde{N}_{t+d}$ as the linear least squares predictor of N_{t+d}. If N_t is a stationary Gaussian process then the unrestricted least squares predictor is, of course, linear in $N_t, N_{t-1}, \ldots$.

Equation (10.4.10) can be re-written to give an explicit expression for U_t in terms of $Y_t, Y_{t-1}, \ldots$. Assuming that N_t is a stationary linear process of the form,

$$N_t = \sum_{u=0}^{\infty} g_u \varepsilon_{t-u} = G(B)\varepsilon_t, \tag{10.4.11}$$

$\{\varepsilon_t\}$ being a white noise process, we have, from (10.1.37),

$$\tilde{N}_{t+d} = \sum_{u=d}^{\infty} g_u \varepsilon_{t+d-u} = \phi(B)\varepsilon_t, \tag{10.4.12}$$

say. Also,

$$Y_t = N_t - \tilde{N}_t = \sum_{u=0}^{d-1} g_u \varepsilon_{t-u} = \psi(B)\varepsilon_t,$$

say, so that

$$\varepsilon_t = \psi^{-1}(B)Y_t. \tag{10.4.13}$$

Substituting (10.4.13) into (10.4.12) and then into (10.4.10) we obtain,

$$U_t = -\{B^d A^{-1}(B)\phi(B)\psi^{-1}(B)\}Y_t \tag{10.4.14}$$

$$= L(B)Y_t, \tag{10.4.15}$$

say. Expanding $L(B)$ as a power series in B, (10.4.15) gives an expression for U_t as a linear combination of $Y_t, Y_{t-1}, \ldots$. Alternatively, writing (10.4.14) as

$$\psi(B)A(B)U_t = -B^d\phi(B)Y_t, \qquad (10.4.16)$$

or

$$L_1(B)U_t = L_2(B)Y_t, \qquad (10.4.17)$$

we obtain an expression for U_t as a linear function of $U_{t-1}, U_{t-2}, \ldots$, $Y_t, Y_{t-1}, \ldots$. Note that if the delay between input and output is one time unit, i.e. $d = 1$, then

$$Y_t = N_t - \tilde{N}_t = \varepsilon_t$$

(assuming $g_0 = 1$), so that the observed output Y_t is identical with the white noise process ε_t.

The three-term controller

The functions ϕ, ψ depend on the sequence $\{g_u\}$ which, in turn, depends on the second order properties of the process N_t. Thus, the form of the minimum variance controller depends not only on the system's transfer function $A(z)$, but also on the model of the disturbance N_t. In particular, if N_t is a non-stationary process whose second differences satisfy the MA(3) model (10.2.15), i.e.

$$\Delta^2 N_t = \varepsilon_t + b_1\varepsilon_{t-1} + b_2\varepsilon_{t-2} + b_3\varepsilon_{t-3},$$

then, with $d = 1$, we may replace ε_t by Y_t in the expression (10.2.18) for $\Delta\tilde{N}_{t+1}$, and, from (10.4.10), ΔU_t takes the form,

$$\Delta U_t = -\{BA^{-1}(B)\}\{\gamma_{-1}\Delta Y_t + \gamma_0 Y_t + \gamma_1 SY_t\}, \qquad (10.4.18)$$

where $\gamma_{-1}, \gamma_0, \gamma_1$ are constants determined by b_1, b_2, b_3 by (10.2.16) and, as before, $S = \Delta^{-1}$ is the summation operator. Equation (10.4.18) has a form which is similar to that of the "*three-term controller*" ("derivative, proportional and integral control"), and this result is due to Box and Jenkins (1962). (Strictly, the expression "three-term controller" is applicable only when U_t (rather than ΔU_t) is of the above form with the factor $\{BA^{-1}(B)\}$ absent.) In fact, if the system has exponential dynamics with unit delay, i.e. if the relationship between U_t and X_t is of the form,

$$(1 + \xi\Delta)X_{t+1} = gU_t,$$

so that

$$A(B) = \frac{gB}{1 + \xi(1 - B)},$$

and if $b_3 = 0$ ($\Rightarrow \gamma_{-1} = 0$), then (10.4.18) reduces to,

$$\Delta U_t = -\frac{1+\xi\Delta}{g}\{\gamma_0 Y_t + \gamma_1 SY_t\}$$

$$= -\frac{1}{g}\{\gamma_0\xi\Delta Y_t + (\gamma_0 + \gamma_1\xi)Y_t + \gamma_1 SY_t\}.$$

This model leads to an exact form of three-term controller for ΔU_t.

Example

Astrom (1970) discusses the following model for the control of a paper-making machine. The quality of a certain type of paper depends on the "dry basis weight", which in turn depends on the "thick stock flow". Regarding the "thick stock flow" as the input (U_t) and the "dry basis weight" (Y_t) as the output, the dynamics of the system is described by the model (with U_t and Y_t measured in suitable units),

$$(1-0\cdot35B)\Delta Y_t = 14\cdot6\Delta U_{t-4} + \lambda(1-0\cdot66B)\varepsilon_t,$$

.r

$$Y_t = \left(\frac{14\cdot6}{1-0\cdot35B}\right)U_{t-4} + \lambda\cdot\frac{(1-0\cdot66B)}{(1-B)(1-0\cdot35B)}\varepsilon_t,$$

where $\lambda = 0\cdot257$. Thus, in this example,

$$d = 4, \qquad A(z) = \frac{14\cdot6z^4}{(1-0\cdot35z)}, \qquad G(z) = \frac{\lambda(1-0\cdot66z)}{(1-z)(1-0\cdot35z)}.$$

Expanding $G(z)$ in the form $G(z) = \sum_{u=0}^{\infty} g_u z^u$, we find

$$g_0 = \lambda, \qquad g_1 = 0\cdot69\lambda, \qquad g_2 = 0\cdot5815\lambda, \qquad g_3 = 0\cdot5434\lambda.$$

Hence

$$\psi(z) = \lambda(1+0\cdot69z + 0\cdot5815z^2 + 0\cdot5434z^3),$$

and

$$\phi(z) = z^{-4}[G(z) - \lambda(1+0\cdot69z + 0\cdot5815z^2 + 0\cdot5434z^3)].$$

From (10.4.14) the minimum variance controller is given by,

$$U_t = -\left[B^4\frac{(1-0\cdot35B)}{14\cdot6}B^{-4}\right]$$

$$\times\left[\frac{\lambda(1-0\cdot66B)}{(1-B)(1-0\cdot35B)} - \lambda(1+0\cdot69B + 0\cdot5815B^2 + 0\cdot5434B^3)\right]$$

$$\times\left[\frac{1}{\lambda(1+0\cdot69B + 0\cdot5815B^2 + 0\cdot5434B^3)}\right]B^{-4}Y_t,$$

or

$$14 \cdot 6(1 - B)(1 + 0 \cdot 69B + 0 \cdot 5815B^2 + 0 \cdot 5434B^3)U_t$$

$$= -[1 - 0 \cdot 66B$$

$$- (1 - B)(1 - 0 \cdot 35B)(1 + 0 \cdot 69B + 0 \cdot 5815B^2 + 0 \cdot 5434B^3)]B^{-4}Y_t$$

$$= -[0 \cdot 53 + 0 \cdot 19B]Y_t.$$

Hence, the controller is

$$U_t = 0 \cdot 31 U_{t-1} + 0 \cdot 11 U_{t-2} + 0 \cdot 04 U_{t-3} + 0 \cdot 54 U_{t-4} - 0 \cdot 03 Y_t - 0 \cdot 01 Y_{t-1}.$$

10.4.2 System Identification

In order to implement the minimum variance controller (10.4.10) it is clearly crucial that we know the system's transfer function $A(z)$. When the transfer function is unknown we may attempt to estimate its structure from observations on U_t and Y_t. This is known as the problem of "*system identification*", and we now indicate briefly some of the standard methods which have been developed. These may be classified broadly into (a) non-parametric approaches, and (b) parametric approaches.

I. *Non-parametric approaches*

Suppose we are given N observations on U_t, together with the corresponding N observations on Y_t, collected while the system is operating in *open loop* form (i.e. without feedback). Assuming that U_t and N_t are jointly stationary processes (so that X_t and Y_t are also stationary), it is reasonable to suppose that U_t and N_t are uncorrelated processes, i.e. $\text{cov}\{U_t, N_t\} = 0$, all t, t'. It then follows from (9.2.18) that,

$$h_{YU}(\omega) = A(e^{-i\omega})h_{UU}(\omega), \tag{10.4.19}$$

where $h_{YU}(\omega)$ denotes the cross-spectral density function between Y_t and U_t and $h_{UU}(\omega)$ denotes the spectral density function of U_t. We may now est mate $h_{YU}(\omega)$ and $h_{UU}(\omega)$ from the data by standard "window techniques" (cf. Section 9.5), leading to estimates $\hat{h}_{YU}(\omega)$, $\hat{h}_{UU}(\omega)$, say, and then a natural estimate of $A(e^{-i\omega})$ is

$$\hat{A}(e^{-i\omega}) = \hat{h}_{YU}(\omega)/\hat{h}_{UU}(\omega). \tag{10.4.20}$$

This estimate of $A(e^{-i\omega})$ is closely related to the *least squares estimate* obtained by minimizing

$$S_1 = \sum_{t=1}^{N} \{Y_t - A(B)U_t\}^2 \tag{10.4.21}$$

with respect to the coefficients $\{a(u)\}$ (cf. the discussion following equation (9.2.25)), the main difference being that (10.4.20) need not necessarily correspond to a physically realizable system—although the corresponding ratio of the true spectra will always be physically realizable if U_t and N_t are uncorrelated.

Closed loop systems

Suppose now that the observations on U_t and Y_t are taken while the system is operating in *closed loop form*, i.e. with a feedback loop already in existence. This situation could arise, for example, when data are collected from an industrial plant (with a view to designing optimal control) while the plant is operating under some form of manual control, or during the operation of some pilot control scheme. The simple estimate (10.4.20) is now no longer valid since (i) the ratio $\{\hat{h}_{YU}(\omega)/\hat{h}_{UU}(\omega)\}$ will no longer be a "backward transform" and hence the estimate (10.4.20) will not correspond to a physically realizable system, and (ii) the presence of feedback induces correlation between the input and output so that U_t and N_t are no longer uncorrelated processes. The first point can be dealt with by using the Wiener–Hopf technique but the second point is more serious and raises problems of "identifiability"; see Akaike (1967), Priestley (1969a). If we consider the asymptotic situation as $N \to \infty$, and replace S_1 by

$$S_1^* = E[Y_t - A(B)U_t]^2, \qquad (10.4.22)$$

then it may be shown (Priestley (1969a)) that if $A(B)$ is restricted to the "one sided" form, S_1^* is minimized by

$$\hat{A}_1(e^{-i\omega}) = \frac{1}{\alpha_{UU}(e^{-i\omega})} \left[\frac{h_{YU}(\omega)}{\alpha_{UU}^*(e^{-i\omega})} \right]_+ , \qquad (10.4.23)$$

where $\alpha_{UU}(e^{-i\omega})$ is obtained from the canonical factorization of $h_{UU}(\omega)$ in the form $h_{UU}(\omega) = \alpha_{UU}(e^{-i\omega})\alpha_{UU}^*(e^{-i\omega})$ and $[\ ]_+$ denotes the "backward transform" of the expression in the brackets: see Section 10.1.3. (The result (10.4.23) follows from the usual Wiener–Hopf type of argument.)

If N_t is a Gaussian process (and the system is in open loop form) the maximum likelihood estimate of $A(e^{-i\omega})$ corresponds to a *weighted least squares* approach. Again considering the asymptotic situation, and writing

$$S_2^* = E[G^{-1}(B)\{Y_t - A(B)U_t\}]^2, \qquad (10.4.24)$$

where $G(z) = \sum_{u=0}^{\infty} g_u z^u$, the $\{g_u\}$ being defined by (10.4.11), it may be

shown (Priestley 1969a) that the restricted weighted least squares estimate of $A(e^{-i\omega})$ is given by

$$\hat{A}_2(e^{-i\omega}) = \frac{G(e^{-i\omega})}{\alpha_{UU}(e^{-i\omega})} \left[\frac{h_{YU}(\omega)}{G(e^{-i\omega})\alpha_{UU}^*(e^{-i\omega})} \right]_+ . \qquad (10.4.25)$$

Suppose, more specifically, that the system is closed by a "linear
feedback plus noise" loop, leading to the schematic representation shown
in Fig. 10.6. Here, the output Y_t is fed back through a linear controller,

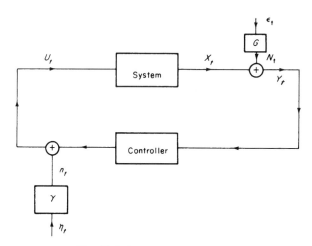

Fig. 10.6. A closed-loop system.

with transfer function $a(e^{-i\omega})$, the output of which is corrupted by a noise
component $n_t = \gamma(B)\eta_t$, ε_t and η_t being uncorrelated processes. The
complete system is described by the equations

$$Y_t = A(B)U_t + G(B)\varepsilon_t, \qquad U_t = a(B)Y_t + \gamma(B)\eta_t.$$

Alternatively, since ε_t and η_t are the only processes "external" to the
system, we may write U_t and Y_t in terms of ε_t and η_t, i.e.

$$(1 - a(B)A(B))U_t = a(B)G(B)\varepsilon_t + \gamma(B)\eta_t,$$

$$(1 - a(B)A(B))Y_t = G(B)\varepsilon_t + A(B)\gamma(B)\eta_t.$$

Using these representations we can easily evaluate $h_{YU}(\omega)$, $h_{UU}(\omega)$, and
it may then be shown (Priestley (1969a)) that $\hat{A}_1(e^{-i\omega})$, $\hat{A}_2(e^{-i\omega})$ are
identical if the function

$$H(z) = \frac{\sigma_\varepsilon^2 a^*(z)|G(z)|^2(1 - a(z)A(z))}{\sigma_\eta^2|\gamma(z)|^2 + \sigma_\varepsilon^2|a(z)|^2|G(z)|^2}$$

is analytic in $|z| \leq 1$. In particular, it follows that $\hat{A}_1$ and $\hat{A}_2$ will be identical if (i) $a(z) \equiv 0$, which corresponds to an *open-loop* system, or (ii) $\gamma(z) \equiv 0$ (so that $H(z) = (a^{-1}(z) - A(z))$), corresponding to a *noise free feedback loop*. The two estimates will be "close" to each other provided $\{\sigma_\varepsilon^2 / \sigma_\eta^2\}$ is small, i.e. provided the noise in the feedback loop has a much larger variance than the system noise; see Priestley (1969a). It should be noted, however, that the presence of correlation between U_t and N_t will mean that both $\hat{A}_1(e^{-i\omega})$ and $\hat{A}_2(e^{-i\omega})$ will be biased estimates of $A(e^{-i\omega})$, the magnitude of the bias depending on the magnitude of $|1 - a(e^{-i\omega})A(e^{-i\omega})|^2$. For further discussion of the closed loop case see Akaike (1967), Box and MacGregor (1974), Caines and Chan (1975), Gustavson, Ljung and Soderstrom (1976), and Harris (1976).

II. *Parametric approach*

The parametric approach to system identification was developed by Box and Jenkins (1962) and Astrom and Bohlin (1966). Here, we assume that $A(z)$ may be represented as a rational function of z involving a finite number of parameters, say

$$A(z) = \frac{z^d(\beta_0 + \beta_1 z + \ldots + \beta_l z^l)}{1 + \alpha_1 z + \ldots + \alpha_k z^k}, \tag{10.4.26}$$

so that the relationship between Y_t and U_t may be expressed as

$$Y_t + \alpha_1 Y_{t-1} + \ldots + \alpha_k Y_{t-k}$$
$$= \beta_0 U_{t-d} + \beta_1 U_{t-d-1} + \ldots + \beta_l U_{t-d-l} + N_t' \tag{10.4.27}$$

or

$$\alpha'(B) Y_t = \beta'(B) U_{t-d} + N_t',$$

say, where $N_t' = (1 + \alpha_1 B + \ldots + \alpha_k B^k) N_t$.

Least squares estimates of the parameters $(\alpha_1, \ldots, \alpha_k, \beta_0, \ldots, \beta_l)$ can be obtained by applying standard regression methods to the observations on Y_t and U_t. However, it is important to note that the "residual", N_t', in (10.4.27) will not, in general, be white noise. Assuming that N_t' can be represented by the model,

$$N_t' = c_0 e_t + c_1 e_{t-1} + \ldots + c_p e_{t-p}$$
$$= C(B) e_t,$$

say, where e_t is a Gaussian white noise process with unit variance, the likelihood function of the output observations (conditional on the input given) is effectively (ignoring "end effects"),

$$L(e_1, \ldots, e_N) = \left(\frac{1}{2\pi}\right)^{N/2} \exp\left(-\frac{1}{2} \sum_{t=1}^{N} e_t^2\right).$$

Hence, the maximum likelihood estimates of $\theta = (\alpha_1, \ldots, \alpha_k, \beta_0, \ldots, \beta_l, c_1, \ldots, c_p)$ are obtained by minimizing

$$V(\theta) = \sum_{t=1}^{N} e_t^2 = \sum_{t=1}^{N} [C^{-1}(B)\{\alpha'(B)Y_t - \beta'(B)U_{t-d}\}]^2. \qquad (10.4.28)$$

Astrom and Bohlin (1966) and Box and Jenkins (1970) have developed iterative techniques for determining the maximum likelihood estimate of θ, given an initial starting value θ_0. (These are essentially "hill climbing" techniques similar in nature to the Newton–Raphson algorithm.) Astrom and Bohlin's algorithm assumes

(1) the roots of $\alpha'(z)$, $\beta'(z)$ and $C(z)$ all lie outside the unit circle (so that, in particular, $C(z)$ is invertible);

(2) there are no common factors in the polynomials $\alpha'(z)$, $\beta'(z)$ and $C(z)$.

Box and Jenkins' algorithm assumes a somewhat more general model for N_t, namely that N_t is an ARMA model of the form,

$$\theta(B)N_t = \phi(B)\varepsilon_t,$$

(where $\theta(B)$ is not necessarily identical with $\alpha'(B)$). For a given set of values of $\alpha_1, \ldots, \alpha_k, \beta_0, \ldots, \beta_l$, and given N values of U_t we can construct the corresponding values of the "true" output X_t, recursively using

$$\alpha'(B)X_t = \beta'(B)U_{t-d},$$

i.e.

$$X_t + \alpha_1 X_{t-1} + \ldots + \alpha_k X_{t-k} = \beta_0 U_{t-d} + \ldots + \beta_l U_{t-d-l}.$$

We can then compute a realization of N_t from

$$N_t = Y_t - X_t,$$

and compute ε_t recursively from the $\{N_t\}$ as in standard ARMA model fitting. Initial estimates of $(\alpha_1, \ldots, \alpha_k, \beta_0, \ldots, \beta_l)$ can be obtained by assuming that N_t' is white noise and applying standard least squares to the model (10.4.27).

There is now a considerable literature on the problem of system identification; for a survey of recent work see Astrom and Eykhoff (1971) and the set of papers in the I.E.E.E. special issue on "System identification and time series analysis" (vol. AC-19, No. 6) December 1974.

10.4.3 Multivariate Systems

In the multivariate case the input will consist of, say, m components, and the true output, noise and observed output will each consist of, say, p components. Thus, writing

$$U'_t = \{U_{1,t}, \ldots, U_{m,t}\}, \qquad X'_t = \{X_{1,t}, \ldots, X_{p,t}\},$$

$$N'_t = \{N_{1,t}, \ldots, N_{p,t}\}, \qquad Y'_t = \{Y_{1,t}, \ldots, Y_{p,t}\},$$

the multivariate analogue of (10.4.2), (10.4.3), becomes

$$X_t = \sum_{u=0}^{\infty} a(u)U_{t-u} = A(B)U_t, \qquad (10.4.29)$$

$$Y_t = X_t + N_t, \qquad (10.4.30)$$

where $a(u) = \{a_{ij}(u)\}$ is a $p \times m$ matrix, $A(z)$ is a $p \times m$ polynomial matrix defined by

$$A(z) = \sum_{u=0}^{\infty} a(u)z^u, \qquad (10.4.31)$$

and $A(e^{-i\omega})$ is the *transfer function matrix*. (Thus, $\{a_{ij}(u)\}$ is the impulse response function of the relationship between the jth input and the ith output, and, similarly if we write $A(z) = \{A_{ij}(z)\}$ then $A_{ij}(e^{-i\omega})$ is the transfer function between the jth input and the ith output.)

If U_t and N_t are jointly stationary *uncorrelated* processes then we have from (9.2.33),

$$h_{YU}(\omega) = A(e^{-i\omega})h_{UU}(\omega),$$

where $h_{YU}(\omega)$, $h_{UU}(\omega)$ are the cross-spectral density matrix between Y_t and U_t and the spectral density matrix of U_t, respectively. Thus, a natural estimate of $A(e^{-i\omega})$ is

$$\hat{A}(e^{-i\omega}) = \hat{h}_{YU}(\omega)\hat{h}_{UU}^{-1}(\omega), \qquad (10.4.32)$$

where $\hat{h}_{YU}(\omega)$, $\hat{h}_{UU}(\omega)$ are estimates of $h_{YU}(\omega)$, $h_{UU}(\omega)$ (see Section 9.5).

Reduction of dimensions

Although (10.4.32) gives a fairly simple non-parametric estimate of $A(e^{-i\omega})$, in some cases it may not be desirable to use this "direct" method of estimation since a purely empirical approach to multivariate model building may involve an excessively large number of input and output variables. (If one does not know which variables are important there may be a tendency to record every conceivable variable which might just possibly influence the system.) Consequently, the transfer function matrix may contain a substantial amount of "redundant" information, and even when it has been estimated it may provide little insight into the basic underlying structure of the system. It may therefore be appropriate, as a first stage of the analysis, to try to reduce the dimensions of the input and output processes by applying linear transformations such that the transformed variables, whilst of lower dimensions, retain as far as possible the basic "input/output" structure of the system. In so doing we must first decide upon suitable criteria for this reduction. In the case of the input we could argue that the total set of variables was "too large" if the output variables could be "explained" in terms of linear relationships on just a few linear combinations of the input variables. In the case of the output variables we could say that the original set was "too large" if the "control" of the system to some set of desired operating conditions could be effected by controlling just a few linear combinations of the output variables. Following the approach of Brillinger (1964, 1969b), Priestley, Subba Rao and Tong (1973, 1974) showed that these considerations lead in a very natural way to the application of principal components analysis on the "Fourier components" of the input and output, $d\mathbf{Z}_U(\omega)$, $d\mathbf{Z}_Y(\omega)$. Specifically, suppose that we wish to control the output vector so that it remains as close as possible to some fixed "set point" which, as in the single input/single output case, may be taken to be zero. We may then introduce a "loss function" $\mathcal{L}$ which measures the cost of deviations of the output from its set point. Suppose that we have a *quadratic loss function* of the form

$$\mathcal{L} = E[\mathbf{Y}_t'\mathbf{Q}\mathbf{Y}_t], \qquad (10.4.33)$$

where $\mathbf{Q} = \{q_{jk}\}$ is a given symmetric positive definite matrix. (Note that since $\mathbf{Y}_t$ is stationary then $\mathcal{L}$ is independent of t.) If we wish to reduce the dimension of $\mathbf{Y}_t$ by transforming to new variables $\mathbf{W}_t$ which retain, as far as possible, those features of the system's performance which we wish to control, we would seek $q(<p)$ linear combinations of the $\{Y_{i,t}\}$ which are orthogonal, and whose total variances approximate as closely as possible the loss function $\mathcal{L}$. Now (10.4.33) may be written in terms of its frequency domain representation by using the spectral representation of $\mathbf{Y}_t$ (cf. (9.5.51)), and making use of the cross-orthogonality of the components of $d\mathbf{Z}_Y(\omega)$ are different frequencies. It is then easy to

show that

$$\mathcal{L} = \int_{-\pi}^{\pi} E[d\mathbf{Z}_Y^{*\prime}(\omega)\mathbf{Q}\,d\mathbf{Z}_Y(\omega)] = \int_{-\pi}^{\pi} \mathcal{L}(\omega)\,d\omega, \qquad (10.4.34)$$

say, where

$$\mathcal{L}(\omega) = E[d\mathbf{Z}_Y^{*\prime}(\omega)\mathbf{Q}(d\omega^{-1})\,d\mathbf{Z}_Y(\omega)].$$

Now consider q linear combinations of the $\{Y_{i,t}\}$, say $W_{1,t}, \ldots, W_{q,t}$ where

$$W_{i,t} = \sum_{j=1}^{p} \sum_{u=-\infty}^{\infty} p_{ij}(u) Y_{j,t-u}, \qquad i = 1, \ldots, q. \qquad (10.4.35)$$

In frequency domain terms (10.4.35) may be re-written as (in an obvious notation),

$$dZ_{W_i}(\omega) = \mathbf{P}_i^{\prime}(\omega)\,d\mathbf{Z}_Y(\omega), \qquad i = 1, \ldots, q, \qquad (10.4.36)$$

where

$$\mathbf{P}_i^{\prime}(\omega) = \{p_{i1}(\omega), \ldots, p_{ip}(\omega)\},$$

and

$$p_{jk}(\omega) = \sum_{u=-\infty}^{\infty} p_{jk}(u) e^{-i\omega u}, \qquad j = 1, \ldots, q, \quad k = 1, \ldots, p. \qquad (10.4.37)$$

Writing $\mathbf{P}(\omega) = \{\mathbf{P}_1(\omega), \ldots, \mathbf{P}_q(\omega)\}$, (10.4.36) is now

$$d\mathbf{Z}_W(\omega) = \mathbf{P}^{\prime}(\omega)\,d\mathbf{Z}_Y(\omega). \qquad (10.4.38)$$

Our problem is to choose the $\{p_{ij}(u)\}$ so that $\sum_{i=1}^{q} \text{var}\{W_i(t)\}$ gives the best approximation (over the class of all q linear orthogonal transformations of $\{Y_i(t)\}$) to $\mathcal{L}$. But we have

$$\text{var}\{W_{i,t}\} = \int_{-\pi}^{\pi} E\{|dZ_{W_i}(\omega)|^2\}\,d\omega,$$

$$= \int_{-\pi}^{\pi} \text{var}\{dZ_{W_i}(\omega)\}, \qquad i = 1, \ldots, q.$$

Treating each frequency component separately, we may obtain an equivalent reformulation of the problem as follows: *for each* ω, choose q vectors, $\mathbf{P}_1(\omega), \ldots, \mathbf{P}_q(\omega)$ so that when the $dZ_{W_i}(\omega)$ are given by (10.4.36), $\sum_{i=1}^{q} \text{var}\{dZ_{W_i}(\omega)\}$ gives the best approximation to

$$\mathcal{L}(\omega) = E[d\mathbf{Z}_Y^{*\prime}(\omega)\mathbf{Q}(d\omega^{-1})\,d\mathbf{Z}_Y(\omega)].$$

Posed in this way the problem becomes identical to that of finding the first q principal components of $X = dZ_Y(\omega)$ when X belongs to an "oblique space" with metric $x'Q_\omega x$, where $Q_\omega = Q(d\omega^{-1})$. Using a result of Rao (1964) it may be shown (see Priestley, Subba Rao and Tong (1973)) that the solution is given by choosing the $P_i(\omega)$ to be the first q eigenvectors of Σ, the variance–covariance matrix of X, with respect to Q_ω^{-1}, i.e. the first q vectors associated with the determinantal equation

$$|\Sigma - \lambda Q_\omega^{-1}| = 0.$$

But here

$$\Sigma = E[dZ_Y^{*\prime}(\omega)\, dZ_Y(\omega)] = h_{YY}(\omega)\, d\omega,$$

and $Q_\omega^{-1} = Q_1^{-1}\, d\omega$. Hence, $P_i(\omega)$ is the ith eigenvector of $h_{YY}(\omega)$ with respect to Q^{-1}, corresponding to the determinantal equation,

$$|h_{YY}(\omega) - \lambda Q^{-1}| = 0, \tag{10.4.39}$$

and the $P_i(\omega)$ satisfy the equations,

$$h_{YY}(\omega)P_i(\omega) = \lambda_i Q^{-1}P_i(\omega), \qquad i = 1, \ldots, q, \tag{10.4.40}$$

where $\lambda_1 \geq \lambda_2 \geq \ldots \geq \lambda_q \geq 0$ are the first q eigenvalues of (10.4.39). When $P_i(\omega)$ has been determined from (10.4.40) the corresponding time domain transformation is given by inverting (10.4.37), yielding

$$p_{jk}(u) = \frac{1}{2\pi} \int_{-\pi}^{\pi} p_{jk}(\omega)\, e^{i\omega u}\, du. \tag{10.4.41}$$

Following Rao (1964) we may test the adequacy of the approximation of $\sum_{i=1}^{q} \text{var}\{W_i(t)\}$ to $\mathscr{L}$ by considering the statistic

$$\Lambda = \frac{\lambda_1 + \lambda_2 + \ldots + \lambda_q}{\lambda_1 + \lambda_2 + \ldots + \lambda_p}.$$

Tong and Sugiyama (1976) have studied the distribution of the above ratio when the eigenvalues λ_i are replaced by the eigenvalues of the *estimated* spectral density matrix, $\hat{h}_{YY}(\omega)$ with respect to Q^{-1}, and have obtained approximations to the distribution function of the statistic, $\hat{\Lambda}$, say.

Now consider the reduction of the dimension of the input vector U_t. Since we aim to describe the system in terms of linear relationships, it is reasonable to seek $r(<m)$ linear combinations of the $\{U_{i,t}\}$, say $V_{1,t} \ldots, V_{r,t}$ such that the "linear predictive efficiency" of $V_t = \{V_{1,t}, \ldots, V_{r,t}\}$ for the output is a maximum. Since at this stage we have already removed "redundant" information from the output we might

consider instead the linear predictive efficiency of V_t for W_t. Let

$$V_j = \sum_{k=1}^{m} \sum_{u=-\infty}^{\infty} m_{jk}(u) U_{k,t-u}, \qquad j = 1, \ldots, r, \qquad (10.4.42)$$

or, in frequency domain terms (with an obvious notation),

$$dZ_{V_j}(\omega) = M_j'(\omega) \, dZ_U(\omega), \qquad j = 1, \ldots, r, \qquad (10.4.43)$$

where

$$M_j'(\omega) = \{m_{j1}(\omega), \ldots, m_{jm}(\omega)\},$$

and

$$m_{jk}(\omega) = \sum_{u=-\infty}^{\infty} m_{jk}(u) \, e^{-i\omega u}. \qquad (10.4.44)$$

Writing $M(\omega) = \{M_1(\omega), \ldots, M_r(\omega)\}$, (10.4.43) becomes,

$$dZ_V(\omega) = M'(\omega) \, dZ_U(\omega). \qquad (10.4.45)$$

Now consider a linear "regression" of $W_{j,t}$ on $V_{1,t}, \ldots, V_{r,t}$ of the form

$$W_{j,t} = \sum_{k=1}^{r} \sum_{u=-\infty}^{\infty} b_{jk}(u) V_{k,t-u} + \text{residual}.$$

The mean square error is

$$M = E\left[W_{j,t} - \sum_{k=1}^{r} \sum_{u=-\infty}^{\infty} b_{jk}(u) V_{k,t-u} \right]^2$$

$$= E\left[\int_{-\pi}^{\pi} e^{i\omega t} \left\{ dZ_{W_j}(\omega) - \sum_{k=1}^{r} B_{jk}(\omega) \, dZ_{V_k}(\omega) \right\} \right]^2,$$

where

$$B_{jk}(\omega) = \sum_{u=-\infty}^{\infty} b_{jk}(u) \, e^{-i\omega u}.$$

Because of the orthogonality of the various $\{dZ(\omega)\}$ we may now write

$$M = \int_{-\pi}^{\pi} E\left[\left| dZ_{W_j}(\omega) - \sum_{k=1}^{r} B_{jk}(\omega) \, dZ_{V_k}(\omega) \right|^2 \right].$$

Thus, the problem of choosing V_t in terms of maximum linear predictive efficiency for W_t is equivalent to choosing $dZ_V(\omega)$ in terms of linear predictive efficiency for $dZ_W(\omega)$. If we now define "maximum linear predictive efficiency" as meaning that the trace of the variance–covariance matrix of the residual is minimized, then using a result of Rao

(1964) it may be shown (see Priestley, Subba Rao and Tong (1973)) that *for each ω, the optimal choice of $M_j(\omega)$ is the jth eigenvector of the matrix* $K'(\omega)K(\omega)$ *with respect to the matrix* $L(\omega)$, where $K(\omega)$ is the covariance matrix between $dZ_W(\omega)$ and $dZ_U(\omega)$, and $L(\omega)$ is the variance–covariance matrix of $dZ_U(\omega)$. Thus,

$$K(\omega) = E[dZ_W(\omega)\, dZ_U^{'*}(\omega)]$$

$$= P'(\omega)E[dZ_Y(\omega)\, dZ_U^{'*}(\omega)]$$

$$= P'(\omega)h_{YU}(\omega)\, d\omega,$$

and

$$L(\omega) = E[dZ_U(\omega)\, dZ_U^{'*}(\omega)] = h_{UU}(\omega)\, d\omega.$$

Hence, $M_j(\omega)$ is the jth eigenvector of

$$h'_{YU}(\omega)P(\omega)P'(\omega)h_{YU}(\omega)$$

with respect to $h_{UU}(\omega)$, corresponding to the determinantal equation

$$|h'_{YU}(\omega)P(\omega)P'(\omega)h_{YU}(\omega) - \lambda h_{UU}(\omega)| = 0. \qquad (10.4.46)$$

When the $M_j(\omega)$ are determined as above the corresponding time domain transformations are given by inverting (10.4.44), i.e.

$$m_{jk}(u) = \frac{1}{2\pi} \int_{-\pi}^{\pi} m_{jk}(\omega)\, e^{i\omega u}\, d\omega. \qquad (10.4.47)$$

In practical applications the spectral matrices $h_{YY}(\omega)$, $h_{YU}(\omega)$, $h_{UU}(\omega)$ will, in general, be unknown, and will have to be estimated numerically from observations on U_t and Y_t. Although in Section 9.5 we discussed standard techniques for estimating these matrices, they can be evaluated only at a discrete set of frequencies, say $\omega_1, \omega_2, \ldots, \omega_K$. This means that the vectors $\{\hat{P}_i(\omega)\}$, $\{\hat{M}_j(\omega)\}$ can similarly be computed only at a discrete set of frequencies, and the corresponding time domain transformations $\{p_{jk}(u)\}$, $\{m_{jk}(u)\}$ have to be determined by computing the "discrete" Fourier transforms of $\hat{P}_i(\omega)$ and $\hat{M}_j(\omega)$. This approach has been applied to a chemical engineering data with considerable success. The data consisted of 4 input and 11 output series, each containing 190 observations, and described temperature and pressures at various points in a distillation column, together with measurements of the concentration of the distillates at the separation points. The analysis described above reduced the output to 2 variables and the input also to 2 variables, these results being in good agreement with those predicted on purely physical considerations. The detailed analysis of this system is given in Haggan (1975) and Haggan and Priestley (1975).

Subba Rao and Tong (1974) have considered an alternative approach based on "factor analysis" in the frequency domain, and Subba Rao (1975, 1976a) has also considered the application of principal components analysis in the time domain. Priestley (1976a) gives a review of these methods.

10.4.4 State Space Representations

In discussing linear systems it is often more convenient to use the so-called "state space" (or "Markovian") representation of the relationship between input and output rather than the explicit form given, for the scalar case, for example, by (10.4.27). The "state-space" form gives a very compact description which is valid provided the relationship between the input and output can be expressed in terms of a finite order linear difference equation. The basic idea rests on the well known result that any finite order linear differential or difference equation can be expressed as a vector first order equation. For example, if we take the (univariate) AR(2) model,

$$X_t + a_1 X_{t-1} + a_2 X_{t-2} = \varepsilon_t, \tag{10.4.48}$$

and write

$$X_t^{(2)} = X_t, \qquad X_t^{(1)} = -a_2 X_{t-1} \quad (= -a_2 X_{t-1}^{(2)}),$$

then (10.4.48) may be re-written as

$$\begin{bmatrix} X_t^{(1)} \\ X_t^{(2)} \end{bmatrix} = \begin{bmatrix} 0 & -a_2 \\ 1 & -a_1 \end{bmatrix} \begin{bmatrix} X_{t-1}^{(1)} \\ X_{t-1}^{(2)} \end{bmatrix} + \begin{bmatrix} 0 \\ 1 \end{bmatrix} \varepsilon_t. \tag{10.4.49}$$

To recover X_t from the vector $(X_t^{(1)}, X_t^{(2)})'$, we now write

$$X_t = (0, 1) \begin{bmatrix} X_t^{(1)} \\ X_t^{(2)} \end{bmatrix}. \tag{10.4.50}$$

The pair of equations (10.4.49), (10.4.50) are completely equivalent to (10.4.48) but whereas (10.4.48) involves a two stage dependence (so that X_t is non-Markovian), (10.4.49) involves only a one stage dependence— so that $(X_t^{(1)}, X_t^{(2)})'$ is a vector Markov process. The device of expressing (10.4.48) in the form (10.4.49) may be viewed alternatively as a special case of the technique of writing certain types of non-Markov processes involving only finite stage dependence as vector Markov processes, and can obviously be extended to general nth order difference equations. Now consider the multivariate system described by (10.4.29), (10.4.30), and suppose that the transfer function matrix takes a "rational" form so that we may write

$$A(z) = [I + a_1 z + a_2 z^2 + \ldots + a_n z^n]^{-1} [b_0 + b_1 z + \ldots + b_{n-1} z^{n-1}],$$

$$(10.4.51)$$

where $|\sum_{i=0}^{n} a_i z^i|$ has no zeros inside or on the unit circle (and the notation $|A|$ means the determinant of A). Then (10.4.29), (10.4.30), reduce to

$$X_t + a_1 X_{t-1} + \ldots + a_n X_{t-n} = b_0 U_t + \ldots + b_{n-1} U_{t-n+1} \quad (10.4.52)$$

$$Y_t = X_t + N_t. \quad (10.4.53)$$

Corresponding to (10.4.49), (10.4.50), we can write (10.4.52), (10.4.53), in "state space" form as

$$x_{t+1} = F x_t + G U_{t+1}, \quad (10.4.54)$$

$$Y_t = H x_t + N_t. \quad (10.4.55)$$

Here, x_t is called the "state vector" (and, in general, is of larger dimension than Y_t), F is called the "system matrix", G the "input matrix", and H the "observation matrix". Akaike (1974a, b) has given a very general treatment of finite order linear systems and has shown that the descriptions (10.4.52), (10.4.53) and (10.4.54), (10.4.55) are indeed equivalent under very general conditions. However, following the analysis of the AR(2) model, one explicit derivation is easily obtained in the following way. Given the model (10.4.52), (10.4.53) (in which we recall that U_t has dimension m and X_t, N_t, Y_t all have dimension p), define the $np \times np$ matrix F by

$$F = \begin{pmatrix} 0_{p \times p} & 0_{p \times p} & \cdots & 0_{p \times p} & -a_n \\ I_{p \times p} & 0_{p \times p} & \cdots & 0_{p \times p} & -a_{n-1} \\ \vdots & \vdots & & \vdots & \vdots \\ 0_{p \times p} & 0_{p \times p} & \cdots & I_{p \times p} & -a_1 \end{pmatrix}, \quad (10.4.56)$$

the $np \times m$ matrix G by

$$G' = (b'_{n-1}, b'_{n-2}, \ldots, b'_0), \quad (10.4.57)$$

the $p \times np$ matrix H by

$$H = (0_{p \times p}, 0_{p \times p}, \ldots, I_{p \times p}), \quad (10.4.58)$$

and partition the $np \times 1$ vector x_t in the form

$$x'_t = (x'_{1,t}, x'_{2,t}, \ldots, x'_{n,t}).$$

It is then easy to check that, with these definitions of F, G, H, (10.4.54),

(10.4.55), become identical with (10.4.52), (10.4.53). (In fact, $x_{n,t}$ is simply the variable X_t in (10.4.52) and $x_{n-1,t}$, $x_{n-2,t}$, ... are defined recursively in terms of $x_{n,t} = X_t$.) Note that, with the above condition on $|\sum_{i=0}^{n} a_i z^i|$, the matrix F is "*stable*", i.e. all its eigenvalues have modulus less than 1, and consequently the process x_t is *stationary*. Now suppose that the input U_t has a finite order MA representation,

$$U_t = \sum_{j=0}^{K} \gamma_j \eta_{t-j} = \gamma(B)\eta_t,$$

say, where η_t is a white noise process. Indeed, η_t is the one step prediction error of U_t, i.e.

$$\eta_t = U_t - \tilde{U}_{t|t-1}$$

where $\tilde{U}_{t|t-1}$ is the linear least squares predictor of U_t, given $U_{t-1}, U_{t-2}, \ldots$; η_t is thus the *innovation* process for U_t. Writing (10.4.52) as

$$\alpha(B)X_t = \beta(B)U_t, \tag{10.4.59}$$

we can now write

$$\alpha(B)X_t = \beta(B)\gamma(B)\eta_t. \tag{10.4.60}$$

Since $\beta(B)$ and $\gamma(B)$ are both finite order polynomial matrices it follows that (10.4.60) has essentially the same finite order difference equation form as (10.4.59), and hence the relationship between X_t and η_t can also be put into a state space form, say,

$$x_{t+1} = Fx_t + G\eta_{t+1}, \tag{10.4.61}$$

$$Y_t = Hx_t + N_t. \tag{10.4.62}$$

(The matrices F, G, in (10.4.61) will, of course, be different from the corresponding matrices in (10.4.54) but for simplicity we retain the same notation.)

Akaike's state-space representation

It should be emphasized that the specific expressions for F, G, H given by (10.4.56), (10.4.57), (10.4.58), provide merely *one* way of writing (10.4.52), (10.4.53), in state-space form. In general, there is a multitude of ways of writing (10.4.52), (10.4.53), in the form (10.4.54), (10.4.55), (or (10.4.57), (10.4.58)). (If we transform the state vector x_t into $x_t^* = Tx_t$, T being an arbitrary non-singular $n \times n$ matrix, then (10.4.54), (10.4.55), can be expressed in the same form in terms of x_t^* by writing $F^* = TFT^{-1}$, $G^* = TG$, $H^* = HT^{-1}$.) Akaike (1974a) has shown, using a geometrical approach, that a *minimal realization* (i.e. a state space representation with x_t having the smallest possible dimension) can be derived by selecting the state vector x_t as any basis of the "predictor space at time t", $R(t+|t-)$, namely the space spanned by the linear least squares predictors of $Y_t, Y_{t+1}, Y_{t+2}, \ldots$, given $U_t, U_{t-1}, U_{t-2}, \ldots$. Akaike describes the space $R(t+|t-)$ as "containing all the information on present and future outputs contained in present and past inputs". The crucial feature of this space is that, for systems described by finite order difference (or differential) equations, *it has a finite dimension, n*. A minimal realization can now be constructed by choosing the state vector (at time t) as *any* basis of $R(t+|t-)$. (Akaike (1974b) discusses a number of special minimal realizations.) This specification of the state vector is in complete accord with the physical interpretation of the "state" of the system at time t_0 as that set of quantities which, together with input for all time points $\geq t_0$, uniquely determines the output for all time points $\geq t_0$. When, for example, the relationship between a scalar input and a scalar output is described by a finite order differential equation, the state of the system at time t_0 is simply the set of "initial conditions" required to determine the solution of the differential equation for all $t \geq t_0$. Intuitively, therefore, we may think of the state at any time point as the totality of information on the future output contained within the past input, and this is precisely the information contained within the set of predictors, $\{\tilde{Y}_t, \tilde{Y}_{t+1|t}, \tilde{Y}_{t+2|t} \ldots\}$, or equivalently, within any set of basis variables of the space spanned by these predictors.

10.5 MULTIVARIATE TIME SERIES MODEL FITTING

As discussed in Section 9.4, a process Y_t is said to follow a multivariate ARMA model if it can be expressed in the form

$$a_0 Y_t + a_1 Y_{t-1} + \ldots + a_k Y_{t-k} = b_0 \varepsilon_t + \ldots + b_l \varepsilon_{t-l}, \qquad (10.5.1)$$

or

$$\alpha(B) Y_t = \beta(B) \varepsilon_t, \qquad (10.5.2)$$

where ε_t is a multivariate white noise process. The ideas discussed above can now be applied to the study of ARMA models by noting that (10.5.1) may be regarded as a special case of (10.4.52), (10.4.53), in which $U_t \equiv \varepsilon_t$ and $N_t \equiv 0$. (The parameter n in (10.4.52) is set as $\max(k+1, l)$.) Thus, Y_t may be regarded as the output of a linear system with a rational transfer function matrix subjected to a white noise input and free from the noise disturbance N_t. The ARMA model (10.5.1) can now be put in the equivalent state space form (setting $N_t \equiv 0$ in (10.4.55)),

$$x_{t+1} = Fx_t + G\varepsilon_{t+1},\tag{10.5.3}$$

$$Y_t = Hx_t.\tag{10.5.4}$$

With the interpretation $U_t = \varepsilon_t$, (10.5.3) follows directly from (10.4.54). However, (10.5.3) can be derived alternatively from (10.4.61) by identifying the input with the output, i.e. by setting $U_t = Y_t$. (Under conditions to be stated below, ε_t is the innovation process of Y_t, and hence, with $U_t = Y_t$, becomes the process η_t in (10.4.61).) In this case, the space $R(t+|t-)$ becomes the space of linear least squares predictors of $Y_t, Y_{t+1}, Y_{t+2}, \ldots$, given its *own present and past*, i.e. given $Y_t, Y_{t-1}, Y_{t-2}, \ldots$.

10.5.1 Identifiability of Multivariate ARMA Models

One of the basic problems associated with the multivariate ARMA model is the *"identification"* of the structure of (10.5.1), given the covariance function of Y_t. By "identification" we mean here the following problem; given that Y_t conforms to some ARMA model (of unspecified orders) can we determine the values of k and l and the matrices $a_0, a_1, \ldots, a_k, b_0, b_1, \ldots, b_l$ *uniquely* from the covariances of Y_t? In its "raw" form (10.5.1) is certainly *not* identifiable in this sense. To see this we need only note, in particular, that if we multiply both sides of (10.5.2) by the same arbitrary matrix polynomial (in B) the covariance structure of Y_t is unaltered. In fact, corresponding to a given covariance structure there will be an "equivalence class" of models, and the problem of identifiability then becomes one of devising a set of rules which select a unique representative model from each equivalence class. This means that we have to impose certain constraints on $\alpha(B)$, $\beta(B)$, and their matrix coefficients which are such that within each equivalence class only one model will satisfy these constraints. This problem has been studied principally by Hannan (1969a, 1970, 1976, 1979b), and in his 1976 paper Hannan lists three different procedures, each of which ensures identifiability. These may be summarized briefly as follows. First of all, we may assume, without loss of generality, that

$$a_0 = b_0 = I.\tag{10.5.5}$$

(With the leading coefficients on the LHS of (10.5.1) set to I, the model is then said to be in "reduced" form; on the other hand, setting $b_0 = I$ involves no loss of generality as long as the variance–covariance matrix of

ε_t is allowed to be of general form.) Next we impose the condition:

$\alpha(B), \beta(B)$ have no common left divisors

(other than unimodular ones),   (10.5.6)

i.e. if $D(B)$ is such that $\alpha(B) = D(B)\alpha_1(B)$, and $\beta(B) = D(B)\beta_2(B)$, ($\alpha_1$, β_2 being polynomials) then $|D(B)|$ is a *constant*. Thirdly (as with the model (10.4.51)), we impose a condition on the zeros of $|\alpha(z)|$ and $|\beta(z)|$, namely:

all the zeros of $|\alpha(z)|$ and of $|\beta(z)|$ lie outside the unit circle.   (10.5.7)

(The condition on the zeros of $|\alpha(z)|$ is required to ensure that Y_t is a stationary process, and the condition on the zeros of $|\beta(z)|$ ensures that the MA operator is "invertible"—see Section 9.4. Without these conditions, the roots of $|\alpha(z)|$ and $|\beta(z)|$ and their reciprocals could be permuted without affecting the covariance structure of Y_t—see, e.g., Quenouille (1957), p. 22, Box and Jenkins (1970), p. 195)—although, in fact, we could allow $|\beta(z)|$ (but not $|\alpha(z)|$) to have zeros *on* the unit circle and still obtain identifiability—see Hannan (1970), p. 370.

The model (10.5.1) may now be identified by any of the following procedures:

(a) From each equivalence class choose that sub-class of models with the smallest value of l; then from this sub-class choose that model with the smallest value of k. If, for such a model, the condition that

the matrix $[b_l : a_k]$ is of full rank   (10.5.8)

is satisfied, then $\alpha(B)$, $\beta(B)$ are uniquely determined.

(b) We may use a "scalar" representation, i.e. a form of (10.5.1) in which the coefficients a_i have the form $a_i = a_i I$. This model is then identifiable provided $a_k \neq 0$.

(c) Alternatively, we may use a "triangular representation", namely, a form of (10.5.1) in which $\alpha(B)$ is lower triangular, and if $\alpha_{ij}(B)$ denotes the (i, j)th element of $\alpha(B)$ then $\alpha_{ij}(B)$ must not be of higher degree than $\alpha_{jj}(B)$. This again identifies the model.

(Note that, in addition to the special conditions stated under (a), (b), (c) the conditions (10.5.5), (10.5.6), (10.5.7) must also hold in each case.) The procedure described under (a) is the simplest, and the usual numerical identification techniques would seem to be most closely related to this approach.

It should, however, be noted that if we are concerned simply with *fitting* a model of the form (10.5.1) to data, *with values of k, l specified a priori*, then all we need are conditions (10.5.5), (10.5.6), (10.5.7) and (10.5.8) to ensure that the matrix coefficients $a_1, \ldots, a_k, b_1, \ldots, b_l$ are statistically identifiable. As an alternative to the condition $b_0 = I$ we could impose a constraint on Σ_ε, the variance–covariance matrix of ε_t, namely, that $\Sigma_\varepsilon = I$. This is easily arranged by writing $\xi_t = K^{-1}\varepsilon_t$, where K is an arbitrary non-singular matrix satisfying $KK' = \Sigma_\varepsilon$, and then re-writing the RHS of (10.5.1) in terms of the white noise process ξ_t. However, the above equation for K does not determine a unique matrix, unless we add an additional constraint, e.g. that K be a lower triangular matrix. In this case Kb_0 will also be a lower triangular matrix. Thus, instead of the condition $b_0 = I$ we can use the alternative conditions that $\Sigma_\varepsilon = I$, and b_0 is lower triangular. Either set of conditions enable the MA coefficients to be estimated, otherwise the MA operator contains redundant parameters. (Note, in particular, that for a p-dimensional pure MA(l) process we have $(l+1)p$ non-zero autocovariances and $(2l+1) \times \frac{1}{2}p(p-1)$ cross-covariances, giving in all $\{(l+1)p^2 - \frac{1}{2}l(l-1)\}$ known quantities. On the other hand, the matrix parameters $b_0, \ldots, b_l$ (with $\Sigma_\varepsilon = I$) involve $(l+1)p^2$ scalar parameters. Thus, there are $\frac{1}{2}l(l-1)$ "redundant" parameters, and these may be removed either by setting $b_0 = I$ and allowing Σ_ε to be an arbitrary (symmetric) matrix or by retaining $\Sigma_\varepsilon = I$ and restricting b_0 to be lower triangular.)

The equivalent state space representation of (10.5.1) (as given by (10.5.3), (10.5.4)) is, of course, equally lacking in identifiability, for, as previously explained, *any* basis of the space $R(t+|t-)$ may serve as a state vector in a minimal realization. However, Akaike (1974a, 1976) has constructed a "canonical form" of the state space representation in which the state vector x_t is of finite dimension np and has as its jth block of p elements $(j = 1, 2, \ldots, n)$ the vector $\tilde{Y}_{t+j-1|t}$ (the linear least squares predictor of Y_{t+j-1}, given $Y_t, Y_{t-1}, Y_{t-2} \ldots$), F takes the form

$$F = \begin{pmatrix} 0 & I & 0 & \cdots & 0 \\ 0 & 0 & I & \cdots & 0 \\ \multicolumn{5}{c}{\dotfill} \\ -a_n & -a_{n-1} & \cdots & \cdots & -a_1 \end{pmatrix} \qquad (10.5.9)$$

(with $a_j = 0$, $j > k$), G has form

$$G' = (I, \ W'(1), \ldots, \ W'(n-1))', \qquad (10.5.10)$$

$\{W(j)\}$ being the impulse response matrices of (10.5.1), (i.e. $W(j)$ is the coefficient of z^j in the expansion of $\alpha^{-1}(z)\beta(z)$), and $H = (I, 0, \ldots, 0)$. Akaike has shown that the parameters $\{a_j\}$, $\{W(j)\}$, in this form of state space representation are uniquely determined provided the variance–covariance matrix of x_t, $E[x_t x_t']$ is of full rank. This means that *the* $\{a_j\}$

and $\{b_i\}$ in (10.5.1) *are uniquely determined provided* $E[x_t x_t']$ *is of full rank*. Akaike has shown further that the condition that $E[x_t x_t']$ be of full rank is equivalent to the condition that the system corresponding to the canonical state space form be *controllable*, i.e. that the matrix

$$[G \vdots FG \vdots \ldots \vdots F^{np-1}G] \tag{10.5.11}$$

be of full rank np. ("Controllability" means, in effect, that starting from an arbitrary initial state, $x^{(0)}$, at time t_0, there exists an input which "steers" the system into any desired state $x^{(1)}$ at some subsequent time t, and (10.5.11) is a well known condition for a linear system to have this property—see, e.g., Rosenbrock (1970).) Hannan (1976) has shown that, in turn, the *condition of controllability is equivalent to the condition* (10.5.8), when the ARMA model for Y_t satisfies condition (10.5.6). Akaike refers to this situation as "*block identifiability*".

10.5.2 Fitting State Space Models via Canonical Correlations Analysis

Given observations on Y_t, Akaike (1974a, 1976) has suggested that we fit the canonical form of state space representation obtained by choosing the state space vector as the *first maximum set of linearly independent elements* within the sequence of predictors,

$$\tilde{Y}^{(1)}_{t|t}, \ \tilde{Y}^{(2)}_{t|t}, \ \ldots, \ \tilde{Y}^{(p)}_{t|t}, \ \tilde{Y}^{(1)}_{t+1|t}, \ \tilde{Y}^{(2)}_{t+1|t}, \ \ldots$$
$$\ldots, \ \tilde{Y}^{(p)}_{t+1|t}, \ \ldots \ \tilde{Y}^{(1)}_{t+j|t}, \ \tilde{Y}^{(2)}_{t+j|t}, \ \ldots, \ \tilde{Y}^{(p)}_{t+j|t}, \ \ldots,$$

where, for the moment, $Y_t^{(r)}$ denotes the rth component of Y_t, $r = 1, \ldots, p$. Writing

$$S(t) = [\tilde{Y}'_{t|t}, \tilde{Y}'_{t+1|t}, \ldots]', \tag{10.5.12}$$

Akaike's method is to test successively the linear dependence of the ith element of S, S_i, on the previous elements, S_{i-1}, S_{i-2}, $\ldots$, S_1. The state vector x_t is then obtained by removing from $S(t)$ all those elements which are linearly dependent on previous elements, leading to a vector with finite dimensions np. Akaike (1976) has given a numerical algorithm for determining this state vector based on the technique of *canonical correlations analysis*. The algorithm is based on a search for linear dependence among the components of the linear predictors of Y_t, Y_{t+1}, $\ldots$ based on a *finite* number of past observations $Y_t, \ldots, Y_{t-M}$, M being a sufficiently large integer. Akaike observes that the number of linearly independent elements among these predictors is the same as the rank of the matrix of regression coefficients of the elements $\{Y^{(r)}_{t+j}\}$ on the variables $\{Y^{(s)}_{t-k}\}$, $s = 1, \ldots, p$, $k = 0, \ldots, M$, and may thus be determined by canonical correlations analysis. Akaike's algorithm is thus based on computing the canonical correlations between the vector,

$$U(t) = [Y'_t, Y'_{t-1}, \ldots, Y'_{t-M}],\qquad(10.5.13)$$

and the vector

$$V(t) = [Y'_t, Y'_{t+1}, Y'_{t+2}, \ldots],\qquad(10.5.14)$$

and determining the dimension np of the state vector as the number of non-zero canonical correlations. In practice, M may be chosen as the order of an AR model fitted to the Y_t using the AIC criterion (see Section 5.4.5), and the canonical correlations analysis proceeds by starting with $V(t) = [Y_t^{(1)}, Y_t^{(2)}, \ldots, Y_t^{(p)}, Y_{t+1}^{(1)}]$ and successively adding new components to $V(t)$ until we fail to obtain an increased number of significant canonical correlations. (Akaike describes a criterion, called "DIC", which can be used to assess whether the addition of a new element to $V(t)$ has increased the number of "significant" canonical correlations.)

As pointed out by Akaike (1974a), the "Hankel matrix" (a block circular matrix whose block elements are the impulse response matrices) can be regarded as the covariance matrix between the present and future output and the present and past input when the input is white noise with the identity matrix as its covariance matrix. Hence, the canonical correlations analysis is, in principle, equivalent to *determining the first np linearly independent (elementary) rows of the Hankel matrix*, and in this sense is related to the Ho–Kalman algorithm for finding minimal realizations of deterministic systems (Ho and Kalman (1966)).

The canonical correlations analysis can be used also to provide estimates of the unknown elements of the F matrix, and the AR model initially fitted to the Y_t observations can be used to provide estimates of Σ_ε, the covariance matrix of the innovations, ε_t, and the impulse response matrices, $W(1), \ldots, W(n-1)$ which enter in the expression for the matrix G. Although these estimates may not be very accurate they may serve as initial estimates in an iterative maximum likelihood procedure. In the scalar case, if the canonical correlations analysis indicates the dimension of the state vector as n, we may then fit an ARMA model of order $(n, n-1)$ by maximum likelihood (assuming, of course, that ε_t is Gaussian). More generally, we may fit a sequence of ARMA models of orders (k, l) where k and l are selected from a "grid" of values in the neighbourhood of the "point" $(n, n-1)$ and select the final model by examining the values of the AIC criterion as (k, l) vary over this grid of points.

Note that when redundant elements have been removed from $S(t)$, the transition matrix F no longer takes the "canonical" form (10.5.9). In fact,

F now has a more complicated structure, but its free elements can be estimated from the canonical correlations analysis, or by maximum likelihood. The matrix F retains its "canonical" form (10.5.9) only when Y_t is *"block identifiable"*.

Once the dimension and structure of the state vector has been determined, we may, as an alternative approach, fit the *"quasi ARMA"* models constructed by Pham-Dinh (1978). These correspond to Akaike's canonical state space representations (i.e. do not involve redundant parameters) but are expressed in an explicit "scalar" time domain form.

Identification of multivariate systems

The discussion above refers to the case where we observe only one process Y_t (the process ε_t being unobservable) and wish to fit an ARMA (or state-space) model. However, in system identifications problems the input process U_t is directly observable, and hence we have observations on both U_t and Y_t. We may, however, reduce this case to the "time series modelling" case by combining U_t and Y_t into a single vector, i.e. by forming the single multivariate process

$$Y_t^* = \left(\frac{Y_t}{U_t} \right),$$

and then fitting an ARMA or state space model to the observations on Y_t^* (assuming, of course, that U_t and Y_t are jointly stationary). This approach is valid for observations taken from both open loop and closed loop systems (in fact, it is ideally suited for the identification of closed loop systems), and it is easy to see that any linear system relationship of the form (10.4.52), (10.4.53) can be "embedded" in an ARMA model for Y_t^*, with the innovations ε_t here playing the role of the "noise disturbance" N_t. This approach has been illustrated by Akaike (1976) with reference to a model relating the "yawing" and "rudder input" of a ship—following Astrom and Kallstrom's (1973) original study of this problem.

Although the ARMA and state-space models are equivalent, the state-space approach has the following advantages;

(a) it makes it easier to deal with the problem of "model identifiability" since canonical forms are more easily constructed for state space models;

(b) it enables a simple recursive technique to be constructed for dealing with prediction or filtering problems.

This recursive technique is based on the celebrated "Kalman filter" algorithm, and we describe the basic ideas in the following section.

10.6 STATE-SPACE APPROACH TO FORECASTING AND FILTERING PROBLEMS: KALMAN FILTERING

In Section 10.3 we discussed the scalar form of the "filtering problem", and derived the Wiener solution. We now consider the Kalman solution, and since this is based on a "state-space" approach we can treat the more general multivariate form of the filtering problem with little additional effort. The multivariate version is as follows: we observe a process Y_t which consists of a "signal" X_t plus a "noise" component N_t, i.e.

$$Y_t = X_t + N_t, \tag{10.6.1}$$

and given observations on Y up to time t, we wish to "estimate" X_{t+m}. We recall that the Wiener solution of the scalar form is based on the assumptions that (i) the "estimate" $\tilde{X}_{t+m}$ of X_{t+m} is a linear function of $\{Y_s; s \le t\}$, (ii) the criterion for determining the optimal linear function is the mean square error $E[X_{t+m} - \tilde{X}_{t+m}]^2$, and (iii) X_t and N_t are jointly stationary with known spectra and cross-spectrum. However, if we assume more specifically that X_t conforms to a finite parameter ARMA model with Gaussian residuals, and that N_t is also Gaussian, then we can evaluate the linear least squares filter $\hat{X}_{t+m}$ by a recursive technique known as the "*Kalman filter*" (Kalman (1960, 1963), Kalman and Bucy (1961)). The theory of the Kalman filter is now very well known and has been discussed by many authors, but the most authoritative derivation is given in the original papers by Kalman and Kalman and Bucy. (For a more expository account see, e.g. Jazwinski (1970).) Nevertheless, it may be of some interest to note that the Kalman filter algorithm may be derived in a very simple way by appealing to a standard result in the statistical technique of "factor analysis". This approach, which was developed by Priestley and Subba Rao (1975) (see also Subba Rao (1976b)) takes the following lines. Under the assumption that X_t is a stationary ARMA process with innovations ε_t, the model for Y_t can be written in the form (10.4.52), (10.4.53), with ε_t replacing U_t in (10.4.52), and thus re-written in the state space form (10.4.61), (10.4.62) (with ε_t replacing η_t), namely,

$$x_{t+1} = Fx_t + G\varepsilon_{t+1}, \tag{10.6.2}$$

$$Y_t = Hx_t + N_t. \tag{10.6.3}$$

(We discuss only the case where X_t follows a stationary ARMA model with constant coefficients, although, as mentioned previously, the Kalman algorithm can be extended to deal with ARMA models involving time-dependent coefficients; see Kalman and Bucy (1961).) We assume that x_t

and Y_t are jointly stationary Gaussian processes, and that ε_t and N_t are also jointly stationary Gaussian zero mean white noise processes, with $E[\varepsilon_t N'_s] = 0$, all s, t. It is then readily shown (see Priestley and Subba Rao (1975)) that $E[x_t] = 0$, all t, and $E[x_t \varepsilon'_s] = 0$, all $s \geq t$, $E[x_t N'_s] = 0$, all s, t, and $E[Y_t \varepsilon'_s] = 0$, all $s \geq t$.

Now let Y^t denote the complete set of variables, $\{Y_t, Y_{t-1}, Y_{t-2}, \ldots\}$, (with a similar convention for the other processes) and let

$$e_t = Y_t - E[Y_t | Y^{t-1}]$$

denote the one step prediction error of Y_t. In view of the assumption that Y_t is Gaussian, the conditional expectation, $E[Y_t | Y^{t-1}]$ is linear in Y_{t-1}, $Y_{t-2}, \ldots$, and hence represents the linear least squares predictor. Taking conditional expectations of both sides of (10.6.3) yields

$$E[Y_t | Y^{t-1}] = H E[x_t | Y^{t-1}]$$

$$= H \tilde{x}_{t|t-1}, \tag{10.6.4}$$

say, where $\tilde{x}_{t|t-1}$ denotes the (linear) least squares predictor of x_t given $Y_{t-1}, Y_{t-2}, \ldots$. Subtracting (10.6.4) from (10.6.3) gives

$$e_t = H f_t + N_t, \qquad t = 1, 2, 3, \ldots, \tag{10.6.5}$$

where

$$f_t = x_t - \tilde{x}_{t|t-1}.$$

Note that $E[f_t] = 0$, all t, so that the covariance matrix of f_t is given by

$$\phi = E[f_t f'_t]. \tag{10.6.6}$$

Since $\tilde{x}_{t|t-1}$ is a linear function of $\{Y_s; s \leq t-1\}$, it follows that f_t and N_t are independent. Since the e_t are also independent, the model (10.6.5) is now exactly the same as that used in standard factor analysis.

Consider now the problem of "estimating" f_t, given observations, Y_t, $Y_{t-1}, \ldots$. Let us define the "optimal" estimate of f_t as

$$\tilde{f}_t = E[f_t | Y^t]. \tag{10.6.7}$$

Since the RHS is a linear function of the $\{Y_t\}$, and in view of the linear relationship between Y_t and e_t, we may equally well write

$$\tilde{f}_t = E[f_t | e^t] = E[\{x_t - \tilde{x}_{t|t-1}\} | e^t]$$

$$= \tilde{x}_{t|t} - \tilde{x}_{t|t-1}. \tag{10.6.8}$$

But since x_t and e_t are jointly stationary and Gaussian, the least squares predictor, $\tilde{x}_{t|t}$, is a linear function of $e_t, e_{t-1}, \ldots$. Hence we may write an equation of the form

$$\tilde{x}_{t|t} = A_0 e_t + A_1 e_{t-1} + A_2 e_{t-2} + \dots$$

Moreover, since the e_t are orthogonal, the coefficients in the above equation will be unaltered if we omit e_t so that we also have

$$\tilde{x}_{t|t-1} = A_1 e_{t-1} + A_2 e_{t-2} + \dots$$

Hence,

$$\tilde{f}_t = \tilde{x}_{t|t} - \tilde{x}_{t|t-1} = A_0 e_t$$

and is thus a function of e_t only. Writing

$$\tilde{f}'_t = [\tilde{f}^{(1)}_t, \tilde{f}^{(2)}_t, \dots, \tilde{f}^{(n)}_t],$$

and

$$A_0 = [a_1, a_2, \dots, a_n]$$

we may write

$$\tilde{f}^{(k)}_t = a'_k e_t$$

and using the well known least squares property of conditional expectations (cf. (10.1.2)), the vector a_k may be determined as that vector for which

$$Q_k = E[\tilde{f}^{(k)}_t - f^{(k)}_t]^2$$

is minimized. In this form the problem becomes identical to that of estimating "factor scores" via the "regression approach", the solution to which is given by Lawley and Maxwell (1971), p. 109. They show that the optimal "estimate" of f_t is given by

$$\tilde{f}_t = \phi H' \Sigma_e^{-1} e_t$$
$$= \phi H' \Sigma_e^{-1} \{ Y_t - H \tilde{x}_{t|t-1} \}, \qquad (10.6.9)$$

where $\Sigma_e = E[e_t e'_t]$ is the covariance matrix of e_t. Combining (10.6.8), (10.6.9) we obtain

$$\tilde{x}_{t|t} = \tilde{x}_{t|t-1} + K \{ Y_t - H \tilde{x}_{t|t-1} \}, \qquad (10.6.10)$$

where

$$K = \phi H' \Sigma_e^{-1}$$

is called the "*Kalman gain*" matrix. But

$$\tilde{x}_{t|t-1} = E[x_t | Y^{t-1}]$$
$$= F E[x_{t-1} | Y^{t-1}]$$
$$= F \tilde{x}_{t-1|t-1}, \qquad (10.6.11)$$

so that (10.6.10), (10.6.11) provide a recursive formula for calculating the optimal "estimate" of the current state vector x_t by means of "updating" the estimate $\tilde{x}_{t-1|t-1}$ of the state vector at the previous time instant.

We may also derive the "m-step" recursive filter as follows. From (10.6.2) we have

$$x_{t+m} = F^m x_t + \sum_{l=1}^{m} F^{m-l} G \varepsilon_{t+l-1}.$$

Since

$$E[\varepsilon_{t+l-1} | Y'] = 0 \qquad \text{for } l \geq 1,$$

we obtain

$$\tilde{x}_{t+m|t} = E[x_{t+m} | Y'] = F^m \tilde{x}_{t|t},$$

where $\tilde{x}_{t|t}$ satisfies (10.6.10).

A simple interpretation of the Kalman one-step recursion may be obtained in the following way. Suppose that we have observed the Y process up to time $(t-1)$ only, and that on the basis of these observations we have computed the optimal "estimate" $\tilde{x}_{t-1|t-1}$ of the state vector at time $(t-1)$. Using only these observations, the best "estimate" of x_t is obviously (from (10.6.2)) $\tilde{x}_{t|t-1} = F \tilde{x}_{t-1|t-1}$, as given by (10.6.11), and the best predictor of the next observation Y_t is obviously (from (10.6.3)) $\tilde{Y}_{t|t-1} = H \tilde{x}_{t|t-1}$. When the value of Y_t becomes available we can compare this with the value predicted from the estimate $\tilde{x}_{t|t-1}$ and "update" our estimate of x_t by taking a linear combination of the previous estimate $\tilde{x}_{t|t-1}$ and the prediction error $\{Y_t - H x_{t|t-1}\}$. This is exactly what (10.6.10) does, and the Kalman gain matrix K may be regarded simply as a "weighting factor". In this sense, the Kalman filter, (10.6.10), is essentially a generalization of the "exponentially weighted moving average" type of predictors.

If we introduce the covariance matrix of the "filter error", i.e. the matrix C_t defined by

$$C_t = E[\{x_t - \tilde{x}_{t|t}\}\{x_t - \tilde{x}_{t|t}\}'] \qquad (10.6.12)$$

then both $\tilde{x}_{t|t}$ and C_t may be evaluated recursively as follows. Compute

$$\tilde{Y}_{t|t-1} = HF \tilde{x}_{t-1|t-1}, \qquad (10.6.13)$$

$$e_t = Y_t - \tilde{Y}_{t|t-1}, \qquad (10.6.14)$$

$$\phi = FC_{t-1}F' + G\Sigma_\varepsilon G', \qquad (10.6.15)$$

$$\Sigma_e = H\phi H' + \Sigma_N, \qquad (10.6.16)$$

$$K = \phi H' \Sigma_e^{-1}, \qquad (10.6.17)$$

then

$$\tilde{x}_{t|t} = F\tilde{x}_{t-1|t-1} + Ke_t, \qquad (10.6.18)$$

and

$$C_t = \phi - K\Sigma_e K'. \qquad (10.6.19)$$

In the above, Σ_ϵ and Σ_N denote respectively the covariance matrices of ϵ_t and N_t, and the recursive calculation requires the specification of H, F, G, Σ_ϵ, Σ_N and $\tilde{x}_{t|t_0}$, C_{t_0} for some "initial" time point t_0. To illustrate the derivation of the above equations we may note that subtracting (10.6.11) from (10.6.2) gives

$$f_t = x_t - \tilde{x}_{t|t-1} = F\{x_{t-1} - \tilde{x}_{t-1|t-1}\} + G\epsilon_t,$$

from which (10.6.15) follows. Similarly, subtracting (10.6.4) from (10.6.3) gives

$$e_t = Y_t - \tilde{Y}_{t|t-1} = H\{x_t - \tilde{x}_{t|t-1}\} + N_t,$$

from which (10.6.16) follows.

It should be noted that the basic concept underlying the Kalman algorithm is the determination of the "estimated" state $\tilde{x}_{t+m}$ as the *conditional expectation* of x_{t+m}, given Y_t, Y_{t-1}, $\ldots$. Although we originally motivated this expression in terms of its least squares property, *we need not regard the least squares property as being of fundamental importance*, since, as observed by Kalman (1978), under the assumption of normality the conditional expectation would still be the optimal solution if we replaced the mean square error criterion $E[\{X_{t+m} - \hat{X}_{t+m}\}^2]$ by a variety of other forms—say, $E[\{X_{t+m} - \hat{X}_{t+m}\}^{2k}]$. We may note further that under the assumption that the conditional distribution of x_{t+m} is multivariate normal, it is completely determined by its mean and variance–covariance matrix.

From (10.6.10), (10.6.11) we may write

$$\tilde{x}_{t+1|t} = F\tilde{x}_{t|t-1} + GK'\{Y_t - \tilde{Y}_{t|t-1}\}, \qquad (K' = G^{-1}FK)$$

$$\tilde{Y}_{t|t-1} = H\tilde{x}_{t|t-1},$$

so that $\tilde{x}_{t|t-1}$, $\tilde{Y}_{t|t-1}$ satisfy exactly the same system equations as x_t, Y_t with $K'\{Y_t - \tilde{Y}_{t|t-1}\}$ acting as the "input".

Thus, if we let Σ denote the linear system determined by the matrices (F, G, H) and write symbolically $\Sigma \equiv (F, G, H)$, the operation of the Kalman filter may be described by the schematic feedback representation of Fig. 10.7 (cf. Kalman (1963)).

Similar results are available for the continuous parameter case (Kalman and Bucy (1961), Kalman (1963)), the recursive relations for $\tilde{x}_{t|t}$ and the covariance matrix C_t now being replaced by differential equations. (The form of the differential equations for the covariance matrix C_t is known in the control theory literature as the "*matrix Riccati equation*".)

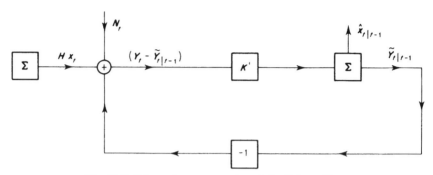

Fig. 10.7. Schematic representation of the Kalman filter.

Although we derived the Kalman filter by treating (10.6.3) as a factor analysis model, the dimension of the state vector x_t will, in general, be considerably larger than the dimension of the observation vector Y_t. Consequently, we cannot pursue this analogy to obtain information on the dimension of x_t by using standard factor analysis techniques purely on the vector Y_t. Indeed, it is obvious from physical considerations that any analysis of the output Y_t at a single time point cannot, in general, tell us anything about the dimensions of x_t, since this quantity depends essentially on the *dynamical* relationship between input and output. Thus, a single observation on Y_t contains no information about the dimensions of x_t, but, in principle, we can extract this information by analysing a *sequence* of observations Y_t, Y_{t+1}, Y_{t+2}, ... More specifically, consider the sequence of predictors $\tilde{Y}_{t+1|t}$, $\tilde{Y}_{t+2|t}$, ... From (10.6.2), (10.6.3), we have

$$\tilde{Y}_{t+m|t} = E[Y_{t+m}|Y^t] = HE[x_{t+m}|Y^t]$$

$$= HF^{m-1}\tilde{x}_{t+1|t}, \qquad \text{all } m \geq 1.$$

Writing out this result for successive values of m, say $m = 1, 2, \ldots, M$, now gives

$$\tilde{Y}_{t+1|t} = H\tilde{x}_{t+1|t},$$

$$\tilde{Y}_{t+2|t} = HF\tilde{x}_{t+1|t},$$

$$\dots\dots\dots\dots\dots\dots$$

$$\tilde{Y}_{t+M|t} = HF^{M-1}\tilde{x}_{t+1|t}.$$

$$(10.6.20)$$

If we now write

$$Y^*_{t+1} = [\tilde{Y}'_{t+1|t}, \ldots, \tilde{Y}'_{t+M|t}]',$$

then (10.6.20) can be written as

$$Y^*_{t+1} = H^*_M \tilde{x}_{t+1|t}, \tag{10.6.21}$$

where the matrix H^*_M is given by

$$H^*_M = [H' \vdots (HF)' \vdots (HF^2)' \vdots \ldots \vdots (HF^{M-1})']'. \tag{10.6.22}$$

Equation (10.6.21) is still of factor analytic form (without any "error" term) but now, if we choose M sufficiently large, the dimension of Y^*_{t+1} can be made much larger than the dimension of x_t. This suggests that, in principle, we can obtain information on the dimension of x_t by applying factor analysis to successive observations on Y^*_{t+1} (corresponding to $t = 1$, 2, 3, . . .). This approach ties in very neatly with the canonical correlations analysis suggested by Akaike, and described in Section 10.5.2.

The form of the matrix H^*_M is very well known in control theory literature, and it plays a central role in the concept of "*observability*". (The concepts of "observability" and "controllability" are both due to Kalman.) The system is said to be "observable" if

$$\text{rank}(H^*_M) = M, \tag{10.6.23}$$

when $M (= np)$ is the dimension of x_t (see, e.g. Rosenbrock (1970)). The idea essentially is that if (10.6.23) holds, then when $N_t \equiv 0$ (i.e. when the observations are "noise free"), it is possible to determine the state vector x_t explicitly from a sequence of observations on Y_t by writing $Y_{t+1}, \ldots,$ Y_{t+M} in terms of x_{t+1} (using equations of exactly the same form as (10.6.20)), and then inverting the matrix H^*_M.

We now discuss briefly two particular forecasting techniques which are based on the Kalman algorithm.

State-space forecasting

Mehra (1977) developed a complete numerical procedure, called "*state-space forecasting*", for predicting future values of a multivariate stationary process Y_t given observations on its past values. This procedure involves basically two main stages:

(a) First fit a canonical state-space model to the given observations using Akaike's canonical correlations technique to determine the dimensions of the state vector and to provide estimates of the non-zero elements of the matrix F. A multivariate AR model is also fitted to the observations (using AIC to determine the order) to provide estimates of Σ_ε and the impulse response matrices $W(1), \ldots,$ $W(n-1)$.

(b) Having fitted the state space model and estimated F, G, Σ_ε, the predictors are computed recursively using Kalman's algorithm. Practical applications of "state-space forecasting" are given in the paper by Mehra cited above.

Bayesian forecasting

Harrison and Stevens (1976) proposed a Bayesian forecasting method based on the assumption that the observations Y_t are described by a model of the form

$$Y_t = H_t \psi_t + N_t, \tag{10.6.24}$$

$$\psi_{t+1} = F\psi_t + \varepsilon_t. \tag{10.6.25}$$

This model is very similar in form to the standard space formulation, (10.6.2), (10.6.3), but here ψ_t represents a vector of time dependent stochastic *parameters* on which Y_t depends, and their evolution over time is described by the "Markov" model (10.6.25). The evolution of Y_t, given ψ_t, is now "concealed" inside the time dependent matrix H_t, which thus replaces the (constant) observation matrix H (see Priestley (1976b)). Assuming that F is known, N_t and ε_t are Gaussian, that at time t, H_t, Σ_N, Σ_ε are known, and that at some initial point (say $t = 0$), ψ has a prior distribution in the form of a multivariate normal distribution with specified mean and covariance matrix, Harrison and Stevens show that the posterior distribution of ψ_t (given observations on Y up to time t) is multivariate normal with mean m_t (say), and covariance matrix C_t, where m_t satisfies exactly the same Kalman filter recursive equations as $\hat{x}_{t|t}$ (viz. (10.6.18)), and C_t satisfies the same equations as previously (viz. (10.6.15), (10.6.16), (10.6.17), (10.6.19) with $G = I$). Thus, the Kalman algorithm can be used to compute the posterior distribution of ψ_{t+1} (say), and then the mean of the posterior distribution of Y_{t+1} is given (assuming that H_{t+1} is known) by

$$\tilde{Y}_{t+1|t} = H_{t+1} \tilde{\psi}_{t+1|t}.$$

Harrison and Stevens (1976) call (10.6.24), (10.6.25) a *"dynamic linear model"*, and show how a wide collection of familiar models (including AR and MA models) can be put in this form.

Although the method of Kalman filtering was first derived within the context of "linear systems", it has now found an extremely wide range of statistical applications well outside this area. One may cite, e.g., its use in Bayesian forecasting (described above), in the analysis of time series models with time dependent coefficients (Young (1974), Bohlin (1976)), and in general regression models (O'Hagan (1978)).

Chapter 11

Non-stationarity and Non-linearity

The classical methods of time series analysis discussed throughout the preceding chapters are all based on two crucial assumptions, namely that:

(a) all series are *stationary* (at least to order 2), or can be reduced to stationarity by some simple transformation such as differencing—as in the case of the ARIMA models;

(b) all models are *linear*, i.e. the observed series can be represented as a linear function of present and past values of an independent "white noise" process.

However, stationarity and linearity are mathematical idealizations which in some cases may be valid only as *approximations* to the real situation. In practical applications the most one could hope for is that, e.g., over the observed time interval the series would not depart "too far" from stationarity for the results of the analysis to be invalid. In order to give this statement a more precise meaning we must first establish some method of characterizing the structure of non-stationary processes, and in the following sections we describe an approach to the spectral analysis of non-stationary processes based on the theory of *"evolutionary spectra"*. This approach was developed by Priestley (1965b, 1966, 1967) and, as we shall see, it provides a convenient framework for interpreting the results of conventional spectral analysis applied to data from non-stationary processes. Moreover, it enables us to construct a theory of linear prediction and filtering for non-stationary processes which parallels very closely the classical Kolmogorov–Wiener theory for stationary processes.

In the latter part of this chapter we discuss the topic of non-linear models, and introduce the reader to some current research work in this area.

816

11.1 SPECTRAL ANALYSIS OF NON-STATIONARY PROCESSES: BASIC CONSIDERATIONS

We have seen that spectral analysis is of fundamental importance in the study of stationary processes. The main reasons for this are: (i) the spectrum has an immediate physical interpretation as a power-frequency distribution; (ii) the spectrum provides information on the stochastic structure of the process, and this is useful as an aid to model-fitting; (iii) the spectral functions play a central role in the theory of linear predictions and filtering; (iv) given a set of observations, the spectrum may be estimated by fairly simple numerical techniques which do not require any specific assumptions on the structure of the process.

Now in generalizing any theory to cover more general situations, an obvious requirement is that the new theory should be consistent with the old one. In the present context this means that any definition of the "spectrum" of a non-stationary process should reduce to the classical definition when, in particular, the process is stationary. However, there are various approaches which we could adopt, all of which satisfy this requirement. The basic motivation underlying the "evolutionary spectra" approach is that it preserves property (i), i.e. *the evolutionary spectra have essentially the same type of physical interpretation as the spectra of stationary processes*, the main distinction being that whereas the spectrum of a stationary process describes the power–frequency distribution for the whole process (i.e. over all time), the evolutionary spectrum is time-dependent and describes the *local* power–frequency distribution *at each instant* of time. As far as we are aware, the theory of evolutionary spectra is the only one so far suggested which preserves this physical interpretation for non-stationary processes. It turns out, however, that in addition to property (i), the evolutionary spectral approach preserves also properties (ii), (iii) and (iv).

To motivate the more general ideas which follow, let us first consider a very simple example of a non-stationary process, namely,

$$X_t = \begin{cases} X_t^{(1)}, & t \le t_0, \\ X_t^{(2)}, & t > t_0, \end{cases}$$

where both $\{X_t^{(1)}\}$, $\{X_t^{(2)}\}$ are stationary processes but with different autocovariance functions. If t_0 is known then a natural way to describe the power–frequency properties of X_t is to specify *two* spectra, one for the interval $t \le t_0$, and one for the interval $t > t_0$. If we consider a slightly more general model, viz., $X_t = X_t^{(i)}$, $t_i < t \le t_{i+1}$, where the $\{X_t^{(i)}\}$ are all different stationary processes, then we would describe the properties of X_t by specifying a different form of spectrum for each of the intervals,

(t_i, t_{i+1}). Generalizing this approach to the case of more general types of non-stationary processes, we are led to the notion of *continuously changing spectrum*, or more precisely, a *time-dependent spectrum*. Clearly, in such cases we could never hope to estimate the spectrum at a particular instant of time, but if we assume that the spectrum is changing *smoothly* over time then, by using estimates which involve only *local* functions of the data, we may attempt to estimate some form of "average" spectrum of the process in the neighbourhood of any particular time instant. We therefore consider the class of non-stationary processes whose statistical characteristics are changing "smoothly" over time.

The notion of "frequency"

If we attempt to define a time-dependent spectrum which possesses a physical interpretation as a "local" power–frequency distribution, the first question we must answer is: what do we mean by "frequency"? This may seem a deceptively simple question, but its study is crucial. For, suppose we are given a real valued continuous parameter non-stationary process $X(t)$ and we have constructed some function, $p(\omega, t)$ say, which is such that, for each t,

$$E[X^2(t)] = \int_{-\infty}^{\infty} p(\omega, t) \, d\omega. \tag{11.1.1}$$

We certainly cannot conclude, on the basis of (11.1.1) alone, that $p(\omega, t)$ represents a decomposition of power over *frequency*, since in the above integral ω is merely a "dummy variable", and there is no reason why it should be related to the physical concept of "frequency". This point is highlighted by the following example. Let $R(s, t) = \text{cov}\{X(s), X(t)\}$ denote the covariance kernel of $\{X(t)\}$. Since $\{X(t)\}$ is a non-stationary process, $R(s, t)$ will no longer be a function of $|s - t|$ only, but $R(t - \tau/2, t + \tau/2)$ measures the covariance between values at time points separated by an interval τ and symmetrically placed about the time point t. If we try (naïvely) to interpret $R(t - \tau/2, t + \tau/2)$ as a form of "local autocovariance function", we may be tempted to define a time-dependent spectral density function by evaluating the Fourier transform of $R(t - \tau/2, t + \tau/2)$ with respect to τ, keeping t fixed, leading to the function

$$\psi_t(\omega) = \frac{1}{2\pi} \int_{-\infty}^{\infty} R(t - \tau/2, t + \tau/2) e^{-i\omega\tau} \, d\tau. \tag{11.1.2}$$

This certainly produces a time-dependent function, and inverting (11.1.2) and setting $\tau = 0$ it is easily seen that

$$E[X^2(t)] = R(t, t) = \int_{-\infty}^{\infty} \psi_t(\omega)\, d\omega.$$

However, ψ_t certainly does not possess a physical interpretation as a power–frequency distribution since $\psi_t(\omega)$ may well take negative values (Priestley (1971c)). Despite the presence of the term $\{e^{-i\omega t}\}$ in the integral in (11.1.2), the variable ω in $\psi_t(\omega)$ has no obvious physical interpretation as far as the properties of $\{X_t\}$ are concerned. (The function $\psi_t(\omega)$ is sometimes referred to as the "*Wigner distribution*", and we will discuss it further in Section 11.3.)

Let us now return for a moment to the case of stationary processes. The reason why we can interpret the spectrum of a stationary process as a power–frequency distribution lies essentially in the fact that, when $X(t)$ is stationary, the *process itself* has a *spectral representation* of the form

$$X(t) = \int_{-\infty}^{\infty} e^{it\omega}\, dZ(\omega). \tag{11.1.3}$$

Heuristically, (11.1.3) tells us that any stationary process can be represented as the "sum" of sine and cosine waves with varying frequencies and (random) amplitudes and phases. We can now identify that *component* in $X(t)$ which has frequency ω, and meaningfully discuss the contribution of this *component* to the total power of the process. In the absence of such a representation we cannot immediately talk about "power distributions over frequency". But a non-stationary process cannot be represented as a "sum" of sine and cosine waves (with orthogonal coefficients)—instead, we have to represent it as a "sum" of other kinds of functions. Since, according to its conventional definition, the term "frequency" is associated specifically with sine and cosine functions, we cannot talk about the "frequency components" of a non-stationary process, unless we first define a more general concept of "frequency" which agrees with our physical understanding. The generalized notion of "frequency" on which the theory of "evolutionary spectra" is based may be seen from the following example. Suppose $X(t)$ is a deterministic function which has the form of a damped sine wave, say,

$$X(t) = A\, e^{-t^2/\alpha^2} \cos(\omega_0 t + \phi). \tag{11.1.4}$$

If we carry out a conventional Fourier analysis of $X(t)$ we see that it contains Fourier components at *all* frequencies—in fact, the Fourier transform of $X(t)$ consists of two Gaussian functions, one centred on ω_0

and the other on $(-\omega_0)$, the "width" of these functions being inversely proportional to the parameter α. In other words, if we represent $X(t)$ as a "sum" of sine and cosine functions with constant amplitudes, we need to include components at all frequencies. However, we can equally well describe $X(t)$ by saying that it consists of just two "frequency" components $(\omega = \omega_0,\ \omega = -\omega_0)$, each component having a *time varying amplitude*, $A \exp(-t^2/\alpha^2)$. Indeed, if we were to examine the *local* behaviour of $X(t)$ in the neighbourhood of the time point t_0, this is precisely what we would observe, i.e. if the interval of observation was small compared with α, $X(t)$ would appear simply as a cosine function with frequency ω_0 and amplitude $A \exp(-t_0^2/\alpha^2)$. However, it would not be physically meaningful to try to assign a "frequency" to a function $X(t)$ of arbitrary form; for example, it would make little sense to talk about the "frequency" of the function $\log \omega t$, and the parameter ω could certainly not be interpreted in this way. For the term "frequency" to be meaningful *the function $X(t)$ must possess what we can loosely describe as an "oscillatory form"*, and we can characterize this property by saying that the Fourier transform of such a function will be concentrated around a particular point ω_0 (or around $\pm \omega_0$ in the real case). Thus, if we have a non-periodic function $X(t)$ whose Fourier transform has an absolute maximum at the point ω_0, we may define ω_0 as "*the frequency*" of this function, the argument being that locally $X(t)$ behaves like a sine wave with (conventional) frequency ω_0, modulated by a "smoothly varying" amplitude function.

It is this type of reasoning which forms the basis of the "evolutionary spectra" approach, and, in fact, this approach rests on a spectral representation of non-stationary processes which is virtually a direct generalization of (11.1.4).

We may note, in passing, that the two different Fourier representations of the function (11.1.4) are equally valid, i.e. we may think of $X(t)$ as consisting either of an infinite number of frequency components with constant amplitudes, or as just two frequency components with time varying amplitudes. These two representations correspond to the use of two different "families" of functions, in the former case the "family" consists of sines and cosines with constant amplitudes, and in the latter case it consists of sines and cosines with time varying amplitude, $A \exp(-t^2/\alpha^2)$. Unless we impose conditions on the form of the "family" of functions which we wish to use we have no way of expressing a preference for one or other of the two alternative representations.

We now present a more formal exposition of the theory of evolutionary spectra.

11.2 THE THEORY OF EVOLUTIONARY SPECTRA

Consider a continuous parameter (complex valued) stochastic process, $\{X(t)\}$, $-\infty < t < \infty$. (Most of the following discussion will, with the usual modifications, apply equally well to discrete parameter processes.) We assume that the process is "trend free", and has zero mean and finite variance, i.e. for all t,

$$E[X(t)] = 0, \qquad E[|X(t)|^2] < \infty.$$

The *covariance kernel* is then defined by

$$R(s, t) = E[X^*(s)X(t)]. \tag{11.2.1}$$

If $\{X(t)\}$ is (weakly) stationary, i.e. if $R(s, t)$ is a function of $|s - t|$ only, then we know from the Wiener–Khintchine theorem (Section 4.8.2) that $R(s, t)$ admits the representation

$$R(s, t) = \int_{-\infty}^{\infty} e^{i\omega(t-s)} \, dH(\omega), \tag{11.2.2}$$

where $H(\omega)$ (the integrated spectrum) has the properties of a distribution function on the interval $(-\infty, \infty)$. Also, we know from Theorem 4.11.1 that corresponding to the representation (11.2.2) for $R(s, t)$, $X(t)$ admits the spectral representation

$$X(t) = \int_{-\infty}^{\infty} e^{it\omega} \, dZ(\omega), \tag{11.2.3}$$

where $Z(\omega)$ is a process with orthogonal increments, and $E[|dZ(\omega)|^2] = dH(\omega)$. Moreover, if we start from the representation (11.2.3) then (11.2.2) follows immediately (cf. the derivation of equation (4.11.15)), so that $R(s, t)$ is a function of $|s - t|$ only, and consequently $X(t)$ is (weakly) stationary. It thus follows that if $X(t)$ is non-stationary it cannot be represented in the form (11.2.3) (with the $\{dZ(\omega)\}$ orthogonal), and similarly $R(s, t)$ cannot be represented in the form (11.2.2). However, we know from the discussion of "general orthogonal expansions" (Section 4.11) that for a fairly general class of stochastic processes $R(s, t)$ can be represented in a form similar to (11.2.2), provided we replace the funct ons $\{e^{i\omega t}\}$ by a more general "family" of functions, $\{\phi_t(\omega)\}$.

We now restrict attention to the class of process for which *there exists a family $\mathcal{F}$ of functions $\{\phi_t(\omega)\}$ defined on the real line, and indexed by the suffix t, and a measure $\mu(\omega)$ on the real line, such that for each s, t the* covariance kernel $R(s, t)$ admits a representation of the form,

$$R(s, t) = \int_{-\infty}^{\infty} \phi_s^*(\omega)\phi_t(\omega)\, d\mu(\omega).$$ (11.2.4)

When the parameter space is limited to a finite interval, say $0 \le t \le T$, it is always possible to obtain a representation of the form (11.2.4) in terms of the eigenfunctions of the covariance kernel $\{R(s, t)\}$ (cf. Parzen (1959)). It should be noted that although we have described $\mathcal{F}$ as a family of functions, each defined on the ω-axis and indexed by the parameter t, we may also think of $\mathcal{F}$ as a family of functions $\phi_\omega(t)$, say, each defined on the t-axis and indexed by the parameter ω. In fact, when we study the properties of various families, it is convenient to adopt the latter description.

In order for var$\{X(t)\}$ to be finite for each t, $\phi_t(\omega)$ must be quadratically integrable with respect to the measure μ, for each t. It then follows from Theorem 4.11.12 that whenever $R(s, t)$ has the representation (11.2.4), the process $\{X(t)\}$ admits a representation of the form,

$$X(t) = \int_{-\infty}^{\infty} \phi_t(\omega)\, dZ(\omega)$$ (11.2.5)

where $Z(\omega)$ is an orthogonal process, with

$$E[|dZ(\omega)|^2] = d\mu(\omega).$$

The measure $\mu(\omega)$ here plays the same role as the integrated spectrum $H(\omega)$ in the case of stationary processes, so that the analogous situation to the case of an absolutely continuous spectrum is obtained by assuming that *the measure $\mu(\omega)$ is absolutely continuous with respect to Lebesgue measure*.

Parzen (1959) has pointed out that if there exists a representation of $\{X(t)\}$ of the form (11.2.5), then there is a multitude of different representations of the process, each representation based on a different family of functions. (The situation is in some ways similar to the selection of a basis for a vector space.) When the process is stationary, we have seen that one valid choice of functions is the complex exponential family,

$$\phi_t(\omega) = e^{i\omega t}.$$ (11.2.6)

This family provides the well known spectral decomposition (cf. (11.2.3)) in terms of sine and cosine "waves", and forms the basis of the physical interpretation of spectral analysis as a "power distribution over frequency". However, if the process is non-stationary this choice of family of functions is no longer valid. This is hardly surprising, since the sine and cosine waves are themselves "stationary" and it is natural that they should form the "basic elements" used in building up models of stationary processes. If we wish to introduce the notion of frequency in the analysis of non-stationary processes, we are led to seek new "basic elements" which, although "non-stationary", have an *oscillatory form*, and in which *the notion of "frequency" is still dominant*. One class of basic elements (or more precisely, family of functions) which possess the required structure may be obtained as follows.

Suppose that, for each fixed ω, $\phi_t(\omega)$ (considered as a function of t) possesses a (generalized) Fourier transform whose modulus has an *absolute maximum* at frequency $\theta(\omega)$, say. Then we may regard $\phi_t(\omega)$ as an *amplitude modulated* sine wave with frequency $\theta(\omega)$, and write $\phi_t(\omega)$ in the form

$$\phi_t(\omega) = A_t(\omega) \, e^{i\theta(\omega)t}, \tag{11.2.7}$$

where the modulating function $A_t(\omega)$ is such that the modulus of its (generalized) Fourier transform has an absolute maximum at *the origin* (*i.e. zero frequency*). We now formalize this approach in the following definition.

Definition 11.2.1 *The function of t, $\phi_t(\omega)$, will be said to be an* oscillatory *function if, for some (necessarily unique) $\theta(\omega)$ it may be written in the form* (11.2.7) *where $A_t(\omega)$ is of the form*

$$A_t(\omega) = \int_{-\infty}^{\infty} e^{it\theta} \, dK_\omega(\theta) \tag{11.2.8}$$

with $|dK_\omega(\theta)|$ having an absolute maximum at $\theta = 0$.

The function $A_t(\omega)$ may now be regarded as the "envelope" of $\phi_t(\omega)$. If, further, the family $\{\phi_t(\omega)\}$ is such that $\theta(\omega)$ is a single-valued function of ω (i.e. if no two distinct members of the family have Fourier transforms whose maxima occur at the same point), then we may transform the variable in the integral in (11.2.4) from ω to $\theta(\omega)$, and by suitably redefining $A_t(\omega)$ and the measure $\mu(\omega)$, write

$$R(s, t) = \int_{-\infty}^{\infty} A_s^*(\omega) A_t(\omega) \, e^{i\omega(t-s)} \, d\mu(\omega) \tag{11.2.9}$$

and correspondingly

$$X(t) = \int_{-\infty}^{\infty} A_t(\omega) \, e^{i\omega t} \, dZ(\omega), \tag{11.2.10}$$

where

$$E[|dZ(\omega)|^2] = d\mu(\omega).$$

Definition 11.2.2 *If there exists a family of oscillatory functions* $\{\phi_t(\omega)\}$ *in terms of which the process* $\{X(t)\}$ *has a representation of the form* (11.2.5), $\{X(t)\}$ *will be termed an "oscillatory process".*

It follows that any oscillatory process also has a representation of the form (11.2.10), where the family $A_t(\omega)$ satisfies the condition of definition (11.2.1), and that, without loss of generality, we may write any family of oscillatory functions in the form

$$\phi_t(\omega) = A_t(\omega) \, e^{i\omega t}.$$

We may note that since (11.2.6) is a particular case of (11.2.7) (with $A_t(\omega) \equiv 1$, all t, ω and $\theta(\omega) \equiv \omega$), the class of *oscillatory processes certainly includes all second order stationary processes.*

Evolutionary (power) spectra

Consider an oscillatory process of the form (11.2.10), with auto-covariance kernel $R(s, t)$ of the form (11.2.9). For any particular process $\{X(t)\}$ there will, in general, be a large number of different families of oscillatory functions in terms of each of which $\{X(t)\}$ has a representation of the form (11.2.10), with each family inducing a different measure $\mu(\omega)$. For a particular family $\mathscr{F}$ of spectral functions $\{\phi_t(\omega)\}$ it is tempting to define the spectrum of $X(t)$ (with respect to $\mathscr{F}$) simply as the measure $\mu(\omega)$. However, such a definition would not have the interpretation of a "power distribution over frequency". For, from (11.2.9) we may write

$$\text{var}\{X(t)\} \equiv R(t, t) = \int_{-\infty}^{\infty} |A_t(\omega)|^2 \, d\mu(\omega). \qquad (11.2.11)$$

Since $\text{var}\{X(t)\}$ may be interpreted as a measure of the "total power" of the process at time t, (11.2.11) gives a decomposition of total power in which the contribution from "frequency" ω is $\{|A_t(\omega)|^2 \, d\mu(\omega)\}$. This result is consistent with the interpretation of equation (11.2.10) as an expression for $X(t)$ as the limiting form of a "sum" of sine waves with different frequencies and time-varying random amplitudes $\{A_t(\omega) \, dZ(\omega)\}$. We are thus led to the following definition:

Definition 11.2.3 *Let* $\mathscr{F}$ *denote a particular family of oscillatory functions,* $\{\phi_t(\omega)\} \equiv \{A_t(\omega) \, e^{i\omega t}\}$, *and let* $\{X(t)\}$ *be an oscillatory process having a representation of the form* (11.2.10) *in terms of the family* $\mathscr{F}$. *We define the evolutionary power spectrum at time* t *with respect to the family* $\mathscr{F}$, $dH_t(\omega)$, *by*

$$dH_t(\omega) = |A_t(\omega)|^2 \, d\mu(\omega). \qquad (11.2.12)$$

Note that when $\{X(t)\}$ is stationary, and $\mathcal{F}$ is chosen to be the family of complex exponentials, $dH_t(\omega)$ reduces to the standard definition of the (integrated) spectrum. The evolutionary spectrum has the same physical interpretation as the spectrum of a stationary process, namely, that it describes a distribution of power over frequency, but whereas the latter is determined by the behaviour of the process over all time, the former represents specifically the spectral content of the process in the neighbourhood of the time instant t.

Although, according to definition (11.2.3), the evolutionary spectrum, $dH_t(\omega)$, depends on the choice of family $\mathcal{F}$, it follows from equation (11.2.11) that

$$\text{var}\{X(t)\} = \int_{-\infty}^{\infty} dH_t(\omega), \qquad (11.2.13)$$

so that the value of the integral of $dH_t(\omega)$ is independent of the particular family $\mathcal{F}$, and, for all families, represents the total power of the process at time t.

It is now convenient to "standardize" the functions $A_t(\omega)$ so that, for all ω,

$$A_0(\omega) = 1, \qquad (11.2.14)$$

i.e. we incorporate $|A_0(\omega)|$ in the measure $\mu(\omega)$. With this convention, $d\mu(\omega)$ represents the evolutionary spectrum at $t = 0$, and $|A_t(\omega)|^2$ represents the change in the spectrum, relative to zero time. We now have, for each ω,

$$\int_{-\infty}^{\infty} dK_\omega(\theta) = 1,$$

so that the Fourier transforms of the $\{A_t(\omega)\}$ are normalized to have unit integrals. (In some respects it may be preferable to specify that the $\{A_t(\omega)\}$ be normalized so that

$$\int_{-\infty}^{\infty} |dK_\omega(\theta)| = 1, \qquad \text{all } \omega,$$

rather than $\int_{-\infty}^{\infty} dK_\omega(\theta) = 1$, as above. For now the function $B_{\mathcal{F}}(\omega)$ (defined by (11.2.27)) is invariant under translations of the t-axis, i.e. if $\mathcal{F}' \equiv \{A_{t-k}(\omega) e^{it\omega}\}$, then $B_{\mathcal{F}'}(\omega) \equiv B_{\mathcal{F}}(\omega)$. This is clearly a desirable property, since if $A_t(\omega)$ is a "slowly varying" function of t so is $A_{t-k}(\omega)$. However, if we use this alternative normalization, $d\mu(\omega)$ is no longer the evolutionary spectrum at time zero.)

There is an interesting alternative interpretation of oscillatory processes in terms of *time varying filters*. Let $\{X(t)\}$ be of the form (11.2.10) and suppose that for each fixed t we may write (formally)

$$A_t(\omega) = \int_{-\infty}^{\infty} e^{i\omega u} k_t(u) \, du.$$

Then from (11.2.10)

$$X_t = \int_{-\infty}^{\infty} S(t-u) k_t(u) \, du, \tag{11.2.15}$$

where $S(t) = \int_{-\infty}^{\infty} e^{i\omega t} dZ(\omega)$ is a *stationary process* with spectrum $d\mu(\omega)$. Thus $X(t)$ may be interpreted as the result of passing a stationary process through a time-varying filter $\{k_t(u)\}$. Conversely, any process of the form (11.2.15) (with $k_t(u)$ chosen so that $A_t(\omega)$ is of the required form) may be written in the form (11.2.10). Thus, the evolutionary spectrum at time t, $|A_t(\omega)|^2 \, d\mu(\omega)$, may be interpreted as the spectrum (in the classical sense) of the stationary process which we *would* have obtained if the filter $\{k_t(u)\}$ was held fixed in the state which it attained at the time instant t.

Later we show how, for a certain class of processes, evolutionary spectra may be estimated from a sample record of $\{X(t)\}$, and by examining the variations of $dH_y(\omega)$ over time we are enabled to study continuously the changing spectral pattern of the process.

The uniformly modulated process

One interesting example of a non-stationary process satisfying the model (11.2.10) is the following,

$$X(t) = C(t) X^{(0)}(t), \tag{11.2.16}$$

where $\{X^{(0)}(t)\}$ is a stationary process with zero mean and spectrum $dH(\omega)$, and the function $C(t)$ (with $C(0) = 1$) has a generalized Fourier transform whose modulus has an absolute maximum at the origin. (For example, $C(t)$ may be any non-negative real-valued function whose Fourier transform exists.) Since $X^{(0)}(t)$ is stationary, we may write,

$$X^{(0)}(t) = \int_{-\infty}^{\infty} e^{i\omega t} dZ(\omega),$$

where $Z(\omega)$ is orthogonal with

$$E[|dZ(\omega)|^2] = dH(\omega),$$

so that

$$X(t) = \int_{-\infty}^{\infty} C(t) e^{i\omega t} dZ(\omega). \tag{11.2.17}$$

We may note that since $\mathcal{F}_0 \equiv \{C(t) e^{i\omega t}\}$ is a family of oscillatory functions, the process defined by (11.2.16) is an oscillatory process, and with respect to $\mathcal{F}_0$, has evolutionary spectrum,

$$dH_t(\omega) = |C(t)|^2 \, dH(\omega). \qquad (11.2.18)$$

It should be observed, however, that the process defined by (11.2.16) is a very special case of the model (11.2.10), in that all the spectral components (w.r.t. $\mathcal{F}_0$) *are varying over time in exactly the same way.* More specifically, for any pair of frequencies ω_1, ω_2 and time instants t_1, t_2,

$$\frac{dH_{t_1}(\omega_1)}{dH_{t_2}(\omega_1)} = \frac{dH_{t_1}(\omega_2)}{dH_{t_2}(\omega_2)}.$$

A process for which there exists a family $\mathcal{F}_0$ such that the evolutionary spectrum (w.r.t. $\mathcal{F}_0$) has the above property will be called a *uniformly modulated process.*

Linear filters

One of the most useful features of the spectral representation of stationary processes is that it enables the effect of linear transformations (i.e. "filters") to be described purely in terms of the effect on individual spectral components. Thus, if we consider a linear transformation of a stationary process $\{X(t)\}$ of the form

$$Y(t) = \int_{-\infty}^{\infty} g(u) X(t-u) \, du, \qquad (11.2.19)$$

then, from (4.12.8), the spectra of $\{X_t\}$ and $\{Y_t\}$ are related by (with an obvious notation),

$$dH^{(Y)}(\omega) = |\Gamma(\omega)|^2 \, dH^{(X)}(\omega), \qquad (11.2.20)$$

where

$$\Gamma(\omega) = \int_{-\infty}^{\infty} g(u) e^{-iu\omega} \, du$$

is the *transfer function* of the filter $\{g(u)\}$. Hence, $dH^{(Y)}(\omega_1)$, say, is determined purely by $dH^{(X)}(\omega_1)$ and $\Gamma(\omega_1)$, and is not affected by $dH^{(X)}(\omega)$ at other frequencies. We now show that this property holds (in an approximate sense) for evolutionary spectra when we consider linear transformations of non-stationary processes.

Suppose that $\{X(t)\}$ satisfies a model of the form (11.2.10), and consider a slightly more general form of the transformation (11.2.19), namely,

$$Y(t) = \int_{-\infty}^{\infty} g(u)X(t-u) e^{i\omega_0(t-u)} du \qquad (11.2.21)$$

where ω_0 is any constant frequency. Then we may write

$$Y(t) = \int_{-\infty}^{\infty} \Gamma_{t,\omega+\omega_0}(\omega)A_t(\omega+\omega_0) e^{i\omega t} dZ(\omega+\omega_0), \qquad (11.2.22)$$

where, for any t, λ, θ,

$$\Gamma_{t,\lambda}(\theta) = \int_{-\infty}^{\infty} g(u)\{A_{t-u}(\lambda)/A_t(\lambda)\} e^{-iu\theta} du. \qquad (11.2.23)$$

The function $\Gamma_{t,\omega}(\omega)$ will be termed the *generalized transfer function of the filter $\{g(u)\}$ with respect to the family* $\mathscr{F}$.

Now the representation of $Y(t)$ given by (11.2.22) is not necessarily of the form (11.2.10), since the modulus of the (generalized) Fourier transform of $\{\Gamma_{t,\omega+\omega_0}(\omega)A_t(\omega+\omega_0)\}$ may not have an absolute maximum at zero frequency. If not, then the function

$$\hat{\phi}_t(\omega) = \Gamma_{t,\omega+\omega_0}(\omega)A_t(\omega+\omega_0) e^{i\omega t}$$

will still, in general, be oscillatory, but its "dominant" frequency will be slightly shifted from ω.

There is, however, an important case where, for each t, λ, the function $\Gamma_{t,\lambda}(\theta)$ reduces approximately to $\Gamma(\theta)$, namely when $A_{t-u}(\lambda)$ is, for each t, λ, *slowly varying* compared with the function $g(u)$. Thus, we assume that $g(u)$ decays rapidly to zero as $|u| \to \infty$, and that $A_{t-u}(\lambda)$ is approximately constant over the range of u for which $g(u)$ is non-negligible. In this case, we may write, heuristically, (for each t, λ)

$$\Gamma_{t,\lambda}(\theta) \sim \Gamma(\theta), \qquad \text{all } \theta, \qquad (11.2.24)$$

so that using (11.2.22) we may write $\{Y(t)\}$ in the form

$$Y(t) \sim \int_{-\infty}^{\infty} A_t(\omega+\omega_0) e^{i\omega t} d\tilde{Z}(\omega), \qquad (11.2.25)$$

where

$$E[|d\tilde{Z}(\omega)|^2] = |\Gamma(\omega)|^2 d\mu(\omega+\omega_0).$$

Thus, we have

$$dH_t^{(Y)}(\omega) \sim |\Gamma(\omega)|^2 dH_t^{(X)}(\omega+\omega_0), \qquad (11.2.26)$$

where the evolutionary spectra $dH_t^{(Y)}(\omega)$ and $dH_t^{(X)}(\omega)$ are both defined with respect to the same family of oscillatory functions $\mathscr{F} \equiv \{A_t(\omega) e^{i\omega t}\}$.

In order to define more precisely the notion of a "slowly varying" function $A_{t-u}(\lambda)$, and to examine in more detail the approximation (11.2.24), we now introduce the notion of "*semi-stationary processes*".

Semi-stationary processes

Let $\{X(t)\}$ be an oscillatory process whose non-stationary characteristics are changing "slowly" over time. Then we may expect that there will exist a family $\mathcal{F}$ of oscillatory functions $\phi_t(\omega) = A_t(\omega) e^{i\omega t}$ in terms of which $\{X(t)\}$ has a representation of the form (11.2.10), and which are such that for each ω, $A_t(\omega)$ is (in some sense) a slowly varying function of t. Now there are, of course, various ways of defining a slowly varying function, but for our purposes the most convenient characterization is obtained by specifying that its Fourier transform must be "highly concentrated" in the region of zero frequency.

For each family $\mathcal{F}$, define the function $B_{\mathcal{F}}(\omega)$ by

$$B_{\mathcal{F}}(\omega) = \int_{-\infty}^{\infty} |\theta|\, |dK_\omega(\theta)|. \tag{11.2.27}$$

(Note that $B_{\mathcal{F}}(\omega)$ is a measure of the "width" of $|dK_\omega(\theta)|$.)

Definition 11.2.4 *A family $\mathcal{F}$ of oscillatory functions will be termed* semi-stationary *if the function $B_{\mathcal{F}}(\omega)$ is bounded for all ω, and the constant $B_{\mathcal{F}}$ defined by*

$$B_{\mathcal{F}} = \left[\sup_\omega \{B_{\mathcal{F}}(\omega)\} \right]^{-1} \tag{11.2.28}$$

will be termed the characteristic width *of the family $\mathcal{F}$.*

Definition 11.2.5 *A* semi-stationary process $\{X(t)\}$ *is now defined as one for which* $\exists$ *a semi-stationary family $\mathcal{F}$ in terms of which $X(t)$ has a representation of the form* (11.2.10).

For example, the uniformly modulated process, defined above, is a semi-stationary process, since the family $\mathcal{F}_0 \equiv \{C(t) e^{i\omega t}\}$ is semi-stationary. (Note that, since $C(t)$ is independent of ω, $B_{\mathcal{F}_0}(\omega)$ is independent of ω.)

For a particular semi-stationary process $\{X(t)\}$ consider the class $\mathcal{C}$ of semi-stationary families $\mathcal{F}$, in terms of each of which $\{X(t)\}$ admits a spectral representation. We define the *characteristic width*, B_X, of the process $\{X(t)\}$ by

$$B_X = \sup_{\mathcal{F} \in \mathcal{C}} \{B_{\mathcal{F}}\}. \tag{11.2.29}$$

Roughly speaking, B_X (or more precisely, $2\pi B_X$) may be interpreted as the maximum interval over which the process may be treated as "approximately stationary". Note that for stationary processes the class $\mathscr{C}$ contains the family of complex exponentials, which has infinite characteristic width. Consequently, all stationary processes have infinite characteristic width.

Now let $\mathscr{C}^* \subset \mathscr{C}$ denote the sub-class of families whose characteristic widths are each equal to B_X, and let $\mathscr{F}^*$ denote any family $\in \mathscr{C}^*$. For example, if $\{X(t)\}$ is stationary, $\mathscr{C}^*$ contains only one family, namely the complex exponentials, so that $\mathscr{F}^*$ is uniquely determined as this family. (However, as far as the theory of evolutionary spectra is concerned, the uniqueness of $\mathscr{F}^*$ is not required.) If $\mathscr{C}^*$ is empty, let $\mathscr{F}^*$ denote any family whose characteristic width is arbitrarily close to B_X.

We now consider the spectral representation of $\{X(t)\}$ in terms of the family $\mathscr{F}^*$. Thus we write

$$X(t) = \int_{-\infty}^{\infty} A_t^*(\omega)\, e^{i\omega t}\, dZ^*(\omega), \qquad (11.2.30)$$

where $E[|dZ^*(\omega)|^2] = d\mu^*(\omega)$, say, and the functions $\phi_t(\omega) = \{A_t^*(\omega)\, e^{i\omega t}\} \in \mathscr{F}^*$. It is now clear that if the evolutionary spectrum of $\{X(t)\}$ is defined with respect to $\mathscr{F}^*$, (11.2.26) will be a valid approximation provided that the "width" of $g(u)$ is much smaller than B_X, i.e. provided that, for each ω, $dK_\omega^*(\theta)$ (the Fourier transform of $A_t^*(\omega)$) "behaves as a δ-function with respect to $\Gamma(\omega)$". To define this notion more precisely, we introduce the following definition.

Definition 11.2.6 *We will say that $u(x)$ is a pseudo δ-function of order ε with respect to $v(x)$, if for any k, $\exists\ \varepsilon(\ll 1)$ independent of k, such that*

$$\left| \int_{-\infty}^{\infty} u(x)v(x+k)\, dx - v(k) \int_{-\infty}^{\infty} u(x)\, dx \right| < \varepsilon.$$

Now suppose that

(a) the filter $\{g(u)\}$ is square integrable and normalized, so that

$$2\pi \int_{-\infty}^{\infty} |g(u)|^2\, du = \int_{-\infty}^{\infty} |\Gamma(\omega)|^2\, d\omega = 1, \qquad (11.2.31)$$

(b)

$$\int_{-\infty}^{\infty} |u|\, |g(u)|\, du = B_g \text{ (say)}. \qquad (11.2.32)$$

Note that B_g is a measure of the "width" of $\{g(u)\}$. The following result holds:

Lemma 11.2.1 *Let* $\{\mathscr{F}\}$ *be a semi-stationary family with characteristic width* $B_{\mathscr{F}}$. *Then, for each* t, ω, $\{e^{it\omega}\,dK_\omega(\theta)\}$ *is a pseudo* δ-*function of order* $(B_g/B_{\mathscr{F}})$ *with respect to* $\Gamma(\theta)$.

Proof. For any k, write

$$\int_{-\infty}^{\infty} e^{it\theta}\Gamma(\theta+k)\,dK_\omega(\theta) = \Gamma(k)\int_{-\infty}^{\infty} e^{it\theta}\,dK_\omega(\theta) + R(k)$$

where

$$R(k) = \int_{-\infty}^{\infty} \theta\,e^{it\theta}\Gamma'\{k+\eta(\theta)\theta\}\,dK_\omega(\theta),$$

where, for each θ, $0 \leqslant \eta(\theta) \leqslant 1$. But

$$|R(k)| \leqslant \sup_\theta |\Gamma'(\theta)| \int_{-\infty}^{\infty} |\theta|\,|dK_\omega(\theta)| \leqslant B_g/B_{\mathscr{F}},$$

in virtue of (11.2.32), and the result follows.

We are now in a position to derive a more exact form of the relation (11.2.24).

Theorem 11.2.2 *Let* $\{g(u)\}$ *be a filter satisfying* (11.2.31), (11.2.32) *and* $\Gamma_{t,\lambda}(\theta)$ *its generalized transfer function with respect to a semi-stationary family* $\mathscr{F}$ *of characteristic width* $B_{\mathscr{F}}$. *If, for any* $\varepsilon\,(>0)$, *we choose* $\{g(u)\}$ *so that*

$$B_g < \varepsilon \cdot B_{\mathscr{F}},$$

then

$$|A_t(\lambda)|\,|\Gamma_{t,\lambda}(\theta) - \Gamma(\theta)| < \varepsilon, \qquad \textit{for all } t, \lambda, \theta.$$

Proof. We have, from (11.2.23),

$$A_t(\lambda)\Gamma_{t,\lambda}(\theta) = \int_{-\infty}^{\infty} g(u)A_{t-u}(\lambda)\,e^{-iu\theta}\,du.$$

Substituting for $A_{t-u}(\lambda)$ in terms of its Fourier transform $dK_\lambda(\theta)$, we obtain

$$A_t(\lambda)\Gamma_{t,\lambda}(\theta) = \int_{-\infty}^{\infty}\int_{-\infty}^{\infty} g(u)\,e^{-iu\theta}\,e^{i(t-u)\phi}\,dK_\lambda(\phi)\,du$$

$$= \int_{-\infty}^{\infty} e^{it\phi}\Gamma(\theta+\phi)\,dK_\lambda(\phi),$$

on interchanging the order of integration. However, according to Lemma

11.2.1, $\{e^{it\phi} dK_\lambda(\phi)\}$ is a pseudo δ-function of order $(B_g/B_{\mathscr{F}})$ with respect to $\Gamma(\phi)$. Thus, if $\{g(u)\}$ is chosen so that, for given ε,

$$B_g < \varepsilon B_{\mathscr{F}},$$

then

$$\left| \int_{-\infty}^{\infty} e^{it\phi} \Gamma(\theta + \phi) \, dK_\lambda(\phi) - \Gamma(\theta) \int_{-\infty}^{\infty} e^{it\phi} \, dK_\lambda(\phi) \right| < \varepsilon.$$

Noting that $\int_{-\infty}^{\infty} e^{it\phi} \, dK_\lambda(\phi) = A_t(\lambda)$, the result follows.

Determination of evolutionary spectra

Let $\{X(t)\}$ be a semi-stationary process with characteristic width B_X, and $\{g(u)\}$ a filter satisfying (11.2.31), (11.2.32), with width B_g. For any frequency ω_0, define the process $\{Y(t)\}$ as in (11.2.21), i.e. write

$$Y(t) = \int_{-\infty}^{\infty} g(u) X(t-u) e^{-i\omega_0(t-u)} \, du. \qquad (11.2.33)$$

Using the representation of $\{X(t)\}$ in terms of the family $\mathscr{F}^*$ given by (11.2.30), it follows from (11.2.22) that we may write

$$Y(t) = \int_{-\infty}^{\infty} \Gamma^*_{t,\omega+\omega_0}(\omega) A_t^*(\omega + \omega_0) e^{i\omega t} \, dZ^*(\omega + \omega_0), \qquad (11.2.34)$$

where $\Gamma^*_{t,\lambda}(\theta)$ is the generalized transfer function of $\{g(u)\}$ with respect to the family $\mathscr{F}^*$. Due to the orthogonality of $Z^*(\omega)$, it follows that

$$E[|Y(t)|^2] = \int_{-\infty}^{\infty} |\Gamma^*_{t,\omega+\omega_0}(\omega)|^2 |A_t^*(\omega + \omega_0)|^2 \, d\mu^*(\omega + \omega_0). \qquad (11.2.35)$$

Now suppose that $\{g(u)\}$ is chosen so that $B_g \leq \varepsilon B_X$. Then according to Theorem 11.2.2 (remembering that $\mathscr{F}^*$ has characteristic width B_X or is arbitrarily close to B_X) we may write

$$\Gamma^*_{t,\omega+\omega_0}(\omega) = \Gamma(\omega) + r(t, \omega_0, \omega),$$

where

$$|r(t, \omega_0, \omega)| < \varepsilon/|A_t^*(\omega + \omega_0)|.$$

Thus we obtain from (11.2.35),

$$E[|Y(t)|^2] = \int_{-\infty}^{\infty} |\Gamma(\omega)|^2 |A_t^*(\omega + \omega_0)|^2 \, d\mu^*(\omega + \omega_0) + I_1 + I_2 + I_3,$$

say, where

$$I_1 = \int_{-\infty}^{\infty} \Gamma(\omega) r^* |A_t^*(\omega + \omega_0)|^2 \, d\mu^*(\omega + \omega_0),$$

$$I_2 = \int_{-\infty}^{\infty} \Gamma^*(\omega) r |A_t^*(\omega + \omega_0)|^2 \, d\mu^*(\omega + \omega_0),$$

$$I_3 = \int_{-\infty}^{\infty} |r|^2 |A_t^*(\omega + \omega_0)|^2 \, d\mu^*(\omega + \omega_0).$$

Now

$$|I_3| \le \varepsilon^2 \int_{-\infty}^{\infty} d\mu^*(\omega) = O(\varepsilon^2)$$

and

$$|I_2| \le \varepsilon \int_{-\infty}^{\infty} |\Gamma(\omega)| \, |A_t^*(\omega + \omega_0)| \, d\mu^*(\omega + \omega_0).$$

To show that $|I_2| = O(\varepsilon)$, it remains to prove that the integral on the right-hand side of the above inequality remains finite as $B_g \to 0$. To demonstrate this fact, let the set Ω be defined by

$$\Omega = \{\omega; |\Gamma(\omega)| \, |A_t^*(\omega + \omega_0)| \le 1\}.$$

Then

$$\int_{-\infty}^{\infty} |\Gamma(\omega)| \, |A_t^*(\omega + \omega_0)| \, d\mu^*(\omega + \omega_0)$$

$$\le \int_{\Omega} d\mu^*(\omega + \omega_0) + \int_{\bar{\Omega}} |\Gamma(\omega)|^2 |A_t^*(\omega + \omega_0)|^2 \, d\mu^*(\omega + \omega_0).$$

The first term is finite, since $d\mu^*(\omega)$ is the evolutionary spectrum at zero time with respect to $\mathscr{F}^*$, and the second term is finite since $\Gamma(\omega)$ is normalized so that $\int_{-\infty}^{\infty} |\Gamma(\omega)|^2 \, d\omega = 1$.

The term I_1 may be treated similarly, so that in terms of

$$dH_t^*(\omega) = |A_t^*(\omega)|^2 \, d\mu^*(\omega),$$

the evolutionary spectrum of $\{X(t)\}$ with respect to the family $\mathscr{F}^*$, we have the following.

Theorem 11.2.3

$$E[|Y(t)|^2] = \int_{-\infty}^{\infty} |\Gamma(\omega)|^2 \, dH_t^*(\omega + \omega_0) + O(\varepsilon),$$

where $O(\varepsilon)$ denotes a term which may be made arbitrarily small by choosing B_g sufficiently small relative to B_X.

Now consider the case where the measure $\mu^*(\omega)$ is absolutely continuous with respect to Lebesgue measure, so that for each t we may write

$$dH_t^*(\omega) = h_t^*(\omega)\, d\omega, \qquad (11.2.36)$$

where $h_t^*(\omega)$, the evolutionary *spectral density function*, exists for all ω. Then, re-writing Theorem 11.2.3 in terms of $h_t^*(\omega)$, we have, to $O(\varepsilon)$,

$$E[|Y(t)|^2] \doteq \int_{-\infty}^{\infty} |\Gamma(\omega)|^2 h_t^*(\omega + \omega_0)\, d\omega. \qquad (11.2.37)$$

So far we have worked with the representation of $\{X(t)\}$ in terms of the family $\mathscr{F}^*$. However, as the validity of (11.2.37) depends only on the condition

$$B_g \ll B_{\mathscr{F}^*},$$

it is clear that, for fixed B_g, (11.2.37) will still be approximately true if instead we work with a representation of $\{X(t)\}$ in terms of any other semi-stationary family $\mathscr{F}$ whose characteristic width $B_{\mathscr{F}} \gg B_g$. Thus, if $dH_t(\omega) = h_t(\omega)\, d\omega$ is the evolutionary spectrum of $\{X(t)\}$ with respect to such a family, then (11.2.37) will still hold approximately if we substitute $h_t(\omega)$ for $h_t^*(\omega)$.

However, it must be remembered that (11.2.37) is only an approximation. In fact, if we work in terms of a general family $\mathscr{F}$, the exact value of $E[|Y(t)|^2]$ is given by (cf. 11.2.35),

$$E[|Y(t)|^2] = \int_{-\infty}^{\infty} |\Gamma_{t, \omega + \omega_0}(\omega)|^2\, dH_t(\omega + \omega_0). \qquad (11.2.38)$$

Thus, the *exact* value of $E[|Y(t)|^2]$ is an average of $dH_t(\omega)$ over *both* frequency and time, and we note that since $E[|Y(t)|^2]$ is independent of the choice of $\mathscr{F}$, *the value of this average of $dH_t(\omega)$ (over time and frequency) must also be independent of $\mathscr{F}$.* Thus, the right-hand side of (11.2.38) has an unambiguous interpretation as *an "average" of the total power of the process contained within a band of frequencies in the region of ω_0 and an interval of time in the neighbourhood of t.*

Now, in writing (11.2.37) we have assumed that the effect of the "time-averaging" is negligible, since the condition $B_g \ll B_{\mathscr{F}}$ implies that $dH_t(\omega)$ is changing very slowly over the effective range of the filter $\{g(u)\}$.

However, the degree of accuracy of (11.2.37) depends on the ratio $(B_g/B_{\mathscr{F}})$. For example, if $B_g = 0$, i.e. $g(u) = \delta(0)$, then (11.2.37) is exact for all $\mathcal{F}$, and reduces to (11.2.13), namely,

$$E[|Y(t)|^2] = \int_{-\infty}^{\infty} dH_t(\omega). \qquad (11.2.39)$$

On the other hand, if $g(u) = \lim_{T\to\infty}\{g_T(u)\}$ where

$$g_T(u) = \begin{cases} \surd(1/T), & |u| \le T/2, \\ 0, & |u| > T/2, \end{cases}$$

so that $B_g = \infty$, then it may be shown that in this case

$$E[|Y(t)|^2] = \lim_{T\to\infty} \int_{-\infty}^{\infty} |G_{T,\omega_0}(\omega)|^2 \, d\mu(\omega + \omega_0) \qquad (11.2.40)$$

where

$$G_{T,\omega_0}(\omega) = \frac{1}{\sqrt{T}} \int_{t-T/2}^{t+T/2} A_u(\omega + \omega_0) e^{iu\omega} \, du.$$

Note that $E[|Y(t)|^2]$ is independent of t, and reduces to the classical definition of the spectrum of $\{X(t)\}$ (if it were stationary), but that (11.2.37) is now invalid for all families.

A comparison of equations (11.2.39) and (11.2.40) is interesting. The right-hand side of (11.2.39) is a function only of the evolutionary spectrum *at time* t, and does not involve its values at other instants of time, but it provides no information on the distribution of $dH_t(\omega)$ over frequency, since it is completely independent of ω_0. However, assuming that, for each T, $|G_{T,\omega_0}(\omega)|^2$ is highly concentrated in the region $\omega = 0$ (as will generally be the case since we have assumed that, for all ω, $A(\omega)$ is a slowly varying function of t), the right-hand side of (11.2.40) will be approximately equal to

$$d\mu(\omega_0)\left\{ \lim_{T\to\infty} \int_{-\infty}^{\infty} |G_{T,\omega_0}(\omega)|^2 \, d\omega \right\},$$

and this quantity, being completely independent of t, may be interpreted as a form of "average" over t of the values of $dH_t(\omega_0)$, for $-\infty \le t \le \infty$.

Thus we see that *the more accurately we try to determine $dH_t(\omega)$ as a function of time, the less accurately we determine it as a function of frequency, and vice-versa.* This feature suggests a form of UNCERTAINTY PRINCIPLE, namely, *"in determining evolutionary spectra, one cannot obtain simultaneously a high degree of resolution in both the time domain and frequency domain".* Daniells (1965) has pointed out that this uncertainty principle is completely analogous to Heisenberg's uncertainty principle in quantum mechanics; the analogy with quantum mechanical concepts is discussed further by Tjøstheim (1976a).

It follows that if we wish to determine empirically the spectrum at a particular instant of time, we must sacrifice some degree of resolvability in the frequency domain, and vice-versa.

Suppose now that we fix the degree of resolution in the frequency domain, i.e. we set a *lower bound* to B_g. Then for a particular family $\mathscr{F}$, the resolution in the time domain will be determined by the value of $B_g/B_\mathscr{F}$. Clearly then, we obtain the maximum possible resolution in time by working in terms of the family with the maximum characteristic width.

Thus, if $\mathscr{C}^*$ contains only one member, i.e. if $\mathscr{F}^*$ is uniquely determined, then $\mathscr{F}^*$ provides the *natural representation* for $\{X(t)\}$, and is the family in terms of which we can most precisely express the time varying spectral pattern of the process. In particular, we now see why the natural representation of stationary processes is given in terms of the complex exponential family—the reason being simply that in this case $\mathscr{F}^*$ is unique and is just this family.

If $\mathscr{C}^*$ contains several families, $\mathscr{F}_1^*, \mathscr{F}_2^*, \mathscr{F}_3^*, \ldots$, then we may say that each $\mathscr{F}_i^*$ $(i = 1, 2, 3, \ldots)$ provides a natural representation for the process. For, let $dH_t^{(i)}(\omega)$ denote the evolutionary spectrum with respect to $\mathscr{F}_i^*$, $(i = 1, 2, \ldots)$, and let $\hat{h}_t^{(i)}(\omega)$ denote the "smoothed" evolutionary spectrum given by

$$\hat{h}_t^{(i)}(\omega_0) = \int_{-\infty}^{\infty} |\Gamma(\omega)|^2 \, dH_t^{(i)}(\omega + \omega_0).$$

Thus we have, for each i,

$$E[|Y(t)|^2] = \hat{h}_t^{(i)}(\omega_0) + O(B_g/B_X).$$

If now we fix the filter width B_g, then to $O(B_g/B_X)$ the "smoothed" evolutionary spectra with respect to each $\mathscr{F}_i^*$ are identical. Consequently, as "smoothed" spectra are the most we can determine, we may regard the representations with respect to each $\mathscr{F}_i^*$ as equivalent—at least as far as their corresponding spectra are concerned.

Finally, consider the case where $\mathscr{C}^*$ is empty, i.e. there is no family $\in \mathscr{C}$ with characteristic width B_X. Then there is no natural representation for the process since now, given any family $\in \mathscr{C}$, there is always another member of $\mathscr{C}$ with larger characteristic width. However, for this case we may redefine the sub-class $\mathscr{C}^*$ so that it includes all families whose characteristic widths lie between $(B_X - \eta)$ and B_X, where η is arbitrarily small. Then, by the argument used above, it follows that all families $\in \mathscr{C}^*$ give rise to almost identical evolutionary spectra.

From now on, we will consider only the representation of $\{X(t)\}$ in terms of a particular family $\mathscr{F}^*$, and when we refer to the *evolutionary spectrum of* $\{X(t)\}$ (without reference to any particular family) we will mean $|A_t^*(\omega)|^2 d\mu^*(\omega)$, the evolutionary spectrum with respect to the family $\mathscr{F}^*$.

The papers by Tjøstheim (1976a,b,c,d) give a very interesting discussion of the way in which evolutionary spectral representations can be derived via the use of "spectral generating operators", and their relationship with quantum theoretic concepts. Mandrekar (1972) gives a characterization of oscillatory processes as "deformed curves" in a Hilbert space.

11.2.1 Estimation of Evolutionary Spectra

Suppose we are given a sample record of $\{X(t)\}$, say for $0 \le t \le T$. We now consider the problem of estimating the evolutionary spectrum $dH_t(\omega)$, for $0 \le t \le T$, from the sample. (We omit the asterisks in $A_t^*(\omega)$, $d\mu^*(\omega)$, and $dF_t^*(\omega)$, it being understood that all functions are now defined with respect to the family $\mathscr{F}^*$.) We will treat here only the case where the measure $\mu(\omega)$ is *absolutely continuous with respect to Lebesgue measure*, so that we may write, for each t,

$$dH_t(\omega) = h_t(\omega)\, d\omega, \quad \text{all } \omega,$$

where $h_t(\omega)$, *the evolutionary spectral density function, exists for all* ω.

Let $\{g(u)\}$ be a filter of width B_g satisfying the conditions (11.2.31), and (11.2.32) and write, for any frequency ω_0,

$$U(t) = \int_{t-T}^{t} g(u)X(t-u)\, e^{-i\omega_0(t-u)}\, du. \tag{11.2.41}$$

We assume that

$$B_g \ll B_X \ll T,$$

so that for $t \gg 0$, the limits in the above integral may be replaced effectively by $(-\infty, \infty)$, when $U(t)$ becomes identical with the process $\{Y(t)\}$ defined in (11.2.33). In fact, the difference between $U(t)$ and $Y(t)$ is due to "transients" (or "end-effects") in the filter output, and we assume that these are negligible for t sufficiently greater than zero. It then follows from Theorem 11.2.3 that

$$E[|U(t)|^2] = \int_{-\infty}^{\infty} |\Gamma(\omega)|^2 h_t(\omega + \omega_0)\, d\omega + O(B_g/B_X). \tag{11.2.42}$$

At this point it is interesting to note an important difference between the estimation of evolutionary spectra and the estimation of spectra for stationary processes. In the latter case we may still employ the technique described above, but the bandwidth of $|\Gamma(\omega)|^2$ is chosen as a function of T which tends to zero as $T \to \infty$. However, in dealing with evolutionary spectra, the bandwidth of $|\Gamma(\omega)|^2$ (which varies inversely with B_g) is limited by the restriction $B_g \ll B_X$. In other words, since we have chosen the filter so that it only operates *locally* on $\{X(t)\}$, thereby assuring a high degree of resolution in the time domain, we must sacrifice some degree of resolution in the frequency domain. Thus, in order to estimate $h_t(\omega)$ we must assume that, for each t, it is *smooth* compared with $|\Gamma(\omega)|^2$, i.e. that its bandwidth is substantially larger than the bandwidth of $|\Gamma(\omega)|^2$ (cf. Section 7.3). In this case, $|\Gamma(\omega)|^2$ is a pseudo δ-function with respect to $h_t(\omega)$ (the order being the ratio of the bandwidths), and we may write

$$E[|U(t)|^2] \sim h_t(\omega_0) \qquad (11.2.43)$$

(remembering that $\int_{-\infty}^{\infty} |\Gamma(\omega)|^2 \, d\omega = 1$). Thus, $E[|U(t)|^2]$ is an (approximately) unbiased estimate of $h_t(\omega_0)$. Now a straightforward calculation (Priestley (1966)) shows that for $X(t)$ a *normal* process,

$$\text{var}\{|U(t)|^2\} \sim \left\{ \int_{-\infty}^{\infty} |\Gamma(\omega)|^2 h_t(\omega + \omega_0) \, d\omega \right\}^2, \qquad (11.2.44)$$

which, being independent of T, means that $|U(t)|^2$ will not be a very useful estimate of $h_t(\omega_0)$ in practice. (This is completely analogous to the behaviour of the periodogram in classical spectral analysis.) However, to reduce sampling fluctuations we may "smooth" the values of $|U(t)|^2$ for neighbouring values of t. In so doing, we increase the precision of our estimates by sacrificing some degree of resolvability in the time domain.

We therefore consider a weight-function $w_{T'}(t)$ depending on the parameter T', which satisfies

(a) $w_{T'}(t) \geq 0$, all t, T';

(b) $w_{T'}(t)$ decays to zero as $|t| \to \infty$, all T';

(c) $\int_{-\infty}^{\infty} w_{T'}(t) \, dt = 1$, all T';

(d) $\int_{-\infty}^{\infty} \{w_{T'}(t)\}^2 \, dt < \infty$, all T'.

Write

$$W_{T'}(\lambda) = \int_{-\infty}^{\infty} e^{-i\lambda t} w_{T'}(t) \, dt. \qquad (11.2.45)$$

We assume that $\exists$ a constant C such that

(e) $\displaystyle \lim_{T' \to \infty} \left\{ T' \int_{-\infty}^{\infty} |W_{T'}(\lambda)|^2 \, d\lambda \right\} = C.$ \hfill (11.2.46)

Now let

$$V(t) = \int_{-\infty}^{\infty} w_{T'}(u) |U(t-u)|^2 \, du. \tag{11.2.47}$$

Again, we assume that $w_{T'}(u)$ decays sufficiently fast so that the above integral may be evaluated from a finite length of $|U(t)|^2$. It follows from (11.2.42) that

$$E[V(t)] \sim \int_{-\infty}^{\infty} \int_{-\infty}^{\infty} w_{T'}(u) h_{t-u}(\omega + \omega_0) |\Gamma(\omega)|^2 \, du \, d\omega$$

$$= \int_{-\infty}^{\infty} \bar{h}_t(\omega + \omega_0) |\Gamma(\omega)|^2 \, d\omega, \tag{11.2.48}$$

where

$$\bar{h}_t(\omega + \omega_0) = \int_{-\infty}^{\infty} w_{T'}(u) h_{t-u}(\omega + \omega_0) \, du.$$

From (11.2.48) we see that $E[V(t)]$ is a "smoothed" form of $h_t(\omega_0)$, smoothed over both time and frequency. If, as before, we assume that $h_t(\omega)$ is "flat" compared with $|\Gamma(\omega)|^2$ (each t), then (11.2.48) may be written

$$E[V(t)] \sim \bar{h}_t(\omega_0), \tag{11.2.49}$$

so that $V(t)$ is an (approximately) *unbiased estimate of the (weighted) average value of $h_t(\omega_0)$ in the neighbourhood of the time-instant t.*

An investigation of the sampling properties of $V(t)$ is straightforward, if lengthy, and we merely summarize here the main result. (A derivation of equations (11.2.44) and (11.2.50) is given in Priestley (1966).) Assuming that $\{(1/T')|W_{T'}(\omega)|^2\}$ is a pseudo δ-function with respect to $|\Gamma(\omega)|^2$, i.e. that the "width" of $|W_{T'}(\omega)|^2$ is much *smaller* than the "width" of $|\Gamma(\omega)|^2$, i.e. that the "width" of $w_{T'}(t)$ is much *larger* than B_g, we may show that

$$\text{var}\{V(t)\} \sim \{\bar{\tilde{h}}_t^2(\omega_0)\}\left\{ \int_{-\infty}^{\infty} |W_{T'}(\omega)|^2 \, d\omega \right\}\left\{ \int_{-\infty}^{\infty} |\Gamma(\omega)|^4 \, d\omega \right\}(1 + \delta_{0,\omega_0}),$$

$$\tag{11.2.50}$$

where

$$\tilde{h}_t^2(\omega_0) = \frac{\displaystyle\int_{-\infty}^{\infty} h_{t-u}^2(\omega_0)\{w_{T'}(u)\}^2\,du}{\displaystyle\int_{-\infty}^{\infty}\{w_{T'}(u)\}^2\,du}.$$

For large T', we obtain from (11.2.50) and condition (e),

$$\mathrm{var}\{V(t)\} \sim \frac{\{\tilde{h}_t^2(\omega_0)\}C}{T'}\left\{\int_{-\infty}^{\infty}|\Gamma(\omega)|^4\,d\omega\right\} \qquad (\omega_0 \neq 0). \quad (11.2.51)$$

Thus the sampling fluctuations of $V(t)$ are of $O(1/\sqrt{T'})$.

The discussion so far has been in terms of a general filter $\{g(u)\}$ and a general weight function $\{w_{T'}(t)\}$. In practice, we must choose a particular form for each of these functions. As far as the form of $\{g(u)\}$ is concerned, we may make use of any of the standard "windows" used in spectral analysis of stationary processes, remembering that $|\Gamma(\omega)|^2$ (and not $\Gamma(\omega)$) corresponds to the spectral window. For example, if we choose

$$g(u) = \begin{cases} \dfrac{1}{2\sqrt{h\pi}}, & |u| \leq h, \\[2mm] 0, & |u| > h, \end{cases} \qquad (11.2.52)$$

then

$$|\Gamma(\omega)|^2 = \frac{1}{\pi}\frac{\sin^2 h\omega}{h\omega^2},$$

corresponding to the Bartlett window with parameter $(2h)$. The weight function $\{w_{T'}(t)\}$ may also be chosen from the customary collection of "windows", with the time and frequency domains interchanged.

For example, we may choose

$$w_{T'}(t) = \begin{cases} 1/T', & -T'/2 \leq t \leq T'/2, \\ 0, & \text{otherwise}, \end{cases} \qquad (11.2.53)$$

(corresponding to the Daniell window) so that

$$W_{T'}(\lambda) = \frac{\sin(T'\lambda/2)}{(T'\lambda/2)}$$

and

$$\lim_{T'\to\infty}\left\{T'\int_{-\infty}^{\infty}|W_{T'}(\lambda)|^2\,d\lambda\right\} = 2\pi.$$

Alternatively, we may choose

$$w_{T'}(t) = \begin{cases} (1/T')\, e^{-t/T'}, & t \geqslant 0, \\ 0, & t < 0, \end{cases} \qquad (11.2.54)$$

then

$$W_{T'}(\lambda) = 1/(1 + i\lambda T')$$

and

$$\lim_{T' \to \infty} \left\{ T' \int_{-\infty}^{\infty} |W_{T'}(\lambda)|^2 \, d\lambda \right\} = \pi.$$

With $\{g(u)\}$ given by (11.2.52) and $W_{T'}(t)$ by (11.2.53) we find (for $\omega_0 \neq 0$),

$$\frac{\text{var}\{V(t)\}}{\bar{h}_t^2(\omega_0)} \sim \frac{2\pi}{T'} \left\{ \int_{-\infty}^{\infty} |\Gamma(\omega)|^4 \, d\omega \right\} = \frac{4h}{3T'},$$

while for $\{g(u)\}$ given by (9.12) and $W_{T'}(t)$ by (9.14) we find

$$\frac{\text{var}\{V(t)\}}{\bar{h}_t^2(\omega_0)} \sim \frac{\pi}{T'} \left\{ \int_{-\infty}^{\infty} |\Gamma(\omega)|^4 \, d\omega \right\} = \frac{2h}{3T'}.$$

Thus, with the same value of T, the estimate corresponding to the weight function (11.2.53) has a variance *twice* as large as that corresponding to (11.2.54).

When $w_{T'}(u)$ has the rectangular form, (11.2.53), the computational form of the evolutionary spectral estimate is very similar to the method of estimating spectra (of stationary processes) by "*averaging of time blocks*"—see Section 7.6. However, the basic difference is that, in the present context, each "block" of time points is used separately to construct an estimate of the evolutionary spectrum corresponding to the mid-point of the interval.

Discrete parameter processes

A spectral theory for discrete parameter non-stationary processes may be developed in exactly the same way as that for continuous parameter processes. Thus, an oscillatory discrete parameter process will have a representation of the form

$$X_t = \int_{-\pi}^{\pi} e^{it\omega} A_t(\omega) \, dZ(\omega), \qquad t = 0, \pm 1, \pm 2, \ldots, \qquad (11.2.55)$$

where, for each ω, the sequence $\{A_t(\omega)\}$ has a generalized Fourier transform whose modulus has an absolute maximum at origin, and $Z(\omega)$

is an orthogonal process on the interval $(-\pi, \pi)$, with $E[|dZ(\omega)|^2] = d\mu(\omega)$. (In view of the factor $e^{it\omega}$ in the integrand in (11.2.55), there is no loss of generality in taking the limits to be $(-\pi, \pi)$.)

The evolutionary spectrum at time t (with respect to the family of sequences $\mathcal{F} \equiv \{e^{it\omega}A_t(\omega)\}$) is defined as

$$dH_t(\omega) = |A_t(\omega)|^2 \, d\mu(\omega), \qquad -\pi \leqslant \omega \leqslant \pi. \qquad (11.2.56)$$

The evolutionary spectral density function $h_t(\omega)$ (with respect to the family with maximum characteristic width) may be estimated by the method described in the previous section, except that the function $\{g(u)\}$ is replaced by a sequence $\{g_u\}$ and the weight function $w_{T'}(t)$ by a sequence $\{w_{T',t}\}$. Thus we write

$$U_t = \sum_{u=-\infty}^{\infty} g_u X_{t-u} \, e^{-i\omega_0(t-u)}, \qquad (11.2.57)$$

$$V_t = \sum_{\nu=-\infty}^{\infty} w_{T',\nu} |U_{t-\nu}|^2, \qquad (11.2.58)$$

and it may be shown that

$$E(V_t) \sim \int_{-\pi}^{\pi} \bar{h}_t(\omega + \omega_0) |\Gamma(\omega)|^2 \, d\omega, \qquad (11.2.59)$$

where

$$\bar{h}_t(\omega + \omega_0) = \sum_{\nu=-\infty}^{\infty} w_{T',\nu} \, h_{t-\nu}(\omega + \omega_0)$$

and

$$\Gamma(\omega) = \sum_{u=-\infty}^{\infty} g_u \, e^{-iu\omega}.$$

We may also show that

$$\mathrm{var}\{V_t\} \sim \{\tilde{h}_t^2(\omega_0)\}\left\{\int_{-\pi}^{\pi} |W_{T'}(\omega)|^2 \, d\omega\right\}\left\{\int_{-\pi}^{\pi} |\Gamma(\omega)|^4 \, d\omega\right\}(1 + \delta_{0, \pm\pi, \omega_0}),$$

$$(11.2.60)$$

where

$$\tilde{h}_t^2(\omega_0) = \frac{\displaystyle\sum_{\nu=-\infty}^{\infty} h_{t-\nu}^2(\omega_0)\{w_{T',\nu}\}^2}{\displaystyle\sum_{\nu=-\infty}^{\infty} \{w_{T',\nu}\}^2}$$

and

$$W_{T'}(\omega) = \sum_{\nu=-\infty}^{\infty} w_{T',\nu} \, e^{-i\nu\omega}.$$

If the covariance kernel $R(s, t)$ were known precisely, then in principle one could determine the measure $d\mu(\omega)$ corresponding to a given family $\{A_t(\omega)\}$ by formally inverting (11.2.9). Having found $d\mu(\omega)$, then $dH_t(\omega)$ would be given by $dH_t(\omega) = |A_t(\omega)|^2 \, d\mu(\omega)$. (This would be analogous to determining the spectrum of a stationary process by inverting the Wiener–Khintchine relationship.) However, as far as the estimation of evolutionary spectra is concerned, it is not necessary to specify—or even to know—the form of a suitable family. In fact, the estimate described above (cf. (11.2.47)) is computed entirely from the observed values of $X(t)$, and may be interpreted as a "smoothed" form of the evolutionary spectrum corresponding to *any* family. Of course, the degree of smoothing varies from family to family, and as we have seen, the family $\mathscr{F}^*$ induces the smallest amount of smoothing. Thus, the evolutionary spectral estimate is most accurate when interpreted as an estimate of $h_t^*(\omega)$, the evolutionary spectral density function with respect to the family $\mathscr{F}^*$. In this case, we may say that the estimation procedure automatically "selects" the family $\mathscr{F}^*$.

In evaluating the evolutionary spectral estimates we have to choose suitable values for the parameters associated with the functions $g(u)$, $w_{T'}(u)$ (corresponding roughly to the *"frequency domain bandwidth"* and *"time domain bandwidth"* of the estimate), and the *"design relations"* governing the choice of these parameters are discussed by Priestley (1966).

A test for stationarity (based on an examination of the "homogeneity" of evolutionary spectral estimates) has been proposed by Priestley and Subba Rao (1969).

An extension of the theory of evolutionary spectra to the case of *non-stationary bivariate processes* has been developed by Priestley and Tong (1973).

Some interesting practical applications of evolutionary spectral analysis are given by Hammond (1968, 1973), Rao and Shapiro (1970), and Tayfun, Yang and Hsiao (1971). In particular, Hammond has used this approach in the analysis of jet engine noise. Hasofer and Petocz (1978) have used evolutionary spectral representations to study the "upcrossings" of oscillatory processes.

Numerical studies

To examine the validity of the suggested methods for estimating evolutionary spectra, realizations were simulated for non-stationary processes generated from known models. The graphs shown below are taken from Priestley (1965b) and refer to a spectral analysis of a uniformly modulated discrete parameter process $\{X_t\}$ generated from the model

$$X_t = \{\exp[-(t - 500)^2/2 \times 200^2]\} Y_t, \qquad t = 0, 1, 2, \ldots,$$

in which Y_t is the second order autoregressive process

$$Y_{t+2} - 0 \cdot 8 \, Y_{t+1} + 0 \cdot 4 \, Y_t = \varepsilon_t,$$

the $\{\varepsilon_t\}$ being independent variables, each having the distribution $N(0, 100^2)$.

The spectral density function of Y_t is given by

$$h^{(Y)}(\omega) = \frac{\sigma_Y^2}{2\pi} \left\{ \frac{0 \cdot 792}{1 \cdot 4 - 3 \cdot 136 \cos \omega + 2 \cdot 24 \cos^2 \omega} \right\},$$

where $\sigma_Y^2 = \operatorname{var}(Y_t) = 133^2$, and we may note that $h^{(Y)}(\omega)$ has a single peak (in the range $0 \le \omega \le \pi$) at (approximately) $\omega = \pi/4$, with bandwidth (approximately) $10\pi/36$. With respect to the family $\mathcal{F}_0$ the evolutionary spectral density function of X_t is given by (cf. (11.2.18))

$$h_t^{(X)}(\omega) = \{\exp[(t - 500)^2/2 \times 200^2]\}^2 h^{(Y)}(\omega).$$

The estimates of $h_t^{(X)}(\omega)$ shown in the graphs were obtained from the discrete time analogue of equation (11.2.47), in which $W_{T'}(u)$ is given by equation (11.2.53) with $T' = 200$, and U_t is given by equation (11.2.41) with $g(u)$ of the form (11.2.52) and $h = 7$. With this value of h the bandwidth of the corresponding filter $|\Gamma(\omega)|^2$ is approximately half of that of $h^{(Y)}(\omega)$ at the peak frequency.

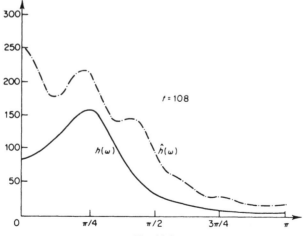

Fig. 11.1

For the family $\mathscr{F}_0$, $B_{\mathscr{F}_0}(\omega)$ is independent of ω and the characteristic width of $\mathscr{F}_0$ is given by

$$B_{\mathscr{F}_0} = 200\sqrt{(\pi/2)}.$$

We find also that the filter "width" corresponding to $g(u)$ of the above form is given by (see equation (11.2.32))

$$B_g = 7^{3/2}/2\sqrt{\pi}$$

so that

$$B_g/B_{\mathscr{F}_0} \doteq 0{\cdot}02.$$

The graphs in Figs. 11.1–11.8 show the estimated and theoretical forms of $h_t^{(X)}(\omega)$ for $t = 108(100)808$.

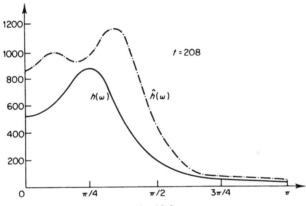

Fig. 11.2

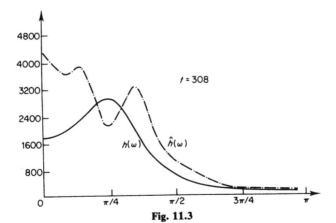

Fig. 11.3

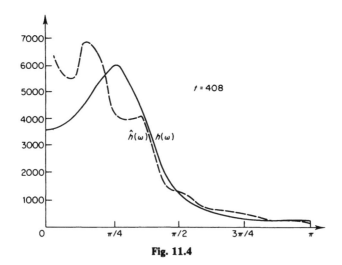

Fig. 11.4

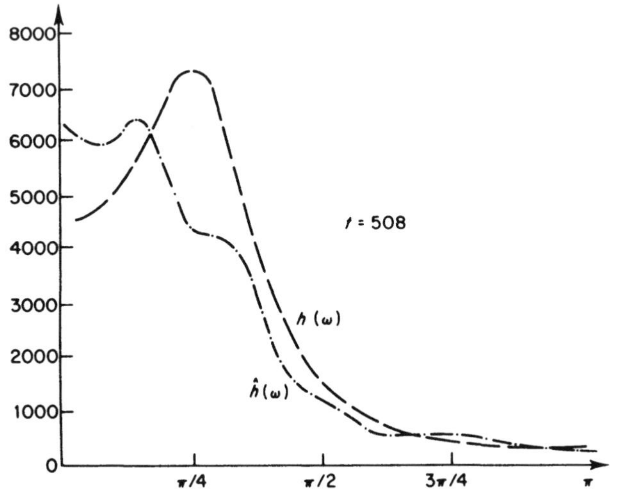

Fig. 11.5

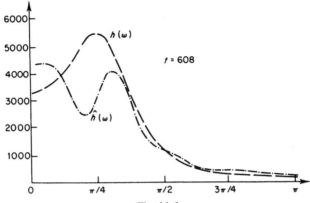

Fig. 11.6

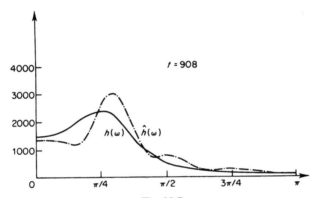

Fig. 11.7

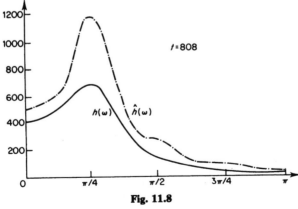

Fig. 11.8

For comparison, Fig. 11.9 shows the estimated spectral density function of the autoregressive process. The form of estimate used is exactly the same as that used for the non-stationary process, and corresponds to $t = 500$. (Further simulation studies of evolutionary spectral estimation are given in Chan and Tong (1976).)

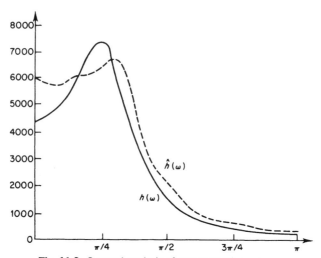

Fig. 11.9. Spectral analysis of autoregressive process.

11.2.2 Complex Demodulation

In our study of mixed spectra (Chapter 8) we considered stationary processes which consist of "signal" plus "noise", the "signal" having a discrete spectrum and the "noise" a continuous spectrum. A typical model for such a process is, in the discrete parameter case (cf. (8.2.1)),

$$X_t = \sum_{r=1}^{k} A_r e^{i\omega_r t} + Z_t, \tag{11.2.61}$$

(Z_t denoting the "noise" term with $E(Z_t) = 0$). If we now allow the "signal" to be non-stationary, the natural generalization of (11.2.61) is

$$X_t = \sum_{r=1}^{k} A_r(t) e^{i\omega_r t} + Z_t, \tag{11.2.62}$$

where, for each r, $A_r(t)$ is a slowly changing function of t. This corresponds to the situation where both the amplitude and phases of the harmonic components "drift" over time, and could represent, e.g., a radio signal which is subject to intermittent "fading". The problem of detecting the presence of such signals, and estimating their amplitudes and phases, can be studied by means of a technique called "*complex demodulation*", due to Tukey (1961). Complex demodulation is, in effect, a "local" form of harmonic analysis, and in the present context it may be regarded as an "evolutionary" version of harmonic analysis.

The basic ideas may be explained as follows. If the $\{\omega_r\}$ are known *a priori*, we can study the time-dependent behaviour of the rth term in the "signal" process by shifting its frequency down to zero, and this is easily accomplished by multiplying X_t by $\exp(-i\omega_r t)$, leading to, say,

$$X'_t = e^{-i\omega_r t}X_t = A_r(t) + \sum_{s \neq r} A_s(t)\, e^{i(\omega_s - \omega_r)t} + e^{-i\omega_r t}Z_t.$$

If we now pass X'_t through a suitably designed low-pass filter, then, provided the frequencies $\omega_1, \omega_2, \ldots, \omega_k$ are spaced sufficiently far apart, the filter will remove all the "signal" terms other than $A_r(t)$. Also, since Z_t has a continuous spectrum, the power due to components with frequencies in the neighbourhood of ω_r will be very small, and hence the contribution from $\{\exp(-i\omega_r t)Z_t\}$ to the output of the filter will also be very small. Thus, the filter output will be essentially just the term $A_r(t)$. The time-dependent amplitude of the rth term is then given by $|X'_t|$, and similarly the time-dependent phase is $\arg\{X'_t\}$. (We must, however, remember that the bandwidth of the filter cannot be made too small since its impulse response function must decay sufficiently fast relative to the rate of variation of the $\{A_r(t)\}$ for the approximation (11.2.26) to be valid, that is, so that we can use standard linear filter theory despite the non-stationary character of the input.) In the more realistic case when the $\{\omega_r\}$ are unknown, we can proceed by using a "guessed" value of ω_r, say $\hat{\omega}_r$, and forming

$$X'_t = e^{-i\hat{\omega}_r t}X_t = A_r(t)\, e^{i(\omega_r - \hat{\omega}_r)t} + \sum_{s \neq r} A_s(t)\, e^{i(\omega_s - \hat{\omega}_r)} + e^{-i\hat{\omega}_r t}Z_t.$$

Provided $\hat{\omega}_r$ is sufficiently close to ω_r, the filter will pass the first term in the above but suppress the remainder. Thus, the output will consist essentially of a low frequency oscillation $\exp(i(\omega_r - \hat{\omega}_r)t)$ modulated by $A_r(t)$. We can then estimate $(\omega_r - \hat{\omega}_r)$ using a local version of the method of "secondary analysis" (see Section 6.1.4), and this, in turn, enables us to estimate ω_r.

To examine the operation of this technique in more detail let $\{g_u\}$ be a real symmetric filter (i.e. $g_u = g_{-u}$), and write

$$U_t = \sum_{u=-\infty}^{\infty} g_u X_{t-u} \cos \omega_0(t-u), \tag{11.2.63}$$

$$V_t = \sum_{u=-\infty}^{\infty} g_u X_{t-u} \sin \omega_0(t-u), \tag{11.2.64}$$

so that

$$W_t = U_t - iV_t = \sum_{u=-\infty}^{\infty} g_u X_{t-u} \, e^{-i\omega_0(t-u)}. \qquad (11.2.65)$$

Now consider the case where the "signal" contains just one term and the "noise" is absent (i.e. $Z_t \equiv 0$), so that X_t may be written in real terms as

$$X_t = A_1(t) \cos(\omega_1 t + \phi_1).$$

Assuming that $A_1(t)$ changes "slowly" over the effective range of $\{g_u\}$—so that $A_1(t)$ may be treated as a constant—we have

$$W_t = \tfrac{1}{2} A_1(t) \sum_{u=-\infty}^{\infty} g_u [\exp(-i\{(\omega_0-\omega_1)(t-u)-\phi_1\})$$

$$+ \exp(-i\{(\omega_0+\omega_1)(t-u)+\phi_1\})]$$

$$= \tfrac{1}{2} A_1(t)\Gamma(\omega_0-\omega_1) \exp(-i\{(\omega_0-\omega_1)t-\phi_1\})$$

$$+ \tfrac{1}{2} A_1(t)\Gamma(\omega_0+\omega_1) \exp(-i\{(\omega_0+\omega_1)t+\phi_1\}),$$

$$(11.2.66)$$

where

$$\Gamma(\omega) = \sum_{u=-\infty}^{\infty} g_u \, e^{-i\omega u} \qquad (11.2.67)$$

is the transfer function corresponding to $\{g_u\}$. (Note that since $\{g_u\}$ is a symmetric filter, $\Gamma(\omega)$ is a real symmetric function of ω.) If ω_0 is chosen so that $\omega_0 \sim \omega_1$, and if the filter is constructed so that $\Gamma(\omega) \sim 0$, $|\omega| > 2\omega_0$, then the second term in (11.2.66) is effectively zero, and we may write

$$U_t \sim \tfrac{1}{2} A_1(t)\Gamma(\omega_0-\omega_1) \cos\{(\omega_0-\omega_1)t-\phi_1\}, \qquad (11.2.68)$$

$$V_t \sim \tfrac{1}{2} A_1(t)\Gamma(\omega_0-\omega_1) \sin\{(\omega_0-\omega_1)t-\phi_1\}. \qquad (11.2.69)$$

Hence we have,

$$A_1(t) \sim 2(U_t^2 + V_t^2)^{1/2} \Big/ \Gamma(\omega_0-\omega_1), \qquad (11.2.70)$$

and

$$\psi_t = \tan^{-1}\left\{\frac{V_t}{U_t}\right\} \sim (\omega_0-\omega_1)t - \phi_1, \quad (\text{mod } 2\pi). \qquad (11.2.71)$$

Thus, if X_t contains a harmonic component with frequency ω we may detect the presence of this component by plotting the "phase" ψ_t as a function of t. *This plot should be approximately linear, with slope* $(\omega_0 - \omega_1)$, *and hence we may estimate both* $(\omega_0 - \omega_1)$ *and* ϕ_1.

The method of "secondary analysis" (discussed in Section 6.1.4) is now seen to be a special form of complex demodulation in which $g_u = 1$; $u = 0$, $1, \ldots, mT_1$ (and zero otherwise), and the phases ψ_t are plotted only at integral multiples of (mT_1). However, in secondary analysis the amplitude is assumed to be contant and the objective is simply to obtain an improved estimate of the frequency. In the present context the amplitude is a (slowly changing) function of t, and now we are interested in determining those intervals where the harmonic is actually present (i.e. those values of t for which $A_1(t) > 0$). Such intervals may be detected by plotting ψ_t against t and seeking those values of t where ψ_t appears to have a "*linear regression*" on t. To pursue this further we need to know the distribution of ψ_t, both in the "null" case (where there is no signal and the observations consist purely of noise), and in the non-null case when a signal is present.

Before discussing these distributions we should note that (11.2.71) defines ψ_t only *mod* 2π. Consequently, when plotting ψ_t we face the same difficulty as in the plotting of phase spectra (see Section 9.5.4). As in the case of phase spectra, we may adopt the convention of taking *the range of* ψ_t *to be* $(-\pi, \pi)$, *and giving* ψ_t *the same sign as* U_t. (Each pair of values of (U_t, V_t) then determines a unique value of ψ_t in $(-\pi, \pi)$). We may then follow the procedure discussed in Section 9.5.4, namely that, for each t, we plot three values of ψ_t, one in the range $(-3\pi, -\pi)$, one in the range $(-\pi, \pi)$ and one in the range $(\pi, 3\pi)$. This should then make it easier to detect visually the presence of linear trends.

Distribution of ψ_t in the null case

Suppose that in (11.2.66), $A_r(t) = 0$, all r, so that $X_t \equiv Z_t$. Then

$$U_t = \sum_{u=-\infty}^{\infty} g_u Z_{t-u} \cos \omega_0(t - u)$$

(with a similar expression for V_t) and

$$W_t = \sum_{u=-\infty}^{\infty} g_u Z_{t-u} e^{-i\omega_0(t-u)}.$$

Let Z_t have spectral density function $h_Z(\omega)$, so that its autocovariance function is

$$R_Z(s) = \int_{-\pi}^{\pi} e^{i\omega s} h_Z(\omega) \, d\omega.$$

Then

$$E(W_t^2) = \sum_u \sum_v g_u g_v e^{-i\omega_0(t-u)} e^{-i\omega_0(t-v)} R_Z(v-u)$$

$$= e^{-2i\omega_0 t} \int_{-\pi}^{\pi} \Gamma(\omega+\omega_0)\Gamma(\omega-\omega_0)h_Z(\omega)\,d\omega.$$

Similarly we find

$$E(|W_t|^2) = \int_{-\pi}^{\pi} \Gamma^2(\omega-\omega_0)h_Z(\omega)\,d\omega.$$

Using results such as $\text{var}(U_t) = \frac{1}{2}\mathcal{R}\{E(W_t^2)+E(|W_t|^2)\}$, we obtain

$$\text{var}(U_t) = \frac{1}{2}\int_{-\pi}^{\pi} \{\cos 2\omega_0 t\,\Gamma(\omega+\omega_0)\Gamma(\omega-\omega_0)+\Gamma^2(\omega-\omega_0)\}h_Z(\omega)\,d\omega$$

$$\text{var}(V_t) = \frac{1}{2}\int_{-\pi}^{\pi} \{\Gamma^2(\omega-\omega_0)-\cos 2\omega_0 t\,\Gamma(\omega+\omega_0)\Gamma(\omega-\omega_0)\}h_Z(\omega)\,d\omega,$$

and

$$\text{cov}(U_t,\,V_t) = \frac{1}{2}\sin 2\omega_0 t\int_{-\pi}^{\pi} \Gamma(\omega+\omega_0)\Gamma(\omega-\omega_0)h_Z(\omega)\,d\omega.$$

If the sequence $\{g_u\}$ is to work successfully as a complex demodulation filter then, as we have previously seen, we must have $\Gamma(\omega)\sim 0$, $|\omega|\geq 2\omega_0$. Suppose now that we impose the stronger condition, $\Gamma(\omega)\sim 0$, $|\omega|\geq\omega_0$. Then integrals of the form

$$\int_{-\pi}^{\pi} \Gamma(\omega+\omega_0)\Gamma(\omega-\omega_0)h_Z(\omega)\,d\omega$$

will be effectively zero. In fact, such expressions will be exactly zero if, e.g., $\Gamma(\omega)$ has the ideal low-pass form

$$\Gamma(\omega) = \begin{cases} 1/2h, & |\omega|\leq h, \quad h\leq\omega_0, \\ 0, & \text{otherwise}. \end{cases}$$

We now have, to $O\{\Gamma(\omega+\omega_0)\Gamma(\omega-\omega_0)\}$,

$$\text{var}(U_t) = \text{var}(V_t) \sim \int_{-\pi}^{\pi} \Gamma^2(\omega)h_Z(\omega+\omega_0)\,d\omega, \qquad (11.2.72)$$

and

$$\text{cov}(U_t,\,V_t) = 0. \qquad (11.2.73)$$

Assuming further that $\Gamma^2(\omega)$ is "highly peaked" relative to $h_Z(\omega + \omega_0)$ (i.e. that the bandwidth of $h_Z(\omega)$ at $\omega = \omega_0$ is much larger than ω_0), we may write

$$\text{var}(U_t) = \text{var}(V_t) \sim \tfrac{1}{2}k_0 h_Z(\omega_0) = \sigma_0^2,$$

say, where

$$k_0 = \int_{-\pi}^{\pi} \Gamma^2(\omega)\, d\omega.$$

If Z_t is a zero-mean Gaussian process, then we may take U_t and V_t to be independent normal variables, and hence the joint distribution of (U_t, V_t) is

$$p(U_t, V_t) = \frac{1}{2\pi\sigma_0^2} \exp\left\{ -\frac{1}{2\sigma_0^2}(u^2 + v^2) \right\}.$$

Writing

$$U_t = r_t \cos \psi_t, \qquad 0 \le r_t < \infty,$$
$$V_t = r_t \sin \psi_t, \qquad -\pi \le \psi_t \le \pi,$$

the distribution of $\psi_t = \tan^{-1}(V_t/U_t)$ is given by

$$p(\psi_t) = \left[\frac{1}{2\pi\sigma_0^2} \int_0^{\infty} r_t e^{-r_t^2/2\sigma_0^2}\, dr_t \right] = \frac{1}{2\pi}, \qquad -\pi \le \psi_t \le \pi.$$

Hence, ψ_t has a rectangular distribution on $(-\pi, \pi)$, and

$$E(\psi_t) = 0, \qquad E(\psi_t^2) = \pi^2/3.$$

(The principal value of ψ_t similarly has a uniform distribution on $(-\pi/2, \pi/2)$, and this is identical with the approximate distribution of the estimated phase spectrum in the case of zero coherency—see Section 9.5.2.)

Distribution of ψ_t in the non-null case

Consider, for simplicity, the case

$$X_t = A_1 \cos(\omega_1 t + \phi_1) + Z_t.$$

Then, with an obvious notation, we may write

$$U_t = \tfrac{1}{2}A_1 \Gamma(\omega_0 - \omega_1) \cos\{(\omega_0 - \omega_1)t - \phi_1\} + U_t(Z), \qquad (11.2.74)$$
$$V_t = \tfrac{1}{2}A_1 \Gamma(\omega_0 - \omega_1) \sin\{(\omega_0 - \omega_1)t - \phi_1\} + V_t(Z). \qquad (11.2.75)$$

Hence, ψ_t may be written in the form

$$\psi_t = \tan^{-1}\left\{\frac{a_t + V_t(Z)}{b_t + U_t(Z)}\right\},$$

where the constants a_t, b_t are given by (11.2.74), (11.2.75). (Here, we treat ϕ_1 as a constant, its value being that appropriate to the particular realization of X_t with which we are dealing.) We note that ψ_t may be written alternatively in the form,

$$\psi_t = \tan^{-1}\left(\frac{a_t}{b_t}\right) + \tan^{-1}\left\{\frac{b_t V_t(Z) - a_t U_t(Z)}{b_t^2 + a_t^2 + (b_t U_t(Z) - a_t V_t(Z))}\right\}. \tag{11.2.76}$$

Hence the distribution of ψ_t may be reduced to the distribution of $\tan^{-1}(P/Q)$, where P and Q are independent normal variables with

$$E(P) = 0, \qquad \operatorname{var}(P) = \sigma_0^2(a_t^2 + b_t^2) = \sigma_1^2, \quad \text{say},$$

$$E(Q) = c', \qquad \operatorname{var}(Q) = \sigma_1^2,$$

where $c' = a_t^2 + b_t^2$.

The joint distribution of P, Q has density function

$$\frac{1}{2\pi\sigma_1^2}\exp\left(-\frac{1}{2\sigma_1^2}\{(Q - c')^2 + P^2\}\right).$$

Write

$$P = \sigma_1 r \sin\theta, \qquad 0 \leqslant r < \infty,$$

$$Q = \sigma_1 r \cos\theta, \qquad -\pi \leqslant \theta \leqslant \pi.$$

If we plot the points ψ_t on a cylinder then we may define the range of ψ_t to be *any* interval of length 2π. For convenience, we now take ψ_t to be defined on the range $(-\pi + \alpha)$ to $(\pi + \alpha)$, where $\alpha = \tan^{-1}(a_t/b_t)$ $(-\pi \leqslant \alpha \leqslant \pi)$, so that θ has the range $-\pi \leqslant \theta \leqslant \pi$. The joint distribution of (r, θ) now has density function

$$(1/2\pi)r[\exp(-\tfrac{1}{2}\{r^2 - 2rc\cos\theta + c^2\})],$$

where $c = c'/\sigma_1 = (a_t^2 + b_t^2)^{1/2}/\sigma_0$. Hence, the distribution of θ has density function

$$p(\theta) = \frac{e^{-c^2/2}}{2\pi}\int_0^\infty r\exp(\{-\tfrac{1}{2}r^2 + rc\cos\theta\})\,dr,$$

or

$$p(\theta) = \frac{e^{-c^2/2}}{2\pi}[1 + \exp(\tfrac{1}{2}c^2\cos^2\theta)\sqrt{2\pi}\,c\cos\theta\,\Phi(c\cos\theta)] \tag{11.2.77}$$

where Φ denotes the distribution function of $N(0, 1)$.

It follows immediately that

$$E(\theta) = 0,$$

so that

$$E(\psi_t) = \alpha = \tan^{-1}(a_t/b_t) = (\omega_0 - \omega_t)t - \phi_1. \qquad (11.2.78)$$

It does not seem possible to evaluate the higher moments of θ analytically, but we may observe that as $c \to \infty$ the distribution of θ becomes concentrated round $\theta = 0$ and tends to normality, i.e. its distribution takes the form

$$p(\theta) \sim \frac{c}{\sqrt{2\pi}} \exp(-\tfrac{1}{2}c^2 \sin^2 \theta) \sim \frac{c}{\sqrt{2\pi}} \exp(-\tfrac{1}{2}c^2 \theta^2).$$

Hence, as $c \to \infty$, $\operatorname{var}(\theta) \sim 1/c^2$.

For small c we may expand $p(\theta)$ in the form

$$p(\theta) = \frac{e^{-c^2/2}}{2\pi} \left[1 + \frac{(c \cos \theta)^2}{1} + \frac{(c \cos \theta)^4}{1 \cdot 3} + \frac{(c \cos \theta)^6}{1 \cdot 3 \cdot 5} + \dots \right.$$

$$+ \tfrac{1}{2} \left\{ c \cos \theta + \tfrac{1}{2} (c \cos \theta)^3 + \frac{1}{1 \cdot 2} \cdot \frac{1}{2^2} (c \cos \theta)^5 \right.$$

$$\left. \left. + \frac{1}{1 \cdot 2 \cdot 3} \cdot \frac{1}{2^3} (c \cos \theta)^7 + \dots \right\} \right]. \qquad (11.2.79)$$

For small values of c the above series may be used to compute the moments of θ numerically.

11.3 SOME OTHER DEFINITIONS OF TIME-DEPENDENT "SPECTRA"

Apart from the theory of evolutionary spectra various other definitions of time-dependent "spectra" for non-stationary processes have been discussed in the literature. We briefly review some of these below.

I. *Instantaneous power spectra*

The first attempt to define time-dependent spectra is due to Page (1952), who, in a pioneering paper, introduced the term "*instantaneous power spectra*". Page first introduces the quantity

$$g_T^*(\omega) = \left| \int_0^T X(t) e^{-i\omega t} dt \right|^2 \qquad (11.3.1)$$

$(g_T^*(\omega)$ is, in fact, proportional to the periodogram of the observations on the interval $(0, T)$—cf. (6.2.186)), and defines the "spectrum" $f^*(\omega)$ of the process as

$$f^*(\omega) = \lim_{T \to \infty} E\{g_T^*(\omega)\}. \qquad (11.3.2)$$

He then defines the "instantaneous power spectrum", $\rho_t(\omega)$, by writing

$$E\{g_T^*(\omega)\} = \int_0^T \rho_t(\omega) \, dt,$$

so that

$$\rho_t(\omega) = \frac{d}{dt} E\{g_t^*(\omega)\}, \qquad (11.3.3)$$

and

$$f^*(\omega) = \int_0^\infty \rho_t(\omega) \, d\omega.$$

Thus, the instantaneous power spectrum $\{\rho_t(\omega) \, dt\}$ represents, roughly speaking, the difference between the power distributions of the process over the interval $(0, t)$ and over the interval $(0, t + dt)$. That is, if we take one record of $X(t)$ over the interval $(0, t)$ and another record over the interval $(0, t + dt)$, pass each through an ideal wave-analyser, and measure the difference between these two power spectra, then the expected value of this difference at frequency ω corresponds to $\{\rho_t(\omega) \, dt\}$. However, this is, in general, quite different from the evolutionary spectrum which represents the power distribution of the process *within* the interval $(t, t + dt)$, i.e. the evolutionary spectrum corresponds roughly to what we would measure if we took just that portion of $X(t)$ lying between t and $t + dt$ and passed it through a wave-analyser. (In the stationary case both definitions reduce to the classical definition of the spectrum.) From the point of view of physical interpretation, we feel that the evolutionary spectrum is the more relevant quantity. An interesting discussion of "instantaneous power spectra" is given by Kharkevich (1960).

II. *The Wigner spectrum*

Let $X(t)$ be a continuous parameter non-stationary process, with covariance kernel $R(s, t)$. As mentioned in Section 11.1, the Wigner spectrum $\psi_t(\omega)$ is defined as the Fourier transform of $R(t - \tau/2, t + \tau/2)$, regarded as a function of τ with t fixed. Thus, $\psi_t(\omega)$ is given by (cf.

(11.1.2)),

$$\psi_t(\omega) = \frac{1}{2\pi} \int_{-\infty}^{\infty} R(t - \tau/2, t + \tau/2) e^{-i\omega\tau} d\tau. \qquad (11.3.4)$$

The function $\psi_t(\omega)$ clearly reduces to the classical definition of the spectrum when the process is stationary, and its mathematical form bears a superficial resemblance to the classical definition. However, as previously noted, $\psi_t(\omega)$ lacks completely a meaningful physical interpretation. This definition of a time-dependent spectrum was discussed by Mark (1970), who pointed out that, e.g., $\psi_t(\omega)$ may take negative values for certain processes. Mark therefore introduced a second type of "spectrum", called the *"physical spectrum"*, $S(\omega, t, W)$, defined as follows. Let $W(t)$ be a suitable real valued function such that $W(0) > 0$, $W(t)$ is "small" outside the neighbourhood of $t = 0$, and $\int_{\infty}^{\infty} W^2(t) dt = 1$. Then $S(\omega, t, W)$ is defined by

$$S(\omega, t, W) = E\left[\left|\int_{-\infty}^{\infty} W(t - u)X(u) e^{-i\omega u} du\right|^2\right]. \qquad (11.3.5)$$

Unfortunately, $S(\omega, t, W)$ is of little use as a *definition* of a time dependent spectrum since it involves an arbitrary function $W(t)$. Thus, each possible choice of $W(t)$ leads to a different expression for $S(\omega, t, W)$, and no particular $S(\omega, t, W)$ is characteristic purely of the process $X(t)$. However, suppose we consider an "idealized" form of $S(\omega, t, W)$ in which the function $W(t)$ becomes in the limit a δ-function. In this case $S(\omega, t, W)$ would describe the behaviour of $X(t)$ at a single time point and would not be "contaminated" by the behaviour of $X(t)$ at other time points. But it is impossible to achieve this type of "idealized" spectrum by considering a limiting form of $S(\omega, t, W)$ since, as $W(t) \to \delta(t)$, the transfer function of the filter corresponding to $W(t)$ becomes uniform, and $S(\omega, t, W) \to E[X^2(t)]$ and so has lost all dependence on the frequency variable ω. On the other hand, if we try to design a filter which has perfect frequency selectivity, then its transfer function must have zero bandwidth and consequently its impulse response function will never die out. This, in turn, means that the corresponding expression for $S(\omega, t, W)$ will now depend on the behaviour of $X(t)$ over all time, and will not, therefore, describe the spectral properties of the process in the neighbourhood of the specific time point t. In other words, $S(\omega, t, W)$ will have lost all form of time dependence. This feature is, of course, a consequence of the basic *"uncertainty principle"* which tells us that we cannot obtain simultaneously an arbitrarily high degree of resolution in both the time and frequency domains.

Nevertheless, if $W(t)$ is not chosen to correspond to either of the two extreme forms, an expression of the form (11.3.5) (modified so that the expectation is replaced by a "time domain" average) can provide a very useful *estimate* of the "idealized" spectrum. In fact, if we identify the function $W(t)$ with the function $g(u)$ in (11.2.41), then we see at once that $S(\omega, t, W)$ *is simply the expression* $E[|U(t)|^2]$ *and by* (11.2.42) *this is just a "smoothed" version of the evolutionary spectral density function,* $h_t(\omega)$. Moreover, if we replace the expectation in (11.3.5) by a time domain average, then $S(\omega, t, W)$ becomes identical with the evolutionary spectral estimate $V(t)$ as given by (11.2.47) (see Priestley (1971c)). For further discussion of the Wigner spectrum, see De Brujin (1967).

III. *Tjøstheim's approach*

Tjøstheim (1976a) has proposed a definition of a time-dependent spectrum based on the representation of a process in terms of its "innovations". Let $\{X_t\}$ be a discrete parameter non-stationary process; Cramer (1961) has shown that if $\{X_t\}$ is "purely non-deterministic" (in the sense of Section 10.1.5) it possesses a one-sided linear representation of the form

$$X_t = \sum_{u=0}^{\infty} a_t(u)\varepsilon_{t-u}, \tag{11.3.6}$$

where $\{\varepsilon_t\}$ is a stationary white noise process, with variance σ_ε^2, say. (Note that since X_t is non-stationary the coefficients $a_t(u)$ in (11.3.6) will be time-dependent.) Tjøstheim now defines the "spectrum at time t" by

$$p_t(\omega) = \frac{\sigma_\varepsilon^2}{2\pi} \left| \sum_{u=0}^{\infty} a_t(u)\, e^{-i\omega u} \right|^2. \tag{11.3.7}$$

Comparing (11.3.6) with the discrete parameter form of (11.2.15) we see that $p_t(\omega)$ is, *in fact, exactly the same as the evolutionary spectrum of* $\{X_t\}$ *defined with respect to the family* $\{A_t^{(1)}(\omega) = \sum_{u=0}^{\infty} a_t(u)\, e^{-i\omega u}\}$ (and with $d\mu(\omega) \propto d\omega$). Thus, Tjøstheim's definition is a special case of the evolutionary spectrum definition, the emphasis being placed on the family of functions generated by the coefficients in (11.3.6). This approach leads to a uniquely defined time-dependent spectrum, and is certainly attractive from a theoretical point of view. However, the major drawback with this definition lies in the fact that unless the family $\{A_t^{(1)}(\omega)\}$ is *oscillatory* the variable ω in (11.3.7) will not possess a physical interpretation in terms of *"frequency"*, and consequently $p_t(\omega)$ will not represent a "power/frequency" distribution. Also, the estimation procedure described in Section 11.2.1 leads inevitably to the evolutionary spectrum defined with respect to the family $\mathcal{F}^*$ of maximum characteristic width. It is difficult to see, therefore, how one could estimate $p_t(\omega)$ empirically from quadratic functions of the observations—unless, of course, one had prior knowledge of the form of the $\{a_t(u)\}$ sequence.

11.4 PREDICTION, FILTERING AND CONTROL FOR NON-STATIONARY PROCESSES

In our introduction to evolutionary spectra we mentioned that it was possible to develop a theory of linear prediction and filtering for non-stationary processes in which the evolutionary spectrum plays almost the same role as does the classical spectrum in the Kolmogorov–Wiener theory for stationary processes. In this section we sketch the main ideas involved in this extension of prediction and filtering theory; a more detailed account is given in Abdrabbo and Priestley (1967, 1969) and Priestley (1971b).

I. *Least square prediction*

Let $\{X_t\}$ be a descrete parameter non-stationary process. We wish to construct a predictor $\hat{X}_{t+m}$ for $X_{t+m}(m>0)$, based on that linear function of $X_t, X_{t-1}, \ldots,$ which minimizes the mean square error,

$$\mathcal{M}(m) = E[\{X_{t+m} - \tilde{X}_{t+m}\}^2]. \qquad (11.4.1)$$

As in the stationary case, the solution may be derived either in the time domain or in the frequency domain, and below we briefly indicate the frequency domain solution. (Continuous parameter processes may be treated similarly.)

Assume that $\{X_t\}$ is an oscillatory process whose evolutionary spectrum is absolutely continuous, with evolutionary spectral density function $h_t(\omega)$. Assume further that, for each t, $h_t(\omega)$ has the canonical factorization

$$h_t(\omega) = |\alpha_t(e^{-i\omega})|^2, \qquad (11.4.2)$$

where the function $\alpha_t(z)$ has no poles or zeros inside the unit circle, $|z| \leq 1$. Then $\alpha_t(e^{-i\omega})$ may be written as a one-sided Fourier series

$$\alpha_t(e^{-i\omega}) = \sum_{u=0}^{\infty} k_t(u)\, e^{-i\omega u}. \qquad (11.4.3)$$

Note that a necessary condition for the validity of (11.4.2) is that

$$\int_{-\pi}^{\pi} \log\{h_t(\omega)\}\, d\omega > -\infty, \qquad \text{all } t. \qquad (11.4.4)$$

We may now re-write the spectral representation (11.2.10) as

$$X_t = \int_{-\pi}^{\pi} e^{it\omega}\alpha_t(\omega)\, dz(\omega), \qquad (11.4.5)$$

where $z(\omega)$ is an orthogonal process with $E[|dz(\omega)|^2] = d\omega$. Let the predictor of X_{t+m} be

$$\tilde{X}_{t+m} = \sum_{u=0}^{\infty} b_t(u)X_{t-u}. \qquad (11.4.6)$$

Using (11.4.5), $\tilde{X}_{t+m}$ may be written as

$$\tilde{X}_{t+m} = \int_{-\pi}^{\pi} e^{it\omega}B_t(\omega)\, dz(\omega),$$

where

$$B_t(\omega) = \sum_{u=0}^{\infty} b_t(u)\alpha_{t-u}(\omega)\, e^{-i\omega u}. \qquad (11.4.7)$$

We also have

$$X_{t+m} = \int_{-\pi}^{\pi} e^{i(t+m)\omega}\alpha_{t+m}(\omega)\, dz(\omega),$$

so that, using the orthogonality of $z(\omega)$,

$$\mathcal{M}(m) = \int_{-\pi}^{\pi} |e^{im\omega}\alpha_{t+m}(\omega) - B_t(\omega)|^2\, d\omega. \qquad (11.4.8)$$

We must now choose $b_t(u)$ or equivalently $B_t(\omega)$, so as to minimize (11.4.8). The solution is obtained by first showing that, for each t, $B_t(\omega)$ is a "backward transform". We then use the Wiener–Hopf technique to decompose $\{e^{i\omega m}\alpha_{t+m}(\omega)\}$ into the sum of its "backward" and "forward" transforms. Specifically, we write, for each t,

$$e^{i\omega m}\alpha_{t+m}(\omega) = C_t^{(1)}(\omega) + C_t^{(2)}(\omega),$$

where

$$C_t^{(1)}(\omega) = \frac{1}{2\pi} \sum_{u=0}^{\infty} e^{-i\omega u}\left[\int_{-\pi}^{\pi} e^{i\theta u}\{e^{i\theta m}\alpha_{t+m}(\theta)\}\, d\theta\right], \qquad (11.4.9)$$

$$C_t^{(2)}(\omega) = \frac{1}{2\pi} \sum_{u=-\infty}^{-1} e^{-i\omega u}\left[\int_{-\pi}^{\pi} e^{i\theta u}\{e^{i\theta m}\alpha_{t+m}(\theta)\}\, d\theta\right]. \qquad (11.4.10)$$

Then (11.4.8) becomes

$$\mathcal{M}(m) = \int_{-\pi}^{\pi} |C_t^{(1)}(\omega) - B_t(\omega)|^2 \, d\omega + \int_{-\pi}^{\pi} |C_t^{(2)}(\omega)|^2 \, d\omega.$$

It follows immediately that $\mathcal{M}(m)$ is minimized by setting, for each t,

$$B_t(\omega) = C_t^{(1)}(\omega), \qquad\qquad (11.4.11)$$

and the prediction variance is then

$$[\mathcal{M}(m)]_{\min} = \int_{-\pi}^{\pi} |C_t^{(2)}(\omega)|^2 \, d\omega.$$

Taking Fourier transforms of both sides of (11.4.11) it may be shown (see Abdrabbo and Priestley (1967)) that we obtain the following equation for $\{b_t(u)\}$,

$$\sum_{u=0}^{v} b_t(u) k_{t-u}(v-u) = k_{t+m}(v+m), \qquad v = 0, 1, 2, \ldots, \quad (11.4.12)$$

where $k_t(u)$ is the sequence defined by (11.4.3). Thus, given the form of the evolutionary spectral density function, $h_t(\omega)$, for all t, we may determine the form of the function $k_t(u)$, for all t, and hence find $b_t(u)$ from (11.4.12). The system of equations (11.4.12) has a "triangular" form, and is easily solved for $\{b_t(u)\}$ by repeated back-substitution.

II. *Linear filtering*

Here, we observe the process

$$Y_t = X_t + N_t \qquad\qquad (11.4.13)$$

up to time t, and wish to "estimate" X_{t+m} by a linear function, $\tilde{X}_{t+m}$, of $Y_t, Y_{t-1}, Y_{t-2}, \ldots$ which minimizes the mean square error

$$\mathcal{M}(m) = E[\{X_{t+m} - \tilde{X}_{t+m}\}^2]. \qquad\qquad (11.4.14)$$

Assume that X_t and N_t are uncorrelated processes, that each is oscillatory, and that they admit evolutionary spectral representations with respect to the same family, $\mathcal{F} = \{e^{i\omega t} A_t(\omega)\}$, with associated measures $d\mu_X(\omega)$, $d\mu_N(\omega)$, each of which is absolutely continuous with respect to $d\omega$. Then we may write X_t as

$$X_t = \int_{-\pi}^{\pi} e^{it\omega} A_t(\omega) \, dZ_X(\omega), \qquad\qquad (11.4.15)$$

where $E[|dZ_X(\omega)|^2] = d\mu_X(\omega)$, and similarly Y_t may be written

$$Y_t = \int_{-\pi}^{\pi} e^{it\omega} A_t(\omega) \, dZ_Y(\omega), \qquad (11.4.16)$$

where $E[|dZ_Y(\omega)|^2] = d\mu_Y(\omega) \doteq d\mu_X(\omega) + d\mu_N(\omega)$. Now let

$$\tilde{X}_{t+m} = \sum_{u=0}^{\infty} b_t(u) Y_{t-u}. \qquad (11.4.17)$$

Using (11.4.16), $\tilde{X}_{t+m}$ can be expressed in the form

$$\tilde{X}_{t+m} = \int_{-\pi}^{\pi} e^{i\omega t} \beta_t(\omega) \, dZ_Y(\omega), \qquad (11.4.18)$$

where

$$\beta_t(\omega) = \sum_{u=0}^{\infty} b_t(u) A_{t-u}(\omega) e^{-i\omega u},$$

while from (11.4.15),

$$X_{t+m} = \int_{-\pi}^{\pi} e^{i\omega(t+m)} A_{t+m}(\omega) \, dZ_X(\omega). \qquad (11.4.19)$$

Substituting (11.4.18), (11.4.19) in (11.4.14) we find

$$\mathcal{M}(m) = E\left[\left|\int_{-\pi}^{\pi} e^{i\omega(t+m)} A_{t+m}(\omega) \, dZ_X(\omega) - \int_{-\pi}^{\pi} e^{i\omega t} \beta_t(\omega) \, dZ_Y(\omega)\right|^2\right].$$

$$(11.4.20)$$

We now assume further that the evolutionary spectral density functions of X_t, Y_t both admit canonical factorizations of the form

$$h_{X,t}(\omega) = |A_t(\omega)|^2 (d\mu_X/d\omega) = |\alpha_t^{(X)}(e^{-i\omega})|^2,$$

$$h_{Y,t}(\omega) = |A_t(\omega)|^2 (d\mu_Y/d\omega) = |\alpha_t^{(Y)}(e^{-i\omega})|^2,$$

say, where both $\alpha_t^{(X)}(e^{-i\omega})$, $\alpha_t^{(Y)}(e^{-i\omega})$ have one-sided Fourier series. Then we may re-write (11.4.20) as

$$\mathcal{M}(m) = \int_{-\pi}^{\pi} |e^{im\omega} \alpha_{t+m}^{(0)}(\omega) - B_t(\omega)|^2 \, d\omega + \int_{-\pi}^{\pi} h_{X,t+m}(\omega)\left\{1 - \left(\frac{d\mu_X/d\omega}{d\mu_Y/d\omega}\right)\right\} d\omega,$$

$$(11.4.21)$$

where

$$\alpha_t^{(0)}(\omega) = h_{X,t}(\omega)/\alpha_t^{(X)}(e^{i\omega}),$$

and

$$B_t(\omega) = \sum_{u=0}^{\infty} b_t(u)\alpha_{t-u}^{(X)} e^{-i\omega u}.$$

We now have to choose $b_t(u)$, or equivalently $B_t(\omega)$, so as to minimize (11.4.21). As in the case of pure prediction we can show that, for each t, $B_t(\omega)$ is a "backward transform", and thus the form of $B_t(\omega)$ which minimizes (11.4.21) can again be found via the Wiener–Hopf technique. Thus, for each t we write

$$e^{im\omega}\alpha_{t+m}^{(0)}(\omega) = C_t^{(1)}(\omega) + C_t^{(2)}(\omega),$$

$(C_t^{(1)}(\omega)$, $C_t^{(2)}(\omega)$ being given by expressions similar to (11.4.9), (11.4.10)) and (11.4.21) now becomes

$$\mathcal{M}(m) = \int_{-\pi}^{\pi} |C_t^{(1)}(\omega) - B_t(\omega)|^2 \, d\omega + \int_{-\pi}^{\pi} |C_t^{(2)}(\omega)|^2 \, d\omega$$

$$+ \int_{-\pi}^{\pi} h_{X,t+m}(\omega)\left\{1 - \frac{d\mu_X/d\omega}{d\mu_Y/d\omega}\right\} d\omega. \qquad (11.4.22)$$

The above expression is clearly a minimum when

$$B_t(\omega) = C_t^{(1)}(\omega). \qquad (11.4.23)$$

Taking Fourier transforms of both sides of (11.4.23) we may show (see Abdrabbo and Priestley (1969)) that

$$\sum_{u=0}^{v} b_t(u)k_{t-u}^{(Y)}(v-u) = l_t(v), \qquad v = 0, 1, 2, \ldots, \qquad (11.4.24)$$

where the $\{k_t^{(Y)}(u)\}$ are the Fourier coefficients of $\alpha_t^{(Y)}(e^{-i\omega})$, and the $\{l_t(v)\}$ are the Fourier coefficients of $C_t^{(1)}(\omega)$. As in the prediction problem, the set of equations (11.4.24) is of "triangular" form and may be solved iteratively.

   With the above choice for $B_t(\omega)$ (equation (11.4.23)), the minimum mean square filtering error is

$$[\mathcal{M}(m)]_{\min} = \int_{-\pi}^{\pi} |C_t^{(2)}(\omega)|^2 \, d\omega + \int_{-\pi}^{\pi} h_{X,t+m}(\omega)\left\{1 - \frac{d\mu_X/d\omega}{d\mu_Y/d\omega}\right\} d\omega. \quad (11.4.25)$$

Since $\mu_Y = \mu_X + \mu_N$, the second term is (11.4.25) may be expressed in the alternative form

$$\int_{-\pi}^{\pi} \{h_{X,t+m}(\omega)h_{N,t+m}(\omega)\}/\{h_{X,t+m}(\omega) + h_{N,t+m}(\omega)\} \, d\omega. \quad (11.4.26)$$

This result is completely analogous to that obtained in the stationary case (compare (11.4.26) with (10.3.17)), with "stationary" spectra replaced by evolutionary spectra. (Note that in (11.4.36) we have, for convenience, expressed all evolutionary spectra in terms of their values at time $(t+m)$. However, apart from the factor in the numerator we could replace $(t+m)$ by any other value of t, say $t=0$.) We have already observed that (10.3.17) represents the minimum mean square error for the case where one observes Y_t for all t (in which case, $B_t(\omega)$ need no longer be restricted to the class of "backward" transforms), and that consequently it is impossible to recover X_t with perfect precision unless the "signal" and "noise" have non-overlapping spectra. It is interesting to note that a corresponding results holds for non-stationary processes, i.e. the expression (11.4.26) vanishes only when the "signal" and "noise" have *non-overlapping evolutionary spectra*. However, we do not require the two sets of evolutionary spectra to be non-overlapping at *all* time instants; in view of the remarks noted above, (11.4.26) will vanish if, for each t, $h_{X,t}(\omega)$ does not overlap with (say) $h_{N,0}(\omega)$. This type of result is obviously a consequence of our initial assumption that X_t and N_t admit representations with respect to the same family of functions, but it seems reasonable to suppose that in a more general situation the expression (11.4.26) would still represent the mean square error for infinite realizations, and that consequently X_t could be determined with perfect precision provided $h_{X,t}(\omega)$ and $h_{N,t}(\omega)$ are non-overlapping for each value of t.

III. *Linear stochastic control*

Once we are able to predict non-stationary processes we may apply the results to control problems in which the "noise" has a general non-stat onary character. Consider the control systems described in Section 10.4.1 (Fig. 10.5), and we will suppose, as before, that the system has linear dynamics with a "dead time" of d units between input and output. Denoting the system's transfer function by $A(e^{-i\omega})$, we saw in Section 10.4.1 that the minimum variance feedback controller is (cf. (10.4.10)),

$$U_t = -\{B^d A^{-1}(B)\}\tilde{N}_{t+d}, \tag{11.4.27}$$

where $\tilde{N}_{t+d}$ is the linear least squares predictor of N_{t+d}, given N_t, N_{t-1}, N_{t-2}, ... Box and Jenkins (1970) have suggested that the noise distribution, N_t, may be described by an ARIMA model of the form

$$\alpha(B)(1-B)^D N_t = \beta(B)\varepsilon_t, \tag{11.4.28}$$

where ε_t is a white noise process, and $\alpha(z)$, $\beta(z)$ are the polynomials

$$\alpha(z) = 1 + \alpha_1 z + \ldots + \alpha_k z^k,$$

$$\beta(z) = 1 + \beta_1 z + \ldots + \beta_l z^l.$$

Here, N_t may be thought of as being generated by passing white noise through an "unstable" filter with transfer function, $\beta(z)/\{\alpha(z)(1-z)^D\}$, having a Dth order pole at $z = 1$. Thus, N_t will be a non-stationary process, but this non-stationary behaviour is of a rather restricted type in that it is controlled by a set of constant (i.e. time invariant) parameters. Thus, if we use the model (11.4.28) and estimate its parameters from an initial length of noise record, all our prediction formulae for the future behaviour of the noise are determined completely by the form of the initial observed record. This demands considerable faith in the assumptions that the above model with the initially fitted parameters will adequately describe the noise process at all future times. To free ourselves from this restriction we may consider a more general model in which the parameters are *time-dependent*, so that the character of the non-stationarity may be quite general. The generalized model may be written as

$$\alpha_t(B)(1-B)^D N_t = \beta_t(B)\varepsilon_t, \tag{11.4.29}$$

where $\alpha_t(z)$, $\beta_t(z)$ denote the polynomials,

$$\alpha_t(z) = 1 + \alpha_{1,t} z + \ldots + \alpha_{k,t} z^k,$$

$$\beta_t(z) = 1 + \beta_{1,t} z + \ldots + \beta_{l,t} z^l,$$

and, for each t, the roots of $\alpha_t(z)$ and $\beta_t(z)$ lie outside the unit circle. Prediction formulae for the model (11.4.29) may be derived by the general method described above, and estimates of the parameters $(\alpha_{1,t} \ldots, \alpha_{k,t}; \beta_{1,t} \ldots, \beta_{l,t})$ can be found by fitting a rational function to the estimated evolutionary spectra of N_t (see Priestley (1969b)). Using this approach we may effect a form of "*parameter adaptive control*", in the sense that the controller sets the input in accordance with the *current* form of the noise process, and will automatically adjust to changes in the noise structure. Thus, at each instant of time, the input is adjusted so as to be optimal for the *local* behaviour of the noise process.

To illustrate this approach, consider the case where N_t conforms to the following model,

$$\Delta^2 N_t = \varepsilon_t + c_{1,t}\varepsilon_{t-1} + c_{2,t}\varepsilon_{t-2} + c_{3,t}\varepsilon_{t-3}, \tag{11.4.30}$$

so that in terms of the model (11.4.29)

$$D = 2; \quad \alpha_t(z) \equiv 1, \quad \text{all } t; \quad \beta_t(z) = 1 + c_{1,t} z + c_{2,t} z^2 + c_{3,t} z^3.$$

Suppose that the system has unit time delay (i.e. $d = 1$); then we require $\tilde{N}_{t+1}$. It may be shown (Priestley (1969b)) that

$$\Delta\tilde{N}_{t+1} = \Delta\{\gamma_{-1}(t)\varepsilon_t\} + \gamma_0(t)\varepsilon_t + S\{\gamma_1(t)\varepsilon_t\}, \qquad (11.4.31)$$

($S = \Delta^{-1}$ denoting the summation operator), where $\gamma_{-1}(t)$, $\gamma_0(t)$, $\gamma_1(t)$ are given by

$$\gamma_{-1}(t) = c_{3,t+3},$$

$$\gamma_0(t) = 1 - 2c_{3,t+3} - c_{2,t+2},$$

$$\gamma_1(t) = c_{1,t+1} + c_{2,t+2} + c_{3,t+3} + 1.$$

The optimum control input is now given by

$$\Delta U_t = -\{BA^{-1}(B)\}\Delta\tilde{N}_{t+1}, \qquad (11.4.32)$$

and with this input, $Y_t = N_t - \tilde{N}_t = \varepsilon_t$, and the above equation may be written

$$\Delta U_t = -\{BA^{-1}(B)\}[\Delta\{\gamma_{-1}(t)Y_t\} + \gamma_0(t)Y_t + S\{\gamma_1(t)Y_t\}]. \quad (11.4.33)$$

This form of control may be described as an *adaptive version of the three-term controller* (10.4.18). The "weights", γ_{-1}, γ_0, γ_1, attached to the "differential", "proportional", and "integral" terms, respectively, are here allowed to vary over time, their values at each time instant being those which are "locally" optimum.

Provided $c_{1,t}$, $c_{2,t}$, $c_{3,t}$ are "smooth" functions of t, so that the family

$$\mathscr{F}_c = \{1 + c_{1,t}e^{-i\omega} + c_{2,t}e^{-2i\omega} + c_{3,t}e^{-3i\omega}\}$$

is *oscillatory*, the evolutionary spectra of $\Delta^2 N_t$ with respect to $\mathscr{F}_c$ are

$$h_t^{(\Delta^2 N)}(\omega) = \frac{\sigma_\varepsilon^2}{2\pi} \cdot |1 + c_{1,t}e^{-i\omega} + c_{2,t}e^{-2i\omega} + c_{3,t}e^{-3i\omega}|^2. \quad (11.4.34)$$

The values of the estimated form of $h_t^{(\Delta^2 N)}(\omega)$ at any three frequencies would be sufficient to obtain crude estimates of $c_{1,t}$, $c_{2,t}$, $c_{3,t}$ but a more efficient procedure is to fit the appropriate model to $\log\{\hat{h}_t^{(\Delta^2 N)}(\omega)\}$ by least squares.

Tong (1974a, 1974b) has discussed the problem of controlling time-dependent linear systems, and Subba Rao and Tong (1972, 1973) have proposed tests for the time-dependence of transfer functions.

11.5 GENERAL NON-LINEAR MODELS

Before considering non-linear models it is natural to enquire as to the degree of generality of linear models as descriptions of stationary processes. Now in Section 10.1.1 we saw that any stationary process X_t with a purely continuous spectrum can be represented in the form (cf. (10.1.97))

$$X_t = \sum_{u=-\infty}^{\infty} g_u \varepsilon_{t-u}, \qquad (11.5.1)$$

where $\{\varepsilon_t\}$ is an uncorrelated process. Moreover, if the spectral density function of X_t satisfies the Paley–Wiener condition (10.1.11), the process $\{\varepsilon_t\}$ can be chosen so that (11.5.1) reduces to the "one-sided" form,

$$X_t = \sum_{u=0}^{\infty} g_u \varepsilon_{t-u}. \qquad (11.5.2)$$

Equation (11.5.2) gives us a representation of X_t as a "general linear process" (in the sense of Section 3.5.7), thus confirming the statement made in Section 3.5.7 that such representations provide a fairly general description for processes with continuous spectra. Assuming that the function $G(z) = \sum_{u=0}^{\infty} g_u z^u$ can be adequately approximated by a rational function, $\alpha(z)/\beta(z)$, (11.5.2) now reduces to the familiar ARMA model,

$$X_t + \alpha_1 X_{t-1} + \ldots + \alpha_k X_{t-k} = \varepsilon_t + \beta_1 \varepsilon_{t-1} + \ldots + \beta_l \varepsilon_{t-l} \qquad (11.5.3)$$

in which ε_t is, as above, an uncorrelated process.

Now throughout our previous discussions we have used the term "ARMA model" to describe any relationship of the form (11.5.3) in which the $\{\varepsilon_t\}$ are merely *uncorrelated*, but we must note that although (11.5.3) completely determines the second order properties of X_t (such as its autocovariance and spectral density functions), it certainly does not determine the full probabilistic properties of X_t. To achieve this we must specify the full probabilistic structure of $\{\varepsilon_t\}$—as we would if we said, for example, that the $\{\varepsilon_t\}$ are *independent*. Thus, strictly speaking, we should not refer to (11.5.3) as a "model" for X_t unless the $\{\varepsilon_t\}$ process is fully specified. In the present discussion we adopt the convention that *the term "linear model" means a representation of the form*

$$X_t = \sum_{u=0}^{\infty} g_u e_{t-u}, \qquad (11.5.4)$$

in which $\{e_t\}$ is a sequence of strictly independent random variables. Of course, as far as second order properties are concerned, ε_t and e_t have identical properties; in particular, both processes have uniform ("white") spectra, but they may differ substantially in other respects. However, if X_t is a *Gaussian* process then (assuming that (11.5.2) is invertible) it follows that ε_t is also Gaussian, and being an uncorrelated process it must now be an independent process. Hence, in this special case the distinction between the processes ε_t and e_t disappears, and any stationary Gaussian process with a purely continuous spectrum which satisfies (10.1.11) can always be described by a linear model of the form (11.5.4).

Returning to the general case of non-Gaussian processes, let us now consider the problem of predicting the future value of a process, given observations up to time t. In the case of the strictly independent process, e_t, the past contains no information on the future, and hence the best predictor of a future value of e_t is simply its (unconditional) mean. For the uncorrelated process, ε_t, it is still true that if we restrict attention to *linear* predictors then, in this sense, the past contains no information on the future. However, the past may well contain useful information on future values if we allow predictors which are non-linear functions of the observations. The following example illustrates this point. Let the process η_t be defined by

$$\eta_t = e_t + \beta e_{t-1} e_{t-2}, \tag{11.5.5}$$

where e_t is an independent process with zero mean and constant variance. It is clear that η_t also has zero mean and constant variance, and its autocovariance function is given by

$$E[\eta_t \eta_{t+s}] = E[e_t e_{t+s} + \beta e_{t-1} e_{t-2} e_{t+s}$$
$$+ \beta e_t e_{t+s-1} e_{t+s-2} + \beta^2 e_{t-1} e_{t-2} e_{t+s-1} e_{t+s-2}].$$

For all $s \neq 0$ each of the above terms has zero expectation. (For example, in the second and third terms there is always at least one e whose time point differs from the other two, and a similar argument holds for the fourth term.) Thus, η_t is an uncorrelated process, and, as far as its second order properties are concerned, it behaves just like an independent process. However, given observations up to time t one can clearly construct a non-trivial predictor of η_{t+1}. Specifically, if we adopt the mean square error criterion, the optimal predictor of η_{t+h} is its conditional expectation, i.e.

$$\tilde{\eta}_{t+h} = E[\eta_{t+h} | \eta_t, \eta_{t-1}, \ldots],$$

and for $h = 1$ we find from (11.5.5)

$$\tilde{\eta}_{t+1} = \beta e_t e_{t-1}. \tag{11.5.6}$$

(To express $\tilde{\eta}_{t+1}$ in terms of past η_t's we would, of course, have to invert (11.5.5) to express each e_t as a function of $\eta_t, \eta_{t-1}, \ldots$.) As noted by Granger and Andersen (1976), if a process η_t of the above form was obtained as the "residual" from a more general model, all the conventional tests for "white noise" based on the behaviour of the autocovariance or autocorrelation function would confirm that the residuals were, in fact, white noise, and hence that there was no further model structure left to fit. However, as we have seen, one could certainly exploit the non-linear structure of the η_t process in order to improve the predictors of the original series. This is, of course, the reason why in the Box–Jenkins algorithm for forecasting ARMA models the residuals are always assumed to be strictly *independent*; see Section 10.2.

The above example shows that one can indeed have a stationary process which does not conform to the general linear model (11.5.4) and it follows *a fortiori* that many types of non-stationary processes would also fall outside the domain of linear models. In the following section we discuss general types of non-linear models, and later we consider some special classes of non-linear models, namely the "*bilinear*", "*threshold autoregressive*", and "*exponential autoregressive*" models.

11.5.1 Volterra Expansions and Polyspectra

The first systematic study of non-linear models is due to Wiener (1958) who considered non-linear relationships between an "input" U_t and an "output" X_t of a physical system. Wiener's model takes the form (in the discrete parameter case),

$$X_t = \sum_{u=0}^{\infty} g_u U_{t-u} + \sum_{u=0}^{\infty} \sum_{v=0}^{\infty} g_{uv} U_{t-u} U_{t-v}$$

$$+ \sum_{u=0}^{\infty} \sum_{v=0}^{\infty} \sum_{w=0}^{\infty} g_{uvw} U_{t-u} U_{t-v} U_{t-w} + \dots \qquad (11.5.7)$$

The RHS of (11.5.7) is now usually referred to as a "*Volterra series expansion*" (see Volterra (1959)). Wiener was concerned mainly with the case where both the input and output are observable, but in the context of time series modelling we may take U_t to be an (unobservable) independent process. With $U_t \equiv e_t$, the first term in (11.5.7) is identical with the general linear model (11.5.4), and the successive terms may thus be called the "quadratic", "cubic", ... components. There is a considerable literature on the theoretical properties of these models (see, e.g., the review by Brillinger (1970)), but from the point of view of statistical estimation these models are extremely unwieldy, and in its "raw" state the estimation problem is almost intractable.

The basic difficulty is that (11.5.7) contains far too many parameters, so that we could not possibly hope to estimate these efficiently from a finite set of observations unless we assume either that the sequences $\{g_u\}$, $\{g_{uv}\}$, $\{g_{uvw}\}$, ... possess some form of "smoothness" properties, or that they may be expressed as known functions of some relatively small number of other parameters. In the case of linear models the "smoothness" condition is imposed on the "transfer function" $\Gamma_1(\omega)$ (the Fourier transform of $\{g_u\}$) and if the input U_t is observable, this leads to the estimation of $\Gamma_1(\omega)$ via cross-spectral analysis (see Section 9.2.1). On the other hand, the "finite parameter" approach in the linear case is usually achieved by assuming that $\Gamma_1(\omega)$ is a rational function leading to an ARMA type of model. In the non-linear case there is no single function which characterizes the model in the same way as the transfer function characterizes a linear model. Rather, we have to consider the infinite sequence of "*generalized transfer functions*" defined by

$$\Gamma_1(\omega_1) = \sum_{u=0}^{\infty} g_u e^{-i\omega_1 u}$$

$$\Gamma_2(\omega_1, \omega_2) = \sum_{u=0}^{\infty} \sum_{v=0}^{\infty} g_{uv} e^{-i\omega_1 u - i\omega_2 v} \qquad (11.5.8)$$

$$\Gamma_3(\omega_1, \omega_2, \omega_3) = \sum_{u=0}^{\infty} \sum_{v=0}^{\infty} \sum_{w=0}^{\infty} g_{uvw} e^{-i\omega_1 u - i\omega_2 v - i\omega_3 w}$$

$$\vdots \qquad \vdots$$

(Note that, without loss of generality, we may take each sequence, $g_{uvw}, \ldots$, to be a symmetric function of u, v, w, ...) To gain insight into the physical interpretation of these functions, let us suppose that U_t is stationary with spectral representation

$$U_t = \int_{-\pi}^{\pi} e^{it\omega} dZ_U(\omega).$$

Then (11.5.7) may be re-written as

$$X_t = \int_{-\pi}^{\pi} e^{it\omega_1} \Gamma_1(\omega_1) dZ_U(\omega_1)$$

$$+ \int_{-\pi}^{\pi} \int_{-\pi}^{\pi} e^{it(\omega_1 + \omega_2)} \Gamma_2(\omega_1, \omega_2) dZ_U(\omega_1) dZ_U(\omega_2)$$

$$+ \int_{-\pi}^{\pi} \int_{-\pi}^{\pi} \int_{-\pi}^{\pi} e^{it(\omega_1 + \omega_2 + \omega_3)} \Gamma_3(\omega_1, \omega_2, \omega_3) dZ_U(\omega_1) dZ_U(\omega_2) dZ_U(\omega_3)$$

$$+ \ldots \qquad (11.5.9)$$

Thus, $\Gamma_1(\omega_1)$ has a similar interpretation to that of the conventional transfer function of a linear model, while $\{\Gamma_2(\omega_1, \omega_2) dZ_U(\omega_1) dZ_U(\omega_2)\}$ represents the contribution of the components with frequencies ω_1, ω_2 in U_t to the component with frequency $(\omega_1 + \omega_2)$ in X_t, etc. In particular, if U_t consists of just one frequency component, say $U_t = A e^{i\omega_0 t}$, then

$dZ_U(\omega) = A$, $\omega = \omega_0$, and is zero otherwise, and hence (11.5.9) gives

$$X_t = A\Gamma_1(\omega_0) e^{i\omega_0 t} + A^2\Gamma_2(\omega_0, \omega_0) e^{2i\omega_0 t} + A^3\Gamma_3(\omega_0, \omega_0, \omega_0) e^{3i\omega_0 t} + \ldots,$$

(11.5.10)

i.e. the output X_t consists of components with frequencies ω_0, $2\omega_0$, $3\omega_0$, $\ldots$, as may be verified by substituting $U_t = A e^{i\omega_0 t}$ directly in (11.5.7). (Note the important difference between linear and non-linear systems: in the linear case an input with frequency ω_0 emerges at the output with the same frequency ω_0, whereas in the non-linear case, the output contains components at frequencies ω_0, $2\omega_0$, $3\omega_0$, $\ldots$. This phenomenon of "frequency *multiplication*" is a well known feature of non-linear systems: in the case of audio amplifiers, for example, the components with frequencies $2\omega_0$, $3\omega_0$, $\ldots$ are referred to as "harmonic distortion".) When we think of the analogy between (11.5.7) and a polynomial regression model, it becomes clear that we cannot hope to determine all the generalized transfer functions using only second order properties of U_t and X_t. In fact, if we wish to "fit" the first m terms in (11.5.7) we will need to consider all the joint moments up to order $(m+1)$ of U_t and X_t, and (following the approach used to determine transfer functions of linear models) this suggests that it would be useful to introduce the *Fourier transforms* of these higher order moments.

Polyspectra

These Fourier transforms—or rather, the Fourier transforms of the corresponding higher order cumulants—are called "*polyspectra*", and are defined formally as follows.

Let $\{X_t\}$ be a process stationary up to order k, and let $C(s_1, s_2, \ldots, s_{k-1})$ denote the joint cumulant of order k of the set of random variables $\{X_t, X_{t+s_1}, \ldots, X_{t+s_{k-1}}\}$, i.e. $C(s_1, s_2, \ldots, s_{k-1})$ is the coefficient of $(z_1 z_2 \ldots z_k)$ in the expansion of the cumulant generating function

$$K(z_1, z_2, \ldots, z_k) = \log[E[\exp\{z_1 X_t + z_2 X_{t+s_1} + \ldots + z_k X_{t+s_{k-1}}\}]].$$

(Note that, by the stationarity condition, $C(s_1, s_2, \ldots, s_{k-1})$ does not depend on t.) The *kth order polyspectrum*, (or *kth order cumulant spectrum*) is defined by

$$h_k(\omega_1, \omega_2, \ldots, \omega_{k-1}) =$$

$$\left(\frac{1}{2\pi}\right)^{k-1} \sum_{s_1=-\infty}^{\infty} \cdots \sum_{s_{k-1}=-\infty}^{\infty} C(s_1, \ldots, s_{k-1}) \exp[-i(\omega_1 s_1 + \ldots + \omega_{k-1} s_{k-1})].$$

(11.5.11)

(We are assuming, of course, that the above Fourier transform exists, a sufficient condition being that

$$\sum_{s_1} \cdots \sum_{s_{k-1}} |C(s_1, \ldots s_{k-1})| < \infty.)$$

Since the second order cumulant $C(s_1)$ is simply $\text{cov}\{X_t, X_{t+s_1}\}$, it follows that the second order polyspectrum is exactly the same as the conventional power spectrum, i.e. $h_2(\omega) \equiv h(\omega)$. Also, the third order cumulant $C(s_1, s_2)$ is identical with the third order moment about the mean, i.e.

$$C(s_1, s_2) = E[\{X_t - \mu_X\}\{X_{t+s_1} - \mu_X\}\{X_{t+s_2} - \mu_X\}] = \mu(s_1, s_2),$$

say (where $\mu_X = E\{X_t\}$), and hence the third order polyspectrum may be written

$$h_3(\omega_1, \omega_2) = \left(\frac{1}{2\pi}\right)^2 \sum_{s_1=-\infty}^{\infty} \sum_{s_2=-\infty}^{\infty} \mu(s_1, s_2) e^{-i(\omega_1 s_1 + \omega_2 s_2)}, \quad (11.5.12)$$

and inverting this transform we obtain

$$\mu(s_1, s_2) = \int_{-\pi}^{\pi} \int_{-\pi}^{\pi} e^{i(s_1 \omega_1 + s_2 \omega_2)} h_3(\omega_1, \omega_2) \, d\omega_1 \, d\omega_2. \quad (11.5.13)$$

The function $h_3(\omega_1, \omega_2)$ is sometimes called the "*bispectrum*" (this terminology being due to Tukey (1959)), and the fourth order polyspectrum, $h_4(\omega_1, \omega_2, \omega_3)$, is called the "*trispectrum*". If we write

$$X_t - \mu_X = \int_{-\pi}^{\pi} e^{it\omega} \, dZ_X(\omega),$$

then $\mu(s_1, s_2)$ can be written as

$$\mu(s_1, s_2) = \int_{-\pi}^{\pi} \int_{-\pi}^{\pi} \int_{-\pi}^{\pi} e^{it(\omega_1 + \omega_2 + \omega_3)} e^{i(s_1 \omega_1 + s_2 \omega_2)}$$

$$\times E[dZ_X(\omega_1) \, dZ_X(\omega_2) \, dZ_X(\omega_3)].$$

Since the LHS is a function of s_1, s_2 only, so is the RHS, and it follows that $E[dZ_X(\omega_1) \, dZ_X(\omega_2) \, dZ_X(\omega_3)]$ must vanish except along the line, $\omega_1 + \omega_2 + \omega_3 = 0$. Hence, we may write the bispectrum in the alternative form

$$h_3(\omega_1, \omega_2) \, d\omega_1 \, d\omega_2 = E[dZ_X(\omega_1) \, dZ_X(\omega_2) \, dZ_X(-\omega_1 - \omega_2)]. \quad (11.5.14)$$

The outstanding property of polyspectra is that *all polyspectra of higher order than the second vanish when* $\{X_t\}$ *is a Gaussian process*. This follows immediately from the well known result that all joint cumulants of higher order than the second vanish for multivariate normal distributions. Hence, the bispectrum, trispectrum, and all higher order polyspectra are identically zero if X_t is Gaussian, and these higher order spectra may thus be regarded as measures of the departure of the process from Gaussianity.

Given two process X_t, Y_t (jointly stationary up to the appropriate order) we may similarly define *"cross-polyspectra"* of various orders. For example, the *third order cross-spectrum* is defined by

$$h_{XXY}(\omega_1, \omega_2) = \left(\frac{1}{2\pi}\right)^2 \sum_{s_1=-\infty}^{\infty} \sum_{s_2=-\infty}^{\infty} \mu_{XXY}(s_1, s_2) \, e^{-i(\omega_1 s_1 + \omega_2 s_2)} \quad (11.5.15)$$

where

$$\mu_{XXY}(s_1, s_2) = E[\{Y_t - \mu_Y\}\{X_{t+s_1} - \mu_X\}\{X_{t+s_2} - \mu_X\}].$$

Writing

$$Y_t - \mu_Y = \int_{-\pi}^{\pi} e^{it\omega} \, dZ_Y(\omega),$$

a similar argument to that used above shows that $h_{XXY}(\omega_1, \omega_2)$ can be written as

$$h_{XXY}(\omega_1, \omega_2) \, d\omega_1 \, d\omega_2 = E[dZ_X(\omega_1) \, dZ_X(\omega_2) \, dZ_Y(-\omega_1 - \omega_2)]. \quad (11.5.16)$$

The second order cross-polyspectrum is, of course, identical with the conventional cross-spectra density function between Y_t and X_t.

The idea of constructing the Fourier transforms of the higher order cumulants was suggested by Kolmogorov, and polyspectra were introduced by Shiryaev (1960). Brillinger (1965) and Brillinger and Rosenblatt (1967) have given a comprehensive treatment of the theoretical properties of polyspectra, and have discussed also the estimation of polyspectra from sample records. (These estimation procedures are based on a generalization of the "window" technique applied to products of the finite Fourier transform of the data, and follow, in effect, the sample analogue of, e.g., (11.5.14).) Bispectra are discussed by Tukey (1959) and Akaike (1966), and an application of bispectral analysis to the study of ocean waves is described by Hasselman, Munk and MacDonald (1963), an application to tides by Cartwright (1968), and applications to turbulence by Lii, Rosenblatt, and Van Atta (1976), and Helland, Lii, and Rosenblatt (1979).

In the special case where the RHS of (11.5.7) contains just one term—say the term of order m—and U_t is Gaussian, we may determine the mth order generalized transfer function in terms of the $(m+1)$th order cross-polyspectrum between U_t and X_t. For example, suppose that (11.5.7) consists of just the quadratic term, with the possible addition of a stationary "noise" term. Then we may write

$$X_t = \sum_{u=0}^{\infty} \sum_{v=0}^{\infty} g_{uv} U_{t-u} U_{t-v} + N_t,$$

where we may assume that $E(U_t) = E(N_t) = 0$ and N_t is independent of U_t. We now have,

$$X_t = \left\{ \int_{-\pi}^{\pi} \int_{-\pi}^{\pi} e^{it(\omega_1 + \omega_2)} \Gamma_2(\omega_1, \omega_2) \, dZ_U(\omega_1) \, dZ_U(\omega_2) \right\} + N_t,$$

so that

$$E[X_t U_{t+s_1} U_{t+s_2}] =$$

$$\int_{-\pi}^{\pi}\!\!\!\int\!\!\!\int\!\!\!\int \exp\{it(\omega_1 + \omega_2) + i(t + s_1)\omega_3 + i(t + s_2)\omega_4\} \Gamma_2(\omega_1, \omega_2)$$

$$\times E[dZ_U(\omega_1) \, dZ_U(\omega_2) \, dZ_U(\omega_3) \, dZ_U(\omega_4)].$$

But, since $dZ_U(\omega)$ is Gaussian, the expectation of the above quadruple product can be written

$$E[dZ_U(\omega_1) \, dZ_U(\omega_2)]E[dZ_U(\omega_3) \, dZ_U(\omega_4)]$$

$$+ E[dZ_U(\omega_1) \, dZ_U(\omega_3)]E[dZ_U(\omega_2) \, dZ_U(\omega_4)]$$

$$+ E[dZ_U(\omega_1) \, dZ_U(\omega_4)]E[dZ_U(\omega_2) \, dZ_U(\omega_3)].$$

Substituting this expression in the multiple integral, and recalling that, e.g.,

$$E[dZ_U(\omega_1) \, dZ_U(\omega_2)] = 0, \qquad \omega_1 \neq -\omega_2;$$

$$E[dZ_U(\omega_1) \, dZ_U(-\omega_1)] = h_U(\omega_1) \, d\omega_1,$$

$h_U(\omega)$ being the (second order) spectral density function of U_t, we obtain,

$$E[X_t U_{t+s_1} U_{t+s_2}] = \int \Gamma_2(\omega_1, -\omega_1) h_U(\omega_1) \, d\omega_1 \int e^{i(s_1 - s_2)\omega_3} h_U(\omega_3) \, d\omega_3$$

$$+ \int\!\!\!\int \Gamma_2(-\omega_3, -\omega_4) \, e^{is_1\omega_3 + is_2\omega_4} h_U(\omega_3) h_U(\omega_4) \, d\omega_3 \, d\omega_4$$

$$+ \int\!\!\!\int \Gamma_2(-\omega_4, -\omega_3) \, e^{is_1\omega_3 + is_2\omega_4} h_U(\omega_3) h_U(\omega_4) \, d\omega_3 \, d\omega_4.$$

$$(11.5.17)$$

We now write

$$\mu_{UUX}(s_1, s_2) = E[X_t U_{t+s_1} U_{t+s_2}] - E[U_{t+s_1} U_{t+s_2}]E[X_t], \quad (11.5.18)$$

and noting that the first term in (11.5.17) is just the second term in (11.5.18), we finally obtain (remembering that, without loss of generality, we may take g_{uv} to be symmetric in u, v so that $\Gamma_2(\omega_1, \omega_2)$ is symmetric in ω_1, ω_2),

$$\mu_{UXX}(s_1, s_2) = 2 \int_{-\pi}^{\pi} \int \Gamma_2(-\omega_3, -\omega_4) e^{is_1\omega_3 + is_2\omega_4} h_U(\omega_3) h_U(\omega_4) \, d\omega_3 \, d\omega_4.$$

It now follows that (writing ω_1 for ω_3 and ω_2 for ω_4),

$$h_{UUX}(\omega_1, \omega_2) = 2\Gamma_2(-\omega_1, -\omega_2) h_U(\omega_1) h_U(\omega_2),$$

or

$$\Gamma_2(\omega_1, \omega_2) = \frac{h_{UUX}(-\omega_1, -\omega_2)}{2h_U(\omega_1) h_U(\omega_2)}. \quad (11.5.19)$$

More generally, if (11.5.7) contains just the mth order term (with additive noise), Shiryaev (1960) has shown that (see also Brillinger (1970)),

$$\Gamma_m(\omega_1, \omega_2, \ldots \omega_m) = \frac{h_{UU\cdots UX}(-\omega_1, -\omega_2, \ldots, -\omega_m)}{m! h_U(\omega_1) h_U(\omega_2) \ldots h_U(\omega_m)}, \quad (11.5.20)$$

where the numerator is the mth order cross polyspectrum between X_t and U_t. This result is a very nice generalization of the classical result for the transfer function of a linear system (see (9.2.19)), to which it reduces when $m = 1$. If (11.5.7) does indeed contain just one term, (11.5.20) enables us to estimate $\Gamma_m(\omega_1, \omega_2, \ldots, \omega_m)$ from a finite number of observations on U_t, X_t by replacing the terms in the numerator and denominator of (11.5.20) by their sample estimates. Unfortunately, (11.5.20) is no longer valid if (11.5.7) contains a mixture of terms of different orders, but we can apply a similar approach if we first adopt Wiener's technique of re-writing the RHS of (11.5.7) as a sum of *orthogonal* terms. This type of orthogonal expansion requires the use of Hermite polynomials (Wiener (1958)) and the "re-parametrized" generalized transfer function can then each be obtained via expressions of the form (11.5.20). (See Brillinger (1970).)

A test for linearity

We have seen that if a process is Gaussian then all its polyspectra (of higher order than the second) are identically zero. Thus, if a process has a non-zero bispectrum this could be due to either,

(i) the process conforms to a linear model of the form (11.5.4) but the $\{e_t\}$ are non-normal; or

(ii) the process conforms to a non-linear model of the form (11.5.7), with $U_t \equiv e_t$ being either normal or non-normal.

Subba Rao and Gabr (1981) have constructed two tests aimed at detecting

(a) whether the process is *Gaussian*—in which case, given that it is stationary with a purely continuous spectrum, it must necessarily conform to a linear model; and

(b) if the process is non-Gaussian, whether it conforms to a *linear* model.

Case (a) is examined by testing the null-hypothesis that the bispectrum is zero at all frequencies; the test statistic is constructed from the values of the estimated bispectrum over a grid of "frequencies", and has a form similar to Hotelling's T^2 statistic. Case (b) is tested by using the result that for the linear model (11.5.4),

$$\mu(s_1, s_2) = \beta \sum_{u=-\infty}^{\infty} g_u g_{u+s_1} g_{u+s_2}, \tag{11.5.21}$$

where $\beta = E(e_t^3)$ is the third moment of the $\{e_t\}$, and $g_u = 0$, $u < 0$. (This result follows immediately from the independence of the $\{e_t\}$.) Hence, for linear models the bispectrum has the form

$$h_3(\omega_1, \omega_2) = \frac{\beta}{4\pi^2} \cdot \Gamma(\omega_1)\Gamma(\omega_2)\Gamma\{-(\omega_1 + \omega_2)\}, \tag{11.5.22}$$

where $\Gamma(\omega) = \sum_{u=0}^{\infty} g_u e^{-i\omega u}$. But the spectral density function of the process is

$$h(\omega) = \frac{\sigma_e^2}{2\pi} |\Gamma(\omega)|^2, \tag{11.5.23}$$

and hence the function

$$T(\omega_1, \omega_2) = \frac{|h_3(\omega_1, \omega_2)|^2}{h(\omega_1)h(\omega_2)h(\omega_1 + \omega_2)} = \text{constant} \qquad (\text{all } \omega_1, \omega_2). \tag{11.5.24}$$

The test for linearity is based on testing the constancy of the sample values of $T(\omega_1, \omega_2)$ (computed by replacing $h_3(\omega_1, \omega_2)$ and $h(\omega)$ by their sample estimates) over a grid of frequencies, and the test statistic again has a form similar to Hotelling's T^2.

These procedures do not, of course, provide complete tests for either Gaussianity or linearity. It is quite possible for a non-Gaussian process to have a zero bispectrum, and a complete test would involve testing *all* the higher order polyspectra. (In fact, it is clear from the above expression for $h_3(\omega_1, \omega_2)$ that this function will vanish for a linear model provided only that e_t has zero third moment.) Similarly, the constancy of $T(\omega_1, \omega_2)$ does not necessarily imply that the process follows a linear model. However, it seems reasonable to suppose that in most practical situations deviations from Gaussianity or linearity would show up in the form of the *bispectrum*, and this is the basis of the above tests.

11.6 SOME SPECIAL NON-LINEAR MODELS: BILINEAR, THRESHOLD, AND EXPONENTIAL MODELS

The principal difficulty in treating non-linear models from a time domain approach has been the lack hitherto of a finite parameter representation of sufficient generality. Recently, however, control theorists have introduced the class of so-called *"bilinear models"* which seem to offer exciting possibilities. A bilinear model is most conveniently described in its state space form, which is typically given by

$$x_{t+1} = Ax_t + B\varepsilon_{t+1} + \varepsilon_t Cx_t, \qquad (11.6.1)$$

$$X_t = Hx_t. \qquad (11.6.2)$$

(We now return to our previous convention of denoting the "residual" process by ε_t, but it should be noted that throughout this section ε_t denotes an *independent* white noise process.) If $C = 0$, (11.6.1) reduces to (10.5.3), the conventional state space description of a linear model, but when C is present the model is non-linear. The bilinear model described above is, in general, a special case of the Volterra series expansion (11.5.7), but it can be shown (Brockett (1976), Sussman (1976)) that, with suitable choices of A, B, C, H, it can, over a finite time interval, approximate with an arbitrary degree of accuracy to any "resonable" model of the form (11.5.7). The outstanding advantage of the bilinear model is that it involves only a finite number of parameters, and hence makes it feasible to consider the problem of fitting such models to data. It is in this sense that the bilinear model may be regarded as a natural non-linear extension of the ARMA type model.

Granger and Andersen (1976, 1978) have considered a more general version of the bilinear scheme which, in "scalar" form, may be expressed as

$$X_t + a_1 X_{t-1} + \ldots + a_k X_{t-k} = \varepsilon_t + b_1 \varepsilon_{t-1} + \ldots + b_l \varepsilon_{t-l} + \sum_{i=1}^{m} \sum_{j=1}^{n} c_{ij} \varepsilon_{t-i} X_{t-j}.$$

(11.6.3)

This is a direct "bilinear" generalization of the linear ARMA model, and clearly (11.6.3) reduced to the ARMA form if $c_{ij} = 0$, all i, j. Although (11.6.3) cannot be put into the simple state space form (11.6.2), it admits the more general state space representation (Granger and Andersen (1978)),

$$x_{t+1} = Ax_t + B\varepsilon_{t+1} + \sum_{i=0}^{m-1} C_{i+1} x_t \varepsilon_{t-i}$$

$$X_t = Hx_t,$$

(11.6.4)

where, writing $p = \max(k, n)$, x_t has dimensions p,

$$A_{p \times p} = \begin{pmatrix} -a_1 & -a_2 & \cdots & -a_p \\ 1 & 0 & \cdots & 0 \\ 0 & 1 & \cdots & 0 \\ \cdots & \cdots & \cdots & \cdots \\ 0 & 0 & 1 & 0 \end{pmatrix} \qquad B_{p \times (l+1)} = \begin{pmatrix} 1 & b_1 & \cdots & b_l \\ 0 & 0 & \cdots & 0 \\ \cdots & \cdots & \cdots & \cdots \\ 0 & 0 & & 0 \end{pmatrix},$$

the $p \times p$ matrix C_i is given by

$$C_i = \begin{pmatrix} c_{i1} & \cdots & c_{ip} \\ 0 & \cdots & 0 \\ 0 & \cdots & 0 \end{pmatrix}, \qquad i = 1, \ldots, m,$$

$$H_{1 \times p} = (1, 0, \ldots 0),$$

and

$$\varepsilon_t = (\varepsilon_t, \varepsilon_{t-1}, \ldots, \varepsilon_{t-l}).$$

(In the above, $a_i = 0$ if $i > k$ and $c_{ij} = 0$ if $i > m$.)

Although bilinear models have been studied in some depth in control theory literature (see, e.g., d'Alessandro (1972), Bruni, di Pillo, and Koch (1974), Mohler (1973), Isidori (1973)) the statistical analysis of these models is still in its infancy. Granger and Andersen (1976, 1978) have considered some special homogeneous models, and Subba Rao (1977, 1981) has discussed the problem of a maximum likelihood estimation for bilinear time series models of specified order. However, the general structural properties of these models (including, in particular, the problem of identifiability) need to be developed further.

In dealing with ARMA models one has, in addition to the time domain description, the equivalent frequency domain characterization, in that the transfer function is known to be a rational function. Thus, one can fit such models either by a "time domain approach" or by a "frequency domain approach". In the frequency domain approach the parameters of the model are estimated by fitting the appropriate rational function to a non-parametric estimate of the transfer function, or, in the "time series" case, by fitting the appropriate rational function to a non-parametric estimate of the spectrum—see, e.g., Hannan (1969b). Before one can generalize the frequency domain approach to bilinear models it is obviously necessary to construct some function of frequency whose form uniquely characterizes the structure of the bilinear model. Now it was noted above that it is not possible to describe the input/output relations of a general non-linear system by a single function of frequency, instead we have the infinite sequence of functions, $\Gamma_1(\omega_1)$, $\Gamma_2(\omega_1, \omega_2)$, $\Gamma_3(\omega_1, \omega_2, \omega_3), \dots$. However, bilinear models involve only a finite number of parameters, and hence for these models it may be possible to construct a single function of frequency which plays a role equivalent to that of the transfer function for linear models. We must note that we cannot, of course, hope to preserve the physical interpretation of the linear transfer function since for non-linear systems, (a) a sine wave input with frequency ω_0 does not emerge at the output simply as sine wave with frequency ω_0, the output containing additional components at frequencies $2\omega_0$, $3\omega_0$, $\dots$, and (b) we lose the principle of "superposition", i.e. if we form a given linear combination of inputs the output is not a linear combination of the corresponding outputs. Thus, we cannot hope to describe a non-linear system purely in terms of its responses to single sine wave inputs of varying frequency, and our objective here is simply to construct a function of frequency which, for a bilinear model, determines the matrices A, B, C, H, and which possesses a physical interpretation so that it can be estimated non-parametrically. The following study of a first order (i.e. scalar) bilinear model (Priestley (1978a, b)) suggests a possible approach to this problem.

Consider the first order model

$$x_{t+1} = ax_t + b\varepsilon_{t+1} + c\varepsilon_t x_t. \tag{11.6.5}$$

Setting $\varepsilon_t = e^{i\omega t}$, it is easy to show that when (11.6.5) is expanded in the form (11.5.7), the "linear" transfer function, $\Gamma_1(\omega_1)$, is given by

$$\Gamma_1(\omega_1) = \frac{b\, e^{i\omega_1}}{(e^{i\omega_1} - a)}. \tag{11.6.6}$$

Similarly, setting $\varepsilon_t = e^{i\omega_1 t} + e^{i\omega_2 t}$, we may show that the "quadratic" transfer function, $\Gamma_2(\omega_1, \omega_2)$, is given by

$$\Gamma_2(\omega_1, \omega_2) = \frac{c}{2\{e^{i(\omega_1 + \omega_2)} - a\}} \{\Gamma_1(\omega_1) + \Gamma_1(\omega_2)\}$$

$$= \frac{\frac{1}{2}cb}{\{e^{i(\omega_1 + \omega_2)} - a\}} \cdot \left\{\frac{e^{i\omega_1}}{e^{i\omega_1} - a} + \frac{e^{i\omega_2}}{e^{i\omega_2} - a}\right\}. \qquad (11.6.7)$$

In particular, along the "diagonal", $\omega_1 = \omega_2 = \omega$,

$$\Gamma_2(\omega, \omega) = \frac{cb\, e^{i\omega}}{(e^{2i\omega} - a)(e^{i\omega} - a)}. \qquad (11.6.8)$$

More generally, one can show that,

$$\Gamma_k(\omega, \omega, \ldots, \omega) = \frac{c^{k-1}b\, e^{i\omega}}{(e^{ik\omega} - a)(e^{i(k-1)\omega} - a) \ldots (e^{i\omega} - a)}. \qquad (11.6.9)$$

The function $\Gamma_1(\omega_1)$ is identical with the conventional transfer function of the linear system obtained by setting $c = 0$ in (11.6.5), and thus (being independent of c) it cannot characterize the bilinear system. However, the "quadratic" transfer function $\Gamma_2(\omega_1, \omega_2)$ involves all three parameters, a, b, and c, and thus the system is, in principle, determined from a knowledge of this function. In fact, all we need to determine the system is a knowledge of the function $\Gamma_2(\omega_1, \omega_2)$ along the diagonal $\omega_1 = \omega_2$, i.e. a knowledge of $\Gamma_2(\omega, \omega)$. If $\Gamma_2(\omega_1, \omega_2)$ is known, and the parameters a, b, c thus determined, it follows that all the generalized transfer functions are then determined. Thus, for the scalar model (11.6.5), *we may regard the single function $\Gamma_2(\omega, \omega)$ as the "transfer function" of the system.*

Estimation of parameters in bilinear models

The problem of estimating the parameters of a bilinear model does not differ, in principle, from that of estimating the parameters of a linear model—provided, of course, that the order of the model (i.e. the total number of parameters involved) is specified *a priori*. Thus, if we assume that ε_t is a Gaussian process, then given observation on the model (11.6.3) for $t = 1, \ldots N$, the likelihood function may, for large N, be written approximately as

$$L(\boldsymbol{\theta}) \propto \exp\left(-\frac{1}{2\sigma_\varepsilon^2} \sum_{t=1}^{N} \varepsilon_t^2\right), \qquad (11.6.10)$$

where $\boldsymbol{\theta}$ denotes the set of parameters $\{a_p, b_q, c_{rs}; p = 1, \ldots, k; q = 1, \ldots, l; r = 1, \ldots, m; s = 1, \ldots, n\}$, and where, for each value of $\boldsymbol{\theta}$, ε_t may be computed recursively from (11.6.3). The maximum likelihood estimate, $\boldsymbol{\theta}$, is therefore obtained by minimizing

$$V(\boldsymbol{\theta}) = \sum_{t=1}^{N} \varepsilon_t^2 \qquad (11.6.11)$$

with respect to each element of $\boldsymbol{\theta}$. Given an initial set of estimates, $\hat{\boldsymbol{\theta}}_0$, a standard Newton–Raphson iterative technique may be used to find that value, $\hat{\boldsymbol{\theta}}$, which minimizes $V(\boldsymbol{\theta})$. This approach has been developed by Subba Rao (1977, 1981), who also proposed an alternative method based on "repeated residuals". For example, if we consider the first order bilinear model,

$$x_{t+1} = ax_t + bx_t\varepsilon_t + \varepsilon_{t+1}, \qquad (11.6.12)$$

then we may write this as

$$\{1 + bx_tB\}\varepsilon_{t+1} = x_{t+1} - ax_t. \qquad (11.6.13)$$

For small b, an approximate inversion gives

$$\begin{aligned}\varepsilon_{t+1} &= \{1 - bx_tB\}\{x_{t+1} - ax_t\} \\ &= x_{t+1} - ax_t - bx_t^2 + bax_tx_{t-1}. \end{aligned} \qquad (11.6.14)$$

Neglecting the product term, x_tx_{t-1}, initial estimates of a and b can be obtained by a standard least squares approach. (Otherwise, if the product term is not neglected, a non-linear least squares approach would be needed to estimate a and b.) Once initial estimates, $\hat{a}_0$ and $\hat{b}_0$, are obtained, $\hat{\varepsilon}_t$ can be computed recursively from the equation,

$$\hat{\varepsilon}_{t+1} = x_{t+1} - \hat{a}_0x_t - \hat{b}_0x_t\hat{\varepsilon}_t, \qquad t = 1, \ldots N$$

(starting with $\hat{\varepsilon}_1 = 0$), and given $\{x_t, \hat{\varepsilon}_t; t = 1, \ldots, N\}$, we can find new estimates of a and b which minimize

$$\sum_{t=1}^{N} \{x_{t+1} - ax_t - bx_t\hat{\varepsilon}_t\}^2. \qquad (11.6.15)$$

This procedure is then repeated until the estimates converge. Subba Rao (1977) studied the first order bilinear model,

$$x_{t+1} = 0\cdot4x_t + 0\cdot4x_t\varepsilon_t + \varepsilon_{t+1},$$

with the ε_t independent $N(0, 1)$ variables, and on the basis of 200 observations obtained the following estimates:
  (1) Newton–Raphson method: $\hat{a} = 0\cdot4978$; $\hat{b} = 0\cdot3935$;
  (2) Repeated residual method: $\hat{a} = 0\cdot5135$; $\hat{b} = 0\cdot4188$.

Analysis of the sunspot series

The Wolf sunspot series has attracted the attention of quite a number of time series analysts, and was, in fact, the series studied by Yule (1927) in his pioneering paper on autoregressive models. Data on the Wolf annual sunspot index are available from 1700 onwards (see Waldmeier (1961)), and the main feature is a cycle of activity varying in duration from about nine to fourteen years, with an average "period" of about eleven years. This feature prompted Yule to fit an AR(2) model, and Moran (1954) also fitted an AR(2) model and later extended it to an AR(5) form. Box and Jenkins (1970) analysed the "short series" of 100 observations (from 1770 to 1869) and tentatively identified the model as either AR(2) or AR(3). (Morris (1977) gives an interesting review of different models which various authors have attempted to fit to this series.)

However, the series exhibits another intriguing feature, namely, different gradients of "ascensions" and "descensions" (i.e. in each cycle the rise to the maximum has a steeper gradient than the fall to the next minimum). This feature suggests the possibility of a non-linear model, and to study this aspect Haggan and Subba Rao fitted both linear and bilinear models and compared their relative fit. The results of their analyses are summarized below. In each case the data consisted of 246 observations (with mean 43·53), and all the linear models were fitted to mean-corrected data. The orders of the linear models were determined according to Akaike's AIC criterion (see Section 5.4.5) which may be written as

$$\text{AIC}(p) = n \log \hat{\sigma}^2 + 2p,$$

where n is the (effective) number of observations used, p is the number of model parameters, and $\hat{\sigma}^2$ the estimated residual variance. The appropriate model order is chosen by seeking the minimum value of $\text{AIC}(p)$.

Linear models

1. *AR Models*: $X_t + a_1 X_{t-1} + \ldots + a_k X_{t-k} = \varepsilon_t$
   The "best" model is AR(9), with estimated coefficients:

$$\hat{a}_1 = -1·21; \quad \hat{a}_2 = 0·47; \quad \hat{a}_3 = 0·12; \quad \hat{a}_4 = -0·14; \quad \hat{a}_5 = 0·16;$$

$$\hat{a}_6 = -0·09; \quad \hat{a}_7 = 0·08; \quad \hat{a}_8 = -0·09; \quad \hat{a}_9 = -0·10.$$

The estimated residual variance is $\hat{\sigma}^2 = 194·43$, and the corresponding AIC value is 1316·44.

2. *Subset AR Models*

"Best" subset AR model (using only three lagged terms) includes lags 1, 2, 9, and the estimated coefficients are:

$$\hat{a}_1 = -1.24; \quad \hat{a}_2 = 0.54; \quad \hat{a}_9 = -0.15.$$

The estimated residual variance is $\hat{\sigma}^2 = 199.2$, and the corresponding AIC value is 1310·39.

3. *ARMA Models*: $X_t + a_1 X_{t-1} + \ldots + a_k X_{t-1} = \varepsilon_t + b_1 \varepsilon_{t-1} + \ldots + b_l \varepsilon_{t-l}$.

The "best" model is ARMA (6, 6) with estimated coefficients:

$$\hat{a}_1 = -0.52; \quad \hat{a}_2 = -0.44; \quad \hat{a}_3 = -0.51; \quad \hat{a}_4 = 1.09;$$

$$\hat{a}_5 = 0.07; \quad \hat{a}_6 = -0.25; \quad \hat{b}_1 = 0.71; \quad \hat{b}_2 = -0.07;$$

$$\hat{b}_3 = -1.09; \quad \hat{b}_4 = -0.08; \quad \hat{b}_5 = 0.41; \quad \hat{b}_6 = 0.42.$$

The residual variance is $\hat{\sigma}^2 = 185.27$, and the corresponding AIC value is 1309·8. For all the linear models the AIC values are calculated with $n = 246$ (which, in the case of the AR models, is essentially equivalent to using starting values $X_0 = X_{-1} = \ldots = X_{-k+1} = 0$). The number of parameters, p, includes the estimated mean, so that, e.g., for the AR(9) model we set $p = 10$.

Bilinear models

The model considered is a special form of (11.6.3) in which the pure MA terms on the RHS are dropped, and $n = k$. The model may thus be written

$$X_t + a_1 X_{t-1} + \ldots + a_k X_{t-k} = d + \sum_{i=1}^{m} \sum_{j=1}^{k} c_{ij} \varepsilon_{t-i} X_{t-j}.$$

(The constant d was included since here the model was fitted to the raw (i.e. non-mean-corrected) data.) Using the AIC criterion, a bilinear model of the above form was chosen with $k = 3$, $m = 4$. The estimated residual variance is $\sigma^2 = 149.60$, and the corresponding AIC value is 1243·93. The estimated parameters are:

$$\hat{d} = 10.9132; \quad \hat{a}_1 = -1.93; \quad \hat{a}_2 = 1.46; \quad \hat{a}_3 = -0.27,$$

the matrix of $\hat{c}_{ij}$ values being (i corresponds to the rows and j to the columns),

$$\begin{array}{ccc} -0.55 \times 10^{-2} & -0.57 \times 10^{-2} & -0.17 \times 10^{-2} \\ 0.32 \times 10^{-2} & -0.56 \times 10^{-2} & 0.71 \times 10^{-2} \\ -0.18 \times 10^{-2} & -0.82 \times 10^{-2} & 1.45 \times 10^{-2} \\ 0.08 \times 10^{-2} & 0.58 \times 10^{-2} & -0.03 \times 10^{-2}. \end{array}$$

The parameters were estimated by minimizing $\sum \varepsilon_t^2$ using a Newton–Raphson iterative technique, the starting values being obtained by using the first stage of the "repeated residuals" method described above. The total number of parameters is 16, but for comparison with the linear models we may ignore the parameter d (which corresponds simply to a "mean correction"), giving the "effective" number of parameters as 15.

Although the bilinear parameters c_{ij} are all quite small, their effect, as measured by the reduction in the residual variance, is quite substantial. (However, care is needed in comparing the AIC values since the AIC values for all the linear models are computed with $n = 246$, whereas the AIC for the bilinear model is computed with $n = 246 - \max(m, k) = 242$.) Some idea of the improvement in fit due to the bilinear terms may be gained by using a crude form of the asymptotic χ^2 test (cf. Bartlett (1955), p. 226), namely, that under the hypothesis that the ARMA model holds

2[maximized log likelihood for the bilinear model
$-$maximized log likelihood for the ARMA model]$\sim \chi_q^2$,

where q is the difference in the number of parameters of the two models. For the comparison of the ARMA and bilinear model this gives

$$2\left[\left(-\frac{n}{2}\log_e \hat{\sigma}_B^2\right) - \left(-\frac{n}{2}\log_e \hat{\sigma}_L^2\right)\right] \sim \chi_3^2,$$

or

$$n \log_e \{\hat{\sigma}_L^2/\hat{\sigma}_B^2\} \sim \chi_3^2,$$

where we may take $n = 246 - \max(k, m) = 242$, and $\hat{\sigma}_L^2$, $\hat{\sigma}_B^2$ denote respectively the estimated residual variances obtained from the ARMA and bilinear models. Using the above values of n, $\hat{\sigma}_L^2$, $\hat{\sigma}_B^2$, we obtain

$$n \log_e \{\hat{\sigma}_L^2/\hat{\sigma}_B^2\} = 51 \cdot 75,$$

which is very highly significant when referred to the χ_3^2 distribution.

Threshold autoregressive models

A different class of finite parameter non-linear models was introduced by Tong (1978a, b, c) who described them as *"threshold models"*. The basic idea is to start with a linear model for a process X_t, and then allow the parameters to vary according to the values of a finite number of past values of X_t, or a finite number of past values of an associated process Y_t. Thus, a "first order threshold autoregressive model" (TAR(1)) would be typically described by

$$X_t = \begin{cases} a^{(1)}X_{t-1} + \varepsilon_t^{(1)}, & \text{if } X_{t-1} < d, \\ a^{(2)}X_{t-1} + \varepsilon_t^{(2)}, & \text{if } X_{t-1} \geq d. \end{cases} \tag{11.6.16}$$

This type of model was first studied by Tong in connection with the analysis of river flow data (Tong (1977)), and it can obviously be extended to a "k-threshold" form, namely

$$X_t = a^{(i)}X_{t-1} + \varepsilon_t^{(i)}, \qquad \text{if } X_{t-1} \in R_i; \quad i = 1, \ldots, k, \tag{11.6.17}$$

where $R_1, \ldots, R_k$ are given subsets of the real line R^1. Looked at in this way, the k-threshold model may be regarded as a "piecewise-linear" approximation to the general non-linear first order model

$$X_t = \lambda(X_{t-1}) + \varepsilon_t, \tag{11.6.18}$$

($\lambda(x)$ being some general function of x) considered by Jones (1978). Higher order threshold autoregressive models may be similarly defined; for example, a TAR(l) model is given by

$$X_t + a_1^{(i)}X_{t-1} + \ldots + a_l^{(i)}X_{t-l} = \varepsilon_t^{(1)},$$

$$\text{if } (X_{t-1}, \ldots, X_{t-l}) \in R^{(i)}; \quad i = 1, \ldots, k, \tag{11.6.19}$$

where $R^{(i)}$ is a given region of the l-dimensional Euclidean space R^l. This model may be viewed as a piecewise linear approximation to

$$X_t = f(X_{t-1}, X_{t-2}, \ldots, X_{t-l}) + \varepsilon_t. \tag{11.6.20}$$

Using the state space representation of AR models (Section 10.4), the TAR(l) model may be re-written as

$$x_{t+1} = A^{(i)}x_t + B\varepsilon_{t+1}^{(i)}, \qquad \text{if } x_t \in R^{(i)}, \tag{11.6.21}$$

$$X_t = Hx_t,$$

and this may be compared with the state space representation of bilinear models. The essential distinction is that in the bilinear case the non-linearity is injected by introducing the "product term", $\varepsilon_t Cx_t$, whereas in the threshold model the non-linearity arises in terms of the functional relationship between x_{t+1} and x_t, with the residual ε_t still entering "linearly". However, one can establish a tenuous link between bilinear and threshold models, as illustrated by the following example. Consider a first order (scalar) bilinear system in which the (physical) input U_t, output X_t, and noise ε_t are related by

$$X_{t+1} = aX_t + cU_tX_t + \varepsilon_{t+1}. \tag{11.6.22}$$

If U_t is determined by a feedback mechanism of the form

$$U_t = \begin{cases} +\alpha, & \text{if } X_t < d, \\ -\alpha, & \text{if } X_t \geqslant d, \end{cases} \qquad (11.6.23)$$

then the model for X_t can obviously be expressed as

$$X_{t+1} = \begin{cases} a^{(1)} X_t + \varepsilon_{t+1}, & \text{if } X_t < d, \\ a^{(2)} X_t + \varepsilon_{t+1}, & \text{if } X_t \geqslant d, \end{cases} \qquad (11.6.24)$$

where $a^{(1)} = a + \alpha c$, $a^{(2)} = a - \alpha c$.

One important property of threshold models is that, under suitable conditions, they can give rise to "*limit cycle*" behaviour—a feature which is well known in the study of deterministic non-linear differential equations. Thus, if in (11.6.19) we "switch off" the process ε_t at time t_0, i.e. if we set $\varepsilon_t \equiv 0$, $t > t_0$, then (11.6.19) may possess a solution which has an *asymptotic* periodic form. This means that, for sufficiently large t, the one-step predictor, $E[X_t | X_{t-1}, X_{t-2}, \ldots]$ will have a periodic structure (when regarded as a function of t), and consequently such models can be used to describe "cyclic" phenomena, as an alternative to the more rigid classical models based on the superposition of harmonic components on a linear stationary residual. (See Tong (1978b), Lim and Tong (1978).)

Estimation of parameters

Under the assumption that the residuals $\varepsilon_t^{(i)}$ are Gaussian, maximum likelihood estimates of the parameters of threshold models can be obtained by an iterative search technique similar to that used for linear models. For a given number of thresholds, given threshold regions, and given sets of parameters $\{a_1^{(i)}, \ldots, a_l^{(i)}; i = 1, \ldots, k\}$, the residuals $\varepsilon_t^{(i)}$ can be computed from (11.6.19), and $\sum_t \{\varepsilon_t^{(i)}\}^2$ evaluated. The orders of the individual models for each of the threshold regions can be determined by applying the AIC criterion, and the AIC criterion can then be applied to the overall model to determine the appropriate number of thresholds and the associated regions. In practice, one would usually start by considering a set of threshold regions defined by just a single past observation X_{t-d} and the AIC criterion can be applied to determine the most appropriate value of d. (See Tong (1978b).)

A threshold autoregressive model for the sunspot series

As an illustration of the above approach, Lim and Tong fitted a TAR model to 216 observations on the Wolf sunspot series (covering the years

1700 to 1915). The model finally selected consists of a fourth order scheme and a tenth order scheme, the "switching" being determined by a single threshold, namely,

$$X_t = 8.6432 + 1.8277X_{t-1} - 1.4342X_{t-2} + 0.4153X_{t-3}$$
$$+ 0.1175X_{t-4} + \varepsilon_t^{(1)}, \qquad \text{if } X_{t-3} \leqslant 36.3,$$

$$X_t = 8.3678 + 0.8337X_{t-1} - 0.1047X_{t-2} - 0.1204X_{t-3}$$
$$+ 0.1209X_{t-4} - 0.2227X_{t-5} + 0.0726X_{t-6}$$
$$+ 0.1242X_{t-7} - 0.2527X_{t-8} + 0.1640X_{t-9}$$
$$+ 0.0855X_{t-10} + \varepsilon_t^{(2)}, \qquad \text{if } X_{t-3} > 36.3,$$

with

$$\hat{\sigma}_{\varepsilon}^2{}_{(1)} = 209.72, \qquad \hat{\sigma}_{\varepsilon}^2{}_{(2)} = 92.42,$$

the "pooled" mean residual sum of squares being 148·79. The above model contains 16 parameters, and the estimated "pooled" residual variance is very close to the residual variance estimated from the bilinear model discussed above. Lim and Tong's model for the sunspot series exhibits an asymmetric limit cycle with an exact eleven year period, the "rise" period from the year of minimum activity to the next year of maximum activity being four years.

Exponential autoregressive models

"Exponential autoregressive" models were introduced by Ozaki (1978) and Haggan and Ozaki (1978) to reproduce certain features of non-linear random vibrations theory. Non-linear random vibrations are typically described by second order differential equations of the form

$$\ddot{x}(t) + f\{\dot{x}(t)\} + g\{x(t)\} = y(t), \qquad (11.6.25)$$

where $f\{\cdot\}$ (the "damping force"), and $g\{\cdot\}$ (the "restoring force") are non-linear functions, and $y(t)$ is a stochastic "input". Two well known examples are *Duffing's equation*,

$$\ddot{x}(t) + c\dot{x}(t) + ax(t) + b\{x(t)\}^3 = y(t),$$

and *Van der Pol's equation*,

$$\ddot{x}(t) + f\{\dot{x}(t)\} + ax(t) = y(t).$$

If $f\{\cdot\}$ and $g\{\cdot\}$ are both linear functions, (11.6.25) reduces to the familiar second order linear differential equation,

$$\ddot{x}(t) + \alpha_1 \dot{x}(t) + \alpha_2 x(t) = y(t),$$

say, and, when $y(t)$ is white noise, $x(t)$ becomes a (continuous parameter) AR(2) process with spectral density function given by (4.10.11). For suitable α_1, α_2 this spectral density function has a peak at frequency ω_1, say. When the system is "driven" by a white noise input the output $x(t)$ then exhibits an "approximate" periodic behaviour with period $2\pi/\omega_1$. Here, the value of ω_1 depends only on α_1, α_2 and remains constant over time. On the other hand, if the input is periodic, i.e. if $y(t) = A \cos \omega_0 t$, say, then the output is of the form

$$x(t) = A' \cos(\omega_0 t + \phi),$$

say, with $A' = A|\Gamma(\omega_0)|$, where $|\Gamma(\omega_0)| = |-\omega_0^2 + i\alpha_1\omega_0 + \alpha_2|$ is a *continuous function* of ω_0. Thus, the amplitude of $x(t)$ changes continuously as we vary ω_0.

However, if a non-linear system is driven by a white noise input the output may still be oscillatory but the "approximate" frequency (ω_1) may vary with the local amplitude of the oscillation, i.e. ω_1 may take different values depending upon whether $x(t)$ is locally "small" or "large". This is known as "*amplitude-dependent frequency*". Also, if a non-linear system is driven by a periodic input, with frequency ω_0, the output may be approximately periodic, with the same frequency ω_0, i.e.

$$x(t) \doteqdot B \cos(\omega_0 t + \phi),$$

but now the amplitude B may be a discontinuous function of ω_0, i.e. as ω_0 passes through certain frequencies B may "jump" from one value to another: this effect occurs, for example, in the solution of Duffing's equation, and is known as a "*jump phenomenon*".

To construct a time series model which reproduces the effect of "amplitude-dependent frequency" Ozaki suggested that one may start with a (discrete parameter) AR(2) model,

$$X_t - a_1 X_{t-1} - a_2 X_{t-2} = \varepsilon_t, \qquad (11.6.26)$$

and then allow the coefficients a_1, a_2 to depend on X_{t-1}. Specifically, he suggests that the coefficients be made *exponential functions* of X_{t-1}, i.e. take the form

$$a_1 = \phi_1 + \pi_1 \exp(-\gamma X_{t-1}^2), \qquad a_2 = \phi_2 + \pi_2 \exp(-\gamma X_{t-1}^2).$$

If we continue to regard (11.6.26) as a "linear" model, its "resonant frequency" will occur at the minimum of

$$|1 - a_1 \exp(-i\omega) - a_2 \exp(-2i\omega)|,$$

and hence will change with the magnitude of X_{t-1}. (For large X_{t-1} we have $a_1 \sim \phi_1$, $a_2 \sim \phi_2$, while for small X_{t-1}, $a_1 \sim \phi_1 + \pi_1$, $a_2 \sim \phi_2 + \pi_2$. In this respect the model behaves rather like the threshold AR model, but here the coefficients change smoothly between the two extreme values.) In addition to exhibiting the "amplitude-dependent frequency" effect, this model can also give rise to "jump phenomena" and "limit cycles" (Ozaki (1978)). With a_1, a_2 chosen as above, (11.2.26) is called a second order "*exponential autoregressive*" model.

More generally, Ozaki has considered kth order exponential models of the form

$$X_t = (\phi_1 + \pi_1 \exp(-\gamma X_{t-1}^2))X_{t-1} + \ldots + (\phi_k + \pi_k \exp(-\gamma X_{t-1}^2))X_{t-k} + \varepsilon_t.$$

$$(11.6.27)$$

Estimation of parameters

Haggan and Ozaki (1978) have proposed the following procedure for estimating the parameters of (11.6.27).
(i) First fix the value of γ; then $\phi_1, \pi_1, \ldots, \phi_k, \pi_k$, may be estimated by standard least squares regression analysis of X_t on X_{t-1}, $X_{t-1} \exp(-\gamma X_{t-1}^2)$, $\ldots$. The order of the model k is determined by minimizing the AIC criterion.
(ii) The above analysis is repeated using a range of values of γ, and the AIC criterion used to select the most suitable value of γ. The values of γ considered are usually chosen so that $\exp(-\gamma X_{t-1}^2)$ varies reasonably widely over the range $(0, 1)$.

Example

Haggan and Ozaki (1978) fitted a model of the form (11.6.27) to the logarithmically transformed (and mean-corrected) Canadian lynx data, covering the years 1821–1934, and, using the above procedure, found the best order to be given by $k = 11$, with $\hat{\gamma} = 3 \cdot 89$. The estimated values of the $\{\phi_i\}$ and $\{\pi_i\}$ were as follows:

i	1	2	3	4	5	6	7	8	9	10	11
$\hat{\phi}_i$	1·09	−0·28	0·27	−0·45	0·41	−0·36	0·22	−0·10	0·22	−0·07	−0·38
$\hat{\pi}_i$	0·01	−0·49	−0·06	0·30	−0·54	0·61	−0·53	0·30	−0·18	0·18	0·16

The estimated residual variance for this "exponential" model is $\hat{\sigma}_\varepsilon^2 = 0 \cdot 0321$, which compares favourably with the model fitted by Campbell and Walker (1977), for which $\hat{\sigma}_{\varepsilon}^2 = 0\cdot0437$. (Campbell and Walker's model was based on a "mixed spectrum" scheme involving one harmonic component superimposed on an AR(2) process.) Lim and Tong (1978) fitted a threshold AR model to the Canadian lynx data and obtained orders of 8, 3 (depending on whether $X_{t-2} \leqslant 3\cdot1163$ or $X_{t-2} > 3\cdot1163$ respectively), with an overall residual variance $\hat{\sigma}_{\varepsilon}^2 = 0\cdot0360$. The above "exponential" model exhibits a limit cycle with an approximate period of ten years; for small X_{t-1} the period is approximately $10\cdot1$ years, while for large X_{t-1} it is approximately $9\cdot4$ years—which agrees well with the behaviour observed in the data. The threshold AR model also shows a limit cycle with an approximate ten-year period, the cycles being asymmetric with the "rise" period roughly double the length of the "fall" period.

The present era of time series analysis—based on stationary linear models—has achieved enormous success in terms of its ability to provide reasonably accurate descriptions for a great variety of observational data, and the current popularity of time series forecasting methods provides ample confirmation of this assertion. However, both the theory and "technology" of stationary linear model fitting has reached a very refined state, and it is doubtful whether there are any major new advances to be made in this area. We know enough from other branches of science to realize that the world is neither "stationary" nor "linear", and the work which has been done so far on non-linear models clearly shows that it is quite feasible to fit such models to real data, and to obtain substantial improvements in fit over that obtained from linear models. The next era of time series analysis must surely lie in this direction: much work remains to be done, but the prospects are indeed exciting.

Appendix

The following tables (A.1–A.5) list the data for the five standard models (white noise, AR(1), AR(2), MA(2), and ARMA(2, 2)) used as illustrative examples throughout the text. Table A.6–A.10 list the theoretical and sample autocovariances for these five sets of data. The specifications of the five models are as follows.

(1) Gaussian white noise process, zero mean, unit variance.

(2) AR(1) model: $X_t - 0 \cdot 6 X_{t-1} = \varepsilon_t$, $\varepsilon_t \sim N(0, 1)$.

(3) AR(2) model: $X_t - 0 \cdot 4 X_{t-1} + 0 \cdot 7 X_{t-2} = \varepsilon_t$, $\varepsilon_t \sim N(0, 1)$.

(4) MA(2) model: $X_t = \varepsilon_t + 1 \cdot 1 \varepsilon_{t-1} + 0 \cdot 2 \varepsilon_{t-2}$, $\varepsilon_t \sim N(0, 1)$.

(5) ARMA(2, 2) model: $X_t + 1 \cdot 4 X_{t-1} + 0 \cdot 5 X_{t-2} = \varepsilon_t - 0 \cdot 2 \varepsilon_{t-1} - 0 \cdot 1 \varepsilon_{t-2}$, $\varepsilon_t \sim N(0, 1)$.

In tables A.1–A.5, the observations should be read consecutively down the columns, starting with the top of the left-hand column and finishing with the bottom of the right-hand column. For example, the first five observations from the white noise process are $-0 \cdot 99$, $0 \cdot 39$, $-0 \cdot 27$, $0 \cdot 64$, $-0 \cdot 29$, while the last five observations are $-0 \cdot 30$, $1 \cdot 73$, $0 \cdot 22$, $-1 \cdot 44$, $0 \cdot 12$.

Table A.1. *500 observations from a white noise process.*

−·99	−2·22	−1·66	·37	−·20	·09	1·09	1·65	1·56	1·32	−·44
·39	1·45	−1·35	·25	−·25	·73	−·79	−1·93	−·16	−·71	·86
−·27	2·33	−·75	−·16	−1·12	−·81	−·28	−1·49	−·79	−·00	−1·46
·64	−1·00	·66	·50	·44	·76	·29	·44	−1·23	·23	·17
−·29	−1·43	−·75	−·15	−1·13	−1·61	−·33	·32	−·80	·44	·26
−·83	−·51	−·02	−1·70	1·23	−·19	·17	1·46	−·47	·57	·05
1·21	−·29	−1·30	−1·61	·06	−1·07	·51	−1·47	−2·49	·07	·52
−1·26	1.09	1·11	·04	−·93	−·09	−·24	−·01	1·02	·97	−·66
−·39	−·67	·14	−1·60	−·17	·19	·12	·90	1·77	·92	·36
1·11	−1·36	·35	1·09	·63	·65	1·24	−·82	1·73	−·42	−·73
1·10	·24	·19	−·19	−·83	1·29	·79	−·10	−·71	−1·18	−·53
−·99	·57	−·31	·74	−1·66	·86	1·93	1·81	·59	·47	1·26
·62	−·43	−·42	−·64	1·21	·69	·37	−·52	1·19	−·99	·22
·02	·39	·45	−1·89	−1·08	−·28	1·00	−1·18	−1·08	−·58	·48
1·22	·34	1·34	1·10	·36	−1·66	−1·24	−1·85	−·16	−·09	−1·17
2·53	−·06	·16	·01	−·83	−·56	·19	·10	−·01	−1·79	−·65
·15	1·15	1·00	−1·80	2·02	−·63	·09	−·60	−·39	−·40	·58
·46	·32	−·09	−·30	·58	−·53	−3·03	−1·04	−·27	−1·60	·34
·67	1·26	·21	·10	−·20	·10	−·68	1·96	−1·22	1·62	−·49
·78	·68	·74	3·10	·31	·20	·87	1·01	−1·12	1·56	−·49
·04	−1·06	·01	·08	1·52	·43	1·64	·36	−·95	−·86	−·09
−1·12	1·08	·56	·24	·91	1·28	−·85	−·50	·39	·72	−1·84
−·37	−·02	−1·03	−·96	−·01	·43	−·32	1·38	·05	1·62	·37
−·41	1·60	−1·45	−·02	1·27	−1·00	−·61	−1·26	−1·31	1·46	−·66
−·34	−·24	·06	·74	−·07	−·29	1·36	−·79	1·59	−·96	1·67
−·35	·22	−1·41	·24	−·57	·73	−·35	−·73	−·77	−1·29	−·19
1·50	1·65	·13	−1·42	−·90	−·18	·31	·31	−·25	−1·21	·40
·23	·15	·76	1·62	−·75	−·99	·40	1·23	·11	−·56	−·46
·06	·74	−·41	·21	·22	·92	−·17	−·93	−2·01	−·70	−·24
−1·53	−·69	−·71	−1·05	−·68	−·44	−·91	·25	−1·52	−·46	−·82
−·16	1·00	1·82	−·37	−1·56	−·48	−·23	−·82	·65	·30	−1·64
·10	·22	−1·22	−·21	·35	·62	·18	·85	−·65	−·01	·54
−2·03	·24	−·15	−1·96	1·56	1·73	−·25	−1·45	−·84	−·37	1·31
1·68	·24	−1·86	1·81	−1·51	−1·45	·53	−·82	·35	−1·72	·04
−1·03	−·67	·21	1·03	·07	−·70	·10	−·76	·39	·27	−·53
·40	·11	−·30	1·01	−·51	−1·18	1·44	−·52	−·24	−·42	−·30
·72	3·07	−·67	−·74	−·51	−·05	−·59	−·47	1·64	−·44	1·73
·67	−1·52	·86	−·37	1·56	−·03	·03	·15	−·24	−·98	·22
−·32	·12	−·07	−·31	1·20	1·53	2·17	−·42	−·64	·46	−1·44
−·76	1·77	2·43	−1·50	·04	·92	·25	·14	−·12	·43	·12
·21	−1·16	−1·42	·08	·14	1·91	−1·15	−1·09	−·98	·60	
−2·23	·73	·90	−·71	−1·07	·51	−·08	2·31	·93	1·33	
1·18	·74	1·58	·71	−1·71	·74	·20	−1·84	1·04	−·17	
−·57	·22	−·18	−·11	−1·26	−1·87	−·18	·52	−·87	−·90	
·56	1·11	−·23	1·87	−·72	1·21	−·03	−2·69	·37	−1·24	
−·29	·88	−1·04	−·97	−2·31	·91	−·81	·06	−·52	−1·11	

Table A.2. *500 observations from an AR(1) process.*

-1·24	-2·34	-·53	-·10	-·15	-2·09	1·94	1·13	·62	1·11	-1·64
-·35	·04	-1·67	·19	-·33	-·52	·37	-1·25	·21	-·04	-·12
-·48	2·36	-1·75	-·04	-1·32	-1·12	-·06	-2·23	-·67	-·03	-1·53
·35	·41	-·39	·47	-·36	·09	·25	-·90	-1·63	·22	-·75
-·08	-1·18	-·98	·13	-1·35	-1·55	-·18	-·22	-1·78	·57	-·19
-·88	-1·22	-·61	-1·62	·42	-1·12	·06	1·33	-1·54	·92	-·06
·68	-1·02	-1·66	-2·59	·31	-1·74	·54	-·68	-3·41	·62	·48
-·85	·48	·11	-1·51	-·75	-1·14	·08	-·42	-1·03	1·35	-·37
-·90	-·39	·20	-2·51	-·62	-·50	·17	·65	1·15	1·73	·14
·57	-1·60	·48	-·42	·26	·36	1·34	-·43	2·42	·61	-·65
1·44	-·72	·48	-·44	-·68	1·50	1·59	-·35	·74	-·81	-·92
-·12	·14	-·02	·47	-2·07	1·76	2·89	1·60	1·04	-·01	·71
·55	-·35	-·44	-·36	-·03	1·75	2·10	·44	1·81	-1·00	·65
·35	·18	·19	-2·11	-1·10	·76	2·26	-·92	·00	-1·18	·87
1·43	·45	1·45	-·17	-·30	-1·20	·11	-2·40	-·16	-·80	-·65
3·39	·21	1·03	-·09	-1·02	-1·28	·25	-1·34	-·11	-2·26	-1·04
2·18	1·28	1·61	-1·85	1·41	-1·40	·24	-1·41	-·46	-1·75	-·05
1·77	1·09	·88	-1·41	1·42	-1·37	-2·89	-1·88	-·55	-2·65	·31
1·73	1·91	·74	-·75	·66	-·72	-2·41	·83	-1·55	·03	-·30
1·82	1·82	1·18	2·65	·71	-·23	-·58	1·50	-2·05	1·58	-·67
1·13	·03	·72	1·66	1·95	·29	1·30	1·27	-2·18	·09	-·49
-·44	1·10	·99	1·23	2·08	1·46	-·07	·26	-·92	·78	-2·13
-·63	·64	-·44	-·22	1·24	1·31	-·37	1·53	-·50	2·09	-·91
-·79	1·98	-1·71	-·15	2·02	-·22	-·83	-·34	-1·61	2·72	-1·21
-·81	·95	-·97	·64	1·14	-·42	·87	-·99	·62	·67	·94
-·83	·78	-1·99	·63	·12	·47	·17	-1·33	-·40	-·89	·37
1·00	2·12	-1·06	-1·04	-·83	·10	·41	-·49	-·49	-1·75	·62
·83	1·42	·12	·99	-1·25	-·92	·64	·93	-·19	-1·61	-·08
·55	1·59	-·34	·81	-·54	·36	·22	-·37	-2·12	-1·67	-·29
-1·19	·26	-·91	-·57	-1·00	-·23	-·78	·03	-2·79	-1·46	-·99
-·88	1·16	1·27	-·71	-2·16	-·62	-·70	-·80	-1·03	-·57	-2·24
-·42	·92	-·46	-·64	-·95	·24	-·24	·37	-1·27	-·35	-·80
-2·29	·79	-·43	-2·35	·99	1·87	-·40	-1·22	-1·60	-·58	·83
·31	·72	-2·11	·40	-·91	-·33	·29	-1·56	-·60	-2·07	·54
-·85	-·24	-1·06	1·27	-·48	-·89	·27	-1·70	·03	-·97	-·21
-·11	-·04	-·94	1·77	-·79	-1·72	1·61	-1·54	-·22	-1·01	-·43
·66	3·05	-1·23	·32	-·99	-1·08	·38	-1·40	1·51	-1·04	1·47
1·07	·31	·12	-·17	·97	-·68	·25	-·68	·66	-1·60	1·10
·32	·31	-·00	-·42	1·78	1·13	2·33	-·83	-·24	-·50	-·78
-·57	1·95	2·43	-1·75	1·11	1·59	1·64	-·36	-·27	·13	-·35
-·13	·01	·04	-·97	·80	2·86	-·16	-1·30	-1·14	·68	
-2·31	·74	·92	-1·30	-·59	2·23	-·18	1·53	·24	1·73	
-·20	1·18	2·13	-·07	-2·06	2·07	·10	-·92	1·18	·87	
-·69	·93	1·10	-·15	-2·50	-·63	-·12	-·03	-·16	-·37	
·15	1·67	·43	1·78	-2·22	·83	-·10	-2·71	·28	-1·47	
-·20	1·88	-·78	·10	-3·64	1·41	-·86	-1·57	-·35	-1·99	

Table A.3. *500 observations from an AR(2) process.*

−·22	−2·60	−1·72	·08	−1·73	−.38	1·98	.85	3.49	2·11	·45
−·31	·82	−2·93	1·77	−·38	1·86	−1·79	−·31	1·39	·26	1·35
−·24	4·48	−·72	·49	−·06	·20	−2·39	−2·20	−2·68	−1·38	−1·23
·76	·22	2.42	−·55	·67	−·46	·59	−·22	−3·27	−·50	−1·27
·18	−4·47	·73	−·72	−·82	−1·93	1·57	1·77	−·23	1·20	·62
−1·29	−2·45	−1·43	−1·61	·43	−·64	·39	2·33	1·73	1·41	1·18
·57	1·86	−2·38	−1·75	·80	·03	−·44	−1·78	−1·64	−·21	·56
−·13	3·55	1·15	·46	−·91	·37	−·69	−2·35	−·84	−·09	−1·26
−·84	−·55	2·27	−·19	−1·09	·32	·15	1·21	2·58	1·03	−·53
·86	−4·07	·45	·69	·84	·52	1·78	1·31	3·35	·05	−·06
2·03	−1·00	−1·21	·22	·27	1·28	1·40	−·42	−1·18	−1·88	−·18
−·78	3·02	−1·11	·34	−2·14	1·01	1·24	·73	−2·23	−·31	1·23
−1·11	1·48	−·02	−·66	·17	·20	−·11	·06	1·12	·20	·84
·12	−1·14	1·22	−2·39	·48	−·91	·08	−1·67	·93	−·28	−·05
2·04	−1·15	1·84	·60	·43	−2·16	−1·13	−2·56	−·58	−·34	−1·78
3·26	·28	·04	1·93	−1·00	−·79	−·32	·24	−·89	−1·72	−1·32
·02	2·07	−·28	−1·45	1·32	·57	·75	1·29	−·35	−·85	1·29
−1·81	·95	−·23	−2·23	1·80	·25	−2·50	−·69	·21	−·73	1·78
−·07	·19	·31	·22	−·40	−·20	−2·21	·78	−·90	1·92	−·67
2·02	·09	1·03	4·75	−1·11	−·05	1·74	1·80	−1·62	2·84	−2·01
·90	−1·16	·20	1·82	1·36	·55	3·88	·54	−·97	−1·07	−·42
−2·18	·56	−·08	−2·36	2·23	1·54	−·51	−1·55	1·14	−1·69	−·60
−1·87	1·02	−1·20	−3·18	−·06	·66	−3·25	·38	1·19	1·69	·42
·36	1·61	−1·88	·36	−·31	−1·81	−1·55	−·02	−1·63	3·32	−·07
1·12	−·31	·15	3·11	−·15	−1·48	3·02	−1·07	·10	−·82	1·34
−·16	−1·04	−·03	1·23	−·41	1·40	1·94	−1·14	·41	−3·95	·39
·66	1·45	·01	−3·10	−·96	1·42	−1·02	·60	−·16	−2·22	−·38
·60	1·46	·79	−·48	−·85	−1·40	−1·37	2·27	−·25	1·31	−·89
−·16	·31	−·11	2·19	·55	−·64	·00	−·44	−2·00	1·38	−·33
−2·01	−1·59	−1·30	·16	·14	·28	·05	−1·51	−2·14	−·83	−·33
−·85	·15	1·37	−1·84	−1·89	·08	−·21	−1·12	1·19	−·99	−1·54
1·17	1·39	·23	−1·06	−·50	·45	·06	1·46	1·33	·18	·15
−·97	·69	−1·02	−1·10	2·68	1·85	−·08	−·08	−1·14	·39	2·45
·48	−·46	−2·43	2·11	−·09	−1·03	·45	−1·88	−1·03	−1·68	·91
−·16	−1·34	−·04	2·65	−1·84	−2·40	·33	−1·46	·78	−·68	−1·88
·00	−·11	1·38	·59	−1·18	−1·42	1·26	·21	·80	·48	−1·69
·84	3·96	−·09	−2·36	·30	1·06	−·32	·63	1·41	·23	2·37
1·00	·14	−·14	−1·72	2·51	1·40	−·98	·26	−·24	−1·22	2·35
−·50	−2·60	−·07	·65	1·99	1·35	2·00	−·76	−1·72	−·19	−2·16
−1·66	·63	2·51	−·04	−·92	·48	1·74	−·35	−·65	1·21	−2·39
−·10	·91	−·37	−·39	−1·63	1·16	−1·86	−·69	−·03	1·21	
−1·11	·65	−1·01	−·84	−1·08	·64	−2·03	2·28	1·37	·96	
·81	·37	1·44	·64	−1·00	·18	·69	−·44	1·61	−·63	
·53	−·09	1·10	·74	−·91	−2·25	1·52	−1·25	−1·18	−1·82	
·21	·81	−·80	1·72	−·38	·18	·10	−2·88	−1·23	−1·53	
−·58	1·27	−2·13	−·80	−1·82	2·56	−1·83	−·22	−·18	−·45	

Table A.4. *500 observations from an MA(2) process.*

-1·39	-2·43	-·47	-·82	-·90	-2·59	2·33	·76	1·09	·83	-1·91
-·77	-1·05	-3·00	·45	-·67	·37	·59	-·27	1·57	·64	·15
-·04	3·48	-2·57	·19	-1·44	·02	-·94	-3·28	-·66	-·52	-·60
·42	1·85	-·44	·37	-·85	·02	-·18	-1·58	-2·14	·09	-1·26
·36	-2·06	-·17	·36	-·88	-·93	-·07	·51	-2·31	·69	·16
-1·02	-2·28	-·71	-1·77	·07	-1·81	-·14	1·90	-1·60	1·10	·37
·23	-1·14	-1·47	-3·52	1·18	-1·60	·63	·20	-3·17	·79	·62
-·10	·67	-·33	-2·08	-·63	-1·31	·35	-1·34	-1·81	1·17	-·08
-1·53	·47	1·10	-1·88	-1·19	-·13	-·05	·59	2·39	2·00	-·26
·43	-1·89	·73	-·66	·26	·84	1·32	·17	3·88	·78	-·47
2·24	-1·39	·61	·69	-·17	2·05	2·17	-·82	1·55	-1·46	-1·26
·44	·57	-·03	·75	-2·45	2·41	3·05	1·54	·15	-·91	·53
-·24	·25	-·72	·13	-·78	1·89	2·65	1·45	1·70	-·70	1·50
·51	·02	-·07	-2·45	-·08	·65	1·79	-1·39	·34	-1·58	·97
1·37	·68	1·75	-1·11	-·59	-1·84	-·07	-3·26	-1·11	-·93	-·60
3·88	·39	1·72	·84	-·66	-2·44	-·98	-2·17	-·41	-2·00	-1·84
3·17	1·16	1·44	-1·57	1·18	-1·57	·05	-·86	-·44	-2·38	-·37
1·13	1·57	1·04	-2·28	2·63	-1·33	-2·90	-1·68	-·71	-2·39	·84
1·21	1·84	·31	-·60	·84	-·61	-4·00	·69	-1·60	-·21	·00
1·61	2·12	·95	3·15	·21	·21	-·49	2·95	-2·52	3·03	-·96
1·04	-·07	·86	3·51	1·83	·67	2·46	1·86	-2·42	1·19	-·73
-·92	·05	·72	·94	2·65	1·80	1·14	·10	-·88	·09	-2·04
-1·59	·96	-·41	-·69	1·30	1·93	-·93	·90	·29	2·25	-1·67
-1·04	1·80	-2·47	-1·03	1·45	-·27	-1·14	·16	-1·17	3·39	-·62
-·86	1·51	-1·74	·52	1·33	-1·31	·63	-1·90	·16	·97	1·01
-·80	·27	-1·63	1·05	-·39	·21	1·03	-1·85	·71	-2·06	1·51
1·05	1·84	-1·41	-1·01	-1·54	·56	·20	-·65	-·79	-2·83	·52
1·81	2·01	·62	·11	-1·86	-1·04	·67	1·42	-·33	-2·16	-·06
·61	1·23	·45	1·71	-·79	-·21	·33	·48	-1·94	-1·56	-·67
-1·42	·15	-1·01	-·50	-·59	·36	-1·02	-·52	-3·71	-1·34	-1·17
-1·83	·39	·96	-1·49	-2·26	-·79	-1·27	-·73	-1·42	-·34	-2·59
-·38	1·19	·63	-·83	-1·50	-·00	-·25	-·00	-·24	·23	-1·43
-1·95	·68	-1·14	-2·27	1·63	2·31	-·10	-·67	-1·42	-·32	1·57
-·53	·55	-2·27	-·39	·27	·57	·28	-2·24	-·70	-2·13	1·59
·41	-·36	-1·86	2·63	-1·28	-1·95	·63	-1·96	·62	-1·70	-·22
-·40	-·58	-·44	2·51	-·73	-2·24	1·66	-1·52	·27	-·47	-·88
·96	3·05	-·96	·58	-1·06	-1·49	1·02	-1·20	1·46	-·85	1·29
1·54	1·88	·06	-·98	·90	-·32	-·33	-·47	1·51	-1·54	2·06
·57	-·94	·73	-·86	2·81	1·49	2·09	-·35	-·58	-·70	-·86
-·98	1·60	2·52	-1·92	1·67	2·59	2·64	-·29	-·88	·75	-1·42
-·69	·81	1·25	-1·64	·42	3·22	-·44	-1·02	-1·25	1·16	
-2·15	-·19	-·18	-·93	-·91	2·79	-1·29	1·14	-·17	2·07	
-1·23	1·32	2·28	-·06	-2·86	1·68	-·11	·49	1·87	1·41	
·29	1·19	1·74	·52	-3·35	-·96	·03	-1·04	·46	-·81	
·18	1·50	-·11	1·90	-2·44	-·70	-·18	-2·49	-·38	-2·26	
·21	2·14	-1·33	1·07	-3·35	1·87	-·87	-2·80	-·28	-2·65	

Table A.5. *500 observations from an ARMA(2, 2) process.*

−·52	3·89	−2·21	·16	6·74	3·66	6·82	2·98	−9·86	3·93	−·08
1·61	−·16	1·87	−·68	−6·19	−3·00	−7·06	−5·95	6·94	−5·25	1·09
−2·24	·54	−1·83	·63	4·24	1·40	6·24	5·58	−5·70	5·40	−3·07
2·99	−2·30	2·58	−·04	−2·15	·39	−4·79	−3·90	3·45	−4·63	4·13
−3·45	1·48	−3·49	−·50	−·21	−2·92	3·22	3·05	−2·46	4·17	−3·87
2·49	−1·05	3·66	−1·00	2·78	3·95	−1·91	−·97	1·54	−3·06	3·33
−·36	·69	−4·60	·39	−3·86	−4·94	1·57	−1·97	−3·24	2·12	−2·25
−2·17	·76	5·98	·48	2·95	5·08	−1·61	3·38	5·33	−·53	·72
2·95	−2·27	−6·02	−2·32	−2·19	−4·33	1·58	−2·69	−4·04	·40	·56
−1·74	1·46	5·66	4·41	2·34	4·15	−·17	1·09	4·26	−1·00	−1·88
1·88	−·33	−4·80	−5·26	−3·13	−2·50	−·02	−·20	−5·19	·01	1·94
−3·07	·39	3·51	5·83	1·65	1·97	1·76	1·65	5·69	1·23	−·34
4·07	−·95	−2·89	−6·30	·88	−1·12	−2·56	−3·09	−4·23	−2·70	−·47
−4·17	1·55	2·86	4·07	−3·22	·07	3·43	2·24	1·70	2·73	1·13
4·96	−1·39	−1·27	−1·00	4·52	−1·21	−5·00	−3·15	−·33	−2·35	−2·63
−2·57	1·01	·19	−·65	−5·52	1·47	5·63	3·88	−·26	·21	2·66
·64	·41	1·20	−·50	7·62	−1·80	−5·20	−4·29	·15	·85	−1·59
·58	−·98	−2·08	1·08	−7·65	1·43	1·40	3·14	−·28	−2·64	1·18
−·56	2·24	2·44	−·93	6·39	−·84	·56	−·03	−·82	5·24	−1·47
1·10	−2·26	−1·66	3·87	−4·83	·68	−·18	−·82	·44	−4·63	1·04
−1·45	·71	·95	−5·51	5·04	−·16	1·50	1·12	−·81	2·52	−·67
·26	1·36	−·01	5·69	−4·07	1·06	−3·28	−1·84	1·61	−·48	−1·36
·21	−2·39	−1·61	−6·23	2·84	−1·27	3·52	3·46	−1·78	·98	2·98
−·65	4·15	·96	6·02	−·76	·03	−3·75	−5·41	·33	−·06	−4·04
·58	−5·18	−·08	−4·48	−·69	·46	5·00	5·16	2·28	−1·82	5·92
−·73	5·28	−1·64	3·36	·66	·23	−5·70	−4·97	−4·31	1·33	−6·73
2·33	−3·18	2·75	−4·00	−1·37	−·85	5·72	4·92	4·64	−1·81	6·73
−2·94	1·61	−2·15	5·80	1·07	·05	−4·79	−3·16	−4·10	1·68	−6·58
2·81	−·12	1·07	−6·10	−·35	1·49	3·57	·76	1·42	−1·91	5·65
−4·03	−1·48	−1·11	4·38	−·69	−2·63	−3·52	·83	−1·06	1·57	−5·34
4·37	3·21	3·03	−3·27	−·30	2·46	3·11	−2·32	1·93	−·78	3·20
−3·82	−3·66	−5·19	2·35	1·49	−1·37	−2·28	3·82	−2·80	·29	−·86
1·13	3·61	5·67	−3·54	−·30	2·34	1·37	−5·72	2·18	−·41	·97
2·41	−3·06	−7·04	6·00	−2·18	−4·45	−·21	5·49	−1·07	−1·21	−1·20
−5·10	1·73	7·62	−5·77	3·42	4·48	−·36	−5·27	·81	2·55	·52
6·38	−·67	−7·31	5·70	−4·07	−4·94	1·99	4·36	−·95	−3·27	−·33
−5·64	3·20	5·80	−6·14	3·56	4·93	−3·48	−3·76	2·57	2·92	2·05
5·19	−6·29	−3·44	5·43	−1·24	−4·33	3·88	3·38	−3·67	−3·30	−2·80
−4·97	7·32	1·74	−4·69	·89	5·14	−1·47	−3·26	3·10	3·87	1·23
3·59	−5·20	1·65	2·45	−·98	−4·42	−·08	3·08	−2·47	−3·32	·06
−2·15	2·09	−5·07	−·67	·93	5·18	−·57	−3·75	1·02	3·18	
−·98	·46	7·21	−·86	−1·92	−5·01	·97	6·23	·95	−1·63	
4·06	−·98	−6·02	2·38	·71	4·87	−·73	−9·03	−·89	·20	
−5·77	1·14	4·23	−3·08	−·84	−6·39	·33	10·19	−·40	−·46	
6·61	−·12	−3·27	4·95	·53	8·02	−·11	*****	1·45	−·51	
−6·72	·24	1·49	−6·73	−2·36	−7·17	−·79	12·75	−2·33	·17	

Table A.6. *Theoretical and sample autocovariances for the white noise series.*

Table A.7. *Theoretical and sample autocovariances for the AR(1) series.*

0	1·0000	1·0402	0	1·5625	1·4006
1	0·0000	−·0055	1	·9375	·7702
2	0·0000	−·0361	2	·5625	·4138
3	0·0000	−·0547	3	·3375	·1997
4	0·0000	−·0451	4	·2025	·0604
5	0·0000	−·0707	5	·1215	−·0513
6	0·0000	−·0380	6	·0729	−·1396
7	0·0000	−·0333	7	·0437	−·0485
8	0·0000	·0391	8	·0262	·0275
9	0·0000	·0922	9	·0157	·0609
10	0·0000	·0569	10	·0094	·0214
11	0·0000	−·0323	11	·0057	·0539
12	0·0000	·0454	12	·0034	·0126
13	0·0000	−·0255	13	·0020	−·0781
14	0·0000	·0439	14	·0012	−·0444
15	0·0000	·0532	15	·0007	·0477
16	0·0000	−·0394	16	·0004	·1326
17	0·0000	−·0567	17	·0003	·1212
18	0·0000	−·0136	18	·0002	·1555
19	0·0000	·0044	19	·0001	·1330
20	0·0000	·0008	20	·0001	·0686
21	0·0000	−·0021	21	·0000	−·0368
22	0·0000	−·0064	22	·0000	−·0852
23	0·0000	−·0110	23	·0000	−·0642
24	0·0000	·0269	24	·0000	−·0968
25	0·0000	·0047	25	·0000	−·0883
26	0·0000	−·0302	26	·0000	−·0458
27	0·0000	·0007	27	·0000	·0444
28	0·0000	−.0428	28	·0000	·0444
29	0·0000	−·0685	29	·0000	·0600

Table A.8. *Theoretical and sample autocovariances for the AR(2) series.*

0	2·0757	1·9838
1	·4884	·4131
2	−1·2576	−1·2341
3	−·8449	−·7131
4	·5424	·6067
5	·8084	·6481
6	−·0563	−·2729
7	−·5884	−·5342
8	−·1960	−·0307
9	·3335	·3176
10	·2706	·1715
11	−·1252	−·0076
12	−·2395	−·1238
13	−·0081	−·1726
14	·1644	−·0036
15	·0715	·1659
16	−·0865	·0655
17	−·0846	−·1364
18	·0267	−·0126
19	·0699	·1890
20	·0093	·1019
21	−·0452	−·1523
22	−·0246	−·1304
23	·0218	·0838
24	·0259	·0628
25	−·0049	−·0789
26	−·0201	−·0734
27	−·0046	·0371
28	·0122	·0272
29	·0081	·0387

Table A.9. *Theoretical and sample autocovariances for the MA(2) series.*

0	2·2500	2·0902
1	1·3200	1·1656
2	·2000	·0809
3	0·0000	−·0761
4	0·0000	−·0931
5	0·0000	−·2332
6	0·0000	−·3155
7	0·0000	−·1233
8	0·0000	·1057
9	0·0000	·1182
10	0·0000	·0634
11	0·0000	·1025
12	0·0000	·0107
13	0·0000	−·1959
14	0·0000	−·1710
15	0·0000	·0558
16	0·0000	·1849
17	0·0000	·1754
18	0·0000	·2067
19	0·0000	·2201
20	0·0000	·0893
21	0·0000	−·1024
22	0·0000	−·1488
23	0·0000	−·0917
24	0·0000	−·1255
25	0·0000	−·1606
26	0·0000	−·0531
27	0·0000	·0816
28	0·0000	·0805
29	0·0000	·0589

Table A.10. *Theoretical and sample autocovariances for the ARMA(2, 2) series.*

0	12·6690	12·2492
1	−11·8510	−11·3820
2	10·1570	9·5321
3	−8·2942	−7·4471
4	6·5334	5·4189
5	−4·9997	−3·6062
6	3·7329	2·1322
7	−2·7262	−1·1106
8	1·9502	·4343
9	−1·3672	−·0151
10	·9390	−·1126
11	−·6310	−·0321
12	·4139	·3517
13	−·2639	−·6667
14	·1626	·8863
15	−·0956	−1·0793
16	·0526	1·2140
17	−·0258	−1·1898
18	·0099	1·0134
19	−·0009	−·7145
20	−·0037	·3685
21	·0056	·0444
22	−·0060	−·4949
23	·0056	·8660
24	−·0048	−1·1248
25	·0040	1·2938
26	−·0031	−1·4156
27	·0024	1·4362
28	−·0018	−1·3276
29	·0013	1·1553

References

Abdrabbo, N. A. and Priestley, M. B. (1967). On the prediction of non-stationary processes. *J. Roy. Statist. Soc. Ser. B*, **29**, 570–585.

Abdrabbo, N. A. and Priestley, M. B. (1969). Filtering non-stationary signals. *J. Roy. Statist. Soc. Ser. B*, **31**, 150–159.

Abraham, B. and Box, G. E. P. (1975). Stochastic difference equation models. Technical Report, Department of Statistics, University of Wisconsin, Madison, Wisconsin, U.S.A.

Abraham, B. and Box, G. E. P. (1978). Deterministic and forecast-adaptive time-dependent models. *Appl. Statist.*, **27**, 120–130.

Akaike, H. (1960). Effect of timing-error on the power spectrum of sampled data. *Ann. Inst. Statist. Math.*, **11**, 145–165.

Akaike, H. (1964). Studies on the statistical estimation of frequency response functions. *Ann. Inst. Statist. Math., Suppl. III*, **15**, 5–17.

Akaike, H. (1965). On the statistical estimation of the frequency response function of a system having multiple input. *Ann. Inst. Statist. Math.*, **17**, 185–210.

Akaike, H. (1966). Note on higher order spectra. *Ann. Inst. Statist. Math.*, **18**, 123–126.

Akaike, H. (1967). Some problems in the application of the cross-spectral method. "Advanced Seminar on Spectral Analysis of Time Series" (Ed. Harris). Wiley, New York.

Akaike, H. (1969a). Fitting autoregressive models for prediction. *Ann. Inst. Statist. Math.*, **21**, 243–247.

Akaike, H. (1969b). Power spectrum estimation through autoregressive model fitting. *Ann. Inst. Statist. Math.*, **21**, 407–419.

Akaike, H. (1971). Information theory and an extension of the maximum likelihood principle. Research Memorandum No. 46, Institute of Statistical Mathematics, Tokyo. Published in 2nd Int. Symp. on Inf. Theory (Eds. B. N. Petrov and F. Csaki), pp. 267–281. Akademiai Kiade, Budapest (1973).

Akaike, H. (1973). Maximum likelihood identification of Gaussian autoregressive moving average models. *Biometrika*, **60**, 255–265.

Akaike, H. (1974a). Markovian representation of stochastic processes and its application to the analysis of autoregressive moving average processes. *Ann. Inst. Statist. Math.*, **26**, 363–387.

Akaike, H. (1974b). Stochastic theory of minimal realisations. *I.E.E.E. Trans. Automatic Control*, **AC-19**, No. 6, 667–673.

Akaike, H. (1974c). A new look at the statistical model identification. *I.E.E.E. Trans. Automatic Control*, **AC-19**, 716–722.

Akaike, H. (1976). Canonical correlations analysis of time series and the use of an information criterion. *In* "Advances and Case Studies in System Identification" (Eds. R. Mehra and D. G. Lainiotis). Academic Press, New York and London.

Akaike, H. (1977). An entropy maximisation principle. *In* Proc. Symp. on Applied Statistics (Ed. P. Krishnaiah). North-Holland, Amsterdam.

Akaike, H. (1978). A Bayesian analysis of the minimum AIC procedure. *Ann. Inst. Statist. Math.*, **30A**, 9–14.

Akaike, H. (1979). A Bayesian extension of the minimum AIC procedure of autoregressive model fitting. *Biometrika*, **66**, 237–242.

Akaike, H. and Yamanouchi, Y. (1962). On the statistical estimation of frequency response function. *Ann. Inst. Statist. Math.*, **14**, 23–56.

Alexander, M. J. and Vok, C. A. (1963). Tables of the cumulative distribution of sample multiple coherence. Research Report 63-67, Rocketdyne Division, North American Aviation Inc., Los Angeles, Cal.

Ali, M. M. (1977). Analysis of autoregressive moving average models: estimation and prediction. *Biometrika*, **64**, 535–545.

Ali, M. M. (1978). Corrections to "Analysis of autoregressive moving average models: estimation and prediction". *Biometrika*, **65**, 677.

d'Allesandro, P. (1972). Structural properties invariance and insensitivity of discrete time bilinear systems. *Ricerche di Automatica*, **3**, 158–169.

Amos, D. E. and Koopmans, L. H. (1963). Tables of the distribution of the coefficient of coherences for stationary bivariate Gaussian processes. Sandia Corporation Monograph SCR-483.

Anderson, N. (1974). On the calculation of fitted coefficients for maximum entropy spectral analysis. *Geophysics*, **39**, 69–72.

Anderson, R. L. (1942). Distribution of the serial correlation coefficient. *Ann. Math. Statist.*, **13**, 1–13.

Anderson, R. L. and Anderson, T. W. (1950). Distribution of the circular serial correlation coefficient for residuals from a fitted Fourier series. *Ann. Math. Statist.*, **21**, 59–81.

Anderson, T. W. (1948). On the theory of testing serial correlation. *Skand. Aktuarietidskr.*, **31**, 88–116.

Anderson, T. W. (1971). "The Statistical Analysis of Time Series". Wiley, New York.

Anderson, T. W. (1975). Maximum likelihood estimation of parameters of an autoregressive process with moving average residuals and other covariance matrices with linear structure. *Ann. Statist.*, **3**, 1283–1304.

Anderson, T. W. (1977). Estimation for autoregressive moving average models in the time and frequency domains. *Ann. Statist.*, **5**, 842–865.

Anderson, T. W. and Walker, A. M. (1964). On the asymptotic distribution of the autocorrelations of a sample from a linear stochastic process. *Ann. Math. Statist.*, **35**, 1296–1303.

Ansley, G. F. (1979). An algorithm for the exact likelihood of a mixed autoregressive moving average process. *Biometrika*, **66**, 59–65.

Astrom, K. J. (1970). "Introduction to Stochastic Control Theory". Academic Press, New York and London.

Astrom, K. J. and Bohlin, T. (1966). Numerical identification of linear dynamic systems from normal operating records. Proc. IFAC Symp. on Theory of Self Adapting Control Systems, Sept. 1965. Plenum, New York.

Astrom, K. J. and Eykhoff, P. (1971). System identification—a survey. *Automatica*, **7**, 123–162.

Astrom, K. J. and Kallstrom, C. G. (1973). Application of system identification techniques to the determination of ship dynamics. *In* "Identification and System Parameter Estimation" (Ed. P. Eykhoff), pp. 415–424. North-Holland, Amsterdam.

Bartlett, M. S. (1946). On the theoretical specification of sampling properties of autocorrelated time series. *J. Roy. Statist. Soc. Suppl.*, **8**, 27–41.

Bartlett, M. S. (1950). Periodogram analysis and continuous spectra. *Biometrika*, **37**, 1–16.

Bartlett, M. S. (1954). Problèmes de l'analyse spectrale des séries temporelles stationnaires. *Publ. Inst. Statist. Univ. Paris*, **III-3**, 119–134.

Bartlett, M. S. (1955). "An Introduction to Stochastic Processes with Special Reference to Methods and Applications," 1st Ed. Cambridge University Press, London.

Bartlett, M. S. (1957). Discussion on "Symposium on spectral approach to time series". *J. Roy. Statist. Soc. Ser. B*, **19**, 1–63.

Bartlett, M. S. (1963). Statistical estimation of density functions. *Sankhyā Ser. A*, **25**, 245–254.

Bartlett, M. S. (1966). "An Introduction to Stochastic Processes with Special Reference to Methods and Applications", 2nd Ed. Cambridge University Press, London.

Bartlett, M. S. (1967). Inference and stochastic processes. *J. Roy. Statist. Soc. Ser. A*, **130**, 457–477.

Bartlett, M. S. (1978). Correlation or spectral analysis? *The Statistician*, **27**, 147–158.

Bartlett, M. S. and Diananda, P. H. (1950). Extensions of Quenouille's test for autoregressive schemes. *J. Roy. Statist. Soc. Ser. B*, **12**, 108–115.

Beamish, N. (1977). The autoregressive (maximum entropy) spectral estimate. M.Sc. Thesis, University of Manchester.

Beamish, N. and Priestley, M. B. (1981). A study of AR and window spectral estimation. *Appl. Statist.*, **30**, No. 1.

Berk, K. N. (1974). Consistent autoregressive spectral estimates. *Ann. Statist.*, **2**, 489–502.

Bhansali, R. J. (1974). Asymptotic properties of the Wiener–Kolmogorov predictor: I. *J. Roy. Statist. Soc. Ser. B*, **36**, 61–73.

Bhansali, R. J. (1977a). Asymptotic properties of the Wiener–Kolmogorov predictor: II. *J. Roy. Statist. Soc. Ser. B*, **39**, 66–72.

Bhansali, R. J. (1977b). An application of Priestley's $P(\lambda)$ test to the Southend tide heights data. *In* "Recent Developments in Statistics" (Ed. J. R. Barra *et al.*), pp. 351–356, North-Holland, Amsterdam.

Bhansali, R. J. (1979). A mixed spectrum analysis of the lynx data. *J. Roy. Statist. Soc. Ser. A*, **142**, No. 2, 199–209.

Bhansali, R. J. (1980). Autoregressive and window estimates of the inverse correlation function. *Biometrika*, **67**, 551–566.

Bingham, C., Godfrey, M. D. and Tukey, J. W. (1967). Modern techniques in power spectrum estimation. *I.E.E.E. Trans. Audio Electroacoustics*, **AU-15**, 56–66.

Blackman, R. B. and Tukey, J. W. (1959). "The Measurement of Power Spectrum from the point of view of Communications Engineering". Dover: New York. (Originally published in *Bell Systems Tech. Journal*, **37**, 185–282 and 485–569 (1958)).

Blanc-Lapierre, A. and Fortet, R. (1946). Deux notes: sur la décomposition spectrale des fonctions aléatoires stationnaires d'ordre deux. *C.R. Acad. Sci. Paris*, **222**, 467 and 713.

Bloomfield, P. (1970). Spectral analysis with randomly missing observations. *J. Roy. Statist. Soc. Ser. B*, **32**, 369–380.

Bochner, S. (1936). "Lectures on Fourier Analysis". Princeton University Press, Princeton.

Bohlin, T. (1976). Four cases of identification of changing systems. *In* "System Identification: Advances and Case Studies" (Eds. R. Mehra and D. G. Laniotis). Academic Press, New York and London.

Boneva, L. I., Kendall, D. G. and Stefanov, I. (1971). "Spline transformations: three new diagnostic aids for the statistical data-analyst". *J. Roy. Statist. Soc. Ser. B*, **33**, 1–70.

Box, G. E. P., Hillmer, S. C. and Tiao, G. C. (1976). Analysis and modelling of seasonal time series. Technical Report No. 465, Department of Statistics, University of Wisconsin.

Box, G. E. P. and Jenkins, G. M. (1962). Some statistical aspects of adaptive optimisation and control. *J. Roy. Statist. Soc. Ser. B*, **24**, 297–343.

Box, G. E. P. and Jenkins, G. M. (1970). "Time Series Analysis, Forecasting and Control". Holden-Day, San Francisco.

Box, G. E. P. and MacGregor, J. F. (1974). The analysis of closed-loop dynamic-stochastic systems. *Technometrics*, **16**, 391–398.

Box, G. E. P. and Newbold, P. (1971). Some comments on a paper of Coen, Gomme and Kendall. *J. Roy. Statist. Soc. Ser. A*, **134**, 229–240.

Box, G. E. P. and Pierce, D. A. (1970). Distribution of residual autocorrelations in autoregressive–integrated moving average time series models. *J. Amer. Statist. Assoc.*, **65**, 1509–1526.

Box, G. E. P. and Tiao, G. C. (1975). Intervention analysis with applications to economic and environmental problems. *J. Amer. Statist. Assoc.*, **70**, 70–79.

Brigham, E. O. (1974). "The Fast Fourier Transform". Prentice-Hall, New Jersey.

Brigham, E. O. and Morrow, R. E. (1967). The fast Fourier transform. *I.E.E.E. Spectrum*, **4**, 63–70.

Brillinger, D. R. (1964). The generalisation of the techniques of factor analysis, canonical correlation and principal components to stationary time series. Paper presented at the Royal Statistical Society Conference, Cardiff.

Brillinger, D. R. (1965). An introduction to polyspectra. *Ann. Math. Statist.*, **36**, 1351–1374.

Brillinger, D. R. (1968). Estimation of the cross-spectrum of a stationary bivariate Gaussian process from its zeros. *J. Roy. Statist. Soc. Ser. B*, **30**, 145–159.

Brillinger, D. R. (1969a). Asymptotic properties of spectral estimates of second order. *Biometrika*, **56**, 375–390.

Brillinger, D. R. (1969b). The canonical analysis of stationary time series. *In* "Multivariate Analysis: II (Ed. P. R. Krishnaiah). Academic Press, New York and London.

Brillinger, D. R. (1970). The identification of polynomial systems by means of higher order spectra. *J. Sound Vib.*, **12**, 301–313.

Brillinger, D. R. (1972). "The spectral analysis of stationary interval functions". Proc. Seventh Berkeley Symp. on Prob. Statist. (Eds. L. Le Cam, J. Neyman and E. L. Scott), pp. 483–513. University of California Press, Berkeley.

Brillinger, D. R. (1975). "Time Series: Data Analysis and Theory". Holt, Rinehart and Winston, New York.

Brillinger, D. R. and Rosenblatt, M. (1967). Asymptotic theory of k-th order spectra. In "Spectral Analysis of Time Series" (Ed. B. Harris), pp. 153–188. Wiley, New York.

Brockett, R. W. (1976). Volterra series and geometric control theory. Automatica, 12, 167–172.

Bruni, C., Di Pillo, G. and Koch, G. (1974). Bilinear systems: an appealing class of "nearly linear" systems in theory and applications. I.E.E.E. Trans. Automatic Control, AC-19, 334–348.

Brunt, D. (1917). "The Combination of Observations". Cambridge University Press, London.

Burg, J. P. (1967). Maximum entropy spectral analysis. Paper presented at the 37th Annual International S.E.G. Meeting, Oklahoma City, Oklahoma.

Burg, J. P. (1972). The relationship between maximum entropy spectra and maximum likelihood spectra. Geophysics, 37, 375–376.

Caines, P. E. (1972). Relationship between Box-Jenkins-Astrom control and Kalman linear regulator. Proc. I.E.E.E., 119, 615–620.

Caines, P. E. and Chan, C. W. (1975). Feedback between stationary stochastic processes. I.E.E.E. Trans. Automatic Control, AC-20, 498–508.

Campbell, M. J. and Walker, A. M. (1977). A survey of statistical work on the MacKenzie River series of annual Canadian lynx trappings for the years 1821–1934, and a new analysis. J. Roy. Statist. Soc. Ser. A, 140, 411–431.

Capon, J. (1969). High resolution frequency-wave number spectrum analysis. Proc. I.E.E.E., 57, 1408–1418.

Cartwright, D. E. (1968). A unified analysis of tides and surges round North and East Britain. Philos. Trans. Roy. Soc. London Ser. A, 263, 1–55.

Chan, W.-Y. T. and Tong, H. (1975). A simulation study of the estimation of evolutionary spectral functions. Appl. Statist., 24, 333–341.

Chanda, K. C. (1964). Asymptotic expansions for tests of goodness of fit for linear autoregressive schemes. Biometrika, 51, 459–465.

Chatfield, C. (1975). "The Analysis of Time Series: Theory and Practice". Chapman and Hall, London.

Chatfield, C. (1979). Inverse autocorrelations. J. Roy. Statist. Soc. Ser. A, 142, No. 3, 363–377.

Chatfield, C. and Prothero, D. L. (1973a). Box–Jenkins seasonal forecasting: problems in a case study. J. Roy. Statist. Soc. Ser. A, 136, 295–336.

Chatfield, C. and Prothero, D. L. (1973b). A reply to some comments by Box and Jenkins. J. Roy. Statist. Soc. Ser. A, 136, 345–352.

Cleveland, W. S. (1972). The inverse autocorrelations of a time series and their applications. Technometrics, 14, 277–293.

Cleveland, W. S. and Parzen, E. (1975). The estimation of coherence, frequency response and envelope delay. Technometrics, 17, 167–172.

Cleveland, W. P. and Tiao, G. C. (1976). Decomposition of seasonal time series: a model for the X-11 program. J. Amer. Statist. Assoc., 17, 581–587.

Clinger, W. and Van Ness, J. W. (1976). On unequally spaced time points in time series. Ann. Statist. 4, 736–745.

Comes, P. and Michel, G. (1974). High resolution Fourier spectra of stars and planets. *Astrophys. J.*, **190**, L29–L32.

Cooley, J. W., Lewis, P. A. W. and Welch, P. D. (1967). Historical notes on the fast Fourier transform. *I.E.E.E. Trans. Electroacoustics*, **AU-15**, 76–79.

Cooley, J. W., Lewis, P. A. W. and Welch, P. D. (1970). The application of the fast Fourier transform algorithm to the estimation of spectra and cross-spectra. *J. Sound Vib.*, **12**, 339–352.

Cooley, J. W. and Tukey, J. W. (1965). An algorithm for the machine calculation of complex Fourier series. *Math. Comp.*, **19**, 297–301.

Cooper, D. M. and Thompson, R. (1977). A note on the estimation of the parameters of the autoregressive moving average process. *Biometrika*, **64**, 625–628.

Cox, D. R. and Miller, H. D. (1965). "The Theory of Stochastic Processes". Methuen, London.

Cramer, H. (1951). "A contribution to the theory of stochastic processes". Proc. 2nd Berkeley Symp. on Math. Statist. and Prob., pp. 329–339. University of California Press.

Cramer, H. (1961). On some classes of non-stationary processes. Proc. 4th Berkeley Symp. on Math. Statist. and Prob., pp. 57–78. University of California Press.

Daniell, P. J. (1946). Discussion on "Symposium on autocorrelation in time series". *J. Roy. Statist. Soc., Suppl.* **8**, 88–90.

Daniells, H. E. (1956). The approximate distribution of serial correlation coefficients. *Biometrika*, **43**, 169–185.

Daniells, H. E. (1957). Discussion on "Symposium on spectral approach to time series". *J. Roy. Statist. Soc. Ser. B*, **19**, 1–63.

Daniells, H. E. (1962). The estimation of spectral densities. *J. Roy. Statist. Soc. Ser. B*, **24**, 185–198.

Daniells, H. E. (1965). Discussion on "Evolutionary spectra and non-stationary processes". *J. Roy. Statist. Soc. Ser. B*, **27**, 234.

Davenport, W. B. and Root, W. L. (1958). "Random Signals and Noise". McGraw-Hill, New York.

Davies, N. and Newbold, P. (1979). "Some power studies of a portmanteau test of time series model specification". *Biometrika*, **66**, 153–155.

Davies, N., Triggs, C. M. and Newbold, P. (1977). Significance levels of the Box-Pierce portmanteau statistic in finite sample. *Biometrika*, **65**, 297–303.

De Bruijn, N. G. (1967). Uncertainty principle in Fourier analysis. *In* "Inequalities", pp. 59–71. Academic Press, New York and London.

Deistler, M., Dunsmuir, W. and Hannan, E. J. (1976). Vector linear time series models: connections and extensions. *Adv. in Appl. Probability*, **10**, 360–372.

Dixon, W. J. (1944). Further contributions to the problem of serial correlation. *Ann. Math. Statist.*, **15**, 119–144.

Doob, J. L. (1949). Time series and harmonic analysis. Proc. 1st Berkeley Symp. on Math. Statist. and Prob., pp. 303–343. University of California Press.

Doob, J. L. (1953). "Stochastic Processes". Wiley, New York.

Dunsmuir, W. (1979). A central limit theorem for parameter estimation in stationary vector time series and its application to models with a signal observed with noise. *Ann. Statist.*, **7**, 490–506.

Dunsmuir, W. and Hannan, E. J. (1976). Vector linear time series models. *Adv. in Appl. Probability*, **8**, 339–364.

Durbin, J. (1959). Efficient estimation of parameters in moving average models. *Biometrika*, **46**, 306–316.

Durbin, J. (1960). The fitting of time series models. *Rev. Inst. Internat. Statist.*, **28**, 233–244.

Durbin, J. (1961). Efficient fitting of linear models for continuous stationary time series from discrete data. *Bull. Inst. Internat. Statist.* **38**, 273–282.

Durbin, J. (1963). Trend elimination for the purpose of estimating seasonal and periodic components in time series. *In* "Time Series Analysis" (Ed. M. Rosenblatt). Wiley, New York.

Durbin, J. and Murphy, M. J. (1975). Seasonal adjustment based on a mixed adaptive-multiplicative model. *J. Roy. Statist. Soc. Ser. A*, **138**, 385–410.

Durbin, J. and Watson, G. S. (1950). Testing for serial correlation in least squares regression—I. *Biometrika*, **37**, 409–428.

Durbin, J. and Watson, G. S. (1951). Testing for serial correlation in least squares regression—II. *Biometrika*, **38**, 159–178.

Durbin, J. and Watson, G. S. (1971). Testing for serial correlation in least squares regression—III. *Biometrika*, **58**, 1–19.

Einstein, A. (1905). "Investigations on the Theory of Brownian Movement". Dover, New York (1956).

Enochson, L. D. and Goodman, N. R. (1965). Gaussian approximations to the distribution of sample coherence. Technical Report AFFDL TR 65-67, Research and Tech. Div., AFSC, Wright-Patterson Air Force Base, Ohio.

Epanechnikov, V. A. (1969). Non-parametric estimation of a multivariate probability density. *Theor. Probability Appl.*, **14**, 153–158.

Fejer, L. (1910). Lebesquesche Konstanten und divergente Fourierreihen. *J. Reine Angew. Math.*, **138**, 22–53.

Feller, W. (1948). On the Kolmogarov–Smirnov theorems for empirical distributions. *Ann. Math. Statist.*, **19**, 177–189.

Feller, W. (1950). "An Introduction to Probability Theory and its Applications", Vol. I. Wiley, New York.

Feller, W. (1966). "An Introduction to Probability Theory and its Applications", Vol. II. Wiley, New York.

Fisher, R. A. (1929). Tests of significance in harmonic analysis. *Proc. Roy. Soc. Ser. A*, **125**, 54–59.

Fraser, D. A. S. (1958). "Statistics: An Introduction". Wiley, New York.

Gasser, T. (1974). A resistant spectrum estimate. Research Report No. 4, Fachgruppe fur Statistik, Eidgenoessische Technische Hochschule, Zurich.

Gaster, M. and Roberts, J. B. (1977). On the spectral analysis of randomly sampled records by a direct transform. *Proc. Roy. Soc. Ser. A.*, **354**, 27.

Gentleman, W. M. and Sande, G. (1966). Fast Fourier transforms—for fun and profit. AFIPS, 1966 Fall Joint Computer Conference, Conf. Proc., **28**, 563–578. Spartan: Washington.

Godfrey, L. G. (1979). Testing the adequacy of a time series model. *Biometrika*, **66**, 67–72.

Godfrey, M. D. (1974). Computational methods for time series. *Bull. Inst. Math. Appl.*, **10**, 224–227.

Godolphin, E. J. (1977). A direct representation for the maximum likelihood estimator of a Gaussian moving average process. *Biometrika*, **64**, 375–384.

Goldolphin, E. J. (1978). Modified maximum likelihood estimation of Gaussian moving averages using a pseudoquadratic convergence criterion. *Biometrika*, **65**, 203–206.

Goodman, N. R. (1957). On the joint estimation of the spectra, cospectrum and quadrative spectrum of a two-dimensional Gaussian process. Scientific Paper No. 10, Engineering Statistics Laboratory, New York University, N.Y.

Goodman, N. R. (1963). Statistical analysis based on a certain multivariate complex Gaussian distribution (an introduction). *Ann. Math. Statist.*, **34**, 152–177.

Goodman, N. R. (1965). Measurement of matrix frequency response functions and multiple coherence functions. Research and Technology Division, AFSC, AFFDL. TR 65-66, Wright–Patterson Air Force Base, Ohio.

Gower, J. C. (1955). A note on the periodigram of the Beveridge Wheat Price Index. *J. Roy. Statist. Soc. Ser. B*, **17**, 228–234.

Granger, C. W. J. and Andersen, A. P. (1976). Introduction to bilinear time series models. Discussion Paper No. 76-5, Department of Economics, University of California, San Diego.

Granger, C. W. J. and Andersen, A. P. (1978). "An Introduction to Bilinear Time Series Models". Vandenhoeck and Ruprecht, Göttingen.

Granger, C. W. J. and Hatanaka, M. (1964). "Spectral Analysis of Economic Time Series". Princeton University Press, New Jersey.

Grenander, U. (1950). Stochastic processes and statistical inference. *Ark. Mat.*, **1**, 195–277.

Grenander, U. (1951). On empirical spectral analysis of stochastic processes. *Ark. Mat.*, **1**, 503–531.

Grenander, U. (1954). On the estimation of regression coefficients in the case of an autocorrelated disturbance. *Ann. Math. Statist.*, **25**, 252–272.

Grenander, U. (1958). Resolvability and reliability in spectral analysis. *J. Roy. Statist. Soc. Ser. B*, **20**, 152–157.

Grenander, U. and Rosenblatt, M. (1953). Statistical spectral analysis arising from stationary stochastic processes. *Ann. Math. Statist.*, **24**, 537–558.

Grenander, U. and Rosenblatt, M. (1957a). "Statistical Analysis of Stationary Time Series". Wiley, New York.

Grenander, U. and Rosenblatt, M. (1957b). Some problems in estimating the spectrum of a time series. Proc. 3rd Berkeley Symp. on Statist. and Prob. University of California Press, Berkeley.

Groves, G. W. and Hannan, E. J. (1968). Time series regression of sea level on weather. *Rev. Geophys.*, **6**, 129–174.

Gudmundsson, G. (1971). Time series analysis of imports, exports and other economic variables. *J. Roy. Statist. Ser. Soc. A*, **134**, 383.

Gustavson, I., Ljung, L. and Soderstrom, T. (1976). Identification of processes in closed loop: identifiability and accuracy aspects. Report No. 7602(c), Department of Automatic Control, Lund Institute of Technology, Sweden.

Haggan, V. (1975). Dimensionality reduction in multivariable stochastic systems: application to a chemical engineering plant. *Int. J. Control*, **22**, 763–772.

Haggan, V. (1977). Discussion on "Some comments on the Canadian lynx data" by H. Tong, *J. Roy. Statist. Soc. Ser. A*, **140**, 458.

Haggan, V. and Ozaki, T. (1978). Amplitude-dependent AR model fitting for non-linear random vibrations. Proc. International Time Series Meeting, University of Nottingham, England, March 1979 (Ed. O. D. Anderson). North-Holland, The Netherlands.

Haggan, V. and Priestley, M. B. (1975). Dimensionality reduction in multivariable stochastic systems. *Int. J. Control*, **22**, 245–259.

Halmos, P. R. (1951). "Introduction to Hilbert Space". Chelsea, New York.

Hamilton, D. C. and Watts, D. G. (1978). Interpreting partial autocorrelation functions of seasonal time series models. *Biometrika*, **65**, 135–140.

Hammond, J. K. (1968). On the response of single and multi degree-of-freedom systems to non-stationary random excitations. *J. Sound Vib.*, **7**, 393.

Hammond, J. K. (1973). Evolutionary spectra in random vibrations. *J. Roy. Statist. Soc. Ser. B*, **35**, 167–188.

Hannan, E. J. (1955). Exact tests for serial correlation. *Biometrika*, **42**, 133–142.

Hannan, E. J. (1958a). The estimation of the spectral density function after trend removal. *J. Roy. Statist. Soc. Ser. A*, **20**, 323–333.

Hannan, E. J. (1958b). The asymptotic powers of certain tests of goodness of fit for time series. *J. Roy. Statist. Soc. Ser. B*, **20**, 143–151.

Hannan, E. J. (1960). "Time Series Analysis". Methuen, London.

Hannan, E. J. (1961). "Testing for a jump in the spectral function". *J. Roy. Statist. Soc. Ser. B*, **23**, 394–404.

Hannan, E. J. (1969a). The identification of vector mixed autoregressive moving average systems. *Biometrika*, **56**, 223–225.

Hannan, E. J. (1969b). The estimation of mixed moving average autoregressive systems. *Biometrika*, **56**, 579–593.

Hannan, E. J. (1970). "Multiple Time Series". Wiley, New York.

Hannan, E. J. (1973). The asymptotic theory of linear time series models. *J. Appl. Probability*, **10**, 130–145.

Hannan, E. J. (1976). The identification and parameterisation of ARMAX and state space forms. *Econometrica*, **44**, 713–723.

Hannan, E. J. (1979a). Estimating the dimension of a linear system. Proc. International Time Series Meeting, University of Nottingham, England, March 1979 (Ed. O. D. Anderson). North-Holland, The Netherlands.

Hannan, E. J. (1979b). The statistical theory of linear systems. *In* "Developments in Statistics" (Ed. P. R. Krishnaiah), Vol. 2, pp. 83–121. Academic Press, New York and London.

Hannan, E. J. (1980). The estimation of the order of an ARMA process. *Ann. Statist.* **8**, 1071–1081.

Hannan, E. J. and Quinn, B. G. (1979). The determination of the order of an autoregression. *J. Roy. Statist. Soc. Ser. B*, **41**, 190–195.

Hannan, E. J. and Thomson, P. J. (1971). The estimation of coherence and group delay. *Biometrika*, **58**, 469–481.

Harris, C. J. (1976). Problems in systems identification and control. *Bull. Inst. Maths. Appl.*, **12**, 139–150.

Harrison, P. J. and Stevens, C. F. (1976). Bayesian forecasting. *J. Roy. Statist. Soc.*, **38**, 205–248.

Hartley, H. O. (1949). Tests of significance in harmonic analysis. *Biometrika*, **36**, 194–201.

Hasofer, A. M. and Petocz, P. (1978). The envelope of an oscillatory process and its upcrossings. *J. Adv. Appl. Probability*, **10**, 711–716.

Hasselman, K., Munk, W. and Macdonald, G. (1963). Bispectrum of ocean waves. *In* "Time Series Analysis" (Ed. M. Rosenblatt), pp. 125–139. Wiley, New York.

Heine, V. (1955). Models for two-dimensional stationary stochastic processes. *Biometrika*, **42**, 170–178.

Helland, K. N., Lii, K. S. and Rosenblatt, M. (1979). Bispectra and energy transfer in grid-generated turbulence. *In* "Developments in Statistics" (Ed. P. R. Krishnaiah), Vol. 2, pp. 125–155. Academic Press, New York.

Helson, H. and Lowdenslager, D. (1958). Prediction theory and Fourier series in several variables. *Acta Math.*, **99**, 165–202.

Hillmer, S. C. and Tiao, G. C. (1979). Likelihood function of stationary multiple autoregressive moving average models. *J. Amer. Statist. Assoc.*, **74**, 652–661.

Ho, B. L. and Kalman, R. E. (1966). Effective construction of linear state-variable models from input/output functions. *Regelungestechnik*, **14**, 545–548.

Hosking, J. R. M. (1978). A unified derivation of the asymptotic distributions of goodness-of-fit statistics for autoregressive time series models. *J. Roy. Statist. Soc. Ser. B*, **40**, 341–349.

Isidori, A. (1973). Direct construction of minimal bilinear realisations from non-linear input/output maps. *I.E.E.E. Trans. Automatic Control*, **AC-18**, No. 6, 626–631.

Isserlis, L. (1918). On a formula for the product moment coefficient of any order of a normal frequency distribution in any number of variables. *Biometrika*, **12**, 134–139.

Jazwinski, A. H. (1970). "Stochastic Processes and Filtering Theory". Academic Press, New York and London.

Jenkins, G. M. (1956). "Tests of hypotheses in the linear autoregressive model: II—null distributions for higher order schemes: non-null distributions". *Biometrika*, **43**, 186–199.

Jenkins, G. M. (1961). General considerations in the analysis of spectra. *Technometrics*, **3**, 133–166.

Jenkins, G. M. (1963). Cross-spectral analysis and the estimation of linear open loop transfer functions. *In* "Time Series Analysis" (Ed. M. Rosenblatt), pp. 267–278. Wiley, New York.

Jenkins, G. M. (1965). A survey of spectral analysis. *Appl. Statist.*, **14**, 2–32.

Jenkins, G. M. and Priestley, M. B. (1957). The spectral analysis of time series. *J. Roy. Statist. Soc. Ser. B*, **19**, 1–12.

Jenkins, G. M. and Watts, D. G. (1968). "Spectral Analysis and its Applications". Holden-Day, San Francisco.

Jones, D. A. (1978). Non-linear autoregressive processes. *Proc. Roy. Soc. Ser. A*, **360**, 71–95.

Jones, R. H. (1962). Spectral analysis with regularly missed observations. *Ann. Math. Statist.*, **33**, 455–461.

Jones, R. H. (1971). Spectrum estimation with missing observations. *Ann. Inst. Statist. Math.*, **23**, 387–398.

Jones, R. H. (1974). Identification and autoregressive spectrum estimation. *I.E.E.E. Trans. Automatic Control*, **AC-19**, 894–897.

Kaiser, J. F. (1966). Digital filters. *In* "System Analysis by Digital Computer" (ed. F. F. Kuo and J. F. Kaiser). Wiley, New York.

Kalman, R. E. (1960). A new approach to linear filtering and prediction problems. *Trans. ASME J. Basic Engrg., Series D*, **82**, 35–45.

Kalman, R. E. (1963). New methods of Wiener filtering theory. *In* Proc. 1st. Symp. on Eng. Appns. of Random Functions Theory and Prob. (Eds. J. L. Bogdanoff and F. Kozin). Wiley, New York.

Kalman, R. E. (1978). A retrospective after twenty years: from the pure to the applied. Chapman Conf. on Appns. of Kalman Filter to Hydrology, Hydraulics and Water Resources, American Geophysical Union, Pittsburgh.

Kalman, R. E. and Bucy, R. S. (1961). New results in linear filtering and prediction problems. *Trans. ASME. J. Basic. Engrg., Series D*, **83**, 95–108.

Kashyap, R. L. (1977). A Bayesian comparison of different classes of dynamic models using empirical data. *I.E.E.E. Trans. on Automatic Control*, **AC-22**, 715–727.

Kendall, M. G. and Stuart, A. (1966). "The Advanced Theory of Statistics", Vols. I, II, III. Griffin, London.

Kharkevich, A. A. (1960). "Spectra and Analysis". Consultants Bureau, New York Enterprises Inc., New York.

Khintchine, A. (1934). "Korrelationtheorie der stationären Prozesse". *Math. Ann.*, **109**, 604–615.

Kingman, J. F. C. and Taylor, S. J. (1966). "Introduction to Measure and Probability". Cambridge University Press, London.

Kleiner, B., Martin, R. D. and Thomson, D. J. (1979). Robust estimation of power spectra. *J. Roy. Statist. Soc. Ser. B*, **41**,

Kohn, R. (1977). Note concerning the Akaike and Hannan estimation procedures for an autoregressive moving average process. *Biometrika*, **66**, 622–625.

Kolmogorov, A. (1933). Grundbegriffe der Wahrscheinlichkeitsrechung, *Ergebnisse der Math. (Berlin)*, **2**, No. 3

Kolmogorov, A. (1941a). Stationary sequences in Hilbert space (in Russian). *Bull. Math. Univ. Moscow*, **2**, No. 6.

Kolmogorov, A. (1941b). Interpolation und extrapolation von stationären Zufälligen Folgen. *Bull. Acad. Sci. (Nauk)*, *U.S.S.R.*, *Ser. Math.*, **5**, 3–14.

Koopmans, L. H. (1964a). On the coefficient of coherence for weakly stationary stochastic processes. *Ann. Math. Statist.*, **35**, 532–549.

Koopmans, L. H. (1964b). On the multivariate analysis of weakly stationary stochastic processes. *Ann. Math. Statist.*, **35**, 1765–1780.

Koopmans, L. H. (1974). "The Spectral Analysis of Time Series". Academic Press, New York and London.

Koopmans, T. (1942). Serial correlation and quadratic forms in normal variates. *Ann. Math. Statist.*, **13**, 14–33.

Kufner, A. and Kadlec, J. (1971). "Fourier Series". Iliffe Books, London.

Lacoss, R. T. (1971). Data adaptive spectral analysis methods. *Geophysics*, **36**, 661–675.

Lanczos, C. (1956). "Applied Analysis". Prentice-Hall, Englewood Cliffs.

Lawley, D. N. and Maxwell, A. E. (1971). "Factor Analysis as a Statistical Method". Butterworth, London.

Lehmann, E. L. (1959). "Testing Statistical Hypotheses". Wiley, New York.

Leipnik, R. B. (1947). Distribution of the serial correlation coefficient in a circularly correlated universe. *Ann. Math. Statist.*, **18**, 80–87.

Lii, K. S., Rosenblatt, M. and Van Atta, C. (1976). Bispectral measurements in turbulence. *J. Fluid. Mech.*, **77**, 45–62.

Lim, K. S. and Tong, H. (1978). Threshold antoregressive modelling of cyclical data. Technical Report No. 102, Department of Mathematics (Statistics), University of Manchester Institute of Science and Technology, England.

Lindley, D. V. (1965). "Introduction to Probability and Statistics from a Bayesian Viewpoint", Vols. I, II. Cambridge University Press, London.

Ljung, G. M. and Box, G. E. P. (1978). On a measure of lack of fit in time series models. *Biometrika*, **65**, 297–303.

Ljung, G. M. and Box, G. E. P. (1979). The likelihood function of stationary autoregressive moving average models. *Biometrika*, **66**, 265–270.

Loève, M. (1945). Sur les fonctions aléatoires stationnaires de second ordre. *Rev. Sci. Paris*, **83**, 297.

Loève, M. (1963). "Probability Theory". Van Nostrand, Princeton.

Lomnicki, Z. A. and Zaremba, S. K. (1957a). On estimating the spectral density function of a stochastic process. *J. Roy. Statist. Soc. Ser. B*, **19**, 13–37.

Lomnicki, Z. A. and Zaremba, S. K. (1957b). On some moments and distributions occurring in the theory of linear stochastic processes—I. *Monatsh. Math.*, **61**, 318–358.

Lomnicki, Z. A. and Zaremba, S. K. (1957c). On the estimation of autocorrelation in time series. *Ann. Math. Statist.*, **28**, 1.

Lomnicki, Z. A. and Zaremba, S. K. (1959a). On some moments and distributions occurring in the theory of linear stochastic processes—II. *Monatsh. Math.*, **63**, 128–168.

Lomnicki, Z. A. and Zaremba, S. K. (1959b). Bandwidth and resolvability in statistical spectral analysis. *J. Roy. Statist. Soc. Ser. B*, **21**, 169–171.

Longuett-Higgins, M. S. (1957). The statistical analysis of a randomly moving surface. *Philos. Trans. Roy. Soc. London Ser. A*, **249**, 287–321.

MacNeill, I. B. (1974). "Tests for periodic components in multiple time series. *Biometrika*, **61**, 57–70.

MacNeill, I. B. (1977). A test of whether several series share common periodicities. *Biometrika*, **64**, 495–508.

McClave, J. T. (1973). On the bias of autoregressive approximation to moving averages. *Biometrika*, **60**, 599–605.

McClave, J. T. (1974). A comparison of moving average estimation procedures. *Comm. Statist.*, **3**, 865–883.

McLeod, A. I. (1977). Improved Box–Jenkins estimators. *Biometrika*, **64**, 531–534.

Madow, W. G. (1945). Note on the distribution of the serial correlation coefficient. *Ann. Math. Statist.*, **16**, 308–310.

Mallows, C. L. (1973). Some comments on C_p. *Technometrics*, **15**, 661–675.

Mandrekar, V. (1972). A characterisation of oscillatory processes and their prediction. *Proc. Amer. Math. Soc.*, **32**, 280–284.

Mann, H. B. and Wald, A. (1943). On the statistical treatment of linear stochastic difference equations. *Econometrica*, **11**, 173–220.

Mark, W. D. (1970). Spectral analysis of the convolution and filtering of non-stationary stochastic processes. *J. Sound. Vib.*, **11**, 19.

Martin, R. D. (1979). Robust estimation for time series. Proc. International Time Series Meeting, University of Nottingham, England, March 1979 (Ed. O. D. Anderson). North-Holland, Amsterdam.

Masani, P. (1966). Recent trends in multivariate prediction theory. *In* "Multivariate Analysis—I" (Ed. P. Krishnaiah), pp. 351–382. Academic Press, New York and London.

Mehra, R. K. (1977). Kalman filters and their application to forecasting. *Management Sciences* (special issue on "Forecasting"). Reprinted in "TIMS Studies in the Management Sciences" (Ed. Makridakis and Wheelwright), Vol. 12. North Holland, Amsterdam.

Mentz, R. P. (1977). Estimation in the first order moving average model through the finite autoregressive approximation. *J. Econometrics*, **6**.

Meyer, P. L. (1965). "Introductory Probability and Statistical Applications". Addison-Wesley, Massachusetts.

Middleton, D. (1960). "Statistical Communication Theory". McGraw-Hill, New York.

Mohler, R. R. (1973). "Bilinear Control Processes". Academic Press, New York and London.

Mood, A. M. and Graybill, F. A. (1963). "Introduction to the Theory of Statistics". McGraw-Hill, New York.

Moran, P. A. P. (1954). Some experiments on the prediction of sunspot numbers. *J. Roy. Statist. Soc. Ser. B*, **16**, 112–117.

Morris, J. (1977). Forecasting the sunspot cycle. *J. Roy. Statist. Soc. Ser. A*, **140**, 437–447.

Murthy, V. K. (1962). Estimation of the spectrum. *Ann. Math. Statist.*, **32**, 730–738.

Nadaraya, E. N. (1970). Remarks on non-parametric estimates for density functions and regression curves. *Theor. Probability Appl.*, **15**, 134.

Neave, H. R. (1970a). Spectral analysis of a stationary time series using initially scarce data. *Biometrika*, **57**, 111–122.

Neave, H. R. (1970b). An improved formula for the asymptotic variance of spectrum estimates. *Ann. Math. Statist.*, **41**, 70–77.

Neave, H. R. (1971). The exact error in spectrum estimates. *Ann. Math. Statist.*, **42**, 961–975.

Neave, H. R. (1972). A comparison of lag window generators, *J. Amer. Statist. Assoc.*, **67**, 152–158.

Nettheim, N. (1966). The estimation of coherence. Technical Report No. 5, Department of Statistics, Stanford University.

Newbold, P. (1974). The exact likelihood function for a mixed autoregressive moving average process. *Biometrika*, **61**, 423–426.

Newbold, P. and Granger, C. W. J. (1974). Experience with forecasting univariate time series and the combination of forecasts. *J. Roy. Statist. Soc. Ser. A*, **137**, 131–165.

Nicholls, D. F. (1967). Estimation of the spectral density function when testing for a jump in the spectrum. *Austral. J. Statist.*, **9**, 103–108.

Nicholls, D. F. (1969). Testing for a jump in co-spectra. *Austral. J. Statist.*, **11**, 7–13.

Nicholls, D. F. (1972). On Hannan's estimation of ARMA models. *Austral. J. Statist.*, **3**, 262–269.

Nicholls, D. F. (1973). "Frequency domain estimation procedures for linear models". *Biometrika*, **60**, 202–205.

Nicholls, D. F. (1976). The efficient estimation of vector linear time series models. *Biometrika*, **63**, 381–390.

Nicholls, D. F. and Hall, A. D. (1979). The exact likelihood function of multivariate autoregressive moving average models. *Biometrika*, **66**, 259–264.

O'Hagan, A. (1978). Curve fitting and optimal design for prediction. *J. Roy. Statist. Soc. Ser. B*, **40**, 1–42.

Osborn, D. R. (1976). Maximum likelihood estimation of moving average processes". *Ann. Econom. Social Measurement*, **5**, 75–87.

Osborn, D. R. (1977). Exact and approximate maximum likelihood estimators for vector moving average processes. *J. Roy. Statist. Soc. Ser. B*, **39**, 114–118.

Ozaki, T. (1978). Non-linear models for non-linear random vibrations. Technical Report No. 92, Department of Mathematics (Statistics), University of Manchester Institute of Science and Technology, England.

Page, C. H. (1952). Instantaneous power spectra. *J. Appl. Phys.*, **23**, 103–106.

Paley, R. E. C. and Wiener, N. (1934). "Fourier Transforms in the Complex Domain". Amer. Math. Soc., Providence.

Pandit, S. M. and Wu, S. M. (1975). Unique estimates of the parameters of a continuous stationary process. *Biometrika*, **62**, 497–502.

Parthasarathy, K. R. (1960). On the estimation of the spectrum of a stationary stochastic process. *Ann. Math. Statist.*, **31**, 568–573.

Parzen, E. (1957a). On consistent estimates of the spectrum of a stationary time series. *Ann. Math. Statist.*, **28**, 329–348.

Parzen, E. (1957b). On choosing an estimate of the spectral density function of a stationary time series. *Ann. Math. Statist.*, **28**, 921–932.

Parzen, E. (1958). On asymptotically efficient consistent estimates of the spectral density function of a stationary time series. *J. Roy. Statist. Soc. Ser. B*, **20**, 303–322.

Parzen, E. (1959). Statistical inference on time series by Hilbert space methods: I. Technical Report No. 23, Department of Statistics, Stanford University. (Published in "Time Series Analysis Papers" by E. Parzen. Holden-Day, San Francisco.)

Parzen, E. (1960). "Modern Probability and its Applications". Wiley, New York.

Parzen, E. (1961a). An approach to time series analysis. *Ann. Math. Statist.*, **32**, 951–989.

Parzen, E. (1961b). Mathematical considerations in the estimation of spectra. *Technometrics*, **3**, 167–190.

Parzen, E. (1962). "Stochastic Processes". Holden-Day, San Francisco.

Parzen, E. (1962a). On the estimation of a probability density function and mode. *Ann. Math. Statist.*, **33**, 1065–1076.

Parzen, E. (1963a). Notes on Fourier analysis and spectral windows. Technical Report No. 48, Department of Statistics, Stanford University. (Published in "Time Series Analysis Papers" by E. Parzen. Holden-Day, San Francisco.)

Parzen, E. (1963b). On spectral analysis with missing observations and amplitude modulation. *Sankhyā, Ser. A*, **25**, 383–392.

Parzen, E. (1967). On empirical multiple time series analysis. Proc. 5th Berkeley Symp. Math. Statist. Prob. (Eds. L. Le Cam and J. Neyman), pp. 305–340. University of California Press, Berkeley.

Parzen, E. (1969). Multiple time series modelling. *In* "Multivariate Analysis, II" (Ed. P. Krishnaiah), pp. 389–409. Academic Press, New York and London.

Parzen, E. (1971). Efficient estimation of stationary time series mixed schemes. *Bull. Inst. Internat. Statist.*, **44**, 315–319.

Parzen, E. (1974). Some recent advances in time series modelling. *I.E.E.E. Trans. Automatic Control*, **AC-19**, 723–729.

Parzen, E. (1977a). Non-parametric statistical data science: a unified approach based on density estimation and testing for white noise. Technical Report No. 47, Statistical Science Division, State University of New York, Buffalo.

Parzen, E. (1977b). Multiple time series modelling: determining the order of approximating autoregressive schemes. *In* "Multivariate Analysis, IV" (Ed. P. Krishnaiah), pp. 283–295. North-Holland, Amsterdam.

Phadke, M. S. and Kedem, G. (1978). Computation of the exact likelihood function of multivariate moving average models. *Biometrika*, **65**, 511–520.

Phadke, M. S. and Wu, S. M. (1969). Modelling of continuous stochastic processes from discrete observations with applications to sunspot data. *J. Amer. Statist. Assoc.*, **69**, 325–329.

Pham-Dinh, T. (1975). Estimation et teste dans les modeles parametriques des proces stationnaires. Ph.D. Thesis, University of Grenoble, France.

Pham-Dinh, T. (1977a). Estimation of parameters in the ARMA model when the characteristic polynomial has a root near the unit circle. *In* "Recent Developments in Statistics" (Ed. J. R. Barra *et al.*). North-Holland, Amsterdam.

Pham-Dinh, T. (1977b). Estimation of parameters of a continuous time Gaussian stationary process with rational spectral density. *Biometrika*, **64**, 385–399.

Pham-Dinh, T. (1978). On the fitting of multivariate processes of the autoregressive moving average type. *Biometrika*, **65**, 99–107.

Pierson, W. J. and Tick, L. J. (1957). Stationary random processes in meteorology and oceanography. *Bull. Inst. Internat. Statist.*

Press, H. and Houlboult, J. C. (1955). Some applications of generalised harmonic analysis to gust loads on airplanes. *J. Aero. Sci.*, **22**, 17–26.

Press, H. and Tukey, J. W. (1956). Power spectral methods of analysis and their application to problems in airplane dynamics. Bell Systems Monograph No. 2606.

Priestley, M. B. (1962a). The analysis of stationary processes with mixed spectra—I. *J. Roy. Statist. Soc. Ser. B*, **24**, 215–233.

Priestley, M. B. (1962b). The analysis of stationary processes with mixed spectra—II. *J. Roy. Statist. Soc. Ser. B*, **24**, 511–529.

Priestley, M. B. (1962c). Basic considerations in the estimation of power spectra. *Technometrics*, **4**, 551–564.

Priestley, M. B. (1962d). Analysis of two-dimensional processes with discontinuous spectra. Technical Report No. 6, Department of Statistics, Stanford University.

Priestley, M. B. (1963). The spectrum of a continuous process derived from a discrete process. *Biometrika*, **50**, 517–520.

Priestley, M. B. (1964a). Estimation of the spectral density function in the presence of harmonic components. *J. Roy. Statist. Soc. Ser. B*, **26**, 123–132.

Priestley, M. B. (1964b). Analysis of two-dimensional processes with discontinuous spectra. *Biometrika*, **51**, 195–217.

Priestley, M. B. (1965a). The role of bandwidth in spectral analysis. *Appl. Statist.*, **14**, 33–47.

Priestley, M. B. (1965b). Evolutionary spectra and non-stationary processes. *J. Roy. Statist. Soc. Ser. B*, **27**, 204–237.

Priestley, M. B. (1966). Design relations for non-stationary processes. *J. Roy. Statist. Soc. Ser. B*, **28**, 228–240.

Priestley, M. B. (1967). Power spectral analysis of non-stationary random processes. *J. Sound Vib.*, **6**, 86–97.

Priestley, M. B. (1969a). Estimation of transfer functions in closed-loop stochastic systems. *Automatica*, **5**, 623–632.

Priestley, M. B. (1969b). Control systems with time dependent parameters. *Bull. Inst. Internat. Statist.*, **37**. (Paper presented at the 37th Session of the I.S.I., London.)

Priestley, M. B. (1971a). Fitting relationships between time series. *Bull. Inst. Internat. Statist.*, **38**, 1–27. (Paper presented at the 38th session of the I.S.I., Washington, U.S.A.)

Priestley, M. B. (1971b). Time-dependent spectral analysis and its applications in prediction and control. *J. Sound Vib.*, **17**, 139–156.

Priestley, M. B. (1971c). Some notes on the physical interpretation of non-stationary stochastic processes. *J. Sound Vib.*, **17**, 51–54.

Priestley, M. B. (1976a). Applications of multivariate techniques in the study of multivariable stochastic systems. Paper presented at the 6th Int. Conf. on Stochastic Processes, Tel-Aviv, Israel. (Summary in *Adv. Appl. Probability*, **9**, 202–205, 1977.)

Priestley, M. B. (1976b). Discussion on "Bayesian Forecasting". *J. Roy. Statist. Soc. Ser. B*, **38**, 205–248.

Priestley, M. B. (1978a). System identification, Kalman filtering and stochastic control. Paper presented at the I.M.S. special topics meeting on Time Series, Ames, Iowa, U.S.A. *In* "Directions in Time Series" (Ed. D. R. Brillinger and G. C. Tiao), I.M.S. Publication (1980).

Priestley, M. B. (1978b). Non-linear models in time series analysis. *The Statistician*, **27**, 159–176.

Priestley, M. B. and Chao, M. T. (1972). Non-parametric function fitting. *J. Roy. Statist. Soc. Ser. B*, **34**, 385–392.

Priestley, M. B. and Gibson, C. H. (1965). Estimation of power spectra by a wave-analyser. *Technometrics*, **7**, 553–559.

Priestley, M. B. and Subba Rao, T. (1969). A test for stationarity of time series. *J. Roy. Statist. Soc. Ser. B*, **31**, 140–149.

Priestley, M. B. and Subba Rao, T. (1975). The estimation of factor scores and Kalman filtering for discrete parameter stationary processes. *Int. J. Control*, **21**, 971–975.

Priestley, M. B., Subba Rao, T. and Tong, H. (1973). Identification of the structure of multivariable stochastic systems. *In* "Multivariate Analysis, III" (Ed. P. Krishnaiah). Academic Press, New York and London.

Priestley, M. B., Subba Rao, T. and Tong, H. (1974). Applications of principal components analysis and factor analysis in the identification of multi-variable systems. *I.E.E.E. Trans. Automatic Control*, **AC-19**, 730–734.

Priestley, M. B. and Tong, H. (1973). On the analysis of bivariate non-stationary processes. *J. Roy. Statist. Soc. Ser. B*, **35**, 153–188.

Prothero, D. L. and Wallis, K. F. (1976). Modelling macroeconomic time series. *J. Roy. Statist. Soc. Ser. A*, **139**, 468–500.

Quenouille, M. H. (1947). A large sample test for the goodness of fit of autoregressive schemes. *J. Roy. Statist. Soc. Ser. A*, **110**, 123–129.

Quenouille, M. H. (1949). Approximate tests of correlation in time series. *J. Roy. Statist. Soc. Ser. B*, **11**, 68–84.

Quenouille, M. H. (1957). "The Analysis of Multiple Time Series". Griffin, London.

Rao, C. R. (1964). The use and interpretation of principal component analysis in applied research. *Sankhyā, Ser. A*, **26**, 329–358.

Rao, A. G. and Shapiro, A. (1970). Adaptive smoothing using evolutionary spectra. *Management Sciences*, **17**, 208–218.

Revfeim, K. J. A. (1969). Iterative techniques for the estimation of parameters in time series models. Ph.D. Thesis, University of Manchester.

Rice, S. O. (1944–5). Mathematical analysis of random noise. *Bell System Tech. J.*, **23**, 282–332; **24**, 46–156. (Reprinted in "Noise and Stochastic Processes" (Ed. N. Wax). Dover, New York, 1954.)

Riesz, F. and Sz. Nagy, B. (1955). "Functional Analysis". Ungar, New York.

Rissanen, J. and Caines, P. E. (1979). The strong consistency of maximum likelihood estimators for ARMA processes. *Ann. Statist.*, **7**, 297–315.

Rosenblatt, M. (1956a). Remarks on non-parametric estimates of a density function. *Ann. Math. Statist.*, **33**, 1065–1076.

Rosenblatt, M. (1956b). Some regression problems in time series analysis. Proc. 3rd Berkeley Symp. on Math. Statist. Prob., pp. 165–186. University of California Press, Berkeley.

Rosenblatt, M. (1959). Statistical analysis of stochastic processes with stationary residuals. *In* "Probability and Statistics" (Ed. U. Grenander), pp. 246–275. Almquist and Wiksell, Stockholm.

Rosenblatt, M. (1971). Curve estimates. *Ann. Math. Statist.*, **42**, 1815–1842.

Rosenbrock, H. H. (1970). "State Space and Multivariable Theory". Nelson, London.

Rozanov, Yu. A. (1967). "Stationary Random Processes". Holden-Day, San Francisco.

Rubin, H. (1945). On the distribution of the serial correlation coefficient. *Ann. Math. Statist.*, **16**, 211–215.

Sargen, J. D. (1953). An approximate treatment of the properties of the correlogram and periodogram. *J. Roy. Statist. Soc. Ser. B*, **15**, 140–152.

Schaerf, M. C. (1964). Estimation of the covariance and autoregressive structure of a stationary time series. Technical Report No. 12, Department of Statistics, Stanford University.

Scheffe, H. (1959). "The Analysis of Variance". Wiley, New York.

Schuster, A. (1898). On the investigation of hidden periodicities with application to a supposed 26-day period of meteorological phenomena, *Terr. Mag. Atmos. Elect.*, **3**, 13–41.

Schwarz, G. (1978). Estimating the dimension of a model. *Ann. Statist.*, **6**, 461–464.

Shapiro, H. S. and Silverman, R. A. (1960). Alias-free sampling of random noise. *J. SIAM*, **8**, 225–248.

Shibata, R. (1976). Selection of the order of an autoregressive model by Akaike's information criterion. *Biometrika*, **63**, 117–126.

Shimshoni, M. (1971). On Fisher's test of significance in harmonic analysis. *Geophys. J. Roy. Astron. Soc.*, **23**, 373–377.

Shiryaev, A. N. (1960). Some problems in the spectral theory of higher order moments—I. *Theor. Probability Appl.*, **5**, 265–284.

Silverman, B. W. (1978). Choosing the window width when estimating a density. *Biometrika*, **65**, 1–11.

Silvey, S. D. (1970). "Statistical Inference". Chapman and Hall, London.

Singleton, R. C. (1969). An algorithm for computing the mixed radix fast Fourier transform. *I.E.E.E. Trans. Audio Electron*, **AU-17**, 93–103.

Sommerfeld, A. J. W. (1949). "Partial Differential Equations in Physics". New York.

Stone, M. (1979). Comments on model selection criteria of Akaike and Schwarz. *J. Roy. Statist. Soc. Ser. B*, **41**, 276–278.

Stuart, R. D. (1961). "An Introduction to Fourier Analysis". Methuen, London.

Subba Rao, T. (1975). An innovations approach to the reduction of the dimensions in a multivariable stochastic system. *Inst. J. Control*, **21**, 673–680.

Subba Rao, T. (1976a). Canonical factor analysis and stationary time series models. *Sankhyā, Ser. B*, **38**, 256–271.

Subba Rao, T. (1976b). A note on the bias in Kalman–Bucy filtering. *Int. J. Control*, **23**, 641–645.

Subba Rao, T. (1977). On the estimation of bilinear time series models. *Bull. Inst. Internat. Statist.*, **41**. (Paper presented at the 41st session of I.S.I., New Delhi, India.)

Subba Rao, T. (1981). On the theory of bilinear models. *J. Roy. Statist. Soc. Ser. B*, **43**.

Subba Rao, T. and Gabr, M. M. (1981). A test for linearity of stationary time series. *J. Time Series Anal.*, **2**.

Subba Rao, T. and Tong, H. (1972). A test for time-dependence of linear open-loop systems. *J. Roy. Statist. Soc. Ser. B*, **34**, 235–250.

Subba Rao, T. and Tong, H. (1973). On some tests for the time-dependence of a transfer function. *Biometrika*, **60**, 589–597.

Subba Rao, T. and Tong, H. (1974). Identification of the covariance structure of state space models. *Bull. Inst. Math. Appl.*, **10**, 201–203.

Sugiyama, T. and Tong, H. (1976). On a statistic useful in dimensionality reduction. *Comm. Statist.*, **A5**(8), 711–721.

Sussman, H. J. (1976). Existence and uniqueness of minimal realisations of non-linear systems—I: initialised systems. *J. Math. Syst. Theory*

Tayfun, M. A., Yang, C. Y. and Hsiao, G. C. (1971). On non-stationary random wave spectra. Int. Symp. on Stochastic Hydraulics, Pittsburgh, U.S.A., May 1971.

Taylor, A. E. (1964). "Introduction to Functional Analysis". Wiley, New York.

Thomson, D. J. (1977). Spectrum estimation techniques for characterisation and development of WTH wave-guide—I, II. *Bell Systems Tech. J.*, **56**, I: 1769–1865; II: 1983–2005.

Tiao, G. C. and Ali, M. M. (1971). Analysis of correlated random effects: linear model with two random components. *Biometrika*, **58**, 37–51.

Tiao, G. C. and Box, G. E. P. (1979). An introduction to applied multiple time series analysis. Tech. Report No. 582, Dept. of Statistics, Univ. of Wisconsin.

Tiao, G. C. and Hillmer, S. C. (1978). Some considerations of decomposition of a time series. *Biometrika*, **65**, 497–502.

Tiao, G. C., Box, G. E. P. and Hamming, W. J. (1975). Analysis of Los Angeles photo-chemical smog data: a statistical overview. *J. Air. Pollution Control Assoc.*, **25**, 260.

Tick, L. J. (1963). Conditional spectra, linear systems and coherency. In "Time Series Analysis" (Ed. M. Rosenblatt), pp. 197–203. Wiley, New York.

Tick, L. J. (1967). Estimation of coherence. In "Advanced Seminar on Spectral Analysis" (Ed. B. Harris), pp. 133–152. Wiley, New York.

Titchmarsh, E. C. (1939). "The Theory of Functions". Oxford University Press, London.

Titchmarsh, E. C. (1948). "Introduction to the Theory of Fourier Integrals". Oxford University Press, London.

Tjøstheim, D. (1976a). Spectral generating operators for non-stationary processes. *Adv. Appl. Probability*, **8**, 831–846.

Tjøstheim, D. (1976b). A commutation relation for wide sense stationary processes. *SIAM J. Appl. Math.*, **30**, 115–122.

Tjøstheim, D. (1976c). On random processes that are almost strict sense stationary. *Adv. Appl. Probability*, **8**, 820–830.

Tjøstheim, D. (1976d). Spectral representations and density operators for infinite dimensional homogeneous random fields. *Z. Wahscheinlichkeitshorie*, **35**, 323–336.

Tong, H. (1974a). On time dependent linear transformations of non-stationary stochastic processes. *J. Appl. Probability*, **11**, 53–62.

Tong, H. (1974b). Frequency domain approach to regulation of linear systems. *Automatica*, **10**, 533–538.

Tong, H. (1977a). Some comments on the Canadian Lynx data. *J. Roy. Statist. Soc. Ser. A*, **140**, 432–436.

Tong, H. (1977b). Discussion on "Stochastic modelling of riverflow time series" by Lawrance and Kottegoda. *J. Roy. Statist. Soc. Ser. A*, **140**,

Tong, H. (1978a). On a threshold model. *In* "Pattern Recognition and Signal Processing" (Ed. C. H. Chen). Sijthoff and Noordhoff, Amsterdam.

Tong, H. (1978b). Threshold autoregressions, limit cycles and cyclical data. Technical Report No. 101, Department of Mathematics (Statistics), University of Manchester Institute of Science and Technology, England.

Tong, H. (1978c). A view on non-linear time series model building. Proc. International Time Series Meeting, University of Nottingham, England, March 1979 (Ed. O. D. Anderson). North-Holland, Amsterdam.

Tukey, J. W. (1949). The sampling theory of power spectrum estimates. Proc. Symp. on Applications of Autocorrelation Analysis to Physical Problems, NAVEXOS— P-735, 47–67. Office of Naval Research, Department of the Navy, Washington, U.S.A.

Tukey, J. W. (1959). An introduction to the measurement of spectra. *In* "Probability and Statistics" (Ed. U. Grenander), pp. 300–330. Wiley, New York.

Tukey, J. W. (1961). Discussion emphasising the connection between analysis of variance and spectrum analysis. *Technometrics*, **3**, 191–219.

Tukey, J. W. (1967). An introduction to the calculations of numerical spectrum analysis. *In* "Advanced Seminar on Spectral Analysis" (Ed. B. Harris), pp. 25–46. Wiley, New York.

Tukey, J. W. (1978). When should which spectrum approach be used? Paper presented at the Institute of Statisticians Conference, King's College, Cambridge, July 1978.

Tunnicliffe-Wilson, G. (1973). The estimation of parameters in multivariate time series models. *J. Roy. Statist. Soc. Ser. B*, 76–85.

Ulrych, T. J. and Bishop, T. N. (1975). Maximum entropy spectral analysis and autoregressive decomposition. *Reviews of Geophysics and Space Physics*, **13**, 183–200.

Volterra, V. (1959). "Theory of Functionals and of Integro-Differential Equations". Dover, New York.

Von Neumann, J. (1941). Distribution of the ratio of the mean-square successive difference to the variance. *Ann. Math. Statist.*, **12**, 367–395.

Wahba, G. (1968). On the distribution of some statistics useful in the analysis of jointly stationary time series. *Ann. Math. Statist.*, **39**, 1849–1862.

Wahba, G. (1978). Automatic smoothing of the log periodogram. Technical Report No. 536, Department of Statistics, University of Wisconsin.

Waldheimer, M. (1961). "The Sun-Spot Activity in the Years 1610–1960". Schulthess, Zürich.

Walker, A. M. (1952). Some properties of the asymptotic power functions of goodness-of-fit tests for linear autoregressive schemes. *J. Roy. Statist. Soc. Ser. B*, **14**, 117–134.

Walker, A. M. (1961). Large sample estimation of parameters for moving average models. *Biometrika*, **48**, 343–357.

Walker, A. M. (1962). Large sample estimation of parameters for autoregressive processes with mvoing average residuals. *Biometrika*, **49**, 117–131.

Walker, A. M. (1964). Asymptotic properties of least squares estimates of parameters of the spectrum of a stationary non-deterministic time series. *J. Austral. Math. Soc.*, **4**, 363–384.

Walker, A. M. (1965). Some asymptotic results for the periodogram of a stationary time series. *J. Austral. Math. Soc.*, **5**, 107–128.

Walker, A. M. (1971). On the estimation of a harmonic component in a time series with stationary independent residuals. *Biometrika*, **58**, 21–36.

Walker, A. M. and Young, A. (1955). The analysis of observations on the variations of latitude. *Monthly Notices of the Roy. Astron. Soc.*, **115**, 443.

Walker, G. (1914). On the criteria for the reality of relationships or periodicities. *Calcutta Ind. Met. Memo.*, **21**, 9.

Watson, G. N. (1944). "A Treatise on the Theory of Bessel Functions", 2nd Edn. Cambridge University Press, London.

Watson, G. S. (1967). Linear least squares regression. *Ann. Math. Statist.*, **28**, 1679–1699.

Whittaker, J. M. (1935). "Interpolatory Function Theory". Cambridge University Press, London.

Whittle, P. (1951). Hypothesis Testing in Time Series Analysis. (Ph.D. Thesis, Uppsala.) Almquist & Wiksell, Uppsala.

Whittle, P. (1952a). The simultaneous estimation of a time series harmonic components and covariance structure. *Trabajos. Estadist.*, **3**, 43–57.

Whittle, P. (1952b). Tests of fit in time series. *Biometrika*, **39**, 309.

Whittle, P. (1953). The analysis of multiple time series. *J. Roy. Statist. Soc. Ser. B*, **15**, 125–139.

Whittle, P. (1954a). Some recent contributions to the theory of stationary processes. Appendix to Second Edition of "A Study in the Analysis of Stationary Time Series" by H. Wold. Almquist & Wiksell, Uppsala.

Whittle, P. (1954b). The statistical analysis of a Seiche record. *J. Marine Research (Sears Foundation)*, **13**, 76–100.

Whittle, P. (1954c). On stationary processes in the plane. *Biometrika*, **41**, 434–449.

Whittle, P. (1957). Curve and periodogram smoothing. *J. Roy. Statist. Soc. Ser. B*, **19**, 38–47.

Whittle, P. (1958). On the smoothing of probability density functions. *J. Roy. Statist. Soc. Ser. B*, **20**, 334–343.

Whittle, P. (1959). Sur la distribution du maximum d'un polynome trigonométrique à coefficients aléatoires. *Colloques Internationaux du Centre Nationale de la Recherche Scientifique*, **87**, 173–184.

Whittle, P. (1963). "Prediction and Regulation". English Universities Press, London.

Wiener, N. (1930). Generalised harmonic analysis. *Acta Math.*, **55**, 117–258.

Wiener, N. (1949). "The Extrapolation, Interpolation and Smoothing of Stationary Time Series with Engineering Applications". Wiley, New York.

Wiener, N. (1958). "Non-linear Problems in Random Theory". M.I.T. Press, Cambridge, Mass.

Wiener, N. and Masani, P. (1957). The prediction theory of multivariate stochastic processes: I. *Acta Math.*, **98**, 111–150.

Wiener, N. and Masani, P. (1958). The prediction theory of multivariate stochastic processes: II. *Acta Math.*, **99**, 93–137.

Wilks, S. S. (1961). "Mathematical Statistics". Wiley, New York.

Wold, H. O. A. (1938). "A Study in the Analysis of Stationary Time Series". Almquist & Wiksell, Uppsala.

Wold, H. O. A. (1949). A large sample test of moving averages. *J. Roy. Statist. Soc. Ser. B*, **11**, 297.

Woodroofe, M. B. and Van Ness, J. W. (1967). The maximum deviation of sample spectral densities. *Ann. Math. Statist.*, **38**, 1558–1570.

Yaglom, A. M. (1955). The correlation theory of processes whose nth differences constitute a stationary process. *Mat. Sb.*, **37** (79), 1, 141–196. (English translation in "American Mathematical Society Translations", Series 2, Vol. 8 (1958), p. 87. Amer. Math. Soc., Providence, R.I.)

Yaglom, A. M. (1962). "An Introduction to the Theory of Stationary Random Functions". Prentice-Hall, Englewood Cliffs.

Young, P. C. (1974). Recursive approaches to time series analysis. *Bull. Inst. Math. Appl.*, **10**, 209–224.

Yule, G. U. (1927). On a method of investigating periodicities in disturbed series with special reference to Wolfer's sunspot numbers. *Philos. Trans. Roy. Soc. London Ser. A*, **226**, 267–298.

Author Index

This index covers the material in both Volume 1 and Volume 2. Page numbers from Volume 1 (pp. 1–653) are printed in bold type, those from Volume 2 (pp. 654–890) are printed in ordinary type.

Subject Index

This index covers the material in both Volume 1 and Volume 2. Page numbers from Volume 1 (pp. 1–653) are printed in bold type, those from Volume 2 (pp. 654–890) are printed in ordinary type.

lvii

Printed and bound by CPI Group (UK) Ltd, Croydon, CR0 4YY

13/05/2025

01869521-0001